Distribution costs: Exp. 447; E 449, 460, 481
Drinking and driving: Exp. 244
Drinking and heart disease: E 335
Drinking and smoking: Exp. 325, 361
Drug dosage: E 623
Drug reaction: E 607, 670

Earnings ratio: E 253
Ecology: E 154, 370, 450, 542
Economic lot size: E 605
Economic order quantity: E 605, 626
Economic rent: E 712
Education and earnings: Exp. 330
Effect of taxation on production: E 605
Effective rate of interest: T 245–46; Exp. 217, 222; E 257
Elasticity of demand: T 646–47; Exp. 648; E 650–52, 736, 774
Electricity cost function: Exp. 177; E 532
Employee classifications: E 327
Employee layoffs: E 361
Employee transfers: E 327
Encyclopedia sales: E 353
Energy use: E 425
Entomology: E 796
Epidemic: E 514, 565, 607, 623, 627, 678, 729, 736, 796
Errors: T 634; Exp. 634; E 635–36
Exponential depreciation: E 232
Exponential growth: E 231–32
Extra cost of production: Exp. 660
Eye color: E 354

Fan cards: E 354
Farming: E 108, 182
Fencing: Exp. 186; E 189, 604
Fick's law of diffusion: E 558
Fish farming: E 542
Fish stocks: E 481
Fuel consumption: E 623
Furniture shipment: E 87

Gas station operations: Exp. 483; E 167
GNP increase: Exp. 218, 258
Gompertz function: E 255
Government contracts: E 349
Grain silo designs: E 607
Graph theory: E 380
Group travel: E 607
Growth of capital: E 711, 729, 795
Growth of GNP: E 258, 490, 514, 795
Growth of sales: E 507, 794

Hair dressing rates: Exp. 106
Handbill design: E 604
Heating costs: E 606
Hiring and promotions: Exp. 341; E 317
House equity: E 284
Housing starts: E 205, 550

Income and car ownership: E 800
Income distribution: E 744
Income elasticity: E 774

Income tax policy: Exp. 529; E 154, 532
Increasing average cost: E 572
Input–output analysis: T 407–10; Exp. 411, 4⸱⸱
Insecticide⸱⸱
Instantaneo⸱⸱ 513, 5⸱⸱
Insufficient⸱⸱
Insurance c⸱⸱
Internationa⸱⸱
Inventory c⸱⸱, E 004–06, 612, 627, 635
Inventory matrix: Exp. 366–67; E 397
Inventory valuation: E 380
Investment of funds: Exp. 69, 83, 216, 222, 244; E 70, 71, 87, 154, 165, 389, 407
Investment return: Exp. 446; E 257

Job applicants: E 317, 339
Job assignments: E 361

Labor requirements: E 550
Land costs: E 606
Land measurements: E 721
Learning curve: T 703; Exp. 704; E 254, 257, 680, 710, 747
Learning model: E 627
Length of telephone calls: E 744, 748
Life of automobile: E 744
Lifetime of light bulbs: Exp. 740
Lifetime of machines: Exp. 261, 269; E 266–67, 270
Limited growth models: E 736
Linear models: T 136, 633; Exp. 136–37; E 142, 165, 636
Loan amortization: T 280; Exp. 281, 282; E 283, 307, 308
Loan repayment: Exp. 264–65, 293, 295; E 257, 260, 266, 298–99, 307
Location of office typists: Exp. 248
Logistic model: T 249–50, 732–34; Exp. 250–51, 734; E 254, 623, 736
Lorentz curve: T 701; E 710, 747

Machinery savings and costs: E 711
Magazine publishing: E 102
Mailing costs: E 143, 163
Management training: E 339
Manufacturer's defects: E 362
Manufacturer's profits: Exp. 100; E 71, 87, 101, 389
Manufacturer's revenue: E 107
Marginal average cost: T 539; Exp. 517, 539; E 523, 542, 549, 681
Marginal average cost rate: E 563
Marginal cost: T 515; Exp. 516; E 522, 534, 549, 558, 565, 572, 663, 669, 680, 730
Marginal cost rate: E 563
Marginal demand: T 769; Exp. 770; E 534, 565, 773
Marginal physical productivity: E 565

Marginal productivity: T 520, 768; E 534, 798
Marginal productivity of money: Exp. 784
Marginal profit: T 519; Exp. 519, 546; E 522, 662, 774, 798
Marginal propensity to save and consume: T 521; E 680
Marginal revenue: T 518; Exp. 518–19, 537; E 522, 534, 541, 547, 558, 565, 663, 680
Marginal revenue analysis: E 572
Marginal revenue rate: E 563
Marginal tax rate: T 521
Marginal yield: T 521
Market equilibrium: T 158; Exp. 158–60, 387; E 163, 166, 389
Market price fluctuation: E 730
Market share: E 424, 441
Market stability: E 163
Marco chains: T 414–23; Exp. 416–17, 420; E 423–25
Married couple survival: E 361
Material distribution: E 449
Material happiness: E 572
Maximizing output: E 780
Maximum profit: Exp. 450, 454, 598; E 188, 213, 605, 607, 612, 626, 747, 787
Maximum profits and tax revenue: Exp. 601
Maximum revenue: E 605–07, 612
Maximum revenue and profits: E 188, 213
Maximum timber yield: E 606
Maximum yield from sales tax: E 606
Medicine: E 359, 490, 558, 729, 781
Memory retention: E 627
Microbiology: E 767
Minimum average cost: E 604, 612
Minimum cost: Exp. 597, 609; E 189, 780, 788
Minimum marginal cost: E 605
Minimum reaction time: E 612
Mixture problems: Exp. 69, 151; E 71, 154, 166, 459, 479
Monthly compounding: Exp. 216; E 224–25
Monthly earnings: Exp. 67; E 212
Mortgage amortization: E 283, 284
Mortgage borrowing limit: Exp. 281
Multiple choice exam: E 349, 353
Mutations: E 327

Natural resources: E 670
Newspaper circulation: E 252
Newspaper profits: E 71
Newton's law of cooling: E 254, 730
Nitrogen expulsion: E 257

Off-shore oil refinery: E 607
Oil consumption: E 681
Oil exploration: Exp. 337–39; E 670
Oil pollution: E 514, 549
Opinion poll: Exp. 352; E 354
Opinion survey: E 425

MATHEMATICAL ANALYSIS

for Business, Economics, and the Life and Social Sciences

FOURTH EDITION

MATHEMATICAL ANALYSIS

for Business, Economics, and the Life and Social Sciences

Jagdish C. Arya
Robin W. Lardner

Department of Mathematics, Simon Fraser University

Prentice Hall, Englewood Cliffs, New Jersey 07632

Library of Congress Cataloging-in-Publication Data

Arya, Jagdish C.
   Mathematical analysis for business, economics, and the life and
social sciences / Jagdish C. Arya, Robin W. Lardner. -- 4th ed.
      p.   cm.
   Includes index.
   ISBN 0-13-564287-6
   1. Mathematical analysis. 2. Business mathematics. 3. Economics,
Mathematical.   I. Lardner, Robin W.   II. Title.
QA300.A75 1993
515--dc20   92-27048
                                                           CIP

Editor-in-chief: Tim Bozik
Senior editor: Steve Conmy
Executive editor: Priscilla McGeehon
Senior managing editor: Jeanne Hoeting
Production editor: Nicholas Romanelli
Design director: Florence Dara Silverman
Interior design: Merle Poweski
Cover design: Patricia McGowan
Prepress buyer: Paula Massenaro
Manufacturing buyer: Lori Bulwin

 © 1993, 1989, 1985, 1981 by Prentice-Hall, Inc.
A Simon & Schuster Company
Englewood Cliffs, New Jersey, 07632

Printed in the United States of America
10 9 8 7 6 5 4

ISBN 0-13-564287-6

Prentice-Hall International (UK) Limited, *London*
Prentice-Hall of Australia Pty. Limited, *Sydney*
Prentice-Hall Inc., *Toronto*
Prentice-Hall Hispanoamericana, S.A., *Mexico*
Prentice-Hall of India Private Limited, *New Delhi*
Prentice-Hall of Japan, Inc., *Tokyo*
Simon & Schuster Asia Pte Ltd., *Singapore*
Editora Prentice-Hall do Brasil, Ltda., *Rio de Janeiro*

To Niki and Shanti

Contents

PREFACE   xiii

PART ONE
ALGEBRA

1 REVIEW OF ALGEBRA   1

1-1   The Real Numbers   2
1-2   Fractions   10
1-3   Exponents   18
1-4   Fractional Exponents   23
1-5   Algebraic Operations   29
1-6   Factors   38
1-7   Algebraic Fractions   46
        Chapter Review   54

2 EQUATIONS IN ONE VARIABLE   57

2-1   Linear Equations   58
2-2   Applications of Linear Equations   66
2-3   Quadratic Equations   71
2-4   Applications of Quadratic Equations   79
        Chapter Review   86

3 INEQUALITIES 88

3-1 Sets and Intervals   89
3-2 Linear Inequalities in One Variable   95
3-3 Quadratic Inequalities in One Variable   102
3-4 Absolute Values   108
Chapter Review   114

4 STRAIGHT LINES 117

4-1 Cartesian Coordinates   118
4-2 Straight Lines and Linear Equations   126
4-3 Applications of Linear Equations   136
4-4 Systems of Equations   144
4-5 Applications to Business and Economics   154
Chapter Review   164

5 FUNCTIONS AND GRAPHS 168

5-1 Functions   169
5-2 Quadratic Functions and Parabolas   183
5-3 More Simple Functions and Graphs   189
5-4 Combinations of Functions   200
5-5 Implicit Relations and Inverse Functions   205
Chapter Review   211

6 LOGARITHMS AND EXPONENTIALS 214

6-1 Compound Interest and Related Topics   215
6-2 Exponential Functions   226
6-3 Logarithms   232
6-4 Applications and Further Properties of Logarithms   243
Chapter Review   255

PART TWO
FINITE MATHEMATICS

7 PROGRESSIONS AND THE MATHEMATICS OF FINANCE 259

7-1 Arithmetic Progressions and Simple Interest   260
7-2 Geometric Progressions and Compound Interest   267
7-3 Mathematics of Finance   274

7-4 Difference Equations   284
7-5 Summation Notation   299
Chapter Review   306

8 PROBABILITY 309

8-1 Sample Spaces and Events   310
8-2 Probability   317
8-3 Conditional Probability   328
8-4 Bayes' Theorem (Optional)   335
8-5 Permutations and Combinations   340
8-6 Binomial Probabilities   349
8-7 The Binomial Theorem (Optional)   354
Chapter Review   359

9 MATRIX ALGEBRA 363

9-1 Matrices   364
9-2 Multiplication of Matrices   370
9-3 Solution of Linear Systems by Row Reduction   381
9-4 Singular Systems   390
Chapter Review   395

10 INVERSES AND DETERMINANTS 399

10-1 The Inverse of a Matrix   400
10-2 Input-Output Analysis   407
10-3 Markov Chains (Optional)   414
10-4 Determinants   425
10-5 Inverses by Determinant   433
Chapter Review   439

11 LINEAR PROGRAMMING 442

11-1 Linear Inequalities   443
11-2 Linear Optimization (Geometric Approach)   450
11-3 The Simplex Tableau   461
11-4 The Simplex Method   470
Chapter Review   480

PART THREE
CALCULUS

12 THE DERIVATIVE   482

12-1 Increments and Rates   483
12-2 Limits   491
12-3 The Derivative   501
12-4 Derivatives of Power Functions   507
12-5 Marginal Analysis   514
12-6 Continuity and Differentiability (Optional)   523
    Chapter Review   532

13 CALCULATIONS OF DERIVATIVES   535

13-1 Derivatives of Products and Quotients   536
13-2 The Chain Rule   542
13-3 Derivatives of Exponential and Logarithmic Functions   550
13-4 Higher Derivatives   559
    Chapter Review   563

14 OPTIMIZATION AND CURVE SKETCHING   566

14-1 The First Derivative and the Graph of the Function   567
14-2 Maxima and Minima   572
14-3 The Second Derivative and Concavity   580
14-4 Curve Sketching for Polynomials   589
14-5 Applications of Maxima and Minima   594
14-6 Absolute Maxima and Minima   608
14-7 Asymptotes   613
    Chapter Review   623

15 MORE ON DERIVATIVES   629

15-1 Differentials   630
15-2 Implicit Differentiation   636
15-3 Logarithmic Differentiation and Elasticity   643
    Chapter Review   651

16 INTEGRATION   654

16-1 Antiderivatives   655
16-2 Method of Substitution   663
16-3 Tables of Integrals   670
16-4 Integration by Parts   674
    Chapter Review   678

17 THE DEFINITE INTEGRAL   682

17-1 Areas under Curves   683
17-2 More on Areas   692
17-3 Applications to Business and Economics   701
17-4 Average Value of a Function   712
17-5 Numerical Integration   715
17-6 Differential Equations: An Introduction   721
17-7 Separable Differential Equations   730
17-8 Applications to Probability   736
     Chapter Review   745

18 SEVERAL VARIABLES   749

18-1 Functions and Domains   750
18-2 Partial Derivatives   760
18-3 Applications to Business Analysis   767
18-4 Optimization   775
18-5 Lagrange Multipliers   781
18-6 Method of Least Squares   789
     Chapter Review   796

Appendices   801

Answers to Odd-Numbered Exercises   819

Index   853

Preface

In this new edition, as in the preceding one, we have striven to present the areas of basic mathematics, algebra, finite mathematics, and differential and integral calculus in a form that will be of maximum use to students whose major area is outside mathematics or the physical sciences. The main direction of the book is toward applications in the fields of business and economics, but a substantial number of exercises are also included that concern applications in other areas of the social and life sciences, which should make the book of use to students in these areas.

In this edition, while the basic framework of the book has been left unchanged, a significant number of revisions have been made. We have added a section in Chapter 7 on difference equations and their applications in financial mathematics and have expanded to two sections the coverage of differential equations in Chapter 17. Chapter 6, on exponential and logarithm functions; the treatment of quadratic inequalities in Chapter 3; and the first four sections in Chapter 14, on optimization, have all been completely revised; and the applications in Chapters 2 and 4 have been divided and placed in closer proximity to the algebra that relates to them. In addition to these major revisions and additions, a great many smaller ones have made throughout the book, consisting of extra worked examples or amplifications of the discussion. Most of the exercise sets have been modified, with the addition of several hundred new exercises.

Several new pedagogical tools are new to this edition. A detailed list of contents and objectives is included at the beginning of each chapter and a chapter review added at the end. An expanded use is made of boxes to emphasize major formulas and results. Perhaps most useful of all for the student is the inclusion in the margin, throughout the text, of "think boxes" that contain generally straightforward questions that are directly linked to the adjacent discussion. Asterisks (*) precede the more challenging exercises.

As before, the orientation of the book is toward the teaching of the applications and use of mathematics rather than toward pure mathematics for its own sake. Proofs of theorems are not stressed and are not given a prominent place in the development of the text. Typically, after stating a particular theorem we have first illustrated it and discussed its significance with a number of examples, and only then has a proof been given. The more difficult proofs are omitted entirely.

This de-emphasis of mathematical detail allows the students who are primarily interested in the applications of mathematics the time to improve their skills in using the various techniques. It is our experience that such students who learn to master the techniques usually develop reasonably sound intuition in the process, and the lack of complete mathematical rigor is not a serious deficiency.

The material in the book has been chosen to consist of those parts of basic mathematics that are of most interest to students majoring in business and economics, and also to students in the social and life sciences. The applications given in these areas has been completely integrated into the development: sometimes a particular application is used to motivate a certain piece of mathematics; elsewhere, a certain mathematical result might be applied either immediately or in a subsequent section, to a particular problem in, say, business analysis. Usually, the applications are given in close proximity to the particular piece of mathematics that is being applied. Having said this, however, it should be added that the mathematics in the book is usually developed "cleanly," that is, not in the context of any particular application. Only after establishing each result at a purely algebraic level is it applied to a practical problem. We believe that it is usually a mistake to teach a mathematical method within the limited context of one particular application, because then the student is not alerted to the possibly wider applications of the method in other problem areas.

The book is divided into three parts. Part One consists of pre-calculus algebra, Part Two covers finite mathematics, and Part Three covers calculus. Parts Two and Three are almost totally independent of each other and can be covered in either order.

The pre-calculus algebra comprises the first six chapters of the book. In the first three of these we have given a fairly detailed review of high-school algebra and of the solution of equations and inequalities in one variable. *Students who are familiar with this material may prefer to begin directly with Chapter 4,* in which linear equations and systems are discussed. The remainder of the first part consists of a chapter on functions and one on exponentials and logarithms.

The finite mathematics part of the book itself consists of three almost independent parts: Chapter 7 on the mathematics of finance, Chapter 8 on discrete probability, and Chapters 9–11 on matrices, determinants, and linear

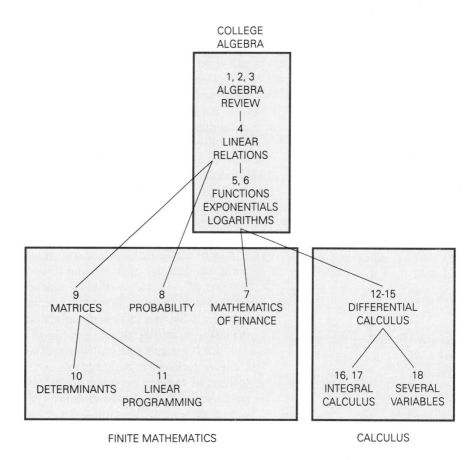

COLLEGE ALGEBRA

1, 2, 3
ALGEBRA
REVIEW
|
4
LINEAR
RELATIONS
|
5, 6
FUNCTIONS
EXPONENTIALS
LOGARITHMS

9
MATRICES

8
PROBABILITY

7
MATHEMATICS
OF FINANCE

12-15
DIFFERENTIAL
CALCULUS

10
DETERMINANTS

11
LINEAR
PROGRAMMING

16, 17
INTEGRAL
CALCULUS

18
SEVERAL
VARIABLES

FINITE MATHEMATICS

CALCULUS

programming. Chapter 11 on linear programming requires some of the material in Chapter 9, but does not require Chapter 10.

Chapters 12–15 cover the differential calculus of one variable. The first two of these chapters are concerned with the definition and calculation of the derivative and with its significance from various points of view. The next chapter applies differential calculus to optimization and curve sketching, and the fourth of these chapters contains certain more advanced topics in differential calculus.

Chapters 16 and 17 concern integral calculus. The first of these covers antiderivatives and some techniques of integration, while the second is concerned with the definite integral and its application to the computation of areas and to business analysis, differential equations, and probability.

The final chapter provides an introduction to the differential calculus of functions of several variables.

By selecting appropriate chapters and sections, the book can be used for a variety of different types of course. For example, courses covering college algebra or algebra and finite mathematics or algebra and calculus or finite mathematics and calculus can all be taught by making use of appropriately selected chapters. The foregoing diagram illustrates the prerequisite structure of the book, and it will be clear from the diagram how each type of course can be put together.

There is available a substantial package of supplements for this new edition that has components to aid both the professor and the student.

An **Instructor's Manual** is available for professors. Written by the authors, this supplement contains complete solutions to every problem. In addition to this, a **Test Item File and Test Generator** is available. These, too, were written by the authors, and they include over 1500 test items which can be accessed with an IBM-PC or compatible computer. A set of **Transparency Acetates** are also part of this package, as well as a **Gradebook/Class Record File Software** which allows an instructor to keep class records, compute class statistics, and a good deal more. All of the above-mentioned ancillary materials are available free upon adoption of the text.

For the student the authors have prepared a **Study Guide and Student Solutions Manual.** This salable item contains additional exercises keyed to sections of the text, detailed solutions to every fourth problem, and chapter tests. *How to Study Calculus* is a small booklet that contains strategies and hints for achieving success in calculus courses. It is available free to students who have purchased the text.

EPIC: Exploration Programs in Calculus 2nd Ed. is an interactive software package for both students and professors that covers all basic calculus topics and requires no computer literacy. This software is available in three versions: (a) a departmental version which is free to adopters and may be copied as needed; (b) a student version which is a salable item, reasonably priced; and (c) a demonstration version. Also included in this supplements package is **Interactive Experiments in Calculus,** an interactive software package for the Apple II series and Macintosh computers.

We wish to express our thanks to the following persons who reviewed the manuscript of the revision and provided helpful comments and suggestions: Michael J. Bradley, Merrimack College; Richard Weimer, Frostburg State University; Karen Mathiason, West Texas State University; Ronald Edwards, Westfield State University; Yoe Itokawa, University of Alabama, Birmingham; and Greg Taylor, Wake Forest University.

J.C.A.
R.W.L.

CHAPTER 1

Review of Algebra

Chapter Objectives

This chapter reviews the fundamental techniques of algebra. It is intended for those students who need, for one reason or another, to brush up on their basic algebraic skills. The following list gives the contents and objectives of each section.

1-1 THE REAL NUMBERS
   (a) A survey of the system of real numbers;
   (b) Fundamental properties of real numbers;
   (c) Elementary algebraic operations involving negatives and reciprocals and expressions with parentheses.

1-2 FRACTIONS
   (a) Multiplication and division of fractions;
   (b) Simplification of a fraction by cancelation of common factors;
   (c) Addition and subtraction of fractions; lowest common denominator.

1-3 EXPONENTS
   (a) Definition of powers with integer exponents;
   (b) The five properties of exponents;
   (c) Simplification of expressions involving integer exponents.

1-4 FRACTIONAL EXPONENTS
   (a) Definition of fractional exponents;
   (b) Simplification of expressions involving fractional exponents and radicals.

ALGEBRAIC OPERATIONS
   (a) Addition and subtraction of algebraic expressions;
   (b) Simplification of an expression by combining like terms;

   (c) Multiplication of multinomial expressions using the distributive property or by the method of arcs;
   (d) Four standard formulas for certain products and squares of binomials;
   (e) Division of an algebraic expression by a monomial;
   (f) Long division of a polynomial by a polynomial.

1-6 FACTORS
   (a) Extraction of monomial factors from an expression;
   (b) Standard formulas for factoring the difference of two squares and the sum and difference of two cubes;
   (c) The grouping method of factoring expressions;
   (d) Factoring quadratic expressions.

1-7 ALGEBRAIC FRACTIONS
   (a) Simplification of algebraic fractions by cancelation of monomial and multinomial common factors;
   (b) Rationalization of a fraction with a binomial denominator containing radicals;
   (c) Addition, subtraction, multiplication and division of fractions with multinomial numerators and denominators.

CHAPTER REVIEW

1-1 THE REAL NUMBERS

We shall begin by giving a brief outline of the structure of the real numbers. The numbers 1, 2, 3, and so on, that we use for counting are called the **natural numbers.** If we add or multiply any two natural numbers, the result is always a natural number. For example, $8 + 5 = 13$ and $8 \times 5 = 40$; the sum, 13, and the product, 40, are natural numbers. But if we subtract or divide two natural numbers, the result is *not* always a natural number. For example, $8 - 5 = 3$ and $8 \div 2 = 4$ are natural numbers, but $5 - 8$ and $2 \div 7$ are not natural numbers. Thus, within the system of natural numbers, we can always add and multiply but cannot always subtract or divide.

To overcome the limitation of subtraction, we extend the natural number system to the system of **integers.** The integers include the natural numbers, the negative of each natural number, and the number zero (0). Thus we may represent the system of integers by

$$\ldots, \quad -3, \quad -2, \quad -1, \quad 0, \quad 1, \quad 2, \quad 3, \quad \ldots.$$

Clearly, all the natural numbers are also integers. If we add, multiply, or subtract any two integers, the result is always an integer. For example, $-3 + 8 = 5$, $(-3)(5) = -15$, and $3 - 8 = -5$ are all integers. But we still cannot always divide one integer by another and get an integer as the result. For instance, we see that: $8 \div (-2) = -4$ is an integer, but $-8 \div 3$ is not. Thus, within the system of integers, we can add, multiply, and subtract, but we cannot always divide.

To overcome the limitation of division, we extend the system of integers to the system of **rational numbers.** This system consists of all the fractions a/b, where a and b are integers with $b \neq 0$.

A number is a rational number if it can be expressed as a ratio of two integers with the denominator nonzero. Thus $\frac{8}{3}$, $-\frac{5}{7}$, $\frac{0}{3}$, and $6 = \frac{6}{1}$, are examples of rational numbers. We can add, multiply, subtract, and divide any two rational numbers (with division by zero excluded)* and the result is always a rational number. Thus all the four fundamental operations of arithmetic, addition, multiplication, subtraction, and division are possible within the system of rational numbers.

When a rational number is expressed as a decimal, the decimal either terminates or develops a pattern that repeats indefinitely. For example, $\frac{1}{4} = 0.25$ and $\frac{93}{80} = 1.1625$ both correspond to decimals that terminate, whereas $\frac{1}{6} = 0.1666\ldots$ and $\frac{4}{7} = 0.5714285714285\ldots$ correspond to decimals with repeating patterns.

There also exist some numbers in common use that are not rational—that is, they cannot be expressed as the ratio of two integers. For example, $\sqrt{2}$, $\sqrt{3}$, and π are not rational numbers. Such numbers are called **irrational num-**

*See the final paragraph of this section.

bers. The essential defference between rational and irrational numbers can be seen through their decimal expressions. When an irrational number is represented by a decimal, the decimal continues indefinitely without developing any recurrent pattern. For example, to ten decimal places $\sqrt{2} = 1.4142135623\ldots$ and $\pi = 3.1415926535\ldots$. No matter to how many decimal places we express these numbers, they would never develop a repeating pattern, in contrast to the patterns that occur with rational numbers.

The term **real number** is used to mean a number that is either rational or irrational. The system of real numbers consists of all possible decimals. Those decimals that are terminating or repeating correspond to the rational numbers, while the rest correspond to the irrational numbers. 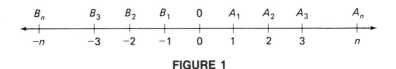 1

Geometrically, the real numbers can be represented by the points on a straight line called a **number line.** In order to do this, we select an arbitrary point O on the line to represent the number zero. The positive numbers are then represented by the points to the right of O and the negative numbers by the points to the left of O. If A_1 is a point to the right of O such that OA_1 is of unit length, then A_1 represents the number 1. The integers $2, 3, \ldots, n, \ldots$ are represented by the points $A_2, A_3, \ldots, A_n, \ldots$, which are on the right of O, such that

$$OA_2 = 2OA_1, \quad OA_3 = 3OA_1, \quad \ldots, \quad OA_n = nOA_1, \quad \ldots.$$

Similarly, if $B_1, B_2, \ldots, B_n, \ldots$ are the points to the left of O such that the distances $OB_1, OB_2, OB_3, \ldots, OB_n, \ldots$ are equal to the distances $OA_1, OA_2, \ldots, OA_n, \ldots$, respectively, then the points $B_1, B_2, B_3, \ldots, B_n, \ldots$ represent the negative integers $-1, -2, -3, \ldots, -n, \ldots$. In this way, all the integers can be represented by the points on a number line. (See Figure 1.)

FIGURE 1

Rational numbers can be represented by points on a number line that lie an appropriate fractional number of units from O. For example, the number $\frac{9}{2}$ is represented by the point which lies four and one-half units to the right of O and $-\frac{7}{3}$ is represented by the point which lies two and one-third units to the left of O. In a similar manner, every rational number can be represented by a point on the line.

It turns out that every *irrational* number can also be represented by a point on a number line. Consequently, all the real numbers, both rational and irrational, can be represented by such points. Furthermore, each point on a number line corresponds to one and only one real number. Because of this, it is quite common to use the word *point* to mean *real number*.

☛ **1.** What types of numbers are the following:

(a) $\dfrac{-3}{2}$;

(b) $(-\sqrt{2})^2$;

(c) $\dfrac{\pi}{2}$.

Answer (a) rational, real;

(b) natural, integer, real;

(c) irrational, real.

Properties of Real Numbers

When two real numbers are added, the result is always a real number; similarly, when two real numbers are multiplied, the result is also always a real number. These two operations of addition and multiplication are fundamental to the system of real numbers, and they possess certain properties which we shall now briefly discuss. These properties by themselves may appear to be rather elementary, perhaps even obvious, but *they are vital to understanding the various algebraic manipulations with which we shall later be involved.*

COMMUTATIVE PROPERTIES If a and b are any two real numbers, then

$$a + b = b + a \quad \text{and} \quad ab = ba.$$

For example, $3 + 7 = 7 + 3$, $3 + (-7) = (-7) + 3$, $3 \cdot 7 = 7 \cdot 3$, and $(3)(-7) = (-7)(3)$. These properties state that it does not matter in which order two numbers are added or multiplied—we get the same answer whichever order we use. They are known as the **commutative properties of addition and multiplication,** respectively.

ASSOCIATIVE PROPERTIES If a, b, and c are any three real numbers, then

$$(a + b) + c = a + (b + c) \quad \text{and} \quad (ab)c = a(bc).$$

For example, $(2 + 3) + 7 = 2 + (3 + 7) = 12$ and $(2 \cdot 3) \cdot 7 = 2 \cdot (3 \cdot 7) = 42$. These properties are known as the **associative properties of addition and multiplication,** respectively. They state that if three numbers are being added (or multiplied) together then it does not matter which two of them are added (or multiplied) together first. We get the same answer in either case.

Because of these properties, it is not necessary to write the parentheses in the above expressions. We can write $a + b + c$ for the sum of a, b, and c and abc for their product without ambiguity.

DISTRIBUTIVE PROPERTIES If a, b, and c are real numbers, then

$$a(b + c) = ab + ac \quad \text{and} \quad (b + c)a = ba + ca.$$

For example, $2(3 + 7) = 2(3) + 2(7) = 6 + 14 = 20$. This is clearly true because $2(3 + 7) = 2 \cdot 10 = 20$. Also, $(-2)[3 + (-7)] = (-2)(3) + (-2)(-7) = -6 + 14 = 8$. We can evaluate the given expression directly, getting the same answer: $(-2)[3 + (-7)] = (-2)(-4) = 8$.

The second form of the distributive property actually follows from the first since, by the commutative property,

$$(b + c)a = a(b + c) \quad \text{and} \quad ba + ca = ab + ac.$$

Since the two right sides are equal to one another by virtue of the first distributive property, the two left sides must also be equal.

The distributive properties are particularly important in algebraic calculations. As we shall see, they underlie many operations involved in simplifying expressions and, if read "backwards," that is, from right to left, they form the basis for factoring methods. ☞ 2

The following examples will illustrate some elementary uses of these properties of real numbers in simplifying algebraic expressions.

EXAMPLE 1

(a) $x(y + 2) = xy + x(2)$ (distributive property)
$\qquad\qquad\quad = xy + 2x$ (commutative property)

(b) $2x + 3x = (2 + 3)x$ (distributive property)
$\qquad\qquad = 5x$

(c) $2(3x) = (2 \cdot 3)x$ (associative property)
$\qquad\quad = 6x$

(d) $(2x)(3x) = [(2x) \cdot 3]x$ (associative property)
$\qquad\qquad = [3 \cdot (2x)]x$ (commutative property)
$\qquad\qquad = [(3 \cdot 2)x]x$ (associative property)
$\qquad\qquad = (6x)x$
$\qquad\qquad = 6(x \cdot x)$ (associative property)
$\qquad\qquad = 6x^2$

where x^2 denotes $x \cdot x$.

This final answer could be obtained by collecting the similar terms in the original product: the numbers 2 and 3 multiplied together give 6 and the two x's multiplied together give x^2. The next part further illustrates this procedure.

(e) $[5(3ab)](2a) = (5 \cdot 3 \cdot 2)(a \cdot a)b = 30a^2b.$

This answer can be justified with a sequence of steps using the associative and commutative laws, as in part (d).

(f) $2x + (3y + x) = 2x + (x + 3y)$ (commutative property)
$\qquad\qquad\qquad = (2x + x) + 3y$ (associative property)
$\qquad\qquad\qquad = (2x + 1x) + 3y$
$\qquad\qquad\qquad = (2 + 1)x + 3y$ (distributive property)
$\qquad\qquad\qquad = 3x + 3y$

☞ **2.** Which properties of real numbers are being used in each of the following?

(a) $2 + 3 \cdot 4 = 2 + 4 \cdot 3$;
(b) $2 + 3 \cdot 4 = 3 \cdot 4 + 2$;
(c) $2 + (3 + 4) = (3 + 4) + 2$;
(d) $2 + (3 + 4) = 4 + (2 + 3)$;
(e) $3x + 3x = (3 + 3)x$;
(f) $3x + xy = x(3 + y)$.

Answer (a) Commutative;
(b) commutative;
(c) commutative;
(d) both commutative and associative;
(e) distributive;
(f) both distributive and commutative.

$$(g)\ 2x(4y + 3x) = (2x)(4y) + (2x)(3x) \qquad \text{(distributive property)}$$

$$= (2 \cdot 4)(x \cdot y) + (2 \cdot 3)(x \cdot x) \qquad \text{(associative and commutative properties as in part (a))}$$

$$= 8xy + 6x^2.$$

The distributive property can be used in the case when more than two quantities are added within the parentheses. That is,

$$a(b + c + d) = ab + ac + ad$$

and so on.

EXAMPLE 2

$$4(x + 3y + 4z) = 4x + 4(3y) + 4(4z) \qquad \text{(distributive property)}$$
$$= 4x + (4 \cdot 3)y + (4 \cdot 4)z \qquad \text{(associative property)}$$
$$= 4x + 12y + 16z$$

IDENTITY ELEMENTS If a is any real number, then

$$a + 0 = a \qquad \text{and} \qquad a \cdot 1 = a.$$

That is, if 0 is added to a, the result is still a and if a is multiplied by 1, the result is again a. For this reason, the numbers 0 and 1 are often called the **identity elements** for addition and multiplication, respectively, because they leave any number unchanged under their respective operations.

INVERSES If a is any real number, then there exists a unique real number called the **negative of a** (denoted by $-a$) such that

$$a + (-a) = 0.$$

If a is *nonzero*, there also exists a unique real number called the **reciprocal of a** (denoted by a^{-1}) such that

$$a \cdot a^{-1} = 1.$$

Observe the similarity between these two definitions: when $-a$ is added to a, the result is the additive identity element and when a^{-1} is multiplied by a, the result is the multiplicative identity element. We often refer to $-a$ as the **additive inverse of a** and to a^{-1} as the **multiplicative inverse of a**. (Sometimes a^{-1} is simply called the **inverse of a.**)

☞ **3.** Which properties of real numbers are being used in each of the following?

(a) $x + 3x = 1x + 3x = (1 + 3)x = 4x$;

(b) $(2 + 1) + (-1) = 2 + [1 + (-1)] = 2 + 0 = 2$

(c) $3 \cdot \frac{1}{3} = 1$.

EXAMPLE 3

(a) The additive inverse of 3 is -3 since $3 + (-3) = 0$. The additive inverse of -3 is 3 since $(-3) + 3 = 0$. Since the additive inverse of -3 is, by definition, $-(-3)$, it follows that $-(-3) = 3$. In fact, a corresponding result holds for any real number a:

$$-(-a) = a.$$

(b) The multiplicative inverse of 3 is 3^{-1} since $3 \cdot 3^{-1} = 1$. The multiplicative inverse of 3^{-1} is by definition denoted by $(3^{-1})^{-1}$ and satisfies the requirement that $3^{-1} \cdot (3^{-1})^{-1} = 1$. But since $3^{-1} \cdot 3 = 1$, it follows that $(3^{-1})^{-1}$ is equal to 3.

Again this result generalizes for any nonzero real number a:

$$(a^{-1})^{-1} = a.$$

The inverse of the inverse of a is equal to a.

Having defined the additive and multiplicative inverses of a, we can define what we mean by the operations of *subtraction* and *division*. We define $a - b$ to mean the number $a + (-b)$, that is, a plus the negative of b. Similarly, we define $a \div b$ to mean the number ab^{-1}, that is, a multiplied by the reciprocal of b. The expression $a \div b$ is defined only when $b \neq 0$. It is also denoted by the fraction a/b, and we have

Definition of $\dfrac{a}{b}$: $\qquad\qquad \dfrac{a}{b} = ab^{-1}.$ $\qquad\qquad$ (1)

Putting $a = 1$ in Equation (1), we have

$$\frac{1}{b} = 1 \cdot b^{-1} = b^{-1}.$$

Hence $1/b$ means the same as the multiplicative inverse b^{-1}. For example, $3^{-1} = \frac{1}{3}$. It therefore follows from Equation (1) that

$$\frac{a}{b} = a\left(\frac{1}{b}\right)$$

Answer (a) Multiplicative identity element and distributive property; (b) associative property, additive inverse, and additive identity; (c) multiplicative identity and definition of $\dfrac{1}{a}$.

since $b^{-1} = 1/b$. ☞ **3**

EXAMPLE 4

(a) $\dfrac{7}{(\frac{1}{3})} = 7\left(\dfrac{1}{3}\right)^{-1}$ (Equation (1), with $a = 7$ and $b = \frac{1}{3}$)

$$= 7(3^{-1})^{-1} = 7(3) = 21$$

This result extends to any pair of real numbers a and b ($b \neq 0$):

$$\frac{a}{1/b} = ab.$$

(b) For any real number, $(-1)b = -b$. This follows because

$$\begin{aligned}
b + (-1)b &= 1 \cdot b + (-1)b \\
&= [1 + (-1)]b \qquad \text{(distributive property)} \\
&= 0 \cdot b = 0
\end{aligned}$$

Therefore, $(-1)b$ must be the additive inverse of b, namely $-b$.

(c) $\begin{aligned}[t]
a(-b) &= a[(-1)b] \qquad \text{(by part (b))} \\
&= (-1)(ab) \qquad \text{(using associative and} \\
&= -(ab) \qquad\qquad \text{commutative properties)}
\end{aligned}$

For example, $3(-7) = -(3 \cdot 7) = -21$.

(d) $\begin{aligned}[t]
3(x - 2y) &= 3[x + (-2y)] \qquad \text{(definition of subtraction)} \\
&= 3x + 3(-2y) \qquad \text{(distributive property)} \\
&= 3x - [3(2y)] \qquad \text{(from part (c))} \\
&= 3x - [(3 \cdot 2)y] \qquad \text{(associative property)} \\
&= 3x - 6y
\end{aligned}$

In general, the distributive property extends to expressions that involve negative signs. For example,

$$a(b - c) = ab - ac.$$

Thus we can solve this example more directly.

$$3(x - 2y) = 3x - 3(2y) = 3x - 6y$$

Observe that when an expression inside parentheses is multiplied by a negative quantity, every term inside the parentheses must change sign.

$$-(a + b) = (-1)(a + b) = (-1)a + (-1)b = -a - b$$

EXAMPLE 5

$$-2(x - 3y) = (-2)x - (-2)(3y)$$
$$= -2x + 6y$$

Note that both the x and $-3y$ inside the parentheses change signs, becoming $-2x$ and $+6y$, respectively.

☞ 4. Are the following defined?

(a) $\dfrac{a}{b + (3b - 4b)}$;

(b) $\dfrac{b + (3b - 4b)}{a}$.

Answer (a) No;
(b) yes, provided $a \neq 0$.

Note on Division by Zero. The statement $a/b = c$ is true if and only if the inverse statement $a = b \cdot c$ is true. Consider a fraction in which the denominator b is zero, such as $\frac{3}{0}$. This cannot equal any real number c because the inverse statement $3 = 0 \cdot c$ cannot be true for any c. Therefore $\frac{3}{0}$ is not well defined. Also, $\frac{0}{0}$ is not a well-defined real number because the inverse statement $0 = 0 \cdot c$ is true for all real numbers c. Thus we conclude that any fraction with denominator zero is not a well-defined real number or, equivalently, *division by zero is a meaningless operation.* For example, $x/x = 1$ is true only if $x \neq 0$. ☞ 4

EXERCISES 1-1

1. State whether each of the following is true or false. Replace each false statement with a corresponding true statement.

a. $3x + 4x = 7x$ **b.** $(3x)(4x) = 7x$

c. $2(5 - 4y) = 10 - 4y$

d. $-(x + y) = -x + y$

e. $5x - (2 - 3x) = 2x - 2$

f. $5 - 2x = 3x$

g. $-3(x - 2y) = -3x - 6y$

h. $(-a)(-b)(-c) \div (-d) = -(abc \div d)$

i. $a \div (b \div c) = (ac) \div b$

j. $a - (b - c) = (a + c) - b$

k. $(-x)(-y) = -xy$

l. $\dfrac{-a}{-b} = \dfrac{a}{b}$

m. $\dfrac{0}{x} = 0$ for all real numbers x

(2–60) Simplify the following expressions.

2. $5 - (-3)$

3. $-7 - (-3)$

4. $5(-3)$

5. $(-3)(-7)$

6. $8 \div (-2)$

7. $(-9) \div (-3)$

8. $-(2 - 6)$

9. $-(-4 - 3)$

10. $(3)(-2)(-4)$

11. $(-5)(-3)(-2)$

12. $3(1 - 4)$

13. $2(-2 - 3)$

14. $-2(-4 - 2)$

15. $-4(3 - 6)$

16. $-6 - 2(-3 - 2)$

17. $3(x + 2y)$

18. $4(2x + z)$

19. $2(2x - y)$

20. $3(4z - 2x)$

21. $-(x - 6)$

22. $-(-x - 3)$

23. $3(x - 4)$

24. $2(-x - 3)$

25. $-2(-x - 2)$

26. $-4(x - 6)$

27. $-x(y - 6)$

28. $-x(-y - 6)$

29. $2(x - y) + 4x$

30. $3y + 4(x + 2y)$

31. $-2z - 3(x - 2z)$

32. $-4x - 2(3z - 2x)$

33. $(x + y) + 4(x - y)$

34. $3(y - 2x) - 2(2x - 2y)$

35. $5(7x - 2y) - 4(3y - 2x)$

36. $4(8z - 2t) - 3(-t - 4z)$

37. $x(-y)(-z)$

38. $(-x)(-y)(-z)$

39. $(-2)(-x)(x + 3)$

40. $(-x)(-y)(2 - 3z)$

41. $2(-a)(3 - a)$

42. $(-37p)(2q)(q - p)$

43. $x(-2)(-x - 4)$

44. $(-2x)(-3)(-y - 4)$

45. $-x(x - 2) + 2(x - 1)$

46. $-2(-3x)(-2y + 1) - (-y)(4 - 5x)$

47. $2x + 5 - 2(x + 2)$

48. $3x - t - 2(x - t)$

49. $2(x - y) - x$

50. $4x(x + y) - x^2$

51. $4[2(x + 1) - 3]$

52. $x[3(x - 2) - 2x + 1]$

53. $x[-3(-4 + 5) + 3]$

54. $4[x(2 - 5) - 2(1 - 2x)]$

55. $x^{-1}(x + 2)$

56. $x^{-1}(2x - 1)$

57. $(-2x)^{-1}(3x - 1)$

58. $(-3x)^{-1}(6 + 2x)$

59. $(xy)^{-1}(x + y)$

60. $(-xy)^{-1}(2x - 3y)$

In Section 1-1, we showed how the fraction a/b is defined as the product of a and the inverse of b:

$$\frac{a}{b} = ab^{-1} \qquad (b \neq 0).$$

In particular,

$$\frac{1}{b} = b^{-1}.$$

5. Evaluate (a) $\dfrac{2}{3} \cdot \dfrac{7}{3}$; (b) $\dfrac{x}{2} \cdot \dfrac{7}{5}$.

From this definition it is possible to derive all the properties which are commonly used in calculating with fractions. In this section, we shall briefly discuss such calculations.*

Multiplication of Fractions

The product of two fractions is obtained by first multiplying the two numerators and then the two denominators.

$$\left(\frac{a}{b}\right)\left(\frac{c}{d}\right) = \frac{ac}{bd}$$

EXAMPLE 1

(a) $\left(\dfrac{2}{3}\right)\left(\dfrac{5}{9}\right) = \dfrac{2 \cdot 5}{3 \cdot 9} = \dfrac{10}{27}$

(b) $\left(\dfrac{2x}{3}\right)\left(\dfrac{4}{y}\right) = \dfrac{(2x)4}{3 \cdot y} = \dfrac{8x}{3y}$

(c) $3x\left(\dfrac{4}{5y}\right) = \left(\dfrac{3x}{1}\right)\left(\dfrac{4}{5y}\right) = \dfrac{(3x) \cdot 4}{1 \cdot (5y)} = \dfrac{12x}{5y}$ ☞ **5**

Division of Fractions

In order to divide one fraction by another, the second fraction is inverted and then multiplied by the first. In other words,

$$\left(\frac{a}{b}\right) \div \left(\frac{c}{d}\right) = \left(\frac{a}{b}\right)\left(\frac{d}{c}\right) = \frac{ad}{bc}.$$

***Answer** (a) $\dfrac{14}{9}$; (b) $\dfrac{7x}{10}$.

---

*Proofs of the quoted properties are given as a series of theorems at the end of this section.

EXAMPLE 2

(a) $\left(\dfrac{3}{5}\right) \div \left(\dfrac{7}{9}\right) = \left(\dfrac{3}{5}\right)\left(\dfrac{9}{7}\right) = \dfrac{27}{35}$

(b) $\left(\dfrac{3x}{2}\right) \div \left(\dfrac{4}{y}\right) = \left(\dfrac{3x}{2}\right)\left(\dfrac{y}{4}\right) = \dfrac{3xy}{8}$

(c) $5y \div \left(\dfrac{6}{5x}\right) = \left(\dfrac{5y}{1}\right)\left(\dfrac{5x}{6}\right) = \dfrac{25xy}{6}$

(d) $\left(\dfrac{3}{2x}\right) \div (2y) = \left(\dfrac{3}{2x}\right) \div \left(\dfrac{2y}{1}\right) = \left(\dfrac{3}{2x}\right)\left(\dfrac{1}{2y}\right) = \dfrac{3}{4xy}$

(e) $\left(\dfrac{a}{b}\right)^{-1} = 1 \div \left(\dfrac{a}{b}\right) = 1 \cdot \dfrac{b}{a} = \dfrac{b}{a}$

☛ **6.** Evaluate
(a) $\dfrac{2}{3} \div \dfrac{3}{2}$; (b) $\dfrac{x}{2} \div \dfrac{7}{5}$.

(That is, the reciprocal of any fraction is obtained by exchanging the numerator and denominator of the fraction.)   ☛ **6**

In view of this last result, we can reword the rule above for division of fractions: *to divide by a fraction, you must multiply by its reciprocal.*

Cancelation of Common Factors

The numerator and denominator of any fraction can be multiplied or divided by any *nonzero* number, without changing the value of the fraction.

$$\dfrac{a}{b} = \dfrac{ac}{bc} \qquad (c \neq 0)$$

EXAMPLE 3

(a) $\dfrac{a}{b} = \dfrac{2a}{2b}$

(b) $\dfrac{3}{5} = \dfrac{6}{10} = \dfrac{9}{15} = \dfrac{-12}{-20} = \cdots$

(c) $\dfrac{5x}{6} = \dfrac{10x^2}{12x}$ (provided that $x \neq 0$)

Answer (a) $\dfrac{4}{9}$; (b) $\dfrac{5x}{14}$.

This property of fractions can be used in order to reduce a fraction to its **lowest terms,** which means dividing the numerator and denominator by all their common factors. (This is also called **simplifying the fraction.**)

EXAMPLE 4

$$\text{(a)} \quad \frac{70}{84} = \frac{2 \cdot 5 \cdot 7}{2 \cdot 2 \cdot 3 \cdot 7} = \frac{\cancel{2} \cdot 5 \cdot \cancel{7}}{\cancel{2} \cdot 2 \cdot 3 \cdot \cancel{7}} = \frac{5}{2 \cdot 3} = \frac{5}{6}$$

Observe that the numerator and denominator are first written in terms of their prime factors and then numerator and denominator are divided by those factors which are common to both numbers, namely 2 and 7. (This process is sometimes called *canceling*.)

$$\text{(b)} \quad \frac{6x^2y}{8xy^2} = \frac{2 \cdot 3 \cdot x \cdot x \cdot y}{2 \cdot 2 \cdot 2 \cdot x \cdot y \cdot y} = \frac{\cancel{2} \cdot 3 \cdot \cancel{x} \cdot x \cdot \cancel{y}}{\cancel{2} \cdot 2 \cdot 2 \cdot \cancel{x} \cdot y \cdot \cancel{y}} = \frac{3x}{4y}$$

$$(xy \neq 0)$$

In this example, the numerator and denominator were divided by $2xy$ in the simplification.

$$\text{(c)} \quad \frac{2x(x+1)}{4y(x+1)} = \frac{x}{2y} \qquad (x + 1 \neq 0)$$

☛ 7. Evaluate

(a) $\dfrac{2}{3} \cdot \dfrac{15}{4}$, (b) $\dfrac{x}{2} \div \dfrac{3x}{8y}$

Here the common factor of $2(x + 1)$ is divided from numerator and denominator. ☛ 7

Addition and Subtraction of Fractions

When two fractions have a common denominator, they may be added by simply adding their numerators.

$$\frac{a}{c} + \frac{b}{c} = \frac{a + b}{c}$$

A similar rule applies for subtraction:

$$\frac{a}{c} - \frac{b}{c} = \frac{a - b}{c}.$$

EXAMPLE 5

$$\text{(a)} \quad \frac{5}{12} + \frac{11}{12} = \frac{5 + 11}{12} = \frac{16}{12} = \frac{4}{3}$$

$$\text{(b)} \quad \frac{3}{2x} - \frac{5}{2x} = \frac{3 - 5}{2x} = \frac{-2}{2x} = -\frac{1}{x}$$

(Note the cancelation of common factors in arriving at the final answers.)

Answer (a) $\dfrac{5}{2}$; (b) $\dfrac{4y}{3}$.

When two fractions with unequal denominators are to be added or subtracted, the fractions must first be rewritten with the same denominator.

EXAMPLE 6 Simplify:

$$\text{(a) } \frac{5}{6} + \frac{1}{2}; \qquad \text{(b) } \frac{5}{6} - \frac{3}{4}$$

Solution

(a) We can write $\frac{1}{2} = \frac{1 \cdot 3}{2 \cdot 3} = \frac{3}{6}$. Then both fractions have the same denominator, so we can add.

$$\frac{5}{6} + \frac{1}{2} = \frac{5}{6} + \frac{3}{6} = \frac{5 + 3}{6} = \frac{8}{6} = \frac{4}{3}$$

(b) In part (a), we multiplied the numerator and denominator of $\frac{1}{2}$ by 3 to give it a denominator equal to that of the other fraction. In this part, both fractions must be changed to obtain a common denominator. We write

$$\frac{5}{6} = \frac{10}{12} \quad \text{and} \quad \frac{3}{4} = \frac{9}{12}.$$

Therefore

$$\frac{5}{6} - \frac{3}{4} = \frac{10}{12} - \frac{9}{12} = \frac{10 - 9}{12} = \frac{1}{12}.$$

☛ **8.** What is the LCD in each case?

(a) $\frac{2}{3}$ and $\frac{5}{6}$; (b) $\frac{1}{2xy}$ and $\frac{x}{8y}$.

In general, when adding or subtracting fractions with different denominators, we first replace each fraction by an equivalent fraction having some common denominator. To keep the numbers as small as possible, we choose the smallest possible common denominator, called the **least common denominator** (LCD). We would still get the right answer by using a larger common denominator, but it is preferable to use the smallest possible denominator. For example, in part (b) of Example 6, we could use 24 as a common denominator:

$$\frac{5}{6} - \frac{3}{4} = \frac{20}{24} - \frac{18}{24} = \frac{20 - 18}{24} = \frac{2}{24} = \frac{1}{12}.$$

The final answer is the same, but we had to work with bigger numbers.

In order to find the LCD of two or more fractions, the two denominators must first be written in terms of their prime factors. The LCD is then formed by taking all of the prime factors that occur in any of the denominators. Each such prime factor is included as many times as it occurs in any one denominator. For example, to find the LCD of $\frac{5}{6}$ and $\frac{3}{4}$, we write the denominators as $6 = 2 \cdot 3$ and $4 = 2 \cdot 2$. The prime factors that occur are 2 and 3, but 2 occurs twice in one denominator. So the LCD is $2 \cdot 2 \cdot 3 = 12$.

As a second example, consider the LCD of $5/12x$ and $7/10x^2y$. We write

$$12x = 2 \cdot 2 \cdot 3 \cdot x \quad \text{and} \quad 10x^2y = 2 \cdot 5 \cdot x \cdot x \cdot y.$$

Taking each factor the greatest number of times it occurs, we have

Answer (a) 6; (b) $8xy$.

$$\text{LCD} = 2 \cdot 2 \cdot 3 \cdot 5 \cdot x \cdot x \cdot y = 60x^2y. \quad ☛ 8$$

EXAMPLE 7 Simplify:

$$\text{(a) } \frac{x}{6} + \frac{3y}{4}; \qquad \text{(b) } \frac{1}{9x} - \frac{1}{6}; \qquad \text{(c) } \frac{a}{c} + \frac{b}{d};$$

$$\text{(d) } \frac{4a}{5b - \dfrac{b}{3}}; \qquad \text{(e) } 3x \div \left(\frac{1}{3x^2} - \frac{3}{4xy} \right)$$

Solution

(a) The LCD is 12.

$$\frac{x}{6} = \frac{2x}{12} \qquad \text{and} \qquad \frac{3y}{4} = \frac{3(3y)}{12} = \frac{9y}{12}$$

Therefore

$$\frac{x}{6} + \frac{3y}{4} = \frac{2x}{12} + \frac{9y}{12} = \frac{2x + 9y}{12}$$

(b) The LCD in this case is $18x$, so

$$\frac{1}{9x} = \frac{2}{18x} \qquad \text{and} \qquad \frac{1}{6} = \frac{3x}{18x}.$$

Then

$$\frac{1}{9x} - \frac{1}{6} = \frac{2}{18x} - \frac{3x}{18x} = \frac{2 - 3x}{18x}.$$

☞ **9.** Evaluate and simplify

(a) $\dfrac{2}{3} + \dfrac{5}{4};$ (b) $\dfrac{x}{2y} - \dfrac{7x}{8y}$

(c) The LCD in this case is cd.

$$\frac{a}{c} + \frac{b}{d} = \frac{ad}{cd} + \frac{bc}{cd} = \frac{ad + bc}{cd} \qquad ☞ 9$$

(d) Here we have a fraction whose denominator itself involves a fraction. First we simplify the denominator:

$$5b - \frac{b}{3} = \frac{15b - b}{3} = \frac{14b}{3}.$$

The given expression is then

$$\frac{4a}{14b/3} = 4a\left(\frac{14b}{3}\right)^{-1} = 4a\left(\frac{3}{14b}\right) = \frac{6a}{7b}.$$

(e) First we simplify the expression in parentheses. The LCD is $12x^2y$.

$$\frac{1}{3x^2} - \frac{3}{4xy} = \frac{4y}{12x^2y} - \frac{9x}{12x^2y} = \frac{4y - 9x}{12x^2y}.$$

Answer (a) $\dfrac{23}{12};$ (b) $-\dfrac{3x}{8y}.$

Therefore the given expression is equal to

$$3x \div \left(\frac{4y - 9x}{12x^2y}\right) = \frac{3x}{1} \cdot \frac{12x^2y}{4y - 9x} = \frac{36x^3y}{4y - 9x}$$

(where $x^3 = x \cdot x^2 = x \cdot x \cdot x$).

Proofs of Theorems

We conclude this section by giving proofs of the basic properties of fractions which have been used in the above examples.

THEOREM 1

$$\left(\frac{1}{b}\right)\left(\frac{1}{d}\right) = \frac{1}{bd}$$

PROOF By definition, $\left(\frac{1}{b}\right) = b^{-1}$ and $\left(\frac{1}{d}\right) = d^{-1}$, so that

$$\left(\frac{1}{b}\right)\left(\frac{1}{d}\right) = b^{-1}d^{-1}.$$

But

$$(b^{-1}d^{-1})(bd) = (b^{-1}b) \cdot (d^{-1}d) \qquad \text{(using the associative and}$$
$$= 1 \cdot 1 = 1. \qquad \qquad \text{commutative properties)}$$

Therefore $b^{-1}d^{-1}$ must be the multiplicative inverse of bd, that is,

$$b^{-1}d^{-1} = \frac{1}{bd}.$$

as required.

Note This result can be rewritten as $(bd)^{-1} = b^{-1}d^{-1}$.

THEOREM 2

$$\left(\frac{a}{b}\right)\left(\frac{c}{d}\right) = \frac{ac}{bd}$$

PROOF

$$\frac{a}{b} = ab^{-1} = a\left(\frac{1}{b}\right)$$

and

$$\frac{c}{d} = c\left(\frac{1}{d}\right).$$

Therefore, using the commutative and associative properties, we can write

$$\left(\frac{a}{b}\right)\left(\frac{c}{d}\right) = a\left(\frac{1}{b}\right) \cdot c\left(\frac{1}{d}\right) = ac \cdot \left(\frac{1}{b} \cdot \frac{1}{d}\right)$$

$$= ac\left(\frac{1}{bd}\right) \qquad \text{(by Theorem 1)}$$

$$= \frac{ac}{bd}$$

as required.

THEOREM 3

$$\left(\frac{a}{b}\right)^{-1} = \frac{b}{a}$$

PROOF By definition, $a/b = ab^{-1}$. Therefore, by Theorem 1,

$$\left(\frac{a}{b}\right)^{-1} = (ab^{-1})^{-1} = a^{-1}(b^{-1})^{-1}.$$

But $(b^{-1})^{-1} = b$, so

$$\left(\frac{a}{b}\right)^{-1} = a^{-1}b = ba^{-1} = \frac{b}{a}$$

as required.

THEOREM 4

$$\left(\frac{a}{b}\right) \div \left(\frac{c}{d}\right) = \left(\frac{a}{b}\right) \cdot \left(\frac{d}{c}\right)$$

PROOF By definition, $x \div y = xy^{-1}$. Therefore we have the following:

$$\left(\frac{a}{b}\right) \div \left(\frac{c}{d}\right) = \left(\frac{a}{b}\right) \cdot \left(\frac{c}{d}\right)^{-1} = \left(\frac{a}{b}\right) \cdot \left(\frac{d}{c}\right) \qquad \text{(by Theorem 3)}$$

THEOREM 5

$$\frac{a}{b} = \frac{ac}{bc} \qquad (c \neq 0)$$

PROOF For any $c \neq 0$, the fraction $c/c = 1$, since, by definition $c/c = cc^{-1}$. Therefore, by Theorem 2,

$$\frac{ac}{bc} = \left(\frac{a}{b}\right) \cdot \left(\frac{c}{c}\right) = \frac{a}{b} \cdot 1 = \frac{a}{b}$$

as required.

THEOREM 6

$$\frac{a}{c} + \frac{b}{c} = \frac{a+b}{c} \qquad (c \neq 0)$$

PROOF By definition,

$$\frac{a}{c} = ac^{-1} \qquad \text{and} \qquad \frac{b}{c} = bc^{-1}.$$

Therefore

$$\frac{a}{c} + \frac{b}{c} = ac^{-1} + bc^{-1} = (a+b)c^{-1} \qquad \text{(by the distributive property)}$$

$$= \frac{a+b}{c}$$

as required.

EXERCISES 1-2

1. State whether each of the following is true or false. Replace each false statement with a corresponding true statement.

a. $\dfrac{3}{x} + \dfrac{4}{x} = \dfrac{7}{x}$

b. $\dfrac{x}{3} + \dfrac{x}{4} = \dfrac{x}{7}$

c. $\dfrac{a}{b} + \dfrac{c}{d} = \dfrac{a+c}{b+d}$

d. $\dfrac{a}{b} \cdot \left(\dfrac{c}{d} \cdot \dfrac{e}{f}\right) = \dfrac{ace}{bdf}$

e. $\left(\dfrac{a}{b} \div \dfrac{c}{d}\right) \div \dfrac{e}{f} = \dfrac{adf}{bce}$

f. $\dfrac{a}{b} \div \left(\dfrac{c}{d} \div \dfrac{e}{f}\right) = \dfrac{adf}{bce}$

g. $\dfrac{1}{a} + \dfrac{1}{b} = \dfrac{1}{a+b}$

h. $\dfrac{\not{x}}{\not{x} + y} = \dfrac{1}{1+y}$

i. $\dfrac{6}{7} \cdot \dfrac{8}{9} = \dfrac{6 \cdot 9 + 7 \cdot 8}{7 \cdot 9}$

j. $\dfrac{1+2+3+4+5}{2+4+6+8+10} = \dfrac{1}{2}$

(2–58) Evaluate the following expressions. Write the answers in simplest terms.

2. $\dfrac{2}{9} \cdot \dfrac{6}{5}$

3. $\left(\dfrac{8}{3}\right)\left(\dfrac{15}{4}\right)$

4. $\dfrac{3}{4} \cdot \dfrac{8}{5} \cdot \dfrac{4}{9}$

5. $\dfrac{2}{5} \cdot \dfrac{3}{6} \cdot \dfrac{10}{7}$

6. $\left(\dfrac{3x}{25}\right)\left(\dfrac{25}{9x}\right)$

7. $\left(\dfrac{14x}{15y}\right)\left(\dfrac{25y}{24}\right)$

8. $7x^2\left(\dfrac{6y}{21x}\right)$

9. $\left(-\dfrac{2x}{3y}\right)(-5xy)$

10. $\left(\dfrac{18}{11}\right) \div \left(\dfrac{8}{33}\right)$

11. $\left(\dfrac{14}{3}\right) \div \left(\dfrac{6}{15}\right)$

12. $\dfrac{4}{9} \div \left(\dfrac{2}{3} \cdot 8\right)$

13. $\left(\dfrac{12}{25} \cdot \dfrac{15}{7}\right) \div \dfrac{20}{7}$

14. $\left(\dfrac{7x}{10}\right) \div \left(\dfrac{21x}{5}\right)$

15. $(2x) \div \left(\dfrac{3xy}{5}\right)$

16. $4 \div \left(\dfrac{8}{9x}\right)$

17. $\left(\dfrac{3}{8x}\right) \div \left(\dfrac{4x}{15}\right)$

18. $\left(\dfrac{3x^2}{20} \cdot 4y\right) \div \left(\dfrac{6xy}{25}\right)$

19. $\left(\dfrac{5x}{2} \cdot \dfrac{3y}{4}\right) \div \left(\dfrac{x^2y}{12}\right)$

20. $8xy \div \left(\dfrac{2x}{3} \cdot \dfrac{2x}{5y}\right)$

21. $6x^2 \div \left(\dfrac{4x}{y} \cdot \dfrac{3y^2}{2}\right)$

22. $\left(\dfrac{8}{9t} \div \dfrac{1}{3st}\right) \cdot \dfrac{s}{4}$

23. $\left(\dfrac{3}{4xy} \div \dfrac{x}{y}\right) \cdot \dfrac{2xy}{9}$

44. $\dfrac{a}{3b} - 2\left(\dfrac{a}{b} - \dfrac{b}{2a}\right)$

45. $\left(\dfrac{x}{2} + \dfrac{2}{x}\right) \div \left(\dfrac{6}{x}\right)$

24. $\left(\dfrac{2}{x} \div \dfrac{z}{2}\right) \div \dfrac{4}{z}$

25. $\left(\dfrac{2xt}{3} \div \dfrac{x}{4t}\right) \div \dfrac{2t}{3}$

46. $\left(\dfrac{x}{9y} + \dfrac{1}{6xy}\right) \div \left(\dfrac{1}{3xy}\right)$

47. $\left(\dfrac{1}{4} - \dfrac{2}{5}\right) \div \left(\dfrac{1}{2} - \dfrac{1}{5}\right)$

26. $\dfrac{2}{z} \div \left(\dfrac{z}{2} \div \dfrac{4}{z}\right)$

27. $\dfrac{2xt}{3} \div \left(\dfrac{x}{4t} \div \dfrac{2t}{3}\right)$

48. $\left(\dfrac{2}{3} + \dfrac{1}{12}\right) \cdot \left(\dfrac{7}{10} + \dfrac{1}{4}\right)$

28. $\dfrac{1}{6} - \dfrac{1}{2}$

29. $\dfrac{1}{10} + \dfrac{1}{15}$

49. $\dfrac{\frac{1}{2} - \frac{1}{3}}{\frac{1}{4} + \frac{1}{5}}$

50. $\dfrac{\frac{8}{5} + \frac{2}{3}}{2 + \frac{4}{7}}$

30. $\dfrac{4x}{5} - \dfrac{x}{10}$

31. $\dfrac{1}{x} + \dfrac{1}{2x}$

51. $\dfrac{\frac{1}{3} - \frac{1}{4}}{\frac{1}{5} - \frac{1}{6}}$

52. $\dfrac{2 - \frac{3}{4}}{3 + \frac{1}{8}}$

32. $\dfrac{x}{2} + \dfrac{x}{3}$

33. $\dfrac{y}{2x} + \dfrac{1}{3x}$

53. $\dfrac{7x - \dfrac{2x}{3}}{15y - \dfrac{y}{3}}$

54. $\dfrac{\dfrac{1}{2x} - \dfrac{1}{3x}}{\dfrac{1}{4y} - \dfrac{1}{5y}}$

34. $\dfrac{a}{6b} - \dfrac{a}{2b}$

35. $\dfrac{a}{6b} + \dfrac{2a}{9b}$

36. $\dfrac{7}{6x} + \dfrac{3}{4x^2}$

37. $\dfrac{3y}{10x^2} - \dfrac{1}{6x}$

55. $\dfrac{\left(\dfrac{2a}{3b}\right)\left(\dfrac{4b}{5}\right) + a}{2b + \dfrac{b}{15}}$

56. $\dfrac{\left(\dfrac{5p}{2q}\right)\left(\dfrac{p}{3}\right) + \dfrac{p^2}{8q}}{4p + \dfrac{p}{12}}$

38. $\dfrac{x}{p^2} + \dfrac{y}{pq}$

39. $\dfrac{x}{y} + \dfrac{y}{z} + \dfrac{z}{x}$

40. $\dfrac{x}{y} - \dfrac{y}{x}$

41. $\dfrac{x^2}{3y} + 4y$

57. $\left(\dfrac{a}{b} + \dfrac{2a}{3b}\right) \div \left[\left(\dfrac{3x}{8}\right) \div \left(\dfrac{x}{9}\right) + \dfrac{1}{4}\right]$

42. $\dfrac{1}{6} + \left(\dfrac{2}{x} + \dfrac{x}{2}\right)$

43. $\dfrac{1}{6} - \left(\dfrac{2}{x} - \dfrac{x}{2}\right)$

58. $\left(\dfrac{xy}{6}\right) \div \left[\left(\dfrac{2}{3}\right) \div \left(\dfrac{x}{6}\right) - \dfrac{3x}{4}\right]$

◣ 1-3 EXPONENTS

If m is a *positive integer*, then a^m (read *a to the power m* or *the mth power of a*) is defined to be the product of m a's multiplied together. Thus

$$a^m = a \cdot a \cdot a \cdots \cdot a.$$

In this product, the factor a appears m times. For example,

$$2^4 = 2 \cdot 2 \cdot 2 \cdot 2 = 16 \qquad \text{(four factors of 2)}$$

$$3^5 = 3 \cdot 3 \cdot 3 \cdot 3 \cdot 3 = 243 \qquad \text{(five factors of 3)}.$$

In the expression a^m, m is called the **power** or **exponent** and a the **base**. Thus, in 2^4 (the fourth power of 2), 2 is the base and 4 is the power or exponent. This definition of a^m when the exponent is a positive integer holds for all real values of a.

Observe the pattern in Table 1, in which several powers of 5 are given in decreasing order. Let us try to complete the table. We notice that every time the exponent is decreased by 1, the number in the right column is *divided* by 5.

☛ 10. Evaluate
(a) $\left(-\frac{1}{2}\right)^0$;   (b) $\left(-\frac{1}{2}\right)^{-3}$.

This suggests that the table should be completed by continuing to divide by 5 with each reduction in the exponent. Thus we are led to the following:

TABLE 1

5^4	625
5^3	125
5^2	25
5^1	5
5^0	?
5^{-1}	?
5^{-2}	?
5^{-3}	?
5^{-4}	?

$$5^1 = 5 \qquad 5^0 = 1 \qquad 5^{-1} = \frac{1}{5} = \frac{1}{5^1}$$

$$5^{-2} = \frac{1}{25} = \frac{1}{5^2} \qquad 5^{-3} = \frac{1}{125} = \frac{1}{5^3} \qquad 5^{-4} = \frac{1}{625} = \frac{1}{5^4}$$

This pattern naturally leads to the following definition of a^m when the exponent m is zero or a negative integer.

DEFINITION If $a \neq 0$, then $a^0 = 1$, and if m is any *positive* integer (so that $-m$ is a *negative* integer), then

$$a^{-m} = \frac{1}{a^m}.$$

For example, $4^0 = 1$, $\left(\frac{3}{7}\right)^0 = 1$, $(-5)^0 = 1$, and so on. Also,

Answer (a) 1;   (b) $-2^3 = -8$.

$$3^{-4} = \frac{1}{3^4} = \frac{1}{81} \qquad \text{and} \qquad (-2)^{-5} = \frac{1}{(-2)^5} = \frac{1}{-32} = -\frac{1}{32}. \quad ☛ \textbf{10}$$

From these definitions, it is possible to establish a series of properties called the **laws of exponents.** These are as follows.

Property 1
$$a^m \cdot a^n = a^{m+n}$$

That is, *to multiply two powers with the same base we must add the two exponents.* This result holds for any real number a, except that if either m or n is negative, we require $a \neq 0$.

EXAMPLE 1

(a) $5^2 \cdot 5^3 = 5^{2+3} = 5^5$

We can verify that this is correct by expanding the two powers in the product.

$$5^2 \cdot 5^3 = (5 \cdot 5) \cdot (5 \cdot 5 \cdot 5) = 5 \cdot 5 \cdot 5 \cdot 5 \cdot 5 = 5^5$$

▶ 11. Simplify
(a) $4^3 \cdot 4^{-5}$; (b) $x^4 \cdot x^{-6} \cdot x^2$.

Answer (a) $\frac{1}{16}$; (b) 1.

(b) $x^5 \cdot x^{-3} = x^{5+(-3)} = x^2$

Again, we can verify this result by expanding the two powers.

$$x^5 \cdot x^{-3} = (x \cdot x \cdot x \cdot x \cdot x)\left(\frac{1}{x \cdot x \cdot x}\right) = x \cdot x = x^2 \quad \text{▶} \; 11$$

Property 2

$$\frac{a^m}{a^n} = a^{m-n} \qquad (a \neq 0)$$

That is, *to divide one power by another with the same base, subtract the exponent in the denominator from the exponent in the numerator.*

EXAMPLE 2

▶ 12. Simplify
(a) $3^3 \div 3^{-2}$; (b) $x^4 \div (x^{-6} \cdot x^2)$.

Answer (a) $3^5 = 243$; (b) x^8.

(a) $\dfrac{5^7}{5^3} = 5^{7-3} = 5^4$

(b) $\dfrac{4^3}{4^{-2}} = 4^{3-(-2)} = 4^{3+2} = 4^5$

(c) $\dfrac{3^{-2}}{3} = \dfrac{3^{-2}}{3^1} = 3^{-2-1} = 3^{-3}$

(d) $\dfrac{x^2 \cdot x^{-4}}{x^{-3}} = \dfrac{x^{2-4}}{x^{-3}} = x^{2-4-(-3)} = x^1 = x \quad \text{▶} \; 12$

Property 3

$$(a^m)^n = a^{mn} \qquad (a \neq 0 \text{ if } m \text{ or } n \text{ is negative or zero})$$

That is, *a power raised to a power is equal to the base raised to the product of the two exponents.*

EXAMPLE 3

(a) $(3^3)^2 = 3^{3 \cdot 2} = 3^6$

We can see that this is correct, since

$$(3^3)^2 = 3^3 \cdot 3^3 = 3^{3+3} = 3^6.$$

▶ 13. Simplify
(a) $3^3 \cdot (3^2)^{-2}$; (b) $(x^4)^4 \div (x^{-3})^{-3}$.

Answer (a) $3^{-1} = \frac{1}{3}$; (b) x^7.

(b) $(4^{-2})^{-4} = 4^{(-2)(-4)} = 4^8$

(c) $x^5(x^{-2})^{-1} = x^5 \cdot x^{(-2)(-1)} = x^5 \cdot x^2 = x^{5+2} = x^7$

(d) $\dfrac{(x^2)^{-2}}{(x^{-2})^{-2}} = \dfrac{x^{(2)(-2)}}{x^{(-2)(-2)}} = \dfrac{x^{-4}}{x^4} = x^{-4-4} = x^{-8}$

(e) $\dfrac{1}{x^{-p}} = (x^{-p})^{-1} = x^{(-p)(-1)} = x^p \quad \text{▶} \; 13$

In an expression such as $3c^5$, the base is c, not $3c$. If we want the base $3c$ we must enclose it in parentheses and write $(3c)^5$. For example, $3 \cdot 2^3 = 3 \cdot 8 = 24$, not the same as $(3 \cdot 2)^3 = 6^3 = 216$. For the case when the base is a product, we have the following property.

> **Property 4**
> $$(ab)^m = a^m b^m \qquad (ab \neq 0 \text{ if } m \leq 0)$$

That is, *the product of two numbers all raised to the mth power is equal to the product of the mth powers of the two numbers.* ☛ **14**

EXAMPLE 4

(a) $6^4 = (2 \cdot 3)^4 = 2^4 \cdot 3^4$

(b) $(x^2 y)^4 = (x^2)^4 y^4 = x^8 y^4$

(c) $(3a^2 b^{-3})^2 = 3^2 (a^2)^2 (b^{-3})^2 = 9a^4 b^{-6}$

(d) $\dfrac{(xy^3)^{-2}}{(x^2 y)^{-4}} = \dfrac{x^{-2}(y^3)^{-2}}{(x^2)^{-4} y^{-4}} = \dfrac{x^{-2} y^{-6}}{x^{-8} y^{-4}} = \dfrac{x^{-2}}{x^{-8}} \cdot \dfrac{y^{-6}}{y^{-4}} = x^{-2-(-8)} y^{-6-(-4)} = x^6 y^{-2}$

> **Property 5**
> $$\left(\frac{a}{b}\right)^m = \frac{a^m}{b^m} \qquad (b \neq 0 \text{ and } a \neq 0 \text{ if } m \leq 0)$$

That is, *the quotient of two numbers all raised to the mth power is equal to the quotient of the mth powers of the two numbers.*

EXAMPLE 5

(a) $\left(\dfrac{3}{2}\right)^4 = \dfrac{3^4}{2^4}$; (b) $\left(\dfrac{x}{y}\right)^5 = \dfrac{x^5}{y^5} = x^5 y^{-5}$;

(c) $x^3 \left(\dfrac{y}{x^2}\right)^{-2} = x^3 \dfrac{y^{-2}}{(x^2)^{-2}} = x^3 \dfrac{y^{-2}}{x^{-4}} = x^{3-(-4)} y^{-2} = x^7 y^{-2}$ ☛ **15**

EXAMPLE 6 Simplify the following, eliminating parentheses and negative exponents.

(a) $\dfrac{(ax)^5}{x^{-7}}$; (b) $\dfrac{(x^{-2})^2}{(x^2 z^3)^3}$; (c) $x^4(2x - 3x^{-2})$;

(d) $(x^{-1} + y^{-1})^{-1}$; (e) $\dfrac{x^{-1} + y^{-1}}{(xy)^{-1}}$

Solution

(a) $\dfrac{(ax)^5}{x^{-7}} = \dfrac{a^5 x^5}{x^{-7}} = a^5 x^{5-(-7)} = a^5 x^{12}$

(b) $\dfrac{(x^{-2})^2}{(x^2z^3)^3} = \dfrac{x^{(-2)(2)}}{(x^2)^3(z^3)^3} = \dfrac{x^{-4}}{x^6z^9} = \dfrac{1}{x^{10}z^9}$

Note that if we wish to avoid negative exponents, both factors must be left in the denominator.

(c) $x^4(2x - 3x^{-2}) = x^4(2x) - x^4(3x^{-2}) = 2x^{4+1} - 3x^{4-2} = 2x^5 - 3x^2$

(d) First we must simplify the expression inside the parentheses. The common denominator is xy.

$$x^{-1} + y^{-1} = \frac{1}{x} + \frac{1}{y} = \frac{y}{xy} + \frac{x}{xy} = \frac{y + x}{xy}$$

■ 16. It would be *completely* *incorrect* in Example 6(d) if we were to write $(x^{-1} + y^{-1})^{-1} = (x^{-1})^{-1} + (y^{-1})^{-1} = x + y$. Can you see why this is wrong? Try giving x and y two values, such as 2 and 4.

Now recall that the reciprocal of a fraction is obtained by interchanging the numerator and denominator. So

$$(x^{-1} + y^{-1})^{-1} = \left(\frac{y + x}{xy}\right)^{-1} = \frac{xy}{y + x}.$$

(e) $\dfrac{x^{-1} + y^{-1}}{(xy)^{-1}} = \dfrac{x^{-1} + y^{-1}}{x^{-1}y^{-1}} = \dfrac{x^{-1}}{x^{-1}y^{-1}} + \dfrac{y^{-1}}{x^{-1}y^{-1}} = \dfrac{1}{y^{-1}} + \dfrac{1}{x^{-1}} = y + x$

Alternative Solution

$$\frac{x^{-1} + y^{-1}}{(xy)^{-1}} = (x^{-1} + y^{-1}) \cdot xy$$

$$= x^{-1} \cdot xy + y^{-1} \cdot xy \qquad \text{(distributive property)}$$

$$= 1 \cdot y + 1 \cdot x = y + x \qquad ☛ 16$$

EXERCISES 1-3

(1–61) Simplify the following, avoiding all parentheses and negative exponents in the final answer.

1. $(2^5)^2$

2. $(3^4)^3$

3. $(a^3)^7$

4. $(x^4)^5$

5. $(-x^2)^5$

6. $(-x^5)^2$

7. $y^2 \cdot y^5$

8. $x^7 \cdot x^4$

9. $a^3 \cdot a^{-5}$

10. $b^{-2} \cdot b^6$

11. $(3x)^2x^{-7}$

12. $(4x)^{-2}x^4$

13. $(2x)^2(2x^{-1})^3$

14. $\dfrac{x^3}{2}(4x^{-1})^2$

15. $(x^2yz)^3(xy)^4$

16. $(3yz^2)^2(y^3z)^3$

17. $(x^{-2}y)^{-2}$

18. $(ab^{-3})^{-1}$

19. $(xy^2z^3)^{-1}(xyz)^3$

20. $(x^2pq^2)^2(xp^2)^{-1}$

21. $\dfrac{(2^4)^2}{4^2}$

22. $\dfrac{(3^3)^2}{3^5}$

23. $\left(\dfrac{1}{3}\right)^{-2} \div 3^{-4}$

24. $\left(\dfrac{1}{5}\right)^3 \div 5^{-2}$

25. $\dfrac{x^5}{x^{-2}}$

26. $\dfrac{y^{-3}}{y^{-7}}$

27. $\dfrac{(x^2)^3}{x^4}$

28. $\dfrac{z^{-8}}{(z^2)^4}$

29. $\dfrac{(a^{-2})^6}{(a^4)^{-3}}$

30. $\dfrac{(b^{-7})^2}{(b^3)^3}$

31. $\dfrac{(-x^3)^2}{(-x)^{-3}}$

32. $\dfrac{(-y^{-1})^{-3}}{(-y^2)^{-2}}$

33. $\dfrac{(x^2y)^{-3}}{(xy)^2}$

34. $\dfrac{(ab^{-2})^{-1}}{a^{-2}b^{-1}}$

35. $\dfrac{(-2xy)^3}{x^3y}$

36. $\dfrac{(-ab^2c)^{-1}}{a^{-2}bc^{-1}}$

37. $\dfrac{(-3x)^2}{-3x^2}$

38. $\dfrac{(2x^2y)^{-1}}{(-2x^2y^3)^2}$

39. $\dfrac{(2a^{-1}b^2)^2}{(a^3b)^3}$

40. $\dfrac{(x^{-3}y^4)^3}{(-3x^2y^{-2})^2}$

41. $x^2(x^4 - 2x)$

42. $x^3(x^{-1} - x)$

43. $2x(x^5 + 3x^{-1})$

44. $3x^2(x^4 + 2x^{-3})$

45. $x^4(2x^2 - x - 3x^{-2})$

46. $2x^{-3}(x^5 - 3x^4 + x)$

47. $(2^{-1} + x^{-1})^{-1}$

48. $[(2x)^{-1} + (2y)^{-1}]^{-1}$

49. $(xy)^{-1}(x^{-1} + y^{-1})^{-1}$

50. $(a^{-2} + b^{-2})^{-1}$

51. $\left(\dfrac{7}{x}\right)\left(\dfrac{3}{14x}\right) + \left(\dfrac{3}{2x}\right)^2$

52. $x^{-3}\left(\dfrac{6}{5x}\right)^{-1} - \left(-\dfrac{1}{2x}\right)^2$

53. $\dfrac{3y}{10x^3} + \dfrac{2}{15xy}$

54. $\dfrac{5}{12x^{-3}} - \dfrac{2}{15x^{-2}}$

55. $\dfrac{1}{2x^{-2}} + \dfrac{1}{3x^{-2}}$

56. $\dfrac{1}{4y^{-4}} - \dfrac{1}{3y^{-4}}$

57. $\left(\dfrac{x^3y}{4}\right) \div \left(\dfrac{4}{x} \div \dfrac{6}{y^3}\right)$

58. $\dfrac{x^{-3}}{4x} - \dfrac{x}{6x^5}$

59. $y^{-5}\left(2xy \div \dfrac{x}{3y^2}\right)$

60. $\left(\dfrac{2}{x} + x^{-1}\right) \div \left(\dfrac{x^2}{2} + \dfrac{1}{5x^{-2}}\right)$

61. $x^{-1} \div (x + x^{-1})^{-1}$

◥ 1-4 FRACTIONAL EXPONENTS

Having defined a^m when m is any integer, we shall now extend the definition to the case when m is any rational number. We should like to make this extension in such a way that Properties 1 to 5 of Section 1-3 continue to hold when m and n are no longer integers.

We shall first consider the definition of $a^{1/n}$ where n is a nonzero integer. If Property 3 is to remain true when $m = 1/n$, then it must be true that

$$(a^{1/n})^n = a^{(1/n)n} = a^1 = a.$$

So, if we set $b = a^{1/n}$, it is necessary that $b^n = a$.

EXAMPLE 1

    (a) $8^{1/3} = 2$ since $2^3 = 8$.

    (b) $(-243)^{1/5} = -3$ since $(-3)^5 = -243$.

When n is an even integer, two difficulties arise with this definition of $a^{1/n}$. For example, let $n = 2$ and $a = 4$. Then $b = 4^{1/2}$ if $b^2 = 4$. But there are *two* numbers whose square is equal to 4, namely $b = 2$ and $b = -2$. Thus we need to decide which we mean when we write $b = 4^{1/2}$. In fact, we shall *define* $4^{1/2}$ to mean $+2$.

Second, suppose that a is negative. Then $b = a^{1/2}$ if $b^2 = a$. But the square of any real number (positive, negative, or zero) is never negative. For example, $4^2 = 16$ and $(-3)^2 = 9$, both of which are positive numbers. Thus b^2 is never negative for any real number b, so when $a < 0$, $a^{1/2}$ does not exist in the real numbers. For example, $(-1)^{1/2}$ or $(-\tfrac{4}{3})^{1/2}$ have no meanings as real numbers. This leads us to adopt the following definition.

DEFINITION If n is a positive even integer (such as 2, 4, or 6) and if a is any *nonnegative* real number, then b is said to be the **principal nth root of a** if $b^n = a$ and $b \geq 0$. Thus the principal nth root of a is the *nonnegative* number which, when raised to the nth power, gives the number a. We denote the principal nth root by $b = a^{1/n}$.

If n is a positive *odd* integer (such as 1, 3, or 5) and if a is *any* real number, then b is the *nth root of a* if $b^n = a$, again denoted by $a^{1/n}$. Thus

$$b = a^{1/n} \text{ if } b^n = a; \qquad b \geq 0 \text{ if } n \text{ is even.}$$

☛ **17.** Evaluate the following, if they exist: (a) $(-27)^{1/3}$; (b) $(64)^{1/6}$; (c) $\sqrt[5]{-32}$; (d) $(-\frac{1}{16})^{1/4}$; (e) $\sqrt[6]{-729}$; (f) $\sqrt[101]{-1}$.

The odd roots are defined for all real numbers a, but the even roots are defined only when a is nonnegative.

EXAMPLE 2

    (a) $32^{1/5} = 2$ because $2^5 = 32$.

    (b) $(-216)^{1/3} = -6$ because $(-6)^3 = -216$.

    (c) $16^{1/4} = 2$ because $2^4 = 16$ and $2 > 0$.

    (d) $(729)^{1/6} = 3$ because $3^6 = 729$ and $3 > 0$.

    (e) $1^{1/n} = 1$ for every positive integer n, because $1^n = 1$.

    (f) $(-1)^{1/n} = -1$ for every positive odd integer n, because $(-1)^n = -1$ when n is odd.

    (g) $(-81)^{1/4}$ does not exist, because negative numbers have nth roots only when n is odd.

The symbol $\sqrt[n]{a}$ is also used instead of $a^{1/n}$. The symbol $\sqrt{}$ is called a **radical sign** and $\sqrt[n]{a}$ is often called a **radical.** When $n = 2$, $a^{1/2}$ is denoted simply by $\sqrt{a}$ rather than $\sqrt[2]{a}$: it is called the **square root** of a. Also, $\sqrt[3]{a} = a^{1/3}$ is the third root of a, usually called the **cube root,** $\sqrt[4]{a} = a^{1/4}$ is the fourth root of a, and so on. The results in Example 2 can be restated using this notation:

    (a) $\sqrt[5]{32} = 2$;   (b) $\sqrt[3]{-216} = -6$;   (c) $\sqrt[4]{16} = 2$;

    (d) $\sqrt[6]{729} = 3$;   (e) $\sqrt[n]{1} = 1$ for n a positive integer;

    (f) $\sqrt[n]{-1} = -1$ for n a positive odd integer;

    (g) $\sqrt[4]{-81}$ does not exist.   ☛ 17

Answer (a) -3; (b) 2; (c) -2; (d) and (e) do not exist; (f) -1.

Now we are in a position to define $a^{m/n}$ with a rational exponent m/n.

DEFINITION Let n be a positive integer, m be a nonzero integer, and a be a real number. Then, if $a^{1/n}$ exists, we define

$$a^{m/n} = (a^{1/n})^m$$

That is, *the (m/n)th power of a is the mth power of the nth root of a.*

   Note If n is even, a must be nonnegative. If m is negative, a must be nonzero.

EXAMPLE 3

   (a) $9^{3/2} = (9^{1/2})^3 = 3^3 = 27$

   (b) $4^{-1/2} = (4^{1/2})^{-1} = 2^{-1} = \frac{1}{2}$

   (c) $16^{-3/4} = (16^{1/4})^{-3} = 2^{-3} = \frac{1}{8}$

From part (b) of Example 3, we can generalize to the following result.

$$a^{-1/n} = \frac{1}{\sqrt[n]{a}}$$

This follows since

$$a^{-1/n} = (a^{1/n})^{-1} = \frac{1}{a^{1/n}}.$$

THEOREM If $a^{m/n}$ exists, then

$$a^{m/n} = (a^m)^{1/n}$$

That is, *the (m/n)th power of a is equal to the nth root of the mth power of a.*

   This theorem, which we shall not prove, offers an alternative method of calculating any fractional power.

EXAMPLE 4

   (a) $16^{3/4} = (16^{1/4})^3 = 2^3 = 8$, or $16^{3/4} = (16^3)^{1/4} = (4096)^{1/4} = 8$

   (b) $36^{3/2} = (36^{1/2})^3 = 6^3 = 216$, or $36^{3/2} = (36^3)^{1/2} = (46{,}656)^{1/2} = 216$

   Note: If m/n is not in its lowest terms, then $(a^m)^{1/n}$ may exist while $a^{m/n}$ does not. For example, let $m = 2$, $n = 4$ and $a = -9$. Then

$$(a^m)^{1/n} = [(-9)^2]^{1/4} = 81^{1/4} = 3$$

but $a^{m/n} = (-9)^{2/4} = [(-9)^{1/4}]^2$ does not exist.

   From Examples 3 and 4, it is clear that when evaluating $a^{m/n}$, it is easier to take the nth root first and then raise to the mth power; we then work with smaller numbers. In other words, in practice, to evaluate $a^{m/n}$ we use the definition $(a^{1/n})^m$ rather than $(a^m)^{1/n}$.

With these definitions, it is possible to show that the laws of exponents, which were stated in Section 1-3, remain valid for fractional exponents. Let us restate these laws since they are so important.

---

1. $a^m \cdot a^n = a^{m+n}$ **2.** $\dfrac{a^m}{a^n} = a^{m-n}$ **3.** $(a^m)^n = a^{mn}$

4. $(ab)^m = a^m b^m$ **5.** $\left(\dfrac{a}{b}\right)^m = \dfrac{a^m}{b^m}$

In writing these laws, we must keep in mind that certain restrictions apply: in any power, if the exponent is negative, the base must not be zero; and if the exponent involves an even root, the base must not be negative.

☛ **18.** Simplify (a) $3^{1/3} \cdot 3^{2/3}$; (b) $3^{1/3} \cdot (3^{2/3})^{-2}$; (c) $(x^{1/2})^3 \cdot \sqrt{x}$; (d) $(x^{1/3})^{1/2} \div x^{7/6}$; (e) $(8x)^{2/5} \cdot \left(\dfrac{x}{4}\right)^{3/5}$.

EXAMPLE 5

(a) $5^3 \cdot 5^{7/2} = 5^{3+7/2} = 5^{13/2}$

(b) $4^{-2} \cdot 4^{7/3} = 4^{-2+7/3} = 4^{1/3}$

(c) $\dfrac{4^{7/2}}{(4)^{3/2}} = 4^{7/2-3/2} = 4^2 = 16$

(d) $\dfrac{9^{1/2}}{9^{-2}} = 9^{1/2-(-2)} = 9^{5/2} = (9^{1/2})^5 = 3^5 = 243$

(e) $\dfrac{x^{9/4}}{x^4} = x^{9/4-4} = x^{-7/4}$

(f) $(5^3)^{7/6} = 5^{3(7/6)} = 5^{7/2}$

(g) $(3^{-4/3})^{-6/5} = 3^{(-4/3)(-6/5)} = 3^{8/5}$

(h) $a^{-m} = (a^m)^{-1} = \dfrac{1}{a^m}$ for any rational number m

(i) $(36)^{1/2} = (4 \cdot 9)^{1/2} = 4^{1/2} \cdot 9^{1/2} = 2 \cdot 3 = 6$

(j) $(x^2 y)^{1/2} = (x^2)^{1/2} y^{1/2} = x^{2(1/2)} y^{1/2} = xy^{1/2}$

(k) $(3a^{2/5}b^{-4})^{-1/2} = 3^{-1/2}(a^{2/5})^{-1/2}(b^{-4})^{-1/2} = 3^{-1/2}a^{-1/5}b^2$

(l) $\sqrt[4]{ab} = (ab)^{1/4} = a^{1/4}b^{1/4} = \sqrt[4]{a}\sqrt[4]{b}$

(m) $\sqrt{x/y} = \left(\dfrac{x}{y}\right)^{1/2} = \dfrac{x^{1/2}}{y^{1/2}} = \dfrac{\sqrt{x}}{\sqrt{y}}$

Answer (a) 3; (b) 3^{-1}; (c) x^2; (d) x^{-1}; (e) x.

(n) $\left(\dfrac{8}{27}\right)^{-2/3} = \dfrac{8^{-2/3}}{27^{-2/3}} = \dfrac{(8^{1/3})^{-2}}{(27^{1/3})^{-2}} = \dfrac{2^{-2}}{3^{-2}} = \dfrac{\frac{1}{4}}{\frac{1}{9}} = \left(\dfrac{1}{4}\right)\left(\dfrac{9}{1}\right) = \dfrac{9}{4}$ ☛ **18**

EXAMPLE 6 Find m such that $\dfrac{\sqrt[3]{9}}{27} = 3^m$.

Solution We express both sides as a power of 3.

$$\frac{\sqrt[3]{9}}{27} = \frac{9^{1/3}}{3^3} = \frac{(3^2)^{1/3}}{3^3} = \frac{3^{2/3}}{3^3} = 3^{(2/3)-3} = 3^{-7/3}$$

Therefore $m = -\frac{7}{3}$.

EXAMPLE 7 Evaluate: (a) $\left(1\dfrac{64}{225}\right)^{1/2}$; (b) $\left(\dfrac{64x^3}{7}\right)^{-2/3}$

Solution

(a) $\left(1\dfrac{64}{225}\right)^{1/2} = \left(\dfrac{289}{225}\right)^{1/2} = \left(\dfrac{17^2}{15^2}\right)^{1/2}$

$\qquad\qquad = \left[\left(\dfrac{17}{15}\right)^2\right]^{1/2} \qquad$ (by Law 5)

$\qquad\qquad = \left(\dfrac{17}{15}\right)^{2\cdot(1/2)} \qquad$ (by Law 3)

$\qquad\qquad = \left(\dfrac{17}{15}\right)^1 = 1\dfrac{2}{15}$

(b) $\left(\dfrac{64x^3}{27}\right)^{-2/3} = \left(\dfrac{4^3x^3}{3^3}\right)^{-2/3} = \left[\left(\dfrac{4x}{3}\right)^3\right]^{-2/3} \qquad$ (by Law 5)

$\qquad\qquad = \left(\dfrac{4x}{3}\right)^{-2} = \dfrac{1}{(4x/3)^2} \qquad$ (by Law 3)

$\qquad\qquad = \dfrac{1}{16x^2/9} = \dfrac{9}{16x^2}$

EXAMPLE 8 Simplify the following expression.

$$\frac{4^p \cdot 27^{p/3} \cdot 125^p \cdot 6^{2p}}{8^{p/3} \cdot 9^{3p/2} \cdot 10^{3p}}$$

Solution In expressions such as this, it is usually a good idea to express all the bases in terms of their prime factors.

$\dfrac{4^p \cdot 27^{p/3} \cdot 125^p \cdot 6^{2p}}{8^{p/3} \cdot 9^{3p/2} \cdot 10^{3p}} = \dfrac{(2^2)^p \cdot (3^3)^{p/3} \cdot (5^3)^p \cdot (2 \cdot 3)^{2p}}{(2^3)^{p/3} \cdot (3^2)^{3p/2} \cdot (2 \cdot 5)^{3p}}$

$\qquad\qquad = \dfrac{2^{2p} \cdot 3^{3p/3} \cdot 5^{3p} \cdot 2^{2p} \cdot 3^{2p}}{2^{3\cdot(p/3)} \cdot 3^{2\cdot(3p/2)} \cdot 2^{3p} \cdot 5^{3p}} \qquad$ (by Laws 3 and 5)

$\qquad\qquad = \dfrac{(2^{2p} \cdot 2^{2p})(3^p \cdot 3^{2p}) \cdot 5^{3p}}{(2^p \cdot 2^{3p})(3^{3p}) \cdot 5^{3p}} \qquad$ (combining terms with like bases)

$\qquad\qquad = \dfrac{2^{4p} \cdot 3^{3p} \cdot 5^{3p}}{2^{4p} \cdot 3^{3p} \cdot 5^{3p}} = 1$

EXAMPLE 9 Simplify $(\sqrt{27} + \sqrt{75})/2\sqrt{12}$.

Solution We observe that the three radicals in this expression can be simplified by factoring out a perfect square from each of the numbers.

$$\sqrt{27} = \sqrt{9 \cdot 3} = \sqrt{9} \cdot \sqrt{3} = 3\sqrt{3}$$
$$\sqrt{75} = \sqrt{25 \cdot 3} = \sqrt{25} \cdot \sqrt{3} = 5\sqrt{3}$$
$$\sqrt{12} = \sqrt{4 \cdot 3} = \sqrt{4} \cdot \sqrt{3} = 2\sqrt{3}$$

Therefore

$$\frac{\sqrt{27} + \sqrt{75}}{2\sqrt{12}} = \frac{3\sqrt{3} + 5\sqrt{3}}{2(2\sqrt{3})} = \frac{8\sqrt{3}}{4\sqrt{3}} = \frac{8}{4} = 2.$$

☛ **19.** Simplify (a) $\sqrt[3]{4} \cdot \sqrt[3]{16}$;
(b) $\sqrt[3]{3} \div (\sqrt[3]{9})^2$;
(c) $\sqrt[4]{x^3} \cdot \sqrt{\sqrt{x}}$;
(d) $\sqrt{x}(\sqrt{x^3} + 3\sqrt{x})$.

EXAMPLE 10 Simplify: (a) $\sqrt{x}(\sqrt{x^3} + \sqrt[3]{x^2})$; (b) $\dfrac{\sqrt{x} + 2x}{\sqrt[3]{x}}$

Solution Express the radicals in terms of fractional exponents and then use the distributive properties and laws of exponents.

(a) $\sqrt{x}(\sqrt{x^3} + \sqrt[3]{x^2}) = x^{1/2}(x^{3/2} + x^{2/3})$
$= x^{1/2} \cdot x^{3/2} + x^{1/2} \cdot x^{2/3}$
$= x^2 + x^{7/6}$

(b) $\dfrac{\sqrt{x} + 2x}{\sqrt[3]{x}} = \dfrac{x^{1/2} + 2x}{x^{1/3}}$
$= (x^{1/2} + 2x)x^{-1/3}$
$= x^{1/2} \cdot x^{-1/3} + 2x^1 \cdot x^{-1/3}$
$= x^{1/6} + 2x^{2/3}$ ☛ **19**

Answer (a) 4; (b) 3^{-1}; (c) x;
(d) $x^2 + 3x$.

EXERCISES 1-4

(1–6) Find m such that the following statements are true.

1. $8\sqrt[3]{2} = 2^m$ **2.** $\dfrac{\sqrt[3]{2}}{8} = 2^m$

3. $\sqrt[3]{\dfrac{2}{8}} = 2^m$ **4.** $3\sqrt{3} \cdot \sqrt[3]{3} = 3^m$

5. $\sqrt{\sqrt{2}} = 4^m$ **6.** $\sqrt[4]{\sqrt[3]{\sqrt{2}}} = 2^m$

(7–26) Evaluate the following expressions.

7. $\sqrt{81}$ **8.** $\sqrt[3]{27}$

9. $\sqrt{1\frac{9}{16}}$ **10.** $\sqrt[3]{3\frac{3}{8}}$

11. $\sqrt[5]{-32}$ **12.** $\sqrt[3]{-0.125}$

13. $\sqrt{(-3)^2}$ **14.** $\sqrt{(-\frac{2}{5})^2}$

15. $(81)^{-3/4}$ **16.** $(\frac{8}{27})^{-4/3}$

17. $(0.16)^{-1/2}$ **18.** $(-0.16)^{3/4}$

19. $0.125^{-2/3}$ **20.** $0.0016^{3/4}$

21. $(9^{-3} \cdot 16^{3/2})^{1/6}$ **22.** $9^{3/4} \cdot 3^{-1/2}$

23. $16^{4/5} \cdot 8^{-2/5}$ **24.** $25^{1/3}(\frac{1}{5})^{-4/3}$

25. $(27)^{-2/3} \div (16)^{1/4}$ **26.** $-(\frac{1}{36})^{1/8} \div (6)^{-5/4}$

(27–56) Simplify the following expressions.

27. $(16x^4)^{3/4}$ **28.** $\left(\dfrac{27x^3}{64}\right)^{2/3}$

29. $(32x^5y^{-10})^{1/5}$ **30.** $\sqrt[3]{\dfrac{8a^3}{27b^3}}$

31. $\sqrt[4]{x^{3/2} \cdot 16x^{1/2}}$ **32.** $(x^{1/3} \cdot x^{-2/5})^3$

33. $(x^{1/2} \cdot x^{-1/3})^2$ **34.** $(16x^{-4})^{-1/2} \div (8x^6)^{1/3}$

35. $\dfrac{x^{3/7}y^{2/5}}{x^{-1/7}y^{1/5}}$

36. $\dfrac{a^{4/9}b^{-3/4}}{a^{2/9}b^{-1/2}}$

50. $a^{2/3}\cdot b^{-5/7}\cdot\left(\dfrac{a}{b}\right)^{7/8}\cdot\dfrac{a^{11/24}}{b^{23/56}}$

37. $\left(\dfrac{p^{-1/5}q^{2/5}}{p^{-3/5}q^{-2/5}}\right)^{10}$

38. $\dfrac{(x^2y)^{-1/3}(xy)^{1/4}}{(xy^{-2})^{1/12}}$

51. $\dfrac{2^{3m}\cdot 3^{2m}\cdot 5^m\cdot 6^m}{8^m\cdot 9^{3m/2}\cdot 10^m}$

52. $\dfrac{(x^{a+b})^2(y^{a+b})^2}{(xy)^{2a-b}}$

39. $\dfrac{2x^{5/2}}{y^{3/4}}\div\dfrac{x^{2/3}}{3y^{2/5}}$

40. $(-2x^2y)^{1/5}(4^{-1}xy^{-2})^{-2/5}$

53. $\left(\dfrac{x^a}{x^b}\right)^c\cdot\left(\dfrac{x^b}{x^c}\right)^a\cdot\left(\dfrac{x^c}{x^a}\right)^b$

54. $\left(\dfrac{x^{a+b}}{x^{2b}}\right)\left(\dfrac{x^{b+c}}{x^{2c}}\right)\left(\dfrac{x^{c+a}}{x^{2a}}\right)$

41. $3\sqrt{45}+\sqrt{20}$

42. $2\sqrt{24}-\sqrt{54}$

55. $\dfrac{(27)^{2n/3}\cdot(8)^{-n/6}}{(18)^{-n/2}}$

56. $\dfrac{28^m\cdot 35^m\cdot 10^{3m}}{8^{5m/3}\cdot 49^m\cdot 25^{2m}}$

43. $2\sqrt{18}-\sqrt{32}$

44. $\dfrac{8\sqrt{2}-4\sqrt{8}}{\sqrt{32}}$

57. State whether the following statements are true or false.

  a. $\sqrt{5}=\sqrt{2}+\sqrt{3}$ **b.** $\sqrt{8}=\sqrt{2}+\sqrt{2}$

45. $\sqrt{63}-\sqrt{175}+4\sqrt{112}$

  c. $\sqrt{21}=\sqrt{7}\cdot\sqrt{3}$ **d.** $\sqrt{(-3)^2}=3$

46. $\sqrt{112}-\sqrt{63}+\dfrac{224}{\sqrt{28}}$

  e. $\sqrt{-9}=-3$ **f.** $\sqrt{a^2}=a$ for all real a

  g. $\sqrt{a^2+b^2}=a+b$ if $a>0$ and $b>0$

47. $\dfrac{20}{\sqrt{5}}-2\sqrt{20}+\dfrac{50}{\sqrt{125}}$

48. $2\sqrt[3]{-16}-\sqrt[3]{-54}$

  h. $a^m\cdot a^n=a^{mn}$ **i.** $\dfrac{a^m}{a^n}=a^{m/n}$

49. $a^{2/3}\cdot a^{-3/4}\cdot(a^2)^{-1/6}\cdot\dfrac{1}{(a^{1/12})^5}$

  j. $\sqrt[3]{\sqrt[3]{a}}=a^{1/6}$ **k.** $\sqrt{a^2}=a$ if $a>0$

◣ 1-5 ALGEBRAIC OPERATIONS

Quantities of the type $2x^2-3x+7$, $5y^3-y^2+6y+2$, and $2x-3/y+4$ are called **algebraic expressions.** The building blocks of an algebraic expression are called its **terms.** For example, the expression $2x^2-3x+7$ has three terms, $2x^2$, $-3x$ and 7. The expression $x^2y/3-y/x$ has two terms, $x^2y/3$ and y/x.

In the term $2x^2$, the factor 2 is called the **numerical coefficient** (or simply the **coefficient**). The factor x^2 is called the **literal part** of this term. In the term $-3x$, the coefficient is -3 and the literal part is x. In the term $x^2y/3$, the coefficient is $\frac{1}{3}$ and the literal part is x^2y. The term 7 has no literal part and is called a **constant term.** The coefficient is 7.

An algebraic expression containing only one term is called a **monomial.** An expression that contains exactly two terms is called a **binomial,** and one containing exactly three terms is called a **trinomial.** The following are a few examples of expressions of these types.

  Monomials: $2x^3$, $-5y^2$, $7/t$, 3, $2xy/z$

  Binomials: $2x+3$, $3x^2-5/y$, $6x^2y-5zt$

  Trinomials: $5x^2+7x-1$, $2x^3+4x-3/x$, $6y^2-5x+t$

In general, an algebraic expression containing more than one term is called a **multinomial.**

Addition and Subtraction of Expressions

When 4 apples are added to 3 apples we get 7 apples. In the same way, $4x + 3x = 7x$. This is simply a consequence of the distributive property, since

$$4x + 3x = (4 + 3)x = 7x.$$

If you compare with Section 1-1 you will see that here we are using the distributive law "backwards," that is, from right to left. In a similar way, we can add any two expressions whose literal parts are the same. We simply add together the two numerical coefficients.

EXAMPLE 1

    (a) $2x + 9x = (2 + 9)x = 11x$

    (b) $4ab + 3ab = (4 + 3)ab = 7ab$

    (c) $\dfrac{2x}{y} + \dfrac{x}{2y} = 2 \cdot \dfrac{x}{y} + \dfrac{1}{2} \cdot \dfrac{x}{y} = \left(2 + \dfrac{1}{2}\right)\dfrac{x}{y} = \dfrac{5}{2} \cdot \dfrac{x}{y} = \dfrac{5x}{2y}$

Two or more terms in an algebraic expression are said to be **like** if they have equal literal parts. For example, the two terms $2x^2y$ and $5yx^2$ are like terms since their literal parts, x^2y and yx^2, are equal. Similarly, the three terms $3x^2yz^3$, $-7x^2z^3y$ and $yz^3x^2/2$ are all like terms. In general, two like terms can differ only in their numerical coefficients or in the order in which the variables appear.

Two or more like terms can be added or subtracted by making use of the distributive property, as in Example 1. Further examples follow.

EXAMPLE 2

    (a) $2x^3 - 7x^3 = (2 - 7)x^3 = -5x^3$

    (b) $5x^2y - 3x^2y + 2yx^2 = (5 - 3 + 2)x^2y = 4x^2y$

Terms that are not like cannot be combined in the above manner. Thus the terms in the expression $2x^2 + 5xy$ cannot be combined into a single term.

When adding or subtracting two or more algebraic expressions, we rearrange the terms in the two expressions so that like terms are grouped together. ☞ 20

☞ **20.** Simplify the following:

(a) $2ab^2 - 4b^2a$;

(b) $x^3 + 2x - (2x^3 - 2x)$.

EXAMPLE 3 Add $5x^2y^3 - 7xy^2 + 3x - 1$ and $6 - 2x + 4xy^2 + 3y^3x^2$.

Solution The required sum is

$$5x^2y^3 - 7xy^2 + 3x - 1 + (6 - 2x + 4xy^2 + 3y^3x^2)$$
$$= 5x^2y^3 - 7xy^2 + 3x - 1 + 6 - 2x + 4xy^2 + 3x^2y^3.$$

Answer (a) $-2ab^2$ (b) $-x^3 + 4x$.

Rearranging the terms, so that like terms are grouped together, we obtain the sum in the following form.

$$5x^2y^3 + 3x^2y^3 - 7xy^2 + 4xy^2 + 3x - 2x - 1 + 6$$

$$= (5 + 3)x^2y^3 \quad + (-7 + 4)xy^2 + (3 - 2)x + (-1 + 6)$$

$$= \qquad 8x^2y^3 \quad + \quad (-3)xy^2 \quad + \quad 1x \quad + 5$$

$$= \qquad 8x^2y^3 \quad - \quad 3xy^2 \quad + \quad x \quad + 5$$

EXAMPLE 4 Subtract $3x^2 - 5xy + 7y^2$ from $7x^2 - 2xy + 4y^2 + 6$.

Solution In this case we want

$$7x^2 - 2xy + 4y^2 + 6 - (3x^2 - 5xy + 7y^2).$$

Upon removal of the parentheses, each term inside changes sign. Thus the above expression is equivalent to the following.

$$7x^2 - 2xy + 4y^2 + 6 - 3x^2 + 5xy - 7y^2$$

$$= 7x^2 - 3x^2 - 2xy + 5xy + 4y^2 - 7y^2 + 6$$

$$= (7 - 3)x^2 + (-2 + 5)xy + (4 - 7)y^2 + 6$$

$$= \qquad 4x^2 \quad + \quad 3xy \quad + (-3)y^2 \quad + 6$$

$$= \qquad 4x^2 \quad + \quad 3xy \quad - \quad 3y^2 \quad + 6$$

Multiplication of Expressions

The expression $a(x + y)$ denotes the product of a and $x + y$. To simplify this expression by removing the parentheses, we multiply each term within the parentheses by the number outside, in this case a:

$$a(x + y) = ax + ay.$$

Once again we are using the distributive property. Similarly, this method is used whenever a multinomial expression is multiplied by a monomial.

EXAMPLE 5

(a) $-2(x - 3y + 7t^2) = (-2)x - (-2)(3y) + (-2)(7t^2)$

$$= -2x + 6y - 14t^2.$$

(b) $x^2y(x^2 + 3x - 5y^3) = x^2y \cdot x^2 + x^2y \cdot 3x - x^2y \cdot 5y^3$

$$= x^4y + 3x^3y - 5x^2y^4 \quad \text{☛ 21}$$

☛ **21.** Simplify the following by removing the parentheses:
(a) $3(x - 2) + x(x - 3)$;
(b) $x^3 - 2x - 2x(x^2 - 1)$.

When multiplying two multinomial expressions together, we must use the distributive property more than once in order to remove the parentheses. Consider the product $(x + 2)(y + 3)$. We can use the distributive property to remove the first parentheses.

$$(x + 2)(y + 3) = x(y + 3) + 2(y + 3)$$

Answer (a) $x^2 - 6$; (b) $-x^3$.

To see this, just set $y + 3 = b$. Then

$$(x + 2)(y + 3) = (x + 2)b = x \cdot b + 2 \cdot b = x(y + 3) + 2(y + 3).$$

In general, the distributive properties of Section 1-1 work with a, b, c replaced by any expressions (as do the other properties of real numbers). We now use this property again to remove the remaining parentheses.

$$x(y + 3) = xy + x \cdot 3 = xy + 3x$$

and

$$2(y + 3) = 2y + 2 \cdot 3 = 2y + 6.$$

Therefore $(x + 2)(y + 3) = xy + 3x + 2y + 6$.

In Figure 2 the four terms (products) on the right can be obtained by multiplying each of the terms in the first parentheses in turn by each of the terms in the second parentheses. Each term in the first parentheses is connected by an arc to each term in the second parentheses and the corresponding product is written down. The four products then give the complete expansion of the given expression. ☞ 22

FIGURE 2

☞ **22.** Use the distributive property (or method of arcs) to remove the parentheses:

(a) $(x + 2)(x + 3)$;

(b) $(x^2 + 2)(x^2 - 2)$.

You may also have come across the FOIL method of multiplying two binomial expressions. (FOIL stands for "First, Outer, Inner, Last.") This is equivalent to the method of arcs described here. However the arc method is much better because you can use it to multiply any two multinomials.

EXAMPLE 6 Expand the product $(3x - 4)(6x^2 - 5x + 2)$. (This means to remove the parentheses.)

Solution We use the distributive property

$$(3x - 4)(6x^2 - 5x + 2) = 3x(6x^2 - 5x + 2) - 4(6x^2 - 5x + 2)$$
$$= (3x)(6x^2) - (3x)(5x) + (3x)(2)$$
$$+ (-4)(6x^2) - (-4)(5x) + (-4)(2)$$
$$= 18x^3 - 15x^2 + 6x - 24x^2 + 20x - 8$$
$$= 18x^3 - 15x^2 - 24x^2 + 6x + 20x - 8$$

(grouping like terms)

$$= 18x^3 - (15 + 24)x^2 + (6 + 20)x - 8$$
$$= 18x^3 - 39x^2 + 26x - 8$$

Answer (a) $x^2 + 5x + 6$;

(b) $x^4 - 4$.

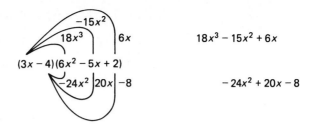

$$18x^3 - 15x^2 + 6x$$

$$-24x^2 + 20x - 8$$

FIGURE 3

Alternatively, we can obtain the answer by drawing arcs connecting each term in the first parentheses to each term in the second. In this case, there are six such arcs, giving six products in the expansion on the right. (See Figure 3.)

EXAMPLE 7 Simplify $3\{5x[2 - 3x] + 7[3 - 2(x - 4)]\}$.

Solution To simplify an expression which involves more than one set of parentheses, we always start with the innermost parentheses.

$$
\begin{aligned}
3\{5x[2 - 3x] + 7[3 - 2(x - 4)]\} &= 3\{5x[2 - 3x] + 7[3 - 2x + 8]\} \\
&= 3\{10x - 15x^2 + 21 - 14x + 56\} \\
&= 3\{-15x^2 + 10x - 14x + 21 + 56\} \\
&= 3\{-15x^2 - 4x + 77\} \\
&= -45x^2 - 12x + 231
\end{aligned}
$$

There are certain special products that are encountered often and may be treated as standard formulas. First we consider the product $(x + a)(x + b)$.

$$
\begin{aligned}
(x + a)(x + b) &= x(x + b) + a(x + b) \\
&= x^2 + bx + ax + ab \\
&= x^2 + (b + a)x + ab
\end{aligned}
$$

Therefore

$$
(x + a)(x + b) = x^2 + (a + b)x + ab. \tag{1}
$$

EXAMPLE 8

(a) Taking $a = 2$ and $b = 7$ in Equation (1), we have

$$(x + 2)(x + 7) = x^2 + (2 + 7)x + 2 \cdot 7 = x^2 + 9x + 14.$$

(b) $(x + 3)(x - 2) = (x + 3)(x + (-2))$

$$= x^2 + [3 + (-2)]x + 3(-2) = x^2 + x - 6$$

In Equation (1), if we replace b by a, we get

$$(x + a)(x + a) = x^2 + (a + a)x + a \cdot a$$

or

$$(x + a)^2 = x^2 + 2ax + a^2. \qquad (2)$$

This result gives an expansion for the square of a binomial. *The square of the sum of two terms is equal to the sum of the squares of the two terms plus twice their product.*

EXAMPLE 9

(a) $(2x + 7)^2 = (2x)^2 + 2(2x)(7) + 7^2 = 4x^2 + 28x + 49$

(b) $\left(3x + \dfrac{4}{y}\right)^2 = (3x)^2 + 2(3x)\left(\dfrac{4}{y}\right) + \left(\dfrac{4}{y}\right)^2 = 9x^2 + \dfrac{24x}{y} + \dfrac{16}{y^2}$

☛ **23.** Use the standard formulas (1)–(4) to remove the parentheses:
(a) $(x + 2)(x - 3)$;
(b) $(x^2 + y)(x^2 - y)$;
(c) $(x + x^{-1})^2$.

If we replace a by $-a$ in Formula (2), we obtain another formula.

$$(x - a)^2 = x^2 - 2ax + a^2 \qquad (3)$$

This expresses *the square of the difference of two terms as the sum of the squares of the two terms minus twice their product.*

Finally, if we replace b by $-a$ in Equation (1), we obtain

$$(x + a)(x - a) = x^2 + (a - a)x + a(-a) = x^2 + 0x - a^2.$$

Thus we have

$$(x + a)(x - a) = x^2 - a^2. \qquad (4)$$

This result states that the *product of the sum and difference of two terms is the difference of the squares of the two terms.*

EXAMPLE 10

(a) $(2x + 3)(2x - 3) = (2x)^2 - 3^2 = 4x^2 - 9$

(b) $(\sqrt{3} + \sqrt{2})(\sqrt{3} - \sqrt{2}) = (\sqrt{3})^2 - (\sqrt{2})^2 = 3 - 2 = 1$

(c) $(3x - 4y)(3x + 4y) = (3x)^2 - (4y)^2 = 9x^2 - 16y^2$ ☛ 23

Division of Expressions

We saw in Theorem 6 of Section 1-2 that the distributive law extends to division, and we have the following general expressions.

$$\frac{a + b}{c} = \frac{a}{c} + \frac{b}{c}$$

Answer (a) $x^2 - x - 6$;
(b) $x^4 - y^2$; (c) $x^2 + 2 + x^{-2}$.

This property is useful when dividing an algebraic expression by a monomial, since it allows us to divide each term separately by the monomial.

EXAMPLE 11

(a) $\dfrac{2x^2 + 4x}{2x} = \dfrac{2x^2}{2x} + \dfrac{4x}{2x} = x + 2$

Observe that we divide each term by the common factor $2x$.

(b) $\dfrac{2x^3 - 5x^2y + 7x + 3}{x^2} = \dfrac{2x^3}{x^2} - \dfrac{5x^2y}{x^2} + \dfrac{7x}{x^2} + \dfrac{3}{x^2}$

$\qquad\qquad\qquad\qquad = 2x - 5y + \dfrac{7}{x} + \dfrac{3}{x^2}$

(c) $\dfrac{25t^3 + 12t^2 + 15t - 6}{3t} = \dfrac{25t^3}{3t} + \dfrac{12t^2}{3t} + \dfrac{15t}{3t} - \dfrac{6}{3t}$

$\qquad\qquad\qquad\qquad = \dfrac{25t^2}{3} + 4t + 5 - \dfrac{2}{t}$

In a fraction, the number or algebraic expression in the numerator is often called the **dividend** (which means the quantity which is being divided) and the number or expression by which it is divided is called the **divisor.** In part (b) of Example 11, $2x^3 - 5x^2y + 7x + 3$ is the *dividend* and x^2 is the *divisor*, for example.

When we want to divide an algebraic expression by a divisor that contains more than one term, we can often use a process called **long division.** We shall describe this process for expressions that contain only positive integral powers of a single variable. (Such expressions are called **polynomials.**)

EXAMPLE 12 Divide $23 - 11x^2 + 2x^3$ by $2x - 3$.

Solution Here $23 - 11x^2 + 2x^3$ is the dividend and $2x - 3$ is the divisor. Before we start division, the terms in the dividend and divisor must be arranged in order of decreasing powers of x and any missing powers filled in with zero coefficients. Thus the dividend must be written as $2x^3 - 11x^2 + 0x + 23$.

$$
\begin{array}{r}
\text{Divisor} \;\rightarrow\; 2x - 3 \overline{\smash{\big)}\,2x^3 - 11x^2 + 0x + 23} \quad
\end{array}
$$

$$
\begin{array}{r}
x^2 - 4x - 6 \quad \leftarrow \textbf{Quotient} \\
2x - 3 \overline{\smash{\big)}\,2x^3 - 11x^2 + 0x + 23} \quad \leftarrow \textbf{Dividend} \\
\underline{2x^3 - 3x^2} \qquad\qquad\qquad\quad \\
-8x^2 + 0x + 23 \quad \\
\underline{-8x^2 + 12x} \qquad\quad \\
-12x + 23 \\
\underline{-12x + 18} \\
5 \quad \leftarrow \textbf{Remainder}
\end{array}
$$

☛ **24.** Using long division, simplify $(3x^2 + 11x + 4) \div (x + 3)$.

The details of the long division are shown above and are explained as follows: First of all, we divide $2x^3$ (the first term in the dividend) by $2x$ (the first term in the divisor), obtaining $2x^3/2x = x^2$. This gives the first term in the quotient. We multiply the divisor, $2x - 3$, by the first term in the quotient, x^2, to obtain $2x^3 - 3x^2$. Subtracting this from the dividend, we have the difference $-8x^2 + 0x + 23$. To obtain the next term in the quotient, we divide the first term in this difference, $-8x^2$, by $2x$, the first term in the divisor. This gives $-8x^2/2x = -4x$, which becomes the second term in the quotient. We multiply the divisor again by this second term, $-4x$, obtaining $-8x^2 + 12x$; we then subtract this from $-8x^2 + 0x + 23$, which gives the next difference, $-12x + 23$. We continue this process until we obtain a difference whose highest power is less than that of the divisor. We call this last difference the **remainder.** The answer may be written as

Answer Quotient = $3x + 2$;
remainder = -2

$$\frac{2x^3 - 11x^2 + 23}{2x - 3} = x^2 - 4x - 6 + \frac{5}{2x - 3}. \qquad ☛ \ 24$$

In general, we have

$$\boxed{\frac{\text{Dividend}}{\text{Divisor}} = \text{Quotient} + \frac{\text{Remainder}}{\text{Divisor}}.}$$

Remark This form of writing the result of the long division is exactly the same as we use in arithmetic. For example, consider the fraction $627/23$, in which the dividend is 627 and the divisor is 23. By ordinary long division we find that the quotient is 27 and the remainder is 6.

$$
\begin{array}{r}
27 \quad \leftarrow \textbf{Quotient} \\
\textbf{Divisor} \rightarrow \ 23\overline{)627} \quad \leftarrow \textbf{Dividend} \\
46 \quad\quad\quad\quad \\
\overline{167} \quad\quad\quad \\
161 \quad\quad\quad \\
\overline{6} \quad \leftarrow \textbf{Remainder}
\end{array}
$$

We therefore write

$$\frac{627}{23} = 27 + \frac{6}{23}.$$

Now, instead of dividing 627 by 23, try dividing $6x^2 + 2x + 7$ by $2x + 3$. When $x = 10$ this amounts to the same thing. You should find a quotient of $2x + 7$ and a remainder of 6. The algebraic long division mirrors the arithmetic one.

If we multiply both sides of this calculation by 23, we obtain the result

$$627 = (27 \cdot 23) + 6.$$

25. Check whether the following long division is correct:
$$\frac{3x^2 - 3x - 10}{x - 2} = 3x + 3 + \frac{4}{x - 2}.$$

Answer You should check that
$3x^2 - 3x - 10 = (3x + 3)(x - 2) + 4.$
This is not correct. (The remainder should be -4.)

This is an example of the general result that

> Dividend = (Quotient)(Divisor) + Remainder.

This is a useful result, because it allows us to check the answer to any long division. We can use the result to verify Example 12.

$$2x^3 - 11x^2 + 23 = (x^2 - 4x - 6)(2x - 3) + 5$$

Dividend = (Quotient)(Divisor) + Remainder ☛ 25

EXERCISES 1-5

(1–56) In the following exercises, perform the indicated operation and simplify.

1. $(5a + 7b - 3) + (3b - 2a + 9)$

2. $(3x^2 - 5x + 7) + (-2 + 6x - 7x^2 + x^3)$

3. $(2\sqrt{a} + 5\sqrt{b}) + (3\sqrt{a} - 2\sqrt{b})$

4. $(4xy + 5x^2y - 6x^3)$
$+ (3y^3 - 6xy^2 + 7xy + x^3 - 2x^2y)$

5. $(7t^2 + 6t - 1) - (3t - 5t^2 + 4 - t^3)$

6. $(x^2 + 3xy + 4y^2) - (2x^2 - xy + 3y^2 - 5)$

7. $(2\sqrt{x} + \sqrt{2y}) - (\sqrt{x} - 2\sqrt{2y})$

8. $(5\sqrt{xy} - 3) - (2 - 4\sqrt{xy})$

9. $4(2x + 3y) + 2(5y + 3x)$

10. $2(x - 4y) + 3(2x + 3y)$

11. $-(x - 7y) - 2(2y - 5x)$

12. $3(x^2 - 2xy + y^2) - (2xy - x^2 + 2y^2)$

13. $x(2x^2 + 3xy + y^2) - y(5x^2 - 2xy + y^2)$

14. $a^2b(a^3 + 5ab - b^3) + 2ab(a^4 - 2a^2b + b^3a)$

15. $(x - 3)(y + 2)$

16. $(x + 4)(y - 5)$

17. $(2x + 1)(3y - 4)$

18. $(5x - 2)(2y - 5)$

19. $(a + 2)(3a - 4)$

20. $(x + 3y)(2x + y)$

21. $(x + 3)(2x^2 - 5x + 7)$

22. $(a - 2b)(a^2 - 2ab + b^2)$

23. $(x + 4)(x - 4)$

24. $(y^2 - 2)(y^2 + 2)$

25. $(2t + 5x)(2t - 5x)$

26. $(\sqrt{a} - \sqrt{b})(\sqrt{a} + \sqrt{b})$

27. $(\sqrt{x} + 3\sqrt{y})(\sqrt{x} - 3\sqrt{y})$

28. $(5\sqrt{x} + 2y)(5\sqrt{x} - 2y)$

29. $(x + y - z)(x + y + z)$

30. $(x - 2y + z)(x + 2y + z)$

31. $(x^2 - 1)(x^3 + 2)$

32. $(y^2 + 2y)(y^3 - 2y^2 + 1)$

33. $\left(x^2 - \frac{1}{x}\right)(x^3 + 2x)$

34. $\left(2xy - \frac{x}{y}\right)\left(xy^2 + \frac{2y}{x}\right)$

35. $(y + 6)^2$ **36.** $(x - 5)^2$

37. $(2x + 3y)^2$ **38.** $(4x - 5y)^2$

39. $(\sqrt{2}x - \sqrt{3}y)^2$ **40.** $(\sqrt{x} + 2\sqrt{y})^2$

41. $(2x + 3y)^2 + (2x - 3y)^2$

42. $3[(x + y)^2 - (x - y)^2]$

43. $xy[(x + y)^2 + (x - y)^2]$

44. $(3a - b)^2 + 3(a + b)^2$

45. $3\{x^2 - 5[x + 2(3 - 5x)]\}$

46. $2\{a^2 - 2a[3a - 5(a^2 - 2)]\} + 7a^2 - 3a + 6$

47. $2a\{(a + 2)(3a - 1) - [a + 2(a - 1)(a + 3)]\}$

48. $(a + 3b)(a^2 - 3ab + b^2) - (a + b)^2(a + 2b)$

49. $\dfrac{4x^3 - 3x^2}{2x}$ **50.** $\dfrac{15x^5 - 25x^3}{5x^2}$

51. $\dfrac{x^3 + 7x^2 - 5x + 4}{x^2}$

52. $\dfrac{y^4 + 6y^3 - 7y^2 + 9y - 3}{3y^2}$

53. $\dfrac{t^2 - 2t + 7}{\sqrt{t}}$ **54.** $\dfrac{t^3 + 2t^2 - 3t + 1}{t\sqrt{t}}$

55. $\dfrac{6x^2y - 8xy^2}{2xy} + \dfrac{x^3y^2 + 2x^2y^3}{x^2y^2}$

56. $\dfrac{3x^4 - 9x^2y^2}{3x^3y} - \dfrac{4x^3 - 8xy^2}{2x^2y}$

(57–64) Simplify the following by long division:

57. $(x^2 - 5x + 6) \div (x - 2)$

58. $(6x^2 + x - 1) \div (3x - 1)$

59. $(t^2 + 1) \div (t - 1)$

60. $(6x^2 - 5x + 1) \div (2x - 3)$

61. $(x^3 + 2x^2 + x + 5) \div (x + 2)$

62. $x^3 \div (x + 1)$

63. $(2x^3 - 3x^2 + 4x + 6) \div (2x + 1)$

64. $(6x^3 + 11x^2 - 19x + 5) \div (3x - 2)$

◥ 1-6 FACTORS

If the product of two integers a and b is c, that is, $c = a \cdot b$, then a and b are called **factors** of c. In other words, an integer a is a factor of another integer c if a divides c exactly with no remainder. For example, 2 and 3 are factors of 6; 2, 3, 4 and 6 are all factors of 12; and so on.

This terminology is also used for algebraic expressions. If two (or more) algebraic expressions are multiplied together, these expressions are said to be *factors* of the expression obtained as their product. For example, the expression $2xy$ is obtained by multiplying 2, x, and y, so 2, x, and y are the factors of $2xy$. Furthermore, for example, $2y$ is a factor of $2xy$ since $2xy$ can be obtained by multiplying $2y$ by x.

Similarly, x is a factor of the expression $2x^2 + 3x$ since we can write $2x^2 + 3x = x(2x + 3)$ and x^2 is a factor of $6x^2 + 9x^3$ since we can write $6x^2 + 9x^3 = x^2(6 + 9x)$.

The process of writing a given expression as the product of its factors is called **factoring** the expression. In this section, we shall discuss certain methods by which multinomial expressions can be factored.

The first step in factoring a multinomial expression is to extract all the monomial factors that are common to all of the terms. The following example illustrates this.

EXAMPLE 1 Factor all the common monomial factors from the following expressions.

(a) $x^2 + 2xy^2$

(b) $2x^2y + 6xy^2$

(c) $6ab^2c^3 + 6a^2b^2c^2 + 18a^3bc^2$

26. Remove all the common factors from

☞ 26. Remove all the common factors from

(a) $12ab - 8a^2b$;

(b) $4xyz - 6x^2z + 12xy^2$;

(c) $x(3x - 1)^2 - y(3x - 1)^2$.

Answer (a) $4ab(3 - 2a)$;

(b) $2x(2yz - 3xz + 6y^2)$;

(c) $(3x - 1)^2(x - y)$.

Solution

(a) Let us write each term in the given expression in terms of its basic factors.

$$x^2 = x \cdot x \qquad 2xy^2 = 2 \cdot x \cdot y \cdot y.$$

Looking at the two lists of basic factors, we see that only the factor x is common to both terms. So we write

$$x^2 + 2xy^2 = x \cdot x + x \cdot 2y^2 = x(x + 2y^2).$$

Observe how the distributive property is used to extract the common factor x.

(b) Expressing each term in terms of basic factors, we have

$$2x^2y = 2 \cdot x \cdot x \cdot y \quad \text{and} \quad 6xy^2 = 2 \cdot 3 \cdot x \cdot y \cdot y.$$

The factors 2, x, and y occur in both lists, so the common factor is $2xy$. This gives

$$2x^2y + 6xy^2 = 2xy \cdot x + 2xy \cdot 3y = 2xy(x + 3y),$$

again using the distributive property.

(c) We first factor the terms.

$$6ab^2c^3 = 2 \cdot 3 \cdot a \cdot b \cdot b \cdot c \cdot c \cdot c$$

$$6a^2b^2c^2 = 2 \cdot 3 \cdot a \cdot a \cdot b \cdot b \cdot c \cdot c$$

$$18a^3bc^2 = 2 \cdot 3 \cdot 3 \cdot a \cdot a \cdot a \cdot b \cdot c \cdot c$$

The common factor of these three terms is $2 \cdot 3 \cdot a \cdot b \cdot c \cdot c = 6abc^2$.

$$6ab^2c^3 + 6a^2b^2c^2 + 18a^3bc^2 = 6abc^2 \cdot bc + 6abc^2 \cdot ab + 6abc^2 \cdot 3a^2$$

$$= 6abc^2(bc + ab + 3a^2) \quad \text{☞ 26}$$

Now let us turn to the question of extracting factors that are themselves binomial expressions from multinomial expressions of various kinds. Some of the formulas established in Section 1-5 are useful for factoring, in particular the following formula.

$$a^2 - b^2 = (a - b)(a + b) \qquad (1)$$

This formula can be used to factor any expression which is reducible to a *difference of two squares*.

EXAMPLE 2 Factor completely: (a) $x^2y^4 - 9$; (b) $5x^4 - 80y^4$.

Solution

(a) The given expression can be written as

$$(xy^2)^2 - 3^2$$

which is the difference of two squares. Using Formula (1) with $a = xy^2$ and $b = 3$, we have

$$x^2y^4 - 9 = (xy^2)^2 - 3^2 = (xy^2 - 3)(xy^2 + 3).$$

Neither of the expressions in parentheses on the right side can be factored any further.

(b) First of all, we check whether we can take out any common monomial factor in $5x^4 - 80y^4$. In this case, because each term is divisible by 5, we take out a common factor of 5.

$$5x^4 - 80y^4 = 5(x^4 - 16y^4)$$

The expression $x^4 - 16y^4$ is a difference of squares.

$$5x^4 - 80y^4 = 5[(x^2)^2 - (4y^2)^2]$$
$$= 5[(x^2 - 4y^2)(x^2 + 4y^2)]$$
$$= 5(x^2 - 4y^2)(x^2 + 4y^2)$$

The factoring is not complete, because $x^2 - 4y^2 = x^2 - (2y)^2$ can be factored further as $(x - 2y)(x + 2y)$. Thus we need one more step.

$$5x^4 - 80y^4 = 5(x^2 - 4y^2)(x^2 + 4y^2)$$
$$= 5(x - 2y)(x + 2y)(x^2 + 4y^2) \quad \text{☛ 27}$$

☛ **27.** Using the difference of squares formula, factor $2x^2 - 4$.

Notes 1. Formula (1) allows us to factor any expression which takes the form of the difference between two squares. There is no corresponding formula expressing the sum $a^2 + b^2$ as the product of two or more factors. An expression that involves the *sum* of two squares, such as $a^2 + b^2$ or $4x^2 + 9y^2$, cannot be factored.

Expressions such as $a^3 + b^3$, $a^4 + b^4$, and so on, that involve the sum of two higher powers can however be further factored. This is discussed below.

2. We can write

$$x^2 - 2 = x^2 - (\sqrt{2})^2 = (x - \sqrt{2})(x + \sqrt{2}).$$

It is usually acceptable to include irrational numbers (such as $\sqrt{2}$) in the factors. However, we would usually not wish to use expressions involving $\sqrt{x}$ as factors. For example, as a rule it would not be useful to write

$$x - 4 = (\sqrt{x})^2 - 2^2 = (\sqrt{x} - 2)(\sqrt{x} + 2).$$

A useful technique in factoring multinomial expressions involving an even number of terms is the **grouping method.** In this method, the terms are grouped in pairs, and the common monomial factors are extracted from each pair of terms. This often reveals a binomial factor common to all the pairs. This method is particularly useful for expressions containing four terms.

Answer $(\sqrt{2}x - 2)(\sqrt{2}x + 2)$ or $2(x - \sqrt{2})(x + \sqrt{2})$.

EXAMPLE 3 Factor $ax^2 + by^2 + bx^2 + ay^2$.

Solution We can group the terms in the given expression into those which have x^2 as a factor and those which have y^2 as a factor:

$$(ax^2 + bx^2) + (ay^2 + by^2).$$

Each term in the first parentheses is divisible by x^2, and each term in the second parentheses is divisible by y^2; therefore we can write this expression as

$$x^2(a + b) + y^2(a + b).$$

We notice that $(a + b)$ is common to both terms. Thus we have

$$x^2(a + b) + y^2(a + b) = (a + b)(x^2 + y^2).$$

Hence the given expression has as its factors $(a + b)$ and $(x^2 + y^2)$.

EXAMPLE 4 Factor the expression $2x^3y - 4x^2y^2 + 8xy - 16y^2$.

Solution We first observe that the terms in this expression have a common monomial factor of $2y$ and we write

$$2x^3y - 4x^2y^2 + 8xy - 16y^2 = 2y(x^3 - 2x^2y + 4x - 8y)$$

☛ 28. By grouping, factor the expression $x^3 + 2x^2 - 9x - 18$.

Inside the parentheses, we group the first two terms together and take out the common factor x^2; we also group the last two terms and take out the common factor of 4.

$$x^3 - 2x^2y + 4x - 8y = x^2(x - 2y) + 4(x - 2y)$$
$$\underbrace{}_{x^2 \text{ common}} \quad \underbrace{}_{4 \text{ common}}$$

$$= (x - 2y)(x^2 + 4)$$

Observe that this same result could also be obtained by grouping the first and third terms and the second and fourth terms together.

$$x^3 + 4x - 2x^2y - 8y = x(x^2 + 4) - 2y(x^2 + 4)$$
$$\underbrace{}_{x \text{ common}} \quad \underbrace{}_{-2y \text{ common}}$$

$$= (x^2 + 4)(x - 2y)$$

Returning to the original expression, we have

$$2x^3y - 4x^2y^2 + 8xy - 16y^2 = 2y(x - 2y)(x^2 + 4).$$

It is not possible to factor the expressions on the right any further, so the factoring is complete. **☛ 28**

An important type of factoring which arises frequently involves finding the factors of expressions of the type

$$x^2 + px + q$$

Answer $(x + 2)(x^2 - 9) = (x + 2)(x - 3)(x + 3)$.

where p and q are constants. Often such expressions can be written as the product of two factors $(x + a)$ and $(x + b)$, where a and b are two real numbers. For example, it is readily verified that the expression $x^2 + 3x + 2$ (in which $p = 3$ and $q = 2$) is equal to the product of $x + 1$ and $x + 2$:

$$x^2 + 3x + 2 = (x + 1)(x + 2).$$

In this case, $a = 1$ and $b = 2$.

In general, with p and q given, we wish to find a and b such that

$$x^2 + px + q = (x + a)(x + b).$$

But we saw in Section 1-5 that

$$(x + a)(x + b) = x^2 + (a + b)x + ab$$

and so

$$x^2 + px + q = x^2 + (a + b)x + ab.$$

These two expressions are the same provided that $a + b = p$ and $ab = q$. So, in order to determine a and b, we must find two numbers whose sum is equal to p and whose product is equal to q. In terms of the original expression $x^2 + px + q$, the sum $a + b$ is equal to the coefficient of x and the product ab is equal to the constant term.

The procedure for finding a and b is to examine all possible pairs of integers whose product is equal to q. We then select the pair (if one exists) whose sum is the given coefficient of x.

EXAMPLE 5 Factor $x^2 + 7x + 12$.

Solution Here $p = 7$ and $q = 12$. We must find two numbers a and b whose product is 12 and whose sum is 7. Let us consider all the possible pairs of factors of 12.

$$a = 1, \qquad b = 12, \qquad a + b = 13$$
$$a = -1, \qquad b = -12, \qquad a + b = -13$$
$$a = 2, \qquad b = 6, \qquad a + b = 8$$
$$a = -2, \qquad b = -6, \qquad a + b = -8$$
$$a = 3, \qquad b = 4, \qquad a + b = 7$$
$$a = -3, \qquad b = -4, \qquad a + b = -7$$

From the above list, we see that the appropriate choice is $a = 3$ and $b = 4$. Therefore

$$x^2 + 7x + 12 = (x + 3)(x + 4).$$

Note The choice $a = 4$ and $b = 3$ gives exactly the same pair of factors.

EXAMPLE 6 Factor (a) $x^2 - 5x + 6$; (b) $3x^2 - 3x - 6$.

Solution

(a) In order to factor $x^2 - 5x + 6$, we have to find two factors of $+6$ (the constant term) whose sum is -5 (the coefficient of x). The possible factors of 6 are $(1)(6)$, $(-1)(-6)$, $(2)(3)$, and $(-2)(-3)$. The two factors of 6 that have the sum -5 are -2 and -3. Thus we let $a = -2$ and $b = -3$.

$$x^2 - 5x + 6 = (x + a)(x + b) = [x + (-2)][x + (-3)] = (x - 2)(x - 3)$$

(b) We first observe that there is a common monomial factor of 3:

$$3x^2 - 3x - 6 = 3(x^2 - x - 2).$$

To factor $x^2 - x - 2$, we have to find two factors of -2 (the constant term) whose sum is -1 (the coefficient of x). The possible factors of -2 are $1(-2)$ and $(-1)(2)$. Only the factors of 1 and -2 have the sum -1, that is, $1 + (-2) = -1$. Thus

$$x^2 - x - 2 = (x + 1)[x + (-2)] = (x + 1)(x - 2).$$

Our original expression therefore factors as follows.

$$3x^2 - 3x - 6 = 3(x^2 - x - 2) = 3(x + 1)(x - 2)$$

EXAMPLE 7 Factor $x^2 + 6x + 9$.

☛ **29.** Factor (a) $4x^2 - 16x + 16$;
(b) $x^2 + x - 12$.

Solution We have $p = 6$ and $q = 9$. Clearly the two factors of 9 whose sum is 6 are 3 and 3. Thus the given expression has factors $x + 3$ and $x + 3$, so

$$x^2 + 6x + 9 = (x + 3)(x + 3) = (x + 3)^2. \quad ☛ \ 29$$

Now let us turn to the problem of factoring an expression of the form

$$mx^2 + px + q$$

where m, p, q are nonzero constants with $m \neq 1$ or -1. In this case, the first step consists of finding two factors of the product mq that have a sum p, the coefficient of x. Then we split p into the sum of these two factors. This changes the given expression into the sum of four terms. These four terms can be considered two by two and factored by the method of grouping. This method is illustrated by Examples 8 and 9.

Answer (a) $4(x - 2)^2$;
(b) $(x - 3)(x + 4)$.

EXAMPLE 8 Factor $3x^2 + 11x + 6$.

Solution In this expression, the coefficients are $m = 3$, $p = 11$, and $q = 6$. The product of the coefficient of x^2 and the constant term is $mq = 3(6) = 18$. We must find two factors of this product 18 that have a sum equal to 11, the coefficient of x. Clearly, two such factors of 18 are 9 and 2. Thus, in the given expression, we split the coefficient of x, 11, into $9 + 2$ and write,

$$3x^2 + 11x + 6 = 3x^2 + (9 + 2)x + 6 = 3x^2 + 9x + 2x + 6.$$

We can take $3x$ as a common factor from the first two terms and 2 as a common factor from the last two terms.

$$3x^2 + 11x + 6 = 3x(x + 3) + 2(x + 3) = (x + 3)(3x + 2).$$

In the last step, we have taken $x + 3$ as common factor from the two terms.

EXAMPLE 9 Factor $6x^2 - 5x - 4$.

Solution The product of the coefficient of x^2 and the constant term is $6(-4) = -24$. We must find two factors of -24 that add up to -5, the coefficient of x. Clearly the two factors of -24 that have the sum -5 are 3 and -8. Therefore we write -5 as $-8 + 3$ in the given expression:

$$6x^2 - 5x - 4 = 6x^2 + (-8 + 3)x - 4$$

$$= 6x^2 - 8x + 3x - 4$$

$$= 2x(3x - 4) + 1(3x - 4)$$

$$= (3x - 4)(2x + 1) \qquad \text{☛ 30}$$

☛ **30.** Factor (a) $4x^2 - 9x + 2$;
(b) $6x^2 - x - 12$.

EXAMPLE 10 Factor $2(x + y)^2 - 5(x + y) + 3$.

Solution Let $z = x + y$. Then the given expression becomes

$$2z^2 - 5z + 3.$$

The product of the outer coefficients is $2 \cdot 3 = 6$. Two numbers whose product is 6 and whose sum is -5 are -2 and -3. So we write

$$2z^2 - 5z + 3 = 2z^2 - 2z - 3z + 3 = 2z(z - 1) - 3(z - 1)$$

$$= (2z - 3)(z - 1) = (2x + 2y - 3)(x + y - 1)$$

after replacing z by $x + y$ in the last step.

The following two formulas are helpful in factoring an expression which can be expressed as either the sum or the difference of two cubes.

$a^3 + b^3 = (a + b)(a^2 - ab + b^2)$	(2)
$a^3 - b^3 = (a - b)(a^2 + ab + b^2)$	(3)

These formulas can be verified by multiplying out the two expressions on the right. Alternatively they can be found by long division of $(a^3 \pm b^3) \div (a \pm b)$. (See Exercises 57–64 in Section 1-5.)

EXAMPLE 11 Factor $8x^3 + 27y^3$.

Solution We use Formula (2).

$$8x^3 + 27y^3 = (2x)^3 + (3y)^3$$

$$= (2x + 3y)[(2x)^2 - (2x)(3y) + (3y)^2]$$

$$= (2x + 3y)(4x^2 - 6xy + 9y^2)$$

Answer (a) $(x - 2)(4x - 1)$;
(b) $(2x - 3)(3x + 4)$.

● **31.** Factor $24x^4 + 3x$.

Answer $3x(2x + 1)(4x^2 - 2x + 1)$.

Note that the expression $4x^2 - 6xy + 9y^2$ cannot be factored further because the product of the coefficient of x^2 and the constant term is $4(9y^2) = 36y^2$, which cannot be expressed as the product of two factors whose sum is $-6y$, the coefficient of x. ● **31**

EXAMPLE 12 Factor the expression

$$(m + n)^4(m - n) - (m - n)^4(m + n).$$

● **32.** Factor $6(x + 2y)^{7/3}(3x - y)^{5/4} - 2(x + 2y)^{4/3}(3x - y)^{9/4}$.

Solution First set $x = m + n$ and $y = m - n$. The given expression is then

$$x^4y - y^4x = xy(x^3 - y^3) = xy(x - y)(x^2 + xy + y^2).$$

Now, $x - y = m + n - (m - n) = 2n$ and

$$
\begin{aligned}
x^2 + xy + y^2 &= (m + n)^2 + (m + n)(m - n) + (m - n)^2 \\
&= (m^2 + 2mn + n^2) + (m^2 - n^2) + (m^2 - 2mn + n^2) \\
&= 3m^2 + n^2.
\end{aligned}
$$

The given expression therefore factors as

Answer $14y(x + 2y)^{4/3}(3x - y)^{5/4}$.

$$xy(x - y)(x^2 + xy + y^2) = 2n(m + n)(m - n)(3m^2 + n^2). \quad ● \ \textbf{32}$$

Note According to Formulas (2) and (3), the sum and difference of two cubes can always be factored. In fact, every expression of the type $a^n + b^n$ or $a^n - b^n$ can be factored for all integers $n \geq 2$ with the single exception of the sum of two squares, $a^2 + b^2$. For example,

$$
\begin{aligned}
a^4 - b^4 &= (a^2 - b^2)(a^2 + b^2) = (a - b)(a + b)(a^2 + b^2) \\
a^5 + b^5 &= (a + b)(a^4 - a^3b + a^2b^2 - ab^3 + b^4) \\
a^4 + b^4 &= (a^2 + \sqrt{2}ab + b^2)(a^2 - \sqrt{2}ab + b^2)
\end{aligned}
$$

and so on.

Summary of Factoring:

1. The first step in factoring an algebraic expression should be to remove all common monomial factors.

2. If the remaining factor is the *difference of two squares*, the *difference of two cubes*, or the *sum of two cubes*, then use Formula (1), (2), or (3) in order to factor further.

3. In order to factor a multinomial expression consisting of four terms, the *grouping method* should be used.

4. A trinomial expression of the type $mx^2 + px + q$ can often be factored into the product of two factors of the type $(ax + b)(cx + d)$, as outlined above.

(1–79) Factor the following expressions completely.

1. $3a + 6b$

2. $2x^2 + 10xy + 4x^3$

3. $4xy - 6yz$

4. $5x^2y + 10xy^2$

5. $2u + av - 2v - au$

6. $px - qy + py - qx$

7. $xy + 4x - 2y - 8$

8. $pq - 6q - 3p + 18$

9. $3x - py - 3y + px$

10. $2px - 3y + py - 6x$

11. $6xz - 16y - 24x + 4yz$

12. $15ac - 9ad - 30bc + 18bd$

13. $x^2 - 16$

14. $4y^2 - 25$

15. $3t^2 - 108a^2$

16. $5x^2 - 20y^2$

17. $x^3y - 25xy^3$

18. $x^5 - 4x^3y^2$

19. $x^2 + 3x + 2$

20. $x^2 + 5x + 6$

21. $x^2 + x - 2$

22. $x^2 - 7x + 12$

23. $x^2 - x - 2$

24. $x^2 - 8x + 12$

25. $x^2 - 15x + 54$

26. $x^2 - 14x + 48$

27. $x^2 - 12x + 11$

28. $x^2 - 9x + 20$

29. $2x^2 + 2x - 12$

30. $3x^2 - 6x + 3$

31. $5y^4 + 25y^3 - 70y^2$

32. $12x - 7x^2 + x^3$

33. $2x^2 + 5x + 3$

34. $6x^2 + 10x - 4$

35. $9 + 12x + 4x^2$

36. $9t^2 - 12t + 4$

37. $5x^2 - 17x + 6$

38. $2t^2 - 3t - 14$

39. $10x^2 - 11x - 6$

40. $2t^2 - 7t + 6$

41. $3q^2 + 20q + 32$

42. $10p^2 + 3p - 18$

43. $6x^3y + 4x^2y - 10xy$

44. $(x^3 - 9x) + (45 - 5x^2)$

45. $x^2 + 6xy + 5y^2$

46. $x^2 - 4xy - 5y^2$

47. $p^2 - pq - 20q^2$

48. $s^2 + 7st - 30t^2$

49. $2t^2 + tu - 6u^2$

50. $2x^2 - 9xy + 10y^2$

51. $6a^2 + ab - 15b^2$

52. $18u^2 + 15uv - 12v^2$

53. $x^3 - 27$

54. $8t^3 + 125$

55. $27u^3 + 8v^3$

56. $128x^3 - 54$

57. $64x^4y^2 - 27xy^5$

58. $x^2y^2 - a^2y^2 - b^2x^2 + a^2b^2$

59. $x^2y^2 - 9y^2 - 4x^2 + 36$

60. $5u^2v^2 - 20v^2 + 15u^2 - 60$

61. $x^2z^2 - 4z^2 + x^4 - 4x^2$

62. $ax^3 + by^3 + bx^3 + ay^3$

63. $x^3 + y^3 + x^2y + xy^2$

64. $x^3y - 8 + 8y - x^3$

65. $(x + y)^3(3x - 2y)^4 + 2(x + y)^4(3x - 2y)^3$

66. $2(a - b)^2(a + b)^3 - 5(a + b)^2(a - b)^3$

67. $(x + y)^2 + 3(x + y) + 2$

68. $2(x + y)^2 + 5(x + y) + 2$

69. $3(a - b)^2 + 5(a - b) + 2$

70. $2(p - q)^2 - (p - q) - 1$

71. $3x^{2n} + 7x^n + 2$

72. $x^6 + y^6$

73. $x^6 - 8y^6$

74. $x^4 - 16y^4$

75. $(2x + 1)^2 - (x + 3)(x + 1)$

76. $5 + (2x + 3)^2 - (3x + 2)(x + 1)$

*77. $x^4 + 4y^4$

*78. $16a^4 + b^4$

*79. $x^5 + y^5$

◤ 1-7 ALGEBRAIC FRACTIONS

The term **algebraic fraction** is used generally for the ratio of two expressions containing one or more variable, such as the following.

$$\frac{x^2 - 7x + 5}{2x + 3} \quad \text{and} \quad \frac{x^2y + xy^2}{x - y}$$

To make an algebraic fraction meaningful, it is understood that the variable or variables do not take values which make the denominator of the fraction zero. Thus, in the fraction on the left, $x \neq -\frac{3}{2}$, because if $x = -\frac{3}{2}$, $2x + 3 = 2(-\frac{3}{2}) + 3 = -3 + 3 = 0$, and the denominator is zero. Similarly, in the fraction on the right, $y \neq x$.

In this section, we shall study methods of simplifying algebraic fractions and of adding, subtracting, multiplying, and dividing two or more such fractions. Factoring plays an important role in such operations, as will be clear from the following examples. *The basic principles involved will be the same as those described for simplifying fractions in Section 1-2.*

Simplification of Fractions

EXAMPLE 1 Simplify $\dfrac{4x^2 - 20x + 24}{6 + 10x - 4x^2}$.

Solution First of all, we completely factor the expressions that appear in the numerator and denominator. In this case, we have

$$4x^2 - 20x + 24 = 4(x^2 - 5x + 6) = 2 \cdot 2(x - 2)(x - 3)$$

and

$$6 + 10x - 4x^2 = -2(2x^2 - 5x - 3) = -2(2x + 1)(x - 3).$$

Note that in factoring the denominator, we first made the coefficient of x^2 positive, so that the x terms in the factors are positive both in the numerator and denominator. Therefore

$$\frac{4x^2 - 20x + 24}{6 + 10x - 4x^2} = \frac{2 \cdot 2(x - 2)(x - 3)}{-2(2x + 1)(x - 3)}$$

$$= \frac{2(x - 2)}{-(2x + 1)} = \frac{-2(x - 2)}{2x + 1}$$

Observe that we have divided the numerator and denominator by the factors 2 and $x - 3$, which appear in both the numerator and denominator. This cancellation of factors was justified in Section 1-2 (see page 00 and Theorem 5). It can be done for binomial factors such as $(x - 3)$ in this example just as well as for monomial factors. (Such factors must always be nonzero; otherwise the original fraction would not be well-defined.) ☛ 33

> ☛ **33.** Simplify $\dfrac{2x^2 - 4x + 2}{x^2 - 4x + 3}$.
> State any values of x at which the given fraction is not equal to your answer.

We sometimes encounter fractions that involve radicals in the denominator, such as

$$\frac{2}{3 - \sqrt{2}} \quad \text{and} \quad \frac{x}{\sqrt{x + 2} - \sqrt{2}}.$$

The first fraction is purely arithmetical, whereas the second is algebraic. In such cases, when the denominator involves only two terms, we can simplify the fraction by an operation called **rationalizing the denominator.** Consider

Answer $\dfrac{2(x - 1)}{x - 3}$, $x \neq 1$.

the first of the two fractions above as an example. If we multiply numerator and denominator by $3 + \sqrt{2}$, this has the effect of putting the radical into the numerator:

$$\frac{2}{3 - \sqrt{2}} = \frac{2(3 + \sqrt{2})}{(3 - \sqrt{2})(3 + \sqrt{2})}.$$

This works since the denominator in this new fraction can be simplified using the formula for the difference of two squares,

$$(a - b)(a + b) = a^2 - b^2.$$

Taking $a = 3$ and $b = \sqrt{2}$, we have

$$(3 - \sqrt{2})(3 + \sqrt{2}) = 3^2 - (\sqrt{2})^2 = 9 - 2 = 7.$$

Therefore

$$\frac{2}{3 - \sqrt{2}} = \frac{2(3 + \sqrt{2})}{7}.$$

In general, to rationalize a fraction involving an expression in the form $A + \sqrt{B}$ in the denominator, we multiply numerator and denominator by $A - \sqrt{B}$. If $A - \sqrt{B}$ occurs, we multiply numerator and denominator by $A + \sqrt{B}$. More generally, if a factor of the type $P\sqrt{A} \pm Q\sqrt{B}$ occurs in the denominator of a fraction, we multiply the numerator and denominator by $(P\sqrt{A} \mp Q\sqrt{B})$. (Note the change in sign of the second term.)

EXAMPLE 2 Rationalize the denominators of the following expressions.

(a) $\dfrac{1}{2\sqrt{5} + 3\sqrt{3}}$; (b) $\dfrac{x - 3}{\sqrt{x + 2} - \sqrt{5}}$

Solution

(a) The factor $2\sqrt{5} + 3\sqrt{3}$ occurs in the denominator so we multiply by $2\sqrt{5} - 3\sqrt{3}$:

$$\frac{1}{2\sqrt{5} + 3\sqrt{3}} = \frac{1 \cdot (2\sqrt{5} - 3\sqrt{3})}{(2\sqrt{5} + 3\sqrt{3})(2\sqrt{5} - 3\sqrt{3})}$$

$$= \frac{2\sqrt{5} - 3\sqrt{3}}{(2\sqrt{5})^2 - (3\sqrt{3})^2}$$

$$= \frac{2\sqrt{5} - 3\sqrt{3}}{4 \cdot 5 - 9 \cdot 3} = \frac{2\sqrt{5} - 3\sqrt{3}}{20 - 27}$$

$$= -\frac{1}{7}(2\sqrt{5} - 3\sqrt{3}).$$

(b) We multiply by $\sqrt{x + 2} + \sqrt{5}$:

$$\frac{x - 3}{\sqrt{x + 2} - \sqrt{5}} = \frac{(x - 3)(\sqrt{x + 2} + \sqrt{5})}{(\sqrt{x + 2} - \sqrt{5})(\sqrt{x + 2} + \sqrt{5})}$$

$$= \frac{(x - 3)(\sqrt{x + 2} + \sqrt{5})}{(\sqrt{x + 2})^2 - (\sqrt{5})^2}$$

$$= \frac{(x - 3)(\sqrt{x + 2} + \sqrt{5})}{x + 2 - 5}$$

$$= \frac{(x - 3)(\sqrt{x + 2} + \sqrt{5})}{x - 3}$$

$$= \sqrt{x + 2} + \sqrt{5}$$

where in the last step we canceled the common factor $(x - 3)$. ☛ 34

☛ **34.** Rationalize the denominator of $\dfrac{5 + \sqrt{2}}{5 - \sqrt{2}}$.

Addition and Subtraction of Fractions

Two or more fractions that have a common denominator can be added or subtracted simply by adding or subtracting their numerators while keeping the denominator unchanged.

EXAMPLE 3

(a) $\dfrac{2x + 3}{x + 1} + \dfrac{x - 1}{x + 1} = \dfrac{(2x + 3) + (x - 1)}{x + 1} = \dfrac{2x + 3 + x - 1}{x + 1}$

$$= \dfrac{3x + 2}{x - 1}$$

(b) $\dfrac{2x + 5}{x - 1} - \dfrac{7}{x - 1} = \dfrac{(2x + 5) - 7}{x - 1} = \dfrac{2x - 2}{x - 1} = \dfrac{2(x - 1)}{x - 1} = 2$

When the fractions to be added or subtracted do not have the same denominator, we first find their least common denominator (LCD) and then replace each of the given fractions by an equivalent fraction having this LCD as its denominator. The method is, in principle, no different from that described in Section 1-2.

To find the LCD of two or more fractions, we factor each denominator completely. The LCD is then obtained by multiplying all the distinct factors that appear in the denominators and raising each factor to the highest power to which it occurs in any one denominator. For example, the LCD of

$$\frac{2x + 1}{x - 3} \quad \text{and} \quad \frac{3x - 1}{2x + 7} \quad \text{is} \quad (x - 3)(2x + 7).$$

The LCD of $\dfrac{x + 1}{(x - 1)^2}, \quad \dfrac{5}{(x - 1)(x + 2)}, \quad \text{and} \quad \dfrac{7}{(x + 2)^3(x + 3)}$ is

$$(x - 1)^2(x + 2)^3(x + 3).$$

Answer $\frac{1}{23}(27 + 10\sqrt{2})$.

EXAMPLE 4 Simplify $\dfrac{2x + 1}{x + 2} + \dfrac{x - 1}{3x - 2}$.

Solution Here the denominators are already completely factored. The LCD in this case is $(x + 2)(3x - 2)$. To replace the first fraction, $(2x + 1)/(x + 2)$, by an equivalent fraction with the LCD $(x + 2)(3x - 2)$ as its denominator, we multiply the numerator and denominator by $3x - 2$. Thus

$$\frac{2x + 1}{x + 2} = \frac{(2x + 1)(3x - 2)}{(x + 2)(3x - 2)}.$$

Similarly,

$$\frac{x - 1}{3x - 2} = \frac{(x - 1)(x + 2)}{(x + 2)(3x - 2)}.$$

Therefore we have the following sum.

$$\frac{2x + 1}{x + 2} + \frac{x - 1}{3x - 2} = \frac{(2x + 1)(3x - 2)}{(x + 2)(3x - 2)} + \frac{(x - 1)(x + 2)}{(x + 2)(3x - 2)}$$

$$= \frac{(2x + 1)(3x - 2) + (x - 1)(x + 2)}{(x + 2)(3x - 2)}$$

$$= \frac{(6x^2 - x - 2) + (x^2 + x - 2)}{(x + 2)(3x - 2)}$$

$$= \frac{7x^2 - 4}{(x + 2)(3x - 2)} \qquad ☛ \ 35$$

☛ **35.** Simplify $\dfrac{4x + 2}{x^2 - 1} - \dfrac{3}{x - 1}$.

EXAMPLE 5 Simplify $\dfrac{5}{x^2 - 3x + 2} - \dfrac{1}{x + 2} + \dfrac{3}{x^2 - 4x + 4}$.

Solution The given expression, after factoring the denominators, is

$$\frac{5}{(x - 1)(x - 2)} - \frac{1}{x + 2} + \frac{3}{(x - 2)^2}.$$

Here the LCD is $(x - 1)(x - 2)^2(x + 2)$.

$$\frac{5}{(x - 1)(x - 2)} - \frac{1}{x + 2} + \frac{3}{(x - 2)^2}$$

$$= \frac{5(x - 2)(x + 2)}{(x - 1)(x - 2)^2(x + 2)} - \frac{(x - 1)(x - 2)^2}{(x + 2)(x - 1)(x - 2)^2}$$

$$+ \frac{3(x - 1)(x + 2)}{(x - 2)^2(x - 1)(x + 2)}$$

$$= \frac{5(x - 2)(x + 2) - (x - 1)(x - 2)^2 + 3(x - 1)(x + 2)}{(x - 1)(x + 2)(x - 2)^2}$$

Answer $\dfrac{1}{x + 1}$, $x \neq 1$.

$$= \frac{5(x^2 - 4) - (x - 1)(x^2 - 4x + 4) + 3(x^2 + x - 2)}{(x - 1)(x + 2)(x - 2)^2}$$

$$= \frac{5x^2 - 20 - (x^3 - 5x^2 + 8x - 4) + 3x^2 + 3x - 6}{(x - 1)(x + 2)(x - 2)^2}$$

$$= \frac{-x^3 + 13x^2 - 5x - 22}{(x - 1)(x + 2)(x - 2)^2}$$

EXAMPLE 6 Simplify $\sqrt{1 - x^2} + \dfrac{1 + x^2}{\sqrt{1 - x^2}}$.

Solution In this case, we write both terms as fractions with an LCD of $\sqrt{1 - x^2}$.

$$\sqrt{1 - x^2} = \frac{\sqrt{1 - x^2}\sqrt{1 - x^2}}{\sqrt{1 - x^2}} = \frac{1 - x^2}{\sqrt{1 - x^2}}$$

Thus we have the following sum.

$$\sqrt{1 - x^2} + \frac{1 + x^2}{\sqrt{1 - x^2}} = \frac{1 - x^2}{\sqrt{1 - x^2}} + \frac{1 + x^2}{\sqrt{1 - x^2}}$$

$$= \frac{1 - x^2 + 1 + x^2}{\sqrt{1 - x^2}} = \frac{2}{\sqrt{1 - x^2}}$$

Multiplication of Fractions

Two or more fractions can be multiplied together simply by multiplying their numerators and denominators, as illustrated in Example 7.

EXAMPLE 7

(a) $\dfrac{2x + 1}{x - 2} \cdot \dfrac{3 - x}{x + 1} = \dfrac{(2x + 1)(3 - x)}{(x - 2)(x + 1)}$

(b) $\dfrac{x^2 - 5x + 6}{6x^2 + 18x + 12} \cdot \dfrac{4x^2 - 16}{2x^2 - 5x - 3}$

$$= \frac{(x^2 - 5x + 6)(4x^2 - 16)}{(6x^2 + 18x + 12)(2x^2 - 5x - 3)}$$

This product can be simplified by factoring the numerator and denominator and dividing numerator and denominator by their common factors.

$$\frac{(x - 2)(x - 3) \cdot 2 \cdot 2(x - 2)(x + 2)}{2 \cdot 3(x + 1)(x + 2)(x - 3)(2x + 1)} = \frac{2(x - 2)(x - 2)}{3(x + 1)(2x + 1)}$$

$$= \frac{2(x - 2)^2}{3(x + 1)(2x + 1)}$$

Division of Fractions

To divide a fraction a/b by another fraction c/d, we invert c/d and multiply. (See page 10 and Theorem 4 of Section 1-2.)

$$\frac{a}{b} \div \frac{c}{d} = \frac{a/b}{c/d} = \frac{a}{b} \cdot \frac{d}{c}$$

This method is illustrated for algebraic fractions in Example 8.

EXAMPLE 8

(a) $\dfrac{2x + 3}{x - 1} \div \dfrac{x + 3}{2x^2 - 2} = \dfrac{2x + 3}{x - 1} \cdot \dfrac{2x^2 - 2}{x + 3}$

$\qquad\qquad = \dfrac{(2x + 3) \cdot 2(x - 1)(x + 1)}{(x - 1)(x + 3)}$

$\qquad\qquad = \dfrac{2(x + 1)(2x + 3)}{(x + 3)}$

(b) $\dfrac{\dfrac{3x - 1}{x - 2}}{\dfrac{x + 1}{1}} = \dfrac{\dfrac{3x - 1}{x - 2}}{\dfrac{x + 1}{1}} = \dfrac{3x - 1}{x - 2} \cdot \dfrac{1}{x + 1} = \dfrac{3x - 1}{(x - 2)(x + 1)}$ ☛ **36**

EXAMPLE 9 Simplify $\dfrac{x + 2 - \dfrac{4}{x - 1}}{\dfrac{x^2 - 5x + 6}{x^2 - 1}}$.

Solution First of all, we simplify the numerator.

$$x + 2 - \frac{4}{x - 1} = \frac{x + 2}{1} - \frac{4}{x - 1} = \frac{(x + 2)(x - 1)}{x - 1} - \frac{4}{x - 1}$$

$$= \frac{(x + 2)(x - 1) - 4}{x - 1} = \frac{x^2 + x - 6}{x - 1}$$

Using this value for the numerator, we complete the division.

$$\frac{\dfrac{x^2 + x - 6}{x - 1}}{\dfrac{x^2 - 5x + 6}{x^2 - 1}} = \frac{x^2 + x - 6}{x - 1} \cdot \frac{x^2 - 1}{x^2 - 5x + 6}$$

$$= \frac{(x^2 + x - 6)(x^2 - 1)}{(x - 1)(x^2 - 5x + 6)}$$

$$= \frac{(x - 2)(x + 3)(x - 1)(x + 1)}{(x - 1)(x - 2)(x - 3)}$$

$$= \frac{(x + 3)(x + 1)}{x - 3}$$ ☛ **37**

☛ **36.** Simplify
$\dfrac{4x + 2}{x^2 - 1} \cdot \dfrac{3x^2 + 4x + 1}{2x + 1} \div \dfrac{3}{x - 1}$.

Answer

$\dfrac{2(3x + 1)}{3}$, $\quad x \neq 1, -1$ *or* $-\frac{1}{2}$.

☛ **37.** Simplify $\dfrac{2x^2 - 3x + 1}{((x^2 - 1)/(2x + 1))}$.

Answer

$\dfrac{(2x - 1)(2x + 1)}{x + 1}$, $\quad x \neq 1$ *or* $-\dfrac{1}{2}$.

EXERCISES 1-7

(1–40) In the following questions, perform the indicated operations and simplify.

1. $\dfrac{4x}{2x+3} + \dfrac{6}{2x+3}$

2. $\dfrac{2x}{x-2} - \dfrac{4}{x-2}$

3. $\dfrac{x^2}{x-3} - \dfrac{5x-6}{x-3}$

4. $\dfrac{2-3x}{x-1} + \dfrac{x^2}{x-1}$

5. $\dfrac{2x+1}{x+2} + 3$

6. $\dfrac{3x-2}{x+1} - 2$

7. $\dfrac{x}{x+2} + \dfrac{3}{2x-1}$

8. $\dfrac{x}{2x-6} + \dfrac{x-2}{x+1}$

9. $\dfrac{2}{x-1} - \dfrac{3x+1}{x+1}$

10. $\dfrac{x}{2x+3} - \dfrac{2x-3}{4x+1}$

11. $\dfrac{2x}{2x-1} - \dfrac{x+2}{x+1}$

12. $\dfrac{2}{5x-6} - \dfrac{4}{10x-2}$

13. $\dfrac{1}{x^2-5x+6} - \dfrac{1}{x^2-3x+2}$

14. $\dfrac{x}{x^2+2x-3} + \dfrac{1}{x^2+x-2}$

15. $\dfrac{x}{x^2+2x-3} + \dfrac{1}{1-2x+x^2}$

16. $\dfrac{2}{9x^2-6x+1} - \dfrac{3}{x+1} + \dfrac{1}{3x^2+2x-1}$

17. $\dfrac{1}{x^2+4x+3} + \dfrac{3}{x^2-1} - \dfrac{2}{x+3}$

18. $\dfrac{x}{2x^2-x-1} - \dfrac{3}{1-2x+x^2} + 2$

19. $\left(\dfrac{x^2-1}{x}\right)\left(\dfrac{x^2+2x}{x+1}\right)$

20. $\left(\dfrac{x^2+4x}{2x+6}\right)\left(\dfrac{2x+4}{x+4}\right)$

21. $\dfrac{2x+4}{1-x} \cdot \dfrac{x^2-1}{3x+6}$

22. $\dfrac{x^2-7x+12}{x^2-x-2} \cdot \dfrac{x^2+4x+3}{2x^2-5x-3}$

23. $\left(\dfrac{x^2+5x+6}{x^2-6x+8}\right)\left(\dfrac{2x^2-9x+4}{2x^2+7x+3}\right)$

24. $\left(\dfrac{2x^4-2x}{2x^2-5x-3}\right)\left(\dfrac{2x^2-3x-2}{x^3+x^2+x}\right)$

25. $\left(3+\dfrac{1}{x-1}\right)\left(1-\dfrac{1}{3x-2}\right)$

26. $\left(x-\dfrac{3}{x-2}\right)\left(\dfrac{9}{x^2-9}-1\right)$

27. $\left(\dfrac{x^2+x}{2x+1}\right) \div \left(\dfrac{x^3-x}{4x+2}\right)$

28. $\left(\dfrac{3x-6}{2x^2+4x+2}\right) \div \left(\dfrac{x^2-4}{x^2+3x+2}\right)$

29. $\dfrac{3x^2-x-2}{x^2-x-2} \div \dfrac{3x^2+5x+2}{2x^2-5x+2}$

30. $\dfrac{2x^2+x-1}{2x^2+10x+12} \div \dfrac{1-4x^2}{4x^2+8x-12}$

31. $\dfrac{\dfrac{x^2+x-2}{2x+3}}{\dfrac{x^2-4}{2x^2+5x+3}}$

32. $\dfrac{1-1/t^2}{t+1-2/t}$

33. $\dfrac{x+2+\dfrac{3}{x-2}}{x-6+\dfrac{7}{x+2}}$

34. $\dfrac{p-\dfrac{2}{p+1}}{1-\dfrac{4p+7}{p^2+4p+3}}$

35. $\dfrac{x^{-1}+y^{-1}}{(x+y)^{-1}}$

36. $\dfrac{(x-y)^{-1}}{(x^{-2}-y^{-2})^{-1}}$

37. $\dfrac{x^{-2}+y^{-2}}{x^{-2}-y^{-2}} \cdot \dfrac{x-y}{x+y}$

38. $\dfrac{y^{-2}-x^{-2}}{xy^{-1}-yx^{-1}}$

39. $\dfrac{1}{h}\left(\dfrac{1}{x+h} - \dfrac{1}{x}\right)$

40. $\dfrac{1}{h}\left[\dfrac{1}{(x+h)^2} - \dfrac{1}{x^2}\right]$

(41–52) Rationalize the denominators of the following expressions.

41. $\dfrac{1}{3+\sqrt{7}}$

42. $\dfrac{3+\sqrt{2}}{2-\sqrt{3}}$

43. $\dfrac{1+\sqrt{2}}{\sqrt{5}+\sqrt{3}}$

44. $\dfrac{6\sqrt{2}}{\sqrt{3}+\sqrt{6}}$

45. $\dfrac{3}{3+\sqrt{3}}$

46. $\dfrac{1}{2\sqrt{3}-\sqrt{6}}$

47. $\dfrac{1}{\sqrt{x} - \sqrt{y}}$

48. $\dfrac{\sqrt{x} - \sqrt{y}}{\sqrt{x} + \sqrt{y}}$

49. $\dfrac{x}{\sqrt{x + 2} - \sqrt{2}}$

50. $\dfrac{x}{\sqrt{x + 1} - \sqrt{x - 1}}$

51. $\dfrac{2x - 2}{\sqrt{x + 3} - 2\sqrt{x}}$

52. $\dfrac{4 - x}{2\sqrt{x + 5} - 3\sqrt{x}}$

(53–56) Rationalize the numerators of the following expressions.

53. $\dfrac{5 - \sqrt{3}}{2}$

54. $\dfrac{\sqrt{x + 4} - \sqrt{x}}{2}$

55. $\dfrac{\sqrt{x + h} - \sqrt{x}}{h}$

56. $\dfrac{\sqrt{x - 2 + h} - \sqrt{x - 2}}{h}$

CHAPTER REVIEW

Key Terms, Symbols, and Concepts

1.1 Natural number, integer, rational number, irrational number, real number.

The real number line.

Commutative, associative, and distributive properties.

Identity elements, additive inverse of a (the negative of a, $-a$), multiplicative inverse of a (the reciprocal of a, a^{-1}).

Subtraction: $a - b \equiv a + (-b)$.
Division: $a \div b \equiv a \cdot b^{-1}$.

1.2 Fraction. Definitions: $\dfrac{1}{b} \equiv b^{-1}$, $\dfrac{a}{b} \equiv a \cdot b^{-1}$

Rules for multiplying and dividing fractions. Cancellation of common factors.

Lowest common denominator (LCD). Addition and subtraction of fractions.

1.3 Power (exponent), base. a^n (a to the power n).

Properties of exponents.

1.4 Principal nth root of a: $b = a^{1/n}$ if $b^n = a$.

Radical, square root, cube root, nth root. $\sqrt{a}$, $\sqrt[3]{a}$, $\sqrt[n]{a}$.

Fractional exponents: $a^{m/n} = (a^{1/n})^m$.

Extension of the five properties of exponents to fractional exponents and radicals.

1.5 Algebraic expression, monomial, binomial, multinomial expressions.

Term, literal part, (numerical) coefficient; constant term.

Like terms, addition and subtraction of like terms.

Multiplication of expressions using the distributive property; method of arcs.

Binomial square formulas. Difference of squares formula.

Division by monomial. Long division of polynomial expressions.

Divisor, dividend, quotient, remainder.

1.6 Factors. Monomial factors. Grouping method.

Factoring by formulas for difference of squares, sum and difference of cubes.

Factoring expressions of the type $x^2 + px + q$ and $mx^2 + px + q$ with $m \neq 1$, -1

1.7 Rationalizing the denominator.

Techniques for addition, subtraction, multiplication, division, and simplification of algebraic fractions.

Formulas

Properties of exponents

$$a^m \cdot a^n = a^{m+n}, \qquad \frac{a^m}{a^n} = a^{m-n}, \qquad (a^m)^n = a^{mn},$$

$$(ab)^m = a^m b^m, \qquad \left(\frac{a}{b}\right)^m = \frac{a^m}{b^m}.$$

Binomial square formulas:

$$(x \pm a)^2 = x^2 \pm 2ax + a^2.$$

Difference of squares formula:

$$(x + a)(x - a) = x^2 - a^2.$$

Sum and difference of cubes formulas:

$$x^3 \pm a^3 = (x \pm a)(x^2 \mp ax + a^2).$$

REVIEW EXERCISES FOR CHAPTER 1

1. State whether each of the following is true or false. Replace each false statement with a corresponding true statement.

a. $a^m \cdot b^n = (ab)^{mn}$

b. $a^m + b^m = (a + b)^m$

c. $(2^0)^m = 1$

d. $(a - b)^2 = a^2 - b^2$

e. $-2(a + b) = -2a + b$

f. $(x + y)^2 = x^2 + y^2$

g. $\sqrt{a - b} = \sqrt{a} - \sqrt{b}$

h. $\dfrac{a + 2b}{a} = 2b$

i. $\sqrt[3]{a^2} = \sqrt[6]{a^4}$

j. $\dfrac{1}{a} - \dfrac{1}{b} = \dfrac{1}{a - b}$

k. $\dfrac{a/b}{c} = \dfrac{a}{b} \cdot \dfrac{c}{1}$

l. $(2a)^5 = 2a^5$

m. $\dfrac{a}{b} \div \dfrac{c}{d} = \dfrac{a \div c}{b \div d}$

n. $\dfrac{a}{b} \cdot \dfrac{c}{d} = \dfrac{a \cdot c}{b \cdot d}$

o. $(-1)^n = -1$ if n is an odd integer.

p. $\left(\dfrac{2}{3}\right)\left(\dfrac{3}{4}\right)\left(\dfrac{4}{5}\right)\left(\dfrac{5}{6}\right)\left(\dfrac{6}{7}\right) = \dfrac{2}{7}$

q. Every terminating decimal number represents a rational number.

r. Every rational number can be expressed as a terminating decimal.

(2–24) In the following expressions, perform the indicated operations and simplify the results.

2. $(125)^{2/3} \div (81)^{-3/4}$

3. $(32)^{-2} \cdot (243)^{1/5}$

4. $(2x^5)^2 \div (2x^3)^3$

5. $\dfrac{(3a^2b^{-1})^4}{(6a^{-1}b)^3}$

6. $\dfrac{(-5p^{-2}q^{-3})^{-2}}{(-10p^2q)^{-3}}$

7. $\dfrac{(-6x^{-1}y^{-2})^3}{(-2x^{-4}y^{-3})^4}$

8. $2\sqrt{\dfrac{3p^4}{q}} - \dfrac{p^2}{\sqrt{3q}}$

9. $(2x^3)^{-1/2}(2x)^{1/2}$

10. $(r^{-2/5})^2(r^{3/10})^3(r^{-2/15})$

11. $\dfrac{2x^{5/2}}{y^{3/4}} \div \dfrac{x^{2/3}}{3y^{2/5}}$

12. $\left(\dfrac{x^a}{x^b}\right)^{a+b}\left(\dfrac{x^b}{x^c}\right)^{b+c}\left(\dfrac{x^c}{x^a}\right)^{c+a}$

13. $\left(\dfrac{x^a}{x^b}\right)^c\left(\dfrac{x^b}{x^c}\right)^a\left(\dfrac{x^a}{x^c}\right)^b$

14. $\dfrac{1}{x + 2} + \dfrac{2}{x - 3}$

15. $\dfrac{1}{x^2 + 1} + 2$

16. $\dfrac{1}{x - 1} + \dfrac{1}{x^2 - 3x + 2} - \dfrac{1}{x^2 - 2x + 1}$

17. $\dfrac{2}{x^2 + 2x + 1} - \dfrac{1}{x^2 + 4x + 3} + \dfrac{3}{x^2 - x - 2}$

18. $\dfrac{x + y}{p^2 - q^2} \div \dfrac{x^2 - y^2}{p + q}$

19. $\dfrac{x^2 + 4x + 4}{y^2 - 9} \div \dfrac{x^2 + 5x + 6}{y^2 - y - 6}$

20. $\dfrac{a^2 - b^2}{2a + 4} \div \dfrac{a^2 - 3ab + 2b^2}{a^2 - 4}$

21. $\dfrac{a^2 + 2ab + b^2}{x^2 + 5x + 6} \cdot \dfrac{x^2 - x - 6}{a^2 - b^2}$

22. $\left(x - \dfrac{2}{x + 1}\right)\left(x + \dfrac{1}{x + 2}\right)$

23. $\left(a + \dfrac{2}{a + 3}\right)\left(a - \dfrac{9}{a}\right)$

24. $\dfrac{x + 1}{x - 2} + \dfrac{3x^2 - 27}{x + 3} - \dfrac{2x + 1}{2x - 1}$

(25–42) Factor the following expressions completely.

25. $3x^2 - 75y^2$

26. $x^2 + 7x + 10$

27. $6x^2 - x - 15$

28. $2p^2 + p - 28$

29. $x^2 + x - 12$

30. $u^2 - 2u - 3$

31. $k^2 + k - 20$

32. $10t^2 + 3tu - u^2$

33. $8x^2 - 18x + 9$

34. $12x^2 + 20x - 25$

35. $y^2 - 3y - 10$

36. $12x^2 + 7xy - 12y^2$

37. $(a + 4)(a - 3) + (2a + 3)(a + 1)$

38. $(x + 2)(x^2 + x - 1) + (2x - 1)(x^2 - 3x - 2)$

39. $4(x + 1)^2 - (2x + 5)^2$

40. $(p + q)^2 + 3(p + q) - 4$

41. $x^3 + \dfrac{8}{x^3}$

42. $(x - 1)(x^2 + 1) + (x + 1)(x^2 - 1)$

43. Prove that $\dfrac{1}{\sqrt{2} - 1} + \dfrac{2}{\sqrt{3} + 1} = \sqrt{2} + \sqrt{3}$.

44. Given $\sqrt{2} \approx 1.414$ and $\sqrt{3} \approx 1.732$, evaluate $\dfrac{1}{\sqrt{3} - \sqrt{2}}$ without using a calculator, tables, or long division. *

---

*The symbol $\approx$ means "is approximately equal to." It should always be used when a number is rounded.

CHAPTER 2

Equations in One Variable

Chapter Objectives

2-1 LINEAR EQUATIONS

  (a) The solutions of an equation;

  (b) Equivalent equations obtained by the addition and multiplication principles;

  (c) Solution of linear equations and equations reducible to linear equations.

2-2 APPLICATIONS OF LINEAR EQUATIONS

  (a) Translation of a word problem into an algebraic equation;

  (b) Solution of applied problems that lead to linear equations; earnings, profits, investment, and mixtures problems.

2-3 QUADRATIC EQUATIONS

Solution of a Quadratic Equation by

  (a) The factoring method;

  (b) The quadratic formula;

  (c) Completing the square.

2-4 APPLICATIONS OF QUADRATIC EQUATIONS

Solution of applied problems that lead to quadratic equations: apartment rental, pricing decision, and investment problems.

CHAPTER REVIEW

2-1 LINEAR EQUATIONS

An **equation** is a statement that expresses the equality of two algebraic expressions. It generally involves one or more variables and the equality symbol, $=$. The following are examples of equations.

$$2x - 3 = 9 - x \tag{1}$$

$$y^2 - 5y = 6 - 4y \tag{2}$$

$$2x + y = 7 \tag{3}$$

$$\frac{a}{1 - r} = s \tag{4}$$

In Equation (1), the variable is the letter x, while in Equation (2), it is y. In Equation (3), we have two variables, x and y. We do not allow the variable in any equation to take a value that would make an expression occurring in the equation undefined. For example, in Equation (4), r cannot be 1 because this would result in division by zero.

The two expressions separated by the equality symbol are called the two **sides** of the equation; individually they are called the *left side* and the *right side*.

Equations involving only constants and no variables are either true or false statements. For example,

$$3 + 2 = 5 \qquad \text{and} \qquad \tfrac{3}{15} = \tfrac{4}{20}$$

are true statements, while

$$2 + 5 = 6 \qquad \text{and} \qquad \tfrac{3}{2} = \tfrac{2}{3}$$

are false statements.

An equation containing a variable usually becomes a true statement for certain values of the variable, whereas it is a false statement for other values of the variable. For example, consider the equation

$$2x - 3 = x + 2.$$

If x takes the value 5, this equation becomes

$$2(5) - 3 = 5 + 2 \quad \text{or} \quad 10 - 3 = 5 + 2$$

which is a true statement. On the other hand, if x takes the value 4, we get

$$2(4) - 3 = 4 + 2 \quad \text{or} \quad 5 = 6$$

which is a false statement.

A value of the variable that makes an equation a true statement is called a **root** or **solution** of the given equation. We say that the equation is *satisfied* by such a value of the variable.

1. Which of the following numbers is a solution of the equation $x^3 - 3x^2 + 4 = 0$: $-2, -1, 0, 1, 2$?

Thus, for example, 5 is a root of the equation $2x - 3 = x + 2$. Similarly, -2 is a root of the equation $y^2 + 3y = 6 + 4y$ because when we substitute -2 for y in the equation we obtain

$$(-2)^2 + 3(-2) = 6 + 4(-2)$$

or $4 - 6 = 6 - 8$, which is a true statement.

Similarly, 5 is *not* a root of the equation $t^2 + 2t = 6 + 3t$ because when t is replaced by 5, we have

$$(5)^2 + 2(5) = 6 + 3(5)$$

or $25 + 10 = 6 + 15$, which is not a true statement. **1**

We are often interested in finding the roots of some given equation—that is, in determining all the values of the variable that make the equation a true statement. The process of finding the roots is called **solving the equation.** In carrying out this process, we usually perform certain operations on the equation which transform it into a new equation that is simpler to solve. Such simplications must be made in such a way that the new equation has the same roots as the original equation. The following two operations give new equations, while at the same time satisfy this requirement of leaving the roots of the equation unchanged.

1. (ADDITION PRINCIPLE) We can add or subtract any constant or any algebraic expression that has a well-defined value to both sides of the equation.

2. (MULTIPLICATION PRINCIPLE) We can multiply or divide both sides of an equation by any *nonzero* constant or any *nonzero* expression involving the variable.

   (**Note** Multiplication by an expression may produce an equation whose roots differ from those of the original equation if the expression becomes zero for certain values of the variable, as illustrated below.)

Observe that according to both these principles we must *do the same thing to both sides of the equation.*

Consider, for example, the equation

$$x - 3 = 2 \tag{5}$$

Let us add 3 to both sides of this equation. By the addition principle, this operation will not change the roots of the equation.

$$x - 3 + 3 = 2 + 3$$

After simplification, this becomes

$$x = 5.$$

We conclude, therefore, that if x satisfies Equation (5) then $x = 5$: this is the one and only solution of Equation (5).

Answer -1 and 2.

As a second example, consider the equation

$$5x = 15. \tag{6}$$

Let us divide both sides of this equation by 5. By the multiplication principle, this operation will not change the roots of the equation since the number by which we are dividing is nonzero. We get

$$\frac{5x}{5} = \frac{15}{5}$$

or

$$x = 3.$$

Thus the one and only solution of Equation (6) is $x = 3$.

Two equations with exactly the same solutions are said to be **equivalent** to one another. Operations 1 and 2 therefore transform a given equation into a new equation that is equivalent to the old one. In solving a given equation, we may have to use these operations several times in succession.

EXAMPLE 1 Solve the equation

$$5x - 3 = 2x + 9. \tag{7}$$

Solution First let us subtract $2x$ from both sides of the equation and simplify.

$$5x - 3 - 2x = 2x + 9 - 2x$$

$$5x - 2x - 3 = 2x - 2x + 9$$

$$3x - 3 = 9 \tag{8}$$

Next we add 3 to both sides and simplify.

$$3x - 3 + 3 = 9 + 3$$

$$3x = 12 \tag{9}$$

Finally, we divide both sides by 3 (which is nonzero).

$$\frac{3x}{3} = \frac{12}{3}$$

$$x = 4$$

Thus the solution of Equation (7) is $x = 4$. ☞ 2

We observe that Equation (8) could be obtained from Equation (7) simply by moving the term $2x$ from the right side to the left side and changing its sign. We would get

$$5x - 3 - 2x = 9$$

or

$$3x - 3 = 9$$

☞ **2.** Are the following pairs of equations equivalent?
(a) $1 - 2x = y$ and $1 - y = 2x$;
(b) $2(x - 1) = 0$ and $x = 1$;
(c) $(x + 1)(x - 1) = 0$ and $x - 1 = 0$;
(d) $x = 1$ and
$$x + \frac{1}{x - 1} = 1 - \frac{1}{1 - x}.$$

Answer (a) Yes; (b) yes;
(c) no ($x = -1$ is a solution of the first equation but not of the second);
(d) no ($x = 1$ is a solution of the first equation but not of the second).

which agrees with Equations (8). Again, we obtain Equation (9) from Equation (8) by simply moving the term -3 from the left side to the right side and changing its sign. We would get

$$3x = 9 + 3$$

or

$$3x = 12.$$

Thus we can see that the addition principle stated earlier implies the following: *We can move any term from one side of an equation to the other side after changing its sign without affecting the roots of the equation.*

According to this principle, the equation $5x + 3 = 2x$ is equivalent to $5x - 2x + 3 = 0$, or $3 = 2x - 5x$.

According to the multiplication principle, any expression by which we multiply or divide must be nonzero, so care must be taken not to multiply or divide the equation by an expression that can be equal to zero. For example, consider the equation

$$x^2 = 5x.$$

Clearly, $x = 0$ is a root of this equation. If we divide both sides by x, we obtain

$$x = 5.$$

We see that $x = 0$ is *not* a root of this resulting equation, although zero was a root of the original equation. The problem is that we have divided both sides by x, which can be zero, and this violates the multiplication principle. In dividing by x, we have lost one root of the equation. To avoid such pitfalls, *care must be exercised not to multiply or divide by an expression that contains the variable unless we are sure that this expression is nonzero.*

An important class of equations consists of those called **polynomial equations.** In a polynomial equation, the two sides may consist of one or several terms added together, each term comprising a nonnegative integral power* of the variable multiplied by a constant coefficient. The **degree** of a polynomial equation is the highest power of the variable that occurs in the equation.

EXAMPLE 2

(a) $\frac{2}{3}x^2 - 1 = 3x + 2$ is a polynomial equation of degree 2.

(b) $x^4 - \frac{3}{2}x^2 - 5x = 4$ is a polynomial equation of degree 4.

(c) $(x^2 + 1)/(x + 1) = 2x$ is not a polynomial equation because of the fraction with x in its denominator.

---

*In other words, each exponent is a whole number.

A polynomial equation of degree 1 is called a **linear equation** and a polynomial equation of degree 2 is called a **quadratic equation.** Linear and quadratic equations will be studied in this and the next three sections of this chapter. We have the following definition.

DEFINITION The *standard form* of a **linear equation** in the variable x is

$$ax + b = 0 \qquad (a \neq 0)$$

where a and b are constants.

EXAMPLE 3

(a) $x - 4 = 0$ is a linear equation. Moving the 4 to the right side and changing its sign, we have $x = 4$. (*Note* This is equivalent to adding 4 to both sides.) Thus the number 4 is the only solution of this equation.

(b) $2x + 3 = 0$ is a linear equation. Moving 3 to the right side, we have $2x = -3$; dividing by 2, we find $x = -\frac{3}{2}$. Thus $-\frac{3}{2}$ is the only solution of the given equation.

(c) In the general case,

$$ax + b = 0$$

we can move the constant b to the right side, which gives

$$ax = -b.$$

Now if we divide by a, we get $x = -b/a$. Thus the linear equation $ax + b = 0$ has one and only one solution, namely $x = -b/a$.

Observe that in solving these equations, we kept the terms involving x on the left side of the equation and moved the constant terms to the right side. This is the general strategy for solving all linear equations. (We used it in solving Example 1 earlier.)

Often equations arise that do not appear at first glance to be linear, but which may be reduced to linear equations by appropriate simplification. In carrying out such a reduction, the following step-by-step procedure is often helpful.

Step 1 Remove any fractions which occur in the equation by multiplying both sides by the common denominator of the fractions involved.
Step 2 Expand any parentheses which occur. (Steps 1 and 2 may be interchanged.)
Step 3 Move all the terms containing the variable to the left side and all other terms to the right side; then simplify, if possible, by combining like terms.

This procedure is amplified in the following examples.

EXAMPLE 4 Solve the equation $3x - 4(6 - x) = 15 - 6x$.

Solution

    Step 1 Since there are no fractions in the equation, we do not need step 1.

    Step 2 Expanding the parentheses gives

$$3x - 24 + 4x = 15 - 6x.$$

    Step 3 Moving all the terms containing the variable to the left and the constant terms to the right, not forgetting to change their signs, we get

$$3x + 4x + 6x = 15 + 24$$

or

$$13x = 39.$$

We now obtain a solution by dividing both sides by 13, the coefficient of x.

$$x = \tfrac{39}{13} = 3$$

EXAMPLE 5 Solve the following equation

$$\frac{5x}{3} - \frac{x - 2}{4} = \frac{9}{4} - \frac{1}{2}\left(x - \frac{2x - 1}{3}\right)$$

Solution After removing the last parentheses, we can write the given equation as

$$\frac{5x}{3} - \frac{x - 2}{4} = \frac{9}{4} - \frac{x}{2} + \frac{2x - 1}{6}.$$

In order to remove the fractions, we multiply both sides by 12, the common denominator, and simplify.

$$12\left(\frac{5x}{3}\right) - 12\left(\frac{x - 2}{4}\right) = 12\left(\frac{9}{4}\right) - 12\left(\frac{x}{2}\right) + 12\left(\frac{2x - 1}{6}\right)$$

$$4(5x) - 3(x - 2) = 3(9) - 6x + 2(2x - 1)$$

$$20x - 3x + 6 = 27 - 6x + 4x - 2$$

Moving the x-terms all to the left and the constant terms to the right, we have

$$20x - 3x + 6x - 4x = 27 - 2 - 6$$

$$19x = 19.$$

Finally, dividing both sides by 19, we obtain $x = 1$, the required solution. ☞ **3**

EXAMPLE 6 Solve the equation

$$\frac{x - 2t}{a} = \frac{3(x - y)}{z}$$

    (a) for x; (b) for t.

☞ **3.** Solve the following equations.

(a) $3 - 2x = 7$;

(b) $4 - x = 3x - 4$;

(c) $3(x + 2) = 2(8 - x)$;

(d) $\tfrac{2}{3}(1 - 2x) = 4 - \tfrac{1}{2}(3x + 4)$.

Answer (a) -2; (b) 2; (c) 2;
(d) 8.

Solution Here the common denominator is az. Multiplying both sides by az to clear the fractions, we have

$$z(x - 2t) = 3a(x - y)$$

$$xz - 2zt = 3ax - 3ay. \qquad (10)$$

(Note that neither a nor z can be zero, since otherwise the given equation would have a fraction with zero denominator. Hence we are allowed to multiply by az.)

(a) Since we are solving for x, all of the other letters involved in the equation are treated as constants. Moving all the terms containing the variable x to the left and all the terms without x to the right, we have

$$xz - 3ax = -3ay + 2zt$$

$$x(z - 3a) = 2zt - 3ay.$$

☞ **4.** Solve for r: $\quad S = \dfrac{a}{1 - r}$.

We divide both sides of the equation by $z - 3a$, assuming this factor to be nonzero.

$$x = \frac{2zt - 3ay}{z - 3a}$$

(b) Since we are solving for t, we shall keep only those terms which contain the variable t on the left and move all other terms to the right. Thus, from Equation (10), we have

$$-2zt = 3ax - 3ay - xz.$$

Dividing both sides by $-2z$, the coefficient of t, which, as noted above, cannot be zero, we have

$$t = \frac{3ax - 3ay - xz}{-2z} = \frac{1}{2z}(-3ax + 3ay + xz)$$

Answer $r = 1 - a/S$.

which is the required solution for the variable t. ☞ **4**

☞ **5.** What is wrong with the following? We are required to solve the equation

$$\frac{1}{x - 2} = 2 + \frac{x - 3}{2 - x}$$

First multiply through by $(x - 2)$:

$$1 = 2(x - 2) - (x - 3).$$

That is,

$$1 = 2x - 4 - x + 3 = x - 1.$$

Therefore $x = 2$ is the solution.

Answer When $x = 2$ the original equation contains undefined terms. There is no solution.

EXAMPLE 7 Solve the equation $(2x + 1)^2 = 4(x^2 - 1) + x - 1$.

Solution At first glance, this equation does not appear to be linear because of the terms involving x^2. However, we shall see that it does reduce to a linear equation. Let us remove the parentheses and move all the terms involving x to the left side of the equation. We obtain

$$4x^2 + 4x + 1 = 4x^2 - 4 + x - 1$$

$$4x^2 + 4x - 4x^2 - x = -4 - 1 - 1.$$

We observe that the two terms involving $4x^2$ cancel one another out (that is, $4x^2 + 4x - 4x^2 - x = (4 - 4)x^2 + (4 - 1)x = 0x^2 + 3x$) and we are left with

$$3x = -6.$$

Hence the solution is $x = -2$. ☞ **5**

EXERCISES 2-1

(1–10) Check whether the given number(s) are solutions of the corresponding equations.

1. $3x + 7 = 12 - 2x$; 1

2. $5t - 3 = 18 + 3(1 - t)$; 3

3. $\dfrac{u + 2}{3u - 1} + 1 = \dfrac{6 - u}{u + 1}$; 2

4. $\dfrac{1 - 2y}{3 - y} + y = \dfrac{1}{y + 2}$; -2

5. $x^2 = 5x - 6$; $2, 5$

6. $y^2 + 12 = 7y$; $4, -3$

7. $\dfrac{5}{x} - \dfrac{3}{2x} = \dfrac{x}{2}$; 3

8. $\dfrac{7}{x + 1} + \dfrac{15}{3x - 1} = 8$; $-\frac{1}{2}, \frac{1}{3}$

9. $\dfrac{3}{x - 1} - \dfrac{5x}{x + 2} = \dfrac{1}{4}$; 1

10. $4x + \dfrac{7}{x} = 3$; 0

(11–14) Reduce the following equations to polynomial equations and state the resulting degree.

11. $x^3 - 7x^2 + 5 = x(x^2 - 1) + 3x^2 - 2$

12. $(y - 2)(y + 5) = (2y - 1)(y + 1) + 7$

13. $y^2 + 7 = (y - 1)^2 + 3y$

14. $(u - 1)^2 = (u + 1)(u + 3) + 5$

(15–32) Solve the following equations.

15. $1 + x = 3 - x$

16. $3x + 7 = 3 + 5x$

17. $2x - 5 = -15 - 3x$

18. $2 - 7x = 3x - 2$

19. $4(x - 3) = 8 - x$

20. $2x - 5(1 - 3x) = 1 - 3(1 - 2x)$

21. $3 - 2(1 - x) = 5 + 7(x - 3)$

22. $6y - 5(1 + 2y) = 3 + 2(1 - y)$

23. $3z - 2 + 4(1 - z) = 5(1 - 2z) - 12$

24. $5[1 - 2(2z - 1)] = -3(3z - 1) + 1$

25. $1 - 2[4 - 3(x + 1)] = 4(x - 5) - 1$

26. $3[2x + 1 - 2(2x - 1)] + 4 = 2[1 + 2(3 - x)]$

27. $\dfrac{3x + 7}{2} = \dfrac{1 + x}{3}$

28. $\dfrac{2x - 7}{3} = 5 - \dfrac{3x - 2}{4}$

29. $1 - \dfrac{2u - 3}{4} = \dfrac{2 - 5u}{3} - 3u$

30. $\dfrac{5y - 6}{2} = y - \dfrac{2 - y}{3}$

31. $\frac{1}{3}(2y + 1) + \frac{1}{2}y = \frac{2}{5}(1 - 2y) - 4$

32. $\dfrac{1}{2}\left[1 + \dfrac{1}{4}(3z - 1)\right] = \dfrac{2z}{3} - \dfrac{1}{2}$

(33–40) Reduce the following equations to linear equations and solve them.

33. $(x - 4)^2 = (x - 2)^2$

34. $(x - 1)(x + 3) = (x + 2)(x - 3) + 1$

35. $x^2 + (x + 1)^2 = (2x - 1)(x + 3)$

36. $(3x - 1)(x + 2) + 5x = (2x + 1)(x - 3) + x^2$

37. $(2x + 1)(x - 1) + x^2 = 3(x - 1)(x + 2) - 3$

38. $(3x + 1)(2x - 1) - 2x^2 = (2x - 3)^2 + 6x + 5$

39. $x(x + 2)(x + 4) + x^3 = 2(x + 1)^3$

40. $(x + 1)^3 + (x - 1)^3 = 2x^3$

(41–44) Solve the following equations for the indicated variables.

41. $ax + by = cz$: **(a)** for x, **(b)** for b

42. $S = \dfrac{a - rl}{1 - r}$: **(a)** for r, **(b)** for l

43. $\dfrac{1}{x} + \dfrac{1}{y} = \dfrac{1}{t}$: **(a)** for x, **(b)** for t

44. $\dfrac{2}{x} + \dfrac{3}{xy} = 1$: **(a)** for x, **(b)** for y

2-2 APPLICATIONS OF LINEAR EQUATIONS

Algebraic methods are very useful in solving applied problems in many different fields. Such problems are generally stated in verbal form; before we can make use of our algebraic tools, it is necessary to translate the verbal statements into corresponding algebraic statements. The following step-by-step procedure will very often be helpful in carrying out this process.

> **Step 1** Represent the unknown quantity—that is, the quantity to be determined—by an algebraic symbol, such as x. In some problems, two or more quantities must be determined; in such cases we choose only one of them to be x.
>
> **Step 2** Express all of the other quantities involved in the problem, if there are any, in terms of x.
>
> **Step 3** Translate verbal expressions occurring in the problem into algebraic expressions involving x. In this context, such as *is* or *was* are translated into the algebraic symbol $=$.
>
> **Step 4** Solve the algebraic statement or statements according to the methods of algebra.
>
> **Step 5** Translate the algebraic solution back into verbal form.

In verbal problems, a number of typical expressions occur involving phrases such as some amount more than or less than a certain value or multiples such as twice or half of a certain quantity. The following examples illustrate how to translate such expressions into algebraic terms.

EXAMPLE 1

(a) If Jack has x dollars and Jill has 5 more than Jack, then Jill has $(x + 5)$ dollars. If Sam has 3 less than Jack, then Sam has $(x - 3)$ dollars.

(b) If Chuck is x years old and his father is 4 years more than twice Chuck's age, then Chuck's father is $(2x + 4)$ years old.

(c) If a certain store sells x refrigerators per month and a second store sells 5 less than one-third as many, then the second store sells $(\frac{1}{3}x - 5)$ refrigerators. ☛ 6

We shall begin with some examples of an elementary nature to illustrate as simply as possible the translation between verbal and algebraic forms.

EXAMPLE 2 Determine two consecutive integers whose sum is 19.

Solution

Step 1 Since we must find two integers, we must decide which of them to call x. Let us denote the smaller integer by x.

☛ **6.** In Example 1(a), Amanda has as many dollars as Jack, Jill and Sam altogether. How many does she have?
In Example 1(c), The first store makes a profit of $30 on each refrigerator and the second store makes a profit of $75. By how much does the monthly profit of the first store exceed that of the second?

Answer (a) $3x + 2$ dollars;
(b) $5x + 375$ dollars.

Step 2 The second integer is then $x + 1$, since the two are consecutive.

Step 3 The expression *sum of the two integers* is translated into the algebraic expression $x + (x + 1)$. The statement that this sum is 19 translates into the equation

$$x + (x + 1) = 19.$$

Step 4 We solve for x.

$$2x + 1 = 19$$
$$2x = 19 - 1 = 18$$
$$x = \tfrac{18}{2} = 9$$

Step 5 The smaller of the integers is therefore 9. The larger, $x + 1$, is 10. ☛ 7

☛ **7.** A triangle has two equal sides and the third is longer by 8 units. If the perimeter exceeds twice the length of the shortest side by 20 units, what are the lengths of the three sides?

EXAMPLE 3 A husband is 7 years older than his wife. Ten years ago he was twice her age. How old is he?

Solution Let x denote the present age of the man in years. Since his wife is 7 years younger than he is, her present age must be $(x - 7)$ years.

Ten years ago, the age of the man was 10 years less than it is now, so his age then was $x - 10$. (For example, if his present age is $x = 38$, then 10 years ago he was $x - 10 = 38 - 10 = 28$ years old.) Similarly, 10 years ago his wife's age was 10 years less than it is now, so her age was $(x - 7) - 10$ or $x - 17$. We are told that at that time the man's age, $x - 10$, was twice his wife's age, $x - 17$. Thus we write

$$x - 10 = 2(x - 17).$$

We simplify and solve for x.

$$x - 10 = 2x - 34$$
$$x - 2x = -34 + 10$$
$$-x = -24$$
$$x = 24$$

The present age of the husband is therefore 24 years. His wife is 17. Ten years ago they were 14 and 7, respectively.

EXAMPLE 4 *(Monthly Earnings)* A salesperson earns a basic salary of $600 per month plus a commission of 10% on the sales she makes. She finds that on average, she takes $1\frac{1}{2}$ hours to make $100 worth of sales. How many hours must she work on the average each month if her monthly earnings are to be $2000?

Solution Suppose that she works x hours per month. Each $\frac{3}{2}$ hours, she makes $100 in sales, so each hour, she averages two-thirds of this or $(200/3)$ in sales. Her commission is 10% or one-tenth of this, so her average commission per hour is $\frac{20}{3}$. In x hours, she will therefore earn a commission of $\left(\frac{20}{3}\right)x$ dollars.

Answer 12, 12, and 20.

Adding on her basic salary, we obtain a total monthly income of $600 + (\frac{20}{3})x$. This must be equal to 2000, so we obtain the equation

$$600 + \tfrac{20}{3}x = 2000.$$

Solving gives the following equations.

$$\tfrac{20}{3}x = 2000 - 600 = 1400$$

$$x = \tfrac{3}{20}(1400) = 210$$

The salesperson must therefore work 210 hours per month, on the average, if she is to reach the desired income level.

EXAMPLE 5 *(Profit)* A cattle dealer bought 1000 steers for $150 each. He sold 400 of them at a profit of 25%. At what price must he sell the remaining 600 if his average profit on the whole lot is to be 30%?

Solution His profit on each steer in the 400 already sold is 25% of the cost price, which is 25% of $150, or $37.50. On 400 steers, his profit was therefore $37.50 × 400 = $15,000. Let his selling price on the remaining 600 steers be x dollars. Then his profit per steer is $x - 150$ and his profit on the remaining 600 is $600(x - 150)$ dollars. Therefore his total profit on the whole purchase is

$$15,000 + 600(x - 150) \text{ dollars.}$$

This profit should be 30% of the price he paid for the 1000 steers, that is, 30% of $150,000. This is equal to $[\frac{3}{10}(150,000)]$, or $45,000. Thus we arrive at the equation

$$15,000 + 600(x - 150) = 45,000.$$

We next solve.

$$15,000 + 600x - 90,000 = 45,000$$

$$600x = 45,000 - 15,000 + 90,000 = 120,000$$

$$x = \frac{120,000}{600} = 200$$

The dealer must sell the remaining steers at $200 each for a 30% average profit.

_____

If a sum of money P dollars is invested for 1 year at an annual rate of interest R percent, the amount of annual interest is given by

$$I = P\left(\frac{R}{100}\right) \text{ dollars.}$$

For example, a sum of $5000 invested at 6% per annum will produce an amount of interest each year given by

$$I = \$5000\left(\frac{6}{100}\right) = \$300.$$

If this interest is withdrawn each year, then both the principal P and the interest I remain the same from year to year. ☛8

EXAMPLE 6 (*Investment*) Ms. Cordero has $70,000 to invest. She wants to receive an annual income of $5000. She can invest her funds in 6% government bonds or, with a greater risk, in 8.5% mortgage bonds. How should she invest her money in order to minimize her risk and yet earn $5000?

Solution Let the amount invested in government bonds be x dollars. Then the amount invested in mortgage bonds is $(70,000 - x)$ dollars. Income received from government bonds at 6% is $\frac{6}{100}x$ dollars. Income received from mortgage bonds at 8.5% is

$$\frac{8.5}{100}(70,000 - x) \text{ dollars} = \frac{85}{1000}(70,000 - x) \text{ dollars}.$$

Since the total income received from the two types of bonds must be $5000,

$$\frac{6}{100}x + \frac{85}{1000}(70,000 - x) = 5000.$$

We multiply both sides by 1000 and solve for x.

$$60x + 85(70,000 - x) = 5,000,000$$
$$60x + 5,950,000 - 85x = 5,000,000$$
$$-25x = 5,000,000 - 5,950,000$$
$$= -950,000$$
$$x = \frac{-950,000}{-25} = 38,000$$

Thus Ms. Cordero should invest $38,000 in government bonds and the remaining $32,000 in mortgage bonds. She could increase her income by investing a bigger proportion of her capital in mortgage bonds, but this would increase her risk.

EXAMPLE 7 (*Mixture Problem*) A winery wishes to make 10,000 liters of sherry by fortifying white wine, which has an alcohol content of 10%, with brandy, which has an alcohol content of 35% by volume. The sherry is to have an alcohol content of 15%. Determine the quantities of wine and brandy which should be mixed together to produce the desired result.

Solution Let x liters of brandy be used in making the 10,000 liters of sherry. Then the volume of white wine used will be $(10,000 - x)$ liters. Since brandy contains 35% alcohol, the amount of alcohol in x liters of brandy is $\frac{35}{100}x$. Similarly, the wine contains 10% alcohol, so $(10,000 - x)$ liters of wine contain $\frac{1}{10}(10,000 - x)$ liters of alcohol. Therefore the total amount of alcohol in the mixture will be

$$\tfrac{35}{100}x + \tfrac{1}{10}(10,000 - x) \text{ liters}.$$

The mixture is to contain 15% alcohol, so in the 10,000 liters there should be $\frac{15}{100}(10,000) = 1500$ liters of alcohol. Therefore we have the equation

$$\frac{35}{100}x + \frac{1}{10}(10,000 - x) = 1500.$$

Solving, we have the following.

$$\frac{35}{100}x + 1000 - \frac{1}{10}x = 1500$$

$$\frac{35}{100}x - \frac{1}{10}x = 1500 - 1000 = 500$$

$$35x - 10x = 50,000$$

$$25x = 50,000$$

$$x = \frac{50,000}{25} = 2000$$

So 2000 liters of brandy and 8000 liters of wine must be mixed together. ☛ 9

Answer 23%.

EXERCISES 2-2

(1–3) If Joe has x dollars, how many dollars does Judy have in each case?

1. She has $4 more than Joe.

2. She has $3 less than twice as much as Joe.

3. She has $2 more than half as much as Joe.

(4–7) If Joe is x years old and Judy is 4 years younger, how old is Fred in each case?

4. Fred is 3 years older than Judy.

5. Fred is 1 year more than the average age of Joe and Judy.

6. Fred is 10 years less than the sum of Joe's and Judy's ages.

7. Fred is 2 years less than five times the difference between Joe's and Judy's ages.

8. Bruce and Jack together have $75. If Jack has $5 more than Bruce, how much does Jack have?

9. In a business mathematics class, there are 52 students. If the number of boys is 7 more than twice the number of girls, determine the number of girls in the class.

10. A father is three times as old as his son. In 12 years, he will be twice the age of his son. How old are the father and the son now?

11. Five years ago, Marlene was twice as old as her brother. Find the present age of Marlene if the sum of their ages today is 40 years.

12. Sue has 3 more nickels than dimes and 5 more dimes than quarters. In all she has $2.10. How many of each coin does she have?

13. I have twice as many dimes in my pocket as I have quarters. If I had 4 fewer dimes and 3 more quarters, I would have $2.60. How many dimes and quarters do I have?

14. *(Investment)* A man invests twice as much at 8% as he invests at 5%. His total annual income from the two investments is $840. How much is invested at each rate?

15. *(Investment)* A college has $60,000 to invest in an endowment fund in order to have an annual income of $5000 for a scholarship. Part of this will be invested in government bonds at 8% and the remainder in long-term fixed deposits of 10.5%. How much should be invested in each to provide the required income?

16. *(Investment)* The trustees of an endowment fund want to invest $18,000 in two kinds of securities paying 9% and 6% annual dividends, respectively. How much must be invested at each rate if the annual income is to be equivalent to a yield of 8% on the total investment?

17. *(Investment)* A man invested $2000 more at 8% than at 10% and received a total interest income of $700 for 1 year. How much did he invest at each rate?

18. *(Investment)* A company invests $15,000 at 8% and $22,000 at 9%. At what rate should the remaining sum of $12,000 be invested so that the combined annual interest income from the three investments is $4500?

19. *(Sale Price)* During a sale an item is marked down 20%. If its sale price is $2, what was its original price?

20. *(Wholesale Price)* An item sells for $12. If the price markup is 50% of the wholesale price, what is the whole-sale price?

21. *(Percentage Discount)* A trader offers a 30% discount on the marked price of an article and yet makes a profit of 10%. If it costs $35 to the trader, what must be the marked price?

22. *(Mixtures)* Ten pounds of peanuts worth 75¢ per pound and 12 pounds of walnuts worth 80¢ per pound are mixed with pecans worth $1.10 per pound to produce a mixture worth 90¢ per pound. How many pounds of pecans should be used?

23. *(Mixtures)* How much of a 10% acid solution must be mixed with 10 ounces of a 15% acid solution to obtain a 12% acid solution?

24. *(Mixtures)* How much water should be added to 15 ounces of 20% acid solution to obtain a 12% acid solution?

25. *(Mixtures)* A sample of seawater has 20% salt content. Fresh water is added to yield 75 ounces of 8% salt solution. How much seawater was in the sample?

26. *(Mixtures)* How much water must be evaporated from 300 ounces of 12% salt solution to obtain a 15% salt solution?

27. *(Mixtures)* Substance A contains 5 milligrams of niacin per ounce, and substance B contain 2 milligrams of niacin per ounce. In what proportions should A and B be mixed so that the resulting mixture contains 4 milligrams of niacin per ounce?

28. *(Agriculture)* A crop of potatoes yields an average 16 metric tons of protein per square kilometer of planted area, while corn yields 24 metric tons per square kilometer. In what proportions must potatoes and corn be planted in order to yield 21 tons of protein per square kilometer from the combined crop?

29. *(Manufacturer's Profit)* It costs a manufacturer $2000 to buy the tools to manufacture a certain household item. If it costs 60¢ for the material and labor for each item produced and if the manufacturer can sell the items for 90¢ each, find how many items should be produced and sold to make a profit of $1000.

30. *(Newspaper Profits)* The cost of publishing each copy of a weekly magazine is 28¢. The revenue from the dealer sales is 24¢ per copy and from advertising is 20% of the revenue obtained from sales in excess of 3000 copies. How many copies must be published and sold each week to earn a weekly profit of $1000?

31. *(Car Sales)* A used-car dealer bought two cars for $2900. He sold one at a gain of 10% and another at a loss of 5% and still made a gain of $185 on the whole transaction. Find the cost of each car.

32. *(Wage Rate)* A businessperson is setting up a small business. His fixed costs are $720 per week, and he plans to employ 48 hours of labor per week. He wishes to ensure that his profit is equal to the cost of labor and that his product is sold at only 40% over total cost. What wage rate should he pay? If he manufactures 70 items per week, for what price should he sell them?

◥ 2-3 QUADRATIC EQUATIONS

An equation of the type

$$ax^2 + bx + c = 0 \qquad (a \neq 0) \qquad (1)$$

where a, b, and c are all constants, is called a **quadratic equation** in the variable x.

There are three methods of solving such an equation: factoring, use of the quadratic formula, and completing the square. Whichever of these methods is used, the first step in solving is to arrange the equation in the standard form of Equation (1). In this form, the right side of the equation is zero; on the left side, the x^2-terms, x-terms, and constant terms are collected together. The procedure in arriving at this standard form is therefore first to remove all fractions that occur by multiplying through by their common denominator, then to remove all parentheses, next to transfer all terms to the left side of the equation, and finally to group all like terms together.

The following examples illustrate this procedure, together with the factoring method.

EXAMPLE 1 Solve the equation $3(x^2 + 1) = 5(1 - x)$.

Solution There are no fractions in this equation. Removing the parentheses, we find

$$3x^2 + 3 = 5 - 5x.$$

After the terms on the right are moved across, the equation becomes

$$3x^2 + 3 - 5 + 5x = 0$$

or

$$3x^2 + 5x - 2 = 0.$$

Thus we have a quadratic equation in which the coefficients are $a = 3$, $b = 5$, and $c = -2$. When using the method of factors, we factor the expression on the left. In this example, we have

$$3x^2 + 5x - 2 = (3x - 1)(x + 2)$$

and so the equation takes the form

$$(3x - 1)(x + 2) = 0.$$

The product of the two factors $(3x - 1)$ and $(x + 2)$ is zero. We now use the following property of real numbers:

Zero Factor Property:

If A and B are real numbers and $AB = 0$, then either $A = 0$ or $B = 0$ or both.*

Thus either $3x - 1 = 0$ or $x + 2 = 0$. In the first case, $3x = 1$ and so $x = \frac{1}{3}$. In the second case, $x + 2 = 0$ implies $x = -2$. Thus either $x = \frac{1}{3}$ or $x = -2$; these two numbers provide us with the two roots of the given equation. ☞ **10**

☞ **10.** Solve each equation:
(a) $(x - 2)(x + 4) = 0$;
(b) $(y + 2)(2y - 5) = 0$.

Answer (a) $x = 2$ or -4;
(b) $y = -2$ or $\frac{5}{2}$.

*The product of two factors cannot be zero unless one of the two factors is equal to zero.

We see then that the crux of the factoring method consists of writing the quadratic expression $ax^2 + bx + c$ that occurs in the standard form of the equation as the product of two linear factors. Since this product is given to be zero, it follows that one of the two factors must also be zero.

EXAMPLE 2 Solve $(2x + 3)(3x - 1) = -4$.

Solution We write the given equation with right side zero and simplify.

$$(2x + 3)(3x - 1) + 4 = 0$$

$$(6x^2 + 7x - 3) + 4 = 0$$

$$6x^2 + 7x + 1 = 0$$

Factoring, we have

$$(6x + 1)(x + 1) = 0.$$

Therefore we have the following.

$$6x + 1 = 0 \quad \text{or} \quad x + 1 = 0$$

$$6x = -1 \qquad\qquad x = -1$$

$$x = -\tfrac{1}{6}$$

The required roots are $-\tfrac{1}{6}$ and -1. ☛ **11**

☛ **11.** Solve by factoring:

$$2x^2 + x - 21 = 0.$$

Quadratic Formula

You should recall from previous work in algebra that the roots of the quadratic equation

$$ax^2 + bx + c = 0 \qquad (a \neq 0)$$

are given by the **quadratic formula**

$$x = \frac{-b \pm \sqrt{b^2 - 4ac}}{2a}$$

This formula is widely used and should be memorized. It will be proved at the end of this section.

To solve a quadratic equation, we can use this formula in the following way. First we reduce the equation to the standard form. Then we identify a, b, and c, the three coefficients that occur in that standard form, and simply substitute these coefficients into the quadratic formula.

EXAMPLE 3 Solve the equation $(2x + 3)(3x - 1) = -4$.

Solution This equation was solved by the method of factors in Example 2; we shall now solve it by using the quadratic formula.

Answer $x = 3$ or $-\tfrac{7}{2}$

The given equation when expressed in standard form (see Example 2) is

$$6x^2 + 7x + 1 = 0.$$

Comparing this with the standard equation $ax^2 + bx + c = 0$, we have $a = 6$, $b = 7$, and $c = 1$. The quadratic formula gives the following.

$$x = \frac{-b \pm \sqrt{b^2 - 4ac}}{2a}$$

$$= \frac{-7 \pm \sqrt{49 - 4(6)(1)}}{2(6)}$$

$$= \frac{-7 \pm \sqrt{25}}{12}$$

$$= \frac{-7 \pm 5}{12}$$

$$= \frac{-7 + 5}{12} \quad \text{or} \quad \frac{-7 - 5}{12}$$

$$= \frac{-2}{12} \quad \text{or} \quad \frac{-12}{12}$$

$$= \frac{-1}{6} \quad \text{or} \quad -1$$

Hence the roots are $-\frac{1}{6}$ and -1, exactly as we found in Example 2.

Remark The method of factors is often a quicker method of solving a quadratic equation than the formula method, but on many occasions it is difficult to spot the factors. Furthermore, many quadratic expressions do not have rational roots; in such cases it is virtually impossible to factor by inspection.

EXAMPLE 4 Solve the equation $2x^2 - x - 2 = 0$.

Solution Comparing the given equation with the standard equation $ax^2 + bx + c = 0$, we see that the coefficients are $a = 2$, $b = -1$, and $c = -2$. Thus we have the following.

$$x = \frac{-b \pm \sqrt{b^2 - 4ac}}{2a}$$

$$= \frac{-(-1) \pm \sqrt{(-1)^2 - 4(2)(-2)}}{2 \cdot 2}$$

$$= \frac{1 \pm \sqrt{1 + 16}}{4}$$

$$= \frac{1 \pm \sqrt{17}}{4}$$

☞ 12. Solve the equation:

$$x^2 - 3x + 1 = 0$$

Answer $x = \frac{1}{2}(3 \pm \sqrt{5})$.

Hence the two roots are $\frac{1}{4}(1 + \sqrt{17}) \approx 1.281$ and $\frac{1}{4}(1 - \sqrt{17}) \approx -0.781$.* **☞ 12**

EXAMPLE 5 Solve the equation $x^4 - 3x^2 - 7 = 0$.

Solution As it stands, this equation is not a quadratic equation. However, if we set $x^2 = z$, we obtain

$$z^2 - 3z - 7 = 0,$$

which is a quadratic equation for z. From the quadratic formula we get the solutions

$$z = \frac{-(-3) \pm \sqrt{(-3)^2 - 4(1)(-7)}}{2 \cdot 1} = \frac{3 \pm \sqrt{37}}{2}.$$

These are $z \approx 4.54$ and $z \approx -1.54$. But since $z = x^2$, then z cannot be negative, so only the first of these roots applies. Taking its square root we then get

☞ 13. Solve the equations:
(a) $x^6 - 7x^3 - 8 = 0$;
(b) $x^4 - 7x^2 - 8 = 0$.

$$x = \pm\sqrt{\frac{1}{2}(3 + \sqrt{37})} \approx \pm\sqrt{4.54} \approx \pm 2.13. \quad \text{☞ 13}$$

Completing the Square

The third method of solving quadratic equations is called **completing the square.** The underlying property of real numbers is the following:

> Square Root Property:
>
> If $X^2 = A$ where $A \geq 0$, then $X = \pm\sqrt{A}$.

For example, if $X^2 = 3$, then either $X = +\sqrt{3} \approx 1.73$ or $X = -\sqrt{3} \approx -1.73$. The aim of this method is to write the quadratic equation in the form $X^2 = A$, where A is some number and X is a linear expression involving the variable x. We shall explain it by using the particular quadratic equation

$$x^2 + 6x - 7 = 0. \tag{2}$$

Let us write this equation in the equivalent form.

$$x^2 + 6x = 7. \tag{3}$$

We observe from the binomial square identity that

$$(x + 3)^2 = x^2 + 2 \cdot x \cdot 3 + 3^2 = x^2 + 6x + 9. \tag{4}$$

Comparing the right side of Equation (4) with the left side Equation (3), we see that they differ only by the constant 9. So if we add 9 to both sides of Equation (3), we get

$$x^2 + 6x + 9 = 7 + 9 = 16$$

Answer (a) $x = 2$ or -1;
(b) $x = \pm\sqrt{8}$.

---

* See the footnote on page 56.

> ☛ **14.** Solve the equations:
> (a) $x^2 - 9 = 0$;
> (b) $(x + 1)^2 = 4$;
> (c) $(x + 1)^2 = -4$.

Answer (a) $x = \pm 3$; (b) $x = 1$, -3; (c) no solution.

or, in other words,

$$(x + 3)^2 = 16.$$

This equation is now easily solved by taking the square root of both sides.

$$x + 3 = 4 \quad \text{or} \quad x + 3 = -4$$

Thus either $x = 4 - 3 = 1$ or $x = -4 - 3 = -7$. The two solutions are $x = 1$ and $x = -7$. ☛ **14**

The question remains why we decided, following Equation (3), to look at the quantity $(x + 3)^2$. Why not consider $(x - 3)^2$ or $(x + 57)^2$ instead? The reasons is that, after expanding this binomial square, we want the result to coincide with the left side of Equation (3) as far as the x^2- or x-terms are concerned. For example, if we took $(x - 3)^2$ instead, we would have $(x - 3)^2 = x^2 - 6x + 9$; although the x^2-term is the same on the left side of Equation (3), the x-term is different. In order to get the same coefficient of x as in Equation (3) we must have $(x + k)^2$, where k is half the coefficient of x as in Equation (3)—that is, k equals half of 6, or 3.

The procedure in solving a quadratic equation by completing the square is outlined in these steps.

Step 1 Divide through by the coefficient of x^2.

Step 2 Move the constant term to the right side.

Step 3 Add k^2 to both sides of the equation, where k is half the coefficient of x occurring on the left side.

Step 4 The left side of the equation will now be the perfect square $(x + k)^2$, so the solution is found by taking the square root of both sides.

EXAMPLE 6 Solve the equation $2x^2 - x - 2 = 0$ by completing the square.

Solution

Step 1 Dividing through by 2, we get

$$x^2 - \tfrac{1}{2}x - 1 = 0.$$

Step 2 $\qquad\qquad\qquad x^2 - \tfrac{1}{2}x = 1.$

Step 3 The coefficient of x is $-\tfrac{1}{2}$. We must take k equal to half of this, namely, $-\tfrac{1}{4}$. So we must add $k^2 = (-\tfrac{1}{4})^2 = \tfrac{1}{16}$ to both sides.

$$x^2 - \tfrac{1}{2}x + \tfrac{1}{16} = 1 + \tfrac{1}{16} = \tfrac{17}{16}$$

Step 4 The left side of this equation is now $(x + k)^2$, that is, $[x + (-\tfrac{1}{4})]^2$. So

$$(x - \tfrac{1}{4})^2 = \tfrac{17}{16}.$$

Taking the square root of both sides, we find that

$$x - \frac{1}{4} = \pm\sqrt{\frac{17}{16}} = \pm\frac{\sqrt{17}}{4}$$

> ☛ **15.** Complete the square in each case:
> (a) $x^2 - 4x = 1$;
> (b) $3x^2 + 2x = 1$;
> (c) $2y^2 + 5y + 2 = 0$.

Answer (a) $(x - 2)^2 = 5$; (b) $(x + \tfrac{1}{3})^2 = \tfrac{4}{9}$; (c) $(y + \tfrac{5}{4})^2 = \tfrac{9}{16}$.

and therefore $x = \tfrac{1}{4} \pm \sqrt{17}/4$. (These agree with the roots found in Example 4.) ☛ **15**

We close this section by deriving the quadratic formula from the quadratic equation $ax^2 + bx + c = 0$, with $a \neq 0$. The method of proof follows the method of completing the square. We begin by moving the constant term to the right:

$$ax^2 + bx = -c.$$

Dividing both sides by a (this is possible because $a \neq 0$), we get

$$x^2 + \frac{b}{a}x = -\frac{c}{a} \qquad (5)$$

According to the method of completing the square, we must divide the coefficient of x (which is b/a) by 2 (giving $b/2a$), square the result, and add this to both sides. We have the following:

$$x^2 + \frac{b}{a}x + \left(\frac{b}{2a}\right)^2 = -\frac{c}{a} + \left(\frac{b}{2a}\right)^2 = \frac{-4ac + b^2}{4a^2}.$$

But the left side here is $(x + b/2a)^2$, as can be seen from the binomial square formula. Therefore we obtain

$$\left(x + \frac{b}{2a}\right)^2 = \frac{b^2 - 4ac}{4a^2}.$$

After taking the square root of both sides, we get

$$x + \frac{b}{2a} = \pm\sqrt{\frac{b^2 - 4ac}{4a^2}} = \pm\frac{\sqrt{b^2 - 4ac}}{2a}$$

Therefore

$$x = \frac{-b \pm \sqrt{b^2 - 4ac}}{2a}$$

as required.

One final remark: The quantity $D = b^2 - 4ac$ is called the **discriminant.** If $D = 0$, the term involving square root in the quadratic formula becomes zero. In this case, the two roots of the equation coincide, so there are not two distinct roots. For example, an equation of this type is the quadratic equation $x^2 - 10x + 25 = 0$, which has just one root, $x = 5$.

If $D < 0$, then the quantity under the square root in the quadratic formula is negative. In this case, the quadratic equation $ax^2 + bx + c = 0$ has no roots that are real numbers. For example, consider the equation $x^2 - 2x + 2 = 0$ (where $a = 1$, $b = -2$, and $c = 2$). From the quadratic formula, we have the following:

$$x = \frac{-b \pm \sqrt{b^2 - 4ac}}{2a}$$

$$= \frac{-(-2) \pm \sqrt{(-2)^2 - 4(1)(2)}}{2(1)}$$

$$= \frac{2 \pm \sqrt{-4}}{2}.$$

But the expression $\sqrt{-4}$ has no meaning as a real number, so we must conclude that the given equation has no real roots.*

EXERCISE 2-3

(1–22) Solve the following equations by factoring.

1. $x^2 + 5x + 6 = 0$ **2.** $x^2 + 3x + 2 = 0$

3. $x^2 + 9x + 14 = 0$ **4.** $x^2 - 5x + 6 = 0$

5. $x^2 + 4x + 4 = 0$ **6.** $x^2 - 6x + 9 = 0$

7. $x^2 - 7x + 12 = 0$ **8.** $x^2 + 2x - 3 = 0$

9. $x^2 - 1 = 0$ **10.** $x^2 - 25 = 0$

11. $x^2 - 8x = 0$ **12.** $4x^2 - 5x = 0$

13. $6x^2 + \frac{5}{2}x + \frac{1}{4} = 0$ **14.** $\frac{x^2}{2} + \frac{10}{3}x + 2 = 0$

15. $2x^2 + 5x + 3 = 0$ **16.** $3x^2 - 11x + 10 = 0$

17. $6x^2 + x - 2 = 0$ **18.** $4x^2 - 4x - 15 = 0$

19. $(x + 3)(x - 3) = x - 9$

20. $6x^2 - \frac{1}{2}x - \frac{1}{4} = 0$ **21.** $x^4 - 5x^2 + 4 = 0$

22. $x^4 - 3x^2 + 2 = 0$

(23–34) Solve the following equations by the quadratic formula.

23. $x^2 + 3x + 1 = 0$ **24.** $x^2 - 4x + 2 = 0$

25. $2x^2 + 3x - 4 = 0$ **26.** $3x^2 + 6x - 2 = 0$

27. $x^2 + x - 3 = 0$ **28.** $4x^2 - 12x + 9 = 0$

29. $4x^2 + 20x + 25 = 0$ **30.** $2x^2 + 5x - 3 = 0$

31. $5x(x + 2) + 6 = 3$

32. $(4x - 1)(2x + 3) = 18x - 4$

33. $(x + 1)^2 = 2(x - 1)^2$ **34.** $(2x + 1)^2 = 3(x + 1)^2$

(35–44) Solve the following equations by completing the square.

35. $x^2 + 6x - 1 = 0$ **36.** $x^2 + 2x - 4 = 0$

37. $x^2 - 3x - 1 = 0$ **38.** $x^2 + 5x + 5 = 0$

39. $4x^2 - 8x - 3 = 0$ **40.** $2x^2 - 14x + 1 = 0$

---

*Quantities that are the square roots of negative numbers are called *imaginary* numbers. In particular, $\sqrt{-1}$ is called the imaginary unit and is denoted by i. Then, for example we can write $\sqrt{-4} = \sqrt{(4)(-1)} = 2\sqrt{-1} = 2i$. In a similar way, every imaginary number can be written in the form iB, where B is some real number.

The solution of the last example can be written in the form

$$x = \tfrac{1}{2}(2 \pm \sqrt{-4}) = \tfrac{1}{2}(2 \pm 2i) = 1 \pm i.$$

We see that these solutions consist of two parts, a **real part,** which is 1, and an **imaginary part,** which is i or $-i$, depending on which root we take. Any number that can be written as the sum of a real number and an imaginary number is called a **complex number.** In general, a complex number has the form $A + iB$, where A and B are real numbers.

Thus when $b^2 - 4ac > 0$, the solutions of a quadratic equation consist of two different real numbers. When $b^2 - 4ac = 0$, there is only one solution and it is a real number. And when $b^2 - 4ac < 0$, there are two different solutions, both of which are complex numbers.

All of the standard operations can be carried out with complex numbers. One simply has to remember that $i^2 = -1$.

41. $7x + 3(x^2 - 5) = x - 3$

42. $2x(4x - 1) = 4 + 2x$

43. $x(x + 1)(x + 3) = (x + 2)^3$

44. $(x + 1)^3 - (x - 1)^3 = 8x$

(45–68) Solve the following equations by any appropriate method.

45. $6x^2 = 11$ **46.** $5x^2 + 7 = 0$

47. $6x^2 = 11x$ **48.** $2(x^2 + 1) = 5x$

49. $15x^2 = 40(x + 2)$ **50.** $(3x + 5)(2x - 3) = -8$

51. $3x(2x - 5) = -4x - 3$

52. $(x + 1)^2 = 2x^2$ **53.** $x^2 = 2(x - 1)(x + 2)$

54. $2x(x + 1) = x^2 - 1$ **55.** $\frac{2}{3}x^2 - \frac{5}{3}x = x - 1$

56. $\frac{x^2}{3} + 2x = 1 + x$ **57.** $\frac{x^2}{3} = \frac{11}{6}x + 1$

58. $5x^2 - \frac{7}{2}x = \frac{1}{2}x + 1$ **59.** $2x^2 - 3x - 1 = 0$

60. $x^2 + 3x - 2 = 0$ **61.** $3x^2 = 5x - 3$

62. $2x^2 = 5x - 2$

63. $(2x + 3)(x + 1) = (x + 2)(x - 1) + 2$

64. $(3x - 1)(x + 2) = (2x - 1)(x + 3) + 5$

65. $x^4 - 3x^2 - 4 = 0$ **66.** $2x^4 - x^2 - 1 = 0$

67. $2x^{2/3} + x^{1/3} - 1 = 0$ **68.** $x^{2/5} - 3x^{1/5} + 2 = 0$

69. Solve $s = ut + \frac{1}{2}gt^2$ for t.

70. Solve $s = \dfrac{2a}{1 + a^2}$ for a.

71. Solve $A = 2\pi R(R + H)$ for R.

72. Solve $A = 2x^2 + 4xy$ for x.

73. If 2 is one root of $x^2 - kx + 2 = 0$, find the other root.

74. If -1 is one root of $2x^2 + 5x + k = 0$, find the other root.

75. Use the quadratic formula to solve the equation

$$x^2 - 2xy + 1 - 3y^2 = 0$$

  a. For x in terms of y;

  b. For y in terms of x.

76. Use the quadratic formula to solve the equation

$$3x^2 - 2y^2 = xy + 1$$

  a. For x in terms of y;

  b. For y in terms of x.

◥ 2-4 APPLICATIONS OF QUADRATIC EQUATIONS

EXAMPLE 1 Sue is 7 years older than Bobby. If the product of their ages is 60, how old is Bobby?

Solution Let x denote Bobby's age. Then Sue is $x + 7$ years old. We are told that the product

$$(\text{Bobby's Age}) \cdot (\text{Sue's Age}) = x(x + 7) = 60.$$

That is,

$$x^2 + 7x - 60 = 0.$$

This factors as $(x - 5)(x + 12) = 0$, so $x = 5$ or -12. But x cannot be negative, so Bobby's age is 5.

EXAMPLE 2 A box with no top is to be formed from a rectangular sheet of tin by cutting out 4-inch squares from each corner and folding up the sides. If the width of the box is 3 inches less than the length and the box is to hold 280 cubic inches, find the dimensions of the sheet of tin.

Solution Let x inches denote the width of the box; then its length is $(x + 3)$ inches and its height 4 inches. (See Figure 1.) The volume of the box is given by

$$(\text{Length})(\text{Width})(\text{Height}) = (x + 3)(x)(4) = 4x(x + 3).$$

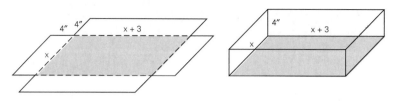

FIGURE 1

But the box is to hold 280 cubic inches, so

$$4x(x + 3) = 280.$$

Dividing both sides by 4, we have

$$x(x + 3) = 70$$

$$x^2 + 3x - 70 = 0. \tag{i}$$

Comparing this with the general quadratic equation $ax^2 + bx + c = 0$ we have $a = 1$, $b = 3$, $c = -70$. Then by the quadratic formula the roots of (i) are given by

$$x = \frac{-b \pm \sqrt{b^2 - 4ac}}{2a}$$

$$= \frac{-3 \pm \sqrt{9 - 4(1)(-70)}}{2(1)}$$

$$= \frac{-3 \pm \sqrt{9 + 280}}{2}$$

$$= \frac{-3 \pm 17}{2}$$

$$= \frac{-3 + 17}{2} \quad \text{or} \quad \frac{-3 - 17}{2}$$

$$= 7 \quad \text{or} \quad -10.$$

But $x = -10$ is unacceptable, because x represents the width of the box, and the width cannot be a negative number. Thus $x = 7$.

The dimensions of the tin sheet before we cut the corners are $x + 8$ and $(x + 3) + 8$. Since $x = 7$, the dimensions are 15 inches and 18 inches. ☞ **16**

☞ **16.** Solve Example 2 if the width is 4 inches less than the length and the volume is 240 cubic inches.

Answer 14 × 18 inches.

EXAMPLE 3 *(Apartment Rental)* Steve owns an apartment building that has 60 suites. He can rent all the suites if he charges a rent of $180 per month. At a higher rent, some of the suites will remain empty; on the average, for each increase of $5 in rent, 1 suite becomes vacant with no possibility of renting it. Find the rent he should charge per suite in order to obtain a total income of $11,475.

☛ **17.** In Example 3, what is the total rental income when the rent is $200 per month?

Solution Let n denote the number of 5-dollar increases. Then the increase in rent per suite is $5n$ dollars, which means that the rent per suite is $(180 + 5n)$ dollars. The number of units not rented will then be $n,$ so that the number rented will be $60 - n$. The total rent he will receive is given by

Rental Income = (Rent per Suite) × (Number of Suites Rented).

Therefore

$$11,475 = (180 + 5n)(60 - n)$$

or

$$11,475 = 5(36 + n)(60 - n).$$

Dividing both sides by 5, we get

$$2295 = (36 + n)(60 - n) = 2160 + 24n - n^2.$$

Therefore

$$n^2 - 24n + 135 = 0$$

$$(n - 9)(n - 15) = 0.$$

Thus $n = 9$ or 15. Hence the rent charged should be $180 + 5n$, which is $180 + 45 = \$225$ or $180 + 75 = \$255$. In the first case, 9 of the suites will be vacant and the 51 rented suites will produce an income of $225 each. In the second case, when the rent is $255, 15 suites will be vacant, and only 45 rented, but the total revenue will be the same. ☛ **17**

Answer $200 × 56.

The *revenue* of a business for a given period of operation is the name given to its total income for that period. The *profit* is equal to this income minus the costs of operation for the period in question. We write this

Profit = Revenue − Costs.

☛ **18.** A firm sells its product for $9 per unit. It costs $$(4x + 3000)$ to produce x units per week. What are the firm's revenue and profit if x units are produced and sold per week?

When revenue comes from selling a particular good, we also have the general equation

Revenue = (Selling Price per Unit) × (Number of Units Sold). ☛ **18**

Answer Revenue = $9x$,
  profit = $5x - 3000$.

EXAMPLE 4 *(Pricing Decision)* The egg marketing board of British Columbia knows from past experience that if it charges p dollars per dozen eggs, the number sold per week will be x million dozens, where $p = 2 - x$. Its total weekly revenue would then be $R = xp = x(2 - x)$ million dollars. The cost to the industry of producing x million dozen eggs per week is given by $C = 0.25 + 0.5x$ million dollars. What price should the marketing board set for eggs to ensure a weekly profit of $0.25 million?

Solution The profit is given by the following equation.

$$P = R - C$$
$$= x(2 - x) - (0.25 + 0.5x)$$
$$= -x^2 + 1.5x - 0.25$$

Setting this equal to 0.25, we obtain the equation

$$-x^2 + 1.5x - 0.25 = 0.25$$

or

$$x^2 - 1.5x + 0.5 = 0.$$

Using the quadratic formula, we find the roots for x.

$$x = \frac{-b \pm \sqrt{b^2 - 4ac}}{2a}$$
$$= \frac{-(-1.5) \pm \sqrt{(-1.5)^2 - 4(1)(0.5)}}{(2)(1)}$$
$$= \frac{1.5 \pm \sqrt{2.25 - 2}}{2}$$
$$= \tfrac{1}{2}(1.5 \pm 0.5)$$
$$= 1 \quad \text{or} \quad 0.5$$

Now $p = 2 - x$. So when $x = 1$, we have $p = 1$, and when $x = 0.5$, $p = 1.5$. Thus the marketing board has a choice of two policies: It can charge $1 per dozen, in which case the sales will be 1 million dozen, or it can charge $1.50 per dozen, when the sales will be 0.5 million dozen per week. In either case the profits to the industry will be $0.25 million per week.

We saw in Example 6 of Section 2-2 that a sum P invested at a rate of interest of $R\%$ earns an amount of interest of $P(R/100)$ in 1 year. At the end of the year, the total value of the investment is

$$\text{Original Principal} + \text{Interest} = P + P\left(\frac{R}{100}\right) = P\left(1 + \frac{R}{100}\right). \quad \text{☞ 19}$$

☞ **19.** A sum of $200 is invested for 2 years at an interest rate of 6% per annum. The first year's interest is not withdrawn and earns interest during the second year. What is the final total value of the investment?

Answer $200(1.06)^2$ or $224.72.

EXAMPLE 5 (*Investment*) A sum of $400 is invested at a rate of interest $R\%$ per annum. At the end of the year both principal and interest are left to earn interest for the second year. Find R if the total value of the investment at the end of the second year is $484.

Solution At the end of the first year, the total value, as discussed above, is

$$P\left(1 + \frac{R}{100}\right) = 400\left(1 + \frac{R}{100}\right) \equiv P_1.$$

This new principal all earns interest during the second year, so the value of the investment at the end of the second year is

$$P_1\left(1 + \frac{R}{100}\right) = 400\left(1 + \frac{R}{100}\right)^2.$$

Thus we have the quadratic equation

$$400\left(1 + \frac{R}{100}\right)^2 = 484$$

to be solved for R. Do not write it in standard form; just take square roots of both sides:

$$\left(1 + \frac{R}{100}\right)^2 = \frac{484}{400} = 1.21 \qquad \text{so} \quad 1 + \frac{R}{100} = \pm 1.1.$$

R cannot be negative, so the meaningful solution is $1 + R/100 = +1.1$ *or* $R = 10$. The rate of interest is 10%.

EXAMPLE 6 (*Investment*) A corporation wishes to set aside a sum of $1 million to be invested at interest and used at a later date to repay two bond issues that will become due. One year after the sum is first invested, $250,000 will be required for the first issue; 1 year later, $900,000 more will be required for the second issue. Determine the rate of interest necessary in order that the investment will be sufficient to cover both repayments.

Solution Let the rate of interest be R percent per annum. When invested at this rate, the value of the investment after 1 year is

$$P\left(1 + \frac{R}{100}\right) = (1 \text{ million})\left(1 + \frac{R}{100}\right) = \left(1 + \frac{R}{100}\right) \text{ million dollars}.$$

At this time, 0.25 million is withdrawn; at the beginning of the second year, the amount still invested is (in millions) therefore

$$P' = \left(1 + \frac{R}{100}\right) - 0.25 = 0.75 + \frac{R}{100}.$$

After a second year at interest, the value of the investment is

$$P'\left(1 + \frac{R}{100}\right) = \left(0.75 + \frac{R}{100}\right)\left(1 + \frac{R}{100}\right).$$

This must be the amount (0.9 million) necessary to pay off the second bond issue. Therefore we arrive at the equation

$$\left(0.75 + \frac{R}{100}\right)\left(1 + \frac{R}{100}\right) = 0.9.$$

Thus

$$0.75 + 1.75\left(\frac{R}{100}\right) + \left(\frac{R}{100}\right)^2 = 0.9.$$

Multiplying both sides by 100^2 to remove the fractions, we arrive at the equation

$$7500 + 175R + R^2 = 9000$$

or

$$R^2 + 175R - 1500 = 0.$$

From the quadratic formula (with $a = 1$, $b = 175$, and $c = -1500$), we find the following value for R.

$$R = \frac{-175 \pm \sqrt{175^2 - 4(1)(-1500)}}{2(1)}$$

$$= \tfrac{1}{2}[-175 \pm \sqrt{30{,}625 + 6000}]$$

$$= \tfrac{1}{2}[-175 \pm \sqrt{36{,}625}]$$

$$\approx \tfrac{1}{2}[-175 \pm 191.4]$$

$$= 8.2 \quad \text{or} \quad -183.2$$

Clearly, the second solution does not make any practical sense—a rate of interest would hardly be negative. The meaningful solution is $R = 8.2$. So the investment must earn 8.2% per annum in order to provide sufficient funds to pay off the bond issues.

EXERCISES 2-4

1. Find two numbers whose sum is 15 and the sum of whose squares is 137.

2. Find two consecutive odd integers whose product is 143.

3. Find two consecutive integers whose product is 132.

4. Find two consecutive even integers the sum of whose squares is 100.

5. The length of the hypotenuse of a right triangle is 13 centimeters. Find the other two sides of the triangle if their sum is 17 centimeters.

6. The diameter of a circle is 8 centimeters. By how much should the radius increase so that the area increases by 33π square centimeters?

7. The perimeter of a rectangle is 20 inches and its area 24 square inches. Find the lengths of its sides.

8. The perimeter of a rectangle is 24 centimeters and its area is 32 square centimeters. Find the lengths of its sides.

9. Equal squares are removed from each corner of a rectangular metal sheet with dimensions 20 by 16 inches. The sides are then folded up to form a rectangular box. If the base of the box has area 140 square inches, find the side of the square that is removed from each corner.

10. A box with square base and no top is to be made from a square piece of metal by cutting 2-inch squares from each corner and folding up the sides. Find the dimensions of the metal sheet if the volume of the box is to be 50 cubic inches.

11. A ball is thrown upward with an initial velocity of 80 feet per second. The height h (in feet) traveled in t seconds is given by the formula

$$h = 80t - 16t^2.$$

a. After how many seconds will the ball reach a height of 64 feet?

b. How long will it be before the ball returns to the ground?

c. Find the maximum height that the ball will reach. (*Hint:* The time of upward travel equals half the time to return to the ground.)

12. A rocket is shot vertically upward from the ground with an initial velocity of 128 feet per second. The rocket is at a height h after t seconds of launching, where $h = 128t - 16t^2$.

  a. After what time will the rocket be at a height of 192 feet above the ground?

  b. When will the rocket return to the ground? Find the maximum height the rocket will reach.

13. *(Cost Problem)* A dealer sold a watch for $75. His percentage profit was equal to the cost price in dollars. Find the cost price of the watch.

14. *(Compound Interest)* For every $100 invested in secured commercial loans, a bank receives $116.64 after 2 years. This amount represents capital and the interest compounded annually. What is the rate of interest?

15. *(Compound Interest)* In two years, the XYZ company will require $1,102,500 to retire some of its bonds. At what rate of interest compounded annually should $1,000,000 be invested over the 2-year period to receive the required amount to retire the bonds?

16. *(Apartment Rental)* Royal Realty has built a new unit of 60 apartments. It is known from past experience that if they charge a monthly rent of $150 per apartment, all the units will be occupied, but for each $3 increase in rent, one apartment unit is likely to remain vacant. What rent should be charged to generate the same $9000 total revenue as is obtained with a rent of $150 and at the same time to leave some vacant suites?

17. *(Apartment Rental)* In Exercise 16, the maintenance, service, and other costs on the building amount to $5000 per month plus $50 per occupied suite and $20 per vacant suite. What rental should be charged if the profit is to be $1225 per month? (The profit is rental revenue minus all costs.)

18. *(Pricing Decision)* If a publisher prices a book at $20, 20,000 copies will be sold. For every dollar by which the price is increased, sales will fall by 500 books. What should the book cost in order to generate a total revenue from sales of $425,000?

19. *(Pricing Decision)* In Exercise 18, the cost of producing each copy is $16. What price should the publisher charge to have a profit of $200,000?

20. *(Pricing Decision)* In Exercise 19, assume that in addition to the cost of $16 per copy, the publisher must pay a royalty to the author of the book equal to 10% of the selling price. What price should now be charged per copy in order to realize a profit of $200,000?

21. *(Investment)* A sum of $100 is invested at interest for 1 year; then, together with the interest earned, it is invested for a second year at twice the first rate of interest. If the total sum realized is $112.32, what are the two rates of interest?

22. *(Investment)* In Exercise 21, $25 is withdrawn after the first year and the remainder is invested at twice the rate of interest. If the value of the investment at the end of the second year is $88, what are the two rates of interest?

23. *(Production and Pricing Decision)* Each week, a company can sell x units of its product at a price of p dollars each, where $p = 600 - 5x$. It costs the company $(8000 + 75x)$ dollars to produce x units.

  a. How many units should the company sell each week to generate a revenue of $17,500?

  b. What price per unit should the company charge to obtain a weekly revenue of $18,000?

  c. How many units should be produced and sold each week to obtain a weekly profit of $5500?

  d. At what price per unit will the company generate a weekly profit of $5750?

24. *(Production and Pricing Decision)* A manufacturer can sell x units of a product each week at a price of p dollars per unit, where $p = 200 - x$. It costs $(2800 + 45x)$ dollars to produce x units.

  a. How many units should be sold each week to generate a revenue of $9600?

  b. At what price per unit will a weekly revenue of $9900 be generated?

  c. How many units should the manufacturer produce and sell each week to obtain a profit of $3200?

  d. At what price per unit will the manufacturer generate a weekly profit of $3150?

25. *(Pricing Policy)* A state liquor board buys whiskey for $2 a bottle and sells it for p dollars per bottle. The volume of sales x (in hundreds of thousands of bottles per week) is given by $x = 24 - 2p$, when the price is p. What value of p gives a total revenue of $7 million per week? What value of p gives a profit to the liquor board of $4.8 million per week?

CHAPTER REVIEW

Key Terms, Symbols, and Concepts

2.1 Equation, solution or root of an equation. Equivalent equations.
The addition and multiplication principles for equations.
Polynomial equation, degree, linear equation, quadratic equation.
Step-by-step procedure for solving a linear equation.

2.2 Step-by-step procedure for handling word problems.
Annual interest formula.

2.3 Standard form of a quadratic equation.
Zero factor property: solution of quadratic equation by factoring.
Quadratic formula. Square root property: completing the square.

2.4 Revenue, costs, profit.

Formulas

Annual interest formula:

$$I = P\frac{R}{100}.$$

$$\text{Value after 1 year} = P\left(1 + \frac{R}{100}\right).$$

Quadratic formula: If $ax^2 + bx + c = 0$, then

$$x = \frac{-b \pm \sqrt{b^2 - 4ac}}{2a}.$$

Profit = Revenue − Costs.

Revenue = (Selling Price per Unit) × (Number of Units Sold).

REVIEW EXERCISES FOR CHAPTER 2

1. State whether each of the following is true of false. Replace each false statement with a corresponding true statement.

a. If both sides of an equation are multiplied by any constant, the roots of the equation remain unchanged.

b. Any expression can be added to both sides of an equation and the roots will remain unchanged.

c. The roots of an equation remain unchanged when both sides are multiplied by an expression containing the variable.

d. It is possible to square both sides of an equation without altering its roots.

e. If $px = q$, then $x = q - p$.

f. A quadratic equation is an equation of the form $ax^2 + bx + c = 0$, where a, b, and c are any constants.

g. The solution of the equation $x^2 = 4$ is given $x = 2$.

h. The roots of the quadratic equation $ax^2 + bx + c = 0$, $a \neq 0$, are given by

$$x = -\frac{b}{2a} \pm \sqrt{b^2 - 4ac}.$$

i. A linear equation always has exactly one root.

j. A quadratic equation always has two different roots.

k. It is possible for a linear equation to have no roots at all.

l. It is possible for a quadratic equation to have no roots at all.

(2–29) Solve the following equations for x.

2. $3(2 - x) + x = 5(2x - 1) + 2$

3. $2(1 - 4x) - 1 = x - 2(2 - 3x)$

4. $4(3x - 1) - 3(2x + 1) = 1 - 7x$

5. $3(2x - 3) - 2(x + 7) = 4(x + 1) - 3$

6. $5x - 2(3x - 1) = x - 2(1 - 5x)$

7. $(3x + 1)^2 - (3x - 1)^2 = 12x + 7$

8. $x^2 + 13x + 40 = 0$

9. $3x^2 - 11x + 10 = 0$

10. $\dfrac{1}{x} + \dfrac{1}{a} = \dfrac{c}{b}$

11. $\dfrac{x}{bc} + \dfrac{x}{ca} + \dfrac{x}{ab} = a + b + c$

12. $(3x - 2)^2 = (3x + 1)^2$

13. $(2x - 1)^2 = 3x^2 + (x - 1)(x - 2)$

14. $(2x + 1)(x - 3) = (2x + 5)(x - 1)$

15. $(x + 2)(x - 3) = 2 + (x - 1)(x - 2)$

16. $(x + 1)(2x - 5) = (x + 2)(x - 3)$

17. $1 + (3x + 4)(x - 2) = (2x + 1)(x - 3)$

18. $(x + 2)(2x - 1) = 1 + (x + 3)(x + 1)$

19. $(x + 1)(2x - 5) = 2(x + 2)(x - 3)$

20. $5x^2 = 13x + 6$

21. $\dfrac{x}{p} + \dfrac{x}{q} + \dfrac{x}{r} = pq + qr + rp$

22. $\dfrac{1}{x + 2} - \dfrac{1}{3} = \dfrac{1}{5}$

23. $28 + (x - 5)(x + 7) = (3x - 1)(x - 2)$

24. $\sqrt{2x + 5} = x + 1$ **25.** $\sqrt{x + 5} = x - 1$

26. $x + 3 = \sqrt{5x + 11}$ **27.** $\sqrt{x - 2} = 2 - x$

28. $2^{x^2} = \dfrac{8}{4^x}$ **29.** $4^x = 8^{3 - x}$

(30–32) Solve the following equations for the indicated variables.

30. $\dfrac{1}{x} + \dfrac{1}{y} = \dfrac{1}{z}$ **a.** for y **b.** for z

31. $S = \dfrac{a - rl}{1 - r}$ **a.** for r **b.** for l

32. $P = P_0(1 + R/100)^2$ **a.** for P_0 **b.** for R

33. *(Investment)* The winner of Western Express Lottery wants to invest his prize money of $100,000 in two investments at 8% and 10%. How much should he invest in each if he wants to obtain an annual income of $8500?

34. *(Furniture Shipment)* The Western Furniture Mart received a shipment of 55 tables, some night tables and some coffee tables. They were billed for $645. If each night table costs $9 and each coffee table costs $15, how many tables of each type were received by the store?

35. *(Manufacturer's Profit)* The manufacturer of a certain product can sell all she can produce at a price of $20 each. It costs her $12.50 to produce each item in materials and labor, and she has additional overhead costs of $7000 per month in order to operate the plant. Find the number of units she should produce and sell to make a profit of $5000 per month.

36. *(Manufacturing Decision)* A television manufacturer wants to decide whether to manufacture his own picture tubes that have been purchased from outside suppliers at $5.70 each. Manufacturing the picture tubes will increase overhead costs by $960 per month and the cost of labor and materials will be $4.20 for each picture tube. How many picture tubes would have to be used by the manufacturer each month to justify a decision to manufacture picture tubes?

37. *(Manufacturer's Profit)* The number of items of a product that a manufacturer can sell each week depends on the price charged for them. Assume that at a price of p dollars, x items per week can be sold, where $x = 300(6 - p)$. Each item costs $3 to manufacture. The profit per item is therefore $(p - 3)$ dollars and the weekly profit is $(p - 3)x$ dollars. Find the value of p that will produce a weekly profit of $600.

38. *(Production Decision)* A manufacturer can sell x units of a product each week at a price of p dollars per unit, where $x = 160(10 - p)$. It costs $(4x + 400)$ dollars to produce x units per week. How many units should be produced and sold to obtain a weekly profit of $1000?

39. *(Business Profits)* A dry-cleaning store opens 8 hours per day from Monday to Friday and is closed for the weekend. The store handles 15 transactions per hour, and the average revenue per transaction is $6. Labor costs are $16 per hour, and there is a weekly rental of $560 for the store and equipment. The only other cost to the operator is for materials: C dollars per transaction.

a. Express the weekly profit P in terms of C.

b. Suppose the store is currently making a weekly profit of $600. The cost of materials, that is, C, is due to go up by 20% next month. Prices to the store's customers are to be raised by 10%. Assuming that nothing else changes, and, in particular, that business does not fall off, what will be the new weekly profit?

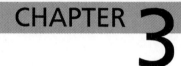

CHAPTER 3

Inequalities

Chapter Objectives

3-1 SETS AND INTERVALS
(a) Definitions of inequality symbols;
(b) Definition of a set and its specification by listing or by rule;
(c) Subsets, equality of sets;
(d) Open and closed intervals of real numbers and their geometric representation.

3-2 LINEAR INEQUALITIES IN ONE VARIABLE
(a) Definition of a linear inequality in one variable;
(b) The solution set of an inequality;
(c) The addition and multiplication rules for inequalities;
(d) The use of these rules to solve linear inequalities.

3-3 QUADRATIC INEQUALITIES IN ONE VARIABLE
(a) Definition of a quadratic inequality;
(b) Solution of a quadratic inequality by factoring the quadratic expression;
(c) Use of the quadratic formula to solve a quadratic inequality.

3-4 ABSOLUTE VALUES
(a) Definition of the absolute value of a real number and its geometrical interpretation in terms of the number line;
(b) Solution of a linear equation involving absolute values;
(c) Solution of a linear inequality involving absolute values.

CHAPTER REVIEW

◥ 3-1 SETS AND INTERVALS

Let us begin by recalling definitions of the symbols $<$, $\leq$, $>$, and $\geq$, called **inequality symbols.**

The real numbers other than zero are divided into the two classes, positive numbers and negative numbers. We write $a > 0$ (*a is greater than zero*) to mean that a is positive and $a < 0$ (*a is less than zero*) to mean that a is negative. The sum $a + b$ and the product $a \cdot b$ of any two positive real numbers a and b are both positive. If a is positive, then $-a$ is negative.

If a and b are two different real numbers, we write $a > b$ if the difference $a - b$ is positive and we write $a < b$ if $a - b$ is negative. For example, $5 > 2$ because $5 - 2 = 3$ is positive and $2 < 8$ because $2 - 8 = -6$ is negative. Geometrically, $a > b$ means that the point on the number line representing the number a lies to the right of the point representing the number b and $a < b$ means that the point representing a lies to the left of the point representing b. (See Figure 1.)

FIGURE 1

We define $a \geq b$ (*a is greater than or equal to b*) to mean that either $a > b$ or $a = b$. Similarly, $a \leq b$ (*a is less than or equal to b*) is used to denote that either $a < b$ or $a = b$. For example, $5 \leq 7$ is true and $5 \geq 5$ because $5 = 5$ is true.

Statements like $a < b$, $a > b$, $a \geq b$, or $a \leq b$ are called **inequalities.** In particular, $a > b$ and $a < b$ are **strict inequalities.** The inequality $a > b$ can be equivalently written in the opposite direction as $b < a$. Thus $5 > 3$ is the same as $3 < 5$.

When a number b lies between two numbers a and c with $a < c$, we write $a < b < c$. The double inequality $a < b < c$ means that both $a < b$ and $b < c$. ☞ 1

Sets

Knowledge of sets and operations on sets is basic to all modern mathematics. Many lengthy statements in mathematics can be written concisely and neatly in terms of sets and set operations.

DEFINITION Any well-defined collection of objects is referred to as a **set.** The objects constituting the set are called the **members** or **elements** of the set.

☞ **1.** Are the following true or false?

(a) $-5 \leq -7$;

(b) $3 > -4$;

(c) If $x > -5$ then $-5 \leq x$;

(d) There exist x such that $-3 \leq x \leq -4$.

Answer (a) False; (b) true; (c) true; (d) false.

By a **well-defined** collection, we mean that given any object, we should be able to decide unambiguously whether or not it belongs to the collection.

A set may be specified in two ways, either by making a list of all of its members or else by stating a rule for membership in the set. Let us examine these two methods in turn.

LISTING METHOD If it is possible to specify all of the elements of a set, the set can be described by listing all the elements and enclosing the list inside braces.

For example, $\{1, 2, 5\}$ denotes the set consisting of the three numbers 1, 2, and 5, and $\{p, q\}$ denotes the set whose only members are the two letters p and q.

In cases where the set contains a large number of elements, it is often possible to employ what is called a **partial listing.** For example, $\{2, 4, 6, \ldots , 100\}$ denotes the set of all the even integers from 2 up through 100. An ellipsis, $\ldots$, is used to show that the sequence of elements continues in a manner that is clear from the first few members listed. The sequence terminates with 100. By the use of the ellipsis, the listing method can be employed in cases in which the set in question contains infinitely many members. For example, $\{1, 3, 5, \ldots \}$ denotes the set of *all* the odd natural numbers. The absence of any number following the ellipsis indicates that the sequence does not terminate, but continues indefinitely.

RULE METHOD There are many cases in which it is not possible or in which it would be inconvenient to list all the members of a particular set. In such a case the set can be specified by stating a rule for membership.

For example, consider the set of all people living in Mexico at the present moment. To specify this set by listing all the members by name would clearly be a prodigious task. Instead we can denote it as follows.

$$\{x \,|\, x \text{ is a person currently living in Mexico}\}.$$

The symbol $|$ stands for *such that,* so this expression is read *the set of all x such that x is a person currently living in Mexico.* The statement that follows the vertical bar inside the braces is the rule specifying the membership in the set.

As a second example, consider the set

$$\{x \,|\, x \text{ is a point on this page}\}$$

which denotes the set of all the points on this page. This is an example of a set that cannot be specified by the listing method even if we wanted to do so.

Many sets can be specified either by listing or by stating a rule, and we can choose whichever of the two methods we like. We shall give several examples of sets, some of which are specified using both methods.

EXAMPLE 1

(a) If N denotes the set of all natural numbers, then we can write

$$N = \{1, 2, 3, \ldots\} = \{k \,|\, k \text{ is a natural number}\}.$$

(b) If P denotes the set of integers between -2 and $+3$ inclusive, then

$$P = \{-2, -1, 0, 1, 2, 3\} = \{x \,|\, x \text{ is an integer}, -2 \leq x \leq 3\}.$$

Observe that the membership rule consists of two conditions separated by a comma. Both conditions must be satisfied by any member of the set.

(c) $Q = \{1, 4, 7, \ldots, 37\}$
$= \{x \,|\, x = 3k + 1, k \text{ is an integer}, 0 \leq k \leq 12\}$

(d) The set of all students currently enrolling at ABC Business College can be represented formally as

$$S = \{x \,|\, x \text{ is a student currently enrolled at ABC Business College}\}.$$

This set could also be specified by listing the names of all the students involved.

(e) The set of all real numbers greater than 1 and less than 2 can be specified by the rule method as

$$T = \{x \,|\, x \text{ is a real number}, 1 < x < 2\}. \quad \bullet 2$$

☛ 2. List the elements in the sets:

(a) $\{x \,|\, x$ is a natural number, $-1 \leq x < 5\}$

(b) $\{x \,|\, x = (k + 4)^{-1}, k$ is an integer, $-2 \leq k \leq 2\}$

A set is said to be **finite** if the number of elements belonging to it is finite, that is, if they can be counted. If the number of elements in a set is not finite, the set is called an **infinite** set.

In Example 1, the sets in parts (b), (c), and (d) are all finite, but those in parts (a) and (e) are infinite.

It is commom to use capital letters to denote sets and lowercase letters to denote the elements in the sets. Observe that we followed this convention in Example 1. If A is any set and x any object, the notation $x \in A$ is used to denote the fact that x is a member of A. The statement $x \in A$ is read x *belongs to A* or x *is an element of A*. The negative statement x *is not an element of A* is denoted by writing $x \notin A$.

In part (b) of Example 1, $2 \in P$ but $6 \notin P$. For the set in part (e), $\sqrt{2} \in T$ and $\frac{3}{2} \in T$, but $2 \notin T$ and $\pi \notin T$.

DEFINITION A set that contains no elements is called an **empty set.** (The terms **null set** and **void set** are also used.)

The symbol $\emptyset$ is used to denote a set that is empty, and the statement $A = \emptyset$ means that the set A contains no members. Examples of empty sets include the following.

$$\{x \,|\, x \text{ is an integer and } 3x = 2\}$$

Answer (a) $\{1, 2, 3, 4\}$;
(b) $\{\frac{1}{2}, \frac{1}{3}, \frac{1}{4}, \frac{1}{5}, \frac{1}{6}\}$.

$\{x \mid x$ is a real number and $x^2 + 1 = 0\}$

The set of all living dragons.

The set of all magnets having only one pole. ☞ 3

DEFINITION A set A is said to be a **subset** of another set B if every element of A is also an element of B. In such a case, we write $A \subseteq B$.

The set A is said to be a **proper subset** of the set B if every element of A is in B but there is at least one element in B which is not in A. In this case, we write $A \subset B$.

EXAMPLE 2

(a) Let $A = \{2, 4, 6\}$ and $B = \{1, 2, 3, 4, 5, 6, 7, 8\}$. Then $A \subset B$.

(b) If N is the set of all natural numbers, I is the set of all integers, Q is the set of all rational numbers, and R is the set of all real numbers, then

$$N \subset I \subset Q \subset R.$$

Answer (a) False; (b) false; (c) false.

(c) The set of all women students at XYZ University is a subset of the set of all students at that university.

(d) Every set is a subset of itself; that is,

$$A \subseteq A \text{ for any set } A$$

However, the statement $A \subset A$ is not true.

(e) An empty set $\emptyset$ is a subset of any set A:

$$\emptyset \subseteq A \text{ for any set } A.$$

In order to explain this last example more fully, let us rephrase the definition of a subset: B is a subset of A if and only if there is no object that belongs to B and does not belong to A. It is clear that there exists no object that belongs to $\emptyset$ and does not belong to A for the simple reason that there exists no object that belongs to $\emptyset$ at all. Hence $\emptyset \subseteq A$. ☞ 4

Answer $\{1, 2, 3\}, \{1, 2\}, \{1, 3\}, \{2, 3\}, \{1\}, \{2\}, \{3\}, \emptyset.$

Two sets are equal to one another if they contain identical elements. More formally, we have the following definition.

DEFINITION Two sets A and B are said to be **equal** if $A \subseteq B$ and $B \subseteq A$. In such a case we write $A = B$.

Thus $A = B$ if there is no object that belongs to A and does not belong to B or that belongs to B and does not belong to A.

EXAMPLE 3

(a) If $A = \{x \mid x^2 = 1\}$ and $B = \{-1, +1\}$, then $A = B$.

(b) If $A = \{y \mid y^2 - 3y + 2 = 0\}$ and $B = \{1, 2\}$, then $A = B$. ☛ 5

Intervals

DEFINITION Let a and b be two real numbers with $a < b$. Then the **open interval** from a to b, denoted by (a, b), is the set of all real numbers x that lie between a and b. Thus

$$(a, b) = \{x \mid x \text{ is a real number and } a < x < b\}.$$

Similarly, the **closed interval** from a to b, denoted by $[a, b]$ is the set of all real numbers that lie between a and b, together with a and b themselves. Thus

$$[a, b] = \{x \mid x \text{ is a real number and } a \le x \le b\}.$$

Semiclosed or **semiopen** intervals are defined as follows.

$$(a, b] = \{x \mid a < x \le b\}$$
$$[a, b) = \{x \mid a \le x < b\}$$

Note The statement that x is a real number has been omitted from the rules defining these sets. This is commonly done to avoid repetition when we are dealing with sets of real numbers. ☛ 6

For all these intervals, (a, b), $[a, b]$, $[a, b)$, and $(a, b]$, a and b are called the **endpoints** of the interval. An open interval does not contain its endpoints, whereas a closed interval contains both its endpoints. A semiclosed interval contains only one of its endpoints. Two methods of representing these intervals are shown in Figure 2.

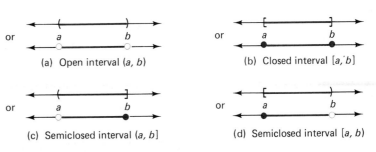

(a) Open interval (a, b)

(b) Closed interval $[a, b]$

(c) Semiclosed interval $(a, b]$

(d) Semiclosed interval $[a, b)$

FIGURE 2

We use the symbol ∞ (*infinity*) and $-\infty$ (*negative infinity*) to describe un-bounded intervals. (See Figure 3.) Note that ∞ and $-\infty$ are not real numbers.

This is enough about sets for our immediate needs. You will find more properties of sets discussed in Section 8-1, in the context of probability theory.

(a) $(a, \infty) = \{x \mid x > a\}$

(b) $[a, \infty) = \{x \mid x \geq a\}$

(c) $(-\infty, a) = \{x \mid x < a\}$

(d) $(-\infty, a] = \{x \mid x \leq a\}$

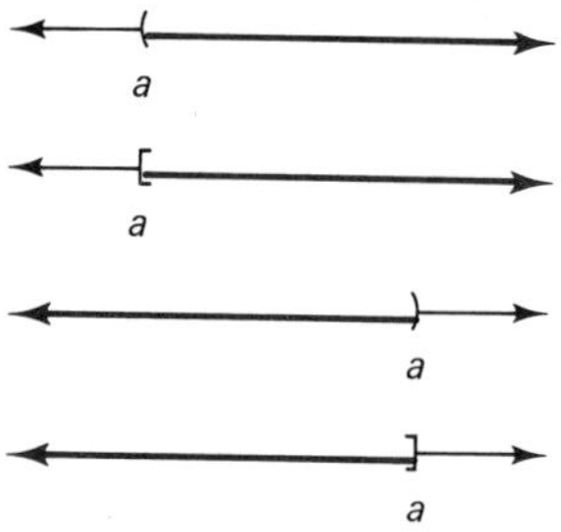

FIGURE 3

EXERCISES 3-1

(1–8) Use the listing method to describe the following sets.

1. The set of all integers less than 5 and greater than -2.

2. The set of all natural numbers less than 50.

3. The set of all integers less than 5.

4. The set of all even numbers greater than 4.

5. The set of all prime numbers less than 20.

6. $\left\{ y \mid y = \dfrac{1}{h + 2}, h \text{ is a natural number} \right\}$

7. $\{x \mid x \text{ is a prime factor of } 36\}$

8. $\left\{ p \mid p = \dfrac{1}{n - 1}, n \text{ is a prime number less than 20} \right\}$

(9–16) Use the rule method to describe the following sets.

9. The set of all even numbers less than 100.

10. The set of all prime numbers less than 30.

11. $\{1, 3, 5, 7, 9, \ldots, 19\}$

12. $\{\ldots, -4, -2, 0, 2, 4, 6, \ldots\}$

13. $\{3, 6, 9, \ldots\}$

14. $\{1, \frac{1}{2}, \frac{1}{3}, \frac{1}{4}, \ldots\}$

15. The interval $[-1, 1]$.

16. The interval $(1, \infty)$.

(17–20) Write the following sets of numbers in the interval form.

17. $3 \leq x \leq 8$ **18.** $-2 < y \leq 7$

19. $-3 > t > -7$ **20.** $t \geq -5$

(21–24) Write the following intervals as inequalities.

21. $[2, 5)$ **22.** $(-3, 7)$

23. $(-\infty, 3)$ **24.** $(-2, \infty)$

25. State whether the following statements are true or false. If false, explain why.

  a. $2 \in \{1, 2, 3\}$ **b.** $3 \subseteq \{1, 2, 3, 4\}$

  c. $4 \in \{1, 2, 5, 7\}$ **d.** $\{a, b\} \subseteq \{a, b, c\}$

  e. $\varnothing = 0$ **f.** $\{0\} = \varnothing$

  g. $0 \in \varnothing$ **h.** $\varnothing \in \{0\}$

  i. $\varnothing \subseteq \{0\}$ **j.** $\{1, 2, 3, 4\} = \{4, 2, 1, 3\}$

  k. $\left\{ x \mid \dfrac{(x - 2)^2}{x - 2} = 0 \right\} = \{x \mid x - 2 = 0\}$

  l. If $A \subseteq B$ and $B \subseteq C$, then $A \subseteq C$.

m. If $A \subseteq B$ and $B \subseteq A$, then $A = B$.

n. The set of all rectangles in a plane is a subset of the set of all squares in a plane.

o. The set of all equilateral triangles is a subset of the set of all triangles.

p. The open interval (a, b) is a subset of the closed interval $[a, b]$.

q. $\{x \mid 2 \leq x \leq 3\} \in \{y \mid 1 \leq y \leq 5\}$

r. $\{x \mid 1 \leq x \leq 2\} = \{y \mid 1 \leq y \leq 2\}$

26. If A is the set of all squares in a plane, B is the set of all rectangles in the same plane, and C is the set of all quadrilaterals in the plane, then which of these sets is a subset of one (or both) of the others?

27. Prove that the set $\{x \mid x^2 - x - 2 = 0\}$ is not a subset of the interval $[0, \infty)$.

28. Is the set $\{x \mid x(x^2 - 1) = 0\}$ a subset of the interval $(0, \infty]$?

29. Is the set $\{x \mid x^2 - x - 6 = 0\}$ a subset of the natural numbers?

30. Is the set $\{x \mid 2x^2 - 3x + 1 = 0\}$ a subset of the integers? Of the rational numbers?

◥ 3-2 LINEAR INEQUALITIES IN ONE VARIABLE

In this section, we shall consider certain inequalities that involve a single variable. The following example gives a simple business problem that results in such an inequality.

Suppose the total cost (in dollars) of production of x units of a certain commodity is given by $C = 3100 + 25x$ and each unit sells for $37. The manufacturer wants to know how many units should be produced and sold to gain a profit of at least $2000.

Suppose x units are produced and sold. The revenue R obtained by selling x units at $37 each is $R = 37x$ dollars. The profit P (in dollars) obtained by producing and selling x units is then given by the following equations.

$$\text{Profit} = \text{Revenue} - \text{Cost}$$

$$P = 37x - (3100 + 25x) = 12x - 3100$$

Since the profit is required to be at least $2000, that is, it should be $2000 or more, we must have

$$P \geq 2000$$

or

$$12x - 3100 \geq 2000. \tag{1}$$

This is an inequality involving the single variable x. We observe that the terms occurring in it are of two types, either constant terms or terms that are constant multiples of the variable x. Any inequality that has terms only of these two types is called a **linear inequality.** If the inequality symbol is either $>$ or $<$, the inequality is called a **strict** inequality; if the symbol is $\geq$ or $\leq$, the inequality is said to be a **weak** inequality.

EXAMPLE 1

(a) $3 - x \leq 2x + 4$ is a weak linear inequality in the variable x.

(b) $\frac{1}{4}z + 3 > 5 - \frac{1}{3}z$ is a strict linear inequality in the variable z.

Any inequality can be written in an equivalent form by switching the two sides and reversing the direction of the inequality sign. For example, $x > 3$ is equivalent to $3 < x$; Example 1(a) is equivalent to $2x + 4 \geq 3 - x$.

DEFINITION The **solution** of an inequality in one variable is the set of all values of the variable for which the inequality is a true statement.

For example, the solution of Equation (1) is the set of all values of x (the number of units sold) that make the profit at least $2000.

As with equations, the solution of an inequality is found by performing certain operations on the inequality to change it to some standard form. There are two basic operations that can be used to manipulate inequalities; we shall now state the rules that govern these operations.

Rule 1

When the same real number is added to or subtracted from both sides of an inequality, the direction of inequality remains unaltered.

In symbols, if $a > b$ and c is any real number, then

$$a + c > b + c \quad \text{and} \quad a - c > b - c.$$

EXAMPLE 2

(a) We see that $8 > 5$ is a true statement. If we add 4 to both sides, we get $8 + 4 > 5 + 4$, or $12 > 9$, which is still true. If we subtract 7 from both sides we get $8 - 7 > 5 - 7$, or $1 > -2$, which is again true.

(b) Let $x - 1 > 3$. Adding 1 to both sides, we get

$$x - 1 + 1 > 3 + 1$$

or

$$x > 4.$$

The set of values of x for which $x - 1 > 3$ is the same set as that for which $x > 4$. ☛ 7

☛ **7.** Add -5 to both sides of the following:
(a) $x + 5 \geq -5$;
(b) $x - 5 < 2$.

Answer (a) $x \geq -10$
(b) $x - 10 < -3$.

In Example 2 we see that the inequality $x > 4$ can be obtained from the given inequality $x - 1 > 3$ by moving the term -1 from the left side to the right side and changing its sign. In general, the above rule allows us to do this kind of operation: *Any term can be moved from one side of the inequality to the*

other side after changing its sign without affecting the direction of inequality.
In symbols, if $a > b + c$, then $a - b > c$, and $a - c > b$.

EXAMPLE 3

(a) If $8 > 5 + 2$, then $8 - 2 > 5$.

(b) If $2x - 1 < x + 4$, then $2x - x < 4 + 1$. Both x and -1 were moved from one side to the other. Simplifying, we then get $x < 5$.

Rule 2

The direction of the inequality is preserved if both sides are multiplied (or divided) by the same positive number and is reversed when multiplied (or divided) by the same negative number.

In symbols, if $a > b$ and c is any positive number, then

$$ac > bc \qquad and \qquad \frac{a}{c} > \frac{b}{c}$$

while if c is any negative number, then

$$ac < bc \qquad and \qquad \frac{a}{c} < \frac{b}{c}.$$

EXAMPLE 4

(a) We know that $4 > -1$ is a true statement. Multiplying both sides by 2, we obtain $8 > -2$, which is still true. If, however, we multiply by (-2), we must reverse the direction of inequality. We get

$$(-2)(4) < (-2)(-1) \qquad or \qquad -8 < 2$$

which is again true.

(b) If $2x \leq 4$, then we can divide both sides by 2 and obtain the equivalent inequality $2x/2 \leq 4/2$, or $x \leq 2$.

(c) If $-3x < 12$, then we divide by -3, which is negative, so we must reverse the inequality:

$$\frac{-3x}{-3} > \frac{12}{-3} \qquad or \qquad x > -4 \quad \text{☛ 8}$$

☛ **8.** Multiply both sides of the following by -2;
(a) $2x > -3$; (b) $-\frac{1}{2}x \leq 3 - x$.

Before doing further examples, let us derive these two basic rules.

PROOF OF RULE 1 We suppose that $a > b$ and we let c be any real number. If $a > b$, then by definition $a - b > 0$. Now consider the difference between $(a + c)$ and $(b + c)$:

$$(a + c) - (b + c) = a + c - b - c = a - b > 0.$$

Answer (a) $-4x < 6$;
(b) $x \geq -6 + 2x$.

But since $(a + c) - (b + c)$ is positive, this means that

$$a + c > b + c$$

which is what we want to prove.

PROOF OF RULE 2 Again we suppose that $a > b$ and we let c be any positive real number. Then, as before, $a - b > 0$. Thus $a - b$ and c are both positive numbers, so their product is also positive:

$$(a - b)c > 0.$$

That is,

$$ac - bc > 0.$$

It follows, therefore, that $ac > bc$, as required. If, on the other hand, c were negative, the product $(a - b)c$ would be negative since one factor is positive and the other negative. It follows that

$$ac - bc < 0$$

and hence that $ac < bc$, as required.

EXAMPLE 5 Find all real numbers that satisfy the inequality

$$3x + 7 > 5x - 1.$$

Solution We move all x-terms to one side of the inequality and all constant terms to the other side. Moving $5x$ to the left side and 7 to the right side, changing their signs, and simplifying gives the following.

$$3x - 5x > -1 - 7 \qquad \text{(Rule 1)}$$
$$-2x > -8$$

We next divide both sides by -2 and reverse the direction of inequality (because -2 is negative).

$$\frac{-2x}{-2} < \frac{-8}{-2} \qquad \text{(Rule 2)}$$
$$x < 4$$

Therefore the solution consists of the set of real numbers in the interval $(-\infty, 4)$. This is illustrated in Figure 4.

FIGURE 4

EXAMPLE 6 Solve the inequality

$$y + \frac{3}{4} \leq \frac{5y - 2}{3} + 1.$$

Solution First of all, we must clear the inequality of fractions. Here the common denominator is 12, so we multiply both sides by 12.

$$12\left(y + \frac{3}{4}\right) \le 12\left(\frac{5y - 2}{3} + 1\right)$$

$$12y + 9 \le 4(5y - 2) + 12$$

$$12y + 9 \le 20y - 8 + 12$$

$$12y + 9 \le 20y + 4$$

Moving y-terms to the left and constant terms to the right, we get

$$12y - 20y \le 4 - 9$$

$$-8y \le -5.$$

We next divide both sides by -8 and reverse the direction of inequality (because -8 is negative).

$$y \ge \frac{-5}{-8} \quad \text{or} \quad y \ge \frac{5}{8}$$

Hence the solution consists of the set of all real numbers greater than or equal to $\frac{5}{8}$, that is, the numbers in the interval $[\frac{5}{8}, \infty)$. This set is illustrated in Figure 5. ☛ **9**

☛ **9.** Find the solutions in interval notation:

(a) $1 - x < 3 - 2x$;

(b) $x + 4 \ge 4x - 2$.

Answer (a) $(-\infty, 2)$;   (b) $(-\infty, 2]$.

FIGURE 5

EXAMPLE 7 Solve the double inequality for x.

$$8 - 3x \le 2x - 7 < x - 13$$

Solution Recall from Section 3-1 that the double inequality $a \le b < c$ means that both $a \le b$ and $b < c$. The given double inequality is therefore equivalent to

$$8 - 3x \le 2x - 7 \quad \text{and} \quad 2x - 7 < x - 13.$$

We solve these two inequalities separately by the methods described above. This gives

$$x \ge 3 \quad \text{and} \quad x < -6.$$

Both of these inequalities must be satisfied by x. But it is clearly impossible for both $x \ge 3$ and $x < -6$ to be satisfied. Thus there is *no solution*; there is no real number which satisfies the given double inequality. ☛ **10**

☛ **10.** Find the solution and draw it on the number line:

$3x - 2 \le 2 - x < x + 6$.

Answer $-2 < x \le 1$.

EXAMPLE 8 Find the solution of the double inequality

$$7 > 5 - 2x \ge 3.$$

Solution In this case, since x appears only in the middle expression, we can manipulate all three parts of the inequality together. First subtract 5 from all three parts:

$$7 - 5 > 5 - 2x - 5 \geq 3 - 5$$

or

$$2 > -2x \geq -2.$$

Next, divide through by -2, reversing both inequality signs:

$$-1 < x \leq 1.$$

The solution consists of the semiclosed interval $(-1, 1]$.

EXAMPLE 9 *(Manufacturer's Profit)* The manufacturer of a certain item can sell all he can produce at the selling price of $60 each. It costs him $40 in materials and labor to produce each item, and he has additional costs (overhead) of $3000 per week in order to operate the plant. Find the number of units he should produce and sell to make a profit of at least $1000 per week.

Solution Let x be the number of items produced and sold each week. Then the total cost of producing x units consists of $3000 plus $40 per item, which is

$$(40x + 3000) \text{ dollars.}$$

The revenue obtained by selling x units at $60 each will be $60x$ dollars. Therefore

$$\text{Profit} = \text{Revenue} - \text{Cost}$$
$$= 60x - (40x + 3000) = 20x - 3000.$$

Since we want a profit of at least $1000 each week, we have the following.

$$\text{Profit} \geq 1000$$
$$20x - 3000 \geq 1000$$
$$20x \geq 4000$$
$$x \geq 200$$

Thus the manufacturer should produce and sell at least 200 units each week. ☞ **11**

☞ **11.** A rectangle has perimeter 24 units. If the difference between the two sides is less than 6 units, find the interval of values for the length of the longest side.

EXAMPLE 10 *(Manufacturing Decision)* The management of a manufacturing firm wants to decide whether they should manufacture their own gaskets, which the firm has been purchasing from outside suppliers at $1.10 each. Manufacturing the gaskets will increase the overhead costs of the firm by $800 per month, and the cost of materials and labor will be 60¢ for each gasket. How many gaskets would have to be used by the firm each month to justify a decision to manufacture their own gaskets?

Answer [6, 9).

Solution Let x be the number of gaskets used by the firm each month. Then the cost of purchasing x gaskets at $1.10 each is $1.10x$ dollars. The cost of manufacturing x gaskets consists of $0.60 per gasket plus an overhead of $800 per month, so the total cost is

$$(0.60x + 800) \quad \text{dollars.}$$

To justify manufacturing gaskets by the firm itself, the following must be true.

$$\text{Cost of Purchasing} > \text{Cost of Manufacturing}$$

$$1.10x > 0.60x + 800$$

$$1.10x - 0.60x > 800$$

$$0.50x > 800$$

$$x > 1600$$

Thus the firm must use at least 1601 gaskets each month to justify manufacturing them.

EXERCISES 3-2

(1–20) Solve the following inequalities.

1. $5 + 3x < 11$

2. $3 - 2y \geq 7$

3. $2u - 11 \leq 5u + 6$

5. $5x + 7 > 31 - 3x$

5. $3(2x - 1) > 4 + 5(x - 1)$

6. $x + \dfrac{4}{3} > \dfrac{2x - 3}{4} + 1$

7. $\dfrac{1}{4}(2x - 1) - x < \dfrac{x}{6} - \dfrac{1}{3}$

8. $\frac{3}{2}(x + 4) \geq 2 - \frac{1}{5}(1 - 4x)$

9. $\dfrac{y + 1}{4} - \dfrac{y}{3} > 1 + \dfrac{2y - 1}{6}$

10. $5 - 0.3t < 2.1 + 0.5(t + 1)$

11. $1.2(2t - 3) \leq 2.3(t - 1)$

12. $2(1.5x - 2.1) + 1.7 \geq 2(2.4x - 3.5)$

13. $5 < 2x + 7 < 13$

14. $4 \geq \dfrac{1 - 3x}{4} \geq 1$

15. $(x + 3)^2 > (x - 2)^2$

16. $(2x + 3)(3x - 1) \leq (6x + 1)(x - 2)$

17. $(3x - 1)(2x + 3) > (2x + 1)(3x + 2)$

18. $(3x + 1)(x - 2) > (x - 3)(3x + 4)$

19. $2x + 1 < 3 - x < 2x + 5$

20. $4 - 2x < x - 2 < 2x - 4$

21. $3x + 7 > 5 - 2x \geq 13 - 6x$

22. $2x - 3 < 1 + x < 3x - 1$

23. $3x - 5 < 1 + x < 2x - 3$

24. $5x - 7 \geq 3x + 1 \geq 6x - 11$

25. *(Investment)* A man has $7000 to invest. He wants to invest some of it at 8% and the rest at 10%. What is the maximum amount he should invest at 8% if he wants an annual interest income of at least $600 per year?

26. *(Investment)* Mrs. K. has $5000 which she wants to invest, some at 6% and the rest at 8%. If she wants an annual interest income of at least $370, what is the minimum amount she should invest at 8%?

27. *(Production Decision)* A manufacturer can sell all units produced at $30 per unit. Fixed costs are $12,000 per month; in addition, it costs $22 to produce each unit. How many units must be produced and sold each month by the company to realize a profit?

28. *(Manufacturer's Profits)* A stereo manufacturer can sell all the units produced at a price of $150 each. Weekly fixed costs are $15,000 and the units cost $100 each in materials and labor. Find the number of stereos which must be manufactured and sold each week to obtain a weekly profit of at least $1000.

29. *(Manufacturing Decision)* A firm manufacturing cars wants to know whether to manufacture their own fan belts, which the firm has been purchasing from outside suppliers at $2.50 for each unit. Manufacturing the belts by the firm will increase its fixed costs by $1500 each month, but it will cost only $1.70 to manufacture each belt. How many belts must be used by the firm each month to justify manufacturing the belts themselves?

30. *(Contracting-out Decision)* A firm can hire a sub-contractor to package each unit of its product at a cost of $2.75. On the other hand, the firm can package its own products by installing a packaging machine. Installing the machine will increase the fixed costs of the firm by $2000 each month, and the packaging itself costs $1.50 per unit.

How many units would have to be made each month to make the installation of the packaging machine worthwhile?

31. *(Magazine Publishing)* The cost of publishing each copy of the weekly magazine *Buy and Sell* is 35¢. The revenue from the dealer sales is 30¢ per copy, and the revenue from advertising is 20% of the revenue obtained from sales in excess of 2000 copies. How many copies must be published and sold each week to earn a weekly profit of at least $1000?

32. *(Magazine Publishing)* The publisher of a monthly magazine has a publishing cost of 60.5¢ per copy. The revenue from dealer sales is 70¢ per copy, and the revenue from advertising is 15% of the revenue obtained from sales in excess of 20,000 copies. How many copies must be published and sold each month to earn a monthly profit in excess of $4000?

◣ 3-3 QUADRATIC INEQUALITIES IN ONE VARIABLE

A **quadratic inequality** in one variable, such as x, is an inequality that contains terms proportional to x and x^2 and constant terms. The standard forms of a quadratic inequality are

$$ax^2 + bx + c > 0 \quad (\text{or} < 0) \qquad \text{or} \qquad ax^2 + bx + c \geq 0 \qquad (\text{or} \leq 0)$$

where a, b, and c are certain constants ($a \neq 0$). ☛ 12

Again we are interested in solving a given inequality, that is, in finding the set of x for which the inequality is true. We can do this by first replacing the inequality sign by an $=$ sign and finding the solutions of the resulting quadratic equation. These solutions divide the number line into intervals. In each interval we choose a point and test whether the inequality is true or false at that point. If it is true at that one point, then it will be true at all points in the interval, and conversely, if it is false at one point in the interval, then it will be false at all points.

EXAMPLE 1 Solve the inequality $x^2 + 3x < 4$.

Solution We first rewrite the inequality in standard form by subtracting 4 from both sides:

$$x^2 + 3x - 4 < 0.$$

Replacing the $<$ by $=$, we obtain the quadratic equation $x^2 + 3x - 4 = 0$. This can be solved by factoring. It becomes $(x - 1)(x + 4) = 0$, so the roots are $x = 1$ and $x = -4$. Plotting these points on the number line, we obtain

☛ **12.** Express in standard form:
$(x + 2)(2x - 1) \leq (3x - 2)^2 + 1$.

Answer $7x^2 - 15x + 7 \geq 0$.

Figure 6. The two points divide the number line into three intervals, $x < -4$, $-4 < x < 1$, and $x > 1$. In each of these intervals the expression $x^2 + 3x - 4$ always keeps the same sign, since it can only change sign by passing through zero, and this happens only when $x = -4$ or 1.

FIGURE 6

Let us take any point in the first interval $x < -4$: we choose $x = -5$. Then $x^2 + 3x - 4 = (-5)^2 + 3(-5) - 4 = 6 > 0$. The inequality is false, so it is false for all points in the interval $x < -4$.

In $-4 < x < 1$ we choose the point $x = 0$. Then $x^2 + 3x - 4 = (0)^2 + 3(0) - 4 = -4 < 0$. The inequality is true, so it is true for all x satisfying $-4 < x < 1$.

In $x > 1$ we choose $x = 2$. Then $x^2 + 3x - 4 = (2)^2 + 3(2) - 4 = 6 > 0$. The inequality is false, so it is false for all $x > 1$.

The solution set is therefore the interval $(-4, 1)$. This is illustrated in Figure 7. ☛ **13**

☛ **13.** Solve the inequalities
(a) $(x - 1)(x - 3) < 0$;
(b) $(x + 1)(x + 4) \geq 0$;
(c) $(x - 3)^2 + 2 \leq 0$.

FIGURE 7

EXAMPLE 2 Solve the inequality $5x \leq 2(x^2 - 6)$.

Solution Moving all the terms to the left, we get

$$5x - 2x^2 + 12 \leq 0.$$

It is always convenient to have the x^2 term with positive coefficient, so that factoring is easier. Thus we multiply both sides by -1 and reverse the direction of inequality.

$$-5x \ \ + 2x^2 - 12 \geq 0$$
$$2x^2 - 5x \ \ - 12 \geq 0.$$

Replacing the $\geq$ sign by $=$ we get the quadratic equation $2x^2 - 5x - 12 = 0$, and by factoring we get

$$(2x + 3)(x - 4) = 0.$$

The roots are $x = -\frac{3}{2}$ and $x = 4$, which divide the number line into the three intervals $(-\infty, -\frac{3}{2})$, $(-\frac{3}{2}, 4)$, and $(4, \infty)$, as shown on Figure 8. Choose any

Answer (a) $1 < x < 3$;
(b) $x \leq -4$ or $x \geq -1$;
(c) no solution.

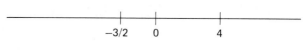

FIGURE 8

point in each interval and test the inequality. In $(-\infty, -\frac{3}{2})$ we choose $x = -2$; in $(-\frac{3}{2}, 4)$ we choose $x = 0$; and in $(4, \infty)$ we choose $x = 5$. It is convenient to set the calculations out as shown in Table 1. The given inequality is therefore true in the whole of the intervals $(-\infty, -\frac{3}{2})$ and $(4, \infty)$, and is false in the interval $(-\frac{3}{2}, 4)$.

TABLE 1

Interval	$(-\infty, -\frac{3}{2})$	$(-\frac{3}{2}, 4)$	$(4, \infty)$
Test Point	-2	0	5
$2x^2 - 5x - 12$	$2(-2)^2 - 5(-2) - 12 = 6 > 0$	$2(0)^2 - 5(0) - 12 = -12 < 0$	$2(5)^2 - 5(5) - 12 = 13 > 0$
Sign	Positive	Negative	Positive

In this case we have a weak inequality, so it is satisfied also where the quadratic expression is zero, namely at $x = -\frac{3}{2}$ and $x = 4$. The end points of the intervals are included in the solution set this time. The solution consists of the two semi-infinite intervals $(-\infty, -\frac{3}{2}]$ and $[4, \infty)$. This solution set is illustrated in Figure 9.

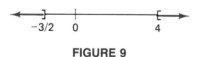

FIGURE 9

Summary of Method of Solution of Quadratic Inequalities:

1. Write the inequality in standard form.

2. Replace the inequality sign by an $=$ sign and solve the resulting quadratic equation. The roots divide the number line into intervals.

3. In each interval choose a point and test the given inequality at that point. If it is true (false) at that point, then it is true (false) at all points in the interval.

4. For a strict inequality the end points of the intervals are not included in the solution set. For a weak inequality they are included.

Sometimes we shall not be able to factor the quadratic expression and it may be necessary to use the quadratic formula to find the dividing points of the intervals.

EXAMPLE 3 Solve the inequality $x^2 - 6x + 6 \le 0$.

Solution The inequality is already in standard form. The corresponding quadratic equation is $x^2 - 6x + 6 = 0$, which does not have rational roots. From the quadratic formula, we have the roots

$$x = \frac{-b \pm \sqrt{b^2 - 4ac}}{2a} = \frac{-(-6) \pm \sqrt{(-6)^2 - 4(1)(6)}}{2 \cdot 1} = \tfrac{1}{2}(6 \pm \sqrt{12})$$

$$= 3 \pm \sqrt{3}.$$

These are approximately 1.27 and 4.73 and as usual divide the real number line into three intervals. We choose a test point in each. (See Table 2 for the details.) The conclusion is that the inequality is false in $(-\infty, 3 - \sqrt{3})$ and $(3 + \sqrt{3}, \infty)$ and is true in $(3 - \sqrt{3}, 3 + \sqrt{3})$.

TABLE 2

Interval	$(-\infty, 3 - \sqrt{3})$	$(3 - \sqrt{3}, 3 + \sqrt{3})$	$(3 + \sqrt{3}, \infty)$
Test point	0	3	5
$f(x) = x^2 - 6x + 6$	$0^2 - 6 \cdot 0 + 6 = 6 > 0$	$3^2 - 6 \cdot 3 + 6 = -3 < 0$	$5^2 - 6 \cdot 5 + 6 = 1 > 0$
Sign	Positive	Negative	Positive

☛ **14.** Solve the inequalities
(a) $x^2 - 2x - 2 \leq 0$;
(b) $x^2 - 2x + 2 > 0$;
(c) $x^2 + 2x + 1 \leq 0$.

Since we have a weak inequality the end points are included, so the solution set is the closed interval $[3 - \sqrt{3}, \ 3 + \sqrt{3}]$, or approximately $[1.27, 4.73]$. This is illustrated in Figure 10. ☛ **14**

1.27   4.73

FIGURE 10

EXAMPLE 4 Solve the inequality $x^2 + 2 > 2x$.

Solution In standard form we have $x^2 - 2x + 2 > 0$. The corresponding quadratic equation is $x^2 - 2x + 2 = 0$, and from the quadratic formula, the roots are

$$x = \frac{-(-2) \pm \sqrt{(-2)^2 - 4(1)(2)}}{2 \cdot 1} = \frac{1}{2}(2 \pm \sqrt{-4}).$$

So, there are not real roots in this case. This means that the expression $x^2 - 2x + 2$ is either positive for all x or negative for all x, since if it changed sign at all it would have to be zero somewhere. All we have to do then is to choose any number as a test point. The simplest is $x = 0$, and we have $0^2 - 2 \cdot 0 + 2 = 2 > 0$. The given inequality is satisfied; hence it is satisfied for all x.

EXAMPLE 5 *(Production and Profits)* The monthly sales x of a certain commodity when its price is p dollars per unit is given by $p = 200 - 3x$. The cost of producing x units per month of the commodity is $C = (650 + 5x)$ dollars.

Answer (a) $1 - \sqrt{3} \leq x \leq 1 + \sqrt{3}$;
(b) $-\infty < x < \infty$ (c) $x = -1$.

How many units should be produced and sold so as to realize a monthly profit of at least $2200?

Solution The revenue R (in dollars) obtained by selling x units at a price p dollars per unit is

$$R = (\text{Units Sold}) \times (\text{Selling Price per Unit})$$

$$= xp$$

$$= x(200 - 3x)$$

$$= 200x - 3x^2.$$

The monthly cost C (in dollars) of manufacturing x units is $C = (650 + 5x)$. The monthly profit P from producing and selling x units is therefore

$$P = R - C$$

$$= (200x - 3x^2) - (650 + 5x)$$

$$= 195x - 3x^2 - 650.$$

Since the profit has to be at least $2200, we have $P \geq 2200$, so

$$195x - 3x^2 - 650 \geq 2200.$$

Writing this in standard form and dividing throughout by -3 (noting the reversal of the inequality sign), we obtain the inequality

$$x^2 - 65x + 950 \leq 0.$$

The roots must be found by the quadratic formula:

$$x = \frac{-b \pm \sqrt{b^2 - 4ac}}{2a}$$

$$= \frac{-(-65) \pm \sqrt{(-65)^2 - 4(1)(950)}}{2 \cdot 1}$$

$$= \tfrac{1}{2}(65 \pm \sqrt{425}).$$

or approximately 22.2 and 42.8. In the three intervals $x < 22.2$, $22.2 < x < 42.8$, and $x > 42.8$ we choose the three points $x = 0$, 40, and 100, respectively. We find that $x^2 - 65x + 950 > 0$ when $x = 0$ and 100 but $x^2 - 65x + 950 < 0$ when $x = 40$. It follows therefore that $x^2 - 65x + 950 < 0$ for all x in the interval $22.2 < x < 42.8$. The solution set of the inequality is therefore the closed interval $[22.2, 42.8]$. ☛ 15

So, to achieve the required goal, the number of units produced and sold per month must lie between 22.2 and 42.8, inclusive.

EXAMPLE 6 *(Pricing Decision)* A hairdresser gets an average of 120 customers per week at a present charge of $8 per haircut. For each increase of 75¢ in the price, the hairdresser will lose 10 customers. What is the maximum price that can be charged so that the weekly earnings will not be less than they are at present?

Solution Let there be x increases of 75¢ in price beyond $8. Then the price per haircut is $(8 + 0.75x)$ dollars, and the number of customers will be

☛ **15.** In Example 5, for what interval of values of x does the monthly profit exceed $2500?

Answer $30 < x < 35$.

$(120 - 10x)$ per week. Then

$$\text{Total Weekly Earnings} = \text{Number of Customers} \times \text{Price per Haircut}$$

$$= (120 - 10x) \times (8 + 0.75x).$$

Earnings from the present 120 customers are $120 \times \$8 = \$960$. Therefore the new earnings should be at least \$960:

$$(120 - 10x)(8 + 0.75x) \geq 960.$$

We simplify:

$$960 + 10x - 7.5x^2 \geq 960$$

$$10x - 0.75x^2 \geq 0.$$

The corresponding equation is $10x - 7.5x^2 = 0$, whose solutions are $x = 0$ and $\frac{4}{3}$. In the three intervals $x < 0$, $0 < x < \frac{4}{3}$, and $x > \frac{4}{3}$ we choose the test points -1, 1, and 2, respectively. We find that $10x - 7.5x^2 < 0$ when $x = -1$ or 2, but $10x - 7.5x^2 > 0$ when $x = 1$. The solution of the inequality is therefore the interval $0 \leq x \leq \frac{4}{3}$. That is, the price of a haircut should be between \$8 and $\$(8 + 0.75 \times \frac{4}{3}) = \$9.00$. The maximum price that can be charged is \$9.00.

EXERCISES 3-3

(1–26) Solve the following inequalities.

1. $(x - 2)(x - 5) < 0$ **2.** $(x + 1)(x - 3) \leq 0$

3. $(2x - 5)(x + 3) \geq 0$ **4.** $(3x - 1)(x + 2) > 0$

5. $x^2 - 7x + 12 \leq 0$ **6.** $9x > x^2 + 14$

7. $x(x + 1) < 2$ **8.** $x(x - 2) \geq 3$

9. $y(2y + 1) > 6$ **10.** $3y^2 \geq 4 - 11y$

11. $(x + 2)(x - 3) > 2 - x$

12. $(2x + 1)(x - 3) < 9 + (x + 1)(x - 4)$

13. $x^2 \geq 4$ **14.** $9x^2 < 16$

15. $x^2 + 3 > 0$ **16.** $x^2 + 1 \leq 0$

17. $x^2 - 6x + 9 \leq 0$ **18.** $x^2 + 4 < 4x$

19. $x^2 + 2x + 1 > 0$ **20.** $x^2 + 9 \geq 6x$

21. $x^2 + 13 < 6x$ **22.** $x^2 + 7 > 4x$

23. $(x - 2)^2 + 5 \geq 0$ **24.** $x^2 + 2x + 4 < 0$

25. $(2x + 3)(x - 3) > (x - 1)(3x + 2)$

26. $(1 - 3x)(x + 2) > (3 - 2x)(x + 3)$

27. *(Manufacturer's Revenue)* At a price of p dollars per unit, x units of a certain commodity can be sold each month in the market, with $p = 600 - 5x$. How many units should be sold each month to obtain a revenue of at least \$18,000?

28. *(Manufacturer's Revenue)* A manufacturer can sell x units of a product each week at a price of p dollars per unit, where $p = 200 - x$. What number of units should be produced and sold each week to obtain a weekly revenue of at least \$9900?

29. *(Production Decision)* In Exercise 27, if it costs $(8000 + 75x)$ dollars to produce x units, how many units should be produced and sold each month to obtain a monthly profit of at least \$5500?

30. *(Pricing Decision)* In Exercise 28, if it costs $(2800 + 45x)$ dollars to produce x units, at what price p will each unit be sold so as to generate a weekly profit of at least \$3200?

31. *(Profitability)* A manufacturer can sell all units of a product at \$25 each. The cost C (in dollars) of producing x units each week is given by $C = 3000 + 20x - 0.1x^2$. How many units should be produced and sold each week to obtain a profit?

32. *(Publisher's Revenue)* A publisher can sell 12,000 copies of a book at $25 each. For every dollar increase in price, sales will fall by 400 copies. What maximum price should be charged for each copy to obtain a revenue of at least $300,000?

33. *(Farming)* A farmer wishes to enclose a rectangular field and has 200 yards of fencing available. Find the possible dimensions of the field if its area must be at least 2100 square yards.

34. One side of the rectangular field is bounded by a river. A farmer has 100 yards of fencing and wants to cover the field's other three sides. If he wants to enclose an area of at least 800 square yards, what are the possible values for the length of the field along the river?

35. An open box is made from a rectangular sheet of metal 16 by 14 feet by cutting equal squares from each corner and folding up the edges. If the area of the base of the box is to be at least 80 square feet, what is the maximum possible height of the box?

36. A rectangular sheet of cardboard is 16 by 10 inches. Squares of equal sizes are cut from each corner and the edges of the cardboard are folded up to form an open

box. What is the maximum height of this box if the base has an area of at least 72 square inches?

37. *(Conservation)* A certain area of water is to be stocked with fish. If n fish are put in, it is known that the average gain in weight of each fish will be $(600 - 3n)$ grams. Find the restrictions on n if the total gain in weight of all the fish stock is to be greater than 28,800 grams.

38. *(Investment)* An investor invests $100 at R percent interest per annum and $100 at $2R$ percent per annum. If the value of the two investments is to be at least $224.80 after 2 years, what restrictions must there be on R?

39. *(Pricing Policy)* A supermarket finds itself with a large stock of apples on hand that must be sold quickly. The manager knows that if the apples are offered at p cents per pound, he will sell x pounds, where $x = 1000 - 20p$. What price should be charged in order to produce a revenue of at least $120?

40. *(Pricing Decision)* A hairdresser gets an average of 120 customers per week at a present charge of $4 per haircut. For each increase of 50¢ in price, the hairdresser will lose 8 customers. What maximum price should be charged to obtain a weekly revenue of at least $520?

◤ 3-4 ABSOLUTE VALUES

If x is a real number, then the **absolute value** of x, denoted by $|x|$, is defined by

> **16.** Evaluate   (a) $-|-5|$;
> (b) $|2 - 3 - 4|$;
> (c) $|2| + |-3| - |4|$.

$$|x| = \begin{cases} x & \text{if } x \geq 0 \\ -x & \text{if } x < 0. \end{cases}$$

For example, $|5| = 5$, $|-3| = -(-3) = 3$, and $|0| = 0$.   ☞ 16

From this definition, it is clear that the *absolute value of a number is always a nonnegative real number;* that is,

$$|x| \geq 0.$$

The absolute value of x is a measure of the "size" of x without regard as to whether x is positive or negative.

EXAMPLE 1 Solve for x.

Answer (a) -5;   (b) 5;   (c) 1.

$$|2x - 3| = 5$$

Solution According to the definition of absolute value, the given equation is satisfied if either

$$2x - 3 = 5 \quad \text{or} \quad 2x - 3 = -5$$

because in either case, the absolute value of $2x - 3$ is 5. If $2x - 3 = 5$, then $2x = 3 + 5 = 8$ and so $x = 4$. Similarly, if $2x - 3 = -5$, then $x = -1$. Thus there are two values of x, $x = 4$ and $x = -1$, that satisfy the given equation.

EXAMPLE 2 Solve for x.

$$|3x - 2| = |2x + 7|$$

☞ **17.** Solve for x:

(a) $|x + 1| = 2$;

(b) $|x - 1| = |3 - 2x|$;

(c) $|x - 1| = (3 - 2x)$.

Solution This equation will be satisfied if either

$$3x - 2 = 2x + 7 \quad \text{or} \quad 3x - 2 = -(2x + 7).$$

Solving these two equations separately, we obtain $x = 9$ and $x = -1$. ☞ **17**

From Examples 1 and 2, it is clear that we have the following general rules for solving equations involving absolute values.

> If $|a| = b$, where $b \geq 0$, then either $a = b$ or $a = -b$.
>
> If $|a| = |b|$, then either $a = b$ or $a = -b$.

  Recall The symbol $\sqrt{a}$ *denotes the nonnegative square root of the real number* $a (a \geq 0)$. For example, $\sqrt{9} = 3$. The negative square root of 9 is denoted by $-\sqrt{9}$. Using the radical symbol, we can give the following alternative definition of absolute value.

> $$|x| = \sqrt{x^2}.$$

For example, $\sqrt{3^2} = \sqrt{9} = 3$, $\sqrt{(-5)^2} = \sqrt{25} = 5 = |-5|$, and $\sqrt{(x - 3)^2} = |x - 3|$.

  We can interpret $|x|$ geometrically. (See Figure 11.) The numbers 3 and 8 on the number line are 5 units apart. Also $|8 - 3| = |5| = 5$ and $|3 - 8| = |-5| = 5$. Thus $|8 - 3| = |3 - 8|$ gives the distance between two points 3 and 8 on a number line. In general, we may interpret $|x - c| = |c - x|$ as the distance between two points x and c on a number line, without

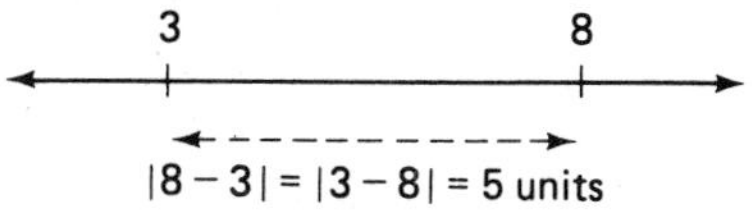

Answer (a) -3 or 1;   (b) $\frac{4}{3}$ or 2;

(c) $\frac{4}{3}$ (right side negative if $x = 2$).

$|8 - 3| = |3 - 8| = 5$ units

FIGURE 11

regard to direction. For example, the equation $|x - 2| = 5$ states that the distance between x and 2 on a number line is 5 units, without regard to direction. Thus x can be either $2 + 5 = 7$ or $2 - 5 = -3$, as shown in Figure 12.

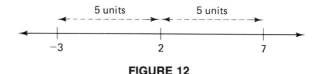

FIGURE 12

Since $|x| = |x - 0|$, $|x|$ represents the distance of the point x on the real number line from the origin O, without regard to direction. (See Figure 13.) Also, since the distance between O and x is the same as the distance between O and $-x$, it follows that

$$|x| = |-x|.$$

For example, $|7| = |-7| = 7$.

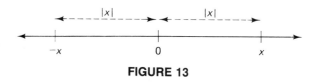

FIGURE 13

In Example 3 a number of statements are re-expressed in terms of absolute values.

EXAMPLE 3

(a) x is at a distance of 3 units from 5: $|x - 5| = 3$.

(b) x is less than 7 units from 4: $|x - 4| < 7$.

(c) x is at least 7 units from -3: $|x - (-3)| \geq 7$ or $|x + 3| \geq 7$.

(d) x is strictly within 3 units of 7: $|x - 7| < 3$.

(e) x is within c units of a: $|x - a| \leq c$. 👉 **18**

Let us now consider some inequalities involving absolute values. The inequality $|x| < 5$ implies that the distance of x from the origin is less than 5 units. Since x can be on either side of O, x lies between -5 and 5, or $-5 < x < 5$. (See Figure 14.) Similarly, $|x| > 5$ implies that x is more than 5 units from the origin on either side, that is, $x < -5$ or $x > 5$. (See Figure 15.) This result is generalized in the following theorem:

👉 **18.** Express the following using absolute values:

(a) x is at most 4 units from 3;

(b) $5 - x$ is 4 units from x.

Answer (a) $|x - 3| \leq 4$;

(b) $|5 - 2x| = 4$.

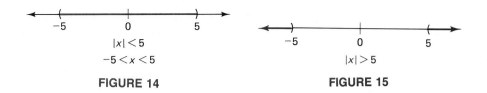

FIGURE 14   FIGURE 15

THEOREM 1 If $a > 0$, then

$$|x| < a \text{ if and only if } -a < x < a; \tag{1}$$
$$|x| > a \text{ if and only if either } x > a \text{ or } x < -a. \tag{2}$$

Figures 16 and 17 illustrate Theorem 1.

FIGURE 16   FIGURE 17

EXAMPLE 4 Solve $|2x - 3| < 5$ for x and express the result in terms of intervals.

Solution Using Statement (1) of Theorem 1, the given inequality implies that

$$-5 < 2x - 3 < 5.$$

Adding 3 to each member of the double inequality and simplifying, we obtain

$$-5 + 3 < 2x - 3 + 3 < 5 + 3$$
$$-2 < 2x < 8.$$

We next divide throughout by 2.

$$-1 < x < 4.$$

Thus the solution consists of the real numbers x that lie in the open interval $(-1, 4)$. (See Figure 18.)

FIGURE 18

EXAMPLE 5 Solve $|2 - 3x| > 7$ for x and express the result in interval notation.

Solution Using Statement (2) of Theorem 1, the given inequality implies that either

$$2 - 3x > 7 \quad \text{or} \quad 2 - 3x < -7.$$

Taking the first inequality, we have

$$2 - 3x > 7.$$

Subtracting 2 from both sides and dividing by -3 (thus reversing the direction of inequality) gives

$$x < -\tfrac{5}{3}.$$

Similarly, solving the second inequality, we obtain

$$x > 3.$$

Thus $|2 - 3x| > 7$ is equivalent to

$$x < -\tfrac{5}{3} \quad \text{or} \quad x > 3.$$

Thus the solution consists of all real numbers *not* in the closed interval $[-\tfrac{5}{3}, 3]$. (See Figure 19.)

-5/3 0   3

FIGURE 19

EXAMPLE 6 Solve $|2x - 3| + 5 \le 0$ for x.

Solution The given inequality can be rewritten as

$$|2x - 3| \le -5.$$

But $|2x - 3|$ can never be negative, so there are no values of x for which the given inequality is true. Thus no solution exists. ☛ 19

EXAMPLE 7 Solve the inequality $|3x - 5| \le x + 1$.

Solution If $(x + 1) < 0$, there can clearly be no solution, since the absolute value on the left side cannot be less than a negative number. Thus the solution set is immediately restricted to $x \ge -1$.

If $x + 1 \ge 0$, we can use Theorem 1 to express the given inequality in the form

$$-(x + 1) \le 3x - 5 \le (x + 1).$$

The left half of this double inequality, $-(x + 1) \le 3x - 5$, leads to $x \ge 1$. The right half, $3x - 5 \le x + 1$, leads to $x \le 3$. The three conditions $x \ge 1$, $x \le 3$, and $x \ge -1$ must all be satisfied. Thus the solution set is $1 \le x \le 3$, or the closed interval $[1, 3]$.

We close this section by stating two basic properties of absolute value. If a and b are two real numbers, then

---

☛ **19.** Solve the inequalities
a) $|1 - x| < 4$;
(b) $|7 - 4x| \ge 3$;
(c) $|x - 1| + |x + 1| < 0$.

Answer (a) $-3 < x < 5$;
(b) $x \le 1$ or $x \ge \tfrac{5}{2}$;
(c) no solution.

$$|ab| = |a| \cdot |b| \tag{3}$$

$$\left| \frac{a}{b} \right| = \frac{|a|}{|b|} \qquad (b \neq 0). \tag{4}$$

EXAMPLE 8

(a) $|(-3)(5)| = |-3||5| = (3)(5) = 15$

(b) $\left| \dfrac{x-2}{1+x} \right| = \dfrac{|x-2|}{|1+x|} \qquad (x \neq -1)$

(c) $\left| \dfrac{x-7}{-3} \right| = \dfrac{|x-7|}{|-3|} = \dfrac{|x-7|}{3}$

Equations (3) and (4) are easily derived from the fact that for any number x, $|x| = \sqrt{x^2}$. For example, Equation (3) is derived as follows.

$$\begin{aligned} |ab| &= \sqrt{(ab)^2} \\ &= \sqrt{a^2 b^2} \\ &= \sqrt{a^2} \cdot \sqrt{b^2} \qquad \text{(using a property of radicals)} \\ &= |a| \cdot |b| \end{aligned}$$

EXERCISES 3-4

(1–4) Evaluate.

1. $\sqrt{2}|-2| + 5|-\sqrt{2}|$ 2. $|\sqrt{3} - 2| + |\sqrt{3} - 1|$

3. $|\pi - 5| - |-2|$ 4. $|3 - \sqrt{5}| - |\sqrt{5} - 2|$

(5–18) Solve the following equations for x.

5. $|3 - 7x| = 4$ 6. $|2x + 5| = 7$

7. $|x + 2| = |3 - x|$ 8. $\left| \dfrac{2x + 1}{3} \right| = |3x - 7|$

9. $|3x - 2| = 4 - x$ 10. $|x + 3| = 5 - x$

11. $|x + 3| = x - 5$ 12. $|3x - 2| = x - 4$

13. $|x - 3| + 7 = 0$ 14. $|2x + 1| + |3x - 2| = 0$

15. $\left| \dfrac{x - 3}{3x - 5} \right| = 6$ 16. $\left| \dfrac{-5x - 2}{x + 3} \right| = 5$

17. $\left| \dfrac{1}{x} - 3 \right| = 4$ 18. $\left| 3 - \dfrac{1}{x - 2} \right| = 7$

(19–36) Solve the following inequalities and express the solution in interval form if possible.

19. $|3x + 7| < 4$ 20. $|2x - 6| \le 3$

21. $|2 - 5x| \ge 3$ 22. $|3 - 4x| < \frac{1}{2}$

23. $5 + 2|3 - 2x| < 7$ 24. $5 - 2|3 - 2x| \le 1$

25. $7 + |3x - 5| \le 5$ 26. $|3x - 13| + 6 \ge 0$

27. $|x + 2| + |2x - 1| \ge 0$

28. $|3x - 2| + |2x - 7| < 0$

29. $\left| \dfrac{5 - x}{3} \right| + 4 \le 2$ 30. $\left| \dfrac{2 - 5x}{4} \right| \ge 3$

31. $|5 - 2x| + 5 \ge 0$ 32. $|2x - 3| + |7 + 3x| < 0$

*33. $|2x - 3| < x - 4$ *34. $|x - 2| < 3 - x$

*35. $|x - 3| < x - 2$ *36. $|3x - 2| > 2x - 3$

(37–38) Express the following statements in terms of absolute values and in terms of interval notation.

37. a. x is less than 5 units from 3.

    b. y is at most 4 units from 7.

    c. t is at a distance of 3 units from 5.

    d. z is strictly within σ (sigma) units from μ (mu).

    e. x differs from 4 by more than 3 units.

    f. $\bar{x}$ differs from μ by more than 3 units.

38. a. x is at least 4 units from -5.

    b. y is at most 7 units from 3.

    c. x is less than 3 units from 9.

    d. x is less than 4 and greater than -4.

e. x is either greater than 3 or less than -3.

f. $\bar{x}$ exceeds μ by more than 2 units.

g. y is less than 7 by more than 3 units.

h. x differs from y by more than 5 units.

39. *(Stocks)* According to a money magazine prediction, the price p of B.C. Tel. stock will not change from its current price of \$22 by more than \$5. Use the absolute notation to express this prediction as an inequality.

40. *(Housing Market)* According to a real estate survey, the price (in dollars) of an average house in Vancouver next year will be given by

$$|x - 210{,}000| \le 30{,}000.$$

Determine the highest and lowest price of the house next year.

CHAPTER REVIEW

Key Terms, Symbols, and Concepts

3.1 The inequality symbols: $<$, $>$, $\le$, $\ge$. Strict inequality. Double inequality.
Set, member, or element of a set. Finite set, infinite set. Empty set.
Listing method, partial listing.
Rule method; the notation $\{x \mid x \text{ satisfies the rule}\}$.
Subset, proper subset. Equality of two sets.
Interval, end points. Closed, open, and semiopen intervals.
Infinite and semi-infinite intervals.
Equivalent notations such as: $\{x \mid a < x \le b\}$, $(a, b]$, or on the number line,

3.2 Linear inequality, solution set of an inequality.
The addition and multipication rules for manipulating inequalities.

Procedure for solution of a linear inequality or a double linear inequality.

3.3 Quadratic inequality.
Step-by-step procedure for solution of a quadratic inequality.

3.4 Absolute value of a real number and its geometrical interpretation.

Formulas

If $|a| = b$ and $b > 0$, then $a = b$ or $a = -b$. If $|a| = |b|$, then $a = b$ or $a = -b$.

If $|a| < b$ and $b > 0$, then $-b < a < b$. If $|a| > b$, then either $a < -b$ or $a > b$.

$$|a| = \sqrt{a^2}; \quad |ab| = |a| \cdot |b|; \quad \left|\frac{a}{b}\right| = \frac{|a|}{|b|}, \quad (b \ne 0).$$

REVIEW EXERCISES FOR CHAPTER 3

1. State whether each of the following is true or false. Replace each false statement with a corresponding true statement.

    a. A linear inequality in one variable has an infinite number of solutions.

b. When two sides of an inequality are multiplied by a nonzero constant, the direction of inequality is preserved.

c. A quadratic inequality has either two solutions, one solution, or no solutions at all.

d. If a negative number is subtracted from both sides of an inequality, the direction of inequality must be reversed.

e. If $|x| = a$, then $x = a$ or $x = -a$ for all values of the constant a.

f. $|x| = |y|$ implies either $x = y$ or $x = -y$.

g. The equation $|x - 2| + |x - 3| = 0$ has *no* solution.

h. $|x + y| = |x| + |y|$ if and only if x and y are of the same sign.

i. If x is any real number, then $|x| \geq x$ and $|x| \geq -x$.

j. $x > y$ implies $|x| > |y|$.

k. If $x^2 > y^2$ then $|x| > |y|$.

l. $|x| < |y|$ if and only if $x^2 < y^2$.

(2–39) Solve the following inequalities.

2. $3(2 - x) + 5 > x - 2(x - 2)$

3. $4x - 2 \leq 3x - 2(2 - 3x)$

4. $(2x + 1)(x + 2) > 2(x + 3)(x - 1)$

5. $x^2 + 3(x - 2) < (x + 3)(x + 2)$

6. $\dfrac{2x - 5}{4} - \dfrac{1 - 2x}{3} > \dfrac{x + 3}{2}$

7. $\dfrac{x + 1}{3} - \dfrac{2x + 1}{6} < \dfrac{1 - 3x}{2}$

8. $(3x - \tfrac{1}{4})^2 < 9(x + \tfrac{1}{2})^2$ **9.** $(2x + \tfrac{1}{3})^2 > 4(x - \tfrac{1}{2})^2$

10. $(3x - 1)(x + 2) > (3x + 2)(x + 1)$

11. $(x + 5)(x + 7) < (x + 9)(x + 3)$

12. $x^2 - 7x + 6 \leq 0$ **13.** $2x^2 < 3x + 5$

14. $5x < 2(x^2 + 1)$ **15.** $3x^2 > 7x - 2$

16. $3(x^2 + 1) \leq 10x$ **17.** $9x + 5 \leq 2x^2$

18. $3 - x > 2x^2$ **19.** $15 - 2x^2 > x$

20. $x^2 + 9 > 4x$ **21.** $x^2 + 12 \geq 6x$

22. $(2x + 1)(x - 2) < (x + 2)(x - 3)$

23. $(3x + 2)(x - 1) \leq (2x - 3)(x + 2)$

*24. $x^3 + 12x > 7x^2$ ***25.** $x^3 \leq 2x^2 + 15x$

26. $2x + 1 \leq 5 - x \leq x - 7$

27. $3x - 1 > x + 3 > 2x - 3$

28. $3x(2 - x) < -9$ **29.** $(x + 1)(2x - 5) \geq -3$

30. $|3 - 4x| < 2$ **31.** $\left| \dfrac{2x - 3}{7} \right| \geq 1$

32. $|2x - 3| \leq 7$ **33.** $|4x - 7| \geq 3$

34. $|2 - 3x| > 7$ **35.** $7 - |x - 3| < 0$

36. $5 + |2x - 5| \geq 0$ **37.** $9 + |2x - 7| \leq 0$

38. $|2x - 3| + |7 + 3x| < 0$

39. $|3x - 5| + |x - 2| \geq 0$

(40–45) Solve the following equations.

40. $|2x - 3| + 7 = 4$ **41.** $|5 - 3x| = x + 2$

42. $|2x - 1| + |3x - 2| = 0$

43. $|3x + 4| - 2|x + 2| = 0$

*44. $|x^2 - 5x| = 4$ ***45.** $|x^2 + 2| = 3x$

46. *(Production and Profits)* A manufacturer can sell all she can produce at a price of $15 per unit. Each unit costs $8 to produce in materials and labor and, in addition, there is an overhead cost of $4000 per week. How many units must be produced if her weekly profit is to be at least $3000?

47. *(Publisher's Profits)* A newspaper costs 25¢ per copy to produce and sell. The publisher receives 20¢ per copy from sales and, in addition, he receives revenue from advertising that is equal to 30% of the revenue from sales in excess of 20,000 copies. How many copies must the publisher sell if:

a. He is at least to break even?

b. He is to make a profit of at least $1000 per issue of the newspaper?

48. *(Apartment Rental)* The owner of an apartment building can rent all the 50 suites if the rent is $150 per month for each suite. For each increase of $5 in the monthly rent, one suite will be vacant with no possibility of renting it. What maximum rent should be charged for each suite to obtain a monthly revenue of at least $8000?

49. *(Pricing Policy)* A state liquor board buys whiskey for $2 a bottle and sells it for p dollars per bottle. The volume of sales x (in hundreds of thousands of bottles per week) is given by $x = 24 - 2p$ when the price is p. What value of p gives a total revenue of $7 million per week? What values of p give a profit to the liquor board of at least $4.8 million per week?

50. *(Revenue from Sales Tax)* A certain luxury item sells for $1000; over a whole state, the sales amount to 20,000 items per year. The state government is considering imposing a sales tax on these items. If the level of tax is set at R percent, the sales will fall by $500R$ items per year. What value of R will provide a total income to the government of $1.68 million per year from the tax? What values of R will provide the government with an income of at least $1.92 million per year?

51. *(Investment Decision)* Mrs. Smith has $60,000 to invest. She can invest her funds in 8% government bonds or with greater risk in 10% mortgage bonds. What minimum amount would she have to invest in mortgage bonds so as to receive an annual income of at least $5500?

52. *(Car Rental)* A firm rents out cars to customers under two plans. In the first plan one can rent a car at $160 per week with unlimited mileage, whereas in the second plan one can rent the same car at $100 per week plus 25¢ for each mile driven. Find the values of weekly mileage for which it is cheaper to rent through the second plan.

53. If x units can be sold each day at a price of $p each, where $p = 60 - x$, how many units must be sold to obtain a daily revenue of at least $800?

54. If x units can be sold at a price of $p each, where $2p + 3x = 200$, what price p per unit must be charged to obtain a revenue of at least $1600?

55. In Exercise 53, it cost $(260 + 12x)$ dollars to produce x units. How many units must be produced and sold each day to obtain a profit of at least $300?

56. In Exercise 54, it costs $(800 + 7x)$ dollars to produce x units. How many units must be produced and sold to obtain a profit of at least $640?

57. *(Pricing and Profits)* In Exercise 53, it costs $(260 + 8x)$ dollars to produce x units. What price p (in dollars) per unit must be charged to obtain a profit of at least $400?

58. *(Pricing and Profits)* In Exercise 54, if it costs $(750 + 10x)$ dollars to produce x units, what price p (in dollars) per unit must be charged to obtain a profit of at least $450?

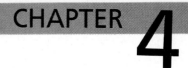

CHAPTER 4

Straight Lines

Chapter Objectives

4-1 CARTESIAN COORDINATES
(a) The Cartesian coordinates of a point in the plane;
(b) The formula for the distance between two points in the plane;
(c) Drawing the graph of an equation in two variables.

4-2 STRAIGHT LINES AND LINEAR EQUATIONS
(a) The rise and run between two points and the slope of a straight line;
(b) The various equations that describe straight lines;
  (1) The point-slope formula,
  (2) The slope-intercept formula,
  (3) The equations of horizontal and vertical lines,
  (4) The general linear equation;
(c) Graphing linear equations.

4-3 APPLICATIONS OF LINEAR EQUATIONS
(a) The linear cost model;
(b) Straight-line depreciation;
(c) The laws of demand and supply;
(d) Trade-off.

4-4 SYSTEMS OF EQUATIONS
(a) Solution of a system of two linear equations in two variables by the substitution method and the addition method;
(b) Geometrical interpretation of a system of linear equations;
(c) Solution by the substitution method of systems of three equations in three variables and some systems of nonlinear equations.

4-5 APPLICATIONS TO BUSINESS AND ECONOMICS
(a) Break-even analysis;
(b) Market equilibrium and its determination;
(c) Effect of sales tax or subsidy on market equilibrium.

CHAPTER REVIEW

◼ 4-1 CARTESIAN COORDINATES

A relationship between two variables is commonly expressed by means of an algebraic equation that involves the two variables. For example, if x is the length (in inches) of one side of a square and if y is the area (in square inches) of the square, then the relation between x and y is expressed by the equation $y = x^2$. For each value of x, the related value of y is obtained by squaring the given value of x.

An algebraic equation of this kind can be represented pictorially, by means of a graph. It is often true that significant features of the given relationship are much more apparent from the graph than they are from the algebraic relation between the variables. When presented with an algebraic relation, it is useful—particularly in applications of mathematics—to develop the habit of asking oneself what its graph looks like.

Graphs are constructed using what are called *Cartesian coordinates*. We draw two perpendicular lines called **coordinate axes,** one horizontal and the other vertical, intersecting each other at a point O. The horizontal line is called the **x-axis,** the vertical line is called the **y-axis,** and O is called the **origin.** A plane with such coordinate axes is called the **Cartesian plane** or simply the **xy-plane.**

We select a unit of length on the two axes. (Usually the units of length on both axes are the same, but it is not necessary that they are the same.) Starting from the origin O as zero, we then mark off number scales on each axis as shown in Figure 1. Positive numbers are marked to the right of O on the x-axis and above O on the y-axis.

Consider any point P in the plane. From P, draw PM perpendicular to the x-axis and PN perpendicular to the y-axis, as shown in Figure 1. If the point M represents the number x on the x-axis and the point N represents the number y on the y-axis, then x and y are called the **Cartesian coordinates** of the point P. We write these two coordinates enclosed in parentheses, in the order (x, y).

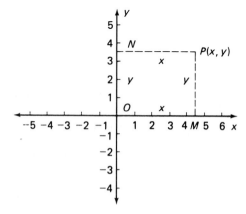

FIGURE 1

In this way, corresponding to each point P in the plane, there is a unique ordered pair of real numbers (x, y), which are the coordinates of the point. And conversely, we can see that corresponding to each ordered pair (x, y) of real numbers, there is a unique point in the plane. This representation of points in the plane by ordered pairs of real numbers is called the **Cartesian coordinate system.**

If the ordered pair (x, y) represents a point P in the plane, then x (the first member) is called the **abscissa** or **x-coordinate** of the point P and y (the second member) is called the **ordinate** or **y-coordinate** of P. The abscissa and ordinate of P together are called the **rectangular Cartesian coordinates** of the point P. The notation $P(x, y)$ is used to denote a point P with coordinates (x, y).

The coordinates of the origin are $(0, 0)$. For each point on the x-axis, the y-coordinate is zero; each point on the y-axis has an x-coordinate of zero. Figure 2 shows several ordered pairs of real numbers and the corresponding points. ☞ 1

☞ **1.** Plot the points $(-4, 0)$, $(0, 4)$, $(2, -1)$, $(-2, -1)$, and $(-3, 3)$.

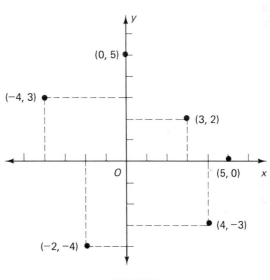

FIGURE 2

Answer

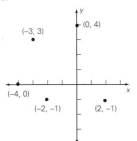

The coordinate axes divide the xy-plane into four parts, called **quadrants.** The quadrants are called the *first, second, third* and *fourth quadrants,* as shown in Figure 3.

(x, y) is in the first quadrant if $x > 0$ and $y > 0$,

(x, y) is in the second quadrant if $x < 0$ and $y > 0$,

(x, y) is in the third quadrant if $x < 0$ and $y < 0$,

(x, y) is in the fourth quadrant if $x > 0$ and $y < 0$.

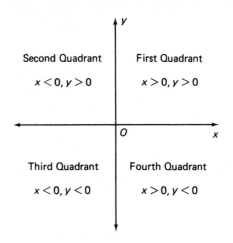

FIGURE 3

THEOREM 1 (DISTANCE FORMULA) If $P(x_1, y_1)$ and $Q(x_2, y_2)$ are any two points in the plane, then the distance d between P and Q is given by

$$d = \sqrt{(x_2 - x_1)^2 + (y_2 - y_1)^2}.$$

Before proving this, let us see how it works for two points on the same horizontal line. Let P be $(4, 3)$ and Q be $(-2, 3)$. (See Figure 4.) Then $y_2 - y_1 = 3 - 3 = 0$ while $x_2 - x_1 = -2 - 4 = -6$. Thus

$$d = \sqrt{(-6)^2 + 0^2} = 6$$

Note that when $y_2 - y_1 = 0$,

$$d = \sqrt{(x_2 - x_1)^2} = |x_2 - x_1|.$$

Similarly, for two points on the same vertical line, $d = \sqrt{(y_2 - y_1)^2} = |y_2 - y_1|$.

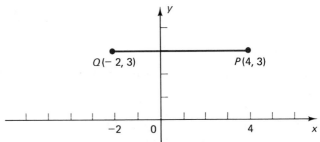

FIGURE 4

PROOF OF THEOREM 1 If PM and QN are perpendiculars from the two points $P(x_1, y_1)$ and $Q(x_2, y_2)$ onto the x-axis, and PA and QB are perpendiculars onto the y-axis, as shown in Figure 5, then the coordinates of the points

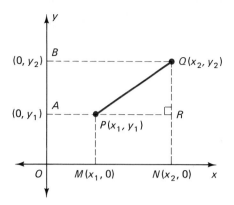

FIGURE 5

M, N, A, and B are as given in Figure 5. Let the point where line PA meets line QN be R, so that PQR is a right triangle with right angle at R.

From Figure 5* we have

$$PR = MN = ON - OM = x_2 - x_1$$

and

$$RQ = AB = OB - OA = y_2 - y_1.$$

We next use the Pythagorean theorem with right triangle PQR.

$$PQ^2 = PR^2 + RQ^2$$

or

$$d^2 = (x_2 - x_1)^2 + (y_2 - y_1)^2.$$

Taking the square root (the nonnegative square root because the distance is always nonnegative), we have

$$d = \sqrt{(x_2 - x_1)^2 + (y_2 - y_1)^2} \tag{1}$$

which proves the result. Equation (1) is known as the **distance formula in the plane.**

EXAMPLE 1 Find the distance between the two points $A(-1, 3)$ and $B(4, 15)$.

---

*Figure 5 has been drawn with P and Q both in the first quadrant. The equations for PR and RQ apply in whatever quadrants the two points lie. However, in Figure 5, Q has been drawn above and to the right of P, so that $x_2 > x_1$ and $y_2 > y_1$. In the more general case—when these conditions are not satisfied—the horizontal and vertical distances between P and Q are given by $PR = |x_2 - x_1|$ and $RQ = |y_2 - y_1|$. It can be seen that Equation (1) for d continues to apply in this general case.

Solution We identify the two given points as

$$(-1, 3) = (x_1, y_1) \quad \text{and} \quad (4, 15) = (x_2, y_2).$$

We now use the distance formula.

$$AB = \sqrt{(x_2 - x_1)^2 + (y_2 - y_1)^2}$$

$$= \sqrt{(4 - (-1))^2 + (15 - 3)^2}$$

$$= \sqrt{5^2 + 12^2} = \sqrt{169} = 13 \quad \text{☛ 2}$$

☛ **2.** What is the distance from $(-3, -2)$ to (a) $(0, 2)$; (b) $(1, 4)$?

EXAMPLE 2 The abscissa of a point is 7 and its distance from the point $(1, -2)$ is 10. Find the ordinate of the point.

Solution Let P be the point whose ordinate is required, and A be the point $(1, -2)$. Let y be the ordinate of the point P. Then the coordinates of P are $(7, y)$, because its abscissa is given to be 7. From the statement of the problem, we are given that

$$PA = 10 \tag{i}$$

Now identifying the two points P and A as

$$(7, y) = (x_1, y_1) \quad \text{and} \quad (1, -2) = (x_2, y_2)$$

Answer (a) 5; (b) $\sqrt{52}$.

and using the distance formula, we have

$$PA = \sqrt{(x_2 - x_1)^2 + (y_2 - y_1)^2} = \sqrt{(1 - 7)^2 + (-2 - y)^2}$$

or

$$10 = \sqrt{36 + (2 + y)^2}$$

from Equation (i). We next square both sides.

$$100 = 36 + (2 + y)^2 = 36 + 4 + 4y + y^2.$$

Therefore

$$y^2 + 4y - 60 = 0$$

or

$$(y + 10)(y - 6) = 0.$$

Therefore one of the following conditions holds:

$$y + 10 = 0 \quad \text{or} \quad y - 6 = 0$$

☛ **3.** Find a if the points $(0, a)$ and $(-1, 1)$ are the same distance from $(1, 2)$.

$$y = -10 \qquad\qquad y = 6$$

The ordinate of the required point P is either 6 or -10. ☛ **3**

DEFINITION The **graph** of an equation involving two variables, such as x and y, is the set of all the points whose coordinates (x, y) satisfy the equation.

Answer $a = 0$ or 4.

Consider, for example, the equation $2x - y - 3 = 0$. One of the points whose coordinates satisfy this equation is $(1, -1)$, since the equation is satisfied when we substitute $x = 1$ and $y = -1$. Other such points are $(0, -3)$ and $(2, 1)$. The graph of this equation is obtained by plotting these points and all of the others which satisfy this equation.

Drawing the *exact* graph of an equation in two variables is usually an impossible task because it would involve plotting infinitely many points. In general practice enough points satisfying the given equation are selected to exhibit the general nature of the graph. These points are plotted and joined by a smooth curve.

When finding the points satisfying a given equation, it is often useful to solve the equation for one variable in terms of the other. For example, if we solve the equation $2x - y - 3 = 0$ for y in terms of x, we have

$$y = 2x - 3.$$

Now if we give values to x, we can calculate the corresponding values for y. For example, if $x = 1$, $y = 2(1) - 3 = -1$; if $x = 5$, $y = 10 - 3 = 7$; and so on.

EXAMPLE 3 Sketch the graph of the quation $2x - y - 3 = 0$.

Solution Solving the given equation for y, we have

$$y = 2x - 3.$$

The values of y corresponding to different values of x are given in Table 1. Plotting these points, we observe that they lie on a straight line. (See Figure 6.) This line is the graph of the given equation.

TABLE 1

x	−2	−1	0	1	2	3	4
y	−7	−5	−3	−1	1	3	5

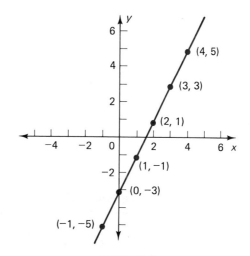

FIGURE 6

EXAMPLE 4 Sketch the graph of the equation $y = 5 - x^2$.

☛ 4. Graph the equations
$y = 1 - x$ and $y = \sqrt{1 - x}$.

Solution Since x appears in the equation only in the second degree and $(x)^2 = (-x)^2$, the table of values of x and y can be abbreviated by combining positive and negative values of x. (See Table 2.)

TABLE 2

x	0	± 1	± 2	± 3
y	5	4	1	-4

We plot these points and join them by a smooth curve, obtaining the graph of the equation $y = 5 - x^2$ shown in Figure 7.   ☛ 4

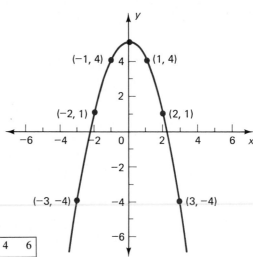

FIGURE 7

Answer

x	-8	-5	-3	-2	-1	0	1	2	4	6
$y = 1 - x$	9	6	4	3	2	1	0	-1	-3	-5
$y = \sqrt{1 - x}$	3	2.45	2	1.73	1.41	1	0	Not defined		

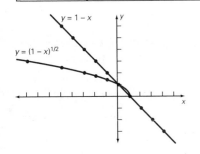

EXAMPLE 5 *(Demand)* If a certain item is offered for sale at a price p per unit, the quantity q demanded in the market is given by the relation $3q + p = 10$. Sketch the graph of this relation. Take q in place of x (horizontal axis) and p in place of y (vertical axis).

Solution Since neither the price p nor the quantity q demanded is negative, only that portion of the graph in the first quadrant is of any practical significance.

Solving the equation for p, we have

$$p = 10 - 3q.$$

Values of p corresponding to a number of different values of q are given in Table 3. For example, when the price is 7, the quantity demanded is only 1 unit. When the price is reduced to 4, 2 units are demanded by the market, and so on.

TABLE 3

q	0	1	2	3
p	10	7	4	1

Plotting these points, we obtain the graph shown in Figure 8.

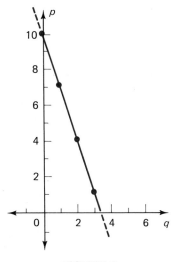

FIGURE 8

Observe that the graph is again a straight line, or rather the portion of a straight line that lies in the first quadrant.

EXERCISES 4-1

1. Plot the following points:

$(2, -5)$; $(-1, 4)$; $(0, 2)$; $(-3, -2)$; $(5, 0)$.

Label each point with its coordinates.

2. Determine the quadrants in which the points in Exercise 1 lie.

(3–6) Find the distance between each pair of points.

3. $(4, -1)$ and $(2, 0)$

4. $(-3, 1)$ and $(-2, -3)$

5. $(\frac{1}{2}, 2)$ and $(-2, 1)$ 6. $(a, 2)$ and $(b, 2)$

7. The ordinate of a point is 6 and its distance from the point $(-3, 2)$ is 5. Find the abscissa of the point.

8. The abscissa of a point is 2 and its distance from the point $(3, -7)$ is $\sqrt{5}$. Find the ordinate of the point.

9. If P is the point $(1, a)$ and its distance from the point $(6, 7)$ is 13, find the value of a.

10. We are given the point $P(x, 2)$. The distance of P from the point $A(9, -6)$ is twice its distance from the point $B(-1, 5)$. Find the value of x.

11. If P is the point $(-1, y)$ and its distance from the origin is half of its distance from the point $(1, 3)$, determine the value of y.

(12–15) Find the equation that the coordinates of the point $P(x, y)$ must satisify in order that the following conditions are met.

12. P is at a distance of 5 units from the point $(2, -3)$.

13. P is at a distance of 3 units from the point $(-1, 3)$.

14. The distance of P from the point $A(2, 1)$ is twice its distance from the point $B(-1, 3)$.

15. The sum of the squares of the distances of the points $A(0, 1)$ and $B(-1, 0)$ from P is 3.

(16–19) Sketch the graph of each equation.

16. $2x + 3y = 6$

17. $3x - 4y = 12$

18. $x^2 - y - 6 = 0$

19. $x = y^2 - 2$

(20–23) Sketch the graph of the following demand relations, where p denotes the price per unit and q is the quantity demanded.

20. $p = -2q + 5$

21. $2p + 3q = 8$

22. $p + q^2 = 14$

23. $p = 25 - q^2$

◼ 4-2 STRAIGHT LINES AND LINEAR EQUATIONS

In this section, we shall examine a number of properties of straight lines. Our first aim will be to investigate the algebraic equation that has a given line as its graph.

    One of the most important properties of a straight line is how steeply it rises or falls, and first we introduce a quantity that will measure this steepness. Let us begin by considering an example. The equation $y = 2x - 4$ has as its graph the straight line shown in Figure 9. Let us choose two points on this line, such as the points $(3, 2)$ and $(5, 6)$, which are denoted, respectively, by P and Q in Figure 9. The difference between the x-coordinates of these two points, denoted by PR, is called the *run* from P to Q:

$$\text{Run} = PR = 5 - 3 = 2.$$

The difference between the y-coordinates of P and Q, equal to the distance QR, is called the *rise* from P to Q:

$$\text{Rise} = QR = 6 - 2 = 4.$$

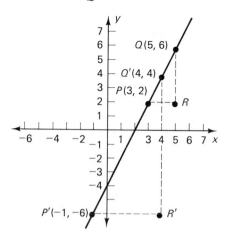

FIGURE 9

We note that the rise is equal to twice the run. This will be the case no matter which pair of points we choose on the given graph. For example, let us take the two points $P'(-1, -6)$ and $Q'(4, 4)$. (See Figure 9.) Then

$$\text{Run} = P'R' = 4 - (-1) = 5 \quad \text{and} \quad \text{Rise} = Q'R' = 4 - (-6) = 10.$$

Again we see that the ratio of rise to run is equal to 2.

The same ratio of rise to run is obtained in the two cases because the two triangles PQR and $P'Q'R'$ are similar. Therefore the ratios of corresponding sides are equal: $QR/PR = Q'R'/P'R'$. This ratio is called the **slope** of the given straight line. The line in Figure 9 has a slope equal to 2.

The slope of a general straight line is defined similarly. Let P and Q be any two points on the given line. (See Figure 10.) Let them have coordinates (x_1, y_1) and (x_2, y_2), respectively. Let R be the intersection of the horizontal line through P and the vertical line through Q. Then we define the **run from P to Q** as $x_2 - x_1$ and the **rise from P to Q** as $y_2 - y_1$:

$$\boxed{\text{Run} = x_2 - x_1 \qquad \text{Rise} = y_2 - y_1.}$$

From Figure 10, the run is the horizontal distance PR and the rise is the vertical distance QR. (If Q turns out to lie below R, which happens if the line slopes downward to the right, the rise is negative. We could also choose Q to lie to the left of P, in which case $x_2 < x_1$ and the run would be negative.)

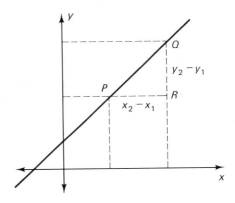

FIGURE 10

The **slope** of the line is defined to be the ratio of rise to run. It is usually denoted by the letter m. Hence

$$m = \frac{\text{rise}}{\text{run}} = \frac{y_2 - y_1}{x_2 - x_1}. \tag{1}$$

Note that Equation (1) for slope is meaningful as long as $x_2 - x_1 \neq 0$; that is, provided that the line is nonvertical. *Slope is not defined for vertical lines.*

It should be noted that the slope of a line remains the same, no matter how we choose the positions of the two points P and Q on the line.

If the slope m of a line is positive, the line ascends to the right. The larger the value of m, the more steeply the line is inclined to the horizontal. If m is negative, then the line descends to the right. If $m = 0$, then the line is horizontal. These properties are illustrated in Figure 11.

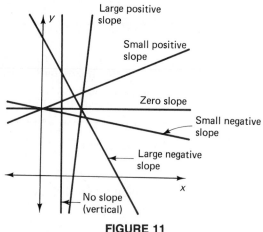

FIGURE 11

☛ **5.** Calculate the rise and run from P to Q and the slope of PQ in each case:

(a) P (1, 0) and Q (−1, −6);

(b) P (−3, 4) and Q (−3, 6);

(c) P (−2, 2) and Q (5, 2).

EXAMPLE 1 Find the slope of the line joining the two points $(1, -3)$ and $(3, 7)$.

Solution Using Equation (1), the slope is

$$m = \frac{7 - (-3)}{3 - 1} = \frac{10}{2} = 5.$$

EXAMPLE 2 The slope of the line joining the two points $(3, 2)$ and $(5, 2)$ is

$$m = \frac{2 - 2}{5 - 3} = 0.$$

Thus the line joining these two points is horizontal.

EXAMPLE 3 The slope of the line joining $P(2, 3)$ and $Q(2, 6)$ is given by

$$m = \frac{6 - 3}{2 - 2} = \frac{3}{0}$$

which is undefined. Thus the line joining P and Q has *no* slope. In this case, line PQ is vertical. ☛ **5**

What information do we need to be given in order to be able to draw a particular straight line? One way in which a line can be specified is by giving two points that lie on it. Once two points are given, the whole line is determined, since there is only one straight line through any two points.

Answer: (a) rise = −6, run = −2, slope = 3; (b) 2, 0, slope not defined; (c) 0, 7, 0.

Through any *one* point, there are of course many different straight lines with slopes ranging from large to small, positive or negative. However if the slope is given, then there is only one line through the point in question. Thus a second way in which a straight line can be specified is by giving one point on it and its slope.

Our immediate task will be to determine the equation of the straight line with given slope m that passes through a given point (x_1, y_1). Let (x, y) be a point on the line different from the given point (x_1, y_1). (See Figure 12.) Then the slope m of the line joining the two points (x_1, y_1) and (x, y) is given by

$$m = \frac{y - y_1}{x - x_1}.$$

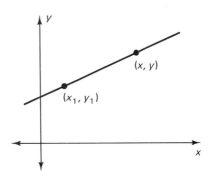

FIGURE 12

It follows therefore that

$$y - y_1 = m(x - x_1). \qquad (2)$$

This is called the **point-slope** formula for the line.

EXAMPLE 4 Find the equation of the line through the point $(5, -3)$ with slope -2.

Solution Using Equation (2) with $m = -2$ and $(x_1, y_1) = (5, -3)$, we find that the required equation of the straight line is as follows.

$$y - (-3) = -2(x - 5)$$
$$y + 3 = -2x + 10$$
$$y = -2x + 7$$

EXAMPLE 5 Find the equation of the straight line passing through the two points $(1, -2)$ and $(5, 6)$.

Solution The slope of the line joining $(1, -2)$ and $(5, 6)$ is

$$m = \frac{6 - (-2)}{5 - 1} = \frac{8}{4} = 2.$$

Using the point-slope formula, the equation of the straight line through $(1, −2)$ with slope $m = 2$ is

$$y − (−2) = 2(x − 1)$$

$$y = 2x − 4.$$

This is the line shown in Figure 9. ● **6, 7**

In the point-slope formula, let (x_1, y_1) be $(0, b)$. (See Figure 13.) Then Equation (2) becomes

$$y − b = m(x − 0)$$

or

$$y = mx + b. \tag{3}$$

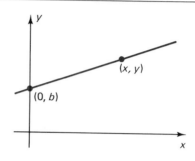

FIGURE 13

The quantity b, which gives the point on the y-axis that is cut off by the straight line, is called the **y-intercept** of the line. Equation (3) is called the **slope-intercept** formula of a line.

If the given line is horizontal, then its slope is $m = 0$ and Equation (3) reduces to

$$y = b. \tag{4}$$

This is the equation of a horizontal line at a distance b from the x-axis. (See Figure 14.) In particular, if we take $b = 0$ in Equation (4), we get $y = 0$, which is the equation of the x-axis itself.

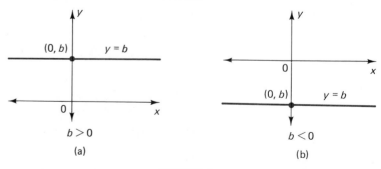

FIGURE 14

Next, suppose that the line in question is *vertical* and let it intersect the x-axis at the point $A(a, 0)$ as shown in Figure 15. If $P(x, y)$ is a general point on the line, the two points P and A have the same abscissa; that is, $x = a$. Every point on the line satisfies this condition, so we can say that the equation of this vertical line is

$$x = a. \tag{5}$$

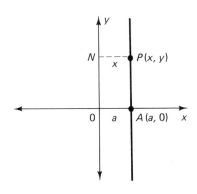

FIGURE 15

☞ **8.** Find the equations of the horizontal and vertical lines through the point $(4, -2)$.

For example, if $a = 0$, we have the equation $x = 0$, which is the equation of the y-axis. Similarly, $x = 2$ is the equation of the vertical line lying 2 units to the right of the y-axis, and $x = -4$ is the equation of the vertical line lying 4 units to the left of the y-axis.   ☞ **8**

EXAMPLE 6 Find the equation of the straight line passing through the two points $(2, 5)$ and $(2, 9)$.

Solution The slope of the line joining $(2, 5)$ and $(2, 9)$ is given by

$$m = \frac{y_2 - y_1}{x_2 - x_1} = \frac{9 - 5}{2 - 2} = \frac{4}{0}$$

which is undefined. Thus the line joining the two given points is vertical. We know that the equation of any vertical line is of the form

$$x = a.$$

The given line passes through the point $(2, 5)$, which has an x-coordinate of 2. Therefore $a = 2$ and the equation of the straight line is

$$x = 2.$$

A **general linear equation** (or a first-degree equation) in two variables x and y is an equation of the form

$$Ax + By + C = 0 \tag{6}$$

Answer Horizontal: $y = -2$, vertical: $x = 4$.

where A, B, and C are constants and A and B are *not both* zero.

In the light of the above discussion, we are in a position to describe the graph of the general linear equation, Equation (6), for different values of A and B.

1. $B \neq 0$, $A \neq 0$. In this case, Equation (6) takes the form

$$y = -\frac{A}{B}x - \frac{C}{B}$$

when solved for y. Comparing with Equation (3), we see that this is the equation of a straight line whose slope is $-A/B$ and y-intercept is $-C/B$.

2. $B \neq 0$, $A = 0$. When solved for y, Equation (6) becomes

$$y = -\frac{C}{B}$$

From Equation (4), we see that this is the equation of a horizontal line whose y-intercept is $-C/B$.

3. $A \neq 0$, $B = 0$. When $B = 0$, Equation (6) can be written in the form

$$x = -\frac{C}{A}.$$

This is the equation of a vertical line that intersects the x-axis at the point $(-C/A, 0)$, from Equation (5).

Thus the graph of the general linear equation (6) is, in each case, a straight line. A linear equation of the form of Equation (6) is often referred to as the **general equation** of a straight line. Table 4 summarizes the various forms taken by the equation of a straight line.

TABLE 4

Formula Name	Equation
1. Point-Slope Formula	$y - y_1 = m(x - x_1)$
2. Slope-Intercept Formula	$y = mx + b$
3. General Formula	$Ax + By + C = 0$, A, B not both zero
4. Horizontal Line	$y = b$
5. Vertical Line	$x = a$

EXAMPLE 7 Given the linear equation $2x + 3y = 6$, find the slope and y-intercept of its graph.

Solution To find the slope and y-intercept of the line, we must express the given equation in the form

$$y = mx + b.$$

That is, we must solve the equation for y in terms of x.

$$2x + 3y = 6$$
$$3y = -2x + 6$$
$$y = -\tfrac{2}{3}x + 2$$

Comparing with the general form $y = mx + b$, we have $m = -\frac{2}{3}$ and $b = 2$. Thus the slope is equal to $-\frac{2}{3}$ and y-intercept is equal to 2. **☛ 9**

Parallel and Perpendicular Lines

Two lines with slopes m_1 and m_2 are *parallel* if $m_1 = m_2$. The two lines are *perpendicular* if $m_1 m_2 = -1$. That is,

---

Two parallel lines have equal slopes.

The product of the slopes of two perpendicular lines is equal to -1*

EXAMPLE 8 Find the equation of a straight line passing through the point $(2, 5)$ and perpendicular to the line $x + 2y - 6 = 0$.

Solution The given line $x + 2y - 6 = 0$, or $y = (-\frac{1}{2})x + 3$, has a slope of $(-\frac{1}{2})$. Let m be the slope of the required line through $(2, 5)$. Since the two lines are perpendicular, the product of their slopes is -1, that is,

$$(-\tfrac{1}{2})m = -1, \quad \text{or} \quad m = 2.$$

From the point-slope formula, the equation of the line through $(2, 5)$ with slope $m = 2$ is

$$y - 5 = 2(x - 2), \quad \text{or} \quad y = 2x + 1. \quad \text{☛ 10}$$

Graphing Linear Equations

EXAMPLE 9 Sketch the graph of the linear equation $3x - 4y = 12$.

Solution We know that the graph of a linear equation in two variables is always a straight line, and a straight line is completely determined by two points. Thus to sketch the graph of the given linear equation, we find two *different* points (x, y) satisfying the given equation, plot them, and then join these points by a straight line. Putting $x = 0$ in the given equation, we get

$$-4y = 12, \quad \text{or} \quad y = -3.$$

Thus one point on the line is $(0, -3)$. Putting $y = 0$ in the given equation, we get

$$3x = 12, \quad \text{or} \quad x = 4.$$

---

*Or equivalently, m_2 is the negative reciprocal of m_1. The proof is outlined in Exercise 41.

11. Find the points where the following lines meet the x- and y-axes:

(a) $2x - 3y + 2 = 0$;

(b) $y - 4x - 5 = 0$;

(c) $\dfrac{x}{a} + \dfrac{y}{b} = 1$.

Hence (4, 0) is a second point on the line. Plotting the two points $(0, -3)$ and $(4, 0)$, which lie on the two coordinate axes, and joining them by a straight line, we obtain the graph of the given equation as shown in Figure 16. ☞ **11**

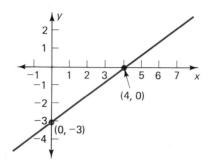

FIGURE 16

When plotting the graph of a linear relation, the simplest procedure in most cases is to find the two points where the graph crosses the coordinate axes, as we did in Example 9. There are occasions, however, when this is inconvenient; for example, one of these points of interesection may be off the graph paper we are using. It is also impossible to use this technique if the graph happens to pass through the origin. In such cases, we can use any pair of points on the graph in order to draw it, choosing the most convenient values of x at which to calculate y.

Alternatively, it is sometimes helpful to use the slope in order to sketch the graph. For example, the equation in Example 9 can be written in the form $y = \frac{3}{4}x - 3$, which shows that the slope is $\frac{3}{4}$ and the y-intercept is -3. Thus we can plot the point $(0, -3)$. In addition, for every 4 units we move along the positive x direction, the graph is going to rise by 3 units, since slope = rise/run = $\frac{3}{4}$. So if we move horizontally 4 units and vertically 3 units from the point $(0, -3)$ we obtain a second point on the graph. Or we can move horizontally 8 units and then vertically 6 units to get a second point, and so on. This is illustrated in Figure 17. From the y-intercept of -3, we have

Answer (a) $(-1, 0)$ and $\left(0, \dfrac{2}{3}\right)$;

(b) $\left(-\dfrac{5}{4}, 0\right)$ and $(0, 5)$;

(c) $(a, 0)$ and $(0, b)$.

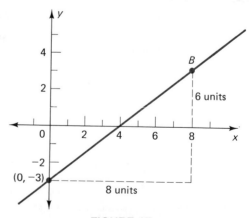

FIGURE 17

moved 8 units horizontally and then 6 units vertically to arrive at point B. A straight line can then be drawn connecting the point $(0, -3)$ and B. The general principle underlying this approach is that on any nonvertical line, if x increases by an amount h, then y increases by h slopes.

EXERCISES 4-2

(1–6) Find the slopes of the lines joining each pair of points.

1. $(2, 1)$ and $(5, 7)$

2. $(5, -2)$ and $(1, -6)$

3. $(2, -1)$ and $(4, -1)$ **4.** $(3, 5)$ and $(-1, 5)$

5. $(-3, 2)$ and $(-3, 4)$ **6.** $(1, 2)$ and $(1, 5)$

(7–24) Find the equation of the straight lines satisfying the conditions in each of the following exercises.

7. Passing through $(2, 1)$ with slope 5.

8. Passing through $(1, -2)$ with slope -3.

9. Passing through $(3, 4)$ with zero slope.

10. Passing through $(2, -3)$ with no slope.

11. Passing through $(3, -1)$ and $(4, 5)$.

12. Passing through $(2, 1)$ and $(3, 4)$.

13. Passing through $(3, -2)$ and $(3, 7)$.

14. With slope -2 and y-intercept 5.

15. With slope $\frac{1}{3}$ and y-intercept -4.

16. With slope 3 and y-intercept 0.

17. Passing through $(2, -1)$ and parallel to the line $3x + y - 2 = 0$.

18. Passing through $(1, 3)$ and parallel to the line $2x - y + 3 = 0$.

19. Passing through $(2, 1)$ and perpendicular to the line $x + y = 0$.

20. Passing through $(-1, 2)$ and perpendicular to $2x - 3y + 4 = 0$.

21. Passing through $(3, 4)$ and perpendicular to $x = 2$.

22. Passing through $(2, -3)$ and parallel to $3y + 2 = 0$.

23. Passing through $(0, -1)$ and parallel to the line through $(2, 2)$ and $(3, 1)$.

24. Passing through $(2, 3)$ and perpendicular to the line through $(-1, -2)$ and $(2, 1)$.

(25–30) Find the slope and the y-intercept for each of the following linear relations.

25. $3x - 2y = 6$ **26.** $4x + 5y = 20$

27. $y - 2x + 3 = 0$ **28.** $\dfrac{x}{3} + \dfrac{y}{4} = 1$

29. $2y - 3 = 0$ **30.** $3x + 4 = 0$

(31–38) Determine whether the following pairs of lines are parallel, perpendicular, or neither.

31. $2x + 3y = 6$ and $3x - 2y = 6$

32. $y = x$ and $x + y = 1$

33. $y = 2x + 3$ and $x = 2y + 3$

34. $4x + 2y = 1$ and $y = 2 - 2x$

35. $x = -2 - 3y$ and $2x + 6y = 5$

36. $3x + 4y = 1$ and $3x - 4y = 1$

37. $y - 3 = 0$ and $x + 5 = 0$

38. $2x - 5 = 0$ and $3 - x = 0$

39. A line passes through the points $(1, 2)$ and $(2, 1)$. Find the coordinates of the points where this line meets the coordinate axes.

40. A line passes through the point $(1, 1)$ and is perpendicular to the line joining $(1, 2)$ and $(1, 4)$. Find the y intercept of this line.

*41. Consider the two perpendicular lines $y = m_1x$ and $y = m_2x$ passing through the origin. Choose any two points P and Q, one on each line. The triangle OPQ is right-angled at O. Use Pythagoras's theorem to prove that $m_1m_2 = -1$.

4-3 APPLICATIONS OF LINEAR EQUATIONS

In this section, we shall discuss some applications of linear equations and straight lines to problems in business and economics.

Linear Cost Model

In the production of any commodity by a firm, there are two types of costs involved; these are known as *fixed costs* and *variable costs*. **Fixed costs** are costs that have to be met no matter how much or how little of the commodity is produced; that is, they do not depend on the level of production. Examples of fixed costs are rents, interest on loans and bonds, and management salaries.

Variable costs are costs that depend on the level of production, that is, on the amount of commodity produced. Material costs and labor costs are examples of variable costs. The total cost is given by

Total Cost = Variable Costs + Fixed Costs.

Consider the case when the *variable costs per unit of commodity is constant*. In this case, the total variable costs are proportional to the amount of commodity produced. If m dollars denotes the variable cost per unit, then the total variable costs of producing x units of commodity is mx dollars. If the fixed costs are b dollars, then the total cost y_c (in dollars) of producing x units is given by

Total Cost = Total Variable Costs + Fixed Costs

$$y_c = mx + b. \tag{1}$$

Equation (1) is an example of a **linear cost model.** The graph of Equation (1) is a straight line whose slope represents the variable cost per unit and whose y-intercept gives the fixed costs.

EXAMPLE 1 *(Linear Cost Model)* The variable cost of processing 1 pound of coffee beans is 50¢ and the fixed costs per day are $300.

(a) Give the linear cost equation and draw its graph.

(b) Find the cost of processing 1000 pounds of coffee beans in one day.

Solution

(a) If y_c represents the cost (in dollars) of processing x pounds of coffee beans per day, then according to the linear model, we have

$$y_c = mx + b$$

where m is the variable cost per unit and b the fixed cost. In our case, $m = 50¢ = \$0.50$ and $b = \$300$. Therefore

$$y_c = 0.5x + 300. \tag{2}$$

To sketch the graph of Equation (2), let us first find two points on it.

Letting $x = 0$ in Equation (2), we have $y_c = 300$; letting $x = 200$ in Equation (2), we have $y_c = 0.5(200) + 300 = 400$. Thus two points satisfying Cost Equation (2) are (0, 300) and (200, 400). Plotting these two points and joining them by a straight line, we obtain the graph shown in Figure 18. Note that the relevant portion of the graph lies totally in the first quadrant because x and y_c are both nonnegative quantities.

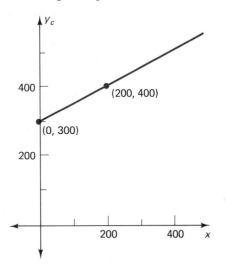

FIGURE 18

(b) Substituting $x = 1000$ in Equation (2), we get

$$y_c = 0.5(1000) + 300 = 800.$$

Thus the cost of processing 1000 pounds of coffee beans per day will be $800. ☛ 12

EXAMPLE 2 *(Cost Model)* The cost of manufacturing 10 typewriters per day is $350, while it costs $600 to produce 20 typewriters per day. Assuming a linear cost model, determine the relationship representing the total cost y_c of producing x typewriters per day and draw its graph.

Solution We are given two points (10, 350) and (20, 600) that lie on the graph of a linear cost model. The slope of the line joining these two points is

$$m = \frac{600 - 350}{20 - 10} = \frac{250}{10} = 25.$$

Using the point-slope formula, the required equation of the (linear cost model) straight line with slope $m = 25$ and passing through the point (10, 350) is

$$y - y_1 = m(x - x_1)$$

$$y_c - 350 = 25(x - 10) = 25x - 250;$$

that is,

$$y_c = 25x + 100. \tag{3}$$

☛ **12.** Find an expression for y_c in a linear cost model if the fixed cost is $4000 per period and it costs $7000 to produce 200 units of output.

Answer $y_c = 15x + 4000$

The graph of Equation (3) in this case is not a continuous straight line, because x cannot take fractional values as it represents the number of typewriters produced. The variable x can take the integer values 0, 1, 2, 3, 4, 5, . . . only. The corresponding values of y_c are given in Table 5.

TABLE 5

x	0	1	2	3	4	5	6	. . .
y_c	100	125	150	175	200	225	250	. . .

Plotting these points and several others, we obtain the graph shown in Figure 19. Note that the graph consists of separate (discrete) points rather than a continuous straight line. ☞ **13**

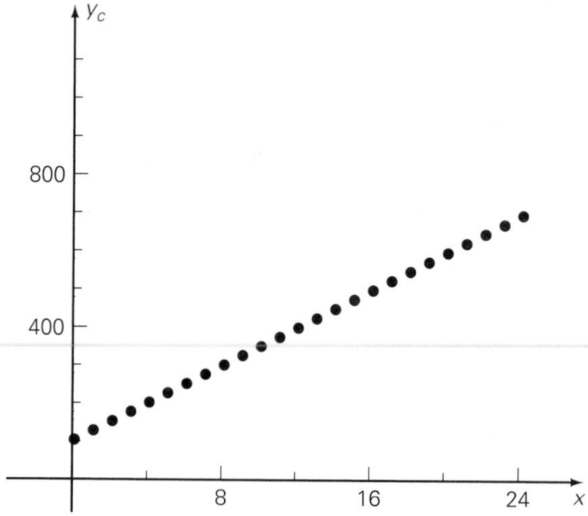

☞ **13.** If it costs \$4500 to produce 75 units per week and \$5200 to produce 100 units per week, what are the fixed weekly costs and the variable cost per unit?

FIGURE 19

Straight-Line Depreciation

When a company buys a piece of equipment or machinery it enters the value of that equipment as one of the assets on its balance sheet. In succeeding years this value must be decreased because the equipment slowly wears out or becomes outdated. This gradual reduction in the value of an asset is called *depreciation*. One common method of calculating the amount of depreciation is to reduce the value by a constant amount each year in such a way that the value is reduced to the scrap value at the end of the estimated lifetime of the equipment. This is called *straight-line depreciation*. We have

Rate of Depreciation (per Year)

= (Initial Value − Scrap Value) ÷ (Lifetime in Years)

Answer \$2400 and \$28 per unit.

EXAMPLE 3 *(Depreciation)* A firm buys a piece of machinery for $150,000. It expects that the lifetime of the machinery will be 12 years with zero scrap value. Find the amount of depreciation per year and a formula for the depreciated value after x years.

Solution

$$\text{Depreciation per Year} = (\text{Initial Purchase Price}) \div (\text{Lifetime in Years})$$
$$= (150,000 \text{ dollars}) \div (12 \text{ years})$$
$$= 12,500 \text{ dollars}$$

$$\text{Value after } x \text{ Years} = (\text{Initial Value}) -$$
$$(\text{Depreciation per Year})(\text{Number of years})$$
$$= (150,000 \text{ dollars})$$
$$- (12,500 \text{ dollars per year})(x \text{ Years})$$
$$= 150,000 - 12,500x \text{ dollars}$$

The graph of this relation is shown in Figure 20. ☞ **14**

☞ **14.** A firm is using straight-line depreciation to compute the value of its recently constructed plant. After 2 years it is valued at $8.8 million and after 6 years at $7.2 million. What was the initial cost and after how many years will the value be depreciated to zero?

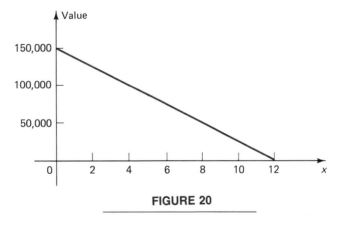

FIGURE 20

Supply and Demand

The laws of demand and supply are two of the fundamental relationships in any economic analysis. The quantity x of any commodity that will be purchased by consumers depends on the price at which that commodity is made available. A relationship that specifies the amounts of a particular commodity that consumers are willing to buy at various price levels is called the **law of demand.** The simplest law is a linear relation of the type

$$p = mx + b \tag{4}$$

where p is the price per unit of the commodity and m and b are constants. The graph of a demand law is called the **demand curve.** Observe that, as given, p is expressed in terms of x. This way of writing the demand law gives the price level at which the quantity x can be sold.

It is a well-recognized fact that if the price per unit of a commodity is increased, the demand for that commodity will decrease because fewer pur-

Answer $9.6 million, 24 years.

chasers can afford it, whereas if the price per unit is decreased—that is, the commodity is made cheaper—the demand will increase. In other words, slope m of the demand relation of Equation (4) is negative. Thus the graph of Equation (4) slopes downward to the right, as shown in part (a) of Figure 21. Since the price p per unit and the quantity x demanded are both nonnegative numbers, the graph of Equation (4) has been drawn only in the first quadrant.

The amount of a particular commodity that its suppliers are willing to make available depends on the price at which they can sell it. A relation specifying the amount of any commodity that manufacturers (or sellers) can make available in the market at various prices is called the **law of supply.** The graph of the law of supply is known as the **supply curve.**

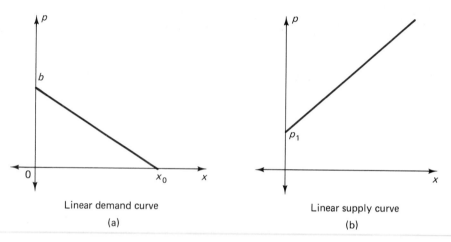

Linear demand curve
(a)

Linear supply curve
(b)

FIGURE 21

In general, suppliers will make a greater quantity available in the market if they can get a high price and a smaller quantity if they get a low price. In other words, the supply increases with increase in price. A typical linear supply curve is shown in part (b) of Figure 21. The price p_1 corresponds to the price below which suppliers will not supply the commodity.

EXAMPLE 4 *(Demand)* A dealer can sell 20 electric shavers per day at $25 per shaver, but he can sell 30 shavers if he charges $20 per shaver. Determine the demand equation, assuming it is linear.

Solution Taking the quantity x demanded as the abscissa (or x-coordinate) and the price p per unit as the ordinate (or y-coordinate), the two points on the demand curve have coordinates

$$x = 20, p = 25 \quad \text{and} \quad x = 30, p = 20.$$

Thus the points are (20, 25) and (30, 20). Since the demand equation is linear, it is given by the equation of the straight line passing through the two points

(20, 25) and (30, 20). The slope of the line joining these two points is

$$m = \frac{20 - 25}{30 - 20} = -\frac{5}{10} = -0.5.$$

From the point-slope formula, the equation of the line through (20, 25) with slope $m = -0.5$ is

$$y - y_1 = m(x - x_1).$$

Since $y = p$, we have

$$p - 25 = -0.5(x - 20)$$

$$p = -0.5x + 35$$

which is the required demand equation. (See Figure 22.) ☛ 15

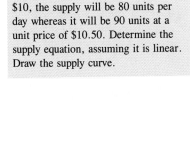
☛ **15.** When the price per unit is $10, the supply will be 80 units per day whereas it will be 90 units at a unit price of $10.50. Determine the supply equation, assuming it is linear. Draw the supply curve.

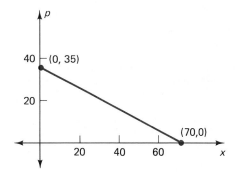

FIGURE 22

Trade-off

Planners frequently have to decide among different ways of allocating limited resources. For example, a manufacturer may have to allocate plant capacity between two different products. If the relation between the amounts of the two products is linear, the slope of its graph can be interpreted as giving the trade-off of one product against the other. Consider the following example.

EXAMPLE 5 *(Transit Decision)* A city government has a budget of $200 million for capital expenditure on transportation, and it intends to use it to construct additional subways or highways. It costs $2.5 million per mile to build highways and $4 million per mile for subways. Find the relationship between the number of miles of highway and of subway that can be built to completely use the available budget. Interpret the slope of the linear relation that is obtained.

Solution Suppose x miles of highway and y miles of subway are built. The cost of constructing x miles of highway at $2.5 million per mile is $2.5x$ million dollars and the cost of building y miles of subway at $4 million per mile is $4y$

Answer $p = \frac{1}{20}x + 6$.

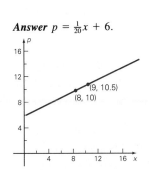

million dollars. Since the total cost has to equal the budget allotted for the purpose,

$$2.5x + 4y = 200.$$

This equation provides the required relation between the numbers of miles that can be constructed within the budget.

Solving the given equation for y, we have

$$y = -\tfrac{5}{8}x + 50.$$

The slope of this line is $-\tfrac{5}{8}$, which expresses the fact that every additional mile of highway construction will be at the cost of $\tfrac{5}{8}$ mile of subway construction. Solving the original equation for x in terms of y, we get

$$x = -\tfrac{8}{5}y + 80.$$

Thus each additional mile of subway construction is traded off against $\tfrac{8}{5}$ miles of highway construction.

EXERCISES 4-3

1. *(Linear Cost Model)* The variable cost of manufacturing a table is $7 and the fixed cost is $150 per day. Determine the total cost y_c of manufacturing x tables per day. What is the cost of manufacturing 100 tables per day?

2. *(Linear Cost Model)* The total cost of manufacturing 100 cameras per week is $700 and of 120 cameras per week is $800.

a. Determine the cost equation, assuming it to be linear.

b. What are the fixed cost and the variable cost per unit?

3. *(Linear Cost Model)* It costs a company $75 to produce 10 units of a certain item per day and $120 to produce 25 units of the same item per day.

a. Determine the cost equation, assuming it to be linear.

b. What is the cost of producing 20 items per day?

c. What are the variable cost per item and the fixed cost?

4. *(Linear Cost Model)* Johnson's Moving Company charges $70 to move a certain machine 15 miles and $100 to move the same machine 25 miles.

a. Determine the relation between total charges and the distance moved, assuming it to be linear.

b. What is the minimum charge for moving this machine?

c. What is the rate for each mile the machine is moved?

5. *(Linear Cost Model)* Fixed costs of manufacturing a certain product are $300 per week and the total cost of manufacturing 20 units per week is $410. Determine the relationship between the total cost and the number of units manufactured, assuming it to be linear. What will it cost to manufacture 30 units per week?

6. *(Linear Cost Model)* A hotel rents a room to one person at the rate of $25 for the first night and $20 for each succeeding night. Express the cost y_c of the bill in terms of x, the number of nights the person stays at the hotel.

7. *(Linear Cost Model)* A catering company will provide banquets for groups of people at a cost of $10 per person plus an overhead charge of $150. Find the cost y_c they would charge for catering for x people.

8. *(Linear Cost Model)* The cost of a bus ticket in Jonesville depends linearly on the distance traveled. A journey of 2 miles costs 40¢, while a journey of 6 miles costs 60¢. Determine the cost of a ticket for a journey of x miles.

9. *(Demand Relation)* A detergent manufacturer finds that weekly sales are 10,000 packets when the price is $1.20 per packet, but that the sales increase to 12,000 when the price is reduced to $1.10 per packet. Determine the demand relation, assuming that it is linear.

10. *(Demand Relation)* A television manufacturer finds that at $500 per television set, sales are 2000 sets per month.

However, at $450 per set, sales are 2400 units. Determine the demand equation, assuming it to be linear.

11. *(Supply Equation)* At a price of $2.50 per unit, a firm will supply 8000 shirts per month; at $4 per unit, the firm will supply 14,000 shirts per month. Determine the supply equation, assuming it to be linear.

12. *(Demand Relation)* A tool manufacturer can sell 3000 hammers per month at $2 each, whereas only 2000 hammers can be sold at $2.75 each. Determine the demand law, assuming it to be linear.

13. *(Supply Equation)* At a price of $10 per unit, a firm will supply 1200 units of its product, and at $15 per unit, 4200 units. Determine the supply relation, assuming it to be linear.

14. *(Apartment Rental)* Georgia Realty owns an apartment complex that has 50 suites. At a monthly rent of $400, all the suites are rented, whereas if the rent is increased to $460 per month, only 47 suites can be rented.

a. Assuming a linear relationship between the monthly rent p and the number of suites x that can be rented, find this relationship.

b. How many suites will be rented if the monthly rent is raised to $500?

c. How many suites will be rented if the rent is dropped to $380 per month?

15. *(Depreciation)* John bought a new car for $10,000. What is the value V of the car after t years, assuming that it is depreciating linearly each year at the rate of 12% of its original value. What is the car worth after 5 years?

16. *(Depreciation)* A firm bought new machinery for $15,000. If it depreciated linearly by $750 per year and if it had a scrap value of $2250, for how long would the machinery be in use? What was the value V of the machinery after t years of use and after 6 years of use?

17. *(Depreciation)* Ms. Kim bought a new television set for $800 and it depreciated linearly each year by 15% of its original value. What is the value of the set after t years and after 6 years of use?

18. *(Depreciation)* Let P be the purchase price, S the scrap value, and N the lifetime in years of a piece of equipment. Show that according to straight-line depreciation, the value of the equipment after t years is given by $V = P - (P - S)(t/N)$.

19. *(Machine Allocation)* A company manfactures two types

of a certain product. Each unit of the first type requires 2 machine-hours and each unit of the second type requires 5 machine-hours. There are 280 machine-hours available each week.

a. If x units of the first type and y units of the second type are made each week, find the relation between x and y if all the machine-hours are used.

b. What is the slope of the equation in part (a)? What does it represent?

c. How many units of the first type can be manufactured if 40 units of the second type are manufactured in a particular week?

20. *(Work Allocation)* The Boss-Toss Company manufactures two products, X and Y. Each unit of X requires 3 work-hours and each unit of Y requires 4 work-hours. There are 120 work-hours available each day.

a. If x units of X and y units of Y are manufactured each day and all the available work-hours are used, find the relationship between x and y.

b. Give the physical interpretation of the slope of the linear relation obtained.

c. How many units of X can be made in one day if 15 units of Y are made the same day?

d. How many units of Y can be made in one day if 16 units of X are made the same day?

21. *(Reduction of Inventory)* Jackson Stores have 650 units of item X in stock and their average sales per day of this item is 25 units.

a. If y represents the inventory (of item X in stock) at a time t (measured in days), determine the linear relation between y and t. (Use $t = 1$ to represent the end of the first day, and so on.)

b. How long would it take to run out of stock?

c. On what sales day is the new order placed if it is placed when the stock level reaches 125 units?

22. *(Political Science)* In an election to the U.S. House of Representatives, it is estimated that if the Democrats win 40% of the popular vote they will obtain 30% of the seats and that for every percentage point by which their share of the vote increases, their share of the seats will increase by 2.5%. Assuming a linear relationship $y = mx + c$ between x, the percentage of the vote, and y, the percentage of the seats, calculate m and c. What percentage of

the seats will the Democrats get if they win 55% of the popular vote?

23. *(Zoology)* The average weight W of the antlers of deer is approximately related to the age A of the deer by the equation $W = mA + c$. For a particular species, it is found that when $A = 30$ months, $W = 0.15$ kilogram, whereas when $A = 54$ months, $W = 0.36$ kilogram. Find m and c and calculate the age at which W reaches 0.5 kilogram.

24. *(Agriculture)* For the last 40 years the average yield y (bushels per acre) of corn in the United States has increased with time t according to the approximate equation $y = mt + c$. In 1950 the average yield was 38 bushels per acre, whereas in 1965 it was 73 bushels per acre. Calculate m and c. (Take $t = 0$ to be 1950.) Estimate what the average yield will be in 1990 assuming the same equation continues to hold.

25. *(Diet Planning)* A patient in the hospital who is on a liquid diet has the choice of prune juice and orange juice to satisfy his daily requirement of thiamine, which is 1 milligram. One ounce of prune juice contains 0.05 milligram of thiamine, and 1 ounce of orange juice contains 0.08 milligram of thiamine. Let his daily consumption be x ounces of prune juice and y ounces of orange juice. What is the relationship between x and y that exactly satisfies his thiamine requirement?

26. *(Diet Planning)* An individual on a strict diet plans to breakfast on cornflakes, milk, and a boiled egg. After allowing for the egg, his diet allows a further 300 calories for this meal. One ounce of milk contains 20 calories and 1 ounce (about one cupful) of cornflakes (plus sugar) contains 160 calories. What is the relation between the number of ounces of milk and of cornflakes that can be consumed?

◥ 4-4 SYSTEMS OF EQUATIONS

Many problems in business and economics lead to what are called *systems of linear equations*. Consider for example, the following situation.

The owner of a television store wants to expand his business by buying and displaying two new types of television sets that have recently appeared on the market. Each television set of the first type costs $300 and each set of the second type costs $400. Each of the first type of set occupies 4 square feet of floor space, whereas each set of the second type occupies 5 square feet. If the owner has only $2000 available for this expansion and 26 square feet of floor space, how many sets of each type should be bought and displayed to make full use of the available capital and space?

Suppose the owner buys x television sets of the first type and y sets of the second type. Then it costs $300x$ to buy the first type of sets and $400y$ to buy the second type of sets. Since the total amount to be spent is $2000, it is necessary that

$$300x + 400y = 2000. \tag{i}$$

Also, the amounts of space occupied by the two types of sets are $4x$ square feet and $5y$ square feet, respectively. The total space available for these two types of sets is 26 square feet. Therefore

$$4x + 5y = 26. \tag{ii}$$

To find the number of sets of each type that can be bought and displayed, we must solve the Equations (i) and (ii) for x and y. That is, we must find the values of x and y that satisfy both Equations (i) and (ii) simultaneously. Observe that each of these equations is a linear equation in x and y.

DEFINITION A **system of linear equations** in two variables x and y consists of two equations of the type

$$a_1 x + b_1 y = c_1 \tag{1}$$

$$a_2 x + b_2 y = c_2 \tag{2}$$

where a_1, b_1, c_1, a_2, b_2, and c_2 are six given constants. The **solution** of the system defined by Equations (1) and (2) is the set of values of x and y that satisfy both equations.

Equations (i) and (ii) form such a system of linear equations. If we identify Equation (i) with Equation (1) and Equation (ii) with Equation (2), the six constants have the values $a_1 = 300$, $b_1 = 400$, $c_1 = 2000$, $a_2 = 4$, $b_2 = 5$, and $c_2 = 26$.

Our main concern in this section is to solve systems of linear equations algebraically. Solution by the use of algebraic methods involves the elimination of one of the variables, either x or y, from the two equations; this allows determination of the value of the other variable. Elimination of one of the variables can be achieved either by substitution or by adding a suitable multiple of one equation to the other. The two processes are illustrated in Example 1.

EXAMPLE 1 Solve the two equations that resulted from the problem posed in the beginning of this section.

$$300x + 400y = 2000 \tag{i}$$

$$4x + 5y = 26 \tag{ii}$$

Solution *(Method of Substitution)* In this case, we solve one of the given equations for x or y (whichever is simplest) and substitute the value of this variable in the other equation. From Equation (ii) (solving for x), we have

$$4x = 26 - 5y$$

$$x = \frac{26 - 5y}{4}. \tag{iii}$$

We substitute this value of x in Equation (i) and solve for y.

$$300\left(\frac{26 - 5y}{4}\right) + 400y = 2000$$

$$75(26 - 5y) + 400y = 2000$$

$$1950 - 375y + 400y = 2000$$

$$25y = 2000 - 1950 = 50$$

$$y = 2$$

Substituting $y = 2$ in (iii) we have

$$x = \tfrac{1}{4}(26 - 10) = 4.$$

☞ 16. Solve the following system by using the first equation to substitute for y in the second: $3x - y = 7$, $2x + 4y = 14$.

Thus the solution of the system of Equations (i) and (ii) is $x = 4$ and $y = 2$. In other words, the dealer should buy and display 4 sets of the first type and 2 sets of the second type to make use of all the available space and capital. **☞ 16**

Alternative Solution *(Method of Addition)*

$$300x + 400y = 2000 \qquad \text{(i)}$$

$$4x + 5y = 26 \qquad \text{(ii)}$$

According to this method, we make the coefficients of either x or y in the two equations exactly the same magnitude and opposite in sign; we then add the two equations to eliminate one of the variables. Observe that if we multiply both sides of Equation (ii) by -80, we make the coefficient of y the same magnitude as the coefficient in Equation (i), but of opposite sign. Equation (ii) becomes

$$-320x - 400y = -2080. \qquad \text{(iv)}$$

Recall that Equation (i) is

$$300x + 400y = 2000.$$

We add these two equations; the y-terms cancel and we obtain

$$(-320x - 400y) + (300x + 400y) = -2080 + 2000 \qquad \text{(v)}$$

or

$$-20x = -80$$

$$x = 4.$$

Answer Substitute $y = 3x - 7$. Solution is $x = 3$, $y = 2$.

Substituting $x = 4$ in one of the given equations (we use Equation (ii)), we have

$$16 + 5y = 26$$

$$5y = 26 - 16 = 10$$

$$y = 2$$

Thus the solution is $x = 4$ and $y = 2$, the same as obtained by the first method. **☞ 17**

☞ 17. Solve the following system by eliminating x by the method of addition: $x + y = 3$, $2x + 3y = 11$.

Note The operations involved in these methods of solutions do not change the set of solutions. For example, any x and y that satisfy Equation (ii) also satisfy Equation (iv); any x and y that satisfy both Equations (iv) and (i) also satisfy Equation (v), and so on. Thus the methods yield values of x and y that are solutions of the original pair of equations.

EXAMPLE 2 Solve the following system.

$$\frac{x - y}{3} = \frac{y - 1}{4}$$

$$\frac{4x - 5y}{7} = x - 7$$

Answer Multiply first equation by -2, then add to the second. Solution is $x = -2$, $y = 5$.

Solution The first step is to get rid of fractions in the given equations. We multiply both sides of the first equation by 12, the common denominator, and simplify.

$$4(x - y) = 3(y - 1)$$
$$4x - 4y = 3y - 3$$
$$4x - 7y = -3$$

Mutiplying both sides of the second equation by 7 and simplifying, we obtain

$$4x - 5y = 7(x - 7) = 7x - 49$$
$$-3x - 5y = -49$$

Multiplying throughout by -1, we get

$$3x + 5y = 49.$$

Thus the given system of equations is equivalent to the following system of linear equations.

$$4x - 7y = -3 \qquad \text{(i)}$$
$$3x + 5y = 49 \qquad \text{(ii)}$$

We use the substitution method. We solve Equation (i) for x.

$$4x = 7y - 3 \quad \text{or} \quad x = \tfrac{1}{4}(7y - 3) \qquad \text{(iii)}$$

Substituting this value of x into Equation (ii), we obtain

$$\tfrac{3}{4}(7y - 3) + 5y = 49.$$

We multiply both sides by 4 and solve for y.

$$3(7y - 3) + 20y = 196$$
$$21y - 9 + 20y = 196$$
$$41y = 196 + 9 = 205$$
$$y = \frac{205}{41} = 5$$

Putting $y = 5$ in Equation (iii), we get

$$x = \tfrac{1}{4}(35 - 3) = 8.$$

Thus the required solution is $x = 8$ and $y = 5$.

A system of linear equations and its solution has an important geometrical interpretation. For example, let us consider the following system.

$$x + y = 3 \qquad \text{(3)}$$
$$3x - y = 1 \qquad \text{(4)}$$

Using either of the above methods of solution, we can easily see that the solution in this case is $x = 1$ and $y = 2$.

Each of Equations (3) and (4) is a linear equation in x and y and so has as its graph a straight line in the xy-plane. In the case of Equation (3), we find the point where the line meets the x-axis by setting $y = 0$. This gives $x = 3$, so the line passes through the point (3, 0). Similarly, setting $x = 0$, we find $y = 3$, so that the line meets the y-axis at the point (0, 3). These two points are shown in Figure 23, and the graph of Equation (3) is drawn as the straight line joining them.

☞ 18. Draw the graphs and find the point of intersection of the lines whose equations are
$3y - 2x = 6$ and $4y + 3x = 24$.

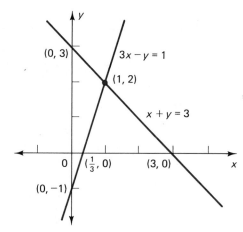

FIGURE 23

Proceeding in a similar way for Equation (4), we find the points $(0, -1)$ and $(\frac{1}{3}, 0)$ on the coordinate axes.

Any pair of values of x and y that satisfies Equation (3) corresponds to a point (x, y) on the first straight line. Any pair of values that satisfies Equation (4) corresponds to a point (x, y) on the second straight line. Thus if x and y satisfy *both* equations, then the point (x, y) must lie on *both* of the lines. In other words (x, y) must be the point at which the lines intersect. From Figure 23, we see that this point is (1, 2), so the solution in this case is $x = 1$ and $y = 2$, as stated earlier. ☞ **18**

Now let us return to the general system of linear equations.

$$a_1x + b_1y = c_1 \tag{1}$$

$$a_2x + b_2y = c_2 \tag{2}$$

The graphs of these two equations consist of two straight lines in the xy-plane, since any linear equation always represents a straight line, as we saw in Section 4-2. Any pair of values of x and y that satisfy both Equations (1) and (2) must correspond to a point (x, y) that lies on both of the lines.

Let us denote the two straight lines by L and M, respectively. Then there are three possibilities.

1. Lines L and M intersect each other. Since the point of intersection,

Answer $x = \frac{48}{17} \approx 2.82$,
$y = \frac{66}{17} \approx 3.88$.

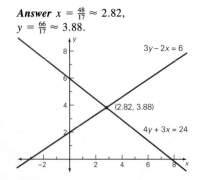

(x_0, y_0), lies on both lines, the coordinates (x_0, y_0) satisfy the equations of both lines and hence they provide a solution of the given system. *This solution is unique,* because if the two straight lines intersect, they intersect each other at only one point. (See part (a) of Figure 24.)

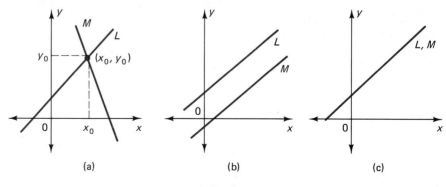

(a)   (b)   (c)

FIGURE 24

2. Lines L and M are parallel. In this case, the lines do not meet each other and there is no point that lies on both the lines. Thus there are no values of x and y that will satisfy both the equations. In other words, the equations have *no solution* in this case. (See part (b) of Figure 24.)

3. Lines L and M coincide. In such a case, every point on line L is also on line M. In this case, the system has an *infinite number of solutions,* namely every ordered pair (x, y) that lies on $L(= M)$. (See part (c) of Figure 24.)

EXAMPLE 3 Solve the following system of equations.

$$x + 2y = 4$$

$$3x + 6y - 8 = 0$$

Solution We solve the first equation for x.

$$x = 4 - 2y$$

We then susbtitute this value of x in the second equation and simplify.

$$3(4 - 2y) + 6y - 8 = 0$$

$$12 - 6y + 6y - 8 = 0$$

$$4 = 0$$

This is impossible. Thus the given equations have *no solution.* This is illustrated graphically in Figure 25. The two straight lines are parallel in this case and do not intersect. We can see this easily by writing the given equations in slope-intercept form.

$$y = -\tfrac{1}{2}x + 2$$

$$y = -\tfrac{1}{2}x + \tfrac{4}{3}$$

19. How many solutions do the following systems have?

(a) $x - 3y = 1$, $y = \dfrac{1}{3}x - 1$

(b) $3y = 5x - 2$,
$\qquad x + y + 2 = 4(y - x + 1)$.

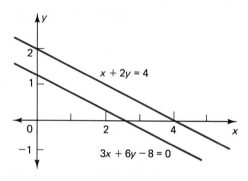

FIGURE 25

The two lines have the *same slope* $(-\tfrac{1}{2})$ but different y-intercepts. Thus the two lines are parallel, with no common point of intersection.

EXAMPLE 4 Solve the system of equations

$$2x - 3y = 6 \qquad \text{(i)}$$

$$\frac{x}{3} - \frac{y}{2} = 1 \qquad \text{(ii)}$$

Solution Multiplying both sides of the second equation by -6, we obtain

$$-2x + 3y = -6.$$

Adding this equation to the first equation, we get

$$0 = 0$$

an equation that is always true.

Writing the two equations in the slope-intercept form, we find that they both reduce to the equation

$$y = \tfrac{2}{3}x - 2.$$

Answer (a) No solution; (b) infinitely many solutions.

Since the two equations are identical, the two lines coincide in this case and the two given equations are equivalent to one another. In fact, Equation (ii) can be obtained from Equation (i) by multiplying the latter by $\tfrac{1}{6}$. We have an infinite number of solutions in this case: any pair of values (x, y) that satisfies Equation (i) provides a solution. One such pair is $(6, 2)$, for example, and another is $(0, -2)$. **19, 20**

20. For what values of c does the following system have no solution or infinitely many solutions?

$$2y + x + 6 = 0, \quad y = c - \frac{1}{2}x.$$

The method of substitution is often useful when we have a system of equations in which one equation is linear and the other is not.

EXAMPLE 5 Solve the following system of equations

$$2x - y = 3$$

$$x^2 + y^2 = 5$$

Answer No solution if $c \neq -3$; infinitely many solutions if $c = -3$.

Solution In this system, one of the two equations is not linear. The method of solution still consists of elimination of one of the variables, x or y, from the two equations. From the first equation, we have

$$y = 2x - 3.$$

We substitute this value of y into the second equation and simplify.

$$x^2 + (2x - 3)^2 = 5$$
$$x^2 + 4x^2 - 12x + 9 = 5$$
$$5x^2 - 12x + 4 = 0$$

This factors to give

$$(x - 2)(5x - 2) = 0.$$

Therefore either

$$x - 2 = 0 \quad \text{or} \quad 5x - 2 = 0$$
$$x = 2 \qquad\qquad x = \tfrac{2}{5}.$$

Next, we substitute these values into the equation we used earlier to substitute for y, namely* $y = 2x - 3$.

$y = 2x - 3$	$y = 2x - 3$
$= 2(2) - 3$	$= 2(\tfrac{2}{5}) - 3$
$= 1$	$= -\tfrac{11}{5}$

Hence there are two solutions,

$$x = 2,\, y = 1 \qquad \text{and} \qquad x = \tfrac{2}{5},\, y = -\tfrac{11}{5}. \quad \text{☞} 21$$

We shall continue by solving an applied problem involving simultaneous equations.

EXAMPLE 6 *(Mixture)* The Britannia Store, which specializes in selling all kinds of nuts, sells peanuts at $0.70 per pound and cashews at $1.60 per pound. At the end of the month, the owner of the store finds that peanuts are not selling well and decides to mix peanuts and cashews to make a mixture of 45 pounds, which could sell for $1.00 per pound. How many pounds of peanuts and cashews should be mixed to keep the same revenue?

Solution Let the mixture contain x pounds of peanuts and y pounds of cashews. Since the total mixture is 45 pounds,

$$x + y = 45.$$

The revenue from x pounds of peanuts at $0.70 per pound is $0.7x$ dollars and

---

☞ 21. Use substitution to solve the system $x + 2y = 8$, $xy = 6$.

Answer Two solutions:
$x = 2$, $y = 3$ and $x = 6$, $y = 1$.

*It would be incorrect to substitute the x-values into the nonlinear equation.

the revenue from y pounds of cashews at \$1.60 per pound is $1.6y$ dollars. The revenue obtained from the mixture of 45 pounds at \$1.00 per pound will be \$45. Since the revenue from the mixture must be the same as from the separate nuts, we have the following equation.

Revenue from Peanuts + Revenue from Cashews = Revenue from Mixture

$$0.7x + 1.6y = 45$$

$$7x + 16y = 450$$

Thus we arrive at the following system of linear equations.

$$x + y = 45$$

$$7x + 16y = 450$$

From the top equation, we have $x = 45 - y$. We then substitute this value of x into the bottom equation and solve for y.

$$7(45 - y) + 16y = 450$$

$$315 - 7y + 16y = 450$$

$$9y = 450 - 315 = 135$$

$$y = 15$$

☛ **22.** In Example 6, find the amounts that must be mixed to make 40 pounds of mixture costing \$1.15 per pound.

Therefore $x = 45 - y = 45 - 15 = 30$.

Thus 30 pounds of peanuts should be mixed with 15 pounds of cashews to form the mixture. ☛ **22**

The method of substitution can also often be used to solve systems of equations with three or more variables. Such systems are covered more fully in Section 9-3, but meanwhile here is one example.

EXAMPLE 7 Solve the following system of equations for x, y, and z.

$$x - y - z = 1$$

$$2x + y + 3z = 6$$

$$-4x - 2y + 3z = 6$$

Solution We solve the first equation for x in terms of y and z.

$$x = 1 + y + z$$

Next we substitute this expression for x into the remaining two equations.

$$2(1 + y + z) + y + 3z = 6$$

$$-4(1 + y + z) - 2y + 3z = 6$$

After simplification these become

$$3y + 5z = 4$$

Answer $x = 20$, $y = 20$.

$$-6y - z = 10$$

We now have two equations to solve for y and z. From the last of these we have $z = -6y - 10$, and substituting this into the first equation, we obtain

$$3y + 5(-6y - 10) = 4$$
$$-27y = 54$$
$$y = -2$$

To complete the solution we calculate z and finally x.

$$z = -6y - 10 = -6(-2) - 10 = 2$$
$$x = 1 + y + z = 1 + (-2) + 2 = 1$$

The solution is therefore $x = 1$, $y = -2$, and $z = 2$.

___

Remark The addition method can also be used to eliminate one of the variables, leaving a system of two equations for the two remaining variables.

EXERCISES 4-4

(1–24) Solve the following systems of linear equations.

1. $x - y = 1$ and $2x + 3y + 8 = 0$

2. $2x - 3y = 1$ and $5x + 4y = 14$

3. $4x - y = -2$ and $3x + 4y = 27$

4. $3u + 2v = 9$ and $u + 3v = 10$

5. $3x + 5t = 12$ and $4x - 3t = -13$

6. $2p - q = 3$ and $p = 5 - 3q$

7. $7x - 8y = 4$ and $\dfrac{x}{2} + \dfrac{y}{3} = 3$

8. $\dfrac{x + y}{2} - \dfrac{x - y}{3} = 8$ and $\dfrac{x + y}{3} + \dfrac{x - y}{4} = 11$

9. $\dfrac{x}{4} + \dfrac{y}{5} + 1 = \dfrac{x}{5} + \dfrac{y}{4} = 23$

10. $\dfrac{x - 2y}{3} = 2 + \dfrac{2x + 3y}{4}$ and

   $\dfrac{3x - 2y}{2} = \dfrac{-y + 5x + 11}{4}$

11. $5x - 7y + 2 = 0$ and $15x - 21y = 7$

12. $2u - 3v = 12$ and $-\dfrac{u}{3} + \dfrac{v}{2} = 4$

13. $x + 2y = 4$ and $3x + 6y = 12$

14. $2p + q = 3$ and $\frac{2}{3}p + \frac{1}{3}q = 1$

15. $x + y = 3$
    $y + z = 5$
    $x + z = 4.$

16. $x + 2y = 1$
    $3y + 5z = 7$
    $2x - y = 7.$

17. $x + y + z = 6$
    $2x - y + 3z = 9$
    $-x + 2y + z = 6.$

18. $x + 2y - z = -3$
    $3y + 4z = 5$
    $2x - y + 3z = 9$

19. $3x_1 + 2x_2 + x_3 = 6$
    $2x_1 - x_2 + 4x_3 = -4$
    $x_1 + x_2 - 2x_3 = 5.$

20. $2u - 3v + 4w = 13$
    $u + v + w = 6$
    $-3u + 2v + w + 1 = 0.$

21. $x + 3y + 4z = 1$
    $2x + 7y + z = -7$
    $3x + 10y + 8z = -3$

22. $3x - 2y + 4z = 3$
    $4x + 3y = 9$
    $2x + 4y + z = 0$

23. $x + y = 3$ and $x^2 + y^2 = 29$

24. $2x + y = 5$ and $xy = 2$

25. *(Ore Refining)* Two metals, A and B, can be extracted from two types of ore, I and II. One hundred pounds of ore I yields 3 ounces of A and 5 ounces of B and 100 pounds of ore II yields 4 ounces of A and 2.5 ounces of B. How many pounds of ore I and II will be required to produce 72 ounces of A and 95 ounces of B?

26. *(Machine Allocation)* A firm manufactures two products, A and B. Each product has to be processed by two machines, I and II. Each unit of type A requires 1 hour of processing by machine I and 1.5 hours by machine II, and each unit of type B requires 3 hours on machine I and 2 hours on machine II. If machine I is available for 300 hours each month and machine II for 250 hours, how many units of each type can be manufactured in one month if the total available time on the two machines is to be utilized?

27. *(Purchasing Decision)* A company is trying to purchase and store two types of items, X and Y. Each item X costs $3 and each item Y costs $2.50. Each item X occupies 2 square feet of floor space and each item Y occupies 1 square foot of floor space. How many units of each type can be purchased and stored if $400 is available for purchasing and 240 square feet of floor space is available for storing these items?

28. *(Blending of Coffee)* A store sells two types of coffee, one at $2.00 per pound and the other at $1.50 per pound. The owner of the store makes 50 pounds of a new blend of coffee by mixing these two types of coffee and sells it at $1.60 per pound. How many pounds of coffee of each type should be mixed to keep the revenue unchanged?

29. *(Mixtures)* A chemical store has two types of acid solutions. One contains 25% acid and the other contains 15% acid. How many gallons of each type should be mixed to obtain 200 gallons of a mixture containing 18% acid?

30. *(Income Tax Policy)* The income tax department has a certain tax rate on the first $5000 of taxable income and a different rate on taxable income over $5000 but less than $10,000. The government wishes to fix the tax rates in such a way that a person with taxable income of $7000 should pay $950 in tax, while one with taxable income of $9000 should pay $1400 in tax. Find the two rates.

31. *(Personnel)* A certain company employs 53 persons in its two branch offices. Of these people, 21 are university graduates. If one-third of those in the first office and three-sevenths of those in the second office are university graduates, how many persons are employed in each office?

32. *(Investment)* A person invested a total of $25,000 in three different investments at 8%, 10%, and 12%. The total annual return was $2440, and the returns from the 8% and 12% investments were equal. How much was invested at each rate?

33. *(Production Decision)* A fertilizer plant produces three types of fertilizer. Type A contains 25% potash, 45% nitrate, and 30% phosphate. Type B contains 15% potash, 50% nitrate, and 35% phosphate. Type C contains no potash, 75% nitrate, and 25% phosphate. The plant has supplies of 1.5 tons per day of potash, 5 tons per day of nitrates, and 3 tons per day of phosphate. How much of each type of fertilizer should be produced so as to use up exactly the supplies of ingredients?

34. *(Ecology)* A fish of species 1 consumes 10 grams of food 1 and 5 grams of food 2 per day. A fish of species 2 consumes 6 grams of food 1 and 4 grams of food 2 per day. If a given environment has 2.2 kilograms of food 1 and 1.3 kilograms of food 2 available daily, what population sizes of the two species will consume exactly all the available food?

35. *(Ecology)* Three species of birds eat aphids from different parts of trees. Species 1 feed half of the time on the top levels and half of the time on the middle levels of the trees. Species 2 feed half on the middle levels and half on the lower levels. Species 3 feed entirely on the lower levels. There are equal numbers of aphids available on the middle and lower levels, but only half this number available on the upper levels. What should be the relative sizes of the populations of the three species in order that the supply of aphids will be entirely consumed?

4-5 APPLICATIONS TO BUSINESS AND ECONOMICS

In this section we shall discuss some important applications of systems of equations.

Break-even Analysis

If the total cost y_c of production exceeds the revenue y_R obtained from the sales, then a business is running at a loss. On the other hand, if the revenue exceeds the costs, there is a profit. If the cost of production equals the revenue obtained from the sales, there is no profit or loss, so the business breaks even. The number of units produced and sold in this case is called the **break-even point.**

EXAMPLE 1 *(Break-even Analysis)* For a watchmaker, the cost of labor and materials per watch is $15 and the fixed costs are $2000 per day. If each watch sells for $20, how many watches should be produced and sold each day to guarantee that the business breaks even?

Solution Let x watches be produced and sold each day. The total cost of producing x watches is

$$y_c = \text{Total Variable Costs} + \text{Fixed Costs} = 15x + 2000.$$

Since each watch sells for $20, the revenue y_R obtained by selling x watches is

$$y_R = 20x.$$

The break-even point is obtained when the revenue equals costs, that is,

$$20x = 15x + 2000.$$

We have $5x = 2000$ or $x = 400$.

Thus 400 watches must be produced and sold each day to guarantee no profit or loss. Figure 26 illustrates the graphical interpretation of the break-even point. When $x < 400$, the cost y_c exceeds the revenue y_R and there is a loss. When $x > 400$, the revenue y_R exceeds the cost y_c so that a profit results.

Note that graphically, the break-even point corresponds to the intersection of two straight lines. One of the lines has the equation $y = 15x + 2000$, corresponding to the cost of production, and the other line has the equation $y = 20x$, corresponding to the revenue. ☞ 23

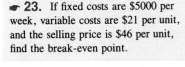

☞ 23. If fixed costs are $5000 per week, variable costs are $21 per unit, and the selling price is $46 per unit, find the break-even point.

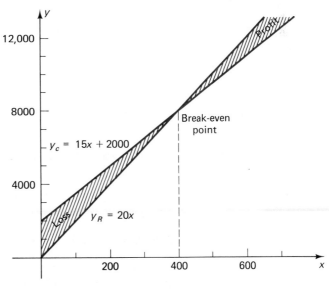

FIGURE 26

Answer 200 units per week.

EXAMPLE 2 *(Break-even Analysis)* Suppose the total daily cost (in dollars) of producing x chairs is given by

$$y_c = 2.5x + 300.$$

(a) If each chair sells for $4, what is the break-even point?

(b) If the selling price is increased to $5 per chair, what is the new break-even point?

(c) If it is known that at least 150 chairs can be sold each day, what price should be charged to guarantee no loss?

Solution The cost is given by

$$y_c = 2.5x + 300.$$

(a) If each chair sells for $4, the revenue (in dollars) obtained by the sale of x chairs is

$$y_R = 4x.$$

At the break-even point, we have $y_c = y_R$; that is,

$$4x = 2.5x + 300.$$

Thus $1.5x = 300$ or $x = 200$. The break-even point is 200 chairs.

(b) If the selling price is increased to $5 per chair, the revenue in this case is

$$y_R = 5x.$$

The break-even point obeys $y_R = y_c$, so

$$5x = 2.5x + 300.$$

Thus $2.5x = 300$, or $x = 120$. With the new selling price, the break-even point is 120 chairs.

(c) Let p dollars be the price charged for each chair. Then the revenue obtained from the sale of 150 chairs is $y_R = 150p$ and the cost of producing 150 chairs is $y_c = 2.5(150) + 300 = 675$. To guarantee a break-even situation, we must have $y_R = y_c$; that is,

$$150p = 675 \quad \text{or} \quad p = 4.50.$$

Thus, $4.50 per chair should be charged to guarantee break-even (at worst) if at least 150 chairs will be sold each day.

When an economist uses a linear relationship to describe a relation between two variables, it is not claimed that the true relationship is linear, but rather that a linear relationship is a good approximation to the observed data over the range of interest. If the observed data are found to lie on or close to a straight line when plotted on a graph, we can use a linear relationship as an approximate representation of the data. The way in which this can be carried out will be described in Section 18-6.

Although observed data often lie close to a straight line, there are also many cases in which they do not, and in such cases it is not reasonable to use a linear equation to approximate the relationship between two given variables. For example, the cost of manufacturing x items of a certain type may not be given by a linear cost model, $y_c = mx + b$, but may depend on x in some more complex way. In principle, a break-even analysis remains unchanged in such cases, but the algebra involved in finding the break-even point becomes more complicated.

EXAMPLE 3 *(Nonlinear Break-even Analysis)* A candy company sells its boxes of chocolates for $2 each. If x is the number of boxes produced per week (in thousands), then the management knows that the production costs are given in dollars by

$$y_c = 1000 + 1300x + 100x^2.$$

Determine the level of production at which the company will break even.

Solution The revenue from selling x thousands of boxes at $2 each is given by

$$y_R = 2000x.$$

In order to break even, revenue must equal costs; so

$$1000 + 1300x + 100x^2 = 2000x.$$

Dividing both sides of this equation by 100 and moving all the terms to the left, we have

$$x^2 - 7x + 10 = 0.$$

If we factor this expression, we get

$$(x - 2)(x - 5) = 0$$

and so $x = 2$ or $x = 5$.

We find, therefore, that there are two break-even points in this problem. The company can decide to make 2000 boxes per week ($x = 2$), with revenues and costs both $4000. Or they can make 5000 boxes per week ($x = 5$), when revenues and costs will again balance at $10,000.

It is instructive in this example to look at the profitability of the company. The weekly profit P is given by revenue minus costs.

$$
\begin{aligned}
P &= y_R - y_c \\
&= 2000x - (1000 + 1300x + 100x^2) \\
&= -1000 + 700x - 100x^2 \\
&= -100(x - 2)(x - 5)
\end{aligned}
$$

● **24.** If a firm's daily costs are $20,000 + 200x - x^2$ when x units are produced per day, and the selling price is $100 per unit, find the break-even point.

When $x = 2$ or 5, the profit is zero, and these are the break-even points. When $2 < x < 5$, we have $x - 2 > 0$ and $x - 5 < 0$. Because the product contains two negative signs, P is positive in this case. Thus the company makes a positive profit when $2 < x < 5$; that is, when it manufactures and sells between 2000 and 5000 boxes per week. ● **24**

Market Equilibrium

If the price of a certain commodity is too high, consumers will not purchase it, whereas if the price is too low, suppliers will not sell it. In a competitive market, when the price per unit depends only on the quantity demanded and the supply available, there is always a tendency for the price to adjust itself so that the quantity demanded by purchasers matches the quantity which suppliers are willing to supply. **Market equilibrium** is said to occur at a price when the quantity demanded is equal to the quantity supplied. This corresponds to the point of intersection of the demand and supply curves. (See Figure 27.)*

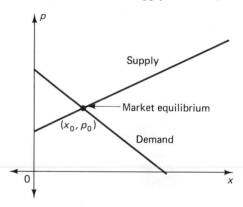

FIGURE 27

Algebraically, the market equilibrium price p_0 and quantity x_0 are determined by solving the demand and supply equations simultaneously for p and x. Note that equilibrium price and quantity are only meaningful if they are not negative.

EXAMPLE 4 Determine the equilibrium price and quantity for the following demand and supply laws.

$$D: \quad p = 25 - 2x \qquad (1)$$

$$S: \quad p = 3x + 5 \qquad (2)$$

Solution Equating the two values of p in Equations (1) and (2), we have

$$3x + 5 = 25 - 2x.$$

---

*Some simple models of the way in which a market adjusts itself to equilibrium are described in Section 17-6.

Answer 200 units per day.

25. If the demand law is
$2p + 3x = 36$ and the supply law is
$2p = x + 12$,
graph the demand and supply curves
and find the point of market
equilibrium.

The solution is readily seen to be $x = 4$. Substituting $x = 4$ in Equation (1), we get

$$p = 25 - 8 = 17.$$

Thus the equilibrium price is 17 and quantity is 4 units. The graphs of the supply and demand curves are shown in Figure 28.

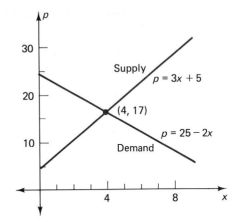

FIGURE 28

EXAMPLE 5 If the demand and supply equations are, respectively,

$$D: \quad 3p + 5x = 22 \tag{3}$$

$$S: \quad 2p - 3x = 2 \tag{4}$$

determine the values of x and p at market equilibrium.

Solution Equations (3) and (4) form a system of linear equations for the two variables x and p. Let us solve them by the addition method. Multiplying both sides of Equation (3) by 3 and both sides of Equation (4) by 5, we obtain

$$9p + 15x = 66$$

$$10p - 15x = 10.$$

We next add these equations and simplify.

$$9p + 15x + 10p - 15x = 66 + 10$$

$$19p = 76$$

Thus $p = 4$. Substituting this value of p into Equation (3), we have

$$3(4) + 5x = 22.$$

Therefore $x = 2$. Market equilibrium thus occurs when $p = 4$ and $x = 2$. **25**

Like most linear relations in economics, linear demand and supply equations provide an approximate representation of the true relations between price

Answer $x = 6$, $p = 9$

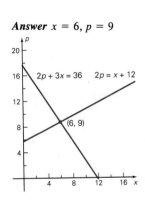

and quantity, and cases do arise in which such linear approximations are not good enough. The determination of market equilibrium when the demand equation or the supply equation (or both) is nonlinear can involve quite complicated calculations.

EXAMPLE 6 *(Market Equilibrium)* The demand for goods produced by an industry is given by the equation $p^2 + x^2 = 169$, where p is the price and x is the quantity demanded. The supply is given by $p = x + 7$. What are the equilibrium price and quantity?

Solution The equilibrium price and quantity are the positive values of p and x that satisfy both the demand and supply equations.

$$p^2 + x^2 = 169 \tag{5}$$

$$p = x + 7 \tag{6}$$

Substituting the value of p from Equation (6) into Equation (5) and simplifying gives the following.

$$(x + 7)^2 + x^2 = 169$$

$$2x^2 + 14x + 49 = 169$$

$$x^2 + 7x - 60 = 0$$

Factoring, we find that

$$(x + 12)(x - 5) = 0$$

which gives $x = -12$ or 5. The negative value of x is inadmissible, so $x = 5$. Substituting $x = 5$ in Equation (6), we get

$$p = 5 + 7 = 12.$$

Thus the equilibrium price is 12 and the quantity is 5.

Additive Tax and Market Equilibrium

Often the government imposes additional taxes on certain commodities in order to raise more revenue or gives subsidies to producers of specified commodities to make these essential commodities available to consumers at reasonable prices. We shall consider the effect of an additional tax or subsidy on market equilibrium under the following assumptions.

1. The quantity demanded by consumers depends on the market price alone. Denote this price paid by consumers by p_c.

2. The quantity made available by suppliers is determined by the price received by them. Denote this price by p_s.

3. The price paid by consumers equals the price received by suppliers plus the tax t per unit: $p_c = p_s + t$. If, instead, a subsidy s per unit is given, then $p_c = p_s - s$.

EXAMPLE 7 *(Subsidy and Market Equilibrium)* The demand law for a certain commodity is $5p + 2x = 200$ and the supply law is $p = \frac{4}{5}x + 10$.

(a) Find the equilibrium price and quantity.

(b) Find the equilibrium price and quantity after a tax of 6 per unit is imposed. Find the increase in price and the decrease in quantity demanded.

(c) What subsidy will cause the quantity demanded to increase by 2 units?

Solution The demand and supply equations are

$$D: \quad 5p + 2x = 200 \tag{7}$$

$$S: \quad p = \tfrac{4}{5}x + 10. \tag{8}$$

(a) Substituting the value of p from Equation (8) into Equation (7) and simplifying gives the following.

$$5(\tfrac{4}{5}x + 10) + 2x = 200$$

$$4x + 50 + 2x = 200$$

$$6x = 150$$

$$x = 25$$

Therefore, from Equation (8),

$$p = \tfrac{4}{5}(25) + 10 = 20 + 10 = 30.$$

Thus the equilibrium price and quantity before taxation are

$$p = 30 \quad \text{and} \quad x = 25.$$

(b) Let p_c be the price paid by consumers and p_s the price received by suppliers. Then Equations (7) and (8) become

$$D: \quad 5p_c + 2x = 200 \tag{9}$$

$$S: \quad p_s = \tfrac{4}{5}x + 10. \tag{10}$$

If a tax of 6 per unit is imposed, then $p_c = p_s + 6$, so the supply equation can be written

$$S: \quad p_c - 6 = \tfrac{4}{5}x + 10$$

or

$$p_c = \tfrac{4}{5}x + 16. \tag{11}$$

Substituting the value of p_c from Equation (11) into Equation (9), we get

$$5(\tfrac{4}{5}x + 16) + 2x = 200.$$

The solution is $x = 20$. Therefore, from Equation (11),

$$p_c = \tfrac{4}{5}(20) + 16 = 32.$$

26. In Example 7, graph the original supply and demand equations (7) and (8) and the modified demand and supply equations (9) and (11) on the same axes. Show the two equilibrium points. Geometrically, what is the effect of the sales tax?

Answer On the plane of x and consumer price, p_c, the effect of the tax is to move the supply curve upward.

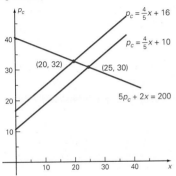

27. Re-solve parts (a) and (b) of Exercise 7 if the supply equation is changed to $p = x + 12$ and if a sales tax of 7 per unit is imposed in part (b).

Comparing with part (a), we see that the effect of the tax is to increase the market price by 2 (from 30 to 32) and to decrease the market demand by 5 (from 25 to 20). ☛ 26 (see p. 161), 27

(c) Again let p_c be the price paid by consumers and p_s the price received by suppliers, so the demand and supply equations are still given by Equations (9) and (10). This time, $p_c = p_s - s$, where s is the subsidy per unit.

We wish to have demand 2 more than the equilibrium demand of 25, that is, $x = 27$. Then, from Equation (9),

$$p_c = \tfrac{1}{5}(200 - 2x) = \tfrac{1}{5}(200 - 54) = 29.2$$

and from Equation (10),

$$p_s = 10 + \tfrac{4}{5}x = 10 + \tfrac{4}{5}(27) = 31.6.$$

Answer (a) $x = 20, p = 32$; (b) $x = 15, p_c = 34$.

Therefore, $s = p_s - p_c = 31.6 - 29.2 = 2.4$. A subsidy of 2.4 per unit will increase the demand by 2 units.

EXERCISES 4-5

1. *(Break-even Analysis)* The variable cost of producing a certain item is 90¢ per item and the fixed costs are $240 per day. The item sells for $1.20 each. How many items should be produced and sold to guarantee no profit and no loss?

2. *(Break-even Analysis)* The fixed costs of producing a certain product are $5000 per month and the variable cost is $3.50 per unit. If the product sells for $6.00 each, find each of the following.

a. The break-even point.

b. The number of units that must be produced and sold each month to obtain a profit of $1000 per month.

c. The loss when only 1500 units are produced and sold each month.

3. *(Break-even Analysis)* The cost of producing x items is given by $y_c = 2.8x + 600$ and each item sells for $4.00.

a. Find the break-even point.

b. If it is known that at least 450 units will be sold, what should be the price charged for each item to guarantee no loss?

4. *(Break-even Analysis)* A manufacturer produces items at a variable cost of 85¢ per item and the fixed costs are

$280 each day. If the item can be sold for $1.10, determine the break-even point.

5. *(Break-even Analysis)* In Exercise 4, if the manufacturer can reduce the variable cost to 70¢ per item by increasing the daily fixed costs to $350, is it advantageous to do so? (Such a reduction might be possible, for example, by purchasing a new machine that would cut production costs but would increase interest charges.)

6. *(Break-even Analysis)* The cost of producing x items per week is given by $y_c = 1000 + 5x$. If each item can be sold for $7, determine the break-even point. If the manufacturer can reduce variable costs to $4 per item by increasing fixed costs to $1200 per week, would it be wise to do so?

7. *(Nonlinear Break-even Analysis)* The cost of producing x items per day is given in dollars by $y_c = 80 + 4x + 0.1x^2$. If each item can be sold for $10, determine the break-even point.

*__8.** *(Nonlinear Break-even Analysis)* The cost of producing x items per day is given in dollars by $y_c = 2000 + 100\sqrt{x}$. If each item can be sold for $10, determine the break-even point.

(9–10) *(Market Equilibrium)* Find the equilibrium price and

162   CH. 4 ∣ STRAIGHT LINES

quantity for the following demand and supply curves:

9. D: $2p + 3x = 100$
   S: $p = \frac{1}{10}x + 2$

10. D: $3p + 5x = 200$
    S: $7p - 3x = 56$

11. D: $4p + x = 50$
    S: $6p - 5x = 10$

12. D: $5p + 8x = 80$
    S: $3x = 2p - 1$

13. D: $p^2 + x^2 = 25$
    S: $p = x + 1$

14. D: $p^2 + 2x^2 = 114$
    S: $p = x + 3$

15. *(Market Equilibrium)* A dealer can sell 200 units of a certain commodity per day at $30 per unit and 250 units at $27 per unit. The supply equation for that commodity is $6p = x + 48$.

   a. Find the demand equation for the commodity, assuming it to be linear.

   b. Find the equilibrium price and quantity.

   c. Find the equilibrium price and quantity if a tax of $3.40 per unit is imposed on the commodity. What is the increase in price and the decrease in quantity demanded?

   d. What subsidy per unit will increase the demand by 24 units?

   e. What additive tax per unit should be imposed on the commodity so that the equilibrium price per unit increases by $1.08?

16. *(Market Equilibrium)* At a price of $2400, the supply of a certain commodity is 120 units while its demand is 560 units. If the price is raised to $2700 per unit, the supply and demand will be 160 units and 380 units, respectively.

   a. Determine the demand and supply equations, assuming them to be linear.

   b. Determine the market equilibrium price and quantity.

   c. If a tax of $110 per unit is imposed on the commodity, what are the new equilibrium price and quantity? What is the increase in price and decrease in quantity?

   d. What subsidy per unit will decrease the market price by $15?

(17–18) *(Insufficient Demand)* Solve the following supply and demand equations. Explain where market equilibrium would be.

17. S: $p = x + 5$
    D: $3p + 4x = 12$

18. S: $2p - 3x = 8$
    D: $3p + 6x = 9$

19. *(Market Equilibrium)* For a certain product, if the price is $4 per unit, consumers will buy 10,000 units per month. If the price is $5 per unit, consumers will buy 9000 units per month.

   a. Assuming the demand curve is a straight line, find its equation.

   b. The supply equation for this product is

   $$p = \begin{cases} 3.2 + \left(\dfrac{x}{2000}\right) & \text{if } 0 \le x \le 6000 \\ 5 + \left(\dfrac{x}{5000}\right) & \text{if } 6000 \le x. \end{cases}$$

   Find the equilibrium point and the total amount spent by consumers on this product at the equilibrium price.

20. *(Multiple Market Equilibria)* A monopoly supplier of a certain commodity is content to supply a sufficient quantity to guarantee a constant revenue. Thus the supply relation has the form $xp =$ constant. If the supply relation is $xp = 3$ and the demand relation is $x + p = 4$, find the points of market equilibrium.

21. *(Multiple Market Equilibria)* In the preceding exercise find the points of market equilibrium for the supply relation $xp = 5$ and the demand relation $3x + 4p = 30$.

22. *(Multiple Market Equilibria)* For a particular commodity the supply relation is

   $$x = \begin{cases} 6p & \text{if } 0 \le p < 1 \\ \dfrac{6}{p} & \text{if } p \ge 1 \end{cases}$$

   and the demand relation is $x + 2p = 7$. Find the points of market equilibrium. Graph the supply and demand equations.

*23. *(Market Stability)* A market equilibrium point is stable if, when the market price deviates slightly from its equilibrium value, there is a tendency for the price to be driven back toward equilibrium. Let the equilibrium price be p_0. Suppose that for $p > p_0$ the quantity supplied, x_P, exceeds the quantity demanded, x_D. Then there will be a tendency for the price to fall. That is, the market will be stable under price increases if $x_S > x_D$ whenever $p > p_0$. Similarly, if $x_S < x_D$ whenever $p < p_0$, the market will be stable under price decreases. Show that the market equilibria in Exercises 9-14 are all stable. Discuss the stability of each of the multiple equilibria in Exercises 20-22.

CHAPTER REVIEW

Key Terms, Symbols, and Concepts

4.1 Cartesian plane or xy-plane, coordinate axes, x-axis, y-axis, origin.

Cartesian coordinates, abscissa or x-coordinate, ordinate or y-coordinate.

First, second, third, and fourth quadrants.

Graph of an equation.

4.2 Rise and run from P to Q. Slope of a line.

Point-slope formula, Slope-intercept formula. General linear equation.

Horizontal and vertical lines.

Parallel and perpendicular lines.

Graphing a linear equation using the intercepts or using a point and the slope.

4.3 Linear cost model, fixed costs, variable costs.

Straight-line depreciation. Trade-off.

Law of demand, demand curve. Law of supply, supply curve.

4.4 System of linear equations. Solution of a system of equations.

Method of substitution. Method of addition.

Geometrical interpretation of a system and its solution.

4.5 Break-even point.

Market equilibrium, equilibrium price, and quantity.

Sales tax, subsidy.

Formulas

The distance formula: $d = \sqrt{(x_2 - x_1)^2 + (y_2 - y_1)^2}$.

Slope of a line: $m = \dfrac{y_2 - y_1}{x_2 - x_1}$.

Point-slope formula: $y - y_1 = m(x - x_1)$.

Slope-intercept formula: $y = mx + b$.

Horizontal line: $y = b$.   Vertical line: $x = a$.

General linear equation: $Ax + By + C = 0$.

Parallel lines: $m_1 = m_2$.

Perpendicular lines: $m_1 \cdot m_2 = -1$.

REVIEW EXERCISES FOR CHAPTER 4

1. State whether the following statements are true or false. If false, replace by the corresponding correct statement.

   a. If a point lies on the x-axis, then its abscissa is zero.

   b. Each point on the y-axis has its y-coordinate zero.

   c. If a point (x, y) is in the first quadrant, then $x \geq 0$, $y \geq 0$.

   d. The origin $(0, 0)$ lies in all the four quadrants.

   e. A horizontal line has no slope.

   f. If a line slopes down to the right, then its slope is negative.

   g. A vertical line has zero slope.

   h. The distance of the point (a, b) from the origin $(0, 0)$ is $a + b$.

   i. The slope of the line joining two points (x_1, y_1), (x_2, y_2) is given by $m = \dfrac{y_2 - y_1}{x_2 - x_1}$ for all values of x_1, y_1, x_2, y_2.

   j. The linear equation $Ax + By + C = 0$ represents a straight line for all values of the constants A, B, and C.

   k. The slope of the line given by $x = my + b$ is m.

   l. If the variable cost per unit of producing a certain commodity is constant, then the commodity obeys a linear cost model.

   m. The equation of the x-axis is $y = 0$.

   n. The equation $x = c$, c a constant, always represents a vertical line.

   o. If the numbers x and y are of the same sign (that is, both positive or both negative), then the point $P(x, y)$ lies in either the first or the third quadrant.

p. If the point $P(x, y)$ lies in the second or fourth quadrant, then x and y are of opposite signs.

(2–8) Find the equation of each line.

2. Passing through $(2, 3)$ and $(3, 1)$.

3. Passing through $(-1, 2)$ and $(5, 2)$.

4. Passing through $(3, 0)$ and $(0, -4)$.

5. Passing through $(2, -1)$ with no slope.

6. Passing through $(-3, 2)$ with zero slope.

7. Passing through $(1, -3)$ and perpendicular to $x/2 + y/3 = 1$.

8. Passing through the point of intersection of the lines $2x - 3 = 0$ and $x + y = 2$ and parallel to $2x - 3y + 7 = 0$.

9. Find the intercepts made by the line $2x - 3y = 6$ on the coordinate axes.

10. A line is parallel to $3x - 4y = 7$ and passes through the point $(1, 1)$. Find the intercepts made by this line on the coordinate axes.

(11–24) Solve the following systems of equations.

11. $x + 2y = 3;\quad 2x - 3y = -1$

12. $\dfrac{x}{2} - \dfrac{y}{3} = -1;\quad \dfrac{2x}{3} + \dfrac{y}{2} = 10$

13. $\dfrac{1}{x} + \dfrac{1}{y} = 12;\quad \dfrac{2}{3x} + \dfrac{3}{2y} = 13$

14. $2x + 3y = 5;\quad \dfrac{x}{3} + \dfrac{y}{2} = 1$

15. $3u - 5v = 8;\quad 2u + 3v + 1 = 0$

16. $2p - q = \dfrac{3}{4};\quad \dfrac{1}{p} + \dfrac{1}{q} = 6$

17. $\dfrac{2}{u} - \dfrac{3}{v} = -15;\quad \dfrac{1}{2u} + \dfrac{1}{3v} = 6$

18. $3x - 4y = 12;\quad \dfrac{x}{4} = 1 + \dfrac{y}{3}$

19. $3x - 4y = 13;\quad 2x + 3y = 3$

20. $\dfrac{2}{x} + \dfrac{3}{y} = 2;\quad \dfrac{5}{x} + \dfrac{8}{y} = 5\dfrac{1}{6}$

$\left(\text{Hint: Let } \dfrac{1}{x} = u \text{ and } \dfrac{1}{y} = v.\right)$

21. $\begin{aligned} x + y - 2z &= -1 \\ 2x - 3y + z &= 13 \\ -3x + 2y + 5z &= -8 \end{aligned}$

22. $\begin{aligned} u + 2v + 3w &= 17 \\ 2u - v + 4w &= 11 \\ 3u + 7v - 8w &= -14 \end{aligned}$

23. $x + y = 9;\quad \dfrac{1}{x} + \dfrac{1}{y} = \dfrac{1}{2}$

*24. $\dfrac{1}{x} + \dfrac{1}{y} = \dfrac{7}{12};\quad xy = 12$

25. *(Work Allocation)* A company manufactures two types of products, A and B. Each product A requires 2.5 work-hours and each product B requires 4 work-hours. If 320 work-hours are available each week, find the relationship between the number of items of each type that can be made to make full use of the available work-hours.

26. *(Plant Allocation)* A chemical plant is used to manufacture two chemicals, A and B. It can operate by two processes. Process 1 produces 2 tons per hour of A and 5 tons per hour of B; process 2 produces 3 tons per hour of A and 4 tons per hour of B. The company can sell 260 tons per week of A and 440 tons per week of B. How many hours per week should the plant be operated on each of the two processes?

27. *(Linear Cost Model)* A manufacturer has fixed costs of $3000 and the variable costs of $25 per unit. Find the equation relating costs to production. What is the cost of producing 100 units?

28. *(Linear Cost Model)* A mining firm finds that it can product 7 tons of ore at a cost of $1500 and 15 tons at a cost of $1800. Assuming a linear cost-output model, find the fixed cost and the variable costs. What will be the cost of producing 20 tons of ore?

29. *(Blending of Tobacco)* A tobacco dealer mixed 8 pounds of one grade of tobacco with 5 pounds of another grade to obtain a blend worth $35. A second blend worth $49 was then made by mixing 10 pounds of the first grade with 8 pounds of the second grade. Find the price per pound of each grade.

30. *(Investment)* An individual receives $1860 in interest on two investments, with the first earning 6% and the second earning 8%. The next year the rates were interchanged and the earnings were $1920. Find the amount of each investment.

31. *(Investment)* A person made two investments, the first being $3750 less than the second. If, during the first year, the rate of the first investment was 2% more than that of the second and the income from each was $900, find the amount of the second investment and the rate that it earned.

32. *(Mixture)* A pharmacist wishes to obtain 30 fluid ounces of a solution containing 70% alcohol. There are two solutions, containing 50% and 80% alcohol, respectively. How much of each solution should be used?

33. *(Market Equilibrium)* At a price of $50 per ton, the demand for a certain commodity is 4500 tons whereas the supply is 3300 tons. If the price is increased by $10 per ton, the demand and supply will be 4400 tons and 4200 tons, respectively.

  a. Assuming linearity, determine the laws of supply and demand.

  b. Find the equilibrium price and quantity.

  c. If the additional tax of $2 per ton is imposed on the supplier, find the increase in equilibrium price and decrease in equilibrium quantity.

  d. What subsidy per ton should be given to the supplier so that the equilibrium quantity increases by 55 tons?

34. *(Market Equilibrium)* At a price of $30 per pair of shoes, a manufacturer can supply 2000 pairs of shoes per month, whereas the demand is 2800 pairs. At a price of $35 per pair, 400 more pairs per month can be supplied. However, at this increased price, the demand reduces by 100 pairs.

  a. Assuming linear relationships, determine the supply and demand relations.

  b. Determine the equilibrium price and quantity.

  c. If the government imposes a tax of $1.50 per pair of shoes, find the new equilibrium price and quantity.

  d. What additive tax per pair should be imposed to raise the equilibrium price to $40?

35. *(Break-even Analysis)* A company has fixed costs of $2500, and the total costs of producing 200 units is $3300.

  a. Assuming linearity, write down the cost-output equation.

  b. If each item produced sells for $5.25, find the break-even point.

  c. How many units should be produced and sold so that a profit of $200 results?

36. *(Break-even Analysis and Costs)* A shoe manufacturing company breaks even if its sales are $180,000 per year. If the yearly fixed costs are $45,000 and each pair of shoes sells for $30, find the average variable cost per pair.

37. *(Contracting-out Decision)* Atlas Motors buys fan belts for its "Atlas" cars for $2.50 each. The management is considering manufacturing its own belts at fixed costs of $6000 per year and a variable cost of $1.30 per belt. How many belts should be required each year to justify manufacturing by the firm itself?

38. *(Break-even Analysis)* The demand equation of a company's product is $4p + x = 50$, where x units can be sold at a price of p dollars each. If it costs $(105 + 1.5x)$ dollars to produce x units, at what price p should each item be sold to break even?

39. *(Market Equilibrium and Revenue)* The supply and demand equation for a certain product are $2p - x = 10$ and $p = 8000/(x + 370)$ where p is the price per unit in thousands of dollars and x is the number of units sold per month.

  a. Find the equilibrium point.

  b. Determine the total revenue received by the manufacturer at the equilibrium point.

40. *(Advertising Allocation Decision)* A company wishes to buy advertising time on both radio and television, so as to reach the same number of potential customers through each form of advertising. Each minute of radio advertising costs $500 and reaches 12,000 potential customers, whereas each minute of television advertising costs $1000 and reaches 16,000 potential customers. How many minutes of each form of advertising time should the company buy, if the total cost is to be $150,000?

41. *(Production and Investment)* The management of a certain firm will not invest money to manufacture a new product unless it receives a 15% return on its fixed costs. The firm can sell all it produces at a price of $10 each. The variable cost per unit is $6 and the fixed costs are $40,000. How many units should be produced and sold so as to obtain the required return?

42. *(Market Equilibrium)* The demand and supply equations of a certain commodity are given by $p + x^2 = 20$ and $3p - 8 = x$ respectively, where p is the price in dollars and x is the quantity sold in units of thousands. Find the equilibrium price and quantity.

43. *(Demand and Supply)* There is no demand for a new brand of video cameras if the price per camera is $1700 or more. For each drop of $100 in the price, the demand increases by 200 units. The manufacturer will not market these cameras if the unit price is $500 or less and will supply 1400 cameras at a price of $850 each. Determine the demand and supply equations, assuming them to be linear. What is the equilibrium price and quantity?

44. *(Investment)* A person invested a total of $40,000 in bonds, mutual funds, and term deposits which yield 8%, 15%, and 10% respectively. The amount invested in bonds and term deposits is three times the amount invested in mutual funds. How much is invested in each if the yearly return on these investments is $4260?

45. *(Management Decision)* A firm can produce its product by using two methods. The cost of producing x units by the first method is $(10x + 20,000)$ dollars, whereas it costs $(15x + 9000)$ dollars if the second method is used. The firm can sell all it produces at $30 each. Which method of production should be used by the firm's management if the projected sales are:

a. 800 units?   **b.** 2500 units?

c. 1500 units?

46. *(Gas Station Operations)* The operator of a self-service gas station pays $150 rent per week and $30 business tax per week. For each liter of gas sold he receives 3¢ from the oil company which owns the station.

a. Assuming that on an average each car buys 25 liters of gas, express the weekly profit P as a function of q, the number of cars visiting the station in a week.

b. How many cars must visit the station in a week for the operator to break even?

47. *(Demand Relation)* A sharp rise in the cost of silver forced a film manufacturing company to increase the price of a 20-exposure roll from $2.25 to $2.50. As a result, monthly sales fell from 3 million to 2.6 million rolls.

a. Find the linear equation (in slope-intercept form) that describes price p in dollars in terms of monthly sales x expressed in millions of rolls.

b. What would the sales have been if the company had decreased its price to $2.00 per roll, assuming the linear equation you have derived applies?

48. *(Refrigerator Sales)* A distributor sells two kinds of refrigerators, an imported model on which federal taxes increase the price by 15% and a domestic model on which the taxes increase the price by only 5%. A pair, consisting of one of each kind of refrigerator, costs (before taxes) $1200 and the difference in their prices (after taxes) is $214. What is the cost of each refrigerator, given that the price of the imported model is the larger?

49. *(Break-even and Profits)* A car dealership has fixed costs of $110,000 per year (covering salaries, equipment, interest charges, rent, etc.). The dealership buys 500 cars a year from the manufacturer at $8500 per unit. These cars are sold to the public for $10,000 each, of which $500 goes to the salesperson as commission. In the last two weeks of the year any cars not yet sold are discounted to $8500, of which $300 goes to the salesperson.

a. Write an expression for P, the annual profit, in terms of n, the number of cars sold at full price during the year. Assume the whole stock is cleared in the year-end sale.

b. What is the break-even value of n?

c. Assuming that $1,300,000 has been invested in the business, what value of n corresponds to a return on investment of 20%?

CHAPTER 5

Functions and Graphs

Chapter Objectives

5-1 FUNCTIONS

(a) Definition of a function; domain and range of a function;
(b) Evaluation of a function;
(c) The graph of a function; the vertical line test;
(d) Construction of functions from verbal information;
(e) Polynomial, rational and algebraic functions.

5-2 QUADRATIC FUNCTIONS AND PARABOLAS

(a) Determination of the vertex of a parabola given by a quadratic function;
(b) Graphing a quadratic function;
(c) Minimization and maximization problems involving quadratic functions.

5-3 MORE SIMPLE FUNCTIONS AND GRAPHS

(a) Power functions and their graphs;
(b) The center-radius formula for a circle;
(c) Determination of the center and radius from the general form of the equation of a circle;
(d) The graph of the absolute value function and other functions involving absolute values.

5-4 COMBINATIONS OF FUNCTIONS

(a) The sum, difference, product, and quotient of two functions and their domains;
(b) The composition of two functions and its domain.

5-5 IMPLICIT RELATIONS AND INVERSE FUNCTIONS

(a) Definition of an implicit relation and implicit function;
(b) Determination of explicit functions from an implicit relation;
(c) The inverse of a function, if necessary with restricted domain;
(d) Connections between the graphs of a function and its inverse.

CHAPTER REVIEW

5-1 FUNCTIONS

The concept of function is one of the basic ideas in all mathematics. Almost every study that concerns the application of mathematics to practical problems or that involves the analysis of empirical data makes use of this mathematical concept.

A function expresses the idea of one quantity depending on or being determined by another. The following are examples of this idea.

1. The area of a circle depends on the length of its radius; if the radius is known, its area can be determined. We say that the area is a function of the radius.

2. The weekly cost of producing any good depends on the number of units produced. We say that cost is a function of number of units.

3. The benefits paid out by the welfare system of a country depend on the unemployment rate.

4. The quantity of any item that will be made available by its suppliers depends on the price they can get for it. Quantity is a function of price.

We begin by giving a formal definition of a function.

DEFINITION Let X and Y be two nonempty sets. Then a **function** from X to Y is a *rule* that assigns to each element $x \in X$ a unique $y \in Y$. If a function assigns y to a particular x, we say that y is the **value** of the function at x.

A function is generally denoted by a single letter such as f, g, F, or G.

Let f denote a given function. The set X for which f assigns a unique $y \in Y$ is called the **domain** of the function f. It is often denoted by D_f. The corresponding set of values $y \in Y$ is called the **range** of the function and is often denoted by R_f.

EXAMPLE 1

(a) Let X be the set of students in a class. Let f be the rule which assigns to each student his or her final grade. Since each student is associated with exactly one final grade, this rule does define a function. In this case, the domain is the set of all students in the class and the range is the set of all the different grades awarded. (For example, R_f might be the set $\{A, B, C, D, F\}$.)

(b) The value of the financial assets of a firm is a function of time. Here the domain is the set of values of time and the range of the function is the set of values of the assets (say in dollars). ☛ 1

If a function f assigns a value y to a certain x in the domain, we write

$$y = f(x).$$

☛ **1.** Do the following define functions?

(a) The rule that assigns to each person the number of his or her children;

(b) the rule that assigns to each person the names of his or her children;

(c) the rule that assigns to each person the name of his or her first child;

(d) a French-to-English dictionary.

Answer (a) Yes; (b) no; (c) yes; (d) no (usually more than one English word corresponds to each French word).

We read $f(x)$ as "f of x"; it is called the *value of f at x*. Note that $f(x)$ is *not* the product of f and x.

If a function f is expressed by a relation of the type $y = f(x)$, then x is called the **independent variable** or **argument of f** and y is called the **dependent variable.**

Generally, we shall encounter functions that are expressed by stating the value of the function by means of an algebraic formula in terms of the independent variable involved. For example, $f(x) = 5x^2 - 7x + 2$ and $g(p) = 2p^3 + 7/(p + 1)$.

EXAMPLE 2 Given $f(x) = 2x^2 - 5x + 1$, find the value of f when $x = a$, $x = 3$, $x = -2$, and $x = -\frac{1}{4}$; that is, find $f(a), f(3), f(-2)$ and $f(-\frac{1}{4})$.

Solution We have

$$f(x) = 2x^2 - 5x + 1. \tag{1}$$

To find $f(a)$, we replace x by a in Equation (1).

$$f(a) = 2a^2 - 5a + 1$$

To evaluate $f(3)$, we substitute 3 for x on both sides of Equation (1).

$$f(3) = 2(3)^2 - 5(3) + 1 = 18 - 15 + 1 = 4$$

Similarly,

$$f(-2) = 2(-2)^2 - 5(-2) + 1 = 19$$

and

☛ **2.** If $f(x) = (x + 1)^{-1}$, evaluate $f(1), f(0)$, and $f(-1)$.

$$f(-\tfrac{1}{4}) = 2(-\tfrac{1}{4})^2 - 5(-\tfrac{1}{4}) + 1 = \tfrac{19}{8}. \quad ☛ 2$$

EXAMPLE 3 Given $\quad g(x) = 3x^2 - 2x + 5, \quad$ evaluate: $\quad$ (a) $g(1 + h)$; (b) $g(1) + g(h)$; (c) $[g(x + h) - g(x)]/h$.

Solution We have

$$g(x) = 3x^2 - 2x + 5. \tag{2}$$

(a) To evaluate $g(1 + h)$, we must replace x in Equation (2) by $1 + h$.

$$g(1 + h) = 3(1 + h)^2 - 2(1 + h) + 5$$
$$= 3(1 + 2h + h^2) - 2 - 2h + 5$$
$$= 3h^2 + 4h + 6$$

(b) Replacing x by 1 and h, respectively, in Equation (2) we get

$$g(1) = 3(1^2) - 2(1) + 5 = 3 - 2 + 5 = 6$$

and

Answer $f(1) = 0.5, f(0) = 1$, and $f(-1)$ does not exist.

$$g(h) = 3h^2 - 2h + 5.$$

Therefore

$$g(1) + g(h) = 6 + 3h^2 - 2h + 5 = 3h^2 - 2h + 11.$$

(c) Replacing the argument x in Equation (2) by $x + h$, we have the following.

$$g(x + h) = 3(x + h)^2 - 2(x + h) + 5$$
$$= 3(x^2 + 2xh + h^2) - 2x - 2h + 5$$
$$= 3x^2 - 2x + 5 + h(3h + 6x - 2)$$

Therefore

$$\frac{1}{h}[g(x + h) - g(x)]$$

$$= \frac{1}{h}[3x^2 - 2x + 5 + h(3h + 6x - 2) - (3x^2 - 2x + 5)]$$

$$= 3h + 6x - 2.$$

☛ **3.** If $f(x) = |x|$, evaluate $\dfrac{f(1 + h) - f(1)}{h}$.

The quantity $[g(x + h) - g(x)]/h$ for a given function g will be very important to us when we study calculus in Chapter 12.   ☛ **3**

EXAMPLE 4 Evaluate $F(0)$, $F(1)$, and $F(4)$ for the function F defined by

$$F(x) = \frac{x - 4}{\sqrt{2 - x}}.$$

Solution First replace x by 0:

$$F(0) = \frac{0 - 4}{\sqrt{2 - 0}} = \frac{-4}{\sqrt{2}} = -2\sqrt{2}.$$

Next, replace x by 1:

$$F(1) = \frac{1 - 4}{\sqrt{2 - 1}} = \frac{-3}{\sqrt{1}} = -3.$$

Finally, replace x by 4:

$$F(4) = \frac{4 - 4}{\sqrt{2 - 4}} = \frac{0}{\sqrt{-2}} \quad \text{not defined.}$$

$F(4)$ does not exist; in other words, 4 is not in the domain of F.

In most cases of interest, the domains and ranges of the functions with which we are concerned are subsets of the real numbers. In such cases, the function is commonly represented by its *graph*. The graph of a function f is obtained by plotting all of the points (x, y), where x belongs to the domain of f and $y = f(x)$, treating x and y as Cartesian coordinates.

Answer 1 if $h \geq -1$, $h \neq 0$;
$\dfrac{-(2 + h)}{h}$ if $h < -1$.

EXAMPLE 5 $f(x) = 2 + 0.5x^2$. The domain of f is the set of all real numbers, since we can evaluate $f(x)$ for any real value of x. Some of the values of this function are shown in Table 1, in which selected values of x are listed in the top row and the values of $y = f(x)$ are given beneath the corresponding values of x. The points corresponding to the values of x and y are plotted as dots in Figure 1. The graph of the function $f(x) = 2 + 0.5x^2$ is shown as the U-shaped curve passing through the heavy dots.

TABLE 1

x	0	1	2	3	4	-1	-2	-3	-4
$y = f(x)$	2	2.5	4	6.5	10	2.5	4	6.5	10

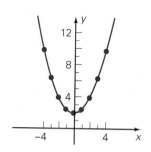

FIGURE 1

EXAMPLE 6 The monthly costs of a small manufacturer are given in thousands of dollars by $C = 10 + 2x$, where x is the number of employees. The average cost per employee is then given by

$$f(x) = \frac{10 + 2x}{x} = \frac{10}{x} + 2.$$

Graph the function f for $1 \le x \le 10$.

Solution In this case x must be a whole number, so $D_f = \{1, 2, 3, \ldots\}$. We find the values shown in Table 2.

TABLE 2

x	1	2	3	4	5	6	7	8	9	10
$f(x)$	12	7	5.33	4.5	4	3.67	3.43	3.25	3.11	3

4. Graph the functions
$y = 4 + x - \frac{1}{4}x^2$ and $y = \frac{1}{2}x - 1$
for $-4 \leq x \leq 4$.

Answer Using the following table of values we obtain the graphs shown below:

x		-4	-3	-2	-1	0	1	2	3	4
$y = 4 + x - \frac{1}{4}x^2$		-4	-1.25	1	2.75	4	4.75	5	4.75	4
$y = \frac{1}{2}x - 1$		-3	-2.5	-2	-1.5	-1	-0.5	0	0.5	1

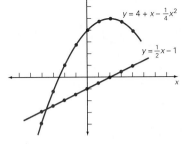

The graph is shown in Figure 2. Note that the graph consists of discrete points, not a continuous curve. **4**

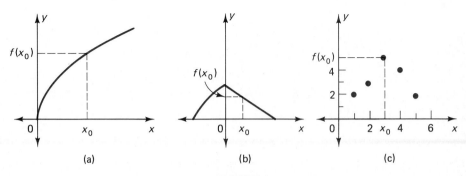

FIGURE 2

Suppose that we are given a certain curve in the xy-plane. How can we tell whether or not it is the graph of some function $y = f(x)$?

Vertical-line test:

Any given curve (or set of points) in the xy-plane is the graph of a function (with y as dependent variable) provided that any vertical line meets the graph in at most one point.

Any vertical line corresponds to some particular value, say $x = x_0$, of the independent variable, and the point where this vertical line meets the graph determines the value of y that corresponds to x_0. That is, the graph itself provides the rule that relates each value of x to some value of y. If the vertical line $x = x_0$ does not meet the graph at all, then x_0 simply does not belong to the domain.

The graphs in Figure 3 represent functions. (Note that in part (c), the domain of the function is the set of integers $\{1, 2, 3, 4, 5\}$, so the graph simply consists of five points rather than a curve.)

(a)   (b)   (c)

FIGURE 3

► 5. Look at the graphs in Figures 6 through 17 in Chapter 4. Are any of them *not* the graph of a function?

On the other hand, the graphs in Figure 4 do not represent functions. These are not functions because there are vertical lines that meet the graphs in more than one point. Thus, corresponding to the value of $x = x_0$ on the first graph, there are two values y_1 and y_2 for y. In such a case, the value of x does not determine a *unique value* of y.   **► 5**

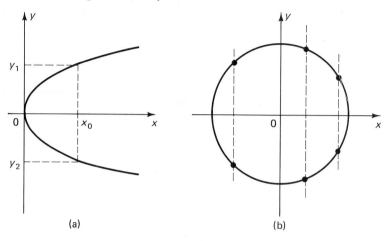

(a)   (b)

FIGURE 4

On the graph of a function, the values along the x-axis at which the graph is defined constitute the domain of the function. Correspondingly, the values along the y-axis at which the graph has points constitute the range of the function. This is illustrated in Figure 5. Here we have

$$D_f = \{x \,|\, -2 \le x \le 3\} \qquad \text{and} \qquad R_f = \{y \,|\, 0 \le y \le 2\}.$$

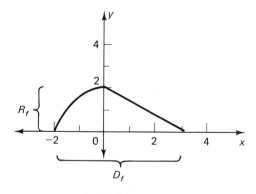

FIGURE 5

Often the domain of a function is not stated explicitly. In such cases, it is understood to be the set of all values of the argument for which the given rule makes sense. For a function f defined by an algebraic expression, the domain of f is the set of all real numbers x for which $f(x)$ is a well-defined real num-

Answer The vertical lines in Figures 11 and 15 are not. All the others are graphs of functions.

ber. For example, the domain of the function $f(x) = \sqrt{x}$ is the set of nonnegative real numbers, since the square root makes sense only if $x \geq 0$. Similarly, in the case of the function $g(x) = x^2/(x - 3)$, the domain is the set of all real numbers except $x = 3$, since when $x = 3$ the denominator becomes zero and $g(3)$ is not defined.

In general, when finding the domain of a function we must bear these two conditions in mind: *Any expression underneath a square root cannot be negative and the denominator of any fraction cannot be zero.* (More generally, any expression underneath a radical of even index such as $\sqrt[4]{}$ or $\sqrt[6]{}$ cannot be negative.)

EXAMPLE 7 Find the domain of g, where

$$g(x) = \frac{x + 3}{x - 2}.$$

Solution Clearly, $g(x)$ is not a well-defined real number for $x = 2$. For any other value of x, $g(x)$ is a well-defined real number. Thus the domain of g is the set of all real numbers except 2.

EXAMPLE 8 Find the domain of f if $f(x) = \sqrt{x - 4}$.

Solution The domain of f is the set of all values of x for which the expression under the radical sign is nonnegative. That is,

$$x - 4 \geq 0 \text{ or } x \geq 4.$$

For $x < 4$, $f(x)$ is not a real number, since the quantity beneath the square root, $x - 4$, is negative. The graph of f is shown in Figure 6, in which a few explicit points are plotted. ☞ 6

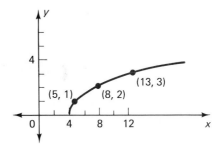

FIGURE 6

EXAMPLE 9 Find the domain of h where

$$h(x) = \frac{x}{(x - 2)\sqrt{x - 1}}.$$

Solution Here the radical is defined only for $x \geq 1$. But the denominator is zero if $x = 1$ or $x = 2$, so these two points must be excluded from the do-

☞ **6.** What are the domain of the functions
(a) $y = (x - 2)^2$;   (b) $y = \sqrt{x - 2}$?

Answer (a) All real numbers;
(b) $\{x \,|\, x \geq 2\}$.

☞ 7. What is the domain in each case?

(a) $y = \dfrac{\sqrt{x-1}}{(x+2)}$; (b) $y = \dfrac{\sqrt{1-x}}{(x+2)}$.

Answer (a) $\{x \mid x \geq 1\}$;
(b) $\{\{x \mid x \leq 1, x \neq -2\}$.

☞ 8. What is the domain of the function in Example 10?

Answer Mathematically, this function is defined for all x. In practice, the domain would be limited to the interval $0 \leq x \leq 2$.

main. Therefore

$$D_h = \{x \mid x > 1, x \neq 2\}. \quad \text{☞ } 7$$

In applied problems, it is often necessary to construct an algebraic function from certain verbal information.

EXAMPLE 10 *(Cost of Telephone Link)* A telephone link is to be constructed between two towns that are situated on opposite banks of a river at points A and B. The width of the river is 1 kilometer and B lies 2 kilometers downstream from A. It costs c dollars per kilometer to construct a line over land and $2c$ dollars per kilometer under water. The telephone line will follow the river bank from A for a distance x (kilometers) and then will cross the river diagonally in a straight line directly to B. Determine the total cost of the line as a function of x.

Solution Figure 7 illustrates this problem. The telephone line proceeds from A to C, a distance x along the bank, then diagonally across from C to B. The cost of the segment AC is cx, whereas the cost of CB is $2c(CB)$. The total cost (call it y) is therefore given by

$$y = cx + 2c(CB).$$

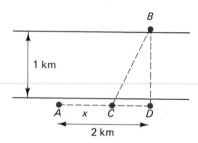

FIGURE 7

In order to complere the problem, we must express CB in terms of x. We apply the Pythagorean theorem to the triangle BCD.

$$BC^2 = BD^2 + CD^2$$

But $BD = 1$ (the width of the river) and

$$CD = AD - AC = 2 - x.$$

Therefore

$$BC^2 = 1^2 + (2-x)^2 = 1 + (4 - 4x + x^2) = x^2 - 4x + 5.$$

Thus the cost is given by

$$y = cx + 2c\sqrt{x^2 - 4x + 5}.$$

This is the required expression, giving y as a function of x. ☞ 8

In the preceding examples, we have been concerned with functions that are defined by a single algebraic expression for all values of the independent variable throughout the domain of the function. It sometimes happens that we need to use functions that are defined by more than one expression.

EXAMPLE 11 *(Electricity Cost Function)* Electricity is charged to consumers at the rate of 10¢ per unit for the first 50 units and 3¢ for amounts in excess of this. Find the function $c(x)$ that gives the cost of using x units of electricity.

Solution For $x \leq 50$, each unit costs 10¢, so the total cost of x units is $10x$ cents. So $c(x) = 10x$ for $x \leq 50$. When $x = 50$, we get $c(50) = 500$; the cost of the first 50 units is equal to 500¢. When $x > 50$, the total cost is equal to the cost of the first 50 units (that is, 500¢) plus the cost of the rest of the units used. The number of these excess units is $x - 50$, and they cost 3¢ each, so their total cost is $3(x - 50)$ cents. Thus the total bill when $x > 50$ comes to

$$c(x) = 500 + 3(x - 50) = 500 + 3x - 150 = 350 + 3x.$$

We can write $c(x)$ in the form

$$c(x) = \begin{cases} 10x & \text{if } x \leq 50 \\ 350 + 3x & \text{if } x > 50. \end{cases}$$

The graph of $y = c(x)$ is shown in Figure 8. Observe how the nature of the graph changes at $x = 50$, where one formula takes over from the other. ☛ 9

☛ **9.** To ship a parcel from Vancouver to Paris, France, a courier service charges $50 for packages weighing up to 2 kilograms and $10 for each additional kilogram or part thereof. Graph the function that expresses the cost of shipping a package of weight x kilograms for $x \leq 8$.

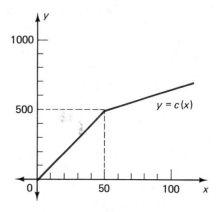

FIGURE 8

Answer

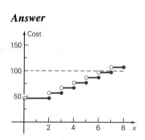

EXAMPLE 12 Consider the function f defined by

$$f(x) = \begin{cases} 4 - x & \text{if } 0 \leq x \leq 4 \\ \sqrt{x - 4} & \text{if } x > 4 \end{cases}$$

The domain of this function is the set of all nonnegative real numbers. For $0 \leq x \leq 4$, the function is defined by the algebraic expression $f(x) = 4 - x$,

10.
Graph the function f where
$$f(x) = \begin{cases} \frac{1}{2}x + 4 & \text{if } x < 2 \\ 8 - x & \text{if } x \geq 2. \end{cases}$$

while for $x > 4$, it is defined by the expression $f(x) = \sqrt{x - 4}$. Some values of $f(x)$ are given in Table 3 and the graph of this function is shown in Figure 9. It consists of two segments: For x between 0 and 4, the graph consists of a straight line segment with equation $y = 4 - x$. ☛ **10**

TABLE 3

x	0	2	4	5	8	13
$y = f(x)$	4	2	0	1	2	3

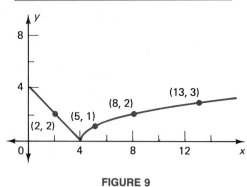

FIGURE 9

In these examples, the function under consideration has been defined by two algebraic expressions. It is sometimes necessary to consider functions defined by three or more different expressions. (For example, see Exercise 50.)

We close this section by discussing some simple functions. A function of the form

$$f(x) = b$$

where b is some constant is called a **constant function.** (See Figure 10.) We have already encountered such functions in Section 4-2. The graph of f is a straight line parallel to the x-axis and at a distance $|b|$ above or below the x-axis depending on whether b is positive or negative. In this case

$$D_f = \text{the set of all real numbers} \quad \text{and} \quad R_f = \{b\}.$$

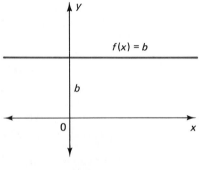

FIGURE 10

Answer

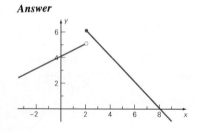

A function f defined by the relation

$$f(x) = a_n x^n + a_{n-1} x^{n-1} + \cdots + a_1 x + a_0 \qquad (a_n \neq 0)$$

where $a_0, a_1, \ldots, a_n$ are constants and n is a nonnegative integer, is said to be a **polynomial function of degree n.** For example, the functions f and g defined by

$$f(x) = 3x^7 - 5x^4 + 2x - 1 \quad \text{and} \quad g(x) = x^3 + 7x^2 - 5x + 3$$

are polynomial functions of degree 7 and 3, respectively.

If the degree of the polynomial function is 1, then the function is a **linear function.** The general form of the linear function is given by

$$f(x) = mx + b \qquad (m \neq 0)$$

where m and b are constants. (See Figure 11.) As we know from Section 4-2, the graph of a linear function is a straight line with slope m and y-intercept of b. Here D_f and R_f equal the set of all real numbers.

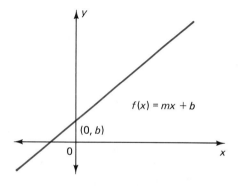

FIGURE 11

If the degree of the polynomial function is 2, then the function is called a **quadratic function.** The general quadratic function may be defined as

$$g(x) = ax^2 + bx + c \qquad (a \neq 0)$$

where a, b, and c are constants. We shall discuss this function in detail in the next section.

Similarly, a polynomial function of degree 3 is called a **cubic function.** For example, the function f defined by

$$f(x) = 2x^3 - 5x^2 + 7x + 1$$

is a cubic function.

If the degree of a polynomial function is zero, then it reduces to a constant function.

If a function can be expressed as the quotient of two polynomial functions, then it is called a **rational function.** Examples of rational functions are

$$f(x) = \frac{x^2 - 9}{x - 4} \quad \text{and} \quad g(x) = \frac{2x^3 - 7x + 1}{5x^2 - 2}.$$

In general, any rational function has the form $f(x) = p(x)/q(x)$, where $p(x)$ and $q(x)$ are polynomials in x.

If the value $f(x)$ of a function f is found by a finite number of algebraic operations, f is called an **algebraic function.** The algebraic operations are addition, subtraction, multiplication, division, raising to powers, and extracting roots. For example, the functions f and g defined by

$$f(x) = \frac{(2x + 1)^2 - \sqrt{x^3 + 1}}{(x^2 + 1)^4} \qquad \text{and} \qquad g(x) = (2x^2 - 1)^{-1/7} + 5x^{3/4}$$

are algebraic functions.

Apart from algebraic functions, there are other functions called **transcendental functions.** Examples of transcendental functions are logarithmic functions, and exponential functions, which we shall discuss in Chapter 6.

EXERCISES 5-1

1. Given $f(x) = 3x + 2$, find $f(1)$, $f(-2)$, $f(x^2)$ and $f(x + h)$.

2. Given $f(x) = 5 - 2x$, find $f(3)$, $f(-1)$, $f(x^2)$ and $f(x + h)$.

3. Given $f(t) = 5t + 7$, find $f(1)$, $f(-3)$, $f(c)$, $f(1 + c)$ and $f(1) + f(c)$.

4. Given $f(x) = 3 - 4x$, find $f(a)$, $f(a + 1)$ and $f(a) + f(1)$.

5. Given $f(x) = x^2$, find $f(3)$, $f(-2)$, $f(a)$, $f(\sqrt{x})$ and $f(x + h)$.

6. Given $f(x) = 3x^2 + 7$, find $f(c)$, $f(c + h)$ and $f(c + h) - f(c)$.

7. Given $f(x) = 3$, find $f(1/x)$, $f(x^2)$, $f(x + 2)$ and $f(x + h)$.

8. Given $f(y) = 5$, find $f(1/y)$, $f(y^2)$, $f(y + 3)$, $f(7)$ and $f(y + h)$.

9. Given $f(x) = \sqrt{x}$, find $f(4)$, $f(x^2)$ and $f(a^2 + h^2)$.

10. Given $f(x) = \sqrt{x - 16}$, find $f(25)$, $f(0)$, and $f(7)$.

11. Given $f(t) = 3t^2 - 5t + 7$, find $f(0)$, $f(1/t)$, $f(c) + f(h)$ and $f(c + h)$.

12. Given $f(u) = 2u^2 + 3u - 5$, find $f(0)$, $f(1/x)$, $f(x + h)$, and $f(x + h) - f(x)$.

13. Given

$$f(x) = \begin{cases} 2x - 3 & \text{if } x \geq 5 \\ 6 - 3x & \text{if } x < 5 \end{cases}$$

find each of the following.

a. $f(0)$ b. $f(7)$ c. $f(-2)$
d. $f(5 + h)$ and $f(5 - h)$, where $h > 0$.

14. Given

$$g(x) = \begin{cases} 4x + 3 & \text{if } -2 \leq x < 0 \\ 1 + x^2 & \text{if } 0 \leq x \leq 2 \\ 7 & \text{if } x > 2 \end{cases}$$

evaluate each of the following.

a. $g(1)$ b. $g(3)$ c. $g(-1)$
d. $g(0)$ e. $g(-3)$
f. $g(2 + h)$ and $g(2 - h)$ if $2 > h > 0$

*15. If $F(t) = t/(1 + t)$ and $G(t) = t/(1 - t)$, show that $F(t) - G(t) = -2G(t^2)$.

*16. If $y = f(x) = (x + 1)/(x - 1)$, show that $x = f(y)$.

17. If $f(x) = x^2 + 1$ and $g(x) = 2x - 1$, find $f[g(2)]$.

18. If $f(x) = g(x) + h(x)$, $g(x) = x^2 + 3$, and $f(x) = x^3$, find $h(2)$.

(19–24) Evaluate $[f(x + h) - f(x)]/h$ where $f(x)$ is as given below.

19. $f(x) = 2x + 5$

20. $f(x) = 3x - 7$

21. $f(x) = 7$

22. $f(x) = x^2$

23. $f(x) = x^2 - 3x + 5$

24. $f(x) = 3x^2 + 5x - 2$

(25–42) Find the domain of the function whose values are given.

25. $f(x) = 2x + 3$

26. $f(t) = 2t^2 - 3t + 7$

27. $h(x) = \dfrac{x - 1}{x - 2}$

28. $D(p) = \dfrac{2p + 3}{p - 1}$

29. $g(x) = \dfrac{x + 1}{x^2 - 3x + 2}$

30. $f(x) = \dfrac{x^2 - 4}{x - 2}$

31. $F(u) = \dfrac{u + 2}{u^2 + 1}$

32. $G(t) = \sqrt{t - 2}$

33. $F(y) = -\sqrt{3y - 2}$

34. $g(t) = \dfrac{1}{\sqrt{2t - 3}}$

35. $G(u) = \dfrac{2}{\sqrt{3 - 2u}}$

36. $f(x) = \sqrt{x^2 + 16}$

37. $f(x) = \begin{cases} 2x - 3 & \text{if } x > 5 \\ 6 - 3x & \text{if } x < 5 \end{cases}$

38. $f(x) = \begin{cases} 4x + 3 & \text{if } -2 \le x < 0 \\ 1 + x^2 & \text{if } 0 \le x \le 2 \\ 7 & \text{if } x > 2 \end{cases}$

39. $f(x) = \begin{cases} \dfrac{1}{x - 2} & \text{if } x < 3 \\ 3x + 5 & \text{if } x > 3 \end{cases}$

40. $f(x) = \begin{cases} \dfrac{1}{5 - x} & \text{if } x < 2 \\ \dfrac{1}{x + 5} & \text{if } x \ge 2 \end{cases}$

41. $g(x) = \begin{cases} \dfrac{2x + 3}{x - 4} & \text{if } x \ge 1 \\ \dfrac{1}{1 - 2x} & \text{if } x < 1 \end{cases}$

42. $F(t) = \begin{cases} 5t - 7 & \text{if } t > 3 \\ \dfrac{1}{t - 4} & \text{if } t < 3 \end{cases}$

(43–46) Sketch the graphs of the following functions:

43. $f(x) = \begin{cases} 1 & \text{if } x > 0 \\ 2 & \text{if } x \le 0 \end{cases}$

44. $f(x) = \begin{cases} 1 & \text{if } x < 0 \\ x & \text{if } x \ge 0 \end{cases}$

45. $f(x) = \begin{cases} x & \text{if } x > 1 \\ -x & \text{if } x < 1 \end{cases}$

46. $f(x) = \begin{cases} x - 3 & \text{if } x < 3 \\ 2x - 6 & \text{if } x > 3 \end{cases}$

47. *(Cost Function)* The cost of producing x units of their product per week is determined by a company to be given by

$$C(x) = 5000 + 6x + 0.002x^2$$

Evaluate the cost of producing:

a. 1000 units per week.

b. 2500 units per week.

c. No units.

48. *(Cost Function)* For the cost function

$$C(x) = 10^{-6}x^3 - (3 \times 10^{-3})x^2 + 36x + 2000$$

calculate the cost of producing:

a. 2000 units.

b. 500 units.

49. *(Physiology)* In a test for blood sugar metabolism, conducted over a time interval, the amount of sugar in the blood was a function of time t (measured in hours) and given by:

$$A(t) = 3.9 + 0.2t - 0.1t^2.$$

Find the amount of sugar in the blood:

a. At the beginning of the test.

b. 1 hour after the beginning.

c. $2\frac{1}{2}$ hours after the beginning.

50. *(Air Pollution)* The average level of air pollution in a certain city varies during the day as follows:

$$p(t) = \begin{cases} 2 + 4t & \text{if } 0 \le t < 2 \\ 6 + 2t & \text{if } 2 \le t < 4 \\ 14 & \text{if } 4 \le t < 12 \\ 50 - 3t & \text{if } 12 \le t < 16 \end{cases}$$

Here t is time in hours, with $t = 0$ corresponding to 6 A.M. and $t = 16$, to 10 P.M. Graph this function. What are the pollution levels at 8 A.M., 12 noon, 6 P.M., and 8 P.M.?

51. *(Cost Function)* A radio manufacturing firm has fixed costs of $3000 and the cost of labor and material is $15 per radio. Determine the cost function, that is, the total cost as a function of number of radios manufactured. If each radio is sold for $25, determine the revenue function and the profit function.

52. *(Revenue Function)* A manufacturer can sell 300 units of its product in a month at a charge of $20 per unit and 500 units at a charge of $15 per unit. Express the market demand x (the number of units that can be sold each month) as a function of the price per unit, assuming it to be a linear function. Express the revenue as:

a. A function of the price.

b. A function of x.

53. *(Farming)* A farmer has 200 yards of fencing to enclose a rectangular field. Express the area A of the field as a function of the length of one side of it.

54. *(Geometry)* A rectangle is inscribed in a circle of radius 3 centimeters. Express the area A of the rectangle as a function of the length of one of its sides.

55. *(Cost Function)* A cistern is constructed to hold 300 cubic feet of water. The cistern has a square base and four vertical sides, all made of concrete, and a square top made of steel. If concrete costs $1.50 per square foot and steel costs $4 per square foot, determine the total cost C as a function of the length of the side of the square base.

56. Repeat Exercise 55 if the cistern is a cylinder with circular base and top. Express the cost C as a function of radius r of the base of the cylinder.

57. *(Cost Function)* Sugar costs 25¢ per pound for amounts up to 50 pounds and 20¢ for amounts over 50 pounds. If $C(x)$ denotes the cost of x pounds of sugar, express $C(x)$ by means of suitable algebraic expressions and sketch its graph.

58. *(Cost Function)* A retailer can buy oranges from the wholesaler at the following prices: 20¢ per pound if 20 pounds or less are purchased; 15¢ per pound for amounts over 20 pounds and up to 50 pounds, and 12¢ per pound for amounts over 50 pounds. Determine the cost $C(x)$ of purchasing x pounds of oranges.

59. *(Revenue Function)* An apartment building has 70 suites which can all be rented if the rent charged is $200 per month. For each increase of $5 in rent, one suite becomes vacant with no possibility of renting it. Express the total monthly revenue R as a function of:

a. x, where x denotes the number of $5 increases in rent.

b. The monthly rent p.

60. *(Profit Function)* The demand equation of a company's product is $2p + 3x = 16$, where x units can be sold at a price of $p each. If it costs $(100 + 2x)$ dollars to produce x units, express the profit P as a function of:

a. The demand x.

b. The price p.

61. *(Quantity Discount)* A travel agent offers a vacation package priced at $500 per person for groups of six or more, with a discount of 10% off this price after the first twelve in the group. Construct the function $C(x)$ giving the average cost per individual in a group of size x ($x \ge 6$).

62. *(Mailing Cost)* The cost of mailing a first-class letter is 35¢ for each 10 grams or part thereof. Construct the function $C(W)$ that gives the cost in cents of mailing a letter of weight W (not exceeding 50 grams).

63. A rectangle has one side of x inches. The perimeter of the rectangle is 20 inches. Express the area A as a function of x and state the domain of this function.

64. Equal squares of side x are cut off from each corner of an 18-inch-square cardboard and then the sides are folded up to form an open box. Express the volume V of the box as a function of x and determine the domain of this function.

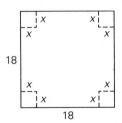

65. Equal squares of side x are cut off from each corner of a 20 cm. $\times$ 16 cm. rectangular sheet and then the sides are folded up to form an open box. If $V = f(x)$ denotes the volume of the box, determine $f(x)$ and state its domain.

66. A rectangle on one side x inches is inscribed in a circle of radius c inches. Express the area A of the rectangle as a function of x and determine the domain of this function.

(67–72) State whether or not the following graphs represent functions.

67.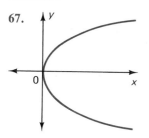

68.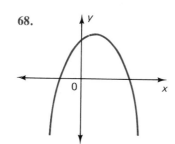

69.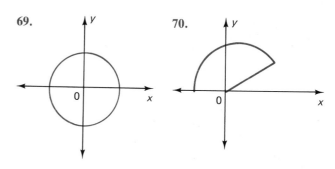

70.

71.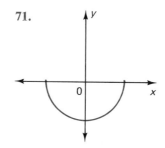

72.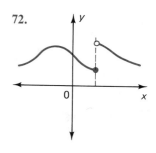

5-2 QUADRATIC FUNCTIONS AND PARABOLAS

A function of the form

$$f(x) = ax^2 + bx + c \qquad (a \neq 0)$$

where a, b, and c are constants, is called a *quadratic function*. The domain of $f(x)$ is the set of all real numbers.

The simplest quadratic function is obtained by setting b and c equal to zero, in which case we obtain $f(x) = ax^2$. Typical graphs of this function in the two cases where a is positive or negative are shown in Figure 12. The low-

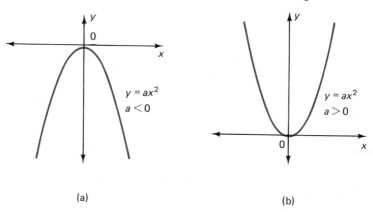

$y = ax^2$
$a < 0$

(a)

$y = ax^2$
$a > 0$

(b)

FIGURE 12

est point on the graph when $a > 0$ occurs at the origin, while the origin is the highest point when $a < 0$. Each of these graphs is called a **parabola.** The origin (which is the lowest or highest point in the two cases) is called the **vertex** of the parabola.

The general quadratic function $f(x) = ax^2 + bx + c$ has a graph identical in shape and size to the graph of $y = ax^2$; the only difference is that the vertex of $f(x) = ax^2 + bx + c$ is shifted away from the origin.

THEOREM 1 The graph of the function $f(x) = ax^2 + bx + c$ $(a \neq 0)$ is a parabola that opens upward if $a > 0$ and downward if $a < 0$. Its vertex (which is the lowest point when $a > 0$ and the highest point when $a < 0$) is at the point

$$x = -\frac{b}{2a} \quad \text{and} \quad y = \frac{4ac - b^2}{4a}.$$

Typical graphs of the quadratic function $y = ax^2 + bx + c$ are shown in Figure 13.

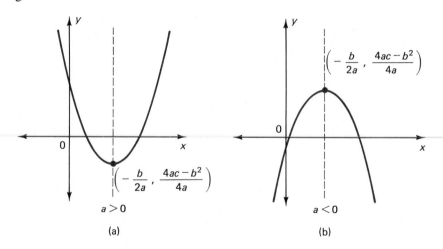

FIGURE 13

Notes **1.** If $b = c = 0$, the quadratic function reduces to $f(x) = ax^2$. The coordinates of the vertex given in Theorem 1 then reduce to $x = y = 0$, which is consistent with our earlier statements.

2. It is not worth remembering the formula in Theorem 1 for the y-coordinate of the vertex. It is easier to substitute the value $x = -b/2a$ into the equation of the parabola, $y = ax^2 + bx + c$ (see Example 1).

3. The parabola is symmetrical about the vertical line through the vertex. This line is called the **axis** of the parabola.

4. Theorem 1 is most easily proved using the method of *completing the square* (see Section 2-3). Can you manage to prove it? ☞ **11**

EXAMPLE 1 Sketch the parabola $y = 2x^2 - 4x + 7$ and find its vertex.

Solution Compare the given equation

$$y = 2x^2 - 4x + 7$$

with the standard quadratic function

$$y = ax^2 + bx + c$$

we have $a = 2$, $b = -4$, and $c = 7$. The x-coordinate of the vertex is

$$x = -\frac{b}{2a} = -\frac{-4}{2(2)} = 1.$$

To find the y-coordinate of the vertex, the simplest way is to substitute $x = 1$ in the given equation for the parabola.

$$y = 2(1)^2 - 4(1) + 7 = 5$$

Thus the vertex is at the point $(1, 5)$.

Alternatively, we could use the formula given in Theorem 1.

$$y = \frac{4ac - b^2}{4a} = \frac{4(2)(7) - (-4)^2}{4(2)} = \frac{56 - 16}{8} = 5$$

Since $a = 2 > 0$, the parabola opens upward; that is, the vertex $(1, 5)$ is the lowest point on the parabola. The graph of this parabola can be sketched by plotting a few points (x, y) lying on it. Values of y corresponding to selected values of x are shown in Table 4. Plotting these points and joining them with a smooth curve, we obtain the graph as shown in Figure 14. (Note that the graph in this example meets the y-axis at the point $(0, 7)$ but does not meet the x-axis at all. It is often helpful when drawing the graph of some given function to find the points where this graph meets the coordinate axes.) ☞ 12

☞ **12.** The parabola that is identical in size with $y = ax^2$ and has vertex at (h, k) is given by the equation
$y - k = a(x - h)^2$.
What is the equation of the parabola identical in size with $y = 2x^2$ with vertex at $(1, 5)$?

Answer $y - 5 = 2(x - 1)^2$ or $y = 2x^2 - 4x + 7$ (cf. Example 1).

TABLE 4

x	-1	0	1	2	3
y	13	7	5	7	13

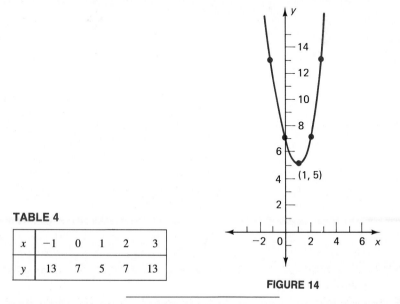

FIGURE 14

► **13.** Let the parabola
$y = ax^2 + bx + c$ cross the x-axis at
$x = p$ and q. If $a > 0$, the graph
lies below the x-axis for x between p
and q. That is, $ax^2 + bx + c < 0$
for x between p and q and
$ax^2 + bx + c > 0$ outside this
interval. If $ax^2 + bx + c$ is never
zero for any real x, then
$ax^2 + bx + c > 0$ for all x. Relate
this to our technique of solving
quadratic inequalities in Section 3.3.
What are the corresponding
conditions if $a < 0$?

As stated before, the vertex of a parabola represents the lowest point when $a > 0$ or the highest point when $a < 0$. It follows, therefore, that for $a > 0$, the function $f(x) = ax^2 + bx + c$ takes its minimum value at the vertex of the corresponding parabola. That is, $f(x)$ is smallest when $x = -b/2a$. Correspondingly, when $a < 0$, the function $f(x) = ax^2 + bx + c$ takes its largest value when $x = -b/2a$. This largest or the smallest value of $f(x) = ax^2 + bx + c$ can be obtained by substituting $x = -b/2a$ in $f(x)$. ► **13**

Problems in which we are required to calculate the maximum and minimum values of certain functions arise very frequently in applications. We shall study them at some length in Chapter 14. However, some of these problems can be solved by making use of the properties of parabolas. The following examples belong to this category.

EXAMPLE 2 *(Fencing)* A farmer has 200 yards of fencing with which to enclose a rectangular field. One side of the field can make use of a fence that already exists. What is the maximum area that can be enclosed?

Solution Let the sides of the field be denoted by x and y, as shown in Figure 15, with the side labeled y parallel to the fence that already exists. Then the length of the new fence is $2x + y$, which must equal the 200 yards available.

$$2x + y = 200$$

The area enclosed is $A = xy$. But $y = 200 - 2x$, so

$$A = x(200 - 2x) = 200x - 2x^2. \tag{1}$$

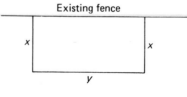

FIGURE 15

Comparing this with $f(x) = ax^2 + bx + c$, we see that A is a quadratic function of x, with $a = -2$, $b = 200$, and $c = 0$. Therefore, since $a < 0$, the quadratic function has a maximum value at the vertex, that is, when

$$x = -\frac{b}{2a} = -\frac{200}{2(-2)} = 50.$$

Answer If $a < 0$, $ax^2 + bx + c > 0$
for x between p and q and
$ax^2 + bx + c < 0$ outside this
interval.
If $ax^2 + bx + c$ is never zero for
any real x, then $ax^2 + bx + c < 0$
for all x.

The maximum value of A is obtained by substituting $x = 50$ in Equation (1).

$$A = 200(50) - 2(50^2) = 10,000 - 5,000 = 5,000$$

The maximum area that can be enclosed is 5000 square yards. The dimensions of this largest area are $x = 50$ yards and $y = 100$ yards. (*Note* $y = 200 - 2x$.)

☛ 14. In Example 3, express the profit P as a function of x instead of p and then find its maximum value. Is the answer the same?

EXAMPLE 3 *(Pricing Decision)* The demand per month, x, for a certain commodity at a price p dollars per unit is given by the relation

$$x = 1350 - 45p$$

The cost of labor and material to manufacture this commodity is $5 per unit and the fixed costs are $2000 per month. What price p per unit should be charged to the consumers to obtain a maximum monthly profit?

Solution The total cost C (in dollars) of producing x units per month is

$$C = \text{Variable Costs} + \text{Fixed Costs} = 5x + 2000.$$

The demand x is given to be

$$x = 1350 - 45p.$$

Using this value of x in C, we have

$$C = 5(1350 - 45p) + 2000 = 8750 - 225p.$$

The revenue R (in dollars) obtained by selling x units at p dollars per unit is

$$R = \text{Price per Unit} \times \text{Number of Units Sold}$$

$$= px = p(1350 - 45p) = 1350p - 45p^2.$$

The profit P (in dollars) is then given by the difference between revenue and cost.

$$P = R - C$$

$$= -45p^2 + 1350p - (8750 - 225p)$$

$$= -45p^2 + 1575p - 8750$$

The profit P is a quadratic function of p. Since $a = -45 < 0$, the graph is a parabola opening downward and the maximum profit is realized at the vertex. In this case, we have

$$a = -45, \qquad b = 1575, \quad \text{and} \quad c = -8750.$$

The vertex of the parabola is given by

$$p = -\frac{b}{2a} = -\frac{1575}{2(-45)} = \frac{1575}{90} = 17.5.$$

Thus a price of $p = \$17.50$ per unit must be charged to consumers to obtain a maximum profit. The maximum profit then will be

$$P = -45(17.5)^2 + 1575(17.5) - 8750 = 5031.25$$

or $5031.25 per month. ☛ **14**

Answer $P = x(30 - \frac{1}{45}x) - (5x + 2000) = -\frac{1}{45}x^2 + 25x - 2000.$
Vertex is at $x = 562.5$. This corresponds to $p = 17.5$ as in the given solution.

EXAMPLE 4 *(Rental-Charge Decision)* Mr. Woolhouse owns an apartment building with 60 suites. He can rent all the suites if he charges a monthly rent

► **15.** In Example 4, suppose that, due to a change in the rental market, all suites are rented when the rent is $240 but still one suite becomes vacant for each $5 increase in rent. What then is the optimum rent?

of $200 per suite. At a higher rent, some suites will remain vacant. On the average, for each increase in rent of $5, one suite remains vacant with no possibility of renting it. Determine the functional relationship between the total monthly revenue and the number of vacant units. What monthly rent will maximize the total revenue? What is this maximum revenue?

Solution Let x denote the number of vacant units. The number of rented apartments is then $60 - x$ and the monthly rent per suite is $(200 + 5x)$ dollars. If R denotes the total monthly revenue (in dollars), then

$$R = (\text{Rent per Unit})(\text{Number of Units Rented})$$
$$= (200 + 5x)(60 - x)$$
$$= -5x^2 + 100n + 12{,}000.$$

The total monthly revenue R is a quadratic function of x with

$$a = -5, \quad b = 100, \quad \text{and} \quad c = 12{,}000.$$

The graph of R is a parabola which opens downward (since $a < 0$) and has a vertex at the maximum point. The vertex is given by

$$x = -\frac{b}{2a} = -\frac{100}{2(-5)} = 10$$
$$R = -5(10)^2 + 100(10) + 12{,}000 = 12{,}500$$

Answer $R = (240 + 5x)(60 - x)$ has vertex when $x = 6$ and the optimum rent is $270.

Thus when 10 units are unoccupied, the revenue is greatest. The rent per suite is then $(200 + 5x)$ dollars, or $250, and the total revenue is $12,500 per month. ► **15**

EXERCISES 5-2

(1–6) Determine the vertices of the following parabolas.

1. $y = 2x^2 - 3$

2. $y = -1 - x^2$

3. $y = x^2 + 2x + 2$

4. $y = x^2 - 3x - 3$

5. $y = 2 - x - 2x^2$

6. $y = -2x - x^2$

(7–10) Sketch the following parabolas and determine their vertices.

7. $y = 2x^2 + 3x - 1$

8. $y = 4x - x^2$

9. $y = 3 - x - 3x^2$

10. $y = 4x^2 + 16x + 4$

(11–14) Determine the maximum or minimum value, as appropriate, of the following functions.

11. $f(x) = x^2 - 3x$

12. $f(x) = 2x - 5x^2$

13. $f(x) = 1 - x - x^2$

14. $f(x) = 3x^2 + x - 1$

15. *(Maximum Revenue)* The monthly revenue from selling x units of a certain commodity is given by $R(x) = 12x - 0.01x^2$ dollars. Determine the number of units that must be sold each month to maximize the revenue. What is the corresponding maximum revenue?

16. *(Maximum Profit)* The profit $P(x)$ obtained by manufacturing and selling x units of a certain product is given by

$$P(x) = 60x - x^2.$$

Determine the number of units that must be produced and sold to maximize the profit. What is the maximum profit?

17. *(Maximum Revenue and Profit)* A firm has a monthly fixed cost of $2000, and the variable cost per unit of its product is $25.

a. Determine the cost function.

b. The revenue R obtained by selling x units is given by $R(x) = 60x - 0.01x^2$. Determine the number of units that must be sold each month so as to maximize the revenue. What is this maximum revenue?

c. How many units must be produced and sold each month to obtain a maximum profit? What is this maximum profit?

18. *(Minimum Cost)* The average cost per unit (in dollars) of producing x units of a certain commodity is $C(x) = 20 - 0.06x + 0.0002x^2$. What number of units produced will minimize the average cost? What is the corresponding minimum cost per unit?

19. *(Fencing)* A farmer has 500 yards of fencing with which to enclose a rectangular paddock. What is the largest area that can be enclosed?

20. *(Planting Decision)* The yield of apples from each tree in an orchard is $(500 - 5x)$ pounds, where x is the density with which the trees are planted (that is, the number of trees per acre). Find the value of x that makes the total yield per acre a maximum.

21. *(Agriculture)* If rice plants are sown at a density of x plants per square foot, the yield of rice from a certain location is $x(10 - 0.5x)$ bushels per acre. What value of x maximizes the yield per acre?

22. *(Planting Decision)* If apple trees are planted at 30 per acre, the value of the crop produced by each tree is $180. For each additional tree planted per acre, the value of the

crop falls by $3. What number of trees must be planted per acre to obtain the maximum value of the crop? What is this maximum value per acre of the crop?

23. *(Book Pricing)* If a publisher prices a book at $20 per book, 10,000 copies can be sold. For every dollar increase in price, sales fall by 400 copies. What should be charged per book to obtain the maximum revenue? What is the value of this maximum revenue?

24. *(Pricing Decision)* In Exercise 23, the cost of producing each copy is $13. What price should the publisher charge for each book to gain a maximum profit?

25. *(Rental-Charge Decision)* Chou-ching Realty has built a new rental unit of 40 apartments. It is known from market research that if a rent of $150 per month is charged, all the suites will be occupied. For each $5 increase in rent, one unit will remain vacant. What monthly rent should be charged for each unit to obtain the maximum monthly revenue? Find this maximum revenue.

26. *(Pricing Decision)* The market demand for a certain product is x units when the price charged to consumers is p dollars, where
$$15p + 2x = 720.$$

The cost (in dollars) of producing x units is given by $C(x) = 200 + 6x$. What price p per unit should be charged to consumers to obtain a maximum profit?

27. Show that the vertex of the parabola whose equation is $y = a(x - h)^2 + k$ is at the point (h, k).

◥ 5-3 MORE SIMPLE FUNCTIONS AND GRAPHS

> ☛ **16.** What is the graph of f in the case $n = 1$?

In this section, we shall discuss a few more simple functions of common use and interest.

Power Functions

A function of the form

$$f(x) = ax^n$$

where a and n are nonzero constants, is called a **power function.** We shall consider some special cases of functions of this type.

1. $n = 2$: In this case $f(x) = ax^2$, and we have a special case of the quadratic functions discussed in Section 2. The graph of $y = ax^2$ is a parabola with vertex at the origin, opening upward if $a > 0$ and downward if $a < 0$. ☛ **16**

Answer A straight line through the origin with slope a.

2. $n = \frac{1}{2}$: In this case, $f(x) = ax^{1/2} = a\sqrt{x}$. The graph of this function is *one-half of a parabola* that opens toward the right. If $a > 0$, the graph is the upper half of the parabola, while if $a < 0$, it is the lower half. Thus the graph rises or falls to the right depending on whether $a > 0$, or $a < 0$. The domain of f is the set of all nonnegative real numbers. (See Figure 16.)

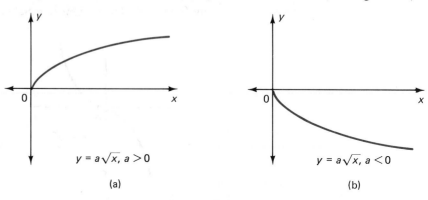

$y = a\sqrt{x}, \, a > 0$

(a)

$y = a\sqrt{x}, \, a < 0$

(b)

FIGURE 16

3. $n = -1$: In this case $f(x) = a/x$. The domain of $f(x)$ consists of all real numbers except zero. Figure 17 shows the graphs of $y = 1/x$ and $y = -1/x$ (that is, corresponding to $a = \pm1$). The graph of $y = a/x$ for $a > 0$ is similar in form to that of $y = 1/x$ and that for $a < 0$ is similar in form to the graph of $y = -1/x$. The graph of $y = a/x$ is called a **rectangular hyperbola**. As x moves closer and closer to zero, the denominator in $f(x) = a/x$ becomes very small, so $f(x)$ becomes numerically very large. It may become large and positive or large and negative, depending on the signs of a and x. These possibilities are clear from Figure 17. The y-axis is said to be a **vertical asymptote** to the graph.

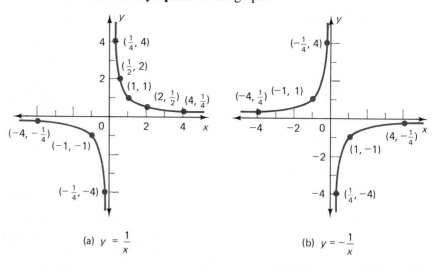

(a) $y = \dfrac{1}{x}$

(b) $y = -\dfrac{1}{x}$

FIGURE 17

☛ **17.** Draw the graphs of
$f(x) = x^{-2}$ and $f(x) = x^{-1/2}$,
corresponding to $n = -2$ and
$n = -\frac{1}{2}$ respectively and to $a = 1$.

It can be seen from the graph that as x becomes very large (positive or negative), $f(x)$ becomes closer and closer to zero; however, it is never quite equal to zero. The x-axis is said to be a **horizontal asymptote** to the graph. ☛ **17**

4. $n = 3$: In this case, $f(x) = ax^3$. The graph of $f(x)$ is the cubic curve shown in Figure 18. The domain is equal to the set of all real numbers.

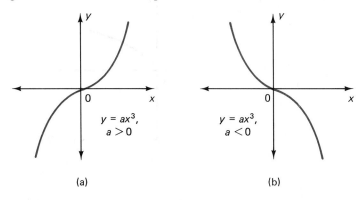

$y = ax^3$, $a > 0$	$y = ax^3$, $a < 0$
(a)	(b)

FIGURE 18

Figure 19 provides a comparison of the graphs of the function $y = ax^n$ for various values of n. The case $a > 0$ is shown, and the graphs are drawn only for the quadrant in which x and y are nonnegative. (In business and economic applications, we are commonly concerned with variables that take only nonnegative values.)

Answer

x	± 3	± 2	± 1	$\pm \frac{1}{2}$	$\pm \frac{1}{3}$
$y = x^{-2}$	$\frac{1}{9}$	$\frac{1}{4}$	1	4	9
x	9	4	1	$\frac{1}{4}$	$\frac{1}{9}$
$y = x^{-1/2}$	$\frac{1}{3}$	$\frac{1}{2}$	1	2	3

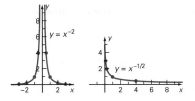

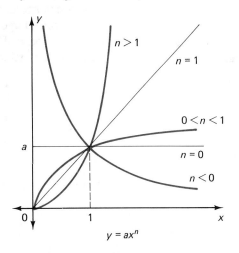

$y = ax^n$

FIGURE 19

We see that all the graphs pass through the point $(1, a)$. When $n > 1$, the graph rises as we move to the right and, moreover, rises more and more steeply as x increases. The functions $y = ax^2$ and $y = ax^3$ encountered previously are examples that fall into this category. The case $n = 1$ corresponds to the straight line $y = ax$ passing through the origin and the point $(1, a)$.

When $0 < n < 1$, the graph of $y = ax^n$ still rises as we move to the right, but it rises less steeply as x increases. The function $y = ax^{1/2}$, which, as we saw before, has a graph that is half of a parabola, is an example of this type.

The case $n = 0$ corresponds to a horizontal straight line. Finally, when $n < 0$, the function $y = ax^n$ has a graph that falls as we move to the right and is asymptotic to the x- and y-axes. The rectangular hyperbola, with equation $y = ax^{-1}$, is an example of such a graph.

EXAMPLE 1 *(Revenue)* A firm has a total revenue of $500 per day regardless of the price it charges for its product. Find the demand relation and graph the demand curve.

Solution Let p denote the price (in dollars) per unit of the product and x is the number of units that can be sold at the price p. Then to obtain $500, we must have

$$500 = \text{Price per Unit} \times \text{Number of Units Sold} = px.$$

That is,

$$p = \frac{500}{x}.$$

TABLE 5

x	25	50	100	125	250	500
p	20	10	5	4	2	1

Selected values of x and $p = 500/x$ are given in Table 5. Plotting these points and joining them by a smooth curve, we obtain the curve shown in Figure 20. We have restricted the graph to the first quadrant because neither the price nor the quantity sold can be negative. The graph is one-half of a rectangular hyperbola.

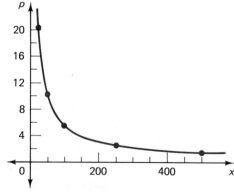

FIGURE 20

Circles

A **circle** is the set of all points that lie at a constant distance (called the **radius**) from a given point (called the **center**).

Let us find the equation of the circle with center at the point (h, k) and radius r. (See Figure 21.) Let (x, y) be any point on the circle. Then the distance between this point (x, y) and the center (h, k) is given by the distance formula to be

$$\sqrt{(x - h)^2 + (y - k)^2}.$$

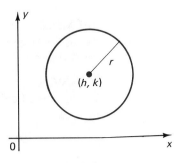

FIGURE 21

Setting this equal to the given radius r, we obtain the equation

$$\sqrt{(x - h)^2 + (y - k)^2} = r$$

which, on squaring, gives the following equation.

$$(x - h)^2 + (y - k)^2 = r^2 \qquad (1)$$

This is called the **center-radius** form of the equation of the circle. In particular, if the center is at the origin, $h = k = 0$ and Equation (1) reduces to

$$x^2 + y^2 = r^2 \qquad (2)$$

EXAMPLE 2 Find the equation of a circle with center at $(2, -3)$ and radius 5.

Solution Here $h = 2$, $k = -3$, and $r = 5$. Using the standard equation of a circle, we have

$$(x - 2)^2 + (y - (-3))^2 = 5^2 \qquad \text{or} \qquad (x - 2)^2 + (y + 3)^2 = 25.$$

We expand the squares and simplify.

$$x^2 + y^2 - 4x + 6y - 12 = 0 \quad ☛ \, 18$$

Answer (a) $x^2 + y^2 + 6x - 8y = 0$;
(b) $x^2 + y^2 - 2x + 4y - 13 = 0$.
(Radius
$= \sqrt{[1 - (-2)]^2 + [(-2) - 1]^2}$
$= \sqrt{18}$.)

Equation (1), when expanded and simplified, can be written as

$$x^2 + y^2 - 2hx - 2ky + (h^2 + k^2 - r^2) = 0.$$

This is of the form

$$x^2 + y^2 + Bx + Cy + D = 0 \qquad (3)$$

where B, C, and D are constants given by

$$B = -2h, \qquad C = -2k, \qquad \text{and} \qquad D = h^2 + k^2 - r^2.$$

Equation (3) is called the **general form** of the equation of a circle.

Given any equation in general form, we can readily determine the center and radius of the circle that it represents. For

$$h = -\frac{B}{2} \qquad \text{and} \qquad k = -\frac{C}{2}$$

gives the coordinates of the center immediately. The radius is then obtained from the equation $r^2 = h^2 + k^2 - D$. However, rather than trying to remember these formulas, it is easier to use the method of completing the square as in the following example.

EXAMPLE 3 Determine whether the graph of

$$2x^2 + 2y^2 - 5x + 4y - 1 = 0$$

is a circle. If the graph is a circle, find its center and radius.

Solution Dividing the equation throughout by 2 (to make the coefficients of x^2 and y^2 equal to 1), we get

$$x^2 + y^2 - \tfrac{5}{2}x + 2y - \tfrac{1}{2} = 0$$

Comparing this with Equation (3), we see that we do have an equation of the correct general form.

We now group the x-terms all together and the y-terms all together and move the constant to the right side:

$$(x^2 - \tfrac{5}{2}x \quad) + (y^2 + 2y \quad) = \tfrac{1}{2}.$$

We wish to arrange this in the same form as Equation (1), which involves completing the square inside each of the two parentheses. We take half the coefficient of x, $-\tfrac{5}{4}$, square it, $\tfrac{25}{16}$, and add this into the first parentheses. Then half of the coefficient of y, 1, square it, 1, and add this into the second parentheses. (See Section 2-3.) The same terms must of course be added to the right side also. We get

$$(x^2 - \tfrac{5}{2}x + \tfrac{25}{16}) + (y^2 + 2y + 1) = \tfrac{1}{2} + \tfrac{25}{16} + 1 = \tfrac{49}{16}.$$

The two parentheses are now perfect squares:

$$(x - \tfrac{5}{4})^2 + (y + 1)^2 = \tfrac{49}{16}.$$

Comparing with Equation (1) we see that $h = \tfrac{5}{4}$, $k = -1$, and $r^2 = \tfrac{49}{16}$, or $r = \tfrac{7}{4}$. The center of the circle is $(\tfrac{5}{4}, -1)$ and the radius is $\tfrac{7}{4}$ units.

Note If, after completing the square we find a negative value for r^2 on the right side, the graph contains no points. ☞ **19**

It sometimes happens that a business firm has a choice between two (or more) ways of using certain of its resources to form different end products. Resources such as the available raw materials, plant and machinery, or labor may, in certain instances, be directed towards the production of several different items, and the company may choose how much of each to produce. For example, a shoe manufacturer can produce either men's or women's shoes from the same resources, an oil refinery can choose a variety of different grades of oil and gasoline to make from its crude oil, and so on.

In general, these different products compete for the use of the available resources—that is, an increase in the amount of one product must be accompanied by a decrease in the amounts of others. These various amounts are related by an equation. When there are only two products involved, this equation can be graphed, and its graph is called the **product transformation curve.**

EXAMPLE 4 *(Product Transformation Curve)* A shoe manufacturing firm can produce men's and women's shoes by varying the production process. The possible amounts x and y (in hundreds of pairs) are related by the equation

$$x^2 + y^2 + 40x + 30y = 975.$$

Plot the product transformation curve for this firm.

Solution The given equation is of the general form of Equation (3) and hence has a circle as its graph. The coefficients are

$$B = 40, \quad C = 30, \quad \text{and} \quad D = -975.$$

The coordinates of the center of the circle are

$$h = -\frac{B}{2} = -\frac{40}{2} = -20 \quad \text{and} \quad k = -\frac{C}{2} = -\frac{30}{2} = -15$$

so the center is the point $(-20, -15)$. The radius is

$$r = \tfrac{1}{2}\sqrt{B^2 + C^2 - 4D} = \tfrac{1}{2}\sqrt{(40)^2 + (30)^2 - 4(-975)} = 40.$$

Figure 22 shows the product transformation curve. Note that since x and y

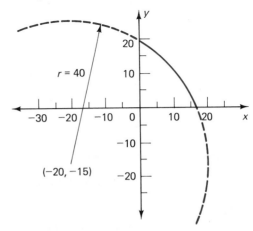

FIGURE 22

must in practice be both nonnegative, only the portion of the curve in the first quadrant is of practical significance. The rest of the curve is shown as a dashed curve in Figure 22.

The set of points (x, y) that satisfy the relation $x^2 + y^2 = a^2$ consists of the points on the circle whose center is the origin and whose radius is a. We may speak of this circle as being the graph of the relation $x^2 + y^2 = a^2$. (See Figure 23.)

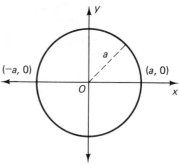

FIGURE 23

By the vertical line test, it is clear that a circle cannot be a graph of a function: for any value of x lying between $-a$ and a (excluding $x = \pm a$), there are two values of y. We can see this algebraically by solving the equation $x^2 + y^2 = a^2$ for y:

$$y = \pm\sqrt{a^2 - x^2}.$$

This shows that there are two values of y for each x, depending on whether we choose the positive or the negative square root.

In fact, the complete circle represents two functions. The upper semicircle is the graph of the function $y = +\sqrt{a^2 - x^2}$. in which the positive square root is taken for y; the lower semicircle is the graph of the function $y = -\sqrt{a^2 - x^2}$, in which the negative square root is taken. (See Figure 24.) ☛ **20**

☛ **20.** Determine the two functions that describe the upper and lower semicircles of (a) the circle with center $(0, 2)$ and radius 1; (b) the circle with center $(2, -1)$ and radius 3.

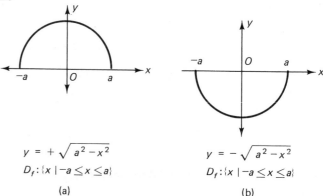

$$y = +\sqrt{a^2 - x^2}$$
$$D_f: \{x \mid -a \le x \le a\}$$

(a)

$$y = -\sqrt{a^2 - x^2}$$
$$D_f: \{x \mid -a \le x \le a\}$$

(b)

FIGURE 24

Answer (a) $y = 2 \pm \sqrt{1 - x^2}$;
(b) $y = -1 \pm \sqrt{9 - (x - 2)^2}$.

☞ **21.** How are the graphs of the following functions related to Figure 26?

(a) $y = |2 - x|$;
(b) $y = |2 - x| - 2$.

Absolute-Value Functions

If x is a real number, the absolute value of x, denoted by $|x|$, is defined as

$$|x| = \begin{cases} x & \text{if } x \geq 0 \\ -x & \text{if } x < 0. \end{cases}$$

Clearly, $|x| \geq 0$; that is, *the absolute value of a real number is always non-negative*.

We call $f(x) = |x|$ the **absolute-value function.** The domain of f is the set of all real numbers and the range is the set of all nonnegative real numbers. The graph of $y = |x|$ is shown in Figure 25.

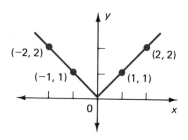

FIGURE 25

EXAMPLE 5 Consider the function

$$f(x) = |x - 2|.$$

The domain of f is the set of all real numbers and the range is the set of all nonnegative real numbers. Let us draw the graph of $f(x)$.

Setting $y = f(x)$, we have

$$y = |x - 2|$$

or, using the above definition of absolute value,

$$y = x - 2 \quad \text{if } x - 2 \geq 0 \quad \text{(that is, if } x \geq 2)$$

and

$$y = -(x - 2) \quad \text{if } x - 2 < 0 \quad \text{(that is, if } x < 2).$$

Therefore the graph of $f(x)$ consists of portions of the two straight lines

$$y = x - 2 \quad \text{and} \quad y = -(x - 2) = 2 - x$$

for $x \geq 2$ and $x < 2$, respectively. The graph is as shown in Figure 26. Note that $y \geq 0$ for all x. ☞ **21**

EXAMPLE 6 Consider the function

Answer (a) Same as Figure 26; (b) graph in Figure 26 must be moved down 2 units.

$$f(x) = \frac{|x|}{x}.$$

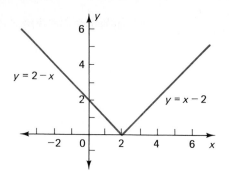

FIGURE 26

Clearly, the function is not defined for $x = 0$, since for this value of x the denominator becomes zero. Thus the domain of f is the set of all real numbers except zero.

$$\text{If } x > 0, \quad f(x) = \frac{|x|}{x} = \frac{x}{x} = 1.$$

$$\text{If } x < 0, \quad f(x) = \frac{|x|}{x} = \frac{-x}{x} = -1.$$

For example,

$$f(-3) = \frac{|-3|}{(-3)} = \frac{3}{(-3)} = -1.$$

Thus the range consists of only two numbers, 1 and -1.

The graph of f consists of two straight lines (one above and one below the x-axis) that are parallel to the x-axis and at a distance 1 from it. This is shown in Figure 27. Note that the endpoints where the two lines meet the y-axis are *not* included in the graph. This is indicated by drawing small, open circles at the ends of the two lines.

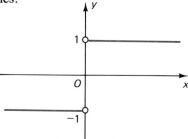

FIGURE 27

EXAMPLE 7 Sketch the graph of the function f where

$$f(x) = 4 - 2|3 - x|$$

and hence find the maximum value of f.

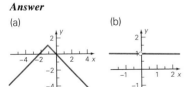

22. Sketch the graphs of

(a) $y = 1 - |x - 1|$;

(b) $y = \left(\dfrac{|x|}{x}\right)^2$.

Solution For $3 - x \geq 0$, that is, for $x \leq 3$, $|3 - x| = 3 - x$ and so

$$f(x) = 4 - 2(3 - x) = 2x - 2.$$

For $3 - x < 0$, that is, for $x > 3$, $|3 - x| = -(3 - x)$ and so

$$f(x) = 4 - 2[-(3 - x)] = 10 - 2x.$$

So the graph of f consists of two half-lines, as shown in Figure 28. The maximum value of f is 4, occurring for $x = 3$. ☞ **22**

Answer

(a)

(b)

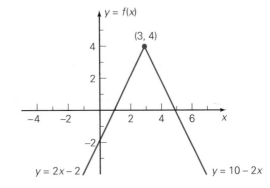

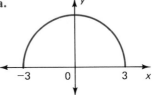

FIGURE 28

EXERCISES 5-3

(1–14) Find the domains of the following functions and sketch their graphs.

1. $f(x) = \sqrt{4 - x^2}$ 2. $f(x) = 2 - \sqrt{9 - x^2}$

3. $g(x) = -\sqrt{3 - x}$ 4. $f(x) = \sqrt{x - 2}$

5. $f(x) = \dfrac{1}{x}$ 6. $f(x) = \dfrac{-3}{x - 2}$

7. $f(x) = x^3$ 8. $f(x) = 1 - x^3$

9. $f(x) = 2 - |x|$ 10. $g(x) = |x| + 3$

11. $f(x) = |x + 3|$ 12. $F(x) = -|x - 2|$

13. $f(x) = \dfrac{|x - 3|}{x - 3}$ 14. $G(x) = \dfrac{2 - x}{|x - 2|}$

15. Which of the following half-circles represent the graphs of functions? In each case where the answer is positive, determine the functional equation for the function from the graph.

a.

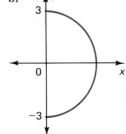

b.

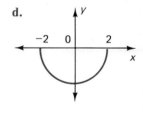

c.

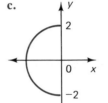

d.

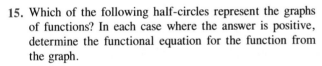

(16–23) Find the equation of each circle.

16. Center $(0, 2)$ and radius 5.

17. Center $(2, 5)$ and radius 3.

18. Center $(-3, 0)$ and radius 4.

19. Center $(0, 0)$ and radius 7.

20. Center $(1, -3)$ and passes through the point $(2, -1)$.

21. Center $(-2, 1)$ and passes through the point $(0, 4)$.

22. Center $(2, 2)$ and touches both coordinate axes.

23. Touches both coordinate axes, lies in the second quadrant, and is of radius 3 units.

(24–29) Determine whether each of the following equations represents a circle. If it does, find its center and radius.

24. $x^2 + y^2 + 2x + 2y + 1 = 0$

25. $x^2 + y^2 - 4x - 8y + 4 = 0$

26. $x^2 + y^2 + 3x - 5y + 1 = 0$

27. $2x^2 + 2y^2 - 5x + 4y - 1 = 0$

28. $3x^2 + 3y^2 - 2x + 4y - (\frac{11}{9}) = 0$

29. $x^2 + y^2 + 4x + 8y + 25 = 0$

30. *(Demand Curve)* At a price of p dollars per unit, a manufacturer can sell x units of his product, where x and p are related by

$$x^2 + p^2 + 400x + 300p = 60,000.$$

Plot the demand curve. What is the highest price above which no sales are possible?

31. *(Demand Relation)* A car dealer can sell x cars of a particular model when he charges p dollars per car, where

$$x^2 + p^2 + 4000x + 2500p = 19,437,500.$$

Plot the demand curve for this model of car. What is the highest price up to which sales are still possible?

32. *(Product Transformation Curve)* The owner of an apple orchard can produce either apples or apple cider. The possible amounts x of apples (in kilograms) and y of apple cider (in liters) are related by the equation

$$x^2 + y^2 + 8x + 250y = 6859.$$

Plot the graph of this relation (called the *product transformation curve*) and determine the maximum amounts of apples or apple cider that can be produced.

33. *(Product Transformation Curve)* The Coronado cycle industries manufactures two types of bicycles called *Coronado* and *Eastern Star*. The possible quantities of x and y (in thousands) that can be produced per year are related by

$$x^2 + y^2 + 6x + 10y = 47.$$

Sketch the product transformation curve for this industry. What are the maximum numbers of cycles of each type that can be produced?

34. *(Minimum Distance)* An airplane flies a distance of 1000 miles over the ocean, its route passing over two islands, one after 200 miles and the other after 700 miles. If x is the distance along the route to a given point $(0 \le x \le 1000)$, determine the function $f(x)$ that is equal to the distance of that point from the nearest land. Sketch its graph.

35. *(Minimum Distance)* In the previous exercise the function $g(x)$ is equal to the distance of the point x from the nearest land *ahead* of the airplane. Write algebraic expressions for $g(x)$.

(36–45) Determine the maximum and/or minimum values of the following functions if they exist, and the values of x at which they occur. (*Hint*: In each case consider the graph of the function.)

36. $f(x) = 2 - |x + 1|$

37. $f(x) = |2x + 1| + 2$

38. $f(x) = x - |x|$

39. $f(x) = \dfrac{|x - 5|}{5 - x}$

40. $f(x) = 1 + \sqrt{4 - x^2}$

41. $f(x) = \sqrt{1 - 9x^2} - 2$

42. $f(x) = 1 - \sqrt{9 - x^2}$

43. $f(x) = \sqrt{16 - x^2} - 3$

44. $f(x) = 2 - \sqrt{3 - 2x}$

45. $f(x) = 2\sqrt{1 - x} + 1$

◣ 5-4 COMBINATIONS OF FUNCTIONS

A variety of situations arise in which we have to combine two or more functions in one of several ways to get new functions. For example, let $f(t)$ and $g(t)$ denote the incomes of a person from two different sources at time t; then the combined income from the two sources is $f(t) + g(t)$. From the two functions f and g, we have in this way obtained a third function, the *sum* of f and g. If $C(x)$ denotes the cost of producing x units of a certain commodity and $R(x)$ is the revenue obtained from the sale of x units, then the profit $P(x)$ obtained by producing and selling x units is given by $P(x) = R(x) - C(x)$. The new function P so obtained is the *difference* of the two functions R and C.

If $P(t)$ denotes population of a country and $I(t)$ is the per capita income at time t, then the national income of that country is given by $P(t)I(t)$. This is an example in which a new function is formed as the *product* of two functions. Correspondingly, we can define the *quotient* of two functions. Let $F(t)$ be the total daily supply of protein available in a certain country at time t and $P(t)$ the population. Then the average per capita supply of protein per day is $F(t)/P(t)$.

These kinds of examples lead us to the following abstract definitions.

DEFINITION Given two functions f and g, the **sum, difference, product,** and **quotient** functions are defined as follows.

Sum:	$(f + g)(x) = f(x) + g(x)$
Difference:	$(f - g)(x) = f(x) - g(x)$
Product:	$(f \cdot g)(x) = f(x) \cdot g(x)$
Quotient:	$\left(\dfrac{f}{g}\right)(x) = \dfrac{f(x)}{g(x)}, \quad$ provided $g(x) \neq 0$.

The domains of the sum, difference, and product fuunctions are all equal to the common part of the domains of f and g, that is, the set of x at which both f and g are defined. In the case of the quotient function, the domain is the common part of the domains of f and g except for those values of x for which $g(x) = 0$.

EXAMPLE 1 Let $f(x) = 1/(x - 1)$ and $g(x) = \sqrt{x}$. Find $f + g$, $f - g$, $f \cdot g$, f/g, and g/f. Determine the domain in each case.

Solution We have:

$$(f + g)(x) = f(x) + g(x) = \frac{1}{x - 1} + \sqrt{x}$$

$$(f - g)(x) = f(x) - g(x) = \frac{1}{x - 1} - \sqrt{x}$$

$$(f \cdot g)(x) = f(x) \cdot g(x) = \frac{1}{x - 1} \cdot \sqrt{x} = \frac{\sqrt{x}}{x - 1}$$

$$\left(\frac{f}{g}\right)(x) = \frac{f(x)}{g(x)} = \frac{1/(x - 1)}{\sqrt{x}} = \frac{1}{\sqrt{x}(x - 1)}$$

$$\left(\frac{g}{f}\right)(x) = \frac{g(x)}{f(x)} = \frac{\sqrt{x}}{1/(x - 1)} = \sqrt{x}(x - 1).$$

Since its denominator becomes zero when $x = 1$, $f(x)$ is not defined for $x = 1$, so the domain of f is the set of all real numbers except 1.

Similarly, $g(x)$ is defined for the values of x for which the expression under the radical sign is nonnegative, that is, for $x \geq 0$. Thus

$$D_f = \{x \mid x \neq 1\} \quad \text{and} \quad D_g = \{x \mid x \geq 0\}.$$

The common part of D_f and D_g is

$$\{x \mid x \geq 0 \quad \text{and} \quad x \neq 1\} \tag{1}$$

This set provides the domain of $f + g, f - g$ and $f \cdot g$.

Since $g(x) = \sqrt{x}$ is zero when $x = 0$, this point must be excluded from the domain of f/g. Thus the domain of f/g is

$$\{x \mid x > 0 \quad \text{and} \quad x \neq 1\}.$$

Since $f(x)$ is never zero, the domain of g/f is again the common part of D_f and D_g, namely Set (1). It appears from the algebraic formula for $(g/f)(x)$ that this function is well-defined when $x = 1$. In spite of this, it is still necessary to exclude $x = 1$ from the domain of this function, as g/f is defined only at points where both g and f are defined. ☛ 23

23. Given $f(x) = \sqrt{1 - x}$ and $g(x) = \sqrt{1 - x^2}$, write down expressions for the values of

(a) $f \cdot g$; (b) $\dfrac{f}{g}$; (c) $\dfrac{g}{f}$.

In each case give the domain.

Another way in which two functions can be combined to yield a third function is called the *composition* of functions. Consider the following situation.

The monthly revenue R of a firm depends on the number x of the units that it produces and sells. In general, we can say $R = f(x)$. Usually, the number x of units it can sell depends on the price p per unit it charges the customers, so that $x = g(p)$. If we eliminate x from the two relations $R = f(x)$ and $x = g(p)$, we have

$$R = f(x) = f(g(p)).$$

This gives R as a function of the price p. Observe how R is obtained as a function of p by using the function $g(p)$ as the argument of the function f. This leads to the following definition.

DEFINITION Let f and g be two functions. Let x belong to the domain of g and be such that $g(x)$ belongs to the domain of f. Then the **composite function** $f \circ g$ (read f circle g) is defined by

$$(f \circ g)(x) = f(g(x)).$$

EXAMPLE 2 Let $f(x) = 1/(x - 2)$ and $g(x) = \sqrt{x}$. Evaluate

(a) $(f \circ g)(9)$;   (b) $(f \circ g)(4)$;   (c) $(f \circ g)(x)$;
(d) $(g \circ f)(6)$;   (e) $(g \circ f)(1)$;   (f) $(g \circ f)(x)$.

Solution

(a) $g(9) = \sqrt{9} = 3$. Therefore

$$(f \circ g)(9) = f(g(9)) = f(3) = 1/(3 - 2) = 1.$$

Answer

(a) $(f \cdot g)(x) = (1 - x)\sqrt{1 + x}$, domain $= \{x \mid -1 \leq x \leq 1\}$;
(b) $(f/g)(x) = 1/\sqrt{1 + x}$, domain $= \{x \mid -1 < x < 1\}$;
(c) $(g/f)(x) = \sqrt{1 + x}$, domain $= \{x \mid -1 \leq x < 1\}$.

(b) $g(4) = \sqrt{4} = 2$. We have
$$(f \circ g)(4) = f(g(4)) = f(2) = 1/(2 - 2).$$

This is not defined. The value $x = 4$ does not belong to the domain of $f \circ g$, so $(f \circ g)(4)$ cannot be found.

(c) $g(x) = \sqrt{x}$
$$(f \circ g)(x) = f(g(x)) = f(\sqrt{x}) = \frac{1}{\sqrt{x} - 2}.$$

(d) $f(6) = 1/(6 - 2) = \frac{1}{4}$;
$$(g \circ f)(6) = g(f(6)) = g(\tfrac{1}{4}) = \sqrt{\tfrac{1}{4}} = \tfrac{1}{2}$$

(e) $f(1) = 1/(1 - 2) = -1$;
$$(g \circ f)(1) = g(f(1)) = g(-1) = \sqrt{-1}$$

which is not a real number. We cannot evaluate $(g \circ f)(1)$ as 1 does not belong to the domain of $g \circ f$.

(f) $f(x) = 1/(x - 2)$
$$(g \circ f)(x) = g(f(x)) = g\left(\frac{1}{x - 2}\right) = \sqrt{\frac{1}{x - 2}} = \frac{1}{\sqrt{x - 2}}.$$

The domain of $f \circ g$ is given by
$$D_{f \circ g} = \{x \mid x \in D_g \quad \text{and} \quad g(x) \in D_f\}.$$

It can be shown that, for the functions in Example 2,
$$D_{f \circ g} = \{x \mid x \geq 0 \quad \text{and} \quad x \neq 4\}$$

and

$$D_{g \circ f} = \{x \mid x > 2\}. \quad \text{☛ 24}$$

☛ **24.** Write down expressions for $f \circ g(x)$ and $g \circ f(x)$ in the following cases. In each case give the domain of the compositions.
(a) $f(x) = \sqrt{1 - x}$ and $g(x) = x + 1$;
(b) $f(x) = x^{-2}$ and $g(x) = \sqrt{x - 1}$.

EXAMPLE 3 *(Revenue)* The monthly revenue R obtained by selling deluxe model shoes is a function of the demand x in the market. It is observed that, as a function of price p per pair, the monthly revenue and demand are
$$R = 300p - 2p^2 \quad \text{and} \quad x = 300 - 2p.$$

How does R depend on x?

Answer (a) $(f \circ g)(x) = \sqrt{-x}$, domain $x \leq 0$;
$(g \circ f)(x) = \sqrt{1 - x} + 1$, domain $x \leq 1$; (b) $(f \circ g)(x) = (x - 1)^{-1}$, domain $x > 1$;
$(g \circ f)(x) = \sqrt{x^{-2} - 1}$, domain $-1 \leq x \leq 1$, $x \neq 0$.

Solution If $R = f(p)$ and $p = g(x)$, then R is obtained as a function of x by means of the composition $R = (f \circ g)(x) = f(g(x))$. The function $f(p)$ is given by $R = f(p) = 300p - 2p^2$. However, in order to obtain $g(x)$, we must solve the demand relation $x = 300 - 2p$ to express p as a function of x. We get
$$p = \tfrac{1}{2}(300 - x).$$

We substitute this value of p in R and simplify.

$$R = 300p - 2p^2$$
$$= 300 \cdot \tfrac{1}{2}(300 - x) - 2 \cdot \tfrac{1}{4}(300 - x)^2$$
$$= (150)(300) - 150x - \tfrac{1}{2}(300^2 - 600x + x^2)$$
$$= 150x - 0.5x^2$$

This is the required result, expressing the monthly revenue R as a function of the demand x in the market.

EXERCISES 5-4

(1–5) Find the sum, difference, product, and quotient of the two functions f and g in each of the following exercises. Determine the domains of the resulting functions.

1. $f(x) = x^2$; $g(x) = \dfrac{1}{x - 1}$

2. $f(x) = x^2 + 1$; $g(x) = \sqrt{x}$

3. $f(x) = \sqrt{x - 1}$; $g(x) = \dfrac{1}{x + 2}$

4. $f(x) = 1 + \sqrt{x}$; $g(x) = \dfrac{2x + 1}{x + 2}$

5. $f(x) = (x + 1)^2$; $g(x) = \dfrac{1}{x^2 - 1}$

(6–13) Given $f(x) = x^2$ and $g(x) = \sqrt{x - 1}$, evaluate each of the following.

6. $(f \circ g)(5)$ **7.** $(g \circ f)(3)$

8. $(f \circ g)(\tfrac{5}{4})$ **9.** $(g \circ f)(-2)$

10. $(f \circ g)(\tfrac{1}{2})$ **11.** $(g \circ f)(\tfrac{1}{3})$

12. $(f \circ g)(2)$ **13.** $(g \circ f)(1)$

(14–21) If $f(x) = 1/(2x + 1)$ and $g(x) = -\sqrt{x}$, evaluate each of the following.

14. $(f \circ g)(1)$ **15.** $(f \circ g)(\tfrac{1}{4})$

16. $(f \circ g)(-1)$ **17.** $(f \circ g)(4)$

18. $(g \circ f)(0)$ **19.** $(g \circ f)(\tfrac{3}{2})$

20. $(g \circ f)(-1)$ **21.** $(g \circ f)(-\tfrac{1}{2})$

(22–32) Determine $(f \circ g)(x)$ and $(g \circ f)(x)$ in the following exercises.

22. $f(x) = x^2$; $g(x) = 1 + x$

23. $f(x) = \sqrt{x} + 1$; $g(x) = x^2$

24. $f(x) = \dfrac{1}{x + 1}$; $g(x) = \sqrt{x} + 1$

25. $f(x) = 2 + \sqrt{x}$; $g(x) = (x - 2)^2$

26. $f(x) = x^2 + 2$; $g(x) = x - 3$

27. $f(x) = \sqrt{x}$, $g(x) = x^2$

28. $f(x) = |x|$; $g(x) = x^2$

29. $f(x) = x - 1$; $g(x) = x^{-1}$

30. $f(x) = 3x - 1$; $g(x) = \dfrac{x + 1}{3}$

31. $f(x) = 3$; $g(x) = 7$

32. $f(x) = 4$; $g(x) = x^2$

(33–36) Find $f(x)$ and $g(x)$ such that each composite function $f \circ g$ is as described. (The answer is not unique. Choose f and g to be as simple as you can.)

33. $(f \circ g)(x) = (x^2 + 1)^3$

34. $(f \circ g)(x) = \sqrt{2x + 3}$

35. $(f \circ g)(x) = \dfrac{1}{x^2 + 7}$

36. $(f \circ g)(x) = \dfrac{1}{\sqrt{x^2 - 5}}$

37. *(Revenue Function)* The demand x for a certain commodity is given by $x = 2000 - 15p$, where p is the price per unit of the commodity. The monthly revenue R obtained from the sales of this commodity is given by $R = 2000p - 15p^2$. How does R depend on x?

38. *(Revenue Function)* A manufacturer can sell q units of a product at a price p per unit, where $20p + 3q = 600$. As a function of quantity q demanded in the market, the total weekly revenue R is given by $R = 30q - 0.15q^2$. How does R depend on the price p?

39. *(Chemical Reaction)* The rate at which a chemical is produced in a certain reaction depends on temperature T according to the formula $R = T^5 + 3\sqrt{T}$. If T varies with time t according to $T = 3(t + 1)$, express R as a function of t and evaluate R when $t = 2$.

40. *(Physics)* The velocity of a falling body varies with distance traveled according to the formula $v = 8\sqrt{y}$ (v = velocity in feet per second, y = distance in feet). The distance fallen varies with time t (in seconds) according to the formula $y = 16t^2$. Express v as a function of t.

41. If $f(x) = ax - 4$ and $g(x) = bx + 3$, find the condition on a and b such that $(f \circ g)(x) = (g \circ f)(x)$ for all x.

42. If $f(x) = x + a$ and $g(t) = t + b$, show that $(f \circ g)(x) = (g \circ f)(x)$.

43. *(Housing Starts)* The number of housing starts per year, N, depends on the mortgage interest rate r percent according to the formula

$$N(r) = \frac{50}{100 + r^2}$$

where N is in millions. The interest rate is currently at 12% and is predicted to decrease to 8% over the next 2 years according to the formula

$$r(t) = 12 - \frac{8t}{t + 24}$$

where t is time measured in months from now. Express N as a function of time t. Calculate the value of N when $t = 6$.

5-5 IMPLICIT RELATIONS AND INVERSE FUNCTIONS

When y is a given function of x, that is, $y = f(x)$, then we often say that y is an **explicit function** of the independent variable x. Examples of explicit functions are $y = 3x^2 - 7x + 5$, and $y = 5x + 1/(x - 1)$.

Sometimes the fact that y is a function of x is expressed indirectly by means of some equation of the type $F(x, y) = 0$, in which both x and y appear as arguments of the function F on the left side. An equation of this type is called an **implicit relation** between x and y.

EXAMPLE 1 Consider $xy + 3y - 7 = 0$. In this equation, we have a function on the left involving both x and y, and the equation provides an implicit relation between x and y. In this case we can solve for y.

$$y(x + 3) = 7$$

$$y = \frac{7}{x + 3}$$

Thus we can express y as an explicit function. In this example, the given implicit relation is equivalent to a certain explicit function. This is not always the case, as the following examples show.

EXAMPLE 2 Consider the implicit relation $x^2 + y^2 = 4$. In this case, we can again solve for y.

$$y^2 = 4 - x^2$$

$$y = +\sqrt{4 - x^2} \quad \text{and} \quad y = -\sqrt{4 - x^2}$$

These last two are explicit functions. Thus the implicit relation $x^2 + y^2 = 4$ leads to the two explicit functions,

$$y = +\sqrt{4 - x^2} \quad \text{and} \quad y = -\sqrt{4 - x^2}.$$

EXAMPLE 3 Consider the implicit relation $x^2 + y^2 + 4 = 0$. If we try to solve for y, we obtain

$$y^2 = -4 - x^2.$$

Whatever the value of x, the right side of the equation is always negative, so we cannot take the square root. In this case, the implicit relation has no solution. (We say that its domain is empty.)

EXAMPLE 4 $y^5 + x^3 - 3xy = 0$. This given relation does imply that y is a function of x, but we cannot solve for y in terms of x; that is, we cannot express y as an explicit function of x by means of any algebraic formula. ☛ **25**

When the fact that y is a function of x is *implied* by some relation of the form $F(x, y) = 0$, we speak of y as an **implicit function** of x. As in Example 4, this does not necessarily mean that we can actually find a formula expressing y as a function of x.

Given an implicit relation $F(x, y) = 0$, we usually are free to choose which of the variables x or y to regard as the independent variable. Consider the demand relation

$$2p + 3x = 12 \tag{1}$$

where x is the quantity demanded at a price p per unit. This equation defines p as an implicit function of x. Solving for p, we get

$$p = 6 - \tfrac{3}{2}x \tag{2}$$

which expresses p as an explicit function of x. The graph of Equation (2) is shown in Figure 29.

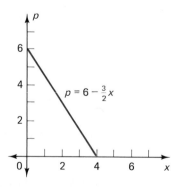

FIGURE 29

☛ **25.** Find the explicit function or functions corresponding to the following relations:

(a) $x(y - 2) = 2y - 1$;

(b) $x^2 - (y + 1)^2 = 2$;

(c) $x^2 + y^2 + 6x - 8y = 0$.

Answer (a) $y = \dfrac{2x - 1}{x - 2}$;

(b) $y = -1 \pm \sqrt{x^2 - 2}$;

(c) $y = 4 \pm \sqrt{25 - (x + 3)^2}$.

Equation (1) could also be viewed as defining x as an implicit function of p. Solving Equation (1) (or (2)) for x in terms of p, we obtain

$$x = 4 - \tfrac{2}{3}p. \tag{3}$$

Equation (3) expresses the demand x as a function of price p. Here p is treated as the independent variable and x as the dependent variable. The graph of Equation (3) is shown in Figure 30.

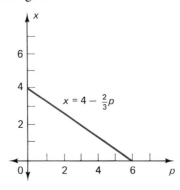

FIGURE 30

Thus the relation in Equation (1) defines not one but two functions:

$$p = f(x) = 6 - \tfrac{3}{2}x \qquad \text{if } p \text{ is regarded as a function of } x$$

and

$$x = g(p) = 4 - \tfrac{2}{3}p \qquad \text{if } x \text{ is regarded as a function of } p.$$

The two functions f and g, where

$$f(x) = 6 - \tfrac{3}{2}x \quad \text{and} \quad g(p) = 4 - \tfrac{2}{3}p$$

are called **inverse functions** of each other.

In general, let $y = f(x)$ be some given function. The equation $y - f(x) = 0$ represents an implicit relation between x and y. If we regard x as the independent variable, we can solve this relation for y, obtaining our original function, $y = f(x)$. On the other hand, we may wish to regard y as the independent variable and to solve for x in terms of y. We will not always be able to do this, but if we can, the solution is written as $x = f^{-1}(y)$ and f^{-1} is called the *inverse function* of f.

Notes **1.** $f^{-1}(y)$ is not to be confused with the negative power

$$[f(y)]^{-1} = \frac{1}{f(y)}.$$

2. If we take the composition of f and its inverse function, we find that

$$(f^{-1} \circ f)(x) = x \qquad \text{and} \qquad (f \circ f^{-1})(y) = y.$$

In other words, the composition of f and f^{-1} gives the identity function, that is, the function which leaves the variable unchanged.

26. Given $y = f(x)$ as follows, find $x = f^{-1}(y)$.
(a) $y = 3x + 1$; (b) $y = x^7$;
(c) $y = x^{1/3} + 1$.

EXAMPLE 5 Find the inverse of the function $f(x) = 2x + 1$.

Solution Setting $y = f(x) = 2x + 1$, we must solve for x as a function of y.

$$2x = y - 1$$

$$x = \frac{y - 1}{2}$$

Therefore the inverse function is given by $f^{-1}(y) = (y - 1)/2$. The graphs of $y = f(x)$ and $x = f^{-1}(y)$ are shown in Figures 31 and 32, respectively. Both graphs in this case are straight lines. Observe that when plotting the graph of $x = f^{-1}(y)$, the y-axis is taken as the horizontal axis and the x-axis as the vertical axis because y is the independent variable.

Answer (a)$x = \frac{1}{3}(y - 1)$;
(b) $x = y^{1/7}$; (c) $x = (y - 1)^3$.

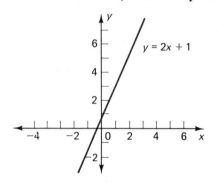

FIGURE 31

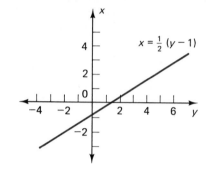

FIGURE 32

27. The graph of f contains just the five points $(1, 2)$, $(2, 3)$, $(3, 1)$, $(4, -1)$ and $(5, 4)$. Find the values of $f(1), f^{-1}(1), f^{-1}(3), f(5), f^{-1}(5)$. What are the domains of f and f^{-1}?

EXAMPLE 6 Find the inverse of the function $f(x) = x^3$ and sketch its graph.

Solution Setting $y = f(x) = x^3$, we solve for x, obtaining $x = f^{-1}(y) = y^{1/3}$. The graphs of $y = f(x)$ and $x = f^{-1}(y)$ are shown in Figures 33 and 34, respectively. **26, 27**

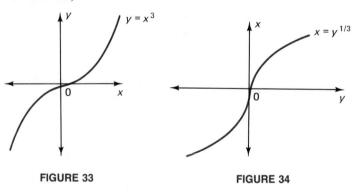

FIGURE 33

FIGURE 34

Answer $f(1) = 2, f^{-1}(1) = 3$,
$f^{-1}(3) = 2, f(5) = 4, f^{-1}(5)$ is not defined.
Domain of $f = \{1, 2, 3, 4, 5\}$.
Domain of $f^{-1} = \{-1, 1, 2, 3, 4\}$.

It can be seen from these two examples that the graphs of a function $y = f(x)$ and its inverse function $x = f^{-1}(y)$ are closely related. In fact, the graph of the inverse function is obtained by flipping over the graph of the original function so that the coordinate axes become interchanged. For example, hold the graph of $y = x^3$ in front of a mirror in such a way that the y-axis is horizontal and the x-axis points vertically upward. The reflection you will see will be the graph $x = y^{1/3}$ that is shown in Figure 34.

The graphs of $y = f(x)$ and $x = f^{-1}(y)$ consist of precisely the same sets of points (x, y). The difference rests only in that the axes are drawn in different directions in the two cases.

However, we sometimes want to consider the graphs of f and f^{-1} on the same set of axes. To do this, we must interchange the variables x and y in the statement $x = f^{-1}(y)$ and express f^{-1} in the form $y = f^{-1}(x)$. For example, in Example 6 we started with $y = f(x)$ as $y = x^3$ and found the inverse $x = f^{-1}(y)$ as $x = y^{1/3}$. But we can equally well write this as $y = f^{-1}(x)$, or $y = x^{1/3}$. ☛ 28

There is an interesting relationship between the graphs of the functions $y = f(x)$ and $y = f^{-1}(x)$ on the same axes. The graph of either of these functions can be obtained by reflecting the other graph about the line $y = x$. Figure 35 shows the graphs of the function $f(x) = x^3$ and its inverse function, $f^{-1}(x) = x^{1/3}$, drawn on the same axes. Clearly, the two graphs are reflections of one another about the line $y = x$.

☛ **28.** Given $f(x)$ as follows, find $f^{-1}(x)$.
(a) $f(x) = 2x - 4$;
(b) $f(x) = 1 + x^{-1}$.

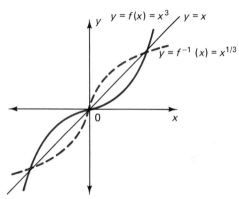

FIGURE 35

In general, let (a, b) be any point on the graph of $y = f(x)$. Then $b = f(a)$. It follows, therefore, that $a = f^{-1}(b)$, so that (b, a) is a point on the graph of $y = f^{-1}(x)$. Now it can be shown* that the two points (a, b) and (b, a) are reflections of one another about the line $y = x$. Consequently, to every point (a, b) on the graph of $y = f(x)$, there exists a point (b, a) on the graph of $y = f^{-1}(x)$ that is the reflection of the first point about the line $y = x$.

Not every function has an inverse. Consider, for example, the function $y = x^2$. Solving for x in terms of y, we obtain

$$x^2 = y \qquad \text{or} \qquad x = \pm\sqrt{y}.$$

Answer (a) $f^{-1}(x) = \frac{1}{2}(x + 4)$;
(b) $f^{-1}(x) = (x - 1)^{-1}$.

*Show that the line joining (a, b) and (b, a) is perpendicular to $y = x$ and the midpoint lies on $y = x$.

☞ 29. By placing a suitable restriction on the domain of f, find $x = f^{-1}(y)$ in each case:

(a) $f(x) = (x - 2)^2$;

(b) $f(x) = (x + 2)^4$;

(c) $f(x) = (x^2 - 4)^2$.

So, for any value of y in the region $y > 0$, there are two possible values of x. Thus we cannot say that x is a function of y.

This example is illustrated graphically in Figures 36 and 37. Figure 36 shows the graph of $y = x^2$, which is a parabola opening upward. Figure 37 shows the same graph, but with the axes flipped over; that is, the y-axis is horizontal and the x-axis is vertical. For each $y > 0$, we have two values of x, $x = +\sqrt{y}$ and $x = -\sqrt{y}$; for example, when $y = 1$, x has the value $+1$ and -1, both of which satisfy the relation $y = x^2$.

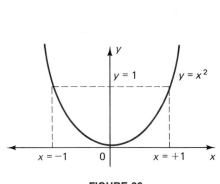

FIGURE 36   **FIGURE 37**

The graph in Figure 37 corresponds to two functions rather than one. The upper branch of the parabola is the graph of $x = +\sqrt{y}$, while the lower branch is the graph of $x = -\sqrt{y}$. Thus we can say that the function $y = x^2$ has two inverse functions, one given by $x = +\sqrt{y}$ and the other by $x = -\sqrt{y}$.

In a case such as this, it is possible to make the definition of f^{-1} unambiguous by restricting the values of x. For example, if x is restricted to the region $x \geq 0$, then $y = x^2$ has the unique inverse $x = +\sqrt{y}$. On the other hand, if x is restricted to the region $x \leq 0$, then the inverse is given by $x = -\sqrt{y}$. Placing a restriction on x in this way means restricting the domain of the original function f. We conclude, therefore, that in cases where a function $y = f(x)$ has more than one inverse function, the inverse can be made unique by placing a suitable restriction on the domain of f.   **☞ 29**

It is worth observing that *a function f(x) has a unique inverse whenever every horizontal line intersects its graph in at most one point.*

Answer

(a) $x = 2 + \sqrt{y}$ if domain is $x > 2$; $x = 2 - \sqrt{y}$ if domain is $x < 2$.

(b) $x = -2 + \sqrt[4]{y}$ if $x > -2$; $x = -2 - \sqrt[4]{y}$ if domain is $x < -2$.

(c) $x = \sqrt{4 + \sqrt{y}}$ if $x > 2$; $x = -\sqrt{4 + \sqrt{y}}$ if $x < -2$; $x = \sqrt{4 - \sqrt{y}}$ if $0 \leq x < 2$; $x = -\sqrt{4 - \sqrt{y}}$ if $-2 < x \leq 0$.

EXERCISES 5-5

(1–14) Find the explicit function or functions corresponding to the following implicit relations.

1. $3x + 4y = 12$

2. $5x - 2y = 20$

3. $xy + x - y = 0$

4. $3xy + 2x - 4y = 1$

5. $x^2 - y^2 + x + y = 0$

6. $x^2y + xy^2 - x - y = 0$

7. $x^2 + y^2 + 2xy = 4$

8. $9x^2 + y^2 = 6xy = 25$

9. $4x^2 + 9y^2 = 36$

10. $4x^2 - 9y^2 = 36$

11. $\sqrt{x} + \sqrt{y} = 1$

12. $xy^2 + yx^2 = 6$

13. $xy^2 + (x^2 - 1)y - x = 0$

14. $y^2 - 3xy + (2x^2 + x - 1) = 0$

(15–24) Find the inverse of each of the following functions. In each case, draw the graphs of the function and its inverse.

15. $y = -3x - 4$

16. $y = x - 1$

17. $p = 4 - \frac{2}{5}x$

18. $q = 3p + 6$

19. $y = \sqrt{3x - 4}$

20. $y = \sqrt{\frac{1}{4}x + 2}$

21. $y = x^5$

22. $y = \sqrt{x}$

23. $y = \sqrt{4 - x}$

24. $y = -\sqrt{2 - x}$

(25–30) By placing a suitable restriction on the domain of each of the following functions, find an inverse function.

25. $y = (x + 1)^2$

26. $y = (3 - 2x)^2$

27. $y = x^{2/3}$

28. $y = \sqrt{x^2 + 1}$

29. $y = |x - 1|$

30. $y = |x| - 1$

CHAPTER REVIEW

Key Terms, Symbols, and Concepts

5.1 Function, value of a function; domain and range of a function.
Independent variable or argument, dependent variable.
Graph of a function. The vertical line test.
Polynomial function, degree. Linear, quadratic, and cubic functions.
Rational function, algebraic function, transcendental function.

5.2 Parabola, vertex, axis.

5.3 Power function, $f(x) = ax^n$. Graphs of power functions for various n.
Circle. Center-radius formula. General equation of a circle.
Product transformation curve.
Absolute value function and its graph.

5.4 Sum, difference, product, and quotient of two functions. Composition of two functions: $f \circ g$.

5.5 Implicit relation, implicit function.
The inverse f^{-1} of a function f.
Relationships between the graphs of f and f^{-1}.

Formulas

If $y = ax^2 + bx + c = f(x)$, then vertex is at $x = -b/2a$, $y = f(-b/2a)$.

If $f(x) = ax^2 + bx + c$, and $a > 0$, then f is minimum when $x = -b/2a$; if $a < 0$, then f is maximum when $x = -b/2a$.

Center-radius formula: $(x - h)^2 + (y - k)^2 = r^2$.

General equation of a circle: $x^2 + y^2 + Bx + Cy + D = 0$.

REVIEW EXERCISES FOR CHAPTER 5

1. State whether each of the following is true or false. Replace each false statement with a corresponding true statement.

a. The domain of $f(x) = |x - 3|$ is the set of all real numbers greater than or equal to 3.

b. The range of $f(x) = x/x$ is $\{1\}$.

c. The graph of $f(x) = (x^2 - 4)/(x - 2)$ is a straight line which is broken at the point $(2, 4)$.

d. If $x \neq 3$, then $|x^2 - 9|/(x - 3) = x + 3$.

e. A given curve is the graph of a function if any vertical line meets the curve in at least one point.

f. A function is a rule that assigns to each value in the domain at least one value in the range.

g. For all real numbers x, $\sqrt{x^2} = |x|$.

h. For all values of a, b, and c, $F(x) = ax^2 + bx + c$ represents a quadratic function.

i. The domain of a polynomial function is the set of all integers.

j. If f and g are two functions, then $f + g, f - g, fg$, and f/g have the same domain.

k. If f and g are two functions such that both the composite functions $f \circ g$ and $g \circ f$ are defined, then $f \circ g = g \circ f$.

l. The graph of a quadratic function is a parabola with the vertex at the origin.

m. In an implicit function of the form $F(x, y) = 0$, x and y are both independent variables.

n. The function $y = f(x)$ has a unique inverse if and only if any horizontal line meets the graph of $f(x)$ in at most one point.

o. The graph of f^{-1} is the reflection of the graph of f about the y-axis.

2. Give an example of a function f that satisfies the stated property for all values of x and y.

a. $f(x) = f(-x)$ (Such a function is called an *even function*.)

b. $f(-x) = -f(x)$ (Such a function is called an *odd function*.)

c. $f(x + y) = f(x) + f(y)$

3. Two functions f and g are said to be equal if $f(x) = g(x)$ for all x in the domain and $D_f = D_g$. Use this criterion to determine which of the following functions are equal to $f(x) = (2x^2 + x)/x$.

a. $g(x) = 2x + 1$

b. $h(x) = \sqrt{1 + 4x + 4x^2}$

c. $F(x) = \dfrac{2x^3 + x^2}{x^2}$

d. $G(x) = \dfrac{(x^3 + 2x)(1 + 2x)}{x(x^2 + 2)}$

4. Find an equation of the circle with center $(-1, 2)$ and that passes through the point $(3, 4)$.

5. A circle of radius 5 units has its center at $(p, -1)$ and passes through the point $(1, 2)$. Determine p.

6. Determine the radius of a circle that passes through the point $(-3, 1)$ and has its center at $(1, -2)$.

7. Find the equation of the circle of radius 3 that lies in the first quadrant and touches both coordinate axes.

8. Determine the equation of the circle that has center $(-3, 4)$ and that passes through the origin.

(9–14) Determine whether each of the following equations represents a circle. If it does, find its center and radius and the points (if any) where it meets the coordinate axes.

9. $x^2 + y^2 + 4x + 4y - 1 = 0$

10. $x^2 + y^2 - 2x + 6y + 6 = 0$

11. $x^2 + y^2 - 3x + 4y + 8 = 0$

12. $x^2 + y^2 - 6x - 4y - 3 = 0$

13. $x^2 + y^2 + 10x - 4y + 4 = 0$

14. $x^2 + y^2 + 2x - 8y + 17 = 0$

15. Find the domain of $f(x) = |x| - 2$. Sketch the graph of f.

16. Find the domain of $g(x) = (x^2 - 4)/(x - 2)$ and sketch its graph.

17. If $f(x) = |x|$ and $g(x) = 1 - x^2$, determine $f \circ g$ and $g \circ f$.

18. Find $f(x)$ and $g(x)$ such that $(f \circ g)(x) = 1/(3x - 1)^2$. (The answer is not unique.)

19. Find $f(x)$ and $g(x)$ such that $(g \circ f)(t) = \sqrt{t}/(t + 1)$.

20. *(Parking Lot Charges)* A downtown parking lot charges \$2.00 for the first hour of parking and \$1.00 for each additional hour or portion thereof to a maximum of \$6.00 per day. Express the total parking charges P (in dollars) as a function of the number of hours the car is parked each day.

21. *(Monthly Earnings)* A salesperson earns a base salary of \$1000 per month plus a commission of 8% on the total sales she makes over \$6000. Express her monthly earnings E as a function of x, where x is the total monthly sales in dollars.

a. What is the domain of this function?

b. What will be her total salary when she makes sales of \$5000 and \$8000?

22. The revenue R for a certain commodity depends on the price p per unit charged and is given by the function $R = f(p) = 300p - 20p^2$. The price p per unit charged is a function of the demand x and is given by $p = g(x) = 15 - 0.05x$. Determine $(f \circ g)(x)$ and interpret your result.

23. A manufacturer can sell x units of its product at a price of p per unit where $2p + 0.05x = 8$. As a function of the quantity x demanded in the market, the revenue R is given by $R = 4x - 0.025x^2$. How does R depend on the price p?

24. *(Cost Function)* Pacific Airways charges $6 to transport each pound of merchandise 900 miles and $10 to transport each pound 1700 miles. Determine the cost function, assuming it to be a linear function of distance.

25. *(Optimum Rent)* The owner of an apartment building can rent all 60 suites if $120 per month is charged for each suite. If the rent is increased by $5, two of the suites remain unoccupied with no possibility of renting them. Assuming the relation between the number of vacant suites and the rent to be linear find:

a. The revenue as a function of monthly rent per unit.

b. The revenue as a function of the number of occupied suites.

c. The rent that maximizes the monthly revenue.

26. *(Optimum Rent)* Royal Trust owns a large apartment building consisting of 200 suites. From past experience, it is known that all 200 suites can be rented out at $120 per month per suite and only 170 suites can be rented if the monthly rent is $150 per month. Assuming that the demand for apartments is a linear function of the monthly rent charged per suite, determine the total revenue as a function of monthly rent per suite. What monthly rent per suite will maximize the total revenue?

27. *(Advertising and Sales)* The number y of units sold each week of a certain product depends on the amount x (in dollars) spent on advertising and is given by $y = 70 + 150x - 0.3x^2$. How much should be spent each week on advertising to obtain maximum sales volume? What is this maximum sales volume?

28. *(Optimum Pay TV Rate)* The Oriental View pay TV has 5000 subscribers when a monthly fee of $16.50 is charged. For each decrease of $0.50 in the monthly fee, it

can get 200 more subscribers. What rate will maximize the monthly revenue?

29. *(Maximum Revenue and Profits)* The weekly fixed costs of a firm for its product are $200 and the variable cost per unit is $0.70. The firm can sell x units at a price of p per unit where $2p = 5 - 0.01x$. How many units should be produced and sold each week so as to obtain:

a. Maximum revenue?

b. Maximum profit?

30. *(Maximum Profit)* There is no demand for a new brand of video cassette tape if the price per tape is $20 or more. For each drop of $1 in the price, the demand increases by 500 tapes. The cost of producing x tapes is $(12x + 2000)$ dollars. How many tapes should be produced and sold to obtain a maximum profit? What is the price charged per tape for the maximum profit?

31. *(Demand Curve)* A manufacturer can sell x units of a product at p dollars per unit, where x and p are related by

$$x^2 + p^2 + 200x + 150p = 49,400.$$

Plot the demand curve. What is the highest price above which no sales are possible?

(32–34) Solve the following implicit relations to express y as an explicit function of x.

32. $x^2 + 4y^2 = 9.$ **33.** $x - y^2 = 1$

34. $x^2y^2 - x^3y + 2y - 2x = 0$

(35–37) Find the inverse of each function.

35. $y = \sqrt{4 - x}$ **36.** $y = x^2 - 2x$

37. *(Population Size)* The size of an insect population at time t (measured in days) is given by

$$p(t) = 3000 - \frac{2000}{1 + t^2}.$$

Determine the initial *population* $p(0)$ and the population sizes after 1 and 2 days. Find the inverse function expressing t as a function of p for $t \geq 0$.

CHAPTER 6

Logarithms and Exponentials

Chapter Objectives

6-1 COMPOUND INTEREST AND RELATED TOPICS

(a) The compound interest formula for annual compounding and compounding many times per year; nominal and effective interest rates;
(b) Application of the compound interest formula to population growth;
(c) The continuous compounding formula and the number e;
(d) The present value of a future income.

6-2 EXPONENTIAL FUNCTIONS

(a) Definition of exponential functions;
(b) Graphs of growing and decaying exponential functions;
(c) The natural exponential function and its graph;
(d) Applications to population growth.

6-3 LOGARITHMS

(a) Definition of the logarithmic function with general base and its graph;
(b) Exponential and logarithmic forms of a statement;
(c) Properties of logarithms and their use;
(d) Natural logarithms; calculating with natural logarithms;
(e) Common logarithms.

6-4 APPLICATIONS AND FURTHER PROPERTIES OF LOGARITHMS

(a) Use of logarithms to solve exponential equations;
(b) Base change formulas for both exponentials and logarithms;
(c) The logistic function and its applications.

CHAPTER REVIEW

6-1 COMPOUND INTEREST AND RELATED TOPICS

Consider a sum of money, say $100, that is invested at a fixed rate of interest, such as 6% per annum. After 1 year, the investment will have increased in value by 6%, to $106. If the interest is compounded, then during the second year, this whole sum of $106 earns interest at 6%. Thus the value of the investment at the end of the second year will consist of the $106 existing at the beginning of that year plus 6% of $106 in interest, giving a total value of

$$\$106 + (0.06)(\$106) = (1 + 0.06)(\$106) = \$100(1.06)^2 = \$112.36.$$

During the third year, the value increases by an amount of interest equal to 6% of $112.36, giving a total value at the end of the year equal to

$$\$112.36 + \$112.36(0.06) = \$100(1.06)^3.$$

In general, the investment increases by a factor of 1.06 with each year that passes, so after n years its value is $\$100(1.06)^n$.

Let us consider the general case of an investment growing at compound interest. Let a sum P be invested at a rate of interest of R percent per annum. Then the interest in the first year is $(R/100)P$, so the value of the investment after 1 year is

$$P + \left(\frac{R}{100}\right)P = P\left(1 + \frac{R}{100}\right) = P(1 + i)$$

where we have let $i = (R/100)$.

The interest in the second year will be R percent of this new value, $P(1 + i)$:

$$\text{Interest} = \left(\frac{R}{100}\right)P(1 + i).$$

Thus the value after 2 years is

$$P(1 + i) + \left(\frac{R}{100}\right)P(1 + i) = P(1 + i)\left(1 + \frac{R}{100}\right) = P(1 + i)^2.$$

We see that each year the value of the investment is multiplied by a factor of $1 + i$ from its value the previous year. After n years, the value is given by the formula

$$\boxed{\text{Value after } n \text{ years} = P(1 + i)^n, \qquad i = \frac{R}{100}.}$$

The expression $(1 + i)^n$ for selected values for i and n can be found in Table A.3.4 or by using a calculator that has a y^x key.

EXAMPLE 1 *(Investment)* A sum of $200 is invested at 5% interest compounded annually. Find the value of the investment after 10 years.

Solution In this case $R = 5$ and $i = R/100 = 0.05$. After n years, the value of the investment is

$$P(1 + i)^n = 200(1.05)^n.$$

When $n = 10$, this is

$$200(1.05)^{10} = 200(1.628895) = 325.78.$$

The value of the investment is therefore $325.78. ☛ **1**

☛ **1.** $4000 is invested at 9% per annum. What is the value after (a) 5 years; (b) 11 years?

Answer
(a) $4000(1.09)^5 = \$6154.50$;
(b) $4000(1.09)^{11} = \$10,321.71$.

In some cases, interest is compounded more than once per year, for example semiannually (2 times per year), quarterly (4 times per year) or monthly (12 times per year). In these cases, the annual rate of interest R percent which is usually quoted is called the **nominal rate.** If compounding occurs k times per year and if the nominal rate of interest is R percent, this means that the interest rate at each compounding is equal to (R/k) percent. In N years, the number of compoundings is kN. ☛ **2**

☛ **2.** If the nominal annual rate is 12%, what is the interest rate at each compounding when compounding is (a) semiannual; (b) quarterly; (c) monthly; (d) daily?

For example, at 8% nominal interest compounded quarterly, an investment is increased by 2% every 3 months. In 5 years, there would be 20 such compoundings.

In the general case, let $n = Nk$ be the number of compounding periods and $i = R/100k$ be the (decimal) interest rate per period. Then the compound interest formula becomes

Answer (a) 6%; (b) 3%; (c) 1%;
(d) (12/365)%.

$$\boxed{\text{Value after } n \text{ Periods} = P(1 + i)^n, \qquad i = \frac{R}{100k}.}$$

EXAMPLE 2 *(Monthly Compounding)* A sum of $2000 is invested at a nominal rate of interest of 9% compounded monthly. Find the value of the investment after 3 years.

Solution Here $k = 12$, the investment is compounded monthly and the interest rate at each compounding is $R/k = \frac{9}{12} = 0.75\%$. Thus at each compounding, the value is increased by a factor

$$(1 + i) = \left(1 + \frac{R}{100k}\right) = \left(1 + \frac{0.75}{100}\right) = 1.0075.$$

☛ **3.** $4000 is invested at a nominal rate of 9% per annum. What is the value after 5 years if compounding is (a) semiannual; (b) quarterly; (c) monthly?

During 3 years, there will be $n = 3 \cdot 12 = 36$ such compoundings. Hence the value will be

$$2000(1.0075)^{36} = 2000(1.308645) = 2617.29 \text{ dollars} ☛ \mathbf{3}$$

Answer
(a) $4000(1.045)^{10} = \$6211.88$;
(b) $4000(1.0225)^{20} = \$6242.04$;
(c) $4000(1.0075)^{60} = \$6262.72$.

☛ 4. In Example 2, what is the effective annual rate if compounding is quarterly instead of monthly?

EXAMPLE 3 *(Quarterly Compounding)* What is the nominal interest rate required to double the value of an investment in 5 years with compounding every 3 months?

Solution In 5 years there are $5 \cdot 4 = 20$ interest periods. An investment of P increases to $2P$, so we have the equation

$$\text{Value after 20 Periods} = P(1 + i)^{20} = 2P$$

or

$$(1 + i)^{20} = 2.$$

Therefore

$$(1 + i) = 2^{1/20} = 1.03526*$$

and so $i = 0.03526$. But $i = R/100k = R/(100 \cdot 4) = R/400$, where R is the nominal annual rate. Therefore $R = 400i = 400(0.03526) = 14.11$. A nominal interest rate of 14.11% per annum is therefore required.

The **effective rate** of interest of an investment is defined as the annual rate that would provide the same growth if compounded once per year. Consider an investment that is compounded k times per year at nominal rate of interest of $R\%$. Then the investment grows by a factor $(1 + i)^k$ in 1 year, where $i = R/100k$. Let R_{eff} be the effective rate of interest and $i_{\text{eff}} = R_{\text{eff}}/100$. Then, by definition, the investment grows by a factor $(1 + i_{\text{eff}})$ each year, so we have

$$\boxed{1 + i_{\text{eff}} = (1 + i)^k}$$

which enables the effective rate to be calculated.

In Example 2, for example, $(1 + i) = 1.0075$ and $k = 12$. Therefore

$$i_{\text{eff}} = (1 + i)^k - 1 = (1.0075)^{12} - 1 = 0.0938.$$

So $R_{\text{eff}} = 100i_{\text{eff}} = 9.38$ and we have an effective annual interest rate of 9.38%. ☛ 4

EXAMPLE 4 Which is better for the investor, 12% compounded monthly or 12.2% compounded quarterly?

Solution We compute the effective rate for each of the two investments. For the first, $i = 0.01$ and $k = 12$, so

$$i_{\text{eff}} = (1 + i)^k - 1 = (1.01)^{12} - 1 = 0.126825.$$

Answer
$100[(1.0225)^4 - 1]\% = 9.308\%$.

---

* Using the y^x key of a calculator: $2^{1/20} = 2^{0.05} = 1.03526$.

For the second, $i = 0.0305$ and $k = 4$, so

$$i_{\text{eff}} = (1 + i)^k - 1 = (1.0305)^4 - 1 = 0.127696.$$

This second has a larger effective rate, so is the better of the two.

Population Growth Problems

The compound interest formula applies to any quantity which grows by a regular percentage each year. For example, a population of initial size P_0 growing by R percent per year will, after n years, have the size $P_0(1 + i)^n$, where $i = R/100$.

EXAMPLE 5 *(Population Growth)* The population of the earth at the beginning of 1976 was 4 billion and it was growing at 2% per year. What will be the population in the year 2000, assuming that the rate of growth remains unchanged?

Solution When a population grows by 2% a year, this means that its size at any time is 1.02 times what it was one year earlier. Thus at the beginning of 1977, the population was $(1.02) \times 4$ billion. Furthermore, at the beginning of 1978, it was (1.02) times the population at the beginning of 1977, that is, $(1.02)^2 \times 4$ billion. Continuing in this way, we conclude that the population at the beginning of the year 2000, that is, after 24 years, will therefore be

$$(1.02)^{24} \times 4 \text{ billion} = 1.608 \times 4 \text{ billion} = 6.43 \text{ billion}.$$

EXAMPLE 6 *(GNP Increase)* The population of a certain developing nation is increasing by 3% per year. By how much per year must the gross national product (GNP) increase if the income per capita is to double in 20 years?

Solution Let the present population be denoted by P_0. Then, since the population increases by a factor (1.03) each year, the size of the population after n years is given by

$$P = P_0(1.03)^n.$$

Let the present GNP be I_0. Then the present income per capita is obtained by dividing this quantity I_0 by the population size, so is equal to I_0/P_0.

If the GNP increases by R percent per year, then it changes by a factor of $1 + i$ for each year, where $i = R/100$. Therefore, after n years, the GNP is given by

$$I = I_0(1 + i)^n.$$

The income per capita after n years is therefore

$$\frac{I}{P} = \frac{I_0(1 + i)^n}{P_0(1.03)^n} = \frac{I_0}{P_0}\left(\frac{1 + i}{1.03}\right)^n.$$

We wish to find the value of R for which the income per capita when $n = 20$ is equal to twice its present value, I_0/P_0. Therefore we have the equation

$$\frac{I}{P} = \frac{I_0}{P_0}\left(\frac{1+i}{1.03}\right)^{20} = 2\frac{I_0}{P_0}.$$

It follows that

$$\left(\frac{1+i}{1.03}\right)^{20} = 2.$$

Therefore

$$\frac{1+i}{1.03} = 2^{1/20}$$

and so $1 + i = (1.03)(2^{1/20}) = (1.03)(1.0353) = 1.066$. Thus $i = 0.066$ and $R = 100i = 6.6$. Thus the GNP would have to increase by 6.6% per annum if the stipulated goal is to be achieved. ☛ 5

☛ **5.** In Example 6, if the GNP grows by 5% per annum, by what factor will the per capita income increase over a 20 year period?

The exponential growth equation must be applied to population problems with a certain amount of caution. A growing population will grow exponentially, provided there are no factors from its environment that limit or otherwise influence the growth. For instance, if the supply of food available to the population of a certain species is limited, then exponential growth must cease eventually as the food supply becomes insufficient to support the ever-growing population. Among other factors that inhibit indefinite growth are the supply of shelter for the species, which typically is limited, the interaction with predator species, sociological factors that might slow the expansion of the population in overcrowded circumstances, and finally, simply the limited availability of physical space for the population. Such factors as these eventually operate to end the exponential growth of a population and cause it to level off at some value that is the maximum population which can be supported by the given habitat.

It is becoming apparent that such constraints are at this time being imposed on the human population, and it seems quite probable that the human population will not grow exponentially during the coming decades. Thus future projections of the human population over long periods of time based on the exponential growth equation are unlikely to turn out to be very accurate. They indicate what will happen if present trends continue, which may be different from what will happen in fact.

Continuous Compounding

Suppose a sum of money, say $100, is invested at a rate of interest of 8% compounded annually. Each year the value of the investment increases by a factor of 1.08, so that after N years, it is equal to $\$100(1.08)^N$. For example, after 4 years, the investment is worth $\$100(1.08)^4 = \$136.05$.

Answer $\left(\dfrac{1.05}{1.03}\right)^{20} = 1.469.$

Now let us suppose instead that the investment of $100 is compounded semiannually and that the nominal rate of interest is still 8% per annum. This means that the rate of interest per half-year is 4%. Then each half-year, the value of the investment increases by a factor of 1.04. In a period of N years, there are $2N$ such semiannual compoundings; thus after N years the investment is worth $100(1.04)^{2N}$. For example, after 4 years the value is $100(1.04)^8 = 136.86.

Next consider the possibility that the investment is compounded every 3 months, again with the nominal annual interest rate of 8%. Then the rate of interest per quarter is equal to $\frac{8}{4}$ or 2%. Each quarter the value increases by a factor of 1.02 and so each year it increases by a factor $(1.02)^4$. In a period of N years, the value increases to $100(1.02)^{4N}$. For example, after 4 years the value is $100(1.02)^{16} = 137.28.

We can continue in this same manner: Let us divide the year into k equal periods and compound the interest at the end of each of these periods at a nominal annual interest rate of 8%. This means that the interest rate for each period is $8/k$ percent and the investment increases in value by a factor of $1 + 0.08/k$ for each of these small periods. During N years there are kN such compounding periods, so the value after N years is given by the formula $100(1 + 0.08/k)^{kN}$ dollars. For example, after 4 years the value is

$$100\left(1 + \frac{0.08}{k}\right)^{4k} \text{ dollars.} \quad \text{☛ 6} \tag{7}$$

Table 1 shows these values for several different values of k. We first give $k = 1$, 2, and 4 and then give $k = 12$, 52, and 365, which correspond, respectively, to monthly compounding, weekly compounding and daily compounding. Finally, we give $k = 1000$ for comparison. It can be seen that as the frequency of compounding is increased, the value of the investment also increases; however, it does not increase indefinitely, but rather approaches closer and closer to a certain value. To the nearest cent, there is no difference between compounding 365 times a year and 1000 times a year: the value of the investment after 4 years would still be $137.71.

Because of this, we can envisage the possibility of what is called **continuous compounding.** By this we mean that the number k is allowed to become arbitrarily large; we say that k is allowed *to approach infinity* and we write this as $k \to \infty$. This corresponds to compounding the interest infinitely often during the year. With our $100 invested at the nominal rate of 8% per annum, continuous compounding gives a value of $137.71 after 4 years, the same value as daily compounding.

Let us write $k = 0.08p$ in Expression (7), which gives the value of the investment after 4 years. Then $4k = 0.32p$ and the value after 4 years takes the form

$$100\left(1 + \frac{0.08}{k}\right)^{4k} = 100\left(1 + \frac{1}{p}\right)^{0.32p} = 100\left[\left(1 + \frac{1}{p}\right)^p\right]^{0.32}.$$

☛ **6.** Suppose that $1 is invested at a nominal rate of 100%, compounded k times per year. What is the value after 1 year? Compute the value to four decimal places when $k = 365$ (daily compounding).

TABLE 1

k	Value after 4 Years
1	$136.05
2	$136.86
4	$137.28
12	$137.57
52	$137.68
365	$137.71
1000	$137.71

Answer Value $= (1 + \frac{1}{k})^k = 2.7146$ when $k = 365$.

TABLE 2

p	$\left(1 + \dfrac{1}{p}\right)^p$
1	2
2	2.25
10	2.594
100	2.705
1,000	2.717
10,000	2.718

The reason for writing it in this form is that as $k \to \infty$, then $p = k/(0.08)$ also becomes arbitrarily large and the quantity inside the square brackets, $(1 + 1/p)^p$, gets closer and closer to a certain constant as $p \to \infty$. This can be seen from Table 2, in which the values of $(1 + 1/p)^p$ are given for a series of increasingly large values of p. The eventual value to which $(1 + 1/p)^p$ approaches as p increases indefinitely is a number denoted by the letter e. This number is irrational and is equal to 2.71828 to five decimal places.

In the example of continuous compounding above, we see that as $p \to \infty$, the value of the investment after 4 years gets closer and closer to $100e^{0.32}$ dollars.

Let us examine continuous compounding in the general case when a sum P is invested. Let the interest be compounded k times a year at the nominal annual interest rate of R percent. The rate of interest at each compounding is then R/k percent. At each compounding, the value increases by a factor $1 + i/k$ where $i = R/100$. After N years, during which there will have been kN such compoundings, the value will be $P(1 + i/k)^{kN}$.

We introduce $p = k/i$, or $k = ip$. Then $kN = piN$ and the value after N years is

$$P\left(1 + \frac{1}{p}\right)^{piN} = P\left[\left(1 + \frac{1}{p}\right)^p\right]^{iN}.$$

For continuous compounding we must let $k \to \infty$; this means that $p = k/i$ also becomes infinitely large. The quantity inside the square brackets here becomes closer and closer to e as $p \to \infty$, so the value of the investment becomes Pe^{iN}.

We have thus shown that *if a sum P is compounded continuously at a nominal annual rate of interest of R percent,*

$$\boxed{\text{Value after } N \text{ Years} = Pe^{iN}; \quad \left(i = \frac{R}{100}\right).}$$

7. Evaluate (a) $e^{2.1}$; (b) $e^{-1.25}$.

Answer (a) 8.1662; (b) 0.2865.

Values of the expression e^{iN} can be found in Appendix III, Table A.3.3. They can also be obtained from many pocket calculators.

EXAMPLE 7 Using Table A.3.3 or a calculator, find the values of the following.

(a) e^2 (b) $e^{3.55}$ (c) $e^{-0.24}$

Solution We use Table A.3.3.

(a) $e^2 = 7.3891$ (b) $e^{3.55} = 34.813$ (c) $e^{-0.24} = 0.7866$

(In part (c), the value is read from the column headed e^{-x} adjacent to the value 0.24 in the x column.)   ☛ 7

EXAMPLE 8 *(Investment)* An investment of $250 is compounded continuously at a nominal annual rate of interest of $7\frac{1}{2}\%$. What will be the value of the investment after 6 years?

Solution We must use the formula Pe^{iN} for the value after N years. In this example, $P = 250$, $N = 6$, and $i = 7.5/100 = 0.075$. Therefore $iN = (0.075)(6) = 0.45$ and the value is

$$Pe^{iN} = 250e^{0.45} \text{ dollars.}$$

The value of $e^{0.45}$ can be found in Table A.3.3, and we get

$$250e^{0.45} = 250(1.5683) = 392.08.$$

Thus the value of the investment after 6 years is $392.08.

An investment compounded continuously grows by a factor e^i in 1 year, where $i = R/100$. As before, we define an effective annual rate of interest, R_{eff} percent, for continuous compounding as the rate that gives the same growth if compounded once per year. The condition is $1 + i_{\text{eff}} = e^i$, where $i_{\text{eff}} = R_{\text{eff}}/100$. Thus we have

$$R_{\text{eff}} = 100(e^i - 1).$$

EXAMPLE 9 *(Investment)* On its savings accounts, the Piggy Bank of New York gives a nominal annual interest rate of 6%, compounded continuously. The bank wishes to calculate an effective annual rate of interest (i.e., the equivalent annual rate) to use in its advertisements.

Solution We have $R = 6$, so $i = 0.06$. The effective rate is then given by $1 + i_{\text{eff}} = e^i$, or

$$i_{\text{eff}} = e^i - 1 = e^{0.06} - 1 = 1.0618 - 1 = 0.0618.$$

Thus the effective percentage rate is $R_{\text{eff}} = 100 i_{\text{eff}} = 6.18$. The bank can advertise an effective annual interest rate of 6.18%. ☛ 8

Present Value

Another application is to the **present value** of a future revenue or a future liability. Let us suppose that by pursuing a certain business activity, a person expects to receive a certain sum of money, P, at a time n years in the future. This future revenue P is less valuable than would be a revenue of the same amount received at the present time since, if the person received P now, it could be invested at interest, and it would be worth more than P in n years

☛ 8. Find the effective annual rate for continuous compounding at nominal rate of 10%

Answer $100(e^{0.1} - 1) = 10.52\%$.

time. We are interested in finding the sum Q which, if received at the present time and invested for n years, would be worth the same as the future revenue P that the person will receive.

Let us suppose that the interest rate that could be obtained on such an investment is equal to R percent. Then, after n years, the sum Q would have increased to $Q(1 + i)^n$, where $i = R/100$. Setting this equal to P, we obtain the equation

$$Q(1 + i)^n = P \quad \text{or} \quad Q = \frac{P}{(1 + i)^n} = P(1 + i)^{-n}.$$

We call Q the **present value** of the future revenue P.

In calculating present value, it is necessary to make some assumption about the rate of interest R that would be obtained over the n years. In such circumstances, R is called the **discount rate** and we say that the future revenue P is **discounted back** to the present time. ☛ 9

☛ 9. What is the present value of $5000 received 3 years from now if the discount rate is 8%?

EXAMPLE 10 (*Real Estate Sales Decision*) A real estate developer owns a piece of property that could be sold right away for $100,000. Alternatively, the property could be held for 5 years. During this time, the developer would spend $100,000 on developing it; it would then sell for $300,000. Assume that the development cost would be spent in a lump sum at the end of 3 years and must be borrowed from a bank at 12% interest per annum. If the discount rate is assumed to be 10%, calculate the present value of this second alternative and hence decide which of these two alternatives represents the developer's best strategy.

Solution Consider first the money which must be borrowed in order to develop the property. Interest must be paid at 12% on this over a period of 2 years, so when the property is sold, this loan has increased to

$$\$100,000(1.12)^2 = \$125,440.$$

The net proceeds from the sale, after paying off this loan, will be

$$\$300,000 - \$125,440 = \$174,560.$$

This revenue is received 5 years in the future. Discounting back at a rate of 10%, we obtain a present value of

$$\$174,560(1.1)^{-5} = \$108,400.$$

Since the present value of an immediate sale is only $100,000, it is somewhat better if the developer holds the property and sells in 5 years.

Observe the way in which decisions can be made between alternative business strategies by comparing their present values.

Answer
$5000 ÷ (1.08)^3 = 3969.16.

(1–2) If $2000 is invested at 6% compound interest per annum, find the following.

1. The value of the investment after 4 years.

2. The value of the investment after 12 years.

(3–4) If $100 is invested at 8% compound interest per annum find the following.

3. The value of the investment after 5 years.

4. The value of the investment after 10 years.

(5–8) A sum of $2000 is invested at a nominal interest rate of 12%. Calculate its value:

5. After 1 year if compounding is quarterly.

6. After 1 year if compounding is monthly.

7. After 4 years if compounding occurs every 6 months.

8. After 6 years with quarterly compounding.

(9–12) Find an effective rate of interest per year that is equivalent to:

9. 6% nominal rate compounded semiannually.

10. 8% nominal rate compounded quarterly.

11. 12% nominal rate compounded monthly.

12. 12% nominal rate compounded 6 times a year.

(13–16) Find the nominal rate of interest per year which corresponds to an effective rate of:

13. 8.16% when the compounding occurs semiannually.

14. 12.55% when the compounding occurs quarterly.

15. 10% when the compounding occurs monthly.

16. 9% when the compounding occurs 6 times a year.

(17–20) Which is better for the investor:

17. Semiannual compounding with a nominal rate of 8.2% or quarterly compounding at 8%?

18. Semiannual compounding with a nominal rate of 6% or annual compounding at 6.1%?

19. Annual compounding at 8.2% or a quarterly compounding with a nominal rate of 8%?

20. Semiannual compounding with a nominal rate of 12.2% or a monthly compounding with a nominal rate of 12%?

21. What rate of compound interest doubles the value of an investment in 10 years?

22. What rate of compound interest triples the value of an investment in 10 years?

23. A sum of money is invested for 5 years at 3% interest per annum and then for a further 4 years at R percent interest. Find R if the money exactly doubles in value over the 9 years.

24. A sum of money is invested at $R\%$ compounded yearly. It amounts to $21,632 at the end of second year and to $22,497.28 at the end of third year. Find the rate of interest R and the sum invested.

25. A sum of money is invested at $R\%$ compounded semiannually. It amounts to $56,275.44 at the end of the second year and to $59,702.62 at the end of the third year. Find the nominal rate of interest R and the sum invested.

26. (*Population Growth*) The population of the earth at the beginning of 1976 was 4 billion. If the growth rate continues at 2% per year, what will be the population in the year 2026?

(27–32) Evaluate the following using Table A.3.3 in the Appendix.

27. $e^{0.41}$ **28.** $e^{2.75}$

29. e^8 **30.** $e^{-1.05}$

31. $e^{-0.68}$ **32.** $e^{-5.2}$

(33–36) (*Continuous Compounding*) Find the value of each of the following investments.

33. $5000 is compounded continuously for 3 years at a nominal rate of interest of 6% per year.

34. $2000 is compounded continuously for 5 years at a nominal rate of interest of 8% per year.

35. $1000 is compounded continuously for 6 years at a nominal rate of interest of 10% per year.

36. $3000 is compounded continuously for 4 years at a nominal rate of interest of 5% per year.

37. (*Continuous Compounding*) An investment of $100 is

compounded continuously for 2 years at a nominal rate of interest of 9% and then for a further 5 years at a nominal rate of interest of 11%. Calculate the value of the investment after the 7-year period.

38. (*Continuous Compounding*) An investment of $2000 is compounded continuously for 3 years at a nominal rate of 6% per annum and then for a further 4 years at a nominal rate of 8% per annum. Find the value of the investment after the 7-year period.

39. (*Continuous Compounding*) An investment is compounded continuously at a nominal rate of 8% per annum. How long does it take the investment to:

a. double in value?

b. triple in value?*

40. Repeat Exercise 39 for a nominal rate of interest of 10% per year.

(41–43) (*Continuous Compounding*) Calculate the nominal rate of interest for each of the following.

41. $100 compounded continuously for 4 years increases in value to $150.

42. An investment compounded continuously for 10 years doubles in value.

43. An investment compounded continuously for 8 years triples in value.

44. (*Continuous Compounding*) An investment is compounded continuously for 2 years at a nominal rate R percent and for a further 4 years at the nominal rate $2R$ percent. Find R if the value exactly doubles.

45. (*Daily Compounding*) If a bank compounds interest daily with a nominal annual rate of $4\frac{1}{2}\%$, what is the "effective annual rate of interest" that it can use in its advertisements?

46. Repeat Exercise 45 for a nominal rate of 8%.

(47–49) (*Continuous Compounding*) Which is better for the investor:

47. Continuous compounding with a nominal rate of 5% or an annual compounding at 5.2%?

48. Continuous compounding at a nominal rate of 6% or a semiannual compounding at a nominal rate of 6.1%?

49. Continuous compounding at a nominal rate of 8% or a quarterly compounding at a nominal rate of 8.2%?

50. (*Sales and Advertising*) In a competitive market, the volume of sales depends on the amount spent on advertising the product in question. If x dollars are spent per month advertising a particular product, it is found that the volume of sales S per month (in dollars) is given by the formula

$$S = 10,000(1 - e^{-0.001x}).$$

Find the volume of sales when $x = 500$ and $x = 1000$. If x is decreased from $500 to $100 per month, what is the resulting percentage decrease in sales?

51. (*Semiannual Compounding*) Calculate the rate of semiannual interest that is equivalent to an annual interest rate of 8%.

52. (*Monthly Compounding*) Calculate the rate of monthly interest that is equivalent to an annual interest rate of 8%.

53. (*Present Value*) A person expects to receive $1000 every year for the next 3 years, the first payment to arrive in 1 year's time. Calculate the present value of this income, assuming a discount rate of 8% per annum.

54. (*Present Value*) A person owes a debt that is to be paid off in three equal annual installments of $5000, the first payment to be made in 1 year's time. If, instead, the person decides to pay the debt off in a lump sum right away, calculate how much must be paid, assuming a discount rate of 8% per annum. $a = \frac{P}{(1+i)} \sim$

55. (*Present Value*) Which is better if the interest rate is 5%: $1000 now or $1100 in 2 years' time?

56. (*Present Value*) A forest product company owns a stand of timber whose value in t years will be $V(t) = 2(1 + 0.3t)$. Assume a discount rate of 10% compounded annually. Calculate the present value of the timber if it is cut and sold:

a. in 1 year's time.

b. in 6 years' time.

c. in 7 years' time.

d. in 8 years' time.

What do your answers suggest?

57. (*Present Value*) Which is better if the interest rate is 10%: $2000 now or $1150 one year from now and another $1150 two years from now?

---

*$e^{0.693\cdots} = 2$; $e^{1.099\cdots} = 3$.

Consider a certain city with population at a given time of 1 million, with the population increasing at a rate of 10% per year. After 1 year, the population will have increased to 1.1 million. During the second year, the increase in population will be 10% of the size at the beginning of that year, that is, 10% of 1.1 million. Therefore the population size after 2 years will be

$$1.1 + (0.1)(1.1) = (1.1)^2 = 1.21 \text{ million.}$$

During the third year, the increase will be 10% of 1.21 million, giving a total population at the end of the third year equal to

$$1.21 + (0.1)(1.21) = (1.1)(1.21) = (1.1)^3 = 1.331 \text{ million.}$$

Continuing in this way, we see that the size of the population after n years is equal to $(1.1)^n$ million. A graph of this function is shown in Figure 1, in which the values $(1.1)^n$ are shown as dots for $n = 0, 1, 2, \ldots, 10$.

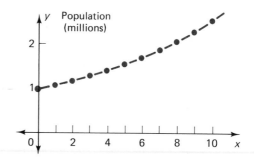

FIGURE 1

The formula $(1.1)^n$ may be used to calculate the size of the population in millions at fractional parts of a year as well as at integer values of n. For example, after 6 months (that is, half a year), the size of the population is $(1.1)^{1/2} = 1.049$ million (to three decimal places). After 2 years and 3 months ($2\frac{1}{4}$ years), the size of the population is $(1.1)^{9/4} = 1.239$ million, and so on.

If all these values of $(1.1)^n$ for fractional values of n are plotted on the graph in Figure 1, it is found that they lie on a smooth curve. This curve is shown in Figure 1, passing, of course, through the heavy dots.

The values of $(1.1)^n$ can only be defined by such elementary means when n is a rational number. For example, when $n = \frac{9}{4}$, $(1.1)^n = (1.1)^{9/4}$ can be defined as the fourth root of 1.1 raised to the ninth power. Similarly, $(1.1)^{7/5}$ can be defined as the fifth root of 1.1 raised to the seventh power. But such a definition in terms of powers and roots cannot be given for $(1.1)^n$ when n is an irrational number; for example $(1.1)^{\sqrt{2}}$ cannot be defined by powers and roots. However, once we have constructed the smooth curve in the preceding figure, we can use that curve to *define* $(1.1)^n$ for irrational values of n. For instance, to find $(1.1)^{\sqrt{2}}$, we simply use the ordinate of the point on the curve that corresponds to the abscissa $n = \sqrt{2}$. In this way, we see that the value of $(1.1)^n$ can be defined for all real values of n, both rational and irrational.

In a similar way we may define the function $y = a^x$ for any positive real number a. First the value of $y = a^x$ is defined for all rational values of x by use of powers and roots. When plotted on a graph, it is found that all such points (x, y) lie on a smooth curve. This curve can then be used to define the value of a^x when x is an irrational number simply by reading off the ordinate of the point on the graph at which x has the given irrational value.

EXAMPLE 1 Construct the graphs of each function.

$$\text{(a) } y = 2^x \qquad \text{(b) } y = \left(\frac{1}{3}\right)^x \qquad \text{(c) } y = 3^x$$

Solution Table 3 gives values of these three functions for a selection of values of x.

TABLE 3

x	-2	-1.5	-1	-0.5	0	0.5	1	1.5	2	3
$y = 2^x$	0.25	0.354	0.5	0.707	1	1.414	2	2.828	4	8
$y = (\frac{1}{3})^x$	9	5.196	3	1.732	1	0.577	0.333	0.192	0.111	0.037
$y = 3^x$	0.111	0.192	0.333	0.577	1	1.732	3	5.196	9	27

For example, when $x = -0.5$,

$$2^x = 2^{-1/2} = \frac{1}{\sqrt{2}} = \frac{1}{1.414} = 0.707$$

to 3 decimal places, and

$$\left(\frac{1}{3}\right)^x = \left(\frac{1}{3}\right)^{-1/2} = 3^{1/2} = \sqrt{3} = 1.732.$$

When plotted, the points indicated by dots on Figure 2 are obtained, and these points can be joined by smooth curves as shown. ☛ 10

A function of the type $y = a^x$ ($a > 0$, $a \neq 1$) is called an **exponential function.** When $a > 1$, the function is called a *growing exponential function,* whereas when $a < 1$, it is called a *decaying exponential function.*

The graphs obtained in Example 1 are characteristic of *exponential functions.* Figure 3 illustrates the graphs of two functions, $y = a^x$ and $y = b^x$, when $a > b > 1$. It is seen that for $x > 0$, these two functions grow at an ever-increasing rate as x increases. Since $a > b$, the graph of $y = a^x$ for positive values of x is above the graph of $y = b^x$ and, moreover, increases more steeply.

On the other hand, for $x < 0$, both functions decrease towards zero as x becomes larger and larger negatively. In this case, the function a^x falls more

☛ **10.** How can the graphs of (a) $y = 3^{-x}$ (b) $y = 2^{-x}$ and (c) $y = 2^{x+1}$ be obtained from those in Figure 2?

Answer (a) $3^{-x} = (\frac{1}{3})^x$ so graph is already given in Figure 2.
(b) Reflecting the graph of 2^x in the y-axis, we obtain the graph of 2^{-x}.
(c) $2^{x+1} = 2 \cdot 2^x$ so graph can be obtained from the graph of 2^x by multiplying each y-value by 2 (or by moving the graph bodily 1 unit to the left).

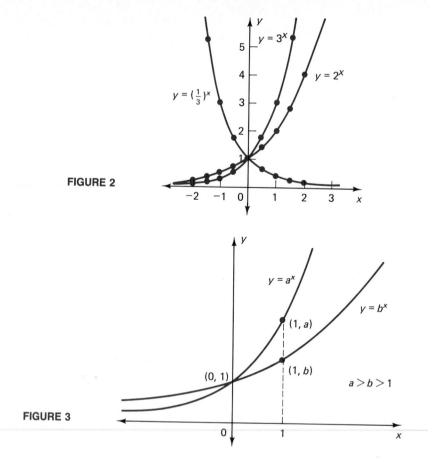

FIGURE 2

FIGURE 3

steeply than b^x and its graph is situated below the graph of $y = b^x$. The two graphs intersect when $x = 0$, since $a^0 = b^0 = 1$.

The graph of $y = a^x$ when $a < 1$ is illustrated in Figure 4. When $a < 1$, a^x decreases as x increases and approaches zero as x becomes larger. As a result, the graph approaches the x-axis more and more closely as x becomes very large.

From the graphs in Figures 3 and 4 it can be seen that the domain of the exponential function, $f(x) = a^x$ is the set of all real numbers and the range is the set of positive real numbers. Thus

> If $a > 0$, $a^x > 0$ for all values of x, positive negative or zero.

The number a that appears in the exponential function a^x is called the **base.** The base can be any *positive real number except* 1^*. It is often useful to use as base the number denoted by e, which is given to five decimal places by $e = 2.71828$. The corresponding exponential function is written e^x and is

---

* If $a = 1$, then $f(x) = a^x = 1^x = 1$ is a constant function.

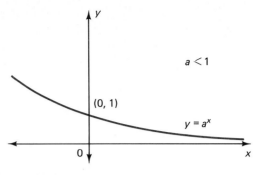

FIGURE 4

called the **natural exponential function.** Since e lies between 2 and 3, the graph of $y = e^x$ lies between the graphs of $y = 2^x$ and $y = 3^x$ shown in Figure 2.

The reason this particular exponential function is so important cannot be explained fully until we know some calculus. However, we have already seen in Section 6-1 that the base e arises in a natural way when considering continuous compounding of interest.

EXAMPLE 2 By using values from Table A.3.3, sketch the graph of the functions $y = e^x$ and $y = e^{-x}$ for $-2 \le x \le 2$.

Solution From Table A.3.3 we get the values of these two functions (rounded to two decimal places) shown in Table 4.

TABLE 4

x	-2	-1.5	-1	-0.5	0	0.5	1	1.5	2
e^x	0.14	0.22	0.37	0.61	1.00	1.65	2.72	4.48	7.39
e^{-x}	7.39	4.48	2.72	1.65	1.00	0.61	0.37	0.22	0.14

Plotting these points and joining them by a smooth curve, we obtain the graphs shown in Figure 5.

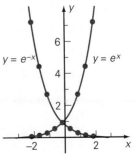

FIGURE 5

The graph of e^x is an example of a growing exponential function, illustrated in Figure 3. The base $e \approx 2.7 > 1$. The function e^{-x} is a decaying exponential function such as illustrated in Figure 4. Note that $e^{-x} = (e^{-1})^x$, so the base is $e^{-1} \approx 0.37 < 1$. Note that the graph of $y = e^{-x}$ is the reflection of the graph of $y = e^x$ in the y-axis. There is, in fact, a general property for any function f that the graph of $y = f(-x)$ is the reflection in the y-axis of the graph of $y = f(x)$. Remember that as with any exponential function, $e^x > 0$ for all values of x. ☛ 11

Answer

x	-4	-3	-2	-1	0	1	2	3	4
$e^{x/2}$	0.14	0.22	0.37	0.61	1.00	1.65	2.72	4.48	7.39

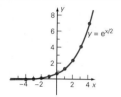

EXAMPLE 3 What are the domain and range of the function $y = 2 - e^{-x}$?

Solution Since e^{-x} is defined for any x, the domain is the set of all real numbers. Since $e^{-x} > 0$ for all x, $2 - e^{-x} < 2$. The range of this function is the set of all real numbers less than 2, since e^{-x} takes all positive values.

EXAMPLE 4 (Population Growth) The population of a certain developing nation is found to be given in millions by the formula

$$P = 15e^{0.02t}$$

where t is the number of years measured from 1960. Determine the population in 1980 and the projected population in 2000, assuming this formula continues to hold until then.

Solution In 1980, $t = 20$ and so

$$P = 15e^{(0.02)(20)} = 15e^{0.4} = 15(1.4918) = 22.4 \quad \text{(to one decimal place)}.$$

So in 1980, the population would be 22.4 million. After a further 20 years, $t = 40$ and so

$$P = 15e^{(0.02)(40)} = 15e^{0.8} = 15(2.2255) = 33.4 \quad \text{(to one decimal place)}.$$

Thus in 2000, the projected population will be 33.4 million. ☛ 12

EXAMPLE 5 (Population Growth) The population of a certain city at time t (measured in years) is given by

$$P(t) = P_0 e^{0.03t} \qquad P_0 = 1.5 \text{ million}.$$

What is the percentage growth per year?

Solution After n years, the population is

13. If $ce^{0.05t} = c(1 + i)^t$ for all values of t, what is the value of i?

$$P_0(1 + i)^n = P_0 e^{0.03n}$$
$$(1 + i)^n = e^{0.03n}$$
$$1 + i = e^{0.03} = 1.0305$$
$$i = 0.0305.$$

Therefore, since $i = R/100$.

$$R = 100(0.0305) = 3.05.$$

The population grows by 3.05% per year. ☞ **13**

Answer $i = e^{0.05} - 1 = 0.05127.$

Note In these two examples of population growth, the population was expressed in terms of a certain natural exponential function. It is in fact very common in the case of variables that are increasing (or decaying) to express them using the base e in the form ce^{kt} (or ce^{-kt}), where c and k are positive constants. It is not essential, however, to use base e, and in Examples 5 and 6 of Section 6-1 we already saw cases where a population growth is expressed with a different base, in the form $c(1 + i)^t$. Both these forms are correct, and as Example 5 here shows, are equivalent.

EXERCISES 6-2

(1–4) Given $f(x) = e^x$, show that:

1. $f(0) = 1$

2. $f(x + y) = f(x) \cdot f(y)$

3. $\dfrac{f(x)}{f(y)} = f(x - y)$

4. $[f(x)]^n = f(nx)$

(5–12) Find the domain and range of the following functions.

5. $y = f(x) = 2^{-x}$

6. $y = f(x) = (0.2)^{-x}$

7. $y = f(t) = -2^t$

8. $y = f(u) = -3e^{-u}$

9. $y = g(x) = 5 + 2e^x$

10. $y = f(t) = 3 - 2e^{-t}$

*11. $y = f(x) = \dfrac{1}{3 + 2^x}$

*12. $y = f(x) = (2 + 3e^{-x})^{-1}$

(13–20) Construct the graphs of the following exponential functions by calculating and plotting a few points.

13. $y = (\frac{3}{2})^x$

14. $y = (\frac{1}{2})^x$

15. $y = (\frac{1}{2})^{-x}$

16. $y = 3^{-x}$

17. $y = (\frac{1}{3})^x$

18. $y = -3^x$

19. $y = (\frac{2}{3})^{-x}$

20. $y = (\frac{3}{4})^x$

(21–32) By using the graphs of $y = e^x$ and $y = e^{-x}$, sketch roughly the graphs of the following exponential functions.

21. $y = -e^x$

22. $y = e^{2x}$

23. $y = 1 + e^{-x}$

24. $y = 3 - 2e^x$

25. $y = e^{|x|}$

26. $y = e^{-|x|}$

27. $y = e^{x+|x|}$

28. $y = e^{x-|x|}$

29. $y = e^{x^2}$

30. $y = e^{-x^2}$

*31. $y = \dfrac{1}{1 + e^x}$

*32. $y = \dfrac{e^x}{1 + 2e^x}$

33. A population of microorganisms doubles every 20 minutes. If 200 organisms are present initially, find a formula for the population size after t hours.

34. A population of microorganisms is doubling every 45 minutes. If 5000 organisms are present initially, how many will there be after:

a. 3 hours?

b. 6 hours?

c. t hours?

35. During the autumn, half of a population of flies dies off on average every three days. If initially the population size is one million, find the number of survivors after:

a. 3 weeks.

b. t weeks.

36. *(Population Growth)* The population of a certain city at time t (measured in years) is given by the formula

$$P = 50,000e^{0.05t}.$$

Calculate the population:

a. When $t = 10$.   **b.** When $t = 15$.

37. *(Population Decline)* A certain depressed economic region has a population which is in decline. In 1970, its population was 500,000, and thereafter its population was given by the formula

$$P = 500,000e^{-0.02t}$$

where t is time in years. Find the population in 1980. Assuming this trend continues, find the projected population in the year 2000.

38. In Exercise 36, calculate the percentage growth of the population per year.

39. In Exercise 37, calculate the percentage decline in the population per year. Is it constant or does it depend on t?

40. *(Profits Growth)* The profits of a certain company have been increasing by an average of 12% per year between 1975 and 1980. In 1980, they were $5.2 million. Assuming that this growth rate continues, find the profits in 1985.

41. *(Exponential Depreciation)* A machine is purchased for $10,000 and depreciates continuously from the date of purchase. Its value after t years is given by the formula

$$V = 10,000e^{-0.2t}.$$

a. Find the value of the machine after 8 years.

b. Find the percentage decline in value each year.

*42. *(Break-even Analysis)* By examining its competitors, a manufacturing company concludes that the number N of its employees increases exponentially with its volume of weekly sales x according to the formula $N = 100e^{0.02x}$. The average wage cost is $6 per hour with a 35-hour workweek. The company's product sells for $2000 each. Plot graphs of the weekly wage bill and the weekly revenue as functions of x for $10 < x < 130$ and estimate graphically the interval of x-values in which the company can make a profit.

◥ 6-3 LOGARITHMS

The inverse of a function $f(x)$ is obtained by solving the equation $y = f(x)$ for x, thus expressing x as a function of y: $x = f^{-1}(y)$. We can consider the possibility of constructing the inverse of the function a^x. In order to do so, we must solve the equation $y = a^x$ for x. Now such an equation cannot be solved in terms of the functions we know so far, so a new name must be invented for the solution. We write the solution in the form $x = \log_a y$, which we call the **logarithm of y with base a.** Thus

$$x = \log_a y \quad \text{if and only if} \quad y = a^x.$$

From the statement $y = a^x$, we see that a must be raised to the power x in order to obtain y. This leads us to an alternative verbal definition (since $x = \log_a y$).

$\log_a y$ is the power to which a must be raised in order to get y.

The function a^x is only defined when $a > 0$. Furthermore, when $a = 1$, then $1^x = 1$ for all x, and this function cannot have an inverse. Therefore, in these definitions *a can be any positive number except 1*. From now on it will always be understood that the base a satisfies the conditions $a > 0$, $a \neq 1$.

EXAMPLE 1 Construct the graph of the logarithm function with base 2. What are the domain and range of this function?

Solution Let us use x as the independent variable and write

$$y = \log_2 x.$$

According to the definition this means the same thing as

$$x = 2^y.$$

(Note that x and y have been interchanged and $a = 2$.) We can now construct Table 5, in which we give a series of values for y and calculate the corresponding values of x. For example, when $y = -2$, $x = 2^{-2} = 1/2^2 = 0.25$, so the point $(x, y) = (0.25, -2)$ lies on the graph. The tabulated points are plotted in Figure 6 and joined by a smooth curve. Note that the x-axis is drawn horizontally since when we write $y = \log_2 x$, we are implying that x is the independent variable.

TABLE 5

y	-2	-1.5	-1	-0.5	0	0.5	1	1.5	2	3
x	0.25	0.354	0.5	0.707	1	1.414	2	2.828	4	8

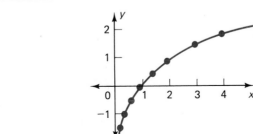

FIGURE 6

The domain of this function is the set of positive real numbers and its range is the set of all real numbers. This is true for $y = \log_a x$ with any base a.

☞ 14

☞ **14.** Graph $y = 3^x$ and $y = \log_3 x$ on the same axes.

Answer In the two tables, values of x and y are interchanged. Observe that the graphs are reflections of each other in the line $y = x$.

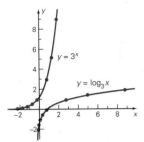

x	-2	-1.5	-1	-0.5	0	0.5	1	1.5	2
$y = 3^x$	0.11	0.19	0.33	0.58	1.00	1.73	3	5.20	9
$y = \log_3 x$	-2	-1.5	-1	-0.5	0	0.5	1	1.5	2
$x = 3^y$	0.11	0.19	0.33	0.58	1.00	1.73	3	5.20	9

As with any other inverse function, the graph of the logarithm function $x = \log_a y$ for a general base a can be obtained from the graph of the exponential function $y = a^x$ simply by flipping the axes. The two graphs are illustrated in Figure 7 for a typical case with $a > 1$. Note that:

> $\log_a y$ is defined if and only if $y > 0$;
>
> when $a > 1$, $\log_a y > 0$ when $y > 1$ and $\log_a y < 0$ when $y < 1$.

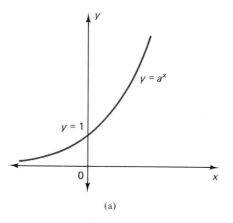

(a)

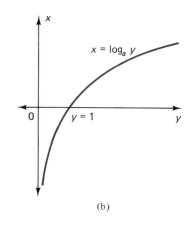

(b)

FIGURE 7

The statement that $x = \log_a y$ means exactly the same thing as the statement that $y = a^x$. For example, the statement $\log_3 9 = 2$ is true because it means the same as the statement $3^2 = 9$. (Here $a = 3$, $y = 9$, and $x = 2$.) Of these two equivalent statements, $x = \log_a y$ is called the **logarithmic form** and $y = a^x$ is called the **exponential form.**

EXAMPLE 2 Write each statement in logarithmic form.

   (a) $2^4 = 16$ (b) $\left(\tfrac{1}{2}\right)^{-3} = 8$

Write each statement in exponential form to verify that each is correct.

   (c) $\log_4 8 = \tfrac{3}{2}$ (d) $\log_{27}\left(\tfrac{1}{9}\right) = -\tfrac{2}{3}$

Solution

   (a) We have $2^4 = 16$. Comparing this with the equation $a^x = y$, we see that $a = 2$, $x = 4$, and $y = 16$. The logarithmic form, $x = \log_a y$, is

$$\log_2 16 = 4.$$

   (b) Comparing $\left(\tfrac{1}{2}\right)^{-3} = 8$ with $a^x = y$, we have $a = \tfrac{1}{2}$, $x = -3$, and $y = 8$. The logarithmic form is

$$\log_{1/2} 8 = -3. \quad \text{☛ 15}$$

☛ **15.** Write in logarithmic form

(a) $5^{-2} = 0.04$;   (b) $\left(\dfrac{2}{5}\right)^{-3} = 15.625$.

Answer (a) $\log_5 (0.04) = -2$;
(b) $\log_{2/5} (15.625) = -3$.

(c) Comparing $\log_4 8 = \frac{3}{2}$ with the equation $\log_a y = x$, we have $a = 4$, $y = 8$, and $x = \frac{3}{2}$. The exponential form, $y = a^x$, is $8 = 4^{3/2}$. Since this is clearly a true statement, it follows that the given logarithmic form must also be true.

(d) Here $a = 27$, $y = \frac{1}{9}$, and $x = -\frac{2}{3}$ and the exponential form is $27^{-3/2} = \frac{1}{9}$. Again, this is easily verified.   **☛ 16**

EXAMPLE 3 Find the values of: (a) $\log_2 16$; (b) $\log_{1/3} 243$.

Solution

(a) Let $x = \log_2 16$. Then from the definition of logarithm, it follows that $16 = 2^x$. But $16 = 2^4$, so $x = 4$. Therefore $\log_2 16 = 4$.

(b) Let $x = \log_{1/3} 243$. Then, from the definition, $243 = \left(\frac{1}{3}\right)^x$. But $243 = 3^5 = \left(\frac{1}{3}\right)^{-5}$. Therefore $x = -5$.   **☛ 17**

EXAMPLE 4 Evaluate $2^{\log_4 9}$.

Solution Let $x = \log_4 9$, so that $9 = 4^x$. The quantity we wish to calculate is therefore

$$2^{\log_4 9} = 2^x = (4^{1/2})^x = (4^x)^{1/2} = 9^{1/2} = 3.$$

Some Properties of Logarithms

In the definition of the logarithm, let us take $x = 0$. Then

$$y = a^x = a^0 = 1.$$

Therefore the statement $\log_a y = x$ takes the form

$$\log_a 1 = 0.$$

Thus we see that *the logarithm of 1 with any base is equal to 0.*
Next, let us take $x = 1$. Then

$$y = a^x = a^1 = a.$$

Therefore the statement $\log_a y = x$ becomes

$$\log_a a = 1.$$

So *the logarithm of any positive number with the same base is always equal to 1.*

Logarithms also have the following four properties.

$$\log_a(uv) = \log_a u + \log_a v \qquad a^x \cdot a^y = a^{x+y} \qquad (1)$$

$$\log_a\left(\frac{u}{v}\right) = \log_a u - \log_a v \qquad \frac{a^x}{a^y} = a^{x-y} \qquad (2)$$

$$\log_a\left(\frac{1}{v}\right) = -\log_a v \qquad a^{-x} = \frac{1}{a^x} \qquad (3)$$

$$\log_a(u^n) = n \log_a u \qquad (a^x)^n = a^{xn} \qquad (4)$$

(Beside each property, we have also given the related property of exponents.) These properties form the basis for the use of logarithms in performing arithmetic calculations. We shall prove them at the end of this section. Meanwhile, we shall illustrate them with two examples.

EXAMPLE 5

(a) Since $8 = 2^3$, it follows that $\log_2 8 = 3$. Since $16 = 2^4$, it follows that $\log_2 16 = 4$. From Property (1), therefore,

$$\log_2(8 \cdot 16) = \log_2 8 + \log_2 16 \qquad \text{or} \qquad \log_2(128) = 3 + 4 = 7.$$

This statement is obviously true since $128 = 2^7$.

(b) We can also calculate $\log_2(128)$ by making use of Property (4) of logarithms.

$$\log_2(128) = \log_2(2^7) = 7 \log_2 2$$

But $\log_a a = 1$ for any base a, so $\log_2 2 = 1$. Therefore $\log_2 128 = 7$.

(c) Since $81 = \left(\frac{1}{3}\right)^{-4}$, it follows that $\log_{1/3} 81 = -4$. Therefore, by Property (3),

$$\log_{1/3}\left(\tfrac{1}{81}\right) = -\log_{1/3} 81 = -(-4) = 4.$$

This result is clearly correct since $\frac{1}{81} = \left(\frac{1}{3}\right)^4$.

EXAMPLE 6 If $x = \log_2 3$, express the following quantities in terms of x.

(a) $\log_2\left(\frac{1}{3}\right)$ (b) $\log_2 \frac{2}{3}$ (c) $\log_2 18$ (d) $\log_2 \sqrt{\frac{27}{2}}$

Solution

(a) $\log_2\left(\frac{1}{3}\right) = -\log_2 3 \qquad$ (Property 3)
$$= -x$$

(b) $\log_2\left(\frac{2}{3}\right) = \log_2 2 - \log_2 3 \qquad$ (Property 2)
$$= 1 - x$$

(Here we used the fact that $\log_a a = 1$ for any base a.)

(c) $\log_2 (18) = \log_2 (2 \cdot 3^2)$
$$= \log_2 2 + \log_2 3^2 \qquad \text{(Property 1)}$$
$$= 1 + 2 \log_2 3 \qquad \text{(Property 4)}$$
$$= 1 + 2x$$

(d) $\log_2 \sqrt{\frac{27}{2}} = \log_2 (\frac{27}{2})^{1/2}$
$$= \tfrac{1}{2} \log_2 (\tfrac{27}{2}) \qquad \text{(Property 4)}$$
$$= \tfrac{1}{2} [\log_2 27 - \log_2 2] \qquad \text{(Property 2)}$$
$$= \tfrac{1}{2} [\log_2 3^3 - 1]$$
$$= \tfrac{1}{2} [3 \log_2 3 - 1] \qquad \text{(Property 4)}$$
$$= \tfrac{1}{2} (3x - 1) \qquad \bullet \ 18$$

☛ **18.** In Example 6, express the following in terms of x.
(a) $\log_2 (\tfrac{1}{6})$; (b) $\log_2 \sqrt{2}$;
(c) $\log_2 \sqrt[3]{\dfrac{3}{4}}$.

Natural Logarithms

We can also form logarithms with base e. These are called **natural logarithms** (or **Napierian logarithms**). They are denoted by the symbol ln. The definition is

$$y = e^x \qquad x = \log_e y = \ln y.$$

That is, the function $x = \ln y$ is the inverse of the function $y = e^x$.

As with all logarithms, $\ln y$ is defined only for $y > 0$. If $y > 1$, then $\ln y$ is positive, while if $y < 1$, $\ln y$ is negative.

The natural logarithm has properties corresponding to those discussed above for a general base. We list them again here:

$$\ln 1 = 0 \qquad \ln e = 1$$

$$\ln(uv) = \ln u + \ln v \qquad \ln\left(\frac{u}{v}\right) = \ln u - \ln v$$

$$\ln\left(\frac{1}{v}\right) = -\ln v \qquad \ln(u^n) = n \ln u.$$

With a suitable pocket calculator, we can find the natural logarithm of any number by pressing the appropriate button. If no calculator is available, a table of natural logarithms is provided in Appendix III (see Table A.3.2); Example 7 illustrates its use.

EXAMPLE 7 Using Table A.3.2, find the values of the following.

(a) $\ln 3.4$ (b) $\ln 100$ (c) $\ln 340$ (d) $\ln 0.34$

Solution

(a) From the table, we find directly that $\ln 3.4 = 1.2238$.

Answer (a) $-1 - x$; (b) $\tfrac{1}{2}$;
(c) $\tfrac{1}{3}(x - 2)$.

(b) At the bottom of the table, we find that $\ln 10 = 2.3026$. Therefore

$$\ln 100 = \ln 10^2 = 2 \ln 10 = 2(2.3026) = 4.6052.$$

(c) $\ln 340 = \ln (3.4 \times 100)$

$$= \ln 3.4 + \ln 100$$

$$= 1.2238 + 4.6052 = 5.8290$$

Observe that Table A.3.2 provides the natural logarithms only of numbers lying between 1 and 10. For numbers lying outside this range, an appropriate power of 10 has to be extracted (in this example, 100).

(d) $\ln(0.34) = \ln (3.4 \times 10^{-1})$

$$= \ln 3.4 + \ln (10^{-1})$$

$$= \ln 3.4 - \ln 10$$

$$= 1.2238 - 2.3026 = -1.0788 \quad \text{☛ 19}$$

☛ **19.** Using Table A.3.2, evaluate (a) $\ln 5.48$; (b) $\ln 0.548$.

EXAMPLE 8 Given that $\ln 2 = 0.6931$ and $\ln 3 = 1.0986$ to four decimal places, evaluate:

(a) $\ln 18$ (b) $\ln\sqrt{54}$ (c) $\ln(\frac{1}{4}e)$

Solution Using the properties of natural logarithms above, we get

(a) $\ln 18 = \ln (2 \cdot 3^2) = \ln 2 + \ln(3^2)$

$$= \ln 2 + 2 \ln 3$$

$$= 0.6931 + 2(1.0986) = 2.8903.$$

(b) $\ln\sqrt{54} = \frac{1}{2} \ln 54 = \frac{1}{2} \ln(2 \cdot 3^3)$

$$= \frac{1}{2}(\ln 2 + \ln 3^3)$$

$$= \frac{1}{2}(\ln 2 + 3 \ln 3)$$

$$= \frac{1}{2}[0.6931 + 3(1.0986)] = 1.9945.$$

(c) $\ln (e/4) = \ln e - \ln 4$

$$= 1 - \ln (2^2) = 1 - 2 \ln 2$$

$$= 1 - 2(0.6931) = -0.3862.$$

EXAMPLE 9 Simplify the following expression without using tables or calculators.

$$E = \ln 2 + 16 \ln \left(\tfrac{16}{15}\right) + 12 \ln \left(\tfrac{25}{24}\right) + 7 \ln \left(\tfrac{81}{80}\right)$$

Solution $E = \ln 2 + 16 \ln \left(\dfrac{2^4}{3 \cdot 5}\right) + 12 \ln \left(\dfrac{5^2}{2^3 \cdot 3}\right) + 7 \ln \left(\dfrac{3^4}{2^4 \cdot 5}\right)$

$$= \ln 2 + 16(\ln 2^4 - \ln 3 - \ln 5)$$

$$+ 12(\ln 5^2 - \ln 2^3 - \ln 3) + 7(\ln 3^4 - \ln 2^4 - \ln 5)$$

$$= \ln 2 + 16(4 \ln 2 - \ln 3 - \ln 5)$$

$$+ 12(2 \ln 5 - 3 \ln 2 - \ln 3) + 7(4 \ln 3 - 4 \ln 2 - \ln 5)$$

Answer (a) 1.7011; (b) −0.6015.

☞ 20. Simplify the following expressions using properties of natural logarithms:

(a) $\ln (4x^2) - \ln (6x)$;

(b) $\ln (e^{2x} \sqrt{x})$;

(c) $\ln (1 - x^2) + \ln \left(\dfrac{1 - x}{1 + x} \right)$;

(d) $\ln (x^3) - \ln (x^2)$;

Answer (a) $\ln (\frac{2}{3}x)$; (b) $2x + \frac{1}{2} \ln x$; (c) $2 \ln (1 - x)$; (d) $\ln x$; (e) $\frac{3}{2}$.

$$= \ln 2(1 + 64 - 36 - 28) + \ln 3(-16 - 12 + 28)$$
$$+ \ln 5(-16 + 24 - 7)$$
$$= \ln 2 + \ln 5 = \ln (2 \cdot 5) = \ln 10 \quad \text{☞ 20}$$

EXAMPLE 10 Solve the following equations for x.

(a) $2 \ln (2x + 2) = \ln \left(1 + \dfrac{12x}{25} \right) + 2 \ln 10$

(b) $\log_x (3 - 2x) = 2$

Solution

(a) Using the properties of logarithms, we can write the given equation in the form

$$\ln(2x + 2)^2 = \ln \left(1 + \frac{12x}{25} \right) + \ln 100$$

$$= \ln \left[100 \left(1 + \frac{12x}{25} \right) \right] = \ln (100 + 48x).$$

Therefore

$$(2x + 2)^2 = 100 + 48x.$$

This is a quadratic equation whose solutions are easily seen to be $x = 12$ and $x = -2$. However, when $x = -2$, the term $\ln(2x + 2) = \ln(-2)$ in the original equation is not defined. So $x = -2$ cannot be a solution and $x = 12$ is the only solution.

(b) The statement $\log_x p = q$ is equivalent to the exponential form $p = x^q$. The given equation is of this form with $p = 3 - 2x$ and $q = 2$, so is equivalent to

$$3 - 2x = x^2.$$

☞ 21. Solve for x:

(a) $\ln (e^{\sqrt{x}}) = 2$;

(b) $\ln (x + 1) - \ln (x - 1) = 1$;

(c) $2 \ln (x - 1) = \ln (x + 3) + \ln 2$.

The roots of this quadratic equation are $x = 1$ and $x = -3$. But in the original equation, x is a base, and cannot equal 1 or be negative. So the given equation has no solution. **☞ 21**

Two important properties of natural logarithms are obtained by eliminating either x or y from the two defining equations, $y = e^x$, $\ln y = x$. Substituting $y = e^x$ into the second equation, we get $\ln y = \ln (e^x) = x$. Alternatively, substituting $x = \ln y$ into the first equation we get $y = e^x = e^{\ln y}$. Thus we have

---

Answer (a) 4; (b) $\dfrac{e + 1}{e - 1}$;

(c) 5. ($x = -1$ is not a solution.)

$$\boxed{e^{\ln y} = y \quad \text{for all } y > 0 \qquad \ln (e^x) = x \quad \text{for all } x}$$

For example, $e^{\ln 2} = 2$, $\ln (e^{-3}) = -3$.

Common Logarithms

At one time, logarithms were used extensively in carrying out arithmetical calculations involving multiplication, division, and the calculation of powers and roots. With the widespread availability of electronic calculators, such usage has diminished considerably, although in some areas logarithms are still used. (For example, a ship's navigator must still learn how to use logarithm tables and other tables to guard against the possibility of electronics failure.)

The logarithms ordinarily used for this purpose are called **common logarithms** and are obtained by using the number 10 as base (that is, $a = 10$). Thus the common logarithm of a number y is $\log_{10} y$; however, to avoid cumbersome notation, the common logarithm is usually denoted by $\log y$, the base being omitted.* So when the base is not written, it should be understood to be 10.

These statements are thus equivalent:

$$x = \log y \quad \text{and} \quad y = 10^x.$$

The following example demonstrates the application of Properties 1 to 4 to common logarithms.

☛ **22.** If $\log 2 = x$, find y such that
(a) $\log y = 2x + 1$;
(b) $\log y = \frac{1}{2}(x - 1)$.

EXAMPLE 11 Let us examine the general relations $y = 10^x$ and $x = \log y$ for certain values of x and y.

(a) $x = 1$:   Then $y = 10^1 = 10$, so $\log 10 = 1$.

(b) $x = 2$:   Then $y = 10^2 = 100$, so $\log 100 = 2$.

(c) $x = -1$:   Then $y = 10^{-1} = 0.1$, so $\log 0.1 = -1$.

(d) To four places, we find from Table A.3.1 that $\log 3 = 0.4771$. This means that $3 = 10^{0.4771}$. Using Property 1 of logarithms stated above, it follows that

$$\log 30 = \log 3 + \log 10 = 1.4771$$

$$\log 300 = \log 3 + \log 100 = 2.4771$$

$$\log(0.3) = \log 3 + \log 0.1 = -1 + 0.4771 = -0.5229.$$

(e) Also from Table A.3.1, we find that $\log 2 = 0.3010$. Thus

$$\log 6 = \log(3 \cdot 2) = \log 3 + \log 2 = 0.4771 + 0.3010 = 0.7781$$

and

$$\log 4 = \log(2^2) = 2 \log 2 = 0.6020. \quad ☛ \mathbf{22}$$

Answer (a) 40;   (b) $\sqrt{0.2}$.

*In some books, the notation $\log y$ is used to mean the natural logarithm of y.

Proof of Basic Properties of Logarithms

1. Let $x = \log_a u$ and $y = \log_a v$. Then from the defintion of logarithm,

$$u = a^x \quad \text{and} \quad v = a^y.$$

It follows, therefore, that

$$uv = (a^x)(a^y) = a^{x+y}$$

after using one of the fundamental properties of exponents. Consequently, from the definition of logarithm, it follows that $x + y$ must be the logarithm of uv with base a:

$$x + y = \log_a(uv).$$

In other words, substituting for x and y we have the required formula

$$\log_a(uv) = \log_a u + \log_a v$$

2. The second result may be obtained by considering u/v.

$$\frac{u}{v} = \frac{a^x}{a^y} = a^x a^{-y} = a^{x-y}$$

Thus $x - y = \log_a(u/v)$, or equivalently, $\log_a(u/v) = \log_a u - \log_a v$.

3. We use Property 2 and let $u = 1$. When $u = 1$, $\log_a u = 0$, and we get

$$\log_a\left(\frac{1}{v}\right) = -\log_a v.$$

4. Fourth, let $x = \log_a u$, so that $u = a^x$. Then $u^n = (a^x)^n = a^{xn}$. Thus $xn = \log_a(u^n)$, or

$$\log_a u^n = n \log_a u.$$

EXERCISES 6-3

(1–6) Verify the following statements and rewrite them in logarithmic form with an appropriate base.

1. $(27)^{-4/3} = \frac{1}{81}$
2. $(16)^{3/4} = 8$
3. $(125)^{2/3} = 25$
4. $8^{-5/3} = \frac{1}{32}$
5. $\left(\frac{8}{27}\right)^{-1/3} = \frac{3}{2}$
6. $\left(\frac{625}{16}\right)^{-3/4} = \frac{8}{125}$

(7–10) Write the following equations in exponential form and hence verify them.

7. $\log_3 27 = 3$
8. $\log_{1/9}\left(\frac{1}{243}\right) = \frac{5}{2}$
9. $\log_4\left(\frac{1}{2}\right) = -\frac{1}{2}$
10. $\log_2\left(\frac{1}{4}\right) = -2$

(11–22) Find the values of the following expressions by using the definition of logarithm.

11. $\log_2 512$
12. $\log_{27} 243$
13. $\log_{\sqrt{2}} 16$
14. $\log_8 128$
15. $\log_2 0.125$
16. $\log_a 32 \div \log_a 4$
17. $10^{\log 100}$
18. $10^{\log 2}$
19. $\log_4(2^p)$
20. $\log_2(4^p)$
21. $2^{\log_{1/2} 3}$
22. $3^{\log_9 2}$

(23–28) Given that $\log 5 = 0.6990$ and $\log 9 = 0.9542$, evaluate the following expressions without using tables or calculators.

23. $\log 2$

24. $\log 3$

25. $\log 12$

26. $\log 75$

27. $\log 30$

28. $\log \sqrt{60}$

(29–36) Write each of the following as the logarithm of a single expression.

29. $\log (x + 1) - \log x$

30. $\log x + \log 5 - \log y$

31. $2 \ln x - 3 \ln y + 4 \ln t$

32. $\ln t - 2 \ln u + 3 \ln v$

33. $2 \ln x + x \ln 3 - \frac{1}{2} \ln (x + 1)$

34. $x \ln 2 + 5 \ln (x - 1) - 2 \ln (x + 3)$

35. $\log x + 2 \log y - 3$

36. $2 + 3 \ln t - 4 \ln x$

(37–50) Solve the following equations for x without using tables or calculators.

37. $\log_2 (x + 3) = -1$

38. $\log_x 4 = 2$

39. $\log_x (5x - 6) = 2$

40. $\log_x (6 - x) = 2$

41. $\log_x (6 - 5x) = 2$

42. $\ln (x + 2) - \ln (x - 1) = \ln 4$

43. $\ln (10x + 5) - \ln (4 - x) = \ln 2$

44. $\log_3 3 + \log_3 (x + 1) - \log_3 (2x - 7) = 4$

45. $\ln x = \ln 3 + 2 \ln 2 - \frac{3}{4} \ln 16$

46. $\ln (4x - 3) = \ln (x + 1) + \ln 3$

47. $\log (2x + 1) - \log (3 - x) = \log 5$

48. $\log (2x + 1) + \log (x + 3) = \log (12x + 1)$

49. $\log (x - 2) = \log (3x - 2) - \log (x - 2)$

50. $\log (x + 3) + \log (x - 1) = \log (1 - x)$

51. If $\ln (x - y) = \ln x - \ln y$, find x in terms of y.

52. Show that
$$x^{\ln y - \ln z} \cdot y^{\ln z - \ln x} \cdot z^{\ln x - \ln y} = 1$$

(53–54) Prove the following without using tables or calculators.

53. $7 \log \left(\frac{16}{15}\right) + 5 \log \left(\frac{25}{24}\right) + 3 \log \left(\frac{81}{80}\right) = \log 2$

54. $3 \log \left(\frac{36}{25}\right) + \log \left(\frac{6}{27}\right)^3 - 2 \log \left(\frac{16}{125}\right) = \log 2$

(55–60) Evaluate the following using Table A.3.2 in Appendix III.

55. $\ln 3.41$ 56. $\ln 2.68$

57. $\ln 84.2$ 58. $\ln 593$

59. $\ln 0.341$ 60. $\ln 0.00917$

61. If $f(x) = \ln x$, show that:

   a. $f(xy) = f(x) + f(y)$ b. $f\left(\frac{x}{y}\right) = f(x) - f(y)$

   c. $f(ex) = 1 + f(x)$ d. $f\left(\frac{e}{x}\right) = 1 - f(x)$

   e. $f(x^n) = nf(x)$ f. $f(x) + f\left(\frac{1}{x}\right) = 0$

62. If $f(x) = \log x$, show that $f(1) + f(2) + f(3) = f(1 + 2 + 3)$.

(63–74) Find the domain of the following functions.

63. $f(x) = \ln (x - 2)$ 64. $f(x) = \ln (3 - x)$

65. $f(x) = \ln (4 - x^2)$ 66. $f(x) = \ln (9 + x^2)$

67. $f(x) = 1 + \ln x$ 68. $f(x) = \log |x - 3|$

69. $f(x) = \dfrac{1}{\ln x}$ 70. $f(x) = \dfrac{1}{1 - \ln x}$

71. $f(x) = \dfrac{1}{1 - e^x}$ 72. $f(x) = \dfrac{e^x}{3 - e^x}$

73. $f(x) = \dfrac{\ln x}{1 + \ln x}$ 74. $\sqrt{\ln x}$

(75–82) Using the graph of $y = \ln x$, roughly sketch the graphs of the following functions.

75. $f(x) = \ln (-x)$ 76. $f(x) = \ln |x|$

77. $f(x) = 1 + \ln x$ 78. $f(x) = -\ln x$

79. $f(x) = 2 \ln x$ 80. $f(x) = \ln x^2$

81. $f(x) = \ln(x - 3)$ 82. $f(x) = \ln |x + 2|$

83. *(Cost Function)* A manufacturing company finds that the

cost of producing x units per hour is given by the formula

$$C(x) = 5 + 10 \log (1 + 2x).$$

Calculate:

a. The cost of producing 5 units per hour.

b. The extra cost of increasing the production rate from 5 units per hour to 10 units per hour.

c. The extra cost of increasing from 10 to 15 units per hour.

84. *(Advertising and Sales)* A company finds that the number of dollars per week y that it must spend on advertising in order to sell x units of its product is given by

$$y = 200 \ln \left(\frac{400}{500 - x} \right).$$

Calculate the amount of advertising expenditure needed to sell:

a. 100 units.

b. 300 units.

c. 490 units.

85. *(Cost Function)* A company is expanding its production facility and has the choice between two designs. The cost functions for the two are $C_1(x) = 3.5 + \log (2x + 1)$ and $C_2(x) = 2 + \log (60x + 105)$, where x is the rate of production. Find the rate x at which the two designs have the same costs. For larger values of x which design is the cheaper?

86. *(Animal Physiology)* If W is the weight of an average animal of a certain species at age t, it is often found that

$$\ln W - \ln (A - W) = B(t - C)$$

where A, B, and C are certain constants. Express W as an explicit function of t.

◼ 6-4 APPLICATIONS AND FURTHER PROPERTIES OF LOGARITHMS

Exponential Equations

One of the most important applications of logarithms is to solve certain types of equations in which the unknown variable appears as an exponent. Consider the following examples.

EXAMPLE 1 *(Population Growth)* In 1980, the population of a certain city was 2 million and was increasing at the rate of 5% each year. When will the population pass the 5 million mark, assuming this growth rate continues?

Solution At a rate of increase of 5%, the population is multiplied by a factor 1.05 each year. After n years, starting from 1980, the population level is therefore

$$2(1.05)^n \text{ million.}$$

We require the value of n for which this level is 5 million, so we have

$$2(1.05)^n = 5 \quad \text{or} \quad (1.05)^n = 2.5.$$

Observe that in this equation, the unknown quantity n appears as an exponent. We can solve it by taking logarithms of both sides. It does not matter which base we use for the logarithms, but common logarithms are usually the most convenient. We obtain

$$\log (1.05)^n = \log 2.5$$

or, using Property 4 of logarithms,

$$n \log 1.05 = \log 2.5$$

Therefore

$$n = \frac{\log 2.5}{\log 1.05} = \frac{0.3979}{0.0212} \quad \text{(from Table A.3.1.)}$$

$$= 18.8$$

It therefore takes 18.8 years for the population to climb to 5 million. This level will be reached during 1998. ☛ 23

EXAMPLE 2 *(Investment)* The sum of $100 is invested at 6% compound interest per annum. How long does it take the investment to increase in value to $150?

Solution At 6% interest per annum, the investment grows by a factor of 1.06 each year. Therefore, after n years, the value is $100(1.06)^n$. Setting this equal to 150, we obtain the following equation for n:

$$100(1.06)^n = 150 \quad \text{or} \quad (1.06)^n = 1.5.$$

We take logarithms of both sides and simplify.

$$\log (1.06)^n = n \log (1.06) = \log (1.5)$$

$$n = \frac{\log (1.5)}{\log (1.06)} = \frac{0.1761}{0.0253} = 6.96$$

Thus it takes almost 7 years for the investment to increase in value to $150. ☛ 24

It will be seen that these two examples lead to an equation of the type

$$a^x = b$$

where a and b are two given positive constants and x is the unknown variable. Such an equation can always be solved by taking logarithms of both sides.

$$\log (a^x) = x \log a = \log b \quad \text{and so} \quad x = \frac{\log b}{\log a}. \quad ☛ 25$$

There is no difference in principle here between problems involving growing exponential functions ($a > 1$) and those involving decaying exponentials $a < 1$). The following example involves a decaying exponential function.

EXAMPLE 3 *(Drinking and Driving)* Shortly after consuming a substantial dose of whiskey, the alcohol level in a person's blood rises to a level of 0.3 milligram per milliliter. Thereafter, this level decreases according to the formula $(0.3)(0.5)^t$, where t is the time measured in hours from the time at which the peak level is reached. How long is it before the person can legally drive her automobile? (In her locality, the legal limit is 0.08 milligram per milliliter of blood alcohol.)

☛ **23.** In Example 1, find the value of n using natural logs instead of common logs.

Answer
$$n = \frac{\ln (2.5)}{\ln (1.05)} = \frac{0.9163}{0.04879} = 18.78.$$

☛ **24.** In Example 2, find the value of n using natural logs instead of common logs.

Answer
$$n = \frac{\ln (1.5)}{\ln (1.06)} = \frac{0.4055}{0.05827} = 6.959.$$

☛ **25.** Solve for x:
(a) $2^x = 5$;   (b) $5^x = 2(3^x)$.

Answer (a) $x = \frac{\log 5}{\log 2} = 2.322$;

(b) $x = \frac{\log 2}{\log 5 - \log 3} = 1.357$.

Solution We wish to find the value of t at which

$$(0.3)(0.5)^t = 0.08.$$

That is,

$$(0.5)^t = \frac{0.08}{0.3} = 0.267.$$

Taking logarithms, we obtain

$$\log (0.5)^t = t \log (0.5) = \log (0.267)$$

and so,

$$t = \frac{\log (0.267)}{\log (0.5)} = \frac{(-0.5735)}{(-0.3010)} \qquad \text{(from Table A.3.1)}$$

$$= 1.91$$

It therefore takes 1.91 hours before the person is legally fit to drive. ☛ 26

☛ **26.** If, in Example 3, the legal limit were 0.05 instead of 0.08, how long would it then take to be legally fit to drive?

In Example 9 of Section 6-1 we showed how to calculate an effective annual rate of interest given the nominal rate. If we wish to go the other way round, for continuous compounding, we need logarithms.

EXAMPLE 4 *(Investment)* What nominal rate of interest, when compounded continuously, gives the same growth over a whole year as a 10% annual rate of interest?

Solution A sum P invested at a nominal rate of interest R percent compounded continuously has a value Pe^i after 1 year, with $i = R/100$. (Take $x = 1$ in the formula for continuous compounding.) If invested at 10% per annum, it would increase by a factor of 1.1 during each year. Therefore we must let

$$Pe^i = (1.1)P \text{ or } e^i = 1.1.$$

If we take natural logarithms of both sides, we get

$$\ln (e^i) = \ln (1.1).$$

But $\ln (e^x) = x$ for any real number x, so

$$i = \ln (1.1) = 0.0953.$$

Therefore $R = 100i = 9.53$.

So 10% interest compounded annually is equivalent to the annual growth provided by a 9.53% nominal rate of interest compounded continuously.

Answer $t = \dfrac{\log (0.05 \div 0.3)}{\log (0.5)}$

$= \dfrac{-0.7782}{-0.3010} = 2.58$ hours.

The following formulas summarize the procedures for changing between nominal and effective rates for continuous compounding.

27. If the effective rate is 15%, what is the nominal rate if compounding is continuous?

$$
\begin{array}{lll}
\text{Nominal to effective:} & i_{\text{eff}} = e^i - 1, & i = \dfrac{R_{\text{nom}}}{100} \\[3mm]
\text{Effective to nominal:} & i = \ln(1 + i_{\text{eff}}), & i_{\text{eff}} = \dfrac{R_{\text{eff}}}{100}
\end{array}
$$

☞ 27

Change of Base

Any exponential function can be written in terms of a natural exponential function. Let $y = a^x$. Then, since we can write $a = e^{\ln a}$, it follows that

$$y = (e^{\ln a})^x = e^{(\ln a)x}.$$

Thus we have

Answer 100 ln (1.15) = 13.98%.

> **Base Change Formula for Exponentials**
> $$a^x = e^{kx} \quad \text{where} \quad k = \ln a.$$

Thus any exponential function $y = a^x$ can be written in the equivalent form $y = e^{kx}$, where $k = \ln a$.

EXAMPLE 5 In Example 3, the alcohol level in a person's blood at a time t was given by the formula $(0.3)(0.5)^t$ mg/ml. We can write this in terms of e,

$$(0.5)^t = e^{kt}$$

where

$$k = \ln(0.5) = \ln 5 - \ln 10 = 1.6094 - 2.3026 = -0.69$$

to two decimal places. Therefore the alcohol level after t hours is $(0.3)e^{-(0.69)t}$.

☞ 28

28. Express the following in the form e^{kt}:
(a) 2^t; (b) 0.2^t; (c) $(1 + i)^t$.

It is the usual practice to write any growing exponential function a^x in the form e^{kx}, where $k = \ln a$. A decaying exponential function, defined by a^x with $a < 1$, would normally be written as e^{-kx}, where k is a positive constant given by $k = -\ln a$. The constant k is known as the **specific growth rate** for the function e^{kx} and as the **specific decay rate** for the decaying exponential e^{-kx}.

When an equation must be solved for an unknown in an exponent and the base is e, it is generally easier to use natural logarithms rather than common logarithms.

Answer (a) $e^{(0.6931)t}$; (b) $e^{-(1.609)t}$;
(c) $e^{[\ln(1+i)]t}$.

EXAMPLE 6 *(Population Growth)* The population of a certain developing nation is given in millions by the formula

$$P = 15e^{0.02t}$$

where t is the time in years measured from 1970. When will the population reach 25 million, assuming this formula continues to hold?

Solution Setting $P = 25$, we obtain the equation

$$15e^{0.02t} = 25 \quad \text{or} \quad e^{0.02t} = \tfrac{25}{15} = 1.667.$$

Again, we have an equation in which the unknown variable appears in the exponent, and we can solve for t by taking the logarithm of both sides. However, because the exponential has base e, it is easiest to take natural logarithms, since $e^{0.02t} = 1.667$ means the same as $0.02t = \ln 1.667$. Therefore

$$t = \frac{\ln 1.667}{0.02} = \frac{0.5108}{0.02} = 25.5.$$

It therefore takes 25.5 years for the population to reach 25 million, which it does therefore midway through 1995.

It is possible to express logarithms with respect to one base in terms of logarithms with respect to any other base. This is done by means of the **base-change formula,** which says that

---

Base-Change Formula for Logarithms

$$\log_a y = \frac{\log_b y}{\log_b a}.$$

Before proving this formula let us look at two important special cases. First take $b = e$, so that $\log_b y = \ln y$ and $\log_b a = \ln a$. Then we get

---

$$\log_a y = \frac{\ln y}{\ln a}.$$

So, *the logarithm of y with base a is equal to the natural logarithm of y divided by the natural logarithm of a.*

Second, take $b = 10$, so that $\log_b y = \log y$ and $\log_b a = \log a$. Then we get

---

$$\log_a y = \frac{\log y}{\log a}$$

---

expressing $\log_a y$ as the ratio of the common logarithms of y and a.

EXAMPLE 7 If $a = 2$, we have the following:

$$\log_2 y = \frac{\ln y}{\ln 2} = \frac{\ln y}{0.6931}.$$

☛ **29.** Express the following in terms of natural logarithms:
(a) $\log_2 e$; (b) $\log_5 6$.

For example,

$$\log_2 3 = \frac{\ln 3}{\ln 2} = \frac{1.0986}{0.6931} = 1.5850. \quad \text{☛} \ 29$$

Alternatively, we can use common logarithms:

$$\log_2 3 = \frac{\log 3}{\log 2} = \frac{0.4771}{0.3010} = 1.5850.$$

Secondly, let $y = b$ in the base-change formula. Then the numerator of the right side becomes $\log_b b$, which equals 1. Therefore we obtain the following results.

$$\log_b a = \frac{1}{\log_a b} \qquad (\log_a b)(\log_b a) = 1$$

EXAMPLE 8 If $b = 10$, we have

$$\log_a 10 = \frac{1}{\log a}.$$

For example,

$$\log_2 10 = \frac{1}{\log 2} = \frac{1}{0.3010} = 3.3219$$

$$\log_3 10 = \frac{1}{\log 3} = \frac{1}{0.4771} = 2.0959$$

$$\ln 10 = \log_e 10 = \frac{1}{\log e}.$$

To four decimal places, the values of these two logarithms are

$$\log e = \log (2.7183) = 0.4343 \quad \text{and} \quad \ln 10 = 2.3026.$$

The two numbers are reciprocals of one another.

The base-change formula allows us to relate a logarithm with a general base a to a common logarithm. In particular, taking $a = e$, we can express the natural logarithm in terms of the common logarithm:

$$\log_e y = \frac{\log y}{\log e} = \frac{\log y}{0.4343}.$$

Thus

Answer (a) $\dfrac{1}{\ln 2}$ (b) $\dfrac{\ln 6}{\ln 5}$.

$$\ln y = 2.3026 \log y.$$

Thus, in order to find the natural logarithm of y, we can determine the common logarithm of y and multiply it by 2.3026. However, this method of finding the natural logarithm of a number is not very convenient when compared to the use of a separate table. The relationship between the two logarithms is of some theoretical importance, however.

EXAMPLE 9 From the Table A.3.1, we find that $\log 2 = 0.3010$, so $\log 0.2 = 0.3010 - 1 = -0.6990$. The natural logarithms of 2 and 0.2 are, therefore,

$$\ln 2 = 2.3026 \log 2 = (2.3026)(0.3010) = 0.6931$$

and

$$\ln 0.2 = 2.3026 \log 0.2 = (2.3026)(-0.6990) = -1.6095.$$

It is not difficult to prove the base-change formula for logarithms. We start with the pair of equivalent statements

$$y = a^x \quad \text{and} \quad x = \log_a y.$$

Similarly, if $a = b^c$, then $c = \log_b a$. But then

$$y = a^x = (b^c)^x = b^{cx}$$

and from this it follows that $cx = \log_b y$. Therefore, substituting for c and x, we obtain the required formula.

$$\log_b y = cx = (\log_b a)(\log_a y)$$

or, $\log_a y = \log_b y / \log_b a$ as required.

The Logistic Model

Earlier, when discussing the growth of populations, we mentioned that an exponential growth function can be used for populations growing without restraint from their environments. However when the habitat imposes limitations on growth, exponential growth does not continue indefinitely, and eventually the population size levels off. The function that is most commonly used to model a restricted growth of this kind is called the **logistic model.** It is based on the following equation for the population size.

$$y = \frac{y_m}{1 + Ce^{-kt}} \tag{1}$$

Here y is the population size at time t and y_m, C, and k are three positive constants.

A typical graph of y against t for this logistic function is shown in Figure 8. We note that when t becomes very large, e^{-kt} becomes very small, so the denominator in Equation (1) becomes closer and closer to 1. Therefore y itself gets closer and closer to y_m as t gets very large. This is apparent from the graph

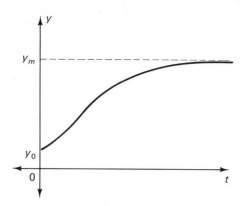

FIGURE 8

in Figure 8, which levels off and approaches the horizontal line $y = y_m$ as t becomes large.

If the initial value y_0 of y is much smaller than the eventual value y_m, then the population size shows a period of growth for small values of t that is approximately exponential. Eventually, however, the growth slows down and finally levels off, approaching y_m as t becomes very large. This final level y_m represents the maximum population size that can be supported by the given environment.

EXAMPLE 10 *(Logistic Population Growth)* A certain population grows according to the logistic equation, with constants $y_m = 275$ million, $C = 54$, and $k = (\ln 12)/100$. The variable t is measured in years. What are the population sizes when $t = 0$, 100, and 200?

Solution When $t = 0$, the size is

$$y_0 = \frac{y_m}{1 + Ce^0} = \frac{275}{1 + 54} = 5 \quad \text{(million)}.$$

We substitute $t = 100$ into Equation (1).

$$y = \frac{y_m}{1 + Ce^{-k(100)}}$$

Now $100k = \ln 12$, so

$$e^{-100k} = e^{-\ln 12} = e^{\ln (1/12)} = \tfrac{1}{12}.$$

Therefore

$$y = \frac{275}{1 + 54(\tfrac{1}{12})} = \frac{275}{1 + (\tfrac{9}{2})} = 50 \quad \text{(million)}.$$

When $t = 200$,

$$y = \frac{y_m}{1 + Ce^{-k(200)}} = \frac{y_m}{1 + C(e^{-100k})^2}$$

$$= \frac{275}{1 + 54(\frac{1}{12})^2} = \frac{275}{1 + (\frac{3}{8})} = 200 \quad \text{(million)}.$$

(This example provides an approximate application of the logistic equation to the population of the United States in the years 1777 ($t = 0$) to 1977 ($t = 200$). For this example, the eventual population size is 275 million.)

The logistic equation is used in many situations other than the growth of populations. The essential qualitative features of the logistic function are that for small values of t, it resembles an exponential function, while for large values of t, it levels off and approaches closer and closer to a certain limiting value. These features occur in a number of phenomena and account for the widespread use of this function.

An example is the spread of information through a population. For example, the information may be a piece of news, a rumor, or knowledge about some new product that has recently come on the market. If p represents the proportion of the population that is aware of the information, then for small values of t, p is small and grows typically in an exponential manner. However, p cannot exceed 1, and as t becomes large, p approaches closer and closer to this value as the information spreads through the whole population. Using the logistic equation, we would model p by means of the expression

$$p = \frac{1}{1 + Ce^{-kt}}.$$

EXAMPLE 11 *(Spread of Information)* At $t = 0$, 10% of all stockbrokers have heard about the impending financial collapse of a large airline. Two hours later, 25% have heard about it. How long will it be before 75% have heard about it?

Solution If $t = 0$, we find that

$$p = \frac{1}{1 + Ce^{-k(0)}} = \frac{1}{1 + C} = 0.1.$$

Therefore $1 + C = 10$, or $C = 9$. Next, when $t = 2$, we have

$$p = \frac{1}{1 + Ce^{-k(2)}} = \frac{1}{1 + 9e^{-2k}} = 0.25.$$

Therefore

$$1 + 9e^{-2k} = 4$$

$$9e^{-2k} = 3$$

$$e^{-2k} = \tfrac{1}{3}.$$

Taking natural logarithms of both sides, we find that

$$-2k = \ln\left(\tfrac{1}{3}\right) = -\ln 3 \qquad \text{and so} \qquad k = \tfrac{1}{2}\ln 3.$$

Having found the values of k and C, we know the precise form of p as a function of t. We wish to calculate the value of t at which $p = 0.75$.

$$p = \frac{1}{1 + 9e^{-kt}} = 0.75 = \tfrac{3}{4}$$

$$1 + 9e^{-kt} = \tfrac{4}{3}$$

$$9e^{-kt} = \tfrac{1}{3}$$

$$e^{-kt} = \tfrac{1}{27}$$

Again, taking natural logarithms of both sides, we get

$$-kt = \ln\left(\tfrac{1}{27}\right) = -\ln 27$$

and so

$$t = \frac{\ln 27}{k} = \frac{\ln 27}{\tfrac{1}{2}\ln 3} = \frac{2\ln(3^3)}{\ln 3} = \frac{6\ln 3}{\ln 3} = 6.$$

Thus it takes 6 hours before 75% of stockbrokers have heard about the collapse of the airline.

EXERCISES 6-4

(1–10) Solve the following equations for x.

1. $10^x = 25$
2. $2^x = 25$
3. $3^x 2^{3x} = 4$
4. $3^x 2^{1-x} = 10$
5. $3^x = 2^{2-x}$
6. $(3^x)^2 = 2\sqrt{2^x}$
7. $(2^x)^x = 25$
8. $(2^x)^x = 3^x$
9. $a^x = cb^x$
10. $(a^x)^2 = b^{x+1}$

11. (Population Growth) The population of the earth in 1976 was 4 billion and was growing at 2% per year. If this rate of growth continues, when will the population reach 10 billion?

12. (Population Growth) The population of China in 1970 was 750 million and was growing at 4% per year. When would this population reach 2 billion, assuming that the same growth rate continued? (The current growth rate is considerably lower.)

13. (Population Growth) Using the data of Exercises 11 and 12, calculate when the population of China would be equal to half the population of the earth.

14. (Profits Growth) The profits of a company had been increasing at an average of 12% per year between 1980 and 1985 and in 1985 they reached the level of $5.2 million. Assuming this rate of growth continues, how long will it be before they reach $8 million per year?

15. (Newspaper Circulation) Two competing newspapers have circulations of 1 million and 2 million, respectively. If the first is increasing its circulation by 2% each month, while the circulation of the second is declining by 1% each month, calculate how long it will be before the two have equal circulations.

(16–17) (Compound Interest) Suppose $1000 is invested at 8% interest compounded annually.

16. How long does it take to increase to $1500?

17. How long does it take to increase to $3000?

18. (Compound Interest) The following rule of thumb is often employed by people in finance: If the rate of interest is R percent per annum, then the number of years, n, for an investment to double is given by dividing R into 70 (that

is, $n = 70/R$). Calculate n *exactly* for the following values of R: 4, 8, 12, 16, and 20. Compare your answers to those obtained from the formula $n = 70/R$ and so assess the accuracy of the rule of thumb.

(19–22) Use the base change formula to prove the following.

19. $(\log_b a)(\log_c b)(\log_a c) = 1$

20. $(\log_b a)(\log_c b)(\log_d c) = \log_d a$

21. $\ln x = (\log x)(\ln 10)$

22. $(\ln 10)(\log e) = 1$

(23–26) Express the following functions in the form $y = ae^{kt}$.

23. $y = 2^t$

24. $y = (1000)2^{t/3}$

25. $y = 5(1.04)^t$

26. $y = 6 \times 10^8(1.05)^t$

27. *(Population Growth)* The earth's population is 4 billion at present and is increasing by 2% each year. Express the population y at a time t years from now in the form $y = ae^{kt}$.

28. *(Depreciation)* A company purchases a machine for $10,000. Each year the value of the machine decreases by 20%. Express the value in the form be^{kt}, where b and k are constants and the time $t = 0$ corresponds to the date of purchase.

29. *(Rise in C.P.I.)* Between January 1975 and January 1980, the consumer price index I rose from 121 to 196.

   a. Calculate the average percentage increase per annum during this period.

   b. Express I in the form be^{kt}, with $t = 0$ corresponding to January 1975.

   c. Assuming this growth rate continues, determine when I will reach 250.

30. *(Population Growth)* A population is growing according to the formula

$$P = 5 \times 10^6 e^{0.06t}$$

where t is in years. Calculate the percentage growth per annum. How long does it take the population to increase by 50%?

31. *(Population Growth)* A population has a size given by the formula

$$P = P_0 e^{kt}$$

Find an expression for the percentage of growth per unit of time and for the length of time necessary for the population to double in size and to triple in size.

32. *(Sales Growth)* The sales volume of a certain product is increasing at 12% per year. If the present volume is 500 units per day, how long will it take to reach 800 units per day?

33. *(Advertising Frequency)* The sales volume of a brand of detergent decreases following an advertising campaign according to the formula $V(t) = 750(1.3)^{-t}$, where t is time in months. The next campaign is planned for when the sales volume has fallen to two-thirds of its initial value. How long should be allowed between successive advertising campaigns?

(34–36) Calculate the nominal rate of interest that compounded continuously is equivalent to:

34. 8% annual interest.

35. 12% annual interest.

36. 15% annual interest.

37. *(Share Price)* It was observed that the price-earnings ratio of a certain share changed between the beginning of 1982 and 1987 according to the formula $R = 4(1.2)^t$, where t is time in years from 1982. What was the value of the ratio in 1987 ($t = 5$)? Assuming the increase is maintained, when will the ratio reach the value 20?

(38–39) *(Radioactivity)* Many isotopes of chemical elements are unstable and change spontaneously into some other isotope; such decay is called *radioactivity* and is generally accompanied by the emission of one of three kinds of radiation, called α-, β-, or γ- rays. If a specimen originally contains an amount y_0 of a radioactive isotope, after time t it will contain an amount $y = y_0 e^{-kt}$, where k is called the decay constant.

38. The decay constant for C^{14} (carbon-14) is 1.24×10^{-4} when t is measured in years.

   a. Calculate the percentage of the original specimen that remains after 2000 years and after 10,000 years.

   b. Calculate the number of years for half the specimen to decay. (This is called the *half-life* of the isotope.)

39. The half-life of radium (see Exercise 38) is 1590 years. Calculate its decay constant. If 10 grams of radium is left for 1000 years, how much will remain?

40. *(Bacteria Population)* After an individual eats some infected food, the population of bacteria in his stomach expands by cell division, doubling every 20 minutes. If initially there are 1000 bacteria present, express the population size after t minutes in the form $y = a \cdot 2^{bt}$ and determine the constants a and b. After how many minutes will the number of bacteria be 10,000?

41. *(Stellar Magnitudes)* The magnitude M of a star or planet is defined as $M = -\left(\frac{5}{2}\right) \log (B/B_0)$, where B is the brightness and B_0 is a constant. The planet Venus has an average magnitude of -3.9, and the star Polaris has a magnitude of 2.1. How many times brighter is Venus on average than Polaris?

42. *(Richter Scale)* The magnitude R of an earthquake on the Richter scale is defined as $R = \log (A/A_o)$, where A is the intensity and A_0 a constant. (A is the amplitude of vibration of a standard seismograph located 100 kilometers from the earthquake's epicenter.) The 1964 earthquake in Alaska measured 8.5 on the Richter scale. The largest earthquake ever recorded measured 8.9. How much more intense was this earthquake than the Alaskan quake?

43. *(Decibel Scale)* The loudness L of a sound is defined in decibels as $L = 10 \log (I/I_0)$, where I is the intensity of the sound (the energy falling on a unit area per second) and I_0 is the intensity of the quietest sound the human ear can hear (called the threshold of hearing). A quiet room has an average sound level of 32 decibels. A loud conversation has a noise level of 65 decibels. The threshold of pain occurs at about 140 decibels. Calculate I/I_0 for each of these three noise levels.

44. *(Population Growth)* A certain population of insects consists of two types: T_1 and T_2. Initially the population contains 90 T_1 insects and 10 T_2 insects. The T_1 population grows at 1% per day, and the T_2 population at 4% per day, on the average. When will the population become equally divided between the two types?

45. *(Population Growth)* A population of bacteria doubles in size every 19 minutes. How long does it take to increase from 10^5 to 10^7 organisms?

46. *(Radiotherapy)* When cancer cells are subjected to radiation treatment, the proportion of cells that survive the treatment is given by

$$P = e^{-kr},$$

where r is the radiation level and k a constant. It is found that 40% of the cancer cells survive when $r = 500$ Roentgens. What should the radiation level be in order to allow only 1% to survive?

47. *(Newton's Law of Cooling)* A body at temperature T higher than its surroundings cools down according to the formula $T = T_0 e^{-kt}$, where T_0 is the initial temperature difference, t is time, and k is a constant. It is found that the temperature difference falls to half its initial values in 20 minutes. If $T_0 = 60°C$ initially, how long does it take for the temperature difference to fall to 10°C?

48. *(Sales Growth)* A new product was introduced onto the market at $t = 0$, and thereafter its monthly sales grew according to the formula

$$S = 4000(1 - e^{-kt})^3.$$

If $S = 2000$ when $t = 10$ (that is, after 10 months), find the value of k.

49. *(Learning Curve)* An individual's efficiency in performing a routine task improves with practice. Let t be the time spent learning the task and y a measure of the individual's output. (For example, y might be the number of times the task can be performed per hour.) Then one function often used to relate y to t is

$$y = A(1 - e^{-kt})$$

where A and k are constants. (The graph of such a relation between y and t is called a **learning curve**.) After one hour of practice, a person on an assembly line can tighten 10 nuts in 5 minutes. After 2 hours, the person can tighten 15 nuts in 5 minutes. Find the constants A and k. How many nuts can the person tighten after 4 hours of practice?

50. *(Logistic Model)* A population grows according to the logistic model, with constants $y_m = \left(\frac{124}{3}\right) \times 10^7$, $C = \frac{245}{3}$, and $k = \ln \left(\frac{35}{4}\right)$. Find the population sizes when $t = 0, 1$, and 2.

51. *(Logistic Model)* The weight of a culture of bacteria is given by

$$y = \frac{2}{1 + 3(2^{-t})}$$

where t is measured in hours. What are the weights when $t = 0, 1, 2$ and 4?

52. *(Spread of Information)* A new improved strain of rice is developed. After t years, the proportion of rice farmers who have switched to the new strain is found to be given by a logistic model

$$p = (1 + Ce^{-kt})^{-1}.$$

At $t = 0$, 2% of farmers are using the new strain. Four

years later, 50% are doing so. Evaluate C and k and calculate how many years it is before 90% have switched to the new strain.

*53. *(Gompertz Function)* Another function sometimes used to describe restricted growth of a population is the *Gompertz function*

$$y = pe^{-ce^{-kt}}$$

where p, c, and k are constants. Show that at $t = 0$, $y = pe^{-c}$, and that as t becomes large, y gets closer and closer to the value p.

CHAPTER REVIEW

Key Terms, Symbols, and Concepts

6.1 Compound interest, annual rate of interest.
Nominal rate and effective rate of interest.
Continuous compounding; the number e.
Present value.

6.2 Exponential function, base; $y = a^x$.
Graphs of growing and decaying exponential functions.
Natural exponential function, $y = e^x$, and its graph.

6.3 Logarithm with base a: $\log_a y$.
Graphs of logarithm functions.
Equivalent exponential and logarithmic forms of a statement.
Properties of logarithms.
Natural logarithms: $x = \ln y$.
Common logarithms: $x = \log y$.

6.4 Exponential equation.
Base-change formulas for exponentials and logarithms.
The logistic model.

Formulas

Compound interest:
Value after n periods $= P(1 + i)^n$.

For k compoundings per year at nominal rate $R\%$,

$$i = \frac{R}{100k}.$$

Nominal and effective rates: $1 + i_{\text{eff}} = (1 + i)^k$.

Continuous compounding; value after N years $= Pe^{iN}$;
$1 + i_{\text{eff}} = e^i$.

Present value; P. V. $= P(1 + i)^{-n}$.

$x = \log_a y$ if $y = a^x$.

$x = \ln y$ if $y = e^x$.

$x = \log y$ if $y = 10^x$.

Properties of logarithms (stated here for natural logarithms only):

$$\ln 1 = 0 \qquad \ln e = 1$$

$$\ln (uv) = \ln u + \ln v \qquad \ln \left(\frac{u}{v}\right) = \ln u - \ln v$$

$$\ln \left(\frac{1}{v}\right) = -\ln v \qquad \ln (u^n) = n \ln u$$

$$e^{\ln y} = y \quad \text{for all } y > 0 \qquad \ln (e^x) = x \quad \text{for all } x.$$

Base-change formula for exponentials:

$$a^x = e^{kx} \qquad \text{where } k = \ln a.$$

Base-change formulas for logarithms:

$$\log_a y = \frac{\log_b y}{\log_b a} \qquad \log_a y = \frac{\ln y}{\ln a} \qquad \log_a y = \frac{\log y}{\log a}.$$

REVIEW EXERCISES FOR CHAPTER 6

1. State whether each of the following is true or false. Replace each false statement with a corresponding true statement.

a. $\log_a a = 1$ for all real numbers a

b. Since $(-2)^2 = 4$, we can say that $\log_{-2} 4 = 2$.

c. $(\log_2 3)(\log_3 4) = 2$

d. $\ln 78 = \ln (7.8) + 1$

e. The function a^x represents exponential growth if $a > 1$ and exponential decay if $0 < a < 1$.

f. The function e^{kx} represents exponential growth if $k > 0$ and exponential decay if $k < 0$.

g. $\log_a (x + y) = \log_a x + \log_a y$.

h. If $\ln x > 1$, then x must be greater than 10.

i. $\ln 10 = (\log e)^{-1}$

j. $\ln (1 + 2 + 3) = \ln 1 + \ln 2 + \ln 3$

k. $\log 4 - \log 2 = \log (4 - 2)$

l. $(\log x)^n = n \log x$

m. $\dfrac{\ln x^3}{\ln x^2} = \ln x^3 - \ln x^2$

(2–3) Evaluate each expression.

2. $\log_{\sqrt{27}} 81$ **3.** $\log_{36}(1/\sqrt{6})$

(4–7) If $\log_{12} 2 = x$, express the following in terms of x.

4. $\log_{12} 3$ **5.** $\log_{12} \sqrt{108}$

6. $\log_{27} 12$ **7.** $\log_3 6$

8. Find the value of n for which $(0.081)^n = 0.24$.

(9–16) Solve the following equations for x.

9. $2^{x+1} = 3^{2-2x}$ **10.** $(2^x)^x = 4^{1-x}$

11. $\log_3 (x + 2) + \log_3 (2x + 7) = 3$

12. $\ln (2x + 3) + \ln (x + 2) = \ln (9 + 12x)$

13. $\log (3x + 1) = 1 + \log (2x - 1)$

14. $\ln 2 - \ln (x - 3) = \ln (x - 4)$

15. $\log_x (8x - 3) - 2 \log_x 2 = 2$

16. $\log_x (1 - x) - \log_x 6 = 2$

17. Prove that

$$a^{\ln(b/c)} \cdot b^{\ln(c/a)} = c^{\ln(b/a)}.$$

18. If $\ln \left[\dfrac{a + b}{3}\right] = \left(\dfrac{1}{2}\right)(\ln a + \ln b)$, show that

$$a^2 + b^2 = 7ab.$$

19. If $\ln(x + y) = \ln x + \ln y$, find y in terms of x.

20. Without using tables or a calculator, show that

$$\frac{\log \sqrt{27} + \log 8 + \log \sqrt{1000}}{\log 120} = \frac{3}{2}$$

(21–26) Find the domains of the following functions.

21. $f(x) = \dfrac{e^x}{1 + e^x}$

22. $f(x) = \sqrt{1 - e^x}$

23. $f(x) = \sqrt{e^{-x^2}}$

24. $f(x) = \dfrac{1}{1 + \ln x}$

25. $f(x) = \ln (9 - x^2)$

26. $f(x) = \log |9 - x^2|$

(27–28) Sketch the graphs of the following functions.

27. $f(x) = -3^x$

28. $f(x) = 2 - \ln x$

29. *(Demand)* The demand equation of a certain product is given by $p = 200e^{-x/50}$, where x denotes the number of units that can be sold at a price of $\$p$ each. Express the revenue R as a function of demand x. What will be the total revenue if 25 units are sold?

30. *(Demand)* The demand equation of a certain commodity is given by $p = 1000e^{-x/20}$, where x denotes the number of units which can be sold at a price of $\$p$ each.

  a. How many units, to the nearest unit, can be sold if the price per unit is $\$10.00$?

  b. At what price p per unit, will 80 units be sold?

31. *(Demand)* The demand equation of a certain product is given by $p \ln (x + 1) = 500$, where x units can be sold at a price of $\$p$ each.

  a. How many units, to the nearest unit, can be sold if the price per unit is $\$62.50$?

  b. At what price p per unit will 5000 units be sold?

32. *(Depreciation)* Two years ago, a firm bought a machine for $\$6000$; its present resale value is $\$4500$. Assuming that the machine is depreciating exponentially, what will it be worth 3 years from now?

33. *(Compound Interest)* If $\$500$ is invested at 7% interest per annum compounded annually, what is its value after 7 years?

34. *(Compound Interest)* If $\$100$ is added to the investment in Exercise 33 after each year, calculate the new value after 3 years.

35. *(Loan Repayment)* The sum of $\$1000$ is borrowed at the rate of interest of 10%, compounded annually. The loan is to be repaid in two equal installments, at the end of 1 year and at the end of 2 years. How much must the installments be?

36. *(Loan Repayment)* Repeat Exercise 35 in the case where the loan is repaid in three equal annual installments.

37. *(Present Value)* A man at the age of 45 purchases a deferred endowment from a life insurance company that will pay him a lump sum of $20,000 at the age of 65. The company charges him $5000 for the policy. What discount rate are they using?

38. *(Return on Investment)* An investment club purchased a high-rise apartment complex for $5.5 million and sold it for $9 million after 4 years. Find the annual rate of return (compounded continuously) on the investment.

39. *(Return on Investment)* A person can deposit his money with the bank A, which pays an interest rate of 12.2% compounded semiannually, or with the bank B, which pays an interest rate of 12% compounded monthly. Where should the money be deposited for a better rate of return?

40. *(Population Growth)* The population of Britain in 1600 is believed to have been about 5 million. Three hundred fifty years later, it had increased to 50 million. What was the average percentage growth per year during that period? (Assume a uniform exponential growth.)

41. *(Population Growth)* If a population increases from 5 million to 200 million over a period of 200 years, what is the average percentage growth per year?

42. *(Compound Interest)* A sum of $100 is invested at the nominal rate of interest of 12% per annum. How much is the investment worth after 5 years if compounded:

  a. Annually?

  b. Quarterly?

  c. Continuously?

43. *(Continuous Compounding)* At what nominal rate of interest does money triple in value in 10 years if compounded continuously?

44. *(Effective Rate)* Show that if a sum is invested at a nominal rate of $R\%$ per year compounded k times a year, then the effective rate is $r\%$ per year, where

$$r = [(1 + R/100k)^k - 1](100).$$

45. *(Effective Rate)* Show that if a sum is invested at a nominal rate of $R\%$ per year compounded continuously, then the effective rate is $r\%$, where

$$r = 100(e^{R/100} - 1).$$

46. Find the effective rate of interest corresponding to a nominal rate of 8% compounded:

  a. Semiannually.

  b. Quarterly.

  c. Monthly.

  d. Continuously.

47. Repeat Exercise 46 if the nominal rate is 6%.

48. Find the nominal rate of interest that corresponds to an effective rate of 8.4% when the compounding occurs:

  a. Semiannually.

  b. Quarterly.

  c. Monthly.

  d. Continuously.

49. *(Plant Pest)* The percentage of trees in a fruit plantation that have become infected with a certain pest is given by

$$P(t) = 100/(1 + 50e^{-0.1t})$$

where t is time in days measured from the time the infestation was first confirmed. Evaluate $P(0)$, $P(20)$, and $P(40)$.

50. *(Spread of Disease)* Assume that the number of diagnosed cases of AIDS is increasing exponentially. In British Columbia, the number of such cases was 88 in 1985 and 330 in 1987. Express this number in the form ae^{bt}, where a and b are constants and t is time measured in years from 1985. On the basis of this model, how many cases will there be in British Columbia in the years 1995 and 2000?

51. *(Learning Curve)* The number of items produced per hour on a production line increases as more and more items are produced. Assume that $P(x) = 30 - 15e^{-0.02x}$, where P is the rate of production after x units have been produced. Calculate $P(0)$ and $P(100)$. To what value does $P(x)$ get closer as x gets larger and larger? Interpret this value.

52. *(Nitrogen Expulsion)* When oxygen is administered to a patient, the amount of nitrogen removed from the patient's lungs increases according to the formula $V = 1 - e^{-kt}$ liters, where t is time in minutes and $k = 0.2$. After how many minutes is 90% of the nitrogen removed?

53. *(Growth of GNP)* The GNP of nation A increased from $0.5 to $1.1 billion between 1970 and 1980.

   a. Calculate the average percentage growth per annum.

   b. Express the GNP at time t in the form be^{kt}.

   c. Assuming this growth rate continues, calculate when the GNP will reach $1.5 billion.

54. *(Growth of GNP)* The GNP of nation B during the same period (see Exercise 53) increased from $1.0 to $1.5 billion.

   a. Calculate the average percentage growth per annum for nation B.

   b. Express the GNP in the form $b'e^{k't}$.

   c. Calculate when the GNP of nation A overtakes that of nation B.

CHAPTER 7

Progressions and the Mathemathics of Finance

Chapter Objectives

7-1 ARITHMETIC PROGRESSIONS AND SIMPLE INTEREST
(a) Definition of arithmetic progesssion (A.P.) and formula for the nth term;
(b) Application of A.P.s to straight-line depreciation and simple interest;
(c) Formulas for the sum of n terms in an A.P. and their applications.

7-2 GEOMETRIC PROGRESSIONS AND COMPOUND INTEREST
(a) Definition of geometric progression (G.P.) and formula for the nth term;
(b) Application of G.P.s to exponential depreciation and compound interest;
(c) Formula for the sum of n terms in a G.P. and its application;
(d) Formula for the sum of an infinite G.P.

7-3 MATHEMATICS OF FINANCE
(a) Savings plans. The future value of an annuity;
(b) Annuities. The present value of an annuity;
(c) Table of values and formulas for $a_{\overline{n}|i}$ and $s_{\overline{n}|i}$.
(d) Amortization of a debt.

7-4 DIFFERENCE EQUATIONS
(a) Discrete time processes;
(b) Definition of a difference equation of order k; solution of a difference equation;
(c) Solution of first order difference equations by numerical iteration;
(d) Linear first-order equations; formulas for the solution;
(e) Application of difference equations in the mathematics of finance: savings plans, amortization, and annuities.

7-5 SUMMATION NOTATION
(a) Definition of the sigma notation and examples of its use;
(b) Formulas for the sum of the first n integers, and the sums of the squares and cubes of the first n integers;
(c) Evaluation of the sums of polynomial functions of integers.

CHAPTER REVIEW

A **sequence** is an ordered list of numbers. For example,

$$2, 5, 8, 11, 14, \ldots \tag{1}$$

$$3, 6, 12, 24, 48, \ldots \tag{2}$$

are examples of sequence. In Sequence (1), the **first term** is 2, the **second term** is 5, and so on. It can be seen that each term is obtained by adding 3 to the preceding term. In Sequence (2), the first term is 3 and the fourth term is 24, and any term can be obtained by doubling the preceding term. Sequences of these types arise in many problems, particularly in the mathematics of finance.

A sequence is said to be **finite** if it contains a limited number of terms, that is, if the sequence has a last term. If there is no last term in the sequence, it is called an **infinite** sequence. The terms of a sequence will be denoted by T_1, T_2, T_3, and so on. Thus, for example, T_7 denotes the seventh term, T_{10} the tenth term, and T_n the nth term. The nth term of a sequence is commonly called its **general term.** ☞ **1**

☞ **1.** For the sequence 1, −1, 2, −2, 3, −3, 4, −4 what are T_2 and T_5? Is the sequence finite?

◥ 7-1 ARITHMETIC PROGRESSIONS AND SIMPLE INTEREST

Suppose Mr. Cernac borrows a sum of $5000 from the bank at 1% interest per month. He agrees to pay $200 toward the principal each month, plus the interest on the balance. At the end of the first month, he pays $200 plus the interest on $5000 at 1% per month, which is $50. Thus his first payment is $250 and he owes only $4800 to the bank. At the end of the second month, he pays $200 toward the principal plus the interest on $4800, which is $48 at 1% for 1 month. Thus his second payment is $248. Continuing in this way, his successive payments (in dollars) are

$$250, 248, 246, 244, \ldots, 202.$$

This sequence is an example of an *arithmetic progression*.

DEFINITION A sequence is said to be an **arithmetic progesssion** (A.P.) if the difference between any term and the preceding term is the same throughout the sequence. The algebraic difference between each term and the preceding term is called the **common difference** and is denoted by d.

Mr. Cernac's sequence of payments is an A.P. because the difference between any term and its preceding term is −2. This A.P. has 250 as its first term and −2(=248 − 250) as its common difference. Similarly,

$$2, 5, 8, 11, 14, \ldots$$

is an A.P. with first term 2 and common difference 3.

If a is the first term and d the common difference of an A.P., then the successive terms of the A.P. are

$$a, \quad a + d, \quad a + 2d, \quad a + 3d, \quad \ldots.$$

Answer −1 and 3. Yes.

The nth term is given by the formula

$$T_n = a + (n - 1)d. \qquad (3)$$

For example, letting $n = 1, 2$, and 3, we find

$$T_1 = a + (1 - 1)d = a$$

$$T_2 = a + (2 - 1)d = a + d$$

$$T_3 = a + (3 - 1)d = a + 2d.$$

Further values can be obtained in a similar manner.

Equation (3) contains four numbers, a, d, n, and T_n. If any three of them are given, we can find the fourth.

EXAMPLE 1 Given the sequence

$$1, 5, 9, 13, \ldots$$

find: (a) the fifteenth term; (b) the nth term.

Solution The given sequence is an A.P. because

$$5 - 1 = 9 - 5 = 13 - 9 = 4.$$

Thus the common difference, d, is 4. Also, $a = 1$.

(a) Using Equation (3) with $n = 5$.

$$T_{15} = a + (15 - 1)d = a + 14d = 1 + (14)(4) = 57.$$

2. For the A.P. -3, -0.5, 2, . . . , find a formula for the nth term and compute the 11th term.

(b) $T_n = a + (n - 1)d = 1 + (n - 1)4 = 4n - 3$

Thus the fifteenth term is 57 and nth term is $4n - 3$. ☛ 2

EXAMPLE 2 *(Depreciation)* A firm installs a machine at a cost of $1700. The value of the machine depreciates annually by $150. Find an expression for the value of the machine after n years. If the scrap value of the machine is $200, what is its lifetime?

Solution Since the value of the machine depreciates each year by $150, its value at the end of the first year, second year, third year, and so on, will be

$$1700-150, \quad 1700-2(150), \quad 1700-3(150), \quad \ldots$$

or

$$1550, \quad 1400, \quad 1250, \quad \ldots$$

This sequence of values forms an A.P. with first term $a = 1550$ and common difference $d = 1400 - 1550 = -150$. Thus the nth term is

Answer $T_n = 2.5n - 5.5; T_{11} = 22.$

$$T_n = a + (n - 1)d = 1550 + (n - 1)(-150) = 1700 - 150n.$$

This quantity T_n gives the value of the machine in dollars at the end of the nth year.

We are interested in the value of n at which this value has reduced to the scrap value, since this gives the lifetime of the machine. Thus we let $T_n = 200$ and solve for n.

$$1700 - 150n = 200$$

$$150n = 1700 - 200 = 1500$$

$$n = 10$$

The lifetime of the machine is 10 years. ☛ 3

☛ **3.** An A.P. with 25 terms has first term 100 and last term 28. Find an expression for the general term and compute the middle term.

EXAMPLE 3 Michelle's monthly payments to the bank toward her loan form an A.P. If her sixth and tenth payments are \$345 and \$333, respectively, what will be her fifteenth payment to the bank?

Solution Let a be the first term and d the common difference of the monthly payments of the A.P. Then successive payments (in dollars) are

$$a, \quad a + d, \quad a + 2d, \quad \ldots .$$

Since the sixth and tenth payments (in dollars) are 345 and 333, $T_6 = 345$ and $T_{10} = 333$. Using Equation (3) for the nth term and the given values for T_6 and T_{10}, we have

$$T_6 = a + 5d = 345$$

$$T_{10} = a + 9d = 333.$$

We subtract the top equation from the second equation and simplify.

$$4d = 333 - 345 = -12$$

$$d = -3$$

Substituting this value of d in the top equation, we obtain

$$a - 15 = 345 \quad \text{or} \quad a = 360.$$

Now

$$T_{15} = a + 14d = 360 + 14(-3) = 308.$$

Thus her fifteenth payment to the bank will be \$308.

Simple Interest

Let a sum of money P be invested at an annual rate of interest of R percent. In 1 year the amount of interest earned is given (see page 68) by

$$I = P\left(\frac{R}{100}\right).$$

Answer $T_n = 103 - 3n$; middle term is $T_{13} = 64$.

If the investment is at *simple interest*, then interest in succeeding years is paid only on the principal P and not on any earlier amounts of earned interest.

Thus, a constant amount I is added to the investment at the end of each year. After 1 year the total value is $P + I$, after 2 years it is $P + 2I$, and so on. The sequence of annual values of the investment,

$$P, \quad P + I, \quad P + 2I, \quad P + 3I, \quad \ldots$$

thus forms an arithmetic progression whose first term is P and whose common difference is I. After t years, the value is given by $P + tI$.

Simple Interest:

$$\text{Value after } t \text{ years} = P + tI, \qquad I = P\left(\frac{R}{100}\right).$$

EXAMPLE 4 *(Simple Interest)* A sum of $2000 is invested at simple interest at a 12% annual rate of interest. Find an expression for the value of the investment t years after it was made. Compute the value after 6 years.

Solution Here $P = 2000$ and $R = 12$. The amount of annual interest is therefore

$$I = 2000\left(\frac{12}{100}\right) = 240.$$

After t years the total interest added is $tI = 240t$, so that the value of the investment is

$$P + tI = 2000 + 240t.$$

After 6 years, this value is

$$2000 + 6(240) = 3440 \text{ dollars.} \quad \blacktriangleright 4$$

Sum of n Terms of an A.P.

If a is the first term and d the common difference of an A.P., then the sequence is

$$a, \quad a + d, \quad a + 2d, \quad \ldots .$$

If the sequence contains n terms and if l denotes the last term (that is, the nth term), then

$$l = a + (n - 1)d. \tag{4}$$

The next-to-last term will be $l - d$, the second from the last term will be $l - 2d$, and so on. If S_n denotes the sum of these n terms, then

$$S_n = a + (a + d) + (a + 2d) + \cdots + (l - 2d) + (l - d) + l.$$

If we write this progression in the reverse order, the sum remains unaltered, so

$$S_n = l + (l - d) + (l - 2d) + \cdots + (a + 2d) + (a + d) + a.$$

► 4. A sum of $400 is invested at simple interest of 8% per annum. Find the value after t years. After 10 years, what is the value, and how much total interest has been earned?

Answer $(400 + 32t)$; $720, $320.

● **5.** In a finite A.P., show that the average of all the terms is equal to the average of the first and last terms.

Adding the two values for S_n, we get

$$2S_n = [a + l] + [a + d + l - d] + [a + 2d + l - 2d] + \cdots$$
$$+ [l - d + a + d] + [l + a].$$

There are n terms on the right side and each is equal to $a + l$. Thus

$$2S_n = n(a + l)$$

or

$$S_n = \frac{n}{2}(a + l). \tag{5}$$

Answer The average of n terms is equal to their sum, S_n, divided by n. From Equation (5), this is $S_n/n = \frac{1}{2}(a + l)$.

Substituting the value of l from Equation (4) into Equation (5), we get

$$S_n = \frac{n}{2}[a + a + (n - 1)d] = \frac{n}{2}[2a + (n - 1)d]. \quad ● 5$$

These results are summarized in the following theorem.

THEOREM 1 The sum of n terms of an A.P. with first term a and common difference d is given by

$$S_n = \frac{n}{2}[2a + (n - 1)d].$$

● **6.** Find the sum of all the even numbers less than 200 and all the odd numbers less than 200.

We can also write this formula as

$$S_n = \frac{n}{2}(a + l) \qquad \text{where } l = a + (n - 1)d.$$

EXAMPLE 5 Find the sum of the first 20 terms of the progression

$$2 + 5 + 8 + 11 + 14 + \cdots .$$

Solution The given sequence is an A.P. because

$$5 - 2 = 8 - 5 = 11 - 8 = 14 - 11 = 3.$$

Thus the common difference is $d = 3$. Also, $a = 2$ and $n = 20$. Therefore

$$S_n = \frac{n}{2}[2a + (n - 1)d]$$

$$S_{20} = \frac{20}{2}[2(2) + (20 - 1)3] = 10(4 + 57) = 610. \quad ● 6$$

Answer $2 + 4 + 6 + \cdots + 198 = 9900$; $1 + 3 + 5 + \cdots + 199 = 10,000.$

EXAMPLE 6 *(Loan Repayment)* Consider Mr. Cernac's loan of $5000 from the bank at 1% interest per month. Each month he pays back $200 toward the principal plus the monthly interest on the outstanding balance. How much will he have paid in all by the time he has repaid the loan?

☞ 7. An A.P. has second term 7 and sixth term 15. Find the first term and common difference. How many terms are required to make a sum of 320?

Solution As discussed at the beginning of this section, the sequence of repayments is

$$250, \quad 248, \quad 246, \quad \ldots, \quad 202.$$

These form an A.P. with $a = 250$ and $d = -2$. Since \$200 of principal is repaid each month, the total number of payments is $n = 5000/200 = 25$. The last term is, therefore,

$$l = T_{25} = a + 24d = 250 + 24(-2) = 202$$

as indicated above.

The total payment is given by the sum of all 25 terms.

$$S_n = \frac{n}{2}(a + l) = \frac{25}{2}(250 + 202) = 5650$$

The total amount paid to the bank is \$5650, which means that the interest paid will amount to \$650.

EXAMPLE 7 *(Loan Repayment)* A man agrees to pay an interest-free debt of \$5800 in a number of installments, each installment (beginning with the second) exceeding the previous one by \$20. If the first installment is \$100, find how many installments will be necessary to repay the loan completely.

Solution Since the first installment is \$100 and each succeeding installment increases by \$20, the installments (in dollars) are

$$100, \quad 120, \quad 140, \quad 160, \quad \ldots.$$

These numbers form an A.P. with $a = 100$ and $d = 20$. Let n installments be necessary to repay the loan of \$5800. Then the sum of the first n terms of this sequence must equal 5800, that is, $S_n = 5800$. Using the formula for the sum of an A.P., we get

$$S_n = \frac{n}{2}[2a + (n - 1)d]$$

$$5800 = \frac{n}{2}[200 + (n - 1)20] = \frac{n}{2}(20n + 180) = 10n^2 + 90n.$$

Therefore

$$10n^2 + 90n - 5800 = 0.$$

Dividing through by 10, we get

$$n^2 + 9n - 580 = 0$$

or

$$(n - 20)(n + 29) = 0$$

which gives $n = 20$ or $n = -29$.

Since a negative value of n is not permissible, we have $n = 20$. Thus 20 installments will be necessary to repay the loan. ☞ 7

Answer $a = 5$, $d = 2$; 16 terms.

(1–4) Find the indicated terms of the given sequences.

1. Tenth and fifteenth terms of 3, 7, 11, 15, 19, . . .

2. Seventh and nth terms of 5, 3, 1, -1, . . .

3. rth term of 72, 70, 68, 66, . . .

4. nth term of 4, $4\frac{1}{3}$, $4\frac{2}{3}$, 5, . . .

5. If the third and seventh terms of an A.P. are 18 and 30, respectively, find the fifteenth term.

6. If the fifth and tenth terms of an A.P. are 38 and 23, respectively, find the nth term.

7. Which term of the sequence 5, 14, 23, 32, . . . is 239?

8. The last term of the sequence 20, 18, 16, . . . is -4. Find the number of terms in the sequence.

(9–14) Find the indicated sum of the following progressions.

9. $1 + 4 + 7 + 10 + \cdots$; 30 terms

10. $70 + 68 + 66 + 64 + \cdots$; 15 terms

11. $2 + 7 + 12 + 17 + \cdots$; n terms

12. $3 + 5 + 7 + 9 + \cdots$; p terms

13. $51 + 48 + 45 + 42 + \cdots + 18$

14. $15 + 17 + 19 + 21 + \cdots + 55$

15. How many terms of the sequence 9, 12, 15, . . . are needed to make the sum 306?

16. How many terms of the sequence -12, -7, -2, 3, 8, . . . must be added to obtain a sum of 105?

17. In an A.P., if 7 times the seventh term is equal to 11 times the eleventh term, show that the eighteenth term is zero.

18. *(Loan Repayment)* A man repays a loan of $3250 by paying $20 in the first month and then increasing the payment by $15 every month. How long will it take to clear his loan?

19. *(Depreciation)* A manufacturing company installs a machine at a cost of $1500. At the end of 9 years, the machine has a value of $420. Assuming that yearly depreciation is a constant amount, find the annual depreciation.

20. *(Depreciation)* It cost $2000 to install a machine which depreciated annually by $160. What was the life of the machine if its scrap value was $400?

21. *(Loan Repayments)* Steve's monthly payments to the bank toward his loan form an A.P. If his eighth and fifteenth payments are $153 and $181, respectively, what will be his twentieth payment?

22. *(Salary Increases)* Carla's monthly salary is increased annually in an A.P. She earned $440 a month during her seventh year and $1160 a month during her twenty-fifth year.

    a. Find her starting salary and the annual increment.

    b. What should her salary be at the time of retirement, on completion of 38 years of service?

23. *(Loan Repayments)* In Exercise 21, suppose Steve paid a total of $5490 to the bank.

    a. Find the number of payments he made to the bank.

    b. How much was his last payment to the bank?

24. *(Loan Repayments)* A debt of $1800 is to be repaid in 1 year by making a payment of $150 at the end of each month, plus interest at the rate of 1% per month on the outstanding balance. Find the total interest paid.

25. *(Simple Interest)* A person deposits $50 at the beginning of every month into a savings account in which interest is allowed at $\frac{1}{2}$% per month on the minimum monthly balance. Find the balance of the account at the end of the second year, calculating at simple interest.

26. *(Tube-well Boring Costs)* The cost of boring a tube-well 600 feet deep is as follows: $15 is charged for the first foot and the cost per foot increases by $2 for every subsequent foot. Find the cost of boring the 500th foot and the total cost.

*27. *(Simple Discount)* Let a sum P be borrowed from a bank and be repaid n months later by a single repayment A. If the bank calculates the repayment using a *simple discount rate* of R percent, then P and A are related by the formula

$$P = A\left(1 - \frac{R}{100} \cdot \frac{n}{12}\right).$$

A man borrows money from a bank that uses a simple discount rate of 12%. He will repay the loan with payments of $100 at the end of each month for the next 12 months. How much can he borrow? (Regard each of the monthly repayments A_1, A_2, . . . as generating its own initial loan, P_1, P_2, . . . , and add all these P's together.)

*28. (*Simple Discount*) Miss Brookfield borrowed money from her credit union, which used a simple discount rate of 10%. She promised to repay $50 a month at the end of each month for the next 24 months. How much was the total interest charged by the credit union?

29. (*Loan Repayments*) A man agrees to repay a debt of $1800 in a number of installments, each installment (beginning with the second) less than the previous one by $10. If his fifth installment is $200, find how many installments will be necessary to repay the debt.

30. (*Savings Bonds*) On November 1 every year, a person buys savings bonds of a value exceeding the previous year's purchase by $50. After 10 years, the total cost of the bonds purchased was $4250. Find the value of the bonds purchased:

   a. In the first year.

   b. In the seventh year.

31. (*Savings Plan*) A man invests $200 in a cooperative fund which pays a simple interest of 10% per annum. What is the value of the investment:

   a. After n years?

   b. After 5 years?

32. (*Savings Plan*) Cindy deposits $1000 at the beginning of

each year into her regular savings plan which earns a simple interest of 8% per annum. How much is the plan worth (including the last payment):

   a. At the end of 5 years?

   b. At the end of n years?

33. (*Depreciation*) Often the method of straight-line depreciation is inappropriate because the asset in question loses value much more during its first 1 or 2 years than in later years. An alternative method is referred to as **sum-of-the-years'-digits** method. Let N be the lifetime of the asset and let d be depreciation during the Nth year (i.e., during the last year). Then according to this method the amount of depreciation during the $(N - 1)$st year is $2d$, during the $(N - 2)$nd year is $3d$, and so on, so that the depreciation during the first year is Nd. Show that the depreciation during the nth year is $(N - n + 1)d$ ($n = 1$, 2, . . . , N) and that the total depreciation D during the N years is $D = \frac{1}{2}N(N + 1)d$. (In practice D is set equal to (Initial Cost $-$ Scrap Value after N Years); hence d is determined.)

34. (*Depreciation*) Using sum-of-the-years'-digits depreciation (see Exercise 33) calculate the first years' depreciation on a computer whose initial cost is $230,000 and whose scrap value after 10 years is $10,000.

35. (*Depreciation*) Using sum-of-the-years'-digits depreciation (see Exercise 33), calculate the annual depreciation during each year for a fleet of cars that has a purchase price of $500,000 and a resale value of $200,000 after 3 years.

◤ 7-2 GEOMETRIC PROGRESSIONS AND COMPOUND INTEREST

Suppose $1000 is deposited with a bank that calculates interest at the rate of 10% compounded annually. The value of this investment (in dollars) at the end of 1 year is equal to

$$1000 + 10\% \text{ of } 1000 = 1000(1 + 0.1) = 1000(1.1) = 1100.$$

If the investment is at compound interest, then during the second year interest is paid on this whole sum of $1100 (see pages 215–217). Therefore, the value of the investment (in dollars) at the end of 2 years is

$$1100 + 10\% \text{ of } 1100 = 1100 + 0.1(1100)$$

$$= 1100(1 + 0.1) = 1100(1.1) = 1000(1.1)^2$$

Similarly, the value of the investment at the end of 3 years will be $1000(1.1)^3$

dollars, and so on. Thus the values of the investment (in dollars) at the end of 0 years, 1 year, 2 years, 3 years, and so on, are

$$1000, \quad 1000(1.1), \quad 1000(1.1)^2, \quad 1000(1.1)^3, \quad \ldots.$$

Note the difference between this example and the case of simple interest discussed in the preceding section. With simple interest a constant amount is added each period. With compound interest the value is multiplied by a constant factor each period (1.1 in this example). This sequence is an example of a *geometric progression*.

DEFINITION A sequence of terms is said to be a **geometric progression** (G.P.) if the ratio of each term to its preceding term is the same throughout. This constant ratio is called the **common ratio** of the G.P.

Thus the sequence 2, 6, 18, 54, 162, . . . is a G.P. because

$$\tfrac{6}{2} = \tfrac{18}{6} = \tfrac{54}{18} = \tfrac{162}{54} = 3.$$

The common ratio is 3.

Similarly, the sequence $\tfrac{1}{3}, -\tfrac{1}{6}, \tfrac{1}{12}, -\tfrac{1}{24}, \ldots$ is a G.P. with common ratio $-\tfrac{1}{2}$.

Each term in a G.P. is obtained by multiplying the preceding term by the common ratio. If a is the first term and r the common ratio, then successive terms of the G.P. are

$$a, \quad ar, \quad ar^2, \quad ar^3, \quad \ldots.$$

In this G.P., we observe that the power of r in any term is one less than the number of the term. Thus the nth term is given by

$$T_n = ar^{n-1}. \tag{1}$$

EXAMPLE 1 Find the fifth and nth terms of the sequence 2, 6, 18, 54, ...

Solution The given sequence is a G.P. because

$$\tfrac{6}{2} = \tfrac{18}{6} = \tfrac{54}{18} = 3.$$

Thus successive terms have constant ratio of 3; that is, $r = 3$. Also, $a = 2$. Therefore

$$T_5 = ar^4 = 2(3^4) = 162 \quad \text{and} \quad T_n = ar^{n-1} = 2 \cdot 3^{n-1}. \quad \text{☞ 8}$$

☞ **8.** Find the sixth and the nth term in the G.P. 3, −6, 12, −24, ...

EXAMPLE 2 The fourth and ninth terms of a G.P. are $\tfrac{1}{2}$ and $\tfrac{16}{243}$. Find the sixth term.

Solution Let a be the first term and r be the common ratio of the given G.P. Then, using our given values, we have

$$T_4 = ar^3 = \tfrac{1}{2} \quad \text{and} \quad T_9 = ar^8 = \tfrac{16}{243}.$$

Answer $T_6 = -96$, $T_n = 3 \cdot (-2)^{n-1}$.

We divide the second equation by the first and solve for r.

$$\frac{ar^8}{ar^3} = \frac{\frac{16}{243}}{\frac{1}{2}}$$

$$r^5 = \frac{16}{243} \cdot \frac{2}{1} = \frac{32}{243} = \left(\frac{2}{3}\right)^5.$$

$$r = \frac{2}{3}$$

Substituting this value of r in the first equation, we have

$$a\left(\frac{2}{3}\right)^3 = \frac{1}{2}.$$

Thus

$$a = \frac{1}{2} \cdot \frac{27}{8} = \frac{27}{16}$$

and

☞ **9.** In a G.P., $T_7 = 2$ and $T_{11} = 8$. Find an expression for T_n and compute T_{15}.

$$T_6 = ar^5 = \frac{27}{16}\left(\frac{2}{3}\right)^5 = \frac{27}{16} \cdot \frac{32}{243} = \frac{2}{9}.$$

Hence the sixth term is $\frac{2}{9}$.   ☞ **9**

In Example 2 of Section 7-1 we looked at an example of depreciation in which the amount of annual depreciation was constant. This method is called straight-line depreciation (see also Section 4-3). An alternative method is to depreciate by a fixed percentage of the value the year before.

EXAMPLE 3 *(Depreciation)* A machine is purchased for $10,000 and is depreciated annually at the rate of 20% of its declining value. Find an expression for the value after n years. If the ultimate scrap value is $3000, what is the effective life of the machine (i.e., the number of years until its depreciated value is less than its scrap value)?

Solution Since the value of the machine depreciates each year by 20% of its value at the beginning of the year, the value of the machine at the end of any year is 80% or four-fifths of its value at the beginning of that year. Thus the value (in dollars) of the machine at the end of the first year is

$$\tfrac{4}{5} \text{ of } 10,000 = 10,000\left(\tfrac{4}{5}\right)$$

and at the end of the second year is

$$\tfrac{4}{5} \text{ of } 10,000\left(\tfrac{4}{5}\right) = 10,000\left(\tfrac{4}{5}\right)^2.$$

Similarly, its value (in dollars) at the end of the third year will be $10,000\left(\frac{4}{5}\right)^3$, and so on. Therefore the value (in dollars) of the machine at the end of the first year, second year, third year, and so on, is

$$10,000\left(\tfrac{4}{5}\right), \quad 10,000\left(\tfrac{4}{5}\right)^2, \quad 10,000\left(\tfrac{4}{5}\right)^3, \ldots.$$

This sequence is clearly a G.P. with first term of $10,000\left(\frac{4}{5}\right)$ and common ratio of $\frac{4}{5}$. The nth term, which gives the value of the machine at the end of the nth year, is therefore

Answer There are two answers: $T_n = (\pm\sqrt{2})^{n-5}$, $T_{15} = 32$.

$$T_n = ar^{n-1} = 10,000\left(\tfrac{4}{5}\right) \cdot \left(\tfrac{4}{5}\right)^{n-1} = 10,000\left(\tfrac{4}{5}\right)^n.$$

● **10.** Re-solve Example 3 if the depreciation rate is 10% per annum.

Setting n successively equal to 1, 2, 3, . . . , we obtain the values in Table 1. We see, therefore, that after 5 years the value of the machine is still a little greater than its scrap value of $3000, but after 6 years, its value is below the scrap value. The lifetime of the machine is therefore 6 years.   ● **10**

TABLE 1

n	1	2	3	4	5	6
T_n	8000	6400	5120	4096	3276.8	2621.44

We began this section with an example of compound interest. The general case of an investment growing at compound interest was discussed in Section 6-1. If a sum P is invested at a rate of interest of R percent per annum compounded k times per year, then the value of the investment at the end of the nth period is given by

$$T_n = P(1 + i)^n \qquad i = \frac{R}{100k}.$$

These values for $n = 1, 2, 3, \ldots$ form a geometric progression. The common ratio is $r = 1 + i$ and the first term is $T_1 = P(1 + i)$.

Further applications related to this will be given in the next section.

THEOREM 1 (SUM OF n TERMS OF A G.P.)   If a is the first term and r the common ratio of a G.P., then the sum S_n of n terms of the G.P. is given by

$$S_n = \frac{a(1 - r^n)}{1 - r}. \qquad (2)$$

PROOF The n terms of the given G.P. are

$$a, \quad ar, \quad ar^2, \quad \ldots, \quad ar^{n-2}, \quad ar^{n-1}.$$

Therefore the sum of these terms is

$$S_n = a + ar + ar^2 + \cdots + ar^{n-2} + ar^{n-1}.$$

We multiply both sides by $-r$.

$$-rS_n = -ar - ar^2 - \cdots - ar^{n-1} - ar^n$$

Adding these two equations, we find that all the terms cancel except the first term in the first equation and the last term in the second equation, giving

$$S_n - rS_n = a - ar^n.$$

We factor and solve for S_n.

$$S_n(1 - r) = a(1 - r^n)$$

Answer Value after n years $= T_n$
$= 10{,}000\left(\dfrac{9}{10}\right)^n$. T_n is less than 3000 after 12 years.

$$S_n = \frac{a(1 - r^n)}{1 - r}$$

This proves the result.

Multiplying numerator and denominator of Equation (2) by -1, we obtain the alternative formula

$$S_n = \frac{a(r^n - 1)}{r - 1}.$$

This formula is generally used when $r > 1$, whereas Equation (2) is more useful when $r < 1$.

Note The above formula for S_n is valid only when $r \neq 1$. When $r = 1$, the given G.P. becomes

$$a + a + a + \cdots + a \quad (n \text{ terms})$$

which has the sum na.

EXAMPLE 4 Find the sum of the first 10 terms of the sequence $2 - 4 + 8 - 16 + \cdots$.

Solution The given sequence is a G.P. with $a = 2$ and $r = -\frac{4}{2} = -2$. Here $n = 10$. Therefore

$$S_n = \frac{a(1 - r^n)}{1 - r}$$

or

$$S_{10} = \frac{2(1 - (-2)^{10})}{1 - (-2)} = \tfrac{2}{3}(1 - 2^{10}) = \tfrac{2}{3}(1 - 1024) = -682. \quad \blacktriangleright 11$$

☛ **11.** Find the sum of 11 terms of the G.P.s

(a) $1 + 2 + 4 + 8 + \cdots$;

(b) $2 + 3 + \dfrac{9}{2} + \dfrac{27}{4} + \cdots$

EXAMPLE 5 *(Savings Plan)* Each year a person invests $1000 in a savings plan that pays interest at the fixed rate of 8% per annum. What is the value of this savings plan on the tenth anniversary of the first investment? (Include the current payment paid into the plan.)

Solution The first $1000 has been invested for 10 years, so it has increased in value to

$$\$1000(1 + i)^{10}, \qquad i = \frac{R}{100} = \frac{8}{100} = 0.08.$$

Thus the value is $1000(1.08)^{10}$.

The second $1000 was invested 1 year later; hence it has been in the plan for 9 years. Its value has therefore increased to $1000(1.08)^9$. The third $1000 has been in the plan for 8 years and has the value $1000(1.08)^8$. We continue in the same fashion until we reach the tenth payment of $1000, which was made 9 years after the first. Its value 1 year later is $1000(1.08)

Answer (a) $2^{11} - 1 = 2047$;

(b) $4\left[\left(\dfrac{3}{2}\right)^{11} - 1\right] = \dfrac{175,099}{512}$.

Thus the total value of the plan on its tenth anniversary is obtained by adding all these amounts together with the current payment of $1000.

$$S = 1000(1.08)^{10} + 1000(1.08)^9 + \cdots + 1000(1.08) + 1000.$$

Writing this in reverse order we have

$$S = 1000 + 1000(1.08) + 1000(1.08)^2 + \cdots + 1000(1.08)^{10}.$$

This is a G.P. with $a = 1000$, $r = 1.08$ and $n = 11$. Therefore

$$s = 1000 \frac{(1.08)^{11} - 1}{1.08 - 1}$$

from the formula for the sum of a G.P. Simplifying, we have

$$S = \frac{1000}{0.08}[(1.08)^{11} - 1] = 12{,}500(2.3316 - 1) = 16{,}645.aaaar$$

☛ 12 Resolve Example 5 if the interest rate is 10% per annum.

The value is thus $16,645.   ☛ 12

The sum of the first n terms of the geometric sequence

$$a + ar + ar^2 + \cdots$$

is given by

$$S_n = \frac{a(1 - r^n)}{1 - r}. \tag{2}$$

Let us consider the behavior of r^n for large n when $-1 < r < 1$. To pick a specific example, let $r = \frac{1}{2}$. Then Table 2 gives the values of r^n for several different values of n. From this table, we observe that as n gets larger and larger, r^n gets smaller and smaller. Ultimately, when n becomes very large, r^n approaches zero. This behavior of r^n—namely, that r^n gets closer and closer to zero as n becomes larger and larger—is true whenever $-1 < r < 1$. Thus from Equation (2), we can say that the sum of an infinite number of terms in a G.P. is given by

$$S_\infty = \frac{a(1 - 0)}{1 - r} = \frac{a}{1 - r}.$$

TABLE 2

n	1	2	3	4	5	6	7
r^n	0.5	0.25	0.125	0.0625	0.03125	0.015625	0.0078125

This leads us to the following theorem.

Answer $S = 1000 \dfrac{(1.1)^{11} - 1}{1.1 - 1}$

$= 18{,}531.$

THEOREM 2 (SUM OF AN INFINITE G.P.) The sum S of an infinite geometric sequence

$$a + ar + ar^2 + \cdots$$

is given by

13. The recurring decimal 0.51515151. . . can be expressed as the sum of an infinite G.P. as

$$0.51\left[1 + \frac{1}{100} + \left(\frac{1}{100}\right)^2 + \left(\frac{1}{100}\right)^3 + \cdots\right].$$

Evaluate this sum and hence express the decimal as a fraction. Similarly express the decimals 1.222222 . . . and 0.279279279 . . . as fractions.

Answer $\frac{17}{33}, \frac{11}{9}, \frac{31}{111}$.

$$\boxed{S = \frac{a}{1 - r}, \qquad \text{provided that } -1 < r < 1. \qquad (3)}$$

In financial mathematics, infinite G.P.'s occur in some situations which involve *perpetuity*. An example would be an annuity which continues forever.

EXAMPLE 6 Find the sum of the infinite sequence $1 - \frac{1}{3} + \frac{1}{9} - \frac{1}{27} + \cdots$.

Solution The given sequence is a G.P. with $a = 1$ and $r = -\frac{1}{3}$. The sum is given by

$$S = \frac{a}{1 - r} = \frac{1}{1 - (-\frac{1}{3})} = \frac{1}{\frac{4}{3}} = \frac{3}{4}. \qquad 13$$

EXERCISES 7-2

(1–4) Find the specified term.

1. Ninth term of the sequence 3, 6, 12, 24, . . .

2. Sixth term of the sequence $\sqrt{3}$, 3, $3\sqrt{3}$, 9, . . .

3. nth term of the sequence $\frac{2}{9}, -\frac{1}{3}, \frac{1}{2}, \ldots$

4. pth term of the sequence $\frac{2}{5}, -\frac{1}{2}, \frac{5}{8}, \ldots$

(5–6) Which term of the sequence is the last given term?

5. 96, 48, 24, 12, . . .; $\frac{3}{16}$

6. 18, 12, 8, . . .; $\frac{512}{729}$

7. The second term of a G.P. is 24 and the fifth term is 81. Find the sequence and the tenth term.

8. The fifth, eighth, and eleventh terms of a G.P. are x, y, and z, respectively. Show that $y^2 = xz$.

9. If $x + 9$, $x - 6$, and 4 are the first three terms of a G.P., find x.

10. In a G.P., if the first term is a, the common ratio r, and the last term K, show that the number of terms in the G.P. is given by

$$n = 1 + \frac{\ln K - \ln a}{\ln r}$$

(11–17) Find the indicated sum of the following sequences.

11. $2 + 6 + 18 + 54 + \cdots$; 12 terms

12. $\sqrt{3} - 3 + 3\sqrt{3} - 9 + \cdots$; 10 terms

13. $1 + 2 + 4 + 8 \cdots$; n terms

14. $3 + 1.5 + 0.75 + 0.375 + \cdots$; p terms

15. $1 + \frac{1}{2} + \frac{1}{4} + \frac{1}{8} + \cdots$

16. $1 - \frac{1}{3} + \frac{1}{9} - \frac{1}{27} + \cdots$

17. $\sqrt{2} - \frac{1}{\sqrt{2}} + \frac{1}{2\sqrt{2}} - \cdots$

18. If $y = 1 + x + x^2 + x^3 + \cdots$ $(-1 < x < 1)$, show that

$$x = \frac{y - 1}{y}.$$

19. If $v = 1/(1 + i)$, show that

$$v + v^2 + v^3 + \cdots = \frac{1}{i}.$$

20. Prove that $9^{1/3} \cdot 9^{1/9} \cdot 9^{1/27} \cdots = 3$.

21. Evaluate $4^{1/3} \cdot 4^{-1/9} \cdot 4^{1/27} \cdot 4^{-1/81} \cdots$.

22. Express 0.85555 . . . as a fraction. (*Hint:* Write $0.85555 = 0.8 + 0.05(1 + 0.1 + 0.01 + \cdots)$.)

23. *(Depreciation)* A machine is depreciated annually at the

rate of 10% on its reducing value. The original cost was $10,000 and the ultimate scrap value was $5314.41. Find the effective life of the machine.

24. *(Depreciation)* A car is purchased for $8300. The depreciation is calculated on the diminshing value at 10% for the first 3 years and at 15% for the next 3 years. Find the value of the car after a period of 6 years.

25. *(Depreciation)* A machine is purchased for $10,000. The depreciation is calculated on the reducing value at 8% for the first 2 years and at 10% for the next 5 years. Find the value after a period of 7 years.

26. *(Compound Interest)* If $2000 is invested in a savings account at 8% interest compounded annually, find its value after 5 years.

27. *(Compound Interest)* In Exercise 26, the rate of interest decreases after 6 years to 6% per annum. Find the value of the investment after a further 6 years.

28. *(Quarterly Compounding)* If $5000 is invested in a savings account on which interest is compounded quarterly at a nominal interest rate of 8% per annum, find its value after 3 years.

29. *(Monthly Compounding)* Suppose $4000 is invested in a fixed deposit at a nominal annual interest rate of 6% compounded monthly. Find its value:

   a. After 1 year.   **b.** After 4 years.

30. *(Compound Interest)* A person wishes to invest a certain sum of money in a fixed deposit earning 10% interest per annum for a period of 4 years. At the end of this time the proceeds from the investment will be used to pay off a debt of $10,000 that will become due then. How much must be invested to have enough to pay off the debt?

31. *(Savings Plan)* Every year Mary invests $2000 in a savings account that earns interest at 10% per annum. Find the value of her investment on the twelfth anniversary of her first deposit. (Include the current payment.)

32. *(Savings Plan)* At the beginning of every month, Joe deposits $200 in a savings account that earns interest at the rate of $\frac{1}{2}$% per month on the minimum monthly balance. How much is his investment worth after 2 years (that is, 25 deposits)?

*33. *(Sinking Fund)* A company will require 1 million dollars in exactly 6 years' time to retire a bond issue. In order to accumulate this amount, the company plans to place a certain sum P each year into a special fund (called a *sinking fund*). The last sum will be placed 1 year before the bonds mature. If the fund will earn interest at 8% per annum, how much must P be?

*34. *(Sinking Fund)* Fred takes a mortgage on his house which will become due for repayment in 5 years' time. At that time, the outstanding debt will be $19,500. Fred plans to save a certain sum each month which he will invest in a savings account which pays interest at a 9% nominal annual rate of interest, compounded monthly. His first investment will be made immediately and his last (the sixty-first) will be made on the date of repayment of the mortgage. How much must he save each month if he is to repay the mortgage completely?

*35. *(Present Value of Annuity)* Today is Art's sixty-fifth birthday and he has just received his check for $1000 from the Veteran's Administration. These checks will continue to arrive on his birthday for the rest of his life. Assuming he dies at the age of 75 after receiving his eleventh check, calculate the present value of all the checks received assuming a discount rate of 10% (see page 222).

*36. *(Present Value of Annuity)* Repeat Exercise 35 assuming Art lives to the age of 80 and the discount rate is 8%.

*37. *(Present Value of Annuity)* Aunt Jane receives an old-age pension of $300 per month. Assuming a nominal discount rate of 12% compounded monthly (see page 222) calculate the present value of the next 48 pension payments if the first payment is received 1 month from now. Also calculate the present value of the next 96 and 144 payments.

◤ 7-3 MATHEMATICS OF FINANCE

The basic problems in the mathematics of finance involving simple and compound interest have been discussed in Chapter 6 and in the first two sections of this chapter. In this section, we will describe briefly some very important further applications of sequences that arise in this area.

Savings Plans

The simplest type of savings plan is one in which regular payments of a fixed amount are made into the plan (for example, at the end of every month or once per year), and the balance invested in the plan earns interest at some fixed rate.

EXAMPLE 1 Every month Jane pays $100 into a savings plan that earns interest at $\frac{1}{2}\%$ per month. Calculate the value of her savings: (a) immediately after making her twenty-fifth payment; (b) immediately after making her nth payment.

Solution

(a) The twenty-fifth payment is made 24 months after the first payment. Each investment increases by a factor of 1.005 per month (0.5% per month). Thus the first $100 that Jane invested is worth $100(1.005)^{24}$ after 24 months. The second $100 she saved will have been in the plan for 23 months, so it will be worth $100(1.005)^{23}$. The third $100 will be worth $100(1.005)^{22}$, and so on. The twenty-fourth payment of $100 will have been in the plan for only 1 month, so will be worth $100(1.005)$. The last payment will not yet have earned any interest. Thus the total value of the plan will be the sum of all these amounts; that is,

$$S = 100(1.005)^{24} + 100(1.005)^{23} + 100(1.005)^{22} + \cdots$$
$$+ 100(1.005)^2 + 100(1.005) + 100.$$

But, in the reverse order, this expression is the sum of 25 terms in a G.P. with the first term $a = 100$ and the common ratio $r = 1.005$. Therefore

$$S = \frac{a(r^n - 1)}{r - 1} = \frac{100[(1.005)^{25} - 1]}{1.005 - 1}$$

$$= \frac{100}{0.005}[(1.005)^{25} - 1]$$

$$= 20{,}000[1.13280 - 1] = 2655.91.$$

After 24 months, Jane's savings plan is therefore worth $2655.91.

(b) The nth payment is made $(n - 1)$ months after the first payment. After $(n - 1)$ months, the first $100 invested will be worth $100(1.005)^{n-1}$. The second $100 will be worth $100(1.005)^{n-2}$ since it will have been in the plan for $(n - 2)$ months. The next-to-last payment will again be worth $100(1.005)$ dollars, and the last (nth) payment will be just $100. Thus the total value of the plan will be the following sum.

$$S = 100(1.005)^{n-1} + 100(1.005)^{n-2} + \cdots + 100(1.005)^2$$
$$+ 100(1.005) + 100$$

Again, in the reverse order, the terms in this sum form a G.P. with $a = 100$

and $r = 1.005$. There are n terms, so

$$S = \frac{a(r^n - 1)}{r - 1} = \frac{100[(1.005)^n - 1]}{1.005 - 1}$$

$$= 20,000[(1.005)^n - 1].$$

From this formula, we can calculate the value of the plan after any number of months. For example, after 59 months (that is, when the sixtieth payment is made), the savings plan is worth

$$20,000[(1.005)^{60} - 1] = 20,000[1.34885 - 1] = 6977.00$$

or $6977.00 ☛ 14

☛ 14 Re-solve Example 1 if the monthly payments are $150 and the interest rate is 0.25% per month.

The argument used in this example can be easily extended to the general case. Let us suppose that an amount P is saved each period, where the period may be a month, a quarter, a year, or any other fixed length of time. Let the interest rate be R percent per period. Then each period the investment increases by a factor of $1 + i$, where $i = R/100$. Now let us ask what will be the value of the savings plan after $n - 1$ periods from the first payment, that is, immediately after the nth payment is made.

The first payment of P will have been invested for the full $n - 1$ periods, so will have increased in value to $P(1 + i)^{n-1}$. The second payment will, however, only have been invested for $n - 2$ periods, so will have increased in value to $P(1 + i)^{n-1}$. (See Figure 1.) The nth or last payment will just have been invested, so it will be worth P (Figure 1).

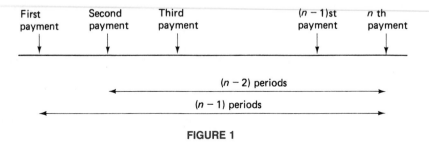

FIGURE 1

The total value of the savings plan will therefore be given by the sum

$$S = P(1 + i)^{n-1} + P(1 + i)^{n-2} + \cdots + P(1 + i) + P.$$

In the reverse order, this is the sum of a G.P. with first term $a = P$ and common ratio $1 + i$. There are n terms, so

$$S = \frac{a(r^n - 1)}{r - 1} = \frac{P[(1 + i)^n - 1]}{(1 + i) - 1} = \frac{P}{i}[(1 + i)^n - 1].$$

If we substitute $P = 100$ and $i = 0.005$, we have the results of Example 1.

Answer (a) $3864.68;
(b) $S = 60,000[(1.0025)^n - 1]$.

15. Using Table A.3.4 find
(a) $s_{\overline{40}|0.02}$; (b) $s_{\overline{25}|0.07}$.

Answer (a) 60.401983;
(b) 63.249038.

16. In Example 1, use Table
A.3.4 to compute the value of the
savings plan immediately after the
50th payment.

In financial mathematics, it is common to use the notation

$$S = P\,s_{\overline{n}|i} \qquad \text{where } s_{\overline{n}|i} = i^{-1}[(1 + i)^n - 1].$$

The quantity $s_{\overline{n}|i}$ is read as 's angle n at i' and represents the value of a savings plan after n regular payments of $1 each. It depends only on the interest rate and the number n. Some values of $s_{\overline{n}|i}$ for different values of i and n are given in Table A.3.4. **15**

For example, the solution to Example 1(a) could be evaluated using the table as

$$S = Ps_{\overline{n}|i} = (100)s_{\overline{25}|0.005} = 100(26.559115) = 2655.91. \quad \text{📌 16}$$

Annuities

An annuity is the term given to a sequence of payments of a certain fixed sum of money at regular intervals of time, for example $2500 on the last day of each quarter, or $2000 on the first of January each year. Example 1 above is an example of an annuity of $100 per month (paid by Jane into her savings plan). The quantity S calculated in part (b) of that example provides the value of the annuity after the nth payment; it is often called the **future value** of the annuity.

More generally, the quantity $s_{\overline{n}|i}$ represents *the future value after n periods of an annuity of $1 per period when $i = R/100$ and the interest rate is R% per period*. (To see this, just put $P = 1$ in the discussion above.)

Another important example of an annuity occurs when a person deposits a sum of money with an insurance company or similar institution and that company repays the money in a series of regular payments of equal amount over a period of time. Between the payments, the balance remaining earns interest at some predetermined rate. The payments continue until the sum deposited has been used up. This is a common method by means of which people secure a pension upon retiring from work.

EXAMPLE 2 On his 65th birthday, Mr. Hoskins wishes to purchase an annuity that will pay him $5000 per year for the next 10 years, the first payment to be made to him on his 66th birthday. His insurance company will give him an interest rate of 8% per annum on the investment. How much must he pay in order to purchase such an annuity?

Solution Consider the first payment, which Mr. Hoskins will receive on his 66th birthday. This payment will be $5000, and it must be paid for by a deposit made 1 year earlier. Let this deposit be A_1. In the intervening year, A_1 will have gained 8% interest, so it will have increased to $(1.08)A_1$. This must be equal to 5000.

$$(1.08)A_1 = 5000$$

$$A_1 = 5000(1.08)^{-1}.$$

Answer $S = 100s_{\overline{50}|0.005} = 5664.52$.

☞ **17.** In Example 2, how much must be paid to purchase the annuity if the interest rate is 6% per annum?

That is, if an amount equal to $5000(1.08)^{-1}$ is deposited on Mr. Hoskins' 65th birthday, it will have increased to $5000 by his 66th birthday.

Now consider the second payment, which he will receive on his 67th birthday. This must be paid by a deposit, A_2, made 2 years earlier. Since A_2 is accumulating interest for 2 years at 8%, we must have

$$A_2(1.08)^2 = 5000$$

if it is to be just enough to grow to $5000 on his 67th birthday. Thus

$$A_2 = 5000(1.08)^{-2}.$$

That is, if an amount equal to A_2 is deposited on Mr. Hoskins' 65th birthday, it will have increased to $5000 by his 67th birthday. Clearly, if Mr. Hoskins deposits the sum $A_1 + A_2$ on his 65th birthday, the investment will be just enough to enable him to collect $5000 on each of his 66th and 67th birthdays.

We can continue in this same way. A deposit $A_3 = \$5000(1.08)^{-3}$ will increase to $5000 after 3 years (on his 68th birthday), and so on. The last $5000 will be paid on his 75th birthday, so it will have been invested for 10 years. Hence it will require a deposit of $\$5000(1.08)^{-10}$.

Thus in order to receive all 10 payments, the total amount A that Mr. Hoskins must deposit on his 65th birthday is given by

$$A = A_1 + A_2 + A_3 + \cdots + A_{10}$$

$$= 5000(1.08)^{-1} + 5000(1.08)^{-2} + 5000(1.08)^{-3} + \cdots + 5000(1.08)^{-10}.$$

This expression represents the sum of 10 terms of a G.P. with first term $a = 5000(1.08)^{-1}$ and common ratio $r = (1.08)^{-1}$. Therefore

$$A = \frac{a(1 - r^n)}{1 - r} = \frac{5000(1.08)^{-1}[1 - (1.08)^{-10}]}{1 - (1.08)^{-1}}.$$

Multiplying numerator and denominator by 1.08, we obtain

$$A = \frac{5000[1 - (1.08)^{-10}]}{1.08[1 - (1.08)^{-1}]} = \frac{5000[1 - (1.08)^{-10}]}{1.08 - 1}$$

$$= \frac{5000}{0.08}[1 - (1.08)^{-10}]$$

$$= \frac{5000}{0.08}(1 - 0.4632)$$

$$= 33{,}550.$$

Thus Mr. Hoskins must deposit $33,550 in order to buy his required annuity of $5000 for 10 years. ☞ **17**

Now let us generalize this example. Suppose an annuity is purchased with a down payment A and the payments made to the annuitant are equal to P. These payments are to be made at regular intervals for n periods starting

Answer $36,800.44.

☞ **18.** Using Table A.3.4 find
(a) $a_{\overline{5}|0.08}$;   (b) $a_{\overline{30}|0.0075}$.

one period after the annuity is purchased. The interest rate is R percent per period.

As in Example 2, the first payment of P—made 1 period after the annuity is purchased—requires a deposit A_1, where

$$A_1(1 + i) = P \qquad \left(i = \frac{R}{100}\right).$$

This is because an investment of A_1 would increase to a value $A_1(1 + i)$ during the intervening period. Therefore

$$A_1 = P(1 + i)^{-1}.$$

In a similar way, an investment of A_2 would increase to a value P after 2 periods if

$$A_2(1 + i)^2 = P$$

or

$$A_2 = P(1 + i)^{-2}.$$

The nth and last payment requires a deposit A_n to be made n periods beforehand, where

$$A_n(1 + i)^n = P \qquad \text{or} \qquad A_n = P(1 + i)^{-n}.$$

Thus if the sum A is to be sufficient to enable all n annuity payments to be made, we must have

$$A = A_1 + A_2 + \cdots + A_n$$
$$= P(1 + i)^{-1} + P(1 + i)^{-2} + \cdots + P(1 + i)^{-n}.$$

Again we have the sum of n terms of a G.P. The first term is $a = P(1 + i)^{-1}$ and common ratio is $r = (1 + i)^{-1}$. Therefore

$$A = \frac{a(1 - r^n)}{1 - r} = \frac{P(1 + i)^{-1}[1 - (1 + i)^{-n}]}{1 - (1 + i)^{-1}}.$$

Multiplying numerator and denominator by $1 + i$, the denominator becomes

$$(1 + i)[1 - (1 + i)^{-1}] = (1 + i) - (1 + i)(1 + i)^{-1} = (1 + i) - 1 = i.$$

Therefore we are left with

$$A = \frac{P}{i}[1 - (1 + i)^{-n}]$$

or

$$A = Pa_{\overline{n}|i} \qquad \text{where } a_{\overline{n}|i} = i^{-1}[1 - (1 + i)^{-n}]. \tag{1}$$

Answer (a) 3.992710;
(b) 26.775080.

Again the values of $a_{\overline{n}|i}$ (read as "a angle n at i") for selected values of n and i are given in Table A.3.4 in Appendix III.   ☞ **18**

For instance, Example 2 can be solved using the table as

$$A = Pa_{\overline{n}|i} = 5000a_{\overline{10}|0.08} = 5000(6.710081) = 33550.40.$$

We often call A the **present value** of an annuity of P per period for n periods. It is the amount that must be paid now to purchase such an annuity. The quantity $a_{\overline{n}|i}$ represents the present value of an annuity of $1 per period for n periods. By contrast, recall that $s_{\overline{n}|i}$ is the future value of such an annuity, that is, the value at the end of all the payments. ☛ 19

When a life insurance company issues a pension policy to a person, it usually does not issue it for a certain specified number of years, but rather for as long as the person concerned remains alive. In such a case, the value of n used is the life expectancy of the person, that is, the number of years (on the average) that people in his or her category survive.

EXAMPLE 3 *(Annuity)* Mrs. Josephs retires at the age of 63 and uses her life savings of $120,000 to purchase an annuity. The life insurance company gives an interest rate of 6% and they estimate that her life expectancy is 15 years. How much annuity (that is, how big an annual pension) will she receive?

Solution In Equation (1), we know that $A = 120,000$ and $i = R/100 = \frac{6}{100} = 0.06$. We wish to calculate P. We have

$$A = Pa_{\overline{n}|i}$$

That is,

$$120,000 = Pa_{\overline{15}|0.06} = P(9.712249) \qquad \text{(from Table A.3.4).}$$

Therefore, $P = 120,000/9.712249 = 12,355.53$ and Mrs. Josephs will receive an annual pension of $12,355.53.

Amortization

When a debt is repaid by regular payments over a period of time, we say that the debt is **amortized.** For example, a person might borrow $5000 from the bank to buy a new car with the arrangement that a specified amount will be paid back each month for the next 24 months. We would like to determine how much the monthly payments should be, given that the bank will charge interest at a certain rate on each outstanding balance. Another example of widespread importance is mortgages, which are again repaid by regular payments typically spread over 20 or 25 years.

Mathematically speaking, the amortization of a debt presents exactly the same problem as paying an annuity. With an annuity, we can view the annuitant as lending a certain amount A to the life insurance company; the company then repays this loan by n regular payments of amount P each. On each outstanding balance, the company adds interest to the credit of the annuitant at the rate of R percent per period. This is identical with the situation arising in the case of a loan. Here the bank lends a specified sum A to the borrower, who then repays this loan by n regular payments of P each. On each outstanding balance, the borrower must add interest at the rate of R percent per period.

☛ **19.** In Example 2, use Table A.3.4 to compute the present value of an annuity of $5000 per year for 20 years.

Answer $A = 5000a_{\overline{20}|0.08}$
$= 49,090.74.$

Consequently, Equation (1) also applies to the amortization of a loan.

$$A = Pa_{\overline{n}|i} \qquad \left(i = \frac{R}{100}\right) \tag{2}$$

EXAMPLE 4 A small construction company wishes to borrow from the bank for expansion of its operations. The bank charges interest at 1% per month, and insists that the loan be repaid within 24 months. The company estimates that they can afford to repay the loan at a rate of $1500 per month. What is the maximum they can borrow?

Solution In the above formula, we can let $P = 1500$, $i = \frac{1}{100} = 0.01$ (since the interest rate is 1% per period, that is, per month in this case), and $n = 24$. Then

$$A = 1500a_{\overline{24}|0.01} = 1500(21.243387) = 31,865.08.$$

Thus the company can borrow up to $31,865.08. ☛ **20**

☛ **20.** In Example 4, if the firm borrows only $20,000, what will be the monthly payments if they arrange to repay the loan over 18 months?

EXAMPLE 5 *(Mortgage Borrowing Limit)* A married couple have a combined income of $45,000. Their mortgage company will allow them to borrow up to an amount at which the repayments are one-third of their income. If the interest rate is 1.2% per month, amortized over 25 years, how much can they borrow?

Solution Here the monthly repayments are $P = (\frac{1}{3} \cdot 45,000) \div 12 = 1250$, and $i = 1.2/100 = 0.012$. In 25 years the number of monthly repayments is $n = 25(12) = 300$. Therefore, the sum which can be borrowed is

$$A = Pa_{\overline{n}|i} = \frac{P}{i}[1 - (1 + i)^{-n}]$$

$$= \frac{1250}{0.012}[1 - (1.012)^{-300}]$$

$$= 104166.67(0.97208)$$

$$= \$101,258.80.$$

Here we have used the formula for $a_{\overline{n}|i}$ because the value of $a_{\overline{n}|i}$ is not given in Table A.3.4 for $n = 300$ and $i = 0.012$.

Usually, we want to use Equation (2) in order to calculate the magnitude of the payments P. Solving for P, we obtain

$$P = \frac{iA}{1 - (1 + i)^{-n}} = \frac{A}{a_{\overline{n}|i}}. \tag{3}$$

Answer $P = \dfrac{20,000}{a_{\overline{18}|0.01}} = 1219.64.$

EXAMPLE 6 During his years as an undergraduate, a student accumulates student loans so that, upon graduation, the debt is $8000. The loan accumulates interest at 8% per annum and is repaid in single installments at the end of each year. How much must the student pay each year to repay the debt in 5 years?

Solution The initial debt is $A = 8000$. The repayment period is $n = 5$. Since the interest rate is $R = 8$, then $i = R/100 = 0.08$. The annual payment is obtained from Equation (3).

$$P = \frac{8000}{a_{\overline{5}|0.08}} = \frac{8000}{3.992710} = 2003.65.$$

The student must therefore repay \$2003.65 at the end of each year in order to amortize the debt within 5 years.

EXERCISES 7-3

(1–8) Use Table A.3.4 to find the following values.

1. $s_{\overline{6}|0.08}$

2. $s_{\overline{30}|0.01}$

3. $s_{\overline{36}|0.0075}$

4. $s_{\overline{40}|0.005}$

5. $a_{\overline{10}|0.01}$

6. $a_{\overline{15}|0.08}$

7. $a_{\overline{20}|0.005}$

8. $a_{\overline{36}|0.0075}$

(In the following, the interest rates quoted are nominal rates.)

(9–14) *(Savings Plan)* Find the value of the savings plan at the end of:

9. Ten years if \$1000 is deposited at the end of each year into an account earning 8% per annum compound interest.

10. Five years if \$2000 is deposited at the end of each year at 6% interest compounded yearly.

11. Four years if \$500 is deposited at the end of every 6 months at 8% interest compounded semiannually.

12. Six years if \$1000 is deposited at the end of every 3 months at 8% interest compounded quarterly.

13. Three years if \$200 is deposited at the end of each month at 12% interest compounded monthly.

14. Five years if \$500 is deposited at the end of every 4 months at 9% interest compounded three times a year.

(15–18) *(Savings Plan)* How much should be deposited at the end of:

15. Each year for 6 years to obtain a sum of \$15,000 at 8% compound interest per annum?

16. Every 6 months to obtain a sum of \$50,000 at the end of

10 years when the interest earned is 6% compounded semiannually?

17. Every 3 months to obtain a sum of \$20,000 at the end of 4 years at 8% annual interest compounded quarterly?

18. Every month for 3 years to obtain a sum of \$8000 at 12% annual interest compounded monthly?

(19–20) *(Savings Plan)* At the end of which year:

19. Will \$100 deposited at the end of each year at 6% annual compound interest amount to \$5486.45?

20. Will \$75 deposited at the end of each month at 6% annual interest compounded monthly amount to \$2950.21?

21. *(Savings Plan)* At the beginning of every year, \$2000 is invested in a savings plan. The interest rate is 8% per annum. Calculate the value of the investment:

a. At the end of the fifth year.

b. At the end of the nth year.

22. *(Savings Plan (Sinking Fund))* Jack is investing money every month in a savings plan which pays interest at $\frac{1}{2}$% per month. Three years (36 months) after starting the plan, he plans to withdraw the money and use it to pay off a second mortgage on his house. If he will require \$8000 to repay the mortgage, how much per month must he save?

23. *(Sinking Fund)* Mr. Smith estimates that it will cost him \$20,000 to send his son to the university 8 years from now. If he deposits a sum of P dollars per month into a savings account paying 6% annual interest compounded monthly, how much should P be so that Mr. Smith will have \$20,000 when his son is ready for his university education?

24. (*Savings Plan*) Mr. Black is due to retire after 5 years and he wants to go on vacation soon after retiring. For this he starts depositing $2500 each year into an account paying 8% annual interest compounded yearly. How much will he have for his vacation?

*25. (*Sinking Fund*) Atlas Industries estimates that it will cost them $200,000 to replace certain machinery 12 years from now, and they set up a sinking fund for this purpose. Initially, they make annual payments of $11,855.41 into an account paying 6% interest compounded yearly. After the eighth payment, the bank increases the interest to 8% per annum compounded yearly. How much should the firm's remaining yearly payments be?

*26. (*Sinking Fund*) Smith's firm estimates that it will cost them $120,000 to replace certain machinery 10 years from now and a sinking fund is set up for the purpose. Initially, they deposit $1986.69 each quarter into an account earning 8% annual interest compounded quarterly. After the thirtieth payment, the bank increases the interest rate to 12% per annum compounded quarterly. How much should be deposited each quarter for the remaining period?

(27–30) Find the present value of an annuity if:

27. $500 is paid annually for 10 years and the interest rate is 7% per annum.

28. $2000 is paid at the end of every 6 months for 5 years and the interest rate is 10% per annum compounded semiannually.

29. $750 is paid quarterly for 3 years and the interest rate is 8% per annum compounded quarterly.

30. $300 is paid monthly for 2 years and the interest rate is 9% per annum compounded monthly.

(31–32) How much will a person receive at the end of:

31. Each year for 8 years if an annuity for $30,000 is purchased and the interest rate is 6% per annum compounded annually?

32. Each month for 3 years if an annuity for $25,000 is purchased and the interest rate is 12% per annum compounded annually?

33. (*Annuity*) An individual wishes to purchase an annuity that will pay $8000 per year for the next 15 years. If the interest rate is 6%, how much will such an annuity cost?

34. (*Annuity*) If the life expectancy of a man is 12 years at retirement, how much will it cost him to buy an annuity for life of $10,000 per annum if the interest rate is 8%?

35. (*Annuity*) Repeat Exercise 34 if the interest rate is:

   a. 7%   b. 5%

36. (*Annuity*) If the man in Exercise 34 has $100,000 with which to buy his annuity, how much annual pension will he receive?

37. (*Annuity*) Mary receives a legacy of $10,000 which she invests at 6% per annum. At the end of every year, she wishes to withdraw a sum P in order to take a trip to Hawaii. How much should P be if the money is to last 10 years?

38. (*Annuity*) When Cy retired, he had $120,000 invested in long-term bonds paying 5% interest. At the beginning of every year, he withdraws a sum P to meet his expenses for the coming year. If the money is to last for 15 years, how much should he withdraw each year?

*39. (*Annuity*) In Exercise 38, calculate how much Cy has remaining in the bonds just after he has made his tenth withdrawal.

40. (*Loan Amortization*) A loan of $5000 is to be repaid by regular monthly installments over 40 months. If the interest rate is 1% per month, how big should the monthly payments be?

41. (*Loan Amortization*) If the loan in Exercise 40 were paid back over 18 months, how big would be the monthly payments then?

42. (*Loan Amortization*) Sam borrows $6000 from the bank at 12% interest per annum. He repays it by regular installments at the end of every year. If the loan is to be paid off in 4 years, how much should his annual payments be?

43. (*Mortgage Amortization*) A mortgage of $40,000 is to be repaid by monthly payments over a period of 25 years (300 payments). If the rate of interest is $\frac{3}{4}$% per month, calculate the monthly payment.

44. (*Mortgage Amortization*) Recalculate Exercise 43 if the mortgage is to be repaid over 20 years.

*45. (*Mortgage Amortization*) In Exercise 43, calculate how much is owed on the mortgage:

   a. After 5 years.   b. After 15 years.

46. (*Auto Loan*) Mr. Brown purchased a new car for $15,000 and made a 10% down payment. The rest was financed by a bank at 12% compounded monthly. If the loan is to

be repaid in 48 monthly installments, how much is his monthly payment?

47. *(House Equity)* Sam's house is worth $90,000, and he still has to make 50 more monthly payments of $450.00 to the bank on his 9% mortgage. How much is his equity in the house? (*Hint:* Equity in the house is the value of the house minus the present value of the loan on the house.)

*48. *(House Equity and Inflation)* Seven years ago, John bought a house for $80,000, which appreciated in value at 9% per year due to inflation. If John has 48 more monthly payments of $500 to make to the bank on his 12% mortgage, find his present equity in the house.

*49. *(Auto Loan)* Ms. Susan pays $300 per month to the bank on her car loan, which is at 12% per annum compounded monthly. If the loan is to be paid off in 48 equal installments, how much of her twenty-fifth payment is interest? (*Hint:* Find Susan's debt at the time of the twenty-fourth payment by calculating the present value of the remaining annuity.)

*50. *(Auto Loan)* Mr. Smith bought a car for $20,000 and made a down payment of 15%. The rest he borrowed from the bank at 9% per annum compounded monthly. If the loan is to be paid off in 36 equal monthly installments, how much is his monthly payment to the bank? How much of his thirtieth payment is interest?

◣ 7-4 DIFFERENCE EQUATIONS

Discrete Time Processes

The sequences that arise in the mathematics of finance are examples of discrete time processes. These are processes that evolve with time but such that the variables involved change at certain definite discrete points in time. For example, the value of an investment compounded monthly changes at the end of each month, and the sequence of values forms a discrete process. (Contrast this with continuous compounding, which is a continuous time process where the value changes from one moment to the next.)

We shall begin by considering the growth of an investment with k compoundings per year and nominal interest rate of $R\%$ per annum. The interest rate at each compounding is then $(R/k)\%$. Let the initial investment be denoted by A_0. After the first compounding, the value, A_1, is given by

$$A_1 = A_0 + \frac{R}{100k}A_0 = (1 + i)A_0 \qquad i = \frac{R}{100k}.$$

The first term here is the original principal and the second is the interest. For the general step we proceed similarly. Let us denote the value after n compoundings by A_n, so that the value after the next compounding will be A_{n+1} and

$$A_{n+1} = A_n + \frac{R}{100k}A_n = (1 + i)A_n.$$

Thus we have a sequence of values, $A_0, A_1, A_2, A_3, \ldots$, and an equation, $A_{n+1} = (1 + i)A_n$, that relates any two successive values. This is an example of a difference equation, which we shall define in a moment.

First, we note that it is not essential that a sequence whose nth term is T_n should actually start with the term denoted by T_1. For example, in the discussion above of the growth of an investment, the successive values were denoted by $A_0, A_1, A_2, A_3, \ldots$ and we have a sequence that starts with a **zeroth term.** It may also be convenient in some cases to include terms such as T_{-1} or in other cases to start a sequence at, for example, T_2. In particular, with sequences that describe processes evolving in time, it is commonly convenient to start the sequence with a zeroth term, corresponding to the initial instant of time at which the process begins. For this reason, we shall in most of our general discussion denote the sequence by

$$y_0, y_1, y_2, y_3, \ldots .$$

DEFINITION Let $y_0, y_1, y_2, y_3, \ldots$ be a sequence of real numbers. A **difference equation of order** k is an equation that relates $y_n, y_{n+1}, \ldots, y_{n+k}$ for every value of n ($n = 0, 1, 2, 3, \ldots$).

EXAMPLE 1

(a) $y_{n+1} = 3y_n$ is a first-order difference equation: it is an equation relating y_n and y_{n+1}.

(b) $ny_{n+1} + (n+1)y_n = n^3$ is also a first-order difference equation since again only y_n and y_{n+1} occur in it.

(c) $y_{n+2} = y_n + y_{n+1}$ is a second-order difference equation: it is an equation relating y_n, y_{n+1}, and y_{n+2}.

(d) $y_{n+2} = \sqrt{ny_n} - n - 1$ is again a second-order difference equation. (Note that y_{n+1} does not occur in this equation.) ☞ **21**

> **21.** Give the order of the following difference equations:
> (a) $y_{n+3} = ny_n$;
> (b) $y_{n+2} - 2y_n = y_{n+1}$;
> (c) $y_{n+1}^2 y_n + 2y_n^3 = 1$.

Remark It is important to realize that if the index n is replaced by any other index that covers the same range of values, the equation so obtained is equivalent to the original difference equation. So, for example, n may be replaced by $n - 1$ or $n - 2$ and the difference equation remains unchanged. For example, in Example 1(a), replacing n by $n - 1$, we get the equivalent difference equation

(a') $$y_n = 3y_{n-1}.$$

For, setting $n = 0$ in (a), we get $y_1 = 3y_0$ and this relation can be obtained from (a') by setting $n = 1$. Setting $n = 1$ in (a), we get $y_2 = 3y_1$ and this can be obtained from (a') by setting $n = 2$, and so on. Any of the relations obtained from (a) by giving n particular values can also be obtained from (a'). Similarly, replacing n by $n - 1$ in Example 1(d), we get the equivalent difference equation

(d') $$y_{n+1} = \sqrt{(n-1)y_{n-1}} - n$$

or, alternatively, replacing n by $n - 2$, we get another equivalent form,

(d'') $$y_n = \sqrt{(n-2)y_{n-2}} - n + 1.$$

DEFINITION A **solution** of a difference equation is a set of values for y_n as a function of n such that when these values are substituted into the difference equation, the latter equation is satisfied identically for all the applicable values of n.

Answer (a) 3; (b) 2; (c) 1.

EXAMPLE 2 Write down the first four terms of the sequence given by

$$y_n = \tfrac{1}{2} n(n + 1) \qquad n = 1, 2, 3, \ldots .$$

Show that this sequence is a solution of the difference equation

$$y_n - y_{n-1} = n.$$

Solution Setting $n = 1, 2, 3,$ and 4 in the given formula, we get the first four terms:

$$y_1 = \tfrac{1}{2}(1)(1 + 1) = 1 \qquad y_2 = \tfrac{1}{2}(2)(2 + 1) = 3$$
$$y_3 = \tfrac{1}{2}(3)(3 + 1) = 6 \qquad y_4 = \tfrac{1}{2}(4)(4 + 1) = 10$$

It is clear that these terms satisfy the given difference equation, for example, $y_2 - y_1 = 2$, $y_3 - y_2 = 3$, and so on. However, we must verify that the sequence is a solution for all n. Since $y_n = \tfrac{1}{2} n(n + 1)$, the preceding term must be

$$y_{n-1} = \tfrac{1}{2}(n - 1)(n - 1 + 1) = \tfrac{1}{2} n(n - 1)$$

and substituting these into the left side of the given difference equation, we get

$$y_n - y_{n-1} = \tfrac{1}{2} n(n + 1) - \tfrac{1}{2} n(n - 1)$$
$$= \tfrac{1}{2} n[(n + 1) - (n - 1)]$$
$$= \tfrac{1}{2} n[2] = n$$

and the difference equation is satisfied, as required. ☛ **22**

☛ **22.** Show that the sequence $y_n = 2^n$ is a solution of the difference equation $y_{n+2} - 5y_{n+1} + 6y_n = 0$.

EXAMPLE 3 Show that the sequence whose nth term is given by

$$y_n = \frac{1}{c + n} \qquad n = 0, 1, 2, \ldots$$

is a solution of the difference equation

$$y_{n+1}(1 + y_n) = y_n$$

for any positive value of the constant c.

Solution If $y_n = \dfrac{1}{c + n}$, then $y_{n+1} = \dfrac{1}{c + n + 1}$. The left side of the difference equation is therefore

$$y_{n+1}(1 + y_n) = \frac{1}{c + n + 1}\left(1 + \frac{1}{c + n}\right)$$
$$= \frac{1}{c + n + 1}\left(\frac{c + n + 1}{c + n}\right)$$
$$= \frac{1}{c + n} = y_n$$

Answer $2^{n+2} - 5 \cdot 2^{n+1} + 6 \cdot 2^n = 0$.

and the difference equation is satisfied. (The reason for restricting c to be positive is simply that if c is a zero or a negative integer, the denominator in one of the terms of the sequence becomes zero.)

Observe that in Example 3 we have a first-order difference equation and a solution that involves an arbitrary constant, c. This is a general feature that we shall return to later.

Solution by Numerical Iteration

In the last two examples, the solution of the difference equation has been given as an algebraic formula for the nth term in the sequence. It is usually very desirable to have such a formula, and later several such solution formulas will be established. However, there are many cases in which a formula cannot be found for the solution of a given difference equation, and in such cases we must be satisfied with a table of values of a sufficient number of terms in the solution sequence. The construction of such a table will be illustrated by the following simple examples.

EXAMPLE 4 A sum of \$1000 is invested at an interest rate of 12% compounded annually. Let $A_0 = 1000$ denote the initial sum and let A_n denote the value of the investment after n years. Write down the difference equation satisfied by this sequence of values and construct the terms up to and including A_8.

Solution This is similar to the problem discussed at the beginning of this section. Here $R = 12$ and $k = 1$, so $1 + i = 1.12$. The sequence satisfies the difference equation $A_n = 1.12\,A_{n-1}$. Since we know the zeroth term, namely $A_0 = 1000$, we can construct as many terms in the sequence as we like, simply by using the difference equation repeatedly for $n = 1, 2, 3$, and so on. We get

$$n = 1: \quad A_1 = 1.12A_0 = 1.12(1000) = 1120$$

$$n = 2: \quad A_2 = 1.12A_1 = 1.12(1120) = 1254.4$$

$$n = 3: \quad A_3 = 1.12A_2 = 1.12(1254.4) = 1404.928$$

and so on. Continuing this process, we get the terms up to A_8 in Table 3. (Values have been rounded off to two decimal places.)

Remark In this example, the solution can also be expressed by means of an algebraic formula. For, using the difference equation with successive values of n, we have

$$A_1 = 1.12A_0$$

$$A_2 = 1.12A_1 = 1.12(1.12A_0) = (1.12)^2A_0$$

$$A_3 = 1.12A_2 = 1.12[(1.12)^2A_0] = (1.12)^3A_0$$

TABLE 3

n	A_n
0	1000.00
1	1254.40
2	1404.93
3	1573.52
4	1762.34
5	1973.82
6	2210.68
7	2475.96
8	2773.08

and so on. Clearly, the general term in the sequence is given by

$$A_n = (1.12)^n A_0.$$

This is, of course, nothing but the usual compound interest formula.

EXAMPLE 5 Given that $y_0 = 1$, construct the solution up to and including y_5 of the difference equation

$$y_{n+1} = \frac{1}{1 + y_n}$$

Solution Setting $n = 0, 1, 2, 3,$ and 4 in the difference equation, we find the solutions successively for $y_1, y_2, y_3, y_4,$ and y_5:

$$n = 0: \qquad y_1 = \frac{1}{1 + y_0} = \frac{1}{1 + 1} = \frac{1}{2}$$

$$n = 1: \qquad y_2 = \frac{1}{1 + y_1} = \frac{1}{1 + \frac{1}{2}} = \frac{2}{3}$$

$$n = 2: \qquad y_3 = \frac{1}{1 + y_2} = \frac{1}{1 + \frac{2}{3}} = \frac{3}{5}$$

$$n = 3: \qquad y_4 = \frac{1}{1 + y_3} = \frac{1}{1 + \frac{3}{5}} = \frac{5}{8}$$

☛ **23.** Compute the first four terms if $y_0 = 2$ and $y_{n+1} y_n + 2y_n^2 = n$.

$$n = 4: \qquad y_5 = \frac{1}{1 + y_4} = \frac{1}{1 + \frac{5}{8}} = \frac{8}{13}.$$ ☛ **23**

EXAMPLE 6 Among the North American population, a number of people suffer from a certain debilitating disease. Each year 1000 new cases of the disease occur and half of the existing cases are cured. At the end of 1991 there were 1200 cases of the disease. Calculate the number of cases to be expected at the end of each subsequent year through 1998.

Solution Let us denote the end of 1991 by $n = 0$ and subsequent years by $n = 1, 2, 3,$ etc. Then $n = 7$ corresponds to the end of 1998. Let y_n denote the number of cases at the end of year n. Then, first, $y_0 = 1200$, the number of cases at the end of 1991.

Second, the number of cases y_n consists of half the y_{n-1} cases from the previous year that are left uncured plus 1000 new cases. So, we can write the difference equation

$$y_n = 0.5y_{n-1} + 1000.$$

We construct the solution by putting n successively equal to 1, 2, 3, ... We find:

$$n = 1: \qquad y_1 = 0.5y_0 + 1000 = 0.5(1200) + 1000 = 1600$$

$$n = 2: \qquad y_2 = 0.5y_1 + 1000 = 0.5(1600) + 1000 = 1800$$

Answer $y_0 = 2$, $y_1 = -4$, $y_2 = \frac{31}{4}$, $y_3 = -\frac{945}{62}$.

$$n = 3: \qquad y_3 = 0.5y_2 + 1000 = 0.5(1800) + 1000 = 1900$$

TABLE 4

n	Year	y_n
0	1991	1200
1	1992	1600
2	1993	1800
3	1994	1900
4	1995	1950
5	1996	1975
6	1997	1988
7	1998	1994

and so on. The complete solution up to $n = 7$ is given in Table 4. We see that at the end of 1998 the number of cases equals 1994.

This technique of calculating the solution by numerical evaluation of successive terms is sometimes called **solution by iteration.** It is a very convenient method if you have a computer, or even a programmable calculator. The program can be designed to loop through successive iterations of the difference equation and so compute as many terms in the sequence as are desired.

Linear First-Order Difference Equations

In the last three examples we constructed the solutions of certain difference equations by direct numerical evaluation of successive members of the sequence. This approach is always possible, and indeed for many difference equations it is the only way of getting a solution. However there are certain types of difference equations for which an algebraic formula can be found for the general term in the sequence, and our purpose in the rest of this section is to examine one class of such equations together with some of their applications, particularly in the mathematics of finance.

In Example 4 we constructed the solution of the difference equation $A_n = 1.12A_{n-1}$ by numerical iteration. Following that example we remarked that the solution can in fact be expressed by means of the formula $A_n = (1.12)^n A_0$. This result generalizes as follows.

THEOREM 1 The general solution of the difference equation

$$y_n = ay_{n-1}$$

where a is a given constant, is

$$y_n = ca^n$$

where c is an arbitrary constant. The value of c is determined if one member of the sequence is given: if y_p is given, then

$$c = y_p a^{-p}.$$

PROOF Writing down the difference equation successively for $n = 1, 2, 3, \ldots$, we find

$$n = 1: \quad y_1 = ay_0$$

$$n = 2: \quad y_2 = ay_1 = a(ay_0) = a^2 y_0$$

$$n = 3: \quad y_3 = ay_2 = a(a^2 y_0) = a^3 y_0$$

and so on. It is obvious that in general $y_n = a^n y_0$. (If you are familiar with the method of induction, you will be able to provide a rigorous justification of this "obvious" fact.) Thus the first part of the theorem is proved, with $c = y_0$. The second part follows immediately by putting $n = p$ in the general formula: $y_p = a^p y_0$. Then

$$c = y_0 = y_p \div a^p = y_p a^{-p}.$$

The solution $y_n = ca^n$ of the difference equation $y_n = ay_{n-1}$ is an exponentially growing sequence if $a > 1$ and is an exponentially decaying sequence if $0 < a < 1$. Note that the general solution of the difference equation involves an arbitrary constant c. In order to fix the value of this constant an extra piece of information is required, namely the value of one term in the sequence. Usually, in practice, the initial term is the one that is given. We shall find this feature repeated for other first-order difference equations: the general solution involves an arbitrary constant and additional information is required to determine that constant.

EXAMPLE 7 Find the solution of the difference equation

$$y_{n+1} = -0.5y_n$$

for which $y_5 = 2$.

Solution We can use Theorem 1 with $a = -0.5$. The general solution is

$$y_n = c(-0.5)^n$$

where $c = y_p a^{-p} = y_5(-0.5)^{-5} = 2(-0.5)^{-5} = -2^6$. Thus, substituting this value of c we get

$$y_n = -2^6(-\tfrac{1}{2})^n = (-1)^{n-1}2^{6-n}. \quad \text{☛ 24}$$

> **☛ 24.** Find the solution of
> $y_n = -5y_{n-1}, \quad y_2 = -3$.

The following theorem deals with a difference equaion that plays a fundamental role in much of the mathematics of finance, as we shall see in later examples.

THEOREM 2 The general solution of the difference equation

$$y_n = ay_{n-1} + b$$

where a and b are given constants (with $a \neq 1$) is

$$y_n = ca^n - \frac{b}{a - 1}$$

where c is an arbitrary constant. The value of c is determined if one member of the sequence is known. If y_p is known, then

$$c = a^{-p}\left[y_p + \frac{b}{a - 1}\right].$$

In particular, if y_0 is known, then $c = y_0 + \dfrac{b}{a - 1}$.

PROOF Define

$$z_n = y_n + \frac{b}{a - 1}.$$

Answer $y_n = -3 \cdot (-5)^{n-2}$.

25. Find the solution for which $y_0 = 2$ of
(a) $y_n + y_{n-1} = 3$;
(b) $y_n + 3y_{n-1} = 3$.

Then

$$z_n - az_{n-1} = \left[y_n + \frac{b}{a-1} \right] - a \left[y_{n-1} + \frac{b}{a-1} \right]$$

$$= \underbrace{y_n - ay_{n-1}}_{=b} + \underbrace{\frac{b}{a-1}[1-a]}_{=-b}$$

$$= b - b = 0.$$

Thus the quantities z_n satisfy the difference equation $z_n - az_{n-1} = 0$. By Theorem 1, the solution for them is therefore $z_n = ca^n$, where c is a constant. Consequently, from the definition of z_n,

$$y_n = z_n - \frac{b}{a-1} = ca^n - \frac{b}{a-1}$$

as required.

The expression given in the statement of the theorem for c in terms of y_p is obtained simply by solving for c the equation

$$y_p = ca^p - \frac{b}{a-1}$$

Note If $a = 1$, the general solution is given by $y_n = y_0 + nb$.

EXAMPLE 8 Find the solution of the difference equation

$$y_n - 2y_{n-1} = 3 \qquad y_1 = 5.$$

Solution This difference equation is of the type in Theorem 2, with $a = 2$ and $b = 3$. The solution is therefore

$$y_n = ca^n - \frac{b}{a-1} = c2^n - \frac{3}{2-1} = c2^n - 3.$$

Setting $n = 1$ and using the given value of y_1 we have

$$y_1 = c2^1 - 3 = 2c - 3 = 5$$

or $c = 4$. The final solution is therefore

$$y_n = 4 \cdot 2^n - 3 = 2^{n+2} - 3. \qquad \text{☞ 25}$$

Applications of Difference Equations in Financial Mathematics

We shall now reexamine the topics from the mathematics of finance discussed in Section 7-3 but now making use of difference equations. The examples will largely be the same as those used earlier so you can compare the two approaches.

Answer (a) $y_n = \frac{3}{2} + \frac{1}{2}(-1)^n$;
(b) $y_n = \frac{3}{4} + \frac{5}{4}(-3)^n$.

EXAMPLE 9 Every month Jane pays $100 into a savings plan that earns interest at $\frac{1}{2}\%$ per month. Calculate the value of her savings (a) immediately after making her nth payment and (b) immediately after making her twenty-fifth payment.

Solution Let y_n denote the value of the plan immediately after the nth payment. Then, first, the initial value is $y_1 = 100$. (An alternative starting value would be $y_0 = 0$.) Second, we can relate y_n to the preceding value y_{n-1} as follows:

Value After nth Payment

$$= \text{Value After } (n-1)\text{th Payment} + \text{Interest} + \text{New Payment}$$

$$y_n = y_{n-1} + 0.005 y_{n-1} + 100.$$

The second term on the right is the interest on the amount y_{n-1} for 1 month at the rate $\frac{1}{2}\%$. Thus we have

$$y_n = 1.005 y_{n-1} + 100$$

which is a difference equation of the same type as in Theorem 2. The two constants are $a = 1.005$ and $b = 100$, and so, from the theorem, the solution is

$$y_n = ca^n - \frac{b}{a-1} = c(1.005)^n - \frac{100}{1.005 - 1} = c(1.005)^n - 20,000.$$

Setting $n = 1$ and using the initial condition, we get

$$y_1 = 1.005c - 20,000 = 100$$

which implies that $c = 20,000$. Thus the solution to part (a) is

$$y_n = 20,000[(1.005)^n - 1].$$

To answer part (b) we simply set $n = 25$ in this formula:

$$y_n = 20,000[(1.005)^{25} - 1] = 20,000(1.13280 - 1) = 2655.91.$$

The value of the savings plan after 25 payments is $2655.91. These answers are the same as we obtained in Example 1 of Section 7-3. ☛ 26

Let us generalize this example. Suppose that an amount P is deposited in a savings plan at the end of each period, where the period may be a month, a quarter, a year, or any other fixed length of time. Let the interest rate be $R\%$ per period. As in the example, let y_n denote the value of the investment immediately after the nth deposit is made. Then, as before,

Value After nth Payment

$$= \text{Value After } (n-1)\text{th Payment} + \text{Interest} + \text{New Payment}$$

$$y_n = y_{n-1} + \frac{R}{100} y_{n-1} + P.$$

☛ **26.** In Example 9, write down the difference equation if the monthly payments are $150 and the interest rate is 1% per month. Find the general solution and the solution that satisfies $y_0 = 0$.

Answer $y_n = y_{n-1} + 0.01 y_{n-1} + 150$,
$y_n = c(1.01)^n - 15,000$,
$y_n = 15,000[(1.01)^n - 1]$.

This difference equation is usually written as

$$y_n = (1 + i)y_{n-1} + P \qquad i = \frac{R}{100}.$$

It has the form given in Theorem 2 with the constants being $a = 1 + i$ and $b = P$. The solution is therefore

$$y_n = ca^n - \frac{b}{a-1} = c(1 + i)^n - \frac{P}{(1 + i) - 1} = c(1 + i)^n - \frac{P}{i}.$$

For the initial condition, we can use either the value after the first payment, $y_1 = P$, or the value before any payments, $y_0 = 0$. We will get the same value of c in either case. Using the former condition, we have

$$y_1 = c(1 + i) - \frac{P}{i} = P$$

which gives $c = P/i$. Substituting this constant into the general solution above gives the final result

$$y_n = \frac{P}{i}[(1 + i)^n - 1].$$

As in Section 7-3, we use the following notation for this solution:

$$y_n = Ps_{\overline{n}|i} \qquad \text{where} \quad s_{\overline{n}|i} = \frac{1}{i}[(1 + i)^n - 1].$$

You can find values of $s_{\overline{n}|i}$ tabulated for different values of n and i in the last column of Table A.3.4 in Appendix III.

Next consider the repayment of a loan, for example repayment of a house mortgage or an automobile loan. Here, a bank or other loan agency lends a certain sum to the customer, who then repays it by means of regular payments, usually every month. Before discussing the general formula for this type of situation, let us consider a specific example.

EXAMPLE 10 A small construction company wishes to borrow from the bank for expansion of its operations. The bank charges interest at 1% per month on the outstanding balance of the loan and insists that the loan be repaid in 24 monthly payments. The company estimates that they can afford to repay the loan at the rate of $1500 per month. What is the maximum amount they can borrow?

Solution Let y_n denote the balance outstanding on the loan immediately after the nth payment. Since the loan is to be repaid in 24 payments, it is necessary that $y_{24} = 0$. We can also derive a difference equation as follows:

Balance After n Payments

$$= \text{Balance After } (n - 1) \text{ Payments} + \text{Interest} - \text{One Payment}$$

$$y_n = y_{n-1} + 0.01y_{n-1} - 1500$$

where the second term on the right is the monthly interest on the preceding outstanding balance, y_{n-1}. This can be written

$$y_n = 1.01y_{n-1} - 1500$$

which has the same general form as the difference equation in Theorem 1 with the constants being $a = 1.01$ and $b = -1500$. The solution is therefore

$$y_n = ca^n - \frac{b}{a-1} = c(1.01)^n + \frac{1500}{1.01-1} = c(1.01)^n + 150{,}000.$$

To find c we must set $n = 24$:

$$y_{24} = c(1.01)^{24} + 150{,}000 = 0$$

and so $c = -150{,}000(1.01)^{-24}$. Substituting this value of c into the solution for y_n above, we get

$$y_n = 150{,}000[1 - (1.01)^{-(24-n)}].$$

We are interested in determining the initial size of the loan, which is the outstanding balance, y_0, after no repayments. Setting $n = 0$, we get

$$y_0 = 150{,}000[1 - (1.01)^{-24}] = 150{,}000[1 - 0.7875661] = 31{,}865.08$$

The maximum loan the company can obtain is therefore \$31,865.08. ☞ 27

☞ **27.** In Example 10, write down the difference equation if the monthly payments are \$900 and the interest rate is $\frac{1}{2}\%$ per month. Find the general solution and the solution that satisfies $y_{24} = 0$.

Now let us generalize this example. Let us suppose that the initial loan is A dollars and is to be repaid in regular installments of P dollars each. Let the interest rate be R percent per period between the payments. As in the example, let y_n denote the balance outstanding on the loan immediately after the nth payment. Then, we have

Balance After n Payments

$$= \text{Balance After } (n - 1) \text{ Payments} + \text{Interest} - \text{One Payment}$$

$$y_n = y_{n-1} + \frac{R}{100}y_{n-1} - P.$$

We rewrite this as

$$y_n = (1 + i)y_{n-1} - P \qquad \text{where} \quad i = \frac{R}{100}.$$

The solution of this difference equation is obtained from Theorem 2 with $a = (1 + i)$ and $b = -P$:

$$y_n = ca^n - \frac{b}{a-1} = c(1 + i)^n + \frac{P}{(1 + i) - 1} = c(1 + i)^n + \frac{P}{i}.$$

Answer $y_n = 1.005y_{n-1} - 900$,
$y_n = c(1.005)^n + 180{,}000$.
$y_n = 180{,}000[1 - (1.005)^{-(24-n)}]$.

Now, the initial loan (which equals the balance y_0 after zero repayments) is given by

$$A = y_0 = c(1 + i)^0 + \frac{P}{i} = c + \frac{P}{i}$$

and so $c = A - P/i$. Substituting this value of c into the solution for the outstanding balance, we get

$$y_n = \left(A - \frac{P}{i}\right)(1 + i)^n + \frac{P}{i}.$$

Now suppose that n is such that the loan has been completely repaid. This means that the outstanding balance has been reduced to zero: $y_n = 0$. This gives the equation

$$\left(A - \frac{P}{i}\right)(1 + i)^n + \frac{P}{i} = 0$$

from which we get

$$A - \frac{P}{i} + \frac{P}{i}(1 + i)^{-n} = 0.$$

This result is commonly written

$$A = Pa_{\overline{n}|i} \qquad \text{where} \qquad a_{\overline{n}|i} = \frac{1}{i}[1 - (1 + i)^{-n}].$$

EXAMPLE 11. During her years as an undergraduate, a student accumulates student loans that total $8000. The loan is to be repaid over the next 5 years by means of a single installment at the end of each year. The interest rate is 8% per annum. Let y_n denote the outstanding balance of the loan after the nth payment. Write down the difference equation satisfied by the sequence y_n and obtain its solution. Find the amount of each repayment. How much will the ex-student require if she is to pay off the loan immediately after the second payment?

Solution In terms of the general notation, $R = 8$, so $i = 0.08$. The initial balance is $y_0 = 8000$. Let the amount of each repayment be P. Then we have the difference equation

Balance After n Payments

$$= \text{Balance After } (n - 1) \text{ Payments} + \text{Interest} - \text{One Payment}$$

$$y_n = y_{n-1} + \frac{8}{100}y_{n-1} - P = 1.08y_{n-1} - P.$$

From Theorem 2, with $a = 1.08$ and $b = -P$, the general solution is

$$y_n = c(1.08)^n + \frac{P}{0.08} = c(1.08)^n + 12.5P.$$

Setting $n = 0$ in this, we find

$$y_0 = c + 12.5P = 8000$$

so that $c = 8000 - 12.5P$. Thus

$$y_n = (8000 - 12.5P)(1.08)^n + 12.5P.$$

If the loan is to be paid off in 5 years, $y_5 = 0$:

$$y_5 = (8000 - 12.5P)(1.08)^5 + 12.5P = 0.$$

Solving this for P, we get

$$P = \frac{8000}{12.5[1 - (1.08)^{-5}]} = 2003.65.$$

Thus the payments are $2003.65 each year. The balance remaining on the loan after two payments is given by

$$y_2 = (8000 - 12.5P)(1.08)^2 + 12.5P$$

$$= [8000 - 12.5(2003.65)](1.08)^2 + 12.5(2003.65) = 5163.61.$$

The balance is therefore $5163.61 after the first two repayments, and this is the amount needed to pay off the loan. ☛ 28

<hr>

Another very similar situation is that of an annuity purchased by a retired person as a pension, usually from a life insurance company.

EXAMPLE 12 After the death of Mr. Josephs, his widow uses part of her capital to purchase an annuity, to be paid in monthly installments, starting 1 month after the purchase. The life insurance company gives an interest rate of 0.5% per month and they estimate her life expectancy to be 10 years. If Mrs. Josephs wishes to receive a monthly income of $1000, how much capital will she be required to invest? If she actually survives for 15 years, what will be the value at the time of her demise of the loss that the insurance company incurs on the deal?

Solution Let y_n denote the balance of capital remaining with the insurance company after the nth payment. Then $y_{120} = 0$, since the plan is designed to make exactly 120 monthly payments. The initial capital that Mrs. Josephs must deposit is y_0.

We can derive a difference equation as in Example 11:

Balance After n Payments

= Balance After $(n - 1)$ Payments + Interest − One Payment

☛ **28.** Repeat Example 11 if the $8000 loan is to be repaid in quarterly payments over five years, with interest rate of 2% per quarter.

Answer $y_n = 1.02y_{n-1} - P,$

$y_n = (8000 - 50P)(1.02)^n + 50P$

where $P = \dfrac{160}{1 - (1.02)^{-20}} = 489.25.$

$$y_n = y_{n-1} + \frac{0.5}{100}y_{n-1} - 1000 = 1.005y_{n-1} - 1000.$$

From Theorem 2 with $a = 1.005$ and $b = -1000$, we get the solution

$$y_n = c(1.005)^n - \frac{-1000}{0.005} = c(1.005)^n + 200,000.$$

Then

$$y_{120} = c(1.005)^{120} + 200,000 = 0$$

which determines the constant c as $c = -200,000(1.005)^{-120}$. Then

$$y_0 = c(1.005)^0 + 200,000 = 200,000[1 - (1.005)^{-120}] = 90,073.45.$$

Thus the initial capital investment required for this annuity is $90,073.45. If Mrs. Josephs survives for 15 years, the company must make 180 monthly payments, so the capital value of the annuity upon her death is y_{180}. This is

$$y_{180} = c(1.005)^{180} + 200,000 = 200,000[1 - (1.005)^{60}] = -69770.02$$

The fact that this is negative represents a loss to the insurance company, the amount of the loss being $69,770.02.

EXERCISES 7-4

(1–6) State the order of the following difference equations.

1. $y_{n+1} + 2y_n = 1/n$

2. $x_{n+1} + (x_n)^2 = 0$

3. $u_{n+3} + \dfrac{1}{u_{n-1}} = nu_{n+2}$

4. $y_{n-2} + \ln(y_n) = 1$

5. $t_{n+4} + 5t_{n-1} = (n + 3)^2$

6. $nx_{n-3} + \sqrt{n}\, x_{n-1} = 4$

(7–14) Show that the following sequences whose nth terms are given below are the solutions of the indicated difference equations (a, b, and c are constants).

7. $y_n = 3n + c$; $y_{n+1} - y_n = 3$

8. $y_n = n(n + 2) + c$; $y_{n+1} - y_n = 2n + 3$

9. $y_n = (-1)^n(n + c)$; $y_n + y_{n-1} = (-1)^n$

10. $y_n = 2^n(an + b)$; $y_{n+2} - 4y_{n+1} + 4y_n = 0$

11. $y_n = (-3)^n(an + b)$; $y_{n+2} + 6y_{n+1} + 9y_n = 0$

12. $y_n = a \cdot 2^n + b \cdot 3^n$; $y_{n+2} - 5y_{n+1} + 6y_n = 0$

13. $y_n = a + b(-1)^n$; $y_{n+2} = y_n$

14. $y_n = a(-1)^n + b(2^n) + c(-3)^n$;
$y_{n+2} + 2y_{n+1} - 5y_n - 6y_{n-1} = 0$

(15–16) Construct a table of values of the solution of each of the following difference equations for the indicated value of n.

15. $y_n - y_{n-1} = (0.6 - 0.01y_{n-1})y_{n-1}$;
$y_0 = 10$, $0 \le n \le 10$

16. $y_n - y_{n-1} = (1.5 - 0.001y_{n-1})y_{n-1}$;
$y_0 = 1800$, $0 \le n \le 10$

17. A certain population has an initial size of 10,000 and is growing by 10% each year. If y_n denotes the population size after n years, write down the difference equation for y_n. Give a table of values of y_n for $n = 0, 1, 2, \ldots, 10$.

18. A sum of $5000 is invested at an interest rate of 15% compounded annually. If A_n denotes the value of the investment after n years, write down the difference equation for A_n. Give a table of values of A_n for $n = 0, 1, 2, \ldots, 8$.

(19–33) Solve the following difference equations.

19. $y_n - 2y_{n-1} = 0$

20. $2y_{n+1} + 3y_n = 0$

21. $y_n - 3y_{n-1} = 0$; $y_0 = 5$

22. $y_n - 1.2y_{n-1} = 0$; $y_0 = 100$

23. $y_n + y_{n-1} = 2$

24. $y_n - \frac{1}{2}y_{n-1} = 3$

25. $y_n - 2y_{n-1} = 1$

26. $y_n - 3y_{n-1} = -5$

27. $y_n - 0.5y_{n-1} = 4; \quad y_3 = 8$

28. $y_n - 3y_{n-1} = 9; \quad y_3 = 5$

29. $y_n - y_{n-1} = 4; \quad y_0 = 6$

30. $y_n = y_{n-1} + 2; \quad y_3 = 25$

31. $y_n - 0.2y_{n-1} = 4; \quad y_0 = 25$

32. $y_n + 2y_{n-1} = -6; \quad y_0 = 0.5$

33. $y_n + y_{n-1} + 6 = 0; \quad y_5 = -7$

(34–36) Solve the following problems by using difference equations.

34. The assets of a certain firm are increasing by 10% each year. If they were $100 million at the end of 1988, how many years will it take the assets to exceed $200 million?

35. Mr. Black has invested $50,000 at an interest rate of 10% compounded annually. If he withdraws $3000 each year on the anniversary of his deposit, in how many years will his investment be over $65,000?

36. A certain population has an initial size of 1000 and grows by 50% each year. If the population is harvested at the rate of 400 per year, find the population size in each of the first 8 years.

37. A sum of $5000 is deposited into a savings account paying 6% interest compounding yearly. If y_n denotes the amount in the account after n years, write down the difference equation of y_n and hence solve it. How much money will be in the account at the end of 8 years?

38. A sum of $2000 is deposited into a savings account paying 5% interest compounding yearly. If y_n denotes the amount in the account after n years, write down the difference equation of y_n and hence solve it. How much money will be in the account at the end of 5 years?

39. A sum of $2000 is invested at a *simple* interest of 8% per year. If y_n denotes the value of investment after n years, write down the difference equation of y_n and hence solve it. What is the value of the investment at the end of 10 years?

40. A sum of $8000 is invested at a *simple* interest of 6% per year. If y_n denotes the value of investment after n years, write down the difference equation of y_n and hence solve it. What is the value of the investment at the end of 5 years?

41. Mr. White took a consumer loan of $5000 at 15% interest compounded yearly and promised to pay $900 at the end of each year. Let y_n denote the amount owing after n years (after the yearly payment is made).

a. Write down the difference equation for y_n and solve it.

b. How much does Mr. White owe after he has paid his tenth yearly payment?

42. Ms. Susan borrowed a sum of $8000 from a bank at 12% interest compounded monthly and promised to pay $600 at the end of each month. Let y_n denote the balance owing after n payments.

a. Write down the difference equation for y_n and solve it.

b. How much does she owe after her sixteenth payment?

43. Every month Steve deposits $200 into a savings plan that earns an interest of 1% per month. Let y_n denote the value of the savings plan immediately after the nth monthly deposit.

a. Write down the difference equation for y_n and solve it.

b. What is the value of his plan immediately after his thirtieth deposit?

44. Every year John deposits $500 into a savings plan that earns an interest of 8% per annum compounded yearly. Let y_n denote the value of the savings plan immediately after the nth deposit.

a. Write down the difference equation for y_n and solve it.

b. What is the value of his plan immediately after his twenty-fifth deposit?

45. Sue wants to borrow some money from the bank to renovate her house. She can afford to pay $200 each month and the bank charges an interest of 1.25% per month. The loan is to be paid off in 3 years. Let y_n denote the balance owed to the bank after n payments.

a. Determine the difference equation satisfied by y_n and solve it.

b. Determine exactly how much she can borrow.

c. How much will she still owe to the bank after she made her twentieth payment?

46. Mr. Brown can afford to pay $500 each month and the bank charges 12% interest compounded monthly. His loan is to be paid off in 25 years. Let y_n denote the balance owed to the bank after n payments.

a. Determine the difference equation satisfied by y_n and solve it.

b. Exactly how much can he borrow from the bank?

c. How much will he still owe to the bank after he has made his 100th payment?

47. Mary borrowed a sum of $10,000 from a bank to buy a new car. The bank charges an interest of 12% per annum compounded monthly and the loan is to be paid off by equal monthly installments of P each. Let y_n denote the amount owed after n monthly payments.

a. Determine the difference equation satisfied by y_n and solve it.

b. Determine the monthly payment of P to·the bank if the loan is to be paid off in 4 years.

48. Bruce borrowed a sum of $15,000 from a bank to renovate his house. The bank charges an interest of 15% per annum compounded monthly and the loan is to be paid off in equal monthly installments of P each. Let y_n denote the amount owed after n monthly payments.

a. Determine the difference equation satisfied by y_n and solve it.

b. Determine the monthly payment of P to the bank if the loan is to be paid off in 5 years.

49. Mr. John wants to purchase an annuity that will give him $500 each month for 10 years and the interest rate is 12% per annum compounded monthly. Let y_n denote the principal remaining in the annuity after n payments.

a. Write down the difference equation satisfied by y_n and solve it.

b. How much should he pay to purchase this annuity?

50. Mr. Tom wants to purchase an annuity that will give him $3000 every year for the next 15 years and the interest rate is 8% per annum compounded yearly. Let y_n denote the principal remaining in the annuity after n payments.

a. Write down the difference equation satisfied by y_n and solve it.

b. How much should he pay to purchase this annuity?

◤ 7-5 SUMMATION NOTATION (OPTIONAL)

The sigma notation is a convenient way of expressing sums that involve large numbers of terms. It is very commonly used in statistics and many other branches of mathematics.

According to the sigma notation (or summation notation) the sum

$$x_1 + x_2 + x_3 + \cdots + x_n$$

is abbreviated by the expression

$$\sum_{i=1}^{n} x_i.$$

This is read as "the sum of x_i as i goes from 1 to n." The Greek letter Σ (capital sigma) corresponds to our S in the English alphabet and suggests the word *sum*. Thus

$$x_1 + x_2 + x_3 + \ldots + x_n = \sum_{i=1}^{n} x_i. \tag{1}$$

The subscript i used in the sigma notation on the right-hand side of (1) above is called the *summation index*. It can be replaced by any other letter, say j or k, r, that is not already being used to stand for something else, and the value of the sum will not be changed. Thus

$$\sum_{i=1}^{n} x_i = \sum_{j=1}^{n} x_j = \sum_{k=1}^{n} x_k, \ldots .$$

In the sum $\sum_{i=1}^{n} x_i$, the index of summation i (which is indicated below Σ) takes the values $1, 2, 3, \ldots, n$. The starting value (1 in this case) is indicated below Σ, and the last value (n in this case) is indicated above Σ. Thus to expand a sum given in the Σ-notation we give all possible integer values to the index of summation in the expression that follows Σ and then add all the terms. For example, in order to expand $\sum_{k=3}^{7} f(x_k)$, we note that the index k takes the values 3, 4, 5, 6, 7 (the starting value 3 is indicated below Σ and the last value 7 is indicated above Σ). Therefore in the expression that follows Σ, that is, in $f(x_k)$, we replace k by 3, 4, 5, 6, 7 and then add all the terms so obtained. Thus

$$\sum_{k=3}^{7} f(x_k) = f(x_3) + f(x_4) + f(x_5) + f(x_6) + f(x_7).$$

The only thing that changes from one term to the next is the numeral in the place indicated by the index of summation (k in this case).

Following are a few other examples of sums given by the Σ-notation.

(a) $\displaystyle\sum_{k=1}^{7} k^3 = 1^3 + 2^3 + 3^3 + 4^3 + 5^3 + 6^3 + 7^3$

(b) $\displaystyle\sum_{k=1}^{5} a^k = a^1 + a^2 + a^3 + a^4 + a^5$

(c) $\displaystyle\sum_{i=2}^{5} \frac{3i}{i-1} = \frac{3(2)}{2-1} + \frac{3(3)}{3-1} + \frac{3(4)}{4-1} + \frac{3(5)}{5-1}$

(d) $\displaystyle\sum_{j=1}^{4} \frac{j+1}{j^2+1} = \frac{1+1}{1^2+1} + \frac{2+1}{2^2+1} + \frac{3+1}{3^2+1} + \frac{4+1}{4^2+1}$

(e) $\displaystyle\sum_{p=1}^{100} \ln p = \ln 1 + \ln 2 + \ln 3 + \cdots + \ln 100$ ☛ 29

EXAMPLE 1 Given $x_1 = 4$, $x_2 = 5$, $x_3 = -1$, and $x_4 = 2$, find

(a) $\displaystyle\sum_{k=1}^{4} (x_k - 2)^2$ (b) $\left[\displaystyle\sum_{k=1}^{4} (x_k - 2) \right]^2$

Solution

(a) $\displaystyle\sum_{k=1}^{4} (x_k - 2)^2 = (x_1 - 2)^2 + (x_2 - 2)^2 + (x_3 - 2)^2 + (x_4 - 2)^2$

$= (4 - 2)^2 + (5 - 2)^2 + (-1 - 2)^2 + (2 - 2)^2$

$= 4 + 9 + 9 + 0$

$= 22$

☛ **29.** Expand the following sums:

(a) $\displaystyle\sum_{j=0}^{3} (j - 1)^2$;   (b) $\displaystyle\sum_{k=1}^{5} \frac{2^k}{k}$;

(c) $\displaystyle\sum_{i=2}^{4} \left(i + \frac{1}{i} \right)$.

Answer (a) $(0 - 1)^2 + (1 - 1)^2 + (2 - 1)^2 + (3 - 1)^2$;

(b) $\dfrac{2^1}{1} + \dfrac{2^2}{2} + \dfrac{2^3}{3} + \dfrac{2^4}{4} + \dfrac{2^5}{5}$;

(c) $(2 + \frac{1}{2}) + (3 + \frac{1}{3}) + (4 + \frac{1}{4})$.

$$\text{(b) } \sum_{k=1}^{4} (x_k - 2) = (x_1 - 2) + (x_2 - 2) + (x_3 - 2) + (x_4 - 2)$$
$$= (4 - 2) + (5 - 2) + (-1 - 2) + (2 - 2)$$
$$= 2 + 3 - 3 + 0$$
$$= 2$$

Thus,

$$\left[\sum_{k=1}^{4} (x_k - 2) \right]^2 = 2^2 = 4. \quad \text{☛ 30}$$

☛ 30. In Example 1, evaluate
(a) $\sum_{j=1}^{4} (x_j - x_j^2)$; (b) $\sum_{k=2}^{4} kx_k$.

It should be observed that the number of terms in the expansion of $\sum_{k=m}^{n}$ is equal to $(n - m + 1)$. Thus $\sum_{k=3}^{7} x_k$ contains $7 - 3 + 1 = 5$ terms, $\sum_{i=1}^{10} x^i$ contains $10 - 1 + 1 = 10$ terms, and so on.

THEOREM 1 If m and n are integers, with $n \geq m$, then

(a) $\sum_{k=m}^{n} c = (n - m + 1)c$, where c is constant.

(b) $\sum_{k=m}^{n} (x_k \pm y_k) = \sum_{k=m}^{n} x_k \pm \sum_{k=m}^{n} y_k$.

(c) $\sum_{k=m}^{n} cx_k = c \sum_{k=m}^{n} x_k$, where c is a constant.

(d) $\sum_{k=m}^{n} (x_k - x_{k-1}) = x_n - x_{m-1}$.

PROOF (a) The number of terms in $\sum_{k=m}^{n} c$ is $(n - m + 1)$, and each term in the expansion is equal to c because the expression that follows Σ, that is c, does not involve the index of summation k. Thus

$$\sum_{k=m}^{n} c = \underbrace{c + c + c + \cdots + c}_{(n - m + 1) \text{ terms}} = (n - m + 1)c.$$

(b) $\sum_{k=m}^{n} (x_k \pm y_m) = (x_m \pm y_m) + (x_{m+1} \pm y_{m+1}) + \cdots + (x_n \pm y_n)$

$$= (x_m + x_{m+1} + \cdots + x_n) \pm (y_m + y_{m+1} + \cdots + y_n)$$

$$= \sum_{k=m}^{n} x_k \pm \sum_{k=m}^{n} y_k$$

(c) $\sum_{k=m}^{n} cx_k = cx_m + cx_{m+1} + \cdots + cx_n$

$$= c(x_m + x_{m+1} + \cdots + x_n)$$

$$= c \sum_{k=m}^{n} x_k$$

Answer (a) -36; (b) 15.

☛ **31.** In Example 3 find

☛ **31.** In Example 3 find

(a) $\sum_{j=1}^{5} (x_j - 3)$; (b) $\sum_{j=1}^{5} (x_j - 3)^2$.

(d) $\sum_{k=m}^{n} (x_k - x_{k-1}) = (x_m - x_{m-1}) + (x_{m+1} - x_m) + (x_{m+2} - x_{m+1})$

$$+ \cdots + (x_{n-1} - x_{n-2}) + (x_n - x_{n-1})$$

$$= x_n - x_{m-1}$$

because all other terms cancel with each other. (Some of these cancellations are indicated by slashes in the preceding expression.)

COROLLARY In particular, when $m = 1$, results (a) and (d) become

(a) $\sum_{k=1}^{n} c = nc.$

(d) $\sum_{k=1}^{n} (x_k - x_{k-1}) = x_n - x_0.$

EXAMPLE 2 Expand the following sums:

(a) $\sum_{k=-4}^{5} (3).$ (b) $\sum_{k=1}^{n} (3).$

Solution

(a) $\sum_{k=-4}^{5} (3)$ contains $5 - (-4) + 1 = 10$ terms, and each of them is equal to 3. Thus

$$\sum_{k=-4}^{5} (3) = 10(3) = 30.$$

(b) $\sum_{k=1}^{n} (3)$ contains $n - 1 + 1 = n$ terms. Thus

$$\sum_{k=1}^{n} 3 = n(3) = 3n.$$

EXAMPLE 3 Given that $\sum_{i=1}^{5} x_i = 13$ and $\sum_{i=1}^{5} x_i^2 = 49$, find $\sum_{i=1}^{5} (2x_i + 3)^2$.

Solution

$$\sum_{i=1}^{5} (2x_i + 3)^2 = \sum_{i=1}^{5} (4x_i^2 + 12x_i + 9)$$

$$= 4 \sum_{i=1}^{5} x_i^2 + 12 \sum_{i=1}^{5} x_i + 5(9)$$

$$= 4(49) + 12(13) + 45$$

$$= 196 + 156 + 45$$

$$= 397 \quad ☛ 31$$

Answer (a) -2; (b) 16.

THEOREM 2

$$\text{(a)} \ \sum_{k=1}^{n} k = 1 + 2 + 3 + \cdots + n = \frac{n(n + 1)}{2}$$

$$\text{(b)} \ \sum_{k=1}^{n} k^2 = 1^2 + 2^2 + 3^2 + \cdots + n^2 = \frac{n(n + 1)(2n + 1)}{6}$$

$$\text{(c)} \ \sum_{k=1}^{n} k^3 = 1^3 + 2^3 + 3^3 + \cdots + n^3 = \left[\frac{n(n + 1)}{2}\right]^2$$

PROOF

$$\text{(a)} \ \sum_{k=1}^{n} k = 1 + 2 + 3 + \cdots + (n - 1) + n$$

The terms in this sum form an arithmetic progression in which there are n terms, the first term being equal to 1 and the common difference also being equal to 1. From the formula on page 264, the sum is given by

$$\sum_{k=1}^{n} k = S_n = \frac{n}{2}[2 \cdot 1 + (n - 1) \cdot 1].$$

Therefore

$$\sum_{k=1}^{n} k = \frac{n(n + 1)}{2}.$$

(b) To prove (b) we shall make use of the following result.

$$k^3 - (k - 1)^3 = k^3 - (k^3 - 3k^2 + 3k - 1)$$

or

$$k^3 - (k - 1)^3 = 3k^2 - 3k + 1$$

This is an identity that is true for all values of k. Putting $k = 1, 2, 3, \ldots, n$, we obtain the following sequence of equations:

$$1^3 - 0^3 = 3 \cdot 1^2 - 3 \cdot 1 + 1$$
$$2^3 - 1^3 = 3 \cdot 2^2 - 3 \cdot 2 + 1$$
$$3^3 - 2^3 = 3 \cdot 3^2 - 3 \cdot 3 + 1$$
$$\vdots$$
$$n^3 - (n - 1)^3 = 3 \cdot n^2 - 3 \cdot n + 1.$$

If these equations are all added vertically we observe that most of the terms on the left-hand side cancel out and we are left with

$$n^3 - 0^3 = 3(1^2 + 2^2 + 3^2 + \cdots + n^2) - 3(1 + 2 + 3 + \cdots + n)$$
$$+ \underbrace{(1 + 1 + \cdots + 1)}_{n \text{ terms}}$$

or,

$$n^3 = 3 \sum_{k=1}^{n} k^2 - 3\frac{n(n+1)}{2} + n$$

where we have made use of Theorems 1(a) and 2(a). Thus

$$3 \sum_{k=1}^{n} k^2 = n^3 - n + \tfrac{3}{2}n(n+1)$$

$$= n(n+1)(n-1) + \tfrac{3}{2}n(n+1)$$

$$= n(n+1)[n-1+\tfrac{3}{2}]$$

$$= n(n+1)\left(\frac{2n+1}{2}\right).$$

Hence

$$\sum_{k=1}^{n} k^2 = \frac{n(n+1)(2n+1)}{6}.$$

which proves the result.

(c) The proof of this part is left as an exercise. (*Hint:* Make use of the identity $k^4 - (k-1)^4 = 4k^3 - 6k^2 + 4k - 1$).

EXAMPLE 4 Evaluate the sum of the squares of the first 100 natural numbers.

Solution
$$1^2 + 2^2 + 3^2 + \cdots + 100^2 = \sum_{k=1}^{100} k^2$$

$$= \frac{100(100+1)(2 \cdot 100+1)}{6}$$

$$= \frac{100(101)(201)}{6} = 338{,}350,$$

where we have made use of Theorem 2(b) for $n = 100$.

EXAMPLE 5 Evaluate the following sum.

$$7^3 + 8^3 + 9^3 + \cdots + 30^3$$

Solution The given sum can be written as

$$7^3 + 8^3 + 9^3 + \cdots + 30^3 = (1^3 + 2^3 + \cdots + 30^3)$$

$$- (1^3 + 2^3 + 3^3 + \cdots + 6^3)$$

$$= \sum_{k=1}^{30} k^3 - \sum_{k=1}^{6} k^3$$

$$= \left[\frac{30(30+1)}{2}\right]^2 - \left[\frac{6(6+1)}{2}\right]^2$$

32. Evaluate (a) $\sum_{j=1}^{20} (j - 1)^2$;

(b) $\sum_{k=20}^{30} k^2$; (c) $\sum_{i=1}^{10} (5i^2 - 3i + 2)$.

$$= (465)^2 - (21)^2$$

$$= 216,225 - 441$$

$$= 215,784.$$

EXAMPLE 6 Evaluate the following sum.

$$\sum_{k=1}^{50} (3k^2 + 2k + 1)$$

Solution First let us find $\sum_{k=1}^{n} (3k^2 + 2k + 1)$. Using Theorems 1 and 2 we have

$$\sum_{k=1}^{n} (3k^2 + 2k + 1) = 3 \sum_{k=1}^{n} k^2 + 2 \sum_{k=1}^{n} k + \sum_{k=1}^{n} 1$$

$$= 3 \left[\frac{n(n + 1)(2n + 1)}{6} \right] + 2 \left[\frac{n(n + 1)}{2} \right] + n(1)$$

$$= \frac{n(n + 1)(2n + 1)}{2} + n(n + 1) + n.$$

If we now replace n by 50 on both sides we obtain

$$\sum_{k=1}^{50} (3k^2 + 2k + 1) = \frac{50(51)(101)}{2} + 50(51) + 50$$

$$= 128,775 + 2550 + 50$$

$$= 131,375. \quad \text{☞ 32}$$

Answer (a) 2470; (b) 6985; (c) 1780.

EXERCISES 7-5

(1–22) Evaluate each sum.

1. $\sum_{k=1}^{4} 2k - 3)$

2. $\sum_{k=0}^{3} (k^2 + 7)$

3. $\sum_{p=2}^{5} (p^2 + p - 1)$

4. $\sum_{i=-3}^{3} (i^2 - i + 2)$

5. $\sum_{i=2}^{4} \frac{i}{i - 1}$

6. $\sum_{q=1}^{4} \left(\frac{q^2 + 1}{q} \right)$

7. $\sum_{n=1}^{3} \frac{1}{n(n + 1)}$

8. $\sum_{k=0}^{5} \left(\frac{1}{k + 1} - \frac{1}{k + 2} \right)$

9. $\sum_{k=1}^{n} (2k - 1)$

10. $\sum_{k=1}^{n} (3k + 2)$

11. $\sum_{j=1}^{n} (j^2 + j + 1)$

12. $\sum_{j=1}^{n} (2j^2 - j + 3)$

13. $\sum_{k=1}^{n} (k + 1)(2k - 1)$

14. $\sum_{k=1}^{n} (k - 1)(k + 1)$

15. $\sum_{k=1}^{n} (k^3 + 7k - 1)$

16. $\sum_{p=1}^{n} (p - 1)(p^2 + p + 1)$

17. $\sum_{p=1}^{20} (p^2 + 7p - 6)$

18. $\sum_{r=1}^{30} (r^3 + 1)$

19. $\sum_{k=1}^{25} (k + 1)(k + 3)$

20. $\sum_{k=1}^{20} (k + 1)(k^2 + 1)$

21. $\sum_{k=11}^{50} k^2$

22. $\sum_{k=6}^{20} (2k^2 + 5k - 3)$

23. Make use of the identity $(k + 1)^2 - k^2 = 2k + 1$ to evaluate the sum $\sum_{k=1}^{n} k$.

24. Make use of the identity $k^2 - (k-1)^2 = 2k - 1$ to evaluate the sum $\sum_{k=1}^{n} k$.

25. Given $x_1 = 1$, $x_2 = -2$, $x_3 = 3$, $x_4 = 7$, $x_5 = 4$, evaluate:

a. $\sum_{p=1}^{5} (2x_p - 3)$ **b.** $\sum_{p=1}^{5} (x_p + 2)^2$

26. Given $x_1 = 1$, $x_2 = 2$, $x_3 = 3$, $x_4 = 4$, $x_5 = 5$, $y_1 = 3$, $y_2 = -1$, $y_3 = 7$, $y_4 = -2$, and $y_5 = -1$, find:

a. $\sum_{p=1}^{5} x_p y_p$ **b.** $\sum_{p=1}^{5} x_p^2 y_p$ **c.** $\sum_{k=1}^{5} (x_k - y_k)^2$

27. Given $\sum_{i=1}^{7} x_i = 13$ and $\sum_{i=1}^{7} x_i^2 = 63$, find:

a. $\sum_{i=1}^{7} (5 - 2x_i)$ **b.** $\sum_{p=1}^{7} (3x_p - 1)^2$

28. Given $\sum_{i=1}^{10} x_i^2 = 15$, $\sum_{i=1}^{10} (x_i + y_i)^2 = 73$, and $\sum_{i=1}^{10} y_i^2 = 26$, find $\sum_{p=1}^{10} x_p y_p$.

CHAPTER REVIEW

Key Terms, Symbols, and Concepts

7.1 Sequence, first term, general term, or nth term (T_n).
Arithmetic progression, common difference.
Formula for T_n.
Sum of n terms, S_n.
Simple interest, Simple interest formulas.

7.2 Geometric progression, common ratio. Formula for T_n.
Sum of n terms, S_n. Sum of an infinite G.P.

7.3 Savings plan, annuity, amortization.
The present value of an annuity, $a_{\overline{n}|i}$.
The future value of an annuity, $s_{\overline{n}|i}$.

7.4 Difference equation of order k.
Solution of a difference equation. General solution.
Solution by numerical iteration.

7.5 Summation notation or sigma notation: $\sum_{i=m}^{n} f(i)$.

Formulas

A.P.: $T_n = a + (n-1)d$.

      $S_n = \frac{1}{2}n[2a + (n-1)d] = \frac{1}{2}n(a+l)$.

Simple interest:

Value after t years $= P + tI$, $I = P\left(\dfrac{R}{100}\right)$.

G.P.: $T_n = ar^{n-1}$. $S_n = \dfrac{a(1-r^n)}{1-r}$.

$S_\infty = \dfrac{a}{1-r}$ if $-1 < r < 1$.

Future value of annuity of \$1: $s_{\overline{n}|i} = \dfrac{1}{i}[(1+i)^n - 1]$.

Present value of annuity of \$1: $a_{\overline{n}|i} = \dfrac{1}{i}[1 - (1+i)^{-n}]$.

Savings plan: Future value, $S = Ps_{\overline{n}|i}$.

Annuity or amortization: Present value, $A = Pa_{\overline{n}|i}$.

If $y_n = ay_{n-1}$, then $y_n = ca^n$.

If $y_n = ay_{n-1} + b$, then $y_n = ca^n - \dfrac{b}{a-1}$ if $a = 1$.

$\sum_{k=m}^{n} c = c(n - m + 1)$ if $a = 1$.

$\sum_{k=m}^{n} (x_k \pm y_k) = \sum_{k=m}^{n} x_k \pm \sum_{k=m}^{n} y_k$.

$\sum_{k=m}^{n} cx_k = c \sum_{k=m}^{n} x_k$.

$\sum_{k=m}^{n} k = 1 + 2 + 3 + \cdots + n = \frac{1}{2}n(n+1)$.

$\sum_{k=m}^{n} k^2 = 1^2 + 2^2 + 3^2 + \cdots + n^2 = \frac{1}{6}n(n+1)(2n+1)$.

$\sum_{k=m}^{n} k^3 = 1^3 + 2^3 + 3^3 + \cdots + n^3 = [\frac{1}{2}n(n+1)]^2$.

1. State whether each of the following is true or false. Replace each false statement with a corresponding true statement.

   a. The nth term of a sequence is given by $T_n = a + (n - 1)d$.

   b. The sum of n terms of an A.P. is given by
   $$S_n = \frac{a(1 - r^n)}{1 - r}.$$

   c. The sum of an infinite G.P. with a as first term and r as common ratio is given by $S = a/(1 - r)$ for all r.

   d. The pth term of a G.P. is given by $T_p = ar^p$.

   e. If a, l, and r are the first term, last term, and the common ratio, respectively, of a G.P., then its sum is given by
   $$\frac{a - rl}{1 - r}.$$

   f. A sequence $T_1, T_2, T_3, \ldots$ is an A.P. if $T_2 - T_1 = T_3 - T_2 = T_4 - T_3 = \cdots$.

   g. A sequence $T_1, T_2, T_3, \ldots$ is a G.P. if $T_2/T_1 = T_3/T_2 = T_4/T_3 = \cdots$.

   h. The sum of n terms of the G.P. a, ar, ar^2, $\ldots$ is given by
   $$S_n = \frac{a(r^n - 1)}{r - 1}.$$

   i. The terms of an arithmetic progression satisfy the equation $T_n - T_{n-1} = d$ for all n, where d is the common difference.

   j. The terms of a geometric progression satisfy the equation $T_n = rT_{n-1}$ for all n, where r is the common ratio.

   k. The sequence 1, $2x$, $3x^2$, $4x^3$, $\ldots$ is a geometric progression.

2. If -18, x, y, z, and 4 form an A.P., find the values of x, y, and z.

3. Find an arithmetic progression of six terms if the first term is $\frac{2}{3}$ and the last term is $\frac{22}{3}$.

4. If $x + 2$, $3x + 1$, and $8 - 3x$ are the first three terms of an A.P., find x.

5. If $x + 1$, $2x + 6$, and $4x + 24$ are the first three terms of a G.P., find x.

6. If 2, p, q, and 54 form a G.P., find the values of p and q.

7. Find a geometric progression of six terms if the third term is 2 and the last term is 0.25.

8. If you save 1¢ today, 2¢ tomorrow, 3¢ the next day, and so on, what will be your savings in 365 days?

(9–14) Find the sum of the following progressions.

9. $3 + 7 + 11 + 15 + \cdots$; 20 terms

10. $20 + 19\frac{1}{3} + 18\frac{2}{3} + 18 + \cdots + 14$

11. $3 + 6 + 12 + 24 + \cdots$; n terms

12. $18 + 6\sqrt{3} + 6 + 2\sqrt{3} + \cdots$; p terms

13. $18 - 12 + 8 - \frac{16}{3} + \cdots$

14. $a + br + ar^2 + br^3 + ar^4 + br^5 + \cdots$; $|r| < 1$

15. The sum of the sequence $25, 22, 19, \ldots$ is 116. Find the number of terms in the sequence and the last term.

16. Find the least value of n for which the sum
   $$1 + 3 + 3^2 + 3^3 + \cdots \text{ to } n \text{ terms}$$
   exceeds 7000.

17. *(Loan Repayment)* A person pays $975 in monthly installments. Each installment is less than the previous one by $5. The amount of first installment is $100. In what time will the entire amount be paid?

18. *(Depreciation)* A machine is depreciated at the rate of 10% on the reducing balance. The original cost was $10,000 and the ultimate scrap value was $3750. Find the effective life of the machine.

19. *(Loan Repayment)* A man secures an interest-free loan of $1530 from a friend and agrees to repay it in 12 installments. He pays $100 as the first installment and then increases each installment by an equal amount over the preceding one. What will be his last installment? How much does each installment differ from the previous one?

20. *(Loan Repayment)* Susan's monthly payments to the bank toward her loan form an A.P. If her fourth and seventh payments are $236 and $242, respectively, what will be her tenth payment?

21. *(Loan Repayment)* Joe borrows $4000 at 1% interest per month. Each month he repays $200 plus the interest on the outstanding balance. Write a formula for his nth payment. Calculate the total amount of interest he pays by the time the loan is repaid.

22. *(Compound Interest)* If $1200 is invested at 9% interest per annum, calculate the value of the investment:

a. After 4 years;   **b.** After n years.

23. *(Annuity)* How much does it cost to buy an annuity of $12,000 per year for 20 years if the interest rate is

a. 6%   **b.** 8% per annum?

24. *(Loan Amortization)* Jake borrowed $5000 from the bank to buy a new car. The bank charges interest on the loan of 1% each month on the outstanding balance at the beginning of the month, and Jake makes regular repayments at the end of each month. If the loan is to be repaid in 24 months, what must the monthly payment be?

25. *(Loan Amortization)* In Exercise 24, how large would Jake's monthly payments need to be in order to repay the loan in 48 months?

26. *(Savings Plan)* Jane regularly saves $250 at the beginning of each month. The bank pays interest of $\frac{1}{2}$% per month on her savings. Calculate the value of her savings account:

a. After 24 months;   **b.** After n months.

27. *(Auto Loan)* Alfred and Judy are planning to buy a new car costing $10,000. They intend to take out a consumer loan for the whole amount to be repaid over 36 months at an interest of $\frac{3}{4}$% per month. What will be their monthly payment?

28. *(Sinking Fund)* In Exercise 27, an alternative plan is to postpone buying the car for 3 years and to save regular amounts at the end of each month for the next 36 months in order to pay for it. Alfred's and Judy's savings plan will earn interest at $\frac{1}{2}$% per month. Unfortunately the price of the car can be expected to increase by 5% per year (compounded) due to inflation. How much would they need to save each month to have enough to buy the car?

(29–30) Show that the following sequences whose nth terms are given below are the solutions of the indicated difference equations (a, b, and c are constants).

29. $y_n = \dfrac{1}{(n+c)^2}$; $y_{n+1} = \dfrac{y_n}{(1+\sqrt{y_n})^2}$ $(c > 0)$

30. $y_n = a + b(2^n) + c(3^n)$;

$y_n - 6y_{n-1} + 11y_{n-2} - 6y_{n-3} = 0$

(31–36) Solve the following difference equations.

31. $y_{n+1} + \frac{1}{4}y_n = 0$ **32.** $y_n - \frac{1}{3}y_{n-1} = 0$

33. $y_n - y_{n-1} = 0$; $y_1 = 3$

34. $y_n + y_{n-1} = 0$; $y_5 = -2$

35. $y_n = y_{n-1} - 3$; $y_0 = 1$

36. $y_n - y_{n-1} = 7$; $y_2 = 5$

37. Ms. Susan has invested $10,000 at 12% per annum compounded annually. If A_n denotes the value of investment after n years, write down the difference equation satisfied by A_n. In how many years will the value of investment exceed $15,000?

38. Mrs. Brown wants to purchase an annuity with her savings of $150,000 that will give her a sum of $P each month for the next 5 years. The insurance company pays an interest of $\frac{1}{2}$% per month. If y_n denotes the principal remaining in the annuity after n months, write down the difference equation satisfied by y_n and hence determine P.

CHAPTER **8**

Probability

Chapter Objectives

8-1 SAMPLE SPACES AND EVENTS
(a) Definitions of sample points and the sample space for an experiment;
(b) Events, impossible event, and certain event; the complement of an event;
(c) Representation of sample spaces using tree diagrams and Venn diagrams;
(d) The union and intersection of two events; mutually exclusive events.

8-2 PROBABILITY
(a) Definition of the probability of an outcome and the probability of an event; interpretation of probability in terms of proportion of occurrences in a large number of repetitions of the experiment;
(b) Calculation of probabilities for sample spaces with equally likely outcomes;
(c) Formula for the probability of a complementary event and its use;
(d) Formula for the probability of the union of two events;
(e) Use of a table of probabilities containing, for two events A and B, the entries $P(A)$, $P(A')$, $P(B)$, $P(B')$, $P(A \cap B)$, $P(A' \cap B)$, $P(A \cap B')$ and $P(A' \cap B')$.

8-3 CONDITIONAL PROBABILITY
(a) Definition of the conditional probability of E_1 given E_2, $P(E_1 | E_2)$;
(b) Formulas for $P(E_1 | E_2)$ and $P(E_2 | E_1)$;
(c) Independent events and the multiplication rule;
(d) Use of the multiplication rule to compute probabilities and to test for the independence of two given events.

8-4 BAYES' THEOREM (OPTIONAL)
(a) Statement of Bayes' theorem for two mutually exclusive events and for n such events;
(b) Use of a tree diagram to solve problems involving Bayes' theorem;
(c) Use of Bayes' theorem to reassess probabilities in light of new evidence.

8-5 PERMUTATIONS AND COMBINATIONS
(a) Fundamental counting principle;
(b) Definition for $_nP_r$, the number of permutations of r objects from among n and $\binom{n}{r}$ the number of combinations of r objects from n;
(c) Formulas for $_nP_r$, and $\binom{n}{r}$ in terms of factorials;
(d) Application of counting techniques and formulas to compute probabilities.

8-6 BINOMIAL PROBABILITIES
(a) Definition and examples of Bernoulli trials;
(b) Formula for the probability of r successes in n trials (binomial probabilities);
(c) Applications of binomial probabilities; opinion surveys.

8-7 THE BINOMIAL THEOREM
(a) The binomial expansion; Pascal's triangle and its construction row by row;
(b) The binomial theorem, formula for the general term, binomial coefficients.

CHAPTER REVIEW

8-1 SAMPLE SPACES AND EVENTS

Historically, the theory of probability originated in investigations conducted by Blaise Pascal (1623–1662) and Pierre de Fermat (1601–1665) in the middle of the seventeenth century at the instigation of certain figures in the gambling world of the time. Today, besides its applications to games of chance, probability theory has become an important tool in such diverse fields as engineering, meteorology, insurance and actuarial work, business operations, and the experimental sciences. Probability theory underlies most of the methods of statistics, a field with widespread applications in almost every area of modern life.

Probability theory is used to handle situations that involve uncertainty. There exist many kinds of observations for which the precise outcome cannot be predicted, even though the set of all possible outcomes can be listed. For example, if a coin is tossed, we cannot predict whether it will fall heads or tails; however, we do know that it will come up one or the other, so that the set of possible outcomes is known. Similarly, when a political election takes place, we cannot predict with certainty how many seats each party will win, but we can make a list of all the possible outcomes. A sales manager may not be able to predict with certainty what sales will be next month, but it may be possible to say, for example, that they will be somewhere between 600 and 1000 units. In order to develop the theory of such observations we introduce the following definition.

DEFINITION The set of all possible outcomes of an experiment is called the **sample space** and is denoted by S. Each element of this set (that is, each outcome) is called a **sample point.** The sample space is said to be **finite** if the number of outcomes is finite. (In this chapter, we shall deal only with finite sample spaces.)

EXAMPLE 1

(a) The experiment consists of tossing a coin. In this case, there are two outcomes, namely heads (H) or tails (T), so that the sample space consists of just two elements. Using set notation, we can write $S = \{H, T\}$.

(b) Two firms, A and B, are competing for two contracts. The set of outcomes can be denoted by $S = \{AA, AB, BA, BB\}$

where, for example AB means that A wins the first contract and B the second, AA means that A wins both contracts, and so on.*

---

*The sample space for a given experiment is not in general unique, since it depends on how we decide to classify the outcomes. For example, in Example 1(b) we may specify the outcomes as simply the number of contracts won by firm A, in which case the sample space is $\{0, 1, 2\}$. However there is usually one particular way of specifying the sample space that is more "fundamental" than the others and that leads to easier calculations.

Tree diagrams are useful for listing the sample spaces of more complicated observations or experiments.

EXAMPLE 2 Suppose we flip two coins. In the tree diagram shown in Figure 1, we first draw branches from the starting point to H and T, the possible outcomes from flipping the first coin.

Then for each of these outcomes, we draw further branches corresponding to the possible outcomes from flipping the second coin: If the first was H, the second could be either H or T and if the first was T, the second could again be either H or T. The elements of the sample space are now obtained by reading from start to the ends of branches (1), (2), (3), and (4). Thus the sample space is

$$S = \{HH, HT, TH, TT\}. \quad \bullet 1$$

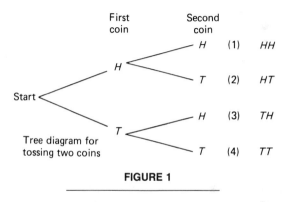

Tree diagram for
tossing two coins

FIGURE 1

We shall, when we come to the question of probability, be concerned with what are called *events*. Let us consider the example of tossing two coins. Then an example of the type of event with which we might be concerned is the event that "both coins fall alike." Looking at the list of elements or points in the sample space, we see that two of the four points satisfy the requirement that both coins fall alike, namely the outcomes HH and TT. We can say, therefore, that the event that both coins fall alike *consists of* the set of outcomes $\{HH, TT\}$.

☛ **1.** List the sample space for flipping three coins. How many outcomes are there? How many outcomes are there for flipping four coins? five coins?

Answer {HHH, HHT, HTH, THH, HTT, THT, TTH, TTT}, 8(=2^3) outcomes; 16(=2^4), 32 (=2^5).

As a second example, the event that "the first coin falls heads" occurs if and only if the outcome is either *HH* or *HT*. Thus this event can be identified with the set of outcomes {*HH*, *HT*}.

The event that "at least one coin falls heads" can be identified with the set of outcomes {*HH*, *HT*, *TH*}.

We see that in each of these examples, an event is associated with a certain set of outcomes of the experiment in question. This leads us to make the following formal definition.

DEFINITION Any subset of a sample space S for a particular experiment is called an **event**. Usually the letter E is used to denote events.

EXAMPLE 3 Consider the experiment of rolling a die with six faces containing 1, 2, 3, 4, 5, and 6 spots, respectively. The sample space is

$$S = \{1, 2, 3, 4, 5, 6\}.$$

(a) $E_1 = \{2, 4, 6\}$ represents the event that the die will show an even number of spots.

(b) $E_2 = \{4, 5, 6\}$ represents the event that the die will show more than 3 spots.

(c) Suppose we are interested in the event E_3 that 7 spots will show up. Clearly this will never happen, because no face on the die has 7 spots. Thus this subset is empty, that is, $E_3 = \emptyset$. Such an event is called an **impossible event**.

☞ **2.** In Example 3, list the events that (a) the number of spots is less than 6;   (b) the number of spots is even and less than 6;   (c) the number of spots is a perfect square.

(d) The event that either an odd or an even number of spots show up is given by the subset $E_4 = \{1, 2, 3, 4, 5, 6\}$. Clearly, this is the whole sample space. An event consisting of all the sample points of the sample space is called a **certain event.**   ☞ 2

It is often useful to represent sample spaces and events by means of what is called a **Venn diagram.** The sample space S itself is represented by a number of points enclosed within a rectangular boundary, each of the points representing one of the outcomes of the experiment. Any event would be represented by a closed region within the rectangle that contains all of the points corresponding to the outcomes in that event. For example, the sample space for rolling a die is shown in part (a) of Figure 2. The events E_1 and E_2 of Example 3 are shown in part (b).

There exist many situations in which a given event can be related to two or more other events. For example, consider the tossing of two coins, for which the sample space consists of the four outcomes $S = \{HH, HT, TH, TT\}$. Then $E_1 = \{HT, TH\}$ represents the event that both coins fall differently and $E_2 = \{TT, TH\}$ is the event that the first coin falls tails. Suppose we are interested in the event that either the first coin falls tails or else both coins fall dif-

Answer (a) {1, 2, 3, 4, 5};
(b) {2, 4};   (c) {1, 4}.

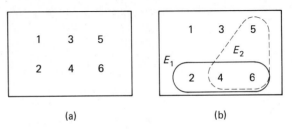

(a)   (b)

FIGURE 2

ferently. This event will occur whenever E_1 happens or E_2 happens or both E_1 and E_2 happen together. That is, this event has the possibilities $\{HT, TH, TT\}$, which are the sample points contained in either of the two events E_1 or E_2. The Venn diagram for this experiment is shown in Figure 3. The event that E_1 or E_2 or both occur consists of all three circled points. More generally, we have the following definition.

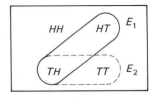

FIGURE 3

DEFINITION Let E_1 and E_2 be two events of a sample space S. Then the **union** of two events E_1 and E_2, denoted by $E_1 \cup E_2$, is the set of all the sample points which are in E_1 or E_2 or both. Thus

$$E_1 \cup E_2 = \{x \mid x \in E_1 \quad \text{or} \quad x \in E_2 \quad \text{or} \quad \text{both}\}.$$

$E_1 \cup E_2$ is read as E_1 *union* E_2. In terms of Venn diagrams, $E_1 \cup E_2$ represents the entire shaded region shown in Figure 4; $E_1 \cup E_2$ represents the event that E_1 *or* E_2 occurs. (Note that we use the inclusive *or*, which means that E_1, E_2, or both E_1 and E_2 occur.)

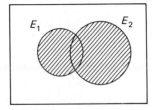

FIGURE 4

EXAMPLE 4 Given a sample space $S = \{1, 2, 3, 4, 5, 6\}$ for the rolling of a die, let E_1 be the event that an even number turns up, so $E_1 = \{2, 4, 6\}$; let E_2 be the event that an odd number turns up, so $E_2 = \{1, 3, 5\}$; and let E_3 be the event that the number turned up is less than 4, so $E_3 = \{1, 2, 3\}$.

(a) $E_1 \cup E_3 = \{1, 2, 3, 4, 6\}$ is the event that either the number turned up is even *or* the number turned up is less than 4.

(b) $E_1 \cup E_2 = \{1, 2, 3, 4, 5, 6\}$ is the event that either the number turned up is even *or* it is odd. Clearly in this case $E_1 \cup E_2 = S$, the whole sample space. ☛ **3**

☛ **3.** In Example 4, what is $E_2 \cup E_3$?

Let us once again consider the sample space $S = \{HH, HT, TH, TT\}$ for flipping two coins. Let $E_1 = \{HT, TH, TT\}$ be the event that at least one coin falls tails and $E_2 = \{HH, HT, TH\}$ be the event that at least one coin falls heads. Consider the event that at least one coin falls tails *and* at least one coin falls heads. This event has the possiblilties $\{HT, TH\}$, which are the sample points common to events E_1 and E_2. This leads us to the following definition.

DEFINITION Let E_1 and E_2 be two events of a sample space S. Then the **intersection** of these events, denoted by $E_1 \cap E_2$, is the set of all the sample points that belong both to E_1 and E_2.

$$E_1 \cap E_2 = \{x \,|\, x \in E_1 \text{ and } x \in E_2\}$$

We read $E_1 \cap E_2$ as E_1 *intersection* E_2. The event $E_1 \cap E_2$ is shaded in Figure 5. The event $E_1 \cap E_2$ occurs when both E_1 *and* E_2 occur.

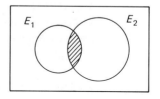

FIGURE 5

EXAMPLE 5 Consider the sample space $S = \{HH, HT, TH, TT\}$ for flipping two coins.

Let $E_1 = \{HH, TT\}$ be the event that both coins fall alike;
$E_2 = \{HT, TH\}$ be the event that both coins fall differently;
$E_3 = \{HH, HT, TH\}$ be the event that at least one coin falls heads;
$E_4 = \{TT, HT\}$ be the event that the second coin falls tails.

Then we have the following events.

$E_1 \cap E_4 = \{TT\}$ is the event that both coins fall alike *and* the second coin falls tails.

Answer $\{1, 2, 3, 5\}$.

$E_2 \cap E_3 = \{HT, TH\}$ is the event that at least one coin falls heads *and* both coins fall differently.

$E_1 \cap E_2 = \emptyset$ is the event that both coins fall alike *and* both coins fall differently.

Note that $E_1 \cap E_2$ is an empty set because the sets E_1 and E_2 have no members in common. $E_1 \cap E_2$ is an impossible event. ☛ 4

☛ **4.** In Example 5, list
(a) $E_1 \cap E_3$; (b) $E_1 \cap E_4$;
(c) $(E_4 \cap E_3) \cup E_1$.

DEFINITION Two events E_1 and E_2 of a sample space are said to be **mutually exclusive** if there is no sample point which is in both E_1 and E_2, that is, if $E_1 \cap E_2 = \emptyset$. In other words, *E_1 and E_2 are mutually exclusive if they cannot occur together in the same experiment.*

Figure 6 represents two mutually exclusive events. The regions representing E_1 and E_2 have no sample points in common.

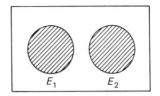

FIGURE 6

EXAMPLE 6 Consider the sample space $\{1, 2, 3, 4, 5, 6\}$. Let $E_1 = \{2, 4, 6\}$, $E_2 = \{1, 5\}$, and $E_3 = \{2, 3\}$. Then E_1 and E_2 are mutually exclusive because $E_1 \cap E_2 = \emptyset$; however, E_1 and E_3 are *not* mutually exclusive because $E_1 \cap E_3 = \{2\}$ is not an empty set. We see that E_2 and E_3 are also mutually exclusive.

EXAMPLE 7 Consider the sample space for drawing a single card from a deck of 52 cards. Then the events *drawing an 8* and *drawing a 10* are mutually exclusive because the card drawn cannot be both an 8 and a 10 at the same time. On the other hand, the events *drawing an ace* and *drawing a heart* are *not* mutually exclusive because it is possible to draw a card that is an ace as well as a heart.

DEFINITION Let E be an event in a sample space S. Then the **complement** E' of the event E with respect to the sample space S is the set of all outcomes in S which are not in E.

$$E' = \{x \mid x \in S \quad \text{and} \quad x \notin E\}$$

Answer (a) {HH}; (b) {TT};
(c) {HT, HH, TT}.

In terms of a Venn Diagram, E' is the region inside the rectangle but outside the region representing event E. (See Figure 7.) The event E' occurs whenever

5. For rolling a die,
$A = \{1, 2, 4, 6\}$ and $B = \{3, 4, 5\}$.
Write down the events
B', $A' \cap B$, and $A' \cup B'$.

Answer $B' = \{1, 2, 6\}$,
$A' \cap B = \{3, 5\}$
and $A' \cup B' = \{1, 2, 3, 5, 6\}$.

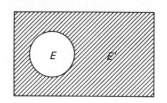

FIGURE 7

6. In the Venn diagram below, what are the two events that are shaded?

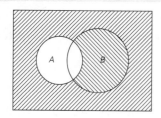

Answer $A' \cap B'$ and $A' \cap B$.

E does not occur. It follows from the definition of E' that the events E and E' are mutually exclusive.

EXAMPLE 8 Consider the sample space $S = \{1, 2, 3, 4, 5, 6\}$ for the roll of a die. Let $E_1 = \{1, 2, 3\}$ and $E_2 = \{4, 6\}$. Then E_1' is the set of all outcomes in S which are not in E_1.

$$E_1' = \{4, 5, 6\}$$

Similarly, $E_2' = \{1, 2, 3, 5\}$. **5, 6**

Final Note The concepts of union, intersection, and complement are used for sets in general, not only for those sets that represent sample spaces.

EXERCISES 8-1

(1–10) Write the sample spaces for the following experiments.

1. Five persons named A, B, C, D, and E contest an election; the observation consists of naming the winner.

2. Three coins are tossed. Each may fall heads or tails with no other possibility.

3. A coin is tossed and at the same time a die is rolled. (The six faces of the die are marked with numbers 1, 2, . . . , 6.)

4. Two dice are rolled and the numbers turned up on the dice are observed.

5. The order in which three competing salesmen end the year in terms of their annual sales is observed.

6. The order in which four competing salesmen end the year is observed.

7. Three rats are chosen from a cage that contains three brown and two white rats. The colors of the selected rats are observed.

8. A card is selected from among the four aces in a deck of playing cards.

9. A spade is selected from a deck of cards.

10. A box contains 3 white balls and 2 black balls that are exactly alike except for color. Two balls are drawn from the box and their colors observed.

(11–14) For the experiment in Exercise 4 above, write the events for which the sum of the numbers turned up on the two dice is equal to each of the following values.

11. 10   **12.** 5   **13.** 7   **14.** 13

(15–22) A card is drawn from a pack of 52 cards. Let the events be defined as follows:

E_1: The card drawn is a heart.

E_2: The card drawn is a black card.

E_3: The card drawn has a denomination less than 7 (ace counts low).

E_4: The card drawn is an ace.

Express the following events in terms of sets, as well as in words.

15. $E_1 \cup E_2$ **16.** $E_1 \cap E_2$

17. $E_1 \cup E_3$ **18.** $E_1 \cap E_3$

19. $E_3 \cap E_4$ **20.** $E_1 \cup E_4$

21. E_2' **22.** E_3'

23. Two coins are tossed. Which of the following events are mutually exclusive?

E_1:   Coins fall alike

E_2:   Coins fall with at least one head.

E_3:   Coins fall with two heads.

E_4:   Coins fall differently.

E_5:   Coins fall with at least one tail.

E_6:   Coins fall with three heads.

(24–27) A card is drawn from a deck of 52 cards. Which of the following pairs of events are mutually exclusive?

24. The card is a 7 and the card is an ace.

25. The card is a 4 and the card is a spade.

26. The card is a face card and the card is a heart.

27. The card is a 5 and the card is a face card.

28. *(Stock Market)* On any given day, a stock on the stock-market may go up, go down, or stay the same in price. An investor holds shares in two companies. List the outcomes for the investor's two stocks after trading on a particular day. How many outcomes would there be for an investor who holds three different stocks?

29. *(Job Applicants)* Among a group of applicants for a certain executive position with an insurance company, let U be the event that a given applicant has a university degree, let E be the event that the applicant has previous experience in insurance, and let F be the event that the applicant is over 40 years old. Express the following events in symbols.

a. The event that an applicant is under 40 and has a university degree.

b. The event that an applicant has neither a degree nor previous experience in insurance.

c. The event that an applicant has a degree but no previous experience and is over 40.

d. The event that an applicant has no degree, has previous experience and is under 40.

30. *(Job Applicants)* The Venn diagram shows the experiment in Exercise 29. Express the four events in parts (a) to (d) of Exercise 29 in terms of the regions I, II, . . . , VIII into which the Venn diagram is divided. Which pairs of those events are mutually exclusive?

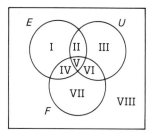

31. *(Job Applicants)* For the Venn diagram in Exercise 30, express in words the events represented by the following regions.

a. Region IV   **b.** Region V

c. Regions I and VIII

d. Regions III and VI

e. Regions VI and VIII

32. *(Promotions)* A corporation plans to promote one of its seven area managers to the position of vice-president of sales. Let S be the event that either Harry or Nick is promoted, T the event that either Harry or Laura is promoted, U the event that either Nick or Sarah is promoted, and V the event that either Fred or Nick is promoted. List the outcomes belonging to each of the following events if the other two managers are Maureen and Joe.

a. $T \cup U$ **b.** $T \cup S$ **c.** S'

d. $V \cup U$ **e.** $V \cap U$ **f.** U'

g. $T \cap U$ **h.** $T \cap V$ **i.** $(T \cup U)'$

◼ 8-2 PROBABILITY

In an experiment or observation for which there are several possible outcomes, we associate with each possible outcome a nonnegative number that we call its **probability.** The probability of any particular outcome indicates how likely that outcome is to occur: the bigger the probability, the more likely it is that the outcome will occur. The sum of the probabilities of all the possible outcomes for an experiment is always equal to 1.

To get an idea of the meaning of probability, suppose that the observation is repeated some large number N times, and let an outcome occur K times. Then the ratio K/N gives the proportion of the times that the given outcome occurs. As we make N larger and larger, this ratio can be expected to become closer and closer to the probability of the outcome in question.

EXAMPLE 1 Let a coin be tossed. The two outcomes H and T are clearly equally likely if the coin is tossed fairly, and each has a probability of $\frac{1}{2}$. (Note that the sum of the probabilities of both possible outcomes is equal to 1.)

Let the coin be tossed N times and let H occur K times in these N tosses. The proportion of heads is then K/N. We know that in general K/N will not be equal to $\frac{1}{2}$; for example, if the coin is tossed twice, it does not have to land heads once and tails once. It can quite easily land heads both times or tails both times. However, if N is very large, the proportion K/N of heads will almost certainly be close to $\frac{1}{2}$. That is K/N gets closer to the probability $\frac{1}{2}$ as N gets larger and larger.

EXAMPLE 2 (Traffic Survey) A traffic survey shows that at a certain intersection, 15% of vehicles turn left, 31% turn right, and 54% make no turn. In this case, the experiment consists of randomly selecting a vehicle approaching the intersection and observing the direction that it turns. The sample space has three points, which can be denoted by

$$\{L, R, N\}$$

indicating left turn, right turn, and no turn, respectively.

The probabilities of these three outcomes are

$$P(L) = 0.15, \qquad P(R) = 0.31, \quad \text{and} \quad P(N) = 0.54.$$

Observe that the sum of these probabilities is 1, as it should be:

$$P(L) + P(R) + P(N) = 0.15 + 0.31 + 0.54 = 1.$$

An event E was defined to be a subset of the outcomes in the sample space of a particular experiment. The event occurs whenever one of the outcomes in this subset occurs. We define the **probability of the event E** to be the sum of the probabilities of all the outcomes that belong to E. It is denoted by $P(E)$.

If the observation is repeated N times, let the event E occur on K of these repetitions. The ratio K/N gives the proportion of the observations in which E occurred. If we make N very large, this proportion will almost certainly be close to the probability $P(E)$.

EXAMPLE 3 In Example 2, let the event E be that the selected vehicle makes a turn of some kind at the intersection. Then E consists of the outcomes

$\{L, R\}$. The probability of E is then the sum of the probabilities of left and right turns:

$$P(E) = P(L) + P(R) = 0.15 + 0.31 = 0.46.$$

This means that if a very large number of vehicles are surveyed, then the fraction of them that make a turn will get closer to 0.46 as the number of vehicles gets larger and larger. ☛ 7

☛ 7. In Example 3, what is the probability of either a left turn or no turn?

Remarks **1.** If $E = \emptyset$, that is, if the event is impossible, then $P(E) = 0$.

2. If $E = S$, that is, the event is certain to happen since it consists of the whole set of possible outcomes, then $P(E) = P(S) = 1$.

3. The sum of probabilities of the outcomes of E cannot exceed the sum for all the possible outcomes, which is 1. Therefore, for any event E, $0 \le P(E) \le 1$.

Very often we find ourselves dealing with observations for which all of the outcomes in the sample space are equally likely to occur. For example, if a coin is tossed, the two outcomes of heads or tails can usually be presumed to be equally likely. Or if a well-balanced die is rolled, the outcomes 1, 2, 3, 4, 5, or 6 should be equally likely to occur. If the die is rolled a large number N of times, each of the six numbers should come up as readily as the rest; that is, each number should appear about one-sixth of the time as long as N is large enough.

DEFINITION The outcomes of an experiment are said to be **equally likely** if their probabilities are all equal to one another. If the sample space contains n sample points, then the probability of each outcome must be $1/n$ if they are equally likely (since the sum of the probabilities of all the outcomes must equal 1).

It follows that if an event E contains k sample points in an experiment in which the outcomes are equally likely, then the probability $P(E) = k/n$:

$$P(E) = \frac{\text{Number of Sample Points in } E}{\text{Number of Sample Points in } S} = \frac{k}{n}.$$

EXAMPLE 4 What is the probability of throwing a number greater than 4 with a standard die?

Solution When the die is rolled, the outcome can be any one of the six numbers 1, 2, 3, 4, 5, or 6, that is, $S = \{1, 2, 3, 4, 5, 6\}$. The event E of throwing a number greater than 4 will consist of the outcomes 5 or 6. Thus $E = \{5, 6\}$. In this case,

$$k = \text{Number of Sample Points in } E = 2$$

$$n = \text{Number of Sample Points in } S = 6.$$

Answer 0.69.

Thus

$$P(E) = \frac{k}{n} = \frac{2}{6} = \frac{1}{3}.$$

EXAMPLE 5 What is the probability of throwing a 7 with a standard die?

Solution As before, the sample space is $S = \{1, 2, 3, 4, 5, 6\}$. The event of throwing a 7 is clearly an impossible event because $7 \notin S$, so $E = \emptyset$. Thus we have $k = 0$, because the empty set has no element. Since $n = 6$,

$$P(E) = \frac{k}{n} = \frac{0}{6} = 0.$$

EXAMPLE 6 Find the probability of throwing at least two heads by tossing three fair coins.

Solution The sample space for tossing three coins is easily obtained from the tree diagram in Figure 8. Thus

$$S = \{HHH, HHT, HTH, HTT, THH, THT, TTH, TTT\}.$$

The event of throwing at least two heads is

$$E = \{HHH, HHT, HTH, THH\}.$$

In this case, k is the number of elements in E, or $k = 4$, and n is the number of

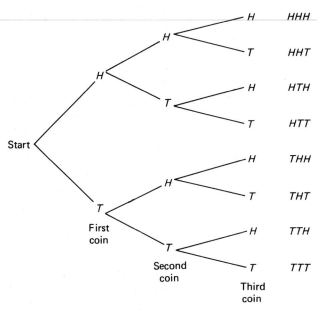

FIGURE 8

> ☛ **8.** If three coins are tossed, what is the probability that (a) they will all land the same way up; (b) there will be more tails than heads; (c) There will be the same number of heads and tails?

Answer (a) $\frac{1}{4}$; (b) $\frac{1}{2}$; (c) 0.

> ☛ **9.** In Example 7, what is the probability that (a) the sum of the two dice is either 4 or 11; (b) the difference between the two dice is not more than 1?

Answer (a) $\frac{5}{36}$; (b) $\frac{4}{9}$.

elements in S, or $n = 8$. Thus

$$P(E) = \frac{k}{n} = \frac{4}{8} = \frac{1}{2}. \quad ☛ 8$$

EXAMPLE 7 Find the probability of throwing a sum of 9 with the roll of two dice.

Solution The sample space can easily be obtained by a tree diagram; it is given in ordered pair notation in Figure 9. In the ordered pair (x, y), x denotes the number on the first die and y denotes the number on the second die. There are 36 ordered pairs in this sample space, so $n = 36$. The sum of 9 on the two dice results if any one of the four pairs (3, 6), (4, 5), (5, 4) or (6, 3) is rolled. (These outcomes are circled in Figure 9.) Thus there are four sample points in the event *a sum of 9 is rolled,* that is, $k = 4$. Hence

$$P\text{ (sum of 9 rolled)} = \frac{k}{n} = \frac{4}{36} = \frac{1}{9}. \quad ☛ 9$$

(1, 1)	(1, 2)	(1, 3)	(1, 4)	(1, 5)	(1, 6)
(2, 1)	(2, 2)	(2, 3)	(2, 4)	(2, 5)	(2, 6)
(3, 1)	(3, 2)	(3, 3)	(3, 4)	(3, 5)	(3, 6)
(4, 1)	(4, 2)	(4, 3)	(4, 4)	(4, 5)	(4, 6)
(5, 1)	(5, 2)	(5, 3)	(5, 4)	(5, 5)	(5, 6)
(6, 1)	(6, 2)	(6, 3)	(6, 4)	(6, 5)	(6, 6)

FIGURE 9

Probability Formulas

$P(E')$ is the probability of the event E', which is complementary to E. Thus $P(E')$ is the probability that the event E does not occur. Similarly, $P(E_1 \cap E_2)$ is the probability of the intersection of E_1 and E_2, or the probability that E_1 *and* E_2 occur and $P(E_1 \cup E_2)$ is the probability that E_1 *or* E_2 occurs. There are certain formulas involving these three probabilities that are often useful in calculating probabilities of events.

Let us begin by considering the following example: Two dice are rolled and we want to find the probability that the sum of the two scores is different from 9.

In Example 7, we calculated the probability that the sum of the two numbers is *equal* to 9. This event corresponds to the four outcomes (3, 6), (4, 5), (5, 4), and (6, 3) and so has probability $\frac{4}{36}$, or $\frac{1}{9}$. The event that the sum of the two scores is different from 9 corresponds to the set of all the outcomes except these four. Since there are 32 other outcomes, the probability of getting a sum different from 9 is equal to $\frac{32}{36}$, or $\frac{8}{9}$.

Observe that in this example we are dealing with two events that are complementary to one another: the event that the sum is equal to 9 and the event that it is different from 9. These two events have probabilities of $\frac{1}{9}$ and $\frac{8}{9}$, respectively. The sum of the two probabilities equals 1. This property of complementary events is generalized in Rule 1.

Rule 1

If E' is the event complementary to E, then

$$P(E') = 1 - P(E).$$

PROOF $P(E)$ is equal to the sum of probabilities of all the outcomes belonging to E. E' consists of all of the outcomes that do not belong to E and $P(E')$ is the sum of probabilities of these outcomes. Therefore the sum $P(E) + P(E')$ is equal to the sum of probabilities of all the possible outcomes in S and must equal $P(S)$, or 1:

$$P(E) + P(E') = 1$$

The stated rule follows from this equation. ☛ **10**

☛ **10.** In Example 7, the probability is $\frac{4}{9}$ that the two dice differ by at most 1. What is the probability that they differ by more than 1?

EXAMPLE 8 In a certain community, the probability of a 70-year-old person living to be 80 is 0.64. What is the probability of a person who is 70 and a member of this community dying sometime in the next 10 years?

Solution If E denotes the event that a 70-year-old person will live for the next 10 years, then the complementary event E' is that a 70-year-old person will die within the next 10 years. We are given that $P(E) = 0.64$ and we want $P(E')$.

$$P(E') = 1 - P(E) = 1 - 0.64 = 0.36$$

Thus the probability of a person who is 70 years old now dying within the next 10 years is 0.36.

In order to introduce the second probability formula, consider the example of rolling two dice. There are 36 outcomes, which are listed in Figure 9. Let E_1 be the event that the sum of the two scores is 8. Then $E_1 = \{(6, 2), (5, 3), (4, 4), (3, 5), (2, 6)\}$ and $P(E_1) = \frac{5}{36}$. Let E_2 be the event that both dice show an odd number of spots. Then

$$E_2 = \{(1, 1), (1, 3), (1, 5), (3, 1), (3, 3), (3, 5), (5, 1), (5, 3), (5, 5)\}.$$

There are 9 outcomes, so $P(E_2) = \frac{9}{36}$.

The event $E_1 \cap E_2$ is the event that both E_1 and E_2 occur and contains only the two outcomes (3, 5) and (5, 3). Therefore $P(E_1 \cap E_2) = \frac{2}{36}$.

Answer $1 - \frac{4}{9} = \frac{5}{9}$.

Similarly, the event $E_1 \cup E_2$ is

$$E_1 \cup E_2 = \{(1, 1), (1, 3), (1, 5), (3, 1), (3, 3), (3, 5), (5, 1), (5, 3),$$
$$(5, 5), (6, 2), (4, 4), (2, 6)\},$$

which has 12 outcomes. Therefore $P(E_1 \cup E_2) = \frac{12}{36}$.

Note that $P(E_1) + P(E_2) = \frac{9}{36} + \frac{5}{36} = \frac{14}{36}$. This is not equal to $P(E_1 \cup E_2)$. The reason that these are not equal to one another is that in forming $P(E_1) + P(E_2)$, the two events (5, 3) and (3, 5) are counted twice—once in E_1 and once in E_2. We see, however, that the difference

$$P(E_1) + P(E_2) - P(E_1 \cup E_2) = \frac{14}{36} - \frac{12}{36}$$

is equal to $\frac{2}{36}$, the probability of the two outcomes (5, 3) and (3, 5). But this is precisely equal to $P(E_1 \cap E_2)$, so we have

$$P(E_1) + P(E_2) - P(E_1 \cup E_2) = P(E_1 \cap E_2).$$

This relation is true in general.

Rule 2:

Let E_1 and E_2 be two events that are subsets of the same sample space S. Then,

$$P(E_1 \cup E_2) = P(E_1) + P(E_2) - P(E_1 \cap E_2).$$

PROOF Figure 10 illustrates the Venn diagram for the experiment in question. The sets of sample points corresponding to E_1 and E_2 are shown. Region I consists of sample points that are in E_1 but outside E_2. Similarly, Region II consists of sample points in E_2 but not in E_1. Region III denotes the sample points that are in both E_1 and E_2, that is, the intersection $E_1 \cap E_2$.

$P(E_1) =$ Sum of Probabilities of the Outcomes in Regions I and III

$P(E_2) =$ Sum of Probabilities of the Outcomes in Regions II and III

Therefore if $P(E_1)$ and $P(E_2)$ are added, we obtain the sum of the probabilities of the outcomes in all three regions, I, II and III, except that those in III are counted twice. But the sum of probabilities in III is equal to $P(E_1 \cap E_2)$,

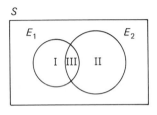

FIGURE 10

and so

$$P(E_1) + P(E_2) - P(E_1 \cap E_2) = \text{Sum of Probabilities of the Outcomes}$$
$$\text{in Regions I, II, and III}$$
$$= P(E_1 \cup E_2)$$

as required.

In particular, if events E_1 and E_2 are mutually exclusive, $E_1 \cap E_2 = \emptyset$, so that $P(E_1 \cap E_2) = 0$. The formula in Rule 2 reduces to

$$P(E_1 \cup E_2) = P(E_1) + P(E_2) \qquad \text{for mutually exclusive events.}$$

EXAMPLE 9 Find the probability of drawing an ace *or* a spade from a deck of 52 cards in a single draw.

Solution A single card can be drawn from a deck of 52 cards in 52 different ways, so that the sample space consists of 52 elements. The event E_1 of drawing an ace can be achieved in 4 ways because there are 4 aces. Thus $P(E_1) = \frac{4}{52}$. The event E_2 of drawing a spade can be achieved in 13 ways because there are 13 spades in the deck. Thus $P(E_2) = \frac{13}{52}$. There is only one card that is both an ace and a spade, so the event $E_1 \cap E_2$ can happen in only one way and $P(E_1 \cap E_2) = \frac{1}{52}$. We are interested in the probability of drawing an ace *or* a spade, that is, $P(E_1 \cup E_2)$. Using Rule 2, we have

$$P(E_1 \cup E_2) = P(E_1) + P(E_2) - P(E_1 \cap E_2)$$
$$= \tfrac{4}{52} + \tfrac{13}{52} - \tfrac{1}{52} = \tfrac{16}{52} = \tfrac{4}{13}. \quad ● \, 11$$

EXAMPLE 10 Find the probability of throwing a sum of 7 or a sum of 9 with the roll of two dice.

Solution The sample space for rolling two dice consists of 36 sample points. (See Figure 9.) Let E_1 be the event that a sum of 7 is obtained and E_2 be the event that a sum of 9 is obtained. Then we have

$$E_1 = \{(6, 1), (5, 2), (4, 3), (3, 4), (2, 5), (1, 6)\}$$

and

$$E_2 = \{(6, 3), (5, 4), (4, 5), (3, 6)\}.$$

Clearly, $E_1 \cap E_2 = \emptyset$, so the events E_1 and E_2 are mutually exclusive. The event of throwing a sum of 7 or 9 is $E_1 \cup E_2$. Thus

$$P(E_1 \cup E_2) = P(E_1) + P(E_2) = \tfrac{6}{36} + \tfrac{4}{36} = \tfrac{10}{36} = \tfrac{5}{18}. \quad ● \, 12$$

A third probability formula will be discussed in Section 8-3.

IMPORTANT NOTE (SHORT METHOD) If a problem on probability involves two events say A and B, then most of the probabilities involving these events can be written down immediately once we complete the rectangular box shown in Table 1. Here A' and B' denote the complements of the events A and B. Note that the vertical columns and horizontal rows add up to their respective end totals, that is,

TABLE 1

	A	A'	
B	$P(A \cap B)$	$P(A' \cap B)$	$P(B)$
B'	$P(A \cap B')$	$P(A' \cap B')$	$P(B')$
	$P(A)$	$P(A')$	1

☛ **13.** Draw a Venn diagram to illustrate the property $P(A \cap B) + P(A \cap B') = P(A)$.

$$P(A \cap B) + P(A \cap B') = P(A); \quad P(A' \cap B) + P(A' \cap B') = P(A')$$
$$P(A \cap B) + P(A' \cap B) = P(B); \quad P(A \cap B') + P(A' \cap B') = P(B')$$
and $\qquad P(A) + P(A') = 1; \qquad P(B) + P(B') = 1 \quad$ ☛ **13**

EXAMPLE 11 In a certain community 30% of the people smoke, 55% of them drink alcohol, and 20% of them smoke as well as drink. Calculate the probability that a randomly selected person (a) smokes but does not drink; (b) neither smokes nor drinks; (c) either smokes or does not drink or both.

Solution If S and D denote the events of smoking and drinking, we are given that

$$P(S) = 0.30, \qquad P(D) = 0.55 \quad \text{and} \quad P(S \cap D) = 0.20.$$

We are asked to calculate

(a) $P(S \cap D')$ (b) $P(S' \cap D')$ (c) $P(S \cup D')$.

We fill in the given information in the probability table (see Table 2). The blank spaces are then filled using the fact that each column and row must add up to its end total (see Table 3). The shaded square is filled last.

For example, 30% smoke, so 70% do not smoke; that is,

$$P(S') = 1 - P(S) = 1 - 0.30 = 0.70.$$

Then of the 30% who smoke, 20% drink so 10% do not drink:

$$P(S \cap D') = P(S) - P(S \cap D) = 0.30 - 0.20 = 0.10.$$

We continue to fill the table in this way, finally obtaining Table 4.

Answer

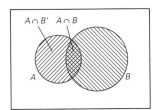

TABLE 2

	S	S'	
D	0.20		0.55
D'			
	0.30		1

TABLE 3

	S	S'	
D	0.20	$0.55 - 0.20 = 0.35$	0.55
D'	$0.30 - 0.20 = 0.10$	$0.45 - 0.10 = 0.35$ or $0.70 - 0.35 = 0.35$	$1 - 0.55 = 0.45$
	0.30	$1 - 0.30 = 0.70$	1

TABLE 4

	S	S'	
D	0.20	0.35	0.55
D'	0.10	0.35	0.45
	0.30	0.70	1

☛ **14.** In Example 11, find the probability that an individual
(a) drinks but does not smoke;
(b) either drinks or does not smoke or both.

Answer (a) $P(D \cap S') = 0.35$
(b) $P(D \cup S') = P(D) + P(S') - P(D \cap S') = 0.9.$

Now from the completed Table 4, the probability that a randomly selected person

(a) smokes but does not drink is $P(S \cap D') = 0.10$
(b) neither smokes nor drinks is $P(S' \cap D') = 0.35$
(c) Either smokes or does not drink is

$$P(S \cup D') = P(S) + P(D') - P(S \cap D')$$
$$= 0.30 + 0.45 - 0.10 = 0.65$$

and so on.   ☛ **14**

EXERCISES 8-2

(1–6) A card is drawn from a well-shuffled pack of 52 cards. Find the probability of drawing each card.

1. A heart.

2. A 2 or a 9.

3. Not a spade.

4. A 6 or a diamond.

5. A 4 or a 5 or a heart.

6. Neither a king nor a queen.

(7–11) A family has three children. Find the probability of each event.

7. There are two boys.

8. There are three girls.

9. There is at least one boy.

10. There is at least one boy and one girl.

11. There is at least one boy or one girl.

(12–16) Two dice are rolled. Find the probability of each event.

12. A sum of 5.

13. A sum of 8 or a sum of 6.

14. Not a sum of 7.

15. At least one 6.

16. Either one 6 (not two) or a sum of 9.

17. If E_1 and E_2 are two mutually exclusive events for which $P(E_1) = 0.3$ and $P(E_2) = 0.45$, find each probability.

a. $P(E_2')$
b. $P(E_1 \cup E_2)$
c. $P(E_1 \cap E_2)$
d. $P(E_1 \cap E_2')$
e. $P(E_1' \cap E_2')$
f. $P(E_1' \cup E_2')$

18. *(Stock Market)* Let E_1 be the event that a stock will rise next year and E_2 the event that its price will remain unchanged. If $P(E_1) = 0.5$ and $P(E_2) = 0.2$, describe the following events in words and calculate the probability of each.

a. E_1'
b. $(E_1 \cup E_2)'$
c. $E_1 \cap E_2'$
d. $E_1' \cap E_2'$
e. $E_1' \cup E_2$

19. *(Personnel)* A person is chosen at random from among a firm's employees. Let E_1 be the event that the person has been an employee for 10 years or more and E_2 be the event that the individual earns a salary of $20,000 or more. If $P(E_1) = 0.25$, $P(E_2) = 0.35$, and $P(E_1' \cup E_2') = 0.8$, describe the following events in words and calculate the probability of each.

a. E_2'
b. $E_2' \cup E_1$
c. $E_1' \cap E_2$
d. $(E_1' \cap E_2')'$

20. If A, B, and C are three events such that $A \cap B = \emptyset$ and

$(A \cup B) \cup C = S$, show that $P(C)$ must be at least equal to $1 - P(A) - P(B)$.

21. The operator of a hot dog stall has found that 70% of all its customers use mustard, 40% use ketchup, and 25% use both. Find the probability that a particular customer will use at least one of these.

22. A business firm has received 150 applications for a vacant position. Of these applicants, 90 had a university degree, 45 a job related experience, and 30 had both. What is the probability that the person selected at random for the job will have neither a degree nor the experience?

23. *(Employee Transfers)* A bank has five junior executives in the head branch. Each year in April, one of these five is selected at random to be transferred as manager to one of the local branches. The selected individual is replaced by a new junior executive. Find the probability that a given junior exective:

a. Stays exactly 2 years in the head branch.

b. Stays more than 2 years in the head branch.

c. Stays 3 years or less in the head branch.

24. *(Employee Classification)* Among a firm's employees, 40% are over 40 years old and 25% are both under 40 and are smokers.

a. What percentage of the employees are under 40 and nonsmokers?

b. What percentage of the employees under 40 are nonsmokers?

25. *(Sales Target)* Seven salespeople are set a monthly sales target. The probability that five or more of them achieve the target is $\frac{7}{10}$ and the probability that five or fewer achieve it is $\frac{2}{5}$. Find the probability that each number of salespeople achieve the target.

a. 6 or 7   **b.** 4 or fewer

c. exactly 5   **d.** 7

26. *(Trainee Retention)* A firm hires five trainee managers per year. The probability that at least one of them will stay on with the firm after the training program is $\frac{7}{10}$ and the probability that they will not all stay on is $\frac{3}{10}$. Find the probability that the number of trainees who stay on is equal to

a. 5   **b.** 0

c. 3   **d.** 1, 2, 3, or 4

27. *(Employee Classification)* A computer firm has 60 college graduates on its staff, 28 of whom had their college training in computer science. Of these employees, 39 are under 40 years old; among these 39, 21 had college training in computer science. If a person is chosen at random from among the 60, what is the probability that he or she will be over 40 and have no training in computer science?

28. *(Supermarket Purchases)* Of the customers at a supermarket, 20% buy 10 items or less, 70% spend more than $10, and 68% spend more than $10 and buy more than 10 items. What is the probability that a customer chosen at random:

a. Spends more than $10 and buys 10 items or less?

b. Spends less than $10 and buys 10 items or less?

29. Prove the following probability statements by appropriate Venn diagrams.

a. $P(E_1 \cap E_2) + P(E_1 \cap E_2') = P(E_1)$

b. $P(E_1' \cap E_2') = 1 - P(E_2) - P(E_1 \cap E_2')$

30. Two cards are drawn from a pack of 52 cards one by one without replacement; that is, the first card is not replaced before the second card is drawn. What is the probability in each case?

a. The first card is an ace and the second is a heart.

b. The first card is a 5 and the second is a 10.

31. *(Mutations)* In a large population of fruit flies, 30% of the flies have a wing mutation, 40% have an eye mutation, and 15% have both eye and wing mutations. A fly is chosen from this population. What is the probability that it has at least one of the mutations?

32. In a certain community 70% of the people smoke, 40% have lung cancer and 25% of the people smoke as well as have lung cancer. If S and L denote the events of smoking and having lung cancer, then (by completing Table 5) find the probability that a randomly selected individual:

TABLE 5

	S	S'	
L			
L'			
			1

a. Does not smoke but has lung cancer.

b. Smokes but does not have lung cancer.

c. Neither smokes nor has lung cancer.

d. Either smokes or does not have lung cancer.

e. Either has lung cancer or does not smoke.

f. Either does not smoke or does not have lung cancer.

33. In a hi-rise apartment building of 200 families, 180 have a television (T.V.) and 150 own a car. There are 14 families who have no T.V. but own a car. If T and C denote the events of having a T.V. and owning a car, then (by completing Table 6) find the probability that a selected family in the hi-rise:

a. Has neither a T.V. nor owns a car.

TABLE 6

	T	T'	
C			
C'			
			1

b. Has a car and a T.V.

c. Has a T.V. but no car.

d. Has either a T.V. or no car.

e. Either does not have a T.V. or has a car.

f. Either does not have a car or does not have a T.V.

8-3 CONDITIONAL PROBABILITY

Additional information about the outcome of an experiment can change the probabilities of associated events. For example, if a card is drawn at random from an ordinary deck of 52 cards, the probability that it is a spade is $\frac{1}{4}$. However, if we are told in addition that the drawn card is black, then the probability that it is a spade is $\frac{1}{2}$. The difference between these two situations is that in the first case, the sample space consists of all 52 cards, while in the second case, the additional information effectively reduces the sample space to the 26 black cards. In each case, 13 of the cards are spades, so that the probabilities of the drawn card being a spade are $\frac{13}{52} = \frac{1}{4}$ and $\frac{13}{26} = \frac{1}{2}$, respectively.

As a second example, let us suppose that an insurance company is choosing a person to fill a certain job from among 50 applicants. Among the applicants, some have a university degree, some have previous experience in the insurance field, and some have both. Suppose that the breakdown is given in Table 7. For example, 5 applicants have both a degree and previous experience in insurance, 20 applicants have neither, and so on. If an applicant is chosen at random from among the 50, the probability that this applicant will have had previous experience is $\frac{15}{50} = 0.3$, since 15 of the applicants altogether have had experience. However, let us suppose that the company considers only those

TABLE 7

	Degree	No Degree
Previous Experience	5	10
No Experience	15	20

applicants who have university degrees and that an applicant is chosen at random from this subset. There are 20 such applicants, 5 of whom have had previous experience, so the probability of selecting one with experience is now $\frac{5}{20}$ or 0.25.

In the second case, the sample space is again effectively reduced by the company's restriction to applicants with degrees. This change of sample space changes the probability of the chosen applicant having previous experience. We can say that for the whole set of applicants, the probability of choosing one with previous experience is 0.3. However, the probability of choosing an applicant with experience given that he or she must also have a degree is 0.25.

This is an example of what is called a *conditional probability*.

DEFINITION Let E_1 and E_2 be two events. Then the **conditional probability of E_2 given E_1,** written $P(E_2 | E_1)$, is the probability that E_2 occurs given that E_1 is known to have occurred.

☛ **15.** Explain verbally what is $P(E_1 | E_2)$. Compute $P(E_1)$ and $P(E_1 | E_2)$.

In the last discussion, if E_1 is the event that the chosen applicant has a degree and E_2 the event that he or she has had previous experience, then $P(E_2) = 0.3$ and the conditional probability $P(E_2 | E_1) = 0.25$. ☛ **15**

EXAMPLE 1 Three coins are tossed. Let E_1 be the event that the first coin falls heads and E_2 the event that two coins fall heads and one falls tails. Calculate $P(E_1)$, $P(E_2)$, $P(E_2 | E_1)$, and $P(E_1 | E_2)$.

Solution The sample space consists of the eight points shown in Figure 11. Here E_1 contains the four outcomes whose first letter is H, and E_2 contains the three outcomes with 2 H's and 1 T. Therefore $P(E_1) = \frac{4}{8} = \frac{1}{2}$ and $P(E_2) = \frac{3}{8}$.

$P(E_2 | E_1)$ is the probability of two heads and one tail given that the first coin falls heads. To calculate this, we must restrict attention to the four outcomes in E_1. Of these, two outcomes also belong to E_2, namely HHT and HTH. Therefore E_2 occurs twice out of the four outcomes in E_1 and we have $P(E_2 | E_1) = \frac{2}{4} = \frac{1}{2}$.

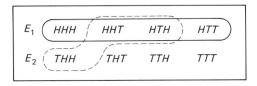

FIGURE 11

Answer $P(E_1 | E_2)$ is the probability that an applicant has a degree given that he or she has previous experience. $P(E_1) = \frac{2}{5}$ (20 out of 50), $P(E_1 | E_2) = \frac{1}{3}$ (5 out of 15).

$P(E_1 | E_2)$ is the probability that the first coin fell heads given that there were two heads and one tail. To calculate it, we must restrict our attention to the three outcomes in E_2. Of these, two also belong to E_1, so E_1 occurs twice out of the three outcomes in E_2. Therefore $P(E_1 | E_2) = \frac{2}{3}$.

Note that in calculating these two conditional probabilities, the two outcomes *HHT* and *HTH*, which belong to both E_1 and E_2, play a key role. These two outcomes form the intersection of E_1 and E_2. We can see that

$$P(E_2|E_1) = \text{(Number of Outcomes in } E_1 \cap E_2)$$
$$\div \text{(Number of Outcomes in } E_1)$$
$$= 2 \div 4$$

and

$$P(E_1|E_2) = \text{(Number of Outcomes in } E_1 \cap E_2)$$
$$\div \text{(Number of Outcomes in } E_2)$$
$$= 2 \div 3. \quad \text{☛} \ 16$$

☛ **16.** Two dice are rolled. Let E_1 be the event that one die shows 6 and E_2 the event that the sum of the two scores is 8. Find $P(E_1|E_2)$ and $P(E_2|E_1)$ and verify Rule 3. (Use Figure 9.)

Alternatively, since the outcomes are all equally likely, we can write the following.

Rule 3:

$$P(E_2|E_1) = \frac{P(E_1 \cap E_2)}{P(E_1)} \quad and \quad P(E_1|E_2) = \frac{P(E_1 \cap E_2)}{P(E_2)}.$$

This result is, in fact, generally true for any pair of events E_1 and E_2. The first relation can be rewritten in the form

---

Rule 3':

$$P(E_1 \cap E_2) = P(E_1)P(E_2|E_1).$$

In words, this says that *the probability that both E_1 and E_2 occur is equal to the probability that E_1 occurs muliplied by the conditional probability that E_2 occurs given that E_1 has already occurred.* In these verbal terms, the equation is intuitively plausible.

EXAMPLE 2 (Education and Earnings) Among the male population between the ages of 30 and 35 living in a certain city, 25% have a university degree, 15% earn more than \$25,000 per year, and 65% have no degree and earn less than \$25,000 per year. What is the probability that a person chosen at random from this group earns more than \$25,000 per year given that: (a) he has a degree; (b) he has no degree?

Solution Let D be the event that the person has a degree and E be the event that he earns more than \$25,000 per year. We are given that $P(D) = 0.25$ and $P(E) = 0.15$.

The complementary events are D', that he has no degree, and E', that he earns less than \$25,000 per year. We are given that the probability of both these events is 0.65, that is, $P(E' \cap D') = 0.65$. The given information is

Answer $P(E_1|E_2) = \frac{2}{5}$,
$P(E_2|E_1) = \frac{2}{11}$, $P(E_1 \cap E_2) = \frac{2}{36}$,
$P(E_1) = \frac{11}{36}$, $P(E_2) = \frac{5}{36}$.

330   CH. 8 | PROBABILITY

summarized in Table 8. In this table, we have used Rule 1 to calculate $P(E') = 1 - P(E) = 1 - 0.15 = 0.85$ and $P(D') = 1 - P(D) = 1 - 0.25 = 0.75$. In verbal terms, 25% of the given age group have a degree, so 75% do not have a degree; 15% earn more than \$25,000, so 85% must earn less.

TABLE 8

	D (Degree)	D' (No Degree)	
E (earns more than \$25,000	$P(E \cap D)$	$P(E \cap D')$	$P(E) = 0.15$
E' (earns less than \$25,000	$P(E' \cap D)$	$P(E' \cap D') = 0.65$	$P(E') = 0.85$
	$P(D) = 0.25$	$P(D') = 0.75$	

In the main part of the table, the sample space is divided into the four segments $E \cap D$, $E \cap D'$, $E' \cap D$, and $E' \cap D'$. For example, the subset $E' \cap D$ corresponds to persons who have a degree and earn less than \$25,000, while $E' \cap D'$ corresponds to those who have no degree and earn less than \$25,000. We are told that this latter group comprises 65% of the whole group, that is, $P(E' \cap D') = 0.65$.

Of the 85% who earn less than \$25,000, 65% have no degree. The remaining 20% must have a degree, so this 20% forms the membership of $E' \cap D$. In terms of probability, we can write

$$P(E' \cap D) + 0.65 = P(E') = 0.85$$

and so $P(E' \cap D) = 0.2$. Similarly, in the column headed by D', we have a total of 75% of the age group. Of these, 65% earn less than \$25,000, so the remaining 10% earn more. That is, $P(E \cap D') = 0.1$. Finally, in the top row, we have $P(E \cap D) + P(E \cap D') = P(E) = 0.15$. Since we now know that $P(E \cap D') = 0.1$, it follows that $P(E \cap D) = 0.05$. The completed table is shown as Table 9.

We wish to calculate the two conditional probabilities

$$P(E|D) \quad \text{and} \quad P(E|D').$$

TABLE 9

	D	D'	
E	0.05	0.1	$P(E) = 0.15$
E'	0.2	0.65	$P(E') = 0.85$
	$P(D) = 0.25$	$P(D') = 0.75$	

From Rule 3,

$$P(E|D) = \frac{P(E \cap D)}{P(D)} = \frac{0.05}{0.25} = 0.2$$

and

$$P(E|D') = \frac{P(E \cap D')}{P(D')} = \frac{0.1}{0.75} = \frac{2}{15} = 0.133.$$

Thus the probability of a person with a degree earning more than \$25,000 is 0.2. The probability of one without a degree doing the same is 0.133. ☛ 17

☛ **17.** In Example 2, find the probability that an individual has no degree given that (a) he earns more than \$25,000 (b) he earns less than \$25,000.

DEFINITION Two events E_1 and E_2 are **independent** if the probability of either one occurring does not depend on whether or not the other occurs. That is, E_1 and E_2 are independent if

$$P(E_1|E_2) = P(E_1) \tag{1}$$

or if

$$P(E_2|E_1) = P(E_2) \tag{2}$$

These two conditions are, in fact, identical. Making use of Rule 3' for independent events, we have from Equation (1) that

$$P(E_1 \cap E_2) = P(E_1|E_2)P(E_2) = P(E_1)P(E_2).$$

If this is substituted into the first part of Rule 3, then Equation (2) is obtained. Thus we have the following rule.

Rule 4:

$$P(E_1 \cap E_2) = P(E_1)P(E_2) \qquad \text{if } E_1 \text{ and } E_2 \text{ are independent.}$$

Thus *if E_1 and E_2 are independent, the probability of E_1 and E_2 both occurring is equal to the product of the separate probabilities of E_1 and E_2.* This result is often called the **multiplication rule** for independent events.

It is often intuitively obvious whether or not two events are independent. For example, let a coin be tossed and a die rolled at the same time and let E_1 be the event that the coin falls heads and E_2 the event that the die shows 5 or 6. Clearly, E_1 and E_2 are independent, since there is no reason why the fall of the coin should influence the number of spots showing on the die.

EXAMPLE 3 Find the probability of tossing a head with a coin and then drawing an ace from a deck of 52 cards.

Solution The probability of getting a head (H) with the flip of a coin is $P(H) = \frac{1}{2}$. The probability of drawing an ace (A) from a pack of 52 cards is $P(A) = \frac{4}{52}$. The two events, tossing a head and drawing an ace, are indepen-

Answer $P(D'|E) = \frac{2}{3}$, $P(D'|E') = \frac{13}{17}$.

dent events. Therefore

$$P(A \text{ and } H) = P(A) \cdot P(H) = \tfrac{4}{52} \cdot \tfrac{1}{2} = \tfrac{1}{26}. \quad \text{☛ } 18$$

EXAMPLE 4 A card is drawn from a deck of 52 cards. It is replaced, the cards are shuffled, and a second card is drawn. Find the probability that the first card is a 7 and the second card is a spade.

Solution These events are independent; the result of the second drawing is not affected by the first drawing since the first card is replaced in the deck before the second card is drawn. We can draw a card out of 52 cards in 52 ways. The card 7 can be drawn in 4 ways, so the probability of drawing a 7 is

$$P(7) = \tfrac{4}{52} = \tfrac{1}{13}.$$

Similarly, the probability of drawing a spade in the second drawing is

$$P(\text{spade}) = \tfrac{13}{52} = \tfrac{1}{4}.$$

Using Rule 4, the probability of drawing a 7 in the first drawing and a spade in the second drawing is

$$P(7 \text{ and spade}) = P(7) \cdot P(\text{spade}) = \tfrac{1}{13} \cdot \tfrac{1}{4} = \tfrac{1}{52}.$$

In cases where the independence of two events is not intuitively obvious, their independence can be verified by demonstrating that the multiplication rule holds or disproved by showing that the rule does not hold.

EXAMPLE 5 *(Return of Defective Automobiles)* Over a certain period of time, an automobile plant turns out 5000 automobiles. Of these, 1000 were built on a Monday, 1000 on a Tuesday, and so on, with 1000 built on a Friday. It was necessary to return 400 of these automobiles for repair of a serious defect during the guarantee period. Of the cars built on a Friday, 150 were returned. Are the two events, a car was built on a Friday and it was defective, independent of one another?

Solution Let F be the event that a car chosen at random was built on a Friday and D be the event that it turned out to be defective. Then the probabilities of these events are

$$P(F) = \tfrac{1000}{5000} = 0.2, \qquad P(D) = \tfrac{400}{5000} = 0.08.$$

Since 150 cars belong to both of these events,

$$P(F \cap D) = \tfrac{150}{5000} = 0.03.$$

But $P(D) \cdot P(F) = (0.08)(0.2) = 0.016 \neq P(F \cap D)$, so the events F and D are not independent.

The probability of a car built on a Friday being defective is the conditional probability $P(D|F)$. It is given by Rule 3:

$$P(D|F) = \frac{P(F \cap D)}{P(F)} = \frac{0.03}{0.2} = 0.15.$$

Answer No. $P(S) = \frac{1}{9}$, $P(D) = \frac{5}{18}$, $P(S \cap D) = \frac{1}{18} \neq P(S) \cdot P(D)$. This probability is not equal to $P(D)$, again showing that the two events are not independent. ☛ **19**

EXERCISES 8-3

(1–2) Two coins are tossed. Find each probability.

1. The second coin falls heads given that both coins fall alike.

2. The second coin falls heads given that the coins fall with at least one head.

(3–4) Two coins are tossed. Are the following events independent?

3. *A*: The coins fall at least one head.
    B: The coins fall both alike.

4. *C*: The first coin falls tails.
    D: The second coin falls either heads or different from the first coin.

(5–10) Two dice are rolled. Find each probability.

5. The sum of the scores is at least 9, given that the first die shows 5 spots.

6. The sum of the scores is at least 9, given that the sum of the scores is at least 8.

7. The difference between the scores is 1, given that the sum of the scores is 7.

8. The second die scores more than the first, given that the sum of the scores is 8.

9. At least one of the dice scores 5 or 6, given that the sum of the scores is 6 or more.

10. The two dice fall alike, given that the sum of the scores is:

a. Even.   **b.** Odd.

(11–14) Three coins are tossed. Find each probability.

11. The coins fall two heads and one tail, given that they do not fall alike.

12. The first two coins fall tails, given that at least two of the coins fall tails.

13. There are more heads than tails, given that the first coin falls heads.

14. The coins do not all fall alike, given that the first two coins fall alike.

15. *(Product Inspection)* A manufacturer's product is examined by two inspectors, A and B. Of the defective products, 20% get by inspector A; and if a defective product gets by A, there is a probability of 0.5 that it will then be passed by inspector B. What is the probability that a defective part gets by both inspectors?

16. *(Product Defects)* Of the cans of meat produced by a firm, 5% are underweight and 1% of the cans are both underweight and have a canning defect. If an underweight can is chosen at random, what is the probability that it will have a canning defect?

17. *(Product Inspection)* A can manufacturer stores its cans in batches of 500. In one particular batch, 150 of the cans are substandard. An inspector selects 2 cans at random from the batch. What is the probability:

a. That both are substandard?

b. That both are good?

18. *(Tea Consumers)* 35% of married men and 50% of married women drink tea regularly. 50% of the men whose wives drink tea also drink tea themselves. Find each probability.

a. That a given married couple both drink tea.

b. That of a married couple, neither drinks tea.

c. That a woman whose husband drinks tea also drinks tea herself.

19. *(Television Advertising)* A company estimates that 30% of the population has seen a television ad for its soap powder. Of those who see the ad, 10% later buy the product. What proportion of the population has both seen the ad and bought the product? If 5% of those who have not seen the ad have also bought the soap powder, find each probability.

a. That a randomly selected individual has bought the soap powder.

b. That a person who has bought the soap powder has not seen the ad.

20. *(Automobile Inspection)* When spot-checked for safety, automobiles are found to have defective tires 15% of the time, defective lights 25% of the time, and both defective tires and lights 8% of the time. Find the probability that a randomly chosen car:

a. Has defective lights, given that its tires have been found defective.

b. Has defective tires, given that its lights have been found defective.

c. Has good tires, given that its lights are good.

21. *(Credit-card Customer Profiles)* A credit-card company finds that 40% of its customers live in urban areas and 60% live in suburban or rural areas. Of the customers, 30% average more than $200 per month on their credit card and 10% both live in urban areas and spend over $200 per month. Find the probability that a given customer lives in an urban area given that the customer spends less than $200 per month on his or her credit card.

22. *(Wealthiness)* In a certain population, 35% are wealthy according to some standard and 30% have wealthy parents. One in six of those with wealthy parents are not wealthy themselves. Find the probability that a person's parents are not wealthy given that the person is not wealthy.

23. *(Component Suppliers)* A manufacturer buys a certain component from two suppliers, A and B. Over a certain period of time, the company used 20,000 of these components, 6000 of which came from supplier A. Of these components, 3% of those supplied by A and $1\frac{1}{2}$% of those supplied by B turned out to be defective. Find the probability that a given defective component came from supplier A.

24. *(Absenteeism and Skill)* A company employs 8 unskilled laborers and 12 skilled tradesmen. Over a period of 100 working days, the company found that it lost 40 workdays of unskilled labor and 40 workdays of skilled labor through absenteeism. Find the probability that a randomly chosen absentee employee is one of the skilled tradesmen.

25. *(Smoking and Cancer)* The probability that a person over 50 years of age is a smoker in a certain community is $\frac{3}{5}$ and the probability that a person over 50 years of age has a cancer is $\frac{1}{20}$. The probability that a person over 50 years of age will be a smoker and have a cancer is $\frac{1}{25}$. Are smoking and cancer disorders independent events?

26. *(Drinking and Heart Disease)* The probability that a person over 60 years old in a certain community drinks alcohol is $\frac{2}{3}$ and the probability that a person over 60 years old has heart disease is $\frac{2}{15}$. The probability that a person over 60 years old both drinks alcohol and has heart disease is $\frac{1}{16}$. Are "drinking alcohol" and "heart disease" independent events?

27. *(Dice Game)* In a certain game, a player rolls two dice and wins if he scores 9 or more. If he fails the first time, he gets a second chance to roll 9 or more. What is the probability of his winning?

28. *(Shooting Game)* Mark has a probability p of hitting the bull's-eye at a firing range. He wins a prize if he hits the bull's-eye, but if he misses he gets a second chance to win. If his probability of getting a prize is $\frac{5}{9}$, what is p?

8-4 BAYES' THEOREM (OPTIONAL)

Consider the following problem. A manufacturer has two assembly lines producing identical models. The first assembly line produces one-third of the total output, whereas the second, more up-to-date line produces two-thirds. Of the items produced on the first line, 5% turn out to be defective, but of those produced on the second line only 2% are defective. If a given unit chosen at random from the whole production is found to be defective, what is the probability that it was made on the first line?

Let B_1 be the event that a randomly chosen unit was produced on the first assembly line and B_2 the event that it was produced on the second line. Let A be the event that the unit turns out to be defective. Then we are given the following probabilities:

$$P(B_1) = \tfrac{1}{3}, \qquad P(B_2) = \tfrac{2}{3}$$

and

$$P(A|B_1) = 0.05, \qquad P(A|B_2) = 0.02$$

We wish to calculate the conditional probability, $P(B_1|A)$, that an item was produced on the first assembly line given that it is defective.

Since $P(A \cap B_1) = P(B_1 \cap A)$, it follows from Rule 3′ in Section 8-3 that

$$P(A|B_1)P(B_1) = P(B_1|A)P(A).$$

Therefore

$$P(B_1|A) = \frac{P(A|B_1)P(B_1)}{P(A)}. \tag{1}$$

On the right side we know the two factors in the numerator, but we do not know $P(A)$ in the denominator.

In order to calculate $P(A)$, it is useful to consider the diagram in Figure 12. In the first branch of the tree, the items are divided into those produced on the first assembly line (B_1) and those produced on the second line (B_2). Then in the second branches of the tree the defective items are indicated (the event A). It is clear from the diagram that defective items must either come through B_1 or through B_2.

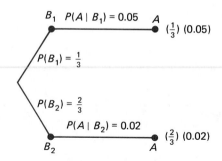

FIGURE 12

One-third of the total production comes from B_1 and of that a fraction 0.05 is defective. So the proportion of the total production which comes from B_1 and is defective is $P(A|B_1)P(B_1)$, or $(\frac{1}{3})(0.05)$. In a similar way the proportion of defective items coming through B_2 is $P(A|B_2)P(B_2)$, or $(\frac{2}{3})(0.02)$ of the total production. Adding these together we obtain the proportion of the total production which is defective, or in other words $P(A)$.

$$P(A) = (\tfrac{1}{3})(0.05) + (\tfrac{2}{3})(0.02) = 0.03$$

Therefore, from Equation (1),

$$P(B_1|A) = \frac{P(A|B_1)P(B_1)}{P(A)} = \frac{(0.05)(\tfrac{1}{3})}{0.03} = \frac{5}{9}.$$

20. Use Bayes' theorem to find the probability that a given defective item came from assembly line 2.

The probability that a given defective item came from assembly line 1 is therefore $\frac{5}{9}$. ☛ **20**

In this example, by means of the tree diagram we break up event A into a branch coming through B_1 and a branch coming through B_2. These two branches have probabilities $P(A|B_1)P(B_1)$ and $P(A|B_2)P(B_2)$, respectively. The probability $P(A)$ is the sum of these,

$$P(A) = P(A|B_1)P(B_1) + P(A|B_2)P(B_2).$$

Therefore Equation (1) takes the form

$$P(B_1|A) = \frac{P(A|B_1)P(B_1)}{P(A|B_1)P(B_1) + P(A|B_2)P(B_2)}.$$

We can express this formula in terms of the tree diagram:

$$P(B_1|A) = \frac{\text{Probability of the Branch through } B_1}{\text{Sum of the Probabilities of both Branches through } B_1 \text{ and } B_2}.$$

This formula is an example of Bayes' theorem. The general theorem, which we shall now state, applies to the case when there are not just two events B_1 and B_2 into which the sample space is divided but n events $B_1, B_2, \ldots, B_n$. For instance, the last example could be generalized to a manufacturer with n assembly lines instead of just two.

THEOREM 1 (BAYES' THEOREM) Let $B_1, B_2, \ldots, B_n$ be n events that are mutually exclusive (i.e., $B_i \cap B_j = \emptyset$ for all i and j, $i \neq j$) and such that any point in the sample space belongs to one and only one of these events. Then if A is any other event, for any B_k we have

$$P(B_k|A) = \frac{P(A|B_k)P(B_k)}{P(A|B_1)P(B_1) + P(A|B_2)P(B_2) + \cdots + P(A|B_n)P(B_n)}.$$

In problems involving more than two events $B_1, \ldots, B_n$, the easiest way to use Bayes' theorem is to draw a tree diagram, as we did earlier.

Bayes' theorem is commonly used as a means of revising our assessment of probabilities in the light of new evidence. As such it forms the basis of the so-called Bayesian approach to statistics.

EXAMPLE 1 (*Oil Exploration*) An oil company estimates that from geological data there is a probability of 0.3 of finding oil in a certain area. It knows from previous experience that if oil is to be found, there is a probability of 0.4 that a positive strike of some kind will be made on the first series of drillings. If the first series of drillings turn out to be negative, what is the probability that oil will eventually be found?

Solution We wish to calculate the probability that oil will be discovered given that the first drilling series is negative. Therefore let B_1 be the event that oil will be discovered and let A be the event that the first series is negative, so that

Answer $P(B_2|A) =$
$\dfrac{P(A|B_2) \cdot P(B_2)}{P(A)} = \dfrac{\frac{2}{3}(0.02)}{0.03} = \frac{4}{9}$.

☛ **21.** In Example 1, use Bayes' theorem to calculate the probability that oil will never be found, given that the first series of drillings is negative.

the probability we want is $P(B_1|A)$. Let B_2 be the complementary event to B_1, namely, the event that oil will never be found. Then

$$P(B_1) = 0.3, \qquad P(B_2) = 1 - P(B_1) = 0.7$$

We know that the probability the first drilling is positive given that oil will be found, that is, $P(A'|B_1)$, is 0.4. Therefore the probability the first drilling is negative given that oil will be found is given by

$$P(A|B_1) = 1 - P(A'|B_1) = 1 - 0.4 = 0.6.$$

We also know that $P(A|B_2) = 1$, since given B_2 (that is, given that oil will never be discovered), the first drillings will certainly be negative.

The tree diagram in Figure 13 shows the two branches leading to the event A, one branch through B_1 the other through B_2. The total probability of A is the sum of the probabilities of the two branches. From Bayes' theorem, the conditional probability $P(B_1|A)$ is given by the probability of the branch leading to A through B_1 divided by the sum of the probabilities of the two branches:

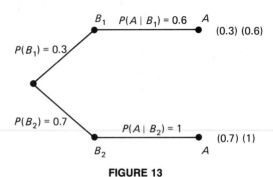

FIGURE 13

$$P(B_1|A) = \frac{(0.6)(0.3)}{(0.6)(0.3) + (1)(0.7)} = \frac{0.18}{0.18 + 0.7} = \frac{9}{44} \approx 0.20.$$

So, given that the first drilling turns out to be negative, the probability of eventually finding oil is 0.20. ☛ **21**

In examples such as this, the probabilities $P(B_1)$ and $P(B_2)$ are called *prior probabilities*. They form our initial estimates of the probabilities of B_1 and B_2 (in this example, based on geological evidence). Subsequently we obtain further evidence through the observation of an event A (a negative drilling) and this causes us to revise our assessment of the probabilities of B_1 and B_2. The new probabilities are $P(B_1|A)$ and $P(B_2|A)$; they are called *posterior probabilities*, that is, probabilities calculated after further evidence comes in.

Let us pursue the preceding example further. Suppose that a second drilling series also turns out to be negative. What then will be the new probability of eventually striking oil?

Answer $P(B_2|A) = \dfrac{P(A|B_2) \cdot P(B_2)}{P(A)}$

$= \dfrac{1 \cdot (0.7)}{[(0.3)(0.6 + 0.7(1)]} = \dfrac{35}{44}.$

☛ 22. A school has equal numbers of boys and girls. 80% of the girls and 30% of the boys have long hair. What is the probability that a randomly chosen child with long hair is a boy?

In order to answer this question we can use Bayes' theorem again, but now as prior probabilities we must use the value

$$P(B_1) = \frac{9}{44} \quad \text{and} \quad P(B_2) = 1 - P(B_1) = \frac{35}{44}.$$

The reason for this is that the observation A (that the first series was negative) must be taken into account in estimating these probabilities.

If C is the event that the second drilling series is negative, then we assume that $P(C|B_1) = 0.6$ and $P(C|B_2) = 1$, as before. Therefore

$$P(B_1|C) = \frac{P(C|B_1)P(B_1)}{P(C|B_1)P(B_1) + P(C|B_2)P(B_2)}$$

$$= \frac{(0.6)(9/44)}{(0.6)(\frac{9}{44}) + (1)(\frac{35}{44})} \approx 0.13.$$

Answer $P(B|LH) =$

$$\frac{P(LH|B) \cdot P(B)}{[P(LH|B) \cdot P(B) + P(LH|G) \cdot P(G)]}$$

$$= \frac{(0.3)(0.5)}{[(0.3)(0.5) + (0.8)(0.5)]} = \frac{3}{11}.$$

Thus the probability of eventually striking oil after the second dry series is 0.13.

Note how each piece of negative evidence reduces the posterior probability of eventually striking oil from 0.3 at first to 0.20 and then to 0.13. **☛ 22**

EXERCISES 8-4

1. *(Management Training)* A large corporation hires management trainees, 30% of whom have a university training in business or economics. Of the trainees, 45% eventually attain management positions with the corporation, but of those trainees with training in business or economics, 70% reach the management level. What is the probability that an individual selected from among those in managerial positions who came through the corporation's trainee program has a university training in business or economics?

2. *(Job Applicants)* After an applicant is interviewed for a job by a certain corporation, the probability that the applicant still wants the job is 0.8, while the probability that the corporation wishes to hire the person is 0.4. Among those applicants the corporation wishes to hire, 90% remain interested after the interview. What is the probability that an applicant who remains interested in a job will be offered one?

3. *(Soda Bottling)* A pop-bottling machine places the caps securely on the bottles 99.5% of the time when it is in correct adjustment but only 80% of the time when it is out of adjustment. If there is a 10% chance that the machine is incorrectly adjusted, what is the probability that a randomly chosen bottle will be securely capped?

4. *(Student Classification)* Among a certain group of students, $\frac{2}{5}$ are male and $\frac{3}{5}$ female. Of the males, $\frac{1}{2}$ are under 21 years of age, whereas $\frac{2}{3}$ of the females are under 21. Find the probability that:

   a. A randomly chosen student is over 21.

   b. A randomly chosen student who is under 21 is also female.

5. *(Production Rejects)* A manufacturer produces identical items on two production lines. Two-fifths of the production is produced on an old production line from which 10% of the output is rejected as substandard, and the other three-fifths is produced on a newer line from which only 4% of the output is substandard. What is the probability that a given rejected item came from the old production line?

6. *(Production Rejects)* In Exercise 5, the manufacturer opened a new production line, which doubled the total output. However, the new line was put into operation in a hurry, and 8% of its output had to be rejected. What is the probability now that a given rejected item came from the oldest of the production lines?

7. *(Production Rejects)* After a while, the manufacturer in Exercise 6 ironed out the bugs in the new production line and reduced the proportion of its substandard output to 3%. Recalculate the stated probability.

8. *(Clothing Inventory)* The women's clothing buyer of a large department store buys 20% of each year's supply of dresses from one manufacturer, A, 30% from a second manufacturer, B, and the remaining 50% from a variety of other sources. Of dresses purchased from A, 80% are sold, 75% of those from B are sold, and 90% of those purchased elsewhere are sold. What is the probability that a given dress that remained unsold at the end of the season came from manufacturer A?

9. *(Ball Game)* A bag contains either 3 black and 5 white balls or 4 black and 4 white balls or 5 black and 3 white balls. These possibilities are known to have equal prior probabilities. A ball is chosen at random from the bag and turns out to be black. Calculate the posterior probabilities for the three possible contents of the bag.

10. *(Ball Game)* In Exercise 9, the selected ball is replaced and a second ball is chosen at random from the bag. If this ball also turns out to be black, recalculate the three posterior probabilities for the initial contents of the bag.

11. *(Coin Toss)* Joe said to Fred, "The penny in my pocket has two heads." Fred, being a skeptic, said "The probability that that statement is true is 1/10." So Joe flipped the penny three times, and it came up heads each time. Fred said, "That makes the probability that the coin is two-headed equal to 8/17." Is he right?

12. *(Repeat Sales)* A company has four salespeople: Joe is responsible for 40% of the sales, Fred, for 25%, Sam, for 20% and Jack, for 15%. Of Joe's customers, 60% return for more purchases from the company, whereas 70% of Fred's customers and 75% of Sam's and Jack's customers

return. What is the probability that a given second-time purchasers from the company is a customer:

a. Of Joe's   b. Of Jack's?

13. *(Shoplifting)* A retail chain has three stores: their main store is responsible for 50% of sales, and two suburban stores are responsible for 30% and 20% of sales. The shoplifting rates at the three stores are 1%, 0.8% and 0.75%, respectively. (The shoplifting rate is the dollar value of stolen goods as a percentage of the sales volume at each store.) What proportion of the chain's total losses from shoplifting occur at the main store?

14. *(Car Insurance)* Of the motorists insured by a certain insurance company, 25% are under 25 years old, 30% are between 25 and 35, 35% are between 35 and 55, and 10% are over 55. The probabilities that a driver in each of these four age groups will be involved in an accident during a given year are 0.02, 0.008, 0.006, and 0.018, respectively. What percentage of the accidents of drivers covered by the company's policies involve drivers who are:

a. Under 25?   b. Over 55?

15. *(Tennis)* A tennis player has two main rivals, A and B, in a certain tournament. If they both enter the tournament, her probability of winning is 0.2; if only A enters, her probability of winning is 0.4; if only B enters, it is 0.5, and if neither of them enters, it is 0.75. The probability that A will enter the tournament is 0.6, and the probability that B will enter is 0.5. (These two events may be assumed to be independent of one another.) If the tennis player does actually win the tournament, what is the probability:

a. That both A and B entered?

b. That neither A nor B entered?

c. That A entered?

8-5 PERMUTATIONS AND COMBINATIONS

It is clear from the examples of the preceding sections that we very often find ourselves dealing with observations for which the outcomes are all equally likely. In these situations, all we need to know to compute the probability of any event is the number of outcomes that lead to the event in question and the total number of outcomes of the observation. One way of evaluating these two numbers is to make a complete list of all of the outcomes and simply count those that belong to the event, as well as the total number of outcomes. This essentially was the method used in Section 8-2. But for some observations, the number of outcomes may be exceedingly large, and listing them all may be a difficult or even impossibly long task. The following example illustrates this point.

EXAMPLE 1

(a) What is the probability of rolling at least two 6's in six rolls of a die? The total number of outcomes when a die is rolled six times is 46,656, and it would clearly be foolish to attempt to list them all.

(b) What is the probability of drawing a full house in a poker hand of five cards? What is the probability of drawing a royal flush? Such questions involve large numbers, since the total number of different poker hands, which is the number of points in the sample space, is about 2.5 million.

(c) A telephone cable containing 10 wires is cut. If the wires in the two broken ends are joined together in a random manner, what is the probability that they will be joined correctly? In this case, there are about 3.6 million ways of joining the two sets of wires together.

(d) In a reorganization of its senior personnel, a large corporation intends to promote 4 of its 20 area managers to positions in its central administration. The number of ways in which the 4 positions can be filled is 116,280. We could ask questions such as, "What is the probability that two of the area managers who are good friends will both be promoted to head office?"

What is needed in situations like this (which, in fact, arise very often) is a method of counting the numbers of outcomes in various events without actually listing them. In this section, we shall briefly outline several techniques that are useful in this situation.

Fundamental Counting Principle

The following fundamental principle underlies many counting methods. Consider an experiment that consists of several (p, for example) independent operations. Let the first operation have n_1 possible outcomes, the second operation have n_2 possible outcomes, and so on, the pth operation having n_p outcomes. Then the total number of outcomes of the whole experiment is given by the product

$$n_1 \cdot n_2 \cdot \cdots \cdot n_p.$$

EXAMPLE 2

(a) Let the experiment consist of flipping 3 coins. We can break this experiment up into three independent operations, each operation being the flip of one of the coins. Each of these has 2 outcomes, heads or tails. The total number of outcomes of the experiment is therefore given by the product $2 \cdot 2 \cdot 2 = 8$. (See Example 6 in Section 8-2.)

☞ 23. How many outcomes are there if (a) 5 coins are flipped; (b) 2 coins are flipped and a die rolled; (c) three people each remove a card from a deck of 52 (leaving 49)?

(b) Let the experiment consist of rolling 2 dice. Each of these rolls has 6 possible outcomes. Therefore the complete experiment, consisting of the 2 rolls, has $6 \cdot 6 = 36$ outcomes. (See Example 7 in Section 8-2.)

(c) A North American corporation is in the process of expanding its business into Europe and intends to send 3 representatives, one to Brussels, one to Frankfurt, and one to Zurich. If there are 10 representatives from which to choose, in how many ways can the 3 be selected?

There are 10 ways of selecting the representative to be sent to Brussels. Having made that choice however, there are only 9 left from which to make the next choice. Thus, if the Frankfurt representative is chosen next, this second choice can be made in 9 ways. Similarly, the Zurich representative can be chosen in only 8 ways once the other two have been selected. Consequently, the number of ways in which the 3 representatives can be chosen is given by the product $10 \cdot 9 \cdot 8 = 720$. ☞ **23**

Part (c) of Example 2 provides an example of what are called **ordered selections,** or **permutations.** We have a group of 10 persons and we wish to select 3 of them for a certain purpose. Since each of the 3 persons is needed for a different purpose, the *order* in which the selection is made is important. On the other hand, if the corporation intended to send all 3 representatives to the same place, then the order would not matter, and we would obtain a different answer.

Consider the general case in which we have n distinguishable objects or persons and we wish to make an ordered selection of r objects from among these n.

Since there are n objects from which to choose, the first object may be chosen in n ways. However, since the first object is not replaced, there are only $n - 1$ objects from which the second can be chosen. Hence the second object can be selected in $n - 1$ ways. Similarly, the third object can be selected in $n - 2$ ways once the first two have been chosen, and so on. The last (r th) object can be chosen in $n - r + 1$ ways. Thus the number of ways in which the whole permutation can be selected is given by the product

$$n(n - 1)(n - 2) \cdots (n - r + 1).$$

This product is denoted by the symbol $_nP_r$ and is called the **number of permutations of r objects from n objects.** The value of $_nP_r$ is given by the product of r consecutive integers, starting with n and decreasing to $n - r + 1$. For example,

$$_8P_3 = 8(8 - 1)(8 - 2) = 8 \cdot 7 \cdot 6 = 336$$

and

$$_{15}P_2 = 15 \cdot 14 = 210.$$

Taking $r = n$, we obtain $_nP_n$, the number of permutations of all n objects, and this is the product of all the integers from n down to 1. For example, $_4P_4 = 4 \cdot 3 \cdot 2 \cdot 1 = 24$.

Answer (a) $2^5 = 32$; (b) $2^2 \cdot 6 = 24$; (c) $52 \cdot 51 \cdot 50 = 132,600$.

☞ **24.** Evaluate (a) $_7P_3$; (b) $_{12}P_3$;
(c) $6! - 5!$

The quantity $n!$, called **n factorial,** is defined to be the product of all the positive integers from 1 up to n. Thus, for example,

$$1! = 1, \quad 2! = 1 \cdot 2 = 2, \quad 3! = 1 \cdot 2 \cdot 3 = 6, \quad 4! = 1 \cdot 2 \cdot 3 \cdot 4 = 24,$$

and $5! = 1 \cdot 2 \cdot 3 \cdot 4 \cdot 5 = 120.$

In general, we can write

$$n! = 1 \cdot 2 \cdot 3 \cdots (n - 1)n.$$

In terms of this notation, we therefore have

$$_nP_n = n!$$

We can express $_nP_r$ in terms of factorials.

EXAMPLE 3

(a) $_8P_3 = 8 \cdot 7 \cdot 6 = \dfrac{8 \cdot 7 \cdot 6 \cdot 5 \cdot 4 \cdot 3 \cdot 2 \cdot 1}{5 \cdot 4 \cdot 3 \cdot 2 \cdot 1} = \dfrac{8!}{5!}$

(b) $_{15}P_2 = 15 \cdot 14 = \dfrac{15 \cdot 14 \cdot 13 \cdot 12 \cdots 2 \cdot 1}{13 \cdot 12 \cdots 2 \cdot 1} = \dfrac{15!}{13!}$ ☞ **24**

Answer (a) 210; (b) 1320;
(c) 600.

☞ **25.** Evaluate $\dfrac{8!}{4!}$ and $\dfrac{12!}{9!}$.
(When calculating with factorials, do not cancel common factors before expanding the factorial. For example, $6! \div 3! \neq 2!$ Instead, $6! \div 3! = (6 \cdot 5 \cdot 4 \cdot 3 \cdot 2 \cdot 1) \div (3 \cdot 2 \cdot 1) = 120.$)

In general,

$$_nP_r = n(n - 1) \cdots (n - r + 1)$$

$$= \frac{n(n - 1) \cdots (n - r + 1)(n - r)(n - r - 1) \cdots 2 \cdot 1}{(n - r)(n - r - 1) \cdots 2 \cdot 1}.$$

We introduced the factors $n - r, n - r - 1, \ldots, 2, 1$ into both numerator and denominator, so the value of the expression has not changed. But the numerator is now equal to $n!$ and the denominator is equal to $(n - r)!$, so we have

$$_nP_r = \frac{n!}{(n - r)!}.$$

Letting $r = n$, we obtain the result that $_nP_n = n!/0!$. This equation is so far meaningless since we have not defined $0!$. However, since $_nP_n = n!$, it becomes a correct statement if we define $0! = 1$. This is the convention used to define $0!$. ☞ **25**

Answer 1680, 1320.

EXAMPLE 4 From a group of 4 people, we are required to select persons to participate in 3 different tests. In how many ways can the selection be made?

Solution Since the tests are *different*, the order in which the 3 are chosen is significant. The number of ways in which the choice can be made is therefore the number of permutations

$$_4P_3 = 4 \cdot 3 \cdot 2 = 24.$$

Now let us consider the case when r objects are selected from n objects, but when the order in which the objects are chosen is of no significance. For instance, in the last example, we could suppose that the 3 persons are chosen from the 4 available people in order to participate in the *same* test. Then the order in which the 3 are chosen would be immaterial, and all we would need to know would be which 3 formed the chosen group.

Let us label the available individuals by the letters a, b, c, and d. If the order in which the choice is made is important, then the number of choices is equal to $_4P_3 = 4 \cdot 3 \cdot 2 = 24$. We can list these 24 choices as follows.

bcd,	*bdc*,	*cbd*,	*cdb*,	*dcb*,	*dbc*
acd,	*adc*,	*cad*,	*cda*,	*dac*,	*dca*
abd,	*adb*,	*bad*,	*bda*,	*dab*,	*dba*
abc,	*acb*,	*bac*,	*bca*,	*cab*,	*cba*

We observe from this list that each group of 3 persons appears 6 times, corresponding to the 6 different ways in which the 3 persons can be ordered. If the order in which the 3 persons are selected is of no importance, all of these 6 permutations of each group are equivalent to one another; for example, *bcd* is equivalent to *bdc*, to *cbd*, and so on. When order is immaterial, there are only 4 different choices, corresponding to the 4 rows in the above table.

Now consider the general problem in which r objects are chosen from among n, where the order of selection is of no significance. Each such choice is called a **combination** of r objects from among n, and the number of combinations is denoted by the symbol $\binom{n}{r}$.

Any permutation of r objects from among n can be formed by first choosing the r objects and then arranging these r objects in an appropriate order. The number of permutations, $_nP_r$, is therefore equal to the number of ways of choosing particular combinations of r objects from among n multiplied by the number of ways in which each combination can be ordered. That is,

$$_nP_r = \binom{n}{r} \times N(r)$$

where $N(r)$ is the number of ordered arrangements of the chosen r objects. But $N(r)$ must be equal to $_rP_r = r!$, the number of permutations of r objects from among r. Therefore

$$_nP_r = \binom{n}{r} r!$$

and so

$$\binom{n}{r} = \frac{{}_nP_r}{r!} = \frac{n!}{(n-r)!\, r!}. \tag{1}$$

We can also write the number of combinations in the form

$$\binom{n}{r} = \frac{n(n-1)\cdots(n-r+1)}{r(r-1)\cdots 2 \cdot 1}$$

Note that both the numerator and denominator in this fraction contain r consecutive integers as factors.

EXAMPLE 5

(a) $\displaystyle \binom{8}{3} = \frac{8(8-1)(8-2)}{3(3-1)(3-2)} = \frac{8\cdot 7\cdot 6}{3\cdot 2\cdot 1} = 56$

(b) $\displaystyle \binom{7}{5} = \frac{7\cdot 6\cdot 5\cdot 4\cdot 3}{5\cdot 4\cdot 3\cdot 2\cdot 1} = \frac{7\cdot 6}{2\cdot 1} = 21$

In the first case, three factors occur in numerator and denominator, while in the second, five factors occur.

Remarks:

$$\binom{n}{n} = \binom{n}{0} = 1, \qquad \binom{n}{r} = \binom{n}{n-r} \quad \text{for all } r.$$

The first two values follow by setting $r = n$ and $r = 0$, respectively, in Equation (1). The last result follows by replacing r by $n - r$ in Equation (1). ☛ **26**

☛ **26.** Evaluate (a) $\binom{9}{3}$ and $\binom{9}{6}$; (b) $\binom{10}{4}$ and $\binom{10}{6}$.

EXAMPLE 6 Compute the number of different poker hands.

Solution A poker hand consists of a set of 5 cards chosen from the available 52 in the deck. The number of different hands is then simply the number of ways of choosing 5 cards from among 52. The order of selection is of no significance, so the number is

$$\binom{52}{5} = \frac{52\cdot 51\cdot 50\cdot 49\cdot 48}{1\cdot 2\cdot 3\cdot 4\cdot 5} = 2{,}598{,}960.$$

EXAMPLE 7 What is the probability of a poker hand of 5 cards containing 4 cards of a kind?

Solution The number of points n in the sample space is the number of differ-

Answer (a) Both 84; (b) both 210.

ent poker hands that are possible. From Example 6, this is

$$n = \binom{52}{5} = \frac{52 \cdot 51 \cdot 50 \cdot 49 \cdot 48}{1 \cdot 2 \cdot 3 \cdot 4 \cdot 5}.$$

The event for which we wish to calculate the probability is that the poker hand contains 4 cards of the same denomination (that is, 4 aces, 4 sevens, and so on). If there are k different ways in which such a hand can be chosen, then the required probability is equal to k/n.

Consider first the possibility that the hand contains 4 aces. The number of hands that have this property is equal simply to the number of ways in which the fifth card can be selected, namely 48 (since there are 48 cards remaining after the four aces have gone).

Similarly, there are 48 poker hands containing four 2's, 48 hands containing four 3's, and so on. Since there are 13 possible denominations, the number of ways of choosing a hand with 4 of a kind, regardless of denomination, is given by

$$k = 13 \cdot 48.$$

The required probability is therefore

$$P = \frac{k}{n} = \frac{13 \cdot 48}{(52 \cdot 51 \cdot 50 \cdot 49 \cdot 48)/(1 \cdot 2 \cdot 3 \cdot 4 \cdot 5)} = \frac{1}{4165}. \quad \text{☛ 27}$$

☛ **27.** What is the probability that a hand of 5 cards contains (a) all diamonds; (b) all five cards of the same suit; (c) 2 aces and 3 kings?

EXAMPLE 8 A box contains 10 electric batteries, 4 of which are defective. If 4 of the batteries are chosen at random from the box, what is the probability that the chosen group contains: (a) 2 defective batteries; (b) at least 2 defective batteries?

Solution The number of points in the sample space is the number of ways of choosing 4 batteries from among 10, that is,

$$n = \binom{10}{4} = \frac{10 \cdot 9 \cdot 8 \cdot 7}{1 \cdot 2 \cdot 3 \cdot 4} = 210.$$

(a) If the selection is to contain 2 defective batteries, it must contain 2 good ones. The number of ways of choosing 2 defective batteries from among the 4 available is $\binom{4}{2} = 6$ ways. The number of ways of choosing the 2 good batteries from among the 6 available is

$$\binom{6}{2} = \frac{6 \cdot 5}{1 \cdot 2} = 15.$$

Therefore the total number of ways of choosing 2 defective and 2 good batteries is $6 \cdot 15 = 90$.

Thus, for this event, the number of sample points is $k = 90$ and so the probability is $k/n = 90/210 = 3/7$.

Answer

(a)
$$\binom{13}{5} \div \binom{52}{5} = \frac{33}{66,640} \approx 0.0005;$$

(b) $4 \cdot \binom{13}{5} \div \binom{52}{5} = \frac{132}{66640}$
$$\approx 0.002;$$

(c) $\binom{4}{2} \cdot \binom{4}{3} \div \binom{52}{5} = \frac{1}{108,290}$
$$\approx 0.000009.$$

(b) If the selection is to contain at least 2 defective batteries, it can have 2 defective and 2 good batteries, 3 defective and 1 good batteries, or 4 defective and 0 good batteries. The first of these selections has a probability 90/210. (See part (a).) Three defective batteries can be chosen in $\binom{4}{3} = 4$ ways and 1 good battery can be chosen in $\binom{6}{1} = 6$ ways. So a group of 3 defective batteries and 1 good one can be chosen in $4 \cdot 6 = 24$ different ways. Hence the probability of such a selection is 24/210.

For the third alternative, there is just 1 way of choosing 4 defective batteries. So the probability of such a choice is 1/210.

The required event is the union of these three events, which are mutually exclusive. The probability of at least 2 defective batteries is therefore the sum of these three probabilities:

☛ 28. Re-solve Example 8 if 6 of the batteries in the box are defective.

$$\frac{90}{210} + \frac{24}{210} + \frac{1}{210} = \frac{23}{42}. \quad ☛ 28$$

(b′) An alternative way of solving part (b) is to consider the complementary event. We have

$$P \text{ (At Least 2 Defective)} = 1 - P \text{ (At Most 1 Defective)}$$

$$= 1 - [P(1 \text{ Defective}) + P(0 \text{ Defective})].$$

Proceeding as above, you can see that

$$P \text{ (1 Defective)} = \frac{1}{210}\binom{4}{1}\binom{6}{3} = \frac{16}{42},$$

$$P \text{ (0 Defective)} = \frac{1}{210}\binom{4}{0}\binom{6}{4} = \frac{3}{42}$$

and so

$$P \text{ (At Least 2 Defective)} = 1 - \left(\frac{16}{42} + \frac{3}{42}\right) = \frac{23}{42}.$$

EXAMPLE 9 (*Location of Office Typists*) An office manager must locate 16 typists into three offices, which hold 8, 5, and 3 typists, respectively. In how many ways can the three groups be chosen to occupy the three offices?

Solution Let us suppose that the manager first selects the 8 typists for the largest office. This choice can be made in $\binom{16}{8}$ ways. Having made this choice, there remain 8 typists from whom to select 5 for the second largest office. This choice can be made in $\binom{8}{5}$ ways. After this, there is no choice left: the remaining 3 typists automatically occupy the remaining office. Thus the number of ways of making the complete choice is given by the product:

Answer (a) $\frac{3}{7}$; (b) $\frac{37}{42}$.

$$\binom{16}{8}\binom{8}{5} = \frac{16!}{8! \, 8!} \cdot \frac{8!}{5! \, 3!} = \frac{16!}{8! \, 5! \, 3!}.$$

You should verify that this same result is obtained if the offices are filled in a different order.

Note The result in this example is an illustration of a general formula. The number of ways in which n different objects can be partitioned into k groups with r_1 objects in the first group, r_2 in the second, . . . , r_k in the kth (where $r_1 + r_2 + \cdots + r_k = n$) is given by

$$\frac{n!}{r_1! \, r_2! \cdots r_k!}.$$

The formula for $\binom{n}{r}$ is a special case of this with $k = 2$.

EXERCISES 8-5

(1–16) Evaluate the following.

1. $_{10}P_2$ **2.** $_6P_4$

3. $_5P_5$ **4.** $_8P_3$

5. $_nP_2$ **6.** $_7P_0$

7. $\binom{10}{2}$ **8.** $\binom{8}{4}$

9. $\binom{15}{0}$ **10.** $\binom{23}{23}$

11. $\binom{50}{48}$ **12.** $\binom{20}{17}$

13. $\dfrac{20!}{18!}$ **14.** $\dfrac{10!}{7!}$

15. $\dfrac{8!}{7! + 6!}$ **16.** $\dfrac{9!}{8! - 7!}$

(17–20) Find the value of n in each case.

17. $(n + 1)! = 20n!$ **18.** $(n + 2)! = 42n!$

19. $_nP_2 = 56$ **20.** $\binom{n}{2} = 10$

21. Three cards are selected one at a time, with replacement, from a deck of 52 cards. What is the probability of selecting an ace, a king, and a queen in that order?

22. Three persons are selected from a group of 5 men and 4 women to fill 3 executive positions in a business firm. Find each probability.

   a. All the selected persons are men.

   b. The selected group consists of 2 men and 1 woman.

   c. The selected group contains more women than men.

23. Two machine parts are randomly selected from a bin containing 10 good parts and 5 defective parts. Find each probability.

   a. Both selected parts are defective.

   b. One good and one defective part are selected.

24. Three balls are selected from a box that contains 5 white, 6 red, and 4 yellow balls. Find each probability.

   a. All 3 balls are red.

   b. The 3 balls are different colors.

   c. The 3 balls are the same color.

25. *(Test-panel Selection)* A sample of 6 persons is selected for a test from a group containing 20 smokers and 10 nonsmokers. What is the probability that the sample contains 4 smokers?

26. *(Play-off Series)* Two teams play off a final series of 3 matches in a sports competition. The first team to win 2 matches wins the competition. How many different outcomes are there for the series? (*Hint:* Use a tree diagram.)

27. *(Play-off Series)* Repeat Exercise 26 for the case when the series consists of 5 matches and the first team to win 3 matches wins the competition.

28. *(Resource Allocation)* A taxicab company has 5 cabs and 4 drivers available when requests for 3 cabs are received. How many different ways are there of selecting the cabs and the drivers to meet these requests? (Assume that any driver can be given any cab to drive.)

29. *(Resource Allocation)* Repeat Exercise 28 for the case when a single request for 3 cabs is received, rather than 3 separate requests.

30. *(Theatre Seating)* Four men and 4 women attend the theater and sit in 8 adjacent seats. How many different seating arrangements are there if men and women alternate along the row?

*31. *(Seating Arrangements)* Four men and 4 women sit around a circular dining table. If men and women alternate around the table, how many different orderings are there? (Two orderings are the same if everyone has the same two neighbors in each.)

*32. *(Seating Arrangements)* Repeat Exercise 31 for the case when the dinner party consists of 4 married couples and no husband and wife sit next to one another.

33. *(Astronaut Selection)* Out of a team of 12 astronauts, 7 have been on a space flight and 5 have not. If 4 of the team are selected at random for a project, what is the probability that at least 2 of those selected have already had experience in space?

34. *(Group Selection)* Six people are chosen from a group consisting of 4 men, 3 women, and 6 children. What is the probability that the selected group contains:

   a. 2 men?

   b. 3 women and 2 children?

35. *(Examination Answers)* A test consists of 10 true-false questions. In how many ways can a student fill in the answer sheet?

36. *(Multiple-choice Examination)* A driving examination contains 20 multiple-choice questions with 3 answers given for each. If an applicant fills out the answer sheet entirely by guesswork, what is the probability that he or she will pass the examination? (At least 19 correct answers are necessary to pass.)

37. *(Monkey Business)* If a monkey playing with a typewriter types out 11 letters at random, what is the probability that it will spell out the word *Shakespeare?*

38. *(Government Contracts)* A government agency must distribute 10 orders among 3 different firms A, B, and C such that A receives 3 orders, B receives 2 orders, and C receives 5 orders. In how many ways can the orders be distributed?

39. *(Government Contracts)* In Exercise 38, 3 of the orders are highly sought after. What is the probability that firm C receives at least two of these orders?

*40. *(Poker Hand)* Find the probability that a poker hand of 5 cards contains a full house (2 cards of one denomination and 3 of another).

41. Three rats are selected from a group of 5 white and 4 brown rats for a certain experiment. What is the probability that:

   a. All the selected rats are white?

   b. The selected rats are 2 white and 1 brown?

   c. All the selected rats are brown?

42. Two flower seeds are randomly selected from a bag containing 10 seeds for red flowers and 5 seeds for white flowers. What is the probability that:

   a. Both result in white flowers?

   b. One of each color is selected?

43. Three flower seeds are selected from a package that contains 5 seeds for white, 6 seeds for red, and 4 for yellow flowers. What is the probability that:

   a. All 3 seeds produce red flowers?

   b. The 3 seeds produce flowers of different colors?

   c. The 3 seeds produce flowers of the same color?

8-6 BINOMINAL PROBABILITIES

In this section we shall be concerned with an experiment that consists of a number of repetitions of a certain basic operation, for example, the experiment of tossing a coin ten times. The individual operations will be called *trials*.

We assume that each trial has two mutually exclusive outcomes that we term success (S) and failure (F). For example, with tossing a coin, we could label getting a head a success and getting a tail a failure (or the other way round, it would make no difference to the results). We assume that the trials are independent; that is, the result of any trial does not depend on the result of any previous trial. Also, we assume that the probabilities of success and failure remain the same from one trial to the next. With these assumptions, the trials are called **Bernoulli trials** after the eighteenth-century Swiss mathematician Jacob Bernoulli.

We denote by p the probability of a success on any trial and by $q = 1 - p$ the probability of failure. The kind of question we are interested in asking is: What is the probability of getting r successes in a sequence of n trials? For example, for tossing a coin, $p = q = 1/2$ and we may be interested in the probability of getting 5 heads, or more than 5 heads, in ten tosses.

As an example, consider an insurance company that issues a policy that pays a certain sum if the insured person dies before his or her 65th birthday but pays nothing if the person lives beyond 65. Then each trial consists of an insured individual and we can define success on any trial to be that the individual lives beyond 65 and failure to be that the person dies before age 65. (From the company's point of view, success and failure are really the other way round, but this seems a little too macabre.) Then p is the probability of any individual living and q is the probability of not living to age 65.

Now let us consider a group of three such persons and let us calculate the probabilities that all three live past 65, that two do so, that only one does and that all three die before 65. The set of outcomes can be listed as follows:

$$SSS, \quad SSF, \quad SFS, \quad FSS, \quad SFF, \quad FSF, \quad FFS, \quad FFF.$$

Here for example, SSF means that the first two people live past 65 (successes) and the third dies before 65 (failure).

The outcome SSS, in which all three survive past 65, has a probability $p \cdot p \cdot p = p^3$, since each Bernoulli trial has a probability p of resulting in a success.

Similarly, the outcome SSF has probability $p \cdot p \cdot q = p^2q$, since the two successes each have probability p and the failure has probability q. In the same way, the other two outcomes SFS and FSS, in which there are two successes and one failure, also have probabilities p^2q. Therefore the total probability of two people surviving and one dying before 65 is equal to the sum of the probabilities of the three outcomes SSF, SFS, and FSS, and hence is given by $3p^2q$.

Continuing in this way, we see that the probability of one person surviving and two dying before 65 is equal to $3pq^2$ and the probability of all three dying before 65 is equal to q^3.

EXAMPLE 1 Let a success be rolling a 6 with a die. Then $p = 1/6$ and $q = 5/6$. The probability of getting two 6's when rolling 3 dice is equal to

$$3p^2q = 3\left(\frac{1}{6}\right)^2\left(\frac{5}{6}\right) = \frac{5}{72}. \quad \text{☞ 29}$$

☞ **29.** In Example 1, what is the probability of getting one six?

Answer $\frac{25}{72}$.

These results are special cases of a general expression for the probability of obtaining r successes in a sequence of n Bernoulli trials ($0 \leq r \leq n$). The general result is given in the following theorem.

THEOREM 1 If p is the probability of success and q the probability of failure in a single Bernoulli trial, then the probability of exactly r successes in a sequence of n independent trials is

$$P(r) = \binom{n}{r} p^r q^{n-r}.$$

EXAMPLE 2 A fair coin is tossed 5 times. What is the probability of getting 3 heads?

Solution Tossing a coin is a Bernoulli trial, because there are two mutually exclusive outcomes, heads (success) and tails (failure). The probability of getting a head in a single trial is $p = 1/2$. Thus $q = 1 - p = 1/2$. Since the coin is tossed 5 times, we have $n = 5$ trials.

In this case $r = 3$, because we want 3 heads (successes). Using Theorem 1, the required probability of 3 heads is

$$P(3) = \binom{n}{r} p^r q^{n-r} = \binom{5}{3} \left(\frac{1}{2}\right)^3 \left(\frac{1}{2}\right)^{5-3} = \frac{5 \cdot 4 \cdot 3}{3 \cdot 2 \cdot 1} \cdot \left(\frac{1}{2}\right)^5 = 10 \cdot \frac{1}{32} = \frac{5}{16}.$$

☛ 30

☛ **30.** In Example 2, what is the probability of getting (a) more than 3 heads in 5 tosses;   (b) 3 heads in 6 tosses?

EXAMPLE 3 *(Airplane Safety)* An airplane with 2 or 4 engines can remain airborne as long as half its engines are functioning. If the probability of any engine breaking down is 10^{-3}, calculate the probability of engine failure causing a crash for each of these types of airplanes.

Solution Let success occur if any engine does not break down. Then the probability of failure, q, is $q = 10^{-3} = 0.001$, and $p = 1 - q = 0.999$. For the two-engine airplane, we have 2 Bernoulli trials (each engine can fail or it cannot fail) and the plane crashes if there are 2 failures. Therefore, P(crashing) $= P$ (2 failures) $= q^2 = 10^{-6}$.

For the four-engine airplane, there are 4 Bernoulli trials and the plane crashes if there are 3 or 4 failures. Therefore

$$P(\text{crashing}) = P(3 \text{ failures}) + P(4 \text{ failures})$$

$$= \binom{4}{3} pq^3 + \binom{4}{4} q^4 = 4pq^3 + q^4$$

$$= 4(0.999)(10^{-3})^3 + (10^{-3})^4$$

$$= 3.997 \times 10^{-9}.$$

(A comparison of these probabilities shows that the four-engine design is much safer from the danger of engine failure than is the two-engine design.)

Answer (a) $P(4) + P(5) = \frac{3}{16}$;
(b) $\frac{5}{16}$.

EXAMPLE 4 *(Opinion Poll)* Just before an election, 40% of the population support the Square Party, 40% support the Round Party, and 20% are undecided. If 6 people are chosen at random from the population and their opinions are sampled, what is the probability that at least half of them will express support for the Squares?

Solution Let a success be the expression of support for the Squares. Then its probability, p, is $p = 0.4$. Correspondingly, $q = 1 - p = 0.6$. There are 6 Bernoulli trials and we want to find the probability of at least 3 successes.

$$P(r \geq 3) = P(3) + P(4) + P(5) + P(6)$$

$$= \binom{6}{3}(0.4)^3(0.6)^3 + \binom{6}{4}(0.4)^4(0.6)^2 + \binom{6}{5}(0.4)^5(0.6) + \binom{6}{6}(0.4)^6(0.6)^0$$

$$= 20(0.01382) + 15(0.00922) + 6(0.00614) + 1(0.00410) = 0.4557$$

As an alternative solution, the event that $r \geq 3$ is complementary to the event that $r \leq 2$. Therefore

$$P(r \geq 3) = 1 - [P(0) + P(1) + P(2)]$$

$$= 1 - \left[\binom{6}{0}(0.4)^0(0.6)^6 + \binom{6}{1}(0.4)(0.6)^5 + \binom{6}{2}(0.4)^2(0.6)^4 \right]$$

$$= 1 - [0.04666 + 6(0.03110) + 15(0.02074)]$$

$$= 1 - 0.5443 = 0.4557. \quad \text{☛ 31}$$

☛ **31.** In Example 4 if the opinions of only four people are sampled, find the probability that (a) 2 of them will express support for the squares; (b) at least 2 of them will do so.

This example illustrates the principle of opinion sampling, commonly carried out prior to elections by numerous newspapers, opinion survey organizations, and so on. Of course, the samples used in such surveys are much bigger than six, typically being about 2000 in size. With such large samples, the use of binomial probabilities becomes quite cumbersome. Suppose, for example, that 2000 people are asked their opinions in Example 4 and we want to calculate the probability that at least half of them express support for the Square party. This is the probability of 1000, 1001, 1002, on up to 2000 successes and is given by the sum of all the probabilities

$$\binom{2000}{r}(0.4)^r(0.6)^{2000-r}$$

with r ranging from 1000 up to 2000. Clearly finding this sum involves a prodigious amount of work (although a modern computer could do it fairly rapidly). We should mention, however, that it is possible to calculate such probabilities as this one relatively easily using the *normal probability distribution*.

Finally, we should mention that the probabilities given in Theorem 1 are called **binomial probabilities.** The reason for this name is connected with the binomial theorem, which we shall discuss in the following section.

Answer (a) 0.3456; (b) 0.5248.

(1–2) A fair coin is tossed four times. What is the probability of getting:

1. One head? **2.** At least one head?

(3–4) A fair die is rolled 5 times. Find the probability of rolling 4 spots each of the following number of times.

3. Exactly 2 times. **4.** At least 2 times.

(5–7) The probability that a couple will have a left-handed child is $\frac{1}{5}$.

5. If the couple has 5 children, what is the probability that exactly 2 are left-handed?

6. If the couple has 4 children, what is the probability that exactly 2 are right-handed?

7. If the couple has 6 children, what is the probability that at least 1 is left-handed?

(8–9) *(Television Viewers)* On the average, 25% of the adult population watch a certain television show. A group of 8 people are chosen at random and independently of one another.

8. What is the probability that exactly two of them watched the show the last time it was on television?

9. What is the probability that more than two of them watched the show?

10. *(Disease Fatalities)* Suppose that 40% of the patients diagnosed as having a certain disease die from it. What is the probability that exactly 1 will die from a group of 4 who have this disease?

11. *(Survival of Rats)* Six rats are administered a dose of poison and the number of rats dying within 24 hours is observed. Suppose that each rat has a probability $\frac{1}{4}$ of dying and that the survival of each rat is independent of the survival of the other rats. What is the probability that:

a. Four rats die?

b. All the rats die?

12. *(Survival of Plants)* Of the trees planted by a landscaping firm, 85% survive. What is the probability that 8 or more out of a group of 10 trees planted will survive?

13. *(Multiple-choice Examination)* A multiple-choice examination contains 10 true-false questions. If a student fills in the answers by random guessing, what is the probability that 5 answers will be correct?

14. *(Multiple-choice Examination)* In Exercise 13, find the probability that the student gets 4 answers correct by guessing.

15. *(Multiple-choice Examination)* A multiple-choice examination contains 10 questions. Five answers are given to each question, only one of which is correct. If a student fills in the answers by random guessing, what is the probability that half of them will be correct?

16. *(Multiple-choice Examination)* In Exercise 15, find the probability that the student gets 4 correct answers.

17. *(Typist Errors)* A typist makes at least one error on average in every fifth letter typed. If eight letters are typed in one afternoon, what is the probability that:

a. None of them has an error?

b. At least two of them have an error?

18. *(Encyclopedia Sales)* An encyclopedia salesperson makes a sale one time out of three once she is admitted into a home. Over one weekend, she is admitted to 6 homes. What is the probability during this weekend that she makes:

a. 1 sale?

b. 2 sales?

c. 3 sales?

19. *(Boat Charters)* A boat charter firm has 8 boats that it hires out by the day. It supplies skippers for those clients who want them; otherwise it charters only the boat. On the basis of long experience, it is known that 1 client in every 5 will want a skipper. If the firm has 3 skippers available, on what fraction of the days will they be unable to meet the demand for skippers? (Assume that all the boats are chartered on each day.)

20. *(Bank Loan Delinquency)* A bank manager knows from experience that, on the average, 10% of the loan customers fall behind in their payments. One day the manager authorizes 7 loans. What is the probability that:

a. None of these 7 borrowers will fall behind in their payments?

b. One of them will fall behind?

c. At least two of them will fall behind?

21. *(Production Inspection)* A defective canning machine seals the lids on the cans incorrectly on 1 can out of 6. If an inspector chooses 2 cans for inspection at random from the output of the machine, what is the probability that the defect will remain unnoticed? If 4 cans are chosen, what is the probability that 2 or more will be found to have defective lids?

22. *(Fan Cards)* A cereal manufacturer states that every box of cereal contains at least 1 card with a photograph of a hockey player and two-thirds of the boxes contain 2 such cards. If Jimmy's mother buys 8 boxes of cereal one week, what is the probability that Jimmy will get from these boxes:

   a. Exactly 12 cards?   **b.** At least 12 cards?

23. *(Opinion Poll)* Just before an election for mayor, 60% of the voters prefer Smith and 40% prefer Jones. If 10 people are chosen at random and asked their preference, find the probability that each of the following numbers of them express a preference for Smith.

   a. 6   **b.** 7   **c.** 5

24. *(Opinion Poll)* In Exercise 23, suppose that 100 voters are asked their opinion and let $P(r)$ denote the probability that exactly r of them express a preference for Smith. Calculate:

   a. $P(60)/P(59)$ **b.** $P(59)/P(58)$

   c. $P(60)/P(61)$ **d.** $P(61)/P(62)$

What conclusion can you draw regarding the outcome $r = 60$?

25. *(Color Blindness)* In a certain population the probability of an individual having color blindness is 0.02. What is the probability that exactly 2 individuals will be color blind out of a group of 10 persons selected from this population?

26. *(Sex Distribution)* Assuming the birth of boys and girls to be equally likely, what proportion of families with exactly 4 children should be expected to have 2 boys and 2 girls?

27. *(Pollen Grains)* The examination of fossilized pollen grains found in the various layers of lake sediment is used to provide information on the type of vegetation that surrounded the lake at the time when the particular layer of sediment was formed. The proportion of pollen grains in the sediment that derive from fir trees of one species or another is 0.6. If 10 grains are examined, what are the probabilities that: (a) 6, (b) 7, and (c) 5 of the grains turn out to be fir pollen?

28. *(Eye Color)* It is known that 60% of the offspring of a certain species of dog are black-eyed. The eye color of one offspring is not related to that of another. What is the probability that there are at least one-third black-eyed pups in a litter of 9?

29. *(Medicine)* There is a 75% probability that penicillin will cure a certain bacterial infection. During a small epidemic a physician treats 8 people with this antibiotic. What is the probability that at least 6 of them will be cured?

◣ 8-7 THE BINOMINAL THEOREM (OPTIONAL)

An expression consisting of only two terms is called a **binomial expression.** For instance, $2x + 3y$, $x^2 - 4a$, and $3a + 5/b$ are all binomial expressions. By actual multiplication, we can show that

$$(x + y)^1 = x + y$$
$$(x + y)^2 = x^2 + 2xy + y^2$$
$$(x + y)^3 = x^3 + 3x^2y + 3xy^2 + y^3 \qquad (1)$$
$$(x + y)^4 = x^4 + 4x^3y + 6x^2y^2 + 4xy^3 + y^4$$
$$(x + y)^5 = x^5 + 5x^4y + 10x^3y^2 + 10x^2y^3 + 5xy^4 + y^5.$$

The expression on the right side is called the **expansion** of the binomial power on the left side. Now, suppose we are interested in the expansion of $(x + y)^{20}$.

To obtain this expansion by actual multiplication of $(x + y)$ by itself 20 times would involve lengthy and tedious calculations. Thus, it would be useful if we had a general formula for the expansion of $(x + y)^n$, where n is any positive integer. The development of such a formula is the purpose of this section.

Let us first make the following observations from the expansions of $(x + y)^n$ for $n = 1, 2, 3, 4,$ and 5 listed above in (1).

1. The number of terms in the expansion on the right is one more than the exponent n. For example, the expansion of $(x + y)^2$ contains 3 terms, $(x + y)^3$ contains 4 terms, and so on. In general, the expansion of $(x + y)^n$ will consist of $(n + 1)$ terms.

2. The first term in the expansion of $(x + y)^n$ is x^n. For example, the first term in the expansion of $(x + y)^3$ is x^3, and so on. Similarly the last term in the expansion of $(x + y)^n$ is y^n.

3. As we move from one term to the next in the expansion, the exponent of x decreases by 1 while the exponent of y increases by 1. The sum of the powers of x and y in any term is equal to the exponent n.

In view of the above observations, we can say that apart from the numerical coefficients, the successive terms of $(x + y)^n$ will consist of

$$x^n, \quad x^{n-1} y, \quad x^{n-2} y^2, \quad x^{n-3} y^3, \quad \ldots, \quad xy^{n-1}, y^n. \qquad (2)$$

Let us now study the pattern of the coefficients of the various terms in the expansion of $(x + y)^n$ in (1). We read off the following coefficients.

BINOMIAL				COEFFICIENTS				
$(x + y)^0$					1			
$(x + y)^1$				1		1		
$(x + y)^2$			1		2		1	
$(x + y)^3$		1		3		3		1
$(x + y)^4$	1		4		6		4	1
$(x + y)^5$	1	5		10		10	5	1

The triangle of numbers formed in this manner is called *Pascal's triangle* after the French mathematician Blaise Pascal (1623–1662). It may be observed that the end numbers in any row are 1 and any number not on the end is the sum of the two closest numbers (one to the left and one to the right) in the preceding row. For example, the 3 in the fourth row is the sum of the 2 and 1 in the third row; and the 10 in the sixth row is the sum of the 6 and 4 in the fifth row, as shown.

This property of the coefficients enables us to write the expansion of $(x + y)^{n+1}$ whenever the expansion of $(x + y)^n$ is known.

EXAMPLE 1 Given the expansion of $(x + y)^5$ in (1), use the Pascal triangle method to write down the expansion of $(x + y)^6$.

Solution The coefficients of $(x + y)^5$ form the bottom row in the triangle printed above. The next row has a 1 at each end, and the numbers in the middle are obtained by adding pairs in the $(x + y)^5$ row:

$$
\begin{array}{cccccccccccc}
& 1 & & 5 & & 10 & & 10 & & 5 & & 1 \\
& \swarrow\searrow & & \swarrow\searrow & & \swarrow\searrow & & \swarrow\searrow & & \swarrow\searrow & & \swarrow\searrow \\
1 & & 1+5 & & 5+10 & & 10+10 & & 10+5 & & 5+1 & & 1 \\
& & =6 & & =15 & & =20 & & =15 & & =6
\end{array}
$$

These numbers are precisely the coefficients of successive terms in the expansion of $(x + y)^6$. Thus,

$$(x + y)^6 = x^6 + 6x^5y + 15x^4y^2 + 20x^3y^3 + 15x^2y^4 + 6xy^5 + y^6. \quad \text{☛ 32}$$

☛ **32.** Following Example 1, write down the expansions of $(x+ y)^7$ and $(x + y)^8$.

From Pascal's triangle, the coefficients of successive terms in the expansion of $(x + y)^4$ are

$$1 \quad 4 \quad 6 \quad 4 \quad 1$$

In combinatorial notation, these coefficients are

$$\binom{4}{0}, \quad \binom{4}{1}, \quad \binom{4}{2}, \quad \binom{4}{3}, \quad \binom{4}{4}$$

because

$$\binom{4}{0} = 1, \quad \binom{4}{1} = 4/1 = 4, \quad \binom{4}{2} = (4 \cdot 3)/(1 \cdot 2) = 6,$$

$$\binom{4}{3} = (4 \cdot 3 \cdot 2)/(3 \cdot 2 \cdot 1) = 4, \quad \text{and} \quad \binom{4}{4} = \frac{4 \cdot 3 \cdot 2 \cdot 1}{4 \cdot 3 \cdot 2 \cdot 1} = 1.$$

In the same manner, the coefficients 1, 5, 10, 10, 5, 1 in the expansion of $(x + y)^5$ are $\binom{5}{0}, \binom{5}{1}, \binom{5}{2}, \binom{5}{3} \binom{5}{4}, \binom{5}{5}$, respectively. These observations lead us to speculate that the coefficients of successive terms in the expansion of $(x + y)^n$, where n is any positive integer, are

$$\binom{n}{0}, \quad \binom{n}{1}, \quad \binom{n}{2}, \quad \binom{n}{3}, \quad \cdots, \quad \binom{n}{n-1}, \quad \binom{n}{n}.$$

This is in fact the case, and combining this sequence with list (2), we can write the expansion of $(x + y)^n$ as stated in the follwing theorem.

THEOREM 1 (THE BINOMIAL THEOREM) If n is a positive integer, then

$$(x + y)^n = \binom{n}{0}x^n + \binom{n}{1}x^{n-1}y + \binom{n}{2}x^{n-2}y^2 + \binom{n}{3}x^{n-3}y^3$$

$$+ \cdots + \binom{n}{n-1}xy^{n-1} + \binom{n}{n}y^n. \tag{3}$$

Answer $x^7 + 7x^6y + 21x^5y^2 + 35x^4y^3$
$+ 35x^3y^4 + 21x^2y^5 + 7xy^6 + y^7.$
$x^8 + 8x^7y + 28x^6y^2 + 56x^5y^3$
$+ 70x^4y^4 + 56x^3y^5 + 28x^2y^6$
$+ 8xy^7 + y^8.$

The right side of (3) is called the **binomial expansion** of $(x + y)^n$. The $(r + 1)$th term in this expansion is called the **general term** and is given by

$$T_{r+1} = \binom{n}{r} x^{n-r} y^r \qquad (4)$$

The coefficients $\binom{n}{0}, \binom{n}{1}, \binom{n}{2}, \cdots, \binom{n}{n}$ of the various terms in the expansion of $(x + y)^n$ are called the **binomial coefficients.**

The binomial expansion may be written in full as follows.

$$(x + y)^n = x^n + \frac{n}{1} x^{n-1} y + \frac{n(n-1)}{1 \cdot 2} x^{n-2} y^2 + \frac{n(n-1)(n-2)}{1 \cdot 2 \cdot 3} x^{n-3} y^3$$

$$+ \cdots + \frac{n(n-1)(n-2) \cdots (n-r+1)}{1 \cdot 2 \cdot 3 \cdot \ldots \cdot r} x^{n-r} y^r + \cdots + y^n.$$

EXAMPLE 2 Find the binomial expansion of $\left(2x + \dfrac{y}{2}\right)^6$.

Solution Using the binomial theorem (3) with $n = 6$ and x replaced by $2x$ and y by $y/2$, we have

$$\left(2x + \frac{y}{2}\right)^6 = \binom{6}{0}(2x)^6 + \binom{6}{1}(2x)^5\left(\frac{y}{2}\right) + \binom{6}{2}(2x)^4\left(\frac{y}{2}\right)^2 + \binom{6}{3}(2x)^3\left(\frac{y}{2}\right)^3$$

$$+ \binom{6}{4}(2x)^2\left(\frac{y}{2}\right)^4 + \binom{6}{5}(2x)\left(\frac{y}{2}\right)^5 + \binom{6}{6}\left(\frac{y}{2}\right)^6$$

$$= 1(64x^6) + \frac{6}{1}(32x^5)\left(\frac{y}{2}\right) + \frac{6 \cdot 5}{1 \cdot 2}(16x^4)\left(\frac{y^2}{4}\right)$$

$$+ \frac{6 \cdot 5 \cdot 4}{1 \cdot 2 \cdot 3}(8x^3)\left(\frac{y^3}{8}\right) + \frac{6 \cdot 5 \cdot 4 \cdot 3}{1 \cdot 2 \cdot 3 \cdot 4}(4x^2)\left(\frac{y^4}{16}\right)$$

$$+ \frac{6 \cdot 5 \cdot 4 \cdot 3 \cdot 2}{1 \cdot 2 \cdot 3 \cdot 4 \cdot 5}(2x)\left(\frac{y^5}{32}\right) + 1\left(\frac{y^6}{64}\right)$$

$$= 64x^6 + 96x^5y + 60x^4y^2 + 20x^3y^3 + \tfrac{15}{4}x^2y^4 + \tfrac{3}{8}xy^5 + \tfrac{1}{64}y^6.$$

EXAMPLE 3 Expand $(2x - 3)^5$.

Solution We can write $(2x - 3)^5 = [2x + (-3)]^5$. Thus, using the binomial theorem (3) with $n = 5$, x replaced by $2x$, and y by -3, we obtain

$$(2x - 3)^5 = \binom{5}{0}(2x)^5 + \binom{5}{1}(2x)^4(-3) + \binom{5}{2}(2x)^3(-3)^2$$

$$+ \binom{5}{3}(2x)^2(-3)^3 + \binom{5}{4}(2x)(-3)^4 + \binom{5}{5}(-3)^5$$

33. Expand (a) $(1 - 2x)^4$;
(b) $\left(x + \dfrac{3}{x^2}\right)^3$.

$= 1(32x^5) + 5(16x^4)(-3) + 10(8x^3)(9)$
$\quad + 10(4x^2)(-27) + 5(2x)(81) + 1(-243)$
$= 32x^5 - 240x^4 + 720x^3 - 1080x^2 + 810x - 243.$ ☛ **33**

Answer
(a) $1 - 8x + 24x^2 - 32x^2 + 16x^4$;
(b) $x^3 + 9 + \dfrac{27}{x^3} + \dfrac{27}{x^6}.$

EXAMPLE 4 Find the seventh term in the expansion of $[x + (2/x^2)]^9$.

Solution We know that the $(r + 1)$th term in $(x + y)^n$ is given by

$$T_{r+1} = \binom{n}{r}x^{n-r}y^r.$$

In this case we want $T_7 = T_{6+1}$, so that $r = 6$. Also y is replaced by $2/x^2$ and $n = 9$. Thus

34. In Example 4, what are the fifth and sixth terms?

$$T_7 = \binom{9}{6}x^{9-6}\left(\frac{2}{x^2}\right)^6 = 84x^3 \cdot \frac{64}{x^{12}} = 5376x^{-9}. \quad ☛ 34$$

The $(r + 1)$th term in the bionomial expansion of $(q + p)^n$ is equal to

$$\binom{n}{r}q^{n-r}p^r$$

Answer $2016x^{-3}$ and $4032x^{-6}$.

But this is the formula for the probability of r successes in n Bernoulli trials. This is the reason why such probabilities are called binomial probabilities.

EXERCISES 8-7

(1–12) Use the binomial theorem to expand the following, and simplify.

1. $(a + b)^7$

2. $(2x + y)^4$

3. $\left(2x + \dfrac{y}{2}\right)^5$

4. $\left(x^2 + \dfrac{1}{x}\right)^6$

5. $\left(\dfrac{2x}{3} + \dfrac{3}{2x}\right)^5$

6. $(x^2 + 3x)^4$

7. $(a - 2b)^6$

8. $\left(2a - \dfrac{b}{2a}\right)^7$

9. $(2p - 3q)^5$

10. $\left(x - \dfrac{1}{x}\right)^7$

11. $\left(p^2 - \dfrac{q}{p}\right)^5$

12. $(1 - x)^8$

(13–22) Find the indicated terms in simplified form in the given binomial expansions.

13. The seventh term in $\left(3x + \dfrac{y}{2}\right)^9$.

14. The eighth term in $\left(\dfrac{x}{2} - \dfrac{4}{x}\right)^9$.

15. The nth term in $\left(x - \dfrac{1}{x^2}\right)^{3n}$.

16. The pth term in $\left(x + \dfrac{1}{2x}\right)^{2p}$.

17. The fifth term from the end in $\left(\dfrac{x^3}{2} + \dfrac{2}{x^2}\right)^9$.

18. The fourth term from the end in $\left(\dfrac{4x}{3} - \dfrac{3}{2x}\right)^7$.

19. The middle term in $\left(x - \dfrac{1}{x}\right)^{10}$.

20. The middle term in $\left(\dfrac{3}{x} - 2x\right)^{12}$.

21. The two middle terms in $\left(3x - \dfrac{x^3}{6}\right)^9$.

22. The two middle terms in $\left(x + \dfrac{2}{x}\right)^7$.

*23. Show that the middle term in the expansion of $(1 + x)^{2n}$ is

$$\frac{1 \cdot 3 \cdot 5 \cdot 7 \cdot \cdots \cdot (2n - 1)}{n!} 2^n x^n$$

where n is a positive integer.

*24. If P is the sum of the odd terms and Q the sum of the even terms in the expansion of $(x + y)^n$, then prove that:

a. $P^2 - Q^2 = (x^2 - y^2)^n$.

b. $4PQ = (x + y)^{2n} - (x - y)^{2n}$.

c. $2(P^2 + Q^2) = (x + y)^{2n} + (x - y)^{2n}$.

*25. The first three terms in the expansion of $(1 + ax)^n$ are $1 + 4x + 7x^2$. Find n and a.

26. Evaluate $(2 + \sqrt{3})^5 + (2 - \sqrt{3})^5$.

27. Evaluate $(\sqrt{2} + 1)^6 - (\sqrt{2} - 1)^6$.

CHAPTER REVIEW

Key Terms, Symbols, and Concepts

8.1 Sample space, sample point (outcome). Venn diagram.
Event, impossible event, certain event.
Union and intersection of two events, $A \cup B$ and $A \cap B$.
Mutually exclusive events.
Complement E' of the event E.

8.2 Probability of an outcome.
Probability of an event, $P(E)$.
Probability formulas.
Probability table.

8.3 Conditional probability of E_1 given E_2, $P(E_1|E_2)$.
Independent events; multiplication rule.

8.4 Bayes' theorem.

8.5 Fundamental counting principle.
Permutation (ordered selection), $_nP_r$.
Combination (unordered selection), $\binom{n}{r}$.

8.6 Bernoulli trial.
Binomial probabilities.

8.7 Binomial expansion. Pascal's triangle.
Binomial coefficients.

Formulas

Equally likely outcomes:

$$P(E) = \frac{\text{Number of Sample Points in } E}{\text{Number of Sample Points in the Sample Space } S}.$$

$P(E') = 1 - P(E)$.

$P(E_1 \cup E_2) = P(E_1) + P(E_2) - P(E_1 \cap E_2)$.

For mutually exclusive events,

$$P(E_1 \cup E_2) = P(E_1) + P(E_2).$$

$P(E_1 \cap E_2) = P(E_1)P(E_2|E_1) = P(E_2)P(E_1|E_2)$.

$$P(E_2|E_1) = \frac{P(E_1 \cap E_2)}{P(E_1)},$$

$$P(E_1|E_2) = \frac{P(E_1 \cap E_2)}{P(E_2)}.$$

For independent events,

$$P(E_1 \cap E_2) = P(E_1)P(E_2).$$

Bayes' theorem for two events:

$$P(B_1|A) = \frac{P(A|B_1)P(B_1)}{P(A|B_1)P(B_1) + P(A|B_2)P(B_2)}.$$

$$_nP_r = \frac{n!}{(n - r)!}, \qquad _nP_n = n!.$$

$$\binom{n}{r} = \frac{n!}{(n - r)!r!} = \frac{n(n - 1) \cdots (n - r + 1)}{r(r - 1) \cdots 2 \cdot 1}.$$

$$\binom{n}{n} = \binom{n}{0} = 1, \qquad \binom{n}{r} = \binom{n}{n - r}.$$

Binomial probabilities: $P(r) = \binom{n}{r}p^r q^{n-r}$.

Binomial theorem:

$$(x + y)^n = \binom{n}{0}x^n + \binom{n}{1}x^{n-1}y + \binom{n}{2}x^{n-2}y^2 + \cdots$$

$$+ \binom{n}{n-1}xy^{n-1} + \binom{n}{n}y^n.$$

General term: $T_{r+1} = \binom{n}{r}x^{n-r}y^r$.

1. State whether each of the following statements is true or false. Replace each false statement with a true statement.

   a. If two events are mutually exclusive, then they are independent.

   b. The probability of any event is a nonnegative real number.

   c. If E_1 and E_2 are two independent events, then $P(E_1 \cap E_2) = P(E_1) + P(E_2)$.

   d. If E_1 and E_2 are two mutually exclusive events, then $P(E_7 \cap E_2) = 0$.

   e. The sample points in a sample space are equally likely outcomes.

   f. The probabilities of complementary events are always equal.

   g. If E_1 and E_2 are independent, they must be mutually exclusive.

   h. If A and B are mutually exclusive, then $P(A) + P(B) \le 1$.

   i. $P(A \mid A) = 1$

   j. $P(A \mid B) = P(B \mid A)$

   k. $P(\emptyset \mid A) = 0$

   l. If $P(B) = 1$, then $P(A \mid B) = P(A)$.

   m. $P(A \mid A') = 1$

   n. If $E_2 \subseteq E_1$, then $P(E_2 \mid E_1) = 1$.

   o. The probability of E_1 given E_2 is equal to one minus the probability of E_2 given E_1.

   p. If E_1 and E_2 are independent, then $P(E_1 \mid E_2) = P(E_1 \cap E_2)$.

   q. For any two events, $P(E_1 \cap E_2) + P(E_1' \cap E_2) = P(E_2)$.

   r. If E_1 and E_2 are independent events, then E_1' and E_2 are independent events.

   s. If A and B are independent events and if A and C are independent events, then B and C are independent events.

(2–13) Evaluate the following.

2. $_7P_4$

3. $_9P_2$

4. $_5P_0$

5. $\binom{8}{6}$

6. $_8P_3 \div \binom{8}{3}$

7. $\binom{10}{2} \div {}_{10}P_2$

8. $\binom{10}{7} \div \binom{9}{3}$

9. $\binom{15}{4} \div \binom{14}{3}$

10. $\binom{n}{r} \div \binom{n-1}{r-1}$

11. $_8P_5 \div {}_7P_4$

12. $_nP_r \div {}_{(n-1)}P_{(r-1)}$

13. $\binom{6}{3}\binom{5}{3} + \binom{6}{2}\binom{5}{4} + \binom{6}{1}\binom{5}{5}$

14. Show that $\binom{10}{6}\binom{6}{2} = \binom{10}{2}\binom{8}{4}$.

15. Show that $\binom{12}{5}\binom{7}{4} = \binom{12}{3}\binom{9}{5}$.

16. Show that $\binom{9}{3} + \binom{9}{4} = \binom{10}{4}$.

17. Of 10 girls in a class, 4 have blue eyes. If two of the girls are chosen at random, what is the probability that:

   a. Both have blue eyes?

   b. Neither has blue eyes;

   c. At least one has blue eyes?

18. Two people are selected at random from a group of 10 married couples. Find each probability.

   a. They are husband and wife.

   b. One is male and one is female.

19. (Smoking and Cancer) The probability that a person in a certain group is a smoker is 0.3, and the probability that a person will have cancer is 0.01. The probability that a person will be a smoker and have a cancer is 0.006. Are smoking and cancer disorders independent?

20. *(Weight and Blood Pressure)* Of the patients examined at the local clinic, 30% have high blood pressure, 35% have excessive weight, and 15% have both. What is the probability that a patient selected at random will have at least one of these characteristics? Are the events excessive weight and high blood pressure independent? Explain.

21. *(Smoking and Drinking)* In a certain community, 40% of the people smoke, 32% of the people drink, and 60% either smoke or drink. What percentage of the people smoke as well as drink?

22. *(Married-couple Survival)* The probability that a husband will live 10 more years is $\frac{1}{5}$ and the probability that his wife will live 10 more years is $\frac{1}{4}$. Find each probability.

   a. Both will live 10 more years.

   b. At least one of them will be alive for 10 more years.

   c. Only the wife will be alive for 10 more years.

23. Let E_1 be the event that a family has children of both sexes and E_2 be the event that a family has at most one boy. Show that the events E_1 and E_2 are:

   a. Independent if the family has 3 children.

   b. Not independent if the family has 2 children.

24. If the events E_1 and E_2 are independent, show that E_1 and E_2' are also independent. (*Hint:* Make a Venn diagram for $E_1 \cap E_2'$.)

25. A family has 8 children. Assuming that the probability that any child is a boy is $\frac{1}{2}$, find the probability that the family will have fewer girls than boys.

26. *(Product Defects)* A batch of 12 new high-grade tires is shipped to a store, which accepts the whole batch if a sample of 3 chosen at random from the batch has no defective tires. If the batch contains 3 defective tires, find the probability that it is accepted.

27. *(Customer Complaints)* Of a firm's customers, 12% subsequently make some kind of complaint about the quality of the service they have received. One salesman named Dick is responsible for one-third of the firm's customers, and 15% of his customers subsequently make a complaint. What proportion of the customers who make complaints are Dick's responsibility? What percentage of the customers who are not Dick's responsibility make complaints?

28. *(Bank Manager's Hours)* A bank manager arrives late for work 45% of the time and leaves early at the end of the day 50% of the time. She does both 15% of the time. What is the probability that she does not leave early on a day on which she arrives late?

29. *(Smoking and Longevity)* In a certain town, 60% of the adults are smokers and 40% nonsmokers. If each smoker has a probability of 25% of living to be 75 and each nonsmoker has a probability of 40%, what proportion of the town's residents aged 75 and over are nonsmokers?

30. *(Lottery)* A person has a probability p of winning a prize in a lottery. What is the probability that the person will win a prize on the first, second, or third try in the lottery? What is the probability of winning a prize sometime in the first n tries at the lottery? (*Hint:* Consider the event of failure n times in a row.)

31. Three dice are rolled. Let E_1 be the event that the scores total 15 or more and E_2 be the event that two 6's are rolled. Find $P(E_1|E_2)$ and $P(E_2|E_1)$.

*32. An event A is independent of each of the events B, C, and $B \cup C$. Prove that A is independent of $B \cap C$.

33. *(Job Assignment)* The coordinator of a cooperative education program has 36 available jobs for the next semester and 30 students in the program. In how many ways can the students be assigned jobs? What would be the answer if only 26 jobs were available?

34. *(Auto Registration)* In a certain country, automobile registration numbers consist of 3 letters followed by 3 digits, where the first digit is nonzero. What is the total number of available registration numbers?

35. *(Employee Lay-offs)* Because of a decline in business, 2 of 6 salespeople employed by a company are to be fired by putting the 6 names in a hat and randomly choosing 2 of them. If Ramon and Carole are two of the salespeople, what is the probability that neither of them will be fired? What is the probability that one and not the other will be fired?

36. *(Air Travel)* Six airlines provide direct flights between two cities. A person makes a round trip between these cities, choosing an airline at random for each leg of the journey (not necessarily flying both ways with the same airline). What is the probability that on at least one leg of the journey the person flies with XYZ Airline, which is one of the six?

37. *(Real Estate Sales)* A real estate salesperson fails to complete a sale one week out of every three, on the average.

What is the probability that she will fail to make a sale in 3 weeks out of the 4 weeks in one month?

38. *(Real Estate Sales)* In Exercise 37, find the probability that the salesperson fails to make sales in at most 2 weeks during a given 5-week period.

39. *(Card Game)* Two cards are dealt from a deck of 52 playing cards. One of the two is chosen at random and turns out to be an ace. What is the probability that the two cards were both aces?

40. *(Tax Audits)* In a certain country, 20% of the income tax returns are subjected to more that routine scrutiny. Of those scrutinized, it is found that 15% are fraudulent; it is estimated that 5% of the returns that are not scrutinized are also fraudulent. What is the probability that a random fraudulent return will be scrutinized?

41. *(Defective Supplies)* A company buys parts from three suppliers: 50% from supplier A, 30% from B, and 20% from C. Of those from A, 1% are defective, of those from B, 1.5%, and of those from C, 1.75%. What is the probability that a given defective part came from A?

42. *(Manufacturing Defects)* Three machines M_1, M_2, and M_3 produce similar engine parts. Of the total output, M_1 produces 45%, M_2 produces 30%, and M_3 produces 25%. Of the items produced by M_1, 8% are defective, as are 6% of those produced by M_2 and 3% of those by M_3. A part is selected at random and is found to be defective. Find the probability:

a. That is was produced by M_1.

b. That it was produced by either M_2 or M_3.

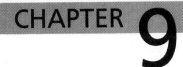

CHAPTER 9

Matrix Algebra

Chapter Objectives

9-1 MATRICES
(a) Definitions of a matrix, the elements, rows and columns, and the size of a matrix;
(b) The rule for equality of two matrices;
(c) The operations of multiplication of a matrix by a number and addition of two matrices;
(d) Some applications of matrices.

9-2 MULTIPLICATION OF MATRICES
(a) Definition of the product of a row matrix multiplied by a column matrix;
(b) Definition of the product of two general matrices, including the condition for such product to exist and the size of the product;
(c) The associative property and failure of the commutative property for matrix multiplication;
(d) The identity matrix and its multiplicative property;
(e) The matrix form of systems of linear equations.

9-3 SOLUTION OF LINEAR SYSTEMS BY ROW REDUCTION
(a) The augumented matrix of a linear system;
(b) The elementary row operations on an augmented matrix;
(c) Row reduction of an augmented matrix $\mathbf{A} \mid \mathbf{B}$ to the form $\mathbf{I} \mid \mathbf{C}$ by systematically reducing the columns from left to right using the elementary row operations.

9-4 SINGULAR SYSTEMS
(a) Examples of the failure of row reduction: systems with an infinite number of solutions;
(b) Inconsistency: systems with no solution;
(c) The general reduced form of an augmented matrix;
(d) Tests using the reduced form for consistency and uniqueness of solution.

CHAPTER REVIEW

A firm produces four products, A, B, C, and D. The manufacture of each product involves specified quantities of two raw materials, X and Y, and also fixed amounts of labor. Suppose the firm wants to compare the numbers of units of X and Y and of labor that are involved in the weekly output of these four products. Sample information for such a case is shown in Table 1. For example, the weekly output of A involves 250 units of X, 160 units of Y, and 80 units of labor.

TABLE 1

Product	A	B	C	D
Units of Material X	250	300	170	200
Units of Material Y	160	230	75	0
Units of Labor	80	85	120	100

Observe that the data in this table naturally form a rectangular array. If the headings are removed, we obtain the following rectangular array of numbers.

$$\begin{bmatrix} 250 & 300 & 170 & 200 \\ 160 & 230 & 75 & 0 \\ 80 & 85 & 120 & 100 \end{bmatrix}$$

This array is an example of a *matrix*.

In this example, it is clear that a rectangular array is a natural way in which to store the twelve given numbers. Each column of three numbers in the array relates to one of the products A, B, C, or D, while each row of four numbers applies to one of the inputs, X, Y, or labor. Thus the number 75 that is in the second row and the third column gives the number of units of the second input (Y) used in the weekly output of the third product (C). The number 80 in the third row and first column represents the number of units of the third input (labor) that is involved in the weekly output of the first product (A), and so on.

Many other sets of tabulated data naturally form rectangular arrays. We shall see later that a number of the calculations that we might wish to make with such data correspond to certain "matrix operations" that are defined in this and following sections.

DEFINITION A **matrix** (plural **matrices**), is a rectangular array of real numbers, which is enclosed in large brackets. Matrices are generally denoted by boldface capital letters such as **A**, **B**, or **C**.

Some examples of matrices are given next.

$$\mathbf{A} = \begin{bmatrix} 2 & -3 & 7 \\ 1 & 0 & 4 \end{bmatrix} \qquad \mathbf{B} = \begin{bmatrix} 3 & 4 & 5 & 6 \\ 7 & 8 & 9 & 1 \\ 5 & 4 & 3 & 2 \end{bmatrix} \qquad \mathbf{C} = \begin{bmatrix} 4 \\ 2 \\ 3 \\ 1 \end{bmatrix}$$

$$\mathbf{D} = \begin{bmatrix} 1 & 2 & 3 & 5 & 6 \end{bmatrix} \qquad \mathbf{E} = \begin{bmatrix} 3 \end{bmatrix}$$

The real numbers which form the array are called the **entries** or **elements** of the matrix. The elements in any horizontal line form a **row** and those in any vertical line form a **column** of the matrix. For example, matrix **B** (above) has three rows and four columns. The elements of the first row are 3, 4, 5, and 6 and those of the third column are 5, 9, and 3.

If a matrix has m rows and n columns, then it is said to be of **size** $m \times n$ (read m by n). Of the matrices given above, **A** is a 2×3 matrix, **B** is a 3×4 matrix, and **C** is a 4×1 matrix.

A matrix of size $1 \times n$ has only one row and a matrix of size $m \times 1$ has only one column. A matrix having only one row is often called a **row matrix** or a **row vector.** Similarly, a matrix having only one column is called a **column matrix** or a **column vector.** In the above examples, **D** is a row vector and **C** is a column vector.

It is often convenient to use a double-subscript notation for the elements of a matrix. In this notation, for example, a_{ij} denotes the element of the matrix **A** that is in the ith row and the jth column. Thus a_{24} denotes the entry in the second row and fourth column of **A**. If **A** is the 2×3 matrix

$$\mathbf{A} = \begin{bmatrix} 2 & -3 & 7 \\ 1 & 0 & 4 \end{bmatrix}$$

☞ **1.** What is the size of the matrix

$$\mathbf{A} = \begin{bmatrix} 0 & 3 \\ 9 & 5 \\ 1 & 7 \end{bmatrix}.$$ Give the elements a_{12}, a_{21}, a_{23}, and a_{32}.

then $a_{11} = 2$, $a_{12} = -3$, $a_{13} = 7$, $a_{21} = 1$, $a_{22} = 0$, and $a_{23} = 4$. ☞ 1

In general, if **A** is an $m \times n$ matrix, we can write the following.

$$\mathbf{A} = \begin{bmatrix} a_{11} & a_{12} & a_{13} & \cdots & a_{1n} \\ a_{21} & a_{22} & a_{23} & \cdots & a_{2n} \\ \cdot & \cdot & \cdot & & \cdot \\ \cdot & \cdot & \cdot & & \cdot \\ \cdot & \cdot & \cdot & & \cdot \\ a_{m1} & a_{m2} & a_{m3} & \cdots & a_{mn} \end{bmatrix}$$

The matrix **A** can be denoted by $[a_{ij}]$ when its size is understood. If the size is also to be specified, then we write $\mathbf{A} = [a_{ij}]_{m \times n}$.

If all the elements of a matrix are zero, we call the matrix a **zero matrix** and denote it by **0.** Thus the following is the zero matrix of size 2×3.

$$\mathbf{0} = \begin{bmatrix} 0 & 0 & 0 \\ 0 & 0 & 0 \end{bmatrix}$$

Answer 3×2. $a_{12} = 3$, $a_{21} = 9$; there is no element a_{23}, $a_{32} = 7$.

A matrix with the same number of rows as columns is called a **square matrix.** The following are examples of square matrices.

$$P = \begin{bmatrix} 1 & 2 \\ 3 & 4 \end{bmatrix}, \qquad Q = \begin{bmatrix} 2 & 1 & 3 \\ 4 & -2 & 1 \\ 3 & 0 & 2 \end{bmatrix}, \qquad R = [2]$$

DEFINITION Two matrices **A** and **B** are said to be **equal** if
  (i) they are of the same size, and
  (ii) their corresponding elements are equal.

For example, let

$$A = \begin{bmatrix} 2 & x & 3 \\ y & -1 & 4 \end{bmatrix} \quad \text{and} \quad B = \begin{bmatrix} a & 5 & 3 \\ 0 & b & 4 \end{bmatrix}.$$

Clearly, **A** and **B** are of the same size and $A = B$ if and only if $a = 2$, $x = 5$, $y = 0$, and $b = -1$.

Scalar Multiplication of a Matrix

Scalar multiplication of a matrix refers to the operation of multiplying the matrix by a real number. If $A = [a_{ij}]$ is an $m \times n$ matrix and c is any real number, then the product $c\mathbf{A}$ is an $m \times n$ matrix obtained by multiplying each element of **A** by the constant c. In other words, $c\mathbf{A} = [ca_{ij}]$. For example, if

$$A = \begin{bmatrix} 1 & 0 & -1 \\ 0 & -2 & 4 \end{bmatrix}$$

then

☛ **2.** Write down $-3\mathbf{A}$.

$$2A = 2\begin{bmatrix} 1 & 0 & -1 \\ 0 & -2 & 4 \end{bmatrix} = \begin{bmatrix} 2(1) & 2(0) & 2(-1) \\ 2(0) & 2(-2) & 2(4) \end{bmatrix} = \begin{bmatrix} 2 & 0 & -2 \\ 0 & -4 & 8 \end{bmatrix}. \qquad ☛ \ 2$$

EXAMPLE 1 A chain of electronics stores has two outlets in Seattle. In May, the sales of televisions, VCRs, and stereos at the two stores were given by the following matrix **A**:

$$\begin{array}{c} \\ \text{Store 1} \\ \text{Store 2} \end{array} \begin{array}{ccc} \text{TVs} & \text{VCRs} & \text{Stereos} \end{array} \\ \begin{bmatrix} 22 & 34 & 16 \\ 14 & 40 & 20 \end{bmatrix} \equiv A.$$

If management sets a sales goal for June of a 50% increase over the May sales, write down the matrix that represents the projected June sales.

Solution Each element in the matrix above must be increased by 50%, that is, multiplied by 1.5. The matrix for June is therefore 1.5**A**, or

$$1.5\begin{bmatrix} 22 & 34 & 16 \\ 14 & 40 & 20 \end{bmatrix} = \begin{bmatrix} 33 & 51 & 24 \\ 21 & 60 & 30 \end{bmatrix}.$$

Answer $\begin{bmatrix} -3 & 0 & 3 \\ 0 & 6 & -12 \end{bmatrix}$

Addition and Subtraction of Matrices

Two matrices **A** and **B** of the *same size* can be added (or subtracted) by adding (or subtracting) their corresponding elements. In other words, if $\mathbf{A} = [a_{ij}]$ and $\mathbf{B} = [b_{ij}]$ are two matrices of the same size, then $\mathbf{A} + \mathbf{B} = [a_{ij} + b_{ij}]$ and $\mathbf{A} - \mathbf{B} = [a_{ij} - b_{ij}]$. Thus

$$
\begin{bmatrix} 2 & 0 & -1 \\ 3 & 4 & 5 \\ 1 & -2 & 3 \end{bmatrix} + \begin{bmatrix} 3 & 1 & 2 \\ 2 & 0 & -3 \\ 3 & 2 & -4 \end{bmatrix} = \begin{bmatrix} 2+3 & 0+1 & -1+2 \\ 3+2 & 4+0 & 5+(-3) \\ 1+3 & -2+2 & 3+(-4) \end{bmatrix}
$$

$$
= \begin{bmatrix} 5 & 1 & 1 \\ 5 & 4 & 2 \\ 4 & 0 & -1 \end{bmatrix}
$$

and

$$
\begin{bmatrix} 2 & 0 & -1 \\ 3 & 4 & 5 \end{bmatrix} - \begin{bmatrix} 4 & 1 & -2 \\ 2 & 6 & 1 \end{bmatrix} = \begin{bmatrix} -2 & -1 & 1 \\ 1 & -2 & 4 \end{bmatrix}. \quad \text{☛ 3}
$$

☛ **3.** If $\mathbf{A} = \begin{bmatrix} 1 & -3 \\ 4 & -2 \end{bmatrix}$, and $\mathbf{B} = \begin{bmatrix} 5 & 2 \\ -4 & 5 \end{bmatrix}$, find $\mathbf{A} + \mathbf{B}$ and $2\mathbf{B} - \mathbf{A}$.

Answer $\mathbf{A} + \mathbf{B} = \begin{bmatrix} 6 & -1 \\ 0 & 3 \end{bmatrix}$ and $2\mathbf{B} - \mathbf{A} = \begin{bmatrix} 9 & 7 \\ -12 & 12 \end{bmatrix}$.

EXAMPLE 2 For the store chain in Example 1, the numbers of televisions, VCRs, and stereos in stock at the two stores at the beginning of May are given by the following matrix **B**:

$$
\mathbf{B} = \begin{bmatrix} 30 & 30 & 20 \\ 18 & 32 & 28 \end{bmatrix}.
$$

(Rows and columns have the same meanings as in Example 1. For example, 32 VCRs were in stock at store 2.) During May, deliveries were made at the stores according to the following matrix **C**:

$$
\mathbf{C} = \begin{bmatrix} 20 & 38 & 12 \\ 10 & 48 & 0 \end{bmatrix}.
$$

Find the matrix that represents the numbers of the three items in stock at the end of May.

Solution For each item at each store we have

Number at End of May =

Number at Beginning of May + Deliveries − Sales

So the matrix we want is given by $\mathbf{B} + \mathbf{C} - \mathbf{A}$, and is

$$
\begin{bmatrix} 30 & 30 & 20 \\ 18 & 32 & 28 \end{bmatrix} + \begin{bmatrix} 20 & 38 & 12 \\ 10 & 48 & 0 \end{bmatrix} - \begin{bmatrix} 22 & 34 & 16 \\ 14 & 40 & 20 \end{bmatrix}
$$

$$
= \begin{bmatrix} 30+20-22 & 30+38-34 & 20+12-16 \\ 18+10-14 & 32+48-40 & 28+0-20 \end{bmatrix}
$$

$$
= \begin{bmatrix} 28 & 34 & 16 \\ 14 & 40 & 8 \end{bmatrix}.
$$

☛ **4.** In Example 2, give the matrix that specifies the minimum deliveries that will be required in June if the sales goal in Example 1 is to be met.

Answer We calculate the sales goal matrix (1.5A) minus the matrix that contains the numbers in store at the end of May:

$$\begin{bmatrix} 33 & 51 & 24 \\ 21 & 60 & 30 \end{bmatrix} - \begin{bmatrix} 28 & 34 & 16 \\ 14 & 40 & 8 \end{bmatrix}$$
$$= \begin{bmatrix} 5 & 17 & 8 \\ 7 & 20 & 22 \end{bmatrix}.$$

☛ **5.** Find **Y** such that
Y + **B** = 2**A**.

Answer $\begin{bmatrix} 5 & 0 & 5 \\ 0 & -11 & 4 \end{bmatrix}$.

For example, store 1 held 28 televisions, 34 VCRs, and 16 stereos in stock at the end of May. ☛ 4

EXAMPLE 3 Given

$$\mathbf{A} = \begin{bmatrix} 3 & 1 & 4 \\ 2 & -3 & 5 \end{bmatrix} \quad \text{and} \quad \mathbf{B} = \begin{bmatrix} 1 & 2 & 3 \\ 4 & 5 & 6 \end{bmatrix}$$

determine the matrix **X** such that **X** + **A** = 2**B**.

Solution We have **X** + **A** = 2**B**, or **X** = 2**B** − **A**.

$$\mathbf{X} = 2\begin{bmatrix} 1 & 2 & 3 \\ 4 & 5 & 6 \end{bmatrix} - \begin{bmatrix} 3 & 1 & 4 \\ 2 & -3 & 5 \end{bmatrix}$$

$$= \begin{bmatrix} 2 & 4 & 6 \\ 8 & 10 & 12 \end{bmatrix} - \begin{bmatrix} 3 & 1 & 4 \\ 2 & -3 & 5 \end{bmatrix} = \begin{bmatrix} -1 & 3 & 2 \\ 6 & 13 & 7 \end{bmatrix} \quad ☛ 5$$

EXERCISES 9-1

1. Give the size of each matrix.

$$\mathbf{A} = \begin{bmatrix} 1 & 0 \\ 2 & 3 \end{bmatrix} \qquad \mathbf{B} = \begin{bmatrix} 2 & 3 & 1 \\ -1 & 2 & 3 \end{bmatrix}$$

$$\mathbf{C} = \begin{bmatrix} 3 \\ 1 \\ 2 \end{bmatrix} \qquad \mathbf{D} = \begin{bmatrix} 1 & 2 & 3 \\ 4 & 5 & 6 \\ 9 & 8 & 7 \end{bmatrix}$$

$$\mathbf{E} = \begin{bmatrix} 3 & 4 & 5 \\ 1 & 0 & 2 \end{bmatrix} \qquad \mathbf{F} = \begin{bmatrix} 2 & -1 \\ -1 & 1 \end{bmatrix}$$

$$\mathbf{G} = \begin{bmatrix} 4 & 1 & 3 \end{bmatrix} \qquad \mathbf{H} = \begin{bmatrix} 1 \end{bmatrix}$$

2. In Exercise 1, if $\mathbf{B} = [b_{ij}]$, find b_{12}, b_{21}, b_{22}, b_{23}, and b_{32}.

3. Give the 2×2 matrix $\mathbf{A} = [a_{ij}]$ for which $a_{ij} = i + j - 2$. (*Hint:* To calculate a_{21}, for example, put $i = 2$ and $j = 1$ in the formula: $a_{21} = 2 + 1 - 2 = 1$.)

4. Give the 3×2 matrix $\mathbf{B} = [b_{ij}]$ for which $b_{ij} = 2i + 3j - 4$.

5. Give an example of a 3×3 matrix $[c_{ij}]$ for which $c_{ij} = -c_{ji}$.

6. Give the 3×4 matrix $\mathbf{A} = [a_{ij}]$ for which

$$a_{ij} = \begin{cases} i + j & \text{if } i \neq j \\ 0 & \text{if } i = j \end{cases}.$$

(7–14) Perform the indicated operations and simplify.

7. $3\begin{bmatrix} 2 & 4 \\ 1 & 3 \end{bmatrix}$

8. $-2\begin{bmatrix} 1 & -2 & 3 \\ -2 & 1 & -4 \\ 3 & 0 & 2 \end{bmatrix}$

9. $\begin{bmatrix} 2 & 1 & 3 \\ -1 & 4 & 7 \end{bmatrix} + \begin{bmatrix} 0 & -1 & 2 \\ 1 & 2 & -8 \end{bmatrix}$

10. $\begin{bmatrix} 3 & 1 & 4 \\ -2 & 5 & -3 \\ 0 & -1 & 2 \end{bmatrix} - \begin{bmatrix} 1 & -2 & 5 \\ 2 & -1 & -4 \\ -3 & 2 & 1 \end{bmatrix}$

11. $2\begin{bmatrix} 1 & 2 \\ -1 & 3 \end{bmatrix} + 3\begin{bmatrix} -2 & 3 \\ 1 & 0 \end{bmatrix}$

12. $3\begin{bmatrix} 2 & 1 \\ -1 & 3 \\ 4 & 7 \end{bmatrix} - 2\begin{bmatrix} 1 & -2 \\ 2 & 3 \\ -3 & 0 \end{bmatrix}$

13. $2\begin{bmatrix} 1 & 2 & 3 \\ 2 & -1 & 0 \\ 4 & 5 & 6 \end{bmatrix} + 3\begin{bmatrix} 0 & -1 & 2 \\ 3 & 2 & -4 \\ -1 & 0 & 3 \end{bmatrix}$

14. $4\begin{bmatrix} 1 & 0 & -3 & 4 \\ 2 & -1 & 5 & 1 \\ 3 & 2 & 0 & -2 \end{bmatrix} - 5\begin{bmatrix} 2 & -1 & 2 & 3 \\ 1 & 0 & -3 & 4 \\ 3 & 1 & 0 & -5 \end{bmatrix}$

(15–24) Determine the values of the variables for which the following matrix equations are true.

15. $\begin{bmatrix} x & 2 \\ 3 & y \end{bmatrix} = \begin{bmatrix} 1 & 2 \\ 3 & 4 \end{bmatrix}$

16. $\begin{bmatrix} 3 & -1 \\ x & 0 \end{bmatrix} = \begin{bmatrix} y+2 & z \\ 4 & t-1 \end{bmatrix}$

17. $\begin{bmatrix} 4 & x & 3 \\ y & -1 & 2 \end{bmatrix} = \begin{bmatrix} y-1 & 2-x & 3 \\ 5 & z+1 & 2 \end{bmatrix}$

18. $\begin{bmatrix} x+2 & 5 & y-3 \\ 4 & z-6 & 7 \end{bmatrix} = \begin{bmatrix} 3 & t+1 & 2y-5 \\ 4 & 2 & z-1 \end{bmatrix}$

19. $\begin{bmatrix} 1 & -2 & x \\ y & 3 & 4 \\ 2 & z & 3 \end{bmatrix} = \begin{bmatrix} 1 & t & 6 \\ 5 & 3 & 4 \\ u & 2 & v \end{bmatrix}$

20. $\begin{bmatrix} x+1 & 2 & 3 \\ 4 & y-1 & 5 \\ u & -1 & z+2 \end{bmatrix} = \begin{bmatrix} 2x-1 & t+1 & 3 \\ v+1 & -3 & 5 \\ -4 & w-1 & 2z-1 \end{bmatrix}$

21. $\begin{bmatrix} x & 3 & 4 \\ 2 & -1 & y \\ 1 & z & -3 \end{bmatrix} + \begin{bmatrix} 1 & t & -1 \\ 3 & 4 & x \\ u & y & 2 \end{bmatrix} = \begin{bmatrix} 2 & 7 & v+1 \\ 5 & w-2 & 3 \\ 0 & 5 & -1 \end{bmatrix}$

22. $\begin{bmatrix} x+1 & -2 & 3 \\ 4 & 1 & z+2 \\ -1 & y & 2 \end{bmatrix} + 2\begin{bmatrix} 3 & -1 & 2 \\ 1 & 2 & -3 \\ 4 & -1 & 0 \end{bmatrix}$

$= \begin{bmatrix} 6 & u+2 & 7 \\ v+1 & 5 & -7 \\ 7 & 0 & w \end{bmatrix}$

23. $3\begin{bmatrix} x & 1 & -1 \\ 0 & -2 & 3 \\ 1 & y & 2 \end{bmatrix} + 2\begin{bmatrix} -2 & t & 0 \\ z & 1 & -1 \\ u & 2 & v \end{bmatrix}$

$= \begin{bmatrix} w-4 & 1 & -v \\ 4 & 2u & 2v+y \\ -1 & x+7 & 12 \end{bmatrix}$

24. $2\begin{bmatrix} 1 & x+1 & 0 \\ 0 & -2 & y-1 \\ z & 1 & 2 \end{bmatrix} - 3\begin{bmatrix} u & -1 & 2 \\ 1 & v+2 & 3 \\ 0 & -3 & 1 \end{bmatrix}$

$= \begin{bmatrix} 8 & 7 & 2v-2z \\ u+y & -7 & 1-7z \\ 4 & w+11 & t \end{bmatrix}$

25. *(Transportation Costs)* A company has plants at three locations, X, Y, and Z, and four warehouses at the locations A, B, C, and D. The cost (in dollars) of transporting each unit of its product from a plant to a warehouse is given by the following matrix.

To	X	Y	Z	←From
A	10	12	15	
B	13	10	12	
C	8	15	6	
D	16	9	10	

a. If the transportation costs are increased uniformly by $1 per unit, what is the new matrix?

b. If the transportation costs go up by 20%, write the new costs in matrix form.

26. *(Supply Costs)* A building contractor finds that the costs (in dollars) of purchasing and transporting specific units of concrete, wood, and steel from three different locations are given by the following matrices (one matrix for each location).

	Concrete	Wood	Steel	
$\mathbf{A} =$	20	35	25	Material Costs
	8	10	6	Transportation Costs
$\mathbf{B} =$	22	36	24	Material Costs
	9	9	8	Transportation Costs
$\mathbf{C} =$	18	32	26	Material Costs
	11	8	5	Transportation Costs

Write the matrix representing the total costs of material and transportation for one unit of concrete, wood, and steel from each of the three locations.

27. *(International Trade)* The trade between three countries I, II, and III during 1986 (in millions of United States dollars) is given by the matrix $\mathbf{A} = [a_{ij}]$, where a_{ij} represents the exports from the ith country to the jth country.

$$\mathbf{A} = \begin{bmatrix} 0 & 16 & 20 \\ 17 & 0 & 18 \\ 21 & 14 & 0 \end{bmatrix}$$

The trade between these three countries during the year 1987 (in millions of U.S. dollars) is given by matrix $\mathbf{B}$.

$$\mathbf{B} = \begin{bmatrix} 0 & 17 & 19 \\ 18 & 0 & 20 \\ 24 & 16 & 0 \end{bmatrix}$$

a. Write a matrix representing the total trade between the three countries for the 2-year period 1986 and 1987.

b. If in 1986 and 1987, 1 U.S. dollar was equal to 5 Hong Kong dollars, write the matrix representing the total trade for the 2 years in Hong Kong dollars.

28. *(Production Matrices)* A firm produces three sizes of recording tapes in two different qualities. The production (in thousands) at its California plant is given by the following matrix.

	Size 1	Size 2	Size 3
Quality 1	27	36	30
Quality 2	18	26	21

The production (in thousands) at its New York plant is given by this matrix:

	Size 1	Size 2	Size 3
Quality 1	32	40	35
Quality 2	25	38	30

a. Write a matrix that represents the total production of recording tapes at both plants.

b. The firm's management is planning to open a third plant at Chicago, which would have one and one-half times the capacity of its plant in California. Write the matrix representing the production at the Chicago plant.

c. What will be the total production of all three plants?

29. *(Production Matrices)* A shoe manufacturer makes black, white, and brown shoes for children, women, and men. The production capacity (in thousands of pairs) at the Seattle plant is given by the following matrix.

	Men's	Women's	Children's
Black	30	34	20
Brown	45	20	16
White	14	26	25

The production at the Denver plant is given by

	Men's	Women's	Children's
Black	35	30	26
Brown	52	25	18
White	23	24	32

a. Give the matrix representing the total production of each type of shoe at both plants.

b. If the production at Seattle is increased by 50% and that at Denver is increased by 25%, give the matrix representing the new total production of each type of shoe.

30. *(Ecology)* Several species in an ecosystem provide the food sources for one another. The element C_{ij} of a consumption matrix equals the number of units of species j consumed daily by an individual of species i. Construct the matrix (C_{ij}) for the following simple ecosystems consisting of just three species.

a. Each species consumes on the average 1 unit of each of the other two species.

b. Species 1 consumes one unit of species 2; species 2 consumes $\frac{1}{2}$ unit each of species 1 and 3; species 3 consumes 2 units of species 1.

c. Species 1 consumes 2 units of species 3; species 2 consumes 1 unit of species 1; species 3 does not consume any other species.

◥ 9-2 MULTIPLICATION OF MATRICES

Suppose a firm manufactures a product using different amounts of three inputs, P, Q, and R (raw materials or labor, for example). Let the number of units of these inputs used for each unit of the product be given by the following row matrix.

$$\begin{array}{ccc} \text{P} & \text{Q} & \text{R} \end{array}$$
$$\mathbf{A} = \begin{bmatrix} 3 & 2 & 4 \end{bmatrix}$$

Then let the cost per unit of each of the three inputs be given by the following column matrix.

$$\mathbf{B} = \begin{bmatrix} 10 \\ 8 \\ 6 \end{bmatrix} \begin{matrix} P \\ Q \\ R \end{matrix}$$

Then the total cost of the three inputs per unit of product is obtained by adding the costs of 3 units of P at a cost of 10 each, 2 units of Q at 8 each, and 4 units of R at 6 each:

$$3 \cdot 10 + 2 \cdot 8 + 4 \cdot 6 = 30 + 16 + 24 = 70$$

We refer to this number as the *product* of the row matrix **A** and the column matrix **B**, written **AB**. Observe that in forming **AB**, the first elements of **A** and **B** are multiplied together, the second elements are multiplied together, the third are multiplied together, and then these three products are added. This method of forming products applies to row and column matrices of any size.

DEFINITION Let **C** be a $1 \times n$ row matrix and **D** be an $n \times 1$ column matrix. Then the **product CD** is obtained by calculating the products of corresponding elements in **C** and **D** and then finding the sum of all n of these products.

EXAMPLE 1 Given the following matrices:

$$\mathbf{K} = [2 \quad 5] \qquad \mathbf{L} = [1 \quad -2 \quad -3 \quad 2]$$

$$\mathbf{M} = \begin{bmatrix} -3 \\ 2 \end{bmatrix} \qquad \mathbf{N} = \begin{bmatrix} 2 \\ 5 \\ -3 \\ 4 \end{bmatrix}$$

Then

$$\mathbf{KM} = 2(-3) + 5(2) = -6 + 10 = 4$$

and

$$\mathbf{LN} = 1(2) + (-2)5 + (-3)(-3) + 2(4) = 9. \quad \text{☛ 6}$$

Notes **1.** The row matrix is always written on the left and the column matrix on the right in such products (for example, **KM,** not **MK**).

2. The row and column matrices *must* have the same number of elements. In Example 1, the products **LM** and **KN** are not defined.

The method of forming products can be extended to matrices in general. Consider the following example. Suppose that a firm manufactures two products, I and II, by using different amounts of the three raw materials, P, Q, and R. Let the units of raw materials used for the two products be given by the following matrix.

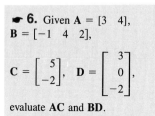

☛ 6. Given $\mathbf{A} = [3 \quad 4]$, $\mathbf{B} = [-1 \quad 4 \quad 2]$,

$$\mathbf{C} = \begin{bmatrix} 5 \\ -2 \end{bmatrix}, \quad \mathbf{D} = \begin{bmatrix} 3 \\ 0 \\ -2 \end{bmatrix},$$

evaluate **AC** and **BD**.

Answer **AC** = 7, **BD** = −7.

$$\begin{array}{ccc} & \text{P} \quad \text{Q} \quad \text{R} \\ \mathbf{A} = & \begin{bmatrix} 3 & 2 & 4 \\ 2 & 5 & 1 \end{bmatrix} & \begin{array}{l} \text{Product I} \\ \text{Product II} \end{array} \end{array}$$

Suppose the firm produces these two products at two plants, X and Y. Let the costs of the raw materials (per unit) at the two locations X and Y be given by matrix **B**.

$$\begin{array}{cc} & \text{X} \quad \text{Y} \\ \mathbf{B} = & \begin{bmatrix} 10 & 12 \\ 8 & 7 \\ 6 & 5 \end{bmatrix} \begin{array}{l} \text{P} \\ \text{Q} \\ \text{R} \end{array} \end{array}$$

The total cost of raw materials for each unit of product I produced at location X is

$$3(10) + 2(8) + 4(6) = 30 + 16 + 24 = 70$$

This is obtained by multiplying the elements of the first row in **A** by the corresponding elements of the first column in **B** and adding them.

Similarly, the total cost of raw materials for each unit of product I produced at plant Y is obtained by multiplying the elements of the first row in **A** by the elements of the *second* column in **B** and adding them.

$$3(12) + 2(7) + 4(5) = 36 + 14 + 20 = 70$$

The total cost of raw materials for each unit of product II produced at plant X is obtained by multiplying the elements of the second row in **A** by the elements in the first column in **B**.

$$2(10) + 5(8) + 1(6) = 20 + 40 + 6 = 66$$

Finally, the total cost of materials for each unit of product II produced at location Y is

$$2(12) + 5(7) + 1(5) = 24 + 35 + 5 = 64.$$

The total costs of raw materials for the two products produced at the two plants X and Y can be arranged in matrix form.

$$\begin{array}{cc} & \text{X} \quad \text{Y} \\ \mathbf{C} = & \begin{bmatrix} 70 & 70 \\ 66 & 64 \end{bmatrix} \begin{array}{l} \text{Product I} \\ \text{Product II} \end{array} \end{array}$$

We say that the matrix **C** is equal to the product **AB** of the original matrices **A** and **B**. This is written as **AB** = **C** or, in full,

$$\begin{bmatrix} 3 & 2 & 4 \\ 2 & 5 & 1 \end{bmatrix} \begin{bmatrix} 10 & 12 \\ 8 & 7 \\ 6 & 5 \end{bmatrix} = \begin{bmatrix} 70 & 70 \\ 66 & 64 \end{bmatrix}.$$

Observe that in forming the product matrix C, each row of A is multiplied by each column of B, just as a row matrix is multiplied by a column matrix. For example, the element c_{21} is obtained by multiplying the second row of A by the first column of B:

$$c_{21} = 2(10) + 5(8) + 1(6) = 66.$$

> In general, if $C = AB$, then *the ij*th *element of the product matrix* C *is obtained by multiplying the i*th *row of* A *by the j*th *column of* B.

In forming the product of two matrices, each row of the first matrix is multiplied in turn by each column of the second matrix. Note that such products can be formed only if the rows of the first matrix have the same number of elements as the columns of the second matrix. In other words, *the product* AB *of two matrices can only be formed if the number of columns in* A *is equal to the number of rows in* B. That is, if A is an $m \times n$ matrix and B a $q \times p$ matrix, then the product AB is defined only if $n = q$.

☛ **7.** Given $A = \begin{bmatrix} 1 & -3 \\ 4 & -2 \end{bmatrix}$, and $B = \begin{bmatrix} 2 \\ 5 \end{bmatrix}$, find AB.

DEFINITION If $A = [a_{ij}]$ is an $m \times n$ matrix and $B = [b_{ij}]$ is an $n \times p$ matrix, then the product AB is an $m \times p$ matrix $C = [c_{ij}]$, where the ijth element c_{ij} is obtained by multiplying the ith row of A and the jth column of B.

☛ 7

EXAMPLE 2 Let

$$A = \begin{bmatrix} 2 & 3 \\ 4 & 1 \end{bmatrix} \quad \text{and} \quad B = \begin{bmatrix} 3 & 1 & 0 \\ 2 & -3 & 4 \end{bmatrix}.$$

Find AB and BA if they exist.

Solution Here A is 2×2 and B is 2×3. Since the number of columns in A is equal to the number of rows in B, the product AB is defined. It is of size 2×3. If $C = AB$, then we can write C as follows.

$$C = \begin{bmatrix} c_{11} & c_{12} & c_{13} \\ c_{21} & c_{22} & c_{23} \end{bmatrix}$$

The element c_{ij} is found by multiplying the ith row of A and the jth column of B. For example, to obtain the element in the first row and second column, that is, c_{12}, we add the products of the elements in the first row of A and the elements in the second column of B.

Row 1 of A	Column 2 of B	Product
2	1	2
3	-3	-9
	Sum	$-7 = c_{12}$

Answer $AB = \begin{bmatrix} -13 \\ -2 \end{bmatrix}$

Thus, in full,

$$AB = \begin{bmatrix} 2 & 3 \\ 4 & 1 \end{bmatrix} \begin{bmatrix} 3 & 1 & 0 \\ 2 & -3 & 4 \end{bmatrix}$$

$$= \begin{bmatrix} 2(3) + 3(2) & 2(1) + 3(-3) & 2(0) + 3(4) \\ 4(3) + 1(2) & 4(1) + 1(-3) & 4(0) + 1(4) \end{bmatrix}$$

$$= \begin{bmatrix} 12 & -7 & 12 \\ 14 & 1 & 4 \end{bmatrix}.$$

In this case, the product **BA** is *not* defined because the number of columns in **B** is not equal to the number of rows in **A**. ☛ 8

8. If $A = \begin{bmatrix} 1 & -3 \\ 4 & -2 \end{bmatrix}$, and

$B = \begin{bmatrix} 5 & 2 \\ -4 & 5 \end{bmatrix}$, find **AB** and **BA**.

EXAMPLE 3 Given

$$A = \begin{bmatrix} 1 & 2 & 3 \\ 4 & 5 & 6 \\ 2 & 1 & 4 \end{bmatrix} \quad \text{and} \quad B = \begin{bmatrix} -2 & 1 & 2 \\ 3 & 2 & 1 \\ 1 & 3 & 2 \end{bmatrix}$$

find **AB** and **BA**.

Solution Here **A** and **B** are both of size 3×3. Thus **AB** and **BA** are both defined and both have size 3×3. We have the following.

$$AB = \begin{bmatrix} 1 & 2 & 3 \\ 4 & 5 & 6 \\ 2 & 1 & 4 \end{bmatrix} \begin{bmatrix} -2 & 1 & 2 \\ 3 & 2 & 1 \\ 1 & 3 & 2 \end{bmatrix}$$

$$= \begin{bmatrix} 1(-2) + 2(3) + 3(1) & 1(1) + 2(2) + 3(3) & 1(2) + 2(1) + 3(2) \\ 4(-2) + 5(3) + 6(1) & 4(1) + 5(2) + 6(3) & 4(2) + 5(1) + 6(2) \\ 2(-2) + 1(3) + 4(1) & 2(1) + 1(2) + 4(3) & 2(2) + 1(1) + 4(2) \end{bmatrix}$$

$$= \begin{bmatrix} 7 & 14 & 10 \\ 13 & 32 & 25 \\ 3 & 16 & 13 \end{bmatrix}$$

$$BA = \begin{bmatrix} -2 & 1 & 2 \\ 3 & 2 & 1 \\ 1 & 3 & 2 \end{bmatrix} \begin{bmatrix} 1 & 2 & 3 \\ 4 & 5 & 6 \\ 2 & 1 & 4 \end{bmatrix}$$

$$= \begin{bmatrix} -2(1) + 1(4) + 2(2) & -2(2) + 1(5) + 2(1) & -2(3) + 1(6) + 2(4) \\ 3(1) + 2(4) + 1(2) & 3(2) + 2(5) + 1(1) & 3(3) + 2(6) + 1(4) \\ 1(1) + 3(4) + 2(2) & 1(2) + 3(5) + 2(1) & 1(3) + 3(6) + 2(4) \end{bmatrix}$$

$$= \begin{bmatrix} 6 & 3 & 8 \\ 13 & 17 & 25 \\ 17 & 19 & 29 \end{bmatrix}$$

Answer $AB = \begin{bmatrix} 17 & -13 \\ 28 & -2 \end{bmatrix}$, and

$BA = \begin{bmatrix} 13 & -19 \\ 16 & 2 \end{bmatrix}.$

Clearly $AB \neq BA$, even though both products are defined.

It is clear from this example that matrix multiplication is not commutative. Even when the two products **AB** and **BA** are both defined for given matrices **A** and **B**, they are usually not equal to one another. (On the other hand, addition of matrices is commutative: **A** + **B** = **B** + **A**.) Matrix multiplication does, however, satisfy the associative property:

If **A**, **B**, and **C** are three matrices of sizes $m \times n$, $n \times p$, and $p \times q$, respectively, then the products **AB**, **BC**, (**AB**)**C** and **A**(**BC**) are all defined. The following property can be shown.

$$(\mathbf{AB})\mathbf{C} = \mathbf{A}(\mathbf{BC}) \qquad \textit{Associative Law}$$

In such products, we can therefore omit the brackets and write simply **ABC**. The product matrix **ABC** is of size $m \times q$. ☛ **9**

If $A = [a_{ij}]$ is a square matrix, then the elements a_{ij} for which $i = j$ (that is, the elements a_{11}, a_{22}, a_{33}, and so on) are called the **diagonal elements** of the matrix.

A square matrix is called an **identity matrix** if all the elements on its diagonal are equal to 1 and all the elements not on the diagonal are equal to zero. The following are identity matrices of sizes 2×2 and 3×3, respectively.

$$\begin{bmatrix} 1 & 0 \\ 0 & 1 \end{bmatrix} \qquad \begin{bmatrix} 1 & 0 & 0 \\ 0 & 1 & 0 \\ 0 & 0 & 1 \end{bmatrix}$$

The identity matrix is usually denoted by **I** when its size is understood without ambiguity.

EXAMPLE 4 Let

$$\mathbf{A} = \begin{bmatrix} a & b \\ c & d \end{bmatrix}.$$

Find **AI** and **IA**, where **I** denotes the identity matrix.

Solution The products **AI** and **IA** are both defined if **A** and **I** are square matrices of the same size. Since **A** is a 2×2 matrix, the identity matrix **I** must also be of the size 2×2; that is,

$$\mathbf{I} = \begin{bmatrix} 1 & 0 \\ 0 & 1 \end{bmatrix}.$$

Thus

$$\mathbf{AI} = \begin{bmatrix} a & b \\ c & d \end{bmatrix}\begin{bmatrix} 1 & 0 \\ 0 & 1 \end{bmatrix} = \begin{bmatrix} a(1) + b(0) & a(0) + b(1) \\ c(1) + d(0) & c(0) + d(1) \end{bmatrix} = \begin{bmatrix} a & b \\ c & d \end{bmatrix} = \mathbf{A}.$$

Similarly,

$$\mathbf{IA} = \begin{bmatrix} 1 & 0 \\ 0 & 1 \end{bmatrix}\begin{bmatrix} a & b \\ c & d \end{bmatrix} = \begin{bmatrix} a & b \\ c & d \end{bmatrix} = \mathbf{A}.$$

☛ **9.** Verify the associative law for

$$\mathbf{A} = [3 \quad 4], \quad \mathbf{B} = \begin{bmatrix} 2 & -1 \\ 3 & -2 \end{bmatrix},$$

$$\mathbf{C} = \begin{bmatrix} 5 \\ -2 \end{bmatrix}.$$

Answer

$$\mathbf{A}(\mathbf{BC}) = [3 \quad 4]\begin{bmatrix} 12 \\ 19 \end{bmatrix} = 112,$$

$$(\mathbf{AB})\mathbf{C} = [18 \quad -11]\begin{bmatrix} 5 \\ -2 \end{bmatrix} = 112.$$

Therefore $AI = IA = A$.

___

We see from this example that when any 2×2 matrix is multiplied by the identity matrix, it remains unchanged. It is easily seen that this result holds for square matrices of any size. In other words, I behaves the same way in matrix multiplication as the number 1 behaves in multiplication of real numbers. This justifies the name *identity matrix* for I. If A is a *square matrix* of any size, then it is always true that

$$AI = IA = A.$$

If A is a square matrix of size $n \times n$, we can multiply it by itself. The resulting product AA is denoted by A^2. It is again of size $n \times n$. Multiplying by A again, we obtain AAA, which is denoted by A^3 and is again of size $n \times n$. We continue multiplying by A, thus defining A^4, A^5, and so on. Note that A^2, A^3, . . . are defined only if A is a square matrix.

Note The product of two matrices can be the zero matrix 0 even though neither matrix is the zero matrix. For example, if

☛ **10.** If A is $m \times n$ and 0_k is the zero matrix of size $k \times k$, evaluate $0_m A$ and $A 0_n$ and in each case give the size.

$$A = \begin{bmatrix} 1 & 0 \\ 0 & 0 \end{bmatrix} \quad \text{and} \quad B = \begin{bmatrix} 0 & 0 \\ 1 & 0 \end{bmatrix}$$

it is easily seen that $AB = 0$ even though $A \neq 0$ and $B \neq 0$. ☛ 10

By using the idea of matrix multiplication, we can write systems of linear equations in the form of matrix equations. Consider, for example, the system

$$2x - 3y = 7$$

$$4x + y = 21$$

consisting of two simultaneous linear equations for the variables x and y. We have the following matrix product.

$$\begin{bmatrix} 2 & -3 \\ 4 & 1 \end{bmatrix} \begin{bmatrix} x \\ y \end{bmatrix} = \begin{bmatrix} 2x - 3y \\ 4x + y \end{bmatrix}$$

But from the given simultaneous equations, we have the following equality.

$$\begin{bmatrix} 2x - 3y \\ 4x + y \end{bmatrix} = \begin{bmatrix} 7 \\ 21 \end{bmatrix}$$

Therefore

$$\begin{bmatrix} 2 & -3 \\ 4 & 1 \end{bmatrix} \begin{bmatrix} x \\ y \end{bmatrix} = \begin{bmatrix} 7 \\ 21 \end{bmatrix}.$$

If we define matrices A, B, and X as

Answer Both are a zero matrix of size $m \times n$.

$$A = \begin{bmatrix} 2 & -3 \\ 4 & 1 \end{bmatrix}, \quad X = \begin{bmatrix} x \\ y \end{bmatrix}, \quad \text{and} \quad B = \begin{bmatrix} 7 \\ 21 \end{bmatrix}$$

then this matrix equation can be written simply as

$$\mathbf{AX} = \mathbf{B}.$$

Observe that the matrices $\mathbf{A}$ and $\mathbf{B}$ have elements whose values are given numbers. Matrix $\mathbf{X}$ contains the unknown quantities x and y. The column matrix $\mathbf{X}$ is commonly called the **variable vector,** $\mathbf{A}$ is called the **coefficient matrix,** and $\mathbf{B}$ is called the **value vector.**

By introducing appropriate matrices $\mathbf{A}$, $\mathbf{B}$, and $\mathbf{X}$, any system of linear equations can be expressed as a matrix equation.

EXAMPLE 5 Express the following system of equations in matrix form.

$$2x + 3y + 4z = 7$$
$$4y = 2 + 5z$$
$$3z - 2x + 6 = 0$$

Solution We first rearrange the equations so that the constant terms are on the right and the variables x, y, and z are aligned in columns on the left.

$$2x + 3y + 4z = 7$$
$$0x + 4y - 5z = 2$$
$$-2x + 0y + 3z = -6$$

Observe that the missing terms are written as $0x$ and $0y$ in the second and third equations. If we define

$$\mathbf{A} = \begin{bmatrix} 2 & 3 & 4 \\ 0 & 4 & -5 \\ -2 & 0 & 3 \end{bmatrix}, \quad \mathbf{X} = \begin{bmatrix} x \\ y \\ z \end{bmatrix}, \quad \text{and} \quad \mathbf{B} = \begin{bmatrix} 7 \\ 2 \\ -6 \end{bmatrix},$$

the given system can be written in the form $\mathbf{AX} = \mathbf{B}$. Again, $\mathbf{A}$ and $\mathbf{B}$ are known matrices of numbers and $\mathbf{X}$ is the matrix whose elements are the unknown variables. ☛ 11

☛ **11.** Express the systems in the form $\mathbf{AX} = \mathbf{B}$:

(a) $x - 4y = 2$, $2x + 6x = 5$;

(b) $-3x + y - 2z = 1$;
$4y - z = 2$, $x + 3z = 4$.

Let us suppose now that we are given a general system of m linear equations involving n variables. We denote the variables by $x_1, x_2, \ldots, x_n$, and suppose that the system takes the following form.

$$
\begin{aligned}
a_{11}x_1 + a_{12}x_2 + &\cdots + a_{1n}x_n = b_1 \\
a_{21}x_1 + a_{22}x_2 + &\cdots + a_{2n}x_n = b_2 \\
a_{31}x_1 + a_{32}x_2 + &\cdots + a_{3n}x_n = b_3 \\
\vdots \qquad \vdots \qquad & \qquad \quad \vdots \qquad \vdots \\
a_{m1}x_1 + a_{m2}x_2 + &\cdots + a_{mn}x_n = b_m
\end{aligned}
\tag{1}
$$

Here the coefficients a_{ij} are certain given numbers, with a_{ij} the coefficient of x_j

Answer (a) $\begin{bmatrix} 1 & -4 \\ 2 & 6 \end{bmatrix}\begin{bmatrix} x \\ y \end{bmatrix} = \begin{bmatrix} 2 \\ 5 \end{bmatrix}$;

(b) $\begin{bmatrix} -3 & 1 & -2 \\ 0 & 4 & -1 \\ 1 & 0 & 3 \end{bmatrix}\begin{bmatrix} x \\ y \\ z \end{bmatrix} = \begin{bmatrix} 1 \\ 2 \\ 4 \end{bmatrix}.$

in the ith equation, and $b_1, b_2, \ldots, b_m$ are the given right sides of the equations.

Let us introduce the $m \times n$ matrix $\mathbf{A}$ whose elements consist of the coefficients of $x_1, x_2, \ldots, x_n$; $\mathbf{A} = [a_{ij}]$.

Note that the first column of $\mathbf{A}$ contains all the coefficients of x_1, the second column contains all the coefficients of x_2, and so on. Let $\mathbf{X}$ be the column vector formed by the n variables $x_1, x_2, \ldots, x_n$, and $\mathbf{B}$ be the column vector formed by the m constants on the right sides of the equations. Thus

$$
\mathbf{X} = \begin{bmatrix} x_1 \\ x_2 \\ x_3 \\ \vdots \\ x_n \end{bmatrix} \quad \text{and} \quad \mathbf{B} = \begin{bmatrix} b_1 \\ b_2 \\ b_3 \\ \vdots \\ b_m \end{bmatrix}.
$$

Now consider the product $\mathbf{AX}$. This product is defined because the number of columns in $\mathbf{A}$ is equal to the number of rows in $\mathbf{X}$. We have

$$
\mathbf{AX} = \begin{bmatrix} a_{11} & a_{12} & \cdots & a_{1n} \\ a_{21} & a_{22} & \cdots & a_{2n} \\ a_{31} & a_{32} & \cdots & a_{3n} \\ \vdots & \vdots & & \vdots \\ a_{m1} & a_{m2} & \cdots & a_{mn} \end{bmatrix} \begin{bmatrix} x_1 \\ x_2 \\ x_3 \\ \vdots \\ x_n \end{bmatrix}
$$

$$
= \begin{bmatrix} a_{11}x_1 + a_{12}x_2 + \cdots + a_{1n}x_n \\ a_{21}x_1 + a_{22}x_2 + \cdots + a_{2n}x_n \\ a_{31}x_1 + a_{32}x_2 + \cdots + a_{3n}x_n \\ \vdots \qquad \vdots \qquad \qquad \vdots \\ a_{m1}x_1 + a_{m2}x_2 + \cdots + a_{mn}x_n \end{bmatrix} = \begin{bmatrix} b_1 \\ b_2 \\ b_3 \\ \vdots \\ b_m \end{bmatrix} = \mathbf{B}
$$

where we have used Equations (1). Thus the system of Equations (1) is again equivalent to the single matrix equation $\mathbf{AX} = \mathbf{B}$.

EXERCISES 9-2

(1–6) If $\mathbf{A}$ is a 3×4 matrix, $\mathbf{B}$ is 4×3, $\mathbf{C}$ is 2×3, and $\mathbf{D}$ is 4×5, find the sizes of the following product matrices.

1. AB

2. BA

3. CA

4. AD

5. CAD

6. CBA

(7–18) Perform the indicated operations and simplify.

7. $\begin{bmatrix} 2 & 3 \end{bmatrix} \begin{bmatrix} 4 \\ 5 \end{bmatrix}$

8. $\begin{bmatrix} 2 & 0 & 1 \end{bmatrix} \begin{bmatrix} 0 & 2 \\ 1 & -1 \\ 3 & 0 \end{bmatrix}$

9. $\begin{bmatrix} 3 & 0 & 1 \\ 2 & 4 & 0 \end{bmatrix} \begin{bmatrix} 4 \\ 5 \\ 6 \end{bmatrix}$

10. $\begin{bmatrix} 1 & -2 \\ -3 & 4 \\ 5 & 6 \end{bmatrix} \begin{bmatrix} 2 \\ 0 \end{bmatrix}$

11.
$$\begin{bmatrix} 1 & 0 & 2 \\ 0 & 2 & -1 \\ -2 & 1 & 0 \end{bmatrix} \begin{bmatrix} 3 & -2 \\ 2 & 1 \\ -1 & 3 \end{bmatrix}$$

12.
$$\begin{bmatrix} 2 & -1 & 0 \\ 1 & 3 & 2 \\ 4 & 0 & -3 \end{bmatrix} \begin{bmatrix} 1 & 0 & 2 \\ 0 & 2 & 1 \\ 2 & 1 & 0 \end{bmatrix}$$

13.
$$\begin{bmatrix} 2 & 3 & 1 \\ -1 & 2 & -3 \\ 4 & 5 & 6 \end{bmatrix} \begin{bmatrix} 1 \\ 2 \\ 3 \end{bmatrix}$$

14.
$$\begin{bmatrix} 2 & 1 & 4 \\ 5 & 3 & 6 \end{bmatrix} \begin{bmatrix} 1 & 0 & 2 & 4 \\ 3 & -1 & 0 & 1 \\ 0 & 2 & 1 & 3 \end{bmatrix}$$

15.
$$\begin{bmatrix} 1 & 2 & 3 \\ 4 & 5 & 6 \end{bmatrix} \begin{bmatrix} -1 & 0 \\ 2 & 4 \\ 0 & 3 \end{bmatrix} \begin{bmatrix} 3 & -1 \\ -2 & 1 \end{bmatrix}$$

16.
$$\begin{bmatrix} 1 & 0 & 2 \\ 0 & 2 & -1 \\ 3 & 1 & 0 \end{bmatrix} \begin{bmatrix} 2 & -1 \\ 1 & 0 \\ 0 & 3 \end{bmatrix} \begin{bmatrix} 0 & 1 & -2 \\ 3 & 0 & 1 \end{bmatrix}$$

17.
$$\begin{bmatrix} 4 & 1 & -2 \\ -3 & 2 & 1 \end{bmatrix} \left(\begin{bmatrix} 5 & 6 \\ 1 & 0 \\ 2 & -3 \end{bmatrix} + \begin{bmatrix} -4 & 2 \\ 3 & 1 \\ -2 & 3 \end{bmatrix} \right)$$

18.
$$\begin{bmatrix} 2 & 1 \\ 0 & 2 \\ 3 & -1 \end{bmatrix} \left(\begin{bmatrix} 1 & -2 \\ 2 & -1 \end{bmatrix} + 3 \begin{bmatrix} 2 & 0 \\ 1 & 2 \end{bmatrix} \right)$$

19. Evaluate $A^2 + 2A - 3I$ for
$$A = \begin{bmatrix} 1 & 2 \\ 2 & 3 \end{bmatrix}.$$

20. Evaluate $A^2 - 5A + 2I$ for
$$A = \begin{bmatrix} 1 & 0 & 0 \\ 0 & 2 & 1 \\ 0 & 0 & 3 \end{bmatrix}.$$

21. Given
$$A = \begin{bmatrix} 1 & 2 \\ 3 & 4 \end{bmatrix} \quad \text{and} \quad B = \begin{bmatrix} 2 & -1 \\ -3 & -2 \end{bmatrix}.$$

  a. Find $(A + B)^2$.

  b. Find $A^2 + 2AB + B^2$.

  c. Is $(A + B)^2 = A^2 + 2AB + B^2$?

22. Given
$$A = \begin{bmatrix} 2 & 3 \\ 1 & 2 \end{bmatrix} \quad \text{and} \quad B = \begin{bmatrix} 1 & 0 \\ 2 & -1 \end{bmatrix}.$$

Compute $A^2 - B^2$ and $(A - B)(A + B)$ and show that $A^2 - B^2 \neq (A - B)(A + B)$.

(23–24) Given
$$A = \begin{bmatrix} p & 1 \\ q & -1 \end{bmatrix} \quad \text{and} \quad B = \begin{bmatrix} 1 & -1 \\ 2 & -1 \end{bmatrix}.$$

Determine p and q so that:

23. $(A + B)^2 = A^2 + B^2$

24. $(A + B)(A - B) = A^2 - B^2$

*(25–28) Determine the matrix A for which each matrix equation is true.

*25.
$$A \begin{bmatrix} 2 & 1 \\ 1 & 0 \end{bmatrix} = \begin{bmatrix} 5 & 3 \end{bmatrix}$$

*26.
$$\begin{bmatrix} 1 & 0 & 2 \\ 2 & -1 & 0 \\ 0 & 1 & 3 \end{bmatrix} A = \begin{bmatrix} 7 \\ 0 \\ 11 \end{bmatrix}$$

*27.
$$\begin{bmatrix} 2 & 0 \\ 1 & -1 \\ 0 & 1 \end{bmatrix} A = \begin{bmatrix} 6 & 0 \\ 3 & -1 \\ 0 & 1 \end{bmatrix}$$

*28.
$$A \begin{bmatrix} 1 & 2 \\ 3 & 4 \end{bmatrix} = \begin{bmatrix} 7 & 10 \\ 15 & 22 \end{bmatrix}$$

(29–34) Express the following systems of linear equations in matrix form.

29. $2x + 3y = 7$
    $x + 4y = 5$

30. $3x - 2y = 4$
    $4x + 5y = 7$

31. $x + 2y + 3z = 8$
    $2x - y + 4z = 13$
    $3y - 2z = 5$

32. $2x - y = 3$
    $3y + 4z = 7$
    $5z + x = 9$

33. $2x + y - u = 0$
    $3y + 2z + 4u = 5$
    $x - 2y + 4z + u = 12$

34. $2x_1 - 3x_2 + 4x_3 = 5$
    $3x_3 + 5x_4 - x_1 = 7$
    $x_1 + x_2 = x_3 + 2x_4$

35. For the following value of A, find a 2×2 nonzero matrix B such that AB is a zero matrix. (There is more than one answer.)
$$A = \begin{bmatrix} 1 & 2 \\ 3 & 6 \end{bmatrix}$$

36. Give an example of two nonzero matrices **A** and **B** of different sizes such that the product **AB** is defined and is a zero matrix. (There are many possible answers.)

(37–40) Determine the matrix $\mathbf{A}^n$ for a general positive integer n, where **A** is as given.

37. $\mathbf{A} = \begin{bmatrix} 1 & 0 \\ 0 & 1 \end{bmatrix}$ **38.** $\mathbf{A} = \begin{bmatrix} 0 & 1 \\ 1 & 0 \end{bmatrix}$

*39. $\mathbf{A} = \begin{bmatrix} 1 & 0 \\ \frac{1}{2} & \frac{1}{2} \end{bmatrix}$ **40.** $\mathbf{A} = \begin{bmatrix} 0 & 1 & 0 \\ 1 & 0 & 0 \\ 0 & 0 & 1 \end{bmatrix}$

41. *(Inventory Valuation)* A dealer in color television sets has five 26-inch sets, eight 20-inch sets, four 18-inch sets, and ten 12-inch sets. The 26-inch sets sell for $650 each, the 20-inch sets sell for $550 each, the 18-inch sets sell for $500 each, and the 12-inch sets sell for $300 each. Express the total selling price of his television stock as the product of two matrices.

42. *(Raw-material Costs)* A firm uses four different raw materials M_1, M_2, M_3, and M_4 in the production of its product. The number of units of M_1, M_2, M_3, and M_4 used per unit of the product are 4, 3, 2, and 5, respectively. The cost per unit of the four raw materials is $5, $7, $6, and $3, respectively. Express the total cost of raw materials per unit of the product as the product of two matrices.

43. *(Raw-material Costs)* A firm uses three types of raw materials M_1, M_2, and M_3 in the production of two products P_1 and P_2. The numbers of units of M_1, and M_2, and M_3 used for each unit of P_1 are 3, 2, and 4, respectively, and for each unit of P_2 are 4, 1, and 3, respectively. Suppose the firm produces 20 units of P_1 and 30 units of P_2 each week. Express the answers to the following questions as matrix products.

a. What is the weekly consumption of the three raw materials?

b. If the unit costs in (dollars) for M_1, M_2, and M_3 are 6, 10, and 12, respectively, what are the costs of raw materials per unit of P_1 and P_2?

c. What is the total amount spent per week on the production of P_1 and P_2?

44. *(Supply Costs)* A building contractor can purchase the required lumber, bricks, concrete, glass, and paint from any of three suppliers. The prices that each supplier charges for each unit of these five materials are given in matrix **A**.

$$\mathbf{A} = \begin{bmatrix} 8 & 5 & 7 & 2 & 4 \\ 9 & 4 & 5 & 2 & 5 \\ 9 & 5 & 6 & 1 & 5 \end{bmatrix}$$

In this matrix, each row refers to one supplier and the columns to the materials, in the order just listed. The contractor has the policy of purchasing all the required materials for any particular job from the same supplier in order to minimize transportation costs. There are three jobs underway at present: job I requires 20 units of lumber, 4 of bricks, 5 of concrete, 3 of glass, and 3 of paint; job II requires 15, 0, 8, 8, and 2 units, respectively; and job III requires 30, 10, 20, 10, and 12 units, respectively. Arrange this information as a 5×3 matrix **B** and form the matrix product **AB**. Interpret the entries in this product and use them to decide which supplier should be used for each job.

45. *(Graph Theory)* A graph consists of a number of points, called vertices, some of which are connected by lines (called edges). Two examples of graphs with four and five vertices are given below.

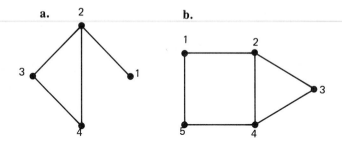

If the vertices are numbered 1, 2, 3, . . . , we define the matrix **A** by setting $a_{ij} = 1$ if there is an edge connecting the vertices i and j and $a_{ij} = 0$ if there is not. Construct **A** for each of the given graphs with the numbering shown. Construct $\mathbf{A}^2$ in each case. Show that the ij-element in $\mathbf{A}^2$ gives the number of routes from vertex i to vertex j that pass through exactly one other vertex. What do you think the elements of $\mathbf{A}^3$ give?

46. *(Applied Graph Theory)* The graph shown represents the number of telephone lines connecting each of four towns.

Let a_{ij} denote the number of lines connecting town i with town j. Construct the matrix $\mathbf{A} = (a_{ij})$. Evaluate $\mathbf{A}^2$ and show that the ij-element in this matrix represents the number of telephone links between town i and town j that pass through exactly one intermediate town. What do the elements of $\mathbf{A} + \mathbf{A}^2$ represent?

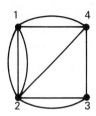

9-3 SOLUTION OF LINEAR SYSTEMS BY ROW REDUCTION

In Section 4-4 we discussed how systems of linear equations arise in certain areas of business and economics. In that section, we solved systems consisting of two linear equations in two unknowns. We shall now develop a method of solving systems of linear equations that can be used regardless of the number of equations involved in the system. Let us illustrate the principles of the method by solving the following simple system of two equations.

$$2x + 3y = 3 \tag{1}$$
$$x - 2y = 5$$

If we interchange the two equations (the reason for this will become clear later), we get the following equivalent system.

$$x - 2y = 5 \tag{2}$$
$$2x + 3y = 3$$

If we multiply the first of these equations by -2, we obtain $-2x + 4y = -10$; we add this equation to the second equation in System (2) and simplify.

$$2x + 3y + (-2x + 4y) = 3 + (-10)$$
$$0x + 7y = -7$$

System (2) then becomes

$$x - 2y = 5 \tag{3}$$
$$0x + 7y = -7.$$

We multiply both sides of the second equation by $\frac{1}{7}$, which gives the equivalent system

$$x - 2y = 5 \tag{4}$$
$$0x + y = -1.$$

From the second equation in System (4), we have $y = -1$. Thus $2y = -2$.

Adding this to the first equation in System (4), we have the following system.

$$x + 0y = 3$$
$$0x + y = -1$$

(5)

Therefore $x = 3$ and $y = -1$ and we have solved the given system of equations.

In the above method, we performed specific operations on the original equations of System (1), transforming them into those of System (5), from which the values of the unknowns x and y can be read directly. With each operation, the system is transformed into a new system that is equivalent to the old one. The operations consisted of the following basic types.

1. Interchanging two equations.

2. Multiplying or dividing an equation by a nonzero constant.

3. Adding (or subtracting) a constant multiple of one equation to (or from) another equation.

If we keep track of the positions of various variables and the equality signs, then a system of linear equations can be written as a matrix with the variables omitted. For example, System (1),

$$2x + 3y = 3$$
$$x - 2y = 5$$

can be abbreviated as

$$\begin{bmatrix} 2 & 3 & | & 3 \\ 1 & -2 & | & 5 \end{bmatrix}.$$

This array of numbers is called the **augmented matrix** for the given system. Note that in writing this augmented matrix, we have written the coefficient matrix elements to the left of the vertical line and the elements of the value vector (that is, the constants on the right sides of the equations) to the right of the vertical line. Thus, if the given system of equations in matrix form is $\mathbf{AX} = \mathbf{B}$, the augmented matrix may be denoted by $\mathbf{A} | \mathbf{B}$. The augmented matrix is simply a way of writing the system of equations without writing down the variables every time.

EXAMPLE 1 For the variables x, y, z, and t, in that order, the augmented matrix

$$\begin{bmatrix} 2 & -1 & 3 & 4 & | & 5 \\ 1 & 3 & -2 & 0 & | & 7 \\ -4 & 0 & 5 & 1 & | & -3 \end{bmatrix}$$

corresponds to the following linear system.

12. Write the augmented matrix for each of the systems:
(a) $x - 4y = 2$, $2x + 6y = 5$;
(b) $-3x + y - 2z = 1$, $4y - z = 2$, $x + 3z = 4$.

$$2x - y + 3z + 4t = 5$$
$$x + 3y - 2z = 7$$
$$-4x + 5z + t = -3 \quad \text{☛ 12}$$

Since each row of the augmented matrix corresponds to an equation in the linear system, the three operations listed earlier correspond to the following three **row operations** on the augmented matrix.

---

1. Interchanging two rows.

2. Multiplying or dividing a row by a nonzero constant.

3. Adding (or subtracting) a constant multiple of one row to (or from) another row.

We shall illustrate the use of row operations on an augmented matrix to solve the following system.

$$3x - 2y = 4$$
$$x + 3y = 5$$

The augmented matrix in this case is

$$\begin{bmatrix} 3 & -2 & | & 4 \\ 1 & 3 & | & 5 \end{bmatrix}.$$

To clarify the use of these operations, we shall also solve the system by operating on the equations, side by side with the corresponding operations on the augmented matrix.

SYSTEM	AUGMENTED MATRIX

$$3x - 2y = 4$$
$$x + 3y = 5$$

$$\begin{bmatrix} 3 & -2 & | & 4 \\ 1 & 3 & | & 5 \end{bmatrix}$$

Interchange the first and second equations:

Interchange the first and second rows:

$$x + 3y = 5$$
$$3x - 2y = 4$$

$$\begin{bmatrix} 1 & 3 & | & 5 \\ 3 & -2 & | & 4 \end{bmatrix}$$

Add -3 times the first equation to the second equation:

Add -3 times the first row to the second row:

$$x + 3y = 5$$
$$0x - 11y = -11$$

$$\begin{bmatrix} 1 & 3 & | & 5 \\ 0 & -11 & | & -11 \end{bmatrix}$$

Divide both sides of the second equation by -11:

Divide the second row by -11:

$$x + 3y = 5$$
$$0x + y = 1$$

$$\begin{bmatrix} 1 & 3 & | & 5 \\ 0 & 1 & | & 1 \end{bmatrix}$$

Answer (a) $\begin{bmatrix} 1 & -4 & | & 2 \\ 2 & 6 & | & 5 \end{bmatrix}$;

(b) $\begin{bmatrix} -3 & 1 & -2 & | & 1 \\ 0 & 4 & -1 & | & 2 \\ 1 & 0 & 3 & | & 4 \end{bmatrix}.$

► **13.** Write down the augmented matrix for the system
$2x + 3y = 4$, $-3x + 5y = 13$.
Obtain the result of the row operations $\frac{1}{2}R_1$ followed by $R_2 + 3R_1$. Give the remaining row operations that complete the reduction.

Subtract 3 times the second equation from the first equation:

$$x + 0y = 2$$
$$0x + \quad y = 1$$

Subtract 3 times the second row from the first row:

$$\begin{bmatrix} 1 & 0 & | & 2 \\ 0 & 1 & | & 1 \end{bmatrix}$$

The solution is therefore $x = 2$ and $y = 1$. Observe that the values of x and y are given by the entries in the last column of this final augmented matrix.

The final augmented matrix from which we read the solution is of the form $\mathbf{I} | \mathbf{C}$, where $\mathbf{I}$ is the identity matrix and $\mathbf{C}$ is a certain column vector. Thus, to obtain the solution of a given system $\mathbf{AX} = \mathbf{B}$, we first write the augmented matrix $\mathbf{A} | \mathbf{B}$ and then use row operations to change it to the form $\mathbf{I} | \mathbf{C}$. This is not always possible;* however, if we succeed, the solution for the variables is given by the entries in the last column $\mathbf{C}$. The final form of the matrix $\mathbf{I} | \mathbf{C}$ that gives the solutions to a system is called the **reduced matrix.** This method of solving linear systems is called the **method of row reduction.**

Before we explain how to select the order of row operations to obtain the reduced matrix from the original augmented matrix, we introduce some notation to avoid repeating lengthy expressions. We shall use the symbol R_p for the p th row of the augmented matrix. Thus R_1 denotes the first row, R_2 the second row, and so on. When we say "apply $R_2 - 2R_1$," this means "subtract twice the first row from the second row," while the operation $R_3 + 4R_2$ consists of adding four times the second row to the third row and $R_2 + R_3$ means adding the third row to the second row (*not* the second row to the third row). Similarly, the operation $2R_3$ means multiplying the third row of the augmented matrix by 2 and $-\frac{1}{2}R_1$ means multiplying the first row by $-\frac{1}{2}$. Finally, notation such as $R_1 \leftrightarrow R_3$ means the operation of interchanging the first and third rows. We shall also use notation such as

$$\text{matrix } \mathbf{A} \quad \xrightarrow{R_1 - 2R_2} \quad \text{matrix } \mathbf{B}$$

which means that matrix $\mathbf{B}$ is obtained by applying the operation $R_1 - 2R_2$ (that is, subtracting twice the seond row from the first row) on matrix $\mathbf{A}$.

► **13**

Now we are in a position to explain the method of row reduction in detail. We shall do this through an example.

EXAMPLE 2 Use the method of row reduction to solve the following system of linear equations.

$$2x - 3y + 4z = 13$$
$$x + \quad y + 2z = 4$$
$$3x + 5y - \quad z = -4$$

Answer $\begin{bmatrix} 2 & 3 & | & 4 \\ -3 & 5 & | & 13 \end{bmatrix} \xrightarrow{\frac{1}{2}R_1}$

$\begin{bmatrix} 1 & \frac{3}{2} & | & 2 \\ -3 & 5 & | & 13 \end{bmatrix} \xrightarrow{R_2+3R_1} \begin{bmatrix} 1 & \frac{3}{2} & | & 2 \\ 0 & \frac{19}{2} & | & 19 \end{bmatrix}.$

$\frac{2}{19}R_2$ followed by $R_1 - \frac{3}{2}R_2$. The solution is $x = -1$, $y = 2$.

*See Section 9-4.

Solution The augmented matrix for this system is

$$\begin{bmatrix} 2 & -3 & 4 & \bigm| & 13 \\ 1 & 1 & 2 & \bigm| & 4 \\ 3 & 5 & -1 & \bigm| & -4 \end{bmatrix}.$$

Our purpose is to apply row operations on this matrix until we obtain its reduced form, that is, until the first three columns form an identity matrix. The best method, generally, is to attack the columns one by one, changing the main diagonal elements to 1 and making the other entries in the columns zero. In the first column of our matrix, the first entry is 2. To change this entry to 1, we could divide R_1 by 2 or, alternatively, we could interchange R_1 and R_2. If we apply $\frac{1}{2}R_1$, we immediately introduce fractions, whereas if we interchange R_1 and R_2 (that is, apply $R_1 \leftrightarrow R_2$), we shall avoid fractions (at least for the time being). Thus it is preferable to apply $R_1 \leftrightarrow R_2$ and obtain

$$\begin{bmatrix} 1 & 1 & 2 & \bigm| & 4 \\ 2 & -3 & 4 & \bigm| & 13 \\ 3 & 5 & -1 & \bigm| & -4 \end{bmatrix}.$$

Now that we have obtained a diagonal entry of 1 in the first column, we use the first row to change all the other entries in that column to zero. First the operation $R_2 - 2R_1$ places a zero in the second element:

$$\begin{bmatrix} 1 & 1 & 2 & \bigm| & 4 \\ 2 - 2(1) & -3 - 2(1) & 4 - 2(2) & \bigm| & 13 - 2(4) \\ 3 & 5 & -1 & \bigm| & -4 \end{bmatrix} = \begin{bmatrix} 1 & 1 & 2 & \bigm| & 4 \\ 0 & -5 & 0 & \bigm| & 5 \\ 3 & 5 & -1 & \bigm| & -4 \end{bmatrix}.$$

Then the operation $R_3 - 3R_1$ makes the third element zero:

$$\begin{bmatrix} 1 & 1 & 2 & \bigm| & 4 \\ 0 & -5 & 0 & \bigm| & 5 \\ 3 - 3(1) & 5 - 3(1) & -1 - 3(2) & \bigm| & -4 - 3(4) \end{bmatrix} = \begin{bmatrix} 1 & 1 & 2 & \bigm| & 4 \\ 0 & -5 & 0 & \bigm| & 5 \\ 0 & 2 & -7 & \bigm| & -16 \end{bmatrix}.$$

We have now reduced the first column to the required form (that is, to the first column of the identity matrix). We now attack the second column. In this column, we must have 1 in the second row and zero in the first and third row. While achieving this goal, *we must be careful not to change the first column.* (This means, for instance, that we cannot add 6 times the first row to the second row because this will change the first column entries.) There are many ways that yield 1 in the second entry in the second column. For example, we can apply $-\frac{1}{5}R_2$ or $R_2 + 3R_3$. Application of $-\frac{1}{5}R_2$ is simpler in this case; it leads to the following matrix.

$$\begin{bmatrix} 1 & 1 & 2 & \bigm| & 4 \\ 0 & 1 & 0 & \bigm| & -1 \\ 0 & 2 & -7 & \bigm| & -16 \end{bmatrix}$$

We now use *the second row* to make the other two entries in the second column zero. The operation $R_1 - R_2$ makes the first element zero and then the operation $R_3 - 2R_2$ makes the third element zero. We can carry out these two operations simultaneously:

$$\begin{bmatrix} 1-0 & 1-1 & 2-0 & 4-(-1) \\ 0 & 1 & 0 & -1 \\ 0-2(0) & 2-2(1) & -7-2(0) & -16-2(-1) \end{bmatrix} = \begin{bmatrix} 1 & 0 & 2 & 5 \\ 0 & 1 & 0 & -1 \\ 0 & 0 & -7 & -14 \end{bmatrix}.$$

Notice that these operations have not changed the first column. Thus we have also reduced the second column to the required form, with 1 on the main diagonal and 0 elsewhere.

Finally we attack the third column. We must make the third entry in this column equal to 1; this can be done by applying $-\frac{1}{7}R_3$. This leads to

$$\begin{bmatrix} 1 & 0 & 2 & 5 \\ 0 & 1 & 0 & -1 \\ 0 & 0 & 1 & 2 \end{bmatrix}.$$

In the third column, the entries in the first and second row must also be zero. We already have zero in the second row. To obtain zero in the first row, we apply the operation $R_1 - 2R_3$. This gives

$$\begin{bmatrix} 1 & 0 & 2-2(1) & 5-2(2) \\ 0 & 1 & 0 & -1 \\ 0 & 0 & 1 & 2 \end{bmatrix} = \begin{bmatrix} 1 & 0 & 0 & 1 \\ 0 & 1 & 0 & -1 \\ 0 & 0 & 1 & 2 \end{bmatrix}.$$

Thus we have attained our goal, that is, we have changed the first three columns of the augmented matrix of the system to an identity matrix. The final matrix represents the system

$$\begin{aligned} 1x + 0y + 0z &= 1 & x &= 1 \\ 0x + 1y + 0z &= -1 & \text{or} \quad y &= -1 \\ 0x + 0y + 1z &= 2 & z &= 2 \end{aligned}$$

from which the required solution can be read directly. ☛ 14

☛ **14.** It is a good idea to substitute your solution back into each of the equations in the original system as a check that it is correct. Try this with the solution to Example 2.

In light of the above example, we may summarize the steps involved in changing the augmented matrix to its reduced form as follows.* Each step is carried out by means of one or more of the row operations given earlier.

---

* The procedure does not always work and must be modified in certain cases. (See Section 9-4.)

☞ **15.** Use the row reduction procedure to solve the following systems:

(a) $2x - 4y + 2 = 0$,
$-x + 3y = 3$; (b) $q + 3r = 1$;
$2p - 5r = 1$, $2p + 2q + 3r = 1$;

EXAMPLE 3 *(Market Equilibrium)* Two products A and B are competitive. The demands x_A and x_B for these products are related to their prices P_A and P_B according to the demand equations

$$x_A = 17 - 2P_A + \tfrac{1}{2}P_B \quad \text{and} \quad x_B = 20 - 3P_B + \tfrac{1}{2}P_A.$$

The supply equations are

$$P_A = 2 + x_A + \tfrac{1}{3}x_B \quad \text{and} \quad P_B = 2 + \tfrac{1}{2}x_B + \tfrac{1}{4}x_A$$

giving the prices at which the quantities x_A and x_B of the two products will be available on the market. For market equilibrium, all four equations must be satisfied (since demand and supply must be equal). Find the equilibrium values of x_A, x_B, P_A, and P_B.

Solution Rearranging the four equations, we obtain the following system.

$$
\begin{aligned}
x_A \qquad\quad + 2P_A - \tfrac{1}{2}P_B &= 17 \\
x_B - \tfrac{1}{2}P_A + 3P_B &= 20 \\
x_A + \tfrac{1}{3}x_B - P_A \qquad\quad &= -2 \\
\tfrac{1}{4}x_A + \tfrac{1}{2}x_B \qquad\quad - P_B &= -2
\end{aligned}
$$

Note that the variables in each equation have been put in the order x_A, x_B, P_A, and P_B. The augmented matrix is as follows.

Answer (a) $x = 3$, $y = 2$
(b) $p = -2$, $q = 4$, $r = -1$.

$$
\left[
\begin{array}{cccc|c}
1 & 0 & 2 & -\tfrac{1}{2} & 17 \\
0 & 1 & -\tfrac{1}{2} & 3 & 20 \\
1 & \tfrac{1}{3} & -1 & 0 & -2 \\
\tfrac{1}{4} & \tfrac{1}{2} & 0 & -1 & -2
\end{array}
\right]
$$

For the first column we apply the operations $R_3 - R_1$ and $R_4 - \frac{1}{4}R_1$ to obtain zeros below the first entry. The result is

$$\begin{bmatrix} 1 & 0 & 2 & -\frac{1}{2} & 17 \\ 0 & 1 & -\frac{1}{2} & 3 & 20 \\ 0 & \frac{1}{3} & -3 & \frac{1}{2} & -19 \\ 0 & \frac{1}{2} & -\frac{1}{2} & -\frac{7}{8} & -\frac{25}{4} \end{bmatrix}$$

Then for the second column, we apply $R_3 - \frac{1}{3}R_2$ and $R_4 - \frac{1}{2}R_2$ to obtain

$$\begin{bmatrix} 1 & 0 & 2 & -\frac{1}{2} & 17 \\ 0 & 1 & -\frac{1}{2} & 3 & 20 \\ 0 & 0 & -\frac{17}{6} & -\frac{1}{2} & -\frac{77}{3} \\ 0 & 0 & -\frac{1}{4} & -\frac{19}{8} & -\frac{65}{4} \end{bmatrix}$$

Before reducing the matrix further, we observe that interchanging the third and fourth rows will help to avoid complicated fractions, since the entry in the third column is $-\frac{1}{4}$ rather than $-\frac{17}{6}$. So making this interchange and multiplying the new R_3 by -4 we have the following.

$$\begin{bmatrix} 1 & 0 & 2 & -\frac{1}{2} & 17 \\ 0 & 1 & -\frac{1}{2} & 3 & 20 \\ 0 & 0 & 1 & \frac{19}{2} & 65 \\ 0 & 0 & -\frac{17}{6} & -\frac{1}{2} & -\frac{77}{3} \end{bmatrix} \begin{array}{c} R_1 - 2R_3 \\ R_2 + \frac{1}{2}R_3 \\ R_4 + \frac{17}{6}R_3 \\ \longrightarrow \end{array} \begin{bmatrix} 1 & 0 & 0 & -\frac{39}{2} & -113 \\ 0 & 1 & 0 & \frac{31}{4} & \frac{105}{2} \\ 0 & 0 & 1 & \frac{19}{2} & 65 \\ 0 & 0 & 0 & \frac{317}{12} & \frac{317}{2} \end{bmatrix}$$

$$\begin{array}{c} \frac{12}{317}R_4 \\ \longrightarrow \end{array} \begin{bmatrix} 1 & 0 & 0 & -\frac{39}{2} & -113 \\ 0 & 1 & 0 & \frac{31}{4} & \frac{105}{2} \\ 0 & 0 & 1 & \frac{19}{2} & 65 \\ 0 & 0 & 0 & 1 & 6 \end{bmatrix}$$

$$\begin{array}{c} R_1 + \frac{39}{2}R_4 \\ R_2 - \frac{31}{4}R_4 \\ R_3 - \frac{19}{2}R_4 \\ \longrightarrow \end{array} \begin{bmatrix} 1 & 0 & 0 & 0 & 4 \\ 0 & 1 & 0 & 0 & 6 \\ 0 & 0 & 1 & 0 & 8 \\ 0 & 0 & 0 & 1 & 6 \end{bmatrix}$$

The solution for market equilibrium is therefore $x_A = 4$, $x_B = 6$, $P_A = 8$, and $P_B = 6$.

EXERCISES 9-3

(1–14) In the following problems, solve the given system (if the solution exists) by using the row reduction method.

1. $2x + 3y = 7$
$\quad 3x - y = 5$

2. $x + 2y = 1$
$\quad 3y + 2x = 3$

3. $u + 3v = 1$
$\quad 2u - v = 9$

4. $3p + 2q = 5$
$\quad p - 3q + 2 = 0$

5. $x + y + z = 6$
$\quad 2x - y + 3z = 9$
$\quad -x + 2y + z = 6$

6. $x + 2y - z = -3$
$\quad 3y + 4z = 5$
$\quad 2x - y + 3z = 9$

7. $3x_1 + 2x_2 + x_3 = 6$
$\quad 2x_1 - x_2 + 4x_3 = -4$
$\quad x_1 + x_2 - 2x_3 = 5$

8.
$$\begin{aligned} 2u - 3v + 4w &= 13 \\ u + v + w &= 6 \\ -3u + 2v + w + 1 &= 0 \end{aligned}$$

9.
$$\begin{aligned} p - q + r &= -1 \\ 3p - 2r &= -7 \\ r + 4q &= 10 \end{aligned}$$

10.
$$\begin{aligned} b &= 3 - a \\ c &= 4 - a - b \\ 3a + 2b + c &= 8 \end{aligned}$$

11.
$$\begin{aligned} x + 2y + z - t &= 0 \\ y - 2z + 2t &= 13 \\ 2x + 4y - z + 2t &= 17 \\ y - z - 3t &= 0 \end{aligned}$$

12.
$$\begin{aligned} p - q - r &= 4 \\ q - r - s &= -5 \\ r - s - p &= -8 \\ p + 2q + 2r + s &= -5 \end{aligned}$$

13.
$$\begin{aligned} x + y + z &= 1 \\ 2x + 3y - w &= 3 \\ -x + 2z + 3w &= 3 \\ 2y - z + w &= 5 \end{aligned}$$

14.
$$\begin{aligned} x_1 + x_2 + x_3 + x_4 &= 2 \\ x_1 - x_2 + x_3 + 2x_4 &= -4 \\ 2x_1 + x_2 - x_3 + x_4 &= 1 \\ -x_1 + x_2 + x_3 - x_4 &= 4 \end{aligned}$$

15. Find x, y, and z such that

$$x[1 \quad 2 \quad -1] - y[2 \quad -1 \quad 3]$$
$$+ z[3 \quad -2 \quad 1] = [9 \quad -1 \quad -2].$$

16. Find a, b, and c such that

$$a[2 \quad 3 \quad -1] + b[1 \quad 2 \quad 3] + c[1 \quad 0 \quad 2]$$
$$= [3 \quad 7 \quad 3].$$

(17–24) Use the method of row reduction to solve the following problems.

17. *(Market Equilibrium)* The demand equation for a certain product is $p + 2x = 25$ and the supply equation is $p - 3x = 5$, where p is the price and x is the quantity demanded or supplied, as the case may be. Find the values of x and p at market equilibrium.

18. *(Market Equilibrium)* The demand and supply equations for a certain commodity are $3p + 5x = 200$ and $7p - 3x = 56$, respectively. Find the values of x and p at market equilibrium.

19. *(Market Equilibrium)* If a sales tax of 11 is imposed on each item in Exercise 18, find the new values of quantity x and the price p_1 paid by consumers. (See Section 4-5.)

20. *(Manufacturer's Profits)* The cost in dollars of producing x items per week of a certain product is given by $C = 3x + 500$. If the items sell for $5 each, how many should be produced to give a weekly profit equal to $300 plus 10% of the production costs?

21. *(Machine Allocation)* A firm produces three products, A, B, and C, which require processing by three machines. The time (in hours) required for processing one unit of each product by the three machines is as follows.

$$\begin{array}{c} \quad\quad\quad\quad \text{A} \quad \text{B} \quad \text{C} \\ \begin{array}{l} \text{Machine I} \\ \text{Machine II} \\ \text{Machine III} \end{array} \begin{bmatrix} 3 & 1 & 2 \\ 1 & 2 & 4 \\ 2 & 1 & 1 \end{bmatrix} \end{array}$$

Machine I is available for 850 hours, machine II, for 1200 hours, and machine III, for 550 hours. How many units of each product should be produced to make use of all the available time on the machines?

22. *(Air Freight)* A shipping company loaded three types of cargo on its light transport plane. The space required by each unit of the three types of cargo was 5, 2, and 4 cubic feet, respectively. Each unit of the three types of cargo weighed 2, 3, and 1 kilograms, respectively, whereas the unit values of the three types of cargo were $10, $40, and $60, respectively. Determine the number of units of each type of cargo loaded if the total value of the cargo was $13,500, it occupied 1050 cubic feet of space, and it weighed 550 kilograms.

23. *(Investment)* A person invested a total of $20,000 in three different investments at 6%, 8%, and 10%. The total annual return was $1624 and the return from the 10% investment was twice the return from the 6% investment. How much was invested in each?

24. A contractor has 5000 work-hours of labor available for three projects. The costs per work-hour of the three projects are $8, $10, and $12, respectively, and the total cost is $53,000. If the number of work-hours for the third project is equal to the sum of the work-hours for the first two projects, find the work-hours that can be used for each project.

(25–26) Solve the following problems by row reduction and comment on the solutions.

25. The demand and supply equations for a certain commodity are $2p + x = 5$ and $3p - 2x = 11$, respectively.

26. In Exercise 21, suppose that machines I, II, and III are available for 1200, 900, and 1100 hours, respectively.

All the systems of linear equations that we solved in the last section had unique solutions. There exist systems of equations that have more than one solution and other systems that have no solutions at all. Such systems are said to be **singular.** Consider the following example.

EXAMPLE 1 Solve the following system.

$$x + y - z = 4$$
$$3x - 2y + 4z = 9$$
$$9x - y + 5z = 30$$

Solution We reduce the augmented matrix for this sytem as follows.

$$\begin{bmatrix} 1 & 1 & -1 & | & 4 \\ 3 & -2 & 4 & | & 9 \\ 9 & -1 & 5 & | & 30 \end{bmatrix} \xrightarrow[R_3 - 9R_1]{R_2 - 3R_1} \begin{bmatrix} 1 & 1 & -1 & | & 4 \\ 0 & -5 & 7 & | & -3 \\ 0 & -10 & 14 & | & -6 \end{bmatrix}$$

$$\xrightarrow{-\frac{1}{5}R_2} \begin{bmatrix} 1 & 1 & -1 & | & 4 \\ 0 & 1 & -\frac{7}{5} & | & \frac{3}{5} \\ 0 & -10 & 14 & | & -6 \end{bmatrix} \xrightarrow[R_3 + 10R_2]{R_1 - R_2} \begin{bmatrix} 1 & 0 & \frac{2}{5} & | & \frac{17}{5} \\ 0 & 1 & -\frac{7}{5} & | & \frac{3}{5} \\ 0 & 0 & 0 & | & 0 \end{bmatrix}$$

So far we have obtained the first two columns in the desired form. However, the third row now consists entirely of zeros, so we are unable to obtain 1 in the third entry of the third column without disturbing the first and second columns. Thus we cannot continue the process of row reduction any further.

The matrix we have obtained corresponds to the following equations.

$$x + \tfrac{2}{5}z = \tfrac{17}{5}$$
$$y - \tfrac{7}{5}z = \tfrac{3}{5} \tag{1}$$

The third equation is $0x + 0y + 0z = 0$, or $0 = 0$, which is true for all values of x, y, and z and can be ignored. We see therefore that the given system of three equations in System (1) can be solved for x and y in terms of z.

$$x = \tfrac{17}{5} - \tfrac{2}{5}z = \tfrac{1}{5}(17 - 2z)$$
$$y = \tfrac{3}{5} + \tfrac{7}{5}z = \tfrac{1}{5}(3 + 7z) \tag{2}$$

The variable z is arbitrary and can take any value. For example, if $z = 1$, then $x = \tfrac{1}{5}(17 - 2) = 3$ and $y = \tfrac{1}{5}(3 + 7) = 2$. Thus $x = 3$, $y = 2$, and $z = 1$ is one solution. By changing the values of z, we get different values of x and y from System (2) and, therefore, different solutions of the given system. Thus the system has an infinite number of solutions. The general form of solution is $x = \tfrac{1}{5}(17 - 2z)$, $y = \tfrac{1}{5}(3 + 7z)$, z, where z is arbitrary. ☛ **16**

☛ 16. Show that the system
$x - 4y + 3z = 4$,
$-3x + 2y - z = -1$,
$-x - 6y + 5z = 7$ has infinitely many solutions. Express x and y in terms of z.

Answer $x = \tfrac{1}{5}z - \tfrac{2}{5}$, $y = \tfrac{4}{5}z - \tfrac{11}{10}$.

The solution in Example 1 is only one form of the general solution. We can, in fact, solve for any two of the variables in terms of the third. For example, if we want to solve for x and z in terms of y, we reduce the matrix to a form containing a second-order identity matrix in the columns corresponding to x and z.

Example 2 illustrates a different situation in which an infinite number of solutions can occur.

EXAMPLE 2 Solve the following system of four equations.

$$x - y + z - t = 5$$
$$2x - 2y + z + 3t = 2$$
$$-x + y + 2z + t = 4$$
$$3x - 3y + z + 3t = 3$$

Solution The augmented matrix for this system is

$$\begin{bmatrix} 1 & -1 & 1 & -1 & 5 \\ 2 & -2 & 1 & 3 & 2 \\ -1 & 1 & 2 & 1 & 4 \\ 3 & -3 & 1 & 3 & 3 \end{bmatrix} \begin{matrix} R_2 - 2R_1 \\ R_3 + R_1 \\ R_4 - 3R_1 \\ \xrightarrow{\hspace{1cm}} \end{matrix} \begin{bmatrix} 1 & -1 & 1 & -1 & 5 \\ 0 & 0 & -1 & 5 & -8 \\ 0 & 0 & 3 & 0 & 9 \\ 0 & 0 & -2 & 6 & -12 \end{bmatrix}$$

At this stage, we observe that the second column contains all zeros below the first row. Thus it is impossible to obtain a 1 in the second position in this column without changing the zeros in the first column. In this kind of predicament, what we do is to forget about the second column and move on to the third. The sequence of row operations $(-1)R_2$ followed by $R_1 - R_2$, $R_3 - 3R_2$, and $R_4 + 2R_2$ gives the matrix in the following form.

$$\begin{bmatrix} 1 & -1 & 1-1 & -1-(-5) & 5-8 \\ 0 & 0 & 1 & -5 & 8 \\ 0 & 0 & 3-3(1) & 0-3(-5) & 9-3(8) \\ 0 & 0 & -2+2(1) & 6+2(-5) & -12+2(8) \end{bmatrix} = \begin{bmatrix} 1 & -1 & 0 & 4 & -3 \\ 0 & 0 & 1 & -5 & 8 \\ 0 & 0 & 0 & 15 & -15 \\ 0 & 0 & 0 & -4 & 4 \end{bmatrix}$$

Having disregarded the second column, we have reduced the third column to the form that the second column would normally have—that is, 1 in the second entry and zeros elsewhere. Applying $\frac{1}{15}R_3$, we now obtain

$$\begin{bmatrix} 1 & -1 & 0 & 4 & -3 \\ 0 & 0 & 1 & -5 & 8 \\ 0 & 0 & 0 & 1 & -1 \\ 0 & 0 & 0 & -4 & 4 \end{bmatrix} \begin{matrix} R_1 - 4R_3 \\ R_2 + 5R_3 \\ R_4 + 4R_3 \\ \xrightarrow{\hspace{1cm}} \end{matrix} \begin{bmatrix} 1 & -1 & 0 & 0 & 1 \\ 0 & 0 & 1 & 0 & 3 \\ 0 & 0 & 0 & 1 & -1 \\ 0 & 0 & 0 & 0 & 0 \end{bmatrix}.$$

As in Example 1, we have obtained an entire row of zeros in the matrix, corresponding to the trivial equation $0 = 0$. The other three rows correspond to the equations

$$x - y = 1, \qquad z = 3, \qquad \text{and} \quad t = -1.$$

17. Show that the system
$2x - 4y + 3z = 4$,
$-x + 2y - z = -1$
$x - 2y + 2z = 3$ has infinitely many solutions. Give the form of the solution.

Thus we see that in this case certain of the variables (z and t) have definite values, while others (x and y) do not. Again the number of solutions is infinite, since we can allow y to have any value whatsoever; x is then given by $x = y + 1$. **17**

_____

There are also systems of equations that have no solution at all.

EXAMPLE 3 Solve the following system.

$$\begin{aligned} x + y + 2z &= 9 \\ 3x - 2y + 7z &= 20 \\ 2x + 7y + 3z &= 27 \end{aligned}$$

Solution We reduce the augmented matrix for the system as follows.

$$\begin{bmatrix} 1 & 1 & 2 & 9 \\ 3 & -2 & 7 & 20 \\ 2 & 7 & 3 & 27 \end{bmatrix} \xrightarrow[R_3 - 2R_1]{R_2 - 3R_1} \begin{bmatrix} 1 & 1 & 2 & 9 \\ 0 & -5 & 1 & -7 \\ 0 & 5 & -1 & 9 \end{bmatrix}$$

Answer $z = 2$, $x = 2y - 1$, y is arbitrary.

$$\xrightarrow{-\frac{1}{5}R_2} \begin{bmatrix} 1 & 1 & 2 & 9 \\ 0 & 1 & -\frac{1}{5} & \frac{7}{5} \\ 0 & 5 & -1 & 9 \end{bmatrix} \xrightarrow[R_3 - 5R_2]{R_1 - R_2} \begin{bmatrix} 1 & 0 & \frac{11}{5} & \frac{38}{5} \\ 0 & 1 & -\frac{1}{5} & \frac{7}{5} \\ 0 & 0 & 0 & 2 \end{bmatrix}$$

The first two columns are in the desired form of an identity matrix. However, we cannot put 1 in the third column and third row without affecting these two columns, so the reduction cannot proceed any further. Let us examine the equation represented by the third row.

18. Reduce the augmented matrix for the system
$2x - 4y + 3z = 4$,
$-3x + 2y - z = 0$,
$5x - 6y + 4z = 3$

and hence show that the system is inconsistent.

$$0x + 0y + 0z = 2, \quad \text{or} \quad 0 = 2$$

Clearly this equation is absurd. Thus the system does not have a solution, that is, there are no values of x, y, and z that satisfy all three equations of the system. **18**

_____

In general, *a system will have no solution if a row is obtained in which all the entries except the last are zero.*

We have seen three possibilities for the solution of a system. It may have a unique solution, infinitely many solutions, or no solution at all. A system is said to be **consistent** if it has at least one solution, whereas it is said to be **inconsistent** if it has no solution. The system in Example 3 is inconsistent, but

Answer Reduced form is
$$\begin{bmatrix} 1 & 0 & -\frac{1}{4} & -1 \\ 0 & 1 & -\frac{7}{8} & -\frac{3}{2} \\ 0 & 0 & 0 & -1 \end{bmatrix}$$

Examples 1 and 2 (as well as all the examples in Section 9-3) involve consistent systems.

It is clear from the examples of this section that the procedure of row reduction outlined in Section 9-3 is not sufficiently general to cope with all cases. We cannot always reduce an augmented matrix to the form $\mathbf{I}\,|\,\mathbf{C}$. More generally, we can reduce it to a form that has the following properties.

---

1. The first nonzero entry in each row is 1.

2. In the column in which such a first 1 appears, all other entries are 0.

3. The first nonzero entry in any row is to the right of the first nonzero entry in every preceding row.

4. Any rows consisting entirely of zeros are below the rows with nonzero entries.

☛ **19** In the Test for Uniqueness, the possibility that $k > n$ is not included. Can you see why this can never happen?
(*Hint:* Look at property 3 of the reduced form.)

☛ **20.** Are the following systems consistent? If so, is the solution unique?
(a) $x - 2y = 4$, $y = \frac{1}{2}x + 2$;
(b) $x - 4y + 3z = 4$,
$-x + 2y - z = -2$, $y - z = -1$;
(c) $2x - y + 3z - 4w = 4$,
$-x - z + 2w = 0$,
$x + 2y - 2w = -3$;
$-2x - y - 2z + 4w = -1$.

The augmented matrix of *any* linear system can be reduced by means of row operations to a form that satisfies these conditions. (The number of equations can be greater or less than the number of variables.) From the final reduced form it is easy to test for consistency and uniqueness of solution.

TEST FOR CONSISTENCY If the final reduced form contains a row in which only the last entry is nonzero, then the system is inconsistent. Otherwise it is consistent.

TEST FOR UNIQUENESS We suppose that the system is consistent. In the final reduced form let the number of rows in which there are non-zero entries be denoted by k (k is called the row-rank of the coefficient matrix A). Let the number of variables be n. Then:

If $k = n$ the system has one and only one solution.

If $k < n$ the system has an infinite number of solutions.   ☛ **19, 20**

These tests are very easy to apply once the reduced form has been obtained.

If a system has fewer equations than the number of variables, the system will always have more than one solution, provided it is not inconsistent. We use the method of Examples 1 and 2 above and try to obtain an identity matrix in the columns corresponding to some of the variables. This then gives the solution for the corresponding variables in terms of the others. This is illustrated in Example 4.

EXAMPLE 4 Solve the following system.

$$3x - 2y + 4z + w = -2$$
$$x + y - 3z + 2w = 12$$

Answer (a) Inconsistent;
(b) consistent, infinitely many solutions; (c) consistent, infinitely many solutions.

Solution The augmented matrix is

$$\begin{bmatrix} 3 & -2 & 4 & 1 & | & -2 \\ 1 & 1 & -3 & 2 & | & 12 \end{bmatrix}.$$

Since there are only two equations in this case, we can obtain an identity matrix of size 2×2 at most. Let us suppose we want to solve for y and w in terms of the remaining variables x and z. Then we must obtain an identity matrix in the two columns corresponding to y and w (that is, the second and fourth columns). Applying $R_1 \leftrightarrow R_2$, we have

$$\begin{bmatrix} 1 & 1 & -3 & 2 & | & 12 \\ 3 & -2 & 4 & 1 & | & -2 \end{bmatrix} \xrightarrow{R_2 + 2R_1} \begin{bmatrix} 1 & 1 & -3 & 2 & | & 12 \\ 5 & 0 & -2 & 5 & | & 22 \end{bmatrix}.$$

Thus we have obtained the y-column, as desired. Now we have to change the entries in the w-column to obtain zero in the top row and 1 in the bottom row. To obtain 1 in R_2, we apply $\frac{1}{5}R_2$.

$$\begin{bmatrix} 1 & 1 & -3 & 2 & | & 12 \\ 1 & 0 & -\frac{2}{5} & 1 & | & \frac{22}{5} \end{bmatrix} \xrightarrow{R_1 - 2R_2} \begin{bmatrix} -1 & 1 & -\frac{11}{5} & 0 & | & \frac{16}{5} \\ 1 & 0 & -\frac{2}{5} & 1 & | & \frac{22}{5} \end{bmatrix}$$

Thus we have obtained an identity matrix in the columns corresponding to y and w, as we set out to do. The system represented by the final matrix is

$$-x + \quad y - \tfrac{11}{5}z + 0w = \tfrac{16}{5}$$
$$x + 0y - \tfrac{2}{5}z + \quad w = \tfrac{22}{5}.$$

After solving for y and w, we have the following equations.

$$y = \tfrac{16}{5} + x + \tfrac{11}{5}z$$
$$w = \tfrac{22}{5} - x + \tfrac{2}{5}z$$

Thus we have expressed y and w in terms of the other two variables, x and z.

EXAMPLE 5 The system

$$2x - 3y + 4z = 7$$
$$6x - 9y + 12z = 22$$

contains two equations involving three variables. It is left as an exercise to verify that this system is inconsistent. (If the process of row reduction is carried out, it will be found that the second row reduces to zeros, except for the last entry.)

EXERCISES 9-4

(1–18) Find the solutions of the following systems where solutions exist.

1.
$$x + y + z = 5$$
$$-x + y + 3z = 1$$
$$x + 2y + 3z = 8$$

2.
$$x + y = 3$$
$$2x + y + z = 4$$
$$2x + 2y - 2z = 5$$

3.
$$x + y + z = 3$$
$$-x - y + z = -1$$
$$3x + 3y + 4z = 8$$

4.
$$6x - 5y + 6z = 7$$
$$2x + y + 6z = 5$$
$$2x - y + 3z = 3$$

5.
$$u - v + 2w = 5$$

6.
$$-x + y + z = 4$$

$$4u + v + 3w = 15$$
$$5u - 2v + 7w = 31$$

$$3x - y + 2z = -3$$
$$4x - 2y + z = 3$$

7.
$$2x + y - z = 2$$
$$3x + 2y + 4z = 8$$
$$5x + 4y + 14z = 20$$

8.
$$a + b - 2c = 3$$
$$2a + 3b + c = 13$$
$$7a + 9b - 4c = 35$$

9.
$$x + 2y - 3z - t = 2$$
$$2x + 4y + z - t = 1$$
$$3x + 6y + 2z + t = -7$$
$$x + 2y + z + t = 6$$

10.
$$p + 2q - r + 2s = 6$$
$$-2p + q + 2r + 3s = 6$$
$$3p + 5q - 3r + s = 0$$
$$p + 2q - r + s = 2$$

11.
$$u + v - w = 4$$
$$3u - v + 2w = -1$$
$$2u + 3v + w = 7$$
$$u + 2v + 3w = 2$$

12.
$$3x + 2y + z = 10$$
$$2x - y + 3z = 9$$
$$x + y - 2z = -3$$
$$2x + 3y + 4z = 20$$

13.
$$x + y - 2z = -3$$
$$2x + 3y + z = 10$$
$$-x + 2y + 3z = 9$$
$$3x + y - z = 4$$
$$x - 2y - z = 2$$

14.
$$x_1 + 2x_2 - x_3 = 2$$
$$3x_1 + x_2 + 4x_3 = 17$$
$$-2x_1 + 3x_2 + 5x_3 = 19$$
$$x_1 + x_2 + 2x_3 = 9$$
$$4x_1 - x_2 + x_3 = 4$$

15.
$$2x - y + 3z = 9$$
$$3y - 6x - 9z = 12$$

16.
$$u - 2v + w = 7$$
$$5u - 10v + 5w = 36$$

17.
$$x + y - z = 2$$
$$2x - 3y + 4z = -3$$

18.
$$2x + y - 3z = 10$$
$$3x + 2y + z = 11$$

19. *(Resource Allocation)* A small construction company makes three types of houses. The first type of house requires 3 units of concrete, 2 units of siding, and 5 units of structural lumber. The second and third types require 2, 3, 5 and 4, 2, 6 units, respectively, of concrete, siding, and lumber. If each month the company has only 150 units of concrete, 100 units of siding, and 250 units of lumber available, find the number of different types of houses which the company can make each month to use up all available concrete, siding, and lumber.

20. *(Production Decision)* A firm produces three products, A, B, and C, which require processing by three machines, I, II, and III. One unit of A requires 3, 1, and 8 hours of processing on the three machines, whereas 1 unit of B requires 2, 3, 3, and 1 unit of C requires 2, 4, and 2 hours on the three machines. The machines I, II, and III are available for 800, 1200, and 1300 hours, respectively. How many units of each should be produced to make use of all the available time on the machines?

21. Repeat Exercise 19 if the number of units of concrete, siding, and lumber available are 100, 80, and 200, respectively.

22. In Exercise 20, how many units of A, B, and C can be produced if the three machines are available for 900, 1200, and 1500 hours, respectively?

CHAPTER REVIEW

Key Terms, Symbols, and Concepts

9.1 Matrix; elements (or entries) in a matrix, row or column in a matrix.
Size of a matrix, row matrix (or vector), column matrix (or vector).
Zero matrix. Square matrix.
Equality of two matrices.
Multiplication of a matrix by a real number (scalar multiplication).
Addition of two matrices of the same size.

9.2 Product of a row matrix and a column matrix.
Multiplication of two matrices, condition for product to exist.
Diagonal elements in a matrix. Identity matrices.
System of equations: variable vector, coefficient matrix, value vector.

9.3 Augmented matrix. Row operations.
Method of row reduction. Reduced form.

9.4 Singular system. General reduced form.
Consistent and inconsistent systems.
Test for consistency. Test for uniqueness.

Formulas

If $\mathbf{P}$ is a $1 \times n$ row matrix an $\mathbf{Q}$ and $n \times 1$ column matrix then the product $\mathbf{PQ}$ is a real number equal to the sum of the n products of corresponding elements in $\mathbf{P}$ and $\mathbf{Q}$.

If $\mathbf{C} = \mathbf{AB}$, then *the ijth element of the product matrix $\mathbf{C}$ is obtained by multiplying the ith row of $\mathbf{A}$ by the jth column of $\mathbf{B}$.*

If $\mathbf{I}$ is the identity matrix of the appropriate size, $\mathbf{AI} = \mathbf{A}$, and $\mathbf{IA} = \mathbf{A}$.

Linear system of equations: $\mathbf{AX} = \mathbf{B}$.

1. Are the following statements true or false? If false, explain why.

   a. The following array of numbers,

$$\begin{bmatrix} 2 & 3 & 4 \\ 0 & 1 & 3 \\ 3 & 2 & \end{bmatrix}$$

      represents a matrix.

   b. If $\mathbf{A} = [a_1 \ \ b_1]$ and $\mathbf{B} = \begin{bmatrix} a_2 \\ b_2 \end{bmatrix}$ then

$$\mathbf{A} + \mathbf{B} = [a_1 + a_2 \ \ b_1 + b_2].$$

   c. If $\mathbf{A}$ and $\mathbf{B}$ are two matrices of the same size, then $\mathbf{A} + \mathbf{B} = \mathbf{B} + \mathbf{A}$.

   d. If $\mathbf{A} + \mathbf{B}$ is defined for two matrices $\mathbf{A}$ and $\mathbf{B}$, then the size of $\mathbf{A} + \mathbf{B}$ is the same as that of $\mathbf{A}$ or $\mathbf{B}$.

   e. The product $\mathbf{AB}$ is defined only if the number of rows in $\mathbf{A}$ is equal to the number of columns in $\mathbf{B}$.

   f. If $\mathbf{A}$ and $\mathbf{B}$ are two matrices of the same size, then $\mathbf{AB}$ and $\mathbf{BA}$ are both defined.

   g. If $\mathbf{A}$ and $\mathbf{B}$ are two matrices such that $\mathbf{AB}$ and $\mathbf{BA}$ are both defined, then $\mathbf{AB}$ is never equal to $\mathbf{BA}$.

   h. If $\mathbf{AB}$ and $\mathbf{BA}$ are both defined, then the size of $\mathbf{AB}$ or $\mathbf{BA}$ is the same as the size of $\mathbf{A}$ or $\mathbf{B}$.

   i. If $\mathbf{A}$ is a matrix of any size and $\mathbf{I}$ is the identity matrix, then $\mathbf{AI} = \mathbf{IA} = \mathbf{A}$.

   j. If $\mathbf{A}$ and $\mathbf{B}$ are two square matrices of the same size, then the size of $\mathbf{AB}$ or $\mathbf{BA}$ is the same as that of $\mathbf{A}$ or $\mathbf{B}$.

   k. If $\mathbf{A} = \mathbf{A} + \mathbf{B}$, then $\mathbf{B}$ is a zero matrix.

   l. If $\mathbf{AB} = \mathbf{0}$, then either $\mathbf{A}$ or $\mathbf{B}$ is a zero matrix.

   m. If a system has the same number of equations as the number of variables, then the system has a unique solution.

   n. If there are more variables than the number of equations, then the system has infinitely many solutions.

   o. A system of linear equations is said to be consistent if it has a unique solution.

2. Give an example of a 2×2 nonzero matrix $\mathbf{A}$ such that $\mathbf{A}^2 = \mathbf{0}$.

(3–8) Perform the indicated matrix operations and simplify.

3. $\begin{bmatrix} 2 & -1 \\ 3 & 4 \end{bmatrix} + 2\begin{bmatrix} 1 & -2 \\ 4 & 3 \end{bmatrix} - 3\begin{bmatrix} 1 & 2 \\ -3 & 0 \end{bmatrix}$

4. $\begin{bmatrix} 1 & 2 & 3 \\ 3 & -1 & 2 \\ -2 & 3 & 1 \end{bmatrix} - 3\begin{bmatrix} 0 & 1 & -1 \\ -2 & 3 & 0 \\ 1 & 0 & 2 \end{bmatrix} + 5\begin{bmatrix} 1 & 0 & 2 \\ -1 & 2 & 3 \\ 0 & -1 & 0 \end{bmatrix}$

5. $\begin{bmatrix} 1 & 0 & -1 \\ 2 & 1 & 0 \end{bmatrix} \begin{bmatrix} 2 & -1 \\ 1 & 3 \\ -3 & 2 \end{bmatrix} + 2\begin{bmatrix} 1 & 2 \\ 3 & 4 \end{bmatrix}$

6. $\begin{bmatrix} 2 & 3 \\ 1 & 0 \\ -3 & 1 \end{bmatrix} - 2\begin{bmatrix} -1 & 2 \\ 0 & -1 \\ 2 & 3 \end{bmatrix} \begin{bmatrix} 2 & 1 \\ 3 & -1 \end{bmatrix}$

7. $\begin{bmatrix} 1 & 2 & 3 \\ 0 & -1 & 2 \end{bmatrix} \begin{bmatrix} 2 & -1 \\ 3 & 4 \\ 1 & 0 \end{bmatrix} - \begin{bmatrix} 3 & 1 \\ -1 & 2 \end{bmatrix} \begin{bmatrix} 0 & 1 \\ 2 & 3 \end{bmatrix}$

8. $\begin{bmatrix} 1 & 0 & -1 \\ 0 & 1 & 2 \end{bmatrix} \left(\begin{bmatrix} 2 & 0 \\ 1 & -1 \\ 0 & 1 \end{bmatrix} + 3\begin{bmatrix} 0 & -1 \\ 1 & 2 \\ 0 & 0 \end{bmatrix} \right) + \begin{bmatrix} 2 & 1 \\ 3 & 0 \end{bmatrix} \begin{bmatrix} -1 & 2 \\ 0 & 1 \end{bmatrix}$

(9–16) Solve the following matrix equations.

9. $x[3 \ \ -1] + y[2 \ \ 1] = [7 \ \ 1]$

10. $x\begin{bmatrix} 2 \\ -1 \\ 3 \end{bmatrix} + y\begin{bmatrix} 1 \\ 2 \\ -1 \end{bmatrix} - z\begin{bmatrix} 3 \\ -1 \\ 2 \end{bmatrix} = \begin{bmatrix} -5 \\ -1 \\ 0 \end{bmatrix}$

11. $x[3 \ \ 1 \ \ 2] + y[2 \ \ -3 \ \ 1] = [1 \ \ 4 \ \ 1]$

12. $x\begin{bmatrix} 1 \\ 2 \\ 3 \end{bmatrix} + 2y\begin{bmatrix} -1 \\ 3 \\ 1 \end{bmatrix} = \begin{bmatrix} 4 \\ 10 \\ 3 \end{bmatrix}$

13. $\begin{bmatrix} 2 & -1 \\ 1 & 3 \end{bmatrix}\begin{bmatrix} x \\ y \end{bmatrix} = \begin{bmatrix} -4 \\ 5 \end{bmatrix}$

14. $\begin{bmatrix} 3 & 2 \\ 4 & -1 \end{bmatrix} \begin{bmatrix} x \\ y \end{bmatrix} + \begin{bmatrix} 5 \\ -3 \end{bmatrix} = \begin{bmatrix} 9 \\ 6 \end{bmatrix}$

15. $\begin{bmatrix} 1 & -1 & 1 \\ 2 & 1 & 3 \\ 1 & 0 & 2 \\ 3 & -2 & 4 \end{bmatrix} \begin{bmatrix} x \\ y \\ z \end{bmatrix} = \begin{bmatrix} 2 \\ 8 \\ 4 \\ 5 \end{bmatrix}$

16. $\begin{bmatrix} 1 & 2 & -1 \\ 3 & 4 & 2 \\ 2 & -1 & 1 \\ 1 & 1 & 3 \end{bmatrix} \begin{bmatrix} x \\ y \\ z \end{bmatrix} = \begin{bmatrix} 4 \\ 5 \\ 0 \\ -1 \end{bmatrix}$

(17–20) Determine the matrix **X** such that each of the following equations is satisfied.

*17. $\begin{bmatrix} 2 & 1 \\ 3 & 4 \end{bmatrix} \mathbf{X} = \begin{bmatrix} 2 & -1 \\ 3 & 1 \end{bmatrix}$

*18. $\begin{bmatrix} 2 & 1 & 3 \\ 1 & 2 & -1 \\ -1 & 1 & 1 \end{bmatrix} \mathbf{X} = \begin{bmatrix} 7 & 14 \\ -3 & 1 \\ 0 & 2 \end{bmatrix}$

*19. $\mathbf{X} \begin{bmatrix} 1 & 2 & 3 \\ 3 & -1 & 2 \end{bmatrix} = \begin{bmatrix} -1 & 5 & 4 \\ 7 & 0 & 7 \end{bmatrix}$

*20. $\mathbf{X} \begin{bmatrix} 1 & 2 & 3 \\ 2 & -1 & 0 \\ 3 & 1 & -1 \end{bmatrix} = \begin{bmatrix} -1 & -2 & 1 \\ 6 & 7 & 8 \end{bmatrix}$

21. Give an example of two 2×2 matrices **A** and **B** such that $\mathbf{A} + \mathbf{B} = 2\mathbf{I}$, where $\mathbf{A} \neq \mathbf{I}$ and $\mathbf{B} \neq \mathbf{I}$. (The answer is not unique.)

22. If $\mathbf{A} = \begin{bmatrix} 3 & 4 \\ -2 & -3 \end{bmatrix}$ then show that $\mathbf{A}^2$ is the identity matrix.

23. *(Politics and Income)* A number of people were interviewed about their political affiliation and their annual income. The following information was obtained:

517 were Liberals earning over $15,000 per year.

345 were Conservatives earning over $15,000 per year.

189 were Democrats earning over $15,000 per year.

257 were Liberals earning under $15,000 per year.

284 were Conservatives earning under $15,000 per year.

408 were Democrats earning under $15,000 per year.

Represent the above information in the form of a matrix. Is this representation unique?

24. *(Inventory Matrix)* The inventory (in gallons) of a small paint store at the beginning of a week is given by matrix **A**.

$$\begin{array}{cccc} & \text{Black} & \text{White} & \text{Red} \\ \mathbf{A} = \begin{bmatrix} 80 & 72 & 45 \\ 50 & 58 & 60 \end{bmatrix} & & & \begin{array}{l} \text{Regular} \\ \text{Deluxe} \end{array} \end{array}$$

Its sales during the week are given by matrix **S**.

$$\begin{array}{cccc} & \text{Black} & \text{White} & \text{Red} \\ \mathbf{S} = \begin{bmatrix} 65 & 70 & 39 \\ 27 & 47 & 35 \end{bmatrix} & & & \begin{array}{l} \text{Regular} \\ \text{Deluxe} \end{array} \end{array}$$

Write the inventory at the end of the week.

25. *(Production Matrix)* The Western Brewery Limited produces three brands of beer in two different sizes. The production (in thousands) per week at its Vancouver plant is

$$\begin{array}{cccc} & & \text{Brand} & \\ & \text{I} & \text{II} & \text{III} \\ \text{Size 1} & \begin{bmatrix} 13 & 27 & 15 \\ \text{Size 2} & 12 & 14 & 24 \end{bmatrix} \end{array}$$

and the weekly production at its Toronto plant is

$$\begin{array}{cccc} & & \text{Brand} & \\ & \text{I} & \text{II} & \text{III} \\ \text{Size 1} & \begin{bmatrix} 20 & 32 & 18 \\ \text{Size 2} & 35 & 24 & 30 \end{bmatrix} \end{array}.$$

a. What is the toal weekly production at the two plants?

b. If the production at the Vancouver plant is increased by 20%, what will the total production be at the two plants now?

26. *(Production Matrix)* A firm produces two types of coffee in three different sizes. The production (in thousands of units) at its plant at location A is given by

$$\begin{array}{cccc} & \text{Size 1} & \text{Size 2} & \text{Size 3} \\ \text{Type 1} & \begin{bmatrix} 20 & 28 & 30 \\ \text{Type 2} & 16 & 22 & 20 \end{bmatrix} \end{array}$$

whereas the production (in thousands) at its plant at location B is given by

$$\begin{array}{cccc} & \text{Size 1} & \text{Size 2} & \text{Size 3} \\ \text{Type 1} & \begin{bmatrix} 30 & 40 & 36 \\ \text{Type 2} & 24 & 20 & 28 \end{bmatrix} \end{array}$$

a. Write the matrix that represents the total production at both plants.

b. The firm's management is planning to open a third plant at a location C, which would have a capacity 20% more than that at location B. Write the matrix representing the production at location C.

c. What will be the total production at all three locations?

27. *(Cost of Purchases)* Steve bought 3 pants, 5 shirts, 2 ties, and 3 jackets from a department store. If the pants are $12 each, the shirts $5 each, the ties $3 each, and the jackets $20 each, use matrix multiplication to represent the total amount Steve spent at the department store.

28. *(Cost of Purchases)* A consulting firm has offices in Miami and Atlanta. The office at Miami has 5 chairs, 7 tables, and 4 typewriters. The Atlanta office has 12 chairs, 16 tables, and 8 typewriters. If the chairs are $10 each, tables $15 each, and typewriters $200 each, express the total amounts spent on these items at the two offices in terms of matrix products.

29. *(Work and Earnings)* Susan earns $5 an hour by tutoring, $6 an hour by typing, and $1.50 an hour by babysitting. The numbers of hours she worked at each type of work over a 4-week period are given by matrix **A**.

$$
\begin{array}{c}
\textit{Week} \\
\begin{array}{cccc}
\textbf{I} & \textbf{II} & \textbf{III} & \textbf{IV}
\end{array} \\
\mathbf{A} = \begin{bmatrix} 15 & 10 & 16 & 12 \\ 6 & 4 & 2 & 3 \\ 2 & 7 & 0 & 4 \end{bmatrix} \begin{array}{l} \text{Tutoring} \\ \text{Typing} \\ \text{Babysitting} \end{array}
\end{array}
$$

If $\mathbf{P} = [5 \quad 6 \quad 1.5]$ denotes her earnings matrix, determine the matrix **PA** and interpret its elements.

30. *(Contracting Charges)* A small contracting firm charges $6 per hour for a truck without a driver, $20 per hour for a tractor without a driver, and $10 per hour for each driver. The firm uses matrix **A** for various types of work.

$$
\begin{array}{c}
\textit{Type of Work} \\
\begin{array}{cccc}
\textbf{I} & \textbf{II} & \textbf{III} & \textbf{IV}
\end{array} \\
\mathbf{A} = \begin{bmatrix} 1 & 1 & 1 & 2 \\ 2 & 0 & 1 & 1 \\ 3 & 1 & 3 & 4 \end{bmatrix} \begin{array}{l} \text{Trucks} \\ \text{Tractors} \\ \text{Drivers} \end{array}
\end{array}
$$

a. If **P** denotes the price matrix that the firm charges, with $\mathbf{P} = [6 \quad 20 \quad 10]$, determine the product **PA** and interpret its elements.

b. Suppose for a small project the firm used 20 hours of type I work and 30 hours of type II work. If **S** denotes the supply matrix,

$$
\mathbf{S} = \begin{bmatrix} 20 \\ 30 \\ 0 \\ 0 \end{bmatrix}
$$

determine and interpret the elements of **AS**.

c. Evaluate and interpret the matrix product **PAS**.

CHAPTER 10

Inverses and Determinants

Chapter Objectives

10-1 THE INVERSE OF A MATRIX
(a) Definition of the inverse of a square matrix; invertible and singular matrices;
(b) Computation of the inverse matrix using row reduction;
(c) Row reduction as a test for singular matrices;
(d) Use of the inverse to solve linear systems.

10-2 INPUT-OUTPUT ANALYSIS
(a) The input-output model of the interacting sectors of an economy;
(b) Definition and construction of the input-output matrix;
(c) The computation of future outputs on the basis of predicted future demands;
(d) The basic assumptions of the input-output model.

10-3 MARKOV CHAINS (OPTIONAL)
(a) Definition of a Markov chain, the transition probabilities, and transition matrix;
(b) Use of tree diagrams to compute future probabilities for a Markov chain;
(c) State vectors; computation of future state vectors using the transition matrix;
(d) Steady-state property of Markov chains; calculation of the steady-state vector.

10-4 DETERMINANTS
(a) Definition and calculation of a 2×2 determinant;
(b) Definition of a 3×3 determinant in terms of its complete expansion;
(c) The minors and cofactors of a 3×3 determinant; calculation of a 3×3 determinant by expansion by any row or column;
(d) Definition of higher-order determinants in terms of expansion by a row or column;
(e) Cramer's rule for systems of three linear equations in three variables; its use for calculation of the solution and as a test for consistency.

10-5 INVERSES BY DETERMINANT
(a) Definition of the transpose of a matrix;
(b) The cofactor matrix and the adjoint matrix;
(c) The condition $|\mathbf{A}| \neq 0$ as a test for consistency;
(d) Use of the adjoint matrix and the determinant to construct the inverse matrix.

CHAPTER REVIEW

10-1 THE INVERSE OF A MATRIX

DEFINITION Let **A** be an $n \times n$ square matrix. Then a matrix **B** is said to be an **inverse** of **A** if it satisfies the two matrix equations

$$\mathbf{AB} = \mathbf{I} \qquad \text{and} \qquad \mathbf{BA} = \mathbf{I}$$

where **I** is the identity matrix of size $n \times n$. In other words, the product of the matrices **A** and **B** in either order is the identity matrix.

It is clear from this definition that **B** must be a square matrix of the same size as **A**; otherwise one or both of the products **AB** and **BA** will not be defined.

EXAMPLE 1 Show that $\mathbf{B} = \begin{bmatrix} -2 & 1 \\ \frac{3}{2} & -\frac{1}{2} \end{bmatrix}$ is an inverse of $\mathbf{A} = \begin{bmatrix} 1 & 2 \\ 3 & 4 \end{bmatrix}$.

Solution To show that **B** is an inverse of **A**, all we need prove is that $\mathbf{AB} = \mathbf{I}$ and $\mathbf{BA} = \mathbf{I}$.

$$\mathbf{AB} = \begin{bmatrix} 1 & 2 \\ 3 & 4 \end{bmatrix} \begin{bmatrix} -2 & 1 \\ \frac{3}{2} & -\frac{1}{2} \end{bmatrix}$$

$$= \begin{bmatrix} 1(-2) + 2(\frac{3}{2}) & 1(1) + 2(-\frac{1}{2}) \\ 3(-2) + 4(\frac{3}{2}) & 3(1) + 4(-\frac{1}{2}) \end{bmatrix} = \begin{bmatrix} 1 & 0 \\ 0 & 1 \end{bmatrix} = \mathbf{I}$$

$$\mathbf{BA} = \begin{bmatrix} -2 & 1 \\ \frac{3}{2} & -\frac{1}{2} \end{bmatrix} \begin{bmatrix} 1 & 2 \\ 3 & 4 \end{bmatrix}$$

$$= \begin{bmatrix} -2(1) + 1(3) & -2(2) + 1(4) \\ \frac{3}{2}(1) - \frac{1}{2}(3) & \frac{3}{2}(2) - \frac{1}{2}(4) \end{bmatrix} = \begin{bmatrix} 1 & 0 \\ 0 & 1 \end{bmatrix} = \mathbf{I}$$

> **1.** Show that $\mathbf{B} = \begin{bmatrix} 1 & 1 \\ 3 & 4 \end{bmatrix}$ is an inverse for $\mathbf{A} = \begin{bmatrix} 4 & -1 \\ -3 & 1 \end{bmatrix}$.

Thus **B** is an inverse of **A**. > **1**

Not every square matrix has an inverse. This is illustrated by Example 2.

EXAMPLE 2 Find an inverse of the matrix **A**, if such an inverse exists, for

$$\mathbf{A} = \begin{bmatrix} 1 & 2 \\ 2 & 4 \end{bmatrix}.$$

Solution Let **B** be an inverse of **A**. If **B** exists, it is a square matrix of the same size as **A** and so must be of the form

$$\mathbf{B} = \begin{bmatrix} a & b \\ c & d \end{bmatrix}$$

where a, b, c, and d are specific entries.

Now the equation $\mathbf{AB} = \mathbf{I}$ implies that

$$\begin{bmatrix} 1 & 2 \\ 2 & 4 \end{bmatrix} \begin{bmatrix} a & b \\ c & d \end{bmatrix} = \begin{bmatrix} 1 & 0 \\ 0 & 1 \end{bmatrix}$$

or

$$\begin{bmatrix} a + 2c & b + 2d \\ 2a + 4c & 2b + 4d \end{bmatrix} = \begin{bmatrix} 1 & 0 \\ 0 & 1 \end{bmatrix}.$$

Therefore, comparing entries in these matrices, we find that

$$\begin{matrix} a + 2c = 1 \\ 2a + 4c = 0 \end{matrix} \quad \text{and} \quad \begin{matrix} b + 2d = 0 \\ 2b + 4d = 1 \end{matrix}.$$

☛ **2.** As in Example 2, show that $\mathbf{A} = \begin{bmatrix} 4 & -6 \\ -2 & 3 \end{bmatrix}$ does not have an inverse.

These systems of equations are inconsistent, as we see if we divide the lower two equations by 2. Thus these systems *have no solution,* so there is no matrix $\mathbf{B}$ that satisfies the condition that $\mathbf{AB} = \mathbf{I}$. Thus $\mathbf{A}$ does not have an inverse. ☛ **2**

DEFINITION A matrix $\mathbf{A}$ is to be **invertible** or **nonsingular** if it has an inverse. If $\mathbf{A}$ does not have an inverse, then it is said to be a **singular matrix**.

It can be shown that the inverse of any nonsingular matrix is unique. That is, if $\mathbf{A}$ has an inverse at all, then it has only one inverse. Because of this, we denote the inverse of $\mathbf{A}$ by $\mathbf{A}^{-1}$ (read $\mathbf{A}$ *inverse*). Thus we have the two equations

$$\mathbf{AA}^{-1} = \mathbf{I} \quad \text{and} \quad \mathbf{A}^{-1}\mathbf{A} = \mathbf{I}.$$

Let us now turn to the problem of finding the inverse of a nonsingular matrix. As an example, suppose we want to find the inverse of

$$\mathbf{A} = \begin{bmatrix} 1 & 3 \\ 2 & 5 \end{bmatrix}.$$

Let

$$\mathbf{B} = \begin{bmatrix} a & b \\ c & d \end{bmatrix}$$

denote the inverse of $\mathbf{A}$. Then $\mathbf{B}$ must satisfy the two equations

$$\mathbf{AB} = \mathbf{I} \quad \text{and} \quad \mathbf{BA} = \mathbf{I}.$$

The matrix equation $\mathbf{AB} = \mathbf{I}$, when written out in full is,

$$\begin{bmatrix} 1 & 3 \\ 2 & 5 \end{bmatrix} \begin{bmatrix} a & b \\ c & d \end{bmatrix} = \begin{bmatrix} 1 & 0 \\ 0 & 1 \end{bmatrix}$$

or

$$\begin{bmatrix} a + 3c & b + 3d \\ 2a + 5c & 2b + 5d \end{bmatrix} = \begin{bmatrix} 1 & 0 \\ 0 & 1 \end{bmatrix}.$$

Therefore

$$\begin{array}{ccc} a + 3c = 1 & & b + 3d = 0 \\ 2a + 5c = 0 & \text{and} & 2b + 5d = 1 \end{array}. \tag{1}$$

Note that the two equations on the left form a system of equations for the unknown entries a and c, while the two equations on the right form a system of equations for b and d. To solve these two systems of equations, we must transform the corresponding augmented matrices

$$\begin{bmatrix} 1 & 3 & | & 1 \\ 2 & 5 & | & 0 \end{bmatrix} \quad \text{and} \quad \begin{bmatrix} 1 & 3 & | & 0 \\ 2 & 5 & | & 1 \end{bmatrix} \tag{2}$$

☛ **3.** Find the sequence of row operations that is needed to reduce each of the two augmented matrices in (2).

to their reduced forms. You can verify that these reduced forms are, respectively,

$$\begin{bmatrix} 1 & 0 & | & -5 \\ 0 & 1 & | & 2 \end{bmatrix} \quad \text{and} \quad \begin{bmatrix} 1 & 0 & | & 3 \\ 0 & 1 & | & -1 \end{bmatrix}. \quad ☛ \mathbf{3}$$

Therefore $a = -5$, $c = 2$, $b = 3$, and $d = -1$. Consequently, the matrix $\mathbf{B}$ that satisfies the equation $\mathbf{AB} = \mathbf{I}$ is

$$\mathbf{B} = \begin{bmatrix} a & b \\ c & d \end{bmatrix} = \begin{bmatrix} -5 & 3 \\ 2 & -1 \end{bmatrix}.$$

It is now easy to verify that this matrix $\mathbf{B}$ also satisfies the equation $\mathbf{BA} = \mathbf{I}$.* Therefore $\mathbf{B}$ is the inverse of $\mathbf{A}$, and we can write

$$\mathbf{A}^{-1} = \begin{bmatrix} -5 & 3 \\ 2 & -1 \end{bmatrix}.$$

In this example, to find the inverse, we solved the two linear systems in Equation (1) by reducing their augmented matrices (Equation (2)). Now it can be observed that the two systems in Equation (1) have the same coefficient matrix $\mathbf{A}$, so it is possible to reduce both augmented matrices in the same calculation since both of them require the same sequence of row operations. The procedure we can use for this simultaneous reduction is to write the coefficient matrix $\mathbf{A}$, draw a vertical line, and write the constants appearing in the right sides of the systems in Equation (1) in two columns, as shown below.

$$\begin{bmatrix} 1 & 3 & | & 1 & 0 \\ 2 & 5 & | & 0 & 1 \end{bmatrix} \tag{3}$$

We now perform row operations in the usual way to reduce the left half of this augmented matrix to an identity matrix.

Answer The *same* sequence works for both matrices:
$R_2 - 2R_1, \ -R_2, \ R_1 - 3R_2$.

*It can be shown (though the proof is quite difficult) that if either one of the two conditions $\mathbf{AB} = \mathbf{I}$ or $\mathbf{BA} = \mathbf{I}$ is satisfied, then the other one is satisfied automatically. This is why we need to use only one of these conditions in order to determine $\mathbf{B}$.

It may be observed that the elements to the right of the vertical line in the augmented matrix (3) form a 2×2 identity matrix. Thus this augmented matrix can be abbreviated as simply $\mathbf{A}|\mathbf{I}$. If we transform $\mathbf{A}|\mathbf{I}$ to reduced form, we shall be simultaneously reducing the two augmented matrices in Equation (2) and therefore solving the two linear systems in Equation (1) at the same time.

In this example, $\mathbf{A}|\mathbf{I}$ is reduced by the following sequence of row operations.

$$\mathbf{A}|\mathbf{I} = \begin{bmatrix} 1 & 3 & | & 1 & 0 \\ 2 & 5 & | & 0 & 1 \end{bmatrix} \xrightarrow{R_2 - 2R_1} \begin{bmatrix} 1 & 3 & | & 1 & 0 \\ 0 & -1 & | & -2 & 1 \end{bmatrix}$$

$$\xrightarrow{-R_2} \begin{bmatrix} 1 & 3 & | & 1 & 0 \\ 0 & 1 & | & 2 & -1 \end{bmatrix}$$

$$\xrightarrow{R_1 - 3R_2} \begin{bmatrix} 1 & 0 & | & -5 & 3 \\ 0 & 1 & | & 2 & -1 \end{bmatrix}$$

This is now the required reduced matrix, since it has an identity matrix to the left of the vertical line. The elements to the right of this line are the solutions of the two systems in Equation (1)—in other words, they form the matrix

$$\begin{bmatrix} a & b \\ c & d \end{bmatrix} = \begin{bmatrix} -5 & 3 \\ 2 & -1 \end{bmatrix}.$$

But this matrix is the inverse of $\mathbf{A}$. We conclude, therefore, that in the reduced form of the augmented matrix $\mathbf{A}|\mathbf{I}$, the inverse of $\mathbf{A}$ appears to the right of the vertical line. To summarize: *Let $\mathbf{A}$ be an invertible square matrix of size $n \times n$, and let $\mathbf{I}$ be the identity matrix of the same size. Then the reduced form of $\mathbf{A}|\mathbf{I}$ is $\mathbf{I}|\mathbf{A}^{-1}$.*

EXAMPLE 3 Find $\mathbf{A}^{-1}$, given

$$\mathbf{A} = \begin{bmatrix} 1 & 2 & 3 \\ 2 & 5 & 7 \\ 3 & 7 & 8 \end{bmatrix}.$$

Solution

$$\mathbf{A}|\mathbf{I} = \begin{bmatrix} 1 & 2 & 3 & | & 1 & 0 & 0 \\ 2 & 5 & 7 & | & 0 & 1 & 0 \\ 3 & 7 & 8 & | & 0 & 0 & 1 \end{bmatrix} \xrightarrow[R_3 - 3R_1]{R_2 - 2R_1} \begin{bmatrix} 1 & 2 & 3 & | & 1 & 0 & 0 \\ 0 & 1 & 1 & | & -2 & 1 & 0 \\ 0 & 1 & -1 & | & -3 & 0 & 1 \end{bmatrix}$$

$$\xrightarrow[R_3 - R_2]{R_1 - 2R_2} \begin{bmatrix} 1 & 0 & 1 & | & 5 & -2 & 0 \\ 0 & 1 & 1 & | & -2 & 1 & 0 \\ 0 & 0 & -2 & | & -1 & -1 & 1 \end{bmatrix}$$

$$\xrightarrow{-\frac{1}{2}R_3} \begin{bmatrix} 1 & 0 & 1 & | & 5 & -2 & 0 \\ 0 & 1 & 1 & | & -2 & 1 & 0 \\ 0 & 0 & 1 & | & \frac{1}{2} & \frac{1}{2} & -\frac{1}{2} \end{bmatrix}$$

☛ 4. Find A^{-1} if $A = \begin{bmatrix} 5 & -4 \\ -4 & 3 \end{bmatrix}$.

$\xrightarrow[\quad]{\begin{array}{c} R_1 - R_3 \\ R_2 - R_3 \end{array}} \begin{bmatrix} 1 & 0 & 0 \\ 0 & 1 & 0 \\ 0 & 0 & 1 \end{bmatrix} \left| \begin{array}{ccc} \frac{9}{2} & -\frac{5}{2} & \frac{1}{2} \\ -\frac{5}{2} & \frac{1}{2} & \frac{1}{2} \\ \frac{1}{2} & \frac{1}{2} & -\frac{1}{2} \end{array} \right.$

This matrix is now in the reduced form $I \mid A^{-1}$.

Thus

$$A^{-1} = \begin{bmatrix} \frac{9}{2} & -\frac{5}{2} & \frac{1}{2} \\ -\frac{5}{2} & \frac{1}{2} & \frac{1}{2} \\ \frac{1}{2} & \frac{1}{2} & -\frac{1}{2} \end{bmatrix} = \frac{1}{2} \begin{bmatrix} 9 & -5 & 1 \\ -5 & 1 & 1 \\ 1 & 1 & -1 \end{bmatrix}.$$

You may verify that this is, in fact, the inverse matrix of A by checking the two equations

$$AA^{-1} \quad \text{and} \quad A^{-1}A = I. \quad \text{☛ 4}$$

Answer $A^{-1} = \begin{bmatrix} -3 & -4 \\ -4 & -5 \end{bmatrix}$.

How do we know whether a given matrix A is invertible or not? If we carry out the process of transforming $A \mid I$ to its reduced form and if at any step we find that one of the rows on the left of the vertical line consists entirely of zeros, then it can be shown that A^{-1} does not exist.

☛ 5. Find A^{-1}, if it exists, if

(a) $A = \begin{bmatrix} 1 & -2 & -1 \\ 0 & 2 & 1 \\ -1 & 1 & 0 \end{bmatrix}$;

(b) $A = \begin{bmatrix} 1 & 4 & 7 \\ 2 & 5 & 8 \\ 3 & 6 & 9 \end{bmatrix}$.

EXAMPLE 4 Find A^{-1} if it exists, given

$$A = \begin{bmatrix} 1 & 2 & 3 \\ 2 & 5 & 7 \\ 3 & 7 & 10 \end{bmatrix}.$$

Solution

$$A \mid I = \begin{bmatrix} 1 & 2 & 3 \\ 2 & 5 & 7 \\ 3 & 7 & 10 \end{bmatrix} \left| \begin{array}{ccc} 1 & 0 & 0 \\ 0 & 1 & 0 \\ 0 & 0 & 1 \end{array} \right. \xrightarrow[\quad]{\begin{array}{c} R_2 - 2R_1 \\ R_3 - 3R_1 \end{array}} \begin{bmatrix} 1 & 2 & 3 \\ 0 & 1 & 1 \\ 0 & 1 & 1 \end{bmatrix} \left| \begin{array}{ccc} 1 & 0 & 0 \\ -2 & 1 & 0 \\ -3 & 0 & 1 \end{array} \right.$$

$$\xrightarrow[\quad]{R_3 - R_2} \begin{bmatrix} 1 & 2 & 3 \\ 0 & 1 & 1 \\ 0 & 0 & 0 \end{bmatrix} \left| \begin{array}{ccc} 1 & 0 & 0 \\ -2 & 1 & 0 \\ -1 & -1 & 1 \end{array} \right.$$

Answer (a) $A^{-1} = \begin{bmatrix} 1 & 1 & 0 \\ 1 & 1 & 1 \\ -2 & -1 & -2 \end{bmatrix}$;

(b) A^{-1} does not exist.

Since the third row to the left of the vertical line consists entirely of zeros, the reduction cannot be completed. We must conclude that A^{-1} does not exist and that A is a *singular* matrix. (See also Example 4 in Section 10-5). **☛ 5**

Inverses of matrices have many uses, one of which is in the solution of systems of equations. In Section 9-3, we solved systems of linear equations by transforming an augmented matrix to its reduced form. In the case when we have n equations in n variables, we can also solve the system by finding the inverse of the coefficient matrix.

A system of equations can be written in matrix form as $AX = B$. If the coefficient matrix A is invertible, then A^{-1} exists. Multiplying both sides of the given matrix by $\mathbf{A}^{-1}$ on the left, we obtain

$$\mathbf{A}^{-1}(\mathbf{AX}) = \mathbf{A}^{-1}\mathbf{B}.$$

Using the asociative property and simplifying, we can write this as follows.

$$(\mathbf{A}^{-1}\mathbf{A})\mathbf{X} = \mathbf{A}^{-1}\mathbf{B}$$

$$\mathbf{IX} = \mathbf{A}^{-1}\mathbf{B}$$

$$\mathbf{X} = \mathbf{A}^{-1}\mathbf{B}$$

Thus we have obtained an expression for the solution $\mathbf{X}$ of the given system of equations.

EXAMPLE 5 Solve the following system of linear equations.

$$x + 2y + 3z = 3$$

$$2x + 5y + 7z = 6$$

$$3x + 7y + 8z = 5$$

Solution The given system of equations in matrix form is

$$\mathbf{AX} = \mathbf{B} \tag{4}$$

where

$$\mathbf{A} = \begin{bmatrix} 1 & 2 & 3 \\ 2 & 5 & 7 \\ 3 & 7 & 8 \end{bmatrix}, \qquad \mathbf{X} = \begin{bmatrix} x \\ y \\ z \end{bmatrix} \quad \text{and} \quad \mathbf{B} = \begin{bmatrix} 3 \\ 6 \\ 5 \end{bmatrix}.$$

Then $\mathbf{A}^{-1}$ (as found in Example 3) is given by

$$\mathbf{A}^{-1} = \tfrac{1}{2}\begin{bmatrix} 9 & -5 & 1 \\ -5 & 1 & 1 \\ 1 & 1 & -1 \end{bmatrix}.$$

It follows that the solution of Equation (4) is given by

$$\mathbf{X} = \mathbf{A}^{-1}\mathbf{B} = \tfrac{1}{2}\begin{bmatrix} 9 & -5 & 1 \\ -5 & 1 & 1 \\ 1 & 1 & -1 \end{bmatrix}\begin{bmatrix} 3 \\ 6 \\ 5 \end{bmatrix}$$

$$= \tfrac{1}{2}\begin{bmatrix} 27 - 30 + 5 \\ -15 + 6 + 5 \\ 3 + 6 - 5 \end{bmatrix} = \tfrac{1}{2}\begin{bmatrix} 2 \\ -4 \\ 4 \end{bmatrix} = \begin{bmatrix} 1 \\ -2 \\ 2 \end{bmatrix}.$$

That is,

$$\begin{bmatrix} x \\ y \\ z \end{bmatrix} = \begin{bmatrix} 1 \\ -2 \\ 2 \end{bmatrix}.$$

Therefore $x = 1$, $y = -2$, and $z = 2$.

● **6.** Solve Example 6 if

$$A = \begin{bmatrix} 0 & 1 & 1 \\ 1 & 2 & -1 \\ -2 & -3 & 1 \end{bmatrix}.$$

At first glance, it may appear that this method of solving a system of equations is much less convenient than the simple method of row reduction described in Section 9-3. The advantage of using the inverse matrix occurs in cases where several systems of equations with the same coefficient matrix must be solved. For such problems as this, the solutions of *all* the systems can be determined immediately once the inverse of the coefficient matrix has been found; it is not necessary to use row reduction over and over again for each system. (See the final remark in the next section.)

EXAMPLE 6 Find the solution of the system **AX** = **B**, where

$$A = \begin{bmatrix} 1 & -1 & 3 \\ 2 & 0 & 1 \\ 4 & 3 & -2 \end{bmatrix}, \qquad X = \begin{bmatrix} x \\ y \\ z \end{bmatrix}, \qquad B = \begin{bmatrix} a \\ b \\ c \end{bmatrix}$$

and a, b, and c are arbitrary real numbers.

Solution We leave it as an exercise for you to compute the inverse matrix. The result is

$$A^{-1} = \frac{1}{7} \begin{bmatrix} -3 & 7 & -1 \\ 8 & -14 & 5 \\ 6 & -7 & 2 \end{bmatrix}.$$

The solution of the system **AX** = **B** is then

$$X = A^{-1}B = \frac{1}{7} \begin{bmatrix} -3 & 7 & -1 \\ 8 & -14 & 5 \\ 6 & -7 & 2 \end{bmatrix} \begin{bmatrix} a \\ b \\ c \end{bmatrix}$$

$$= \frac{1}{7} \begin{bmatrix} -3a + 7b - c \\ 8a - 14b + 5c \\ 6a - 7b + 2c \end{bmatrix}.$$

Answer $x = -\frac{1}{2}(a + 4b + 3c)$,
$y = \frac{1}{2}(a + 2b + c)$, $z = \frac{1}{2}(a - 2b - c)$.

Therefore

$$x = \frac{1}{7}(-3a + 7b - c) \qquad y = \frac{1}{7}(8a - 14b + 5c) \qquad z = \frac{1}{7}(6a - 7b + 2c). \quad ● 6$$

EXERCISES 10-1

(1–16) In the following problems, find the inverse of the given matrix (if it exists).

1. $\begin{bmatrix} 2 & 5 \\ 3 & 4 \end{bmatrix}$

2. $\begin{bmatrix} 3 & 1 \\ 4 & 2 \end{bmatrix}$

3. $\begin{bmatrix} 1 & -2 \\ -3 & 4 \end{bmatrix}$

4. $\begin{bmatrix} 1 & -3 \\ -2 & 6 \end{bmatrix}$

5. $\begin{bmatrix} 3 & -2 \\ -6 & 4 \end{bmatrix}$

6. $\begin{bmatrix} 1 & 2 \\ 0 & 0 \end{bmatrix}$

7. $\begin{bmatrix} 1 & 0 & 2 \\ 0 & 3 & 1 \\ 2 & -1 & 0 \end{bmatrix}$

8. $\begin{bmatrix} 2 & 1 & 0 \\ 1 & 0 & 3 \\ 0 & 2 & 1 \end{bmatrix}$

9. $\begin{bmatrix} 2 & 3 & 4 \\ 1 & 2 & 0 \\ 4 & 5 & 6 \end{bmatrix}$ 10. $\begin{bmatrix} 1 & 2 & 3 \\ -2 & 1 & 4 \\ -3 & -4 & 1 \end{bmatrix}$

11. $\begin{bmatrix} 2 & 1 & -1 \\ 3 & 2 & 0 \\ 4 & 3 & 1 \end{bmatrix}$ 12. $\begin{bmatrix} 3 & -4 & 5 \\ 4 & -3 & 6 \\ 6 & -8 & 10 \end{bmatrix}$

13. $\begin{bmatrix} -1 & 2 & -3 \\ 2 & -1 & 1 \\ 3 & 1 & 2 \end{bmatrix}$ 14. $\begin{bmatrix} -3 & 2 & 1 \\ 2 & -1 & 3 \\ 1 & -3 & 2 \end{bmatrix}$

15. $\begin{bmatrix} 1 & -1 & 1 & 2 \\ 2 & -3 & 0 & 3 \\ 1 & 1 & 1 & 1 \\ 3 & 0 & -1 & 2 \end{bmatrix}$ 16. $\begin{bmatrix} 2 & 1 & 3 & 4 \\ 1 & 1 & 1 & -1 \\ -1 & 1 & -1 & 0 \\ 3 & 0 & 1 & 2 \end{bmatrix}$

(17–24) Solve the following systems of equations by finding the inverse of the coefficient matrix.

17. $2x - 3y = 1$
$3x + 4y = 10$

18. $3x_1 + 2x_2 = 1$
$2x_1 - x_2 = 3$

19. $4u + 5v = 14$
$2v - 3u = 1$

20. $3y - 2z = -4$
$5z + 4y = -13$

21. $2x - y + 3z = -3$
$x + y + z = 2$
$3x + 2y - z = 8$

22. $x + 2y - z = 1$
$2z - 3x = 2$
$3y + 2z = 5$

23. $2u + 3v - 4w = -10$
$w - 2u - 1 = 0$
$u + 2v = 1$

24. $p + 2q - 3r = 1$
$q - 2p + r = 3$
$2r + p - 2 = 0$

25. *(Ore Refining)* Two metals, X and Y, can be extracted from two types of ores, P and Q. One hundred pounds of ore P yield 3 ounces of X and 5 ounces of Y, and 100 pounds of ore Q yield 4 ounces of X and 2.5 ounces of Y. How many pounds of ores P and Q will be required to produce 72 ounces of X and 95 ounces of Y?

26. *(Investment)* A person invested a total of $20,000 in three different investments that yield 5%, 6%, and 8%, respectively. The income from the 8% investment is twice the income from the 5% investment and the total income per year from all three investments is $1296. Find the amount invested in each investment.

27. If A is a nonsingular matrix and $AB = AC$, show that $B = C$.

28. If $AB = A$ and A is nonsingular, show that $B = I$.

29. Given

$$A = \begin{bmatrix} 1 & 3 \\ 2 & 4 \end{bmatrix} \quad \text{and} \quad B = \begin{bmatrix} 2 & -1 \\ -3 & 1 \end{bmatrix}.$$

Verify the result $(AB)^{-1} = B^{-1}A^{-1}$.

30. Use the matrices A and B in Exercise 29 to verify that $(A^{-1}B)^{-1} = B^{-1}A$.

*31. Show that $(A^{-1})^{-1} = A$ for any invertible matrix A.

*32. Show that for any two noninvertible $n \times n$ matrices A and B,

$$(AB)^{-1} = B^{-1}A^{-1}.$$

*33. Show that if two matrices B and C are both inverses of a matrix A, then $B = C$. (*Hint:* Consider BAC.)

10-2 INPUT-OUTPUT ANALYSIS

The input-output model was first introduced in the late forties by Leontief, the recipient of a 1973 Nobel prize, in a study of the U.S. economy. The main feature of this model is that it incorporates the interactions between different industries or sectors which make up the economy. The aim of the model is to allow economists to forecast the future production levels of each industry in order to meet future demands for the various products. Such forecasting is complicated by the linkages between the different industries, through which a change in the demand for the product of one industry can induce a change in the production levels of other industries. For example, an increase in the demand for automobiles leads not only to an increase in the production levels of automobile manufacturers, but also in the levels of a variety of other industries in the economy, such as the steel industry, the rubber industry, and so on. In

Leontief's original model, he divided the U.S. economy into 500 interacting sectors of this type.

In order to describe the model in the simplest possible terms, we consider an economy that consists of only two industries, P and Q. To clarify our ideas, suppose that the interactions between these two industries are those given in Table 1. The first two columns in this table give the *inputs* of the two industries, measured in suitable units. (For example, the units might be millions of dollars per year.) From the first column, we see that in its annual production, industry P uses 60 units of its own product and 100 units of the product of industry Q. Similarly, Q uses 64 units of P's product and 48 units of its own product. In addition, from the last row we see that P uses 40 units of *primary inputs*, which include inputs such as labor, land, or raw materials, while Q uses 48 units of primary inputs.

TABLE 1

	Industry P Inputs	Industry Q Inputs	Final Demands	Gross Output
Industry P Outputs	60	64	76	200
Industry Q Outputs	100	48	12	160
Primary	40	48		
Total Inputs	200	160		

Totaling the columns, we see that the total inputs are 200 units for P and 160 units for Q. It is an assumption of the model that whatever is produced is consumed, or, in other words, the output of each industry must equal the sum of all the inputs (measured in common units). Thus the gross output of the two industries must be 200 units for P and 160 units for Q.

Now consider the first two rows in Table 1, which show how the outputs of each industry are used. Of the 200 units produced by P, 60 are used by that industry itself and 64 are used by Q. This leaves 76 units available to meet the *final demand,* that is, the goods that are not used internally by the producing industries themselves. These would consist primarily of goods produced for household consumption, government consumption, or export. Similarly, of the 160 units produced by Q, 100 are used by P, 48 are used by Q itself, and 12 units are left to meet the final demand.

Suppose that market research predicts that in 5 years, the final demand for P will decrease from 76 to 70 units, whereas for Q, it will increase considerably from 12 to 60 units. The question arises concerning how much each industry should adjust its production level in order to meet these projected final demands.

It is clear that the two industries do not operate independently of one another—for example, the gross output of P depends on the final demand for Q's product, and vice versa. Thus the output of one industry is linked to the outputs of the other industry or industries. Let us suppose that in order to meet the projected final demands in 5 years, P must produce x_1 units and Q must product x_2 units.

From Table 1, we see that to produce 200 units, industry P uses 60 units of its own product and 100 units of Q's product. Thus to produce x_1 units, industry P must use $\frac{60}{200}x_1$ units of its own product and $\frac{100}{200}x_1$ units of Q's product. Similarly, to produce x_2 units, industry Q should use $\frac{64}{160}x_2$ units of P's product and $\frac{48}{160}x_2$ units of its own product. But we have the following equation:

$$\begin{array}{l}\text{Gross Output} \\ \text{of Industry P}\end{array} = \begin{array}{l}\text{Units Consumed} \\ \text{by P}\end{array} + \begin{array}{l}\text{Units Consumed} \\ \text{by Q}\end{array} + \text{Final Demands.}$$

That is

$$x_1 = \tfrac{60}{200}x_1 + \tfrac{64}{160}x_2 + 70$$

since the new final demand is 70 units.

Similarly, out of x_2 units produced by industry Q, $\frac{100}{200}x_1$ units are used up by P and $\frac{48}{160}x_2$ units are used by Q itself. Thus we have,

$$\begin{array}{l}\text{Gross Output} \\ \text{of Industry Q}\end{array} = \begin{array}{l}\text{Units Consumed} \\ \text{by P}\end{array} + \begin{array}{l}\text{Units Consumed} \\ \text{by Q}\end{array} + \text{Final Demands.}$$

That is,

$$x_2 = \tfrac{100}{200}x_1 + \tfrac{48}{160}x_2 + 60.$$

These two equations can be written in matrix form as

$$\begin{bmatrix} x_1 \\ x_2 \end{bmatrix} = \begin{bmatrix} \frac{60}{200} & \frac{64}{160} \\ \frac{100}{200} & \frac{48}{160} \end{bmatrix} \begin{bmatrix} x_1 \\ x_2 \end{bmatrix} + \begin{bmatrix} 70 \\ 60 \end{bmatrix}.$$

Thus

$$\mathbf{X} = \mathbf{AX} + \mathbf{D} \qquad (1)$$

where

$$\mathbf{X} = \begin{bmatrix} x_1 \\ x_2 \end{bmatrix}, \quad \mathbf{A} = \begin{bmatrix} \frac{60}{200} & \frac{64}{160} \\ \frac{100}{200} & \frac{48}{160} \end{bmatrix} \quad \text{and} \quad \mathbf{D} = \begin{bmatrix} 70 \\ 60 \end{bmatrix}.$$

We call **X** the **output matrix,** **D** the **demand matrix,** and **A** the **input-output matrix.** The entries in matrix **A** are called the **input-output coefficients.**

Let us consider the interpretation of the elements of the input-output matrix. As usual, let a_{ij} denote the general element in **A**. We note that of the 200 units of total input into industry P, 60 consist of units of its own product and 100 consist of units of Q's product. Thus the entries $\frac{60}{200}$ and $\frac{100}{200}$ in the first column of the input-output matrix represent the proportion of P's inputs which come from the industries P and Q, respectively. In general, a_{ij} represents the fractional part of the input of industry j that is the product of industry i.

Each entry in the input-output matrix is between 0 and 1, and the sum of the entries in any column is never greater than 1. Note that the input-output matrix

$$\mathbf{A} = \begin{bmatrix} \frac{60}{200} & \frac{64}{160} \\ \frac{100}{200} & \frac{48}{160} \end{bmatrix} = \begin{bmatrix} 0.3 & 0.4 \\ 0.5 & 0.3 \end{bmatrix}$$

☛ 7. What are the outputs if the prediction of future demands is changed to 70 and 50 for P and Q respectively?

Answer $D = \begin{bmatrix} 70 \\ 50 \end{bmatrix}$, $X = \begin{bmatrix} \frac{6900}{29} \\ \frac{7000}{29} \end{bmatrix}$.

☛ 8. The headings in the following table are the same as in Table 1.

20	40	40	100
80	80	40	200
0	80		
100	200		

Construct the input-output matrix and find the outputs if the final demands are changed to 30 and 50 respectively.

Answer $A = \begin{bmatrix} 0.2 & 0.2 \\ 0.8 & 0.4 \end{bmatrix}$.
Outputs 87.5 and 200.

in the preceding example can be obtained directly from Table 1 by dividing each number in the interior rectangle of the table by the gross output of the industry that heads the column. For example, in the first column headed by P, we divide each entry by 200, which is the gross output of industry P. Thus we obtain $\frac{60}{200}$ and $\frac{100}{200}$ as the elements of the first column in the input-output matrix.

Equation (1), $X = AX + D$, is called the **input-output equation**. To find the output matrix X that will meet the projected new final demands, we must solve Equation (1) for X. We have

$$X = AX + D$$

$$X - AX = D.$$

We can write this as

$$IX - AX = D \quad \text{or} \quad (I - A)X = D.$$

We have a system of linear equations whose coefficient matrix is $(I - A)$. We can solve this system by row reduction or alternatively by using the inverse of the coefficient matrix. Suppose $(I - A)^{-1}$ exists. Then, as in Section 10-1, we have

$$(I - A)^{-1}(I - A)X = (I - A)^{-1}D$$

$$X = (I - A)^{-1}D.$$

Thus we see that the output matrix X is determined once the inverse of the matrix $(I - A)$ has been found. This inverse can be calculated using the methods of Section 10-1.

In our example, we have

$$I - A = \begin{bmatrix} 1 & 0 \\ 0 & 1 \end{bmatrix} - \begin{bmatrix} 0.3 & 0.4 \\ 0.5 & 0.3 \end{bmatrix} = \begin{bmatrix} 0.7 & -0.4 \\ -0.5 & 0.7 \end{bmatrix}.$$

Using the methods of Section 10-1, we find that

$$(I - A)^{-1} = \tfrac{1}{29}\begin{bmatrix} 70 & 40 \\ 50 & 70 \end{bmatrix}.$$

Therefore

$$X = (I - A)^{-1}D$$

$$= \tfrac{1}{29}\begin{bmatrix} 70 & 40 \\ 50 & 70 \end{bmatrix}\begin{bmatrix} 70 \\ 60 \end{bmatrix} = \begin{bmatrix} \frac{7300}{29} \\ \frac{7700}{29} \end{bmatrix} = \begin{bmatrix} 251.7 \\ 265.5 \end{bmatrix}.$$

Thus industry P must produce 251.7 units and Q should produce 265.5 units to meet the projected final demands in 5 years. **☛ 7**

It may happen that an economist is uncertain about his or her prediction of future final demands. Thus he or she may wish to calculate the output matrix X for a number of different demand matrices D. In such a case, it is much more convenient to use the formula $X = (I - A)^{-1}D$, involving the inverse matrix, than to use row reduction to solve for X for each different D. **☛ 8**

EXAMPLE 1 *(Input-Output Model)* Suppose that in a hypothetical economy with only two industries, I and II, the interaction between the industries is as shown in Table 2.

(a) Determine the input-output matrix **A**.

(b) Determine the output matrix if the final demands change to 312 units for industry I and 299 units for industry II.

(c) What will then be the new primary inputs for the two industries?

TABLE 2

	Industry I	Industry II	Final Demands	Gross Output
Industry I	240	750	210	1200
Industry II	720	450	330	1500
Primary Inputs	240	300		

Solution

(a) Dividing the first column (headed by industry I) by the gross output of industry I, 1200, and the second column (headed by industry II) by the gross output of industry II, 1500, we obtain input-output matrix **A**.

$$\mathbf{A} = \begin{bmatrix} \frac{240}{1200} & \frac{750}{1500} \\ \frac{720}{1200} & \frac{450}{1500} \end{bmatrix} = \begin{bmatrix} 0.2 & 0.5 \\ 0.6 & 0.3 \end{bmatrix}$$

(b) If **I** denotes the 2×2 identity matrix, then

$$\mathbf{I} - \mathbf{A} = \begin{bmatrix} 1 & 0 \\ 0 & 1 \end{bmatrix} - \begin{bmatrix} 0.2 & 0.5 \\ 0.6 & 0.3 \end{bmatrix} = \begin{bmatrix} 0.8 & -0.5 \\ -0.6 & 0.7 \end{bmatrix}.$$

Using the methods of Section 10-1, we obtain

$$(\mathbf{I} - \mathbf{A})^{-1} = \begin{bmatrix} \frac{35}{13} & \frac{25}{13} \\ \frac{30}{13} & \frac{40}{13} \end{bmatrix} = \frac{5}{13} \begin{bmatrix} 7 & 5 \\ 6 & 8 \end{bmatrix}.$$

If **D** denotes the new demand vector, that is,

$$\mathbf{D} = \begin{bmatrix} 312 \\ 299 \end{bmatrix}$$

and **X** is the new output matrix, then we have

$$\mathbf{X} = (\mathbf{I} - \mathbf{A})^{-1}\mathbf{D}$$

$$= \frac{5}{13} \begin{bmatrix} 7 & 5 \\ 6 & 8 \end{bmatrix} \begin{bmatrix} 312 \\ 299 \end{bmatrix} = \frac{5}{13} \begin{bmatrix} 3679 \\ 4264 \end{bmatrix} = \begin{bmatrix} 1415 \\ 1640 \end{bmatrix}.$$

Thus industry I must produce 1415 units and industry II must produce 1640 units to meet the final new demands.

☛ 9. A two-sector economy is described in the following table:

	Primary Industry	Secondary Industry	Final Demands	Gross Output
Primary	10	75	15	100
Secondary	50	60	40	150
Primary Inputs	40	15		

Construct the input-output matrix and find the outputs if the final demands are changed to 40 and 40 respectively.

Answer $A = \begin{bmatrix} 0.1 & 0.5 \\ 0.5 & 0.4 \end{bmatrix}$,

$X = \begin{bmatrix} \frac{4400}{29} \\ \frac{5600}{29} \end{bmatrix}$.

(c) For industry I, 240 units of primary inputs are needed to produce a gross output of 1200 units. That is, primary inputs are $\frac{240}{1200} = 0.2$ of the gross output. Thus 0.2 of the new output, 1415, gives the new primary inputs for industry I. The primary inputs for industry I are $0.2(1415) = 283$ units. Similarly, the primary inputs for industry II are $\frac{300}{1500} = 0.2$ of the gross output, and so equal $0.2(1640) = 328$ units. The new primary inputs for the two industries will thus be 283 units and 328 units, respectively. ☛ 9

The basic assumptions of the input-output model can be seen in these simple examples involving only two interacting sectors. In a realistic model of a national economy, it is necessary to consider a much larger number of sectors. Such an enlargement of the model introduces great complications in the calculations, and it becomes necessary to use a computer to solve the system of equations. However, the principles involved in the model remain essentially the same as those in our two-sector examples.

We can summarize these basic assumptions as follows.

1. Each industry or sector of the economy produces a single commodity and no two industries produce the same commodity.

2. For each industry, the total value of the output is equal to the total value of all the inputs, and all the outputs are consumed either by other producing sectors or as final demands.

3. The input-output matrix remains constant over the period of time under consideration. Over long periods, technological advances do cause changes in the input-output matrix and this means that predictions based on this model must be relatively short-term to be reliable.

EXERCISES 10-2

1. *(Input-Output Model)* Table 3 gives the interaction between two sectors in a hypothetical economy.

TABLE 3

	Industry I	Industry II	Final Demands	Gross Output
Industry I	20	56	24	100
II	50	8	22	80
Primary Inputs	30	16		

a. Find the input-output matrix **A**.

b. If in five years the final demands change to 74 for industry I and 37 for industry II, how much should each industry produce to meet this projected demand?

c. What will be the new primary input requirements in 5 years for the two industries?

2. *(Input-Output Model)* The interaction between the two sectors of a hypothetical economy is given in Table 4.

a. Find the input-output matrix **A**.

b. Suppose that in 3 years the demand for agricultural products decreases to 63 units and increases to 105

TABLE 4

	Agri-culture	Manu-facturing	Final Demands	Total Output
Agriculture	240	270	90	600
Manufacturing	300	90	60	450
Labor	60	90		

units for manufactured goods. Determine the new output vector to meet these new demands.

c. What will the new labor requirements be for each sector?

3. *(Input-Output Model)* Table 5 gives the interaction between two sectors of a hypothetical economy.

TABLE 5

	Industry P	Q	Final Demands	Total Output
Industry				
P	60	75	65	200
Q	80	30	40	150
Labor	60	45		

a. Determine the input-output matrix **A**.

b. Determine the output matrix if the final demands change to 104 for P and 172 for Q.

c. What are the new labor requirements?

4. *(Input-Output Model)* The interaction between two industries P and Q that form a hypothetical economy is given in Table 6.

a. Determine the input-output matrix A.

TABLE 6

	Industry P	Q	Consumer Demands	Total Output
Industry				
P	46	342	72	460
Q	322	114	134	570
Labor Inputs	92	114		

b. Determine the output matrix if the consumer demands change to 129 for P and 213 for Q.

c. What will the new labor requirements be for the two industries?

*5. For the economy of Exercise 3, it is anticipated that the final demand for the output of industry Q will increase by twice as much over the next few years as the final demand for the output of industry P. During the next 5 years, the total labor pool available to the two industries will increase from 105 units to 150 units. How much must the two final demands increase during this period if this whole labor pool is to be employed?

6. *(Input-Output Model)* The interaction between three industries P, Q, and R is given in Table 7.

TABLE 7

	Industry P	Q	R	Final Demands	Total Output
Industry					
P	20	0	40	40	100
Q	40	40	100	20	200
R	0	80	40	80	200
Primary Input	40	80	20		

a. Construct the input-output matrix.

b. Determine the new outputs of P, Q, and R if the final demands change in the future to 70, 50, and 120, respectively.

c. What will then be new primary inputs for the three industries?

7. Repeat Exercise 6 for the three-sector economy given in Table 8, if the new final demands are 68, 51, and 17 for P, Q, and R respectively.

TABLE 8

	Industry P	Q	R	Final Demands	Total Output
Industry					
P	22	80	76	42	220
Q	88	40	38	34	200
R	66	60	57	7	190
Primary Input	44	20	19		

❦ 10-3 MARKOV CHAINS (OPTIONAL)

A process or a sequence of events developing in time in which the outcome at any stage contains some element of chance is called a **random process,** or a **stochastic process**. For example, the sequence might be the weather conditions in Vancouver on a series of successive days: the weather changes from day to day in an apparently somewhat random manner. Or the sequence might consist of the daily closing prices of a certain stock, again involving a certain degree of randomness. A sequence of elections is another example of a stochastic process.

A very simple example of a stochastic process is a sequence of Bernoulli trials (see Section 8-6), for example, a sequence of tosses of a coin. In this case, the outcome at any stage is independent of all previous outcomes. (In fact, this condition of independence is part of the definition of Bernoulli trials.) However, in most stochastic processes each outcome does depend on what has happened at earlier stages of the process. For example, the weather on any day is not completely random but is affected to some extent by the weather on previous days. Or the closing price of a stock on any one day depends to some extent on the way in which the stock has been behaving on previous days.

The simplest case of a stochastic process in which the outcomes are dependent on one another occurs when the outcome at each stage depends only on the outcome at the immediately preceding stage and not on any of the earlier outcomes. Such processes are called *Markov processes* or *Markov chains* (a chain of events, each event linked to the preceding one).

DEFINITION A **Markov** *chain* is a sequence of similar trials or observations in which each trial has the same finite number of possible outcomes and in which the probabilities of each outcome for a given trial depend only on the outcome of the immediately preceding trial and not on any earlier outcomes.

For example, consider the daily weather conditions at a certain location, and let us suppose that we are concerned simply with whether it is wet or dry on each day. Then each trial (that is, each daily weather observation) has the same two outcomes: wet or dry. If we assume that the probabilities of its being wet or dry tomorrow are determined completely by whether it is wet or dry today, then this sequence of trials would form a Markov chain.

In dealing with Markov chains, it is often useful to think of the sequence of trials as experiments performed on a certain physical system, each outcome leaving this system in a certain *state*. For example, consider a sequence of political elections in a certain country: the system could be taken as the country itself and each election leaves it in a certain state, namely in the control of the winning party. If there are only two strong political parties, call them A and B, which normally control the government, then we can say that the country is in state A or B according to whether the party A or B wins the election.* Each trial (that is, each election) places the country in one of the two states A or B. A sequence of 10 elections might produce results such as the following:

A, B, A, A, B, B, B, A, B, B

The first election in the sequence left party A in power, the second was won by party B, and so on, the tenth election being won by party B.

Let us assume that the probabilities of party A or B winning the next election are completely determined by which party is in power now. For example we might have the following probabilities:

1. If party A is in power, then there is a probability of $\frac{1}{4}$ that party A will win the next election and a probability of $\frac{3}{4}$ that party B will win the next election.

2. If party B is in power, then there is a probability of $\frac{1}{3}$ that party A will win the next election and a probabilty of $\frac{2}{3}$ that party B will stay in power.

In such a case, the sequence of elections forms a Markov chain, since the probabilities of the two outcomes of each election are determiend by the outcome of the preceding election.

The probability information given above can be conveniently represented by the following matrix:

Result of Next Election

A B

$$\begin{array}{cc} \textit{Result of Last} & \text{A} \\ \textit{Election} & \text{B} \end{array} \begin{bmatrix} \frac{1}{4} & \frac{3}{4} \\ \frac{1}{3} & \frac{2}{3} \end{bmatrix}$$

This matrix is called the *transition matrix*. The elements in the transition matrix represent the probabilities that at the next trial the state of the system will change from the party indicated on the left of the matrix to the party indicated above the matrix.

DEFINITION Consider a Markov process in which the system possesses n possible states, denoted by the numbers 1, 2, 3, . . . , n. Let p_{ij} denote the probability that the system will end up in state j after any trial given that it is in state i before the trial. The numbers p_{ij} are called the **transition probabilities** and the $n \times n$ matrix $P = [p_{ij}]$ is called the **transition matrix** for the system.

Notes 1. The sum $p_{i1} + p_{i2} + \cdots + p_{in}$ represents the probability that the system will end up in one of the states 1, 2, . . . , n given that it starts in state i. But since the system has to be in one of these n states, this sum of prob-

---

* In the United States we could think of a sequence of presidential elections, for example, whose outcomes determine control of the executive.

☛ **10.** Are the following possible transition matrices? If the answer is yes, give the probabilities that the system, if in state 3, will change to state 1 and that it will remain in state 3.

(a) $\begin{bmatrix} 0.1 & 0.2 & 0.7 \\ 0.5 & 0.5 & 0.1 \\ 0.4 & 0.3 & 0.2 \end{bmatrix}$;

(b) $\begin{bmatrix} 0 & 0.5 & 0.5 \\ 0.5 & 0 & 0.5 \\ 0.5 & 0.5 & 0 \end{bmatrix}$;

(c) $\begin{bmatrix} 0.4 & 0.3 & 0.3 \\ 0.7 & 0.7 & -0.4 \\ 0.1 & 0.6 & 0.3 \end{bmatrix}$.

abilities must equal 1. This means that the elements in any row of the transition matrix must add up to 1.

2. Each element must be nonnegative: $p_{ij} \geq 0$.

EXAMPLE 1 Given the transition matrix

$$\mathbf{P} = \begin{bmatrix} 0.3 & 0.5 & 0.2 \\ 0.4 & 0.2 & 0.4 \\ 0.5 & 0.4 & 0.1 \end{bmatrix}$$

what is the probability that at the next trial the system will change: (a) from state 2 to state 1? (b) from state 1 to state 3?

Solution By definition, in the transition matrix $\mathbf{P}$, p_{ij} denotes the probability that the system changes from state i to state j. Thus for part (a) we are interested in p_{21}, the probability of a change from state 2 to state 1. This is equal to 0.4 (the element in the second row and first column).

Similarly for part (b), we require $p_{13} = 0.2$. ☛ **10**

EXAMPLE 2 *(Party Government)* In a certain nation, there are three main political parties: Liberal (L), Conservative (C), and Democratic (D). The following transition matrix gives the probabilities that the nation will be controlled by each of the three political parties after an election, given the various possibilities for the result of the preceding election:

$$\begin{array}{c} \\ L \\ C \\ D \end{array} \begin{array}{ccc} L & C & D \\ \begin{bmatrix} 0.7 & 0.2 & 0.1 \\ 0.5 & 0.3 & 0.2 \\ 0.3 & 0.4 & 0.3 \end{bmatrix} \end{array}$$

Given that the Liberal party is in control now, use a tree diagram to determine the probability that the Conservative party will be in power after the next two elections.

Solution The possible results of the next two elections are shown in Figure 1, when we start with the Liberal party in office. Since we want the probability of the Conservative party being in office after the two elections, we are interested only in those branches of the tree diagram which end in C. The numbers along the branches are the appropriate probabilities given in the transition matrix. For example, the number 0.1 on the branch from L to D is the probability that Liberals will be replaced by Democrats at the next election. The number 0.4 on the branch from D to C is the probability that Conservatives will be elected at the second election given that the Democrats are in power after the first. The product $(0.1)(0.4)$ gives the probability that Liberals will be replaced by Democrats and then the Democrats will be replaced by the Conservatives.

The other sequences (or branches) that end with Conservatives in office have the probabilities $(0.7)(0.2)$ and $(0.2)(0.3)$, respectively. Thus the proba-

Answer (a) No   (b) Yes; 0.5, 0. (c) No.

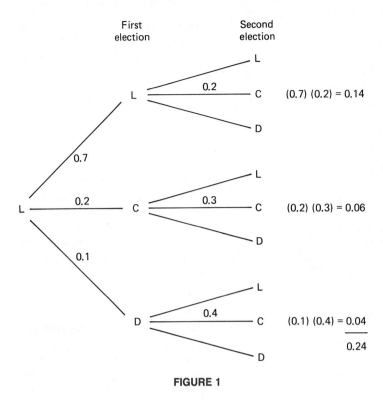

L

0.2 — C   (0.7) (0.2) = 0.14

D

0.7

L

0.2   C   0.3 — C   (0.2) (0.3) = 0.06

D

0.1

L

D   0.4 — C   (0.1) (0.4) = 0.04

0.24

D

FIGURE 1

bility that the Conservatives will be in office after two elections is the sum of probabilities of these three sequences:

$$(0.7)(0.2) + (0.2)(0.3) + (0.1)(0.4) = 0.24.$$

EXAMPLE 3 *(Stock Market Fluctuations)* The price of a stock fluctuates from day to day. When the stock market is stable, an increase one day tends to be followed with a decrease the next day, and a decrease tends to be followed by an increase. We can model these changes in price as a Markov process with two states, the first state being that the price increased on a given day, the second state being that decreased. (The possibility that the price remained unchanged is ignored.) Suppose that the transition matrix is as follows:

Tomorrow's Change

Increase   Decrease

$$\begin{array}{cc} & \text{Increase} \quad \text{Decrease} \\ \begin{array}{c} \textit{Today's} \\ \textit{Change} \end{array} \begin{array}{c} \text{Increase} \\ \text{Decrease} \end{array} & \begin{bmatrix} 0.1 & 0.9 \\ 0.8 & 0.2 \end{bmatrix} \end{array}$$

If the stock went down in price today, find the probability that it will increase 3 days from now.

Solution The possible states of the stock (up or down) for the next 3 days are given in Figure 2, in which the initial state is down. We denote the up and

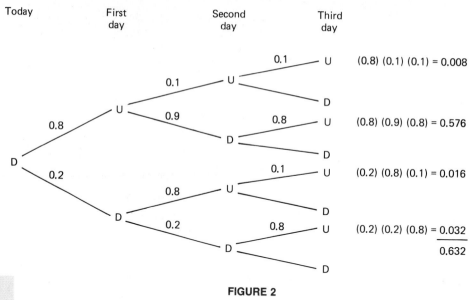

	Today	First day	Second day	Third day	
				0.1 — U	(0.8) (0.1) (0.1) = 0.008
		0.1 — U			
	0.8 — U		0.9	0.8 — U	(0.8) (0.9) (0.8) = 0.576
D			D		
	0.2		0.1 — U	0.1 — U	(0.2) (0.8) (0.1) = 0.016
		0.8 — U			
		D	0.8	0.8 — U	(0.2) (0.2) (0.8) = 0.032
		0.2	D		0.632

FIGURE 2

down states (that is, increase or decrease) by the letters U and D, respectively. We are interested only in the four branches that end up in state U on the third day. The required probability that the stock will go up on the third day when it went down today is obtained by adding the probabilities for all four of these branches and is equal to 0.632. ☛ 11

☛ **11.** Given the transition matrix
$$\begin{bmatrix} 0.3 & 0.7 \\ 0.5 & 0.5 \end{bmatrix},$$
find the probability that the system, starting in state 1, will be (a) in state 1 after 2 trials; (b) in state 2 after 3 trials.

In Example 3 we calculated the probability of the stock going up on the third day. Suppose we want to calculate the probability of the stock going up or down on the tenth day. In this case, the use of a tree diagram would be very cumbersome. In such a situation, matrix algebra can be used to avoid drawing a large tree diagram.

Let a given system have n possible states, so that each trial has n possible outcomes. At any stage in the future we cannot say what state the system will be in, but we might be in a position to give the probabilities that it will be in each of the states 1, 2, . . . , n. (For example, in Example 3, we cannot say whether the stock will be up or down in 3 days' time, but we can say that the probability is 0.632 that it will be up and therefore is $1 - 0.632 = 0.368$ that it will be down.) In general, if $p_1, p_2, \ldots, p_n$ are the probabilities that the system will be in states 1, 2, . . . , n, respectively, then the $1 \times n$ row matrix

$$[p_1 \quad p_2 \quad \cdots \quad p_n]$$

is called a **state matrix** or a **state vector** for the system. Note that $p_1 + p_2 + \cdots + p_n = 1$. We shall denote the initial state matrix by $\mathbf{A}_0$ and the state matrix after k trials (or steps) by $\mathbf{A}_k$.

Consider Example 3. On each day the system (stock) is in one of the two states: state 1 (up) and state 2 (down). In the example, the stock was down ini-

Answer (a) 0.44; (b) 0.588.

tially, so that initially the probability p_1 that the system is in state 1 is zero and the probability p_2 that the system is in state 2 is 1. Thus the initial state matrix $\mathbf{A}_0$ for the system is

$$\mathbf{A}_0 = [p_1 \quad p_2] = [0 \quad 1].$$

As shown in Figure 2, after 1 day, the stock is up (state 1) with probability $p_1 = 0.8$ and is down (state 2) with probability $p_2 = 0.2$. Thus the state matrix $\mathbf{A}_1$ after 1 day is given by

$$\mathbf{A}_1 = [p_1 \quad p_2] = [0.8 \quad 0.2].$$

From Figure 2, the probability that the stock will be up after 2 days is

$$p_1 = (0.8)(0.1) + (0.2)(0.8) = 0.08 + 0.16 = 0.24.$$

Similarly, the probability that the stock will be down after 2 days is

$$p_2 = (0.8)(0.9) + (0.2)(0.2) = 0.72 + 0.04 = 0.76.$$

Thus the state matrix $\mathbf{A}_2$ after 2 days is given by

$$\mathbf{A}_2 = [p_1 \quad p_2] = [0.24 \quad 0.76].$$

After 3 days the state matrix is

$$\mathbf{A}_3 = [p_1 \quad p_2] = [0.632 \quad 0.368].$$

The following theorem shows how to calculate the state matrix for the system at any stage if the state matrix after the previous trial is known.

THEOREM 1 If $\mathbf{P}$ denotes the transition matrix for a Markov chain and $\mathbf{A}_k$ is the state matrix after k trials, then the state matrix $\mathbf{A}_{k+1}$ after the next trial is given by

$$\mathbf{A}_{k+1} = \mathbf{A}_k \mathbf{P}.$$

Consider again the problem in Example 3. The initial state matrix is $\mathbf{A}_0 = [0 \quad 1]$ and the transition matrix is

$$\mathbf{P} = \begin{bmatrix} 0.1 & 0.9 \\ 0.8 & 0.2 \end{bmatrix}.$$

The state matrix after one step (day) is given by

$$\mathbf{A}_1 = \mathbf{A}_0 \mathbf{P} = [0 \quad 1] \begin{bmatrix} 0.1 & 0.9 \\ 0.8 & 0.2 \end{bmatrix} = [0.8 \quad 0.2].$$

The state matrix after 2 days is given by

$$\mathbf{A}_2 = \mathbf{A}_1 \mathbf{P} = [0.8 \quad 0.2] \begin{bmatrix} 0.1 & 0.9 \\ 0.8 & 0.2 \end{bmatrix} = [0.24 \quad 0.76].$$

These results agree with our results calculated earlier directly from the tree diagram. ☛ 12

☛ **12.** Given the transition matrix

$$\begin{bmatrix} 0.3 & 0.7 \\ 0.5 & 0.5 \end{bmatrix},$$

find the state matrices after one and two trials if the initial state matrix is $\mathbf{A}_0 = [0.2 \quad 0.8]$.

Answer $\mathbf{A}_1 = [0.46 \quad 0.54]$,
$\mathbf{A}_2 = [0.408 \quad 0.592]$.

EXAMPLE 4 *(Weather Forecasting)* The variation in weather from one day to the next is assumed to form a Markov chain with the following transition matrix.

Tomorrow

$$
\begin{array}{c}
 \\
 \\
Today \quad
\begin{array}{c}
Sunny \\
Cloudy \\
Rainy
\end{array}
\end{array}
\begin{array}{ccc}
Sunny & Cloudy & Rainy \\
\end{array}
\begin{bmatrix}
0.6 & 0.2 & 0.2 \\
0.2 & 0.5 & 0.3 \\
0.1 & 0.4 & 0.5
\end{bmatrix} = \mathbf{P}
$$

Given that it is cloudy today (Sunday), what is the probability that it will be sunny on Wednesday?

Solution Let the states 1, 2, and 3 stand for sunny, cloudy, and rainy, respectively. Since it is cloudy today (Sunday), we have $p_1 = 0$, $p_2 = 1$, $p_3 = 0$ and the initial state matrix is

$$\mathbf{A}_0 = [p_1 \quad p_2 \quad p_3] = [0 \quad 1 \quad 0].$$

The state matrix after one day (on Monday) is given by

$$\mathbf{A}_1 = \mathbf{A}_0\mathbf{P} = [0 \quad 1 \quad 0]\begin{bmatrix} 0.6 & 0.2 & 0.2 \\ 0.2 & 0.5 & 0.3 \\ 0.1 & 0.4 & 0.5 \end{bmatrix} = [0.2 \quad 0.5 \quad 0.3].$$

The state matrix on Tuesday (after 2 days) is

$$\mathbf{A}_2 = \mathbf{A}_1\mathbf{P} = [0.2 \quad 0.5 \quad 0.3]\begin{bmatrix} 0.6 & 0.2 & 0.2 \\ 0.2 & 0.5 & 0.3 \\ 0.1 & 0.4 & 0.5 \end{bmatrix}$$

$$= [0.25 \quad 0.41 \quad 0.34].$$

The state matrix on Wednesday (after 3 days) is

$$\mathbf{A}_3 = \mathbf{A}_2\mathbf{P} = [0.25 \quad 0.41 \quad 0.34]\begin{bmatrix} 0.6 & 0.2 & 0.2 \\ 0.2 & 0.5 & 0.3 \\ 0.1 & 0.4 & 0.5 \end{bmatrix}$$

$$= [0.266 \quad 0.391 \quad 0.343].$$

Thus the probability that it will be sunny on Wednesday is 0.266.

People who work in downtown Vancouver and do not live in the immediate neighborhood can commute to and from their jobs either by public transportation (buses) or by private automobiles. Currently, 25% of such people use the buses and 75% use their automobiles. Because of a shortage of parking space in the downtown area, the city has increased considerably the number of buses to and from downtown in the hope that more and more people will use the buses and so will alleviate the parking problem. City officials expect on the basis of a survey of the working population, that with the additional buses, each year 60% of those using automobiles will switch to buses, whereas 20% of those using buses will switch back to automobiles. The officials are interested in the long-term effect of the extra buses, that is, in the percentage of commuters who will in the long run use buses or automobiles.

In any year we take the state matrix to be $[p_1 \quad p_2]$, where p_1 is the proportion of commuters using buses and p_2 is the proportion using automobiles. The initial state matrix is

$$\mathbf{A}_0 = [0.25 \quad 0.75].$$

The probability that a person who uses the bus one year will continue to use it the following year is 0.8, while the probability that he will change to automobile is 0.2. The corresponding probabilities for commuters using automobiles are 0.6 and 0.4, so the transition matrix $\mathbf{P}$ is as follows:

People Using

	Buses Next Year	Automobiles Next Year
People Using Buses This Year	0.8	0.2
Automobiles This Year	0.6	0.4

Then the state matrices after 1, 2, 3, . . . , 6 years are as follows:

$$\mathbf{A}_1 = \mathbf{A}_0\mathbf{P} = [0.25 \quad 0.75]\begin{bmatrix} 0.8 & 0.2 \\ 0.6 & 0.4 \end{bmatrix} = [0.65 \quad 0.35]$$

$$\mathbf{A}_2 = \mathbf{A}_1\mathbf{P} = [0.65 \quad 0.35]\begin{bmatrix} 0.8 & 0.2 \\ 0.6 & 0.4 \end{bmatrix} = [0.73 \quad 0.27]$$

$$\mathbf{A}_3 = \mathbf{A}_2\mathbf{P} = [0.73 \quad 0.27]\begin{bmatrix} 0.8 & 0.2 \\ 0.6 & 0.4 \end{bmatrix} = [0.746 \quad 0.254]$$

$$\mathbf{A}_4 = \mathbf{A}_3\mathbf{P} = [0.746 \quad 0.254]\begin{bmatrix} 0.8 & 0.2 \\ 0.6 & 0.4 \end{bmatrix} = [0.749 \quad 0.251]$$

$$\mathbf{A}_5 = \mathbf{A}_4\mathbf{P} = [0.749 \quad 0.251]\begin{bmatrix} 0.8 & 0.2 \\ 0.6 & 0.4 \end{bmatrix} = [0.750 \quad 0.250]$$

$$\mathbf{A}_6 = \mathbf{A}_5\mathbf{P} = [0.75 \quad 0.25]\begin{bmatrix} 0.8 & 0.2 \\ 0.6 & 0.4 \end{bmatrix} = [0.75 \quad 0.25]$$

where we have rounded the decimals to at most three decimal places. We observe that after 5 years, the percentage of people using buses stabilizes at 75% and the percentage using automobiles at 25%.

Suppose now that the initial state matrix is

$$\mathbf{A}_0 = [0.2 \quad 0.8]$$

instead of [0.25 0.75]. In other words, 20% of commuters currently use buses and 80% use automobiles. In this case, the state matrices after 1, 2, 3, 4, and 5 years are as follows:

$$\mathbf{A}_1 = [0.2 \quad 0.8]\begin{bmatrix} 0.8 & 0.2 \\ 0.6 & 0.4 \end{bmatrix} = [0.64 \quad 0.36]$$

$$\mathbf{A}_2 = [0.64 \quad 0.36]\begin{bmatrix} 0.8 & 0.2 \\ 0.6 & 0.4 \end{bmatrix} = [0.728 \quad 0.272]$$

$$\mathbf{A}_3 = [0.728 \quad 0.272]\begin{bmatrix} 0.8 & 0.2 \\ 0.6 & 0.4 \end{bmatrix} = [0.746 \quad 0.254]$$

$$\mathbf{A}_4 = [0.746 \quad 0.254]\begin{bmatrix} 0.8 & 0.2 \\ 0.6 & 0.4 \end{bmatrix} = [0.749 \quad 0.251]$$

$$\mathbf{A}_5 = [0.749 \quad 0.251]\begin{bmatrix} 0.8 & 0.2 \\ 0.6 & 0.4 \end{bmatrix} = [0.750 \quad 0.250]$$

☛ **13.** Repeat the calculation of $\mathbf{A}_1, \ldots, \mathbf{A}_5$ if $\mathbf{A}_0 = [0.3 \quad 0.7]$

Once again we see that the percentage of people using buses stabilizes at 75% and those who will use automobiles at 25%. ☛ **13**

Thus the state matrix stabilizes at [0.75 0.25], no matter whether the initial state matrix is [0.25 0.75] or [0.2 0.8]. These results are not accidental; in many Markov chains the state matrix stabilizes over a large number of trials irrespective of the initial state matrix. This follows from the following theorem, which we state without proof.

THEOREM 2 A transition matrix **P** is said to be **regular** if for some positive integer k the matrix $\mathbf{P}^k$ has no entries equal to zero. If **P** is a regular transition matrix, then regardless of the initial state matrix, the successive state matrices approach some fixed state matrix **B** where $\mathbf{BP} = \mathbf{B}$. The matrix **B** is called the **steady-state matrix** for the system.

In the preceding discussion of an urban transit problem, we found that the steady-state matrix (up to three decimal places) is

$$\mathbf{B} = [0.750 \quad 0.250].$$

Let us show that this same steady-state matrix can be obtained using Theorem 2. The transition matrix is

$$\mathbf{P} = \begin{bmatrix} 0.8 & 0.2 \\ 0.6 & 0.4 \end{bmatrix}.$$

Let $\mathbf{B} = [p_1 \quad p_2]$ be the required steady-state matrix. Since by definition, the sum of probabilities in a state matrix is 1, we must have

$$p_1 + p_2 = 1. \tag{1}$$

Now the equation $\mathbf{BP} = \mathbf{B}$ implies that

$$[p_1 \quad p_2]\begin{bmatrix} 0.8 & 0.2 \\ 0.6 & 0.4 \end{bmatrix} = [p_1 \quad p_2]$$

Answer $\mathbf{A}_1 = [0.66 \quad 0.34],$
$\quad \mathbf{A}_2 = [0.732 \quad 0.268],$
$\quad \mathbf{A}_3 = [0.7464 \quad 0.2536],$
$\quad \mathbf{A}_4 = [0.74928 \quad 0.25072],$
$\quad \mathbf{A}_1 = [0.749856 \quad 0.250144].$

14. Given the transition matrix
$$\begin{bmatrix} 0.3 & 0.7 \\ 0.5 & 0.5 \end{bmatrix},$$
find the steady state matrix by solving the equation $\mathbf{BP} = \mathbf{B}$.

Answer $\begin{bmatrix} \frac{5}{12} & \frac{7}{12} \end{bmatrix} \approx [0.4167 \quad 0.5833]$.

that is,

$$[0.8p_1 + 0.6p_2 \quad 0.2p_1 + 0.4p_2] = [p_1 \quad p_2]$$

which gives

$$0.8p_1 + 0.6p_2 = p_1$$
$$0.2p_1 + 0.4p_2 = p_2.$$

These equations are identical with one another and solution is $p_1 = 3p_2$. Substituting this into Equation (1), we get

$$3p_2 + p_2 = 1 \quad \text{or} \quad p_2 = \tfrac{1}{4} = 0.25.$$

Therefore from Equation (1),

$$p_1 = 1 - p_2 = 1 - 0.25 = 0.75.$$

Hence

$$\mathbf{B} = [p_1 \quad p_2] = [0.75 \quad 0.25].$$

This agrees with the result found earlier. ☞ **14**

EXERCISES 10-3

(1–6) Determine which of the following matrices are transition matrices. If a matrix is not a transition matrix, explain why.

1. $\begin{bmatrix} \frac{1}{4} & \frac{3}{4} \\ \frac{2}{3} & \frac{1}{3} \end{bmatrix}$

2. $\begin{bmatrix} \frac{2}{5} & \frac{3}{5} & 0 \\ 0 & \frac{1}{2} & \frac{1}{2} \\ \frac{1}{7} & \frac{2}{7} & \frac{4}{7} \end{bmatrix}$

3. $\begin{bmatrix} \frac{1}{6} & \frac{2}{3} \\ \frac{3}{4} & \frac{1}{4} \end{bmatrix}$

4. $\begin{bmatrix} \frac{1}{3} & \frac{2}{5} & \frac{4}{15} \\ \frac{1}{4} & \frac{1}{2} & \frac{1}{4} \end{bmatrix}$

5. $\begin{bmatrix} \frac{1}{6} & \frac{1}{3} & \frac{1}{2} \\ -\frac{1}{4} & \frac{3}{4} & \frac{1}{2} \end{bmatrix}$

6. $\begin{bmatrix} \frac{1}{3} & \frac{3}{4} & -\frac{1}{12} \\ \frac{1}{2} & \frac{1}{3} & \frac{1}{6} \\ \frac{2}{5} & 0 & \frac{3}{5} \end{bmatrix}$

(7–12) Which of the following transition matrices are regular?

7. $\begin{bmatrix} 0 & 1 \\ \frac{1}{3} & \frac{2}{3} \end{bmatrix}$

8. $\begin{bmatrix} \frac{1}{4} & \frac{3}{4} \\ 1 & 0 \end{bmatrix}$

9. $\begin{bmatrix} 1 & 0 \\ 0 & 1 \end{bmatrix}$

10. $\begin{bmatrix} 1 & 0 & 0 \\ 0 & 1 & 0 \\ 0 & 0 & 1 \end{bmatrix}$

11. $\begin{bmatrix} \frac{1}{3} & 0 & \frac{2}{3} \\ \frac{1}{2} & \frac{1}{3} & \frac{1}{6} \\ \frac{1}{4} & \frac{1}{2} & \frac{1}{4} \end{bmatrix}$

12. $\begin{bmatrix} \frac{1}{7} & \frac{4}{7} & \frac{2}{7} \\ \frac{1}{3} & \frac{1}{6} & \frac{1}{2} \\ 0 & \frac{1}{5} & \frac{4}{5} \end{bmatrix}$

(13–16) Which of the following are state matrices?

13. $\begin{bmatrix} \frac{1}{5} & 0 & \frac{4}{5} \end{bmatrix}$

14. $\begin{bmatrix} 1 & \frac{1}{2} & \frac{1}{3} \end{bmatrix}$

15. $\begin{bmatrix} 3 & 2 & 1 & 4 \end{bmatrix}$

16. $\begin{bmatrix} \frac{1}{2} & \frac{5}{6} & -\frac{1}{3} & 0 \end{bmatrix}$

17. Suppose that the transition matrix for a certain Markov chain is given by the following.

$$\mathbf{P} = \begin{bmatrix} \frac{2}{3} & \frac{1}{3} \\ \frac{1}{4} & \frac{3}{4} \end{bmatrix} \begin{matrix} \text{State 1} \\ \text{State 2} \end{matrix}$$

with the columns labeled State 1 and State 2.

a. What does the entry $\frac{1}{4}$ in the matrix represent?

b. Assuming that the system is in State 1 initially, use a tree diagram to find the state matrix after two trials.

c. Use Theorem 1 to find the answer to part (b).

d. What is the steady-state matrix for the system?

18. Consider a Markov process with this transition matrix:

$$\begin{array}{cc} & \text{State 1} \quad \text{State 2} \\ \begin{array}{c} \text{State 1} \\ \text{State 2} \end{array} & \begin{bmatrix} 0.6 & 0.4 \\ 0.3 & 0.7 \end{bmatrix} \end{array}$$

a. What does the entry 0.4 in this matrix represent?

b. Assuming the system to be in State 2 initially, use a tree diagram to determine the state matrix after three trials.

c. Use Theorem 1 to determine the state matrix in part (b).

d. Find the steady-state matrix for the system.

19. The transition matrix for a certain Markov process is

$$\begin{bmatrix} 0.3 & 0.5 & 0.2 \\ 0.1 & 0.6 & 0.3 \\ 0.4 & 0.1 & 0.5 \end{bmatrix}.$$

a. If the system is initially in State 1, determine the state matrix after two stages of the process.

b. If the system is in State 2 initially, determine the state matrix after two stages.

c. Determine the steady-state matrix.

20. Repeat Exercise 19 for the transition matrix.

$$\begin{bmatrix} \frac{1}{4} & \frac{1}{2} & \frac{1}{4} \\ \frac{1}{5} & \frac{3}{5} & \frac{1}{5} \\ \frac{3}{10} & \frac{1}{5} & \frac{1}{2} \end{bmatrix}$$

21. (Party Government) The probabilities that a certain country will be governed by one of three political parties X, Y, or Z after the next election are given by the transition matrix.

$$\begin{array}{c} \\ P = \end{array} \begin{array}{c} \text{X} \quad \text{Y} \quad \text{Z} \\ \begin{bmatrix} \frac{1}{2} & \frac{1}{3} & \frac{1}{6} \\ \frac{1}{4} & \frac{3}{4} & 0 \\ \frac{1}{5} & \frac{2}{5} & \frac{2}{5} \end{bmatrix} \begin{array}{c} \text{X} \\ \text{Y} \\ \text{Z} \end{array} \end{array}$$

a. What is the probability that party Z will win the next-but-one election if party X is in power now?

b. What is the probability that party X will be in power after two elections if party Y is in power now?

c. If party Z is in power now, what is the probability that it will be in power after two elections?

d. Find the steady-state matrix. How is this matrix to be interpreted?

22. (Stock-price Fluctuations) The price of a certain stock can go up, down, or stay the same on any day. The probability that the stock price will go up (state 1), go down (state 2), or stay the same (state 3) on the next day is given by the transition matrix:

Tomorrow's Change

$$\begin{array}{c} \\ \begin{array}{c} \textit{Today's} \\ \textit{Change} \end{array} \end{array} \begin{array}{c} \quad \text{Up} \quad \text{Down} \quad \text{Same} \\ \begin{array}{c} \text{Up} \\ \text{Down} \\ \text{Same} \end{array} \begin{bmatrix} 0.2 & 0.7 & 0.1 \\ 0.6 & 0.2 & 0.2 \\ 0.2 & 0.5 & 0.3 \end{bmatrix} \end{array}$$

a. What is the probability that the stock will go down after 2 days if it went up today?

b. What is the steady-state matrix for the Markov process?

23. The probability that a short person will father a short son is 0.75, wheras the probability that a tall man will father a tall son is 0.60. (The possibility of a medium-height offspring is ignored.)

a. What is the probability that a tall man will have a short grandson?

b. What is the probability that a short man will have a tall grandson?

c. Find the steady-state matrix for the process and give its interpretation.

24. (Market Share) Presently, three food-processing firms X, Y, and Z hold 50%, 30% and 20% of the market in coffee. The three firms simultaneously introduce new blends of coffee. With the introduction of these new blends in one year the following occurs:

a. X retains 60% of its customers and loses 20% to Y and 20% to Z.

b. Y retains 50% of its customers and loses 30% to X and 20% to Z.

c. Z retains 70% of its customers and loses 10% to Z and 20% to Y.

Assuming that this trend continues, what share of market will each firm have at the end of 2 years? What share of market will each firm have in the long run?

25. (Agriculture) In a certain region the farms can be classified as arable farms, dairy farms, or mixed. At present 30% are arable, 40% are dairy, and 30% are mixed. The transition matrix from one year to the next is

$$\begin{array}{c c c} & \text{A} & \text{D} & \text{M} \\ \begin{array}{c} \text{A} \\ \text{D} \\ \text{M} \end{array} & \left[\begin{array}{c c c} 0.8 & 0.1 & 0.1 \\ 0.2 & 0.8 & 0 \\ 0.1 & 0.1 & 0.8 \end{array}\right] \end{array}$$

Find the percentage of the three types of farms:

a. Next year.

b. The following year.

c. In the long run.

*26. *(Opinion Survey)* In an opinion poll regarding a certain TV program, 60% of those surveyed said they enjoyed the program, whereas 40% said they did not. The same group was surveyed again 1 week later, and this time 65% said they enjoyed the program, but 35% did not. A further week later 68% said they enjoyed the program. Calculate the transition matrix that represents this change of opinion. If the group is repeatedly surveyed on the same program, what will be the eventual percentage that express liking for it?

*27. *(Energy Use)* In a certain country 90% of the energy was generated from oil, gas, or coal and 10% came from atomic power. Five years later the percentages were 80% and 20%, respectively while after a further 5 years they are 75% and 25%. Assuming the process is Markovian, so that

$$[0.8 \quad 0.2] = [0.9 \quad 0.1]\mathbf{P}$$
$$[0.75 \quad 0.25] = [0.8 \quad 0.2]\mathbf{P},$$

calculate the 2×2 transition matrix $\mathbf{P}$. Find the steady-state matrix and interpret it.

◥ 10-4 DETERMINANTS

Corresponding to any square matrix there is a real number called its *determinant*. The determinant is denoted by enclosing the matrix in vertical bars. For example if $\mathbf{A}$ is the 2×2 matrix given by

$$\mathbf{A} = \begin{bmatrix} 2 & 3 \\ 4 & 5 \end{bmatrix}$$

then its determinant is denoted by $|\mathbf{A}|$, or, in full, by

$$\begin{vmatrix} 2 & 3 \\ 4 & 5 \end{vmatrix}.$$

The determinant of an $n \times n$ matrix is said to be a *determinant of order n*. For example, $|\mathbf{A}|$ just given is a determinant of order 2.

The symbol Δ (delta) is often used to denote a given determinant.

We shall begin by defining determinants of order 2, discussing higher orders later.

DEFINITION A **determinant of order 2** is defined by the following expression.

$$\begin{vmatrix} a_1 & b_1 \\ a_2 & b_2 \end{vmatrix} = a_1 b_2 - a_2 b_1$$

In other words, the determinant is given by the product of the elements a_1 and b_2 on the main diagonal minus the product of the elements a_2 and b_1 on the cross-diagonal. We can indicate these two diagonals by means of arrows.

$$\begin{matrix} (+) & (-) \end{matrix}$$
$$\begin{vmatrix} a_1 & b_1 \\ a_2 & b_2 \end{vmatrix} = a_1 b_2 - a_2 b_1$$

The $(+)$ and $(-)$ signs indicate the signs associated with the two products.

EXAMPLE 1 Evaluate the following determinants.

(a) $\begin{vmatrix} 2 & -3 \\ 4 & 5 \end{vmatrix}$ (b) $\begin{vmatrix} 3 & 2 \\ 0 & 4 \end{vmatrix}$

Solution

(a)

$$\begin{matrix} (+) & (-) \end{matrix}$$
$$\Delta = \begin{vmatrix} 2 & -3 \\ 4 & 5 \end{vmatrix} = 2(5) - 4(-3) = 10 + 12 = 22$$

(b)

$$\begin{matrix} (+) & (-) \end{matrix}$$
$$\Delta = \begin{vmatrix} 3 & 2 \\ 0 & 4 \end{vmatrix} = 3(4) - 0(2) = 12 - 0 = 12 \quad \text{☛ 15}$$

☛ 15. Evaluate the following determinants:

(a) $\begin{vmatrix} 3 & 4 \\ -2 & 2 \end{vmatrix}$; (b) $\begin{vmatrix} 0 & 4 \\ 2 & -5 \end{vmatrix}$;

(c) $\begin{vmatrix} 2 & 10 \\ -1 & 8 \end{vmatrix}$.

DEFINITION A **determinant of order 3** is defined by the following expression.

$$\Delta = \begin{vmatrix} a_1 & b_1 & c_1 \\ a_2 & b_2 & c_2 \\ a_3 & b_3 & c_3 \end{vmatrix} = a_1 b_2 c_3 + a_2 b_3 c_1 + a_3 b_1 c_2 - a_1 b_3 c_2 - a_3 b_2 c_1 - a_2 b_1 c_3$$

The expression on the right is called the **complete expansion** of the third-order determinant Δ. Observe that it contains six terms, three positive and three negative. Each term consists of a product of three elements in the determinant.

When evaluating a third-order determinant, we usually do not use the complete expansion; instead, we use what are called *cofactors*. Each element in the determinant has a cofactor, which is denoted by the corresponding capital letter. For example, A_2 denotes the cofactor of a_2, B_3 denotes the cofactor of b_3, and so on. They are defined as follows.

DEFINITION The **minor** of an element in a determinant Δ is equal to the determinant obtained by deleting the row and column in Δ that contain the given element. If the given element occurs in the ith row and the jth column of Δ, then its **cofactor** is equal to $(-1)^{i+j}$ times its minor.

EXAMPLE 2

Answer (a) 14; (b) −8; (c) 26.

(a) In the above determinant Δ, a_2 occurs in the second row and first column ($i = 2$ and $j = 1$), so its cofactor is

$$A_2 = (-1)^{2+1}\begin{vmatrix} a_1 & b_1 & c_1 \\ a_2 & b_2 & c_2 \\ a_3 & b_3 & c_3 \end{vmatrix} = (-1)^3 \begin{vmatrix} b_1 & c_1 \\ b_3 & c_3 \end{vmatrix} = -\begin{vmatrix} b_1 & c_1 \\ b_3 & c_3 \end{vmatrix}.$$

(b) Since c_3 occurs in the third row and third column ($i = j = 3$), its cofactor is

$$C_3 = (-1)^{3+3}\begin{vmatrix} a_1 & b_1 & c_1 \\ a_2 & b_2 & c_2 \\ a_3 & b_3 & c_3 \end{vmatrix} = (-1)^6 \begin{vmatrix} a_1 & b_1 \\ a_2 & b_2 \end{vmatrix} = \begin{vmatrix} a_1 & b_1 \\ a_2 & b_2 \end{vmatrix}. \quad \text{☞ 16}$$

The connection between a determinant and its various cofactors is given by Theorem 1.

THEOREM 1 The value of a determinant can be found by multiplying the elements in any row (or column) by their cofactors and adding the products for all elements in the given row (or column).

Let us verify that this theorem holds for expansion by the first row. The theorem states that

$$\Delta = a_1 A_1 + b_1 B_1 + c_1 C_1. \tag{1}$$

The three cofactors here are as follows.

$$A_1 = (-1)^{1+1}\begin{vmatrix} b_2 & c_2 \\ b_3 & c_3 \end{vmatrix} = (b_2 c_3 - b_3 c_2)$$

$$B_1 = (-1)^{1+2}\begin{vmatrix} a_2 & c_2 \\ a_3 & c_3 \end{vmatrix} = -(a_2 c_3 - a_3 c_2)$$

$$C_1 = (-1)^{1+3}\begin{vmatrix} a_2 & b_2 \\ a_3 & b_3 \end{vmatrix} = (a_2 b_3 - a_3 b_2)$$

Substituting these into Equation (1), we obtain

$$\Delta = a_1(b_2 c_3 - b_3 c_2) - b_1(a_2 c_3 - a_3 c_2) + c_1(a_2 b_3 - a_3 b_2).$$

It is easily seen that this expression agrees with the complete expansion given in the definition of the third-order determinant.

EXAMPLE 3 Evaluate the determinant

$$\Delta = \begin{vmatrix} 2 & 3 & -1 \\ 1 & 4 & 2 \\ -3 & 1 & 4 \end{vmatrix}.$$

Solution Expanding by the first row, we have

$$\Delta = a_1 A_1 + b_1 B_1 + c_1 C_1$$

$$= 2 \begin{vmatrix} 4 & 2 \\ 1 & 4 \end{vmatrix} - 3 \begin{vmatrix} 1 & 2 \\ -3 & 4 \end{vmatrix} + (-1) \begin{vmatrix} 1 & 4 \\ -3 & 1 \end{vmatrix}$$

$$= 2(4 \cdot 4 - 1 \cdot 2) - 3[1 \cdot 4 - (-3)2] + (-1)[1 \cdot 1 - (-3)4]$$

$$= 2(16 - 2) - 3(4 + 6) - 1(1 + 12) = -15$$

Let us now return to Theorem 1 and verify that it gives the determinant when we expand by the second column. In this case, the theorem states that

$$\Delta = b_1 B_1 + b_2 B_2 + b_3 B_3. \tag{2}$$

Here, the cofactors are

$$B_1 = (-1)^{1+2} \begin{vmatrix} a_2 & c_2 \\ a_3 & c_3 \end{vmatrix} = -(a_2 c_3 - a_3 c_2)$$

$$B_2 = (-1)^{2+2} \begin{vmatrix} a_1 & c_1 \\ a_3 & c_3 \end{vmatrix} = (a_1 c_3 - a_3 c_1)$$

$$B_3 = (-1)^{3+2} \begin{vmatrix} a_1 & c_1 \\ a_2 & c_2 \end{vmatrix} = -(a_1 c_2 - a_2 c_1).$$

Therefore, from Equation (2),

$$\Delta = -b_1(a_2 c_3 - a_3 c_2) + b_2(a_1 c_3 - a_3 c_1) - b_3(a_1 c_2 - a_2 c_1).$$

Again, we can verify that the terms here are the same as the six terms given in the complete expansion.

In a similar way, we can verify that a determinant can be evaluated by expanding by any row or column. ☛ 17

EXAMPLE 4 Evaluate the determinant

$$\Delta = \begin{vmatrix} 2 & -3 & 0 \\ 1 & 4 & 3 \\ -5 & 6 & 0 \end{vmatrix}$$

(a) by expanding by the second column: (b) by expanding by the third column.

Solution

(a) Expanding by the second column, we get

$$\Delta = b_1 B_1 + b_2 B_2 + b_3 B_3$$

$$= -(-3) \begin{vmatrix} 1 & 3 \\ -5 & 0 \end{vmatrix} + 4 \begin{vmatrix} 2 & 0 \\ -5 & 0 \end{vmatrix} - 6 \begin{vmatrix} 2 & 0 \\ 1 & 3 \end{vmatrix}$$

$$= 3[1(0) - 3(-5)] + 4[2(0) - (-5)(0)] - 6[2(3) - 1(0)]$$

$$= 3(15) + 4(0) - 6(6) = 9.$$

☛ **17.** Evaluate the determinant

$$\begin{vmatrix} 2 & 0 & -1 \\ 4 & 3 & 5 \\ -1 & 1 & 0 \end{vmatrix}$$

by expanding
(a) by the first column;
(b) by the second row:

Answer

(a) $2 \cdot \begin{vmatrix} 3 & 5 \\ 1 & 0 \end{vmatrix} - 4 \cdot \begin{vmatrix} 0 & -1 \\ 1 & 0 \end{vmatrix}$

$+ (-1) \cdot \begin{vmatrix} 0 & -1 \\ 3 & 5 \end{vmatrix} = -17;$

(b) $-4 \cdot \begin{vmatrix} 0 & -1 \\ 1 & 0 \end{vmatrix} + 3 \cdot \begin{vmatrix} 2 & -1 \\ -1 & 0 \end{vmatrix}$

$- 5 \cdot \begin{vmatrix} 2 & 0 \\ -1 & 1 \end{vmatrix} = -17.$

● **18.** Evaluate the determinant
$$\begin{vmatrix} -2 & 0 & -3 \\ 4 & 0 & 5 \\ 3 & 11 & 13 \end{vmatrix}.$$

(b) Expanding by the third column, we get

$$\Delta = c_1 C_1 + c_2 C_2 + c_3 C_3$$

$$= (0)\begin{vmatrix} 1 & 4 \\ -5 & 6 \end{vmatrix} - 3\begin{vmatrix} 2 & -3 \\ -5 & 6 \end{vmatrix} + (0)\begin{vmatrix} 2 & -3 \\ 1 & 4 \end{vmatrix}.$$

In this expansion, two of the terms are zero, so $\Delta = -3[2(6) - (-3)(-5)] = 9.$

In Example 4, the same answer was obtained using both methods, but the calculations involved in the second method were a little easier because the third column had two zeros, and two of the three terms in the expansion could immediately be set equal to zero. *It is generally easier to choose a row or column with the maximum number of zeros when expanding a determinant.* ● **18**

From the above discussion, it will be clear that we can expand a determinant of order 3 by any row or column. In such an expansion, the terms alternate in sign and each element in the given row or column multiplies the 2×2 determinant (the minor) obtained by deleting from Δ the row and column that contain that element.

We observe that sometimes the sign of the first term in such an expansion is positive (as in Example 3 and part (b) of Example 4) and sometimes it is negative (as in part (a) of Example 4). In fact, *the first term in an expansion is positive when expanding by the first or third row (or column) and is negative when expanding by the second row (or column).*

These rules extend in a natural way to determinants of orders higher than 3. Any such determinant can be evaluated by expansion along any row or column. The determinant is obtained by multiplying each element in the row (or column) by its cofactor and adding all the products so obtained. The rule for evaluting the cofactors is exactly the same as for 3×3 determinants: the cofactor of the element in the ith row and jth column is equal to $(-1)^{i+j}$ multiplied by the determinant obtained by deleting the ith row and the jth column.

For example, consider the following determinant of order 4.

$$\Delta = \begin{vmatrix} a_1 & b_1 & c_1 & d_1 \\ a_2 & b_2 & c_2 & d_2 \\ a_3 & b_3 & c_3 & d_3 \\ a_4 & b_4 & c_4 & d_4 \end{vmatrix}$$

Its expansion by the first row is given by

$$\Delta = a_1\begin{vmatrix} b_2 & c_2 & d_2 \\ b_3 & c_3 & d_3 \\ b_4 & c_4 & d_4 \end{vmatrix} - b_1\begin{vmatrix} a_2 & c_2 & d_2 \\ a_3 & c_3 & d_3 \\ a_4 & c_4 & d_4 \end{vmatrix} + c_1\begin{vmatrix} a_2 & b_2 & d_2 \\ a_3 & b_3 & d_3 \\ a_4 & b_4 & d_4 \end{vmatrix} - d_1\begin{vmatrix} a_2 & b_2 & c_2 \\ a_3 & b_3 & c_3 \\ a_4 & b_4 & c_4 \end{vmatrix}$$

$$= a_1 A_1 + b_1 B_1 + c_1 C_1 + d_1 D_1$$

where $A_1, B_1, C_1,$ and D_1 denote the cofactors of $a_1, b_1, c_1,$ and d_1, respectively, in Δ.

Answer Expand by the second column. The only nonzero term is
$$-11 \cdot \begin{vmatrix} -2 & -3 \\ 4 & 5 \end{vmatrix} = -22.$$

One important application of determinants is to the solution of systems of linear equations when the number of equations is equal to the number of unknowns. In fact, the concept of determinants originated from the study of such systems of equations. The main result, known as *Cramer's rule,* is stated in Theorem 2 for systems of three equations. The theorem generalizes in a straightforward way to systems of n equations in n unknowns.

THEOREM 2 (CRAMER'S RULE) Consider the following system of three equations in three unknowns, x, y, and z.

$$a_1x + b_1y + c_1z = k_1$$

$$a_2x + b_2y + c_2z = k_2$$

$$a_3x + b_3y + c_3z = k_3$$

Let

$$\Delta = \begin{vmatrix} a_1 & b_1 & c_1 \\ a_2 & b_2 & c_2 \\ a_3 & b_3 & c_3 \end{vmatrix}$$

be the determinant of coefficients and let Δ_1, Δ_2, and Δ_3 be obtained by replacing the first, second, and third columns in Δ, respectively, by the constant terms. In other words,

$$\Delta_1 = \begin{vmatrix} k_1 & b_1 & c_1 \\ k_2 & b_2 & c_2 \\ k_3 & b_3 & c_3 \end{vmatrix}, \qquad \Delta_2 = \begin{vmatrix} a_1 & k_1 & c_1 \\ a_2 & k_2 & c_2 \\ a_3 & k_3 & c_3 \end{vmatrix}, \quad \text{and} \quad \Delta_3 = \begin{vmatrix} a_1 & b_1 & k_1 \\ a_2 & b_2 & k_2 \\ a_3 & b_3 & k_3 \end{vmatrix}.$$

Then *if $\Delta \neq 0$, the given system has a unique solution given by*

$$x = \frac{\Delta_1}{\Delta}, \qquad y = \frac{\Delta_2}{\Delta}, \qquad z = \frac{\Delta_3}{\Delta}.$$

if $\Delta = 0$ and $\Delta_1 = \Delta_2 = \Delta_3 = 0$, then the system has an infinite number of solutions. If $\Delta = 0$ and either $\Delta_1 \neq 0$ or $\Delta_2 \neq 0$ or $\Delta_3 \neq 0$, then the system has no solution.

EXAMPLE 5 Use determinants to solve the following system of equations.

$$2x - 3y + z = 5$$

$$x + 2y - z = 7$$

$$6x - 9y + 3z = 4$$

Solution The determinant of coefficients is

$$\Delta = \begin{vmatrix} 2 & -3 & 1 \\ 1 & 2 & -1 \\ 6 & -9 & 3 \end{vmatrix} = 0$$

as can be seen by expanding by the first row. Replacing the first column elements in Δ by the constant terms, we have

$$\Delta_1 = \begin{vmatrix} 5 & -3 & 1 \\ 7 & 2 & -1 \\ 4 & -9 & 3 \end{vmatrix} = -11.$$

Since $\Delta = 0$ and $\Delta_1 \neq 0$, the given system has *no* solution.

☞ **19.** Use Cramer's rule to solve the systems
(a) $q + 3r = 1$, $2p - 5r = 1$, $2p + 2q + 3r = 1$;
(b) $x - 4y + 3z = 4$, $-x + 2y - z = -2$, $y - z = -1$.

EXAMPLE 6 Use determinants to solve the following system of equations.

$$3x - y + 2z = -1$$
$$2x + y - z = 5$$
$$x + 2y + z = 4$$

Solution The determinant of coefficients is

$$\Delta = \begin{vmatrix} 3 & -1 & 2 \\ 2 & 1 & -1 \\ 1 & 2 & 1 \end{vmatrix}.$$

Expanding by the first row, we obtain

$$\Delta = 3 \begin{vmatrix} 1 & -1 \\ 2 & 1 \end{vmatrix} - (-1) \begin{vmatrix} 2 & -1 \\ 1 & 1 \end{vmatrix} + 2 \begin{vmatrix} 2 & 1 \\ 1 & 2 \end{vmatrix}$$
$$= 3(1 + 2) + 1(2 + 1) + 2(4 - 1) = 18.$$

Since $\Delta \neq 0$, the system has a unique solution given by

$$x = \frac{\Delta_1}{\Delta}, \qquad y = \frac{\Delta_2}{\Delta}, \qquad z = \frac{\Delta_3}{\Delta}.$$

Replacing the first, second, and third columns in Δ, respectively, by the constant terms, we have the following values.

$$\Delta_1 = \begin{vmatrix} -1 & -1 & 2 \\ 5 & 1 & -1 \\ 4 & 2 & 1 \end{vmatrix} = 18$$

$$\Delta_2 = \begin{vmatrix} 3 & -1 & 2 \\ 2 & 5 & -1 \\ 1 & 4 & 1 \end{vmatrix} = 36$$

$$\Delta_3 = \begin{vmatrix} 3 & -1 & -1 \\ 2 & 1 & 5 \\ 1 & 2 & 4 \end{vmatrix} = -18$$

Answer (a) $\Delta = -4$, and $\Delta_1 = 8$, $\Delta_2 = -16$, $\Delta_3 = 4$, so $p = -2$, $q = 4$, $r = -1$;
(b) $\Delta = 0$, and $\Delta_1 = \Delta_2 = \Delta_3 = 0$, so the system is consistent with infinitely many solutions.

Therefore we have the following values for x, y, and z:

$$x = \frac{\Delta_1}{\Delta} = \frac{18}{18} = 1, \qquad y = \frac{\Delta_2}{\Delta} = \frac{36}{18} = 2, \qquad z = \frac{\Delta_3}{\Delta} = -\frac{18}{18} = -1$$

Hence the required solution is $x = 1$, $y = 2$, and $z = -1$. ☞ **19**

It should be pointed out that Cramer's rule is not usually the most efficient method of solving a system of equations. As a rule, the method of row reduction described in Chapter 9 involves shorter calculations. Cramer's rule is important mainly from a theoretical standpoint. One of the most significant results arising from it is that *a system of n linear equations in n unknowns has a unique solution if and only if the determinant of coefficients is nonzero.*

EXERCISES 10-4

(1–4) Write each of the following, given

$$\Delta = \begin{vmatrix} a & b & c \\ p & q & r \\ l & m & n \end{vmatrix}$$

1. The minor of q.

2. The minor of n.

3. The cofactor of r.

4. The cofactor of m.

(5–28) Evaluate the following determinants.

5. $\begin{vmatrix} 3 & -1 \\ 4 & 7 \end{vmatrix}$

6. $\begin{vmatrix} 5 & 8 \\ 3 & -2 \end{vmatrix}$

7. $\begin{vmatrix} -6 & -7 \\ -8 & -3 \end{vmatrix}$

8. $\begin{vmatrix} 2 & x \\ 0 & 1 \end{vmatrix}$

9. $\begin{vmatrix} a & -2 \\ b & 3 \end{vmatrix}$

10. $\begin{vmatrix} 5 & a \\ -a & 4 \end{vmatrix}$

11. $\begin{vmatrix} 32 & 2 \\ 64 & 5 \end{vmatrix}$

12. $\begin{vmatrix} 274 & 3 \\ 558 & 7 \end{vmatrix}$

13. $\begin{vmatrix} 59 & 3 \\ 64 & 0 \end{vmatrix}$

14. $\begin{vmatrix} a+1 & 2-a \\ 2a+3 & 5-2a \end{vmatrix}$

15. $\begin{vmatrix} a & b \\ a+b & b+c \end{vmatrix}$

16. $\begin{vmatrix} x+2 & x-1 \\ 3 & x \end{vmatrix}$

17. $\begin{vmatrix} 2 & 1 & 4 \\ 3 & 5 & -1 \\ 1 & 0 & 0 \end{vmatrix}$

18. $\begin{vmatrix} 1 & -2 & 3 \\ 4 & 0 & 5 \\ 6 & 0 & 7 \end{vmatrix}$

19. $\begin{vmatrix} 1 & 3 & -2 \\ 0 & 2 & 1 \\ 4 & -1 & 3 \end{vmatrix}$

20. $\begin{vmatrix} 3 & 1 & 0 \\ 0 & -2 & 1 \\ 2 & 0 & 5 \end{vmatrix}$

21. $\begin{vmatrix} 2 & 3 & 4 \\ 5 & 6 & 7 \\ 8 & 9 & 10 \end{vmatrix}$

22. $\begin{vmatrix} 5 & 10 & 1 \\ 8 & 5 & 4 \\ 1 & 4 & 2 \end{vmatrix}$

23. $\begin{vmatrix} 1 & 2 & 4 \\ 2 & 5 & 1 \\ 3 & 8 & 4 \end{vmatrix}$

24. $\begin{vmatrix} 7 & 9 & 3 \\ 8 & 2 & 0 \\ 0 & 5 & 4 \end{vmatrix}$

25. $\begin{vmatrix} a & b & c \\ 0 & d & e \\ 0 & 0 & f \end{vmatrix}$

26. $\begin{vmatrix} x & 0 & 0 & 0 \\ a & y & 0 & 0 \\ b & c & z & 0 \\ d & e & f & w \end{vmatrix}$

27. $\begin{vmatrix} 1 & 0 & -1 & 0 \\ 2 & 1 & 0 & -3 \\ 0 & -2 & 1 & 1 \\ 1 & 2 & 0 & -1 \end{vmatrix}$

28. $\begin{vmatrix} 2 & \cdot3 & 4 & 5 \\ 1 & 0 & -1 & 2 \\ 0 & -2 & 1 & 0 \\ 3 & 0 & 2 & 1 \end{vmatrix}$

(29–32) Determine x in each case.

29. $\begin{vmatrix} x & 3 \\ 2 & 5 \end{vmatrix} = 9$

30. $\begin{vmatrix} x+3 & 2 \\ x & x+1 \end{vmatrix} = 3$

31. $\begin{vmatrix} 1 & 0 & 0 \\ x^2 & x-2 & 3 \\ x & x+1 & x \end{vmatrix} = 3$

32. $\begin{vmatrix} x+1 & 2 & x \\ x & x^2 & 2 \\ 0 & 1 & 0 \end{vmatrix} = 1$

(33–50) Use Cramer's rule to solve the following systems of equations.

33. $3x + 2y = 1$
    $2x - y = 3$

34. $2x - 5y = 8$
    $3y + 7x = -13$

35. $4x + 5y - 14 = 0$
    $3y = 7 - x$

36. $2(x - y) = 5$
    $4(1 - y) = 3x$

37. $\frac{1}{3}x + \frac{1}{2}y = 7$
    $\frac{1}{2}x - \frac{1}{5}y = 1$

38. $\frac{2}{3}u + \frac{3}{4}v = 13$
    $\frac{5}{2}u + \frac{1}{3}v = 19$

39. $2x + 3y = 13$
    $6x + 9y = 40$

40. $3x = 2(2 + y)$
    $4y = 7 + 6x$

41. $x + y + z = -1$
    $2x + 3y - z = 0$
    $3x - 2y + z = 4$

42. $2x - y + z = 2$
    $3x + y - 2z = 9$
    $-x + 2y + 5z = -5$

43. $\begin{aligned} 2x + y + z &= 0 \\ x + 2y - z &= -6 \\ x + 5y + 2z &= 0 \end{aligned}$ **44.** $\begin{aligned} x + 3y - z &= 0 \\ 3x - y + 2z &= 0 \\ 2x - 5y + z &= 5 \end{aligned}$ **47.** $\begin{aligned} 2x - y + 3z &= 4 \\ x + 3y - z &= 5 \\ 6x - 3y + 9z &= 10 \end{aligned}$ **48.** $\begin{aligned} 4x + 2y - 6z &= 7 \\ 3x - y + 2z &= 12 \\ 6x + 3y - 9z &= 10 \end{aligned}$

45. $\begin{aligned} x + 2y &= 5 \\ 3y - z &= 1 \\ 2x - y + 3z &= 11 \end{aligned}$ **46.** $\begin{aligned} 2v + 5w &= 3 \\ 4u - 3w &= 5 \\ 3u - 4v + 2w &= 12 \end{aligned}$ **49.** $\begin{aligned} x + 2y - z &= 2 \\ 2x - 3y + 4z &= -4 \\ 3x + y + z &= 0 \end{aligned}$ **50.** $\begin{aligned} 2p - r &= 5 \\ p + 3q &= 9 \\ 3p - q + 5r &= 12 \end{aligned}$

◼ 10-5 INVERSES BY DETERMINANT

In Section 10-1, we used row operations to find the inverse of a nonsingular matrix. It is also possible to calculate inverses by the use of determinants and, in fact, for small matrices (2×2 or 3×3), this method is usually more convenient than using row operations.

DEFINITION Let $\mathbf{A} = [a_{ij}]$ be a matrix of any size. The matrix obtained by interchanging the rows and columns of $\mathbf{A}$ is called the **transpose** of $\mathbf{A}$ and is denoted by $\mathbf{A}^T$. The first, second, third, . . . , rows of $\mathbf{A}$ become the first, second, third, . . . , columns of $\mathbf{A}^T$.

EXAMPLE 1

> ☛ **20.** Give the transposes of
>
> $\mathbf{A} = \begin{vmatrix} -2 & -3 \\ 4 & 5 \end{vmatrix}$, and $\mathbf{B} = \begin{bmatrix} 2 \\ 0 \\ 1 \end{bmatrix}$.

(a) If $\mathbf{A} = \begin{bmatrix} 2 & 3 \\ 5 & 7 \end{bmatrix}$, then $\mathbf{A}^T = \begin{bmatrix} 2 & 5 \\ 3 & 7 \end{bmatrix}$.

(b) If $\mathbf{A} = \begin{bmatrix} a & b \\ p & q \\ u & v \end{bmatrix}$, then $\mathbf{A}^T = \begin{bmatrix} a & p & u \\ b & q & v \end{bmatrix}$. ☛ **20**

DEFINITION Let $\mathbf{A} = [a_{ij}]$ be a square matrix and let A_{ij} denote the cofactor of the element a_{ij} in the determinant of $\mathbf{A}$. (That is, A_{11} denotes the cofactor of a_{11}, the first element in the first row of $\mathbf{A}$; A_{32} denotes the cofactor of a_{32}, the second element in the third row of $\mathbf{A}$; and so on.) The matrix $[A_{ij}]$ whose ij-element is the cofactor A_{ij} is called the **cofactor matrix** of $\mathbf{A}$. The transpose of the cofactor matrix is called the **adjoint** of $\mathbf{A}$ and is denoted by adj $\mathbf{A}$.

EXAMPLE 2 Find the adjoint of the matrix $\mathbf{A}$, where

$$A = \begin{bmatrix} 1 & 2 & 3 \\ 4 & 5 & 6 \\ 3 & 1 & 2 \end{bmatrix}$$

> *Answer* $\mathbf{A}^T = \begin{bmatrix} -2 & 4 \\ -3 & 5 \end{bmatrix}$, and
> $\mathbf{B}^T = [2 \quad 0 \quad 1]$.

Solution Let us first find the cofactors A_{ij} of the various elements a_{ij} of $\mathbf{A} = [a_{ij}]$.

$$A_{11} = (-1)^{1+1} \begin{vmatrix} 1 & 2 & 3 \\ 4 & 5 & 6 \\ 3 & 1 & 2 \end{vmatrix} = \begin{vmatrix} 5 & 6 \\ 1 & 2 \end{vmatrix} = 10 - 6 = 4$$

$$A_{12} = (-1)^{1+2} \begin{vmatrix} 1 & 2 & 3 \\ 4 & 5 & 6 \\ 3 & 1 & 2 \end{vmatrix} = -\begin{vmatrix} 4 & 6 \\ 3 & 2 \end{vmatrix} = -(8 - 18) = 10$$

Similarly,

$$A_{13} = (-1)^{1+3} \begin{vmatrix} 4 & 5 \\ 3 & 1 \end{vmatrix} = 4 - 15 = -11$$

$$A_{21} = (-1)^{2+1} \begin{vmatrix} 2 & 3 \\ 1 & 2 \end{vmatrix} = -(4 - 3) = -1$$

and so on, the other cofactors being $A_{22} = -7$, $A_{23} = 5$, $A_{31} = -3$, $A_{32} = 6$, and $A_{33} = -3$.

Thus the cofactor matrix is

$$[A_{ij}] = \begin{bmatrix} A_{11} & A_{12} & A_{13} \\ A_{21} & A_{22} & A_{23} \\ A_{31} & A_{32} & A_{33} \end{bmatrix} = \begin{bmatrix} 4 & 10 & -11 \\ -1 & -7 & 5 \\ -3 & 6 & -3 \end{bmatrix}.$$

☛ **21.** Give the adjoints of

$$A = \begin{bmatrix} -2 & -3 \\ 4 & 5 \end{bmatrix}, \text{ and}$$

$$B = \begin{bmatrix} 2 & 1 & 0 \\ -1 & 0 & 3 \\ 0 & 2 & 2 \end{bmatrix}.$$

Then adj A is the transpose of $[A_{ij}]$ and so is given by

$$\text{adj } A = \begin{bmatrix} 4 & -1 & -3 \\ 10 & -7 & 6 \\ -11 & 5 & -3 \end{bmatrix}. \quad ☛ \textbf{21}$$

The importance of the adjoint matrix is shown in Theorem 1, which we state without proof.

THEOREM 1 The inverse of a square matrix A exists if and only if $|A|$ is nonzero; in such a case, it is given by the formula

$$A^{-1} = \frac{1}{|A|} \cdot \text{adj } A.$$

Answer adj $A = \begin{bmatrix} 5 & 3 \\ -4 & -2 \end{bmatrix}$,

adj $B = \begin{bmatrix} -6 & -2 & 3 \\ 2 & 4 & -6 \\ -2 & -4 & 1 \end{bmatrix}$.

In the case of a 2×2 matrix, this result takes the following explicit form:

$$\text{If } A = \begin{bmatrix} a_{11} & a_{12} \\ a_{21} & a_{22} \end{bmatrix} \quad \text{then} \quad A^{-1} = \frac{1}{|A|} \begin{bmatrix} a_{22} & -a_{12} \\ -a_{21} & a_{11} \end{bmatrix}.$$

EXAMPLE 3 Find $\mathbf{A}^{-1}$ for the matrix of Example 2,

$$\mathbf{A} = \begin{bmatrix} 1 & 2 & 3 \\ 4 & 5 & 6 \\ 3 & 1 & 2 \end{bmatrix}.$$

Solution We find by expanding by the first row that $|\mathbf{A}| = -9$. Since $|\mathbf{A}| \neq 0$, $\mathbf{A}^{-1}$ exists and is given by

$$\mathbf{A}^{-1} = \frac{1}{|\mathbf{A}|} \cdot \text{adj } \mathbf{A}.$$

Taking the matrix adj $\mathbf{A}$ from Example 2, we obtain

$$\mathbf{A}^{-1} = \frac{1}{-9} \begin{bmatrix} 4 & -1 & -3 \\ 10 & -7 & 6 \\ -11 & 5 & -3 \end{bmatrix} = \begin{bmatrix} -\frac{4}{9} & \frac{1}{9} & \frac{1}{3} \\ -\frac{10}{9} & \frac{7}{9} & -\frac{2}{3} \\ \frac{11}{9} & -\frac{5}{9} & \frac{1}{3} \end{bmatrix}.$$

It is readily verified using matrix multiplication that $\mathbf{A}\mathbf{A}^{-1} = \mathbf{I}$ and $\mathbf{A}^{-1}\mathbf{A} = \mathbf{I}$.

EXAMPLE 4 Show that the matrix

$$\mathbf{A} = \begin{bmatrix} 1 & 2 & 3 \\ 2 & 5 & 7 \\ 3 & 7 & 10 \end{bmatrix}$$

is not invertible. (See Example 4 in Section 10-1).

☛ 22 Use determinants to compute the inverses of the following matrices, if they exist:

(a) $\begin{bmatrix} 2 & 1 \\ 4 & 6 \end{bmatrix}$; (b) $\begin{bmatrix} -3 & 2 \\ 9 & -6 \end{bmatrix}$;

(c) $\begin{bmatrix} 1 & 2 & 3 \\ 4 & 4 & 6 \\ 3 & 2 & 3 \end{bmatrix}$; (d) $\begin{bmatrix} 0 & 1 & 0 \\ -1 & 2 & 3 \\ 1 & 2 & 0 \end{bmatrix}$.

Solution The simplest way of showing this is to demonstrate that the determinant of $\mathbf{A}$ is zero. By expansion, we readily verify that

$$|\mathbf{A}| = \begin{vmatrix} 1 & 2 & 3 \\ 2 & 5 & 7 \\ 3 & 7 & 10 \end{vmatrix} = 0$$

as required. ☛ 22

EXAMPLE 5 *(Input-Output Model)* Table 9 gives the interaction between various sectors of a hypothetical economy.

TABLE 9

	Industry I	Industry II	Industry III	Final Demands	Gross Output
Industry I	20	48	18	14	100
Industry II	30	12	54	24	120
Industry III	30	36	36	72	180
Labor Inputs	20	24	72		

Answer (a) $\frac{1}{8} \begin{bmatrix} 6 & -1 \\ -4 & 2 \end{bmatrix}$;

(d) $\frac{1}{3} \begin{bmatrix} -6 & 0 & 3 \\ 3 & 0 & 0 \\ -4 & 1 & 1 \end{bmatrix}$;

(b) and (c) not invertible ($\Delta = 0$).

(a) Determine the input-output matrix $\mathbf{A}$.

(b) Suppose that in three years, the final demands are anticipated to change to 24, 33, and 75 for the three industries, I, II, and III, respectively. How much should each industry produce to meet this projected demand?

Solution (a) Dividing each column in the inner rectangle by the gross output of the corresponding industry, we obtain the input-output matrix

$$\mathbf{A} = \begin{bmatrix} \frac{20}{100} & \frac{48}{120} & \frac{18}{180} \\ \frac{30}{100} & \frac{12}{120} & \frac{54}{180} \\ \frac{30}{100} & \frac{36}{120} & \frac{36}{180} \end{bmatrix} = \begin{bmatrix} 0.2 & 0.4 & 0.1 \\ 0.3 & 0.1 & 0.3 \\ 0.3 & 0.3 & 0.2 \end{bmatrix}$$

(b) If $\mathbf{I}$ denotes the 3×3 identity matrix, then

$$\mathbf{I} - \mathbf{A} = \begin{bmatrix} 1 & 0 & 0 \\ 0 & 1 & 0 \\ 0 & 0 & 1 \end{bmatrix} - \begin{bmatrix} 0.2 & 0.4 & 0.1 \\ 0.3 & 0.1 & 0.3 \\ 0.3 & 0.3 & 0.2 \end{bmatrix}$$

$$= \begin{bmatrix} 0.8 & -0.4 & -0.1 \\ -0.3 & 0.9 & -0.3 \\ -0.3 & -0.3 & 0.8 \end{bmatrix}.$$

Let $\mathbf{B} = \mathbf{I} - \mathbf{A}$. Then, in order to calculate the future outputs, we need to find the inverse of $\mathbf{B}$. (See Section 10-2.) We can use the method of determinants. We have

$$|\mathbf{B}| = \begin{vmatrix} 0.8 & -0.4 & -0.1 \\ -0.3 & 0.9 & -0.3 \\ -0.3 & -0.3 & 0.8 \end{vmatrix} = 0.336.$$

Since $|\mathbf{B}| = 0.336 \neq 0$, $\mathbf{B}^{-1}$ exists. The cofactors B_{ij} in the determinant $|\mathbf{B}|$ are as follows.

$$B_{11} = (-1)^{1+1} \begin{vmatrix} 0.9 & -0.3 \\ -0.3 & 0.8 \end{vmatrix} = 0.72 - 0.09 = 0.63$$

$$B_{12} = (-1)^{1+2} \begin{vmatrix} -0.3 & -0.3 \\ -0.3 & 0.8 \end{vmatrix} = -(-0.24 - 0.09) = 0.33$$

Continuing in the same manner, we have $B_{13} = 0.36$, $B_{21} = 0.35$, $B_{22} = 0.61$, $B_{23} = 0.36$, $B_{31} = 0.21$, $B_{32} = 0.27$, and $B_{33} = 0.60$.

Taking the transpose of the cofactor matrix, we obtain

$$\text{adj } \mathbf{B} = \begin{bmatrix} 0.63 & 0.35 & 0.21 \\ 0.33 & 0.61 & 0.27 \\ 0.36 & 0.36 & 0.60 \end{bmatrix}.$$

Consequently, the inverse of $\mathbf{B}$ (or $\mathbf{I} - \mathbf{A}$) is given by

$$(\mathbf{I} - \mathbf{A})^{-1} = \mathbf{B}^{-1} = \frac{1}{|\mathbf{B}|} \operatorname{adj} \mathbf{B} = \left(\frac{1}{0.336} \right) \begin{bmatrix} 0.63 & 0.35 & 0.21 \\ 0.33 & 0.61 & 0.27 \\ 0.36 & 0.36 & 0.60 \end{bmatrix}.$$

If $\mathbf{D}$ denotes the new demand vector, that is,

$$\mathbf{D} = \begin{bmatrix} 24 \\ 33 \\ 75 \end{bmatrix}$$

and $\mathbf{X}$ is the new output matrix, then we showed in Section 10-2 that $\mathbf{X} = (\mathbf{I} - \mathbf{A})^{-1} \mathbf{D}$.

Therefore

$$\mathbf{X} = \left(\frac{1}{0.336} \right) \begin{bmatrix} 0.63 & 0.35 & 0.21 \\ 0.33 & 0.61 & 0.27 \\ 0.36 & 0.36 & 0.60 \end{bmatrix} \begin{bmatrix} 24 \\ 33 \\ 75 \end{bmatrix} = \begin{bmatrix} 126.25 \\ 143.75 \\ 195 \end{bmatrix}.$$

Thus industry I should produce 126.25 units, industry II should produce 143.75 units, and industry III should produce 195 units to meet the projected final demands in 3 years.

EXERCISES 10-5

(1–6) Write the transposes of the following matrices.

1. $\begin{bmatrix} 2 & 5 \\ 3 & -7 \end{bmatrix}$

2. $\begin{bmatrix} 3 & 2 & 1 \\ -5 & 7 & 6 \\ 0 & 3 & 2 \end{bmatrix}$

3. $\begin{bmatrix} a_1 & a_2 & a_3 \\ b_1 & b_2 & b_3 \end{bmatrix}$

4. $\begin{bmatrix} 1 & 2 & 3 \\ 3 & 1 & 2 \\ 5 & 4 & 6 \\ 6 & 5 & 4 \end{bmatrix}$

5. $\begin{bmatrix} 1 & 0 \\ 0 & 1 \end{bmatrix}$

6. $[2]$

(7–16) Use the method of determinants to find the inverses of the following matrices (when they exist).

7. $\begin{bmatrix} 3 & 2 \\ -1 & 1 \end{bmatrix}$

8. $\begin{bmatrix} -2 & 5 \\ 1 & -3 \end{bmatrix}$

9. $\begin{bmatrix} -\frac{5}{2} & \frac{3}{2} \\ 2 & -1 \end{bmatrix}$

10. $\begin{bmatrix} 0.3 & -0.1 \\ -0.2 & 0.4 \end{bmatrix}$

11. $\begin{bmatrix} 2 & 1 & -1 \\ 1 & 2 & 3 \\ -1 & 1 & 2 \end{bmatrix}$

12. $\begin{bmatrix} 1 & -1 & 1 \\ -1 & 1 & 1 \\ 1 & 1 & -1 \end{bmatrix}$

13. $\begin{bmatrix} 1 & 0 & 2 \\ 0 & 2 & 1 \\ 2 & 1 & 0 \end{bmatrix}$

14. $\begin{bmatrix} 1 & -1 & 2 \\ 2 & 1 & 0 \\ -1 & 2 & 1 \end{bmatrix}$

15. $\begin{bmatrix} 1 & 2 & 3 \\ 4 & 5 & 6 \\ 7 & 8 & 9 \end{bmatrix}$

16. $\begin{bmatrix} 2 & 1 & 3 \\ 5 & 3 & 7 \\ 7 & 4 & 10 \end{bmatrix}$

17. (*Input-Output Model*) Table 10 gives the interaction between various sectors of a hypothetical economy:

a. Determine the input-output matrix $\mathbf{A}$.

TABLE 10

	Industry I	Industry II	Industry III	Final Demands	Gross Output
Industry					
I	20	40	30	10	100
II	30	20	90	60	200
III	40	100	60	100	300
Primary Inputs	10	40	120		

b. Suppose that in 5 years, the final demands change to 150, 280, and 420 for industries I, II, and III, respectively. How much must each industry produce to meet these projected demands?

c. What will be the new primary input requirements for the three industries in 5 years?

18. *(Input-Output Model)* The interaction between various sectors of a hypothetical economy is given by Table 11.

a. What is the input-output matrix **A**?

b. Suppose that in 3 years, the consumer demands change to 20 for industry I, 50 for industry II, and 70 for industry III. How much must each industry produce in 3 years to meet this projected demand?

c. What will be the new labor input requirements for the three industries in 3 years time?

TABLE 11

	Industry I	Industry II	Industry III	Consumer Demands	Total Output
Industry					
I	16	30	20	14	80
II	32	15	80	23	150
III	24	75	40	61	200
Labor Inputs	8	30	60		

19. *(Input-Output Model)* An economy consists of three sec-

tors, A, B, and C, whose interactions are given in Table 12.

TABLE 12

	A	B	C	Final Demands	Gross Output
A	60	16	80	44	200
B	60	48	20	32	160
C	40	32	60	68	200
Primary Inputs	40	64	40		

a. Determine the input-output matrix.

b. If the final demands change to 50, 60, and 80 units for the products of A, B, and C, respectively, what will be the required production levels to meet these new demands?

20. *(Input-Output Model)* The interaction between three sectors in an economy is given in Table 13.

a. Determine the input-output matrix.

b. If the final demand for secondary industrial products increases to 10 units, determine the new output levels for the three sectors.

c. If the final demand for primary industrial products falls to zero, determine the new output levels for the three sectors.

TABLE 13

	Primary Industry	Secondary Industry	Agriculture	Final Demands	Gross Output
Primary Industry	4	12	3	1	20
Secondary Industry	8	9	6	7	30
Agriculture	2	3	3	7	15
Primary Inputs	6	6	3		

CHAPTER REVIEW

Key Terms, Symbols, and Concepts

10.1 The inverse of a square matrix $\mathbf{A}$, $\mathbf{A}^{-1}$.
Invertible (or nonsingular) matrix, singular matrix.
Computation of $\mathbf{A}^{-1}$ by row reduction.

10.2 Input-output model. Primary inputs, final demands.
Output matrix, demand matrix, input-output matrix and coefficients.

10.3 Markov process (or chain).
Transition probabilities, transition matrix.
State matrix (or vector). Steady-state matrix.

10.4 Determinant of order 2, of order 3, of higher order.
Complete expansion of a determinant of order 3.
The ijth-minor and ijth-cofactor in a determinant.
Expansion of a determinant by the ith row or column.
Cramer's rule; condition for uniqueness of solution.

10.5 Transpose of a matrix.
Cofactor matrix, adjoint matrix.

Formulas

$\mathbf{AA}^{-1} = \mathbf{I}$, $\mathbf{A}^{-1}\mathbf{A} = \mathbf{I}$.

Solution of linear system: If $\mathbf{AX} = \mathbf{B}$ and $\mathbf{A}$ is square, then $\mathbf{X} = \mathbf{A}^{-1}\mathbf{B}$.

Input-output model: $\mathbf{X} = \mathbf{AX} + \mathbf{D}$, $\mathbf{X} = (\mathbf{I} - \mathbf{A})^{-1}\mathbf{D}$.

Markov chain: $\mathbf{A}_{k+1} = \mathbf{A}_k\mathbf{P}$.
For steady-state matrix $\mathbf{B}$: $\mathbf{BP} = \mathbf{B}$.

Determinant of order 2: $\begin{vmatrix} a_1 & b_1 \\ a_2 & b_2 \end{vmatrix} = a_1b_1 - a_2b_1$.

Expansion of a determinant by any row (or column): multiply each element in the row (or column) by the corresponding cofactor and form the sum of the products.

Cramer's rule for a 3×3 system of equations:
$$x = \frac{\Delta_1}{\Delta}, \quad y = \frac{\Delta_2}{\Delta}, \quad z = \frac{\Delta_3}{\Delta}.$$

$\mathbf{A}^{-1}$ exists if and only if $|\mathbf{A}| \neq 0$.

$\mathbf{A}^{-1} = \dfrac{1}{|\mathbf{A}|}$ adj $\mathbf{A}$.

If $\mathbf{A} = \begin{bmatrix} a_{11} & a_{12} \\ a_{21} & a_{22} \end{bmatrix}$ then $\mathbf{A}^{-1} = \dfrac{1}{|\mathbf{A}|} \begin{bmatrix} a_{22} & -a_{12} \\ -a_{21} & a_{11} \end{bmatrix}$

REVIEW EXERCISES FOR CHAPTER 10

1. State whether each of the following is true or false. Replace each false statement by a true statement.

   a. If $\mathbf{A}$ is invertible, then the size of $\mathbf{A}^{-1}$ is the same as that of $\mathbf{A}$.

   b. The identity matrix is its own inverse.

   c. The zero matrix has as its inverse a zero matrix.

   d. If $\mathbf{A}$, $\mathbf{B}$, and $\mathbf{C}$ are three matrices such that $\mathbf{AB} = \mathbf{AC}$, then $\mathbf{B} = \mathbf{C}$.

   e. If a matrix $\mathbf{A}$ contains a zero element, then it is not invertible.

   f. A matrix is invertible as long as at least one of its elements is nonzero.

   g. The square matrix $\mathbf{A}$ is invertible if and only if the determinant $|\mathbf{A}| \neq 0$.

   h. The adjoint of $\mathbf{A}$ is the cofactor matrix of $\mathbf{A}^T$.

   i. If $\mathbf{A}$ is an $n \times n$ square matrix, then the determinant of the matrix $k\mathbf{A}$ $(k \neq 0)$ is equal to $k|\mathbf{A}|$.

   j. If $\mathbf{A}$ is a square matrix and $\mathbf{A}^T$ its transpose, then $|\mathbf{A}| = |\mathbf{A}^T|$.

   k. The cofactor of an element in a determinant Δ is the determinant obtained by deleting the row and column in Δ in which the element occurs.

   l. The cofactor and minor of an element are equal in absolute value but differ in sign.

m. If $\mathbf{A}$ and $\mathbf{B}$ are two square matrices of the same size and if their inverses exist, then $(\mathbf{AB})^{-1} = \mathbf{A}^{-1}\mathbf{B}^{-1}$.

n. If $\mathbf{A}$ and $\mathbf{B}$ are two invertible square matrices of the same size, then $(\mathbf{A}^{-1}\mathbf{B})^{-1} = \mathbf{B}^{-1}\mathbf{A}$.

o. If $\mathbf{P}$ and $\mathbf{Q}$ are two transition matrices of the same size for a Markov chain, then $\mathbf{PQ}$ is also a transition matrix.

p. If $\mathbf{A}$ and $\mathbf{P}$ denote, respectively, the state matrix and transition matrix of a Markov process, then $\mathbf{AP}$ is also a state matrix.

q. If $\mathbf{A}_0$ denotes the initial state matrix, $\mathbf{A}_k$ the state matrix after k trials, and $\mathbf{P}$ the transition matrix, then $\mathbf{A}_k = \mathbf{A}_0\mathbf{P}^k$.

2. Show that if $\mathbf{A}$ and $\mathbf{B}$ are 2×2 matrices such that $\mathbf{A} = k\mathbf{B}$, where k is a constant, then $|\mathbf{A}| = k^2|\mathbf{B}|$. Does this still hold if $\mathbf{A}$ and $\mathbf{B}$ are 3×3? $n \times n$?

(3–14) Find the inverses of the matrices given below, when they exist.

3. $\begin{bmatrix} 1 & -3 \\ 2 & 5 \end{bmatrix}$

4. $\begin{bmatrix} 2 & 4 \\ 5 & -3 \end{bmatrix}$

5. $\begin{bmatrix} a & b \\ -b & a \end{bmatrix}$ $(a, b \neq 0)$

6. $\begin{bmatrix} 1 & 1 \\ a & b \end{bmatrix}$ $(a \neq b)$

7. $\begin{bmatrix} 1 & 0 & 0 \\ 0 & 2 & 0 \\ 0 & 0 & 3 \end{bmatrix}$

8. $\begin{bmatrix} 1 & 0 & 2 \\ 0 & 2 & 1 \\ 2 & 1 & 0 \end{bmatrix}$

9. $\begin{bmatrix} 1 & 2 & 3 \\ 2 & 3 & 4 \\ 3 & 1 & 2 \end{bmatrix}$

10. $\begin{bmatrix} 2 & -1 & 1 \\ 1 & 2 & -1 \\ -1 & 1 & 2 \end{bmatrix}$

11. $\begin{bmatrix} 1 & -2 & 3 \\ 4 & 1 & 6 \\ 7 & 4 & 9 \end{bmatrix}$

12. $\begin{bmatrix} 2 & -1 & -3 \\ 4 & 1 & -1 \\ 7 & -2 & -8 \end{bmatrix}$

13. $\begin{bmatrix} 2 & 1 & 3 \\ 1 & 4 & 2 \end{bmatrix}$

14. $\begin{bmatrix} 1 & 2 & 3 & 4 \\ 5 & 6 & 7 & 8 \\ 2 & 1 & 4 & 3 \end{bmatrix}$

(15–18) Solve the following systems of equations by using the inverse of the coefficient matrix.

15. $4x - 3y = 1$
$3x + 2y = 5$

16. $2u - 5v = 11$
$3u + 4v = 5$

17. $x + y + z = 1$
$2x - y + 3z = -2$
$3x + 2y - z = 6$

18. $3p - 2q + 4r = 13$
$p + q - 2r = 1$
$2p - 3q + 5r = 11$

(19–24) Use determinants to solve the following systems of equations.

19. $3x + 5y = 1$
$2x - 4y = -3$

20. $3u - 2v = 4$
$3v = -5 + 4u$

21. $x - y + z = 2$
$-x + y + z = 4$
$x + y - z = 0$

22. $2x - y - z = 3$
$x - 2y + z = 6$
$x + y - 2z = -3$

23. $2x - 6y + 4z = 9$
$3x + y - 2z = 5$
$3x - 9y + 6z = 13$

24. $6x - 3y + 12z = 15$
$2x + 3y + 5z = 10$
$4x - 2y + 8z = 21$

(25–30) Factor the following determinants.

25. $\begin{vmatrix} a & 2 \\ 8 & a \end{vmatrix}$

26. $\begin{vmatrix} a & -b \\ b & a + 2b \end{vmatrix}$

27. $\begin{vmatrix} x + 1 & 7 \\ 6 & x + 2 \end{vmatrix}$

28. $\begin{vmatrix} x + 4 & x + 14 \\ x & 2x + 1 \end{vmatrix}$

29. $\begin{vmatrix} 1 & 1 & 1 \\ x & 0 & 1 \\ 1 & 1 & x \end{vmatrix}$

30. $\begin{vmatrix} 1 & -1 & 9 \\ x + 1 & x & 2 \\ x + 2 & x + 1 & x - 1 \end{vmatrix}$

31. *(Input-Output Model)* Table 14 gives the interaction between various sectors of a hypothetical economy.

TABLE 14

	Industry I	Industry II	Final Demands	Gross Outputs
Industry I	8	52	20	80
Industry II	56	26	48	130
Primary Inputs	16	52		

a. Find the input-output matrix $\mathbf{A}$.

b. Suppose in 2 years the final demands change to 58 for industry I and 79 for industry II. How much should each industry produce to meet these projected demands?

c. What will be the new primary input requirements for the two industries in 2 years?

32. *(Input-Output Model)* The interaction between various sectors of a hypothetical economy is given in Table 15.

a. Give the input-output matrix **A**.

b. Suppose that in 5 years the final demands change to 140 for industry I and 287 for industry II. How much must each industry produce to meet these projected demands?

TABLE 15

	Industry		Final	Gross
	I	II	Demands	Outputs
Industry				
I	120	384	96	600
II	420	288	252	960
Primary				
Inputs	60	288		

c. What will be the new primary input requirements for the two industries in 5 years?

(33–36) *(City Intersections)* Suppose a small city has all the streets running in north-south or east-west directions. If you are walking in this city, at each intersection the transition matrix will be of the form

Final Direction

$$
\begin{array}{c}
\text{Initial} \\
\text{Direction}
\end{array}
\begin{array}{c}
\quad\quad \text{N} \;\; \text{E} \;\; \text{S} \;\; \text{W} \\
\begin{array}{c} \text{N} \\ \text{E} \\ \text{S} \\ \text{W} \end{array}
\left[
\begin{array}{cccc}
- & - & - & - \\
- & - & - & - \\
- & - & - & - \\
- & - & - & -
\end{array}
\right]
\end{array}
$$

where the letters N, E, S, and W stand for north, east, south, and west directions. Write down the transition matrix for each of the following types of decisions at the intersections.

33. Go straight ahead at every intersection.

34. At each intersection, flip a coin to decide whether to go straight on or make a right turn.

35. Roll an unbiased die to decide whether to turn left (1, 2, 3, 4) or turn right (5, 6).

36. Roll an unbiased die to decide whether to go straight (1, 2, or 3), turn left (4) or turn right (5, 6).

37. *(Shooting Game)* Tom and Dick are playing a shooting game. Tom hits the target 80% of the time and Dick hits 60% of the time. Each man keeps his turn firing if he hits the target and loses if he misses and then it is the other person's turn. Write down the transition matrix for this shooting game.

38. *(Ball Game)* Sandy, Jean, and Sally are throwing a ball to each other. Sandy always throws the ball to Sally; Sally throws to Sandy with probability 0.2 and to Jean with probability 0.8; whereas Jean is as likely to throw the ball to Sandy as she is to Sally. Assume that none of them keeps the ball once it is thrown to her. Write the transition matrix for this Markov process.

39. *(Market Share)* A particular product is being produced by two firms, X and Y. At present, the firm X enjoys 70% of the market and the firm Y enjoys 30%. A survey analysis over the past few years indicates that from one year to another, 20% of firm X's customers switch to firm Y and 10% of firm Y's customers switch to firm X. Assume that this trend continues.

a. Write the transition matrix for this Markov process.

b. What percentage of the market will the firm Y have in 2 years from now?

c. What percentage of the market will the firm X have in the long run?

40. *(Condition of Operating Machines)* At the start of each day, a particular operating machine is either in working order or is broken down. If the machine is in working order at the start of any day, there is a 10% chance that it will be broken at the start of the next day. However, if the machine is broken at the start of any day, there is a 70% chance that it will be repaired and in working order the next day.

a. Write the transition matrix for this Markov process.

b. If the machine is initially in working order, determine the probability that it will be in working order 3 days later.

c. Find the long-run probability that the machine will be found broken at the start of the day.

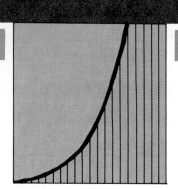

CHAPTER 11

Linear Programming

Chapter Objectives

11-1 LINEAR INEQUALITIES

(a) The graph of a linear inequality in two variables; representation of weak and strict inequalities;
(b) Construction of the graph of several linear inequalities in two variables;
(c) Formulation of systems of inequalities from verbal information.

11-2 LINEAR OPTIMIZATION (GEOMETRIC APPROACH)

(a) Reformulation of verbal optimization problems as linear programming problems; definitions of objective function and constraints;
(b) Feasible solutions; the feasible region and its graph;
(c) Use of indifference lines to solve a linear programming problem; demonstration that the optimum solution occurs at a vertex of the feasible region;
(d) Examples of the geometric method of solution.

11-3 THE SIMPLEX TABLEAU

(a) Introduction of slack variables for a system of inequalities;
(b) Basic feasible solutions (BFS) and their relation to the slack and structural variables and to the vertices of the feasible region;
(c) Representation of a BFS by a simplex tableau;
(d) Departing variable, entering variable and row operations involved in pivoting.

11-4 THE SIMPLEX METHOD

(a) Extension of the tableau and pivoting to include the objective function;
(b) The indicators and the largest indicator criterion for entering variable;
(c) The criterion for departing variable in terms of smallest positive ratio of entry in final column to entry in the column of the entering variable;
(d) The termination criterion for the simplex method;
(e) Artificial variables and the penalty method.

CHAPTER REVIEW

◼ 11-1 LINEAR INEQUALITIES

The inequality $y > 2x - 4$, relating the two variables x and y, is an example of what are called *linear inequalities*. Let us begin by examining this particular example in terms of a graph.

The equation $y = 2x - 4$ has as its graph a straight line whose slope is 2 and whose y-intercept is -4. It is shown as a dashed line in Figure 1. As an example, when $x = 4$, $y = 2(4) - 4 = 4$, so the point $(4, 4)$ lies on the line, as shown in Figure 1.

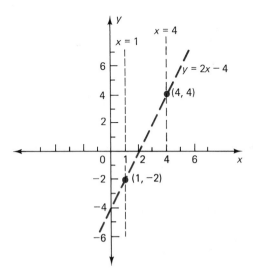

FIGURE 1

Now consider the inequality

$$y > 2x - 4.$$

When $x = 4$, this takes the form $y > 2(4) - 4$, or $y > 4$. Thus the inequality is satisfied at all the points $(4, y)$ where $y > 4$. Graphically, this means that on the vertical line $x = 4$, the inequality $y > 2x - 4$ is satisfied at all points that lie *above* the point $(4, 4)$.

Similarly, we can take the vertical line, $x = 1$. On this line, the inequality $y > 2x - 4$ becomes $y > -2$. It is satisfied by the points $(1, y)$ that lie on this vertical line above the point $(1, -2)$. (See Figure 1.)

It can be seen in a similar way that the inequality $y > 2x - 4$ is satisfied at all of the points (x, y) that lie *above* the straight line $y = 2x - 4$. This region in the xy-plane is said to be the **graph** of the given inequality.

A linear inequality between two variables x and y is any relationship of the form $Ax + By + C > 0$ (or < 0) or $Ax + By + C \geq 0$ (or ≤ 0). The graph of a linear inequality consists of all those points (x, y) that satisfy the inequality. It consists of a region in the xy-plane, not simply a line or a curve.

The graph of the inequality $Ax + By + C > 0$ is a half-plane bounded by the straight line whose equation is $Ax + By + C = 0$. Figure 2 illustrates some linear inequalities. In each case, the half-plane of points that satisfy the inequality is shaded.

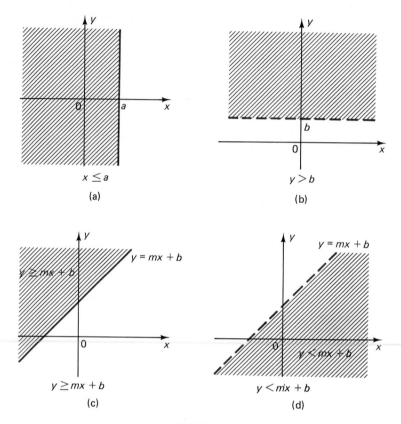

FIGURE 2

The graph of $y > mx + b$ is the half-plane above the line $y = mx + b$, and the graph of $y < mx + b$ is the half-plane below the line $y = mx + b$. If the graph includes the line, we show it by a solid line; otherwise we use a dashed line. A dashed line always corresponds to a strict inequality ($>$ or $<$) and a solid line corresponds to a weak inequality ($\geq$ or $\leq$).

EXAMPLE 1 Sketch the graph of the linear inequality $2x - 3y < 6$.

Solution First we solve the given inequality for y in terms of x (that is, we express it in one of the forms $y > mx + b$ or $y < mx + b$).

$$2x - 3y < 6$$
$$-3y < -2x + 6$$

We now divide both sides by -3. (Recall that when we divide the terms of an inequality by a negative number, the direction of the inequality changes. See Section 3-2.)

$$y > \tfrac{2}{3}x - 2$$

Next we graph the line $y = \tfrac{2}{3}x - 2$. For $x = 0$, we have $y = -2$. Thus $(0, -2)$ is a point on this line. Again, when $y = 0$, we have $\tfrac{2}{3}x - 2 = 0$ or $x = 3$. Thus $(3, 0)$ is another point on the line. We plot these two points and join them by a dashed line (because we have a strict inequality). Since the given inequality, when solved for y, involves the *greater than* sign, the graph is the half-plane *above* the dashed line. (See Figure 3.) ☛ 1

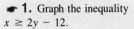

☛ **1.** Graph the inequality $x \geq 2y - 12$.

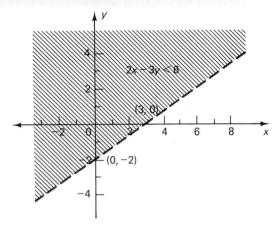

FIGURE 3

Linear inequalities arise in many problems of practical interest. This will become clear in the subsequent sections of this chapter, where we shall study an area of mathematics called *linear programming*. The following examples will provide some typical illustrations of situations that give rise to linear inequalities.

EXAMPLE 2 *(Stocks Investment)* An investor plans to invest up to $30,000 in two stocks, A and B. Stock A is currently priced at $165 and stock B at $90 per share. If the investor buys x shares of A and y shares of B, graph the region in the xy-plane that corresponds to possible investment strategies.

Solution The x shares of stock A at $165 per share cost $165x$ dollars. Similarly, y shares of stock B at $90 cost $90y$ dollars. The total sum invested is therefore

$$(165x + 90y) \text{ dollars}$$

and this cannot exceed $30,000. Thus

$$165x + 90y \leq 30,000.$$

We solve for y.

Answer

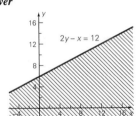

$$90y \le 30{,}000 - 165x$$

$$y \le -\frac{165}{90}x + \frac{30{,}000}{90}$$

$$y \le -\frac{11}{6}x + \frac{1000}{3}$$

The graph of this inequality is shown in Figure 4. In this example, only the region for which $x \ge 0$ and $y \ge 0$ has any significance, so the shaded region is a triangular region rather than a half-plane.

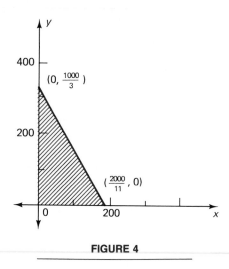

FIGURE 4

In many practical situations, problems arise involving more than one linear inequality. Example 3 illustrates a case in which two such inequalities occur.

EXAMPLE 3 *(Investment Return)* In the preceding example, stock A is currently paying a dividend of $6 per share and stock B is paying $5 per share. If the investor requires that the investment should pay more than $1400 in dividends, draw the graph of the allowed region.

Solution Again let x and y be the numbers of shares of stocks A and B, respectively. The previous inequality, $165x + 90y \le 30{,}000$, still applies. In addition, the dividend payments are $6x$ dollars from stock A and $5y$ dollars from stock B, giving a total of $(6x + 5y)$ dollars. Since this must exceed $1400, we have the second condition that

$$6x + 5y > 1400.$$

This can be rewritten as

$$y > -\tfrac{6}{5}x + 280.$$

► 2. Graph the following set of inequalities:
$$0 \le x \le 4, \ y \ge 0, \ x + 2y < 6.$$

This inequality is satisfied by the points above the line $6x + 5y = 1400$. Since it is a strict inequality, the line is not included and is drawn as a dashed line in Figure 5.

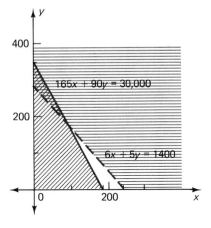

FIGURE 5

Answer

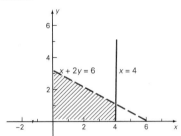

The allowed values of x and y must satisfy both the inequalities $165x + 90y \le 30,000$ and $6x + 5y > 1400$. Hence any allowed point must be *below or on* the line $165x + 90y = 30,000$ and at the same time *above* the line $6x + 5y = 1400$. These two regions are shown shaded differently in Figure 5. The allowed region is the one shaded both ways. Again, negative values of x or y are not allowed. **► 2, 3**

The general procedure for graphing several linear inequalities is as follows. First, graph each inequality separately and shade each allowed region in a different way. The allowed region for all the inequalities together will be where the shaded regions all overlap.

When more than two variables are involved in a system of inequalities, graphical techniques are much less helpful. With three variables, the graphs can still be drawn, but often at some inconvenience; with four or more variables, it becomes impossible to use graphs. Example 4 illustrates a simple problem involving three variables.

► 3. Graph the following set of inequalities:
$$x \ge 0, \ y \ge 0, \ 3x + 2y \le 6,$$
$$x - y \ge -1.$$

Answer

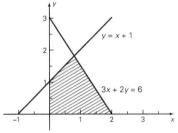

EXAMPLE 4 (*Distribution of TV Sets*) An electronics company makes television sets at two factories, F_1 and F_2. Factory F_1 can produce up to 100 sets per week and F_2 can produce up to 200 sets per week. The company has three distribution centers, X, Y, and Z. Center X requires 50 television sets per week, Y requires 75 sets per week, and Z requires 125 sets per week in order to meet the demands in their respective areas. If factory F_1 supplies x sets per week to distribution center X, y sets to Y, and z sets to Z, write the inequalities satisifed by x, y, and z.

☛ **4.** How are the inequalities in Example 4 modified if
(a) the capacity of factory F_2 is suddenly reduced to 150 sets per week or,
(b) the demands from X, Y, and Z increase respectively to 100,150, and 200 sets per week and the plant capacities are increased to 250 at F_1 and 200 at F_2?

Solution The situation is illustrated in Figure 6. If factory F_1 supplies x sets to center X, then F_2 must supply $(50 - x)$ sets since a total of 50 sets are required at this distribution center. Similarly, F_2 must supply $(75 - y)$ sets to center Y and $(125 - z)$ sets to center Z.

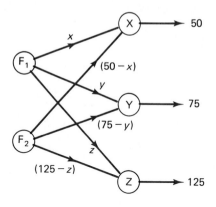

FIGURE 6

The total number of sets supplied by factory F_1 to all three distribution centers is $x + y + z$; this cannot exceed the productive capacity of this factory, which is 100 sets per week. Thus we arrive at the condition

$$x + y + z \leq 100.$$

Similarly, the total number of sets supplied by factory F_2 is equal to

$$(50 - x) + (75 - y) + (125 - z) = 250 - x - y - z.$$

This number cannot exceed 200, which is the most this factory can produce.

$$250 - x - y - z \leq 200$$

that is,

$$x + y + z \geq 50.$$

Since the number of sets supplied by any factory to any distribution center cannot be negative, each of the six quantities x, y, z, $(50 - x)$, $(75 - y)$, and $(125 - z)$ must be greater than or equal to zero. Hence x, y, and z must satisfy the following system of inequalities.

Answer (a) $x + y + z = 100$, so z can be removed from the problem and we are left with $0 \leq x \leq 50$, $0 \leq y \leq 75$, and $x + y \leq 100$.
(b) Again z can be eliminated since $x + y + z = 250$, and we are left with $0 \leq x \leq 100$, $0 \leq y \leq 150$, and $50 \leq x + y \leq 250$.

$$x \geq 0 \qquad y \geq 0 \qquad z \geq 0 \qquad x \leq 50 \qquad y \leq 75 \qquad z \leq 125$$
$$x + y + z \geq 50 \qquad x + y + z \leq 100$$

In order to represent these inequalities geometrically, we must use three-dimensional coordinates (x, y, z). It is possible to draw a suitable figure, but a certain degree of drawing skill is needed to obtain an accurate representation. ☛ **4**

(1–6) Sketch the graphs of the following inequalities in the xy-plane.

1. $x + y > 1$ **2.** $2x + 3y < 6$

3. $2x - y \leq 4$ **4.** $3x \geq y - 6$

5. $2x + 3 > 0$ **6.** $4 - 3y < 0$

(7–18) Sketch the graphs of the following sets of inequalities.

7. $x + y > 2$, $3x + y < 3$

8. $2x + y > 4$, $x + 2y < 4$, $2x - 3y < 3$

9. $0 \leq x \leq 10$, $0 \leq y \leq 15$, $5 \leq x + y \leq 12$

10. $2 \leq x \leq 5$, $1 \leq y \leq 5$, $x + y > 4$,
    $2x + y < 10$

11. $x \geq 0$, $y \geq 0$, $x + 3y \leq 4$, $2x + y \leq 6$

12. $1 \leq x + y \leq 4$, $y - x \geq 0$, $y - 2x \leq 1$

13. $x \geq 0$, $y \geq 0$, $x + y \leq 1$, $2x + 3y \leq 3$

14. $x \geq 0$, $y \geq 0$, $x + y \geq 1$, $2x + 3y \leq 3$

15. $x \geq 0$, $y \geq 0$, $4x + y \leq 4$,
    $2x + 3y \leq 3$, $x + y \geq 3$

16. $x \geq 0$, $y \geq 0$, $x - y \leq 2$, $2x + y \leq 2$

17. $x \geq 0$, $y \geq 0$, $x + 3y \leq 3$, $3x + y \leq 3$,
    $x + 2y \leq 3$

18. $x \geq 0$, $y \geq 0$, $y \geq 2x + 1$, $y + 2x \leq 3$,
    $3y + 2x \leq 3$

19. (*Material Distribution*) A company has 100 tons of sheet aluminum stored at one location and 120 tons stored at a second location. Some of this material must be delivered to two construction projects. The first project requires 70 tons and the second project requires 90 tons. Let x and y denote the amounts delivered from the first storage location to the two projects, respectively. Write the inequalities that must be satisfied by x and y and represent them graphically.

20. (*Distribution Costs*) In Exercise 19, suppose that it costs $10 and $15 per ton to deliver the aluminum from the first storage location to the first and second construction projects, respectively, and $15 and $25 per ton, respectively, to deliver it from the second storage location. If the company requires that the total delivery cost should

not exceed $2700, write the further condition on x and y and represent the allowed region graphically.

21. (*Distribution Costs*) Repeat Exercise 20 if the four delivery costs are $15 and $10, respectively, from the first storage location and $10 and $20, respectively, from the second storage location.

22. (*Warehouse Storage*) A storage company wishes to store up to 120 television sets in its warehouse. It keeps two models in stock, a table model and a floor model. The number of table models must not be less than 40 and the number of floor models must not be less than 30. Represent graphically the possible numbers of sets that can be stored.

23. (*Warehouse Storage*) In Exercise 22, suppose that the floor model requires 12 cubic feet of storage space and the table model requires 8 cubic feet. If the company has 1200 cubic feet of space available for storing the sets, represent the new allowed numbers of sets by a graph.

24. (*Machine Allocation*) A company makes two products, A and B. These products each require a certain amount of time on two machines in their manufacture. Each unit of product A requires 1 hour on machine I and 2 hours on machine II; each unit of product B requires 3 hours on machine I and 2 hours on machine II. The company has 100 hours per week available on each machine. If x units of product A and y units of product B are produced per week, give the inequalities satisfied by x and y and represent them graphically.

25. (*Allocation and Profits*) In Exercise 24, suppose that the company makes profits of $20 on each item A and $30 on each item B. If it is required that the total weekly profit should be at least $1100, represent the allowed values of x and y graphically.

26. (*Allocation and Profits*) In Exercise 25, represent the allowed region graphically if at least 15 of each type of product must be produced per week in order to fulfill contracts.

27. (*Diet Planning*) Sirloin steak costs 15¢ per ounce, and each ounce contains 110 calories and 7 grams of protein. Roast chicken costs 8¢ per ounce, and each ounce contains 83 calories and 7 grams of protein. Represent algebraically the combinations of x ounces of steak and y ounces of chicken that do not exceed $1.00 in cost and that contain less than 900 calories and at least 60 grams of protein.

28. *(Diet Planning)* Miss X has been informed by her doctor that she would be less depressed if she obtained at least the minimum adult requirement of thiamine, which is 1 milligram per day. The doctor suggests that she get half of this from breakfast cereal. Cereal *A* contains 0.12 milligram of thiamine per ounce and cereal *B* contains 0.08 milligram of thiamine per ounce. Determine the possible amounts of these cereals to provide her with at least one-half of the adult daily requirement of thiamine.

29. *(Storage Space)* The storeroom of a chemistry department stocks at least 300 beakers of one size and at least 400 beakers of a second size. It is decided that the total number of beakers stored should not exceed 1200. Determine the possible numbers of the two kinds of beakers that can be stored, and show this by a graph.

30. *(Storage Space)* In Exercise 29, assume that beakers of the first size occupy 9 square inches of shelf space and those of the second size occupy 6 square inches. The total

area of shelf space available for storage is at most 62.5 square feet. Determine the possible numbers of the two beakers and show this by a graph.

31. *(Diet Planning)* A person is considering replacing part of the meat in his diet by soybeans. One ounce of meat contains on average about 7 grams of protein, while 1 ounce of soybeans (undried) contains about 3 grams of protein. If he demands that his daily protein intake from meat and soybeans together should be at least 50 grams, what combination of these two would form an acceptable diet?

32. *(Ecology)* A fish pool is stocked each spring with two species of fish, S and T. The average weight of the fish stocked is 3 pounds for S and 2 pounds for T. Two foods F_1 and F_2 are available in the pool. The average daily requirement of a fish of species S is 2 units of F_1 and 3 units of F_2, whereas for species T, it is 3 units of F_1 and 1 unit of F_2. If at most 600 units of F_1 and 300 units of F_2 are available each day, how should the pool be stocked so that the total weight of the fish in the pool is at least 400 pounds?

◣ 11-2 LINEAR OPTIMIZATION (GEOMETRIC APPROACH)

A linear programming problem is one that involves finding the maximum or minimum value of some linear algebraic expression when the variables in this expression are subject to a number of linear inequalities. The following simple example is typical of such problems.

EXAMPLE 1 *(Maximum Profit)* A company manufactures two products, X and Y. Each of these products requires a certain amount of time on the assembly line and a further amount of time in the finishing shop. Each item of type X needs 5 hours for assembly and 2 hours for finishing, and each item of type Y needs 3 hours for assembly and 4 hours for finishing. In any week, the firm has available 105 hours on the assembly line and 70 hours in the finishing shop. The firm can sell all it can produce of each item and makes a profit of $200 on each item of X and $160 on each item of Y. Find the number of items of each type that should be manufactured per week to maximize the total profit.

Solution It is usually convenient when handling problems of this type to summarize the information in the form of a table. Table 1 shows the information in Example 1.

TABLE 1

	Assembly	Finishing	Profit
X	5	2	200
Y	3	4	160
Available	105	70	

☞ **5.** In Example 1, write down the inequalities if each item of type X requires 3 hours for assembly and 2 for finishing and each item of type Y requires 4 and 3 hours for assembly and finishing respectively.

Suppose that the firm produces x items of type X per week and y items of type Y per week. Then the time needed on the assembly line will be $5x$ hours for product X and $3y$ hours for product Y, or $(5x + 3y)$ hours in all. Since only 105 hours are available, we must have $5x + 3y \leq 105$.

Similarly, it requires $2x$ hours in the finishing shop to finish x items of product X, and $4y$ hours to finish y items of product Y. The total number of hours, $2x + 4y$, cannot exceed the 70 which are available, so we have the second condition, $2x + 4y \leq 70$. ☞ **5**

Each item of type X produces a profit of \$200, so x items produce $200x$ dollars in profit. Similarly, y items of type Y produce $160y$ dollars in profit. Hence the total weekly profit P (in dollars) is given by

$$P = 200x + 160y.$$

We can therefore restate the problem in the following terms: Find the values of x and y that maximize the quantity $P = 200x + 160y$ when x and y are subject to the conditions

$$5x + 3y \leq 105, \qquad 2x + 4y \leq 70, \qquad x \geq 0, \quad \text{and} \quad y \geq 0. \qquad (1)$$

(Observe the conditions that x and y must be nonnegative. These are added for completeness.)

This example is a typical linear programming problem. We have an expression $P = 200x + 160y$, which is linear in the variables x and y, and we wish to find the maximum value of P when x and y satisfy inequalities (1). A more general problem might involve more than two variables and a larger number of inequalities than the four in this example, but otherwise the example is quite representative of problems in the linear programming area.

When investigating any problem in linear programming, when only two variables are involved, a graphical approach is often helpful. Consider inequalities (1). The set of points (x, y) that satisfy all of these inequalities is shown shaded in Figure 7. This shaded region represents the set of *feasible solutions*, that is, the set of values of x and y that the firm is able to adopt. Any point (x, y) that lies outside this shaded region cannot be adopted.

For example, consider the point $x = 12$, $y = 14$, which lies outside the feasible region. To produce 12 items of type X and 14 items of type Y would require $12(5) + 14(3) = 102$ hours on the assembly line and $12(2) + 14(4) = 80$ hours in the finishing shop. Although this would not exceed the available hours on the assembly line, it does exceed those available for finishing; so it does not represent a possible production schedule.

Answer $x \geq 0$, $y \geq 0$,
$3x + 4y \leq 105$, $2x + 3y \leq 70$.

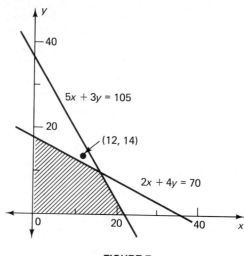

FIGURE 7

Now consider the set of values of x and y that lead to some fixed profit. For example, giving P the value 4000, we see that x and y must satisfy the equation

$$200x + 160y = 4000. \tag{2}$$

All values of x and y satisfying this equation produce a profit of \$4000 per week. This is the equation of a straight line that meets the x-axis at the point $(20,0)$ and the y-axis at the point $(0,25)$, as shown in Figure 8. This line passes through part of the region of feasible solutions. Because of this, we conclude that it is possible for the firm to achieve a profit of 4000 dollars per week. It can do this by choosing any value of (x, y) that lies on the segment AB shown in Figure 8.

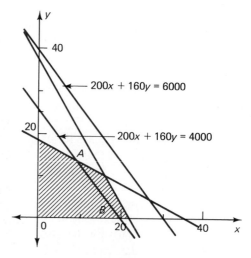

FIGURE 8

On the other hand, consider $P = 6000$. The corresponding values of x and y must satisfy $200x + 160y = 6000$, which again is the equation of a straight line, this time meeting the coordinate axes at the points $(30,0)$ and $(0,37.5)$ This straight line does not cross through the shaded region of feasible solutions (see Figure 8) and hence it is not possible for the firm to make a profit as large as $6000 per week. The maximum possible profit must lie somewhere between $4000 and $6000 per week.

The set of points (x, y) that lead to a given profit P satisfy the equation $200x + 160y = P$. This equation, for a fixed P, has as its graph a straight line in the xy-plane called a **constant profit line** or an **indifference line.** The two lines shown in Figure 8 are constant profit lines corresponding to the values $P = 4000$ and $P = 6000$. ☛ 6

The equation of a constant profit line can be written in the form

$$160y = P - 200x \qquad \text{or} \qquad y = -\frac{5}{4}x + \frac{P}{160}.$$

The line therefore has a slope of $-\frac{5}{4}$ and a y-intercept of $P/160$. It is an important feature that the slope of any constant profit line is the same regardless of the value of P. This means that all the constant profit lines are parallel to one another. As the value of P is increased, the corresponding line of constant profit moves farther from the origin (the y-intercept increases), always with the same slope.

In order to obtain the maximum profit, we must move the constant profit line away from the origin until it just touches the edge of the region of feasible solutions. It is clear from Figure 9 that the line of maximum profit is the one that passes through the corner C on the boundary of the feasible region. The values of x and y at C provide the production volumes of the two products X and Y that lead to the maximum profit.

☛ **6.** In Figures 7 or 8, draw the indifference lines that correspond to a profit of $3000 and $5000 per week. Are these profit levels feasible?

Answer $3000 is feasible, but the $5000 line does not intersect the feasible region.

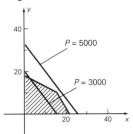

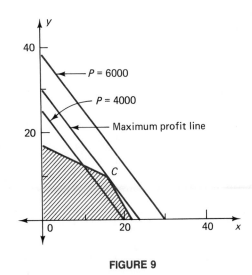

FIGURE 9

☞ 7. Use the geometrical method to find the maximum value of $Z = x + y$ when x and y are restricted by the inequalities $0 \le x \le 4$, $y \ge 0$, $x + 2y \le 6$.

Answer $Z_{max} = 5$, at the vertex $x = 4$, $y = 1$.

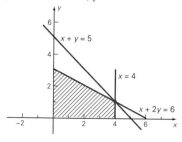

☞ 8. Find the maximum and miminum values of $Z = 2y - x$, when x and y are restricted by the conditions $x \ge 0$, $y \ge 0$, $3x + 2y \le 6$, $x - y \ge -1$.

Answer $Z_{max} = 2.8$ at $x = 0.8$, $y = 1.8$; $Z_{min} = -2$ at $x = 2$, $y = 0$.

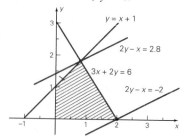

C is the point of intersection of the two straight lines which bound the feasible region. Its coordinates are obtained by solving the equations of these two lines, $5x + 3y = 105$ and $2x + 4y = 70$. Solving these equations, we find that $x = 15$ and $y = 10$. Thus the profit is maximum when the firm manufactures 15 items of type X and 10 items of type Y per week. The maximum weekly profit is given by

$$P_{max} = 200x + 160y = 200(15) + 160(10) = 4600.$$

The maximum profit is thus $4600.

The procedure used to solve this problem can also be used when a greater number of inequalities occur.   **☞ 7, 8**

EXAMPLE 2 A chemical firm makes two brands of fertilizer. Their regular brand contains nitrates, phosphates, and potash in the ratio $3:6:1$ (by weight) and their super brand contains these three ingredients in the ratio $4:3:3$. Each month the firm can rely on a supply of 9 tons of nitrates, 13.5 tons of phosphates, and 6 tons of potash. Their manufacturing plant can produce at most 25 tons of fertilizer per month. If the firm makes a profit of $300 on each ton of regular fertilizer and $480 on each ton of the super grade, what amounts of each grade should be produced in order to yield the maximum profit?

Solution The information given is summarized in Table 2.

TABLE 2

	Nitrates	Phosphates	Potash	Profit
Regular Grade	0.3	0.6	0.1	300
Super Grade	0.4	0.3	0.3	480
Available Supply	9	13.5	6	

Let the firm manufacture x tons of regular grade and y tons of super grade fertilizer per month. Then, since each ton of regular contains 0.3 tons of nitrates and each ton of super contains 0.4 tons of nitrates, the total amount of nitrates used is $0.3x + 0.4y$. This cannot exceed the available supply of 9 tons, so we have the condition $0.3x + 0.4y \le 9$.

Proceeding similarly with the phosphates and potash, we obtain the two further conditions, $0.6x + 0.3y \le 13.5$ and $0.1x + 0.3y \le 6$.

In addition to these inequalities, there is also the condition that the total production of fertilizer, $x + y$, cannot exceed the plant capacity of 25 tons, so $x + y \le 25$. After removing the decimals, we obtain the following system of inequalities that must be satisfied by x and y.

$$3x + 4y \le 90 \qquad 6x + 3y \le 135$$

$$x + 3y \le 60 \qquad x + y \le 25$$

$$x \ge 0 \qquad y \ge 0$$

The feasible region satisfying all these inequalities is shown in Figure 10. It is the interior of polygon *ABCDEO*, which is shaded.

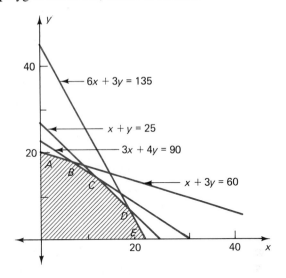

FIGURE 10

Each ton of fertilizer yields a profit of $300 for the regular grade and $480 for the super grade. When the production volumes are x and y tons per month, respectively, the total monthly profit P is

$$P = 300x + 480y.$$

By setting P at some fixed value, this equation again determines a straight line in the xy-plane, a constant profit line. A number of these lines are shown in Figure 11. The lines corresponding to different values of P are all parallel to one another and lie farther from the origin as the values of P increase. For example, we see that the line corresponding to $P = 7200$ passes across the feasible region, whereas the line for $P = 12,000$ does not. It is geometrically obvi-

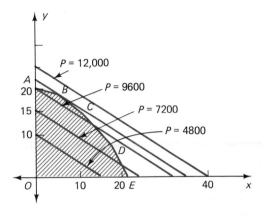

FIGURE 11

ous that the constant profit line with the largest value of P that still intersects the feasible region is the one which passes through corner point B. Point B is the point of intersection of the two straight lines

$$x + 3y = 60 \qquad \text{and} \qquad 3x + 4y = 90.$$

Its coordinates are $x = 6$ and $y = 18$.

We conclude therefore that the maximum profit is obtained by manufacturing 6 tons of regular and 18 tons of super grade of fertilizer per month. The maximum profit is given by

$$P_{max} = 300x + 480y = 300(6) + 480(18) = 10,440 \text{ dollars.}$$

It is worth noting that the production schedule that maximizes the profit uses all the available nitrates and potash, but does not use all of the available phosphates and does not make full use of the plant's capacity. ☛ 9

☛ **9.** In Example 2, how should the firm modify its production strategy if (a) more phosphate becomes available; (b) more potash becomes available? (*Hint:* In answering this type of question (called sensitivity analysis) it is helpful to see how the boundaries of the feasible region change.)

DEFINITION The inequalities that must be satisfied by the variables in a linear programming problem are called the **constraints.** The linear function that is to be maximized or minimized is called the **objective function.**

In the applications in business analysis, the objective function is very often either a profit function (which must be maximized) or a cost function (which must be minimized). It is usual to denote the objective function by the letter Z, and we shall do this from now on.

The following example illustrates a linear programming problem involving the minimization of a cost.

EXAMPLE 3 (*Production Decision*) A chemical company is designing a plant for producing two types of polymer, P_1 and P_2. The plant must be capable of producing at least 100 units of P_1 and 420 units of P_2 per day. There are two possible designs for the basic reaction chambers which are to be included in the plant: each chamber of type A costs \$600,000 and is capable of producing 10 units of P_1 and 20 units of P_2 per day; type B is a cheaper design costing \$300,000 and capable of producing 4 units of P_1 and 30 units of P_2 per day. Because of operating costs it is necessary to have at least 4 chambers of each type in the plant. How many chambers of each type should be included to minimize the cost of construction and still meet the required production schedule?

Solution The given information is summarized in Table 3.

Answer (a) The line *DE* moves to the right. This does not affect the vertex *B* so the optimum solution is unchanged.
(b) The line *AB* moves upward. The vertex *B* moves upward and to the left, so the optimum value of x will decrease and of y will increase. Profit will increase.

TABLE 3

	P_1	P_2	Cost
Chamber A	10	20	6
Chamber B	4	30	3
Required	100	420	

(Costs are given in hundreds of thousands of dollars.) Let the design include x chambers of type A and y chambers of type B. Then the following inequalities must be satisfied.

$$x \geq 4, \qquad y \geq 4;$$

$$10x + 4y \geq 100 \qquad \text{(production of } P_1\text{)}$$

$$20x + 30y \geq 420 \qquad \text{(production of } P_2\text{)}$$

The total cost of the chambers is given by

$$Z = 6x + 3y$$

and Z must be minimized subject to the above constraints. The feasible region (that is, the region satisfying the constraints) is shaded in Figure 12. Observe that this region is unbounded in this example.

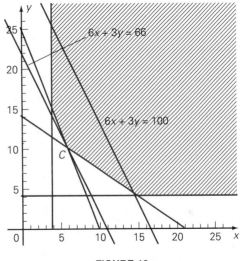

FIGURE 12

The lines of constant cost are obtained by setting Z equal to different constants. Two of these lines are shown in the figure. As Z is decreased, the corresponding line moves closer to the origin, always keeping the same slope; the line of minimum cost is the one that passes through the vertex C of the feasible region.

At C we have two simultaneous equations

$$10x + 4y = 100 \qquad \text{and} \qquad 20x + 30y = 420.$$

Their solution is $x = 6$ and $y = 10$. Therefore the optimal design for the plant is to include 6 reaction chambers of type A and 10 of type B. The minimum cost is

$$Z = 6x + 3y = 6(6) + 3(10) = 66$$

that is $6.6 million.

Consider a general linear programming problem with two variables, x and y. The feasible region, that is, the points (x, y) that satisfy the given set of linear inequalities, will consist of a polygon in the xy-plane. The equation obtained by setting the objective function equal to a constant will always represent a straight line in the xy-plane (for example, the constant profit line). It is intuitively obvious that the extreme (maximum or minimum) values of the objective function within the feasible region will be obtained when this straight line passes through a vertex of the polygon, since as we move the straight line parallel to itself in the direction of increasing (or of decreasing) the value of the objective function, the last point of contact with the feasible region must occur at one of the vertices.

This is illustrated in parts (a) and (b) of Figure 13, which show a series of lines of constant Z (where Z denotes the objective function). As Z increases, the line is moved across the feasible region. The largest and smallest values of Z occur when the line makes its first and last contacts with the feasible region.*

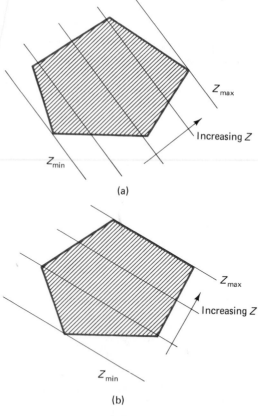

(a)

(b)

FIGURE 13

---

* When the feasible region is unbounded, Z may not have a finite maximum or minimum value.

In part (b), the lines of constant Z are parallel to one of the sides of the feasible region. In this case the largest value of Z occurs when the line of constant Z coincides with that side. Note, however, that it is still true that the maximum value of Z occurs when the line passes through a vertex of the polygon. In fact it passes through two vertices.

This suggests that instead of using the graphical technique of solving a linear programming problem, all we need do is to work out the value of the objective function at each of the vertices of the feasible region. The largest of these vertex values will give the maximum value of the objective function and the smallest of them will give its minimum value. This method of solving such problems can be used quite easily when there are only two variables, although it has no real computational advantage over the graphical method. For more than two variables, neither of these methods offers a practical tool for optimization. Fortunately an alternative method does exist, called the simplex method, and we shall devote the rest of this chapter to introducing it.

EXERCISES 11-2

(1–8) Find the maximum value of the objective function Z subject to the given constraints.

1. $Z = 3x + 2y$; $x \geq 0$, $y \geq 0$, $x + y \leq 5$

2. $Z = 3x + 4y$; $x \geq 0$, $y \geq 0$, $2x + y \leq 3$

3. $Z = 3x + 2y$; $x \geq 0$, $y \geq 0$, $2x + y \leq 4$, $x + 2y \leq 5$

4. $Z = 2(x + y)$; $x \geq 0$, $y \geq 0$, $6x + 5y \leq 17$, $4x + 9y \leq 17$

5. $Z = 5x + y$; $x \geq 0$, $y \geq 0$, $3x + y \leq 7$, $x + y \leq 3$, $x + 2y \leq 5$

6. $Z = x + 3y$; $x \geq 0$, $y \geq 0$, $2x + 3y \leq 6$, $2x + y \leq 5$, $x + 4y \leq 6$

7. $Z = 2x - y$; $x \geq 0$, $y \geq 0$, $x + y \leq 4$, $y - x \geq 3$, $3x + y \geq 6$

8. $Z = x + 3y$; $x \geq 0$, $y \geq 0$, $y \leq x + 1$ $x + y \geq 2$, $2y \geq x - 1$

(9–16) Find the minimum values of the objective function Z subject to the given constraints.

9. $Z = x + y$; $x \geq 0$, $y \geq 0$, $x + 3y \geq 6$, $2x + y \geq 7$

10. $Z = x + 2y$; $x \geq 0$, $y \geq 0$, $x + y \geq 5$, $x + 4y \geq 8$

11. $Z = x - 2y$; $x \geq 0$, $y \geq 0$, $y \leq x + 1$, $x + y \geq 2$

12. $Z = x - 3y$; $0 \leq x \leq 3$, $y \geq 0$, $x + 2y \leq 6$, $x + y \geq 5$

13. $Z = x + 4y$; $0 \leq x \leq 4$, $0 \leq y \leq 4$, $5 \leq x + y \leq 7$

14. $Z = x - y$; $x \geq 0$, $y \geq 0$, $x + y \geq 4$, $x + 2y \leq 10$

15. $Z = x + 2y$; $x \geq 0$, $y \geq 0$, $2x + y \geq 7$, $2y - x \geq -1$, $2x - y \geq -3$

16. $Z = x + y$; $-\frac{1}{2} \leq y - x \leq 2$, $y + 2x \leq 8$, $y + 4x \geq 7$

17. *(Whiskey Blending)* A distilling company has two grades of raw (unblended) whiskey, I and II, from which it makes two different blends. Regular blend contains 50% each of grades I and II, while super blend consists of two-thirds of grade I and one-third of grade II. The company has 3000 gallons of grade I and 2000 gallons of grade II available for blending. Each gallon of regular blend produces a profit of $5, whereas each gallon of super blend produces a profit of $6. How many gallons of each blend should the company produce in order to maximize their profits?

18. *(Mixtures)* A nut company sells two different mixtures of nuts. The cheaper mixture contains 80% peanuts and 20% walnuts, while the more expensive one contains 50% of each type of nut. Each week the company can obtain up to 1800 pounds of peanuts and up to 1200 pounds of walnuts from its sources of supply. How many pounds of each mixture should be produced in order to maximize profits if the profit is 10¢ from each pound of the cheaper mix and 15¢ from each pound of the more expensive mix?

19. *(Production Decision)* A company produces two products, A and B. Each unit of A requires 2 hours on one machine and 5 hours on a second machine. Each unit of B requires 4 hours on the first machine and 3 hours on the second. There are 100 hours available per week on the first machine and 110 hours on the second machine. If the company makes a profit of $70 on each unit of A and $50 on each unit of B, how many of each unit should be produced to maximize the total profit?

20. *(Production Decision)* In Exercise 19, suppose that a single order is received for 16 units of A per week. If it is decided that this order must be filled, determine the new value of the maximum profit.

21. *(Production Decision)* A manufacturer makes two products, A and B, each of which requires time on three machines. Each unit of A requires 2 hours on the first machine, 4 hours on the second machine, and 3 hours on the third machine. The corresponding numbers for each unit of B are 5, 1, and 2, respectively. The company makes profits of $250 and $300 on each unit of A and B respectively. If the numbers of machine hours available per month are 200, 240, and 190 for the first, second, and third machines, respectively, determine how many units of each product must be produced to maximize the total profit.

22. *(Product Decision)* In Exercise 21, suppose that there is a sudden shortage of product A on the market so that the company is able to increase its price of that product. If the profit on each unit of A is increased to $600, determine the new production schedule that maximizes the total profit.

*23. *(Production Decision)* In Exercise 21, suppose that the manufacturer is forced by competition to reduce the profit margin on product B. How low can the profit per unit of B become before the manufacturer is obliged to change his production schedule? (The production schedule should always be chosen to maximize the total profit.)

24. *(Investment Decision)* An investment manager has $1 million of a pension fund, part or all of which is to be

invested. The manager has two investments in mind, a conservative corporate bond that yields 6% per annum and a more risky mortgage that yields 10% per annum. According to government regulations, no more than 25% of the amount invested can be in mortgages. Furthermore, the minimum that can be put into a mortgage is $100,000. Determine the amounts of the two investments that will maximize the total yield.

25. *(Crop-planting Decision)* A farmer has 100 acres on which to plant two crops. The cost of planting the first crop is $20 per acre and the second crop is $40 per acre, and up to $3000 is available to cover the cost of planting. Each acre of the first crop will require 5 work-hours for harvesting and each acre of the second crop will require 20 work-hours. The farmer can rely on having only a total of 1350 work-hours for harvesting the two crops. If the profit is $100 per acre for the first crop and $300 per acre for the second crop, determine the acreage that should be planted with each crop in order to maximize the total profit.

26. *(Crop-planting Decision)* In Exercise 25, determine the acreage that should be planted with each crop if the profit from the second crop rises to $450 per acre.

27. *(Diet Planning)* A hospital dietician wishes to find the cheapest combination of two foods, A and B, that contains at least 0.5 milligram of thiamine and at least 600 calories. Each ounce of A contains 0.12 milligram of thiamine and 100 calories, while each ounce of B contains 0.08 milligram of thiamine and 150 calories. If each food costs 10¢ per ounce, how many ounces of each should be combined?

28. *(Ore Refining)* A mining company has two mines, P and Q. Each ton of ore from the mine P yields 50 pounds of copper, 4 pounds of zinc, and 1 pound of molybdenum. Each ton of ore from Q yields 25 pounds of copper, 8 pounds of zinc, and 3 pounds of molybdenum. The company must produce at least 87,500, 16,000, and 5000 pounds per week of these three metals, respectively. If it costs $50 per ton to obtain ore from P and $60 per ton from Q, how much ore should be obtained from each mine in order to meet the production requirements at minimum cost?

29. *(Distribution Costs)* An automobile manufacturer has two plants located at D and C with capacities of 5000 and 4000 cars per day. These two plants supply three distribution centers, W, E, and N, which require 3000, 4000, and 2000 cars per day, respectively. The shipping costs

per car from each plant to each distribution center are given in Table 4. Let x and y denote the numbers of cars per day shipped from plant D to W and E, respectively; determine the values of x and y that minimize the total shipping cost.

TABLE 4

	W	E	N
D	45	15	25
C	60	10	50

◣ 11-3 THE SIMPLEX TABLEAU

The geometric method and the method of inspection of vertices become impractical as methods of solution of linear programming problems when the number of variables is more than two, and particularly when the number of inequalities is large. For these more complex problems, an alternative, called the **simplex method,** does exist and provides a straightforward and economical way of finding the extrema. We shall describe the simplex method in Section 11-4; in this section, we shall outline certain constructions and operations that are basic to the method.

Suppose we are given the inequality $x + 3y \leq 2$ satisfied by two variables x and y. We can write the inequality in the form

$$2 - x - 3y \geq 0.$$

If we define a new variable t by the equation

$$t = 2 - x - 3y$$

then the inequality takes the form $t \geq 0$. In this way, the original inequality $x + 3y \leq 2$ is replaced by the following equation and inequality.

$$x + 3y + t = 2, \qquad t \geq 0$$

The variable t introduced in this way is called a **slack variable.** The reason for this name is that t is equal to the amount by which $x + 3y$ is less than 2, that is, t measures the *amount of slack* in the given inequality $x + 3y \leq 2$. The original variables in a linear programming problem, such as x, y, are called the **structural variables.**

The first step in using the simplex method is to introduce slack variables so that each inequality in the problem is changed to an equality and in such a way that all slack variables are nonnegative.

EXAMPLE 1 Suppose that a linear programming problem leads to the system of inequalities

$$x \geq 0, \qquad 0 \leq y \leq 1.5, \qquad 2x + 3y \leq 6, \qquad x + y \leq 2.5.$$

We introduce the slack variables

$$t = 1.5 - y, \qquad u = 6 - 2x - 3y, \qquad v = 2.5 - x - y.$$

Then the five variables (x, y, t, u, and v) satisfy the inequalities

$$x \geq 0, \qquad y \geq 0, \qquad t \geq 0, \qquad u \geq 0, \qquad v \geq 0$$

and the linear equations

$$y + t = 1.5 \qquad 2x + 3y + u = 6, \qquad x + y + v = 2.5.$$

☛ 10. Introduce slack variables for the following sets of inequalities

(a) $x \geq 0$, $y \geq 0$, $x + 3y \geq 3$,
$2x + y \leq 2$;
(b) $x \geq 0$, $y \geq 0$, $z \geq 0$,
$x + y + z \leq 30$,
$2x + 3y + 2z \leq 12$.

Observe that in this example, the original set of inequalities has been replaced by three linear equations together with the condition that all of the five variables which occur in these equations are nonnegative. We say that the linear programming problem has been reduced to **standard form.** In general a linear programming problem is said to be in *standard form if it consists of finding the maximum value of an objective function Z which is a linear function of a number of variables such as* x_1, x_2, ..., x_k, *where* x_1, x_2, ..., x_k *are all nonnegative and satisfy a certain number of linear equations.* ☛ 10

When a linear programming problem is changed to standard form, the solution remains unchanged. That is, the values of the variables that optimize the objective function for the new problem are the same as the values which optimize the objective function in the original problem. (Of course, the new problem has extra variables, too.)

EXAMPLE 2 Reduce the problem given in Example 2 of Section 11-2 to standard form.

Solution The given problem concerned a manufacturer of fertilizer who makes x tons of regular grade fertilizer and y tons of super grade. The profit function $Z = 300x + 480y$ is to be maximized subject to the following conditions.

$$x \geq 0, \qquad y \geq 0, \qquad 0.3x + 0.4y \leq 9$$
$$0.6x + 0.3y \leq 13.5, \qquad 0.1x + 0.3y \leq 6, \qquad x + y \leq 25$$

We define slack variables t, u, v, and w in such a way that the last four of these inequalities become equalities:

$$0.3x + 0.4y + t = 9 \qquad 0.6x + 0.3y + u = 13.5$$
$$0.1x + 0.3y + v = 6 \qquad x + y + w = 25. \tag{1}$$

Then the linear programming problem can be stated in standard form in the following way: Maximize the linear function

$$Z = 300x + 480y$$

where x, y, t, u, v, and w are nonnegative variables satisfying the Equations (1).

Answer (a) $x + 3y + t = 3$,
$2x + y + u = 2$,
$x \geq 0$, $y \geq 0$, $t \geq 0$, $u \geq 0$;
(b) $x + y + z + t = 30$,
$2x + 3y + 2z + u = 12$,
$x \geq 0$, $y \geq 0$, $z \geq 0$, $t \geq 0$,
$u \geq 0$.

Let us consider the significance of the slack variables in the context of this example. The manufacture of x tons of regular and y tons of super grade fertilizer uses $0.3x + 0.4y$ tons of nitrates. The condition $0.3x + 0.4y \leq 9$ states that this amount cannot exceed the available supply of 9 tons. The slack variable $t = 9 - (0.3x + 0.4y)$ equals the amount of nitrates that are left over, or unused. The condition $t \geq 0$ has the simple interpretation that the amount of nitrates left over can be zero or positive but cannot be negative.

Similarly, the slack variables u and v represent the amounts of phosphates and potash, respectively, left over when x tons of regular and y tons of super grade fertilizer are produced. The variable w represents the unused plant capacity, that is, the number of additional tons of fertilizer that could be produced if the plant were working at full capacity. As mentioned before, the slack variables measure the amount of slack in the corresponding inequalities.

In the examples so far, the inequalities have all involved the symbol $\leq$ (except for those that state that the variables x and y themselves are nonnegative). The introduction of slack variables also can be done when inequalities of the $\geq$ variety occur.

EXAMPLE 3 Introduce slack variables for the system of inequalities

$$x \geq 0, \qquad y \geq 0, \qquad 3 \leq x + y \leq 9, \qquad 2y - x \geq -6, \qquad y - x \leq 6.$$

Solution Define $t = x + y - 3$ and $u = 9 - x - y$. Then the condition $3 \leq x + y \leq 9$ implies that $t \geq 0$ and $u \geq 0$.

Similarly, if we define $v = 2y - x + 6$, then the condition $2y - x \geq -6$ implies that $v \geq 0$. Finally, setting $w = 6 + x - y$, we also have $w \geq 0$.

Then the six variables x, y, t, u, v, and w are nonnegative and satisfy the following four linear equations.

$$x + y - t = 3 \qquad x + y + u = 9$$
$$x - 2y + v = 6 \qquad -x + y + w = 6$$

Observe that the slack variables are always introduced into the inequalities in such a way that they are nonnegative. *This is done by defining each slack variable to be the high side of the associated inequality minus the low side.*

The number of slack variables that have to be introduced is equal to the number of inequalities in the original problem (not counting the conditions that the structural variables must be nonnegative). Consider, for example, the system of inequalities in Example 1 above. There are two structural variables x and y and three inequalities apart from the conditions $x \geq 0$ and $y \geq 0$. It is therefore necessary to introduce three slack variables (t, u, and v), bringing the total number of variables to $2 + 3 = 5$. These five variables are all nonnegative and satisfy the three linear equations

$$y + t = 1.5, \qquad 2x + 3y + u = 6, \qquad x + y + v = 2.5.$$

The feasible region for this problem is shown in Figure 14 in terms of the original variables x and y. We know that the optimum value of any linear objective function must be attained at one (or more than one) of the vertices of this region. But at each vertex, two of the five variables in the standard problem are always zero: at O, $x = y = 0$; at A, $x = 0$ and $y = 1.5$ so $t = 0$; at C, $x + y = 2.5$ and $2x + 3y = 6$ so that u and v are both zero; at D, $v = y = 0$; and at B, $t = u = 0$.

We conclude that the optimum value of any objective function for this ex-

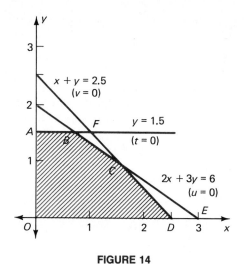

FIGURE 14

☞ **11.** Introduce slack variables for the set of inequalities
$x \geq 0$, $y \geq 0$, $3x + 2y \leq 6$,
$x - y \geq -1$.
Draw the feasible region and give the pair of variables that are zero at each vertex.

ample occurs when two of the five variables x, y, t, u, and v are equal to zero. ☞ **11**

This result generalizes. Suppose we have a linear programming problem with n structural variables and m slack variables. The optimum value of any linear objective function occurs when n out of the total set of $(n + m)$ variables are zero. A point at which n variables are zero is called a **basic solution,** and if this point is also feasible it is called a **basic feasible solution** (BFS). Each BFS corresponds to a vertex of the feasible region, and the solution of any linear programming problem occurs at one (or more) of the BFSs.

We cannot arbitrarily select the n variables to set equal to zero since some of these selections will not correspond to vertices of the feasible region. For example, in Figure 14, the point E corresponds to $y = 0$, $u = 0$, but this is not a BFS since E lies outside the feasible region. (It is easily seen that v is negative at E: E has coordinates $(3,0)$ and so $v = 2.5 - x - y = -0.5$.) Similarly, point F, which corresponds to $t = v = 0$, is not a BFS since $u < 0$ there.

The essence of the simplex method consists first of choosing a particular BFS as a starting point and then of transforming from this to another BFS in such a way that the objective function becomes closer to being optimal. This transformation process is called **pivoting** and is continued until the optimal basic solution is determined. The criterion used to choose the particular pivot that will be made will be the subject of the next section. In this section, we shall simply discuss the transformations themselves.

Let us take an elementary example. Suppose there are two variables, x and y, which satisfy the constraints $x \geq 0$, $y \geq 0$, $2x + 3y \leq 12$, and $4x + y \leq 14$. We introduce slack variables t and u such that $t \geq 0$ and $u \geq 0$ and the inequalities become

$$2x + 3y + t = 12$$

$$4x + y + u = 14.$$

Answer $3x + 2y + t = 6$,
$y - x + u = 1$, $x \geq 0$, $y \geq 0$,
$t \geq 0$, $u \geq 0$.
O: $x = y = 0$; A: $x = u = 0$;
B: $t = u = 0$; C: $t = y = 0$.
The points D: $u = y = 0$ and
E: $t = x = 0$ are not vertices of the feasible region.

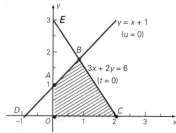

These equations can be summarized by means of the following augmented matrix.

$$\begin{array}{c} \\ t \\ u \end{array} \begin{array}{cccc} x & y & t & u \\ \left[\begin{array}{cccc|c} 2 & 3 & 1 & 0 & 12 \\ 4 & 1 & 0 & 1 & 14 \end{array}\right] \end{array}$$

This matrix is called the **simplex tableau.**

Observe that the variables are listed in the tableau at the head of the column of coefficients corresponding to that variable. Certain of the variables are listed on the left side of the tableau, in this case t and u. Suppose that all of the other variables except t and u are set equal to zero (that is, $x = 0$ and $y = 0$). Then the two equations reduce to

$$2(0) + 3(0) + t = 12 \quad \text{and} \quad 4(0) + 0 + u = 14$$

or $t = 12$ and $u = 14$. So the values of t and u are given by the elements in the augmented matrix that lie in the last column. This is why t and u are set next to the corresponding rows of the tableau.

Looking at the columns headed by t and u in the above tableau, we see that they form a 2×2 unit matrix. It is for this reason that the values of t and u can be read off directly from the last column when $x = y = 0$. The variables t and u are said to form the **basis** for this feasible solution. ☞ 12

In using the simplex method, we move from one BFS to another (that is, from one vertex to another) by replacing the variables in the basis one at a time by variables outside the basis. The variable that is removed from the basis is called the **departing variable** and the variable that replaces it is called the **entering variable.** For example, we might change from the basis (t, u) in the above tableau to the basis (y, u). Then the departing variable would be t and the entering variable would be y.

Figure 15 illustrates this example. The BFS with basis (t, u) corresponds to vertex O and the BFS with basis (y, u) corresponds to vertex A. A pivot from the one BFS to the other corresponds to moving from O to A.

☞ **12.** Construct the simplex tableau for the constraints
$x + y + z + t = 30,$
$2x + 3y + 2z + u = 12,$
$5y + 2z + v = 6, x \geq 0, y \geq 0,$
$z \geq 0, t \geq 0, u \geq 0, v \geq 0.$

Answer
$$\begin{array}{c} \\ t \\ u \\ v \end{array} \begin{array}{ccccccc} x & y & z & t & u & v \\ \left[\begin{array}{cccccc|c} 1 & 1 & 1 & 1 & 0 & 0 & 30 \\ 2 & 3 & 2 & 0 & 1 & 0 & 12 \\ 0 & 5 & 2 & 0 & 0 & 1 & 6 \end{array}\right] \end{array}.$$

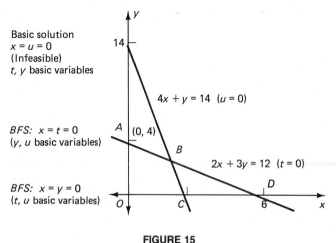

FIGURE 15

☛ 13. Construct the simplex tableau for the constraints $3x + 2y + t = 6$, $y - x + u = 1$, $x \geq 0$, $y \geq 0$, $t \geq 0$, $u \geq 0$. Use row operations to transform from the basis (t, u) to the basis (x, u) and then to the basis (x, y). Draw the feasible region and indicate the vertices involved.

When the basis is (y, u), we require that the values of y and u can be read off from the last column of the tableau if $x = t = 0$. This means that the tableau must be transformed to the form

$$
\begin{array}{c} y \\ u \end{array}
\begin{array}{cccc} x & y & t & u \end{array}
\left[\begin{array}{cccc|c} - & 1 & - & 0 & - \\ - & 0 & - & 1 & - \end{array}\right]
$$

where the dashes denote unknown entries. This transformation is accomplished by means of elementary row operations. For instance, the operation $R_2 - \frac{1}{3}R_1$ (subtracting one-third of the first row from the second row) changes the tableau to

$$
\left[\begin{array}{cccc|c} 2 & 3 & 1 & 0 & 12 \\ \frac{10}{3} & 0 & -\frac{1}{3} & 1 & 10 \end{array}\right].
$$

This places a zero in the y-column, as required. Dividing the first row by 3 reduces the tableau to the desired form.

$$
\begin{array}{c} y \\ u \end{array}
\begin{array}{cccc} x & y & t & u \end{array}
\left[\begin{array}{cccc|c} \frac{2}{3} & 1 & \frac{1}{3} & 0 & 4 \\ \frac{10}{3} & 0 & -\frac{1}{3} & 1 & 10 \end{array}\right]
$$

From this tableau, we conclude that for the BFS in which $x = t = 0$, the values of y and u are 4 and 10, respectively. This second tableau corresponds to the equations

$$
\frac{2}{3}x + y + \frac{1}{3}t = 4
$$
$$
\frac{10}{3}x - \frac{1}{3}t + u = 10
$$

and it is readily seen that setting $x = t = 0$ gives $y = 4$ and $u = 10$. These values are positive, showing that this solution is a feasible one. In general, *a tableau can represent a feasible solution only if the entries in the final column are all nonnegative.*

Let us continue this example and transform again from the basis (y, u) to the basis (x, u). Then the departing variable is y and the entering variable is x. We must transform the x-column to the form currently held by the y-column, namely $\begin{bmatrix} 1 \\ 0 \end{bmatrix}$. The appropriate row operations are $R_2 - 5R_1$ followed by $\frac{3}{2}R_1$ and the result is the tableau

$$
\begin{array}{c} x \\ u \end{array}
\begin{array}{cccc} x & y & t & u \end{array}
\left[\begin{array}{cccc|c} 1 & \frac{3}{2} & \frac{1}{2} & 0 & 6 \\ 0 & -5 & -2 & 1 & -10 \end{array}\right]
$$

This tableau corresponds to the values $y = t = 0$, $x = 6$, $u = -10$ and refers to the point D in Figure 15. This point is a basic solution but is not a feasible one. We can tell this immediately from the negative entry in the final column of the tableau. **☛ 13**

Answer

$$
\begin{array}{c} t \\ u \end{array}
\begin{array}{cccc} x & y & t & u \end{array}
\left[\begin{array}{cccc|c} 3 & 2 & 1 & 0 & 6 \\ -1 & 1 & 0 & 1 & 1 \end{array}\right].
$$

This corresponds to vertex O on the figure.

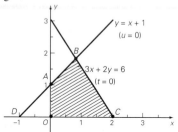

First t departs and x enters. After the operations $\frac{1}{3}R_1$, $R_2 + R_1$ we get the tableau corresponding to vertex C.

$$
\begin{array}{c} x \\ u \end{array}
\begin{array}{cccc} x & y & t & u \end{array}
\left[\begin{array}{cccc|c} 1 & \frac{2}{3} & \frac{1}{3} & 0 & 2 \\ 0 & \frac{5}{3} & \frac{1}{3} & 1 & 3 \end{array}\right].
$$

Next u departs and y enters. After the operations $\frac{3}{5}R_2$, $R_1 - \frac{2}{3}R_2$ we get the tableau corresponding to vertex B:

$$
\begin{array}{c} x \\ y \end{array}
\begin{array}{cccc} x & y & t & u \end{array}
\left[\begin{array}{cccc|c} 1 & 0 & \frac{1}{5} & -\frac{2}{5} & \frac{4}{5} \\ 0 & 1 & \frac{1}{5} & \frac{3}{5} & \frac{9}{5} \end{array}\right].
$$

EXAMPLE 4 A linear programming problem involves finding the maximum value of $Z = x + 4y + 2z$ when x, y, and z are nonnegative variables satisfying the constraints

$$3x + y + 2z \leq 6 \quad \text{and} \quad 2x + 3y + z \leq 6.$$

We introduce the nonnegative slack variables t and u so that

$$3x + y + 2z + t \phantom{{}+u} = 6$$
$$2x + 3y + z \phantom{{}+t} + u = 6.$$

The simplex tableau then has the following form.

$$
\begin{array}{c}
 \\
t \\
u
\end{array}
\begin{array}{cccccc}
x & y & z & t & u & \\
\left[\begin{array}{ccccc|c}
3 & 1 & 2 & 1 & 0 & 6 \\
2 & 3 & 1 & 0 & 1 & 6
\end{array}\right]
\end{array}
$$

Let us transform this tableau to one in which t and y form the basis. This means that u will be the departing variable and y the entering variable, so we must perform row operations in such a way as to change the second column to

$$\begin{bmatrix} 0 \\ 1 \end{bmatrix}.$$

The operation $R_1 - \frac{1}{3}R_2$ (subtracting one-third of the second row from the first row) gives

$$
\left[\begin{array}{ccccc|c}
\frac{7}{3} & 0 & \frac{5}{3} & 1 & -\frac{1}{3} & 4 \\
2 & 3 & 1 & 0 & 1 & 6
\end{array}\right]
$$

and then the operation $\frac{1}{3}R_2$ (dividing the second row by 3) gives the required form.

$$
\begin{array}{c}
 \\
t \\
y
\end{array}
\begin{array}{cccccc}
x & y & z & t & u & \\
\left[\begin{array}{ccccc|c}
\frac{7}{3} & 0 & \frac{5}{3} & 1 & -\frac{1}{3} & 4 \\
\frac{2}{3} & 1 & \frac{1}{3} & 0 & \frac{1}{3} & 2
\end{array}\right]
\end{array}
$$

From this, we conclude that the basic feasible solution for which $x = z = u = 0$ has the values $t = 4$ and $y = 2$ for these variables.

The preceding examples have involved tableaux with two rows. The number of rows in a tableau is equal to the number of slack variables, which in turn is equal to the number of inequalities in the original problem (not counting those of the type $x \geq 0$).

EXAMPLE 5 In Example 2 of this section, we considered the problem of maximizing the function $Z = 300x + 480y$, where the nonnegative variables x, y, t, u, v, and w satisfy the following equations.

$$0.3x + 0.4y + t \qquad\qquad\qquad = 9$$

$$0.6x + 0.3y \qquad + u \qquad\qquad = 13.5$$

$$0.1x + 0.3y \qquad\qquad + v \qquad = 6$$

$$x + \quad y \qquad\qquad\qquad + w = 25$$

Write the simplex tableau for this problem. Transform it first to the basis (t, u, y, w) and then to the basis (x, u, y, w).

Solution The tableau is as follows.

$$
\begin{array}{c}
 t \\
u \\
\to v \\
w
\end{array}
\begin{array}{c}
\begin{array}{cccccc}
x & y & t & u & v & w
\end{array} \\
\left[
\begin{array}{cccccc|c}
0.3 & 0.4 & 1 & 0 & 0 & 0 & 9 \\
0.6 & 0.3 & 0 & 1 & 0 & 0 & 13.5 \\
0.1 & 0.3 & 0 & 0 & 1 & 0 & 6 \\
1 & 1 & 0 & 0 & 0 & 1 & 25
\end{array}
\right] \\
\uparrow
\end{array}
$$

In the first transformation, v is the departing variable and y the entering variable, as indicated by the two arrows. This means that the second column of the tableau must be transformed to the form

$$
\begin{bmatrix}
0 \\
0 \\
1 \\
0
\end{bmatrix}
$$

by means of elementary row operations. This is accomplished by the sequence of operations $R_1 - \frac{4}{3}R_3$, $R_2 - R_3$, $R_4 - \frac{10}{3}R_3$, and $\frac{10}{3}R_3$. The result is the following tableau.

$$
\begin{array}{c}
\to t \\
u \\
y \\
w
\end{array}
\begin{array}{c}
\begin{array}{cccccc}
x & y & t & u & v & w
\end{array} \\
\left[
\begin{array}{cccccc|c}
\frac{1}{6} & 0 & 1 & 0 & -\frac{4}{3} & 0 & 1 \\
\frac{1}{2} & 0 & 0 & 1 & -1 & 0 & \frac{15}{2} \\
\frac{1}{3} & 1 & 0 & 0 & \frac{10}{3} & 0 & 20 \\
\frac{2}{3} & 0 & 0 & 0 & -\frac{10}{3} & 1 & 5
\end{array}
\right] \\
\uparrow
\end{array}
$$

From the last column, we see that in the BFS for which $x = v = 0$, the other variables are $t = 1$, $u = \frac{15}{2}$, $y = 20$, and $w = 5$.

Figure 10 shows the feasible region for this example. The first tableau corresponds to the vertex O $(x = y = 0)$, while the second tableau corresponds to A $(x = v = 0)$.

In the next step, we move to the basis (x, u, y, w), which corresponds to B $(t = v = 0)$. In this step, x is the entering variable and t the departing variable. The sequence of operations $R_2 - 3R_1$, $R_3 - 2R_1$, $R_4 - 4R_1$, and $6R_1$ result in the following tableau.

$$
\begin{array}{c}
 \\
x \\
u \\
y \\
w
\end{array}
\begin{array}{c}
\begin{array}{cccccc}
x & y & t & u & v & w
\end{array} \\
\left[
\begin{array}{cccccc|c}
1 & 0 & 6 & 0 & -8 & 0 & 6 \\
0 & 0 & -3 & 1 & 3 & 0 & \frac{9}{2} \\
0 & 1 & -2 & 0 & 6 & 0 & 18 \\
0 & 0 & -4 & 0 & 2 & 1 & 1
\end{array}
\right]
\end{array}
$$

Again we see that for the basic feasible solution $t = v = 0$, $x = 6$, $u = \frac{9}{2}$, $y = 18$, and $w = 1$. (This basic feasible solution is, in fact, the optimal one for this problem.)

EXERCISES 11-3

(1–12) Introduce slack variables and write the simplex tableau for Exercises 1–6, 17–21, and 25 of Section 11-2.

(13–14) Introduce slack variables and write the simplex tableau for each of the following problems.

13. Maximize $Z = x + 3y + 2z$ subject to $x \geq 0$, $y \geq 0$, $z \geq 0$, $2x + y + z \leq 5$, $x + 2y + z \leq 4$.

14. Maximize $Z = x + y + z$ subject to $x \geq 0$, $y \geq 0$, $z \geq 0$, $4x + 2y + z \leq 11$, $2x + 2y + 3z \leq 15$, $x + 2y + 2z \leq 11$.

(15–22) For the simplex tableaux given below, perform the appropriate row operations to make the indicated change of basis. In each case decide whether the new basis gives a feasible solution. In Exercises 15, 16, and 19–22, illustrate the change of basis with a diagram showing the corresponding change of vertex of the feasible region.

15.
$$
\begin{array}{c}
 \\
t \\
u
\end{array}
\begin{array}{c}
\begin{array}{cccc}
x & y & t & u
\end{array} \\
\left[
\begin{array}{cccc|c}
2 & 3 & 1 & 0 & 8 \\
7 & 6 & 0 & 1 & 19
\end{array}
\right]
\end{array}
\quad (t, u) \rightarrow (y, u) \rightarrow (y, x)
$$

16.
$$
\begin{array}{c}
 \\
s \\
t
\end{array}
\begin{array}{c}
\begin{array}{cccc}
x & y & s & t
\end{array} \\
\left[
\begin{array}{cccc|c}
2 & 1 & 1 & 0 & 10 \\
2 & 5 & 0 & 1 & 18
\end{array}
\right]
\end{array}
\quad (s, t) \rightarrow (s, x) \rightarrow (y, x)
$$

17.
$$
\begin{array}{c}
 \\
t \\
u
\end{array}
\begin{array}{c}
\begin{array}{ccccc}
x & y & z & t & u
\end{array} \\
\left[
\begin{array}{ccccc|c}
1 & 2 & 1 & 1 & 0 & 5 \\
3 & 2 & 4 & 0 & 1 & 16
\end{array}
\right]
\end{array}
\quad (t, u) \rightarrow (y, u) \rightarrow (y, x)
$$

18.
$$
\begin{array}{c}
 \\
t \\
u
\end{array}
\begin{array}{c}
\begin{array}{ccccc}
x & y & z & t & u
\end{array} \\
\left[
\begin{array}{ccccc|c}
3 & 1 & 1 & 1 & 0 & 4 \\
2 & 2 & 4 & 0 & 1 & 10
\end{array}
\right]
\end{array}
\quad (t, u) \rightarrow (y, u) \rightarrow (y, z)
$$

19.
$$
\begin{array}{c}
 \\
s \\
t \\
u
\end{array}
\begin{array}{c}
\begin{array}{ccccc}
x & y & s & t & u
\end{array} \\
\left[
\begin{array}{ccccc|c}
6 & 5 & 1 & 0 & 0 & 17 \\
4 & 9 & 0 & 1 & 0 & 17 \\
2 & 3 & 0 & 0 & 1 & 6
\end{array}
\right]
\end{array}
$$
$$(s, t, u) \rightarrow (s, x, u) \rightarrow (s, x, y) \rightarrow (u, x, y)$$

20.
$$
\begin{array}{c}
 \\
s \\
t \\
u
\end{array}
\begin{array}{c}
\begin{array}{ccccc}
x & y & s & t & u
\end{array} \\
\left[
\begin{array}{ccccc|c}
4 & 1 & 1 & 0 & 0 & 17 \\
1 & 1 & 0 & 1 & 0 & 5 \\
2 & 3 & 0 & 0 & 1 & 12
\end{array}
\right]
\end{array}
$$
$$(s, t, u) \rightarrow (y, t, u) \rightarrow (y, x, u)$$

21.
$$
\begin{array}{c}
 \\
p \\
q \\
r
\end{array}
\begin{array}{c}
\begin{array}{ccccc}
x & y & p & q & r
\end{array} \\
\left[
\begin{array}{ccccc|c}
3 & 2 & 1 & 0 & 0 & 5 \\
1 & 2 & 0 & 1 & 0 & 3 \\
1 & 5 & 0 & 0 & 1 & 6
\end{array}
\right]
\end{array}
$$
$$(p, q, r) \rightarrow (x, q, r) \rightarrow (x, y, r) \rightarrow (x, y, q)$$

22.
$$
\begin{array}{c}
 \\
p \\
q \\
r
\end{array}
\begin{array}{c}
\begin{array}{ccccc}
x & y & p & q & r
\end{array} \\
\left[
\begin{array}{ccccc|c}
4 & 1 & 1 & 0 & 0 & 6 \\
3 & 3 & 0 & 1 & 0 & 9 \\
2 & 5 & 0 & 0 & 1 & 15
\end{array}
\right]
\end{array}
$$
$$(p, q, r) \rightarrow (p, y, r) \rightarrow (p, y, x) \rightarrow (q, y, x)$$

◥ 11-4 THE SIMPLEX METHOD

The procedure used in the simplex method is to continue making changes in the basis variables of the type discussed in the last section until the set of variables that optimizes the objective function is obtained. Each change of variables is made in such a way as to improve the value of the objective function.

Let us consider the method with reference to a particular example. Suppose that we wish to maximize $Z = 2x + 3y$ subject to the constraints $x \geq 0$, $y \geq 0$, $x + 4y \leq 9$, and $2x + y \leq 4$. As usual, we introduce slack variables t and u such that

$$x + 4y + t = 9, \qquad 2x + y + u = 4 \tag{1}$$

where the four variables x, y, t, and u are nonnegative. The simplex tableau is

$$
\begin{array}{c c}
 & \begin{array}{cccc} x & y & t & u \end{array} \\
\begin{array}{c} t \\ u \\ \end{array} &
\left[\begin{array}{cccc|c}
1 & 4 & 1 & 0 & 9 \\
2 & 1 & 0 & 1 & 4 \\
2 & 3 & 0 & 0 & Z
\end{array} \right].
\end{array}
$$

Observe that we have now added an additional row to the tableau that contains the coefficients in the objective function

$$2x + 3y + 0 \cdot t + 0 \cdot u = Z.$$

We start with the BFS in which $x = y = 0$. For this solution, $t = 9$ and $u = 4$. The objective function has the value zero for this BFS. Our aim is to replace one of the variables t or u with either x or y in such a way that Z is increased. Looking at the last row of the tableau, we see that if x is increased by 1, Z increases by 2, whereas if y is increased by 1, Z increases by 3. That is, any increase in y has a bigger effect on Z than the same increase in x. It therefore appears reasonable to take y as the entering variable in forming the new basis.

The elements in the bottom row of the tableau are called the **indicators.** At each stage of the simplex procedure, *the entering variable is the one with the largest positive indicator*. (If the largest indicator occurs twice, we can choose arbitrarily between the two variables.)

We must next decide whether to take t or u as the departing variable. Let us consider these two possibilities in turn.

t **departing:** In this case the basis will consist of (y, u), since y enters and t departs. The BFS for this basis will be obtained by setting $x = t = 0$. From Equations (1), we have $0 + 4y + 0 = 9$ and $2(0) + y + u = 4$. Thus, $y = \frac{9}{4}$ and $u = 4 - y = 4 - \frac{9}{4} = \frac{7}{4}$. This solution is acceptable since y and u are both positive.

u **departing:** In this case, the basis will consist of (t, y) and the BFS corresponds to setting $x = u = 0$. From Equations (1), we have $0 + 4y + t = 9$ and $2(0) + y + 0 = 4$. Therefore $y = 4$ and $t = 9 - 4y = 9 - (4)4 = -7$.

The second solution is not acceptable because t is negative. It follows therefore that we must take t as the departing variable.

This method of deciding on the departing variable can be shortened quite appreciably. Suppose that the tableau has the general form

$$
\begin{array}{c} \\ t \\ u \end{array}
\begin{array}{cccc}
x & y & t & u \\
\end{array}
\left[
\begin{array}{cccc|c}
p_1 & q_1 & 1 & 0 & b_1 \\
p_2 & q_2 & 0 & 1 & b_2
\end{array}
\right]
$$

where p_i, q_i, and b_i denote the indicated entries in the tableau. The corresponding equations would be

$$
\begin{aligned}
p_1 x + q_1 y + t \quad &= b_1 \\
p_2 x + q_2 y \quad + u &= b_2.
\end{aligned}
\tag{2}
$$

Let us suppose that it has already been decided that y is the entering variable, and let us consider the two possibilities that t or u could be the departing variable.

t departing: In this case, the basis will consist of (y, u). The BFS will be obtained by setting $x = t = 0$, in which case Equations (2) give

$$
q_1 y = b_1 \qquad \text{and} \qquad q_2 y + u = b_2.
$$

Therefore

$$
y = \frac{b_1}{q_1} \qquad \text{and} \qquad u = b_2 - q_2 y = b_2 - q_2 \frac{b_1}{q_1}.
$$

Since y and u must both be nonnegtive if this is to be a feasible solution, we require that

$$
\frac{b_1}{q_1} \geq 0 \qquad \text{and} \qquad b_2 - q_2 \frac{b_1}{q_1} \geq 0.
$$

Since b_1 is nonegative (the elements in the last column must always be nonnegative), the first condition is met as long as $q_1 > 0$. Now q_1 is the element in the tableau that lies in the row of the departing variable t and the column of the entering variable y. It is called the **pivot element** for this change of basis. We conclude that in any change of basis variables *the pivot element must be positive*.

The second of the conditions will automatically be satisfied if $q_2 \leq 0$, since then the term $q_2(b_1/q_1)$ will be negative or zero. (Note that $b_2 \geq 0$). If $q_2 > 0$, this second condition can be written as $b_2 \geq q_2(b_1/q_1)$ or

$$
\frac{b_2}{q_2} \geq \frac{b_1}{q_1}.
$$

u departing: By a similar analysis, we conclude that a valid BFS will be obtained with (t, y) as basis provided that the pivot element $q_2 > 0$ and provided that either $q_1 \leq 0$ or, if $q_1 > 0$, then $(b_1/q_1) \geq (b_2/q_2)$.

Observe that the two ratios b_1/q_1 and b_2/q_2 are obtained by dividing the element in the last column of the tableau by the corresponding element in the column of the entering variable. (See Figure 16.) Thus we can summarize*:

If $q_1 > 0$ and $q_2 \leq 0$, t is the departing variable.

If $q_2 > 0$ and $q_1 \leq 0$, u is the departing variable.

If both $q_1 > 0$ and $q_2 > 0$,
  t is the departing variable if $b_1/q_1 \leq b_2/q_2$
  u is the departing variable if $b_2/q_2 \leq b_1/q_1$.

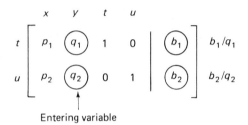

FIGURE 16

Thus *the departing variable is the one whose row in the tableau corresponds to the smallest nonnegative ratio b_i/q_i.*

Let us return to the earlier example. The first two rows of the tableau are shown in Figure 17. Since y is to be the entering variable, we divide each element in the last column by the corresponding element in the column headed by y. The ratios are given at the right of the tableau. Both ratios are positive, and the smaller is $9 \div 4 = 2.25$, which belongs to the t-row in the tableau. So we must take t as the departing variable and the pivot element is 4.

$$
\begin{array}{c}
\begin{array}{cccc} x & y & t & u \end{array} \\
\begin{array}{c} t \\ u \end{array}
\left[
\begin{array}{cccc|c}
1 & \boxed{4} & 1 & 0 & \boxed{9} \\
2 & \boxed{1} & 0 & 1 & \boxed{4}
\end{array}
\right]
\begin{array}{l} 9 \div 4 = 2.25 \\ 4 \div 1 = 4 \end{array}
\end{array}
$$

Entering variable

FIGURE 17

---

* If both $q_1 \leq 0$ and $q_2 \leq 0$, then the problem is unbounded—that is, Z does not have a finite maximum value.

The two row operations $R_2 - \frac{1}{4}R_1$ and $\frac{1}{4}R_1$ then reduce the tableau to the following form.

$$\begin{array}{c} \\ y \\ u \\ \\ \end{array} \begin{array}{c} \begin{array}{cccc} x & y & t & u \end{array} \\ \left[\begin{array}{cccc|c} \frac{1}{4} & 1 & \frac{1}{4} & 0 & \frac{9}{4} \\ \frac{7}{4} & 0 & -\frac{1}{4} & 1 & \frac{7}{4} \\ 2 & 3 & 0 & 0 & Z \end{array}\right] \end{array}$$

In this form, the values of the basis variables y and u can be read directly from the last column for the BFS in which $x = t = 0$. ☞ **14**

We see that Z is still expressed in terms of x and y. We would like to express it in terms of x and t so that when x and t are set equal to zero, the value of Z can be read immediately from the tableau. We can do this by the operation $R_3 - 3R_1$.

$$\begin{array}{c} \\ y \\ u \\ \\ \end{array} \begin{array}{c} \begin{array}{cccc} x & y & t & u \end{array} \\ \left[\begin{array}{cccc|c} \frac{1}{4} & 1 & \frac{1}{4} & 0 & \frac{9}{4} \\ \frac{7}{4} & 0 & -\frac{1}{4} & 1 & \frac{7}{4} \\ \frac{5}{4} & 0 & -\frac{3}{4} & 0 & Z - \frac{27}{4} \end{array}\right] \end{array}$$

The last row in this new tableau stands for the equation

$$Z - \tfrac{27}{4} = \tfrac{5}{4}x - \tfrac{3}{4}t. \tag{3}$$

When $x = t = 0$, this becomes $Z - \frac{27}{4} = 0$, or $Z = \frac{27}{4}$. Thus for the basic feasible solution at which $x = t = 0$, the objective function has the value $\frac{27}{4}$. This certainly represents an improvement over the previous value of zero.

In Equation (3), we observe that if t is made positive, Z would actually decrease. The corresponding indicator (namely $-\frac{3}{4}$) is negative. Therefore we do not want to allow t to enter the basis. The largest positive indicator (in fact, the only positive indicator) is $\frac{5}{4}$, which belongs to x, so x will be the entering variable for the next step in the simplex procedure.

To determine the departing variable, we again divide the last column by the corresponding elements of the column headed by the entering variable. The results are given in Figure 18. The smallest of these quotients is 1, which comes from the u-row, so u will be the departing variable.

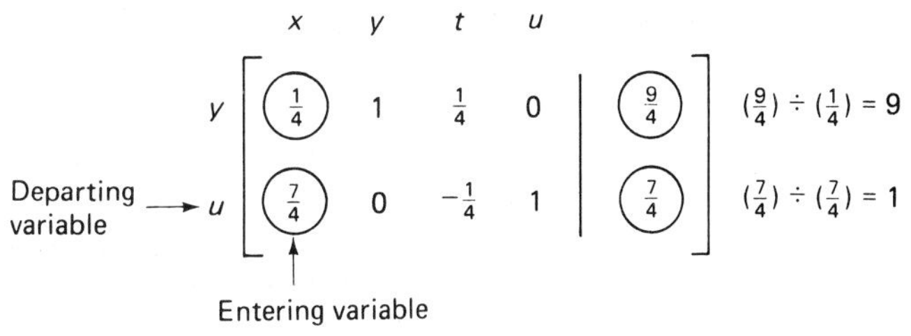

FIGURE 18

The sequence of row operations $R_1 - \frac{1}{7}R_2$, $R_3 - \frac{5}{7}R_2$ and $\frac{4}{7}R_2$ then reduces the tableau to

SEC. 11-4 | THE SIMPLEX METHOD   **473**

14. For the following linear programming problem, write down the initial tableau and perform the first pivot.
Maximize $Z = x + 2y$ subject to $x + y \leq 2$, $2x + y \leq 3$, $x, y \geq 0$.

Answer

$$\begin{array}{c} \\ t \\ u \\ \\ \end{array} \begin{array}{c} \begin{array}{cccc} x & y & t & u \end{array} \\ \left[\begin{array}{cccc|c} 1 & 1 & 1 & 0 & 2 \\ 2 & 1 & 0 & 1 & 3 \\ 1 & 2 & 0 & 0 & Z \end{array}\right] \end{array}$$

y enters (largest indicator) and t departs (ratio 2/1 smaller than 3/1). After row operations $R_2 - R_1$, $R_3 - 2R_1$ we get

$$\begin{array}{c} \\ y \\ u \\ \\ \end{array} \begin{array}{c} \begin{array}{cccc} x & y & t & u \end{array} \\ \left[\begin{array}{cccc|c} 1 & 1 & 1 & 0 & 2 \\ 1 & 0 & -1 & 1 & 1 \\ -1 & 0 & -2 & 0 & Z - 4 \end{array}\right] \end{array}.$$

$$\begin{array}{c} \quad\;\; x \quad\; y \quad\;\; t \quad\;\;\; u \\ \begin{array}{c} y \\ x \\ {} \end{array} \left[\begin{array}{cccc|c} 0 & 1 & \frac{2}{7} & -\frac{1}{7} & 2 \\ 1 & 0 & -\frac{1}{7} & \frac{4}{7} & 1 \\ 0 & 0 & -\frac{4}{7} & -\frac{5}{7} & Z-8 \end{array} \right] \end{array}$$

The BFS for this tableau corresponds to $t = u = 0$. Observe that the last row in the tableau corresponds to the equation

$$Z - 8 = -\tfrac{4}{7}t - \tfrac{5}{7}u,$$

so that when t and u are zero, the value of Z can immediately be determined: $Z - 8 = 0$, or $Z = 8$. The corresponding values of x and y can be read from the last column: $y = 2$ and $x = 1$.

The indicators are now all negative. This means that if either of the variables t or u were given a positive value, Z would decrease. Hence the maximum value of Z is obtained by setting $t = u = 0$, that is, by taking the BFS in which $x = 1$ and $y = 2$. In general, *the simplex procedure should be stopped when there remain no positive indicators.*

EXAMPLE 1 *(Production Decision)* A company makes two types of electronic calculators, a standard model, on which its profit is $5, and a deluxe model, on which its profit is $8. The company estimates that at most 1000 calculators per week can be handled by its distribution network. Because of the rapid growth of the calculator industry, there is a shortage of both the parts and the skilled labor necessary to assemble the calculators. The company can obtain a regular weekly supply of only 5000 electronic circuitry units (chips) necessary for the calculators; each regular calculator needs 3 of these chips and each deluxe calculator needs 6. Furthermore, the company has only 2500 work-hours of skilled labor available per week; each regular calculator requires 3 work-hours to assemble and each deluxe calculator needs 2. How many calculators of each type should be made each week in order to maximize the total profit?

Solution Let x regular calculators and y deluxe calculators be made each week. This requires $3x + 6y$ chips and $3x + 2y$ work-hours of labor. Thus x and y must satisfy the constraints $x \geq 0$, $y \geq 0$, $x + y \leq 1000$, $3x + 6y \leq 5000$, and $3x + 2y \leq 2500$. The weekly profit is then

$$Z = 5x + 8y.$$

Introducing the slack variables t, u, and v, the constraints can be written in the following form.

$$
\begin{aligned}
x + y + t &\phantom{{}+ u + v} = 1000 \\
3x + 6y &\phantom{{}+ t}+ u \phantom{{}+ v} = 5000 \\
3x + 2y &\phantom{{}+ t + u}+ v = 2500
\end{aligned}
$$

where x, y, t, u, and v are all greater than or equal to zero. We then have the simplex tableau below.

15. Solve the following linear programming problem:
Maximize $Z = x + 2y$ subject to
$x + y \le 4$, $x + 5y \le 8$, $x, y \ge 0$.

$$
\begin{array}{c}
\\
t \\
\rightarrow u \\
v \\
\\
\end{array}
\begin{array}{ccccc}
x & y & t & u & v \\
\end{array}
\left[
\begin{array}{ccccc|c}
1 & 1 & 1 & 0 & 0 & 1000 \\
3 & 6 & 0 & 1 & 0 & 5000 \\
3 & 2 & 0 & 0 & 1 & 2500 \\
5 & 8 & 0 & 0 & 0 & Z
\end{array}
\right]
\begin{array}{l}
1000 \div 1 = 1000 \\
5000 \div 6 = 833.3 \\
2500 \div 2 = 1250 \\
\\
\end{array}
$$

The largest of the indicators is 8, in the y-column, so that y becomes the entering variable. To decide on the departing variable, we take the ratios of the entries in the last column to those in the y-column: the smallest of these ratios, $5000 \div 6$, occurs in the u-row, so that u is the departing variable.

We must therefore transform the y-column to the form

$$
\begin{bmatrix}
0 \\
1 \\
0 \\
0
\end{bmatrix}
$$

leaving the t- and v-columns unchanged. The sequence of row operations $R_1 - \frac{1}{6}R_2$, $R_3 - \frac{1}{3}R_2$, $R_4 - \frac{4}{3}R_2$, and $\frac{1}{6}R_2$ achieve this.

$$
\begin{array}{c}
\rightarrow t \\
y \\
v \\
\\
\end{array}
\begin{array}{ccccc}
x & y & t & u & v \\
\end{array}
\left[
\begin{array}{ccccc|c}
\frac{1}{2} & 0 & 1 & -\frac{1}{6} & 0 & \frac{500}{3} \\
\frac{1}{2} & 1 & 0 & \frac{1}{6} & 0 & \frac{2500}{3} \\
2 & 0 & 0 & -\frac{1}{3} & 1 & \frac{2500}{3} \\
1 & 0 & 0 & -\frac{4}{3} & 0 & Z - \frac{20,000}{3}
\end{array}
\right]
\begin{array}{l}
\frac{500}{3} \div \frac{1}{2} \approx 333 \\
\frac{2500}{3} \div \frac{1}{2} \approx 1667 \\
\frac{2500}{3} \div 2 \approx 417 \\
\\
\end{array}
$$

The largest positive indicator is now 1, in the x-column, so x is the entering variable for the next step. Finding the ratios involving the last column and the x-column, we find the smallest ratio occurs in the t-row, so t is the departing variable. Consequently we perform the row operations $R_2 - R_1$, $R_3 - 4R_1$, $R_4 - 2R_1$, and $2R_1$.

$$
\begin{array}{c}
x \\
y \\
v \\
\\
\end{array}
\begin{array}{ccccc}
x & y & t & u & v \\
\end{array}
\left[
\begin{array}{ccccc|c}
1 & 0 & 2 & -\frac{1}{3} & 0 & \frac{1000}{3} \\
0 & 1 & -1 & \frac{1}{3} & 0 & \frac{2000}{3} \\
0 & 0 & -4 & \frac{1}{3} & 1 & \frac{500}{3} \\
0 & 0 & -2 & -1 & 0 & Z - 7000
\end{array}
\right]
$$

At this stage, all the indicators are negative or zero, so we cannot improve the value of Z by further change of basis. The optimum value of Z is 7000, and is achieved by taking $x = \frac{1000}{3}$ and $y = \frac{2000}{3}$. Thus the company should make 333 regular calculators and 667 deluxe calculators per week. ☛ **15**

Answer

$$
\begin{array}{c}
t \\
u \\
\\
\end{array}
\begin{array}{cccc}
x & y & t & u \\
\end{array}
\left[
\begin{array}{cccc|c}
1 & 1 & 1 & 0 & 4 \\
1 & 5 & 0 & 1 & 8 \\
1 & 2 & 0 & 0 & Z
\end{array}
\right]
\begin{array}{l}
4/1 \\
8/5 \leftarrow \\
\\
\end{array}
$$

y enters and u departs:

$$
\begin{array}{c}
t \\
y \\
\\
\end{array}
\begin{array}{cccc}
x & y & t & u \\
\end{array}
\left[
\begin{array}{cccc|c}
\frac{4}{5} & 0 & 1 & -\frac{1}{5} & \frac{12}{5} \\
\frac{1}{5} & 1 & 0 & \frac{1}{5} & \frac{8}{5} \\
\frac{3}{5} & 0 & 0 & -\frac{2}{5} & Z - \frac{16}{5}
\end{array}
\right]
\begin{array}{l}
\frac{12}{5}/\frac{4}{5} = 3 \leftarrow \\
\frac{8}{5}/\frac{1}{5} = 8 \\
\\
\end{array}
$$

Now x enters and t departs:

$$
\begin{array}{c}
x \\
y \\
\\
\end{array}
\begin{array}{cccc}
x & y & t & u \\
\end{array}
\left[
\begin{array}{cccc|c}
1 & 0 & \frac{5}{4} & -\frac{1}{4} & 3 \\
0 & 1 & -\frac{1}{4} & \frac{1}{4} & 1 \\
0 & 0 & -\frac{3}{4} & -\frac{1}{4} & Z - 5
\end{array}
\right].
$$

The optimal solution is then
$x = 3$, $y = 1$, $Z_{\text{mas}} = 5$.

The simplex method can be summarized by the following sequence of steps.

> **Step 1** Introduce nonnegative slack variables to turn the inequalities into equations.
> **Step 2** Construct the simplex tableau.
> **Step 3** Select the entering variable on the basis of the largest positive indicator.
> **Step 4** Compute the ratios of the entries in the last column of the tableau to the entries in the column of the entering variable. The smallest nonnegative quotient determines the departing variable.
> **Step 5** Perform row operations on the tableau to transform the column headed by the entering variable into the form that the column of the departing variable had previously. This should be done without changing the columns headed by the other basic variables.
> **Step 6** Repeat Steps 3, 4, and 5 until none of the indicators are positive. The maximum value of the objective function will then be given in the lower right entry of the tableau.

The simplex method can be used for problems involving more than two variables and any number of inequalities. When these numbers are large, it is necessary to use a computer to perform the calculations, but problems with three variables can generally be computed by hand without too much difficulty.

EXAMPLE 2 Use the simplex method to find the maximum value of the objective function $Z = 4x + y + 3z$, where x, y, and z are nonnegative variables satisfying the constraints $x + y + z \leq 4$, $3x + y + 2z \leq 7$, and $x + 2y + 4z \leq 9$.

Solution We introduce t, u, and v as nonnegative slack variables such that

$$
\begin{aligned}
x + y + z + t &= 4 \\
3x + y + 2z + u &= 7 \\
x + 2y + 4z + v &= 9.
\end{aligned}
$$

The simplex tableau is given below. The largest indicator is 4, belonging to the x-column, so x becomes the entering variable. The quotients of entries in the last column to those in the x-column are calculated on the right. The smallest quotient belongs to the u-row, so u becomes the departing variable.

$$
\begin{array}{c}
\\
\\
\\
Departing \\
Variable
\end{array}
\begin{array}{c}
\\
t \\
\rightarrow u \\
v \\
\\
\end{array}
\begin{array}{c}
\quad x \quad y \quad z \quad t \quad u \quad v \\
\left[\begin{array}{cccccc|c}
1 & 1 & 1 & 1 & 0 & 0 & 4 \\
3 & 1 & 2 & 0 & 1 & 0 & 7 \\
1 & 2 & 4 & 0 & 0 & 1 & 9 \\
4 & 1 & 3 & 0 & 0 & 0 & Z
\end{array}\right]
\end{array}
\begin{array}{l}
4 \div 1 = 4 \\
7 \div 3 = 2.33 \\
9 \div 1 = 9
\end{array}
$$

$\uparrow$
Entering
Variable

The row operations $R_1 - \frac{1}{3}R_2$, $R_3 - \frac{1}{3}R_2$, $R_4 - \frac{4}{3}R_2$, and $\frac{1}{3}R_2$ transform the x-column into the form currently held by the u-column:

$$
\begin{array}{c}
\\
t \\
x \\
\text{Departing} \\
\text{Variable} \to v \\
\\
\end{array}
\begin{array}{cccccc}
x & y & z & t & u & v \\
\end{array}
\left[
\begin{array}{cccccc}
0 & \frac{2}{3} & \frac{1}{3} & 1 & -\frac{1}{3} & 0 \\
1 & \frac{1}{3} & \frac{2}{3} & 0 & \frac{1}{3} & 0 \\
0 & \frac{5}{3} & \frac{10}{3} & 0 & -\frac{1}{3} & 1 \\
0 & -\frac{1}{3} & \frac{1}{3} & 0 & -\frac{4}{3} & 0 \\
\end{array}
\left|
\begin{array}{c}
\frac{5}{3} \\
\frac{7}{3} \\
\frac{20}{3} \\
Z - \frac{28}{3} \\
\end{array}
\right.
\right]
\quad
\begin{array}{l}
\frac{5}{3} \div \frac{1}{3} = 5 \\
\frac{7}{3} \div \frac{2}{3} = 3.5 \\
\frac{20}{3} \div \frac{10}{3} = 2 \\
\\
\end{array}
$$

$$\uparrow$$

Entering
Variable

The only positive indicator now belongs to z, so this variable enters the basis. According to the quotients calculated on the right, v is the departing variable. We perform the sequence of operations $R_1 - \frac{1}{10}R_3$, $R_2 - \frac{1}{5}R_3$, $R_4 - \frac{1}{10}R_3$ and $\frac{3}{10}R_3$ to transform the z-column into the current form of the v-column:

$$
\begin{array}{c}
\\
t \\
x \\
z \\
\\
\end{array}
\begin{array}{cccccc}
x & y & z & t & u & v \\
\end{array}
\left[
\begin{array}{cccccc}
0 & \frac{1}{2} & 0 & 1 & -\frac{3}{10} & -\frac{1}{10} \\
1 & 0 & 0 & 0 & \frac{2}{5} & -\frac{1}{5} \\
0 & \frac{1}{2} & 1 & 0 & -\frac{1}{10} & \frac{3}{10} \\
0 & -\frac{1}{2} & 0 & 0 & -\frac{13}{10} & -\frac{1}{10} \\
\end{array}
\left|
\begin{array}{c}
1 \\
1 \\
2 \\
Z - 10 \\
\end{array}
\right.
\right]
$$

The indicators are now all negative, showing that the maximum value of Z is attained for the corresponding BFS. This is given by $y = u = v = 0$ and the values of t, x, and z can be read from the last column. They are $t = 1$, $x = 1$, and $z = 2$. Thus the maximum value of Z is 10 and is achieved when $x = 1$, $y = 0$, and $z = 2$.

We have described the simplex method for a maximization problem. The easiest way of using it to solve a *minimization* problem is to convert the given problem into one involving maximization. For example, suppose we want to find the values of x and y subject to certain constraints that minimize a cost C given by $C = 2x + 6y + 3$. Then we define $Z = -2x - 6y$, so that $C = 3 - Z$. Then when C has its minimum value, Z must be maximum. We can thus replace the objective in the given problem by the new objective: Maximize $Z = -2x - 6y$. The constraints remain unchanged, and we can proceed by the simplex method as described above because we now have a maximization problem.

In our examples of the simplex method, we have started with a BFS in which the slack variables form the basis and the structural variables are all zero. Sometimes, however, such a solution is not a feasible one, and the procedure must be modified. We shall not go into detail regarding the resolution of this difficulty, but the following example will indicate the basic ideas involved.

EXAMPLE 3 Minimize $C = 10 + x - 2y$ subject to the constraints $x \geq 0$, $y \geq 0$, $x + y \leq 5$ and $2x + y \geq 6$.

Solution First define $Z = -x + 2y$. Then $C = 10 - Z$, and we shall maximize Z, which will be equivalent to minimizing C.

Introducing slack variables in the usual way, the linear programming problem becomes, in standard form,

Maximize $\quad Z = -x + 2y$

Subject to $\quad x + y + t = 5, \qquad 2x + y - u = 6, \qquad x, y, t, u \geq 0$.

Now let us try to find a BFS by setting $x = y = 0$, in order to start the simplex method. We get $t = 5$ and $u = -6$, and this is not feasible because $u < 0$.

It is possible to get around this difficulty by the simple trick of introducing another variable v, called an **artificial variable,** into the second constraint so that the constraints become

$$x + y + t = 5 \qquad 2x + y - u + v = 6 \qquad x, y, t, u, v \geq 0.$$

We can now obtain a BFS by putting $x = y = u = 0$, and the basic variables are $t = 5$ and $v = 6$, both positive.

Introducing v has of course changed the problem. But when $v = 0$, the new set of constraints is the same as the old set. Therefore, if we make sure that v is zero in the final solution of the new problem, this solution must also solve the given problem.

We can make sure that v is driven to zero by changing the objective function to $Z = -x + 2y - Mv$, where M is a very large number, for example, a million. M is called the **penalty** associated with the artificial variable, and its effect is to ensure that any nonzero value of v produces a large negative value of the objective function, which must therefore be less than the maximum value. At the maximum of this new Z, v must be zero.

The tableau for our new problem is then

$$
\begin{array}{c}
 \\
t \\
v \\

\end{array}
\begin{array}{c}
\begin{array}{ccccc}
x & y & t & u & v
\end{array} \\
\left[
\begin{array}{ccccc|c}
1 & 1 & 1 & 0 & 0 & 5 \\
2 & 1 & 0 & -1 & 1 & 6 \\
-1 & 2 & 0 & 0 & -M & Z
\end{array}
\right].
\end{array}
$$

This tableau is not quite in the usual form, however, because the indicator is nonzero in the v-column, and v is a basic variable. The operation $R_3 + MR_2$ fixes that little problem, and leaves

$$
\begin{array}{c}
 \\
t \\
\rightarrow v \\

\end{array}
\begin{array}{c}
\begin{array}{ccccc}
x & y & t & u & v
\end{array} \\
\left[
\begin{array}{ccccc|c}
1 & 1 & 1 & 0 & 0 & 5 \\
2 & 1 & 0 & -1 & 1 & 6 \\
2M-1 & M+2 & 0 & -M & 0 & Z+6M
\end{array}
\right]
\begin{array}{l}
\frac{5}{1} = 5 \\
\frac{6}{2} = 3
\end{array}
\end{array}
$$

$$\uparrow$$

We now proceed with the usual simplex method. The largest indicator is

Use the simplex method to maximize $Z = x$ subject to the constraints $y \geq x + 1$, $x + 2y \leq 8$, $x, y \geq 0$.

$2M - 1$, in the x-column, so x enters the basis, and the usual ratios on the right show that v departs. The row operations $R_1 - \frac{1}{2}R_2$ and $R_3 - (M - \frac{1}{2})R_2$ followed by $\frac{1}{2}R_2$ then produce the tableau

$$
\begin{array}{c}
\rightarrow t \\
x \\
\\
\end{array}
\begin{bmatrix}
x & y & t & u & v & \\
0 & \frac{1}{2} & 1 & \frac{1}{2} & -\frac{1}{2} & 2 \\
1 & \frac{1}{2} & 0 & -\frac{1}{2} & \frac{1}{2} & 3 \\
0 & \frac{5}{2} & 0 & -\frac{1}{2} & -M + \frac{1}{2} & Z + 3
\end{bmatrix}
\begin{array}{l}
2/\frac{1}{2} = 4 \\
3/\frac{1}{2} = 6 \\
\\
\end{array}
$$

$\uparrow$

This time y enters and t departs. The row operations $R_2 - R_1$ and $R_3 - 5R_1$ followed by $2R_1$ produce the tableau

$$
\begin{array}{c}
y \\
x \\
\\
\end{array}
\begin{bmatrix}
x & y & t & u & v & \\
0 & 1 & 2 & 1 & -1 & 4 \\
1 & 0 & -1 & -1 & 1 & 1 \\
0 & 0 & -5 & -3 & -M + 3 & Z - 7
\end{bmatrix}.
$$

Answer After eliminating the artificial variable from the objective function, the initial tableau is

$$
\begin{array}{c}
t \\
v \\
\\
\end{array}
\begin{bmatrix}
x & y & t & u & v & \\
1 & 2 & 1 & 0 & 0 & 8 \\
-1 & 1 & 0 & -1 & 1 & 1 \\
1 - M & M & 0 & -M & 0 & Z + M
\end{bmatrix}.
$$

After two pivots, the final tableau is

$$
\begin{array}{c}
x \\
y \\
\\
\end{array}
\begin{bmatrix}
x & y & t & u & v & \\
1 & 0 & \frac{1}{3} & \frac{2}{3} & -\frac{2}{3} & 2 \\
0 & 1 & \frac{1}{3} & -\frac{1}{3} & \frac{1}{3} & 3 \\
0 & 0 & -\frac{1}{3} & -\frac{2}{3} & -M + \frac{2}{3} & Z - 2
\end{bmatrix}
$$

giving the optimal solution $x = 2$, $y = 3$, and $Z_{max} = 2$.

Now all the indicators are negative, so this solution is optimal: $y = 4$, $x = 1$, and the maximum value of Z is 7. (You can easily verify by the geometrical method that this solution is the correct one.) Finally, the minimum value of $C = 10 - Z$ is 3. ☛ **16**

EXERCISES 11-4

(1–16) Use the simplex method to solve the linear programming problems given in Exercises 1–6, 17–22, 25, and 26 of Section 11-2 and Exercises 13 and 14 of Section 11-3.

17. *(Mixtures)* A nut company sells three different assortments of nuts. The regular assortment contains 80% peanuts, 20% walnuts, and no pecans; the super assortment contains 50% peanuts, 30% walnuts, and 20% pecans; and the deluxe assortment contains 30% peanuts, 30% walnuts, and 40% pecans. The firm has available supplies of up to 4300 pounds of peanuts, 2500 pounds of walnuts, and 2200 pounds of pecans per week. If the profit per pound is 10¢ for each assortment, how many pounds of each should be produced in order to maximize the total profit?

(18–26) Use the simplex method to find the maximum value of the given objective function subject to the constraints stated.

18. $Z = x + y + z$; $x, y, z \geq 0$, $x \leq 6$,
$x + 2y + 3z \leq 12$, $2x + 4y + z \leq 16$

19. $Z = x + 2y - z$; $x, y, z \geq 0$,
$2x + y + z \leq 4$, $x + 4y + 2z \leq 5$

20. $Z = 2x - y + 3z$; $x, y, z \geq 0$,
$x + 3y + z \leq 5$, $2x + 2y + z \leq 7$

21. $Z = x + y + z$; $x, y, z \geq 0$,
$x + 2y + z \leq 5$, $2x + y + 2z \leq 7$,
$2x + 3y + 4z \leq 13$

22. $Z = 3x + y + 4z$; $x, y, z \geq 0$,
$x + 2y + 2z \leq 9$, $2x + y + 3z \leq 13$,
$3x + 2y + z \leq 13$

23. $Z = 4x + 5y$; $x, y \geq 0$, $x + 2y \leq 10$,
$-x + 2y \leq 4$, $3x - y \leq 9$

24. $Z = x$; $x, y \geq 0$, $x \leq 2y$, $x + 2y \leq 4$,
$3x + y \leq 9$

25. $Z = 3x - y + 2z$; $x, y, z \geq 0$, $4x - 3y + 2z \leq 4$,
$3x + 2y + z \leq 1$, $-x + y - 3z \leq 0$

26. $Z = x + z$; $x, y, z \geq 0$, $2x - y + z \leq 6$,
$4x + y + 3z \leq 20$, $-x + z \leq 2$

(27–30) By introducing artificial variables as necessary, use the simplex method to solve Exercises 9, 10, 27, and 29 in Section 11-2.

CHAPTER REVIEW

Key Terms, Symbols, and Concepts

11.1 Linear inequality, graph of a linear inequality.

11.2 Linear programming problem. Constraints, objective function.
Feasible solution, feasible region.
Indifference line.

11.3 Simplex method. Slack variable, structural variable.
Standard form of a linear programming problem.

Basic solution, basic feasible solution (or vertex).
Simplex tableau. Basis.
Pivoting, entering variable, departing variable.

11.4 Indicators. Pivot element.
Step-by-step procedure for the simplex method.
The conditions for selecting the entering and departing variables.
Termination condition.
Artificial variable, penalty method.

REVIEW EXERCISES FOR CHAPTER 11

1. State whether each of the following is true or false. Replace each false statement by a corresponding true statement.

a. The graph of a linear inequality is a dotted line if the inequality is weak and a solid line if the inequality is a strict one.

b. If $y - 2x \geq 1$, then $2x - y \leq 1$.

c. If $y - 3x \leq 2$, then $3x - y > -2$.

d. If $y > a$ and $x > b$, then $y - x > a - b$.

e. If $y > a$ and $x < b$, then $y - x > a - b$.

f. If $y - x > a - b$, then $y > a$ and $x < b$.

g. If $x < a$ and $y < b$, then $x + y < a + b$.

h. $4x - 2y > 6$ is equivalent to $-2x + y > -3$.

(2–4) Draw the graphs of the following sets of inequalities.

2. $x \geq 0, \quad y \geq 0, \quad x + y \leq 4, \quad x + 2y \leq 6$

3. $1 \leq x \leq 5, \quad 2 \leq y \leq 5, \quad 2x + y \geq 5,$
$3x + 2y \leq 20$

4. $0 \leq y - x \leq 6, \quad x + 2y \geq 4, \quad x + y \leq 10,$
$x \geq 0$

(5–12) Solve each of the following linear programming problems:

a. By the geometric approach.

b. Using the simplex method.

5. Maximize $Z = 5x + 7y$ subject to the conditions $x \geq 0, \quad y \geq 0, \quad 3x + 2y \leq 7$, and $2x + 5y \leq 12$.

6. Maximize $Z = 2y - x$ subject to the conditions $x \geq 0$, $y \geq 0$, $x + y \leq 5$, and $x + 2y \leq 6$.

7. Find the maximum and minimum values of $Z = x - y$ subject to the conditions in Exercise 5.

8. Minimize $Z = 4y - 3x$ subject to the conditions $x \geq 0$, $y \geq 0$, $3x + 4y \leq 4$, and $x + 6y \leq 8$.

9. Maximize $Z = 3x - y$ subject to the conditions $2 \leq x \leq 5, \quad y \geq 0,$ and $x + y \leq 6$. (*Hint:* Set $x - 2 = z$).

10. Maximize $Z = x + 2y$ subject to the conditions $x \geq 0$, $y \geq 0$, $2y - x \geq -2$, and $4y + x \leq 9$.

11. Minimize $Z = 2y + x$ subject to the conditions $x \geq 0$, $y \geq 0$, $-y + x \geq -1$, and $3y - x \geq -2$.

12. Maximize $Z = 3y + x$ subject to the conditions $x \geq 0$, $y \geq 0$, $5y - x \geq -5$, $y - x \leq 2$, and $y + 2x \leq 4$.

(13–15) Solve each of the following linear programming problems by the simplex method.

13. Maximize $Z = x + 3y + 4z$ subject to the conditions $x, y, z \geq 0$, $x + y + z \leq 4$, $2x + y + 2z \leq 6$, and $3x + 2y + z \leq 8$.

14. Maximize $Z = x - 2y + 2z$ subject to the conditions $x, z \geq 0, 2 \leq y \leq 5, x + 2y + z \leq 14$, and $2x + y + 3z \leq 14$. (See hint in Exercise 9.)

15. Find the maximum and minimum values of $Z = x +$

$2y - z$ subject to the conditions $x + y + z \le 8$, $x - y + 2z \le 6$, $2x + y - 3z \le 4$, and $x, y, z \ge 0$.

16. (*Distribution Costs*) In Exercises 19 and 20 of Section 11-1, find the values of x and y that minimize the total delivery cost for the aluminum.

17. (*Distribution Costs*) For Exercise 21 of Section 11-1, find the values of x and y that minimize the total delivery cost.

18. (*Machine Allocation*) In Exercises 24 and 25 of Section 11-1, find the values x and y that maximize the total weekly profit.

19. (*Fish Stocks*) In Exercise 32 of Section 11-1, find the numbers of the two species of fish that produce the maximum weight of fish.

20. (*Fish Stocks*) In Exercise 19, suppose that a third species of fish, U, is introduced into the pool. This species consumes 3 units of the food F_1 and 3 units of F_2 per day; the average weight of each fish of species U is 4 pounds. Find the numbers of the three species that produce the maximum weight of fish in the pool.

CHAPTER 12

The Derivative

Chapter Objectives

12-1 INCREMENTS AND RATES

(a) Definition of increments and computation of increment in a dependent variable;
(b) The average rate of change of one variable with respect to another;
(c) Geometrical interpretation of increments and average rate of change;
(d) Average speed over an interval of time.

12-2 LIMITS

(a) Instantaneous speed defined as a limit;
(b) Definition of a limit in general; examples and geometrical significance of limits;
(c) Theorems on limits and their uses to compute limits;
(d) Continuous functions and computation of limits by substitution;
(e) Examples involving cancellation of common factors and rationalization of numerators or denominators of fractions.

12-3 THE DERIVATIVE

(a) Instantaneous rate of change of a function of time;
(b) Definition and formula for the derivative in general;
(c) Calculation of derivatives directly from the definition;
(d) Geometrical interpretation; the equations of tangent lines.

12-4 DERIVATIVES OF POWER FUNCTIONS

(a) The power formula for differentiation;
(b) Theorems on the derivatives of a constant times a function and the sum of two or more functions;
(c) Calculation of derivatives of polynomials and similar functions.

12-5 MARGINAL ANALYSIS

(a) Marginal cost and its significance;
(b) Marginal revenue and marginal profit;
(c) Other examples of marginal rates: marginal productivity, marginal yield, marginal tax rate, marginal propensities to save and consume.

12-6 CONTINUITY AND DIFFERENTIABILITY (OPTIONAL)

(a) One-sided limits and two-sided limits;
(b) Continuous and discontinuous functions; jump discontinuity;
(c) Differentiable and nondifferentiable functions;
(d) Examples of discontinuous and nondifferentiable functions in practice.

CHAPTER REVIEW

12-1 INCREMENTS AND RATES

Differential calculus is the study of the changes that occur in a quantity when changes occur in other quantities on which the original quantity depends. The following are examples of such situations.

1. The change in the total cost of operation of a manufacturing plant that results from each additional unit produced.

2. The change in the demand for a certain product that results from an increase of one unit (for example, $1) in the price.

3. The change in the gross national product of a country with each additional year that passes.

DEFINITION Let a variable x have a first value x_1 and a second value x_2. Then the change in the value of x, which is $x_2 - x_1$, is called the **increment** in x and is denoted by Δx.

We use the Greek letter Δ (delta) to denote a change or increment in any variable.

Δx denotes the change in the variable x.

Δp denotes the change in the variable p.

Δq denotes the change in the variable q.

Let y be a variable dependent on x such that $y = f(x)$ is defined for *all* x between x_1 and x_2. When $x = x_1$, y has the value $y_1 = f(x_1)$. Similarly, when $x = x_2$, y has the value $y_2 = f(x_2)$. The increment in y is then

$$\Delta y = y_2 - y_1$$
$$= f(x_2) - f(x_1)$$

EXAMPLE 1 The volume of gasoline sales from a certain service station depends on the price per gallon. If p is the price per gallon in cents, it is found that the sales volume q (in gallons per day) is given by

$$q = 500(150 - p).$$

Find the increment in sales volume that corresponds to an increase in price from 120¢ to 130¢ per gallon.

Solution Here p is the independent variable and q is the function of p. The first value of p is $p_1 = 120$ and the second value is $p_2 = 130$. The increment in p is

$$\Delta p = p_2 - p_1 = 130 - 120 = 10.$$

The corresponding values of q are as follows.

$$q_1 = 500(150 - p_1) = 500(150 - 120) = 15,000$$

$$q_2 = 500(150 - p_2) = 500(150 - 130) = 10,000$$

Hence the increment in q is given by

$$\Delta q = q_2 - q_1 = 10,000 - 15,000 = -5000.$$

The increment in q measures the *increase* in q, and the fact that it is negative means that q actually decreases. The sales volume decreased by 5000 gallons per day if the price is increased from 120¢ to 130¢. ☛ **1**

☛ **1.** Given $y = 2 - 3x + x^2$, calculate Δx and Δy if
(a) $x_1 = 1$, $x_2 = 2$;
☛(b) $x_1 = -1$, $x_2 = 1$.

Let P be the point (x_1, y_1) and Q be the point (x_2, y_2), both of which lie on the graph of the function $y = f(x)$. (See Figure 1.) Then the increment Δx is equal to the horizontal distance from P to Q, whereas Δy is equal to the vertical distance from P to Q. In other words, Δx is the *run* and Δy is the *rise* from P to Q.

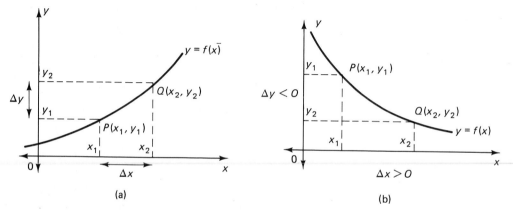

(a)   (b)

FIGURE 1

In the case illustrated in part (a) of Figure 1, both Δx and Δy are positive. It is possible for either or both Δx and Δy to be negative, and Δy can also be zero. A typical example of a case when $\Delta x > 0$ and $\Delta y < 0$ is illustrated in part (b) of Figure 1.

In some of the applications we shall make later on, we shall want to think of the increment Δx as being small—that is, we shall want to consider only small changes in the independent variable. It is, in fact, often understood that Δx means a small increment in x rather than just any increment. In this section, however, no restriction will be placed on the size of increments considered; they may be as small or as large as we like.

Solving the equation $\Delta x = x_2 - x_1$ for x_2, we have $x_2 = x_1 + \Delta x$. Using this value of x_2 in the definition of Δy, we get

$$\Delta y = f(x_1 + \Delta x) - f(x_1).$$

Since x_1 can be any arbitrary value of x, we can drop the subscript and write

Answer (a) $\Delta x = 1$, $\Delta y = 0$;
(b) $\Delta x = 2$, $\Delta y = -6$.

$$\Delta y = f(x + \Delta x) - f(x).$$

Alternatively, since $f(x) = y$, we can write

$$y + \Delta y = f(x + \Delta x).$$

EXAMPLE 2 Given $f(x) = x^2$, find Δy if $x = 1$ and $\Delta x = 0.2$.

Solution Substituting the values of x and Δx in the formula for Δy, we have the following.

$$\Delta y = f(x + \Delta x) - f(x) = f(1 + 0.2) - f(1) = f(1.2) - f(1)$$
$$= (1.2)^2 - (1)^2 = 1.44 - 1 = 0.44$$

Thus a change of 0.2 in the value of x results in a change in y of 0.44. This is illustrated graphically in Figure 2.

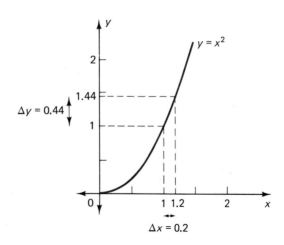

FIGURE 2

EXAMPLE 3 For the function $y = x^2$, find Δy when $x = 1$ for any increment Δx.

Solution

$$\Delta y = f(x + \Delta x) - f(x) = f(1 + \Delta x) - f(1) = (1 + \Delta x)^2 - (1)^2$$
$$= (1 + 2\Delta x + (\Delta x)^2) - 1 = 2\Delta x + (\Delta x)^2$$

Since the expression for Δy in Example 3 holds for all increments Δx, we may solve Example 2 by substituting $\Delta x = 0.2$ in this result. We get

$$\Delta y = 2(0.2) + (0.2)^2 = 0.4 + 0.04 = 0.44$$

as before.

► **2.** Calculate Δy for general values of x and Δx if

(a) $y = 3 - 2x$; (b) $y = 4x - x^2$.

EXAMPLE 4 For the function $y = x^2$, find Δy for general values of x and Δx.

Solution $\Delta y = f(x + \Delta x) - f(x) = (x + \Delta x)^2 - x^2 = 2x\Delta x + (\Delta x)^2$

It is again clear that we recover the result of Example 3 by substituting $x = 1$ in the expression of Example 4. ► **2**

When stated in absolute terms (as in the above examples), changes in the dependent variable are less informative than they would be if stated in relative terms. For example, absolute statements such as "The temperature dropped by 10°C" or "The revenue will increase by $3000" are less informative than relative statements such as, "The temperature dropped by 10°C in the last 5 hours" or "The revenue will increase by $3000 dollars if 60 extra units are sold." From these last statements, we not only know by how much the variable (temperature or revenue) changes, but also we can calculate the average *rate* at which it is changing with respect to a second variable. Thus the average drop in temperature during the last 5 hours is $\frac{10}{5} = 2$°C per hour; and the average increase in revenue if 60 more units are sold is $\frac{3000}{60} = 50$ dollars per unit.

DEFINITION The **average rate of change** of a function f over an interval x to $x + \Delta x$ is defined by the ratio $\Delta y / \Delta x$. Thus the average rate of change of y with respect to x is

$$\frac{\Delta y}{\Delta x} = \frac{f(x + \Delta x) - f(x)}{\Delta x}.$$

Note It is necessary that the whole interval from x to $x + \Delta x$ belong to the domain of f.

Graphically, if P is the point $(x, f(x))$ and Q the point $(x + \Delta x, f(x + \Delta x))$ on the graph of $y = f(x)$, then $\Delta y = f(x + \Delta x) - f(x)$ is the rise and Δx is the run from P to Q. From the definition of slope, we can say that $\Delta y / \Delta x$ is the slope of the straight line segment PQ. Thus the average rate of change of y with respect to x is equal to the slope of the chord PQ joining the two points P and Q on the graph of $y = f(x)$. (See Figure 3.) These points correspond to the values x and $x + \Delta x$ of the independent variable.

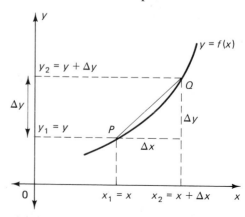

FIGURE 3

Answer (a) $\Delta y = -2\Delta x$;

(b) $\Delta y = 4\Delta x - 2x\Delta x - (\Delta x)^2$.

☞ **3.** Given $y = x^2 + 2x$, calculate the average rate of change of y with respect to x over the interval

(a) from $x_1 = 1$, $x_2 = 3$;

(b) from $x_1 = -1$, $x_2 = 2$.

EXAMPLE 5 *(Cost, Revenue, and Profits)* A chemical manufacturer finds that the cost per week of making x tons of a certain fertilizer is given by $C(x) = 20,000 + 40x$ dollars and the revenue obtained from selling x tons per week is given by $R(x) = 100x - 0.01x^2$. The company is currently manufacturing 3100 tons per wek, but is considering an increase in production to 3200 tons per week. Calculate the resulting increments in cost, revenue, and profit. Find the average rate of change of profit per extra ton produced.

Solution The first value of x is 3100 and $x + \Delta x = 3200$.

$$\Delta C = C(x + \Delta x) - C(x)$$
$$= C(3200) - C(3100)$$
$$= [20,000 + 40(3200)] - [20,000 + 40(3100)]$$
$$= 148,000 - 144,000 = 4000$$

$$\Delta R = R(x + \Delta x) - R(x)$$
$$= R(3200) - R(3100)$$
$$= [100(3200) - 0.01(3200)^2] - [100(3100) - 0.01(3100)^2]$$
$$= 217,600 - 213,900 = 3700$$

So the costs increase by \$4000 under the given increment in production, whereas the revenue increases by \$3700.

It is clear from these results that the profit must decrease by \$300. We can see this more fully if we consider that the profits made by the firm are equal to its revenues minus its costs, so the profit $P(x)$ from the sale of x tons of fertilizer is

$$P(x) = R(x) - C(x)$$
$$P(x) = 100x - 0.01x^2 - (20,000 + 40x)$$
$$= 60x - 0.01x^2 - 20,000.$$

Consequently, the increment in profit when x changes from 3100 to 3200 is

$$\Delta P = P(3200) - P(3100)$$
$$= [60(3200) - 0.01(3200)^2 - 20,000] - [60(3100) - 0.01(3100)^2 - 20,000]$$
$$= 69,600 - 69,900 = -300.$$

The profit therefore decreases by \$300. The average rate of change of profit per extra ton is

$$\frac{\Delta P}{\Delta x} = \frac{-300}{100} = -3$$

where $\Delta x = 3200 - 3100 = 100$. So the profit decreases by an average of \$3 per ton under the given increase in production. ☞ **3**

EXAMPLE 6 When any object is released from rest and allowed to fall freely under the force of gravity, the distance s (in feet) traveled in time t (in seconds)

Answer (a) 6; (b) 3.

is given by

$$s(t) = 16t^2.$$

Determine the average speed of the object during the following intervals of time.

(a) The time interval from 3 to 5 seconds.

(b) The fourth second (from $t = 3$ to $t = 4$ seconds).

(c) The interval between the times 3 and $3\frac{1}{2}$ seconds.

(d) The time interval from t to $t + \Delta t$.

Solution The average speed of any moving object is equal to the distance traveled divided by the interval of time involved. During the interval of time from t to $t + \Delta t$, the distance traveled is the increment Δs, and so the average speed is the ratio $\Delta s / \Delta t$.

(a) Here $t = 3$ and $t + \Delta t = 5$

$$\frac{\Delta s}{\Delta t} = \frac{s(t + \Delta t) - s(t)}{\Delta t} = \frac{s(5) - s(3)}{5 - 3}$$

$$= \frac{16(5^2) - 16(3^2)}{2} = \frac{400 - 144}{2} = \frac{256}{2} = 128$$

Thus during the interval of time $t = 3$ to $t = 5$, the body falls a distance of 256 feet and has an average speed of 128 feet/second.

(b) Here $t = 3$ and $t + \Delta t = 4$.

$$\frac{\Delta s}{\Delta t} = \frac{s(t + \Delta t) - s(t)}{\Delta t} = \frac{s(4) - s(3)}{4 - 3}$$

$$= \frac{16(4^2) - 16(3^2)}{1} = 256 - 144 = 112$$

The body has an average speed of 112 feet/second during the fourth second of fall.

(c) Here $t = 3$ and $\Delta t = 3\frac{1}{2} - 3 = \frac{1}{2}$.

$$\frac{\Delta s}{\Delta t} = \frac{s(t + \Delta t) - s(t)}{\Delta t} = \frac{16(3\frac{1}{2})^2 - 16(3)^2}{\frac{1}{2}}$$

$$= \frac{196 - 144}{\frac{1}{2}} = \frac{52}{\frac{1}{2}} = 104$$

Thus the body has an average speed of 104 feet/second during the time interval 3 to $3\frac{1}{2}$ seconds.

4. If the distance traveled in t seconds is $s = 96t - 16t^2$, calculate the average speed during
(a) the interval from $t = 0$ to $t = 3$;
(b) the interval from $t = 2$ to $t = 4$;
(c) the interval from $t = 3$ to $t = 5$;
(d) the interval from t to $t + \Delta t$.

(d) In the general case,

$$\frac{\Delta s}{\Delta t} = \frac{s(t + \Delta t) - s(t)}{\Delta t} = \frac{16(t + \Delta t)^2 - 16t^2}{\Delta t}$$

$$= \frac{16[t^2 + 2t \cdot \Delta t + (\Delta t)^2] - 16t^2}{\Delta t}$$

$$= \frac{32t \cdot \Delta t + 16(\Delta t)^2}{\Delta t} = 32t + 16\Delta t$$

which is the required average speed during the interval from t to $t + \Delta t$.

All the specific results in Example 6 can be obtained as special cases of part (d) by putting in the appropriate values for t and Δt. For example, the result of part (a) is obtained by setting $t = 3$ and $\Delta t = 2$:

Answer (a) 48; (b) 0; (c) −32; (d) $96 - 32t - 16\Delta t$.

$$\frac{\Delta s}{\Delta t} = 32t + 16\Delta t = 32(3) + 16(2) = 96 + 32 = 128. \quad \text{☛ 4}$$

EXERCISES 12-1

(1–8) Find the increments of the following functions for the given increments in the independent variables.

1. $f(x) = 2x + 7; x = 3, \Delta x = 0.2$

2. $f(x) = 2x^2 + 3x - 5; x = 2, \Delta x = 0.5$

3. $g(x) = \dfrac{x^2 - 4}{x - 2}; x = 1, \Delta x = 2$

4. $f(t) = \dfrac{900}{t}; t = 25; \Delta t = 5$

5. $p(t) = 2000 + \dfrac{500}{1 + t^2}; t = 2, \Delta t = 1$

6. $h(x) = ax^2 + bx + c; x$ to $x + \Delta x$

7. $F(x) = x + \dfrac{2}{x}; x$ to $x + \Delta x$

8. $G(t) = 300 + \dfrac{5}{t + 1}; t$ to $t + \Delta t$

(9–16) Determine the average rate of change of each function for the given interval.

9. $f(x) = 3 - 7x; x = 2, \Delta x = 0.5$

10. $f(x) = 3x^2 - 5x + 1; x = 3, \Delta x = 0.2$

11. $g(x) = \dfrac{x^2 - 9}{x - 3}; x = 2, \Delta x = 0.5$

12. $h(x) = \dfrac{3x^2 + 1}{x}; x = 5, \Delta x = 0.3$

13. $f(t) = \sqrt{4 + t}; t = 5, \Delta t = 1.24$

14. $F(x) = \dfrac{3}{x}; x$ to $x + \Delta x$

15. $G(t) = t^3 + t; t = a$ to $a + h$

16. $f(x) = \dfrac{3}{2x + 1}; x$ to $x + \Delta x$

17. *(Population Change)* The size of the population of a certain mining town at time t (measured in years) is given by

$$p(t) = 10{,}000 + 1000t - 120t^2.$$

Determine the average rate of growth between each pair of times.

a. $t = 3$ and $t = 5$ years

b. $t = 3$ and $t = 4$ years

c. $t = 3$ and $t = 3\frac{1}{2}$ years

d. $t = 3$ and $t = 3\frac{1}{4}$ years

e. t and $t + \Delta t$ years

18. *(Cost Function)* A manufacturer finds that the cost of producing x items is given by

$$C = 0.001x^3 - 0.3x^2 + 40x + 1000.$$

a. Find the increment in cost when the number of units is increased from 50 to 60.

b. Find the average cost per additional unit in increasing the production from 50 to 60.

19. *(Cost Function)* For the cost function in Exercise 18, find the average cost per additional unit in increasing the production from 90 to 100 units.

20. *(Demand Relation)* When the price of a certain item is equal to p, the number of items which can be sold per week (that is, the demand) is given by the formula

$$x = \frac{1000}{\sqrt{p} + 1}.$$

Determine the increment in demand when the price is increased from \$1 to \$2.25.

21. *(Revenue Function)* For the demand function in Exercise 20:

a. Determine the increment in gross revenue when the price per item is increased from \$4 to \$6.25.

b. Determine the average increase in total revenue per dollar of increase in price which occurs with this increment in p.

22. *(Growth of GNP)* During the period from 1970 to 1990, the gross national product of a certain nation was found to be given by the formula $I = 5 + 0.1x + 0.01x^2$ in billions of dollars. (Here the variable x is used to measure years, with $x = 0$ being 1970 and $x = 20$ being 1990.) Determine the average growth in GNP per year between 1975 and 1980.

23. *(TV Ownership)* After television was introduced in a certain developing country, the proportion of households owning a television set t years later was found to be given by the formula $p = 1 - e^{-0.1t}$.

a. Find the increment in p between $t = 3$ and $t = 6$.

b. Find the average rate of change of p per year.

24. *(Population Growth)* The population of a certain island as a function of time t is found to be given by the formula

$$y = \frac{20,000}{1 + 6(2)^{-0.1t}}.$$

Find

a. The increment in y between $t = 10$ and $t = 30$.

b. The average population growth per year during that period.

25. *(Projectile)* A body thrown upwards with a velocity of 100 feet/second reaches a height s after t seconds, where $s = 100t - 16t^2$. Find the average upward velocity in each case.

a. Between $t = 2$ and $t = 3$ seconds.

b. Between $t = 3$ and $t = 5$ seconds.

c. Between t and $t + \Delta t$.

26. *(Revenue Function)* The total weekly revenue R (in dollars) obtained by producing and marketing x units of a certain commodity is given by

$$R = f(x) = 500x - 2x^2.$$

Determine the average change of revenue per extra unit as the number of units produced and marketed per week is increased from 100 to 120.

27. *(Medicine)* When a certain drug is given to a person, its reaction is measured by noting the change in blood pressure, change in body temperature, pulse rate, and other physiological changes. The strength S of the reaction depends on the amount x of drug administered, and is given by

$$S(x) = x^2(5 - x).$$

Determine the average rate of change in the strength of reaction when the amount of drug used changes from $x = 1$ unit to $x = 3$ units.

28. *(Agriculture)* The number of pounds of good quality peaches, P, produced by an average tree in a certain orchard depends on the number of pounds of insecticide x with which the tree is sprayed, according to the formula

$$P = 300 - \frac{100}{1 + x}.$$

Calculate the average rate of increase of P when x changes from 0 to 3.

◼ 12-2 LIMITS

In Example 6 of Section 12-1, we discussed the average speeds of a falling body during a number of different time intervals. However, in many instances in both science and everyday life, the average speed of a moving object does not provide the information of most importance. For example, if a person traveling in an automobile hits a concrete wall, it is not the average speed but the speed at the *instant of collision* that determines whether the person will survive the accident.

What do we mean by the speed of a moving object at a certain instant of time (or *instantaneous speed,* as it is usually called)? Most people would accept that there is such a thing as instantaneous speed—it is precisely the quantity which is measured by the speedometer of an automobile—but the definition of instantaneous speed presents some difficulty. Speed is defined as the distance traveled in a certain interval of time divided by the length of time. But if we are concerned with the speed at a particular instant of time, we ought to consider an interval of time of zero duration. However, the distance traveled during such an interval would be zero, and we would obtain: Speed = Distance ÷ Time = 0 ÷ 0, a meaningless value.

In order to define the instantaneous speed of a moving object at a certain time t, we proceed as follows. During any interval of time from t to $t + \Delta t$, an increment of distance Δs is traveled. The average speed is $\Delta s/\Delta t$. Now let us imagine that the increment Δt becomes smaller and smaller, so that the corresponding interval of time is very short. Then it is reasonable to suppose that the average speed $\Delta s/\Delta t$ over such a very short interval will be very close to the instantaneous speed at time t. Furthermore, the shorter the interval Δt, the better the average speed will approximate the instantaneous speed. In fact, we can imagine that Δt is allowed to get arbitrarily close to zero, so that the average speed $\Delta s/\Delta t$ can be made as close as we like to the instantaneous speed.

In Example 6 of Section 12-1, we saw that the average speed during the time interval from t to $t + \Delta t$, for a body falling under gravity, is given by

$$\frac{\Delta s}{\Delta t} = 32t + 16\Delta t.$$

Setting $t = 3$, we obtain the average speed during a time interval of length Δt following 3 seconds of fall.

$$\frac{\Delta s}{\Delta t} = 96 + 16\Delta t$$

Some values of this velocity are given in Table 1 for different values of the increment Δt. For example, the average velocity between 3 and 3.1 seconds is obtained by setting $\Delta t = 0.1$: $\Delta s/\Delta t = 96 + 16(0.1) = 96 + 1.6 = 97.6$ feet/second.

It is clear from the values in Table 1 that as Δt gets smaller and smaller,

TABLE 1

Δt	0.5	0.25	0.1	0.01	0.001
$\Delta s/\Delta t$	104	100	97.6	96.16	96.016

the average velocity gets closer and closer to 96 feet/second. We can reasonably conclude therefore that 96 feet/second is the instantaneous speed at $t = 3$.

This example is typical of a whole class of problems in which we need to examine the behavior of a certain function as its argument gets closer and closer to a particular value.* In this case, we are concerned with the behavior of the average speed $\Delta s/\Delta t$ as Δt gets closer and closer to zero. In general, we can be interested in the behavior of a function $f(x)$ of a variable x as x approaches a particular value, say c. When we say that x approaches c, we mean that x takes a succession of values that get arbitrarily close to the value c, although x may never be exactly equal to c. (Note that the average speed $\Delta s/\Delta t$ is not defined for $\Delta t = 0$. We can only take a very, very small value of Δt, never a zero value.) We write $x \rightarrow c$ to mean x *approaches* c; for example, we would write $\Delta t \rightarrow 0$ in the above example.

Let us examine the behavior of the function $f(x) = 2x + 3$ as $x \rightarrow 1$. We shall allow x to assume the succession of values 0.8, 0.9, 0.99, 0.999, and 0.9999, which clearly are getting closer and closer to 1. The corresponding values of $f(x)$ are given in Table 2.

TABLE 2

x	0.8	0.9	0.99	0.999	0.9999
$f(x)$	4.6	4.8	4.98	4.998	4.9998

It is clear from the table that as x gets closer to 1, $f(x)$ gets closer to 5. We write $f(x) \rightarrow 5$ as $x \rightarrow 1$.

The values of x considered in Table 2 were all less than 1. In such a case, we say that x approaches 1 from below. We can also consider the alternative case in which x approaches 1 from above, that is, x takes a succession of values getting closer and closer to 1 but always greater than 1. For example, we might allow x to take the sequence of values 1.5, 1.1, 1.01, 1.001, and 1.0001. The corresponding values of $f(x)$ are given in Table 3.

TABLE 3

x	1.5	1.1	1.01	1.001	1.0001
$f(x)$	6	5.2	5.02	5.002	5.0002

---

* The term *argument* was defined on page 170.

Again it is clear that $f(x)$ gets closer and closer to 5 as x approaches 1 from above.

Thus as x approaches 1 either from below or from above, $f(x) = 2x + 3$ approaches 5. We say that the *limit* (or *limiting value*) of $f(x)$ as x approaches 1 is equal to 5. This is written

$$\lim_{x \to 1}(2x + 3) = 5.$$

We now give a formal definition of a limit.

DEFINITION Let $f(x)$ be a function that is defined for all values of x close to c, except possibly at the point c itself. Then L *is said to be the* **limit** *of* $f(x)$ *as* x *approaches* c, *if the difference between* $f(x)$ *and* L *can be made as small as we wish simply by restricting* x *to be sufficiently close to* c. In symbols, we write

$$\lim_{x \to c} f(x) = L \qquad \text{or} \qquad f(x) \to L \quad \text{as} \quad x \to c.$$

In our earlier example, $f(x) = 2x + 3$, $c = 1$, and $L = 5$. We can make the value of the function $2x + 3$ as close as we like to 5 by choosing x sufficiently close to 1. ☞ **5**

In this example, the limiting value of the function $f(x) = 2x + 3$ as $x \to 1$ can be obtained simply by substituting $x = 1$ into the formula $2x + 3$ that defines the function. The question arises as to whether limits can always be found by substituting the value of x into the given expression. The answer to this question is: Sometimes, but not always. The discussion of instantaneous speed on page 491 made this point already. In terms of limits, instantaneous speed was defined as

$$\text{Instantaneous Speed} = \lim_{\Delta t \to 0} \frac{\Delta s}{\Delta t}$$

and if we try directly substituting $\Delta t = 0$, we get $0/0$.

Example 1 illustrates another case when direct substitution does not work.

EXAMPLE 1 If $f(x) = (x^2 - 9)/(x - 3)$, evaluate $\lim_{x \to 3} f(x)$.

Solution If we substitute $x = 3$ into $f(x)$, we obtain $\frac{0}{0}$, and so we conclude that $f(x)$ is not defined for $x = 3$. However, $\lim_{x \to 3} f(x)$ does exist, since we can write

$$f(x) = \frac{x^2 - 9}{x - 3} = \frac{(x - 3)(x + 3)}{x - 3} = x + 3.$$

Dividing out the factor $x - 3$ is valid for all $x \neq 3$, but, of course, is not valid for $x = 3$. It is readily seen that as x approaches 3, the function $x + 3$ gets closer and closer to 6 in value. (You can easily convince yourself of this by

☞ **5.** By computing a few values, as in Tables 2 and 3, evaluate the limits

(a) $\lim_{x \to 3} (2x + 3)$;   (b) $\lim_{x \to -1} 2x^2$;

(c) $\lim_{x \to 1} \dfrac{x - 1}{x + 1}$.

Answer (a) 9;   (b) 2;   (c) 0.

computing a few values as in Tables 2 and 3.) Consequently,

$$\lim_{x \to 3} f(x) = \lim_{x \to 3}(x + 3) = 3 + 3 = 6. \quad ■ 6$$

When evaluating $\lim\limits_{x \to c} f(x)$, it is quite legitimate to divide numerator and denominator by a common factor of $x - c$, as we did in Example 1, in spite of the fact that when $x = c$, these factors are zero. This is because *the limit is concerned with the behavior of $f(x)$ close to $x = c$, but is not concerned at all with the value of f at $x = c$ itself.* As long as $x \neq c$, factors of $x - c$ can be divided out. In fact, Example 1 illustrates a case where $f(x)$ is not even defined for $x = c$ yet $\lim\limits_{x \to c} f(x)$ exists.

Let us examine the idea of limits from the point of view of the graph of the function involved. We shall first consider our initial example in which $f(x) = 2x + 3$. The graph of this function is a straight line of slope 2 and intercept 3. When $x = 1$, $y = 5$.

Consider any sequence of points $P_1, P_2, P_3, \ldots$, on the graph (see Figure 4) such that the x-coordinates of the points are getting closer to 1. Then clearly the points themselves must get closer to the point $(1, 5)$ on the graph, and their y-coordinates approach the limiting value 5. This corresponds to our earlier statement that $\lim\limits_{x \to 1}(2x + 3) = 5$.

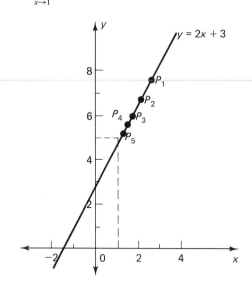

FIGURE 4

The example $f(x) = (x^2 - 9)/(x - 3)$ is a little different. We saw before that as long as $x \neq 3$, we can write $f(x) = x + 3$. So this function also has a straight line as its graph, with slope of 1 and intercept of 3. However $f(x)$ is not defined for $x = 3$, so that the point $(3, 6)$ is missing from the graph. This fact is indicated in Figure 5 by the use of a small circle at this point on the straight line. Again, if we consider a sequence of points $P_1, P_2, P_3, \ldots$, on the graph with x-coordinates aproaching 3, then the points themselves must approach the point (3, 6), even though this point is missing from the graph. Thus, in spite of the fact that $f(3)$ does not exist, the limit of $f(x)$ as $x \to 3$ does exist and is equal to 6.

In the first of these two examples, we have a function $f(x) = 2x + 3$ for which the limit as $x \to 1$ exists and is equal to the value of the function at $x = 1$. In the second example, we have a function $f(x) = (x^2 - 9)/(x - 3)$ for which the limit as $x \to 3$ exists, but this limit is not equal to $f(3)$—in fact, $f(3)$ does not even exist in this case. The first function is said to be *continuous* at $x = 1$; the second function is *discontinuous* at $x = 3$. Roughly speaking, a function is continuous at $x = c$ if its graph passes through this value of x without a jump or a break. For example, the graph in Figure 5 does not pass through $x = 3$ without a break because the point (3, 6) is missing from the graph. More precisely, we have the following definition.

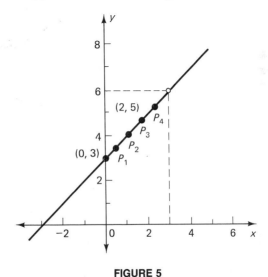

FIGURE 5

DEFINITION A function $f(x)$ is **continuous** at $x = c$ if $f(c)$ is defined and $\lim\limits_{x \to c} f(x)$ exists and is equal to $f(c)$.

We shall discuss continuous and discontinuous functions at greater length in Section 12-6.

The calculation of the limiting values of functions in more complicated cases rests on a number of theorems concerning limits. We shall now state these theorems and illustrate their significance with a number of examples, but shall not give proofs of them.

THEOREM 1 If m, b, and c are any three constants, then

$$\lim_{x \to c} (mx + b) = mc + b.$$

☛ **7.** Using Theorems 1 and 2, evaluate the following limits:

(a) $\lim\limits_{x \to 3} (2x + 3)^2$; (b) $\lim\limits_{x \to 4} 2\sqrt{x}$;

(c) $\lim\limits_{x \to 1} \dfrac{4}{x + 1}$.

We note that the function $y = mx + b$ has as its graph a straight line with slope m and y-intercept b. When $x = c$, y is always defined and $y = mc + b$. As x approaches c, the point (x, y) on the graph of this function gets closer and closer to the point $(c, mc + b)$. That is, the value of y gets closer and closer to $mc + b$, as stated in the theorem.

EXAMPLE 2

(a) Taking $m = 2$, $b = 3$, and $c = 1$, we get the result

$$\lim\limits_{x \to 1} (2x + 3) = 2(1) + 3 = 5$$

that we discussed earlier.

(b) Taking $m = 1$, $b = 3$, and $c = 3$, we get

$$\lim\limits_{x \to 3} (x + 3) = 3 + 3 = 6$$

again reproducing a result stated earlier.

THEOREM 2

> (a) $\lim\limits_{x \to c} bf(x) = b \lim\limits_{x \to c} f(x)$
>
> (b) $\lim\limits_{x \to c} [f(x)]^n = [\lim\limits_{x \to c} f(x)]^n$ if $[f(x)]^n$ is defined for x close to $x = c$

EXAMPLE 3

(a) $\lim\limits_{x \to 3} x^2 = [\lim\limits_{x \to 3} x]^2$ (by Theorem 2(b))

$\qquad\qquad = 3^2$ (by Theorem 1)

$\qquad\qquad = 9$

(b) $\lim\limits_{x \to 1} 5(2x + 3)^{-1} = 5 \lim\limits_{x \to 1} (2x + 3)^{-1}$ (by Theorem 2(a))

$\qquad\qquad = 5[\lim\limits_{x \to 1} (2x + 3)]^{-1}$ (by Theorem 2(b))

$\qquad\qquad = 5[2(1) + 3]^{-1}$ (by Theorem 1))

$\qquad\qquad = 5(5)^{-1} = 1$

(c) $\lim\limits_{x \to 3} \dfrac{(x^2 - 9)^3}{12(x - 3)^3} = \dfrac{1}{12} \lim\limits_{x \to 3} \left(\dfrac{x^2 - 9}{x - 3}\right)^3$ (by Theorem 2(a))

$\qquad\qquad = \dfrac{1}{12} \left[\lim\limits_{x \to 3} \left(\dfrac{x^2 - 9}{x - 3}\right)\right]^3$ (by Theorem 2(b))

$\qquad\qquad = \dfrac{1}{12} (6)^3$ (by the result of Example 1)

$\qquad\qquad = 18$ ☛ **7**

Answer (a) 81; (b) 4; (c) 2.

-----

8. Using Theorems 1, 2, and 3, evaluate the following limits:

(a) $\lim\limits_{x \to 3} [(x + 3)(1 - x)]$;

(b) $\lim\limits_{x \to -1} \sqrt{\dfrac{3 - x}{x^3 + 2}}$;

(c) $\lim\limits_{x \to 2} (x + 2x^{-1})$.

THEOREM 3

(a) $\lim\limits_{x \to c} [f(x) + g(x)] = \lim\limits_{x \to c} f(x) + \lim\limits_{x \to c} g(x)$

(b) $\lim\limits_{x \to c} [f(x) - g(x)] = \lim\limits_{x \to c} f(x) - \lim\limits_{x \to c} g(x)$

(c) $\lim\limits_{x \to c} [f(x)g(x)] = [\lim\limits_{x \to c} f(x)][\lim\limits_{x \to c} g(x)]$

(d) $\lim\limits_{x \to c} \left[\dfrac{f(x)}{g(x)}\right] = \dfrac{[\lim\limits_{x \to c} f(x)]}{[\lim\limits_{x \to c} g(x)]}$

provided that the limits on the right exist and in case (d) that the denominator on the right side is different from zero.

EXAMPLE 4

(a) $\lim\limits_{x \to 3} (x^2 + 2x) = \lim\limits_{x \to 3} x^2 + \lim\limits_{x \to 3} 2x$ (by Theorem 3(a))

$= 3^2 + 2(3)$ (by Example 3(a)

$= 9 + 6 = 15$ and Theorem 1)

(b) $\lim\limits_{x \to -1} \left(2x^3 - \dfrac{3}{x - 1}\right) = \lim\limits_{x \to -1} (2x^3) - \lim\limits_{x \to -1} \left(\dfrac{3}{x - 1}\right)$ (by Theorem 3(b))

$= 2 \lim\limits_{x \to -1} x^3 - 3 \lim\limits_{x \to -1} (x - 1)^{-1}$ (by Theorem 2(a))

$= 2[\lim\limits_{x \to -1} x]^3 - 3[\lim\limits_{x \to -1} (x - 1)]^{-1}$ (by Theorem 2(b))

$= 2(-1)^3 - 3(-1 - 1)^{-1}$ (by Theorem 1)

$= -2 + \tfrac{3}{2} = -\tfrac{1}{2}$

(c) $\lim\limits_{x \to 3} \dfrac{(x - 1)(x^2 - 9)}{x - 3} = \lim\limits_{x \to 3} (x - 1)\lim\limits_{x \to 3} \left(\dfrac{x^2 - 9}{x - 3}\right)$ (by Theorem 3(c))

$= \lim\limits_{x \to 3} (x - 1) \lim\limits_{x \to 3} (x + 3)$

$= (3 - 1)(3 + 3) = 12$

(d) $\lim\limits_{x \to -2} \left(\dfrac{x^2}{x - 1}\right) = \dfrac{\lim\limits_{x \to -2} x^2}{\lim\limits_{x \to -2} (x - 1)}$ (by Theorem 3(d))

$= \dfrac{\left(\lim\limits_{x \to -2} x\right)^2}{(-2 - 1)}$ (by Theorems 2(b) and 1)

$= \dfrac{(-2)^2}{-3}$ (by Theorem 1)

$= -\tfrac{4}{3}$ **8**

Answer (a) -12;   (b) 2;   (c) 3.

It probably will not have escaped your notice that in most of these examples the limiting value of the function involved could have been obtained simply by substituting the limit value of x into the given function. This method of simple substitution will in fact always produce the right answer when the function whose limit is being evaluated is continuous. This follows directly from the definition of a continuous function. All polynomials are continuous functions, and any rational function is continuous except at points where the denominator vanishes. So in the case of a rational function, we can always evaluate a limiting value by substitution, provided that the result after substitution is a well-defined number and is not of the form $\frac{0}{0}$ or constant/0. This same remark also applies to algebraic functions of x provided that they are defined on some interval which includes the limiting value of x. ☞ 9

☞ **9.** Evaluate the following limits by substituting the limiting value of x whenever this is allowed:

(a) $\lim_{x \to 3} [(x + 3)(1 - x)]$;

(b) $\lim_{x \to -2} \dfrac{x^2 - 4}{x + 2}$; (c) $\lim_{x \to 2} \dfrac{x^2 - 4}{x + 2}$

(d) $\lim_{x \to 2} \dfrac{x + 2}{x^2 - 4}$.

In the examples that follow, we shall evaluate limits by simple substitution. However, we recommend that you do a number of exercises using the above limit theorems in the manner illustrated by the preceding examples. The reason for this is that we shall encounter cases in later chapters in which use of the theorems plays an essential role, and the limits will not be capable of evaluation by substitution. (See Exercises 47 and 48 for examples.) Only after having mastered the use of the theorems should you adopt the method of substitution as a means of evaluating limits.

It may happen that upon substitution $x = c$ into $f(x)$, we obtain a result of the type constant/0. For example, suppose we tried to evaluate $\lim_{x \to 0} (1/x)$. Substituting $x = 0$, we obtain the result $1/0$, which is not defined. In such a case we would say that *the limit does not exist*. The function $1/x$ becomes indefinitely large as x approaches zero, and does not approach any limiting value. This can be seen from Table 4, which shows a series of values of $1/x$ as x takes succession of smaller and smaller values. It is clear that the corresonding values of $1/x$ get larger and larger and cannot approach any finite limiting value.

TABLE 4

x	1	0.5	0.1	0.02	0.002	0.0002
$\dfrac{1}{x}$	1	2	10	50	500	5000

A further, very important case that can arise is that we obtain the result $0/0$, which is undefined, upon substitutiong $x = c$ into $f(x)$. Limits of this type can often be evaluated by dividing out common factors of $(x - c)$ from numerator and denominator of fractions that occur in $f(x)$. This technique was illustrated earlier in this section, and other examples will now be given.

EXAMPLE 5 Evaluate $\lim_{x \to 1} f(x)$ for the following function.

$$f(x) = \begin{cases} \dfrac{x^2 + 3x + 2}{1 - x^2} & (x \neq -1) \\ 0 & (x = -1) \end{cases}$$

Answer (a) -12; (b) Substitution not allowed; (c) 0; (d) Limit does not exist.

Solution Letting $x = -1$ in the formula holding for $f(x)$ when $x \neq -1$, we have

$$\frac{(-1)^2 + 3(-1) + 2}{1 - (-1)^2} = \frac{1 - 3 + 2}{1 - 1} = \frac{0}{0}.$$

Consequently, we factor the numerator and denominator and cancel the factor $x + 1$ before substituting $x = -1$.

$$\lim_{x \to -1} \frac{x^2 + 3x + 2}{1 - x^2} = \lim_{x \to -1} \frac{(x + 1)(x + 2)}{(1 - x)(1 + x)} = \lim_{x \to -1} \frac{x + 2}{1 - x} = \frac{-1 + 2}{1 - (-1)} = \frac{1}{2}.$$

The fact that $f(-1) = 0$ is irrelevant to the limit. (Recall that the value of the limit is determined by the behavior of the function near the limit point c, but is not at all influenced by the value of f at c itself.)

EXAMPLE 6 Evaluate

$$\lim_{x \to 0} \frac{\sqrt{1 + x} - 1}{x}.$$

Solution When we substitute 0 for x, we get

$$\frac{\sqrt{1 + 0} - 1}{0} = \frac{0}{0}.$$

In this case, we cannot factor the numerator directly to get the factor x that is needed to cancel the x in the denominator. We overcome this difficulty by rationalizing the numerator, that is, by multiplying the numerator and denominator by $(\sqrt{1 + x} + 1)$.*

$$\lim_{x \to 0} \frac{\sqrt{1 + x} - 1}{x} = \lim_{x \to 0} \frac{\sqrt{1 + x} - 1}{x} \cdot \frac{\sqrt{1 + x} + 1}{\sqrt{1 + x} + 1}$$

$$= \lim_{x \to 0} \frac{(\sqrt{1 + x})^2 - 1^2}{x(\sqrt{1 + x} + 1)} = \lim_{x \to 0} \frac{(1 + x) - 1}{x(\sqrt{1 + x} + 1)}$$

$$= \lim_{x \to 0} \frac{x}{x(\sqrt{1 + x} + 1)}$$

$$= \lim_{x \to 0} \frac{1}{\sqrt{1 + x} + 1}$$

$$= \frac{1}{\sqrt{1 + 0} + 1} = \frac{1}{2} \quad \text{☛ 10}$$

Note that in these examples, the final limit has been evaluated by substitution. In reality, the theorems on limits underlie this substitution procedure.

☛ 10. Following Examples 5 and 6, evaluate the limits

(a) $\lim_{x \to -2} \dfrac{x^2 - 4}{x^2 + 5x + 6}$;

(b) $\lim_{x \to 1} \dfrac{1 - \sqrt{2 - x}}{x - 1}$;

(c) $\lim_{x \to 4} \dfrac{x^2 - 3x - 4}{2 - \sqrt{8 - x}}$.

Answer (a) -4; (b) $\frac{1}{2}$; (c) 20.

*See Section 1-5.

(1–30) Evaluate the following limits.

1. $\lim\limits_{x \to 2} (3x^2 + 7x - 1)$

2. $\lim\limits_{x \to -1} (2x^2 + 3x + 1)$

3. $\lim\limits_{x \to 3} \dfrac{x + 1}{x - 2}$

4. $\lim\limits_{x \to 3} \dfrac{x^2 + 1}{x + 3}$

5. $\lim\limits_{x \to 5} \dfrac{x^2 - 25}{\sqrt{x^2 + 11}}$

6. $\lim\limits_{x \to 4} \dfrac{x^2 - 16}{x - 4}$

7. $\lim\limits_{x \to -2} \dfrac{x^2 - 4}{x^2 + 3x + 2}$

8. $\lim\limits_{x \to 1} \dfrac{x^2 - 1}{x^2 + x - 2}$

9. $\lim\limits_{x \to 3} \dfrac{x^2 - 5x + 6}{x - 3}$

10. $\lim\limits_{x \to 2} \dfrac{x^2 - 5x + 6}{x^2 - x - 2}$

11. $\lim\limits_{x \to -1} \dfrac{x^2 + 3x + 2}{x^2 - 1}$

12. $\lim\limits_{x \to 3} \dfrac{x^2 - 9}{x^2 - 5x + 6}$

13. $\lim\limits_{x \to -2} \dfrac{x^2 + 4x + 4}{x^2 - 4}$

14. $\lim\limits_{x \to -1} \dfrac{x^2 + 4x + 3}{x^2 + 3x + 2}$

15. $\lim\limits_{x \to 1} \dfrac{x^2 + x - 2}{x^2 - 3x + 2}$

16. $\lim\limits_{x \to 4} \dfrac{x - 4}{\sqrt{x} - 2}$

17. $\lim\limits_{x \to 9} \dfrac{9 - x}{\sqrt{x} - 3}$

18. $\lim\limits_{x \to 9} \dfrac{\sqrt{x} - 3}{x^2 - 81}$

19. $\lim\limits_{x \to 1} \dfrac{x^3 - 1}{x^2 - 1}$

20. $\lim\limits_{x \to -2} \dfrac{x^3 + 8}{x^2 - 4}$

*21. $\lim\limits_{x \to 4} \dfrac{\sqrt{x} - 2}{x^3 - 64}$

*22. $\lim\limits_{x \to 9} \dfrac{x^3 - 729}{\sqrt{x} - 3}$

23. $\lim\limits_{x \to 2} \dfrac{x + 1}{x - 2}$

24. $\lim\limits_{x \to 0} \dfrac{2x^2 + 5x + 7}{x}$

25. $\lim\limits_{x \to 0} \dfrac{\sqrt{4 + x} - 2}{x}$

26. $\lim\limits_{x \to 2} \dfrac{\sqrt{x + 7} - 3}{x - 2}$

27. $\lim\limits_{x \to 1} \dfrac{\sqrt{x + 3} - 2}{x^2 - 1}$

28. $\lim\limits_{x \to 0} \dfrac{\sqrt{9 + x} - 3}{x^2 + 2x}$

29. $\lim\limits_{x \to 0} \dfrac{\sqrt{1 + x} - 1}{\sqrt{4 + x} - 2}$

30. $\lim\limits_{x \to 1} \dfrac{\sqrt{2 - x} - 1}{2 - \sqrt{x + 3}}$

(31–36) Evaluate $\lim\limits_{x \to c} f(x)$, where $f(x)$ and c are given below.

31. $f(x) = \begin{cases} 3x - 4 & \text{for } x \neq 2 \\ 5 & \text{for } x = 2 \end{cases}$; $c = 2$

32. $f(x) = \begin{cases} x^2 - 3x + 1 & \text{for } x \neq 1 \\ 7 & \text{for } x = 1 \end{cases}$; $c = 1$

33. $f(x) = \begin{cases} \dfrac{x^2 - 4}{x - 2} & \text{for } x \neq 2 \\ 3 & \text{for } x = 2 \end{cases}$; $c = 2$

34. $f(x) = \begin{cases} \dfrac{x^2 - 9}{x + 3} & \text{for } x \neq -3 \\ -5 & \text{for } x = -3 \end{cases}$; $c = -3$

35. $f(x) = \begin{cases} \dfrac{x^2 - 1}{x - 1} & \text{for } x \neq 1 \\ 3 & \text{for } x = 1 \end{cases}$; $c = 1$

36. $f(x) = \begin{cases} \dfrac{x - 9}{\sqrt{x} - 3} & \text{for } x \neq 9 \\ 7 & \text{for } x = 9 \end{cases}$; $c = 9$

(37–41) The functions $f(x)$ and the values of a are given below. Evaluate

$$\lim_{h \to 0} \frac{f(a + h) - f(a)}{h}$$

in each case.

37. $f(x) = 2x^2 + 3x + 1, \quad a = 1$

38. $f(x) = 3x^2 - 5x + 7, \quad a = 2$

39. $f(x) = x^2 - 1, \quad a = 0$

40. $f(x) = x^2 + x + 1, \quad a = x$

41. $f(x) = 2x^2 + 5x + 1, \quad a = x$

42. A body falls from rest under gravity. What is the instantaneous velocity after $1\frac{1}{2}$ seconds?

43. A ball is thrown vertically upwards with velocity 40 feet/second. The distance traveled in feet after t seconds is given by the formula $s = 40t - 16t^2$. Find the instantaneous velocity:

    a. After 1 second.   **b.** After 2 seconds.

44. In Exercise 43, find the instantaneous velocity after t seconds. What occurs when $t = \frac{5}{4}$? What is the instantaneous velocity when $t = \frac{5}{2}$?

45. In this exercise, use your hand calculator to evaluate the function

$$f(x) = \frac{x^4 - 1}{x^3 - 1}$$

for $x = 1.2$, 1.1, 1.05, 1.01, 1.005, and 1.001. Show that $\lim_{x \to 1} f(x) = \frac{4}{3}$. Are your calculated values getting closer to this limit?

46. Use a hand calculator to evaluate

$$f(x) = \frac{\sqrt{x + 3} - 2}{x - 1}$$

for $x = 0.9$, 0.99, 0.999 and 0.9999 and for $x = 1.1$, 1.01, 1.001, and 1.0001. Show that $\lim_{x \to 1} f(x) = \frac{1}{4}$. Are your calculated values getting closer and closer to this limit?

47. Use a hand calculator to evaluate the function

$$f(x) = \frac{e^x - 1}{x}.$$

for $x = 0.1, 0.01, 0.001, 0.0001$, and 0.00001. Are your calculated values getting closer to some number? What do you think is $\lim_{x \to 0} f(x)$?

48. Repeat Exercise 45 for the function

$$f(x) = \frac{\ln x}{x - 1}.$$

What do you think is $\lim_{x \to 1} f(x)$?

12-3 THE DERIVATIVE

In Section 12-2, we saw how the definition of the instantaneous velocity of a moving object leads naturally to a limiting process. The average velocity $\Delta s / \Delta t$ is first found for an interval of time from t to $t + \Delta t$, and then its limiting value is calculated as $\Delta t \to 0$. We might describe $\Delta s / \Delta t$ as the average rate of change of position, s, with respect to time, and its limit is the instantaneous rate of change of s with respect to t.

Now there are many examples of processes that develop in time and we can give corresponding definitions of the instantaneous rate of change of the associated variables.

EXAMPLE 1 *(Population Growth)* During the 10-year period from 1970 to 1980, the population of a certain city was found to be given by the formula

$$P(t) = 1 + 0.03t + 0.001t^2$$

where P is in millions and t is time measured in years from the beginning of 1970. Find the instantaneous rate of growth at the beginning of 1975.

Solution We want the rate of growth at $t = 5$. The increment in P between $t = 5$ and $t = 5 + \Delta t$ is

$$\Delta P = P(5 + \Delta t) - P(5)$$
$$= [1 + 0.03(5 + \Delta t) + 0.001(5 + \Delta t)^2]$$
$$\quad - [1 + 0.3(5) + 0.001(5)^2]$$
$$= 1 + 0.15 + 0.03 \, \Delta t + 0.001(25 + 10 \, \Delta t + (\Delta t)^2)$$
$$\quad - [1 + 0.15 + 0.001(25)]$$
$$= 0.04 \, \Delta t + 0.001 \, (\Delta t)^2.$$

The average rate of growth during this time interval is therefore given by

$$\frac{\Delta P}{\Delta t} = 0.04 + 0.001 \, \Delta t.$$

In order to obtain the instantaneous rate of growth, we must take the limit as $\Delta t \to 0$.

$$\lim_{\Delta t \to 0} \frac{\Delta P}{\Delta t} = \lim_{\Delta t \to 0} [0.04 + 0.001 \, \Delta t] = 0.04$$

☛ **11.** In Example 1, find the instantaneous rate of growth when (a) $t = 0$; (b) $t = 10$.

Thus at the beginning of 1975, the population of the city was growing at the rate of 0.04 million per year (that is, 40,000 per year). ☛ **11**

The instantaneous rate of change of a function such as the one in Example 1 is one case of what we call the *derivative* of a function. We shall now give a formal definition of the derivative.

DEFINITION Let $y = f(x)$ be a given function. Then the **derivative of y with respect to x,** denoted by dy/dx, is defined to be

$$\frac{dy}{dx} = \lim_{\Delta x \to 0} \frac{\Delta y}{\Delta x}$$

or

$$\frac{dy}{dx} = \lim_{\Delta x \to 0} \frac{f(x + \Delta x) - f(x)}{\Delta x}$$

provided that this limit exists.

The derivative is also given the name **differential coefficient,** and the operation of calculating the derivative of a function is called **differentiation.**

If the derivative of a function f exists at a particular value of x, then we say that f is **differentiable** at that point.

The derivative of $y = f(x)$ with respect to x is also denoted by any one of the following symbols.

$$\frac{d}{dx}(y), \quad \frac{df}{dx}, \quad \frac{d}{dx}(f), \quad y', \quad f'(x), \quad D_x y, \quad D_x f$$

Every one of these notations means exactly the same thing as dy/dx.

Note dy/dx represents a single symbol and should not be interpreted as the ratio of two quantities dy and dx. To amplify the notation further, note that dy/dx denotes the derivative of y with respect to x if y is a function of the independent variable x; dC/dq denotes the derivative of C with respect to q if C is a function of the independent variable q; dx/du denotes the derivative of x with respect to u if x is a function of the independent variable u. From the definition,

Answer (a) 0.03; (b) 0.05.

$$\frac{dy}{dx} = \lim_{\Delta x \to 0} \frac{\Delta y}{\Delta x}, \qquad \frac{dC}{dq} = \lim_{\Delta q \to 0} \frac{\Delta C}{\Delta q}, \quad \text{and} \quad \frac{dx}{du} = \lim_{\Delta u \to 0} \frac{\Delta x}{\Delta u}.$$

In order to calculate the derivative dy/dx, we can proceed as follows:

1. Calculate $y = f(x)$ and $y + \Delta y = f(x + \Delta x)$.

2. Subtract the first from the second to get Δy and simplify the result.

3. Divide Δy by Δx and then take the limit of the resulting expression as $\Delta x \to 0$.

The value of dy/dx depends on the choice of x. This is emphasized when we use the notation $f'(x)$, which indicates the derivative $f'(x)$ is a function of x. The value of the derivative at a particular point, say $x = 2$, is then $f'(2)$. For example, in Example 1 we evaluated dP/dt at $t = 5$, or equivalently, $P'(5)$.

☛ **12.** Find $f'(x)$ when $f(x) = 2 - x^2$. Evaluate $f'(3)$ and $f'(-2)$.

EXAMPLE 2 Find $f'(x)$ if $f(x) = 2x^2 + 3x + 1$. Evaluate $f'(2)$ and $f'(-2)$.

Solution Let $y = f(x) = 2x^2 + 3x + 1$. Then

$$y + \Delta y = f(x + \Delta x) = 2(x + \Delta x)^2 + 3(x + \Delta x) + 1$$
$$= 2[x^2 + 2x \cdot \Delta x + (\Delta x)^2] + 3x + 3\Delta x + 1$$
$$= 2x^2 + 4x \cdot \Delta x + 2(\Delta x)^2 + 3x + 3\Delta x + 1$$
$$= 2x^2 + 3x + 1 + \Delta x(4x + 3 + 2\Delta x).$$

Subtracting y from $y + \Delta y$, we have

$$\Delta y = \Delta x(4x + 3 + 2\Delta x)$$

Answer $f'(x) = -2x$, $f'(3) = -6$, $f'(-2) = 4$.

and so $\Delta y/\Delta x = 4x + 3 + 2\Delta x$. Thus

$$\frac{dy}{dx} = \lim_{\Delta x \to 0} \frac{\Delta y}{\Delta x} = \lim_{\Delta x \to 0} (4x + 3 + 2\Delta x), = 4x + 3.$$

☛ **13.** Find $f'(x)$ when $f(x) = x^3$. Evaluate $f'(2)$ and $f'(-2)$.

That is, $f'(x) = 4x + 3$.
When $x = 2$, $f'(2) = 4(2) + 3 = 11$;
when $x = -2$, $f'(-2) = 4(-2) + 3 = -5$.

Note To find $f'(c)$ we must not first find $f(c)$ and then differentiate it: $f'(c) \neq (d/dx)f(c)$. ☛ **12, 13**

EXAMPLE 3 Find $f'(x)$ if $f(x) = \sqrt{x}$.

Solution Let $y = f(x) = \sqrt{x}$. Then $y + \Delta y = f(x + \Delta x) = \sqrt{x + \Delta x}$, so

$$\Delta y = \sqrt{x + \Delta x} - \sqrt{x}.$$

Thus

Answer $f'(x) = 3x^2$, $f'(2) = 12$, $f'(-2) = 12$.

$$\frac{\Delta y}{\Delta x} = \frac{\sqrt{x + \Delta x} - \sqrt{x}}{\Delta x}.$$

We wish to take the limit as $\Delta x \to 0$; before doing so, we must rationalize the numerator. We do this by multiplying numerator and denominator by $(\sqrt{x + \Delta x} + \sqrt{x})$.

$$\frac{\Delta y}{\Delta x} = \frac{(\sqrt{x + \Delta x} - \sqrt{x})(\sqrt{x + \Delta x} + \sqrt{x})}{\Delta x(\sqrt{x + \Delta x} + \sqrt{x})} = \frac{(\sqrt{x + \Delta x})^2 - (\sqrt{x})^2}{\Delta x(\sqrt{x + \Delta x} + \sqrt{x})}$$

$$= \frac{(x + \Delta x) - x}{\Delta x(\sqrt{x + \Delta x} + \sqrt{x})} = \frac{1}{\sqrt{x + \Delta x} + \sqrt{x}}$$

Therefore

$$\frac{dy}{dx} = \lim_{\Delta x \to 0} \frac{\Delta y}{\Delta x} = \lim_{\Delta x \to 0} \frac{1}{\sqrt{x + \Delta x} + \sqrt{x}} = \frac{1}{\sqrt{x} + \sqrt{x}} = \frac{1}{2\sqrt{x}}.$$

Hence $f'(x) = 1/2\sqrt{x}$.

EXAMPLE 4 Evaluate dy/dx for the cubic function

$$y = Ax^3 + Bx^2 + Cx + D$$

where A, B, C, and D are four constants.

Solution Replacing x by $x + \Delta x$, we find that

$$y + \Delta y = A(x + \Delta x)^3 + B(x + \Delta x)^2 + C(x + \Delta x) + D$$
$$= A[x^3 + 3x^2 \Delta x + 3x(\Delta x)^2 + (\Delta x)^3]$$
$$+ B[x^2 + 2x \Delta x + (\Delta x)^2] + C(x + \Delta x) + D.$$

If we now subtract the given expression for y, we find that

$$\Delta y = (y + \Delta y) - y$$
$$= A[x^3 + 3x^2 \Delta x + 3x(\Delta x)^2 + (\Delta x)^3]$$
$$+ B[x^2 + 2x \Delta x + (\Delta x)^2]$$
$$+ C(x + \Delta x) + D - (Ax^3 + Bx^2 + Cx + D)$$
$$= A[3x^2 \Delta x + 3x(\Delta x)^2 + (\Delta x)^3] + B[2x \Delta x + (\Delta x)^2] + C \Delta x.$$

Therefore

$$\frac{\Delta y}{\Delta x} = A[3x^2 + 3x \Delta x + (\Delta x)^2] + B(2x + \Delta x) + C.$$

Allowing Δx to approach zero, we see that the three terms on the right that involve Δx as a factor all approach zero in the limit. The remaining terms give the following result.

$$\frac{dy}{dx} = \lim_{\Delta x \to 0} \frac{\Delta y}{\Delta x} = 3Ax^2 + 2Bx + C \qquad (1)$$

From the result of this example, it is possible to retrieve some of the results in preceding examples. For instance, if we set $A = 0$, $B = 2$, $C = 3$, and $D = 1$, the cubic function in Example 4 becomes $y = 0x^3 + 2x^2 + 3x + 1 = 2x^2 + 3x + 1$, which was discussed in Example 2. From Equation (1), we obtain

$$\frac{dy}{dx} = 3Ax^2 + 2Bx + C = 3(0)x^2 + 2(2)x + 3 = 4x + 3$$

which agrees with the result of Example 2.

Geometric Interpretation

We have already seen that in the case where the independent variable in a function $y = f(t)$ represents time, the derivative dy/dt gives the instantaneous rate of change of y. For example, if $s = f(t)$ represents that distance traveled by a moving object, then ds/dt provides the instantaneous velocity. Apart from this kind of application of derivatives, however, they also have a very great geometrical significance.

If P and Q are two points $(x, f(x))$ and $(x + \Delta x, f(x + \Delta x))$ on the graph of $y = f(x)$, then, as stated in Section 12-1, the ratio

$$\frac{\Delta y}{\Delta x} = \frac{f(x + \Delta x) - f(x)}{\Delta x}$$

represents the slope of the line segment PQ. As Δx becomes smaller and smaller, the point Q moves closer and closer to P and the chord segment PQ becomes more and more nearly a tangent. As $\Delta x \to 0$, the slope of the chord PQ approaches the slope of the tangent line at P. Thus

$$\lim_{\Delta x \to 0} \frac{\Delta y}{\Delta x} = \frac{dy}{dx}$$

represents the slope of the tangent line to $y = f(x)$ at the point $P(x, f(x))$. (See Figure 6.) As long as the curve $y = f(x)$ is "smooth" at P; that is, as long as we can draw a nonvertical tangent at P, the limit will exist.

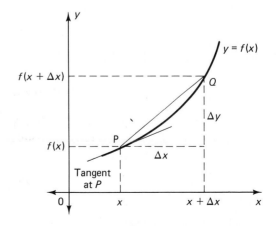

FIGURE 6

14. In Example 4, find the equation of the tangent line at
(a) (1, 1);   (b) (9, 3)

EXAMPLE 5 Find the slope of the tangent and the equation of the tangent line to the graph $y = \sqrt{x}$ at the point (4, 2) and the point $(\frac{1}{4}, \frac{1}{2})$.

Solution In Example 3, we showed that if $f(x) = \sqrt{x}$ then $f'(x) = 1/2\sqrt{x}$. When $x = 4$, $f'(4) = 1/2\sqrt{4} = \frac{1}{4}$. Thus the slope of the tangent at (4, 2) is $\frac{1}{4}$.

To obtain the equation of the tangent line, we can use the point-slope formula

$$y - y_1 = m(x - x_1)$$

with slope $m = \frac{1}{4}$ and $(x_1, y_1) = (4, 2)$. (See Figure 7.) We obtain

$$y - 2 = \tfrac{1}{4}(x - 4)$$

$$y = \tfrac{1}{4}x + 1$$

which is the required equation.

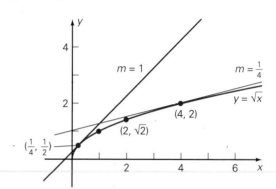

FIGURE 7

Answer (a) $y = \frac{1}{2}x + \frac{1}{2}$;
(b) $y = \frac{1}{6}x + \frac{3}{2}$.

15. Find the equation of the tangent line to the graph of $y = x^2 + x$ at the point (a) (1, 2); (b) (−1, 0).

Answer (a) $y = 3x - 1$;
(b) $y = -x - 1$.

When $x = \frac{1}{4}$, $f'(\frac{1}{4}) = 1/2\sqrt{\frac{1}{4}} = 1$. Thus the slope of the tangent at $(\frac{1}{4}, \frac{1}{2})$ is 1. (See Figure 7.) From the point-slope formula, its equation is

$$y - \tfrac{1}{2} = 1 \cdot (x - \tfrac{1}{4}) \quad \text{or} \quad y = x + \tfrac{1}{4}. \quad \text{☛ 14, 15}$$

EXERCISES 12-3

(1–14) Find the derivatives of the following functions with respect to the independent variables involved.

1. $f(x) = 2x - 5$

2. $f(x) = 2 - 5x$

3. $g(x) = 7$

4. $g(t) = -3$

5. $f(x) = x^2$

6. $g(t) = 3t^2 + 1$

7. $f(u) = u^2 + u + 1$

8. $g(x) = x^2 - 3x + 7$

9. $h(x) = 7 - 3x^2$

10. $f(x) = 1/x$

11. $g(x) = \dfrac{1}{x + 1}$

12. $h(u) = \dfrac{2}{1 - u}$

13. $f(t) = \dfrac{1}{2t + 3}$

14. $g(u) = \dfrac{u}{u + 1}$

15. Find dy/dx if:

   a. $y = 3 - 2x^2$;

   b. $y = 3x + 7$.

16. Find du/dt if:

    a. $u = 2t + 3$;   **b.** $u = 1/(2t + 1)$.

17. Find dx/dy if:

    a. $x = \sqrt{y}$;   **b.** $x = (y + 1)/y^2$.

18. Find dp/dq if:

    a. $p = 1/(3 + 2q)$;   **b.** $p = 1/\sqrt{q}$.

19. Find $f'(2)$ if $f(x) = 5 - 2x$.

20. Find $g'(4)$ if $g(x) = (x + 1)^2$.

21. Find $F'(3)$ if $F(t) = t^2 - 3t$.

22. Find $G'(1)$ if $G(u) = u^2 - u + 3$.

23. Find $h'(0)$ if $h(y) = y^2 + 7y$.

24. Find $H'(2)$ if $H(t) = 1/(t - 1)$.

(25–32) Find the slope of the tangent to the graphs of the following functions at the indicated points. Determine the equation of the tangent line in each case.

25. $y = 3x^2 - 4$ at $x = 2$

26. $y = x^2 + x + 2$ at $x = -2$

27. $f(x) = \dfrac{1}{x}$ at $x = 3$ **28.** $g(x) = \dfrac{1}{x - 1}$ at $x = 2$

29. $y = \dfrac{x + 1}{x}$ at $x = 1$ **30.** $f(x) = \sqrt{x - 1}$ at $x = 5$

31. $f(x) = \dfrac{x + 1}{x - 1}$ at $x = 2$

32. $g(t) = 5t^2 + 1$ at $t = -3$

33. *(Sales Growth)* The sales volume of a particular phonograph record is given as a function of time t by the formula

$$S(t) = 10{,}000 + 2000t - 200t^2$$

where t is measured in weeks and S is the number of records sold per week. Determine the rate at which S is changing when:

    a. $t = 0$.   **b.** $t = 4$.   **c.** $t = 8$.

34. *(Population Growth)* A certain population grows according to the formula

$$p(t) = 30{,}000 + 60t^2$$

where t is measured in years. Find the growth rate when:

    a. $t = 2$.   **b.** $t = 0$.   **c.** $t = 5$.

35. *(Chemical Reaction)* During a chemical reaction in which a compound A is broken down, the mass (in grams) of A remaining at time t is given by $m(t) = 9 - 3t + \frac{1}{4}t^2$. Find $m'(t)$ and interpret this quantity. Evaluate $m(0)$, $m'(0)$, $m(6)$, and $m'(6)$.

◣ 12-4 DERIVATIVES OF POWER FUNCTIONS

It is clear from Section 12-3 that finding derivatives of functions by direct use of the definition of the derivative is not always easy and is generally time consuming. This task can be appreciably lightened by the use of certain standard formulas. In this section, we shall develop formulas for finding the derivatives of power functions and combinations of power functions.

Let us begin by going back to Example 4 in Section 12-3. By taking special cases for the coefficients A, B, C, and D in that example, we obtain the following results.

THEOREM 1

    (a) The derivative of a constant function is zero.

    (b) If $y = x$, then $dy/dx = 1$.

    (c) If $y = x^2$, then $dy/dx = 2x$.

    (d) If $y = x^3$, then $dy/dx = 3x^2$.

PROOF

(a) In the function $y = Ax^3 + Bx^2 + Cx + D$, let A, B, and C equal zero. Then $y = D$, a constant function. The general expression for dy/dx is $3Ax^2 + 2Bx + C$ (from Example 4 of Section 12-3) and this is zero when $A = B = C = 0$.

(b) If we let $A = B = D = 0$ and $C = 1$, we get $y = x$ and $dy/dx = 1$, as required.

(c) and (d) are proved in a similar manner.

Geometrically, part (a) of Theorem 1 asserts that the slope of the line $y = c$ is zero at every point on it. This is obviously true because the graph of $y = c$ is a horizontal line, and any horizontal line has zero slope.

EXAMPLE 1 $\dfrac{d}{dx}(6) = 0$ and $\dfrac{d}{dt}\left(\dfrac{3}{2}\right) = 0$

From the results in parts (a)–(d) of Theorem 1, we can observe a certain pattern developing for the derivatives of powers of x, $y = x^n$. We have the following result that holds for any real value of n.

$$\text{If } y = x^n, \quad \text{then } \frac{dy}{dx} = nx^{n-1}. \qquad \text{(Power Formula)}$$

In words, *to find the derivative of any constant power of x, we decrease the power of x by 1 and multiply by the original exponent of x.*

At the end of this section, we shall prove this formula for the derivative of x^n for the case when n is a positive integer. It is, however, true for all real values of n.

EXAMPLE 2

(a) $\dfrac{d}{dx}(x^7) = 7x^{7-1} = 7x^6$

(b) $\dfrac{d}{dy}(y^{3/2}) = \tfrac{3}{2}y^{3/2-1} = \tfrac{3}{2}y^{1/2}$

(c) $\dfrac{d}{dt}\left(\dfrac{1}{\sqrt{t}}\right) = \dfrac{d}{dt}(t^{-1/2}) = -\tfrac{1}{2}t^{-1/2-1} = -\tfrac{1}{2}t^{-3/2}$

(d) $\dfrac{d}{du}\left(\dfrac{1}{u^2}\right) = \dfrac{d}{du}(u^{-2}) = -2u^{-2-1} = -2u^{-3} = -\dfrac{2}{u^3}$

◆ 16. Use the power formula to find

(a) $\dfrac{d}{dx}(x^4)$; (b) $\dfrac{d}{dt}(\sqrt{t})$.

(c) $\dfrac{d}{du}(u^e)$; (d) $\dfrac{d}{dx}(x^x)$.

(e) $\dfrac{d}{dx}(x) = \dfrac{d}{dx}(x^1) = 1 \cdot x^{1-1} = x^0 = 1$ (because $x^0 = 1$)

(f) $\dfrac{d}{dx}(x^{\sqrt{2}}) = \sqrt{2}\,x^{\sqrt{2}-1}$ ◆ **16**

THEOREM 2 If $u(x)$ is a differentiable function of x and c is a constant, then

$$\frac{d}{dx}(cu) = c\,\frac{du}{dx}.$$

That is, *the derivative of the product of a constant and a function of x is equal to the product of the constant and the derivative of the function.*

EXAMPLE 3

(a) $\dfrac{d}{dx}(cx^n) = c\,\dfrac{d}{dx}(x^n) = c(nx^{n-1}) = ncx^{n-1}$

(b) $\dfrac{d}{dt}\left(\dfrac{4}{t}\right) = \dfrac{d}{dt}(4t^{-1}) = 4\,\dfrac{d}{dt}(t^{-1}) = 4(-1\cdot t^{-2}) = -\dfrac{4}{t^2}$

(c) $\dfrac{d}{du}(2\sqrt{u}) = \dfrac{d}{du}(2u^{1/2}) = 2\,\dfrac{d}{du}(u^{1/2}) = 2\cdot\tfrac{1}{2}u^{-1/2} = u^{-1/2}$

THEOREM 3 If $u(x)$ and $v(x)$ are two differentiable functions of x, then

$$\frac{d}{dx}(u + v) = \frac{du}{dx} + \frac{dv}{dx}.$$

In other words, *the derivative of the sum of two functions is equal to the sum of the derivatives of the two functions.*

EXAMPLE 4 Find dy/dx if $y = x^2 + \sqrt{x}$.

Solution The given function is the sum of x^2 and $x^{1/2}$. Therefore, by Theorem 3, we can differentiate these two powers separately.

$$\frac{dy}{dx} = \frac{d}{dx}(x^2) + \frac{d}{dx}(x^{1/2}) = 2x + \tfrac{1}{2}x^{-1/2}$$

This theorem can be readily extended to the sum of any number of functions and also to the differences between functions. For example,

Answer (a) $4x^3$; (b) $\tfrac{1}{2}t^{-1/2}$; (c) eu^{e-1} (d) The power formula cannot be used when the exponent is not a constant. The answer is *not* $x \cdot x^{x-1}$.

$$\frac{d}{dx}(u - v) = \frac{du}{dx} - \frac{dv}{dx}$$

$$\frac{d}{dx}(u + v - w) = \frac{du}{dx} + \frac{dv}{dx} - \frac{dw}{dx}$$

and so on.

EXAMPLE 5 Find the derivative of $3x^4 - 5x^3 + 7x + 2$ with respect to x.

Solution Let $y = 3x^4 - 5x^3 + 7x + 2$. Then

$$\frac{dy}{dx} = \frac{d}{dx}(3x^4 - 5x^3 + 7x + 2) = \frac{d}{dx}(3x^4) - \frac{d}{dx}(5x^3) + \frac{d}{dx}(7x) + \frac{d}{dx}(2).$$

We have used Theorem 3 to express the derivative of the sum $3x^4 - 5x^3 + 7x + 2$ as the sum of the derivatives of $3x^4$, $-5x^3$, $7x$, and 2. Evaluating these four derivatives, we obtain

$$\frac{dy}{dx} = 3(4x^3) - 5(3x^2) + 7(1x^0) + 0 = 12x^3 - 15x^2 + 7$$

because $x^0 = 1$. ● **17**

Let us reconsider Example 2 of Section 12-3. Using the methods of this section we can obtain the answer much more easily than before. For if $f(x) = 2x^2 + 3x + 1$, then

$$f'(x) = 2 \cdot 2x + 3 \cdot 1 + 0 = 4x + 3.$$

Finished.

It is often necessary to rearrange the algebraic form of a function before the theorems can be applied. Expressions involving parentheses can be differentiated after removing the parentheses. For example, if we wish to calculate dy/dx when $y = x^2(2x - 3)$, we first write $y = 2x^3 - 3x^2$. In this form, we can differentiate y as in Example 5, and we obtain $dy/dx = 6x^2 - 6x$. Or if $y = (x + 2)(x^2 - 3)$, then we begin by expanding the parentheses, obtaining $y = x^3 + 2x^2 - 3x - 6$. From this stage, we again proceed as in Example 5, and obtain $dy/dx = 3x^2 + 4x - 3$.

Similarly, we can simplify fractions with monomial denominators before we differentiate. For example, if

$$y = \frac{5t^4 + 7t^2 - 3}{2t^2}$$

we first write $y = \frac{5}{2}t^2 + \frac{7}{2} - \frac{3}{2}t^{-2}$. After differentiating the three terms separately, we obtain

$$\frac{dy}{dt} = 5t + 3t^{-3}. \quad ● \ \mathbf{18}$$

Proofs of Theorems

PROOF OF THEOREM 2 Let $y = cu(x)$. Then if x is replaced by $x + \Delta x$, u becomes $u + \Delta u$ and $y + \Delta y$, so that

$$y + \Delta y = cu(x + \Delta x) = c(u + \Delta u).$$

Subtracting y, we have $\Delta y = c(u + \Delta u) - cu = c\Delta u$. Dividing both sides by Δx gives

$$\frac{\Delta y}{\Delta x} = c\,\frac{\Delta u}{\Delta x}.$$

Taking the limit as $\Delta x \to 0$, we have

$$\lim_{\Delta x \to 0} \frac{\Delta y}{\Delta x} = \lim_{\Delta x \to 0} \left(c\,\frac{\Delta u}{\Delta x} \right) = c \lim_{\Delta x \to 0} \frac{\Delta u}{\Delta x}.$$

That is

$$\frac{dy}{dx} = c\,\frac{du}{dx},$$

as required.

PROOF OF THEOREM 3 Let $y = u(x) + v(x)$. Let x be given an increment Δx. Since y, u, and v are all functions of x, they become $y + \Delta y$, $u + \Delta u$, and $v + \Delta v$, where

$$y + \Delta y = u(x + \Delta x) + v(x + \Delta x) = (u + \Delta u) + (v + \Delta v).$$

Subtracting y from $y + \Delta y$ gives

$$\Delta y = (u + \Delta u + v + \Delta v) - (u + v) = \Delta u + \Delta v.$$

Dividing by Δx, we have

$$\frac{\Delta y}{\Delta x} = \frac{\Delta u}{\Delta x} + \frac{\Delta v}{\Delta x}.$$

If we allow Δx to approach zero, we obtain

$$\lim_{\Delta x \to 0} \frac{\Delta y}{\Delta x} = \lim_{\Delta x \to 0} \frac{\Delta u}{\Delta x} + \lim_{\Delta x \to 0} \frac{\Delta v}{\Delta x} \qquad \text{(by Theorem 3(a), Section 12-2)}$$

That is,

$$\frac{dy}{dx} = \frac{du}{dx} + \frac{dv}{dx}$$

which is the required result.

Finally, let us prove the power formula when n is a positive integer. The proof to be given makes use of the following result from algebra.

If n is a positive integer,

$$a^n - b^n = (a - b)(a^{n-1} + a^{n-2}b + a^{n-3}b^2 + \cdots + ab^{n-2} + b^{n-1}).$$

This result is easy to verify by multiplying the two expressions on the right side term by term. It should be noted that the number of terms in the second parentheses on the right is equal to n, the power of a and b on the left side. Consider the following examples.

$$n = 2: \quad a^2 - b^2 = (a - b)\underbrace{(a + b)}_{\text{2 terms}}$$

$$n = 3: \quad a^3 - b^3 = (a - b)\underbrace{(a^2 + ab + b^2)}_{\text{3 terms}}$$

$$n = 4: \quad a^4 - b^4 = (a - b)\underbrace{(a^3 + a^2b + ab^2 + b^3)}_{\text{4 terms}}, \quad \text{etc.}$$

THEOREM 4 The derivative of x^n with respect to x is nx^{n-1}, when n is a positive integer.

PROOF Let $y = x^n$. When x changes to $x + \Delta x$, y changes to $y + \Delta y$, where

$$y + \Delta y = (x + \Delta x)^n.$$

Subtracting the value for y from that for $y + \Delta y$, we have

$$\Delta y = (x + \Delta x)^n - x^n.$$

In order to simplify this expression for Δy, we make use of the algebraic identity given above, letting $a = x + \Delta x$ and $b = x$. Then $a - b = (x + \Delta x) - x = \Delta x$, and so

$$\Delta y = \Delta x[(x + \Delta x)^{n-1} + (x + \Delta x)^{n-2} \cdot x + (x + \Delta x)^{n-3} \cdot x^2 + \cdots$$
$$+ (x + \Delta x) \cdot x^{n-2} + x^{n-1}].$$

Dividing both sides by Δx and taking the limit as $\Delta x \to 0$, we have

$$\frac{dy}{dx} = \lim_{\Delta x \to 0} \frac{\Delta y}{\Delta x} = \lim_{\Delta x \to 0} [(x + \Delta x)^{n-1} + (x + \Delta x)^{n-2} \cdot x + (x + \Delta x)^{n-3} \cdot x^2$$
$$+ \cdots + (x + \Delta x) \cdot x^{n-2} + x^{n-1}].$$

Now as $\Delta x \to 0$, each term in the brackets approaches the limit x^{n-1}. For example, the second term $(x + \Delta x)^{n-2} \cdot x$ approaches $x^{n-2} \cdot x = x^{n-1}$ as $\Delta x \to 0$. Furthermore, there are n such terms added together, so

$$\frac{dy}{dx} = \underbrace{x^{n-1} + x^{n-1} + x^{n-1} + \cdots + x^{n-1} + x^{n-1}}_{n \text{ terms}} = nx^{n-1}$$

as required.

EXERCISES 12-4

(1–46) Differentiate the following expressions.

1. x^5

2. $x^{\sqrt{3}}$

3. $\dfrac{1}{t^3}$

4. $\dfrac{4}{u^4}$

5. $\dfrac{1}{5u^5}$

6. $\dfrac{x^7}{7}$

7. $\dfrac{1}{\sqrt[3]{x^2}}$

8. $2x - x^3$

9. $4x^3 - 3x^2 + 7$

10. $5 - 2x^2 + x^4$

11. $3x^4 - 7x^3 + 5x^2 + 8$

12. $4x^3 + 2 + \dfrac{1}{x}$

13. $3u^2 + \dfrac{3}{u^2}$

14. $\dfrac{x^6}{6} + \dfrac{6}{x^6}$

15. $x^{1.2} + \dfrac{1}{x^{0.6}}$

16. $x^{0.4} - x^{-0.4}$

17. $2\sqrt{x} + 2/\sqrt{x}$

18. $x^7 + \dfrac{1}{x^7} + 7x + \dfrac{7}{x} + 7$

19. $2\sqrt{x^3} + \dfrac{2}{\sqrt{x^3}}$

20. $2\sqrt{t} - \dfrac{3}{\sqrt[3]{t}}$

21. $2x^{3/2} + 4x^{5/4}$

22. $\sqrt[3]{x} - \dfrac{1}{\sqrt[3]{x}}$

23. $3x^4 + (2x - 1)^2$

24. $(y - 2)(2y - 3)$

25. $(x - 7)(2x - 9)$

26. $\left(x + \dfrac{1}{x}\right)^2$

27. $(u + 1)(2u + 1)$

28. $\left(\sqrt{x} + \dfrac{1}{\sqrt{x}}\right)^2 - \left(\sqrt{x} - \dfrac{1}{\sqrt{x}}\right)^2$

29. $(t + 1)(3t - 1)^2$

30. $(u - 2)^3$

31. $(x + 2)^3$

32. $(x + 1)(x - 1)^2$

33. $\left(\dfrac{x + 1}{x}\right)^3$

34. $\left(\dfrac{2t - 1}{2t}\right)^3$

35. $\left(\dfrac{y + 2}{y}\right)^3 + \left(\dfrac{y - 2}{y}\right)^3$

36. $\dfrac{2y^2 + 3y - 7}{y}$

37. $\dfrac{(x + 1)^2}{x}$

38. $\dfrac{x^2 - 3x + 1}{\sqrt{x}}$

39. $\dfrac{t + 3/t}{\sqrt{t}}$

40. $\dfrac{(x + 1)^2 + (x - 1)^2}{x^2}$

41. $\dfrac{(2t + 3)^2 - (2t - 3)^2}{4t}$

42. $x^3 - \dfrac{x^{1.6}}{x^{2.3}}$

43. $\sqrt{2y} + (3y)^{-1}$

44. $(8y)^{2/3} + (8y)^{-2/3}$

45. $(16t)^{3/4} - (16t)^{-3/4}$

46. $\sqrt[3]{27t^2} - \dfrac{1}{\sqrt[3]{27t^2}}$

47. Find dy/dx if $y = x^3 + 1/x^3$.

48. Find du/dx if $u = x^2 - 7x + 5/x$.

49. Find dy/du if $y = u^3 - 5u^2 + \dfrac{7}{3u^2} + 6$.

50. Find dx/dt if $x = (t^3 - 5t^2 + 7t - 1)/t^2$.

51. If $y = \sqrt{x}$, prove that $2y(dy/dx) = 1$.

52. If $u = 1/\sqrt{x}$, prove that $2u^{-3}(du/dx) + 1 = 0$.

(53–56) Determine the equation of the tangent line to the graph of the following functions at the indicated points.

53. $f(x) = x^2 - 3x + 4$ at $(1, 2)$

54. $f(x) = x^2 + \dfrac{1}{x^2}$ at $(-1, 2)$

55. $f(x) = \dfrac{2}{x}$ at $x = -2$

56. $f(x) = x^3 - \dfrac{1}{x^3}$ at $x = 1$

57. Find all the points on the graph of $y = x^2 - 3x + 7$ where the tangent line is parallel to the line $x - y + 4 = 0$.

58. Find all the points on the graph of $f(x) = x^3 - 5x + 2$ where the tangent line is perpendicular to the line $x + 7y + 4 = 0$.

59. *(Moving Object)* The distnce traveled by a moving object at a time t is equal to $2t^3 - t^{1/2}$. Find the instantaneous velocity:

a. At time t. **b.** At time 4.

60. *(Projectile)* A ball is thrown vertically upward with an initial velocity of 60 feet/second. After time t seconds, its height above the ground is given by $s = 60t - 16t^2$. Find its instantaneous velocity after t seconds. What is special about $t = \frac{15}{8}$?

61. *(Growth of GNP)* In Exercise 22 of Section 12-1, find the instantaneous rates of growth of the GNP in:

a. 1970. **b.** 1980. **c.** 1990.

(The answer will be in billions of dollars per year.)

62. *(Population Growth)* At the start of an experiment, a culture of bacteria is found to contain 10,000 individuals. The growth of the population was observed, and it was found that at any subsequent time t (hours) after the start of the experiment, the population size $p(t)$ could be expressed by the formula

$$p(t) = 2500(2 + t)^2.$$

Determine the formula for the rate of growth of the population at any time t, and in particular calculate the growth rate for $t = 15$ minutes and for $t = 2$ hours.

63. *(Botany)* The proportion of seeds of a certain species of tree that scatter farther than distance r from the base of the tree is given by

$$p(r) = \frac{3}{4}\left(\frac{r_0}{r}\right)^{1/2} + \frac{1}{4}\left(\frac{r_0}{r}\right)$$

where r_0 is a constant. Find the rate of change of the proportion with respect to distance and calculate $p'(2r_0)$.

64. *(Physics)* During rapid (adiabatic) changes in pressure, the pressure p and density ρ of a gas vary according to the law $p\rho^{-\gamma} = c$, where γ and c are constants. Calculate $dp/d\rho$.

65. *(Biochemistry)* According to the Schütz-Borisoff law, the amount y of substrate transformed by an enzyme in a time interval t is given by $y = k\sqrt{cat}$, where c is the concentration of the enzyme, a is the initial concentration of the substrate, and k is a constant. What is the rate at which the substrate is being transformed?

66. *(Projectile)* A ball is thrown into the air with a velocity of 40 feet per second at an angle of 45° to the horizontal. If the x-axis is taken horizontal and the y-axis vertical, the origin being at the initial point of the ball's flight, then the position of the ball at time t is given by $x = 20\sqrt{2}\,t$, $y = 20\sqrt{2}\,t - 16t^2$. Calculate the slope of the path t seconds after the ball is thrown. At what value of t is the slope zero? (*Hint:* Express y in terms of x by eliminating t.)

67. *(Cell Growth)* The mass of a unicellular organism increases with time t according to the formula $m(t) = 2 + 6t + 3t^2$. Find $m'(t)$ and evaluate $m(2)$ and $m'(2)$. Interpret these values.

68. *(Epidemics)* An infectious and debilitating disease is spreading slowly through a population. The number of individuals who have caught the diease after t months is found to be given by the formula

$$N(t) = 1000(t^{3/2} + t^2)$$

Find $N'(t)$. Evaluate $N(9)$ and $N'(9)$ and interpret these values.

69. *(Fay/Lehr Formula)* It is observed that the shape of a spreading oil slick is roughly an ellipse, with its major axis in the direction of the wind. The area of the ellipse in square meters is $A = \pi ab$, where

$$a(t) = b(t) + c_1 t^{3/4}, \qquad b(t) = c_2 t^{1/4}.$$

Here t is time in minutes, c_1 is a constant depending on the wind speed, and c_2 a constant depending on the volume of the spill. If $c_1 = 0.2$ and $c_2 = 15$, calculate the values of $A(t)$ and $A'(t)$ after 15 minutes and after 30 minutes.

◣ 12-5 MARGINAL ANALYSIS

Derivatives have a number of applications in business and economics in constructing what are called *marginal rates*. In this field, the word "marginal" is used to indicate a derivative, that is, a rate of change. A selection of examples will be given.

Marginal Cost

Suppose that the manufacturer of a certain item finds that in order to make x of these items per week, the total cost in dollars is given by $C = 200 + 0.03x^2$. For example, if 100 items are produced per week, the cost is given by $C = 200 + 0.03(100)^2 = 500$. The average cost per item of producing 100 items is therefore $\frac{500}{100} = \$5$.

Now let us suppose that the manufacturer is considering changing the weekly production from 100 to $(100 + \Delta x)$ units per week, where Δx represents the increment in weekly production. The cost becomes

$$C + \Delta C = 200 + 0.03(100 + \Delta x)^2$$

$$= 200 + 0.03[10,000 + 200\Delta x + (\Delta x)^2]$$

$$= 500 + 6\Delta x + 0.03(\Delta x)^2.$$

Therefore the extra cost involved in producing the additional items is

$$\Delta C = (C + \Delta C) - C = 500 + 6\Delta x + 0.03(\Delta x)^2 - 500$$

$$= 6\Delta x + 0.03(\Delta x)^2.$$

The average cost per item of the extra items is therefore

$$\frac{\Delta C}{\Delta x} = 6 + 0.03\Delta x.$$

For example, if the production is increased from 100 to 150 per week (so $\Delta x = 50$), then the average cost of the additional 50 items is equal to $6 + 0.03(50) = \$7.50$ each. If the increase is from 100 to 110 (so $\Delta x = 10$), then the average cost of the extra 10 items is equal to $\$6.30$ each.

We define the **marginal cost** to be the limiting value of the average cost per extra item as the number of extra items approaches zero. Thus we can think of the marginal cost as the average cost per extra item when a very small change is made in the amount produced. In the above example,

$$\text{Marginal Cost} = \lim_{\Delta x \to 0} \frac{\Delta C}{\Delta x} = \lim_{\Delta x \to 0} (6 + 0.03\Delta x) = 6.$$

In the case of a general cost function $C(x)$ representing the cost of producing an amount x of a certain item, the marginal cost is similarly given by

$$\text{Marginal Cost} = \lim_{\Delta x \to 0} \frac{\Delta C}{\Delta x} = \lim_{\Delta x \to 0} \frac{C(x + \Delta x) - C(x)}{\Delta x}$$

Clearly, the marginal cost is nothing but the derivative of the cost function with respect to the amount produced.

$$\text{Marginal Cost} = \frac{dC}{dx}$$

☛ **19.** Find the marginal cost if $C(x) = 4 + 3x - 0.1x^2$. Evaluate $C'(5)$ and state its significance.

The marginal cost measures the rate at which the cost is increasing with respect to increases in the amount produced.

EXAMPLE 1 *(Marginal Cost)* For the cost function

$$C(x) = 0.001x^3 - 0.3x^2 + 40x + 1000$$

determine the marginal cost as a function of x. Evaluate the marginal cost when the production is given by $x = 50$, $x = 100$, and $x = 150$.

Solution We wish to calculate $C'(x)$. The given function $C(x)$ is a combination of powers of x and so can be differentiated by means of the power formula discussed in the last section. We obtain

$$C'(x) = \frac{d}{dx}(0.001x^3 - 0.3x^2 + 40x + 1000)$$

$$= 0.001(3x^2) - 0.3(2x) + 40(1) + 0$$

$$= 0.003x^2 - 0.6x + 40.$$

This function, the marginal cost, gives the average cost per item of increasing the production by a small amount given that x items are already being produced. When 50 units are being produced, the marginal cost of extra items is given by

$$C'(50) = (0.003)(50)^2 - (0.6)(50) + 40 = 7.5 - 30 + 40 = 17.5.$$

When $x = 100$, the marginal cost is

$$C'(100) = (0.003)(100)^2 - (0.6)(100) + 40 = 30 - 60 + 40 = 10.$$

When $x = 150$, the marginal cost is

$$C'(150) = (0.003)(150)^2 - (0.6)(150) + 40 = 67.5 - 90 + 40 = 17.5.$$

Roughly speaking, we can say that the 51st item costs $17.50 to produce, the 101st item costs $10, and the 151st item costs $17.50. (Such statements as these are not *quite* accurate, since the derivative gives the rate for an infinitesimally small increment in production, not for a unit increment.) ☛ **19**

In Example 1, we observe that the marginal cost decreases as the production is increased from 50 to 100 units and then increases again as the production is further increased from 100 to 150. The graph of $C'(x)$ as a function of x is shown in Figure 8.

This type of behavior is quite typical of marginal cost. When the production x increases from small values, the marginal cost decreases—that is, the average cost of the next small increase in production becomes lower. The reason for this lies in the economies of scale, which make the manufacture of small quantities of goods relatively more expensive per unit than the manufacture of larger quantities. However, when x becomes very large, costs start to

Answer $C'(x) = 3 - 0.2x$. $C'(5) = 2$. When 5 units are being produced, the cost rises by 2 per additional unit when an infinitesimally small increment is made in production level.

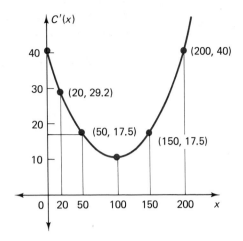

FIGURE 8

increase again as the capacity of the existing production units becomes exhausted and it becomes necessary to invest in new plant or machinery or to pay overtime rates for labor, and so on. This causes an eventual rise in the marginal cost. Thus it is the usual case that the marginal cost first decreases as the production increases, then rises again.

It is worth comparing this type of behavior with the simple linear cost model (see Section 4-3). In that case, $C(x) = mx + b$ (m and b constants) and the marginal cost $C'(x) = m$ is constant for all x. Thus the cost of each additional unit of production is constant, independent of the level of production.

It is important not to confuse the marginal cost with the average cost. If $C(x)$ is the cost function, then the **average cost** of producing x items is the total cost, $C(x)$, divided by the number of items produced.

$$\text{Average Cost per Item} = \frac{C(x)}{x}$$

This is quite different from the marginal cost, which is given by the derivative $C'(x)$. The marginal cost represents the average cost per *additional* unit of a small increase in production. The average cost is commonly denoted by $\overline{C}(x)$.

EXAMPLE 2 For the cost function $C(x) = 1000 + 10x + 0.1x^2$, the marginal cost is $C'(x) = 10 + 0.2x$. The average cost of producing x items is

$$\overline{C}(x) = \frac{C(x)}{x} = \frac{1000}{x} + 10 + 0.1x.$$

These two functions are quite distinct from one another.

Marginal Revenue and Profit

Now consider the revenues derived from the sale of a firm's products or services. If $R(x)$ denotes the revenue in dollars from the sale of x items, we define the **marginal revenue** to be the derivative $R'(x)$.

$$\text{Marginal Revenue} = R'(x) = \lim_{\Delta x \to 0} \frac{\Delta R}{\Delta x}.$$

If the number of items sold increases from x to $x + \Delta x$, there is a corresponding increment in revenue given by

$$\Delta R = \text{New Revenue} - \text{Old Revenue} = R(x + \Delta x) - R(x).$$

The average increase in revenue per additional item sold is obtained by dividing ΔR by the number of additional items, giving $\Delta R / \Delta x$. The limiting value of this average as $\Delta x \to 0$ gives the marginal revenue. Thus the marginal revenue represents the additional income to a firm per additional item sold when a very small increase is made in the number of items sold. That is, it is the rate at which revenue is increasing with respect to increase in the volume of sales.

EXAMPLE 3 *(Marginal Revenue)* If the revenue function is given by

$$R(x) = 10x - 0.01x^2$$

where x is the number of items sold, determine the marginal revenue. Evaluate the marginal revenue when $x = 200$.

Solution We need to evaluate $R'(x)$. Since $R(x)$ is a combination of powers of x, we can use the power formula, obtaining the result

$$R'(x) = \frac{d}{dx}(10x - 0.01x^2) = 10(1) - (0.01)(2x) = 10 - 0.02x.$$

This provides the marginal revenue when a general number x items are sold. When $x = 200$, we get a marginal revenue of

$$R'(200) = 10 - (0.02)(200) = 10 - 4 = 6.$$

Thus when 200 items are sold, any small increase in sales provides an increase of revenue of $6 per item.

The revenue function can be written in the form

$$R(x) = xp$$

where p is the price per item and x is the number of items sold. We saw in Section 4-5 that in many cases there exists a relationship between x and p characterized by the demand equation. The more items the firm wishes to sell, the lower it must set its price; the higher the price is set, in general, the smaller the volume of sales will be.

EXAMPLE 4 *(Marginal Revenue)* Determine the marginal revenue when $x = 300$ if the demand equation is

$$x = 1000 - 100p.$$

Solution We must first write the demand equation in a form in which p is expressed as a function of x.

$$100p = 1000 - x$$

$$p = 10 - 0.01x$$

Then the revenue function is given by

$$R(x) = xp = x(10 - 0.01x) = 10x - 0.01x^2.$$

We observe that this revenue function is the same as the one in the previous example, so we can make use of the result for the marginal revenue:

$$R'(x) = 10 - 0.02x.$$

When the volume of sales is 300, the marginal revenue is therefore given by

$$R'(300) = 10 - (0.02)(300) = 10 - 6 = 4.$$

The profit that a business firm makes is given by the difference between its revenue and its costs. If the revenue function is $R(x)$ when x items are sold and if the cost function is $C(x)$ when x items are produced, then the profit $P(x)$ obtained by producing and selling x items is given by

$$P(x) = R(x) - C(x).$$

☛ **20.** Calculate the marginal revenue for the demand equation $p = 4 - \sqrt{x}$. If the cost function is $C(x) = 1 + x$, find the marginal cost and marginal profit. Evaluate $R'(4)$, $P'(4)$, $R'(6)$, and $P'(6)$.

The derivative $P'(x)$ is called the **marginal profit.** It represents the additional profit per item if the production changes by a small increment. ☛ **20**

EXAMPLE 5 *(Marginal Profit)* The demand equation for a certain item is

$$p + 0.1x = 80$$

and the cost function is

$$C(x) = 5000 + 20x.$$

Compute the marginal profit when 150 units are produced and sold and when 400 units are produced and sold.

Solution The revenue function is given by

$$R(x) = xp = x(80 - 0.1x) = 80x - 0.1x^2.$$

Therefore the profit from producing and selling x items is

$$P(x) = R(x) - C(x)$$

$$= (80x - 0.1x^2) - (5000 + 20x)$$

$$= 60x - 0.1x^2 - 5000.$$

Answer $R'(x) = 4 - \frac{3}{2}\sqrt{x}$, $C'(x) = 1$, $P'(x) = 3 - \frac{3}{2}\sqrt{x}$. $R'(4) = 1$, $P'(4) = 0$, $R'(6) = 4 - \frac{3}{2}\sqrt{6} \approx 0.33$, $P'(6) = 3 - \frac{3}{2}\sqrt{6} \approx -0.67$.

The marginal profit is given by the derivative $P'(x)$. Since $P(x)$ is a combination of powers, we use the power formula to calculate its derivative.

$$P'(x) = \frac{d}{dx}(60x - 0.1x^2 - 5000) = 60 - 0.2x$$

When $x = 150$, we get $P'(x) = 60 - (0.2)(150) = 30$. Thus when 150 items are being produced, the marginal profit, that is, the extra profit per additional item when the production is increased by a small amount, is $30.

When $x = 400$, the marginal profit is $P'(400) = 60 - (0.2)(400) = -20$. Thus when 400 units are being produced, a small increase in production results in a loss (that is, a negative profit) of $20 per additional unit. ☛ **21**

☛ **21.** For the cost function in Example 1, find the marginal profit if the items can be sold for $130 each. Evaluate $P'(200)$, $P'(300)$, and $P'(400)$ and interpret their values.

The use of marginal rates is widespread in the fields of business and economics. Aside from the above examples of marginal cost, marginal revenue and marginal profit, a number of other uses occur. A few of these are summarized briefly below.

Marginal Productivity

Consider a manufcturer with a fixed amount of manufacturing capacity available but with a variable number of employees. Let u denote the amount of labor employed (for example, u might be the number of work-hours per week by the firm's employees) and let x be the amount of output (for example, the total number of items produced per week). Then x is a function of u and we can write $x = f(u)$.

If the amount of labor is given an increment Δu, then the production x changes to $x + \Delta x$ where, as usual, the increment in production is given by

$$\Delta x = f(u + \Delta u) - f(u).$$

The ratio

$$\frac{\Delta x}{\Delta u} = \frac{f(u + \Delta u) - f(u)}{\Delta u}$$

then gives the average additional production per extra unit of labor corresponding to the given increment Δu. If we allow Δu to approach zero, this ratio approaches the derivative dx/du, which is called the **marginal productivity of labor.** Thus

$$\text{Marginal Productivity} = \frac{dx}{du} = \lim_{\Delta u \to 0} \frac{\Delta x}{\Delta u} = \lim_{\Delta u \to 0} \frac{f(u + \Delta u) - f(u)}{\Delta u}.$$

So the marginal productivity of labor measures the increase in production per additional unit of labor, for instance, per additional work-hour, when a small change is made in the amount of labor employed. It is given by the derivative $f'(u)$.

Answer
$P'(x) = -0.003x^2 + 0.6x + 90.$
$P'(200) = 90,$
$P'(300) = 0$ and $P'(400) = -150.$
For a very small increases in production level, profit increases by $90 per unit when $x = 200$, is unchanged when $x = 300$ and decreases by $150 per unit when $x = 400$.

Marginal Yield

Suppose that an investor is faced with the question of how much capital to invest in a business or financial enterprise. If an amount S is invested, the investor will receive a certain return in the form of an income of, let us say, Y dollars per year. In general, the yield Y will be a function of the capital S which is invested: $Y = f(S)$. In a typical case, if S is small, the yield will be small or even zero since the enterprise would not have enough capital to operate efficiently. As S increases, the efficiency of operation improves and the yield increases rapidly. However, when S becomes very large, the efficiency may again deteriorate if the other resources necessary to the operation, such as labor or supplies, cannot increase sufficiently to keep pace with the extra capital. Thus for large S, the yield Y may again level off as S continues to increase.

The **marginal yield** is defined as the derivative dY/dS. It is obtained as the limiting value of $\Delta Y/\Delta S$ and it represents the yield per additional dollar invested when a small increase in capital is made.

Marginal Tax Rate

Let T be the amount of taxes paid by an individual or by a corporation when the income is I. Then we can write $T = f(I)$. If everything else is fixed, then a small increase ΔI in I leads to an increment in T given by $\Delta T = f(I + \Delta I) - f(I)$. The ratio $\Delta T/\Delta I$ represents the fraction to the increment of income that disappears in the form of taxation. If we allow ΔI to approach zero, this ratio approaches the derivative dT/dI, which is called the **marginal rate of taxation.** It represents the proportion of an infinitesimally small increment in income that must be paid in tax.

The marginal rate of taxation is determined by the graduated tax scales. Individuals with very low incomes pay no income tax, and below a certain income level the marginal rate is zero. As income rises, the marginal rate of taxation rises until it reaches a maximum level equal to the maximum proportion payable according to the scale. (See Example 8 in Section 12-6).

Marginal Propensities to Save and to Consume

Let I be the total income (gross national product) of a nation. Each individual among the population who receives part of this income makes a decision to spend part of his or her income on consumable goods and services and to save the rest. Let C be the total amount spent by the population on consumables and S be the total amount of savings. Then $S + C = I$.

In general, the amount of savings is determined by the national income, and we can write $S = f(I)$. The amount of consumption is then given by $C = I - f(I)$.

If the national income receives an increment ΔI, then the savings and consumption also obtain increments ΔS and ΔC, respectively, where

$$\Delta S + \Delta C = \Delta I \qquad \text{and} \qquad \Delta S = f(I + \Delta I) - f(1).$$

The ratio $\Delta S / \Delta I$ represents the fraction of the increment of income that is saved and $\Delta C / \Delta I$ represents the fraction that is consumed. Since

$$\frac{\Delta S}{\Delta I} + \frac{\Delta C}{\Delta I} = \frac{\Delta S + \Delta C}{\Delta I} = \frac{\Delta I}{\Delta I} = 1$$

the sum of these two fractions is equal to 1.

In the limit as $\Delta I \to 0$, these fractions become the corresponding derivatives. We call dS/dI the **marginal propensity to save** and dC/dI the **marginal propensity to consume.** They represent the proportions of a small increment in national income that are saved and consumed, respectively. They are related by the equation

$$\frac{dS}{dI} + \frac{dC}{dI} = 1.$$

EXERCISES 12-5

(1–4) *(Marginal Cost)* Calculate the marginal cost for the following cost functions.

1. $C(x) = 100 + 2x$

2. $C(x) = 40 + (\ln 2)x^2$

3. $C(x) = 0.0001x^3 - 0.09x^2 + 20x + 1200$

4. $C(x) = 10^{-6}x^3 - (3 \times 10^{-3})x^2 + 36x + 2000$

(5–8) *(Marginal Revenue)* Calculate the marginal revenue for the following revenue functions.

5. $R(x) = x - 0.01x^2$

6. $R(x) = 5x - 0.01x^{5/2}$

7. $R(x) = 0.1x - 10^{-3}x^2 - 10^{-5}x^{5/2}$

8. $R(x) = 100x - (\log 5)x^3(1 + \sqrt{x})$

9. *(Marginal Revenue)* If the demand equation is $x + 4p = 100$, calculate the marginal revenue, $R'(x)$.

10. *(Marginal Revenue)* If the demand equation is $\sqrt{x} + p = 10$, calculate the marginal revenue.

11. *(Marginal Revenue)* If the demand equation is $x^{3/2} + 50p = 1000$, calculate the marginal revenue when $p = 16$.

12. *(Marginal Revenue)* If the demand equation is $10p + x + 0.01x^2 = 700$, calculate the marginal revenue when $p = 10$.

13. *(Marginal Profit)* If, in Exercise 9, the cost function is $C(x) = 100 + 5x$, calculate the marginal profit.

14. *(Marginal Profit)* If, in Exercise 10, the cost function is $C(x) = 60 + x$, calculate the marginal profit.

15. *(Marginal Profit)* If, in Exercise 11, the cost function is $C(x) = 50 + x^{3/2}$, evaluate the marginal profit when:

    a. $p = 16$,   **b.** $x = 25$.

16. *(Marginal Profit)* If, in Exercise 12, the cost function is $C(x) = 1000 + 0.1x^2$, evaluate the marginal profit when:

    a. $x = 100$.   **b.** $p = 10$.

17–18. *(Maximum Profit)* In Exercises 13 and 14, find the value of x that makes $P'(x) = 0$ and calculate the corresponding profit. This represents the maximum profit that can be obtained from sale of the item in question. Find the price p that gives this maximum profit.

19. *(Marginal Revenue)* When a barber charged $4 for a haircut, she found that she gave 100 haircuts a week, on the average. When she raised her price to $5, the number of customers per week fell to 80. Assuming a linear demand equation relating price and number of customers, find the marginal revenue function. Then find the price that makes the marginal revenue zero.

20. *(Marginal Profits)* A magazine publisher finds that if he charges $1 for his magazine, he will sell 20,000 copies per month; however, if he charges $1.50, his sales will be only 15,000 copies. It costs him $0.80 to produce each issue and he has fixed overhead costs of $10,000 per month. Assuming a linear demand equation, calculate this marginal profit function and find the price of the magazine that makes the marginal profit equal to zero. Evaluate the profit itself when the price is:

a. $1.80.   b. $1.90.   c. $2.

21. *(Marginal Cost and Average Cost)* Show that if the cost function is of the form $C(x) = ax^2 + bx + c$, then at the value of x for which the marginal cost is equal to the average cost $\bar{C}(x)$, the derivative $(d/dx)\bar{C}(x)$ is zero.

*22. *(Marginal Cost and Average Cost)* Show that the result in Exercise 21 is true for any cost function $C(x)$ that is a polynomial function of x. (That is, $C(x)$ consists of a sum of powers of x, with each power multiplied by a constant.)

23. The consumption function of a certain nation is given by $C(I) = 4 + 0.36I + 0.48I^{3/4}$. Find the marginal propensities to consume and save if the national income is $I = 16$ billion.

◼ 12-6 CONTINUITY AND DIFFERENTIABILITY (OPTIONAL)

In considering the limiting value of a function $f(x)$ as x approaches c, we must consider values of x both less than and greater than c. However, in some cases the behavior of a given function is different for $x < c$ than for $x > c$. In such a case, we may wish to consider separately the possibilities that x might approach c from above or from below.

We say that x **approaches c from above** and write $x \to c^+$ if x takes a sequence of values that get closer and closer to c but always remain greater than c. (See page 492.) We say that x **approaches c from below** and write $x \to c^-$ if x takes a sequence of values that get closer and closer to c, but remain less than c. If $f(x)$ approaches the limiting value L as $x \to c^+$, we write

$$\lim_{x \to c^+} f(x) = L.$$

If $f(x)$ approaches the limiting value M as $x \to c^-$, we write

$$\lim_{x \to c^-} f(x) = M.$$

Limits of this kind are called **one-sided limits**.

EXAMPLE 1 Investigate the limiting values of $f(x) = \sqrt{x - 1}$ as x approaches 1 from above and from below.

Solution As $x \to 1^+$, $x - 1$ approaches zero through positive values. Therefore

$$\lim_{x \to 1^+} \sqrt{x - 1} = 0.$$

On the other hand, as $x \to 1^-$, $x - 1$ still approaches zero, but it is always negative. Hence $\sqrt{x - 1}$ is not defined for $x < 1$, so $\lim_{x \to 1^-} \sqrt{x - 1}$ does not exist.

The graph of $y = \sqrt{x - 1}$ is shown in Figure 9. The domain of this function does not extend to values of x less than 1, so the limit from below cannot exist.

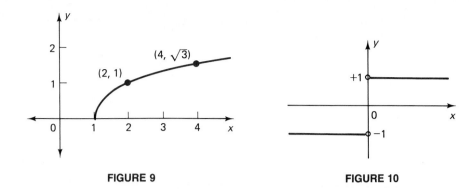

FIGURE 9

FIGURE 10

☞ **22.** Evaluate
$\lim\limits_{x\to1^+} f(x)$ and $\lim\limits_{x\to1^-} f(x)$
in the following cases:
(a) $f(x) = \sqrt{1-x}$;
(b) $f(x) = \dfrac{x-1}{|1-x|}$.

EXAMPLE 2 Investigate the limiting values of $f(x) = |x|/x$ as x approaches 0 from above and from below.

Solution For $x > 0$, $|x| = x$, and so

$$f(x) = \frac{|x|}{x} = \frac{x}{x} = 1.$$

The given function has the value 1 for all $x > 0$ and so must have the limiting value 1 as x approaches 0 from above:

$$\lim_{x\to0^+} \frac{|x|}{x} = 1.$$

For $x < 0$, $|x| = -x$, and so

$$f(x) = \frac{|x|}{x} = \frac{-x}{x} = -1.$$

(For example, when $x = -6$, $f(-6) = |-6|/(-6) = 6/(-6) = -1$.) Hence $f(x)$ is identically equal to -1 for all $x < 0$ and therefore

$$\lim_{x\to0^-} \frac{|x|}{x} = -1.$$

The graph of $y = f(x)$ is shown in Figure 10. Note that $f(x)$ is not defined for $x = 0$ and that the graph makes a jump from -1 to $+1$ as x passes from below zero to above zero. ☞ **22**

The preceding examples illustrate two basic types of behavior. In the first case, only one of the two limits from above and from below existed. In the second case, both limits existed, but their values were different from one another. In both cases, the relevant two-sided limit, $\lim\limits_{x\to c} f(x)$, does not exist. For a general $f(x)$, as illustrated in Figure 11, if the graph of $f(x)$ makes a jump at $x = c$, then the two limits from above and from below are not equal to one another. Note that $\lim\limits_{x\to c} f(x)$ exists if both $\lim\limits_{x\to c^-} f(x)$ and $\lim\limits_{x\to c^+} f(x)$ exist and are equal to one another.

Answer (a) No limit, 0 respectively;
(b) 1 and -1 respectively.

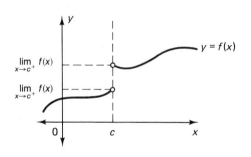

FIGURE 11

☛ **23.** Given
$$f(x) = \begin{cases} 3 - 4x & \text{if } x \geq 2 \\ x^2 - 2 & \text{if } 1 < x < 2 \\ 3 - 4x & \text{if } x < 1 \end{cases}$$
find $\lim_{x \to 1} f(x)$ and $\lim_{x \to 2} f(x)$ if they exist.

EXAMPLE 3 Given

$$f(x) = \begin{cases} 2x + 5 & \text{for } x > 3 \\ x^2 + 2 & \text{for } x \leq 3 \end{cases}$$

find $\lim_{x \to 3} f(x)$.

Solution In this case, $f(x)$ is defined by two different formulas, one for $x \leq 3$ and one for $x > 3$. So we must find the limits separately from above and below.

Since $f(x) = 2x + 5$ for $x > 3$, for the limit from above we find

$$\lim_{x \to 3^+} f(x) = \lim_{x \to 3} (2x + 5) = 2(3) + 5 = 11.$$

Similarly, for $x < 3$, we have $f(x) = x^2 + 2$ and therefore, for the limit from below,

$$\lim_{x \to 3^-} f(x) = \lim_{x \to 3} (x^2 + 2) = 3^2 + 2 = 11.$$

Since $\lim_{x \to 3^+} f(x) = \lim_{x \to 3^-} f(x) = 11$, it follows that $\lim_{x \to 3} f(x)$ exists and is equal to 11.

The graph of $f(x)$ in this case is shown in Figure 12. Note that the graph changes type at $x = 3$, but it does not make a jump at this point. ☛ **23**

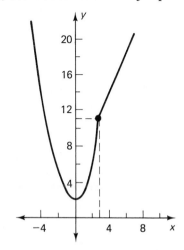

FIGURE 12

Answer $\lim_{x \to 1} f(x) = -1$; $\lim_{x \to 2} f(x)$ does not exist as $\lim_{x \to 2^+} f(x) \neq \lim_{x \to 2^-} f(x)$.

Let us recall the definition of the continuity of a function from Section 12-2.

DEFINITION A function $f(x)$ is said to be **continuous** at a point $x = c$ if the following three conditions are met.

1. $f(x)$ is defined at $x = c$. That is, $f(c)$ is well defined.

2. $\lim\limits_{x \to c} f(x)$ exists.

3. $\lim\limits_{x \to c} f(x) = f(c)$.

If any one of these three conditions is not satisfied, then the function is said to be **discontinuous** at $x = c$. If the two limits of $f(x)$ as x approaches c from above and below are different from one another, we say that $f(x)$ has a **jump discontinuity** at $x = c$.

EXAMPLE 4 The function $f(x) = |x|$ is continuous at $x = 0$. We note that $f(0) = |0| = 0$, so that condition 1 is satisfied. Also $\lim\limits_{x \to 0} f(x)$ exists since, as x approaches zero, $|x|$ approaches the limit zero. Finally, condition 3 is met, since $\lim\limits_{x \to 0} f(x)$ and $f(0)$ are equal to one another, both being zero. The graph of $y = |x|$ is shown in Figure 13. The graph clearly passes through $x = 0$ without a break. It does have a corner (or change of slope) at $x = 0$, but this does not make it discontinuous.

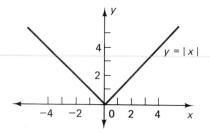

FIGURE 13

In Example 2, we discussed the function $f(x) = |x|/x$. This function is discontinuous at $x = 0$ because $\lim\limits_{x \to 0} f(x)$ does not exist: the limits from above and from below are not equal to one another. The graph makes a jump from -1 to $+1$ as x passes through 0. Another example of a discontinuous function is given in Example 5.

EXAMPLE 5 Given

$$f(x) = \begin{cases} \dfrac{x^2 - 9}{x - 3} & \text{if } x \neq 3 \\ 5 & \text{if } x = 3. \end{cases}$$

Is $f(x)$ continuous at $x = 3$?

Solution

Condition (1) Clearly, $f(x)$ is defined at $x = 3$ and $f(3) = 5$.

Condition (2) $\displaystyle \lim_{x \to 3} f(x) = \lim_{x \to 3} \frac{x^2 - 9}{x - 3} = \lim_{x \to 3} \frac{(x - 3)(x + 3)}{x - 3}$

$$= \lim_{x \to 3} (x + 3) = 3 + 3 = 6$$

Condition (3) $\displaystyle \lim_{x \to 3} f(x) = 6$ and $f(3) = 5$ are not equal.

In this case, the first two conditions are satisfied, but the third condition is *not* met, so the given function is discontinuous at $x = 3$. This is shown graphically in Figure 14. The graph of $f(x)$ has a break at $x = 3$ and the isolated point $(3, 5)$ on the graph is not joined continuously to the rest of the graph. ☞ **24**

☞ **24.** For what values of h and k is the following function continuous at $x = 2$?

$$f(x) = \begin{cases} \dfrac{(x - 2)^2}{x^2 - 4} & \text{if } x > 2 \\ h & \text{if } x = 2 \\ 2x + k & \text{if } x < 2 \end{cases}$$

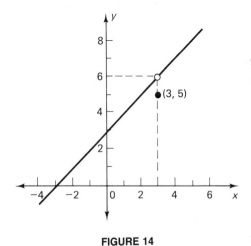

FIGURE 14

It may seem at first sight that discontinuous functions would be of little relevance to practical problems. This is not the case however, as the following example illustrates.

EXAMPLE 6 *(Sugar-cost Function)* A wholesaler sells sugar at 50¢ a pound for amounts up to 100 pounds. For orders greater than 100 and up to 200 pounds, the charge is 45¢ a pound, and for orders over 200 pounds the charge is 40¢ a pound. Let $y = f(x)$ denote the cost in dollars of x pounds of sugar. Then for $x \le 100$, $y = (0.5)x$. For $100 < x \le 200$, the cost is $0.45 per pound, so $y = 0.45x$. Finally, for $x > 200$, $y = 0.4x$. The graph of this function is shown in Figure 15. Clearly, the function is discontinuous at $x = 100$ and 200.

Answer $h = 0$, $k = -4$.

In Section 12-3, we defined the term differentiability: a function $f(x)$ is said to be differentiable at the point x if the derivative

25. *(More difficult)* Using the definition of $f'(0)$ as a limit, show that the function $f(x) = x|x|$ is differentiable at $x = 0$.

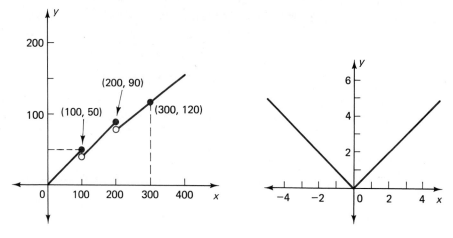

FIGURE 15

FIGURE 16

$$f'(x) = \lim_{\Delta x \to 0} \frac{f(x + \Delta x) - f(x)}{\Delta x}$$

exists at that point.

EXAMPLE 7 Show that the function $f(x) = |x|$ is not differentiable at $x = 0$.

Solution We must take $x = 0$, so that $f(x) = f(0) = 0$ and $f(x + \Delta x) = f(0 + \Delta x) = f(\Delta x) = |\Delta x|$. So

$$\Delta y = f(x + \Delta x) - f(x) = |\Delta x| - 0 = |\Delta x|.$$

Then

$$\frac{dy}{dx} = \lim_{\Delta x \to 0} \frac{\Delta y}{\Delta x} = \lim_{\Delta x \to 0} \frac{|\Delta x|}{\Delta x}.$$

But in Example 2, we discussed this limit and we showed that the limit does not exist. In fact, the two limits from above and below exist but are unequal.

$$\lim_{\Delta x \to 0^+} \frac{|\Delta x|}{\Delta x} = 1, \qquad \lim_{\Delta x \to 0^-} \frac{|\Delta x|}{\Delta x} = -1$$

Therefore f is not differentiable at $x = 0$.

The graph of $y = |x|$ is shown in Figure 16. For $x > 0$, the graph has a constant slope of 1, while $x < 0$ it has a constant slope of -1. For $x = 0$, there is no slope since the graph has a corner at this value of x. This is the geometrical reason why $|x|$ is not differentiable at $x = 0$. **25**

Answer $\lim_{\Delta x \to 0} \dfrac{f(0 + \Delta x) - f(0)}{\Delta x}$

$= \lim_{\Delta x \to 0} \dfrac{\Delta x |\Delta x| - 0|0|}{\Delta x}$

$= \lim_{\Delta x \to 0} \dfrac{\Delta x |\Delta x|}{\Delta x}$

$= \lim_{\Delta x \to 0} |\Delta x|$, which exists and equals zero.

A function $y = f(x)$ is differentiable at a certain value of x if its graph is "smooth" at the corresponding point (x, y) by which we mean that the graph has a well-defined tangent line with a well-defined slope. If the graph has a corner at the point (x, y), then $f(x)$ is not differentiable at that value of x. The preceding example illustrates such a function.

EXAMPLE 8 *(Income Tax)* In the mythical country of Erehwon, the fortunate* inhabitants pay no income tax on their first $10,000 of taxable earnings. The graduated tax rates for higher income levels are given in Table 5. Let I denote taxable earnings and T denote the amount of taxes. Express T as a function of I, draw the graph of this function and discuss its differentiability.

TABLE 5

Taxable Earnings	Tax Rate
$10,001–$20,000	20%
$20,001–$30,000	30%
Over $30,000	40%

Solution For $0 \le I \le 10,000$, $T = 0$. When $10,000 < I \le 20,000$, the amount by which I exceeds 10,000 is taxed at 20%. Therefore, in this range

$$T = 0.2(I - 10,000) = 0.2I - 2000.$$

When $I = 20,000$, $T = 0.2(20,000 - 10,000) = 2000$, so the tax on $20,000 is $2000.

When $20,000 < I \le 30,000$, the amount by which I exceeds 20,000 is taxed at 30%. Thus, in this range,

$$T = 2000 + 0.3(I - 20,000) = 0.3I - 4000.$$

When $I = 30,000$, $T = 0.3(30,000) - 4000 = 5000$, so the tax is $5000.

Continuing this way, we construct a table of values of T as a function of I (see Table 6) and the graph as shown in Figure 17. ☛ 26

TABLE 6

I	T
$I \le 10,000$	0
$10,000 < I \le 20,000$	$0.2I - 2000$
$20,000$	2000
$20,000 < I \le 30,000$	$0.3I - 4000$
$30,000$	5000
$I > 30,000$	$0.4I - 7000$

The graph consists of a number of line segments. Clearly, the amount of tax is a continuous function of taxable earnings, but it is not differentiable at

---

*1 Erehwon dollar = 5 U.S. dollars.

☛ **26.** There is a move to "rationalize" the tax structure in Erehwon by taxing all income above $10,000 and up to and including $30,000 at 25% and all income above $30,000 at 40%. Construct the new version of Table 6 in this case.

Answer

I	T
$I \le 10,000$	0
$10,000 < I \le 30,000$	$0.25I - 2500$
$30,000$	5000
$I > 30,000$	$0.4I - 7000$

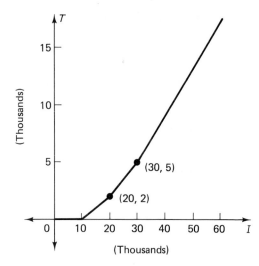

FIGURE 17

27. *(More difficult)* Show that $f(x) = x^{1/3}$ is not differentiable at $x = 0$. (*Hint:* Go back to the definition of the derivative, $f'(0)$.)

Answer $\lim_{\Delta x \to 0} \dfrac{f(0 + \Delta x) - f(0)}{\Delta x}$

$= \lim_{\Delta x \to 0} \dfrac{(\Delta x)^{1/3} - 0^{1/3}}{\Delta x}$

$= \lim_{\Delta x \to 0} (\Delta x)^{-2/3}$,

which does not exist. (It is *not* sufficient to say that f is not differentiable at $x = 0$ because $f'(x) = \frac{1}{3}x^{-2/3}$, which does not exist when $x = 0$. All this shows is that $\lim_{x \to 0} f'(x)$ does not exist and this is not the same as $f'(0)$.)

the points where the graph has corners. These occur at the values of I that mark the divisions on the graduated tax scale. In between these dividing points, T is differentiable, and its derivative gives the marginal tax rate.

Another case in which a function is not differentiable arises when the tangent line at a certain point becomes vertical. In such a case, the slope of the tangent line is not defined at the point in question, so the function is not differentiable at that value of x. For example, we leave it as an exercise to show that the function $f(x) = x^{1/3}$ is not differentiable at $x = 0$. **27**

We observe that in Example 7 we have a function that is defined and is continuous for all values of x, but that is not differentible for all x. At $x = 0$, $f(x) = |x|$ is continuous but not differentiable. Clearly, therefore, *the fact that a function is continuous does not imply that it is differentiable*. However, the converse of this implication is true: *If $f(x)$ is differentiable at a point $x = c$, then it is continuous at $x = c$.* Thus differentiability implies continuity, but not conversely. We shall not give a proof of this result, although it is an important one.

EXERCISES 12-6

(1–4) Use the graph of $f(x)$ on the facing page to estimate the following limits.

1. a. $\lim_{x \to 2^+} f(x)$ **b.** $\lim_{x \to 2^-} f(x)$ **c.** $\lim_{x \to 2} f(x)$

2. a. $\lim_{x \to -3^+} f(x)$ **b.** $\lim_{x \to -3^-} f(x)$ **c.** $\lim_{x \to -3} f(x)$

3. a. $\lim_{x \to 1^+} f(x)$ **b.** $\lim_{x \to 1^-} f(x)$ **c.** $\lim_{x \to 1} f(x)$

4. a. $\lim_{x \to 3^+} f(x)$ **b.** $\lim_{x \to 3^-} f(x)$ **c.** $\lim_{x \to 3} f(x)$

(5–16) Evaluate the following one-sided limits.

5. $\lim_{x \to 1^+} \sqrt{x - 1}$ **6.** $\lim_{x \to 1/2^-} \sqrt{1 - 2x}$

7. $\displaystyle\lim_{x \to 4/3^+} \sqrt{4 - 3x}$

8. $\displaystyle\lim_{x \to -1^-} \sqrt{x + 1}$

9. $\displaystyle\lim_{x \to 1^+} \frac{|x - 1|}{x - 1}$

10. $\displaystyle\lim_{x \to -1^-} \frac{x + 1}{|x + 1|}$

11. $\displaystyle\lim_{x \to 3^-} \frac{|x - 3|}{9 - x^2}$

12. $\displaystyle\lim_{x \to 2^-} \frac{x^2 - x - 2}{|x - 2|}$

13. $\displaystyle\lim_{x \to 0^-} \frac{1}{x^2}$

14. $\displaystyle\lim_{x \to -1^+} \frac{x - 1}{x + 1}$

15. $\displaystyle\lim_{x \to 1^-} \frac{2 - x^2}{|x - 1|}$

16. $\displaystyle\lim_{x \to 2^-} \frac{x^2 - 6}{(x - 2)^3}$

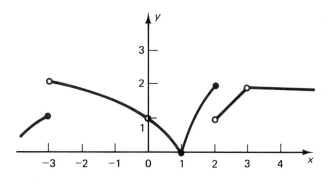

(17–22) Discuss the continuity of the following functions at $x = 0$ and sketch their graphs.

17. $f(x) = \dfrac{x^2}{x}$

18. $g(x) = \sqrt{x^2}$

19. $h(x) = \begin{cases} |x| & \text{for } x \neq 0 \\ 1 & \text{for } x = 0 \end{cases}$

20. $F(x) = \begin{cases} \dfrac{|x|}{x} & \text{if } x \neq 0 \\ 0 & \text{if } x = 0 \end{cases}$

21. $G(x) = \begin{cases} 0 & \text{if } x < 0 \\ 1 & \text{if } x > 0 \end{cases}$

22. $H(x) = \begin{cases} 0 & \text{if } x < 0 \\ x & \text{if } x > 0 \end{cases}$

(23–28) Discuss the continuity of the following functions at the indicated points and sketch their graphs.

23. $f(x) = x^2 + 4x + 7, \quad x = 1$

24. $g(x) = \dfrac{2x - 1}{x - 1}; \quad x = 1$

25. $f(x) = \begin{cases} \dfrac{|x - 3|}{x - 3} & \text{if } x \neq 3 \\ 0 & \text{if } x = 3 \end{cases}; \quad x = 3$

26. $G(x) = \begin{cases} \dfrac{x^2 - 4}{x - 2} & \text{for } x \neq 2 \\ 4 & \text{for } x = 2 \end{cases}; \quad x = 2$

27. $f(x) = \begin{cases} 5x + 7 & \text{for } x > 2 \\ 2x + 3 & \text{for } x \leq 2 \end{cases}; \quad x = 2$

28. $f(x) = \begin{cases} 3x + 5 & \text{for } x < 1 \\ 10 - 2x & \text{for } x > 1 \end{cases}; \quad x = 1$

(29–35) Find the values of x (if any) for which the following functions are not continuous.

29. $f(x) = \dfrac{x + 1}{x - 2}$

30. $f(x) = \dfrac{x^2 + 1}{x^2 - 9}$

31. $f(x) = \dfrac{x^2 + 5}{x^2 - x - 6}$

32. $f(x) = \dfrac{x^2 - 4}{x^2 + 4}$

33. $f(x) = \dfrac{x^2 + 2x + 3}{x^2 + 2x + 4}$

34. $f(x) = \begin{cases} \dfrac{1}{x - 2} & \text{if } x > 3 \\ 2x - 5 & \text{if } x \leq 3 \end{cases}$

35. $f(x) = \begin{cases} \dfrac{1}{x - 2} & \text{if } x \leq 3 \\ 2x + 1 & \text{if } x > 3 \end{cases}$

(36–37) Find the value of h in the following exercises so that $f(x)$ is continuous at $x = 1$.

36. $f(x) = \begin{cases} x^2 - 3x + 4 & \text{if } x \neq 1 \\ h & \text{if } x = 1 \end{cases}$

37. $f(x) = \begin{cases} hx + 3 & \text{if } x \geq 1 \\ 3 - hx & \text{if } x < 1 \end{cases}$

(38–41) Find the values of x for which the following functions are not differentiable.

*38. $f(x) = x^{2/3}$

*39. $f(x) = \begin{cases} 0 & \text{for } x \leq 0 \\ x & \text{for } x > 0 \end{cases}$

*40. $f(x) = \begin{cases} 0 & \text{for } x \leq 0 \\ x^2 & \text{for } x > 0 \end{cases}$

*41. $f(x) = (x - 1)^{1/2}$

42. *(Electricity Cost Function)* A power company charges 10¢ per unit of electricity for the first 50 units used by a household each month and 3¢ per unit for amounts over this. If $c(x)$ denotes the cost of x units per month, discuss the continuity and differentiability of $c(x)$ and draw this graph.

43. *(Cost of an Employee)* Let $f(x)$ denote the cost per week to a firm of hiring an employee who works x hours per week. This consists of (1) an overhead cost of $20, (2) a wage of $6 per hour for the first 35 hours, (3) an overtime wage of $9 per hour for hours worked over 35 hours but not more than 45 hours, and (4) a double overtime wage of $12 per hour for hours worked in excess of 45. Discuss the continuity and differentiability of $f(x)$ and draw its graph.

44. *(Income Tax)* In a certain country the graduated tax rates are as follows: 10% on the first 2000 dinars (the unit of currency); 25% on the next 4000 dinars; and 40% on any additional income. Express the amount of income tax as a function of the income and draw the graph of this function.

45. *(Income Tax)* In the country of Exercise 44, it is proposed to change the tax brackets to the following: no tax on the first 2000 dinars, 30% on the next 4000, and 50% on any additional income. Express the change in an individual's income tax as a function of his or her income and draw the graph of this function.

46. *(Discontinuous Cost Function)* For production levels up to 1000 units per week, a company's cost function is $C(x) = 5000 + 8x$, where x is the production level. If $x > 1000$, a new assembly line must be opened and the cost function becomes $C(x) = 9000 + 6x$. If the units are sold at $16 each, construct the firm's profit function. Graph this function and discuss its continuity.

47. *(Postage Costs)* A first-class letter costs 12¢ per ounce or fraction thereof. Let $f(x)$ denote the cost of mailing a letter weighing x ounces. Discuss the continuity and differentiability of $f(x)$ and draw its graph for $0 < x \leq 8$.

CHAPTER REVIEW

Key Terms, Symbols, and Concepts

12.1 Increment, Δx, Δy.
Average rate of change of y with respect to x: $\Delta y / \Delta x$.
Average speed.

12.2 Instantaneous speed.
Limit (or limiting value): $\lim_{x \to c} f(x)$.
Continuous function.

12.3 Derivative: For $y = f(x)$: $\dfrac{dy}{dx}, \dfrac{df}{dx}, \dfrac{d}{dx} y, y', f'(x)$.
Differentiable, differentiation.
Slope of the tangent line.

12.4 Power formulas for derivatives.

12.5 Marginal cost, $C'(x)$. Average cost, $\overline{C}(x) = C(x)/x$.
Marginal revenue, $R'(x)$. Marginal profit, $P'(x)$.
Marginal productivity, marginal yield, marginal tax rate.
Marginal propensities to save and to consume.

12.6 One-sided limits:
limits from above (from the right), $\lim_{x \to c^+} f(x)$;
limit from below (from the left), $\lim_{x \to c^-} f(x)$;
Continuous, discontinuous, jump discontinuity.

Formulas

$\Delta x = x_2 - x_1$.

If $y = f(x)$, then $\Delta y = f(x + \Delta x) - f(x)$.

Average speed $= \dfrac{\Delta s}{\Delta t}$. Instantaneous speed $= \lim_{\Delta t \to 0} \dfrac{\Delta s}{\Delta t}$.

Limit theorems:

$$\lim_{x \to c} (mx + b) = mc + b.$$

$$\lim_{x \to c} bf(x) = b \lim_{x \to c} f(x).$$

$$\lim_{x \to c} [f(x)]^n = [\lim_{x \to c} f(x)]^n.$$

$$\lim_{x \to c} [f(x) \pm g(x)] = \lim_{x \to c} f(x) \pm \lim_{x \to c} g(x).$$

$$\lim_{x \to c} [f(x)g(x)] = \lim_{x \to c} f(x) \cdot \lim_{x \to c} g(x).$$

$$\lim_{x \to c} \frac{f(x)}{g(x)} = \frac{\lim_{x \to c} f(x)}{\lim_{x \to c} g(x)}$$

For $y = f(x)$: $\quad \dfrac{dy}{dx} = f'(x) = \lim\limits_{\Delta x \to 0} \dfrac{f(x + \Delta x) - f(x)}{\Delta x}$.

Power formula: $\quad$ If $y = x^n \quad \dfrac{dy}{dx} = nx^{n-1}$.

Differentiation theorems:

$$\dfrac{d}{dx}(cu) = c\,\dfrac{du}{dx}, \quad \text{where } c \text{ is a constant.}$$

$$\dfrac{d}{dx}(u + v) = \dfrac{du}{dx} + \dfrac{dv}{dx}.$$

$$P(x) = R(x) - C(x), \; P'(x) = R'(x) - C'(x).$$

REVIEW EXERCISES FOR CHAPTER 12

1. State whether each of the following is true or false. Replace each false statement with a corresonding true statement.

   a. An increment in the independent variable must be positive.

   b. An increment in the dependent variable can be positive, negative, or zero.

   c. A function must be defined at a point if the limit of the function exists at this point.

   d. $x/x = 1$ for all x.

   e. If $g(x) \to 0$ as $x \to c$ and if $\lim\limits_{x \to c} [f(x)/g(x)]$ exists, then $f(x)$ must approach zero as $x \to c$.

   f. A function must be defined at a point if the derivative of the function exists at this point.

   g. The derivative of y with respect to x represents the average rate of change of y with respect to x.

   h. A function $f(x)$ is continuous at $x = c$ if and only if $\lim\limits_{x \to c} f(x)$ exists.

   i. If a function is continuous at a point then it is differentiable at this point.

   j. The derivative of the sum of any number of functions is equal to the sum of their derivatives.

   k. If $f(x) = |x|$, then $f'(0) = 0$.

2. Find Δy when $x = 2$ and $\Delta x = 0.5$ for the function $y = x^3 - 2x^{-1}$.

3. (Cost Function) For the cost function $C(x) = 3000 + 10x + 0.1x^2$, find the increment in cost when production is increased from 100 units to 120 units. Find the average cost per additional unit.

4. (Falling Object) For an object falling under gravity, find the average velocity between $t = 4$ and $t = 5$ seconds ($t = 0$ is the instant at which the object is released).

(5–16) Evaluate the following limits.

5. $\lim\limits_{x \to 5} \dfrac{x^2 - 25}{x + 15}$

6. $\lim\limits_{x \to 2} \dfrac{x^2 + x - 6}{x^2 - 4}$

7. $\lim\limits_{x \to 3} \dfrac{x^2 - 9}{x^2 - 6x + 9}$

8. $\lim\limits_{x \to -1} \dfrac{x^2 - 1}{x^2 + 2x + 1}$

9. $\lim\limits_{x \to 1} \dfrac{x^2 - 2x + 1}{x^2 - 1}$

10. $\lim\limits_{x \to -2} \dfrac{x^2 + 5 + 6}{x^2 - x - 6}$

11. $\lim\limits_{x \to 2} \dfrac{\sqrt{x + 7} - 3}{x + 2}$

12. $\lim\limits_{x \to 1} \dfrac{\sqrt{x + 3} + 2}{x - 2}$

13. $\lim\limits_{x \to 3} \dfrac{\sqrt{x + 1} - 2}{\sqrt{x + 6} + 3}$

14. $\lim\limits_{x \to a} \dfrac{\sqrt{x} - \sqrt{a}}{x - a}$

15. $\lim\limits_{h \to 0} \dfrac{\sqrt{x + h} - \sqrt{x}}{h}$

16. $\lim\limits_{x \to 1} \dfrac{\sqrt{x + 3} - 2}{\sqrt{2 - x} - x}$

(17–18) Find the derivatives of the following functions by using the definition of the derivative as a limit.

17. $f(x) = (x + 1)^{-2}$

18. $g(x) = (x + 1)^{-1/2}$

(19–30) Differentiate the following functions with respect to the argument given.

19. $x^{3/2} - 2x^{-3/2}$

20. $\dfrac{(x + 1)(2x - 1)}{\sqrt{x}}$

21. $x^3\sqrt{x}$

22. $\dfrac{x^2 + 1}{x^{\sqrt{2}}}$

23. $\dfrac{\sqrt{x}}{\sqrt[3]{x}}$

24. $\dfrac{\sqrt[6]{x^5}}{\sqrt{x}}$

25. $\sqrt{x} \cdot \sqrt[3]{x}$

26. $\sqrt[3]{x^2} \cdot \sqrt{x}$

27. $(t^2 + 1)(3t - 2)$

28. $\dfrac{(u + 2)(u - 3)}{u^2}$

29. $\dfrac{(2y^2 - 1)(y^2 + 2)}{2y^3}$

30. $(3x + 1)^2(x + 2)$

(31–32) *(Marginal Cost)* Find the marginal cost for the following cost functions.

31. $C(x) = 500 + 10x^2$

32. $C(x) = (2 \times 10^{-4})x^3 - 0.1x^2 + 25x + 2000$

33. *(Marginal Revenue)* If the demand equation is $x + 20p = 1000$, calculate the marginal revenue.

34. *(Marginal Revenue)* If the demand equation is $x^2 + 400p = 10{,}000$, calculate the marginal revenue.

35–36. *(Marginal Profit)* Calculate the marginal profit in Exercises 33 and 34 if the cost function is $C(x) = 1000 + 10x$.

37. *(Marginal Price)* If the demand function is given by $p = f(x)$, then dp/dx is called *marginal price function*. The demand equation of a certain product is given by $p = 2024 - 2x - x^2$. Determine the marginal price at a demand level of 30 units.

38. *(Marginal Price)* The demand equation of a certain commodity is $p = 50/(x + 1)$. Determine the marginal price function.

39. *(Marginal Demand)* If the demand relation is given by $x = f(p)$, then dx/dp is called the *marginal demand*. If the demand equation of a certain product is $p^2 + x = 20$, find the marginal demand at a price level of $p = 2$. Interpret your result.

40. *(Physical Productivity)* The *physical productivity* ρ is defined as the total physical output by a given number of workers or machines, and is thus a function of the number x of workers or machines. For a certain manufacturing firm, $\rho = 500(x + 1)^2 - 500$. Determine the marginal physical productivity $d\rho/dx$ when $x = 3$.

(41–43) Determine whether the following functions are continuous at the indicated points.

41. $f(x) = \dfrac{x^3 + 3x}{x}; \quad x = 0$

42. $f(x) = \begin{cases} \dfrac{|x^2 - 9|}{x - 3} & \text{if } x \neq 3 \\ 6 & \text{if } x = 3 \end{cases}; \quad x = 3$

43. Determine the value of h if

$$f(x) = \begin{cases} x^2 + h & \text{for } x \neq 1 \\ 3 & \text{for } x = 1 \end{cases}$$

is continuous at $x = 1$.

44. If $f(x) = \dfrac{x^2 - 4}{x - 2}$ for $x \neq 2$ and $f(x)$ is continuous at $x = 2$, find $f(2)$.

45. *(Quantity Discount)* A supplier of bulk sugar sells it at $2 per kilogram for the first 20 kilograms in an order, $1.75 per kilogram for the next 20 kilograms, and $1.50 per kilogram for amounts in excess of 40 kilograms. Give the function $\overline{C}(x)$ that provides the average cost per kilogram for an order of size x kilograms. Is this function continuous and/or differentiable at all values of x?

46. *(Discontinuous Interest Rate)* A bank quotes different interest rates depending on the minimum monthly balance in the account. For amounts less than $1000 their interest rate is 4%; for amounts of $1000 or more and up to $5000 the rate is 6%; for amounts of $5000 or more the rate is 7%. Graph the interest rate as a function of the minimum monthly balance in the account and discuss its continuity and differentiability.

CHAPTER **13**

Calculation of Derivatives

Chapter Objectives

13-1 DERIVATIVES OF PRODUCTS AND QUOTIENTS
  (a) The product rule and its use;
  (b) Application to marginal revenue and the demand law;
  (c) The quotient rule and its use;
  (d) Application to marginal average cost.

13-2 THE CHAIN RULE
  (a) The chain rule and its use; formula for derivative of $[u(x)]^n$;
  (b) The decomposition of functions into "inside" and "outside" parts;
  (c) Combinations of the chain rule with the product and/or quotient rules;
  (d) Related rates.

13-3 DERIVATIVES OF EXPONENTIAL AND LOGARITHMIC FUNCTIONS
  (a) Geometrical definition of the number e;
  (b) Derivatives of e^x and $\ln x$;
  (c) Combination of product and quotient rules with exponentials and logarithms;
  (d) Chain rule formulas for derivatives of $e^{u(x)}$ and $\ln u(x)$;
  (e) Simplification of expressions involving logarithms prior to differentiation.

13-4 HIGHER DERIVATIVES
  (a) Definition of second, third, and higher derivatives in general;
  (b) Computation of higher derivatives.

CHAPTER REVIEW

⬛ 13-1 DERIVATIVES OF PRODUCTS AND QUOTIENTS

In this section, we shall prove and explain the use of two important theorems that provide useful techniques for differentiating complicated functions.

THEOREM 1 (PRODUCT RULE) If $u(x)$ and $v(x)$ are any two differentiable functions, of x, then

$$\frac{d}{dx}(u \cdot v) = u\frac{dv}{dx} + v\frac{du}{dx}.$$

POWER RULE

$1d2 + 2d1$

$v\underline{du} + u\underline{dv}$

That is,

$$(uv)' = uv' + vu'.$$

In words, *the derivative of the product of two functions is equal to the first function times the derivative of the second plus the second function times the derivative of the first*.

EXAMPLE 1 Find y' if $y = (5x^2 - 3x)(2x^3 + 8x + 7)$.

Solution The given function y can be written as a product $y = uv$ if we let

$$u = 5x^2 - 3x \qquad \text{and} \qquad v = 2x^3 + 8x + 7.$$

Then, by the methods of Section 12-4, we see that

$$u' = 10x - 3 \qquad \text{and} \qquad v' = 6x^2 + 8.$$

Therefore, by the product rule,

$$y' = uv' + vu'$$
$$= (5x^2 - 3x)(6x^2 + 8) + (2x^3 + 8x + 7)(10x - 3)$$
$$= \underline{50x^4 - 24x^3 + 120x^2 + 22x - 21.}$$

Notice the procedure here:

1. Identify u and v such that $y = uv$.

2. Calculate u' and v'.

3. Use the product rule to find y'.

In Example 1, we do not actually need the product rule in order to differentiate the given function. We could have calculated y' by multiplying out the right side and expressing y as a sum of powers of x.

$$y = (5x^2 - 3x)(2x^3 + 8x + 7)$$
$$= 10x^5 - 6x^4 + 40x^3 + 11x^2 - 21x$$

$$y' = 10(5x^4) - 6(4x^3) + 40(3x^2) + 11(2x) - 21(1)$$

$$= 50x^4 - 24x^3 + 120x^2 + 22x - 21$$

The examples that we shall now give on the use of the product rule can likewise be solved using the methods of Chapter 12. However, later we shall come across functions for which such an alternative method does not exist. For these functions, it will be essential to use the product rule in order to find their derivatives.

EXAMPLE 2 Given $f(t) = (2\sqrt{t} + 1)(t^2 + 3)$, find $f'(t)$.

Solution We use the product rule with $u = 2\sqrt{t} + 1 = 2t^{1/2} + 1$ and $v = t^2 + 3$. Then $u'(t) = 2 \cdot \frac{1}{2}t^{-1/2} = t^{-1/2}$ and $v'(t) = 2t$. Therefore

$$f'(t) = uv' + vu'$$

$$= (2t^{1/2} + 1)(2t) + (t^2 + 3)(t^{-1/2})$$

$$= 4t^{3/2} + 2t + t^{3/2} + 3t^{-1/2}$$

$$= 5t^{3/2} + 2t + \frac{3}{\sqrt{t}}. \quad \text{☞} \, 1$$

The demand equation gives the price p at which a quantity x of a certain item can be sold during a given span of time. In general, we can write $p = f(x)$. The revenue from the sale of this number of items is then

$$R = xp.$$

Since R is given as a product of two quantities, the marginal revenue, which is the derivative of R with respect to x, can be obtained from the product rule.

$$\frac{dR}{dx} = p\frac{d}{dx}(x) + x\frac{d}{dx}(p) = 1 \cdot p + x\frac{dp}{dx} = p + x\frac{dp}{dx}$$

The derivative dp/dx can be computed from the demand relation. It is the change in price per unit increase in demand that is needed to produce a very small change in demand.

EXAMPLE 3 *(Marginal Revenue)* If the demand equation is linear, we have

$$p = a - bx$$

where a and b are two positive constants. Then $dp/dx = -b$ and the marginal revenue is

$$\frac{dR}{dx} = p + x\frac{dp}{dx} = a - bx + x(-b) = a - 2bx.$$

Note that the marginal revenue in this example can be calculated directly.

$$R = xp = x(a - bx) = ax - bx^2$$

Therefore $R'(x) = a - 2bx$, as before. ☞ **2**

Note. The product rule extends in a straightforward way to the product of more than two functions. For the product of three functions it becomes

$$(uvw)' = u'vw + uv'w + uvw'.$$

☛ 3. Use the quotient rule to differentiate the following functions:

(a) $\dfrac{x}{x - 1}$; (b) $\dfrac{2t + 5}{2t - 5}$;

(c) $\dfrac{1 - u^3}{1 + u^3}$.

THEOREM 2 (QUOTIENT RULE) If $u(x)$ and $v(x)$ are differentiable functions of x, then

$$\frac{d}{dx}\left(\frac{u}{v}\right) = \frac{v\,\dfrac{du}{dx} - u\,\dfrac{dv}{dx}}{v^2}$$

or

$$\left(\frac{u}{v}\right)' = \frac{vu' - uv'}{v^2}.$$

That is, *the derivative of a quotient of two functions is equal to the denominator times the derivative of the numerator minus the numerator times the derivative of the denominator all divided by the square of the denominator.*

EXAMPLE 4 Find y' if

$$y = \frac{x^2 + 1}{x^3 + 4}.$$

Solution First we must identify u and v such that $y = u/v$. This is easy: $u = x^2 + 1$ and $v = x^3 + 4$. Then we have $u' = 2x$ and $v' = 3x^2$. Finally, from the quotient rule

$$y' = \frac{u'v - uv'}{v^2} = \frac{(x^3 + 4)(2x) - (x^2 + 1)(3x^2)}{(x^3 + 4)^2}$$

$$= \frac{2x^4 + 8x - (3x^4 + 3x^2)}{(x^3 + 4)^2} = \frac{-x^4 - 3x^2 + 8x}{(x^3 + 4)^2} \qquad ☛ 3$$

Answer

(a) $\dfrac{(x - 1)\cdot 1 - x\cdot 1}{(x - 1)^2}$

$= \dfrac{-1}{(x - 1)^2}$;

(b) $\dfrac{(2t - 5)\cdot 2 - (2t + 5)\cdot 2}{(2t - 5)^2}$

$= \dfrac{-20}{(2t - 5)^2}$;

(c) $\dfrac{(1 + u^3)\cdot(-3u^2) - (1 - u^3)(3u^2)}{(1 + u^3)^2}$

$= \dfrac{-6u^2}{(1 + u^3)^2}$.

EXAMPLE 5 *(Per Capita Income)* The gross national product of a certain country is increasing with time according to the formula $I = 100 + t$ (billions of dollars). The population at time t is $P = 75 + 2t$ (millions). Find the rate of change of per capita income at time t.

Solution The per capita income, which we denote by y is equal to the GNP divided by the population size:

$$y = \frac{I}{P} = \frac{100 + t}{75 + 2t} \quad \text{(thousands of dollars).}$$

☛ 4. In Example 5, calculate the rate of growth of per capita income if the population growth is reduced to $P = 75 + t$ millions at time t.

To differentiate this we use the quotient rule with $y = u/v$, where $u = 100 + t$ and $v = 75 + 2t$. Then $du/dt = 1$ and $dv/dt = 2$. From the quotient rule,

$$\frac{dy}{dt} = \frac{u'v - uv'}{v^2} = \frac{(75 + 2t) \cdot 1 - (100 + t) \cdot 2}{(75 + 2t)^2} = \frac{-125}{(75 + 2t)^2}. \qquad ☛ 4$$

Answer $\dfrac{dy}{dt} = \dfrac{-25}{(75 + t)^2}$.

EXAMPLE 6 Find dy/dx if $y = \dfrac{(x+1)(x^3-2x)}{x-1}$.

Solution First write $y = u/v$, as a quotient, with $u = (x+1)(x^3-2x)$ and $v = x - 1$. Then from the quotient rule,

$$y' = \frac{u'v - uv'}{v^2}.$$

We have $v' = 1$ immediately, but to find u' we use the product rule. We write $u = u_1 v_1$ where $u_1 = x + 1$ and $v_1 = x^3 - 2x$. Then $u_1' = 1$ and $v_1' = 3x^2 - 2$, so that

$$u' = u_1 v_1' + v_1 u_1' = (x+1)(3x^2-2) + (x^3-2x)\cdot 1$$

$$= 4x^3 + 3x^2 - 4x - 2.$$

Then we have

$$y' = \frac{(x-1)(4x^3+3x^2-4x-2)-(x+1)(x^3-2x)\cdot 1}{(x-1)^2}$$

$$= \frac{3x^4 - 2x^3 - 5x^2 + 4x + 2}{(x-1)^2}.$$

Let $C(x)$ be the cost function for a certain item—that is, $C(x)$ is the cost of manufacturing and marketing a quantity x of the items in question. The derivative $C'(x)$ gives the marginal cost. The ratio $C(x)/x$ equals the total cost divided by the quantity produced and so represents the average cost per unit of producing these items. The derivative of this ratio with respect to x is called the **marginal average cost.** It gives the increase in the average cost per item for each increase of one unit in the amount produced.

In order to compute the marginal average cost from the cost function, we must differentiate the ratio $C(x)/x$. For this, we can use the quotient rule.

$$\text{Marginal Average Cost} = \frac{d}{dx}\left(\frac{C(x)}{x}\right) = \frac{x\dfrac{d}{dx}C(x) - C(x)\dfrac{d}{dx}x}{x^2}$$

$$= \frac{xC'(x) - C(x)}{x^2} = \frac{1}{x}\left[C'(x) - \frac{C(x)}{x}\right]$$

Observe that in this final expression the square brackets contain the difference between the marginal cost, $C'(x)$, and the average cost, $C(x)/x$. Thus we conclude that *the marginal average cost is equal to marginal cost minus average cost all divided by the quantity produced.* In particular, the marginal average cost is zero when marginal cost and average cost are equal to one another.

EXAMPLE 7 *(Marginal Average Cost)* Compute the marginal average cost for the cost function

$$C(x) = 0.001x^3 - 0.3x^2 + 40x + 1000$$

when $x = 100$.

Solution $C'(x) = 0.003x^2 - 0.6x + 40$ and so

$$C'(100) = 0.003(100)^2 - 0.6(100) + 40 = 10$$

$$C(100) = 0.001(100)^3 - 0.3(100)^2 + 40(100) + 1000 = 3000.$$

Therefore the marginal average cost when $x = 100$ is

$$\frac{1}{x}\left[C'(x) - \frac{C(x)}{x}\right] = \frac{1}{100}\left[10 - \frac{3000}{100}\right] = -0.2.$$

So when $x = 100$, the average cost per unit decreases by 0.2 for each additional unit produced. (We can also calculate this answer by forming $\overline{C}(x) = C(x)/x$, then differentiating the resulting expression.) ☛ 5

☛ **5.** Find the marginal cost, average cost, and marginal average cost for the cost function $C(x) = 5 + x + 2x^2$.
Verify that
$\overline{C}'(x) = x^{-1}[C'(x) - \overline{C}(x)]$.

Proofs of Theorems

PROOF OF THEOREM 1 Let $y = u \cdot v$. Then

$$y + \Delta y = (u + \Delta u) \cdot (v + \Delta v)$$

$$= uv + u \cdot \Delta v + v \cdot \Delta u + \Delta u \cdot \Delta v$$

$$= y + u\,\Delta v + v\,\Delta u + \Delta u\,\Delta v.$$

We subtract y from both sides.

$$\Delta y = u\,\Delta v + v\,\Delta u + \Delta u\,\Delta v$$

$$\frac{\Delta y}{\Delta x} = u\frac{\Delta v}{\Delta x} + v\frac{\Delta u}{\Delta x} + \Delta u \cdot \frac{\Delta v}{\Delta x}$$

Taking limits as $\Delta x \to 0$, we have

$$\lim_{\Delta x \to 0}\frac{\Delta y}{\Delta x} = u\lim_{\Delta x \to 0}\frac{\Delta v}{\Delta x} + v\lim_{\Delta x \to 0}\frac{\Delta u}{\Delta x} + \lim_{\Delta x \to 0}\Delta u \lim_{\Delta x \to 0}\frac{\Delta v}{\Delta x}.$$

(Note that parts (a) and (c) of Theorem 3, in Section 12-2 have been used.) The last term on the right is zero because $\Delta u \to 0$ as $\Delta x \to 0$, so we get

$$\frac{dy}{dx} = u \cdot \frac{dv}{dx} + v \cdot \frac{du}{dx}$$

as required.

PROOF OF THEOREM 2 Let $y = u/v$. When x changes to $x + \Delta x$, then y changes to $y + \Delta y$, u to $u + \Delta u$, and v to $v + \Delta v$, so that

$$y + \Delta y = \frac{u + \Delta u}{v + \Delta v}.$$

We subtract $y = u/v$ from both sides.

$$\Delta y = \frac{u + \Delta u}{v + \Delta v} - \frac{u}{v} = \frac{v(u + \Delta u) - u(v + \Delta v)}{v(v + \Delta v)} = \frac{v\,\Delta u - u\,\Delta v}{v(v + \Delta v)}$$

Answer
$C'(x) = 1 + 4x$,
$\overline{C}(x) = 5x^{-1} + 1 + 2x$,
$\overline{C}'(x) = -5x^{-2} + 2$.

Dividing by Δx, we obtain

$$\frac{\Delta y}{\Delta x} = \frac{v\dfrac{\Delta u}{\Delta x} - u\dfrac{\Delta v}{\Delta x}}{v(v + \Delta v)}.$$

If we now take the limits as $\Delta x \to 0$, so that $\Delta y/\Delta x \to dy/dx$, $\Delta u/\Delta x \to du/dx$, and $\Delta v/\Delta x \to dv/dx$, we have

$$\frac{dy}{dx} = \frac{v\dfrac{du}{dx} - u\dfrac{dv}{dx}}{v(v + 0)} = \frac{vu' - uv'}{v^2}$$

since the extra Δv in the denominator tends to zero. Thus we have proved the result stated in the theorem.

EXERCISES 13-1

(1–12) Using the product rule, find the derivatives of the following functions with respect to the variable involved.

1. $y = (x + 1)(x^3 + 3)$

2. $y = (x^3 + 6x^2)(x^2 - 1)$

3. $u = (7x + 1)(2 - 3x)$

4. $u = (x^2 + 7x)(x^2 + 3x + 1)$

5. $f(x) = (x^2 - 5x + 1)(2x + 3)$

6. $g(x) = (x^2 + 1)(x + 1)^2$

7. $f(x) = (3x + 7)(x - 1)^2$

8. $y = (t^2 + 1)\left(t - \dfrac{1}{t}\right)$

9. $u = \left(y + \dfrac{3}{y}\right)(y^2 - 5)$

10. $g(t) = \left(t + \dfrac{1}{t}\right)\left(5t^2 - \dfrac{1}{t^2}\right)$

11. $g(x) = (x^2 + 1)(3x - 1)(2x - 3)$

12. $f(x) = (2x + 1)(3x^2 + 1)(x^3 + 7)$

(13–16) (*Marginal Revenue*) Using the product rule, calculate the marginal revenue for the following demand relations.

13. $x = 1000 - 2p$

14. $p = 40 - \frac{1}{2}\sqrt{x}$

15. $x = 4000 - 10\sqrt{p}$

16. $p = 15 - 0.1x^{0.6} - 0.3x^{0.3}$

17. (*Rate of Change of GNP*) The average income per capita in a certain country at time t is equal to $W = 6{,}000 + 500t + 10t^2$. ($W$ is in dollars and t is in years.) The population size at time t (in millions) is $P = 10 + 0.2t + 0.01t^2$. Calculate the rate of change of GNP at time t. (*Hint:* GNP = population size $\times$ per capita income.)

18. (*Rate of Change of GNP*) Repeat Exercise 17 for the case when $W = 1000 + 60t + t^2$ and $P = 4 + 0.1t + 0.01t^2$.

(19–34) Use the quotient rule to find the derivatives of the following functions with respect to the independent variable involved.

19. $y = \dfrac{3}{2x + 7}$

20. $f(t) = \dfrac{5t}{2 - 3t}$

21. $y = \dfrac{u}{u + 1}$

22. $f(x) = \dfrac{x + 1}{x + 3}$

23. $f(x) = \dfrac{x + 2}{x - 1}$

24. $g(x) = \dfrac{3 - x}{x^2 - 3}$

25. $y = \dfrac{t^2 - 7t}{t - 5}$

26. $y = \dfrac{u^2 - u + 1}{u^2 + u + 1}$

27. $x = \dfrac{\sqrt{u} + 1}{\sqrt{u} - 1}$

28. $t = \dfrac{x^2 - 1}{x^2 + 1}$

29. $y = \dfrac{1}{x^2 + 1}$

30. $y = \dfrac{1}{(t + 1)^2}$

31. $f(x) = \dfrac{(x^2 + 1)(2x + 3)}{3x - 1}$ **32.** $g(t) = \dfrac{(t + 3)(3t^2 + 5)}{2 - 3t}$

33. $y = \dfrac{(2u^3 + 7)(3u^2 - 5)}{u^2 + 1}$ **34.** $y = \dfrac{(t + 1/t)(t^2 + 7t)}{3t + 4}$

(35–38) Find the equation of the tangent line to the graphs of the following functions at the indicated point.

35. $y = (3x^2 + 7)(x + 2)$ at $(-1, 10)$

36. $y = (x + 1/x)(x^2 - 1)$ at $x = 1$

37. $y = \dfrac{2x - 3}{x - 2}$ at $(3, 3)$ **38.** $f(x) = \dfrac{x^2 - 4}{x^2 + 1}$ at $x = -2$

39. Find the points on the curve $f(x) = \dfrac{1}{x^2 + 1}$ where the tangent lines are horizontal.

40. Find the points on the curve $y = \dfrac{x + 2}{x^2 + 5}$ where the tangent lines are horizontal.

41. Find the points on the curve $y = \dfrac{x - 3}{x + 3}$ where the tangent lines have a slope of $\frac{1}{6}$.

42. Find the points on the curve $y = (x + 1/x)(x^2 + 6x)$ where the tangent lines have a slope of -8 units.

(43–44) (*Marginal Average Cost*) Compute the marginal average cost for the following cost functions (a, b, and n are constants).

43. $C(x) = a + bx$ **44.** $C(x) = a + bx^n$

45. (*Per Capita Income*) If the GNP of a nation at time t is $I = 10 + 0.4t + 0.01t^2$ (in billions of dollars) and the population size (in millions) is $P = 4 + 0.1t + 0.01t^2$, find the rate of change of per capita income.

46. Use the quotient rule to *prove* that $(d/dx)(x^{-7}) = -7x^{-8}$. (*Hint:* Write $x^{-7} = 1/x^7$.)

*47. Generalize Exercise 46 to prove that $(d/dx)(x^n) = nx^{n-1}$ when n is any negative integer. (*Hint:* Write $x^n = 1/x^m$, where $m = -n$.)

48. (*Real Wage Rate*) The hourly wage rate of a certain group of workers increased according to the formula $W(t) = 3 + \frac{1}{2}t$ between 1970 and 1980, where t is time in years with $t = 0$ corresponding to 1970. During that time the consumer price index was given by $I(t) = 100 + 3t + \frac{1}{2}t^2$. The real wage is equal to $100\, W(t)/I(t)$ when adjusted for inflation. Calculate the rate of change of this real wage rate in 1970, 1975, 1980.

49. (*Fish Farming*) The weight of a certain stock of fish is given by $W = nw$, where n is the size of the stock and w the average weight of each fish. If n and w change with time t according to the formulas $n = (2t^2 + 3)$ and $w = (t^2 - t + 2)$, find the rate of change of W with respect to time.

50. (*Physics*) The absolute temperature T of a gas is given by $T = cPV$, where P is its pressure, V its volume, and c is some constant depending on the mass of gas. If $P = (t^2 + 1)$ and $V = (2t + t^{-1})$ as functions of time t, find the rate of change of T with respect to t.

51. (*Biology*) The density of algae in a water tank is equal to n/V, where n is the number of algae and V the volume of water in the tank. If n and V vary with time t according to the formulas $n = \sqrt{t}$ and $V = \sqrt{t} + 1$, calculate the rate of change of the density.

52. (*Ecology*) Let x be the size of a certain population of predators and y the size of the population of prey upon which they feed. As functions of time t, $x = t^2 + 4$ and $y = 2t^2 - 3t$. Let u be the number of prey to each predator. Find the rate of change of u.

◼ 13-2 THE CHAIN RULE

Let $y = f(u)$ be a function of u and $u = g(x)$ be a function of x. Then we can write

$$y = f[g(x)]$$

representing y as a function of x, called the *composite function of* f and g. It is denoted by $(f \circ g)(x)$. (See Section 5-4.)

The derivatives of composite functions can be found by the use of the following theorem. A proof will be given at the end of this section.

THEOREM 1 (CHAIN RULE) If y is a function of u and u is a function of x, then

$$\frac{dy}{dx} = \frac{dy}{du} \cdot \frac{du}{dx}.$$

The chain rule provides what is probably the most useful of all the aids to differentiation, as will soon become apparent. It is a tool that is constantly in use when you work with the differential calculus, and you should master its use as soon as possible. When using it to differentiate a complicated function, it is necessary at the start to spot how to write the given function as the composition of two simpler fuctions. The following examples provide some illustrations.

EXAMPLE 1 Find dy/dx when $y = (x^2 + 1)^5$.

Solution We could solve this problem by expanding $(x^2 + 1)^5$ as a polynomial in x. However, it is much simpler to use the chain rule.

Observe that y can be written as a composite function in the following way.

$$y = u^5 \quad \text{where} \quad u = x^2 + 1$$

Then

$$\frac{dy}{du} = 5u^4 \quad \text{and} \quad \frac{du}{dx} = 2x.$$

From the chain rule, we have the following.

☛ 6

$$\frac{dy}{dx} = \frac{dy}{du} \cdot \frac{du}{dx} = 5u^4 \cdot 2x = 5(x^2 + 1)^4 \cdot 2x = 10x(x^2 + 1)^4$$

Another way of writing the chain rule is that if $y = f(u)$, then

$$\frac{dy}{dx} = f'(u)\frac{du}{dx}$$

(since $f'(u) = dy/du$). In particular, if $f(u) = u^n$, then $f'(u) = nu^{n-1}$. Thus we have the following special case of the chain rule.

$$\text{If} \quad y = [u(x)]^n, \quad \text{then} \quad \frac{dy}{dx} = nu^{n-1}\frac{du}{dx}.$$

Think of a composite function as having different layers that you peel off one by one. The outside layer of the function corresponds to the part you would compute last if you were evaluating it. For example, if $y = (x^2 + 1)^5$,

☛ **6.** Differentiate the following functions. Indicate how each function is decomposed.
(a) $y = (1 - x^2)^3$;
(b) $y = \sqrt{2x + 1}$

Answer
(a) $y = u^3, \quad u = 1 - x^2.$
$\dfrac{dy}{dx} = -6x(1 - x^2)^2.$
(b) $y = \sqrt{u} = u^{1/2}, \quad u = 2x + 1.$
$\dfrac{dy}{dx} = \dfrac{1}{\sqrt{2x + 1}}.$

the *outside* part of the function is the fifth power and the *inside* part is $(x^2 + 1)$. If you were evaluating y for a particular value of x, you would first evaluate the inside part, $x^2 + 1$, and then you would raise it to the fifth power. For example, if $x = 2$, then *inside* $= x^2 + 1 = 2^2 + 1 = 5$ and $y = (inside)^5 = 5^5 = 3125$.

When differentiating a composite function, you first differentiate the outside layer of the function, then multiply by the derivative of the inside. In these verbal terms we can rephrase the chain rule as follows.

If $y = f(inside)$, then $\dfrac{dy}{dx} = f'(inside) \cdot$ (derivative of *inside* with respect to x).

If $y = (inside)^n$, then $\dfrac{dy}{dx} = n(inside)^{n-1} \cdot$ (derivative of *inside* with respect to x).

Here *inside* stands for any differentiable function of x.

For example, in the example in which $y = (x^2 + 1)^5$, we would take *inside* to be $x^2 + 1$ and $y = f(inside) = (inside)^5$. Then immediately,

$$\frac{dy}{dx} = 5(inside)^4 \cdot \frac{d}{dx}(inside)$$

$$= 5(x^2 + 1)^4 \cdot \frac{d}{dx}(x^2 + 1)$$

$$= 5(x^2 + 1)^4 \cdot 2x = 10x(x^2 + 1)^4$$

giving the same answer as before.

EXAMPLE 2 Given $f(t) = 1/\sqrt{t^2 + 3}$ find $f'(t)$.

Solution Let $u = t^2 + 3$, so that $y = f(t) = 1/\sqrt{u} = u^{-1/2}$. Then

$$\frac{du}{dt} = 2t \quad \text{and} \quad \frac{dy}{du} = -\frac{1}{2}u^{-3/2} = -\frac{1}{2}(t^2 + 3)^{-3/2}.$$

Thus, by the chain rule,

$$\frac{dy}{dt} = \frac{dy}{du} \cdot \frac{du}{dt}$$

$$= -\frac{1}{2}(t^2 + 3)^{-3/2} \cdot 2t = -t(t^2 + 3)^{-3/2}.$$

Alternatively, we can solve directly.

$$f(t) = \frac{1}{\sqrt{t^2 + 3}} = (t^2 + 3)^{-1/2}$$

Here *inside* is $(t^2 + 3)$ and *outside* is the power $-\frac{1}{2}$. Using the power formula to differentiate this outside part, we have

$$f'(t) = -\frac{1}{2}(t^2 + 3)^{-1/2-1} \cdot \frac{d}{dt}(t^2 + 3)$$

$$= -\frac{1}{2}(t^2 + 3)^{-3/2} \cdot 2t = t(t^2 + 3)^{-3/2}.$$

EXAMPLE 3 Given $y = (x^2 + 5x + 1)(2 - x^2)^4$, find dy/dx.

Solution First we write y as a product, $y = uv$, where $u = x^2 + 5x + 1$ and $v = (2 - x^2)^4$. We have $u' = 2x + 5$ immediately, but to find v' we must use the chain rule. For this, *inside* $= (2 - x^2)$ and the *outside* part of v is the fourth power. Thus

$$v' = \frac{d}{dx}(2 - x^2)^4 = 4(2 - x^2)^3 \cdot \frac{d}{dx}(2 - x^2)$$

$$= 4(2 - x^2)^3 \cdot (-2x) = -8x(2 - x^2)^3.$$

Then finally, from the product rule,

$$y' = uv' + vu' = (x^2 + 5x + 1)[-8x(2 - x^2)^3] + (2 - x^2)^4(2x + 5).$$

Factoring then yields

$$\frac{dy}{dx} = (2 - x^2)^3[-8x(x^2 + 5x + 1) + (2x + 5)(2 - x^2)]$$

$$= (2 - x^2)^3[10 - 4x - 45x^2 - 10x^3]. \quad ☛ 7$$

EXAMPLE 4 Find dy/dx if $y = \left(\dfrac{x - 1}{x + 1}\right)^3$.

Solution Here we have a choice as to how we break this function down. We can write y as a composite function,

$$y = u^3, \qquad u = \frac{x - 1}{x + 1} \tag{1}$$

and then use the chain rule. Alternatively, we can write $y = u/v$ where $u = (x - 1)^3$ and $v = (x + 1)^3$ and then use the quotient rule. Or a third alternative is to write $y = uv$ where $u = (x - 1)^3$ and $v = (x + 1)^{-3}$ and use the product rule. We shall use the first of these methods, but you might like to check that the other methods give the same answer.

From Equations (1), by the chain rule,

$$\frac{dy}{dx} = 3u^2\frac{du}{dx} = 3\left(\frac{x - 1}{x + 1}\right)^2\frac{du}{dx}.$$

To find du/dx we write $u = u_1/v_1$ where $u_1 = x - 1$ and $v_1 = x + 1$. By the

☛ **7.** Differentiate the following functions:
(a) $y = x\sqrt{2x + 1}$;
(b) $y = \dfrac{x}{\sqrt{2x + 1}}$.

Answer
(a) $\dfrac{dy}{dx} = \dfrac{3x + 1}{\sqrt{2x + 1}}$;
(b) $\dfrac{dy}{dx} = \dfrac{x + 1}{(2x + 1)^{3/2}}$.

quotient rule then

$$\frac{du}{dx} = \frac{v_1 u_1' - u_1 v_1'}{v_1^2} = \frac{(x+1)\cdot 1 - (x-1)\cdot 1}{(x+1)^2} = \frac{2}{(x+1)^2}.$$

Thus, finally,

$$\frac{dy}{dx} = 3\left(\frac{x-1}{x+1}\right)^2 \frac{2}{(x+1)^2} = 6\frac{(x-1)^2}{(x+1)^4}. \qquad ☛ 8$$

☛ **8.** Solve Example 4 using the quotient rule or the product rule.

EXAMPLE 5 (*Marginal Profit*) A shoe manufacturer can use his plant to make either men's or women's shoes. If he makes x (in thousands of pairs) men's shoes and y (in thousands of pairs) women's shoes per week, then x and y are related by the equation

$$2x^2 + y^2 = 25.$$

(This equation is the product transformation equation; see Section 5-3.) If the profit is \$10 on each pair of shoes, calculate the marginal profit with respect to x when $x = 2$.

Solution The weekly profit P in thousands of dollars is given by

$$P = 10x + 10y$$

since each thousand pairs of shoes brings in ten thousand dollars in profit, and so $(x + y)$ thousand pairs give $10(x + y)$ thousand dollars profit. But

$$y^2 = 25 - 2x^2$$

or

$$y = \sqrt{25 - 2x^2}.$$

Therefore, we can express P in terms of x alone as

$$P = 10x + 10\sqrt{25 - 2x^2}.$$

The marginal profit with respect to x is just the derivative dP/dx. It measures the increase in profit per unit increase in x when x, the production of men's shoes, is given a small increment. It is

$$\frac{dP}{dx} = \frac{d}{dx}[10x + 10(25 - 2x^2)^{1/2}].$$

In order to differentiate the second term, we must use the chain rule with *inside* = $(25 - 2x^2)$.

$$\frac{d}{dx}(25 - 2x^2)^{1/2} = \tfrac{1}{2}(25 - 2x^2)^{-1/2} \cdot \frac{d}{dx}(25 - 2x^2)$$

$$= \tfrac{1}{2}(25 - 2x^2)^{-1/2}(-4x)$$

$$= -2x(25 - 2x^2)^{-1/2}$$

Answer The first step is:
Quotient rule:
$$\frac{dy}{dx} = \frac{(x+1)^3 \cdot 3(x-1)^2 - (x-1)^3 \cdot 3(x+1)^2}{[(x+1)^3]^2}.$$
Product rule:
$$\frac{dy}{dx} = (x+1)^{-3} \cdot 3(x-1)^2 + (x-1)^3 \cdot [-3(x+1)^{-4}].$$

Therefore

$$\frac{dP}{dx} = 10 + 10\frac{d}{dx}(25 - 2x^2)^{1/2}$$

$$= 10 + 10[-2x(25 - 2x^2)^{-1/2}]$$

$$= 10 - 20x(25 - 2x^2)^{-1/2}.$$

When $x = 2$, the value of y is

$$y = \sqrt{25 - 2x^2} = \sqrt{25 - 2(4)} = \sqrt{17} = 4.1.$$

Thus the firm is producing 2000 pairs of men's shoes and 4100 pairs of women's shoes per week. Its weekly profit is

$$P = 10(x + y) = 10(2 + 4.1) = 61$$

(or $61,000). The marginal profit

$$\frac{dP}{dx} = 10 - 20(2)[25 - 2(4)]^{-1/2} = 10 - \frac{40}{\sqrt{17}} = 0.30.$$

Thus an increase of Δx thousands of pairs of men's shoes gives an approximate increase of $(0.30)\,\Delta x$ thousand dollars in the profit.

Related Rates

Let $y = f(x)$ and suppose that x varies as a function of time t. Then since y is a function of x, y also will vary with time. By using the chain rule, it is possible to find an expression for the rate at which y varies in terms of the rate at which x varies. For we have

$$\frac{dy}{dt} = \frac{dy}{dx} \cdot \frac{dx}{dt} = f'(x)\frac{dx}{dt}$$

and we have a direct relation between the two rates dy/dt and dx/dt. This is called the equation of **related rates.** ☛ 9

☛ **9.** Suppose that $y = \sqrt{x + 2}$. Find dy/dt if $x = 2$ and $dx/dt = 0.5$.

EXAMPLE 6 *(Related Rates)* A firm has the cost function $C(x) = 25 + 2x - \frac{1}{20}x^2$, where x is the production level. If the production level is equal to 5 at present and is increasing at the rate of 0.7 per year, find the rate at which the production costs are rising.

Solution We are given that $dx/dt = 0.7$ (when time is measured in years). The marginal cost is given by

$$\frac{dC}{dx} = 2 - \frac{x}{10}.$$

Therefore

$$\frac{dC}{dt} = \frac{dC}{dx}\frac{dx}{dt} = \left(2 - \frac{x}{10}\right)\frac{dx}{dt}.$$

Answer 0.125.

Substituting $x = 5$, the current production level, we obtain

$$\frac{dC}{dt} = \left(2 - \frac{5}{10}\right)(0.7) = 1.05.$$

Thus the production costs are increasing at the rate of 1.05 per year. ☛ 10

☛ **10.** Repeat Example 6 for the cost function
$C(x) = 12 + 5\sqrt{x} + 3x.$

PROOF OF CHAIN RULE The proof of the chain rule, if given in complete detail, would be a little more complicated than we wish to include. We shall therefore provide a proof which, although covering most cases that arise, does have certain restrictions on its range of applicability.

Let Δx be an increment in x. Since u and y are functions of x, they will change whenever x changes, so we denote their increments by Δu and Δy. Then, as long as $\Delta u \neq 0$, we have

$$\frac{\Delta y}{\Delta x} = \frac{\Delta y}{\Delta u} \frac{\Delta u}{\Delta x}.$$

We now let $\Delta x \to 0$. In this limit, we also have that $\Delta u \to 0$ and $\Delta y \to 0$, and so

$$\lim_{\Delta x \to 0} \frac{\Delta y}{\Delta x} = \lim_{\Delta x \to 0} \left(\frac{\Delta y}{\Delta u} \frac{\Delta u}{\Delta x}\right)$$

$$= \left(\lim_{\Delta x \to 0} \frac{\Delta y}{\Delta u}\right)\left(\lim_{\Delta x \to 0} \frac{\Delta u}{\Delta x}\right)$$

$$= \left(\lim_{\Delta u \to 0} \frac{\Delta y}{\Delta u}\right)\left(\frac{du}{dx}\right)$$

$$= \frac{dy}{du} \cdot \frac{du}{dx}$$

as required.

The reason that this proof is incomplete lies in the assumption that $\Delta u \neq 0$. For most functions $u(x)$, it will be the case that Δu never vanishes when Δx is sufficiently small (but $\Delta x \neq 0$). However, it is conceivable that the function $u(x)$ could be so peculiar in its behavior that Δu vanishes repeatedly as $\Delta x \to 0$. For such an unusual function, the above proof would then break down. It is possible to modify the proof to cover such cases as this, but we shall not do so here.

Answer $(\frac{1}{2}\sqrt{5} + 3)(0.7) \approx 2.88.$

EXERCISES 13-2

(1–36) Find the derivatives of the following functions with respect to the independent variable involved.

1. $y = (3x + 5)^7$

2. $y = \sqrt{5 - 2t}$

3. $u = (2x^2 + 1)^{3/2}$

4. $x = (y^3 + 7)^6$

5. $f(x) = \dfrac{1}{(x^2 + 1)^4}$

6. $g(x) = \dfrac{1}{(x^2 + x + 1)^3}$

7. $h(t) = \sqrt{t^2 + a^2}$

8. $F(x) = \sqrt[3]{x^3 + 3x}$

9. $x = \dfrac{1}{\sqrt[3]{t^3 + 1}}$

10. $y = \left(t + \dfrac{1}{t}\right)^{10}$

11. $y = \left(t^2 + \dfrac{1}{t^2}\right)^5$

12. $y = \dfrac{1}{\sqrt{u^2 + 9}}$

13. $y = (x^2 + 1)^{0.6}$

14. $y = \sqrt{x^2 + \dfrac{1}{x^2}}$

15. $u = \sqrt[3]{t^3 - \dfrac{1}{t^3}}$

16. $y = \sqrt{1 + x \ln 2}$

17. $f(x) = \dfrac{\sqrt{x^2 + 1}}{\sqrt[3]{x^2 + 1}}$

18. $g(x) = (x^4 + 16)^{1/4}$

19. $G(u) = (u^2 + 1)^3(2u + 1)$

20. $H(y) = (2y^2 + 3)^6(5y + 2)$

21. $f(x) = (x + 1)^3(2x + 1)^4$

22. $g(x) = (3x - 1)^5(2x + 3)^4$

23. $f(x) = x^3(x^2 + 1)^7$

24. $u = x^2\sqrt{x^3 + a^3}$

25. $y = [(x + 1)(x + 2) + 3]^4$

26. $u = [(y - 1)(2y + 3) + 7]^5$

27. $y = \left(\dfrac{3x + 2}{x - 1}\right)^7$

28. $y = \left(\dfrac{t}{t + 1}\right)^6$

29. $y = \left(\dfrac{u^2 + 1}{u + 1}\right)^3$

30. $y = \sqrt{\dfrac{3x + 7}{5 + 2x}}$

31. $y = \dfrac{(x^2 + 1)^2}{x + 1}$

32. $x = \dfrac{t^2 + 1}{(t + 1)^3}$

33. $z = \dfrac{x}{\sqrt{x^2 - 1}}$

34. $y = \dfrac{z^2 + 1}{z^2 - 1}$

35. $x = \dfrac{t^2}{\sqrt{t^2 + 4}}$

36. $Z = \dfrac{\sqrt{2x + 1}}{x + 2}$

37. Find $f'(0)$ if $f(x) = (2x + 1)^4(2 - 3x)^3$.

38. Find $f'(1)$ if $f(x) = (x - 1)^7(x^2 + 3)^4$.

(39–42) Find the equation of the tangent line to the graphs of the following functions at the indicated point.

39. $f(x) = \sqrt{x^2 + 9}$ at (4, 5)

40. $f(x) = x\sqrt{x^2 - 16}$ at $x = 5$

41. $(x) = (x - 2/x)^4$ at $x = 2$

42. $y = \dfrac{5}{\sqrt{x^2 + 16}}$ at $x = -3$

(43-44) (*Marginal Cost*) Find the marginal cost for the following cost functions.

43. $C(x) = \sqrt{100 + x^2}$

44. $C(x) = 20 + 2x - \sqrt{x^2 + 1}$

(45–46) (*Marginal Average Cost*) Calculate the marginal average cost for the cost functions of Exercises 43 and 44.

(47–48) (*Marginal Revenue*) Calculate the marginal revenue for the following demand relations.

47. $p = \sqrt{100 - 0.1x - 10^{-4}x^2}$

48. $x = 1000(8 - p)^{1/3}$

49. (*Rate of Cost Increase*) A manufacturer's cost function is given by

$$C(x) = 2000 + 10x - 0.1x^2 + 0.002x^3.$$

If the current production level is $x = 100$ and is increasing at the rate of 2 per month, find the rate at which the production costs are increasing.

50. (*Rate of Revenue Increase*) The manufacturer in Exercise 49 has a revenue function given by $R(x) = 65x - 0.05x^2$. Calculate the rate at which the revenue is increasing and the rate at which the profit is increasing.

51. (*Rate of Revenue Change*) The demand equation for a company's product is $2p + x = 300$, where x units can be sold at a price of $\$p$ each. If the demand is changing at a rate of 2 units per year when the demand has reached 40 units, at what rate is the revenue changing if the company adjusts its price to the changing demand?

52. (*Rate of Profit Change*) In Exercise 51, it costs the company $(225 + 60x)$ dollars to produce x units. When the demand level has reached 40 units and the demand is increasing at a rate of 2 units per year, find the rate at which the profit is changing.

53. (*Oil Pollution*) The area of a circular oil slick from a ruptured pipeline is growing at the rate of 30 square kilometers per day. How fast is the radius increasing when the radius is 5 kilometers?

54. A spherical balloon is being inflated. When the radius is 10 inches and is increasing at the rate of 2 inches every 5 seconds, at what rate is the volume increasing?

55. (*Productivity*) The unit productivity of labor P (i.e., output per work-hour) is a function of the capital K invested

in plant and machinery. Assume that $P = 0.5K^2 + K - 5$, where K is measured in millions of dollars and P in dollars per work-hour. If K is 10 and is increasing at the rate of 2 per year, how fast is P increasing?

*56. (*Labor Requirements*) When the volume of its weekly production is x thousand units, a company observes that the number of its employees is $N = 500(1 + 0.01x + 0.00005x^2)$. If the weekly production is increasing by 5% per year, at what rate is the number of employees increasing when 100,000 units per week are being produced? When 200,000 are being produced?

57. (*Chemical Reaction*) The rate R at which a chemical reaction progresses is equal to $\sqrt{T}$, where T is the temperature. If T varies with time t according to the formula $T = (3t + 1)/(t + 2)$, find the rate of change of R with respect to t.

58. (*Seed Germination*) The proportion P of seeds that germinate depends on the soil temperature T. Suppose that under certain conditions, $P = T^7$, and that T varies with depth x below the surface as $T = (x^2 + 3)/(x + 3)$. Find the rate of change of P with respect to depth.

59. (*Housing Starts*) The number of housing starts per year, N (millions) depends on the annual mortgage interest rate r (percent) according to the formula

$$N(r) = \frac{50}{100 + r^2}$$

a. If r is currently 10 and is increasing at the rate of 0.25 per month, what is the rate of change of N?

b. If $r(t) = 12 - \dfrac{8t}{t + 24}$, where t is time in months, calculate the rate of change of N at $t = 6$.

◣ 13-3 DERIVATIVES OF EXPONENTIAL AND LOGARITHMIC FUNCTIONS

Figure 1 shows the graph of the exponential function $f(x) = a^x$ in a typical case when $a > 1$. When $x = 0$, $y = a^0 = 1$, so the graph passes through the point $(0, 1)$ for any value of a. The slope of the graph as it crosses the y-axis at this point varies, depending on a: the bigger the value of a, the greater the slope when $x = 0$.

Let us select the particular value of a for which the slope of the graph at $x = 0$ is equal to 1. For this value of a, the graph slopes upwards at an angle of $45°$ with the horizontal as it crosses the y-axis. The condition that must be satisfied is that the derivative $f'(0)$ must equal 1. Thus, since in general

$$f'(x) = \lim_{\Delta x \to 0} \frac{f(x + \Delta x) - f(x)}{\Delta x}$$

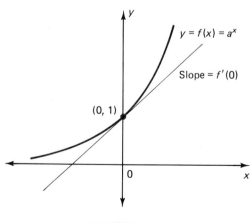

FIGURE 1

we have

$$f'(0) = \lim_{\Delta x \to 0} \frac{f(0 + \Delta x) - f(0)}{\Delta x} = \lim_{\Delta x \to 0} \frac{f(\Delta x) - f(0)}{\Delta x} = \lim_{\Delta x \to 0} \frac{a^{\Delta x} - a^0}{\Delta x}.$$

Since $a^0 = 1$, the condition $f'(0) = 1$ therefore reduces to

$$\lim_{\Delta x \to 0} \frac{a^{\Delta x} - 1}{\Delta x} = 1. \tag{1}$$

This condition determines the value of a for us. It turns out that the value of a that satisfies this condition is $a = e = 2.71828 \ldots$, the base of the natural exponential and logarithmic functions that were introduced in Chapter 6. The proof of this statement is beyond the scope of this book; however, Table 1 provides quite convincing evidence of its accuracy. We know that $e = 2.7183$ to four decimal places, and in the table we have computed the values of the quantity $[(2.7183)^{\Delta x} - 1]/\Delta x$ for a series of values of Δx starting with $\Delta x = 1$ and decreasing to $\Delta x = 0.0001$. It is clear that as Δx becomes smaller and smaller, the quantity in question becomes closer and closer to 1. Therefore Equation (1) is satisfied to a very close approximation by taking $a = 2.7183$. A more exact calculation would show that the quantity $[(2.7183)^{\Delta x} - 1]/\Delta x$ does, in fact, approach the limiting value 1.00000668 (to eight decimal places) as $\Delta x \to 0$. ☛ 11

☛ **11.** Use your calculator with $\Delta x = 0.001$ or 0.0001 to find the approximate values of the limit in Equation (1) when $a = 2$, when $a = 2.5$ and when $a = 3$.

Instead of taking $a = 2.7183$, we could take an even better approximation to the irrational number e—for example, we could take $a = 2.718282$, which is correct to seven significant figures. Then by constructing a table similar to the one above, we could convince ourselves that the limiting value of $(a^{\Delta x} - 1)/\Delta x$ as $\Delta x \to 0$ is even closer to 1. (In fact, with $a = 2.718282$, this limiting value is equal to 1.0000000631 to 10 decimal places.) We can then be confident that Condition (1) is satisfied by choosing the base of the exponential expression to be e.

Let us now evaluate the derivative of the function e^x for a general x. Setting $y = e^x$, we have

$$\frac{dy}{dx} = \lim_{\Delta x \to 0} \frac{e^{x+\Delta x} - e^x}{\Delta x}.$$

TABLE 1

Δx	$\dfrac{(2.7183)^{\Delta x} - 1}{\Delta x}$
1	1.7183
0.1	1.0517
0.01	1.0050
0.001	1.0005
0.0001	1.000057

Answer 0.693, 0.916, and 1.099.

But using a basic property of exponents, $e^{x+\Delta x} = e^x \cdot e^{\Delta x}$, and so

$$\frac{dy}{dx} = \lim_{\Delta x \to 0} e^x \frac{e^{\Delta x} - 1}{\Delta x} = e^x \lim_{\Delta x \to 0} \frac{e^{\Delta x} - 1}{\Delta x} = e^x \cdot 1 = e^x$$

after using the Equation (1) (with a replaced by e).

Thus we have the important result that the derivative of the function e^x is the function itself.

$$\boxed{\text{If} \quad y = e^x, \quad \text{then} \quad \frac{dy}{dx} = e^x.}$$

The reason that the natural exponential function is so important rests in this property that its derivative is everywhere equal to the function itself. It is, apart from a constant factor, the only function that possesses this property. It is this fact that accounts for our interest in the number e and in exponential expressions and logarithms that have e as their bases.

EXAMPLE 1 Evaluate dy/dx if $y = xe^x$.

Solution In order to differentiate the function xe^x, we must use the product rule since we can write $y = uv$ with $u = x$ and $v = e^x$. Then

$$\frac{du}{dx} = 1 \quad \text{and} \quad \frac{dv}{dx} = e^x.$$

Therefore

$$\frac{dy}{dx} = u\frac{dv}{dx} + v\frac{du}{dx} = (x)(e^x) + (e^x)(1) = (x + 1)e^x. \quad ☛ 12$$

EXAMPLE 2 Find dy/dx if $y = e^{x^2}$.

Solution Here we break y up as a composite function, $y = e^u$ where $u = x^2$. (Again, to see this, think of evaluating y. The last thing you would compute would be the exponential function, so this is the outside part.) Then

$$\frac{dy}{du} = e^u, \quad \frac{du}{dx} = 2x$$

so from the chain rule,

$$\frac{dy}{dx} = \frac{dy}{du}\frac{du}{dx} = e^u \cdot 2x = 2xe^{x^2}.$$

By the same method as we used in Example 2, the chain rule allows us to differentiate composite functions of the type $e^{u(x)}$, where $u(x)$ is any differentiable function of x. We obtain the following:

$$\boxed{\text{If} \quad y = e^{u(x)}, \quad \text{then} \quad \frac{dy}{dx} = e^{u(x)}u'(x).}$$

☛ **12.** Differentiate

(a) $y = x^3 e^x$; (b) $y = \dfrac{e^x}{x + 1}$.

Answer (a) $x^2(x + 3)e^x$;

(b) $\dfrac{xe^x}{(x + 1)^2}$.

In verbal form we can say

$$\frac{d}{dx} e^{inside} = e^{inside} \frac{d}{dx} \ (inside)$$

where *inside* is any differentiable function of x. ☞ **13**

A particular case that is worth remembering is $u(x) = kx$, where k is a constant. For this we have

$$\text{If} \quad y = e^{kx} \quad \text{then} \quad \frac{dy}{dx} = ke^{kx}. \tag{2}$$

☞ **13.** Differentiate

(a) $y = e^{3x}$; (b) $y = e^{x^3 - 3x^2}$;

(c) $y = xe^{1/x}$.

EXAMPLE 3 *(Profits and Advertising)* A certain item can be manufactured and sold at a profit of $10 each. If the manufacturer spends x dollars on advertising the product, then the number of items that can be sold will be equal to $1000(1 - e^{-0.001x})$. If P denotes the net profit from sales, calculate dP/dx and interpret this derivative. Evaluate dP/dx when $x = 1000$ and when $x = 3000$.

Solution Since each item produces a profit of $10, the total gross profit from sales is obtained by multiplying the number of sales by $10. The net profit is then obtained by subtracting the advertising costs:

$$P = 10,000(1 - e^{-0.001x}) - x.$$

Therefore

$$\frac{dP}{dx} = -10,000 \frac{d}{dx}(e^{-0.001x}) - 1 = -10,000(-0.001e^{-0.001x}) - 1$$

where we have used Equation (2) with k replaced by -0.001. Then

$$\frac{dP}{dx} = 10e^{-0.001x} - 1.$$

The interpretation of this derivative is that it measures the rate of change of net profit with respect to advertising expenditure. In other words, dP/dx gives the number of dollars increase in net profit produced by an additional dollar spent on advertising.

When $x = 1000$,

$$\frac{dP}{dx} = 10e^{-1} - 1 = 10(0.3679) - 1 = 2.679.$$

So when $1000 is spent on advertising, each additional dollar yields an increase of $2.68 in the net profit.

When $x = 3000$,

Answer (a) $3e^{3x}$;

(b) $(3x^2 - 6x)e^{x^3 - 3x^2}$;

(c) $(1 - x^{-1})e^{1/x}$.

$$\frac{dP}{dx} = 10(e^{-3}) - 1 = 10(0.0498) - 1 = -0.502.$$

Thus when $3000 is spent on advertising, an additional dollar spent this way yields a decrease of $0.50 in the net profit. It is clear in this case that the manufacturer should not do more advertising—the cost of extra advertising would outweigh the value of the additional sales that it would generate. (In fact, when $x = 3000$, more is already being spent on advertising than is desirable.)

EXAMPLE 4 *(Population Growth)* A population grows according to the logistic model (see Section 6-4) such that at time t its size y is given by

$$y = y_m(1 + Ce^{-kt})^{-1}$$

with y_m, C, and k constants. Find the rate of increase of population at time t.

Solution The required rate of increase is dy/dt. We observe that y is a composite function of t of the form

$$y = y_m(inside)^{-1}, \qquad inside = 1 + Ce^{-kt}.$$

Therefore

$$\frac{dy}{dt} = y_m(-1)(inside)^{-2}\frac{d}{dt}(inside)$$

$$= -y_m(1 + Ce^{-kt})^{-2}\frac{d}{dt}(1 + Ce^{-kt}).$$

$$= -y_m(1 + Ce^{-kt})^{-2}(-kCe^{-kt})$$

$$= \frac{ky_m Ce^{-kt}}{(1 + Ce^{-kt})^2}$$

Again Equation (2) has been used to differentiate e^{-kt}.

Let us now evaluate the derivative of the function $y = \ln x$, the natural logarithm function.

If $y = \ln x$, then $x = e^y$. Let us differentiate this second equation with respect to x.

$$\frac{d}{dx}(e^y) = \frac{d}{dx}(x) = 1$$

But from the chain rule, we see that

$$\frac{d}{dx}(e^y) = \frac{d}{dy}(e^y)\cdot\frac{dy}{dx} = e^y\frac{dy}{dx},$$

since $(d/dy)(e^y) = e^y$. Therefore $e^y(dy/dx) = 1$, and so

$$\frac{dy}{dx} = \frac{1}{e^y} = \frac{1}{x}.$$

Thus

$$\boxed{\text{If} \quad y = \ln x, \quad \text{then} \quad \frac{dy}{dx} = \frac{1}{x}.}$$

14. Differentiate

(a) $y = x \ln x$;

(b) $y = x \ln (x + 1)$;

(c) $y = \dfrac{x}{\ln x}$.

EXAMPLE 5 Find dy/dx if $y = \ln (x + c)$ where c is a constant.

Solution We have y a composite function, with $y = \ln u$ and $u = x + c$. Therefore, from the chain rule,

$$\frac{dy}{dx} = \frac{dy}{du} \cdot \frac{du}{dx} = \frac{d}{du}(\ln u) \cdot \frac{du}{dx} = \left(\frac{1}{u}\right) \cdot (1) = \frac{1}{x + c}. \quad \blacktriangleright 14$$

In general, the chain rule allows us to differentiate any composite function of the form $y = \ln u(x)$ in the following way:

$$\frac{dy}{dx} = \frac{dy}{du} \cdot \frac{du}{dx} = \frac{d}{du}(\ln u) \cdot u'(x) = \frac{1}{u} u'(x).$$

$$\boxed{\text{So, if } y = \ln u(x), \quad \text{then} \quad \frac{dy}{dx} = \frac{u'(x)}{u(x)}.}$$

Alternatively, in verbal form,

$$\boxed{\frac{d}{dx} \ln \, (inside) = \frac{1}{inside} \frac{d}{dx}(inside)}$$

where *inside* stands for any differentiable function of x.

EXAMPLE 6 Differentiate $\ln (x^2 + x - 2)$.

Solution Here we take $inside = (x^2 + x - 2)$.

$$\frac{d}{dx} \ln (x^2 + x - 2) = \frac{1}{(x^2 + x - 2)} \frac{d}{dx}(x^2 + x - 2)$$

$$= \frac{1}{(x^2 + x - 2)}(2x + 1)$$

$$= \frac{2x + 1}{x^2 + x - 2}$$

EXAMPLE 7 If $f(x) = \ln x / x^2$, find $f'(1)$.

Solution Write $f(x) = u/v$ where $u = \ln x$ and $v = x^2$. Then $u' = 1/x$ and $v' = 2x$. From the quotient rule,

$$f'(x) = \frac{vu' - uv'}{v^2} = \frac{x^2(1/x) - (\ln x) \cdot 2x}{(x^2)^2} = \frac{x - 2x \ln x}{x^4} = \frac{1 - 2 \ln x}{x^3}.$$

Answer　(a) $1 + \ln x$;

(b) $\dfrac{x}{x + 1} + \ln (x + 1)$;

(c) $\dfrac{\ln x - 1}{(\ln x)^2}$.

Therefore

$$f'(1) = \frac{1 - 2 \ln 1}{1^3} = 1$$

since $\ln 1 = 0$.

When we want to differentiate the logarithm of a product or quotient of various expressions, it is often useful to simplify the given function first by making use of the properties of logarithms. Refer back to Section 6-3 if you would like to review these.

EXAMPLE 8 Find dy/dx when $y = \ln (e^x/\sqrt{x + 1})$.

Solution We first simplify y by using properties of logarithms

$$y = \ln\left(\frac{e^x}{\sqrt{x + 1}}\right) = \ln (e^x) - \ln (\sqrt{x + 1}) = x \ln e - \frac{1}{2} \ln (x + 1)$$

Therefore (since $\ln e = 1$)

$$\frac{dy}{dx} = 1 - \frac{1}{2} \frac{d}{dx} \ln (x + 1) = 1 - \frac{1}{2(x + 1)}.$$

An alternative solution to this problem is to write $y = \ln u$, where $u = e^x/\sqrt{x + 1}$, then use the chain rule to write $y' = (1/u)u'$. We leave it to you to convince yourself that this approach leads to *much* harder calculations than those we just did. **15**

EXAMPLE 9 Find dy/dx if $y = \log x$.

Solution In order to differentiate the common logarithm, we express it in terms of the natural logarithm using the base-change formula on page 247.

$$y = \log x = \log_{10} x = \frac{\ln x}{\ln 10}.$$

Therefore

$$\frac{dy}{dx} = \frac{1}{\ln 10} \frac{d}{dx} \ln x = \frac{1}{\ln 10} \cdot \frac{1}{x} = \frac{0.4343}{x}$$

since $1/\ln 10 = 1/2.3026 \ldots = 0.4343 \ldots$.

Observe that in this example, the common logarithm had to be expressed in terms of a natural logarithm before it could be differentiated. This is equally true of logarithms with respect to any other base, such as $\log_a x$: Such functions must first be expressed as natural logarithms. Similarly, a general exponential function a^x must be expressed as e^{kx} ($k = \ln a$) before it can be differentiated. **16**

Now that we have introduced the derivatives of the exponential and logarithmic functions, let us summarize the three forms of the chain rule that we shall use most. In Table 2, *inside* represents any differentiable function of x.

TABLE 2 *Summary of Chain Rule*

$f(x)$	$f'(x)$		$f(x)$	$f'(x)$
$[u(x)]^n$	$n[u(x)]^{n-1}u'(x)$		$(inside)^n$	$n(inside)^{n-1}\dfrac{d}{dx}(inside)$
$e^{u(x)}$	$e^{u(x)}u'(x)$	OR	e^{inside}	$e^{inside}\dfrac{d}{dx}(inside)$
$\ln u(x)$	$\dfrac{1}{u(x)}u'(x)$		$\ln(inside)$	$\dfrac{1}{inside}\dfrac{d}{dx}(inside)$

EXERCISES 13-3

(1–66) Find dy/dx in the following questions.

1. $y = 7e^x$

2. $y = e^7$

3. $y = e^{3x}$

4. $y = \dfrac{1}{e^x}$

5. $y = e^{x^2}$

6. $y = e^{x^3+1}$

7. $y = e^{\sqrt{x}}$

8. $y = e^{1/x}$

9. $y = xe^x$

10. $y = xe^{-x^2}$

11. $y = x^2 e^{-x}$

12. $y = \dfrac{e^x}{x}$

13. $y = \dfrac{x+1}{e^x}$

14. $y = \dfrac{e^{x^3}}{e^{x^2}}$

15. $y = \dfrac{e^{x^2}}{e^x}$

16. $y = e^{x^2} + (x^2)^e$

17. $y = \dfrac{e^x}{x+2}$

18. $y = \dfrac{1+e^x}{1-e^x}$

19. $y = \dfrac{e^x}{e^x+1}$

20. $y = 3\ln x$

21. $y = \dfrac{\ln x}{7}$

22. $y = \ln 2$

23. $y = \ln 3 + \sqrt{\log 4}$

24. $y = (\ln 3)(\ln x)$

25. $y = \dfrac{\ln x}{\ln 7}$

26. $y = \ln(3x + 7)$

27. $y = \ln(x^2 + 5)$

28. $y = \ln(1 + e^x)$

29. $y = (\ln x)^5$

30. $y = \sqrt{\ln x}$

31. $y = \dfrac{1}{\ln x}$

32. $y = \dfrac{1}{1+\ln x}$

33. $y = \dfrac{1}{\sqrt{\ln x}}$

34. $y = x^2 \ln x$

35. $y = x \ln x^2$

36. $y = x(\ln x - 1)$

37. $y = x^2 \ln(x^2 + 1)$

38. $y = x \ln(x + 1)$

39. $y = e^x \ln x$

40. $y = e^x \ln(x^2 + 1)$

41. $y = \dfrac{\ln x}{x}$

42. $y = \dfrac{1+\ln x}{1-\ln x}$

43. $y = \dfrac{\ln(x+1)}{x+1}$

44. $y = \dfrac{x+2}{\ln(x+2)}$

45. $y = \ln(3^{x^2})$

46. $y = \log(e^x)$

47. $y = x^2 \log(e^x)$

48. $y = \dfrac{\log(e^x)}{x}$

49. $y = \dfrac{x^2}{\ln 3^x}$

50. $y = \dfrac{\ln x}{e^x}$

51. $y = \ln\left(\dfrac{x^2+1}{x+1}\right)$

52. $y = \ln\left[\dfrac{(x+2)e^{3x}}{x^2+1}\right]$

53. $y = \ln\left[\dfrac{\sqrt{x+1}}{x^2+4}\right]$

54. $y = \dfrac{\ln x^3}{\ln x^2}$

(*Hint:* Use base-change formula for Exercises 55–66.)

55. $y = a^x$

56. $y = 3^{x^2+1}$

57. $y = \log_a x$

58. $y = \log_3 (x + 1)$

59. $y = \dfrac{\ln x}{\log x}$

60. $y = \dfrac{\log_2 x}{\log_3 x}$

61. $y = (\log_3 x)(\log_x 2)$

62. $y = \log_x x^2$

63. $y = \log_x (x + 1)$

64. $y = xa^{x^2}$

65. $y = x^2 \log x$

66. $y = \dfrac{\log x}{x}$

67. Find $f'(1)$ if $f(x) = e^x \ln x$

68. Find $f'(0)$ if $f(x) = e^{2x} \ln (x + 1)$

(69–72) Find the equation of the tangent line to the graphs of the following functions at the indicated point.

69. $y = \dfrac{1 - e^x}{1 + e^x}$ at $(0, 0)$

70. $y = x \ln x$ at $x = 1$

71. $y = \ln (x^2 + 1)$ at $x = 0$

72. $y = \ln\left(\dfrac{e^x}{\sqrt{x^2 + 1}}\right)$ at $x = 0$

(73–76) (*Marginal Revenue*) Compute the marginal revenue for the following demand relations.

73. $p = 5 - e^{0.1x}$

74. $p = 4 + e^{-0.1x}$

75. $x = 1000(2 - e^p)$

76. $x = 100 \ln (16 - p^2)$

(77–78) (*Marginal Costs*) Calculate the marginal cost and the marginal average cost for the following cost functions.

77. $C(x) = 100 + x + e^{-0.5x}$

78. $C(x) = \sqrt{25 + x + \ln (x + 1)}$

79. (*Advertising and Sales*) In order to sell x units of its product per week, a company must spend A dollars per week on advertising, where

$$A = 200 \ln \left(\frac{400}{500 - x}\right).$$

The items sell for \$5 each. The net revenue is then $R = 5x - A$. Calculate the rate of change of R with respect to A.

80. (*Marginal Revenue*) A company finds that its revenue is given by $R = 2pe^{-0.1p}$ when its product is priced at p

dollars per unit. Find the marginal revenue with respect to price when p is

a. \$5.

b. \$10.

c. \$15.

81. (*Fick's Law of Diffusion*) According to Fick's law, the diffusion of a solute across the wall membrane of a cell is governed by the equation $c'(t) = k[c_s - c(t)]$, where $c(t)$ is the concentration of the solute in the cell, c_s the concentration in the surrounding medium, and k is a constant depending on the size of the cell and the membrane properties. Show that the function

$$c(t) = c_s + Ce^{-kt}$$

satisfies this equation for any constant C. Relate C to the initial concentration $c(0)$.

*82. (*Survival Function*) During winter, the percentage of the bees in a certain group of hives that have been killed is a function of the average temperature. Assume that $p = 100e^{-0.1e^{0.1T}}$, where T is temperature (degrees Celsius) and p is the percentage of bees killed. If T is decreasing at the rate of 2°C per week, calculate the rate at which p is changing when $T = -10°C$.

83. (*Acidity*) The pH of a solution is defined as

$$\mathrm{pH} = -\log_{10} [\mathrm{H}],$$

where $[\mathrm{H}]$ is the concentration of hydrogen ions. It is a measure of acidity, with pH $= 7$ being a neutral solution. Calculate the values of $d\mathrm{pH}/d[\mathrm{H}]$ when $[\mathrm{H}] = 10^{-4}$, 10^{-7}, and 10^{-10}.

84. (*Medicine*) After injection, the concentration of a certain drug in the blood of a patient changes according to the formula $c = pt^2e^{-kt}$, where p and k are constants. Calculate the rate of increase of concentration at time t.

*85. (*Population Growth*) A certain population is growing according to the formula

$$y = y_m(1 - Ce^{-kt})^3$$

where y_m, C, and k are constants. Find the rate of growth at time t and show that

$$\frac{dy}{dt} = 3ky^{2/3}(y_m^{1/3} - y^{1/3}).$$

*86. (*Spread of Information*) The proportion p of physicians who have heard of a new drug t months after it has become available satisfies the equation

$$\ln p - \ln (1 - p) = k(t - C)$$

where k and C are constants. Express p as a function of t and calculate dp/dt. Show that

$$\frac{dp}{dt} = kp(1 - p).$$

*87. Prove that $(d/dx)(x^n) = nx^{n-1}$ for n any real number and $x > 0$. (*Hint:* Write $x^n = e^{n \ln x}$.)

◤ 13-4 HIGHER DERIVATIVES

If $y = f(t)$ is a function of time t, then, as we have seen, the derivative $dy/dt = f'(t)$ represents the rate at which y changes. For example, if $s = f(t)$ is the distance traveled by a moving object, then $ds/dt = f'(t)$ gives the rate of change of distance or, in other words, the instantaneous *velocity* of the object. Let us denote this velocity by v. Then v is also a function of t, and—as a rule—may be differentiated to give the derivative dv/dt.

When the velocity of an object increases, we say that it *accelerates*. For example, when we press the gas pedal of a car, we cause it to accelerate, that is, to go faster. Let us suppose that over a period of 5 seconds, the car accelerates from a speed of 20 feet/second (which is about 14 miles per hour) to 80 feet/second (55 miles per hour). The increment in velocity is $\Delta v = 60$ feet/second and the time increment $\Delta t = 5$ seconds, so the average acceleration is given by

$$\frac{\Delta v}{\Delta t} = \frac{60}{5} = 12 \text{ feet/second/second.}$$

For a moving body, we are often interested in the *instantaneous acceleration,* which is defined as the limit of the average acceleration $\Delta v/\Delta t$ as $\Delta t \to 0$. In other words, instantaneous acceleration is the derivative dv/dt. It gives the instantaneous rate at which the velocity is increasing.

So, to calculate the acceleration, we must differentiate s and then differentiate the result once more. We have

$$\text{Acceleration} = \frac{dv}{dt} = \frac{d}{dt}\left(\frac{ds}{dt}\right).$$

Acceleration is called the *second derivative* of s with respect to t and is usually denoted by $f''(t)$ or by d^2s/dt^2.

In problems involving moving objects, the second derivative, acceleration, is a quantity of prime importance. For example, the degree of safety of the braking system of an automobile depends on the maximum deceleration it can give (deceleration is just a negative acceleration). Or the medical effects of rocket launching on an astronaut depend on the level of acceleration to which he is subjected. More fundamentally, it is one of the basic laws of mechanics that when an object is acted on by a force, it is caused to accelerate, and the magnitude of the acceleration is directly proportional to the size of the force. Thus acceleration enters into the basic laws of motion in an essential way.

We shall now examine the higher order derivatives in a more abstract context. Let $y = f(x)$ be a given function of x with derivative $dy/dx = f'(x)$. In full, we call this the **first derivative** of y with respect to x. If $f'(x)$ is a differentiable function of x, its derivative is called the **second derivative** of y with respect to x. If the second derivative is a differentiable function of x, its derivative is called the **third derivative** of y, and so on.

The first and all higher-order derivatives of y with respect to x are generally denoted by one of the following types of notation.

$$\frac{dy}{dx}, \qquad \frac{d^2y}{dx^2}, \qquad \frac{d^3y}{dx^3}, \qquad \cdots, \qquad \frac{d^ny}{dx^n}.$$

$$y', \qquad y'', \qquad y''', \qquad \cdots, \qquad y^{(n)}.$$

$$f'(x), \qquad f''(x), \qquad f'''(x), \qquad \cdots, \qquad f^{(n)}(x)$$

From the definition of higher-order derivatives, it is clear that

$$\frac{d^2y}{dx^2} = \frac{d}{dx}\left(\frac{dy}{dx}\right), \qquad \frac{d^3y}{dx^3} = \frac{d}{dx}\left(\frac{d^2y}{dx^2}\right)$$

and so on.

EXAMPLE 1 Find the first and higher-order derivatives of $3x^4 - 5x^3 + 7x^2 - 1$.

Solution Let $y = 3x^4 - 5x^3 + 7x^2 - 1$. Then

$$\frac{dy}{dx} = \frac{d}{dx}(3x^4 - 5x^3 + 7x^2 - 1) = 12x^3 - 15x^2 + 14x.$$

The second derivative of y is obtained by differentiating the first derivative.

$$\frac{d^2y}{dx^2} = \frac{d}{dx}\left(\frac{dy}{dx}\right) = \frac{d}{dx}(12x^3 - 15x^2 + 14x) = 36x^2 - 30x + 14$$

Differentiating again, we obtain the third derivative.

$$\frac{d^3y}{dx^3} = \frac{d}{dx}\left(\frac{d^2y}{dx^2}\right) = \frac{d}{dx}(36x^2 - 30x + 14) = 72x - 30$$

Continuing this process, we have

$$\frac{d^4y}{dx^4} = \frac{d}{dx}\left(\frac{d^3y}{dx^3}\right) = \frac{d}{dx}(72x - 30) = 72$$

$$\frac{d^5y}{dx^5} = \frac{d}{dx}\left(\frac{d^4y}{dx^4}\right) = \frac{d}{dx}(72) = 0$$

17. Calculate the derivatives up to third order:

(a) $y = x^6$; (b) $y = x^{-2}$;
(c) $y = x^2 \ln x$.

Answer

(a) $y' = 6x^5$, $y'' = 30x^4$,
    $y''' = 120x^3$

(b) $y' = -2x^{-3}$, $y'' = 6x^{-4}$,
    $y''' = -24x^{-5}$;

(c) $y' = 2x \ln x + x$,
    $y'' = 2 \ln x + 3$, $y''' = 2x^{-1}$.

$$\frac{d^6 y}{dx^6} = \frac{d}{dx}\left(\frac{d^5 y}{dx^5}\right) = \frac{d}{dx}(0) = 0$$

and so on. ☞ **17**

In this particular example, all derivatives higher than the fourth derivative are zero. This occurs because the fourth derivative is a constant.

EXAMPLE 2 Find the second derivative of $f(t) = e^{t^2+1}$.

Solution To find the first derivative, we use the chain rule. Thus

$$f'(t) = e^{t^2+1} \cdot \frac{d}{dt}(t^2 + 1) = e^{t^2+1} \cdot 2t = 2t e^{t^2+1}.$$

Now $f'(t)$ is the product of two functions $u = 2t$ and $v = e^{t^2+1}$. To find $f''(t)$, we shall use the product rule.

$$f''(t) = 2t\frac{d}{dt}(e^{t^2+1}) + e^{t^2+1}\frac{d}{dt}(2t) = 2t\left[e^{t^2+1}\frac{d}{dt}(t^2 + 1)\right] + e^{t^2+1}(2)$$

where we have used the chain rule to differentiate $v = e^{t^2+1}$.

Therefore

$$f''(t) = 2t[e^{t^2+1} \cdot 2t] + 2e^{t^2+1} = 2e^{t^2+1}(2t^2 + 1).$$

EXAMPLE 3 *(Falling Body)* A body falling under gravity from a position of rest falls a distance $s = 16t^2$ in t seconds. Find its acceleration.

Solution The velocity after t seconds is

$$\frac{ds}{dt} = \frac{d}{dt}(16t^2) = 32t \text{ feet/second.}$$

We obtain the acceleration by differentiating again.

$$\text{Acceleration} = \frac{d^2 s}{dt^2} = \frac{d}{dt}(32t) = 32 \text{ feet/second}^2$$

Note that this is independent of t: A body falling under gravity has a constant acceleration of 32 feet/second². ☞ **18**

18. If the distance traveled in t seconds is $s = 12t - t^3$, evaluate the distance, velocity and acceleration when $t = 1$, $t = 2$, and $t = 3$.

Answer

11, 9, and -6 at $t = 1$;
16, 0, and -12 at $t = 2$;
9, -15, and -18 at $t = 3$.

If $C(x)$ is a manufacturer's cost function—the cost of producing x items—then the first derivative $C'(x)$ gives the marginal cost, that is, the cost per additional item of a small increase in production. The second derivative $C''(x)$ gives the rate of increase of marginal cost with respect to an increase in production. We shall have more to say on the interpretation of this quantity in the next chapter, but meanwhile the following example will illustrate certain aspects of its significance.

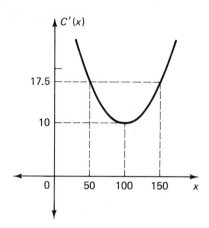

FIGURE 2

EXAMPLE 4 *(Cost Function Analysis)* For the cost function

$$C(x) = 0.001x^3 - 0.3x^2 + 40x + 1000,$$

the marginal cost is

$$C'(x) = 0.003x^2 - 0.6x + 40.$$

The second derivative is

$$C''(x) = 0.006x - 0.6 = 0.006(x - 100).$$

When $x = 150$, the marginal cost is $C'(150) = 17.5$. Furthermore,

$$C''(150) = 0.006(150 - 100) = 0.3.$$

We can interpret this result as signifying that each additional unit produced leads to an increase of 0.3 in the marginal cost.

Observe that in this example, $C''(x) < 0$ when $x < 100$. This means that when $x < 100$, increases in production lead to decreases in marginal cost. The graph of $C'(x)$ as a function of x slopes downward when $x < 100$. (See Figure 2.) However, when $x > 100$, the graph of $C'(x)$ slopes upward, so its slope, $C''(x)$, is positive. In this case, increases in production lead to increases in the marginal cost.

EXERCISES 13-4

(1–4) Find the first and higher-order derivatives of the following functions with respect to the independent variable involved.

1. $y = 3x^5 + 7x^3 - 4x^2 + 12$

2. $u = (t^2 + 1)^2$

3. $f(x) = x^3 - 6x^2 + 9x + 16$

4. $y(u) = (u^2 + 1)(3u - 2)$

5. Find y'' if $y = \dfrac{x^2}{x^2 + 1}$.

6. Find $f'''(t)$ if $f(t) = \dfrac{t - 1}{t + 1}$.

7. Find $g^{(4)}(u)$ if $g(u) = \dfrac{1}{3u + 1}$.

8. Find $\dfrac{d^2y}{dt^2}$ if $y = \sqrt{t^2 + 1}$.

9. Find $\dfrac{d^2u}{dx^2}$ if $u = \dfrac{1}{x^2 + 1}$.

10. Find $\dfrac{d^3y}{dx^3}$ if $y = \dfrac{x^3 - 1}{x - 1}$, $(x \neq 1)$.

11. Find y''' if $y = \ln x$.

12. Find $y^{(4)}$ if $y = x \ln x$.

13. Find $y^{(4)}$ if $y = xe^x$.

14. Find y'' if $y = e^{x^2}$.

15. Find y'' if $y = \ln [(x + 1)(x + 2)]$.

16. Find y''' if $y = x^3 + e^{2x}$.

17. Find y'' if $y = (x + 1)e^{-x}$.

18. Find y'' if $y = \dfrac{x^2 + 1}{e^x}$.

19. If $y = ae^{mx} + be^{-mx}$, where a, b, m are constants, then show that $d^2y/dx^2 = m^2y$.

20. If $y = x + 1/x$, then show that $x^2y'' + xy' - y = 0$

21. *(Velocity and Acceleration)* Find the velocity and the acceleration of a moving object for each given distance s traveled in time t.

a. $s = 9t + 16t^2$

b. $s = 3t^3 + 7t^2 - 5t$

22. *(Velocity and Acceleration)* Suppose the distance s traveled in time t is given by $s = t(3 - t)$.

a. At what times is the velocity zero?

b. What is the value of acceleration when the velocity equals zero?

(23–24) *(Marginal Cost Rate)* Find the marginal cost and the rate of change of marginal cost with respect to volume of production for the following cost functions.

23. $C(x) = 500 + 30x - 0.1x^2 + 0.002x^3$

24. $C(x) = 500 + 20x - 2x \ln x + 0.01x^2$

25. *(Marginal Average Cost Rate)* If $\overline{C}(x)$ is the average cost function, show that

$$\overline{C}''(x) = \frac{C''(x)}{x} - \frac{2C'(x)}{x^2} + \frac{2C(x)}{x^3}.$$

26. *(Marginal Revenue Rate)* If $R(x)$ is the revenue function, show that

$$R''(x) = 2p'(x) + xp''(x)$$

where $p = p(x)$ is the price as a function of demand x.

*27. *(Population Growth)* A population grows according to the logistic equation $y = y_m/(1 + Ce^{-kt})$, where y_m, C, and k are constants. Calculate the rate at which the population growth rate is changing.

CHAPTER REVIEW

Key Terms, Symbols, and Concepts

13.1 Product rule. Quotient rule. Marginal average cost.

13.2 Chain rule. Related rates.

13.4 Second derivative; acceleration. Third, fourth, and higher derivatives.

Formulas

Product rule:

$$\frac{d}{dx}(uv) = u\frac{dv}{dx} + v\frac{du}{dx} \quad \text{or} \quad (uv)' = uv' + vu'.$$

Quotient rule:

$$\frac{d}{dx}\left(\frac{u}{v}\right) = \frac{v\dfrac{du}{dx} - u\dfrac{dv}{dx}}{v^2} \quad \text{or} \quad \left(\frac{u}{v}\right)' = \frac{vu' - uv'}{v^2}.$$

Marginal revenue:

$$\frac{dR}{dx} = \frac{d}{dx}(px) = p + x\frac{dp}{dx}.$$

Marginal average cost:

$$\frac{d\overline{C}}{dx} = \frac{1}{x}[C'(x) - \overline{C}(x)], \quad \text{where} \quad \overline{C}(x) = \frac{C(x)}{x}.$$

Chain rule: $\dfrac{dy}{dx} = \dfrac{dy}{du}\dfrac{du}{dx}$.

Chain rule forms:

If $y = [u(x)]^n$, then $\dfrac{dy}{dx} = n[u(x)]^{n-1}\dfrac{du}{dx}$.

Related rates: If $y = f(x)$, then $\dfrac{dy}{dt} = f'(x)\dfrac{dx}{dt}$.

If $y = e^{kx}$, then $\dfrac{dy}{dx} = ke^{kx}$.

$\dfrac{d}{dx}(e^x) = e^x$.

If $y = e^{u(x)}$, then $\dfrac{dy}{dx} = e^{u(x)}\dfrac{du}{dx}$.

$\dfrac{d}{dx}(\ln x) = \dfrac{1}{x}$.

If $y = \ln u(x)$, then $\dfrac{dy}{dx} = \dfrac{1}{u(x)}\dfrac{du}{dx} = \dfrac{u'(x)}{u(x)}$.

REVIEW EXERCISES FOR CHAPTER 13

1. State whether each of the following is true or false. Replace each false statement with a corresponding true statement.

a. The derivative of the product of two functions is equal to the product of their derivatives.

b. The derivative of the quotient of two functions is equal to the derivative of the numerator divided by the denominator all plus the numerator multiplied by the derivative of 1 over the denominator.

c. If $y = [u(x)]^n$, then $\dfrac{dy}{dx} = n[u(x)]^{n-1}$.

d. If $y = \ln\left[\dfrac{1}{u(x)}\right]$, then $\dfrac{dy}{dx} = u(x)\dfrac{d}{dx}\left[\dfrac{1}{u(x)}\right]$.

e. If $y = e^{\ln u(x)}$, then $\dfrac{dy}{dx} = u'(x)$.

f. The second derivative of any quadratic function is zero.

g. If the acceleration of a moving object is zero, then its velocity is also zero.

h. If $y = [u(x)]^n$, then $\dfrac{d^2y}{dx^2} = n[u(x)]^{n-1}u''(x)$.

i. $\dfrac{d}{dx}(e^{x^2}) = e^{2x}$

j. $\dfrac{d}{dx}[\ln(x^2 + 1)] = \dfrac{1}{x^2 + 1}$

k. $\dfrac{d}{dx}(\ln 2) = \dfrac{1}{2}$

l. $\dfrac{d}{dx}(e^x) = xe^{x-1}$

m. $\dfrac{d}{dx}\left(\dfrac{1}{x^3}\right) = \dfrac{1}{3x^2}$

(2–23) Find dy/dx for the following functions.

2. $y = (3x + 7)\ln x$

3. $y = x^2 \ln x^3$

4. $y = (x + 2)\ln(x + 2)$

5. $y = x^3 e^{-x}$

6. $y = \dfrac{x^2}{e^{3x}}$

7. $y = e^x \ln(x + 3)$

8. $y = (3x - 5)^2(x^2 + 1)^2$

9. $y = (2x + 1)^3(3x - 1)^4$

10. $y = \left(\dfrac{x^2 + 1}{x^2 + 4}\right)^3$

11. $y = \dfrac{(x^2 + 1)^3}{(x - 1)^4}$

12. $y = \dfrac{x}{\sqrt{x^2 + 1}}$

13. $y = x\sqrt{x^2 + 4}$

14. $y = (x + 1)\sqrt{x + 3}$

15. $y = \dfrac{\sqrt{x + 1}}{\sqrt[3]{x + 1}}$

16. $y = x^2 e^{x^2}$

17. $y = x^{\sqrt{2}}\ln x$

18. $y = \ln\left(\dfrac{x^3}{\sqrt{x^2 + 7}}\right)$

19. $y = \ln\left(\dfrac{2^x}{\sqrt{x^3 + 1}}\right)$

20. $y = \ln\left[\dfrac{x^5\sqrt{x^2 + 1}}{(2x - 3)^4}\right]$

21. $y = \ln\left(\dfrac{xe^{x^2}}{\sqrt{3x + 5}}\right)$

22. $y = \ln e^x + \ln x^e$

*23. $y = x^x$

(24–27) Find the equation of the tangent line to the graphs of the following functions at the indicated point.

24. $y = x\sqrt{x^2 + 9}$ at $x = 0$

25. $f(x) = xe^{-x}$ at $x = 0$

26. $f(x) = \dfrac{\ln x}{x}$ at $x = 1$

27. $y = \dfrac{x}{\sqrt{x + 2}}$ at $x = 2$

(28–31) Find d^2y/dx^2 for the following functions.

28. $y = \sqrt[3]{x^3 + a^3}$

29. $y = (3x - 7)^6(x + 1)^4$

30. $y = x^n \ln x$ **31.** $y = \ln(\ln x)$

(32–33) *(Marginal Revenue)* Calculate the marginal revenue for the following demand relations.

32. $p = a - b \ln x$ **33.** $x = a - b \ln p$

(34–35) *(Marginal Cost)* Calculate the marginal cost and marginal average cost for the following cost functions.

34. $C(x) = 600 + 25x - 4(x + 1) \ln(x + 1) + 0.02x^2$

35. $C(x) = 100 + 0.5x + 0.01xe^x$

36. *(Marginal Price)* The demand equation of a certain commodity is $p = 300/(x^2 + 1)$. Find the marginal price at a demand level of 3 units.

37. *(Marginal Price)* If x units can be sold at a price of $\$p$ each where $x + \ln(p + 1) = 50$, $(0 \le x \le 50)$, find the marginal price.

38. *(Marginal Demand)* The demand equation of a certain product is given by $2x + 3 \ln(p + 1) = 60$. Find the marginal demand at a price level of $p = 2$. Interpret your result.

39. *(Marginal Demand)* The demand of a certain product is given by the equation $p^2 + x^2 = 2500$, where x units can be sold at a price of $\$p$ each. Determine the marginal demand at a price level of 40 dollars. Interpret your result.

40. *(Marginal Physical Productivity)* The physical productivity of a certain firm is given by $\rho = 500(x + 4)^{3/2} - 4000$, where x is the number of machines at work. Determine the marginal physical productivity when 5 machines are working. Interpret your result.

41. *(Marginal Revenue and Profits)* The demand equation of a certain product is $p = 300e^{-x/20}$ where x units are sold at a price of $\$p$ each. If the manufacturer has a fixed cost of $\$500$ and the variable cost of $\$20$ per unit, find the marginal revenue and marginal profit functions.

42. *(Marginal Revenue)* If x units can be sold at a price of p each, where $2p + \ln(x + 1) = 70$, find the marginal revenue function.

43. *(Moving Object)* The distance traveled by a moving object up to time t is given by $y = (3t + 1)\sqrt{t + 1}$. Find the instantaneous velocity at time t.

44. *(Population Growth)* The size of a certain population at time t is

$$\left(\frac{t^2 + 3t + 1}{t + 1}\right)^6$$

Find the rate of change of the population size.

45. *(Epidemic)* During an epidemic, the number of infected individuals is given by $f(t) = at^p e^{-t}$ at time t (a and p are constants). Find the value of t for which $f'(t) = 0$.

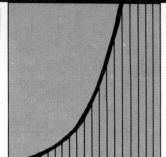

CHAPTER 14

Optimization and Curve Sketching

Chapter Objectives

14-1 THE FIRST DERIVATIVE AND THE GRAPH OF THE FUNCTION

(a) Definition of increasing and decreasing functions and test based on the first derivative;
(b) Determination of intervals where a given function is increasing or decreasing;

14-2 MAXIMA AND MINIMA

(a) Critical points of a function and their determination.
(b) Definition of local maximum and local minimum of a function;
(c) Relationship between local extrema and critical points;
(d) Statement and application of the first derivative test for determining local maxima and minima.

14-3 THE SECOND DERIVATIVE AND CONCAVITY

(a) Relationship between the sign of the second derivative and the concavity properties of the graph;
(b) Determination of intervals in which the graph of a function is concave upwards or concave downwards;
(c) Definition and determination of points of inflection;
(d) The second-derivative test for local maxima and minima and its application.

14-4 CURVE SKETCHING FOR POLYNOMIALS

(a) Examples of the use of the first and second derivatives of a function to obtain a sketch of its graph.

14-5 APPLICATIONS OF MAXIMA AND MINIMA

(a) Examples of the application of the theory of maxima and minima to word problems and practical optimization problems:
(1) An optimal conservation problem,
(2) A number theory problem,
(3) A minimal cost problem,
(4) Three problems of maximizing profit by optimizing the production, schedule or the advertizing expenditure,
(5) A problem of maximizing profits and revenue from taxation,
(6) An inventory cost model.

14-6 ABSOLUTE MAXIMA AND MINIMA

(a) Definition and determination of the absolute maximum and minimum values of a function in an interval;
(b) Examples of determination of absolute maxima and minima.

14-7 ASYMPTOTES

(a) Horizontal asymptotes and the limits of the function at $\pm\infty$;

(b) Determination of limits at $\pm\infty$ for rational functions;

(c) Graph sketching for functions with horizontal asymptotes;

(d) Vertical asymptotes; determination on which side of a vertical asymptote the graph lies;

(e) Graph sketching for functions with vertical asymptotes;

CHAPTER REVIEW

14-1 THE FIRST DERIVATIVE AND THE GRAPH OF THE FUNCTION

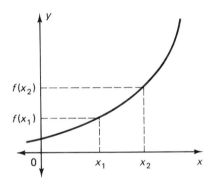

(a) $x_2 > x_1$; $f(x_2) > f(x_1)$

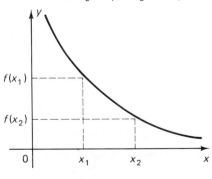

(b) $x_2 > x_1$; $f(x_2) < f(x_1)$

FIGURE 1

In this section we shall consider the significance of the first derivative of a function as it relates to the graph.

DEFINITION A function $y = f(x)$ is said to be an **increasing function** over an interval of values of x if y increases with increase of x. That is, if x_1 and x_2 are any two values in the given interval with $x_2 > x_1$, then $f(x_2) > f(x_1)$.

A function $y = f(x)$ is said to be a **decreasing function** over an interval of its domain if y decreases with increase of x. That is, if $x_2 > x_1$ are two values of x in the given interval, then $f(x_2) < f(x_1)$.

Parts (a) and (b) of Figure 1 illustrate an increasing and a decreasing function, respectively. The graph rises or falls, respectively, as we move from left to right.

THEOREM 1

(a) If $f(x)$ is an increasing function that is differentiable, then $f'(x) \geq 0$.

(b) If $f(x)$ is a decreasing function that is differentiable, then $f'(x) \leq 0$.

PROOF (a) Let x and $x + \Delta x$ be two values of the independent variable, with $y = f(x)$ and $y + \Delta y = f(x + \Delta x)$ the corresponding values of the dependent variable. Then

$$\Delta y = f(x + \Delta x) - f(x).$$

There are two cases to consider, depending on whether $\Delta x > 0$ or $\Delta x < 0$. They are illustrated in Figures 2 and 3.

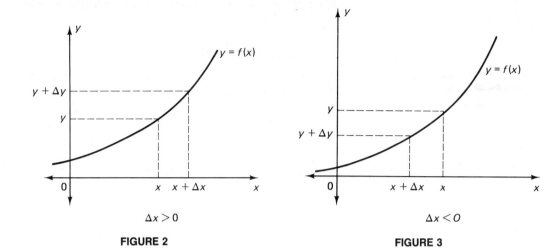

$$\Delta x > 0$$

FIGURE 2

$$\Delta x < 0$$

FIGURE 3

If $\Delta x > 0$, then $x + \Delta x > x$. Therefore, since $f(x)$ is an increasing function, $f(x + \Delta x) > f(x)$, so that $\Delta y > 0$. Consequently, both Δx and Δy are positive, so that $\Delta y/\Delta x > 0$.

The second possibility is that $\Delta x < 0$. Then $x + \Delta x < x$ and so $f(x + \Delta x) < f(x)$. Hence $\Delta y < 0$. In this case, Δx and Δy are both negative, so that again $\Delta y/\Delta x > 0$.

So, in both cases, $\Delta y/\Delta x$ is positive. The derivative $f'(x)$ is the limit of $\Delta y/\Delta x$ as $\Delta x \to 0$, and since $\Delta y/\Delta x$ is always positive, it is clearly impossible for it to approach a negative number as limiting value. Therefore, $f'(x) \geq 0$, as stated in the theorem.

The proof of part (b), when $f(x)$ is a decreasing function, is quite similar and is left as an exercise.

This theorem has a converse, which may be stated as follows.

THEOREM 2

(a) If $f'(x) > 0$ for all x in an interval, then f is an increasing function in that interval.

(b) If $f'(x) < 0$ for all x in an interval, then f is a decreasing function in that interval.

Note   Observe that in Theorem 2, the inequalities are strict.

The proof of this theorem will not be given. However, it is an intuitively obvious result. In part (a), for example, the fact that $f'(x) > 0$ means, geometrically, that the tangent to the graph at any point has positive slope. If the graph of $f(x)$ always slopes upward to the right, then clearly y must increase as x increases. Correspondingly, in part (b), if $f'(x) < 0$, then the graph slopes downward to the right and y decreases as x increases.

These theorems are used to determine the intervals in which a function is increasing or decreasing—that is, where the graph is rising or falling.

EXAMPLE 1 Find the values of x for which the function

$$f(x) = x^2 - 2x + 1$$

is increasing or decreasing.

☛ **1.** By examining the sign of f', decide for what values of x the following functions are increasing or decreasing

(a) $f(x) = x^2$;

(b) $f(x) = x^2 + 4x$;

(c) $f(x) = x^3$.

Solution Since $f(x) = x^2 - 2x + 1$, we have $f'(x) = 2x - 2$. Now $f'(x) > 0$ implies that $2x - 2 > 0$, that is, $x > 1$. Thus $f(x)$ is increasing for all values of x in the interval $x > 1$. Similarly, $f'(x) < 0$ implies that $2x - 2 < 0$, that is, $x < 1$. The function is decreasing for $x < 1$.

The graph of $y = f(x)$ is shown in Figure 4. (Note that $f(1) = 0$, so the point $(1, 0)$ lies on the graph.) For $x < 1$, the graph slopes negatively and for $x > 1$, it slopes positively. ☛ **1**

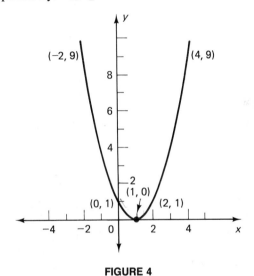

FIGURE 4

EXAMPLE 2 Find the values of x for which the function

$$f(x) = x^3 - 3x$$

is increasing or decreasing.

Solution We have $f'(x) = 3x^2 - 3 = 3(x - 1)(x + 1)$. To find the interval in which $f(x)$ is increasing, we set $f'(x) > 0$, that is,

$$3(x - 1)(x + 1) > 0.$$

This type of quadratic inequality was examined in Section 3-3. The procedure is first to replace the inequality sign by an equals sign and solve for x. We get $3(x - 1)(x + 1) = 0$, giving $x = -1$ and $x = +1$.

These two numbers divide the number line into three intervals: $(-\infty, -1)$, $(-1, 1)$, and $(1, \infty)$. In each of these intervals, $f'(x)$ has constant sign since it only changes sign at $x = \pm 1$, where it is zero. So we simply choose a test point in each interval and calculate the sign of $f'(x)$ at each test point. The results are given in Table 1.

Answer (a) Increasing for $x > 0$, decreasing for $x < 0$;

(b) increasing for $x > -2$, decreasing for $x < -2$;

(c) increasing for $x < 0$ and for $x > 0$.

TABLE 1

Interval	$(-\infty, -1)$	$(-1, 1)$	$(1, \infty)$
Test Point	-2	0	2
$f'(x) = 3x^2 - 3$	$3(-2)^2 - 3 = 9 > 0$	$3(0)^2 - 3 = -3 < 0$	$3(2)^2 - 3 = 9 > 0$
f	Increasing	Decreasing	Increasing

☞ **2.** For what values of x are the following functions increasing or decreasing?

(a) $f(x) = x^3 - 3x^2$;

(b) $f(x) = x^{-1} + x$;

(c) $f(x) = 2 \ln x - x^2$.

We see that $f'(x) > 0$ in $(-\infty, -1)$ and in $(1, \infty)$, so f is an increasing function of x in each of those intervals. In $(-1, 1)$, $f'(x) < 0$, so f is a decreasing function in that interval. The graph of f is shown in Figure 5. ☞ **2**

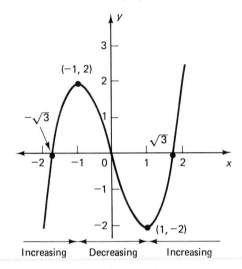

FIGURE 5

EXAMPLE 3 *(Analysis of Cost, Revenue, and Profit Functions)* For the cost function $C(x) = 500 + 20x$ and the demand relation $p = 100 - x$, find the regions in which the cost function, revenue function, and profit function are increasing and decreasing functions of x.

Solution Since $C(x) = 500 + 20x$, $C'(x) = 20$ is always positive. Hence the cost function is an increasing function of x for all values of x. The revenue function is

$$R(x) = xp = x(100 - x) = 100x - x^2.$$

Therefore the marginal revenue is

$$R'(x) = 100 - 2x.$$

So $R'(x) > 0$ when $100 - 2x > 0$, that is, when $x < 50$. When $x > 50$, $R'(x) < 0$. So the revenue function is an increasing function of x for $x < 50$ and a decreasing function of x for $x > 50$.

The profit function is

$$P(x) = R(x) - C(x) = 100x - x^2 - (500 + 20x) = 80x - x^2 - 500.$$

Answer (a) Increasing for $x < 0$ and $x > 2$, decreasing for $0 < x < 2$;

(b) increasing for $x < -1$ and $x > 1$, decreasing for $-1 < x < 0$ and $0 < x < 1$;

(c) increasing for $0 < x < 1$ and decreasing for $x > 1$. (Domain is $x > 0$ only.)

3. Re-solve Example 3 if the demand equation is changed to $p = 120 - 2x$.

Then $P'(x) = 80 - 2x$ and $P'(x) > 0$ when $80 - 2x > 0$, or $x < 40$; alternatively, $P'(x) < 0$ when $x > 40$. So the profit function is an increasing function of x for $x < 40$, and it is a decreasing function of x for $x > 40$. The graphs of the three functions are shown in Figure 6. ☞ **3**

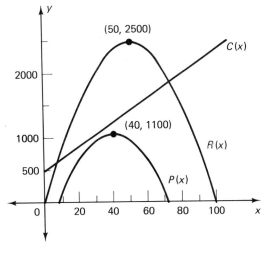

FIGURE 6

The type of behavior found for these three functions is quite typical of general cost, revenue, and profit functions. The cost function is usually an increasing function of the quantity of goods produced (it usually costs more to produce more, although exceptions do occur with certain pricing policies for the raw materials.) Similarly, the revenue function is, in general, an increasing function for small volumes of sales, but usually becomes a decreasing function for large volumes of sales. The profit function commonly has this same qualitative behavior of increasing for small x and decreasing for large x.

Answer R increasing for $0 < x < 30$, decreasing for $x > 30$. P increasing for $0 < x < 25$, decreasing for $x > 25$.

EXERCISES 14-1

(1–24) Find the values of x for which the following functions are: **(a)** increasing; **(b)** decreasing.

1. $y = x^2 - 6x + 7$

2. $y = x^3 - 12x + 10$

3. $f(x) = x^3 - 3x + 4$

4. $f(x) = 2x^3 - 9x^2 - 24x + 20$

5. $f(x) = x + \dfrac{1}{x}$

6. $f(x) = x^2 + \dfrac{1}{x^2}$

7. $f(x) = \dfrac{x}{x + 1}$

8. $f(x) = \dfrac{x + 1}{x - 1}$

9. $y = x + \ln x$

10. $y = x - e^x$

11. $y = x \ln x$

12. $y = xe^{-x}$

13. $y = x^5 - 5x^4 + 1$

14. $y = x^7 - 7x^6$

15. $y = x^2 - 4x + 5$

16. $y = x^3 - 3x + 2$

17. $y = 5x^6 - 6x^5 + 1$

18. $y = x^4 - 2x^2$

19. $y = x^{2/3}$

20. $y = x^{1/5}$

21. $y = \ln x$

22. $y = e^{-2x}$

23. $y = \dfrac{2}{x}$

24. $y = \dfrac{-1}{x}$

(25–28) *(Analysis of Cost, Revenue, and Profit Functions)* For the following cost functions and demand relations, determine the regions in which **(a)** the cost function, **(b)** the revenue function, and **(c)** the profit function are increasing and decreasing.

25. $C(x) = 2000 + 10x$; $\quad p = 100 - \frac{1}{2}x$

26. $C(x) = 4000 + x^2$; $\quad p = 300 - 2x$

27. $C(x) = C_0 + kx$; $\quad p = a - bx$ (a, b, k, and C_0 are positive constants.)

28. $C(x) = \sqrt{100 + x^2}$;
$p = a - (b/x)\sqrt{100 + x^2}$
(Assume $b > a > 0$.)

29. *(Marginal Cost Analysis)* The cost of producing x thousands of a certain product is given by $C(x) = 2500 + 9x - 3x^2 + 2x^3$. At what production level x is the marginal cost:

a. increasing

b. decreasing?

30. Repeat Exercise 29, if $C(x) = 2000 + 15x - 6x^2 + x^3$.

31. *(Marginal Revenue Analysis)* Given the demand relation is $p = 600 - x^2$, where x units can be sold at a price of p each. Find when the marginal revenue is:

a. increasing

b. decreasing.

32. Repeat Exercise 31 for the demand relation $p = 50e^{-x/20}$.

33. *(Marginal and Average Costs)* For the cost function $C(x) = 6 + 2x(x + 4)/(x + 1)$, show that the marginal and average costs are always decreasing for $x > 0$.

34. *(Marginal Revenue)* For the demand relation $p = 50 - \ln(x + 1)$, show that the marginal revenue is always decreasing for $x > 0$.

35. *(Increasing Average Cost)* Show that the average cost function $\overline{C}(x)$ is an increasing function when the marginal cost exceeds the average cost.

36. *(Material Happiness)* Let $H(x)$ be the amount of happiness that an individual derives from the ownership of x units of some material possession. One model sometimes used for this quantity is $H(x) = A \ln(1 + x) - Bx$, where A and B are positive constants with $A > B$. Evaluate $H(0)$. Show that $H(x)$ is an increasing function for small values of x but eventually becomes a decreasing function. Find the value of x at which $H(x)$ is greatest.

◤ 14-2 MAXIMA AND MINIMA

Many of the important applications of derivatives involve finding the maximum or minimum values of a particular function. For example, the profit a manufacturer makes depends on the price charged for the product, and the manufacturer is interested in knowing the price which makes his profit maximum. The **optimum** price (or **best** price) is obtained by a process called **maximization** or **optimization** of the profit function. In a similar way, a real estate company may be interested in knowing the rent to charge for the offices or apartments it controls to generate the maximum rental income; a railway company may want to know the average speed at which trains should run in order to minimize the cost per mile of operation; or an economist may wish to know the level of taxation in a country that will promote the maximum rate of growth of the economy. Before we look at applications such as these, however, we shall discuss the theory of maxima and minima.

DEFINITIONS (a) A function $f(x)$ is said to have a **local maximum** at $x = c$ if $f(c) > f(x)$ for all x sufficiently near to c.

Thus the points P and Q in the graphs in Figure 7 correspond to local maxima of the corresponding functions.

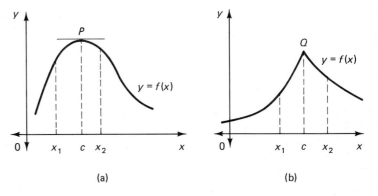

(a)

(b)

FIGURE 7

(b) A function $f(x)$ is said to have a **local minimum** at $x = c$ if $f(c) < f(x)$ for all x sufficiently close to c.

The points A and B in the graphs in Figure 8 correspond to local minima.

(c) The term **extremum** is used to denote either a local maximum or a local minimum. Extrema is the plural of extremum.

A function may have more than one local maximum and more than one local minimum, as is shown in Figure 9. The points A, C, and E on the graph

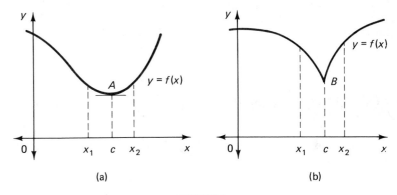

(a)

(b)

FIGURE 8

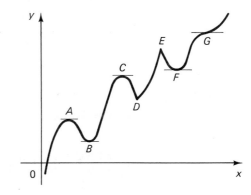

FIGURE 9

4. Give the values of x at which the following graphs have local maxima or minima.

(a)

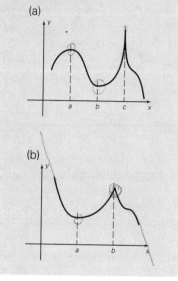

(b)

Answer (a) Local maxima at a and c, local minimum at b; (b) local minimum at a, local maximum at b.

5. What are the critical points of the function f if
(a) $f(x) = x^2$;
(b) $f(x) = |x - 1|$?

Answer (a) $x = 0$; (b) $x = 1$.

correspond to points where the function has local maxima, and the points B, D, and F correspond to points where the function has local minima. **4**

A (local) *maximum or minimum value* of a function is the y-coordinate at the point at which the graph has a local maximum or minimum. *A local minimum value of a function may be greater than a local maximum value.* This can be easily seen from the above graph, where the y-value at F is greater than the y-value at A.

DEFINITION The value $x = c$ is called a **critical point** for a continuous function f if $f(c)$ is well-defined and if *either $f'(c) = 0$ or $f'(x)$ fails to exist* at $x = c$.

In the case when $f'(c) = 0$, the tangent to the graph of $y = f(x)$ is horizontal at $x = c$. This possibility is illustrated in part (a) of Figure 10. The second case, when $f'(c)$ fails to exist, occurs when the graph has a corner at $x = c$ (see part (b) of Figure 10) or when the tangent to the graph becomes vertical at $x = c$ (so that $f'(x)$ becomes infinitely large as $x \to c$). (See part (c) of Figure 10.) **5**

We emphasize the fact that for c to be a critical point, $f(c)$ must be well-defined. Consider, for example, $f(x) = x^{-1}$, whose derivative is $f'(x) = -x^{-2}$.

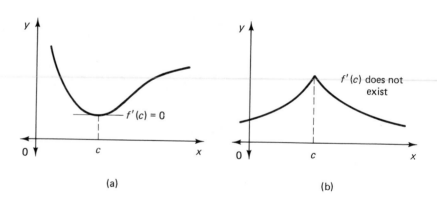

(a)

(b)

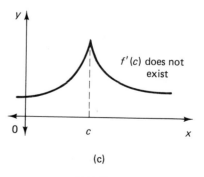

(c)

FIGURE 10

Clearly, $f'(x)$ becomes unbounded as $x \to 0$. However, $x = 0$ is not a critical point for this function since $f(0)$ does not exist.

It is clear from the graphs in Figure 10 that the local extrema of a function f occur only at critical points. But not every critical point of a function corresponds to a local minimum or a local maximum. The point P in part (a) of Figure 11, where the tangent is horizontal, is a critical point but is neither a local maximum nor a local minimum point. The points Q and R in parts (b) and (c) are critical points at which $f'(c)$ fails to exist, but are not extrema of $f(x)$.

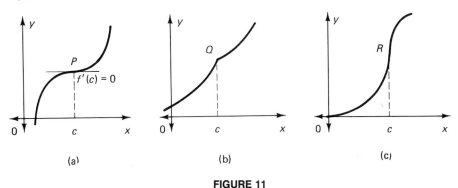

FIGURE 11

Shortly, we shall develop certain tests that will enable us to distinguish those critical points which are local extrema from those which are not. First let us examine critical points themselves through some examples.

EXAMPLE 1 Determine the critical points of the function

$$f(x) = x^3(2x^3 - 3x).$$

Solution We have $f(x) = 2x^6 - 3x^4$. Differentiating, we get

$$f'(x) = 12x^5 - 12x^3 = 12x^3(x^2 - 1).$$

It is clear that $f'(x)$ exists for all x, so the only critical points are those at which $f'(x)$ vanishes:

$$f'(x) = 12x^3(x^2 - 1) = 0$$

so

$$x^3 = 0 \quad \text{or} \quad x^2 - 1 = 0.$$

So the critical points are $x = 0, \pm 1$.

EXAMPLE 2 Determine the critical points of the function

$$f(x) = x^4(x - 1)^{4/5}.$$

Solution Differentiating, using the product rule, we get

$$f'(x) = 4x^3(x - 1)^{4/5} + x^4(\tfrac{4}{5})(x - 1)^{-1/5}$$

6. What are the critical points of the function f if
(a) $f(x) = x^3 + 3x^2$;
(b) $f(x) = x^4 - 8x^2$;
(c) $f(x) = x(x - 4)^{1/3}$.?

$$= \tfrac{4}{5}x^3(x - 1)^{-1/5}[5(x - 1) + x]$$
$$= \tfrac{4}{5}x^3(x - 1)^{-1/5}(6x - 5).$$

Now $f'(x) = 0$ when either $x^3 = 0$ or $6x - 5 = 0$, so we have critical points at $x = 0$ and $x = \tfrac{5}{6}$. However, we observe that $f'(x)$ becomes infinitely large as $x \to 1$ because of the negative power. Since $f(1)$ is well-defined (in fact $f(1) = 0$), $x = 1$ must be a critical point of the type at which $f'(x)$ fails to exist.

EXAMPLE 3 Find the critical points of the function
$$f(x) = x^3 e^{-x^2}.$$

Solution We use the product rule.
$$f'(x) = 3x^2 e^{-x^2} + x^3(-2xe^{-x^2})$$
$$= x^2 e^{-x^2}(3 - 2x^2)$$

The factor e^{-x^2} is never zero. Therefore $f'(x) = 0$ either when $x^2 = 0$ or when $3 - 2x^2 = 0$; that is, either when $x = 0$ or when $x = \pm\sqrt{\tfrac{3}{2}}$. So the given function has three critical points: $x = 0, \pm\sqrt{\tfrac{3}{2}}$. **6**

Answer (a) $x = 0, -2$;
(b) $x = 0, 2, -2$; (c) $x = 3, 4$.

The First-Derivative Test

Not all critical points are local extrema; several examples of critical points that are not local extrema were illustrated in Figure 11. The following theorem provides the first of the two tests that can be used to decide whether a given critical point is a local maximum or minimum or neither.

7. The following functions have a critical point at $x = 0$. Apply the first derivative test to determine the nature of this point.
(a) $f(x) = x^3$;
(b) $f(x) = x^4$;
(c) $f(x) = x^{1/3}$
(d) $f(x) = x^{4/3}$.

THEOREM 1 (FIRST-DERIVATIVE TEST) Let $x = c$ be a critical point of the function f. Then:

(a) If $f'(x) > 0$ for x just below c and $f'(x) < 0$ for x just above c, then c is a local maximum of f. (See part (a) of Figure 12. The $(+)$, $(-)$, or (0) next to each portion of the graph indicates the sign of f'.)

(b) If $f'(x) < 0$ for x just below c and $f'(x) > 0$ for x just above c, then c is a local minimum of f. (See part (b) of Figure 12.)

(c) $f'(x)$ has the same sign for x just below c and for x just above c, then c is not a local extremum of f. (See part (c) of Figure 12.)

Remark In part (a) of the theorem, f changes from increasing to decreasing as x moves to the right through c. In part (b), f changes from decreasing to increasing as x passes through c. In part (c), f is either increasing on both sides of c or decreasing on both sides. **7**

Answer (a) Not a local extremum;
(b) local minimum;
(c) not a local extremum;
(d) local minimum.

EXAMPLE 4 Find the local extrema of f, where $f(x) = x^4 - 4x^3 + 7$.

Solution In this case,
$$f'(x) = 4x^3 - 12x^2 = 4x^2(x - 3).$$

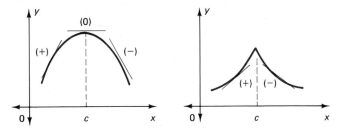

Local maximum at $x = c$

(a)

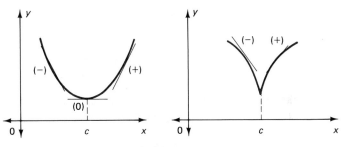

Local minimum at $x = c$

(b)

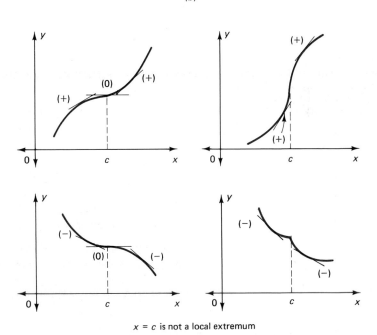

$x = c$ is not a local extremum

(c)

FIGURE 12

f' exists for all x, so the critical points are given by $f'(x) = 0$. That is, $4x^2(x - 3) = 0$, or $x = 0$ and $x = 3$. These critical points divide the number line into the three intervals $(-\infty, 0)$, $(0, 3)$, and $(3, \infty)$. As usual, we determine the sign of f' in each interval by choosing a test point. The results are in Table 2.

TABLE 2

Interval	$(-\infty, 0)$	$(0, 3)$	$(3, \infty)$
Test Point	-1	1	4
$f'(x) = 4x^2(x - 3)$	$4(-1)^2(-1 - 3)$	$4(1)^2(1 - 3)$	$4(4)^2(4 - 3)$
	$= -16 < 0$	$= -8 < 0$	$= 64 > 0$
f	Decreasing	Decreasing	Increasing

At $x = 0$, f' is negative on both sides, so $x = 0$ is not a local extremum. For $x = 3$, f' is negative to the left (f is decreasing) and positive to the right (f is increasing). Therefore, by part (b) of Theorem 1, $x = 3$ is a local minimum of f.

EXAMPLE 5 Find the local maxima and minima of the function $f(x) = x^{2/3}(x - 5)$.

Solution First we find the critical points. From the product rule we have

$$f'(x) = \tfrac{2}{3}x^{-1/3}(x - 5) + x^{2/3} \cdot 1 = \tfrac{5}{3}x^{-1/3}(x - 2).$$

$f' = 0$ when $x = 2$ and f' is undefined when $x = 0$. Thus there are two critical points, namely $x = 0$ and $x = 2$. These points divide the number line into the three intervals, $(-\infty, 0)$, $(0, 2)$, and $(2, \infty)$. Choosing a test point as usual in each of these intervals, we obtain the results shown in Table 3.

TABLE 3

Interval	$(-\infty, 0)$	$(0, 2)$	$(2, \infty)$
Test Point	-1	1	8
$f'(x) = \tfrac{5}{3}x^{-1/3}(x - 2)$	$\tfrac{5}{3}(-1)^{-1/3}(-3) = 5 > 0$	$\tfrac{5}{3}(1)^{-1/3}(-1) = -\tfrac{5}{3} < 0$	$\tfrac{5}{3}(8)^{-1/3}(6) = 5 > 0$
f	Increasing	Decreasing	Increasing

So, just below $x = 0$, f' is positive while just above $x = 0$ it is negative. By part (a) of Theorem 1, therefore, $x = 0$ is a local maximum of f. Just below $x = 2$, f' is negative, while just above $x = 2$ it is positive. By part (b) of Theorem 1, therefore, $x = 2$ is a local minimum of f.

EXAMPLE 6 Find the local maxima and minima of the function $f(x) = x^4/(x - 1)$.

Solution First we find the critical points. From the quotient rule we have

$$f'(x) = \frac{(x - 1) \cdot 4x^3 - x^4 \cdot 1}{(x - 1)^2} = \frac{x^3(3x - 4)}{(x - 1)^2}.$$

For a critical point, $f'(x) = 0$; thus $x = 0$ or $\frac{4}{3}$. (Note that $x = 1$ is not a critical point since $f(1)$ is not defined.)

In this case, we must be a little careful since the domain of the function is not the whole real line. We must consider the four intervals $(-\infty, 0)$, $(0, 1)$, $(1, \frac{4}{3})$, and $(\frac{4}{3}, \infty)$ since $x = 1$ is not in the domain. Choosing a test point as usual in each of these intervals, we obtain the result shown in Table 4.

TABLE 4

Interval	$(-\infty, 0)$	$(0, 1)$	$(1, \frac{4}{3})$	$(\frac{4}{3}, \infty)$
Test Point	-1	$\frac{1}{2}$	$\frac{7}{6}$	2
$f'(x)$	$\dfrac{(-1)^3(-7)}{(-2)^2} > 0$	$\dfrac{(\frac{1}{2})^3(-\frac{5}{2})}{(-\frac{1}{2})^2} < 0$	$\dfrac{(\frac{7}{6})^3(-\frac{3}{6})}{(\frac{1}{6})^2} < 0$	$\dfrac{(2)^3(2)}{(1)^2} > 0$
f	Increasing	Decreasing	Decreasing	Increasing

So, just below $x = 0$, f' is positive, while just above $x = 0$ it is negative. By part (a) of Theorem 1, therefore, $x = 0$ is a local maximum of f. Just below $x = \frac{4}{3}$, f' is negative while just above $x = \frac{4}{3}$ it is positive. By part (b) of Theorem 1, therefore, $x = \frac{4}{3}$ is a local minimum of f.

It is very important in this type of example to use different test points to examine f' just above 0 and just below $\frac{4}{3}$ since the whole interval between these two critical points is not in the domain of the function. ☞ **8**

☞ **8.** Find the local extrema of
(a) $f(x) = 12x - x^3$;
(b) $f(x) = 2x^4 - x^2$;
(c) $f(x) = x^{2/3}(x - 10)$.

Summary of Determining Local Extrema by the First-Derivative Test

Step 1. Find $f'(x)$ and determine the critical points, that is, the points where $f'(x)$ is zero or fails to exist.

Step 2. The critical points divide the domain of f into a number of intervals. In each interval choose a test point and compute $f'(x)$ at that point. If the value is positive, then f is an increasing function in the whole of the corresponding interval. If the value of $f'(x)$ at the test point is negative, then f is decreasing in the whole interval.

Step 3. If f' is positive to the left and negative to the right of a critical point, then that point is a local maximum. If f' is negative to the left and positive to the right of a critical point, then that point is a local minimum. If f' has the same sign on both sides of a critical point, then that point is not a local extremum.

Answer (a) Local minimum at -2, local maximum at $x = 2$;
(b) local minima at $x = \pm\frac{1}{2}$, local maximum at $x = 0$
(c) local maximum at $x = 0$, local minimum at $x = 4$.

EXERCISES 14-2

(1–20) Find the critical points for the following functions.

1. $x^2 - 3x + 7$

2. $3x + 5$

3. $2x^3 - 6x$

4. $2x^3 - 3x^2 - 36x + 7$

5. $x^4 - 2x^2$

6. $x^4 - 4x^3 + 5$

7. $x^2(x - 1)^3$

8. $(x - 1)^2(x - 2)^3$

9. $\dfrac{3x + 1}{3x}$

10. $x^2 + x^{-2}$

11. $\dfrac{x^2}{x-1}$

12. $x^{2/3} - x^{1/3}$

13. $x^{4/5} - 2x^{2/5}$

14. $x(x-1)^{1/3}$

15. $\dfrac{(x-1)^{1/5}}{x+1}$

16. xe^{-3x}

17. $x^3 \ln x$

18. $\dfrac{\ln x}{x}$

19. $2 + |x-3|$

20. $|x^2 - 3x + 2|$

(21–36) Find the values of x at the local maxima and minima of the following functions.

21. $f(x) = x^2 - 12x + 10$

22. $f(x) = 1 + 2x - x^2$

23. $f(x) = x^3 - 6x^2 + 7$

24. $f(x) = x^3 - 3x + 4$

25. $y = 2x^3 - 9x^2 + 12x + 6$

26. $y = 4x^3 + 9x^2 - 12x + 5$

27. $y = x^3 - 18x^2 + 96x$

28. $y = x^3 - 3x^2 - 9x + 7$

29. $y = x^5 - 5x^4 + 5x^3 - 10$

30. $y = x^4 - 4x^3 + 3$

31. $f(x) = x^3(x-1)^2$

32. $f(x) = x^4(x+2)^2$

33. $f(x) = x^{4/3}$

34. $f(x) = x^{1/3}$

35. $f(x) = x \ln x$

36. $f(x) = xe^{-x}$

(37–52) Find the local maximum and minimum values of the following functions.

37. $f(x) = 2x^3 + 3x^2 - 12x - 15$

38. $f(x) = \frac{1}{3}x^3 + ax^2 - 3xa^2 \quad (a > 0)$

39. $f(x) = xe^x$

40. $f(x) = xe^{-2x}$

41. $f(x) = x^3(x-1)^{2/3}$

42. $f(x) = x^4(x-1)^{4/5}$

43. $f(x) = \dfrac{\ln x}{x^2}$

44. $f(x) = \dfrac{(\ln x)^2}{x}$

45. $f(x) = |x-1|$

46. $f(x) = 2 - |x|$

47. $f(x) = |x^2 - 3x + 2|$

48. $f(x) = |6 + x - x^2|$

49. $f(x) = e^{|x|}$

50. $f(x) = e^{x-|x|}$

51. $f(x) = (x-2)^{4/3}$

52. $f(x) = (x+1)^{7/5} + 3$

53. Show that $f(x) = x^3 - 3x^2 + 3x + 7$ has neither a local maximum nor a minimum at $x = 1$.

54. Show that $f(x) = x + 1/x$ has a local maximum and a local minimum value, but the maximum value is less than the minimum value.

◥ **14-3** THE SECOND DERIVATIVE AND CONCAVITY

We have seen in the preceding sections that the sign of the first derivative has a geometrical significance that is extremely useful. We shall now go on to consider the second derivative, which, as we shall see, also has an important geometrical interpretation.

Consider a function f whose graph has the general shape indicated in Figure 13. The slope of the graph is positive, $f'(x) > 0$, so f is an increasing function. Furthermore, the graph shown has the property that as we move to the right the slope becomes greater. That is, the derivative f' must also be an increasing function of x. The graph of f' must have the form shown qualitatively in Figure 14.

Now, by Theorem 2 of Section 14-1, f' is an increasing function of x if its derivative is positive, that is, if $f''(x) > 0$. So, if $f'(x) > 0$ and $f''(x) > 0$,

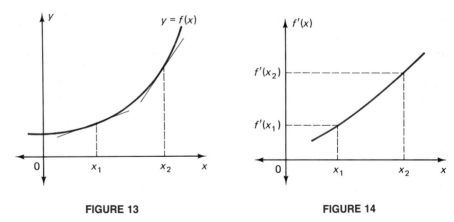

FIGURE 13

FIGURE 14

then the graph of f must have the general form shown in Figure 13. It must slope upward to the right and the slope becomes steeper as x increases.

Now consider a function f whose graph has the general shape indicated in Figure 15. The slope of the graph is negative, $f'(x) < 0$, so f is a decreasing function. Furthermore, the graph shown has the property that as we move to the right the slope becomes less steep. That is, as x increases, the derivative f' increases from large negative values toward zero, as indicated in Figure 16.

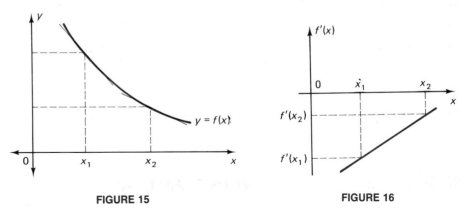

FIGURE 15

FIGURE 16

Again, f', although negative, is an increasing function of x, and this is guaranteed if $f''(x) > 0$. So, if $f'(x) < 0$ and $f''(x) > 0$, then the graph of f must have the general form shown in Figure 15. It must slope downward to the right and the slope becomes less steep as x increases.

The geometrical property that characterizes both of these types of graph is that they are **concave upward.*** We conclude therefore that if $f''(x) > 0$ in some interval, then the graph of f is concave upward in that interval.

---

*A curve is *concave upward* if, given any two points on the curve, the straight-line segment connecting them lies entirely above the curve. A curve is *concave downward* if such a straight-line segment always lies entirely below the curve.

9. Find the intervals where $f''(x)$ is positive and negative in the following cases:

(a) $f(x) = x^3$;

(b) $f(x) = x^4$;

(c) $f(x) = x^3 + 3x^2$.

Now let us consider the alternative possibility that the graph of f is **concave downward.** The two cases corresponding to the two types discussed above are shown in Figure 17. Part (a) illustrates a case when $f'(x) > 0$, but the slope is becoming less steep as x increases. Part (b) shows a case in which $f'(x) < 0$ and the slope is becoming steeper (more negative) as x increases.

In both cases, f' is a decreasing function of x. By Theorem 2 of Section 14-1, f' is decreasing if $f''(x) < 0$ and this therefore is a sufficient condition for the graph of f to be concave downwards. ☛ **9**

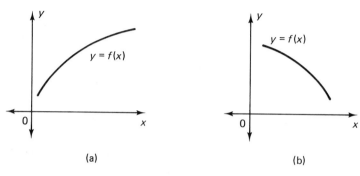

(a)   (b)

FIGURE 17

EXAMPLE 1 Find the values of x for which the graph of

$$y = \tfrac{1}{6}x^4 - x^3 + 2x^2$$

is concave upward or concave downward.

Solution

$$y = \tfrac{1}{6}x^4 - x^3 + 2x^2$$

$$y' = \tfrac{4}{6}x^3 - 3x^2 + 4x$$

$$y'' = 2x^2 - 6x + 4 = 2(x^2 - 3x + 2) = 2(x - 1)(x - 2)$$

We must determine the points where $y'' > 0$ (concave upward) and $y'' < 0$ (concave downward). First, setting $y'' = 0$ we obtain $2(x - 1)(x - 2) = 0$, giving $x = 1$ and $x = 2$. These points divide the number line into the three intervals $(-\infty, 1)$, $(1, 2)$, and $(2, \infty)$. In each of these intervals y'' has constant sign, so we choose a convenient test point and compute the sign of y'' at that point. This determines the sign of y'' in the whole interval. The results are in Table 5.

Answer (a) $f''(x)$ is positive for $x > 0$, negative for $x < 0$;

(b) positive for all $x \neq 0$;

(c) Positive for $x > -1$, negative for $x < 1$.

TABLE 5

Interval	$(-\infty, 1)$	$(1, 2)$	$(2, \infty)$
Test Point	0	$\tfrac{3}{2}$	3
$y'' = 2(x - 1)(x - 2)$	$2(-1)(-2) > 0$	$2(\tfrac{1}{2})(-\tfrac{1}{2}) < 0$	$2(2)(1) > 0$
Concavity	Upwards	Downwards	Upwards

10. Find the intervals where the following functions are concave up and concave down:

(a) $f(x) = 24x^2 - x^4$;

(b) $f(x) = e^{-2x^2}$.

Thus the given function is concave upward for $x < 1$ and for $x > 2$ and is concave downward for $1 < x < 2$. These properties are shown on the graph in Figure 18.

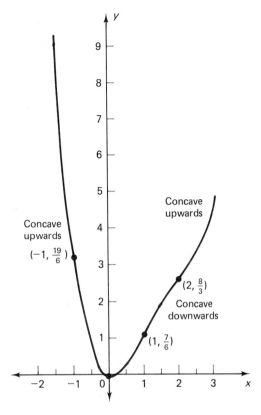

FIGURE 18

EXAMPLE 2 Examine the concavity properties of the cost function

$$C(x) = 2000 + 10x - 0.03x^2 + 10^{-4}x^3.$$

Solution

$$C'(x) = 10 - 0.06x + (3 \times 10^{-4})x^2$$

$$C''(x) = -0.06 + (6 \times 10^{-4})x = (6 \times 10^{-4})(x - 100).$$

We observe that for $x < 100$, $C''(x)$ is negative, which means that the graph of the cost function is concave downward. For $x > 100$, $C''(x) > 0$ and the graph is therefore concave upward. The graph of $C(x)$ has the form shown in Figure 19. ☛ **10**

Answer (a) Concave up for $-2 < x < 2$, concave down for $x < -2$ or $x > 2$;

(b) concave up for $x < -\frac{1}{2}$ or $x > \frac{1}{2}$, concave down for $-\frac{1}{2} < x < \frac{1}{2}$.

The cost function in Example 2 has a qualitative shape that is quite typical of such functions. For small values of x, the cost function is commonly concave downward. This property is due to the fact that increases of produc-

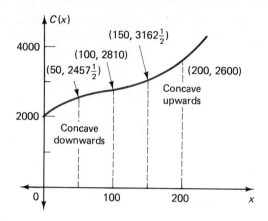

FIGURE 19

tion introduce economies of scale, so the marginal cost $C'(x)$ decreases. Past a certain level of production, however, it becomes more costly to increase production because, for example, new machinery must be purchased and workers paid overtime. At this stage, the marginal cost starts to increase and the cost function becomes concave upward.

Note that the graph of $C(x)$ always slopes upward to the right $(C'(x) > 0)$.

EXAMPLE 3 Find the values of x for which the function

$$f(x) = xe^{2x}$$

is increasing or decreasing and concave upward or downward.

Solution

$$f(x) = xe^{2x}$$
$$f'(x) = (2x + 1)e^{2x}$$
$$f''(x) = 4(x + 1)e^{2x}$$

Since e^{2x} is always positive, the sign of $f'(x)$ is the same as the sign of $(2x + 1)$. Thus $f(x)$ is increasing when $2x + 1 > 0$, that is, when $x > -\frac{1}{2}$, and $f(x)$ is decreasing when $2x + 1 < 0$, that is, when $x < -\frac{1}{2}$.

Similarly, the sign of $f''(x)$ is the same as the sign of $(x + 1)$. Thus $f(x)$ is concave upward when $x > -1$ and concave downward when $x < -1$.

DEFINITION A **point of inflection** on a curve is a point where the curve changes from concave upward to concave downward, or vice versa.

If $x = x_1$ is a point of inflection of the graph of $y = f(x)$, then on one side of x_1 the graph is concave upward, that is, $f''(x) > 0$; on the other side of x_1, the graph is concave downward, that is, $f''(x) < 0$. Thus on passing from one side to the other of $x = x_1$, $f''(x)$ changes sign. At $x = x_1$ itself, it is necessary either that $f''(x_1) = 0$ or that $f''(x_1)$ fails to exist ($f''(x)$ may become infinitely large as $x \to x_1$)

In Example 1, the graph of $y = \frac{1}{6}x^4 - x^3 + 2x^2$ has points of inflection at $x = 1$ and $x = 2$. For example, for $x < 1$, the graph is concave upward, while for x just above 1, the graph is concave downward. So $x = 1$ is a point where the concavity changes, that is, a point of inflection. This is also true for $x = 2$.

Example 1 corresponds to a point of inflection at which $y'' = 0$. Example 4 illustrates the alternative possibility.

EXAMPLE 4 Find the points of inflection of $y = x^{1/3}$

Solution We have

$$y' = \frac{1}{3}x^{-2/3}$$

$$y'' = \frac{1}{3}\left(-\frac{2}{3}\right)x^{-5/3} = -\frac{2}{9}x^{-5/3}.$$

11. Is $x = 0$ a point of inflection for the following functions?

(a) $f(x) = x^5$;

(b) $f(x) = x^8$;

(c) $f(x) = x^{1/5}$;

(d) $f(x) = x^{4/3}$.

Now for $x > 0$, $x^{5/3}$ is positive, so $y'' < 0$. For $x < 0$, $x^{5/3}$ is negative, so $y'' > 0$. Thus the graph is concave upward for $x < 0$ and concave downward for $x > 0$. The value $x = 0$, at which y is also 0, is therefore a point of inflection. (See Figure 20). In this case, y'' becomes indefinitely large as $x \to 0$, so we have a point of inflection at which the second derivative fails to exist. (Note also that as $x \to 0$, y' becomes infinite, so that the slope of the graph becomes vertical at the origin for this particular function.) ☛ **11**

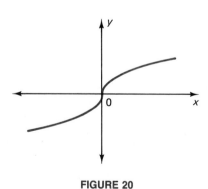

FIGURE 20

Observe that the tangent to a graph at a point of inflection always crosses the graph at that point. This is an unusual property for a tangent—as a rule the graph lies entirely on one side of the tangent line in the neighborhood of the point of tangency.

Answer (a) Yes;

(b) no;

(c) yes;

(d) no.

The Second-Derivative Test

In Section 14-2, we introduced the first-derivative test to distinguish among those critical points that are local maxima or minima or neither. The second derivative provides an alternative test that can be used in certain cases. When it can be used, this test is often easier than the first-derivative test.

Consider the case when the local extremum occurs at a critical point given by $f'(x) = 0$, that is, when the tangent line is horizontal at the point on the graph of f corresponding to the extremum. Then if the point is a local maximum, the graph is concave downward, and if the point is a local minimum, the graph is concave upward. But we know that whenever $f''(x) < 0$, the graph of f is concave downward and whenever $f''(x) > 0$, the graph is concave upward. This leads to the following theorem.

THEOREM 1 (SECOND-DERIVATIVE TEST) Let $f(x)$ be twice differentiable at the critical point $x = c$. Then

(a) $x = c$ is a local maximum of f whenever $f'(c) = 0$ and $f''(c) < 0$;

(b) $x = c$ is a local minimum of f whenever $f'(c) = 0$ and $f''(c) > 0$.

EXAMPLE 5 Find the local maximum and minimum values of

$$x^3 + 2x^2 - 4x - 8.$$

Solution Let $f(x) = x^3 + 2x^2 - 4x - 8$.

$$f'(x) = 3x^2 + 4x - 4.$$

To find the critical points, we set $f'(x) = 0$.

$$3x^2 + 4x - 4 = 0$$

$$(3x - 2)(x + 2) = 0$$

This gives $x = \frac{2}{3}$ or -2. Now

$$f''(x) = 6x + 4.$$

At $x = \frac{2}{3}$,

$$f''\left(\frac{2}{3}\right) = 6\left(\frac{2}{3}\right) + 4 = 8 > 0.$$

Hence, since $f''(x)$ is positive when $x = \frac{2}{3}$, $f(x)$ has a local minimum when $x = \frac{2}{3}$. The local minimum value is given by

$$f\left(\frac{2}{3}\right) = \left(\frac{2}{3}\right)^3 + 2\left(\frac{2}{3}\right)^2 - 4\left(\frac{2}{3}\right) - 8 = -\frac{256}{27}.$$

When $x = -2$, $f''(-2) = 6(-2) + 4 = -8 < 0$. Hence, since $f''(x)$ is negative when $x = -2$, $f(x)$ has a local maximum when $x = -2$. The local maximum value is given by

$$f(-2) = (-2)^3 + 2(-2)^2 - 4(-2) - 8 = 0.$$

Thus the only local maximum value of $f(x)$ is 0, and it occurs when $x = -2$; the only local minimum value is $-\frac{256}{27}$, and it occurs when $x = \frac{2}{3}$.

EXAMPLE 6 Determine the local maxima and minima for $f(x) = (\ln x)/x$.

Solution Using the quotient rule, we have

$$f'(x) = \frac{x \cdot \dfrac{1}{x} - \ln x \cdot 1}{x^2} = \frac{1 - \ln x}{x^2}.$$

For a critical point, $f'(x) = 0$, or

$$\frac{1 - \ln x}{x^2} = 0.$$

That is, $1 - \ln x = 0$. Thus $\ln x = 1 = \ln e$ and so $x = e$.

In this case, we have only one critical point, $x = e$. (Note that $f'(x)$ becomes infinite as $x \to 0$. However, $x = 0$ is not a critical point because $f(0)$ is not defined.)

We use the quotient rule again.

$$f''(x) = \frac{x^2(1 - \ln x)' - (1 - \ln x) \cdot (x^2)'}{(x^2)^2}$$

$$= \frac{x^2(-1/x) - (1 - \ln x)(2x)}{x^4} = \frac{2 \ln x - 3}{x^3}$$

When $x = e$,

☛ **12.** Use the second derivative test to determine the local extrema:
(a) $f(x) = 1 - 2x^2 + x^4$;
(b) $f(x) = x \ln x$;
(c) $f(x) = x^2 - 6x^{4/3}$.

$$f''(e) = \frac{2 \ln e - 3}{e^3} = \frac{2 - 3}{e^3} = -\frac{1}{e^3} < 0$$

where we have used the fact that $\ln e = 1$. Hence $f(x)$ has a local maximum when $x = e$. In this case, there are no local minima. ☛ **12**

The second-derivative test can be used for all local extrema at which $f'(c) = 0$ and $f''(c)$ is nonzero. When $f''(x) = 0$ at a critical point $x = c$, or when $f''(c)$ fails to exist, then the second-derivative test cannot be used to ascertain whether $x = c$ is a local maximum or minimum. In such cases, we must use the first derivative test. The first-derivative test must also be used for all critical points where $f'(c)$ fails to exist.

The following example illustrates several simple cases where the second-derivative test fails.

Answer (a) Local minima at ± 1, local maximum at 0;
(b) local minimum at $x = e^{-1}$;
(c) local minima at $x = \pm 8$; second derivative test fails for $x = 0$.

EXAMPLE 7

(a) Consider $f(x) = x^3$. Then $f'(x) = 3x^2$ and $f'(x) = 0$ when $x = 0$. The only critical point is $x = 0$. Now, $f''(x) = 6x$, so $f''(0) = 0$ and the test fails. (In fact, $f'(x) > 0$ for all $x \neq 0$, so the function is increasing for all $x \neq 0$, and $x = 0$ is not a local extremum.)

(b) Consider $f(x) = x^{6/5}$. Then $f'(x) = \frac{6}{5}x^{1/5}$ and $f'(x) = 0$ at $x = 0$. This is the only critical point. Now $f''(x) = \frac{6}{25}x^{-4/5}$, so $f''(0)$ is undefined. The second-derivative test fails. (In fact, the first-derivative test shows that $x = 0$ is a local minimum.)

(c) Consider $f(x) = x^{2/5}$. Then $f'(x) = \frac{2}{5}x^{-3/5}$ and there is a critical point at $x = 0$ at which f' is undefined. The second-derivative test cannot be applied to this type of critical point. (Actually, $x = 0$ is a local minimum.)

(d) In Example 4 of Section 14-2 we have $f'(x) = 4x^2(x - 3)$ with critical points at $x = 0$ and $x = 3$. Then $f''(x) = 12x^2 - 24x$. We have $f''(0) = 0$ and $f''(3) = 12(3)^2 - 24(3) = 36 > 0$. So the second-derivative test shows that $x = 3$ is a local minimum but this test fails at $x = 0$. ☛ 13

☛ **13.** Does the second derivative test succeed or fail for the critical point at $x = 1$ for each of the following functions?

(a) $f(x) = x^3 - 3x^2 + 3x - 1$;
(b) $f(x) = (x - 1)^{4/7}$;
(c) $f(x) = x(x - \frac{4}{3})^{1/3}$;
(d) $f(x) = x(x - 1)^{1/3}$.

Answer (a) Fails;
(b) fails;
(c) succeeds (minimum);
(d) fails.

Summary of Determining Local Extrema by the Second-Derivative Test:

Step 1. Find $f'(x)$ and determine the critical points. Let $x = c$ be a critical point at which $f'(c) = 0$. The second-derivative test cannot be used for a point where $f'(x)$ fails to exist.

Step 2. Find $f''(x)$ and evaluate it when $x = c$.

Step 3. If $f''(c) < 0$, then f has a local maximum at $x = c$. If $f''(c) > 0$, then f has a local minimum at $x = c$. If $f''(c) = 0$ or $f''(c)$ is undefined, then the test fails.

EXERCISES 14-3

(1–12) Find the values of x for which the following functions are **(a)** concave upward or **(b)** concave downward. Also, find the points of inflection, if any.

1. $x^2 - 4x + 7$

2. $5 + 3x - x^2$

3. $x^3 - 3x + 4$

4. $x^3 + 3x^2 - 9x + 1$

5. $x^4 - 18x^2 + 5$

6. $x^7 - 7x^6 + 2$

7. $x + \dfrac{1}{x}$

8. $\dfrac{1}{x - 2}$

9. $(x - 5)^{3/4}$

10. xe^{-x}

11. $\dfrac{x}{e^{x/2}}$

12. $x - 2\ln x$

(13–20) Find the values of x for which the following functions are **(a)** increasing; **(b)** decreasing; **(c)** concave upward; **(d)** concave downward. Also, find the points of inflection, if any.

13. $3x^2 - 15x + 2$

14. $x^3 - 6x^2 - 15x + 7$

15. $x^4 - 4x^3$

16. $(x - 1)^{1/5}$

17. $x - \ln x$

18. $\dfrac{x^2}{e^x}$

19. $x^2 - 18\ln x$

*20. $|x^2 - 5x - 6|$

(21–24) *(Cost Functions)* Discuss the concavity of the following cost functions.

21. $C(x) = a + bx$

22. $C(x) = \sqrt{100 + x^2}$

23. $C(x) = 1500 + 25x - 0.1x^2 + 0.004x^3$

24. $C(x) = 1000 + 40\sqrt{x} - x + 0.02x^{3/2}$

(25–45) Use second-derivative test to find the local maximum and minimum values of the following functions. If the second-derivative test fails, use the first-derivative test.

25. $x^2 - 10x + 3$ **26.** $x^3 - 27x + 5$

27. $2x^3 - 3x^2 - 36x + 7$ **28.** $x^4 - 8x^2 + 15$

29. $x^4 + 4x^3 - 8x^2 + 3$ **30.** $1 + 3x^2 - x^6$

31. xe^x **32.** x^2/e^{2x}

33. e^{-x^2} **34.** $x^2 \ln x$

35. $x^2 - \ln x$ **36.** $x^5 - 15x^3 + 2$

37. $x \ln x$ **38.** $x \ln x - x$

39. $(x - 1)^3(x - 2)^4$ **40.** $(x + 1)^2(x - 2)^3$

41. $(x - 1)^{4/3}$ ***42.** $x^2(x - 1)^{2/3}$

▨ 14-4 CURVE SKETCHING FOR POLYNOMIALS

It often happens that we would like to get a rough qualitative picture of the graph of a given function without going to the length of plotting a large number of points. The first and second derivatives are effective tools for this purpose. In this section, we shall study their use for graphing polynomial functions. The graphs already shown in Figures 5 and 18 were obtained using the procedures to be described. In Section 14-7, these techniques are extended to other classes of functions.

The basic properties we need have already been established and are summarized in Table 6.

TABLE 6

Signs of $f'(x)$ and $f''(x)$	Properties of graph of f	Shape of the graph
$f'(x) > 0$ and $f''(x) > 0$	Increasing and concave up	
$f'(x) > 0$ and $f''(x) < 0$	Increasing and concave down	
$f'(x) < 0$ and $f''(x) > 0$	Decreasing and concave up	
$f'(x) < 0$ and $f''(x) < 0$	Decreasing and concave down	

EXAMPLE 1 Sketch the graph of the function $y = 2x^3 - 9x^2 + 12x - 2$.

Solution We first find where the function is increasing and decreasing:

$$y' = 6x^2 - 18x + 12 = 6(x - 1)(x - 2).$$

14. Find the intervals in which the graph of each of the following functions belongs to each of the four types.

(a) $f(x) = 2x - x^2$;

(b) $f(x) = 2x^2 - \frac{2}{3}x^3$.

Analyzing the sign of y' as in Section 14-1, we find that $y' > 0$ for $x < 1$ and for $x > 2$ while $y' < 0$ for $1 < x < 2$. Thus the graph is increasing for $x < 1$, decreasing for $1 < x < 2$, and increasing again for $x > 2$. (See part (a) of Figure 21.)

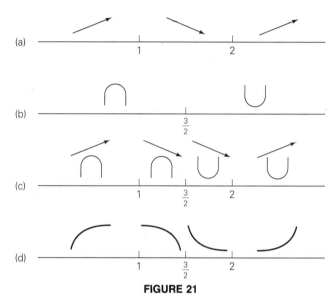

FIGURE 21

By the first-derivative test, y has a local maximum when $x = 1$ and a local minimum when $x = 2$. It is readily found that when $x = 1$, $y = 3$ and when $x = 2$, $y = 2$. Thus the coordinates of the local extrema are $(1, 3)$ and $(2, 2)$.

Now we examine the concavity of the function. We find

$$y'' = 12x - 18 = 12(x - \tfrac{3}{2})$$

and so $y'' > 0$ when $x > \frac{3}{2}$ (concave up), while $y'' < 0$ when $x < \frac{3}{2}$ (concave down). This information is pictured in part (b) of Figure 21. When $x = \frac{3}{2}$, $y = \frac{5}{2}$, so the point $(\frac{3}{2}, \frac{5}{2})$ is a point of inflection on the graph.

Combining parts (a) and (b) of Figure 21, we get part (c); that is, y is increasing and concave down for $x < 1$, y is decreasing and concave down for $1 < x < \frac{3}{2}$, and so on. This information is translated into a qualitative shape for the graph in part (d) of Figure 21.

Finally, we compute the coordinates of the point where the graph crosses the y-axis. When $x = 0$, $y = -2$, so that the point is $(0, -2)$.

To sketch the graph, first we plot the points $(1, 3)$, $(\frac{3}{2}, \frac{5}{2})$, $(2, 2)$, where $f(x)$ changes its nature (increasing to decreasing or concave upward to concave downward) and the point $(0, -2)$ where the graph meets the y-axis. Then using the information of Figure 21 we draw curves of the appropriate type joining these points. This gives the graph as shown in Figure 22. **☛ 14**

Answer

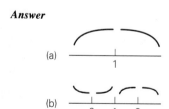

(a)

(b)

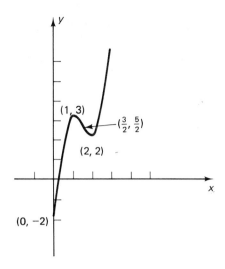

FIGURE 22

The steps involved in sketching the graph of a polynomial function can be summarized in the following procedure.

Step 1: Find $f'(x)$ Find the intervals in which $f'(x)$ is positive or negative: These give the intervals where $f(x)$ is increasing or decreasing, respectively. Find the xy-coordinates of the points dividing these intervals.

Step 2: Find $f''(x)$ Find the intervals in which $f''(x)$ is positive or negative: These give the intervals where $f(x)$ is concave upward or downward, respectively. Find the xy-coordinates of the points separating these intervals.

Step 3: Combine Combine the information of Steps 1 and 2 as in Figure 21.

Step 4: Find a Few Explicit Points For example the intersection with the y-axis is obtained by putting $x = 0$, so that $y = f(0)$. The intersections with the x-axis are obtained by putting $y = 0$. This gives the equation $f(x) = 0$ which must be solved for the values of x at the points of intersection. Sometimes this equation turns out to be too complicated to solve and we must manage without this information.

The present methods can be used instead of those in Section 5-2 for graphing quadratic functions.

EXAMPLE 2 Sketch the graph of $y = 3 + 5x - 2x^2$

Solution

Step 1 $y' = 5 - 4x$. Thus $y' > 0$ when $x < \frac{5}{4}$ and $y' < 0$ when $x > \frac{5}{4}$. When $x = \frac{5}{4}$,

$$y = 3 + 5(\tfrac{5}{4}) - 2(\tfrac{5}{4})^2 = \tfrac{49}{8}$$

Thus the graph is increasing for $x < \frac{5}{4}$ and decreasing for $x > \frac{5}{4}$, and the dividing point on the graph is $(\frac{5}{4}, \frac{49}{8})$. This point is a local maximum.

Step 2 $y'' = -4$. Thus the graph is concave downward for all x.

Step 3 Combining the information of Steps 1 and 2, we have Figure 23(a).

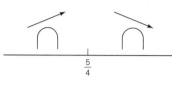

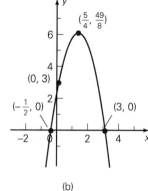

☞ **15.** Sketch the graphs of

(a) $f(x) = \frac{1}{2}x^2 - x$;

(b) $f(x) = x^2 - \frac{2}{3}x^3$;

(c) $f(x) = \frac{1}{4}x^2 - \frac{3}{2}x^{4/3}$

(a)

(b)

FIGURE 23

Answer

(a)

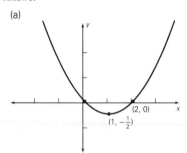

(b)

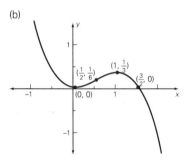

(c)

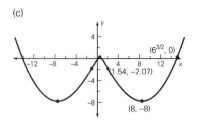

Step 4 When $x = 0$, $y = 3$ giving the point $(0, 3)$. When $y = 0$ we obtain the equation $2x^2 - 5x - 3 = 0$. This quadratic function factors:

$$(2x + 1)(x - 3) = 0$$

and the roots are $x = -\frac{1}{2}$ and $x = 3$. Hence the graph crosses the x-axis at $(-\frac{1}{2}, 0)$ and $(3, 0)$.

Putting all this information together, we can draw a reasonably accurate sketch of the graph, as shown in Figure 23(b). ☞ **15**

EXAMPLE 3 If the number of items produced per week is x (measured in thousands), a manufacturer's cost function is

$$C = 2 + x - \frac{1}{4}x^2 + \frac{1}{24}x^3$$

(in thousands of dollars). Sketch the graph of C as a function of x.

Solution For obvious reasons we are only interested in the region $x \geq 0$.

Step 1 $C'(x) = 1 - \frac{1}{2}x + \frac{1}{8}x^2$. First of all, we set $C'(x) = 0$ to obtain the points where the graph has horizontal tangents.

$$1 - \frac{1}{2}x + \frac{1}{8}x^2 = 0$$

$$x^2 - 4x + 8 = 0$$

From the quadratic formula,

$$x = \frac{-(-4) \pm \sqrt{(-4)^2 - 4 \cdot 1 \cdot 8}}{2 \cdot 1} = \frac{4 \pm \sqrt{-16}}{2}$$

and because of the negative number under the radical, x is not a real number. We conclude that $C'(x)$ is never zero. Thus $C'(x)$ is either positive for all x or negative for all x. But $C'(0) = 1 > 0$, so $C'(x) > 0$ for all x. Therefore C is an increasing function for all x.

Step 2 $C''(x) = -\frac{1}{2} + \frac{1}{4}x = \frac{1}{4}(x - 2)$.   Therefore   when $x > 2$, $C''(x) > 0$ and the graph is concave upward. When $x < 2$, $C''(x) < 0$ and the graph is concave downward.

When $x = 2$,

$$C(2) = 2 + 2 - \tfrac{1}{4}(2)^2 + \tfrac{1}{24}(2)^3 = \tfrac{10}{3}$$

$$C'(2) = 1 - \tfrac{1}{2}(2) + \tfrac{1}{8}(2)^2 = \tfrac{1}{2}$$

So the dividing point is $(2, \frac{10}{3})$ (point of inflection).

Step 3 Combining the information in Steps 1 and 2, we have Figure 24(a).

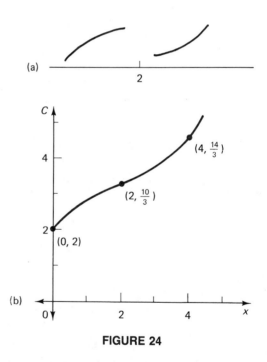

FIGURE 24

Step 4 When $x = 0$, $C = 2$, giving the point $(0, 2)$. Setting $C = 0$ we obtain a cubic equation for x, which we are not in a position to solve. Hence we must dispense whith this information.

It is helpful to have one further point on the graph to the right of $x = 2$, so we compute the value of C for $x = 4$ and find the point $(4, \frac{14}{3})$.

Putting all these results together, we obtain the graph shown in Figure 24(b).

(1–12) Sketch the graphs of the following functions.

1. $y = x^2 - 6x + 7$

2. $y = x^2 - 4x + 5$

3. $y = x^3 - 3x + 4$

4. $y = x^3 - 12x + 10$

5. $y = x^3 - 3x + 2$

6. $y = 2x^3 - 9x^2 - 24x + 20$

7. $y = x^4 - 2x^2$

8. $y = \frac{1}{4}x^4 - x^3 + x^2$

9. $y = x^5 - 5x^4 + 1$

10. $y = x^7 - 7x^6$

11. $y = 5x^6 - 6x^5 + 1$

12. $y = \frac{1}{4}x^4 - 3x^2$

(13–14) Sketch the graphs of the two cost functions in Exercises 23 and 24 of Section 14-3 ($x \geq 0$ only).

14-5 APPLICATIONS OF MAXIMA AND MINIMA

Many situations arise in practice when we want to maximize or minimize a certain quantity. The following example represents a typical case in point.

EXAMPLE 1 (*Optimal Conservation*) A conservationist is stocking a lake with fish. The more fish put in, the more competition there will be for the available food supply, and so the fish will gain weight more slowly. In fact, it is known from previous experiments that when there are n fish per unit area of water, the average amount that each fish gains in weight during one season is given by $w = 600 - 30n$ grams. What value of n leads to the maximum total production of weight in the fish?

Solution The gain in weight of each fish is $w = 600 - 30n$. Since there are n fish per unit area, the total production of fish weight per unit area, P, is equal to nw. Therefore

$$P = n(600 - 30n) = 600n - 30n^2.$$

To find the value of n at which P is maximum, we differentiate and set the derivative dP/dn equal to zero.

$$\frac{dP}{dn} = 600 - 60n$$

and $dP/dn = 0$ when $600 - 60n = 0$, that is, when $n = 10$. Thus the density of 10 fish per unit area gives the maximum total production. The maximum value of P is given by

$$P = 600(10) - 30(10)^2 = 3000$$

that is, 3000 grams per unit area. We can check that this is a local maximum by using the second derivative:

$$\frac{d^2P}{dn^2} = -60$$

The second derivative is negative (for all values of n in fact) so that the critical value $n = 10$ corresponds to a maximum of P.

The graph of P against n is shown in Figure 25. P is zero when n is zero since there are then no fish to produce. As n increases, P increases to a maximum value, then decreases to zero again when $n = 20$. When n gets large, P decreases because for large values of n the fish put on very little weight; even though there are lots of them, the total production is small. ☛ 16

☛ **16.** Re-solve Example 1 if the average weight gain per fish is $w = 800 - 25n$.

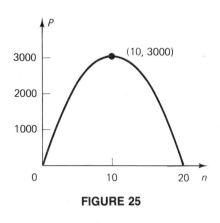

FIGURE 25

Answer $n = 16$.

Let us consider another example of a purely mathematical nature.

EXAMPLE 2 Find two numbers whose sum is 16 and whose product is as large as possible.

☛ **17.** Find two numbers whose product is 64 and whose sum is minimum.

Solution Let the two numbers be x and y, so that $x + y = 16$. If $P = xy$ denotes the product of the two numbers, then we are required to find the values of x and y that make P a maximum.

We cannot differentiate P immediately, since it is a function of two variables, x and y. These two variables are not independent, however, but are related through the condition $x + y = 16$. We must use this condition to eliminate one of the variables from P, thus leaving P as a function of a single variable. We have $y = 16 - x$, and so

$$P = xy = x(16 - x) = 16x - x^2.$$

We must find the value of x that makes P a maximum.

$$\frac{dP}{dx} = 16 - 2x$$

Thus $dP/dx = 0$ when $16 - 2x = 0$, that is, when $x = 8$. The second derivative $d^2P/dx^2 = -2 < 0$, and so $x = 8$ corresponds to a maximum of P.

When $x = 8$, $y = 8$ also, so the maximum value of $P = xy$ is then equal to 64. ☛ 17

Answer 8 and 8.

The solution of optimization problems of the type given above is often found to be one of the most difficult areas of the differential calculus. The main difficulty arises at the level of translating the given word problem into the necessary equations. Once the equations have been constructed, it is usually much more straightforward to complete the solution by using the appropriate bit of calculus. This task of phrasing word problems in terms of mathematical equations is one that occurs repeatedly in all branches of applied mathematics, and it is something which the applied student should master if his or her calculus courses are ever to be useful.

Unfortunately, it is not possible to give hard and fast rules by means of which any word problem can be translated into equations. However here are a few guiding principles that are useful to bear in mind.*

Step 1 Identify all of the variables involved in the problem and denote each by a symbol.

In Example 1, the variables were n, the number of fish per unit area, w, the average gain in weight per fish, and P, the total production of fish weight per unit area. In Example 2, the variables were the two numbers x and y and P, their product.

Step 2 Identify the variable that is to be maximized or minimized and express it in terms of the other variables in the problem.

In Example 1, the total production P is maximized, and we wrote $P = nw$, expressing P in terms of n and w. In Example 2, the product P of x and y is maximized, and of course $P = xy$.

Step 3 Identify all of the relationships between the variables. Express these relationships mathematically.

In the first example, the relationship $w = 600 - 3n$ was given. In the second example, the relationship between x and y is that their sum is equal to 16, so we wrote the mathematical equation $x + y = 16$.

Step 4 Express the quantity to be maximized or minimized in terms of just one of the other variables. In order to do this, the relationships obtained in Step 3 are used to eliminate all but one of the variables.

In Example 1, we have $P = nw$ and $w = 600 - 3n$, so, eliminating w, we obtain P in terms of n: $P = n(600 - 3n)$. In Example 2, we have $P = xy$ and $x + y = 16$, so, eliminating y, we obtain $P = x(16 - x)$.

Step 5 Having expressed the required quantity as a function of one variable, calculate its critical points and test each for local maximum or minimum.

---

* Steps 1 and 3 apply not only to optimization problems but to word problems in general.

Let us follow through these steps in another example.

EXAMPLE 3 *(Minimizing Cost)* A tank is to be constructed with a horizontal, square base and vertical, rectangular sides. There is no top. The tank must hold 4 cubic meters of water. The material of which the tank is to be constructed costs $10 per square meter. What dimensions for the tank minimize the cost of material?

Solution

Step 1 The variables in the problem are the dimensions of the tank and the cost of construction materials. The cost depends upon the total area of the base and sides, which determines the amount of material used in the construction. We let x denote the length of one side of the base and y denote the height of the tank. (See Figure 26.) The quantity to be minimized is the total cost of materials, which we denote by C.

Step 2 C is equal to the area of the tank multiplied by $10, which is the cost per unit area. The base is a square with side x, so has an area equal to x^2. Each side is a rectangle with dimensions x and y, and has an area xy. The total area of base plus the four sides is therefore $x^2 + 4xy$. Consequently, we can write

$$C = 10(x^2 + 4xy).$$

Step 3 We observe that the quantity to be minimized is expressed as a function of two variables, so we need a relationship between x and y to eliminate one of these variables. This relationship is obtained from the requirement (stated in the problem) that the volume of the tank must be 4 cubic meters. The volume equals the area of the base times the height, that is, x^2y, and so we have the condition

$$x^2y = 4.$$

Step 4 From Step 3, $y = 4/x^2$, and so

$$C = 10\left[x^2 + 4x\left(\frac{4}{x^2}\right)\right] = 10\left[x^2 + \frac{16}{x}\right].$$

Step 5 We can at last differentiate to find the critical points of C.

$$\frac{dC}{dx} = 10\left(2x - \frac{16}{x^2}\right) = 20\left(x - \frac{8}{x^2}\right) = 0$$

Thus, $x - 8/x^2 = 0$ and so $x^3 = 8$; that is, $x = 2$.

The base of the tank should therefore have a side 2 meters in length. The height of the tank is then given by

$$y = \frac{4}{x^2} = 4/(2)^2 = 1.$$

It is easily verified that $d^2C/dx^2 > 0$ when $x = 2$, so this value of x provides a local minimum of C. ☛ **18**

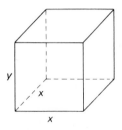

FIGURE 26

☛ **18.** Re-solve Example 3 if the tank has a lid that cost $30 per square meter.

Answer $x = \sqrt[3]{2} \approx 1.26$,
$y = \sqrt[3]{16} \approx 2.52$.

One of the most important applications of the theory of maxima and minima is to the operations of business firms. This occurs for a very simple reason, namely that a business firm selects its mode and level of operation in such a way as to maximize its profit. Thus if the management knows how the profit depends on some variable that can be adjusted, then they will choose the value of that variable that makes the profit as large as possible.

Let us consider the case in which the variable to be adjusted is simply the level of production, x (the number of units of the firm's product produced per week or per month). If the units are sold for price p each, then the revenue is $R(x) = px$. The cost of producing x items depends on x and is denoted by $C(x)$, the cost function. Then the profit as a function of x is given by

$$P(x) = R(x) - C(x) = px - C(x).$$

We wish to choose the value of x that makes this a maximum.

Consider first the case of a small firm that is selling its product in a competitive market. In this case, the volume x of sales by this particular firm will not affect the market price for the item in question. We can assume that the price p is constant, independent of x, being determined by market forces outside the control of our small firm. The following example illustrates a problem of this kind.

EXAMPLE 4 *(Maximizing Profits)* A small manufacturing firm can sell all the items it can produce at a price of $6 each. The cost of producing x items per week (in dollars) is

$$C(x) = 1000 + 6x - 0.003x^2 + 10^{-6}x^3.$$

What value of x should be selected in order to maximize the profits?

Solution The revenue from selling x items at $6 each is $R(x) = 6x$ dollars. Therefore the profit per week is

$$P(x) = R(x) - C(x)$$
$$= 6x - (1000 + 6x - 0.003x^2 + 10^{-6}x^3)$$
$$= -1000 + 0.003x^2 - 10^{-6}x^3.$$

To find the maximum value of P, we find the critical values in the usual way and then test them for local extrema. Differentiating we get

$$P'(x) = 0.006x - (3 \times 10^{-6})x^2$$

and setting $P'(x) = 0$, we find that either $x = 0$ or $x = 2000$. We can test each of these values using the second-derivative test:

$$P''(x) = 0.006 - (6 \times 10^{-6})x$$

so that

$$P''(0) = 0.006 > 0 \qquad \text{and} \qquad P''(2000) = -0.006 < 0.$$

So $x = 0$ is a local minimum of $P(x)$, while $x = 2000$ is a local maximum.

This latter value provides us with the level of production at which the profit is greatest. The maximum profit is given by

$$P(2000) = -1000 + 0.003(2000)^2 - 10^{-6}(2000)^3 = 3000$$

or $3000 per week.

A different situation occurs in the case of a large business firm that is essentially the only supplier of a particular product. In such a case, the firm controls or monopolizes the market and can choose the price at which it wishes to sell the product. The volume of sales is then determined by the price at which the product is offered (through the demand equation). If the demand equation is written in the form $p = f(x)$, then the revenue function is $R = xp = x f(x)$. The profit function is then

$$P(x) = \text{Revenue} - \text{Cost} = x f(x) - C(x)$$

and x must be chosen to maximize this function.

EXAMPLE 5 *(Pricing Decision)* The cost of producing x items per week is

$$C(x) = 1000 + 6x - 0.003x^2 + 10^{-6}x^3.$$

For the particular item in question, the price at which x items can be sold per week is given by the demand equation

$$p = 12 - 0.0015x.$$

Determine the price and volume of sales at which the profit is maximum.

Solution The revenue per week is

$$R(x) = px = (12 - 0.0015x)x.$$

The profit is therefore given by

$$\begin{aligned} P(x) &= R(x) - C(x) \\ &= (12x - 0.0015x^2) - (1000 + 6x - 0.003x^2 + 10^{-6}x^3) \\ &= -1000 + 6x + 0.0015x^2 - 10^{-6}x^3. \end{aligned}$$

To find the maximum value of $P(x)$, we set $P'(x) = 0$.

$$P'(x) = 6 + 0.003x - (3 \times 10^{-6})x^2 = 0$$

We change signs throughout this equation, divide through by 3, and multiply by 10^6 to obtain $x^2 - 1000x - 2 \times 10^6 = 0$. We can factor the left side as

$$(x - 2000)(x + 1000) = 0$$

and so the solutions are $x = 2000$ or -1000. (These solutions could also have been obtained using the quadratic formula.)

The negative root has no practical significance, so we need only consider $x = 2000$. To verify that this does indeed represent a local maximum of the profit function, we can check that $P''(2000) < 0$. This is easily done.

$$P''(x) = 0.003 - (6 \times 10^{-6})x$$

$$P''(2000) = 0.003 - (6 \times 10^{-6})(2000) = -0.009$$

Thus the sales volume of 2000 items per week does give the maximum profit. The price per item corresponding to this value of x is

$$p = 12 - 0.0015x = 12 - 0.0015(2000) = 9. \quad \text{☛ 19}$$

For any firm, the profit is the difference between revenue and costs:

$$P(x) = R(x) - C(x).$$

Therefore, assuming that all functions are differentiable,

$$P'(x) = R'(x) - C'(x).$$

When the profit is maximum, $P'(x) = 0$, and so it follows that $R'(x) = C'(x)$.

This result provides an important general conclusion regarding the operation of any firm: *At the level of production at which the profit is maximum, the marginal revenue equals the marginal cost.*

In a competitive market in which many firms produce similar products at about the same price, the volume of sales can be increased by advertising. However, if too much money is spent on advertisement, the expenditure will outweigh the gain in revenue through increased sales. Again the criterion that must be used in deciding how much to spend on advertising is that the profit should be maximum.

EXAMPLE 6 *(Advertising and Profits)* A company makes a profit of $5 on each item of its product it sells. If it spends A dollars per week on advertising, then the number of items per week it sells is given by

$$x = 2000(1 - e^{-kA})$$

where $k = 0.001$. Find the value of A that maximizes the net profit.

Solution The gross profit from selling x items is $5x$ dollars, and from this we must subtract the cost of advertising. This leaves a net profit given by

$$P = 5x - A = 10{,}000(1 - e^{-kA}) - A. \tag{1}$$

We differentiate to find the maximum value of P.

$$\frac{dP}{dA} = 10{,}000(ke^{-kA}) - 1 = 10e^{-kA} - 1$$

since $k = 0.001$. Setting this to zero, we get

$$10e^{-kA} = 1 \qquad \text{or} \qquad e^{kA} = 10$$

and taking natural logarithms, we obtain

$$kA = \ln 10 = 2.30$$

☛ **19.** Find the value of x that maximizes the profit and the maximum profit if the cost function is $C(x) = (1 + x)^2$ and the demand equation is $p = 10 - x$.

Answer $x = 2$, $P_{\max} = 7$.

to three significant figures. Therefore

$$A = \frac{2.30}{k} = \frac{2.30}{0.001} = 2300.$$

The optimum amount to be spent on advertising is therefore $2300 per week.

The maximum profit is found by substituting this value of A into Equation (1). Since $e^{-kA} = \frac{1}{10}$, it follows that the maximum weekly profit is

$$P_{max} = 10,000(1 - \tfrac{1}{10}) - 2300 = 6700 \text{ dollars}.$$

EXAMPLE 7 *(Maximum Profits and Tax Revenue)* A firm's cost function and the demand functions are $C(x) = 5x$ and $p = 25 - 2x$, respectively.

(a) Find the output level that will maximize the firm's profits. What is the maximum profit?

(b) If a tax of t per unit is imposed which the firm adds to its cost, find the output level that will maximize the firm's profits. What is the maximum profit?

(c) Determine the tax t per unit that must be imposed to obtain a maximum tax revenue.

Solution We have:

$$\text{Revenue} = \text{Price} \times \text{Quantity}$$

or

$$R = px = x(25 - 2x) = 25x - 2x^2.$$

(a) If P denotes the profit function, then

$$P = R - C = 25x - 2x^2 - 5x = 20x - 2x^2, \qquad \frac{dP}{dx} = 20 - 4x$$

For maximum profits, $dP/dx = 0$, or $20 - 4x = 0$, or $x = 5$. Also, $d^2P/dx^2 = -4 < 0$. Thus the profits are maximum at the output level of $x = 5$ units. $P_{max} = 20(5) - 2(5^2) = 50$.

(b) After a tax of t per unit is imposed, the new cost function will be

$$C_N = 5x + tx$$

and the profits are given by

$$P = R - C_N = 25x - 2x^2 - (5x + tx) = (20 - t)x - 2x^2.$$

$$\frac{dP}{dx} = 20 - t - 4x \qquad \text{and} \qquad \frac{d^2P}{dx^2} = -4$$

For optimal profits, $dP/dx = 0$, which gives

$$x = \frac{20 - t}{4} = 5 - \frac{t}{4}.$$

The maximum profit is

$$P_{\text{max}} = (20 - t)\left(\frac{20 - t}{4}\right) - 2\left(\frac{20 - t}{4}\right)^2 = \frac{1}{8}(20 - t)^2.$$

(Notice that any positive tax level t decreases the firm's profits, whereas a negative value of t, which means a subsidy, increases the profits.)

(c) If T denotes the total revenue obtained from the taxation, then

$$T = tx = t\left(5 - \frac{t}{4}\right) = 5t - \frac{t^2}{4}.$$

We wish to maximize T. Now,

$$\frac{dT}{dt} = 5 - \frac{t}{2} \qquad \text{and} \qquad \frac{d^2T}{dt^2} = -\frac{1}{2}.$$

☛ **20.** Repeat parts (b) and (c) of Example 7 if the cost function is $C(x) = (1 + x)^2$ and the demand equation is $p = 10 - x$.

For maximum T, we must have $dT/dt = 0$ and $d^2T/dt^2 < 0$. $dT/dt = 0$ gives $t = 10$. Thus a tax of 10 per unit will produce a maximum tax revenue. ☛ **20**

We shall conclude this section by describing the application of maxima and minima to an **inventory cost model.** Let us consider a particular example. Suppose that a manufacturer has to make 50,000 units of a certain item during a year. There is a choice of a number of different production schedules. All the required units could be made at the beginning of the year in one production run. Because of the economies of mass production, this would minimize the cost of production. However, it would mean that large numbers of items would have to be held in storage until they were needed for sale, and so storage costs would be high and could outweigh the advantage of lower production costs. Perhaps if the annual production were to be manufactured in several smaller batches, the lower storage costs would outweigh the higher costs of production.

Let us suppose that it costs $400 to prepare the manufacturing plant for each production run, that each item then costs $4 to manufacture, and that it costs 40¢ per year for each item carried in storage. We assume that the same number of items is produced in each production run, and denote that number by x. We shall assume that after a batch has been made, the x units are placed in storage and are sold at a uniform rate so that the units stored are exactly used up when the next production run is made. Then the number of units in storage as a function of time is illustrated in Figure 27. At each production run, the number leaps from 0 to x, then steadily decreases at a constant rate to zero. When it reaches zero, the next batch is produced and the number in storage goes up to x again.

It is clear from Figure 27 that the average number of units in storage is $x/2$. Since it costs $0.40 to store each item per year, the storage costs over the year will be $(0.4)(x/2)$ dollars, or $x/5$ dollars.

Since the necessary 50,000 items are produced in batches of size x, the number of production runs per year must be $50,000/x$. The cost of preparing

Answer (b) $x = 2 - \frac{1}{4}t$, $P_{\text{max}} = 2(2 - \frac{1}{4}t)^2 - 1$; (c) $t = 4$, $T_{\text{max}} = 4$.

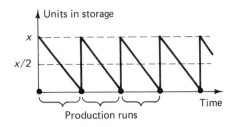

Units in storage

Production runs

FIGURE 27

☛ **21.** Suppose it costs $200 to prepare each production run, $2 per item to manufacture and $4 per year to store each item. Write down the annual cost function if N items are required each year and find the batch size that minimizes it.

the plant for these runs is therefore $(400)(50,000/x) = (2 \times 10^7/x)$ dollars. The manufacturing cost of producing 50,000 items at $4 each is $200,000. Therefore the total manufacturing and storage costs over one year are given (in dollars) by

$$C = \frac{2 \times 10^7}{x} + 200,000 + \frac{x}{5}.$$

We wish to find the value of x that makes C a minimum. Differentiating gives

$$\frac{dC}{dx} = -\frac{2 \times 10^7}{x^2} + \frac{1}{5}.$$

Setting this derivative equal to zero, we obtain the following result.

$$\frac{2 \times 10^7}{x^2} = \frac{1}{5}$$

$$x^2 = (2 \times 10^7)(5) = 10^8$$

$$x = 10^4 = 10,000$$

(The negative root has no practical significance.) Furthermore, we see that

$$\frac{d^2C}{dx^2} = \frac{4 \times 10^7}{x^3}$$

which is positive when $x = 10,000$. So this value of x provides a local minimum of C.

Consequently, the minimum cost is obtained by making

$$50,000/10,000 = 5$$

production runs per year, each run producing 10,000 units. ☛ **21**

This type of inventory cost model also applies to businesses such as warehouses or retail outlets that maintain stocks of items to be sold either to the public or to other businesses. The question is how great a quantity should be ordered each time any item is restocked. If a very large quantity is ordered, the firm will be faced with substantial storage costs, although it will have the advantage of not having to reorder for a long time. On the other hand, if only a small quantity is ordered each time the item is restocked, the storage costs

Answer $C = \dfrac{200N}{x} + 2N + 2x,$

$x = 10\sqrt{N}.$

will be low but the costs of placing the orders will be high, since orders must be placed frequently. Somewhere between these extremes we can expect there to be an optimum size for each order that will make the total cost of storage plus ordering a minimum. This optimum is called the *economic order quantity*. It can be determined, at least for the simplest model, by a method similar to the one just given. (See Exercise 29 in this section and Exercise 31 in the Review Exercises.)

EXERCISES 14-5

1. *(Number Theory)* Find two numbers with sum 10 and with maximum product.

2. *(Number Theory)* Find two numbers with sum 8, such that the sum of their squares is a minimum.

3. *(Number Theory)* Find two positive numbers with sum 75, such that the product of one times the square of the other is a maximum.

4. *(Number Theory)* Find two positive numbers with sum 12, such that the sum of their cubes is a minimum.

5. *(Geometry)* Show that among all the rectangles of area 100 square centimeters, the one with the smallest perimeter is the square of side 10 centimeters.

6. *(Geometry)* What is the area of the largest rectangle that can be drawn inside a circle of radius a?

7. *(Geometry)* What is the area of the largest rectangle that can be drawn inside a semicircle of radius a?

8. *(Fencing Costs)* A farmer wishes to enclose a rectangular paddock of area 900 square yards. Fencing costs $15 per yard. What should be the dimensions of the paddock so as to minimize the cost of the fencing? How is your answer changed if the cost of fencing rises to $20 per yard?

9. *(Fencing Costs)* Repeat Exercise 8 for the case in which one side of the paddock makes use of an existing fence and only three new sides need to be constructed.

10. *(Handbill Design)* A handbill is to contain 48 square inches of printed matter with 3-inch margins at the top and bottom and 1-inch margins on each side. What dimensions for the handbill will consume the least amount of paper?

11. *(Cistern Design)* A cistern is to be constructed to hold 324 cubic feet of water. The cistern has a square base and four vertical sides, all made of concrete, and a square top

made of steel. If the steel costs twice as much per unit area as the concrete, determine the dimensions of the cistern which minimize the total cost of construction.

12. *(Cistern Design)* Repeat Exercise 11 if the shape of the cistern is a cylinder with circular base and top.

13. *(Minimum Average Cost)* The average cost of manufacturing a certain article is given by

$$\overline{C} = 5 + \frac{48}{x} + 3x^2$$

where x is the number of articles produced. Find the minimum value of $\overline{C}$.

14. *(Inventory Cost Model)* The cost of the annual production of an item is

$$C = 5000 + \frac{80,000,000}{x} + \frac{x}{20}$$

where x is the average batch size per production run. Find the value of x which makes C a minimum.

15. *(Minimum Average Cost)* The cost of producing x items of a certain product is

$$C(x) = 4000 + 3x + 10^{-3}x^2 \quad \text{(dollars)}.$$

Find the value of x that makes the average cost per item a minimum.

16. *(Minimum Average Cost)* Repeat Exercise 15 for the cost function $C(x) = 16,000 + 3x + 10^{-6}x^3$ (dollars).

17. *(Minimum Marginal Cost and Average Cost)* The cost function of a firm is given by

$$C(x) = 300x - 10x^2 + x^3/3.$$

Calculate the output x at which:

a. the marginal cost is minimum.

b. the average cost is minimum.

18. (Minimum Marginal Cost) A firm produces x tons of a valuable metal per month at a total cost C given by $C(x) = 10 + 75x - 5x^2 + x^3/3$ dollars. Find the level of output x at which the marginal cost attains its minimum.

19. (Maximum Revenue) The demand function for a certain commodity is given by $p = 15e^{-x/3}$ for $0 \le x \le 8$, where p is the price per unit and x the number of units demanded. Determine the price p and the quantity x for which the revenue is maximum.

20. (Maximum Revenue) Repeat Exercise 19 for the demand law $p = 10e^{-x^2/32}$ for $0 \le x \le 6$.

21. (Maximum Profit) A firm sells all units it produces at $4 per unit. The firm's total cost C of producing x units is given in dollars by

$$C = 50 + 1.3x + 0.001x^2.$$

a. Write the expression for total profit P as a function of x.

b. Find the production volume x so that the profit P is maximum.

c. What is the value of maximum profit?

22. (Maximum Profit) A company finds that it can sell all of a certain product that it produces at the rate of $2 per unit. It estimates the cost function of the product to be $(1000 + \frac{1}{2}(x/50)^2)$ dollars for x units produced.

a. Find an expression for the total profit if x units are produced and sold.

b. Find the number of units produced that will maximize profit.

c. What is the amount of the maximum profit?

d. What would be the profit if 6000 units were produced?

23. (Maximum Profit) In Exercise 15, the items in question are sold for $8 each. Find the value of x that maximizes the profit and calculate the maximum profit.

24. (Maximum Profit) In Exercise 16, the items are sold for $30 each. Find the value of x that maximizes the profit and calculate the maximum profit.

25. (Maximum Profit) For a certain item, the demand equation is $p = 5 - 0.001x$. What value of x maximizes the revenue? If the cost function is $C = 2800 + x$, find the value of x that maximizes the profit. Calculate the maximum profit.

26. (Maximum Profit) Repeat Exercise 25 for the demand equation $p = 8 - 0.02x$ and the cost function $C = 200 + 2x$.

27. (Effect of Taxation on Production) The total cost function of a manufacturer is given by

$$C(x) = 10 + 28x - 5x^2 + \frac{x^3}{3}$$

and the market demand of the product is given by $p = 2750 - 5x$, where p and x denote the price in dollars and the quantity, respectively. A tax at the rate of $222 per unit of the product is imposed, which the manufacturer adds to the cost. Determine the level of output after the tax is imposed for maximum profits. Show that the post-tax output is less than the pretax output for optimal profits.

28. (Effect of Taxation on Production) Repeat Exercise 27 for

$$C(x) = 30 + 12x - 0.5x^2 \quad \text{and} \quad p = 60 - 2x$$

when a tax of $3 per unit is imposed.

29. (Economic Order Quantity) Material is demanded at the rate of 10,000 units per year; the cost price of the material is $2 per unit; the cost of replenishing the stock of the material per order, regardless of the size of the order (x), is $40 per order; and the cost of storing material for one year is 10% of the value of the average inventory $(x/2)$ on hand. C is the annual cost of ordering and storing the material.

a. Show that

$$C = 20,000 + \frac{400,000}{x} + \frac{x}{10}.$$

b. Find the economic order quantity.

30. (Inventory Cost Model) A factory has to produce 96,000 units of an item per year. The cost of material is $2 per unit and the cost of replenishing the stock of material per order regardless of the size x of the order is $25 per order. The cost of storing the material is 30¢ per item per year on the inventory $(x/2)$ in hand. Show that the total cost C is given by

$$C = 192,000 + \frac{2,400,000}{x} + \frac{3x}{20}.$$

Also find the economic lot size (that is, the value of x for which C is minimum).

31. (Inventory Cost Model) A motor car wholesaler sells

100,000 cars per year and reorders them in batches of size x. It costs \$1000 to place each order and \$200 per year in storage charges for each car. Calculate the optimum size of each order in order to minimize the sum of order costs and storage costs.

32. *(Inventory Cost Model)* A manufacturer requires N of a certain part per year. It costs K dollars to place each order for new parts, regardless of the size of the order and it costs I dollars per year to store each item held in inventory. Show that the optimal reorder size is equal to $\sqrt{2NK/I}$. Calculate the minimum total cost of ordering plus storage.

33. *(Land Cost)* A company is seeking a rectangular plot of land on which to construct a new warehouse. The area of the warehouse will be 6400 square meters. It is required to have a 40-meter-wide loading zone along one side of the building and a 10-meter-wide strip along the front for parking. What is the smallest area of land the company must purchase?

34. *(Maximum Revenue)* A steakhouse determines that at a price of \$5 per steak they have an average of 200 customers per night, whereas if the price is \$7 per steak, the average number of customers falls to 100. Determine the demand relation, assuming it is linear. Find the price that maximizes the revenue.

35. *(Profit and Customer Satisfaction)* By reducing the number of its tellers, a bank can cut down on labor costs, but it can expect to lose business because of customer dissatisfaction at having to wait. Assume that the wage rate for tellers is \$80 per day and that the loss of profit from having only n tellers is $5000/(n + 1)$ dollars per day. Determine the value of n that minimizes the sum of this loss plus the wage cost.

36. *(Maximum Volume)* A square of side x is cut from each corner of a rectangular piece of cardboard of size 12 inches $\times$ 18 inches, and the four edges are then bent up to form a box of depth x. Find the value of x that gives the box of greatest volume.

*37. *(Maximum Area)* A square of side x is cut from each corner of a square piece of cardboard of size $y \times y$ and the four edges then bent up to form a box of depth x. The box is required to have a volume of 128 cubic centimeters. Find the values of x and y that minimize the area of the original piece of cardboard.

38. *(Optimal Can Shape)* It is desired to manufacture cylindrical cans with a given volume V. Show that for the

shape of can that minimizes the amount of material used (i.e., minimizes the total area of the sides, top, and bottom), the height is equal to twice the radius (Why do you think most cans are not made like that?)

39. *(Maximum Timber Yield)* A forest company plans to log a certain area of fir trees after a given number of years. The average number of board feet obtained per tree over the given period is known to be equal to $50 - 0.5x$, where x is the number of trees per acre, when x lies between 35 and 80. What density of trees should be maintained in order to maximize the amount of timber per acre?

40. *(Crop Yield)* The yield y (bushels per acre) of a certain crop of wheat is given by $y = a(1 - e^{-kx}) + b$, where a, b, and k are constants and x is the number of pounds per acre of fertilizer. The profit from sale of the wheat is given by $P = py - c_0 - cx$ where p is the profit per bushel, c is the cost per pound of fertilizer, and c_o is an overhead cost. Determine how much fertilizer must be used in order to maximize the profit P.

41. *(Maximum Yield from Sales Tax)* The quantity of an item x that can be sold per month at price p is given by $x = 100(5 - p)$. The quantity that suppliers will make available at a price p_1 is given by $x = 200(p_1 - 1)$. If there is a tax t on each item (so $p_1 = p - t$), determine the quantity x that is sold per month if the market is in equilibrium. Find the value of t that provides the maximum total tax per month for the government.

42. *(Maximum Yield from Sales Tax)* Repeat Exercise 41 if the demand equation is $x = 400(15 - p)$ and the supply equation is $x = 400(2p_1 - 3)$. Calculate the monthly tax yield to the government.

43. *(Building Costs)* The cost of erecting a building containing n floors can often be taken to be of the form $a + bn + cn^2$, where a, b and c are constants. (Here a represents fixed costs such as land costs, b represents a cost that is the same for every floor, such as interior walls, windows, floor covering, and cn^2 represents costs such as structural members, which increase as the square of the number of floors.) Calculate the value of n that makes the average cost per floor a minimum. Show that as land costs increase, this optimum value of n increases.

44. *(Heating Costs)* An individual is planning to insulate a house. At present the annual heating cost is \$3000, but if x inches of insulation are added, the cost will reduce to $3000e^{-0.1x}$ dollars. For each inch of insulation, the owner must borrow $1000 from the bank at 10% interest rate. How many inches should be added to minimize the total of heating cost plus interest?

45. *(Optimal Selling Time)* A speculator purchases a consignment of rare wine whose value will increase according to the formula $V(t) = S(1 + 0.2t)$, where t is time in years. If the wine is sold after t years, the proceeds must be discounted back to give a present value of $P(t) = V(t)e^{-rt}$, where $r = R/100$ and R is the nominal discount rate. After how many years should the wine be sold in order to optimize its present value?

*46. *(Plant Pest)* The percentage of trees in a fruit plantation that have become infected with a certain pest is given by

$$P(t) = \frac{100}{1 + 50e^{-0.1t}}$$

where t is time in days. Calculate the time at which $P'(t)$ is maximum. What is the significance of this time?

47. *(Grain Silo Design)* A grain silo is to be built in the form of a vertical cylinder with a hemispherical roof. The silo is to be capable of storing 10,000 cubic feet of grain. (Assume the grain will be stored only in the cylindrical part, not in the roof.) The hemispherical roof costs twice as much per unit area to manufacture as the cylindrical sides cost. What dimensions should the silo have in order to minimize its total cost?

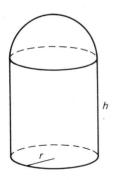

48. *(Group Travel)* A travel agent provides charter holidays for group travel on the following basis: For groups of size up to 50, the fare is $400 per person, while for larger groups, the fare per person is reduced for the whole group by $2 for each person traveling in excess of 50. Find the group size that maximizes the travel agent's revenue.

*49. *(Maximum Revenue and Profit)* A manufacturer prices a

product at $10 each for orders less than 200 and offers a reduction in price of 2¢ for each item by which an order exceeds 200, the reduction applying to the whole order. Find the order size that maximizes the manufacturer's revenue. If the items cost $5 each to manufacture, find the order size that maximizes the manufacturer's profit. How is this last result changed if the manufacturing cost rises to $7 per item?

50. *(Cost of Telephone Link)* It is desired to construct a telephone link between two towns A and B situated on opposite banks of a river. The width of the river is 1 kilometer, and B lies 2 kilometers downstream from A. It costs $c per killometer to construct a line overland and $2c per kilometer underwater. The telephone line will follow the river bank from A for a distance x (kilometers) and then will cross the river diagonally in a straight line directly to B. Determine the value of x which minimizes the total cost.

51. *(Offshore Oil Refinery)* An oil company requires a pipeline from an offshore oilfield to a refinery which is to be built on the neighboring coast. The distance from the oilfield to the nearest point P on the coast is 20 kilometers and the distance along the coast from P to the refinery is 50 kilometers. From the refinery, the pipeline will go a distance x kilometers along the coast, then will follow a straight line underwater to the oilfield. The cost per kilometer of underwater pipeline is three times that of the overland section. Find the value of x which minimizes the total cost of the pipeline.

52. *(Epidemic)* During the course of an epidemic, the proportion of the population infected after a time t is equal to

$$\frac{t^2}{5(1 + t^2)^2}$$

(t is measured in months, and the epidemic starts at $t = 0$). Find the maximum proportion of population that becomes infected. Find also the time at which the proportion of infected individuals is increasing most rapidly.

53. *(Drug Reaction)* The reaction as a function of time (measured in hours) to two drugs is given by

$$R_1(t) = te^{-t}, \qquad R_2(t) = te^{-2t^2}.$$

Which drug has the larger maximum reaction?

54. *(Drug Reaction)* The reaction of a drug at time t after it has been administered is given by $R(t) = t^2e^{-t}$. At what time is the reaction maximum?

14-6 ABSOLUTE MAXIMA AND MINIMA

☛ 22. By inspection of the following graphs, give the values of x at which each function takes its absolute maximum and absolute minimum values in the given interval.

(a)

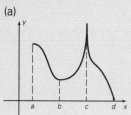

(b)

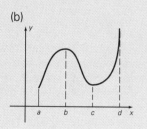

Answer (a) Absolute maximum at c, absolute minimum at d.
(b) absolute maximum at d, absolute minimum at both a and c.

☛ 23. Find the absolute maximum and minimum values of the following functions in the given intervals:
(a) $f(x) = 3 + 4x - x^2$ in $[0, 3]$;
(b) $f(x) = x^3 - 12x + 5$ in $[-3, 4]$.

Answer (a) Absolute maximum 7 at $x = 2$,
absolute minimum 3 at $x = 0$;
(b) absolute maximum 21 at $x = -2$, and $x = 4$,
absolute minimum -11 at $x = 2$.

In some problems, it happens that the independent variable x is restricted to some interval of values, say $a \leq x \leq b$, and we need to find the largest or smallest value of some function $f(x)$ over this set of values of x. In fact, most of our problems in the last section were of this type, although we did not emphasize the fact there. For example, if x is the level of production by some manufacturing firm, then x is restricted to the interval $x \geq 0$ and we are interested in the maximum value of the profit function in this interval. Any local maximum that might occur for a negative value of x is of no significance. This restriction on x does not affect any of the results we obtained, but cases do arise in which similar restrictions can affect the conclusions regarding the optimum.

DEFINITION The **absolute maximum** value of $f(x)$ over an interval $a \leq x \leq b$ of its domain is the largest value of $f(x)$ as x takes all the values from a to b. Similarly, the **absolute minimum** value of $f(x)$ is the smallest value of $f(x)$ as x increases from a to b.

It is intuitively obvious that if $f(x)$ is continuous in $a \leq x \leq b$, the point at which $f(x)$ attains its absolute maximum must be either a local maximum of $f(x)$ or else one of the endpoints a or b. A similar statement holds for the absolute minimum. Thus to find the absolute maximum and absolute minimum values of $f(x)$ over $a \leq x \leq b$, we simply select the largest and the smallest values from among the values of $f(x)$ at the critical points lying in $a \leq x \leq b$ and at the end points a and b. This is illustrated in Example 1. ☛ 22

EXAMPLE 1 Determine the absolute maximum and minimum values of

$$f(x) = 1 + 12x - x^3 \quad \text{in } 1 \leq x \leq 3.$$

Solution We have $f'(x) = 12 - 3x^2$.

Since $f'(x)$ is defined for all x, the critical points of f are given by $f'(x) = 0$, or $x^2 = 4$; that is, $x = \pm 2$. But $x = -2$ is *not* within the given interval $1 \leq x \leq 3$. Thus we consider only the critical point $x = 2$, plus the endpoints $x = 1$ and $x = 3$. The values of $f(x)$ at these points are

$$f(1) = 1 + 12 - 1 = 12$$

$$f(2) = 1 + 24 - 8 = 17$$

$$f(3) = 1 + 36 - 27 = 10.$$

Thus the absolute maximum value of $f(x)$ is 17, which occurs at $x = 2$, and the absolute minimum is 10, which occurs at the endpoint $x = 3$. The graph of $y = 1 + 12x - x^3$ is shown in Figure 28. Within the interval $1 \leq x \leq 3$, the graph has a single local maximum occurring at $x = 2$. The absolute minimum value occurs at the endpoint $x = 3$. ☛ 23

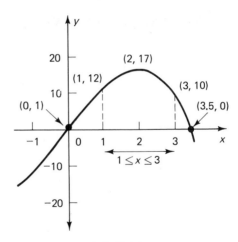

FIGURE 28

EXAMPLE 2 *(Minimum Cost)* An underground cistern is to be constructed to hold 100 cubic feet of radioactive waste. The cistern is to be a circular cylinder in shape. The circular base and vertical sides, which are all underground, cost \$100 per square foot and the lid, at ground level, costs \$300 per square foot because of the necessary shielding. Furthermore, the depth of the tank cannot exceed 6 feet because of a hard rock layer beneath the surface, which would increase the excavation costs enormously if it were to be penetrated. Finally, the radius of the tank cannot exceed 4 feet because of space limitations. What dimensions for the tank will make its cost a minimum?

Solution Let the radius be r feet and the depth be x feet. (See Figure 29.) Then the volume is $\pi r^2 x$, which must be the required 100 cubic feet.

$$\pi r^2 x = 100 \tag{1}$$

The area of the vertical sides is $2\pi rx$ and of the base is πr^2, and all of these cost \$100 per square foot. So

$$\text{Cost (in dollars) of Base and Sides} = (2\pi rx + \pi r^2)(100).$$

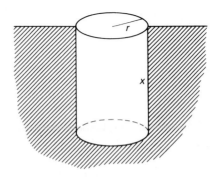

FIGURE 29

The top costs $(\pi r^2)(300)$ dollars. Therefore the total cost C (in dollars) is

$$C = (2\pi rx + \pi r^2)(100) + (\pi r^2)(300) = 200\pi rx + 400\pi r^2.$$

But from Equation (1), $x = 100/\pi r^2$, so substituting for x we find that

$$C = \frac{20{,}000}{r} + 400\pi r^2.$$

To find the local minimum value of C we set $dC/dr = 0$ and solve for r.

$$\frac{dC}{dr} = -\frac{20{,}000}{r^2} + 800\pi r = 0$$

$$800\pi r = \frac{20{,}000}{r^2}$$

$$r^3 = \frac{20{,}000}{800\pi} = \frac{25}{\pi}$$

Therefore $r = \sqrt[3]{25/\pi} \approx 2.00$.

The corresponding value of x is

$$x = \frac{100}{\pi r^2} \approx \frac{100}{\pi (2.00)^2} = 7.96.$$

Thus the dimensions that give the cheapest construction are of radius of 2 feet and a depth of 7.96 feet. However we are not allowed to have a value of x in excess of 6 feet. So although the value $x = 7.96$ gives the minimum value of C, it does not provide the solution to the problem as posed.

Let us calculate the interval of allowed values of r. The largest value of r is given as 4. The smallest occurs when the depth is greatest, that is, when $x = 6$. In that case, $r^2 = 100/\pi x = 100/6\pi$, so

$$r = \sqrt{100/6\pi} \approx 2.30.$$

Thus r is restricted to the interval $2.30 \le r \le 4$.

Within this interval, C has no critical points, so we need to evaluate it at the end points only:

$$C(2.30) = \frac{20{,}000}{2.30} + 400\pi (2.30)^2 \approx 15{,}300$$

$$C(4) = \frac{20{,}000}{4} + 400\pi (4)^2 \approx 25{,}100$$

The absolute minimum value of C is therefore \$15,300 and occurs when $r = 2.30$ (that is, when $x = 6$). The graph of C as a function of r is shown in Figure 30. ☛ **24**

☛ **24.** Re-solve Example 2 if the lid cost \$100 per square foot.

Answer $r = 2.515$, $x = 5.031$.

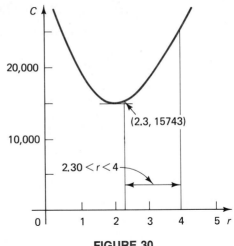

FIGURE 30

Summary of Method of Finding Absolute Extrema

Suppose we want the absolute maximum and/or the absolute minimum of $f(x)$ in the interval $a \leq x \leq b$.

Step 1. Find the critical points of f and *reject* those (if there are any) that are outside the interval $a \leq x \leq b$.

Step 2. Evaluate the given function f at the critical points found in Step 1 and at the end points a and b.

Step 3. The largest and smallest of the values of f found in Step 2 are then respectively the absolute maximum and minimum values of f in $a \leq x \leq b$.

EXERCISES 14-6

(1–14) Find the absolute extrema of the following functions in the indicated intervals.

1. $f(x) = x^2 - 6x + 7; \quad 1 \leq x \leq 6$

2. $f(x) = 9 + 6x - x^2; \quad 1 \leq x \leq 5$

3. $f(x) = x^3 - 75x + 1; \quad -1 \leq x \leq 6$

4. $f(x) = x^3 - 3x + 4; \quad -2 \leq x \leq 2$

5. $f(x) = x^3 - 18x^2 + 60x; \quad -1 \leq x \leq 5$

6. $f(x) = \dfrac{(x - 1)(x - 3)}{x^2}; \quad \frac{1}{2} \leq x \leq 2$

7. $f(x) = \dfrac{(x + 1)(x - 6)}{x^2}; \quad \frac{1}{2} \leq x \leq 2$

8. $f(x) = x + 1/x; \quad -2 \leq x \leq -\frac{1}{2}$

9. $f(x) = xe^{-x}; \quad \frac{1}{2} \leq x \leq 2$

10. $f(x) = x^2 e^{-x^2}; \quad -2 \leq x \leq 2$

11. $f(x) = x - \ln x; \quad e^{-1} \le x \le e$

12. $f(x) = x^{-1} \ln x; \quad \frac{1}{2} \le x \le 10$

*13. $f(x) = x \ln x; \quad 0 < x \le 0.9$

*14. $f(x) = (2x + 1)e^{-x}; \quad 0 \le x < \infty$

15. *(Water Pollution)* When deposited in a lake, organic waste decreases the oxygen content of the water. If t denotes the time in days after the waste is deposited, then it is found experimentally in one instance that the oxygen content is given by

$$y = t^3 - 30t^2 + 6000$$

for $0 \le t \le 25$. Find the maximum and minimum values of y during the first 25 days following the depositing of the waste.

16. *(Minimum Average Cost)* The cost function for a manufacturer is

$$C(x) = 1000 + 5x + 0.1x^2$$

when x items are produced per day. If at most 80 items can be produced per day, determine the value of x that gives the lowest average cost per item.

17. *(Maximum Revenue and Profit)* The cost of producing x items per week is

$$C(x) = 1000 + 6x - 0.003x^2 + 10^{-6}x^3$$

but not more than 3000 items can be produced each week. If the demand equation is

$$p = 12 - 0.0015x$$

determine the level of production that maximizes the revenue and the level that maximizes the profit.

18. *(Production Decision)* The cost function in thousands of dollars is

$$C(x) = 2 + x - \tfrac{1}{4}x^2 + \tfrac{1}{24}x^3$$

where the production level x is in thousands of units per week. The available plant capacity limits x to the range $0 \le x \le 4$. If each item produced can be sold for $2.50, find the production levels that maximize:

a. The revenue.   b. The profit.

How do your conclusions change if the plant capacity is increased to $x = 8$ with the same cost function?

19. *(Production Decision)* The demand equation for a company's product is $p = 200 - 1.5x$, where x units can be sold at a price of p each. It costs the company $(500 + 65x)$ dollars to produce x units each week. How many units should the company produce and sell each week to maximize profit if the production capacity is at most:

a. 60 units?   b. 40 units?

20. *(Minimum Reaction Time)* In a test of airline pilots' speed of reaction to a simulated crisis, it was found that the total time required to react to the crisis varied with the pilot's age x according to the formula $T = 0.04(1700 - 80x + x^2)^{1/2}$ over the age range $30 \le x \le 55$. Within this range, what age gave the smallest reaction time?

21. *(Tank Design)* A company manufactures open water tanks that are required to hold 50 cubic feet. The base must be square. Because of storage and transportation limitations, the side of the base must not exceed 5 feet and the height must not exceed 5 feet. Find the dimensions that minimize the amount of material used (i.e., minimize the surface area.)

22. *(Tank Design)* Repeat Exercise 21 for the case of a tank with circular base whose diameter must not exceed 5 feet.

23. *(Inventory Cost Model)* A computer retailer sells 30,000 per year of a particular brand of personal computer. It costs $1200 to reorder from the manufacturer, regardless of the size of the order, and it costs $2 per year to store each computer. Furthermore, there is storage available for a maximum of 5000 computers at any time. How many times per year should he reorder in order to minimize his total cost?

24. *(Photosynthesis)* If an intensity x of light falls on a plant, the rate of photosynthesis y, measured in appropriate units, is found experimentally to be given by $y = 150x - 25x^2$ as long as $0 \le x \le 5$. Find the maximum and minimum values of y when x lies in the interval $1 \le x \le 5$.

*25. *(Population Measurement)* The size of a certain bacterial population at time t (in hours) is given by $y = a(1 + \tfrac{1}{2}e^t)^{-1}$, where a is a constant. A biologist plans to observe the population over the 2-hour period from $t = 0$ to $t = 2$. What will be the largest and the smallest growth rates that he observes?

In the first part of this section we shall be concerned with the way in which certain functions behave as their argument increases to very large values or decreases to very large negative values.

Consider, for example, $f(x) = 1/x$. A table of values of this function for $x = 1, 10, 100, 1000$, and so on, is given in Table 7. It is clear from these values that as x increases, $f(x)$ gets closer and closer to zero. The behavior is shown also by the graph of $y = 1/x$ in Figure 31. We use the notation $x \to \infty$ (which is read as "*x approaches infinity*") to indicate that x increases without bound. The fact that $1/x$ gets closer and closer to zero as $x \to \infty$ is then written in the form of a limit:

$$\lim_{x \to \infty} \frac{1}{x} = 0$$

TABLE 7

x	1	10	100	1,000	10,000	. . .
$f(x)$	1	0.1	0.01	0.001	0.0001	. . .

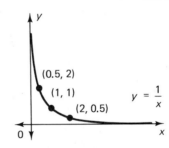

FIGURE 31

As a second example consider the function

$$f(x) = \frac{2x - 1}{x}$$

The values of this function for an increasingly large set of values of x are given in Table 8. It is clear that as x increases, $f(x)$ gets closer and closer to 2. This is also shown by the graph of $y = (2x - 1)/x$ in Figure 32. As x increases, the graph gets closer and closer to the horizontal line $y = 2$. Using the limit notation, we write

$$\lim_{x \to \infty} \frac{2x - 1}{x} = 2.$$

TABLE 8

x	1	10	100	1000	10,000	. . .
$f(x)$	1	1.9	1.99	1.999	1.9999	. . .

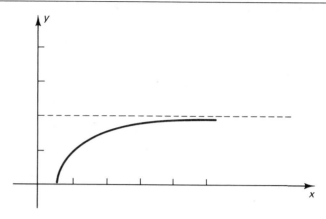

FIGURE 32

We use the notation $x \to -\infty$ to indicate that x becomes increasingly large in magnitude through negative values. This is read as "*x approaches minus infinity.*" The formal definition of the limit notation is then as follows.

DEFINITION A function $f(x)$ approaches the **limiting value** L as $x \to \infty$ if the value of $f(x)$ can be made as close as we like to L simply by taking x large enough. We write

$$\lim_{x \to \infty} f(x) = L.$$

☞ **25.** By calculating a few values as in Table 1 and 2, find $\lim_{x \to \infty} \dfrac{1}{x+1}$ and $\lim_{x \to \infty} \dfrac{3x+1}{x}$.

DEFINITION A function $f(x)$ approaches the **limiting value** L as $x \to -\infty$ if the value of $f(x)$ can be made as close to L as we like by taking x to be a negative number of sufficiently large absolute value. We write

$$\lim_{x \to -\infty} f(x) = L. \quad ☞ 25$$

As we have seen, the function $1/x$ approaches the limit zero as $x \to \infty$. This function also approaches the same limit zero as $x \to -\infty$. These results generalize to inverse powers.

THEOREM 1

$$\lim_{x \to \infty} \frac{1}{x^n} = 0 \text{ for all } n > 0.$$

$$\lim_{x \to -\infty} \frac{1}{x^n} = 0 \text{ for all } n > 0, \text{ provided that } \frac{1}{x^n} \text{ is defined for } x < 0.$$

Answer 0 and 3.

☛ **26.** Evaluate

(a) $\lim\limits_{x \to -\infty} \dfrac{1}{\sqrt[3]{x}}$;

(b) $\lim\limits_{x \to \infty} \dfrac{\sqrt[4]{x}}{\sqrt[6]{x}}$;

(c) $\lim\limits_{x \to -\infty} \left(2 + \dfrac{\sqrt{-x}}{x} \right)$.

Answer (a) 0;
(b) does not exist;
(c) 2.

☛ **27.** Evaluate

(a) $\lim\limits_{x \to -\infty} \dfrac{2x - 1}{1 - x}$;

(b) $\lim\limits_{x \to \infty} \dfrac{2x - x^2}{1 - 3x^2}$;

(c) $\lim\limits_{x \to -\infty} \dfrac{2x + 3x^2}{2x^2 - x - 1}$.

Answer (a) -2;
(b) $\frac{1}{3}$;
(c) $\frac{3}{2}$.

EXAMPLE 1

$$\lim_{x \to \infty} 1/x^2 = 0, \qquad \lim_{x \to \infty} 1/\sqrt{x} = 0, \qquad \lim_{x \to -\infty} x^{-4/3} = 0.$$

Note that limits such as $\lim\limits_{x \to -\infty} 1/\sqrt{x}$ do not exist because $\sqrt{x}$ is not defined for $x < 0$. ☛ **26**

EXAMPLE 2 Calculate $\lim\limits_{x \to \infty} f(x)$ for the following functions.

(a) $f(x) = \dfrac{2x^2 + 2}{x^2 + x + 3}$

(b) $f(x) = \dfrac{x + 1}{x^3 + 3x}$

Solution To calculate the limiting value of such rational functions the general rule is to *divide numerator and denominator by the highest power of x in the denominator* and then use Theorem 1.

(a) Divide by x^2.

$$f(x) = \frac{(2x^2 + 2) \div x^2}{(x^2 + x + 3) \div x^2} = \frac{2 + \dfrac{2}{x^2}}{1 + \dfrac{1}{x} + \dfrac{3}{x^2}}$$

As $x \to \infty$, the inverse powers all approach zero, and

$$f(x) \to \frac{2 + 0}{1 + 0 + 0} = 2.$$

That is, $\lim\limits_{x \to \infty} f(x) = 2$.

(b) Divide by x^3.

$$f(x) = \frac{(x + 1) \div x^3}{(x^3 + 3x) \div x^3} = \frac{\dfrac{1}{x^2} + \dfrac{1}{x^3}}{1 + \dfrac{3}{x^2}} \to \frac{0 + 0}{1 + 0} = 0$$

Thus $\lim\limits_{x \to \infty} f(x) = 0$. ☛ **27**

EXAMPLE 3 *(Advertising and Profits)* According to the estimates of a firm, the profits P from the sale of its new product are related to advertising expenditure x by the formula

$$P(x) = \frac{23x + 15}{x + 4}$$

where P and x are both in millions of dollars. (a) show that $P(x)$ is an increasing function of x. (b) Find the upper limit to the profits, if any exists.

Solution

(a) We have

$$P(x) = \frac{23x + 15}{x + 4}.$$

By the quotient rule, we have

$$P'(x) = \frac{(x + 4)(23) - (23x + 15)(1)}{(x + 4)^2} = \frac{77}{(x + 4)^2}.$$

Since $P'(x) > 0$ for all $x > 0$, $P(x)$ is an increasing function of x, that is, the profits increase with an increase in the amount spent on advertising.

(b) $\lim\limits_{x \to \infty} P(x) = \lim\limits_{x \to \infty} \dfrac{23x + 15}{x + 4} = \lim\limits_{x \to \infty} \dfrac{(23x + 15)/x}{(x + 4)/x}$

$$= \lim\limits_{x \to \infty} \frac{23 + 15/x}{1 + 4/x} = \frac{23 + 0}{1 + 0} = 23.$$

Thus the upper limit on the profits is $23 million. The graph of $P(x)$ is shown in Figure 33. From the graph, it follows that after a while, large increases in advertising spending (x) will increase the profits by only a very small amount. This is an example of what is known as the *law of diminishing returns*.

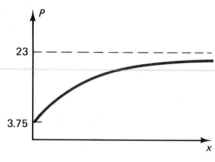

FIGURE 33

If $f(x) \to L$ as $x \to \infty$, then the graph of $y = f(x)$ gets closer and closer to the line $y = L$ as x moves further towards the right. We say that the line $y = L$ is a **horizontal asymptote** of the graph at $+\infty$.

If the graph of $y = f(x)$ has $y = L$ as a horizontal asymptote at $+\infty$, the graph may stay entirely on one side of the line $y = L$ or may cross the asymptote as x increases. Typical examples are shown in Figure 34.

Similarly, if $f(x) \to L'$ as $x \to -\infty$, then the graph of $y = f(x)$ gets closer and closer to the horizontal line $y = L'$ as x moves further to the left. This line is a horizontal asymptote of the graph at $-\infty$.

EXAMPLE 4 Sketch the graph of the function $y = e^{-x^2}$.

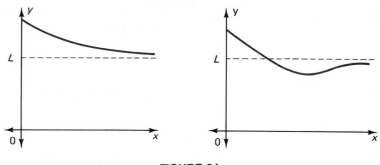

FIGURE 34

Solution We first follow the steps given in Section 14-4.

Step 1 We have the following.

$$f(x) = e^{-x^2}$$

$$f'(x) = -2xe^{-x^2} \qquad \text{(from the chain rule)}$$

The factor e^{-x^2} is never zero, so $f'(x) = 0$ only when $x = 0$. At this value $y = e^0 = 1$. That is, the graph has a horizontal tangent at the point $(0, 1)$ where it crosses the y-axis.

Since e^{-x^2} is always positive, we see that when $x < 0, f'(x) > 0$, so the graph is increasing for $x < 0$. Correspondingly, when $x > 0$, $f'(x) < 0$ and the graph decreases.

Step 2 We use the product rule and differentiate a second time.

$$f''(x) = -2\frac{d}{dx}(xe^{-x^2}) = 2(2x^2 - 1)e^{-x^2}$$

The points of inflection, where $f''(x) = 0$, are given by $2x^2 - 1 = 0$, that is, $x = \pm 1/\sqrt{2}$.

The corresponding values of y are $y = e^{-(\pm 1/\sqrt{2})^2} = e^{-0.5}$. So the points of inflection are $(\pm 1/\sqrt{2}, e^{-0.5}) \approx (\pm 0.71, 0.61)$. The second derivative changes sign at these points of inflection. The factor e^{-x^2} is always positive, so the sign of $f''(x)$ is the same as the sign of $2x^2 - 1$. When $x < -1/\sqrt{2}$, $x^2 > \frac{1}{2}$ so $2x^2 - 1 > 0$. Thus $f''(x) > 0$ in this region and the graph is concave upward. When $-1/\sqrt{2} < x < 1/\sqrt{2}$, $x^2 < \frac{1}{2}$, so $2x^2 - 1 < 0$. Thus $f''(x) < 0$ in this region and the graph is concave downward. When $x > 1/\sqrt{2}$, $x^2 > \frac{1}{2}$ and $2x^2 - 1 > 0$. Thus again $f''(x) > 0$ and the graph is concave upward.

Step 3 We have already found the point $(0, 1)$ where the graph crosses the y-axis. The graph never crosses the x-axis since e^{-x^2} is always positive.

Step 4 Now a new step: we examine the behavior of $f(x)$ as $x \to \pm\infty$. In either case, the exponent $-x^2$ becomes increasingly large and negative. Hence e^{-x^2} gets closer and closer to zero. So the graph has the x-axis ($y = 0$) as a horizontal asymptote as both $x \to \infty$ and $x \to -\infty$.

☛ 28. For the function

$$y = \frac{1}{1 + x^2}, \text{ find}$$

(a) the intervals where it is increasing and decreasing;

(b) the intervals where it is concave up and concave down;

(c) the points of intersection with the coordinate axes;

(d) the horizontal asymptote.

Bringing all the information together we can draw a reasonably accurate sketch as shown in Figure 35. The graph is related to the familiar bell-shaped curve of probability theory. ☛ 28

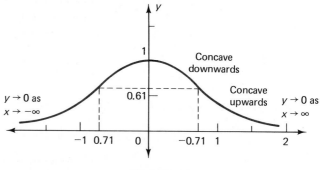

FIGURE 35

Vertical Asymptotes

Consider the behavior of the function $y = 1/x$ as x approaches 0. If x becomes smaller and smaller, its reciprocal becomes larger and larger, as illustrated by the succession of values in Table 9. This feature is illustrated by the graph of $y = 1/x$ since the graph climbs to arbitrarily high values as x decreases toward 0. (See Figure 36.) We write this as

$$\lim_{x \to 0^+} \frac{1}{x} = \infty.$$

TABLE 9

x	1	0.1	0.01	0.001
$y = \dfrac{1}{x}$	1	10	100	1000

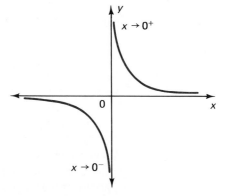

FIGURE 36

Answer (a) Increasing for $x < 0$, decreasing for $x > 0$;
(b) Concave up for $x < -1/\sqrt{3}$ and $x > 1/\sqrt{3}$,
concave down for $-1/\sqrt{3} < x < 1/\sqrt{3}$;
(c) y-axis at $(0, 1)$, does not cross the x-axis;
(d) $y = 0$. (The graph is similar in shape to Figure 35.)

It should be pointed out that this notation is purely a matter of convention, and does not imply that $\lim_{x \to 0^+} (1/x)$ exists in the ordinary sense of a limit. It simply means that as x approaches zero from above, the function $1/x$ increases without bound.

As x approaches zero from below, the values of $1/x$ become larger and larger in a negative direction, as illustrated by Table 10. This is written as

$$\lim_{x \to 0^-} \frac{1}{x} = -\infty,$$

indicating that as x approaches 0 from below, the function $1/x$ decreases without bound.

TABLE 10

x	-1	-0.1	-0.01	-0.001
$\dfrac{1}{x}$	-1	-10	-10	-1000

► **29.** Find $\lim_{x \to c^-} f(x)$ and $\lim_{x \to c^+} (f)x$ for the following functions with the given c:

(a) $f(x) = \dfrac{1}{x^2}$, $c = 0$;

(b) $f(x) = \dfrac{x}{x - 2}$, $c = 2$;

(c) $f(x) = \dfrac{x}{x + 2}$, $c = -2$;

(d) $f(x) = \dfrac{x - 2}{(1 - x)^2}$, $c = 1$.

The line $x = 0$ is called a **vertical asymptote** of the graph of $y = 1/x$. The graph approaches the vertical asymptote more and more closely, y becoming indefinitely large, as x approaches 0. This compares with the line $y = 0$ (the x-axis) which is a *horizontal* asymptote of the graph. The *horizontal* asymptote is obtained from the limit as x approaches $\pm\infty$: $\lim_{x \to \pm\infty} 1/x = 0$. The *vertical* asymptote is obtained at the limiting values of x that make y approach $\pm\infty$. ► **29**

EXAMPLE 5 Find the horizontal and vertical asymptotes of the function

$$y = \frac{2x - 9}{x - 2}$$

and sketch its graph.

Solution First of all, we divide numerator and denominator by x and write

$$y = \frac{2x - 9}{x - 2} = \frac{2 - 9/x}{1 - 2/x} \to \frac{2 - 0}{1 - 0} = 2 \quad \text{as } x \to \pm\infty.$$

Therefore the graph of the given function approaches the line $y = 2$ as its horizontal asymptote both as $x \to +\infty$ and as $x \to -\infty$.

The domain of the given function is the set of all real numbers except $x = 2$. As x approaches 2, the denominator $x - 2$ approaches zero and so y becomes very large. The line $x = 2$ must therefore be a vertical asymptote.

In order to complete the sketch of the graph, we must decide on which sides of the asymptotes the graph lies. As $x \to 2^+$ (x approaches 2 from above), the factor $2x - 9$ in the numerator approaches the limit -5. The de-

Answer (a) ∞, ∞;
(b) $-\infty$, ∞;
(c) ∞, $-\infty$;
(d) $-\infty$, $-\infty$.

nominator $x - 2$ approaches zero through positive values. Therefore y becomes very large and negative, since its numerator is negative and its denominator small but positive.

On the other hand, as $x \to 2^-$ (x approaches 2 from below) the numerator still approaches the limit -5, but the denominator is small and negative. Hence y becomes large and positive. We therefore conclude that

$$y \to -\infty \text{ as } x \to 2^+ \qquad \text{and} \qquad y \to +\infty \text{ as } x \to 2^-.$$

We are thus able to place the graph in relation to the vertical asymptote at $x = 2$. (See Figure 37.) The sections of the graph between the asymptotes can now be filled in. In doing this, it is helpful to find where the graph crosses the two coordinate axes. We note that when $x = 0$, $y = \frac{9}{2}$, so that the graph crosses the y-axis at the point $(0, \frac{9}{2})$. To find where the graph crosses the x-axis, we must set $y = 0$, which means that $2x - 9 = 0$, or $x = \frac{9}{2}$; the crossing occurs at $(\frac{9}{2}, 0)$. These two points are marked on Figure 37. ☞ 30

☞ **30.** Sketch the graph of the function
$$y = \frac{2x}{x + 1}.$$

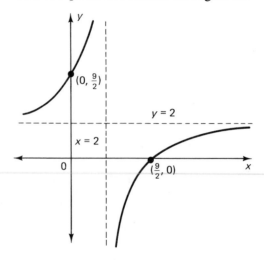

FIGURE 37

EXAMPLE 6 A manufacturer's cost function for producing x thousands of items per week is given in dollars by

$$C = 3000 + 2000x$$

If $\overline{C}(x)$ denotes the average cost per item, sketch the graph of $\overline{C}$ as a function of x.

Solution The number of items produced is $1000x$, so their average cost is

$$\overline{C}(x) = \frac{C(x)}{1000x} = \frac{3000 + 2000x}{1000x} = \frac{3}{x} + 2.$$

We are only interested in the region $x \geq 0$. As $x \to 0$ from above, the first term, $3/x$, becomes unbounded and positive; that is $\overline{C}(x) \to +\infty$ as $x \to 0^+$. The graph therefore has a vertical asymptote at $x = 0$.

Answer

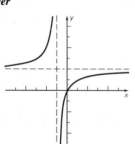

As x becomes large, the term $3/x$ approaches zero, so that $\overline{C}(x) \to 2$ as $x \to \infty$. Thus the line $\overline{C} = 2$ is a horizontal asymptote.

Differentiating, we obtain

$$\overline{C}'(x) = -\frac{3}{x^2} < 0 \qquad \text{for all } x;$$

$$\overline{C}''(x) = +\frac{6}{x^3} > 0 \qquad \text{for all } x > 0.$$

Thus $\overline{C}$ is a decreasing function for all x and its graph is concave upward for all $x > 0$.

The graph is shown in Figure 38. In order to locate the graph better we have computed two explicit points on it, namely when $x = 1$, $\overline{C} = 5$ and when $x = 3$, $\overline{C} = 3$.

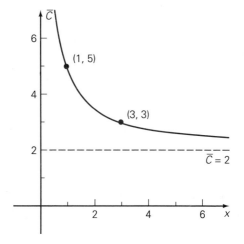

FIGURE 38

Summary of Methods of Finding Asymptotes:

Suppose we want the asymptotes of $y = f(x)$.

1. **Vertical asymptotes.** Find the values of x for which the denominator of any fraction occurring in $f(x)$ becomes zero. If a is any such value, then the line $x = a$ is a vertical asymptote.

2. **Horizontal asymptotes.** Find $\lim\limits_{x \to \infty} f(x)$ and $\lim\limits_{x \to -\infty} f(x)$. If these limits exist and are equal to b and c, respectively, then $y = b$ is a horizontal asymptote at $+\infty$ and $y = c$ is a horizontal asymptote at $-\infty$.

Note A polynomial function possesses no asymptotes, vertical or horizontal.

(1–36) Evaluate the following limits, if they exist.

1. $\lim\limits_{x \to \infty} \left(1 \div \dfrac{2}{x}\right)$

2. $\lim\limits_{x \to -\infty} \left(3 + \dfrac{1}{3x^2}\right)$

3. $\lim\limits_{x \to \infty} \dfrac{x + 1}{2x - 3}$

4. $\lim\limits_{x \to \infty} \dfrac{3x - 5}{5x - 2}$

5. $\lim\limits_{x \to -\infty} \dfrac{5 - 2x}{3x + 7}$

6. $\lim\limits_{x \to -\infty} \dfrac{3 + 2x}{2 + 3x}$

7. $\lim\limits_{x \to \infty} \dfrac{x^2 + 2x + 4}{x^2 + 1}$

8. $\lim\limits_{x \to -\infty} \dfrac{2x^2 - 3}{3x^2 - 2}$

9. $\lim\limits_{x \to -\infty} \dfrac{5 + 3x - 2x^2}{3x^2 + 4}$

10. $\lim\limits_{x \to \infty} \dfrac{\sqrt{4x + 7}}{\sqrt{x + 1}}$

11. $\lim\limits_{x \to \infty} \dfrac{x + 1}{x^2 + 1}$

12. $\lim\limits_{x \to -\infty} \dfrac{1 - x^2}{4 + x^3}$

13. $\lim\limits_{x \to \infty} \dfrac{x^2 + 4}{3x + 7}$

14. $\lim\limits_{x \to -\infty} \dfrac{2 + 3x^2}{x - 2}$

15. $\lim\limits_{x \to -\infty} \dfrac{3 - 4x^2}{2 + 3x}$

16. $\lim\limits_{x \to \infty} \dfrac{x + 1}{(2x + 3)^2}$

17. $\lim\limits_{x \to -\infty} \dfrac{3x^2 + 5}{(2x - 3)^3}$

18. $\lim\limits_{x \to -\infty} \dfrac{1}{(x^2 + 1)^3}$

19. $\lim\limits_{x \to -\infty} \dfrac{1}{\sqrt{1 + x}}$

20. $\lim\limits_{x \to \infty} \dfrac{1}{\sqrt{3 - 2x}}$

21. $\lim\limits_{x \to \infty} \left(x - \dfrac{x^2}{x + 1}\right)$

22. $\lim\limits_{x \to -\infty} \left(x - \dfrac{2x^2}{2x + 3}\right)$

*23. $\lim\limits_{x \to \infty} \dfrac{|x + 2|}{x + 1}$

*24. $\lim\limits_{x \to -\infty} \dfrac{3 + |x - 2|}{2x + 5}$

25. $\lim\limits_{x \to \infty} (1 + 2e^{-x})$

26. $\lim\limits_{x \to -\infty} (3 + 4e^x)$

*27. $\lim\limits_{x \to \infty} \dfrac{2e^x + 3}{3e^x - 1}$

28. $\lim\limits_{x \to -\infty} \dfrac{2e^x + 3}{3e^x + 1}$

*29. $\lim\limits_{x \to -\infty} \dfrac{3e^{-x} + 4}{2e^{-x} - 1}$

30. $\lim\limits_{x \to \infty} \dfrac{3e^{-x} + 4}{2e^{-x} - 1}$

31. $\lim\limits_{x \to \infty} \left(5 + \dfrac{2}{\ln x}\right)$

*32. $\lim\limits_{x \to \infty} \dfrac{3 + \ln x}{2 \ln x - 1}$

*33. $\lim\limits_{x \to \infty} \dfrac{5 - 2 \ln x}{2 + 3 \ln x}$

34. $\lim\limits_{x \to \infty} e^{-|x|}$

35. $\lim\limits_{x \to -\infty} e^{-|x|}$

*36. $\lim\limits_{x \to \infty} \dfrac{2 + 3e^x}{3 + 2e^x}$

(37–46) Evaluate (a) $\lim\limits_{x \to c^+} f(x)$ and (b) $\lim\limits_{x \to c^-} f(x)$ for the following functions $f(x)$ and points c.

37. $f(x) = \dfrac{1}{x - 2}, \quad c = 2$

38. $f(x) = \dfrac{1}{x + 1}, \quad c = -1$

39. $f(x) = \dfrac{1}{4 - x^2}, \quad c = 2$

40. $f(x) = \dfrac{1}{4 - x^2}, \quad c = -2$

41. $f(x) = \dfrac{1}{x^2}, \quad c = 0$

42. $f(x) = \dfrac{x}{(x + 1)^2}, \quad c = -1$

43. $f(x) = \dfrac{x}{x + 1}, \quad c = -1$

44. $f(x) = \dfrac{\sqrt{x}}{x + 1}, \quad c = -1$

45. $f(x) = \dfrac{\sqrt{x^2 - 1}}{x - 1}, \quad c = 1$

46. $f(x) = \dfrac{\sqrt{4 - x^2}}{x - 2}, \quad c = 2$

(47–66) Find the horizontal and vertical asymptotes of the following curves and sketch their graphs.

47. $y = \dfrac{1}{x - 1}$

48. $y = \dfrac{-2}{x + 2}$

49. $y = \dfrac{x + 1}{x - 2}$

50. $y = \dfrac{x - 2}{x + 2}$

51. $y = \dfrac{2x + 1}{x + 1}$

52. $y = \dfrac{3x - 6}{x - 1}$

53. $y = \dfrac{x^2 + 1}{x^2}$

54. $y = \dfrac{x^2}{x^2 + 1}$

55. $y = \dfrac{x^2 - 1}{x^2 + 1}$

56. $y = \dfrac{x}{x^2 + 1}$

*57. $y = \dfrac{x^2 + 1}{x^2 - 1}$

*58. $y = \dfrac{x^2 - 2}{x^2 - 3x + 2}$

59. $y = \ln x$

60. $y = \ln |x|$

61. $y = -\ln x$

62. $y = 2 - \ln x$

63. $y = e^{-2x}$

64. $y = xe^{-x}$

*65. $y = \dfrac{e^x}{x}$

*66. $y = e^{|x|}$

67. *(Epidemics)* During an influenza epidemic, the percentage of the population of Montreal that is infected at time t (measured in days from the start of the epidemic) was given by

$$p(t) = \frac{200t}{t^2 + 100}.$$

Find the time at which $p(t)$ is maximum and sketch the graph of $p(t)$.

*68. *(Fuel Consumption)* A trucking firm finds that at a speed v kilometers per hour, its trucks consume fuel at the rate $(25 + 0.2v + 0.01v^2)$ liters per hour. Construct the function $C(v)$ that provides the number of liters consumed per kilometer at speed v. Graph this function and compute the speed at which it is minimum.

69. *(Reservoir Engineering)* The rate of production of oil from a newly opened reservoir initially increases and then tails off as the reserve is depleted. In a particular case the rate of production is given by $P(t) = 5000te^{-0.2t}$ barrels per day, where t is in years. Sketch the graph of $P(t)$ as a function of t.

70. *(Weather)* The temperature in Vancouver during an average summer day varies approximately according to the formula

$$T(t) = 24e^{-(1 - t/8)^2} \qquad (0 \le t \le 16)$$

where t is time in hours measured form 6 A.M. Sketch the graph of this function.

71. *(Drug Dosage)* The following formula is sometimes used to calculate the dose of a drug to be given to a child of age t.

$$y(t) = \frac{Dt}{t + c}$$

where D is the adult dose and $c > 0$ is a constant. What are the horizontal and vertical asymptotes for this function? Sketch the graph in the following two cases:

a. $c = 10$. **b.** $c = 15$.

72. *(Product Transformation Curve)* (See p. 195.) An automobile company can use its plant to manufacture either compact or full-size motor cars, or both. If x and y are the number of compact and full-size cars produced (in hundreds per day), then the product transformation relation is $xy + 2x + 3y = 6$. Express y as a function of x, find the horizontal and vertical asymptotes, and sketch its graph.

(73–76) *(Average Cost Functions)* For the following cost functions $C(x)$, sketch the graphs of the corresponding average cost $\overline{C}(x) = C(x)/x$.

73. $C(x) = 2 + 3x$

74. $C(x) = 3 + x$

*75. $C(x) = 2 + x - \frac{1}{4}x^2 + \frac{1}{24}x^3$.

*76. $C(x) = 3 + 2x + \frac{1}{6}x^2$

*77. *(Advertising and Profits)* If an amount A (in thousands of dollars) is spent on advertising per week, a company finds that the weekly sales volume is given by

$$x = 2000(1 - e^{-A}).$$

The items are sold at a profit of $2 each. If P denotes the net profit (that is, profit from sales minus advertising costs), express P as a function of A and sketch its graph.

*78. *(Logistic Model)* Sketch the graph of the logistic function

$$y = \frac{y_m}{1 + ce^{-t}} \qquad (y_m, c > 0).$$

CHAPTER REVIEW

Key Terms, Symbols, and Concepts

14.1 Increasing function, decreasing function.

14.2 Local maximum, local minimum, local extremum.

Local maximum (or minimum) value.
Critical point. The first derivative test.

14.3 Concave upward, concave downward.

Point of inflection.
Second-derivative test.

14.4 Step-by-step procedure for sketching graphs of polynomial functions.

14.5 Step-by-step procedure for translating verbal optimization problems into algebraic problems.
Inventory cost model. Economic batch size (or order quantity).

14.6 Absolute maximum and absolute minimum values of a function.

14.7 Limit at infinity; $\lim_{x \to \infty} f(x)$ and $\lim_{x \to -\infty} f(x)$.
Horizontal asymptote.
Vertical asymptote; the notation $\lim_{x \to c} f(x) = \infty$ or $\lim_{x \to c} f(x) = -\infty$.
Procedure for sketching graphs with asymptotes.

Formulas

Test for increasing or decreasing property:
If $f'(x) > 0$ [alternatively, < 0] for all x in an interval,

then f is an increasing function [decreasing function] in that interval.

First-derivative test:
Let $x = c$ be a critical point of the function f. Then:

a. If $f'(x) > 0$ for x just below c and $f'(x) < 0$ for x just above c, then c is a local maximum of f.
b. If $f'(x) < 0$ for x just below c and $f'(x) > 0$ for x just above c, then c is a local minimum of f.
c. If $f'(x)$ has the same sign for x just below c and for x just above c, then c is not a local extremum of f.

Test for concavity:
If $f''(x) > 0$ [alternatively, < 0] for all x in an interval, then f is concave up [concave down] in that interval.

Second-derivative test:
Let $x = c$ be a critical point of the function f at which $f'(c) = 0$ and $f''(c)$ exists. Then $x = c$ is a local maximum if $f''(c) < 0$ and a local minimum if $f''(c) > 0$.

REVIEW EXERCISES FOR CHAPTER 14

1. State whether each of the following is true or false. Replace each false statement with a corresponding true statement.

   a. The function $f(x)$ is increasing for values of x at which $f(x) > 0$ and is decreasing for values of x at which $f(x) < 0$.

   b. If $f(x)$ is increasing for all values of x, then $f'(x)$ is never zero.

   c. The graph of $y = [f(x)]^2$ is concave upward whenever $f(x)f''(x) + [f'(x)]^2 > 0$.

   d. If the graph of $y = f(x)$ is increasing at $x = a$, then the graph of $y = -f(x)$ is decreasing at $x = a$.

   e. At a point of inflection, $f''(x) = 0$.

   f. Every local maximum occurs at a critical point of the function concerned.

   g. If a function f has a maximum or a minimum at $x = c$, then $f'(c)$ must be zero.

   h. If $f'(c) = 0$, then the function f has either a maximum or a minimum at $x = c$.

   i. The tangent to the graph of a function at a point where it is maximum or minimum is either horizontal or vertical.

   j. The tangent to the graph of a function at a point of inflection is always horizontal.

   k. A local maximum value of a function is always greater than a local minimum value of the same function.

   l. Any quadratic function has exactly one local extremum.

   m. Any cubic function has two local extrema.

   n. A firm is operating optimally if it maximizes its revenues.

   o. If $f(x)$ has no local extrema for $a \le x \le b$ and is differentiable in that interval, then its absolute maximum value in the interval is either $f(a)$ or (b).

p. "Advertising always pays, and the more advertising the better."

q. If $f'(x) \leq 0$ for $a \leq x \leq b$, then the absolute maximum value of $f(x)$ in this interval is $f(a)$.

r. If $f(x) \rightarrow L$ as $x \rightarrow +\infty$, then $f(x) \rightarrow L$ as $x \rightarrow -\infty$.

s. As $x \rightarrow -\infty$, $x^{-1/2} \rightarrow 0$.

t. The graph of a function can cross a horizontal asymptote but can never cross a vertical asymptote.

(2–7) Find the values of x for which the following functions are: (a) increasing; (b) decreasing; (c) concave upward; (d) concave downward. Sketch their graphs.

2. $y = -2x^2 + 9x - 4$

3. $y = x^3 - 9x^2 + 24x - 18$

4. $y = 1 - e^{-x^2/2}$ **5.** $y = \frac{1}{6}x^6 - x^4$

6. $y = x^5 - \frac{5}{3}x^3$ **7.** $y = xe^x$

8. If x units can be sold at a price of \$$p$ each, where $2p + 3x = 50$, show that the marginal revenue is never increasing.

9. The demand equation of a certain commodity is given by $p = 200e^{-x/30}$. For what sales level x will the marginal revenue be increasing?

10. For the demand relation $p = 50 - \ln(x + 1)$, show that the marginal demand is always increasing.

11. The industrial production Y of a certain country t years after 1930 was found to be given by $Y = 375/(1 + 215e^{-0.07t})$.

  a. Has the production been increasing or decreasing?

  b. Use natural logarithms to solve for t in terms of Y.

(12–17) Find the critical points of the following functions and ascertain which of them are local maxima or minima.

12. $3 - 2x - 4x^2$ **13.** $2t^3 - 3t^2 + 1$

14. $x^2 e^{3x}$ **15.** $2x^2 - \ln x$

*16. $\sqrt{|x|}$ **17.** $x^{2/5}(1 - x)^2$

18. Determine two values of the constant c so that $f(x) = x + c/x$ may have:

  a. a local maximum at $x = -1$;

  b. a local minimum at $x = 2$.

19. Determine the constant k in such a way that the function $f(x) = x^3 + k/x^2$ may have:

  a. a local minimum at $x = 1$;

  b. a local maximum at $x = -2$;

  c. a point of inflection at $x = 2$.

20. Determine the constants A and B so that the function $f(x) = x^3 + Ax^2 + Bx + C$ may have:

  a. a maximum at $x = -2$ and a minimum at $x = 1$;

  b. a maximum at $x = -3$ and a point of inflection at $x = -2$.

21. Find the restrictions on the constants A, B, and C in order that $f(x) = Ax^2 + Bx + C$ may have a local minimum.

22. Find the absolute extrema of $g(x) = \sqrt{x^2 - 4}$ in $2 \leq x \leq 3$.

23. Find the absolute extrema of $f(x) = x^2(x - 2)^{2/3}$ in $-2 \leq x \leq \frac{5}{2}$.

24. *(Optimal Crop Spraying)* The value of a certain fruit crop (in dollars) is given by

$$V = A(1 - e^{-KI})$$

where A and K are constants and I is the number of pounds per acre of insecticide with which the crop is sprayed. If the cost of spraying is given by $C = BI$, where B is a constant, find the value of I that makes $V - C$ a maximum. What is the interpretation of your result when $AK < B$?

25. *(Maximum Profit)* A company determined that the cost C, the total revenue R, and the number of units produced x are related by

$$C = 100 + 0.015x^2 \quad \text{and} \quad R = 3x.$$

Find the production rate x that will maximize the profits of the firm. Find that profit and also the profit when $x = 120$.

26. *(Maximum Revenue)* A company finds that its total revenue is described by the relation

$$R = 4,000,000 - (x - 2000)^2$$

where R is the total revenue and x the number of units sold.

  a. Find the number of units sold that maximizes total revenue.

b. What is the amount of this maximum total revenue?

c. What would be the total revenue if 2500 units were sold?

27. *(Maximum Profit)* The total cost C of producing x units of a commodity is given by $C = 50 + 2x + 0.5x^2$ and the total revenue R received from the sales is $R = 20x - x^2$. Find the production rate x that will maximize the profits.

28. *(Maximum Profit)* Solve Exercise 27 when

$$C = 300 + 0.075x^2 \quad \text{and} \quad R = 3x.$$

29. *(Maximum Profit)* A radio manufacturer finds that x instruments per week can be sold at p dollars each, where $5x = 375 - 3p$. The cost of production is $(500 + 13x + \frac{1}{5}x^2)$ dollars. Show that the maximum profit is obtained when the production is 30 instruments per week.

30. *(Inventory Cost Model)* A radio manufacturer has to produce 144,000 units of an item per year. The cost of material is \$5 per unit and it costs \$160 to make the factory ready for the production run of the item, regardless of the number of units x produced in a run. The cost of storing the material is 50¢ per item per year on the inventory $(x/2)$ in hand. Show that the total cost C is given by

$$C = 720,000 + \frac{23,040,000}{x} + \frac{x}{4}.$$

Find also the economic lot size, that is, the value of x for which C is minimum.

31. *(Economic Order Quantity)* Let the quantity Q, when purchased in each order, minimize the total cost T incurred in obtaining and storing material for a certain time period to fulfill a given rate of demand for the material during the time period. The material demanded is 10,000 units per year; the cost price of material is \$1 per unit; the cost of replenishing the stock of the material per order, regardless of the size Q of the order, is \$25; and the cost of storing the material is $12\frac{1}{2}\%$ per year of the value of the average inventory $(Q/2)$ on hand.

a. Show that $T = 10,000 + \dfrac{25,000}{Q} + \dfrac{Q}{16}$.

b. Find the economic order quantity and the total cost T corresponding to that value of Q.

c. Find the total cost when each order is placed for 2500 units.

32. *(Agriculture)* The yield of fruit from each tree of an apple

orchard decreases as the density of the trees planted increases. When there are n trees per acre, the average number of apples per tree is known to be equal to $(900 - 10n)$ for a particular variety of apple (when n lies between 30 and 60.) What value of n gives the maximum total yield of apples per acre?

33. *(Optimal Allocation of Production)* A shoe manufacturer can use her plant to make either men's or women's shoes. If she makes x and y thousand pairs of each, respectively, per week, then x and y are related by the product transformation equation

$$2x^2 + y^2 = 25.$$

The manufacturer's profit is \$10 on each pair of men's shoes and \$8 on each pair of women's shoes. Determine how many pairs of each she should make in order to maximize her weekly profit.

34. *(Warehouse Location)* The cost of owning and operating a warehouse is $200,000 + 100,000/x^2$ dollars per year and the cost of delivering goods from the warehouse is $200x + 15,000$ dollars per year, where x is the distance in kilometers from the warehouse to the center of the city. How far from the center of the city should the warehouse be located in order to minimize the total of these costs?

35. *(Taxation and Production)* The demand and the total cost function of a monopolist are $p = 12 - 4x$ and $C(x) = 8x + x^2$. If the tax of t per unit is imposed, find:

a. The quantity x and the price p that corresponds to the maximum profit.

b. The maximum profit.

c. The tax t per unit that maximizes the tax revenue of the government.

36. *(Taxation and Production)* Repeat Exercise 35 if the demand and the total cost function of a monopolist are

$$p = 20 - 0.01x \quad \text{and} \quad C(x) = 500 + 2x - 0.005x^2.$$

37. The production function of a commodity is given by $Q = 40F + 3F^2 - F^3/3$, where Q is the total output and F the units of input.

a. Find the number of units of input that maximizes the output.

b. Find the maximum values of the marginal product dQ/dF.

c. Verify that when the average product Q/F is maximum, it is equal to the marginal product.

38. *(Maximum Marginal Product)* Repeat Exercise 37 for the production function $Q = 96F + 6F^2 - F^3$.

39. *(Optimal Selling Time)* A company produces broiler chickens. If the chickens are kept for t months, the profit from the sale of each chicken is $P(t) = 0.2e^{0.06\sqrt{t}}$ dollars. The present value of this profit is $P(t)e^{-0.01t}$ if the nominal discount rate is 1% per month. After how many months should the chickens be sold in order to maximize this present value?

*40. *(Maximum Population Growth)* A population has size $p(t)$ at time t given by the logistic function

$$p(t) = \frac{A}{1 + Be^{-t}}$$

where A and B are constants. Find the value of t at which the growth rate is maximum. What is the maximum growth rate?

41. *(Memory Retention)* While cramming for an exam, a student learns a large number of facts. At a time t weeks after the exam, the percentage of those facts that the student is able to recall is given by

$$p(t) = \frac{180 + 20e^{0.5t}}{1 + e^{0.5t}}.$$

Calculate $p(0)$, $p(2)$, and $\lim_{x \to \infty} p(t)$. Show that $p'(t) < 0$ and $p''(t) > 0$ for all $t > 0$ and sketch the graph.

*42. *(von Bertalanffy Model)* A function originally proposed to model population growth and sometimes used in other limited growth situations is

$$y(t) = (a - be^{-kt})^p$$

where a, b, k, and p are positive constants. Calculate $y(0)$ and $\lim_{x \to \infty} y(t)$, and show that $y'(t) > 0$ for all $t > 0$. Sketch the graph when $p = 3$, $a = 4$, $k = 1$, and

a. $b = 1$;   **b.** $b = 2$.

43. *(Learning Model)* When a repetitive task (for example, solving calculus problems) is performed a number of times, an individual's probability of doing it correctly increases. One model sometimes used for this probability of success is $p = AN/(N + B)$ where A and B are constants and N is the number of times the task has been performed. By calculating $\lim_{x \to \infty} p$, interpret A. Calculate dp/dN at $N = 0$.

44. *(Sales and Advertising)* The volume of weekly sales is given by

$$x = 1000(5 - 3e^{-A})$$

where A is the amount spent in thousands of dollars per week on advertising. Calculate $\lim_{A \to \infty} x$. If the items are sold for $3 each find the value of A that maximizes Revenue − Advertising Cost.

45. *(Maximum Fruit Yield)* When there are n trees per acre, the average number of peaches per tree is equal to $(840 - 6n)$ for a particular variety of peach. What value of n gives the maximum total yield of peaches per acre?

46. *(Soil Moisture Content)* In a certain location the concentration of water in the soil is given in terms of depth x by the formula

$$c = 1 - e^{-x^2}.$$

Find the depth at which c is increasing most rapidly.

47. *(Epidemic)* During a certain influenza epidemic, the proportion of the population in the lower mainland who are infected is denoted by $y(t)$, where t is the time in weeks since the start of the epidemic. It is found that

$$y(t) = \frac{t}{4 + t^2}.$$

a. What is the physical interpretation of dy/dt?

b. For what value of t is y maximum?

c. For what values of t is y increasing and decreasing?

48. *(Biochemistry)* According to Michaelis and Menton, the intital rate of reaction v in an enzyme-catalyzed reaction depends on the concentration x of the substrate and is given by

$$v = \frac{Vx}{x + M}$$

where V and M are constants. Show that $v \to V$ as $x \to \infty$ and that for any finite value of x, v is always less than V. Find dv/dx, that is, the rate of change of the initial velocity of reaction with respect to the concentration of the substrate.

49. *(Optimal Design)* A home-improvement firm has decided to market an inexpensive rectangular shed with no floor and one open side. If the material for the three sides and the *square* roof costs $2 per square foot, find the dimensions of the shed that minimize the total cost of material if the volume of the shed is 486 cubic feet.

(50–53) Sketch the graphs of the following functions.

50. $y = x - \ln x$

51. $y = xe^{-2x}$

52. $y = e^{-|x|}$

53. $y = e^{x-|x|}$

(54–55) Find the horizontal and vertical asymptotes of the following functions and sketch their graphs.

54. $y = \dfrac{3x + 1}{x - 2}$

55. $y = \ln|x|$

56. (*Marginal Revenue Analysis*) It is given that a demand curve $p = f(x)$ is concave up at all points, that is, $f''(x) > 0$ for all x. Show that the marginal revenue curve is also concave up if either $f'''(x) > 0$ or $f'''(x) < 0$ and numerically less than $(3/x)f''(x)$.

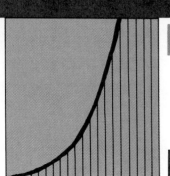

CHAPTER 15

More on Derivatives

Chapter Objectives

15-1 DIFFERENTIALS
(a) Definition and calculation of differentials;
(b) Geometrical interpretation of differentials;
(c) Use of differentials to compute approximate values of a function;
(d) Linear models: use of differentials to approximate a function by a linear function;
(e) Errors: use of differentials to compute approximate values of errors.

15-2 IMPLICIT DIFFERENTIATION
(a) The calculation of derivatives from an implicit relation;
(b) Tangent lines for an implicit relation; points with horizontal or vertical tangents;
(c) Calculation of second derivatives from an implicit relation.

15-3 LOGARITHMIC DIFFERENTIATION AND ELASTICITY
(a) The technique of logarithmic differentiation for functions that consist of the product and/or quotient of a number of factors or of one function raised to the power of another function;
(b) Elasticity of demand and its significance. Calculation of elasticity of demand;
(c) General elasticity and its relation to logarithmic derivatives;
(d) Elasticity of demand and marginal revenue.

CHAPTER REVIEW

■ 15-1 DIFFERENTIALS

Let $y = f(x)$ be a differentiable function of the independent variable x. Up to now, we have used dy/dx to denote the derivative of y with respect to x and treated dy/dx as a single symbol. Now we shall define the new concept of a *differential* so that dx and dy will have separate meanings; this will permit us to think of dy/dx either as the symbol for the derivative of y with respect to x or as the ratio of dy and dx.

DEFINITION Let $y = f(x)$ be a differentiable function of x. Then

(a) dx, called the **differential of the independent variable** x is simply an arbitrary increment of x; that is,

$$dx = \Delta x;$$

(b) dy, the **differential of the dependent variable** y is a function of x and dx given by

$$dy = f'(x)\, dx.$$

The differential dy is also denoted by df.

The following are obvious from the above definition of differentials dx and dy.

1. If $dx = 0$ then $dy = 0$.

2. If $dx \neq 0$, then the ratio of dy divided by dx is given by

$$\frac{dy}{dx} = \frac{f'(x)\, dx}{dx} = f'(x)$$

and so is equal to the derivative of y with respect to x.

There is nothing strange about the last result, because we deliberately defined dy as the product of $f'(x)$ and dx in order that the result in statement 2 would be true.

EXAMPLE 1 If $y = x^3 + 5x + 7$, find dy.

Solution Let $y = f(x)$ so that $f(x) = x^3 + 5x + 7$. Then $f'(x) = 3x^2 + 5$ and, by definition,

$$dy = f'(x)\, dx = (3x^2 + 5)\, dx. \quad \text{☞ 1}$$

It should be noted that dx (or Δx) is another independent variable and the value of dy depends on the *two* independent variables x and dx.

While $dx = \Delta x$, the differential dy of the dependent variable is not equal to the increment Δy. If dx is sufficiently small, however, dy and Δy are approximately equal to one another.

☞ **1.** For the function
$y = 4x - 2x^2$,
find dy when $x = 2$ and when
$x = -2$.

Answer $dy = -4dx$ when $x = 2$;
$dy = 12dx$ when $x = -2$.

EXAMPLE 2 If $y = x^3 + 3x$, find dy and Δy when $x = 2$ and $\Delta x = 0.01$.

Solution If $y = f(x) = x^3 + 3x$, then

$$f'(x) = 3x^2 + 3.$$

Therefore

$$dy = f'(x)\,dx = (3x^2 + 3)\,dx.$$

When $x = 2$ and $dx = 0.01$ then $dy = (12 + 3)(0.01) = 0.15$.
By definition,

$$\Delta y = f(x + \Delta x) - f(x)$$
$$= f(2.01) - f(2)$$
$$= [(2.01)^3 + 3(2.01)] - [2^3 + 3(2)]$$
$$= [8.120601 + 6.03] - 14 = 0.150601.$$

► **2.** For the function $y = x^2$, find dy and Δy when $x = 3$ and when (a) $\Delta x = 0.2$; (b) $\Delta x = 0.05$.

Thus $dy = 0.15$ and $\Delta y = 0.150601$, showing that the differential and increment of y are not quite equal to one another. ► **2**

Geometrical Interpretation of Differentials

Let P be the point whose abscissa is x on the graph of $y = f(x)$, and let Q be the point on the graph whose abscissa is $x + \Delta x$. The increment Δy is the rise from P to Q, or the vertical distance QR on Figure 1.

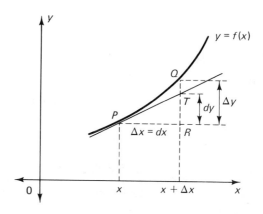

FIGURE 1

Let T be the point with abscissa $x + \Delta x$ on the tangent to the graph at P (see Figure 1). The slope of PT is the derivative $f'(x)$ and is equal to the rise from P to T divided by the run:

Answer
(a) $dy = 1.2$, $\Delta y = 1.24$;
(b) $dy = 0.3$, $\Delta y = 0.3025$.

$$\text{Slope} = f'(x) = \frac{\text{Rise from } P \text{ to } T}{\text{Run from } P \text{ to } T} = \frac{TR}{PR} = \frac{TR}{\Delta x}.$$

Therefore, since $\Delta x = dx$, we have $TR = f'(x)\, dx = dy$. Thus the differential dy is equal to the rise TR along the tangent line at P.

> The differentials dx and $dy = f'(x)\, dx$ are therefore increments in x and y along the tangent line to the graph of f at the point $(x, f(x))$.

From Figure 1, the difference between Δy and dy is equal to the distance QT. It is intuitively clear that if we allow Δx to become smaller, so that Q moves around the curve toward P, then the distance QT rapidly shrinks to zero. Because of this, we can use dy as an approximation to Δy provided that Δx is sufficiently small:

$$\Delta y \approx dy = f'(x)\, dx.$$

Thus, since $f(x + \Delta x) = y + \Delta y$,

> $$f(x + \Delta x) \approx f(x) + f'(x)\, \Delta x.$$

Replacing x by a and Δx by h, we have the following alternative form:

> If h is sufficiently small, then $f(a + h) \approx f(a) + hf'(a)$.   (1)

This approximation is useful because it is often easier to calculate the right side than to calculate $f(a + h)$, particularly if the calculation has to be done for several different value of h. The reason is that the right side is a *linear* function of h. The following example illustrates how this approximation can be used to replace a complicated function by a linear one.

EXAMPLE 3 Find an approximation to the value of $\sqrt{16 + h}$ when h is small.

Solution We are asked to approximate the square root function $\sqrt{x}$ near $x = 16$. So in Equation (1) we take $f(x) = \sqrt{x}$ with $a = 16$. The result is

$$f(16 + h) \approx f(16) + hf'(16). \tag{2}$$

But if $f(x) = \sqrt{x}$, then $f'(x) = 1/(2\sqrt{x})$. In particular,

$$f'(16) = \frac{1}{2\sqrt{16}} = \frac{1}{8}.$$

Substituting $f(16) = 4$ and $f'(16) = \tfrac{1}{8}$ into Equation (2), we get

$$f(16 + h) \approx 4 + \tfrac{1}{8}h$$

that is,

$$\sqrt{16 + h} \approx 4 + \tfrac{1}{8}h.$$

(For example, taking $h = 0.1$, we find that $\sqrt{16.1} \approx 4 + \tfrac{1}{8}(0.1) = 4.0125$.

This compares with the exact value, which is 4.01248 to five decimal places.) ☛ **3**

The usefulness of this kind of approximation is apparent from this example: It is much easier to calculate with the approximate expression $(4 + \frac{1}{8}h)$ than with the expression $\sqrt{16 + h}$, because *the approximation is a linear function of h.*

EXAMPLE 4 A section of land consists of a square with sides of one mile (5280 feet) in length. If a strip of 20 feet is removed along each side for a roadway, how much area is lost from the section?

Solution If x denotes the length of a side, then the area is

$$A = f(x) = x^2.$$

If the side is changed to $x + \Delta x$, then the change in area ΔA is given approximately by the differential

$$\Delta A \approx f'(x)\,\Delta x = 2x\,\Delta x.$$

In this example, $x = 5280$ feet and $\Delta x = -40$ feet (a strip of 20 feet removed from each side, so the width and breadth are reduced by 40 feet). Then

$$\Delta A \approx 2(5280)(-40) = -422,400.$$

Thus the loss of area is approximately 422,400 square feet.

Linear Models

In the approximation formula (1), let us set $a + h = x$. Then $h = x - a$ and we get

$$f(x) \approx f(a) + (x - a)f'(a).$$

But the right side here is a linear function of x. If we define $m = f'(a)$ and $b = f(a) - af'(a)$, then the approximation becomes simply $f(x) \approx mx + b$. Thus we have established the following important result:

> Over a sufficiently short interval of x, any differentiable function of x can be approximated by a linear function.

Referring back to Figure 1, we see that the geometrical basis of this result is that, close to the point P, the graph of f is approximately the same as the tangent line at P.

Use is often made of this result in devising mathematical models of complex phenomena. Suppose that x and y are two economic variables that are related in some complex and not very well-understood way. Then, regardless of the degree of complexity of the relationship (as long as it is smooth), we may approximate it by a linear model $y = mx + b$ for certain constants m and b

☛ 4. A company has an exact cost function given by
$$C(x) = 25 + 11x - x^2.$$
The current production level is $x = 3$. Find a linear cost model that approximates the exact cost function when x is close to 3.

Answer
$$C(x) \approx C(3) + C'(3)(x - 3)$$
$$= 34 + 5x.$$

provided that the range of variation of x is sufficiently restricted. Linear models of this kind are frequently used in economics and elsewhere as a starting point in the analysis of complex phenomena. **☛ 4**

Errors

Differentials are useful in assessing the effects of errors in the measured or estimated values of quantities. Let x be a variable whose value is measured or estimated with a certain possible error and let $y = f(x)$ be some other variable that is calculated from the measured value of x. If the value of x that is used in calculating y is in error, then of course the calculated value of y will also be incorrect.

Let x be the true value of the measured variable and $x + dx$ be the measured value. Then dx is now the **error** in this variable. The true value of the calculated variable is $y = f(x)$, but the actual calculated value is $f(x + dx)$. Thus the error in y is equal to $f(x + dx) - f(x)$. If dx is small, which can usually be presumed to be the case, we can approximate this error in y by the differential dy. Thus we arrive at the result that the error in y is given approximately by $dy = f'(x)\, dx$.

The ratio dx/x is called the **relative error** in x. Correspondingly, the relative error in y is dy/y. If the relative error is multiplied by 100, we obtain what is called the **percentage error** in the variable in question. Often the sign is ignored in stating the percentage error, so we might speak of a percentage error of 2% and mean that the error is either $\pm 2\%$. **☛ 5**

☛ 5. If x is measured with a percentage error of 2%, what is the percentage error in y if
(a) $y = x^2$; (b) $y = \sqrt{x}$?

Answer (a) 4%; 1%.

EXAMPLE 5 *(Error in Estimated Profits)* A sales manager estimates that his staff will sell 10,000 units during the next month. He believes his estimate is accurate to within a maximum percentage error of 3%. If the profit function is

$$P(x) = x - (4 \times 10^{-5})x^2 \quad \text{(dollars per month)}$$

(where x = number of units sold per month), find the maximum percentage error in the estimated profit.

Solution If $x = 10,000$, the profit will be

$$P = 10,000 - (4 \times 10^{-5})(10,000)^2 = 6000.$$

The maximum percentage error in the estimated value of x is 3%, so the maximum error dx is given by

$$dx = 3\% \text{ of } 10,000 = \tfrac{3}{100}(10,000) = 300.$$

The corresponding error in profit is given approximately by

$$dP = P'(x)\, dx$$
$$= (1 - 8 \times 10^{-5}x)\, dx$$
$$= [1 - 8 \times 10^{-5}(10,000)](300)$$
$$= 0.2(300) = 60.$$

6. Re-solve Example 5 if the profit function is
$$P(x) = -9000 + 2x - (6 \times 10^{-5})x^2.$$

Answer 4.8%.

So the maximum error in the estimated profit is $60. The percentage error is therefore given by

$$100 \frac{dP}{P} = 100 \frac{60}{6000} = 1.$$

The maximum percentage error in profit is 1%. **6**

EXERCISES 15-1

(1–10) Find dy for the following functions.

1. $y = x^2 + 7x + 1$

2. $y = (t^2 + 1)^4$

3. $y = t \ln t$

4. $y = ue^{-u}$

5. $y = \ln(z^2 + 1)$

6. $y = \dfrac{x + 1}{x^2 + 1}$

7. $y = \dfrac{e^u}{u + 1}$

8. $y = \dfrac{e^u + 1}{e^u - 1}$

9. $y = \sqrt{x^2 - 3x}$

10. $y = \sqrt{\ln x}$

11. Find dy for $y = x^2 - 1$ when $x = 1$.

12. Find dx for $x = \sqrt{t + 1}$ when $t = 3$.

13. Find dt for $t = \ln(1 + y^2)$ when $y = 0$.

14. Find du for $u = e^{0.5 \ln(1 - t^2)}$ when $t = \frac{1}{2}$.

15. Find dy for $y = x^3$ when $x = 2$ and $dx = 0.01$.

16. Find du for $u = t^2 + 3t + 1$ when $t = -1$ and $dt = 0.02$.

17. Find dx for $x = y \ln y$ when $y = 1$ and $dy = 0.003$.

18. Find df for $df = xe^x$ when $x = 0$ and $dx = -0.01$.

(19–22) Find dy and Δy for the following functions.

19. $y = 3x^2 + 5$ when $x = 2$ and $dx = 0.01$.

20. $y = \sqrt{t}$ when $t = 4$ and $dt = 0.41$.

21. $y = \ln u$ when $u = 3$ and $du = 0.06$.

22. $y = \sqrt{x + 2}$ when $x = 2$ and $dx = 0.84$.

23. Use differentials to approximate the cube root of 9.

24. Use differentials to approximate the 4th root of 17.

25. Use differentials to approximate the 5th root of 31.

26. Use differentials to approximate the value of $(4.01)^3 + \sqrt{4.01}$.

27. *(Errors)* The radius of a sphere is equal to 8 centimeters, with a possible error of ± 0.002 centimeter. The volume is calculated assuming that the radius is exactly 8 centimeters. Use differentials to estimate the maximum error in the calculated volume.

28. *(Percentage Error)* If the volume of a sphere is to be determined to within a percentage error that does not exceed 2%, what is the maximum percentage error allowable in the measured value of the radius?

29. *(Percentage Error)* A manufacturer estimates that sales will be 400 units per week with a possible percentage error of 5%. If the revenue function is $R(x) = 10x - 0.01x^2$, find the maximum percentage error in the estimated revenue.

30. *(Percentage Error)* The cost function for the manufacturer in Exercise 29 is $C(x) = 1000 + x$.

a. Find the maximum percentage error in the estimated costs.

b. Find the maximum percentage error in the estimated profit.

31. *(Approximate Price)* The demand equation for a certain product is $p = 100/\sqrt{x + 100}$. Use differentials to find the approximate price at which 2500 units are demanded.

32. *(Approximate Cost)* The cost function for a certain manufacturer is $C(x) = 400 + 2x + 0.1x^{3/2}$. Use differentials to estimate the change in cost if the production level is increased from 100 to 110.

33. *(Inventory Cost Model)* For the inventory-cost model (see Section 14-5), let D be the total annual demand, s the storage cost per unit per year, a the setup cost for each production run, and b the production cost per item. Then the optimum batch size per production run is given by $x = \sqrt{2aD/s}$. The minimum cost per year of producing the items is $C = bD + \sqrt{2aDs}$. If $D = 10,000$, $s = 0.2$, $a = 10$, and $b = 0.1$, evaluate x and C. Use differentials to estimate the errors in x and C if the true value of s is 0.22.

34. (*Physical Measurement*) The acceleration due to gravity, g, is determined by measuring the period of swing of a pendulum. If the length of the pendulum is l and the measured period is T, then g is given by the formula

$$g = \frac{4\pi^2 l}{T^2}.$$

Find the percentage error in g if:

a. l is measured accurately but T has an error of 1%.

b. T is measured accurately but l has an error of 2%.

35. (*Linear Cost Modeling*) A company is currently producing 200 units per day and its daily cost is $5000. If the marginal cost is $20 per unit, obtain a linear cost model that approximates the cost function $C(x)$ for x close to 200.

36. (*Linear Revenue Modeling*) A company is currently producing 1500 units per month and it sells all the units it produces. Its monthly revenue is $30,000. If the current marginal revenue is $180, obtain a linear formula that approximates the revenue function $R(x)$ for x close to 1500.

37. (*Linear Profit Modeling*) A company is currently producing 50 units per week and its weekly profit is $2000. If the current marginal profit is $15, obtain a linear formula that approximates the weekly profit function $P(x)$ for x close to 50.

38. (*Linear Cost Modeling*) The monthly cost of producing x units of its product for a certain company is given by $C(x) = 2000 + 16x - 0.001x^2$. The company is currently producing 3000 units per month. Obtain a linear cost model that approximates the monthly cost function $C(x)$ for x close to 3000.

39. (*Linear Revenue Modeling*) The weekly demand function of a certain product is $p = 50 - 0.2x$. Currently, the demand is 200 units per week. Obtain a linear formula that approximates the weekly revenue function $R(x)$ for x close to 200.

40. (*Linear Profit Modeling*) The daily demand function of a company's product is $p = 45 - 0.03x$. The cost of producing x units per day is given by $C(x) = 1500 + 5x - 0.01x^2$. The company is currently producing 500 units per day. Obtain a linear formula that approximates the daily profit function $P(x)$ for x close to 500.

15-2 IMPLICIT DIFFERENTIATION

As we discussed in Section 5-5, a relationship between two variables is sometimes expressed through an implicit relation rather than by means of an explicit function. Thus instead of having y given as a function $f(x)$ of the independent variable x, it is possible to have x and y related through an equation of the form $F(x, y) = 0$, in which both variables occur as arguments of some function F. For example, the equation

$$F(x, y) = x^3 + 2x^2y + 3xy^2 + y^3 - 1 = 0$$

expresses a certain relationship between x and y, but y is not given explicitly in terms of x.

The question we wish to consider in this section is how to calculate the derivative dy/dx when x and y are related by an implicit equation. In certain cases, it is possible to solve the implicit equation $F(x, y) = 0$ and to obtain y explicitly in terms of x. In such cases, the standard techniques of differentiation enable the derivative to be calculated in the usual way. However, in many cases it is not possible to solve for the explicit function, and to cope with such situations it is necessary that we use a new technique which is called **implicit differentiation.**

When using this technique, *we differentiate each term in the given implicit relation with respect to the independent variable*. This involves differentiating expressions involving y with respect to x, and to do this, we make use of the chain rule. For example, suppose we wish to differentiate y^3 or $\ln y$ with respect to x. We write the following.

$$\frac{d}{dx}(y^3) = \frac{d}{dy}(y^3) \cdot \frac{dy}{dx} = 3y^2 \frac{dy}{dx}$$

$$\frac{d}{dx}(\ln y) = \frac{d}{dy}(\ln y) \cdot \frac{dy}{dx} = \frac{1}{y}\frac{dy}{dx}$$

In general,

$$\frac{d}{dx}(f(y)) = f'(y)\frac{dy}{dx}. \quad \text{☛ 7}$$

☛ **7.** Find (a) $\dfrac{d}{dx}(\sqrt{y})$;

(b) $\dfrac{d}{dx}(e^y)$; (c) $\dfrac{d}{dy}(x^4)$.

Answer (a) $\dfrac{1}{2\sqrt{y}}\dfrac{dy}{dx}$; (b) $e^y\dfrac{dy}{dx}$;

(c) $4x^3\dfrac{dx}{dy}$.

EXAMPLE 1 Find dy/dx if $x^2 + y^2 = 4$.

Solution Differentiate each term with respect to x.

$$\frac{d}{dx}(x^2) = 2x,$$

$$\frac{d}{dx}(y^2) = \frac{d}{dy}(y^2)\frac{dy}{dx} = 2y\frac{dy}{dx}, \qquad \frac{d}{dx}(4) = 0.$$

We differentiate the given relation and solve for dy/dx:

$$2x + 2y\frac{dy}{dx} = 0, \qquad 2y\frac{dy}{dx} = -2x, \qquad \frac{dy}{dx} = -\frac{2x}{2y} = -\frac{x}{y}.$$

Check Let us check this result using one of the explicit functions associated with the implicit relation $x^2 + y^2 = 4$, namely*

$$y = \sqrt{4 - x^2} = (4 - x^2)^{1/2}.$$

Using the chain rule,

$$\frac{dy}{dx} = \frac{1}{2}(4 - x^2)^{1/2-1} \cdot (-2x) = -\frac{x}{\sqrt{4 - x^2}} = -\frac{x}{y}$$

which is the same result as in Example 1. ☛ 8

☛ **8.** Take the other explicit function associated with $x^2 + y^2 = 4$, namely $y = -\sqrt{4 - x^2}$, and verify that it is still true that $\dfrac{dy}{dx} = \dfrac{-x}{y}$.

EXAMPLE 2 Find dy/dx if $xy + \ln(xy^2) = 7$.

Solution We first simplify the logarithm: $\ln(xy^2) = \ln x + 2\ln y$. Then the relation takes the form

$$xy + \ln x + 2\ln y = 7.$$

---

*Recall that $\sqrt{a}$, or $a^{1/2}$, denotes the positive square root of a. (See page 24.)

Differentiating with respect to x, we obtain

$$\frac{d}{dx}(xy) + \frac{d}{dx}(\ln x) + 2\frac{d}{dx}(\ln y) = \frac{d}{dx}(7) = 0.$$

From the product rule,

$$\frac{d}{dx}(xy) = \frac{d}{dx}(x) \cdot y + x \cdot \frac{d}{dx}(y) = 1 \cdot y + x\frac{dy}{dx} = y + x\frac{dy}{dx}.$$

Also,

$$\frac{d}{dx}(\ln x) = \frac{1}{x} \qquad \text{and} \qquad \frac{d}{dx}(\ln y) = \frac{d}{dy}(\ln y) \cdot \frac{dy}{dx} = \frac{1}{y}\frac{dy}{dx}.$$

Therefore

$$\left(y + x\frac{dy}{dx}\right) + \frac{1}{x} + \frac{2}{y}\frac{dy}{dx} = 0.$$

We group all the derivative terms on the left and move the other terms to the right and solve for dy/dx.

$$\left(x + \frac{2}{y}\right)\frac{dy}{dx} = -\left(y + \frac{1}{x}\right)$$

$$\frac{dy}{dx} = -\frac{y + 1/x}{x + 2/y} = -\frac{y(xy + 1)}{x(xy + 2)} \qquad \text{☛ 9}$$

☛ 9. Find $\dfrac{dy}{dx}$ if

(a) $2x^2 - 3y^2 = 2$;

(b) $x^2 + 4xy + y^2 = 1$.

EXAMPLE 3 Find the equation of the tangent line at the point $(2, -\frac{1}{2})$ on the graph of the implicit relation

$$xy^2 - x^2y + y - x = 0.$$

Solution The slope of the tangent line is equal to the derivative dy/dx evaluated at $x = 2$ and $y = -\frac{1}{2}$. Differentiating through the implicit relation with respect to x, we obtain

$$\frac{d}{dx}(xy^2) - \frac{d}{dx}(x^2y) + \frac{dy}{dx} - 1 = 0.$$

The first two terms must be evaluated using the product rule. We obtain

$$\left(y^2 \cdot 1 + x \cdot 2y\frac{dy}{dx}\right) - \left(x^2 \cdot \frac{dy}{dx} + y \cdot 2x\right) + \frac{dy}{dx} - 1 = 0$$

and so $(2xy - x^2 + 1)(dy/dx) = 2xy - y^2 + 1$. Therefore

$$\frac{dy}{dx} = \frac{2xy - y^2 + 1}{2xy - x^2 + 1}.$$

Setting $x = 2$ and $y = -\frac{1}{2}$, we obtain the slope of the tangent line at the required point to be

$$\frac{dy}{dx} = \frac{2(2)(-\frac{1}{2}) - (-\frac{1}{2})^2 + 1}{2(2)(-\frac{1}{2}) - (2)^2 + 1} = \frac{1}{4}.$$

Answer (a) $\dfrac{dy}{dx} = \dfrac{2x}{3y}$;

(b) $\dfrac{dy}{dx} = \dfrac{-(x + 2y)}{2x + y}$.

10. Find the equation of the tangent line at the point $(2, 1)$ on the graph of the relation $x^3 + y^3 = 9$.

The equation of the tangent line is obtained from the point-slope formula:

$$y - y_1 = m(x - x_1)$$

$$y - (-\tfrac{1}{2}) = \tfrac{1}{4}(x - 2)$$

$$y = \tfrac{1}{4}x - 1. \quad \text{☛ 10}$$

When we evaluate dy/dx from an implicit relation $F(x, y) = 0$, we are assuming that x is the independent variable and y the dependent variable. However, given the implicit relation $F(x, y) = 0$, we could instead regard y as the independent variable with x a function of y. In that case, we should evaluate the derivative dx/dy.

EXAMPLE 4 Given $x^2 + y^2 = 4xy$, find dx/dy.

Solution Here x is an implicit function of y. We differentiate both sides with respect to y.

$$\frac{d}{dy}(x^2) + \frac{d}{dy}(y^2) = 4\frac{d}{dy}(xy)$$

$$2x\frac{dx}{dy} + 2y = 4\left(x \cdot 1 + y\frac{dx}{dy}\right)$$

$$2\frac{dx}{dy}(x - 2y) = 2(2x - y)$$

$$\frac{dx}{dy} = \frac{2x - y}{x - 2y}.$$

Consider the implicit relation $x^2 + y^2 = 4$. The graph of this relation is a circle of radius 2 and center at $(0, 0)$ as shown in Figure 2. In Example 1, we calculated that $dy/dx = -x/y$. If $dy/dx = 0$, then $-x/y = 0$, or $x = 0$. When $x = 0$, $y = \pm 2$. Thus at each of the points $(0, \pm 2)$, the slope dy/dx of the tangent line to the circle is zero and therefore the tangent line is horizontal.

If in Example 1 we take y as the independent variable and differentiate with respect to y instead of x, we find the result that

$$\frac{dx}{dy} = -\frac{y}{x}.$$

If $dx/dy = 0$, then $-y/x = 0$, or $y = 0$. When $y = 0$, $x = \pm 2$. Thus at the points $(\pm 2, 0)$ we have $dx/dy = 0$. But at these points the tangent lines are vertical, as shown in the figure. This result can be generalized as follows.

---

1. If $dy/dx = 0$ at a point, then the tangent line is **horizontal** at that point.

2. If $dx/dy = 0$ at a point, then the tangent line is **vertical** at that point.

Answer $y = -4x + 9$.

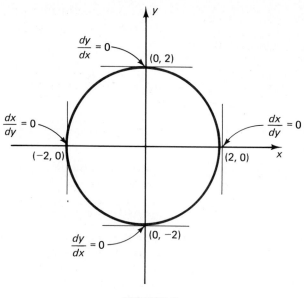

FIGURE 2

EXAMPLE 5 Find the points on the curve $4x^2 + 9y^2 = 36y$ where the tangent line is: (a) horizontal; (b) vertical.

Solution We have

$$4x^2 + 9y^2 = 36y. \qquad (i)$$

(a) Differentiating both sides with respect to x, we obtain

$$\frac{d}{dx}(4x^2) + \frac{d}{dx}(9y^2) = \frac{d}{dx}(36y)$$

$$8x + 18y\frac{dy}{dx} = 36\frac{dy}{dx}$$

$$\frac{dy}{dx} = \frac{-8x}{18y - 36} = \frac{-4x}{9(y - 2)}$$

For the tangent line to be horizontal, we have $dy/dx = 0$, which gives $x = 0$. When $x = 0$ relation (i) gives $0 + 9y^2 = 36y$, so that $y = 0$ or 4. Thus the two points on the curve where the tangent line is horizontal are $(0, 0)$ and $(0, 4)$.

(b) If we differentiate (i) with respect to y instead, we find the result that

$$\frac{dx}{dy} = \frac{-9(y - 2)}{4x}.$$

Setting $dx/dy = 0$ to find the vertical tangents, we get $y = 2$. Putting

$y = 2$ in (i), we find $x = \pm 3$. Thus the tangent lines are vertical at the points $(3, 2)$ and $(-3, 2)$. ☛ **11**

_____

It will probably not have escaped your notice that in the last example, dy/dx and dx/dy were reciprocals of one another. This property is generally true*:

$$\frac{dx}{dy} = \frac{1}{dy/dx}.$$

In other words, _the derivative of the inverse of a function f is the reciprocal of the derivative of f._ Using this property, we can shortcut much of the work in part (b) of Example 5, once we have found dy/dx in part (a).

Higher-order derivatives can also be calculated from an implicit relation. The method consists of first finding the first derivative in the manner outlined above and then differentiating the resulting expression with respect to the independent variable.

EXAMPLE 6 Find d^2y/dx^2 when $x^3 + y^3 = 3x + 3y$.

Solution Here x is the independent variable since we are required to find derivatives with respect to x. So, differentiating implicitly with respect to x, we obtain

$$3x^2 + 3y^2\frac{dy}{dx} = 3 + 3\frac{dy}{dx}.$$

Therefore

$$\frac{dy}{dx} = \frac{x^2 - 1}{1 - y^2}.$$

We differentiate again with respect to x and use the quotient rule.

$$\frac{d^2y}{dx^2} = \frac{d}{dx}\left(\frac{x^2 - 1}{1 - y^2}\right)$$

$$= \frac{(1 - y^2)(d/dx)(x^2 - 1) - (x^2 - 1)(d/dx)(1 - y^2)}{(1 - y^2)^2}$$

_____

*By definition,

$$\frac{dx}{dy} = \lim_{\Delta y \to 0}\frac{\Delta x}{\Delta y} = \left[\lim_{\Delta y \to 0}\frac{\Delta y}{\Delta x}\right]^{-1} = \left[\lim_{\Delta x \to 0}\frac{\Delta y}{\Delta x}\right]^{-1} = \left[\frac{dy}{dx}\right]^{-1}.$$

In the second step we used Limit Theorem 2(b) of Section 12-2.

■ **12.** An alternative way of finding y'' is to use implicit differentiation twice. In Example 6, we have already obtained that $x^2 + y^2 y' = 1 + y'$. Write down the result of differentiating this again with respect to x. Hence calculate y''.

Answer $2x + (2yy' \cdot y' + y^2 y'')$
$$= 0 + y''$$

or $y'' = \dfrac{2[x + y(y')^2]}{1 - y^2}$.

Final result is the same as before.

$$= \frac{2x(1 - y^2) - (x^2 - 1)[-2y(dy/dx)]}{(1 - y^2)^2}$$

At this stage, we observe that the expression for the second derivative still involves the first derivative. Hence, to complete the solution, we must substitute $dy/dx = (x^2 - 1)/(1 - y^2)$.

$$\frac{d^2y}{dx^2} = \frac{2x(1 - y^2) + 2y(x^2 - 1)[(x^2 - 1)/(1 - y^2)]}{(1 - y^2)^2}$$

$$= \frac{2x(1 - y^2)^2 + 2y(x^2 - 1)^2}{(1 - y^2)^3}$$

In the last step, we have multiplied numerator and denominator of the fraction by $1 - y^2$. ■ **12**

EXERCISES 15-2

(1–14) Find dy/dx in each case.

1. $x^2 + y^2 + 2y = 15$

2. $\sqrt{x} + \sqrt{y} = 1$

3. $x^3 + y^3 = a^3$ (a is constant)

4. $x^2 - xy + y^2 = 3$

5. $(y - x)(y + 2x) - 12 = 0$

6. $x^4 + y^4 = 2x^2y^2 + 3$

7. $xy^2 + yx^2 = 6$

8. $x^2y^2 + x^2 + y^2 = 3$

9. $x^5 + y^5 = 5xy$

10. $\dfrac{x^2}{a^2} - \dfrac{y^2}{b^2} = 1$ ($a; b$ are constants)

11. $xy + e^y = 1$

12. $\dfrac{x}{y} + \ln\left(\dfrac{y}{x}\right) = 6$

13. $xy + \ln(xy) = -1$

14. $x^2 + y^2 = 4e^{x+y}$

15. Find dx/dt if $3x^2 + 5t^2 = 15$.

16. Find du/dy if $u^2 + y^2 + u - y = 1$.

17. Find dx/dy if $x^3 + y^3 = xy$.

18. Find dt/dx if $x^3 + t^3 + x^3t^3 = 9$.

(19–22) Find the equation of the tangent to the following curves at the given points.

19. $x^3 + y^3 - 3xy = 3$;   (1, 2)

20. $x^2 + y^2 = 2x + y + 15$;   (−3, 1)

21. $\dfrac{2y}{x} - \dfrac{x}{y} = 1$ at $(2, -1)$

22. $(x - y)(x + 2y) = 4$ at $(2, 1)$

(23–26) Find the points where each given curve has:
a. A horizontal tangent.   b. A vertical tangent.

23. $(x - 1)^2 + (y - 2)^2 = 9$

24. $9x^2 - 4y^2 = 36$

25. $x^2 + y^2 = xy + 12$

26. $x^2 + 3y^2 = 2xy + 48$

27. Find d^2y/dx^2 if $x^2 + y^2 = 4xy$.

28. Find d^2u/dt^2 when $u = 1$ and $t = -1$ if $u^5 + t^5 = 5ut + 5$.

29. Find d^2x/dy^2 when $x = 2$ and $y = 1$ if $x^3 + y^3 - 3xy = 3$.

30. Find d^2y/dx^2 if $(x + 2)(y + 3) = 7$.

31. Find d^2x/dy^2 if $x + y + \ln(xy) = 2$.

32. Find d^2y/dx^2 if $x^2 + y^2 + e^{3y} = 4$.

(33–36) Find dy for the following implicit relations.

33. $xy + y^2 = 3$

34. $y^2 + z^2 - 4yz = 1$

35. $\ln(yz) = y + z$

36. $xe^y + ye^x = 1$

(37–40) Find the rate of change of x with respect to p for the following demand relations.

37. $p = \sqrt{100 - 9x^2}$

38. $p = \dfrac{500}{x^3 + 4}$

39. $2pe^x = 3e^{x/2} - 7p$

40. $7x + x\ln(p + 1) = 2$

41. *(Price and Profit)* The relationship between the price p at which its product is sold and a firm's profit P is $P = 6p - p^2$. Express this relation as an explicit function $p = f(P)$. Evaluate the derivatives dP/dp and dp/dP and show they are reciprocals of one another.

42. *(Product Transformation Function)* A manufacturer can make x thousand pairs of men's shoes or y thousand pairs of women's shoes per week, where x and y are related by

$$2x^3 + y^3 + 5x + 4y = \text{Constant.}$$

At the moment the manufacturer is making 2000 pairs of men's and 5000 pairs of women's shoes per week. Calculate dy/dx for current production levels. What is its significance?

43. *(Predator-Prey Model)* Let x and y be the sizes of two populations, one of which preys on the other. At any time x and y satisfy the implicit relation

$$(x + ty - h)^2 + (y - tx - k)^2 = a^2$$

where a, h, k, and t are certain constants. Calculate dy/dx.

44. *(Physiology)* According to A. V. Hill, the relation between the load F acting on a muscle and the speed V of contraction or shortening of the muscle is given by

$$(F + a)V = (F_0 - F)b$$

where a, b, F_0 are constants that depend on the particular species and type of muscle. Prove that the speed V approaches zero as $F \to F_0$, so that F_0 represents the maximum load under which the muscle can contract. Find dV/dF and dF/dV. Show that each of these derivatives is the reciprocal of the other.

*45. By writing $y = x^{p/q}$ in the form $y^q - x^p = 0$, use implicit differentiation to prove that $(d/dx)(x^n) = nx^{n-1}$ when n is a rational number p/q.

15-3 LOGARITHMIC DIFFERENTIATION AND ELASTICITY

With certain types of functions, a technique known as *logarithmic differentiation* can be used to ease the calculation of the derivative. One situation in which this technique can be used arises when the given function consists of the product or quotient of a number of factors, where each factor may be raised to some power. The method is perhaps best demonstrated through an example.

EXAMPLE 1 Calculate dy/dx if

$$y = \frac{(x + 1)\sqrt{x^2 - 2}}{(x^2 + 1)^{1/3}}$$

Solution We could differentiate this function by use of the product and quotient rules. However, let us instead take the natural logarithm of both sides. Then using the properties of logarithms, we proceed as follows.

$$\ln y = \ln\left[\frac{(x + 1)\sqrt{x^2 - 2}}{(x^2 + 1)^{1/3}}\right]$$

$$= \ln(x + 1) + \ln\sqrt{x^2 - 2} - \ln(x^2 + 1)^{1/3}$$

$$= \ln(x + 1) + \tfrac{1}{2}\ln(x^2 - 2) - \tfrac{1}{3}\ln(x^2 + 1)$$

Now, let us differentiate both sides with respect to x. We use the chain rule as usual with implicit differentiation.

$$\frac{d}{dx}(\ln y) = \frac{d}{dy}(\ln y)\frac{dy}{dx} = \frac{1}{y}\frac{dy}{dx}$$

After differentiating the terms on the right, we find that

$$\frac{1}{y}\frac{dy}{dx} = \frac{1}{x+1} + \frac{1}{2}\cdot\frac{1}{(x^2-2)}\cdot 2x - \frac{1}{3}\cdot\frac{1}{x^2+1}\cdot 2x.$$

We next simplify and multiply by y.

$$\frac{dy}{dx} = y\left[\frac{1}{x+1} + \frac{x}{x^2-2} - \frac{2x}{3(x^2+1)}\right]$$

We can, if we wish, substitute $y = [(x+1)\sqrt{x^2-2}]/(x^2+1)^{1/3}$ in this expression in order to obtain dy/dx entirely in terms of x. ☞ 13

☞ **13.** Use logarithmic differentiation to find $\dfrac{dy}{dx}$ for the following functions:

(a) $y = \sqrt{\dfrac{x-1}{x+1}}$;

(b) $y = (x+1)2^x$.

Another situation in which logarithmic differentiation is useful arises when we must differentiate one function raised to the power of another function.* We shall give two examples, the first an elementary one and the second more complicated.

EXAMPLE 2 Find dy/dx if $y = x^x$ $(x > 0)$.

Solution First of all, let us observe that this derivative can be found using the ordinary techniques of differentiation if we first write y in the form

$$y = x^x = e^{x\ln x}.$$

The chain rule combined with the product rule then allows us to determine dy/dx. However, the method of logarithmic differentiation can be used as an alternative. Taking logarithms of both sides, we get

$$\ln y = \ln (x^x) = x \ln x.$$

We then differentiate with respect to x and use the product rule on the right.

$$\frac{d}{dx}(\ln y) = (1)\ln x + x\left(\frac{1}{x}\right)$$

$$\frac{1}{y}\frac{dy}{dx} = \ln x + 1$$

Therefore

$$\frac{dy}{dx} = y(\ln x + 1) = x^x(\ln x + 1).$$

Answer

(a) $\dfrac{dy}{dx} = \dfrac{y}{2}\left(\dfrac{1}{x-1} - \dfrac{1}{x+1}\right)$

$\qquad\quad = \dfrac{y}{x^2-1}$;

(b) $\dfrac{dy}{dx} = y\left(\ln 2 + \dfrac{1}{x+1}\right)$.

---

*A function of the type $[f(x)]^{g(x)}$ is generally defined only if $f(x) > 0$, and this restriction is assumed in the examples that follow.

EXAMPLE 3 Find dy/dx if

$$y = (x^2 + 1)^{\sqrt{x^3-1}}.$$

Solution Again the calculation is considerably simplified if we take logarithms before differentiating.

$$\ln y = \ln [(x^2 + 1)^{\sqrt{x^3-1}}] = \sqrt{x^3 - 1} \ln (x^2 + 1)$$

We can now differentiate with respect to x (using the product rule on the right).

$$\frac{d}{dx}(\ln y) = \frac{d}{dx}\sqrt{x^3 - 1} \cdot \ln (x^2 + 1) + \sqrt{x^3 - 1}\frac{d}{dx}\ln (x^2 + 1)$$

$$\frac{1}{y}\frac{dy}{dx} = \frac{1}{2}(x^3 - 1)^{-1/2}(3x^2) \cdot \ln (x^2 + 1) + \sqrt{x^3 - 1}\left(\frac{1}{x^2 + 1}\right)2x$$

Simplifying and multiplying by y, we obtain finally

$$\frac{dy}{dx} = y\left[\frac{3x^2}{2\sqrt{x^3 - 1}} \ln (x^2 + 1) + \frac{2x\sqrt{x^3 - 1}}{x^2 + 1}\right]. \quad \text{☛ 14}$$

☛ **14.** Use logarithmic differentiation to find $\dfrac{dy}{dx}$ for the following functions:
(a) $y = x^{x^2}$;
(b) $y = (x^2 + 1)(x^2 + 1)^{x^2+1}$.

It can be seen from these examples that the essence of this method consists of the following steps:

1. Take the logarithm of y and simplify the right side using properties of logarithms.

2. Differentiate and solve for dy/dx. In step 2 we obtain the expression

$$\frac{d}{dx}\ln y = \frac{1}{y}\frac{dy}{dx}.$$

This is called the **logarithmic derivative** of y with respect to x.

EXAMPLE 4 Calculate the logarithmic derivatives of the following functions.

    (a) ax^n (b) $u(x)v(x)$

Solution

    (a) If $y = ax^n$, then $y' = anx^{n-1}$. The logarithmic derivative is therefore

$$\frac{y'}{y} = \frac{anx^{n-1}}{ax^n} = \frac{n}{x}.$$

In the special case when $n = 1$, $y = ax$ and the logarithmic derivative is $1/x$.

    (b) If $y = u(x)v(x)$, then from the product rule it follows that

$$y' = u'v + uv'.$$

Answer (a) $\dfrac{dy}{dx} = yx(2 \ln x + 1)$;

(b) $\dfrac{dy}{dx} = 2xy[1 + \ln (x^2 + 1)]$.

Therefore the logarithmic derivative is

$$\frac{y'}{y} = \frac{u'v + uv'}{uv} = \frac{u'v}{uv} + \frac{uv'}{uv} = \frac{u'}{u} + \frac{v'}{v}.$$

This result can be summarized as follows: *The logarithmic derivative of the product uv is the sum of the logarithmic derivatives of u and v.* ☛ 15

☛ **15.** Find the logarithmic derivatives of the following functions: (a) x; (b) e^x (c) $\ln x$.

EXAMPLE 5 Find dy/dx if $y = x^x + (1 + x)^{1+x}$.

Solution Examples of this type are traps for the unwary student. It is a great temptation to take logarithms immediately and write

$$\ln y = x \ln x + (1 + x) \ln (1 + x).$$

Of course, a moment's thought will reveal the error in doing this. What we must do is to write $y = u + v$ where $u = x^x$ and $v = (1 + x)^{1+x}$. Then

$$\frac{dy}{dx} = \frac{du}{dx} + \frac{dv}{dx}$$

and the two derivatives du/dx and dv/dx can be found separately by logarithmic differentiation. The first of these can be obtained from Example 2.

$$\frac{du}{dx} = x^x(\ln x + 1)$$

For dv/dx, we have

$$\ln v = (1 + x) \ln (1 + x)$$

and therefore, after differentiation with respect to x,

$$\frac{1}{v}\frac{dv}{dx} = \ln (x + 1) + 1$$

Consequently, $dv/dx = (1 + x)^{1+x}[\ln (1 + x) + 1]$. Adding the values for du/dx and dv/dx, we obtain dy/dx as required.

Elasticity

A concept widely used in economics and closely related to logarithmic differentiation is that of elasticity. We shall introduce this idea via the so-called **elasticity of demand.**

For a given item, let p be the price per unit and x the number of units that will be purchased during a specified time interval at the price p, and let $x = f(p)$. The elasticity of demand is usually represented by the Greek letter η (eta) and is defined as follows.*

Answer (a) $\dfrac{1}{x}$; (b) 1; (c) $\dfrac{1}{x \ln x}$.

---

* Take care, some books define η with an extra minus sign.

$$\eta = \frac{p}{x}\frac{dx}{dp} = \frac{pf'(p)}{f(p)}$$

Before doing any examples, let us discuss the significance of η. Suppose that the price is increased from p to $p + \Delta p$. Then, of course, the quantity demanded will change, say to $x + \Delta x$, where $x + \Delta x = f(p + \Delta p)$. Thus $\Delta x = f(p + \Delta p) - f(p)$.

The increment in price is Δp; this increase is a fraction $\Delta p/p$ of the original price. We can also say that the percentage increase in price is $100(\Delta p/p)$. For example, let the original price per unit be $p = \$2$ and let the new price be $\$2.10$. Then $\Delta p = \$0.10$. This increase is a fraction $\Delta p/p = 0.10/2 = 0.05$ of the original price. Multiplying by 100, we see that the percentage increase in price is $100(\Delta p/p) = 100(0.05) = 5\%$.

Similarly, the change Δx in demand is a fraction $(\Delta x/x)$ of the original demand. The percentage change in demand is $100(\Delta x/x)$. (Note that with an increase in price, the demand will actually decrease, so this percentage change in demand will be negative.)

Consider the ratio of these two percentage increases:

$$\frac{\text{Percentage Change in Demand}}{\text{Percentage Change in Price}} = \frac{100(\Delta x/x)}{100(\Delta p/p)} = \frac{p}{x}\frac{\Delta x}{\Delta p}.$$

Comparing this with the definition of η, we see that

$$\eta = \frac{p}{x}\frac{dx}{dp} = \lim_{\Delta p \to 0} \frac{p}{x}\frac{\Delta x}{\Delta p}.$$

Thus the elasticity of demand is equal to the limiting value of the ratio of percentage change in demand to percentage change in price in the limit as the change in price approaches zero.

When the change in price is small, the ratio $\Delta x/\Delta p$ of the two increments is approximately equal to the derivative dx/dp. Thus for Δp small,

$$\frac{p}{x}\frac{\Delta x}{\Delta p} \approx \frac{p}{x}\frac{dx}{dp} = \eta$$

and so the ratio of percentage change in demand to percentage change in price is approximately equal to η. Alternatively, we can say that, when the price change is small,

Percentage Change in Demand $\approx \eta$ (Percentage Change in Price).

For example, if a 2% increase in price causes the demand to decrease by 3%, then the elasticity of demand is approximately equal to $(-3)/(2) = -1.5$. Or, another example, if the elasticity of demand is -0.5, then a 4% increase in price would lead to a change in demand of approximately $(-0.5)(4\%) = -2\%$.

EXAMPLE 6 Calculate the elasticity of demand if the demand equation is $x = k/p$, where k is some positive constant.

Solution Since $x = k/p$, $dx/dp = -k/p^2$. Therefore

$$\eta = \frac{p}{x}\frac{dx}{dp} = \frac{p}{(k/p)}\left(-\frac{k}{p^2}\right) = -1.$$

The elasticity of demand is therefore constant in this case and is equal to -1.

This means that a small percentage increase in price will always lead to an equal percentage decrease in demand.

EXAMPLE 7 Calculate the elasticity of demand if $x = 500(10 - p)$ for each value of p.

(a) $p = 2$ (b) $p = 5$ (c) $p = 6$

Solution In this case, $dx/dp = -500$. Therefore

$$\eta = \frac{p}{x}\frac{dx}{dp} = \frac{p}{500(10 - p)}(-500) = -\frac{p}{10 - p}.$$

We see that the elasticity of demand varies, depending on the price p.

(a) $p = 2$: $\eta = \dfrac{-2}{10 - 2} = -0.25$

Thus when $p = 2$, the percentage decrease in demand is one-fourth of the percentage increase in price.

(b) $p = 5$: $\eta = \dfrac{-5}{10 - 5} = -1$

When $p = 5$, a small percentage increase in price gives an equal percentage decrease in demand.

(c) $p = 6$: $\eta = \dfrac{-6}{10 - 6} = -1.5$

The percentage decrease in demand is one and one-half times the percentage increase in price when $p = 6$. ☛ 16

The demand can be described in terms of elasticity as follows.

Demand is **elastic** if $\eta < -1$; percentage change in demand is greater than percentage change in price. Demand is **inelastic** if $-1 < \eta < 0$; percentage change in demand is less than percentage change in price. If $\eta = -1$, there is **unit elasticity;** percentage change in demand is equal to percentage change in price. ☛ 17

The idea of elasticity can be used for any pair of variables that are func-

☛ **16.** Find the elasticity of demand for the demand relations
(a) $x = 12 - 2p$;
(b) $3x + 4p = 12$.

Answer

(a) $\eta = \dfrac{-2p}{x} = -p/(6 - p)$;

(b) $\eta = \dfrac{-4p}{3x} = \dfrac{-p}{3 - p}$.

☛ **17.** For the demand relation $x = 16 - 2p$, for what values of p is demand elastic and for what values is it inelastic?
(Assume $p < 8$ of course.)

Answer Elastic for $p > 4$, inelastic for $0 \le p < 4$.

tionally related. If $y = f(x)$, then the *elasticity of y with respect to x is defined as*

$$\eta = \frac{x}{y}\frac{dy}{dx}$$

(again denoted by η). It is approximately equal to the ratio of the percentage change in y to the percentage change in x, provided that these changes are small. For example, we can talk about the elasticity of supply with respect to price, which is the percentage change in the supply of an item divided by the percentage change in its price (strictly, in the limit as the price change approaches zero).

Elasticity is closely related to logarithmic derivatives. The logarithmic derivative of y with respect to x is

$$\frac{d}{dx}(\ln y) = \frac{1}{y}\frac{dy}{dx}.$$

The logarithmic derivative of x itself is similarly given by

$$\frac{d}{dx}(\ln x) = \frac{1}{x}\frac{dx}{dx} = \frac{1}{x}.$$

It follows therefore that *elasticity of y with respect to x is equal to the logarithmic derivative of y divided by the logarithmic derivative of x:*

$$\eta = \frac{(d/dx)(\ln y)}{(d/dx)(\ln x)}.$$

Returning to the elasticity of demand, we can establish a connection between this quantity and marginal revenue. The revenue function is given by

$$R(x) = (\text{quantity sold}) \times (\text{price}) = xp$$

Let us think of R as a function of unit price p. The derivative dR/dp is called the **marginal revenue with respect to price** and it gives the increase in revenue per unit increase in price when these increments are small. From $R = xp$, we have, by the product rule,

$$\frac{dR}{dp} = \frac{d}{dp}(px) = x + p\frac{dx}{dp} = x\left(1 + \frac{p}{x}\frac{dx}{dp}\right) = x(1 + \eta). \tag{1}$$

If the demand is elastic, that is, $\eta < -1$, then $1 + \eta < 0$, and from (1) it follows that $dR/dp < 0$. In this case the total revenue R is a decreasing function of price p. Similarly, if the demand is inelastic, that is, $-1 < \eta < 0$, then $1 + \eta > 0$ and from (2), $dR/dp > 0$, so that the revenue R is an increasing function of p. Thus:

> *If demand is elastic, a price increase causes revenue to decrease.*
> *If demand is inelastic, a price increase causes revenue to increase.*
> *For unit elasticity, a price increase causes no change in revenue.*

18. For the demand relation $x = 12 - p^2$, determine the elasticity of demand when (a) $x = 6$; (b) $x = 8$; (c) $x = 9$. In each case, will revenue increase or decrease if the unit price is raised?

EXAMPLE 8 The demand function of a certain product is $p = 10 - 0.2\sqrt{x}$, where x units are sold at a price of p each. Use the elasticity of demand to determine whether a rise in price will increase or decrease the total revenue if the demand is: (a) 900 units, (b) 1600 units.

Solution We first calculate η. $p = 10 - 0.2\sqrt{x}$ gives $dp/dx = -0.1/\sqrt{x}$, so that

$$\eta = \frac{p}{x}\frac{dx}{dp} = \frac{(10 - 0.2\sqrt{x})}{x(-0.1/\sqrt{x})} = \frac{10 - 0.2\sqrt{x}}{-0.1\sqrt{x}} = 2 - \frac{100}{\sqrt{x}}.$$

(a) When $x = 900$, we have

$$\eta = 2 - \frac{100}{30} = \frac{-4}{3}.$$

Since $\eta = -\frac{4}{3} < -1$, the demand is elastic and an increase in price will result in a decrease in total revenue.

(b) When $x = 1600$, we have

$$\eta = 2 - \frac{100}{40} = -\frac{1}{2}.$$

Answer (a) $\eta = -2$. R will decrease;
(b) $\eta = -1$ (R will be unchanged);
(c) $\eta = -\frac{2}{3}$ (R will increase).

Since $\eta = -1/2 > -1$, the demand is inelastic and, therefore, an increase in price will cause the revenue to rise. ☞ **18**

EXERCISES 15-3

(1–16) Use logarithmic differentiation to evaluate dy/dx for the following functions.

1. $y = (x^2 + 1)(x - 1)^{1/2}$

2. $y = (3x - 2)(3x^2 + 1)^{1/2}$

3. $y = (x^2 - 2)(2x^2 + 1)(x - 3)^2$

4. $y = (x^3 - 1)(x + 1)^3(x + 2)^2$

5. $y = \dfrac{(x^2 + 1)^{1/3}}{x^2 + 2}$

6. $y = \dfrac{(2x^2 - 3)^{1/4}}{x(x + 1)}$

7. $y = \left(\dfrac{2x^2 + 5}{2x + 5}\right)^{1/3}$

8. $y = \sqrt{\dfrac{(x + 1)(x + 2)}{x^3 - 3}}$

9. $y = x^{x^2}$

10. $y = x^{\sqrt{x}}$

11. $y = e^{e^x}$

12. $y = x^{e^x}$

13. $y = x^{\ln x}$

14. $y = (\ln x)^x$

15. $y = x^x + x^{1/x}$

16. $y = (x^2 + 1)^x - x^{x^2 + 1}$

(17–20) *(Elasticity of Demand)* Find the elasticity of demand for the following demand relations.

17. $x = k/p^n$ $(k, n$ constants$)$

18. $x = 100(5 - p)$

19. $x = 50(4 - \sqrt{p})$

20. $x = 200\sqrt{9 - p}$

21. *(Elasticity of Demand)* If the demand relation is $x = 400 - 100p$, find the elasticity of demand when:
a. $p = 1$; b. $p = 2$; c. $p = 3$.

22. *(Elasticity of Demand)* For the demand relation $x/1000 + p/8 = 1$, find the elasticity of demand when:
a. $p = 2$; b. $p = 4$; c. $p = 6$.

(23–26) *(Elastic and Inelastic Demand)* For the following demand relations, determine the values of p for which the demand is: a. Elastic. b. Inelastic.

23. $x = 100(6 - p)$ **24.** $x = 800 - 100p$

25. $x = 100(2 - \sqrt{p})$

26. $x = k(a - p)$ (k, a positive constants)

27. *(Elasticity)* The demand relation for a product is $x = 250 - 30p + p^2$, where x units can be sold at a price of p each. Determine the elasticity of demand when $p = 12$. If this price p is increased by 8.5%, find the approximate percentage change in demand.

28. *(Elasticity)* The demand equation for a product is $p = \sqrt{2500 - x^2}$, where x units can be sold at a price of p each. Find the elasticity of demand when $p = 40$. If the price of 40 is decreased by 2.25%, find the approximate percentage increase in demand.

29. *(Elasticity)* For the demand relation $p = 250 - 0.5x$, verify that the demand of x is elastic and the total revenue is an increasing function of x if $0 < x < 250$. Also show that the demand is inelastic and the total revenue is decreasing if $250 < x < 500$.

30. *(Elasticity)* For any linear demand function $p = mx + b$ ($m < 0$ and $b > 0$), show that the demand is elastic if $p > b/2$, inelastic if $p < b/2$, and has a unit elasticity if $p = b/2$.

31. Prove that $\eta = p/(R_m - p)$, where p is average revenue and R_m is marginal revenue. Verify this for the demand equation $p = b + mx$ ($m < 0$, $b > 0$).

*32. *(Elasticity)* The elasticity of demand for a demand function $p = f(x)$ is given by

$$\eta = f(x)/xf'(x).$$

Show that the elasticity of demand ζ for the demand function $p = xf(x)$ is given by $\zeta = \eta/(1 + \eta)$.

33. *(Price Change and Elasticity)* The demand equation of a product is $p = 300 - 0.5x$. Will a rise in price increase or decrease the total revenue if the demand per week is:

a. 200 units **b.** 400 units?

34. *(Price Change and Elasticity)* The demand equation of a certain product is $x = \sqrt{4100 - p^2}$. Will a rise in price increase or decrease the total revenue at a demand level of:

a. 40 units? **b.** 50 units?

35. *(Population Growth)* A population grows according to the Gompertz function $y = pe^{-ce^{-kt}}$. Show that the logarithmic derivative of y is an exponentially decreasing function of t.

CHAPTER REVIEW

Key Terms, Symbols, and Concepts

15.1 Differential, dx and dy.
Errors, relative error, percentage error.

15.2 Implicit differentiation.

15.3 Logarithmic differentiation. Logarithmic derivative.
Elasticity of demand; elasticity of y with respect to x.
Elastic and inelastic demand; unit elasticity.

Formulas

$dy = f'(x)\, dx.$

$f(x + \Delta x) \approx f(x) + f'(x)\Delta x$
or $f(a + h) \approx f(a) + hf'(a).$

$$\frac{d}{dx} f(y) = f'(y)\frac{dy}{dx}.$$

Horizontal tangent: $dy/dx = 0$. Vertical tangent: $dx/dy = 0$.

$$\frac{dx}{dy} = \frac{1}{dy/dx}.$$

$$\eta = \frac{p}{x}\frac{dx}{dp}.$$

Percentage Change in Demand
$$\approx \eta \ (\text{Percentage Change in Price}).$$

Marginal revenue with respect to price: $\dfrac{dR}{dp} = x(1 + \eta).$

If demand is elastic [alternatively inelastic], a price increase causes revenue to decrease [increase].

1. State whether each of the following is true or false. Replace each false statement with a corresponding true statement.

   a. The differential of x^2 is $2x$.

   b. For a linear function f, $df = \Delta f$.

   c. The derivative dy/dx is equal to the ratio of differentials $(dy) \div (dx)$.

   d. In an implicit relation $f(x, y) = 0$, x and y are both independent variables.

   e. $\left(\dfrac{dy}{dx}\right)\left(\dfrac{dx}{dy}\right) = -1$

   f. If y is an increasing function of x, then x is a decreasing function of y.

   g. The logarithmic derivative of y with respect to x is the derivative of $\ln y$ with respect to x.

   h. The logarithmic derivative of x with respect to y is the reciprocal of the logarithmic derivative of y with respect to x.

   i. The logarithmic derivative of x^n is nx^{n-1}.

   j. The logarithmic derivative of x^x is $\ln(ex)$.

   k. When the elasticity of demand is -1, the marginal revenue is zero.

   l. When the revenue is maximum, the elasticity of demand is -1.

   m. For the demand relation $x = kp^{-\alpha}$ (k, α constants), the elasticity of demand is constant.

2. Find dy if $y = \sqrt{x^2 - 2}$.

3. Evaluate dy when $x = 1$ and $dx = \frac{1}{4}$ for the function $y = \ln(3 - x^2)$.

4. Find dy and Δy when $t = 7$ and $dt = 1.03$ for the function $y = (t + 1)^{1/3}$.

5. Use differentials to find the approximate value of $(4.1)^2\sqrt{4.1}$.

6. Find dy/dx if $e^{xy} + x + y = 1$.

7. Find dx/dt if $x^2t + xt^2 + \ln(x + t) = 0$.

8. Find the equation of the tangent to the graph of the relation $x^3y^2 + xy^3 - x^3 - xy - x^2 - y = 0$ at the point $(\frac{1}{3}, 2)$.

9. Find dy if $x = -1$, $y = 0$, and $dx = \frac{1}{3}$, where x and y satisfy the relation in Exercise 8.

10. Find d^2y/dx^2 at $x = -1$, $y = 1$ if $x^3 + y^3 + xy = -1$.

11. Find d^2x/dy^2 at $x = 1$, $y = 0$ if $e^{xy} + e^x + e^y = 2 + e$.

(12-15) Use logarithmic differentiation to find dy/dx for the following functions.

12. $y = (x + 1)(x^3 - 3)^{1/2}(2 - x^2)^{-1/4}$

13. $y = \left[\dfrac{x^4 - 4}{x(x + 1)^2}\right]^{1/3}$

14. $y = x^{x^{-2}}$

15. $y = x^{1/(\ln x)}$

16. (Elasticity of Demand) If the demand relation is $x = 1000 - 50p$, find the elasticity of demand when: **a.** $p = 5$ **b.** $p = 10$ **c.** $p = 15$

17. (Elasticity of Demand) For the demand relation $x/600 + p/12 = 1$, find the values of p for which: **a.** $\eta = -1$ **b.** $\eta = -2$ **c.** $\eta = -\frac{1}{3}$.

18. (Elasticity) For the demand relation $x = k(1 - p - p^2)$, find the value of p for which $\eta = -1$. Find the values of p for which demand is: **a.** elastic; **b.** inelastic.

19. (Elasticity) If the demand relation is $x^2 + p^2 = 25$, find the elasticity of demand when:

    a. $p = 3$ **b.** $p = 4$

20. (Elasticity) For the demand relation $x^2 + px + p^2 = 7$, find the elasticity of demand when $p = 2$.

*21. (Elasticity) If the marginal revenue of a product at a certain level is \$25 and the elasticity of demand at this price is $\eta = -2$, determine the average revenue, that is, the price p.

22. (Elasticity) The demand function of a certain product is $p^2 + x^2 = 6100$, where x units are sold at a price of p dollars each. Will a fall in demand increase or decrease the total revenue at a price level of:

    a. \$50? **b.** \$60?

23. *(Elasticity)* For the demand function $p = f(x)$, show that at a level of production that maximizes the total revenue, the elasticity of demand is -1.

24. *(Elasticity)* The demand function of a certain product is $p + 10x = 5000$. Find the demand level at which it makes no difference to the revenue whether the price p is increased or decreased.

25. Find dy when $x^y = e^x$.

26. Find du if $u^x = x^u$.

CHAPTER 16

Integration

Chapter Objectives

16-1 ANTIDERIVATIVES

(a) Definition and notation for indefinite integrals (antiderivatives);
(b) The power formula for integrals and its application;
(c) The integral of the inverse first power and the natural exponential;
(d) Theorems on the integral of a constant times a function and the sum of two or more functions;
(e) Calculation of cost and revenue functions from marginal cost and marginal revenue.

16-2 METHOD OF SUBSTITUTION

(a) The transformation of an integral by the substitution $u = g(x)$;
(b) The evaluation of integrals by means of an appropriate substitution;
(c) Special case of linear substitutions.

16-3 TABLES OF INTEGRALS

(a) Elementary use of a table of integrals;
(b) Cases where the table must be used more than once;
(c) Substitution combined with use of a table of integrals.

16-4 INTEGRATION BY PARTS

(a) The basic formula for integration by parts;
(b) Application to the case of a polynomial times an exponential function;
(c) Application to the case of a polynomial times a logarithm function;
(d) Cases where integration by parts must be used more than once.

CHAPTER REVIEW

◣ 16-1 ANTIDERIVATIVES

So far in our study of calculus, we have been concerned with the process of differentiation—that is, the calculation and use of the derivatives of functions. This part of the subject is called **differential calculus.** We shall now turn to the second area of study within the general area of calculus, called **integral calculus,** in which we are concerned with the process opposite to differentiation.

We have seen before that, if $s(t)$ is the distance traveled in time t by a moving object, then the instantaneous velocity is $v(t) = s'(t)$, the derivative of $s(t)$. In order to calculate v, we simply differentiate $s(t)$. However, it may happen that we already know the velocity function $v(t)$ and wish to calculate the distance s traveled. In such a situation, we know the derivative $s'(t)$ and require the function $s(t)$, a step opposite to that of differentiation. As another example, we may be concerned with a cost model in which the marginal cost is a known function of production level, and we may wish to calculate the total cost of producing x items. Or, we might know the rate of production of an oil well as a function of time and wish to calculate the total production over a certain period of time.

The process of finding the function when its derivative is given is called **integration,** and the function to be found is called the **antiderivative** or the **integral** of the given function.

To evaluate the antiderivative of some function $f(x)$, we must find a function $F(x)$ whose derivative is $f(x)$. For example, suppose that $f(x) = 3x^2$. Since we know that $(d/dx)(x^3) = 3x^2$, we conclude that we can take $F(x) = x^3$. An antiderivative of $3x^2$ is therefore x^3.

However, this answer is not unique: the functions $x^3 + 4$ and $x^3 - 2$ also have derivative equal to $3x^2$. In fact, for any constant C, $x^3 + C$ has derivative $3x^2$, and therefore $x^3 + C$ is an antiderivative of $3x^2$ for any C. The constant C, which can have any arbitrary value is called the **constant of integration.**

The nonuniqueness aspect is common to all antiderivatives: Any constant can be added to them without destroying their property of being the antiderivative of a given function. However, this is the only ambiguity that there is: if $F(x)$ is any antiderivative of $f(x)$, then any other antiderivative of $f(x)$ differs from $F(x)$ only by a constant. Therefore if $F'(x) = f(x)$, then the general antiderivative of $f(x)$ is given by $F(x) + C$, where C is an arbitrary constant. ◣ 1

Since the constant of integration is arbitrary—that is, it can be any real number—the integral so obtained is given the more complete name **indefinite integral.** Sometimes different methods of evaluating an integral can give different forms for the answer, but it will always be the case that the two answers differ only by a constant.

The expression

$$\int f(x)\, dx$$

◣ **1.** (a) What is the antiderivative of $2x$?
(b) Of what function is $\ln x$ an antiderivative?

Answer (a) $x^2 + C$ where C is an arbitrary constant; (b) x^{-1}.

► **2.** Find (a) $\int x^4\,dx$;

(b) $\int x^{1/2}\,dx$.

is used to denote an arbitrary member of the set of antiderivatives of f. It is read as the integral of $f(x)$, dx. In such an expression, the function $f(x)$ is called the **integrand** and the symbol $\int$ is the **integral sign.** The symbol

$$\int \ldots dx$$

stands for *integral, with respect to x, of* . . . . It is the inverse of the symbol

$$\frac{d}{dx} \ldots$$

which means *derivative, with respect of x, of* . . . . The integral sign and dx go together; the integral sign means operation of integration and the dx specifies that the *variable of integration* is x. The integrand is always put between the integral sign and the differential of the variable integration.

If $F(x)$ is one particular antiderivative of $f(x)$, then

$$\int f(x)\,dx = F(x) + C$$

where C is an arbitrary constant. For example,

$$\int 3x^2\,dx = x^3 + C. \tag{1}$$

From the definition of the integral, it is clear that

$$\frac{d}{dx}\left[\int f(x)\,dx\right] = f(x).$$

That is, the operation of differentiation neutralizes the effect of the operation of integration.

We shall establish a number of simple and standard formulas for integration. The first of these is known as the **power formula;** it gives us the rule to integrate any power of x except the inverse of x.

Consider first $\int x^2\,dx$. We must look for a function whose derivative is x^2. As we saw above, the derivative of x^3 is $3x^2$. Therefore, the derivative of $\frac{1}{3}x^3$ is $\frac{1}{3}(3x^2) = x^2$. Thus $\int x^2\,dx = \frac{1}{3}x^3 + C$.

Next, consider $\int x^3\,dx$, which means a function whose derivative is x^3. But the derivative of x^4 is $4x^3$ and therefore, the derivative of $\frac{1}{4}x^4$ is $\frac{1}{4}(4x^3) = x^3$. Thus $\int x^3\,dx = \frac{1}{4}x^4 + C$. ► **2**

It is now easy to see how this generalizes:

$$\int x^n\,dx = \frac{x^{n+1}}{n+1} + C \quad (n \neq -1) \qquad \text{(Power Formula)}$$

Answer (a) $\frac{1}{5}x^5 + C$;
(b) $\frac{2}{3}x^{3/2} + C$.

Thus *to integrate any power of x except the inverse first power, we must increase the power by 1, then divide by the new exponent* and, finally, add the arbitrary constant of integration.

This formula is obtained by reversing the corresponding formula for differentiation. We observe that

$$\frac{d}{dx}\left(\frac{x^{n+1}}{n+1}\right) = \frac{1}{n+1}\frac{d}{dx}(x^{n+1}) = \frac{1}{n+1}(n+1)x^n = x^n.$$

Therefore, since the derivative of $x^{n+1}/(n+1)$ is x^n, an antiderivative of x^n must be $x^{n+1}/(n+1)$. The general antiderivative is then obtained by adding the constant of integration.

EXAMPLE 1

(a) $\displaystyle\int x^3\,dx = \frac{x^{3+1}}{3+1} + C = \frac{x^4}{4} + C$ $(n = 3)$

(b) $\displaystyle\int \frac{1}{x^2}\,dx = \int x^{-2}\,dx = \frac{x^{-2+1}}{-2+1} + C = \frac{x^{-1}}{-1} + C$

         $= -\dfrac{1}{x} + C$ $(n = -2)$

(c) $\displaystyle\int \frac{1}{\sqrt{t}}\,dt = \int t^{-1/2}\,dt = \frac{t^{-1/2+1}}{\left(-\frac{1}{2}+1\right)} + C = 2\sqrt{t} + C$ $(n = -\frac{1}{2})$

(d) $\displaystyle\int dx = \int 1\,dx = \int x^0\,dx = \frac{x^{0+1}}{0+1} + C = x + C$ $(n = 0)$

☛ 3

☛ **3.** Using the power formula, find
(a) $\displaystyle\int x^{-4}\,dx$;   (b) $\displaystyle\int u^{3/4}\,du$.

A number of formulas giving antiderivatives of simple functions are given in Table 1. Each formula is stated a second time with the variable changed from x to u. All these results are obtained by simply reversing corresponding results for derivatives. Formula 2 requires some comment. For $x > 0$, this formula is straightforward, since then $|x| = x$, and we know that

$$\frac{d}{dx}\ln x = \frac{1}{x}.$$

TABLE 1   *Standard Elementary Integrals*

1. $\displaystyle\int x^n\,dx = \frac{x^{n+1}}{n+1} + C \quad (n \neq -1)$ or $\displaystyle\int u^n\,du = \frac{u^{n+1}}{n+1} + C$	
2. $\displaystyle\int \frac{1}{x}\,dx = \ln|x| + C$ or $\displaystyle\int \frac{1}{u}\,du = \ln|u| + C$	
3. $\displaystyle\int e^x\,dx = e^x + C$ or $\displaystyle\int e^u\,du = e^u + C$	

Answer
(a) $-\frac{1}{3}x^{-3} + C$;   (b) $\frac{4}{7}u^{7/4} + C$.

Since $1/x$ is the derivative of $\ln x$, it follows that the antiderivative of $1/x$ must be $\ln x$, plus the constant of integration.

When $x < 0$, we have $|x| = -x$. Therefore

$$\frac{d}{dx}\ln|x| = \frac{d}{dx}\ln(-x) = \frac{1}{(-x)}(-1) = \frac{1}{x},$$

where the chain rule has been used in carrying out the differentiation. Thus $1/x$ is the derivative of $\ln|x|$ for $x < 0$, as well as for $x > 0$. Therefore the antiderivative of $1/x$ must be $\ln|x| + C$, as given in the table, for all $x \neq 0$.

Now we shall prove two theorems that will simplify the algebra of integration.

THEOREM 1 The integral of the product of a constant and a function of x is equal to the constant times the integral of the function. That is, if c is a constant,

$$\int c f(x)\, dx = c \int f(x)\, dx.$$

EXAMPLE 2

(a) $\displaystyle\int 3x^2\, dx = 3\int x^2\, dx = 3 \cdot \frac{x^3}{3} + C = x^3 + C$

(b) $\displaystyle\int 2e^x\, dx = 2\int e^x\, dx = 2e^x + C$

(c) $\displaystyle\int 5\, dx = 5\int 1\, dx = 5x + C$ ☛ 4

_____

☛ **4.** Find (a) $\displaystyle\int \frac{2}{x}\, dx$;

(b) $\displaystyle\int 4\sqrt[3]{t}\, dt$.

PROOF OF THEOREM 1 We have

$$\frac{d}{dx}\left[c \int f(x)\, dx \right] = c\frac{d}{dx}\left[\int f(x)\, dx \right] = c f(x).$$

Therefore, $cf(x)$ is the derivative of $c \int f(x)\, dx$, and so from the definition of the antiderivative, it follows that $c \int f(x)\, dx$ must be the antiderivative of $cf(x)$. In other words,

$$\int c f(x)\, dx = c \int f(x)\, dx$$

which proves the result.

From this theorem, it follows that we can move any multiplicative constant across the integral sign.

Answer (a) $2\ln|x| + C$; (b) $3t^{4/3} + C$.

Caution Variables must not be moved across the integral sign. For example,

$$\int xe^{-x}\, dx \neq x \int e^{-x}\, dx.$$

THEOREM 2 The integral of the sum of two functions is equal to the sum of their integrals.

$$\int [f(x) + g(x)]\, dx = \int f(x)\, dx + \int g(x)\, dx$$

Note This result may be extended to the difference of two functions or an algebraic sum of any finite number of functions.

☞ **5.** Find (a) $\displaystyle\int \frac{2x^2 + 3}{x}\, dx$;

(b) $\displaystyle\int (1 + \sqrt{v})^2\, dv.$

EXAMPLE 3 Find the integral of $(x - 3/x)^2$.

Solution We expand $(x - 3/x)^2$ to express the integrand as a sum of power functions.

$$\int \left(x - \frac{3}{x}\right)^2 dx = \int \left(x^2 - 6 + \frac{9}{x^2}\right) dx$$

$$= \int x^2\, dx - \int 6\, dx + \int 9x^{-2}\, dx$$

$$= \int x^2\, dx - 6\int 1\, dx + 9\int x^{-2}\, dx$$

$$= \frac{x^{2+1}}{2 + 1} - 6x + 9\frac{x^{-2+1}}{-2 + 1} + C$$

$$= \frac{x^3}{3} - 6x - \frac{9}{x} + C$$

EXAMPLE 4 Find the antiderivative of $\dfrac{3 - 5t + 7t^2 + t^3}{t^2}$.

Solution

$$\int \frac{3 - 5t + 7t^2 + t^3}{t^2}\, dt = \int \left(\frac{3}{t^2} - \frac{5}{t} + 7 + t\right) dt$$

$$= 3\int t^{-2}\, dt - 5\int \frac{1}{t}\, dt + 7\int 1\, dt + \int t\, dt$$

$$= 3\frac{t^{-2+1}}{-1} - 5\ln|t| + 7t + \frac{t^{1+1}}{2} + C$$

$$= -\frac{3}{t} - 5\ln|t| + 7t + \frac{t^2}{2} + C \qquad ☞ 5$$

Answer (a) $x^2 + 3\ln|x| + C$
(b) $v + \frac{4}{3}v^{3/2} + \frac{1}{2}v^2 + C.$

PROOF OF THEOREM 2

$$\frac{d}{dx}\left[\int f(x)\,dx + \int g(x)\,dx\right] = \frac{d}{dx}\left[\int f(x)\,dx\right] + \frac{d}{dx}\left[\int g(x)\,dx\right]$$

$$= f(x) + g(x)$$

Therefore $f(x) + g(x)$ is the derivative of $\int f(x)\,dx + \int g(x)\,dx$, and so by the definition of the antiderivative,

$$\int [f(x) + g(x)]\,dx = \int f(x)\,dx + \int g(x)\,dx.$$

EXAMPLE 5 Find $f(x)$ if $f'(x) = (x^2 + 1)(4x - 3)$ and $f(1) = 5$.

Solution Expanding the parentheses, we get $f'(x) = 4x^3 - 3x^2 + 4x - 3$. Using the theorems above, the antiderivative of this is

$$f(x) = 4(\tfrac{1}{4}x^4) - 3(\tfrac{1}{3}x^3) + 4(\tfrac{1}{2}x^2) - 3x + C$$

$$= x^4 - x^3 + 2x^2 - 3x + C$$

where C is an unknown constant. But in this case we are given the information that $f(1) = 5$, and this allows us to determine the value of C. For, $f(1) = 1^4 - 1^3 + 2 \cdot 1^2 - 3 \cdot 1 + C = C - 1 = 5$. Therefore, $C = 6$, and so

$$f(x) = x^4 - x^3 + 2x^2 - 3x + 6. \quad \blacktriangleright 6$$

☛ **6.** Find $g(x)$ if $g'(x) = 1 - 2x$ and $g(0) = 4$.

EXAMPLE 6 *(Extra Cost of Production)* A company is at present producing 150 units per week of its product. From its past experience, it is known that the cost of producing the xth unit per week (that is, the marginal cost) is given by

$$C'(x) = 25 - 0.02x.$$

Assuming that this marginal cost continues to apply, find the extra cost per week that would be involved in raising the output from 150 to 200 units per week.

Solution The marginal cost is the derivative of the cost function. Therefore the cost function is obtained by integrating the marginal cost function.

$$C(x) = \int C'(x)\,dx = \int (25 - 0.02x)\,dx$$

$$= 25x - (0.02)\frac{x^2}{2} + K = 25x - 0.01x^2 + K$$

where K is the constant of integration. We do not have enough information to determine the value of K. However, we want to calculate the increase in cost resulting from increasing x from 150 to 200—that is, $C(200) - C(150)$.

$$C(200) = 25(200) - 0.01(200)^2 + K = 4600 + K$$

$$C(150) = 25(150) - 0.01(150)^2 + K = 3525 + K$$

Answer $g(x) = x - x^2 + 4$.

7. Find the cost function, $C(x)$, in dollars, given that the marginal cost is $C'(x) = 200 + 2x - 0.003x^2$ and the fixed costs are $22,000.

and therefore

$$C(200) - C(150) = (4600 + K) - (3525 + K) = 1075.$$

The increase in weekly cost would therefore be $1075. Notice that the unknown constant K is not in the final answer. ☛ 7

EXAMPLE 7 *(Revenue and Demand)* The marginal revenue of a firm is given by

$$R'(x) = 15 - 0.01x.$$

(a) Find the revenue function.

(b) Find the demand relation for the firm's product.

Solution

(a) The revenue function $R(x)$ is the integral of the marginal revenue function. Thus

$$R(x) = \int R'(x)\,dx = \int (15 - 0.01x)\,dx$$

$$= 15x - 0.01\frac{x^2}{2} + K = 15x - 0.005x^2 + K$$

where K is the constant of integration. To determine K, we use the fact that the revenue must be zero when no units ares sold. That is, when $x = 0$, $R = 0$. Putting $x = 0$ and $R = 0$ into our expression for $R(x)$, we get

$$0 = 15(0) - 0.005(0^2) + K$$

which gives $K = 0$. Therefore the revenue function is

$$R(x) = 15x - 0.005x^2.$$

(b) If each item the firm produces sells at a price p, then the revenue obtained by selling x items is given by $R = px$. Thus

$$px = 15x - 0.005x^2 \qquad \text{or} \qquad p = 15 - 0.005x$$

which is the required demand relation.

Answer
$C(x) = 22{,}000 + 200x + x^2 - 0.001x^3.$

EXERCISES 16-1

(1–52) Write the integrals of the following.

1. x^7

2. $\sqrt[3]{x}$

3. $1/x^3$

4. $1/\sqrt{x}$

5. $7x$

6. $\ln 2$

7. $\dfrac{e^3}{x}$

8. $x \ln 3$

9. $\dfrac{1}{x \ln 2}$

10. $3x + \dfrac{1}{3x}$

11. $\dfrac{e}{x} + \dfrac{x}{e}$

12. $xe^{-2} + ex^{-2}$

13. $(e^2 - 2^e)e^x$

14. $\sqrt{3x}$

15. $\dfrac{\ln 2}{x^2}$

16. exe^{+1}

17. $x^7 + 7x + \dfrac{7}{x} + 7$

18. $e^x + x^e + e + x$

19. $7x^2 - 3x + 8 + \dfrac{1}{x} + \dfrac{2}{x^2}$

20. $3x^2 - 5x + \dfrac{7}{x} + 2e^3$

21. $(x + 2)(x + 3)$

22. $(x - 2)(2x + 3)$

23. $(x + 1)(3x - 2)$

24. $(x + 3)(2x - 1)$

25. $(x + 2)^2$

26. $(2x - 3)^2$

27. $\left(x + \dfrac{1}{x}\right)^2$

*28. $\left(x - \dfrac{1}{x}\right)^3$

29. $\left(2x - \dfrac{3}{x}\right)^2$

30. $x^2(x + 1)^2$

31. $x^2\left(x + \dfrac{2}{\sqrt{x}}\right)$

32. $\left(\sqrt{x} + \dfrac{3}{\sqrt{x}}\right)^2$

33. $x^3(x + 1)(x + 2)$

34. $\left(\sqrt{x} + \dfrac{1}{\sqrt{x}}\right)\left(x - \dfrac{2}{x}\right)$

35. $(x + 2)\left(3x - \dfrac{1}{x}\right)$

36. $\dfrac{(x + 2)(x + 3)}{x^2}$

*37. $\dfrac{\ln x^3}{\ln x^2}$

*38. $\dfrac{\ln x^2}{\ln x}$

*39. $\dfrac{\ln x}{\ln \sqrt{x}}$

40. $e^x \ln 3$

41. $\dfrac{e^x}{\ln 2}$

*42. $\dfrac{e^{x+2}}{e^{x+1}}$

43. $e^{\ln(x^2+1)}$

44. $e^{3\ln x}$

45. $(\sqrt{x} + 3)^2 \, dx$

46. $\dfrac{3\sqrt{x} + 7}{\sqrt[3]{x}} \, dx$

47. $\dfrac{3x^4 - 12}{x^2 + 2} \, dx$

48. $e^{2\ln x} \, dx$

49. $x \, e^{\ln(x+1)} \, dx$

50. $\dfrac{4 - x}{\sqrt{x} + 2} \, dx$

51. $\dfrac{2x - 18}{\sqrt{x} + 3} \, dx$

52. $\dfrac{x - 8}{2 - \sqrt[3]{x}} \, dx$

(53–58) Find the antiderivatives of the following functions with respect to the independent variable involved.

53. $4x^3 + 3x^2 + 2x + 1 + \dfrac{1}{x} + \dfrac{1}{x^3}$

54. $3e^t - 5t^3 + 7 + \dfrac{3}{t}$

55. $\sqrt{u}(u^2 + 3u + 7)$

56. $\dfrac{2y^3 + 7y^2 - 6y + 9}{3y}$

57. $\sqrt{x}(x + 1)(2x - 1)$

58. $\dfrac{(t - t^2)^2}{t\sqrt{t}}$

(59–62) Evaluate the following integrals.

59. $\displaystyle\int \dfrac{1 + 3x + 7x^2 - 2x^3}{x^2} \, dx$

60. $\displaystyle\int \dfrac{(2t + 1)^2}{3t} \, dt$

61. $\displaystyle\int \left(3\theta^2 - 6\theta + \dfrac{9}{\theta} + 4e^\theta\right) d\theta$

62. $\displaystyle\int (\sqrt{2}y + 1)^2 \, dy$

63. Find $f(x)$ if $f'(x) = (x + 2)(2x - 3)$ and $f(0) = 7$.

64. Find $f(e)$ if $f'(t) = \dfrac{2t + 3}{t}$ and $f(1) = 2e$.

65. *(Velocity and Distance)* The velocity of motion at time t is $(t + \sqrt{t})^2$. Find the distance traveled in time t.

66. *(Acceleration)* The acceleration of a moving object at time t is $3 + 0.5t$.

  a. Find the velocity at any time t if the initial velocity at $t = 0$ is given to be 60 units.

  b. Find the distance traveled by the object at time t if the distance is zero when $t = 0$.

67. *(Marginal Cost)* The marginal cost function of a firm is $C'(x) = 30 + 0.05x$.

a. Determine the cost function, $C(x)$, if the firm's fixed costs are $2000 per month.

b. How much will it cost to produce 150 units in a month?

c. If the items can be sold at $55 each, how many should be produced to maximize the profit. (*Hint:* See page 600.)

68. *(Marginal Cost)* The marginal cost of a certain firm is given by $C'(x) = 24 - 0.03x + 0.006x^2$. If the cost of producing 200 units is $22,700, find:

a. The cost function;

b. The fixed costs of the firm;

c. The cost of producing 500 units.

d. If the items can be sold at $90 each, determine the production level that maximizes the profit.

69. *(Marginal Cost)* The marginal cost of ABC Products Limited is $C'(x) = 3 + 0.001x$ and the cost of manufacturing 100 units is $1005. What is the cost of producing 200 units? The items sell for $5 each. Determine the increment in profit if the sales volume is increased from 1000 to 2000.

70. *(Marginal Cost)* The marginal cost of a certain firm is $C'(x) = 5 + 0.002x$. What are the total variable costs of manufacturing x units?

71. *(Marginal Revenue)* The marginal revenue function of a certain firm is
$$R'(x) = 4 - 0.01x.$$

a. Determine the revenue obtained by selling x units of its product.

b. What is the demand function for the firm's product?

72. *(Marginal Revenue)* The marginal revenue function of a certain firm is
$$R'(x) = 20 - 0.02x - 0.003x^2.$$

a. Find the revenue function.

b. How much revenue will be obtained by selling 100 units of the firm's product?

c. What is the demand function for the firm's product?

73. *(Marginal Profit)* The marginal profit function of a firm is $P'(x) = 5 - 0.002x$ and the firm made a profit of $310 when 100 units were sold. What is the profit function of the firm?

*74. *(Water Consumption)* In a certain city during summer the consumption of water (millions of gallons per hour) is given by the following function.

$$f(t) = \begin{cases} 1 & \text{if } 0 \leq t < 6 \\ t - 5 & \text{if } 6 \leq t < 9 \\ 4 & \text{if } 9 \leq t < 21 \\ 25 - t & \text{if } 21 \leq t < 24 \end{cases}$$

where t is time in hours during the day (24-hour clock). Determine the total consumption between 6 A.M. and 9 A.M. and the total consumption during the whole day.

*75. *(Telephone Demand)* During business hours (8 A.M.– 5 P.M.) the number of telephone calls per minute passing through a switchboard vary according to the formula

$$f(t) = \begin{cases} 5t & \text{if } 0 \leq t < 1 \\ 5 & \text{if } 1 \leq t < 4 \\ 0 & \text{if } 4 \leq t < 5 \\ 3 & \text{if } 5 \leq t < 8 \\ 27 - 3t & \text{if } 8 \leq t < 9 \end{cases}$$

where t is time in hours measured from 8 A.M. Calculate the total number of calls during the business day. How many calls are there between 8 and 11 A.M.?

76. *(Population Growth)* A population of insects grows from an initial size of 3000 to a size $p(t)$ after time t (measured in days). If the growth rate at time t is $5(t + 2t^2)$, determine $p(t)$ and $p(10)$.

◣ 16-2 METHOD OF SUBSTITUTION

Not all integrals can be evaluated directly by the use of the standard integrals and theorems discussed in the preceding section. Many times, however, the given integral can be reduced to a standard integral already known by a change of the variable of integration. Such a method is called the **method of substitution** and corresponds to the chain rule in differentiation.

Suppose F is an antiderivative of f, so that

$$\int f(x)\, dx = F(x) + C.$$

In this equation we can change the name of the variable from x to u:

$$\int f(u)\, du = F(u) + C.$$

Now the basic theorem of the method of substitution states that we can replace u by $g(x)$, where g is any differentiable function, not identically constant, and this equation remains true. In this replacement, du is treated as a differential, in other words, $du = g'(x)\, dx$. Thus we have:

THEOREM 1 If $\int f(u)\, du = F(u) + C$, then

$$\int f[g(x)]g'(x)\, dx = F[g(x)] + C$$

for any differentiable function g that is not a constant function.

Let us illustrate this theorem with some examples before proving it. We start with the power formula

$$\int u^n\, du = \frac{u^{n+1}}{n+1} + C \qquad (n \neq -1)$$

which corresponds to taking $f(u) = u^n$ and $F(u) = u^{n+1}/(n+1)$. Then according to Theorem 1 we must replace the argument u in these two functions by $g(x)$:

$$f[g(x)] = [g(x)]^n \qquad \text{and} \qquad F[g(x)] = \frac{[g(x)]^{n+1}}{n+1}.$$

In this particular case, the theorem then states that

$$\int [g(x)]^n\, g'(x)\, dx = \frac{[g(x)]^{n+1}}{n+1} + C \qquad (n \neq -1).$$

In this result, $g(x)$ can be any differentiable function that is not constant. For example, let us take $g(x) = x^2 + 1$ and $n = 4$. Then $g'(x) = 2x$ and we obtain

$$\int (x^2 + 1)^4 \cdot 2x\, dx = \frac{(x^2 + 1)^{4+1}}{4+1} + C.$$

After division by 2, this becomes

$$\int (x^2 + 1)^4 x\, dx = \frac{(x^2 + 1)^5}{10} + C_1$$

8. State the results that are obtained from the power formula by taking (a) $g(x) = x^2 + 1$ and $n = \frac{1}{2}$; (b) $g(x) = \ln x$ and $n = -2$.

where $C_1 = C/2$. (Note that C_1 can still be any arbitrary constant, since dividing by 2 does not remove the arbitrariness.)

As a further example, let us take $g(x) = \ln x$ and $n = 2$. Since $g'(x)$ then is $1/x$, we get the result

$$\int \frac{(\ln x)^2}{x} \, dx = \frac{(\ln x)^3}{3} + C. \qquad \text{☛ 8}$$

It is clear that by choosing different functions $f(x)$ and $g(x)$, a great variety of different integrals can be evaluated. When actually using this substitution method to evaluate a given integral, it is necessary to spot how to choose these functions in such a way that the given integral is expressed in the form $\int f(u) \, du$ when we substitute $u = g(x)$, with the function f being a sufficiently simple function that the new integral can be easily evaluated. We shall elaborate on this later, but first let us pause to prove the theorem.

PROOF OF THEOREM 1 Set $u = g(x)$. Then from the chain rule,

$$\frac{d}{dx} F[g(x)] = \frac{d}{dx} F(u) = \frac{d}{du} F(u) \cdot \frac{du}{dx} = f(u) g'(x) = f[g(x)] g'(x).$$

Therefore, from the definition of the antiderivative, it follows that

$$\int f[g(x)] g'(x) \, dx = F[g(x)] + C,$$

as required.

EXAMPLE 1 Evaluate $\int (x^2 + 3x - 7)^5 (2x + 3) \, dx$.

Solution We observe that the differential of $x^2 + 3x - 7$ is equal to $(2x + 3) \, dx$, which appears in the integral. Therefore we set $x^2 + 3x - 7 = u$. Then $(2x + 3) \, dx = du$. Using this substitution, the given integral reduces to

$$\int (x^2 + 3x - 7)^5 (2x + 3) \, dx = \int u^5 \, du = \frac{u^6}{6} + C$$

$$= \frac{1}{6} (x^2 + 3x - 7)^6 + C$$

where we have substituted the value of u again.

EXAMPLE 2 Evaluate $\int \frac{1}{x \ln x} \, dx$.

Solution The given integral is

$$\int \frac{1}{x \ln x} \, dx = \int \frac{1}{\ln x} \cdot \frac{1}{x} \, dx.$$

Answer

(a) $\displaystyle\int \sqrt{x^2 + 1} \cdot 2x \, dx$

$= \dfrac{2}{3} (x^2 + 1)^{3/2} + C;$

(b) $\displaystyle\int \frac{1}{x (\ln x)^2} \, dx = -\frac{1}{\ln x} + C.$

Note that we have separated the integrand in such a way that the combination $(1/x) \, dx$ occurs as a distinct factor. This combination is the differential of $\ln x$, and moreover the rest of the integrand is also a simple function of $\ln x$. So we let $\ln x = u$. Then $(1/x) \, dx = du$. The given integral now reduces to

$$\int \frac{1}{x \ln x} \, dx = \int \frac{1}{\ln x} \cdot \frac{1}{x} dx = \int \frac{1}{u} \cdot du = \ln |u| + C$$

$$= \ln |\ln x| + C$$

after substituting $u = \ln x$.

We observe from these examples that the appropriate technique in using the substitution method is to *look for a function* $u = g(x)$ *with a differential* $g'(x) \, dx$ *that occurs in the original integral. The rest of the integrand should be a simple function of* u. The choice of substitution is by nature ambiguous, but you will soon learn from experience to spot the right one to make.

EXAMPLE 3 Evaluate $\int e^{x^2 - 5x}(2x - 5) \, dx$.

Solution We observe that $(2x - 5) \, dx$ occurs in the integral, and this quantity is the differential of $x^2 - 5x$. Therefore we set $u = x^2 - 5x$. Then $du = (2x - 5) \, dx$ and the integral becomes

$$\int e^{x^2 - 5x}(2x - 5) \, dx = \int e^u \, du = e^u + C = e^{x^2 - 5x} + C.$$

Sometimes the appropriate exact differential may not appear in the integral itself, but the function that does appear must be multiplied or divided by a certain constant. This is illustrated by the following examples.

EXAMPLE 4 Evaluate $\int x^2 e^{x^3 + 1} \, dx$.

Solution The derivative of $x^3 + 1$ is $3x^2$. Since the combination $x^2 \, dx$ occurs in the integrand, this suggests setting $u = x^3 + 1$. Then $du = 3x^2 \, dx$, and so $x^2 \, dx = \frac{1}{3} du$. Thus

$$\int x^2 e^{x^3 + 1} \, dx = \int e^u (\tfrac{1}{3} du) = \tfrac{1}{3} \int e^u \, du = \tfrac{1}{3} e^u + C = \tfrac{1}{3} e^{x^3 + 1} + C. \quad \text{☞ 9}$$

EXAMPLE 5 Evaluate $\int \sqrt{2x + 3} \, dx$.

Solution Writing $u = 2x + 3$, we find that $du = 2dx$, that is, $dx = \frac{1}{2} du$.

Then

$$\int \sqrt{2x + 3} \; dx = \int \sqrt{u} \cdot \tfrac{1}{2} du = \tfrac{1}{2} \int u^{1/2} \; du$$

$$= \tfrac{1}{2} \cdot \tfrac{2}{3} u^{3/2} + C = \tfrac{1}{3}(2x + 3)^{3/2} + C.$$

Example 5 is an example of a special type of substitution called a **linear substitution.** In Theorem 1 let us choose $u = ax + b$, where a and b are constants $(a \neq 0)$. That is, $g(x) = ax + b$ and $g'(x) = a$. The statement of the theorem then becomes

$$\int f(ax + b) \cdot a \; dx = F(ax + b) + C_1.$$

Dividing through by a and denoting $C_1/a = C$, we have the following:

THEOREM 2

> If $\int f(x) \; dx = F(x) + C$, then $\int f(ax + b) \; dx = \dfrac{1}{a} F(ax + b) + C$

where a and b are any two constants $(a \neq 0)$. In other words, in order to integrate $f(ax + b)$, we treat $(ax + b)$ as if it were a single variable, then divide the resulting integral by a, the coefficient of x.

Theorem 2 is a powerful tool and can be used to generalize each integral in Table 1 (see Section 16-1) by replacing x by $ax + b (a \neq 0)$. This leads to the types of integrals listed in Table 2.

TABLE 2

1. $\int x^n \; dx = \dfrac{x^{n+1}}{n + 1} + C \quad (n \neq -1)$	1. $\int (ax + b)^n \; dx = \dfrac{1}{a} \cdot \dfrac{(ax + b)^{n+1}}{n + 1} + C$
	$(a \neq 0, n \neq -1)$
2. $\int \dfrac{1}{x} \; dx = \ln |x| + C$	2. $\int \dfrac{1}{ax + b} \; dx = \dfrac{1}{a} \cdot \ln |ax + b| + C \; (a \neq 0)$
3. $\int e^x \; dx = e^x + C$	3. $\int e^{ax+b} \; dx = \dfrac{e^{ax+b}}{a} + C \quad (a \neq 0)$

EXAMPLE 6 Evaluate $\int (3x - 7)^5 \; dx.$

Solution From the first general result in Table 2,

$$\int (ax + b)^n \; dx = \dfrac{(ax + b)^{n+1}}{a(n + 1)} + C.$$

We must set $a = 3$, $b = -7$, and $n = 5$ in this general formula in order to evaluate the required integral.

$$\int (3x - 7)^5\, dx = \frac{(3x - 7)^{5+1}}{(3)(5 + 1)} + C = \frac{1}{18}(3x - 7)^6 + C.$$

EXAMPLE 7 Evaluate $\displaystyle\int e^{5-3x}\, dx.$

Solution Setting $a = -3$ and $b = 5$ in Formula 3 of the table, we obtain

$$\int e^{5-3x}\, dx = \frac{e^{5-3x}}{(-3)} + C = -\frac{1}{3}e^{5-3x} + C.$$

EXAMPLE 8 Evaluate $\displaystyle\int x\sqrt{1 - x}\, dx.$

Solution This example can again be solved by a linear substitution, though it is not quite as straightfoward as Examples 6 and 7. We set $u = 1 - x$, so $du = -dx$. The factor $\sqrt{1 - x}$ in the integrand becomes $\sqrt{u}$ while the factor $x = 1 - u$. Thus

$$\int x\sqrt{1 - x}\, dx = \int (1 - u)\sqrt{u}(-du) = -\int (u^{1/2} - u^{3/2})\, du$$

$$= -\frac{2}{3}u^{3/2} + \frac{2}{5}u^{5/2} + C$$

$$= -\frac{2}{3}(1 - x)^{3/2} + \frac{2}{5}(1 - x)^{5/2} + C. \quad ▪ 10$$

EXERCISES 16-2

(1–14) Make use of a linear substitution or use of Theorem 1 to evaluate the following integrals.

1. $\displaystyle\int (2x + 1)^7\, dx$

2. $\displaystyle\int \sqrt{3x - 5}\, dx$

3. $\displaystyle\int \frac{1}{(2 - 5t)^2}\, dt$

4. $\displaystyle\int \frac{1}{\sqrt{5 - 2x}}\, dx$

5. $\displaystyle\int \frac{1}{2y - 1}\, dy$

6. $\displaystyle\int \frac{1}{1 - 3t}\, dt$

7. $\displaystyle\int \frac{2u - 1}{4u^2 - 1}\, du$

8. $\displaystyle\int \frac{2x + 3}{9 - 4x^2}\, dx$

9. $\displaystyle\int e^{3x+2}\, dx$

10. $\displaystyle\int e^{5-2x}\, dx$

11. $\displaystyle\int \frac{e^5}{e^x}\, dx$

12. $\displaystyle\int \frac{e^{2x}}{e^{5-x}}\, dx$

13. $\displaystyle\int \frac{e^{2x+3}}{e^{1-x}}\, dx$

14. $\displaystyle\int \left(\frac{e^3}{e^{x-1}}\right)^2 dx$

(15–64) Use an appropriate substitution to evaluate the following antiderivatives.

15. $\displaystyle\int (x^2 + 7x + 3)^4(2x + 7)\, dx$

16. $\displaystyle\int (x + 2)(x^2 + 4x + 2)^{10}\, dx$

17. $\displaystyle\int \frac{2x + 3}{(x^2 + 3x + 1)^3}\, dx$

18. $\displaystyle\int \frac{4x-1}{2x^2-x+1}\,dx$

19. $\displaystyle\int x\sqrt{x^2+1}\,dx$

20. $\displaystyle\int x\sqrt{3x^2+4}\,dx$

21. $\displaystyle\int \frac{x}{x^2+1}\,dx$

22. $\displaystyle\int \frac{x^2}{x^3+7}\,dx$

23. $\displaystyle\int \frac{t^2}{\sqrt[3]{t^3+8}}\,dt$

24. $\displaystyle\int t^2\sqrt{1+t^3}\,dt$

25. $\displaystyle\int \frac{(\sqrt{x}+7)^5}{\sqrt{x}}\,dx$

26. $\displaystyle\int \frac{1}{\sqrt{x}\,(1+\sqrt{x})}\,dx$

27. $\displaystyle\int \sqrt{x}\,(2+x\sqrt{x})^5\,dx$

28. $\displaystyle\int x\sqrt{x}\,(1+x^2\sqrt{x})^4\,dx$

29. $\displaystyle\int te^{t^2}\,dt$

30. $\displaystyle\int x^3 e^{x^4}\,dx$

31. $\displaystyle\int \frac{e^{x^n}}{x^{1-n}}\,dx$

32. $\displaystyle\int \frac{e^{\sqrt{x}}}{\sqrt{x}}\,dx$

33. $\displaystyle\int \frac{e^{3/x}}{x^2}\,dx$

34. $\displaystyle\int \frac{x^2}{e^{x^3}}\,dx$

35. $\displaystyle\int \frac{\sqrt{x}}{e^{x\sqrt{x}}}\,dx$

36. $\displaystyle\int \frac{x^2+x^{-2}}{x^3-3x^{-1}}\,dx$

37. $\displaystyle\int x^2 e^{x^3}\,dx$

38. $\displaystyle\int x^{n-1}e^{x^n}\,dx$

39. $\displaystyle\int \frac{(2x-1)e^{x^2}}{e^x}\,dx$

40. $\displaystyle\int \frac{1}{e^x e^{1/x^2}}\,dx$

41. $\displaystyle\int \frac{e^x}{(e^x+1)^2}\,dx$

42. $\displaystyle\int \frac{e^x}{e^x+1}\,dx$

43. $\displaystyle\int \frac{e^{3x}}{3-e^{3x}}\,dx$

44. $\displaystyle\int \frac{e^{x/2}}{1-e^{x/2}}\,dx$

45. $\displaystyle\int \frac{e^x+e^{-x}}{e^x-e^{-x}}\,dx$

46. $\displaystyle\int \frac{e^x-e^{-x}}{e^x+e^{-x}}\,dx$

47. $\displaystyle\int \frac{\ln x}{x}\,dx$

48. $\displaystyle\int \frac{1}{x\sqrt{\ln x}}\,dx$

49. $\displaystyle\int \frac{\sqrt{\ln x}}{x}\,dx$

50. $\displaystyle\int \frac{1}{x(1+\ln x)^4}\,dx$

51. $\displaystyle\int \frac{1}{x}(\ln x)^3\,dx$

52. $\displaystyle\int \frac{\ln(x+1)}{x+1}\,dx$

53. $\displaystyle\int \frac{1}{x(1+\ln x)}\,dx$

54. $\displaystyle\int \frac{1}{(x+3)\ln(x+3)}\,dx$

55. $\displaystyle\int \frac{3t^2+1}{t(t^2+1)}\,dt$

56. $\displaystyle\int \frac{y}{\sqrt{1+y^2}}\,dy$

57. $\displaystyle\int (x+2)\sqrt{x^2+4x+1}\,dx$

58. $\displaystyle\int \frac{x+1}{\sqrt{x^2+2x+7}}\,dx$

59. $\displaystyle\int \frac{\ln(2x)}{x}\,dx$

60. $\displaystyle\int e^{x^2+\ln x}\,dx$

61. $\displaystyle\int \frac{t^2}{t-1}\,dt$

62. $\displaystyle\int \frac{t^3}{t-1}\,dt$

63. $\displaystyle\int x\sqrt{x+1}\,dx$

(*Hint:* Put $\sqrt{x+1}=u$ or $x+1=u^2$.)

64. $\displaystyle\int x^2\sqrt{x-3}\,dx$

65. Find $g(x)$ if $g'(x)=x/\sqrt{x^2+1}$ and $g(0)=2$

66. Find $f(e)$ if $f'(x)=(x+x\ln x)^{-1}$ and $f(1)=0$

(67–72) If $f'(x)=g(x)$, calculate the following integrals.

67. $\displaystyle\int g(3x)\,dx$

68. $\displaystyle\int x\,g(x^2)\,dx$

69. $\displaystyle\int \frac{g(\sqrt{x})}{\sqrt{x}}\,dx$

70. $\displaystyle\int e^x g(e^x)\,dx$

71. $\displaystyle\int x^{-1}g(\ln x)\,dx$

72. $\displaystyle\int x^2 g(x^3)\,dx$

73. (*Marginal Cost*) The marginal cost (in dollars) of a company manufacturing shoes is given by

$$C'(x)=\frac{x}{1000}\sqrt{x^2+2500}$$

where x is the number of pairs of shoes produced. If the fixed costs are \$100, find the cost function.

74. (*Marginal Cost*) A fabric manufacturer has a marginal cost (in dollars) per roll of a particular cloth given by $C'(x)=20xe^{0.01x^2}$, where x is the number of rolls of the cloth produced. If the fixed costs are \$1500, find the cost function.

75. (Unemployment Rate) During a recent downturn in the economy, the percentage of unemployed increased at the rate

$$P'(t) = \frac{0.4e^{-0.1t}}{(1 + e^{-0.1t})^2}$$

where t is time in months. Given that at $t = 0$ there were 4% unemployed, what percentage were unemployed:
a. 10 months later? **b.** 20 months later?

76. (Natural Resource) At present a logging company has a reserve of 100 million board feet of timber. The rate at which the company is cutting and selling the timber is $R(t) = 3e^{0.06t}$ million board feet per year, where t is time in years measured from the present. Calculate the reserve that will remain after t years. How many years will the reserve last without any replanting?

77. (Oil Production) The rate of production in barrels per day

of an oil well varies according to the formula

$$P'(t) = \frac{1,200,000}{(t + 1600)^{3/2}}$$

where t is time (in days) from the start-up of production. Calculate the total production up to time t. Also find the total possible production, that is, $\lim_{t \to \infty} P(t)$.

78. (Population Growth) A bacteria population is growing in such a way that the growth rate at time t (measured in hours) is equal to $1000(1 + 3t)^{-1}$. If the population size is 1000 at $t = 0$, what will be its size after 4 hours?

79. (Drug Reaction) The rate at which antibodies are produced t hours after an injection of serum is given by $f(t) = 10t/(t^2 + 9)$. Find the value of t at which $f(t)$ is maximum and the total number of antibodies produced up to that time.

■ 16-3 TABLES OF INTEGRALS

In the previous section, we introduced the method of substitution, by means of which certain complex integrals can be reduced to one of the three standard integrals listed in Section 16-1. Apart from the substitution method, there are other techniques that are useful when it comes to evaluating integrals, and one of these will be discussed in Section 16-4.

In general, the evaluation of integrals is a matter that requires considerable skill and often ingenuity. The variety of methods that are available for the purpose is an indication of this fact. Moreover, it is not possible to give hard and fast rules about which method or which substitution will work in a given situation, but it is necessary to develop through experience an intuition for which method is likely to work best.

In face of these difficulties, by far the most convenient way to evaluate integrals is by use of a table of integrals. A table of integrals consists simply of a list of a large number of integrals, together with their values. In order to evaluate a certain integral, it is necessary only to extract the answer from the table, substituting the values of any constants as necessary. A number of such tables exist, some more complete than others; there is a fairly short table of integrals in Appendix II; it will, however, be sufficiently complete to allow the evaluation of all integrals appearing in our examples and exercises.

The integrals in this table are classified under certain headings to facilitate the use of the table. For example, all integrals involving a factor of the type $\sqrt{ax + b}$ are listed together and integrands involving $\sqrt{x^2 + a^2}$ are listed together, as are those involving exponential functions, and so on.

EXAMPLE 1 Evaluate $\int \frac{1}{(4 - x^2)^{3/2}} \, dx$.

Solution We must look through the table of integrals until we find an integral of the same form as that given above. The section titled "Integrals Containing $\sqrt{a^2 - x^2}$" is the appropriate place to look, and in Formula 33 we find:

$$\int \frac{1}{(a^2 - x^2)^{3/2}} \, dx = \frac{1}{a^2} \frac{x}{\sqrt{a^2 - x^2}}.$$

This holds for any nonzero value of the constant a, so if we set $a = 2$, we obtain the required integral.

$$\int \frac{1}{(4 - x^2)^{3/2}} \, dx = \frac{1}{4} \frac{x}{\sqrt{4 - x^2}} + C$$

Observe that we must add the constant of integration. ☞ 11

11. Use the table to find
$$\int \frac{x}{3x - 7} \, dx.$$

EXAMPLE 2 Evaluate $\displaystyle\int \frac{1}{2x^2 - 7x + 4} \, dx$.

Solution If we compare this with the standard integral

$$\int \frac{1}{ax^2 + bx + c} \, dx$$

that appears in the table of integrals, we have $a = 2$, $b = -7$, and $c = 4$. Therefore

$$b^2 - 4ac = (-7)^2 - 4(2)(4) = 49 - 32 = 17 > 0.$$

When $b^2 - 4ac > 0$, we have (see Formula 66)

$$\int \frac{1}{ax^2 + bx + c} \, dx = \frac{1}{\sqrt{b^2 - 4ac}} \ln \left| \frac{2ax + b - \sqrt{b^2 - 4ac}}{2ax + b + \sqrt{b^2 - 4ac}} \right|$$

Substituting the values of a, b, and c, we have

$$\int \frac{1}{2x^2 - 7x + 4} \, dx = \frac{1}{\sqrt{17}} \ln \left| \frac{4x - 7 - \sqrt{17}}{4x - 7 + \sqrt{17}} \right| + C$$

where C is the constant of integration that should always be included.

Sometimes the use of tables is not quite so straightforward, and it may be necessary to use the table two or more times in evaluating an integral. The following example illustrates this point.

EXAMPLE 3 Evaluate $\displaystyle\int \frac{1}{x^2\sqrt{2 - 3x}} \, dx$.

Solution If we look up the integrals involving $\sqrt{ax + b}$ in the table, then Formula 23 states that

Answer $\frac{1}{3}x + \frac{7}{9} \ln|3x - 7| + C.$

$$\int \frac{1}{x^n\sqrt{ax + b}} \, dx = -\frac{\sqrt{ax + b}}{(n - 1)bx^{n-1}} - \frac{(2n - 3)a}{(2n - 2)b} \int \frac{1}{x^{n-1}\sqrt{ax + b}} \, dx; \quad (n \neq 1).$$

In our example, $n = 2$, $a = -3$, and $b = 2$. Therefore

$$\int \frac{1}{x^2\sqrt{2 - 3x}}\, dx = -\frac{\sqrt{2 - 3x}}{2x} + \frac{3}{4}\int \frac{1}{x\sqrt{2 - 3x}}\, dx. \qquad (1)$$

To evaluate the integral on the right in Equation (1), we again look up the part of the table of integrals that involves $\sqrt{ax + b}$; Formula 22 gives

$$\int \frac{1}{x\sqrt{ax + b}}\, dx = \frac{1}{\sqrt{b}} \ln \left| \frac{\sqrt{ax + b} - \sqrt{b}}{\sqrt{ax + b} + \sqrt{b}} \right|, \quad \text{if } b > 0.$$

Putting $a = -3$ and $b = 2$ in this, we have

$$\int \frac{1}{x\sqrt{2 - 3x}}\, dx = \frac{1}{\sqrt{2}} \ln \left| \frac{\sqrt{2 - 3x} - \sqrt{2}}{\sqrt{2 - 3x} + \sqrt{2}} \right|.$$

Using this value on the right side in Equation (1),

☞ **12.** Use the table to find $\int (\ln x)^2\, dx$.

$$\int \frac{1}{x^2\sqrt{2 - 3x}}\, dx = -\frac{\sqrt{2 - 3x}}{2x} + \frac{3}{4\sqrt{2}} \ln \left| \frac{\sqrt{2 - 3x} - \sqrt{2}}{\sqrt{2 - 3x} + \sqrt{2}} \right| + C$$

where we have again added the constant of integration C. ☞ **12**

Sometimes, before the table of integrals can be used, it is necessary to make a change of variable by means of substitution to reduce the given integral to one that appears in the table.

*EXAMPLE 4 Evaluate $\displaystyle\int \frac{e^x}{(e^x + 2)(3 - e^x)}\, dx$.

Solution In this case, we do not find the integral in the tables. Let us change the variable of integration first. Clearly, $e^x\, dx$, the differential of e^x, appears in the integral, so we let $e^x = y$. Then $e^x\, dx = dy$ and the given integral now becomes

$$\int \frac{e^x}{(e^x + 2)(3 - e^x)}\, dx = \int \frac{1}{(y + 2)(3 - y)}\, dy.$$

A general integral of this form is given in the tables (Formula 15):

$$\int \frac{1}{(ax + b)(cx + d)}\, dx = \frac{1}{bc - ad} \ln \left| \frac{cx + d}{ax + b} \right| \quad (bc - ad \neq 0).$$

In our example, $a = 1$, $b = 2$, $c = -1$, and $d = 3$, and x is replaced by y. Thus

$$\int \frac{1}{(y + 2)(3 - y)}\, dy = \frac{1}{(2)(-1) - (1)(3)} \ln \left| \frac{-y + 3}{y + 2} \right| + C$$

$$= -\frac{1}{5} \ln \left| \frac{3 - y}{y + 2} \right| + C$$

Answer $x(\ln x)^2 - 2x \ln x + 2x + C$.

13. Use a substitution and then the table to find

$$\int x\sqrt{x^4 + 1}\, dx.$$

where C is the constant of integration. Substituting $y = e^x$, we have

$$\int \frac{e^x}{(e^x + 2)(3 - e^x)}\, dx = -\frac{1}{5} \ln \left| \frac{3 - e^x}{e^x + 2} \right| + C. \quad \bullet 13$$

Whenever an integral is evaluated using a table of integrals, we may readily verify that the answer obtained is correct by differentiating it: The result of the differentiation should be the original integrand. For instance, it is readily verified by standard methods of differentiation that

$$\frac{d}{dx}\left(-\frac{1}{5} \ln \left| \frac{3 - e^x}{e^x + 2} \right| \right) = \frac{e^x}{(e^x + 2)(3 - e^x)}.$$

This provides a check on the answer to Example 4.

You may wonder how tables of integrals are constructed in the first place. There are in fact, a number of techniques (apart from the general substitution method) that are useful in evaluating integrals and that are used in constructing tables of the type given in the Appendix. In the following section, we shall provide an introduction to one of the most important of these techniques.

Provided that you have developed sufficient skill at using tables of integrals, the technique to be given in the following section will not be used that frequently. However, it will still be of some use, since sometimes an integral will be encountered that is not listed in your table. In such a case, this technique may be useful in transforming the given integral into one that is listed.

Answer Substitution $u = x^2$:
$\frac{1}{4}x^2\sqrt{x^4 + 1} + \frac{1}{4} \ln |x^2 + \sqrt{x^4 + 1}| + C.$

EXERCISES 16-3

(1–26) Make use of tables of integrals to evaluate the following integrals.

1. $\displaystyle\int \frac{1}{x^2 - 3x + 1}\, dx$

2. $\displaystyle\int \frac{1}{2x^2 + 5x - 3}\, dx$

3. $\displaystyle\int \frac{x}{(2x - 3)^2}\, dx$

4. $\displaystyle\int \frac{y}{(3y + 7)^5}\, dy$

5. $\displaystyle\int \frac{\sqrt{3x + 1}}{x}\, dx$

6. $\displaystyle\int \frac{t}{(2t + 3)^{5/2}}\, dt$

7. $\displaystyle\int \frac{1}{t\sqrt{16 + t^2}}\, dt$

8. $\displaystyle\int \frac{u^2}{\sqrt{u^2 + 25}}\, du$

9. $\displaystyle\int \frac{y^2}{\sqrt{y^2 - 9}}\, dy$

10. $\displaystyle\int \frac{\sqrt{x^2 + 9}}{x}\, dx$

11. $\displaystyle\int \frac{1}{x\sqrt{3x + 4}}\, dx$

12. $\displaystyle\int \frac{1}{x^2\sqrt{25 - x^2}}\, dx$

13. $\displaystyle\int \frac{1}{x(2x + 3)^2}\, dx$

14. $\displaystyle\int \frac{x^2}{\sqrt{x^2 - 9}}\, dx$

15. $\displaystyle\int x^2(x^2 - 1)^{3/2}\, dx$

16. $\displaystyle\int x^3(\ln x)^2\, dx$

17. $\displaystyle\int x^3 e^{2x}\, dx$

18. $\displaystyle\int y^2 e^{-3y}\, dy$

19. $\displaystyle\int \sqrt{\frac{2x + 3}{4x - 1}}\, dx$

20. $\displaystyle\int \frac{x^2}{3x - 1}\, dx$

*21. $\displaystyle\int \frac{e^x}{(1 - e^x)(2 - 3e^x)}\, dx$

*22. $\displaystyle\int \frac{1}{x \ln x (1 + \ln x)}\, dx$

*23. $\displaystyle\int \frac{x}{(x^2 + 1)(2x^2 + 3)}\, dx$

*24. $\displaystyle\int \frac{y}{(2y^2 + 1)(3y^2 + 2)}\, dy$

*25. $\displaystyle\int \frac{\ln x}{x(3 + 2 \ln x)}\, dx$

*26. $\displaystyle\int x^5 e^{-x^2}\, dx$

◣ 16-4 INTEGRATION BY PARTS

The method of integration by parts can often be used to evaluate an integral with an integrand that consists of a product of two functions. It is analogous to the product formula of differential calculus, and is in fact derived from it.

From differential calculus, we know that

$$\frac{d}{dx}[u(x)v(x)] = u'(x)v(x) + u(x)v'(x)$$

or

$$u(x)v'(x) = \frac{d}{dx}[u(x)v(x)] - u'(x)v(x).$$

Integrating both sides with respect to x, we get

$$\int u(x)v'(x)\, dx = u(x)v(x) - \int u'(x)v(x)\, dx. \tag{1}$$

This equation is commonly written in the form

$$\int u\, dv = uv - \int v\, du$$

after introducing the differentials $du = u'(x)\, dx$ and $dv = v'(x)\, dx$. An alternative way of writing it, however, is as follows.

Let $u(x) = f(x)$ and $v'(x) = g(x)$. Then we can write $v(x) = G(x)$, where $G(x)$ denotes the integral of $g(x)$, and then Equation (1) becomes

$$\int f(x)g(x)\, dx = f(x)G(x) - \int f'(x)G(x)\, dx.$$

This formula expresses the integral of the product $f(x)g(x)$ in terms of the integral of the product $f'(x)G(x)$. It is useful because in many cases the integral of $f'(x)G(x)$ is easier to evaluate than the integral of the original product $f(x)g(x)$. The following example illustrates this point.

EXAMPLE 1 Evaluate $\displaystyle\int xe^{2x}\, dx$.

Solution Choose $f(x) = x$ and $g(x) = e^{2x}$, so that the given integral is equal to $\int f(x)g(x)\,dx$. Then $f'(x) = 1$ and $G(x)$, the integral of $g(x)$, is given by $G(x) = \frac{1}{2}e^{2x} + C_1$, where C_1 is a constant of integration. Substituting these values into the formula for integration by parts, we obtain

$$\int f(x)g(x)\,dx = f(x)G(x) - \int f'(x)G(x)\,dx.$$

$$\int xe^{2x}\,dx = x(\tfrac{1}{2}e^{2x} + C_1) - \int (1)(\tfrac{1}{2}e^{2x} + C_1)\,dx$$

$$= \tfrac{1}{2}xe^{2x} + C_1x - \tfrac{1}{2}\int (e^{2x} + 2C_1)\,dx$$

$$= \tfrac{1}{2}xe^{2x} + C_1x - \tfrac{1}{4}e^{2x} - C_1x + C$$

$$= \tfrac{1}{4}(2x - 1)e^{2x} + C$$

where C again is a constant of integration.

The integral in this example could also be found using Formula 69 in Appendix II. You can verify that the answer obtained is the same as that in Example 1.

Note It should be observed that the first constant of integration C_1 in the above example, which arises in integrating $g(x)$ to obtain $G(x)$, cancels from the final answer. This is always the case when integrating by parts. Therefore, in practice, we never bother to include a constant of integration in $G(x)$, but simply take $G(x)$ to be any particular antiderivative of $g(x)$. ☛ **14**

When using this method, it is important to make the right selection of $f(x)$ and $g(x)$ in expressing the original integrand as a product. Otherwise, the integral of $f'(x)G(x)$ may turn out to be no easier to evaluate than the integral of $f(x)g(x)$. For example, if we reverse the choices in Example 1, setting $f(x) = e^{2x}$ and $g(x) = x$, then $f'(x) = 2e^{2x}$ and $G(x) = \frac{1}{2}x^2$, so that the formula for integration by parts becomes

$$\int e^{2x}x\,dx = e^{2x} \cdot \tfrac{1}{2}x^2 - \int 2e^{2x} \cdot \tfrac{1}{2}x^2\,dx.$$

This equation is quite correct, but it is not much help since the integral on the right is more complicated than our original integral.

One obvious criterion in choosing f and g is that we must be able to integrate $g(x)$ in order to find $G(x)$. Usually we should choose $g(x)$ in such a way that its antiderivative $G(x)$ is a fairly simple function. The following guidelines will be helpful in deciding the choice of f and g.

1. If the integrand is the product of a polynomial in x and an exponential function, it is often useful to take $f(x)$ as the given polynomial. The preceding example illustrates this type of choice.

2. If the integrand contains a logarithmic function as a factor, it is often useful

☛ **14.** Use integration by parts to find

$$\int xe^{-3x}\,dx.$$

Answer

$$-\tfrac{1}{3}xe^{-3x} + \tfrac{1}{3}\int e^{-3x}\,dx$$

$$= -\tfrac{1}{3}xe^{-3x} - \tfrac{1}{9}e^{-3x} + C.$$

to choose this function as $f(x)$. If the integrand consists entirely of a logarithmic function, we can take $g(x) = 1$. The following examples illustrate this.

EXAMPLE 2 Evaluate $\displaystyle\int x^2 \ln x \, dx \quad (x > 0)$.

Solution Choose $f(x) = \ln x$ and $g(x) = x^2$. Then $f'(x) = 1/x$ and $G(x) = \frac{1}{3}x^3$. Substituting into the formula for integration by parts, we obtain

$$\int f(x)g(x) \, dx = f(x)G(x) - \int f'(x)G(x) \, dx$$

$$\int \ln x \cdot x^2 \, dx = \ln x \cdot \frac{1}{3}x^3 - \int \frac{1}{x} \cdot \frac{1}{3}x^3 \, dx.$$

Therefore

$$\int x^2 \ln x \, dx = \tfrac{1}{3}x^3 \ln x - \tfrac{1}{3}\int x^2 \, dx = \tfrac{1}{3}x^3 \ln x - \tfrac{1}{9}x^3 + C. \quad \blacktriangleright \textbf{15}$$

EXAMPLE 3 Evaluate $\displaystyle\int \ln(2x - 1) \, dx$.

Solution In this case, we can express the integrand as a product by writing $f(x) = \ln(2x - 1)$ and $g(x) = 1$. Then

$$f'(x) = \frac{1}{2x - 1} \cdot 2 = \frac{2}{2x - 1} \quad \text{and} \quad G(x) = x.$$

Integration by parts gives

$$\int f(x)g(x) \, dx = f(x)G(x) - \int f'(x)G(x) \, dx$$

or

$$\int \ln(2x - 1) \, dx = \ln(2x - 1) \cdot x - \int \frac{2}{2x - 1} \cdot x \, dx$$

$$= x \ln(2x - 1) - \int \frac{2x}{2x - 1} \, dx$$

$$= x \ln(2x - 1) - \int \left(1 + \frac{1}{2x - 1}\right) dx \text{ by long division*}$$

$$= x \ln(2x - 1) - x - \frac{\ln|2x - 1|}{2} + C$$

$$= (x - \tfrac{1}{2}) \ln(2x - 1) - x + C.$$

---

*Alternatively, you can substitute $n = 2x - 1$.

▶ 15. Use integration by parts to find
$$\int x^{-3} \ln x \, dx.$$

Answer $-\dfrac{\ln x}{2x^2} - \dfrac{1}{4x^2} + C.$

16. Find $\int x^5 e^{x^2}\, dx$.

[*Hint*: First substitute $u = x^2$.]

In the last step we have written $\ln|2x - 1| = \ln(2x - 1)$, since the given integral is defined only when $2x - 1 > 0$.

EXAMPLE 4 Evaluate $\int x^2 e^{mx}\, dx \quad (m \neq 0)$.

Solution Here we choose $f(x) = x^2$ and $g(x) = e^{mx}$. Then $f'(x) = 2x$ and $G(x) = e^{mx}/m$. Using the formula for integration by parts, we have

$$\int f(x)g(x)\, dx = f(x)G(x) - \int f'(x)G(x)\, dx \tag{2}$$

or

$$\int x^2 e^{mx}\, dx = x^2 \cdot \frac{e^{mx}}{m} - \int 2x \cdot \frac{e^{mx}}{m}\, dx = \frac{1}{m}x^2 e^{mx} - \frac{2}{m}\int xe^{mx}\, dx. \tag{3}$$

(Compare this with formula 70 in Appendix II). To evaluate the integral on the right, we again use integration by parts, with $f(x) = x$ and $g(x) = e^{mx}$. Then $f'(x) = 1$ and $G(x) = e^{mx}/m$. Using Equation (2), we get

$$\int xe^{mx}\, dx = x \cdot \frac{e^{mx}}{m} - \int 1 \cdot \frac{e^{mx}}{m}\, dx = \frac{x}{m}e^{mx} - \frac{1}{m} \cdot \frac{e^{mx}}{m}.$$

Substituting the value of this integral into Equation (3), we get

$$\int x^2 e^{mx}\, dx = \frac{1}{m}x^2 e^{mx} - \frac{2}{m}\left(\frac{x}{m}e^{mx} - \frac{1}{m^2}e^{mx}\right) + C$$

$$= \frac{1}{m^3}e^{mx}(m^2 x^2 - 2mx + 2) + C$$

Answer
Integral =
$\frac{1}{2}e^{x^2}(x^4 - 2x^2 + 2) + C$.

where we have finally added the constant of integration C. **16**

EXERCISES 16-4

(1–34) Evaluate the following integrals.

1. $\int x \ln x\, dx$

2. $\int x^3 \ln x\, dx$

3. $\int x^n \ln x\, dx$

4. $\int \ln(x + 1)\, dx$

5. $\int \ln x\, dx$

6. $\int \dfrac{\ln(x^2)}{x^2}\, dx$

7. $\int \sqrt{x} \ln x\, dx$

8. $\int x^3 \ln(x^3)\, dx$

9. $\int \dfrac{\ln \sqrt{x}}{\sqrt{x}}\, dx$

10. $\int (x^2 + 5) \ln x\, dx$

11. $\int (x + 1)^2 \ln(x + 1)\, dx$

12. $\int (x - 2)^3 \ln(x - 2)\, dx$

13. $\int \ln(ex)\, dx$

14. $\int \ln(2x)\, dx$

15. $\int x^2 \ln(ex)\, dx$

16. $\int x^3 \ln(3x)\, dx$

*17. $\int \log x\, dx$

*18. $\int \log_2 x\, dx$

*19. $\int x \log x\, dx$

*20. $\int x^3 \log x\, dx$

21. $\displaystyle\int x\,e^x\,ax$

22. $\displaystyle\int x\,e^{-x}\,dx$

23. $\displaystyle\int x\,e^{mx}\,dx$

24. $\displaystyle\int \frac{x}{e^{2x}}\,dx$

25. $\displaystyle\int (2x+1)e^{3x}\,dx$

*26. $\displaystyle\int e^{x+\ln x}\,dx$

27. $\displaystyle\int \ln(x^x)\,dx$

28. $\displaystyle\int \ln(xe^x)\,dx$

29. $\displaystyle\int x^2 e^x\,dx$

30. $\displaystyle\int y^2 e^{3y}\,dy$

31. $\displaystyle\int x^3 e^{x^2}\,dx$ (*Hint:* Let $x^2 = u$.)

32. $\displaystyle\int e^{\sqrt{x}}\,dx$ (*Hint:* Let $\sqrt{x} = u$.)

33. $\displaystyle\int \ln(x^{x^2})\,dx$

34. $\displaystyle\int e^{2x}\ln(e^x)\,dx$

35. Use integration by parts to verify Formula 74 in Appendix II.

36. Verify Formula 64 in Appendix II.

37. *(Marginal Cost)* A firm has a marginal cost per unit of its product given by

$$C'(x) = \frac{5000\,\ln(x+20)}{(x+20)^2}$$

where x is the level of production. If the fixed costs are $2000, determine the cost function.

38. *(Epidemic)* During the course of an epidemic the rate of arrival of new cases at a certain hospital is equal to $5te^{-(t/10)}$, where t is measured in days, $t = 0$ being the start of the epidemic. How many cases has the hospital handled in total when $t = 5$ and when $t = 10$?

CHAPTER REVIEW

Key Terms, Symbols and Concepts

16.1 Differential calculus, integral calculus. Integration. Antiderivative or indefinite integral. Constant of integration. Integral sign, integrand, variable of integration;

$$\int f(x)\,dx.$$

Power formula for integration.

16.2 Method of substitution. Linear substitution.

16.3 Table of integrals.

16.4 Integration by parts.

Formulas

If $F'(x) = f(x)$, then $\displaystyle\int f(x)\,dx = F(x) + C$.

$$\int x^n\,dx = \frac{x^{n+1}}{n+1} + C \quad (n \neq 1).$$

$$\int \frac{1}{x}\,dx = \ln|x| + C.$$

$$\int e^x\,dx = e^x + C.$$

Properties of integrals:

$$\int cf(x)\,dx = c\int f(x)\,dx,$$

$$\int [f(x) + g(x)]\,dx = \int f(x)\,dx + \int g(x)\,dx,$$

$$\frac{d}{dx}\int f(x)\,dx = f(x),$$

$$\int \frac{d}{dx}F(x)\,dx = F(x) + C.$$

If $\displaystyle\int f(u)\,du = F(u) + C$, then we can substitute $u = g(x)$ and $du = g'(x)\,dx$. That is,

$$\int f[g(x)]g'(x)\,dx = F[(g(x)] + C.$$

If $\displaystyle\int f(u)\,du = F(u) + C$, then

$$\int f(ax+b)\,dx = \frac{1}{a}F(ax+b) + C.$$

Integration by parts:

$$\int f(x)g(x)\,dx = f(x)G(x) - \int f'(x)G(x)\,dx,$$

where $G(x) = \int g(x)\,dx;$

or $\int u\,dv = uv - \int v\,du.$

1. State whether each of the following is true or false. Replace each false statement with a corresponding true statement.

a. The antiderivative of an integrable function is unique.

b. The integral of the sum of two functions is equal to the sum of their integrals.

c. The integral of the product of two functions is equal to the product of their integrals.

d. $\displaystyle\int \frac{d}{dx}[f(x)]\,dx = f(x)$

e. $\displaystyle\frac{d}{dx}\left(\int f(t)\,dt\right) = f(t)$

f. If $f'(x) = g'(x)$, then $f(x) = g(x)$.

g. $\displaystyle\int \ln x\,dx = \frac{1}{x} + C$

h. $\displaystyle\int e^x\,du = e^x + C$

i. $\displaystyle\int \frac{1}{e^t}\,dt = \frac{1}{e^t} + C$

j. $\displaystyle\int [f(x)]^n\,dx = \frac{[f(x)]^{n+1}}{n+1} + C \quad (n \neq -1)$

k. $\displaystyle\int x^n\,dx = \frac{x^{n+1}}{n+1} + C \quad \text{for all } n$

l. $\displaystyle\int x f(x)\,dx = x \int f(x)\,dx$

m. $\displaystyle\int \frac{1}{x^2}\,dx = \ln x^2 + C$

n. $\displaystyle\int e^{x^2}\,dx = e^{x^3/3} + C$

o. $\displaystyle\int e^t\,dt = \frac{e^{t+1}}{t+1} + C$

(2–11) Evaluate the following integrals:

2. $\displaystyle\int (2x + 3)(3x - 2)\,dx$

3. $\displaystyle\int (t + 2/\sqrt{t})^2\,dt$

4. $\displaystyle\int \sqrt{e^x}\,dx$

5. $\displaystyle\int \frac{e^{x^2+1}}{e^{x^2-1}}\,dx$

6. $\displaystyle\int (\log 3)\,dt$

7. $\displaystyle\int \sqrt{\ln(e^x)}\,dx$

8. $\displaystyle\int \sqrt{\log(e^x)}\,dx$

*9. $\displaystyle\int \frac{\log x}{\ln x}\,dx$

*10. $\displaystyle\int \frac{\ln x}{\log x}\,dx$ (*Hint*: Use the base-change formula in Exercises 9 and 10.)

11. $\displaystyle\int \frac{\log e}{\ln 3}\,dx$

(12–19) Make use of an appropriate substitution to evaluate the following integrals.

12. $\displaystyle\int e^x \sqrt{e^x + 1}\,dx$

13. $\displaystyle\int \frac{1}{x + x \ln x}\,dx$

14. $\displaystyle\int \frac{e^{1/x}}{x^2}\,dx$

15. $\displaystyle\int \frac{e^{1+\sqrt{x}}}{\sqrt{x}}\,dx$

16. $\displaystyle\int \frac{x}{\sqrt{1 + x^2}}\,dx$

17. $\displaystyle\int \frac{x^2}{\sqrt[3]{1 + x^3}}\,dx$

18. $\displaystyle\int \frac{1}{(x + 2) \ln(x + 2)}\,dx$

19. $\displaystyle\int \frac{1}{x\sqrt{2 + \ln x}}\,dx$

(20–39) Make use of tables or otherwise to evaluate the following integrals.

20. $\int \dfrac{3x + 2}{(x - 1)(2x + 1)}\, dx$

21. $\int \dfrac{3x - 1}{(x - 1)(x + 2)}\, dx$

22. $\int \dfrac{1}{x(4 - x^2)}\, dx$

23. $\int \dfrac{1}{x^2(9 - x^2)^{3/2}}\, dx$

24. $\int \sqrt{4x^2 - 9}\, dx$

25. $\int \sqrt{25t^2 + 9}\, dt$

26. $\int x^3 e^{2x}\, dx$

27. $\int \dfrac{1}{1 + 2e^{3x}}\, dx$

28. $\int x \ln |x + 1|\, dx$

29. $\int x^5 \log_x x^3\, dx$

*30. $\int \log_3 x\, dx$

31. $\int x^2 (\ln x)^2\, dx$

*32. $\int \dfrac{1}{x(x^4 + 1)}\, dx$

*33. $\int \dfrac{1}{x(x^3 - 1)}\, dx$

34. $\int \dfrac{\ln (x^{2x})}{e^x}\, dx$

35. $\int \dfrac{e^x}{(e^x + 1)(e^x - 4)^2}\, dx$

36. $\int \dfrac{e^{\sqrt{x} - 1}}{\sqrt{x}}\, dx$

37. $\int \dfrac{e^{(1/x^2)}}{x^3}\, dx$

38. $\int x \ln (3x)\, dx$

39. $\int 7 \ln (x/2)\, dx$

40. Find $g(t)$ if $g'(x) = \ln (x^x)$ and $g(1) = 0$.

41. Find $f(e)$ if $f'(t) = \dfrac{\ln t}{t^2}$ and $f(1) = 1$.

42. *(Marginal Cost)* The marginal cost function of a certain firm at a production level x is $C'(x) = 5 - 2x + 3x^2$ and the cost of manufacturing 30 units is $29,050. Determine the cost of manufacturing 50 units.

43. *(Marginal Revenue)* The marginal revenue function of a firm is $R'(x) = 12 - 0.2x - 0.03x^2$.

  a. Determine the revenue function.

  b. How much revenue will be obtained when 20 units are sold?

c. What is the demand function for the firm's product?

d. How many units will the firm be able to sell if it charges $3 per unit?

44. *(Demand Relation)* A firm finds that a price increase of $1 causes a drop in sales of its product by 4 units. In addition, the firm can sell 50 units at a price of $8 each. Find the firm's demand function. (*Hint:* If x is the demand at a price p, then $dx/dp = -4$.)

45. *(Learning Curve)* After a person has been working for t hours on a particular machine, x units will have been produced, where the rate of production (number of units per hour) is given by

$$\frac{dx}{dt} = 10(1 - e^{-t/50}).$$

How many units are produced during the person's first 50 hours on the machine? How many are produced during the second 50 hours?

46. *(Marginal Revenue)* The marginal revenue of a firm is given by $R'(x) = x^2/\sqrt{x^3 + 3600}$.

  a. Find the revenue function.

  b. Find the demand relation.

47. *(Profit Function)* The daily marginal profit of a firm is given by $P'(x) = -2 + x/\sqrt{x^2 + 900}$. If the firm loses $130 per day when it sells only 40 units per day, find the firm's profit function.

48. *(Profit Function)* The marginal profit of a firm is given by $P'(x) = 125 - 2x$. If the firm's fixed costs are $300, determine the profit function.

49. *(Revenue and Demand)* The marginal revenue of a firm for its product is $R'(x) = 10(20 - x)e^{-x/20}$. Find the revenue function and the demand equation for the product.

50. *(Total Sales)* The manufacturer of video games determines that its new game is selling in the market at the rate of $4000t\, e^{-0.2t}$ games per week, where t is the number of weeks since the release of this game. Express the total sales S as a function of t. How many games were sold during the first four weeks?

51. *(Consumption Function)* For a certain developing nation, the marginal propensity to consume is given by $dC/dI = 0.25 + 0.3/\sqrt[3]{I}$ (in billions of dollars). If the consumption is equal to the national income when I is $8 billion, find the consumption function $C(I)$. (Refer to page 521 for marginal propensity to consume.)

52. *(Consumption Function)* The marginal propensity to save is given by $dS/dI = 0.6 - 0.3/\sqrt{I}$ and the consumption C is $7 billion when the income I is $9 billion. Find the consumption function $C(I)$.

53. *(Physical Productivity)* The marginal physical productivity dp/dx for a toy industry is given by $dp/dx = 250(x + 4)^{3/2}$. Find the physical productivity p when 5 machines are working.

54. *(Marginal Average Cost)* The marginal average cost of a certain product is given by $C'(x) = 0.01 - 500/x^2$. If it costs $2300 to produce 200 units, find the cost function $C(x)$.

55. *(Revenue Rate)* The rates at which a company obtains net revenue from one of its mining operations is given by $R'(t) = 20t - t^2$ million dollars per year, where t is time in years measured from when the mine began operation. Find $R(t)$, the total revenue obtained during the first t years of operation. When does $R(t)$ reach a maximum? What is the maximum value of $R(t)$?

*56. *(Traffic Density)* The traffic density on a bridge during the rush hour varies according to the formula

$$f(t) = \begin{cases} 2 + 8t & 0 \le t < 1.5 \\ 23 - 6t & 1.5 \le t < 3 \end{cases}$$

where t is measured in hours from the start of the rush hour and $f(t)$ is measured in thousands of vehicles per hour. How many vehicles cross the bridge during:

a. The first 1.5 hours?

b. The whole rush hour?

*57. *(Oil Consumption)* Since 1970 the rate of oil consumption in a certain country has been given in millions of barrels per day by the following function:

$$R(t) = \begin{cases} 1 + 0.1t & \text{if } 0 \le t < 4 \\ 1.68 - 0.07t & \text{if } 4 \le t < 12 \\ 0.24 + 0.05t & \text{if } 12 \le t < 18 \end{cases}$$

where t is time in years from 1970. Calculate the total consumption:

a. Between 1970 and 1975.

b. Between 1980 and 1985.

c. Between 1970 and 1988.

Note: Don't forget to multiply $R(t)$ by 365.

58. *(Velocity and Distance)* The velocity of a braking car t seconds after the brakes are applied is equal to $u - kt$, where u is the velocity before braking and k is a constant. Find the distance traveled during the t seconds and hence find the distance required to come to a complete stop.

59. At the point x on the graph of $y = f(x)$, the slope of the tangent line is $f'(x) = 4x^3 - 3\sqrt{x}$. If the point $(1, 0)$ lies on the graph, find $f(x)$.

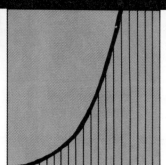

CHAPTER 17

The Definite Integral

Chapter Objectives

17-1 AREAS UNDER CURVES
(a) Definition and notation for the definite integral;
(b) Evaluation of definite integrals; evaluation when a substitution is used;
(c) Area under the graph of a positive continuous function; the fundamental theorem of calculus;
(d) Theorems on the integral of a derivative and the derivative of an integral.

17-2 MORE ON AREAS
(a) Formula for the area between the graph of a negative function and the x-axis;
(b) Evaluation of areas when the function changes sign;
(c) Evaluation of the area between two curves;
(d) The area between a graph and the y-axis;
(e) Improper integrals in which one or both of the limits extends to $\pm\infty$.

17-3 APPLICATIONS TO BUSINESS AND ECONOMICS
(a) The Lorentz curve and coefficient of inequality for income distribution;
(b) Learning curves and the estimation of future production schedules;
(c) Profit maximizing over time;
(d) Present value of a future income;
(e) Consumers' and producers' surplus.

17-4 AVERAGE VALUE OF A FUNCTION
(a) Definition of the average value of a continuous function over an interval;
(b) Examples of evaluation of the average value.

17-5 NUMERICAL INTEGRATION (OPTIONAL)
(a) Formula for the approximation of a definite integral by the trapezoidal rule;
(b) Formula for the approximation of a definite integral by Simpson's rule;
(c) Examples of the approximate evaluation of integrals by these two rules.

17-6 DIFFERENTIAL EQUATIONS: AN INTRODUCTION
(a) Definition and examples of differential equations and solutions of differential equations;
(b) General solution, initial condition and particular solution;
(c) General solution and applications of the differential equation $dy/dt = ky$;
(d) General solution of the differential equation $dy/dt = ky + b$; applications to growth of populations with immigration or harvesting.

17-7 SEPARABLE DIFFERENTIAL EQUATIONS
(a) Definition of a separable first-order differential equation; procedure for solving such equations;
(b) Examples of solution of separable differential equations;
(c) General solution of the logistic differential equation; application to restricted population growth.

17-8 APPLICATIONS TO PROBABILITY

    (a) Continuous random variable and probability
        density function;
    (b) Calculation of probabilities for a continuous
        random variable;
    (c) Uniform probability distribution and applications;
    (d) Exponential probability distribution; applications
        to reliability;
    (e) Calculation of the expectation (average value) of
        a continuous random variable;

CHAPTER REVIEW

◣ 17-1 AREAS UNDER CURVES

In this and the following sections, we shall be concerned with the calculation of the areas of regions that have curved boundaries. Such areas can be evaluated by using what are called definite integrals.

DEFINITION Let $f(x)$ be a function with an antiderivative that we denote by $F(x)$. Let a and b be any two real numbers such that $f(x)$ and $F(x)$ exist for all values of x in the closed interval with endpoints a and b. Then the **definite integral of $f(x)$ from $x = a$ to $x = b$** is denoted by $\int_a^b f(x)dx$ and is defined by

$$\int_a^b f(x)dx = F(b) - F(a).$$

The numbers a and b are called the **limits of integration,** a the **lower limit** and b the **upper limit.** Usually, $a < b$, but this is not essential.

When evaluating a definite integral, it is usual as a matter of convenience to use square brackets on the right side in the following way:

$$\int_a^b f(x)dx = \left[F(x) \right]_a^b = F(b) - F(a).$$

We read *the definite integral of $f(x)$ from $x = a$ to $x = b$ is $F(x)$ at b minus $F(x)$ at a*. The bracket notation in the middle means that the function inside the bracket must be evaluated at the two values of the argument indicated after the bracket. The difference between these two values of the function is then taken in order, value at the top argument minus the value at the bottom argument.

    In evaluating a definite integral, we drop the constant from the antiderivative of $f(x)$ because this constant of integration cancels in the final answer. Let $F(x) + C$ be any antiderivative of $f(x)$, where C is a constant of integration. Then by the above definition

● 1. Evaluate (a) $\int_{-2}^{2} x^2 \, dx$;

(b) $\int_{-3}^{3} x^5 \, dx$; (c) $\int_{1}^{3} \ln t \, dt$.

$$\int_a^b f(x)dx = \Big[F(x) + C\Big]_a^b = [F(b) + C] - [F(a) + C] = F(b) - F(a)$$

and C has disappeared from the answer.

EXAMPLE 1 Evaluate the following definite integrals.

(a) $\int_a^b x^4 \, dx$ (b) $\int_1^3 \frac{1}{t} \, dt$ (c) $\int_0^2 e^{3x} \, dx$

Solution

(a) We have $\int x^4 \, dx = x^5/5$. Thus

$$\int_a^b x^4 \, dx = \Big[\frac{x^5}{5}\Big]_a^b = \frac{b^5}{5} - \frac{a^5}{5} = \frac{1}{5}(b^5 - a^5).$$

(b) $\int_1^3 \frac{1}{t} \, dt = \Big[\ln|t|\Big]_1^3 = \ln|3| - \ln|1| = \ln 3$

(Note that $\ln 1 = 0$.)

(c) $\int_0^2 e^{3x} \, dx = \Big[\frac{e^{3x}}{3}\Big]_0^2 = \frac{e^6}{3} - \frac{e^0}{3} = \frac{1}{3}(e^6 - 1)$ ● 1

When evaluating definite integrals where the antiderivative is found by the method of substitution, it is important to note that the limits of integration also change when the variable of integration changes. This is illustrated in the following example.

EXAMPLE 2 Evaluate $\int_1^2 xe^{x^2} \, dx$.

Solution Let $I = \int_1^2 xe^{x^2} \, dx$. To find an antiderivative of xe^{x^2}, we can make use of the substitution method. We write the given integral as

$$I = \frac{1}{2} \int_1^2 e^{x^2} \cdot 2x \, dx.$$

Since $2x \, dx$, the differential of x^2, occurs in the integral, we let $x^2 = u$ so that $2x \, dx = du$. Therefore

$$I = \frac{1}{2} \int_{x=1}^{x=2} e^u \, du.$$

We have so far left the limits of integration unchanged and have emphasized that they are still limits for the original variable x. We can change them to limits for u: when $x = 1$, $u = 1^2 = 1$ and when $x = 2$, $u = 2^2 = 4$, so the limits are $u = 1$ and $u = 4$. Thus

$$I = \frac{1}{2} \int_1^4 e^u \, du.$$

Answer (a) $\frac{16}{3}$; **(b)** 0;
(c) $3 \ln 3 - 2$.

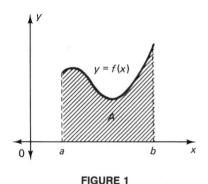

2. Evaluate **(a)** $\displaystyle\int_0^2 x^2 e^{x^3}\, dx$;

(b) $\displaystyle\int_{-1}^3 \frac{x}{x^2 + 1}\, dx$.

Here, it is understood that the limits refer to the new variable of integration u. Then, finally,

$$I = \tfrac{1}{2}\left[e^u \right]_1^4 = \tfrac{1}{2}(e^4 - e^1) = \tfrac{1}{2}e(e^3 - 1). \quad \text{☞ 2}$$

Our main concern in this section will be the calculation of areas bounded by curves. Let $f(x)$ be some given function defined and continuous in an interval $a \le x \le b$ and taking *nonnegative* values in that interval. The graph of $y = f(x)$ then lies entirely above the x-axis, as illustrated in Figure 1. We wish to find a formula for the area A that lies between such a graph and the x-axis and between the vertical lines at $x = a$ and $x = b$. This area is shaded in Figure 1.

There exists a close connection between the area A and the antiderivative of the function $f(x)$. This connection is contained in a theorem called the *fundamental theorem of calculus*, perhaps the most remarkable theorem in the whole of calculus.

THEOREM 1 (FUNDAMENTAL THEOREM OF CALCULUS) Let $f(x)$ be a continuous *nonnegative* function in $a \le x \le b$ and let $F(x)$ be an antiderivative of $f(x)$. Then A, the area between $y = f(x)$ and the x-axis and the vertical lines $x = a$ and $x = b$, is given by the definite integral

$$A = \int_a^b f(x)\, dx = F(b) - F(a).$$

Before discussing the proof of this theorem, let us illustrate its use with some examples.

EXAMPLE 3 Evaluate the area between the graph of $y = x^2$ and the x-axis from $x = 0$ to $x = 2$.

Solution The required area is shaded in Figure 2. Since $f(x) = x^2$ is nonnegative, this area is given by the definite integral $\int_a^b f(x)dx$, where $f(x) = x^2$,

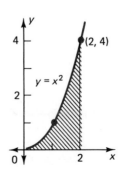

FIGURE 1

FIGURE 2

Answer **(a)** $\tfrac{1}{3}(e^8 - 1)$; **(b)** $\tfrac{1}{2}\ln 5$.

3. Find the area between the x-axis and

(a) the graph of $y = x^2$ for $-2 \le x \le 1$;

(b) the graph of $y = 16 - x^2$ for $0 \le x \le 4$.

$a = 0$, and $b = 2$. Thus the area is

$$\int_0^2 x^2 \, dx = \left[\frac{x^3}{3}\right]_0^2 = \frac{2^3}{3} - \frac{0^3}{3} = \frac{8}{3}. \quad \text{☛ 3}$$

EXAMPLE 4 Evaluate the area bounded by the curve $y = 3x^2 + 2x + 5$, the x-axis, and the lines $x = 1$ and $x = 3$.

Solution Clearly $f(x) = 3x^2 + 2x + 5$ is nonnegative for values of x in $1 \le x \le 3$. Thus the required area is given by the following definite integral.

$$\int_1^3 (3x^2 + 2x + 5) \, dx = \left[x^3 + x^2 + 5x\right]_1^3$$

$$= [3^3 + 3^2 + 5(3)] - [1^3 + 1^2 + 5(1)]$$

$$= 51 - 7 = 44 \text{ square units.} \quad \text{☛ 4}$$

Answer **(a)** 3; **(b)** $\frac{128}{3}$.

If $C(x)$ denotes the total cost of producing x units of a certain commodity, then $C'(x)$ represents the marginal cost function. Now, by the definition of the definite integral,

$$\int_a^b C'(x) \, dx = \left[C(x)\right]_a^b = C(b) - C(a).$$

☛ 4. In Example 4, convince yourself that $f(x) \ge 0$ for $1 \le x \le 3$.

But $C(b) - C(a)$ represents the change in total cost when the production level is changed from a units to b units. It follows that $\int_a^b C'(x) \, dx$ also represents this same change in total cost.

Thus we have the following important result: the change in production costs in increasing the level of production from a units to b units is equal to the area underneath the graph of the marginal cost function ($y = C'(x)$) between $x = a$ and $x = b$.

Similarly, if $R'(x)$ is the marginal revenue function, then the change in revenue when the sales level changes from a units to b units is given by $\int_a^b R'(x) \, dx$. A similar interpretation may be given to $\int_a^b P'(x) \, dx$ where $P'(x)$ is the marginal profit function; it is the change in profit when x changes from a to b.

EXAMPLE 5 The marginal cost function for a firm at the production level x is given by

$$C'(x) = 23.5 - 0.01x.$$

Find the increase in total cost when the production level is increased from 1000 to 1500 units.

Solution The increase in total cost is given by

$$\int_{1000}^{1500} C'(x) \, dx = \int_{1000}^{1500} (23.5 - 0.01x) \, dx$$

Answer Each of the three terms in $f(x)$ is positive when $x > 0$.

$$= \left[23.5x - 0.01\left(\frac{x^2}{2}\right) \right]_{1000}^{1500}$$

$$= 23.5(1500) - 0.005(1500^2)$$

$$- [23.5(1000) - 0.005(1000^2)]$$

$$= 35{,}250 - 11{,}250 - (23{,}500 - 5000) = 5500.$$

The cost increase is thus $5500. ☛ 5

☛ **5.** The marginal revenue function is $(25 - 3x)$. What will be the change in revenue if the sales level is increased from $x = 2$ to $x = 4$?

Theorems 2 and 3 provide some simple properties of definite integrals.

THEOREM 2 If $f(t)$ is continuous in $a \leq t \leq x$, then

$$\frac{d}{dx}\left[\int_a^x f(t)\, dt \right] = f(x).$$

PROOF Let $F(t)$ be an antiderivative of $f(t)$; then by the definition of definite integrals,

$$\int_a^x f(t)\, dt = \left[F(t) \right]_a^x = F(x) - F(a).$$

This is a function of x and can be differentiated with respect to x. Thus

$$\frac{d}{dx}\left[\int_a^x f(t)\, dt \right] = \frac{d}{dx}[F(x) - F(a)] = F'(x).$$

But since $F(t)$ is an antiderivative of $f(t)$, $F'(t) = f(t)$, and so

$$\frac{d}{dx}\left[\int_a^x f(t)\, dt \right] = f(x).$$

EXAMPLE 6 Evaluate $\dfrac{d}{dx}\left[\displaystyle\int_1^x te^t\, dt \right]$.

Solution By Theorem 2, we have

$$\frac{d}{dx}\left[\int_1^x te^t\, dt \right] = xe^x.$$

It would be much longer to go to the length of evaluating the integral, but of course the answer is the same:

$$\int_1^x te^t\, dt = \left[(t - 1)e^t \right]_1^x = (x - 1)e^x \quad \text{(using Formula 69 in Appendix II)}$$

$$\frac{d}{dx}\left[\int_1^x te^t\, dt \right] = \frac{d}{dx}[(x - 1)e^x] = (x - 1)e^x + 1 \cdot e^x = xe^x$$

Answer $\displaystyle\int_2^4 (25 - 3x)\, dx = 32.$

EXAMPLE 7 Evaluate each of the following:

$$\text{(a) } \frac{d}{dx}\left[\int_0^1 x^3 e^{\sqrt{x}}\, dx\right] \qquad \text{(b) } \int_0^1 \frac{d}{dx}(x^3 e^{\sqrt{x}})\, dx$$

Solution

(a) In this case, it is important to realize that the definite integral $\int_0^1 x^3 e^{\sqrt{x}}\, dx$ has a constant value and does not depend on x. Thus its derivative is zero:

$$\frac{d}{dx}\left[\int_0^1 x^3 e^{\sqrt{x}}\, dx\right] = 0.$$

(b) From the definition of antiderivatives, if $F'(x) = f(x)$,

$$\int f(x)\, dx = \int F'(x)\, dx = F(x) + C.$$

Thus

$$\int \frac{d}{dx}(x^3 e^{\sqrt{x}})\, dx = x^3 e^{\sqrt{x}} + C$$

and so, disregarding C,

$$\int_0^1 \frac{d}{dx}(x^3 e^{\sqrt{x}})\, dx = \left[x^3 e^{\sqrt{x}}\right]_0^1 = 1^3 e^{\sqrt{1}} - 0 \cdot e^{\sqrt{0}} = e.$$

Note It is worthwhile to notice the difference between the parts (a) and (b) of Example 7. The positions of the integral sign and the differentiation operator d/dx are reversed. ☛ 6

THEOREM 3

$$\text{(a) } \int_a^a f(x)\, dx = 0$$

$$\text{(b) } \int_a^b f(x)\, dx = -\int_b^a f(x)\, dx$$

$$\text{(c) } \int_a^b f(x)\, dx = \int_a^c f(x)\, dx + \int_c^b f(x)\, dx \text{ where } c \text{ is any other number.}$$

PROOF Let $F(x)$ be any antiderivative of $f(x)$. Then, from the definition of the definite integral, we have the following.

(a) $\int_a^a f(x)\, dx = [F(x)]_a^a = F(a) - F(a) = 0$

(b) We have

$$\int_a^b f(x)\, dx = \left[F(x)\right]_a^b = F(b) - F(a)$$

☛ **6.** Evaluate

(a) $\displaystyle\int_0^1 \frac{d}{dx}(\ln{(x^3 + 1)})\, dx$;

(b) $\displaystyle\frac{d}{dx}\int_0^1 \ln{(t^3 + 1)}\, dt$;

(c) $\displaystyle\frac{d}{dx}\int_0^x \ln{(t^3 + 1)}\, dt$.

Answer (a) ln 2; (b) 0; (c) ln $(x^3 + 1)$.

and

$$\int_b^a f(x)\,dx = \left[F(x)\right]_b^a = F(a) - F(b)$$

so that

$$\int_a^b f(x)\,dx = -\int_b^a f(x)\,dx.$$

(c) The proof of this part is left as an exercise.

EXAMPLE 8

(a) $\displaystyle\int_2^2 x^3 e^{\sqrt{x}}\,dx = 0$

(b) $\displaystyle\int_0^2 x^2\,dx = \int_0^3 x^2\,dx + \int_3^2 x^2\,dx$ by Theorem 3(c)

$$= \int_0^3 x^2\,dx - \int_2^3 x^2\,dx \quad \text{by Theorem 3(b).}$$

(You may like to verify this by evaluating the three integrals.)   ☛ 7

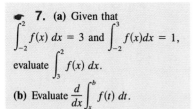

☛ 7. (a) Given that
$$\int_{-2}^2 f(x)\,dx = 3 \text{ and } \int_{-2}^3 f(x)dx = 1,$$
evaluate $\displaystyle\int_3^2 f(x)\,dx.$

(b) Evaluate $\displaystyle\frac{d}{dx}\int_x^b f(t)\,dt.$

We shall conclude this section by giving a proof of the fundamental theorem of calculus. The proof we shall give will necessarily be nonrigorous since we have not given a proper mathematical definition of the area under a curve. Nevertheless, the proof will be a convincing one, and we can assure the more skeptical readers that rigorous proofs do exist.

PROOF OF THE FUNDAMENTAL THEOREM OF CALCULUS We shall prove the theorem for the particular case when $f(x)$ is a nonnegative *increasing* function in $a \le x \le b$, although the proof can quite easily be extended to all continuous functions.

When $f(x) \ge 0$, we want to find an expression for A, the total area under the curve $y = f(x)$. Let us define the area function $A(x)$, which represents the area under the curve $y = f(x)$ from the value a to the value x of the abscissa, where x is any number such that $a \le x \le b$.

$A(x)$ is the shaded area in Figure 3. Thus $A(a) = 0$, because the area obviously shrinks to zero as x approaches a. Further, $A(b)$ is clearly the area under the curve from a to b, that is, the quantity A that we require: $A(b) = A$.

When x is changed to $x + \Delta x$ ($\Delta x > 0$), the area $A(x)$ also increases to $A(x) + \Delta A$, which is the area under the curve between the values a and $x + \Delta x$ of the abscissa. (See Figure 4.) It is reasonable to expect that ΔA is equal to the area that is shaded in this figure. (We cannot prove this rigorously here since we do not have a rigorous definition of area.)

The area ΔA is greater than the area of the inscribed rectangle with height $f(x)$ and width Δx; and ΔA is less than the area of the circumscribed

Answer (a) 2;   (b) $-f(x)$.

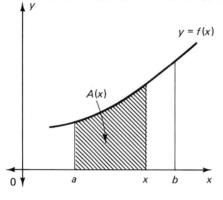

| FIGURE 3 | FIGURE 4 |

rectangle with height $f(x + \Delta x)$ and width Δx. Thus

$$f(x) \cdot \Delta x < \Delta A < f(x + \Delta x) \cdot \Delta x.$$

Dividing throughout by Δx, we obtain

$$f(x) < \frac{\Delta A}{\Delta x} < f(x + \Delta x).$$

Since f is continuous, $f(x + \Delta x) \to f(x)$ as $\Delta x \to 0$. Upon taking the limits of the above inequalities as $\Delta x \to 0$, it follows that $\Delta A/\Delta x$ has a limit, and

$$\lim_{\Delta x \to 0} \frac{\Delta A}{\Delta x} = f(x) \quad \text{or} \quad A'(x) = f(x).$$

Because $F(x)$ is an antiderivative of $f(x)$, it follows also that $F'(x) = f(x)$. Thus $F(x)$ and $A(x)$ are both antiderivatives of $f(x)$ and hence they can differ from each other by at most a constant; that is

$$A(x) = F(x) + C \tag{1}$$

where C is some constant.

Letting $x = a$ and remembering that $A(a) = 0$, we have

$$A(a) = F(a) + C = 0,$$

and so $C = -F(a)$. Replacing C by $-F(a)$ in Equation (1) above, we get

$$A(x) = F(x) - F(a).$$

Finally, putting $x = b$ in this value, we arrive at the result that

$$A(b) = F(b) - F(a).$$

But $A(b) = A$, the required total area under the curve, and we have proved that

$$A = F(b) - F(a) = \int_a^b f(x) \, dx$$

(from the definition of the definite integral).

(1–26) Evaluate the following definite integrals.

1. $\displaystyle\int_0^1 x^2\,dx$

2. $\displaystyle\int_{-1}^3 x^3\,dx$

3. $\displaystyle\int_{-1}^1 t^5\,dt$

4. $\displaystyle\int_0^1 x^{3/2}\,dx$

5. $\displaystyle\int_0^8 \sqrt[3]{x}\,dx$

6. $\displaystyle\int_0^5 (u^2 + u + 1)\,du$

7. $\displaystyle\int_1^2 (3x^2 - 5x + 7)\,dx$

8. $\displaystyle\int_0^3 (x + 1)(2x - 3)\,dx$

9. $\displaystyle\int_1^2 \frac{(2x + 1)(x - 2)}{x}\,dx$

10. $\displaystyle\int_1^3 \left(2x - \frac{1}{x}\right)^2 dx$

11. $\displaystyle\int_0^1 t^4 \ln (e^t)\,dt$

12. $\displaystyle\int_1^4 \frac{(\sqrt{x} + 1)^2}{\sqrt{x}}\,dx$

13. $\displaystyle\int_0^1 x\sqrt{x^2 + 1}\,dx$

14. $\displaystyle\int_0^1 x\,e^{x^2}\,dx$

15. $\displaystyle\int_1^0 (e^x + e^{-x})\,dx$

16. $\displaystyle\int_0^1 \frac{e^{5x} + e^{6x}}{e^{3x}}\,dx$

17. $\displaystyle\int_e^{e^2} \frac{\ln t}{t}\,dt$

18. $\displaystyle\int_1^e \frac{1}{y(1 + \ln y)}\,dy$

19. $\displaystyle\int_{-1}^2 \frac{x}{x^2 + 1}\,dx$

20. $\displaystyle\int_{-1}^1 x\sqrt{x^2 + 4}\,dx$

21. $\displaystyle\int_2^2 (x + 1)(x^2 + 2x + 7)^5\,dx$

22. $\displaystyle\int_3^3 (e^x - \sqrt{\ln x})\,dx$

23. $\displaystyle\int_0^1 \frac{d}{dt}\left[\frac{e^t + 2t - 1}{3 + \ln (1 + t)}\right] dt$

24. $\displaystyle\int_0^1 \frac{d}{dx}\left(\frac{1}{e^{2x} + e^x + 1}\right) dx$

25. $\displaystyle\int_1^1 \frac{d}{dx}\left(\frac{x^2 + 1}{1 + e^x}\right) dx$

26. $\displaystyle\int_2^2 \frac{d}{du}\left(\frac{\ln u}{u + 7}\right) du$

(27–40) Evaluate the areas beneath the graphs of the following functions between the given values of x.

27. $y = 3x + 2$, $x = 1$ and $x = 3$

28. $y = 5x^2$, $x = 0$ and $x = 2$

29. $y = 4 - x^2$, $x = 0$ and $x = 2$

30. $y = 2x^2 + 3x - 1$, $x = 1$ and $x = 4$

31. $y = x^3$, $x = 0$ and $x = 3$

32. $y = 1 + x^3$, $x = 0$ and $x = 2$

33. $y = 2 + x - x^2$, $x = -1$ and $x = 2$

34. $y = x^3 - x$, $x = -1$ and $x = 0$

35. $y = \dfrac{1}{x + 1}$, $x = 0$ and $x = 1$

36. $y = \dfrac{2x}{x^2 + 4}$, $x = 1$ and $x = 3$

37. $y = xe^x$, $x = 0$ and $x = 1$

38. $y = xe^{x^2}$, $x = 0$ and $x = 1$

39. $y = \ln x$, $x = 1$ and $x = e$

40. $y = \dfrac{\ln x}{x}$, $x = 1$ and $x = e^2$

(41–46) Evaluate the following.

41. $\displaystyle\frac{d}{dx}\left(\int_2^x \frac{e^t \ln t}{1 + t^2}\,dt\right)$

42. $\displaystyle\frac{d}{du}\left(\int_u^3 [e^x (\ln x)^4]\,dx\right)$

43. $\displaystyle\frac{d}{dx}\left(\int_1^2 \frac{e^t \ln (t^2 + 1)}{1 + t^3}\,dt\right)$

44. $\displaystyle\frac{d}{dt}\left(\int_2^5 \frac{e^{x^2}}{x + 1}\,dx\right)$

45. $\displaystyle\int_1^2 \frac{d}{dx}(x^2 e^{\sqrt{x}} \ln x)\,dx$

46. $\displaystyle\int_e^1 \frac{d}{dx}\left(\frac{\ln x}{x^2 + 1}\right) dx$

47. *(Change in Revenue)* The marginal revenue function for a firm is given by $R'(x) = 12.5 - 0.02x$. Find the increase in the total revenue of the firm when the sales level increases from 100 units to 200 units.

48. *(Increase in Profits)* The marginal cost of a certain firm is given by $C'(x) = 15.7 - 0.002x$, whereas its marginal revenue is $R'(x) = 22 - 0.004x$. Determine the increase in the profits of the firm when the sales are increased from 500 units to 600 units.

49. *(Change in Revenue)* In Exercise 47, the sales level is first decreased from 100 units to 80 units and then is increased to 150 units. Find the overall change in total revenue.

50. *(Change in Profits)* In Exercise 48, find the change in profits when the sales decrease from 500 units to 400 units.

51. *(Car Repairs)* If the average repair costs on an automobile of age t years are $10(6 + t + 0.6t^2)$ dollars per year, calculate the total repair cost during the first 2 years and during the period between $t = 4$ and $t = 6$.

◥ 17-2 MORE ON AREAS

In Section 17-1, we proved that the area under the curve $y = f(x)$ bounded by the lines $x = a$, $x = b$, and $y = 0$ (the x-axis) is given by the definite integral $\int_a^b f(x)\, dx$ in the case when $f(x) \geq 0$ in $a \leq x \leq b$.

Consider now the corresponding area bounded by the curve $y = f(x)$, the lines $x = a$, $x = b$, and the x-axis in the case when $f(x) \leq 0$ for $a \leq x \leq b$. The area in question clearly lies below the x-axis, as shown in Figure 5.

Let us define $g(x) = -f(x)$ so that $g(x) \geq 0$ for $a \leq x \leq b$. The area bounded by $y = g(x)$ (or $y = -f(x)$), the lines $x = a$, $x = b$, and the x-axis lies above the x-axis. (See Figure 6.) The latter area, as in the last section, is given by the definite integral $\int_a^b g(x)\, dx$. Now

$$\int_a^b g(x)\, dx = G(b) - G(a)$$

where $G(x)$ is the antiderivative of $g(x)$. But since $g(x) = -f(x)$, it must follow that $F(x) = -G(x)$ is an antiderivative of $f(x)$. Thus $G(b) - G(a) =$

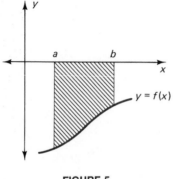

FIGURE 5

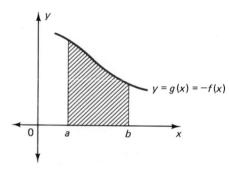

FIGURE 6

$-F(b) + F(a)$, or

$$\int_a^b g(x)\, dx = -[F(b) - F(a)] = -\int_a^b f(x)\, dx.$$

Comparing Figures 5 and 6, it is clear that the two shaded areas are equal in magnitude since one area can be obtained simply by reflecting the other about the x-axis. Thus the area below the x-axis, bounded by the curve $y = f(x)$ and the lines $x = a$ and $x = b$, is given by the definite integral

$$-\int_a^b f(x)\, dx.$$

EXAMPLE 1 Find the area bounded by $y = x^2 - 9$, $x = 0$, $x = 2$, and the x-axis.

Solution The graph of $y = x^2 - 9$ lies below the x-axis for $0 \le x \le 2$. The required area (shaded in Figure 7) is given by

$$-\int_0^2 (x^2 - 9)\, dx = \int_0^2 (9 - x^2)\, dx$$

$$= \left[9x - \frac{x^3}{3} \right]_0^2$$

$$= \left(9(2) - \frac{2^3}{3} \right) - \left(9(0) - \frac{0^3}{3} \right) = \frac{46}{3} \text{ square units.}$$

Let us now consider the area bounded by the curve $y = f(x)$ and the lines $x = a$, $x = b$, and the x-axis in the case when $f(x)$ is sometimes positive and sometimes negative in the interval $a \le x \le b$. (See Figure 8.) Such an area has parts below the x-axis and parts above the x-axis. We shall assume

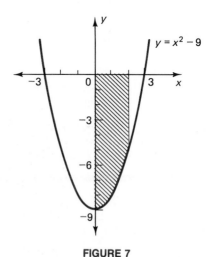

FIGURE 7

that we can find the points where the graph of $y = f(x)$ crosses the x-axis, that is, the values of x for which $f(x) = 0$. In Figure 8, we have illustrated the case when there are two such points, denoted by $x = p$ and $x = q$. In this case,

$$f(x) \geq 0 \quad \text{for} \quad a \leq x \leq p,$$
$$f(x) \leq 0 \quad \text{for} \quad p \leq x \leq q,$$

and

$$f(x) \geq 0 \quad \text{for} \quad q \leq x \leq b.$$

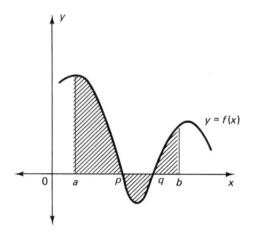

FIGURE 8

In a problem of this type, we calculate the area in each subinterval separately; the required area is then the sum of all these areas. In Figure 8, the areas between $x = a$ and $x = p$ and between $x = q$ and $x = b$ lie above the x-axis, whereas the area between $x = p$ and $x = q$ lies below the x-axis. Therefore the required area is equal to

$$\int_a^p f(x) \, dx + \left[-\int_p^q f(x) \, dx \right] + \int_q^b f(x) \, dx.$$

EXAMPLE 2 Find the area bounded by the x-axis, the curve $y = x^2 - 9$, and the lines $x = 0$ and $x = 4$.

Solution The graph of $y = x^2 - 9$ is shown in Figure 7. For $0 \leq x < 3$, it lies below the x-axis, while for $3 < x < 4$, it lies above. Therefore the required area is given by

$$\int_0^3 -(x^2 - 9) \, dx + \int_3^4 (x^2 - 9) \, dx$$

$$= \left[-\frac{x^3}{3} + 9x \right]_0^3 + \left[\frac{x^3}{3} - 9x \right]_3^4$$

☛ **8.** Find the area between the
x-axis and the graph of $y = 4 - x^2$
for

(a) $2 \le x \le 3$;

(b) $-2 \le x \le 4$.

$$= \left(-\frac{3^3}{3} + 9 \cdot 3 \right) - \left(-\frac{0^3}{3} + 9 \cdot 0 \right) + \left(\frac{4^3}{3} - 9 \cdot 4 \right) - \left(\frac{3^3}{3} - 9 \cdot 3 \right)$$

$$= 18 - 0 + (-\tfrac{44}{3}) - (-18) = \tfrac{64}{3} \text{ square units.} \quad ☛ \, 8$$

Area Between Two Curves

Let us now consider the area bounded by the two curves $y = f(x)$ and $y = g(x)$ and the lines $x = a$ and $x = b$. We shall suppose initially that $f(x) > g(x) \ge 0$ in $a \le x \le b$ so that both curves lie above the x-axis and the curve $y = f(x)$ lies above the curve $y = g(x)$. The area in question is shaded in Figure 9. Clearly this area is the difference between the area bounded by $y = f(x)$ and the x-axis and the area bounded by $y = g(x)$ and the x-axis; that is, the area of the region $CDEF$ between the two curves is equal to the area of $ABEF$ minus the area of $ABDC$.

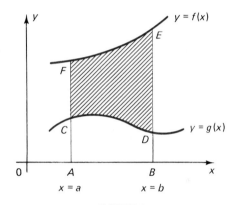

FIGURE 9

Thus the required area is given by

$$\int_a^b f(x) \, dx - \int_a^b g(x) \, dx = \int_a^b [f(x) - g(x)] \, dx.$$

Note that in the integrand $[f(x) - g(x)]$, the first term $f(x)$ relates to the top curve and the second term $g(x)$ relates to the bottom curve. A convenient way of remembering this formula is, therefore,

$$\int_a^b (y_{\text{upper}} - y_{\text{lower}}) \, dx.$$

In this form, it can also be used to calculate the area between two curves when one or both of them lie below the x-axis and also when the curves cross one another.

Answer (a) $\frac{7}{3}$; (b) $\frac{64}{3}$.

EXAMPLE 3 Find the area between the curves $y = x^2 + 5$ and $y = x^3$ and the lines $x = 1$ and $x = 2$.

Solution The graph of $y = x^2 + 5$ lies above the curve $y = x^3$ in the interval $1 \leq x \leq 2$. Thus the required area (shaded in Figure 10) is given by

$$A = \int_1^2 (y_{\text{upper}} - y_{\text{lower}}) \, dx = \int_1^2 [(x^2 + 5) - x^3] \, dx$$

$$= \left[\frac{x^3}{3} + 5x - \frac{x^4}{4} \right]_1^2 = (\tfrac{8}{3} + 10 - 4) - (\tfrac{1}{3} + 5 - \tfrac{1}{4}) = 3\tfrac{7}{12}$$

or $3\tfrac{7}{12}$ square units. ☛ 9

☛ **9.** Evaluate the area between the graphs of $y = x^2$ and $y = x$ for
(a) $0 \leq x \leq 1$.
(b) $1 \leq x \leq 2$.

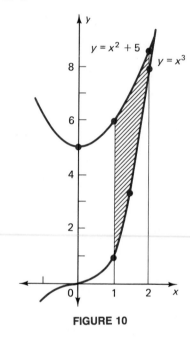

FIGURE 10

EXAMPLE 4 Find the area of the region which is enclosed by the curves $y = -x^2$ and $y = x^2 - 8$.

Solution In this case, we are not given the limits of integration. The first step is to sketch the graphs of the two curves in order to determine the required area that they enclose and the limits of integration. Sketches of the two curves are shown in Figure 11, in which the enclosed area is shaded.

To find the points of intersection of two curves, we must treat the two equations of the curves as simultaneous equations and solve them for x and y. In this particular example, equating the two values of y gives:

$$-x^2 = x^2 - 8 \quad \text{or} \quad 2x^2 - 8 = 0.$$

Therefore, $x = \pm 2$. For the area shown in Figure 11, x thus varies from -2

Answer (a) $\tfrac{1}{6}$; (b) $\tfrac{5}{6}$.

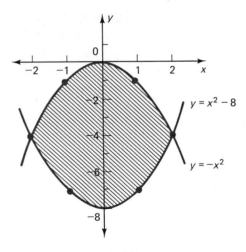

FIGURE 11

to $+2$:

$$\text{Area} = \int_{-2}^{2} (y_{\text{upper}} - y_{\text{lower}})\, dx.$$

This formula still applies, even though both graphs lie below the x-axis. The easiest way to see this is to add a sufficiently large constant to both y's to move both graphs above the x-axis (adding 8 will do). When we form the difference between y_{upper} and y_{lower}, this added constant disappears. In this case, the upper curve is $y = -x^2$, so

$$\text{Area} = \int_{-2}^{2} [(-x^2) - (x^2 - 8)]\, dx$$

$$= \int_{-2}^{2} (8 - 2x^2)\, dx = \left[8x - \tfrac{2}{3}x^3 \right]_{-2}^{2}$$

$$= (16 - \tfrac{16}{3}) - (-16 + \tfrac{16}{3}) = \tfrac{64}{3} \text{ square units.} \quad \text{☛ 10}$$

☛ **10.** Find the area enclosed between $y = 3 - x^2$ and $y = x^2 - 5$.

EXAMPLE 5 Find the area bounded by the curves $y = 1/x$ and $y = x^2$ between $x = \tfrac{1}{2}$ and $x = 2$.

Solution The two curves $y = 1/x$ and $y = x^2$ intersect where $1/x = x^2$, or $x^3 = 1$; that is, when $x = 1$. (See Figure 12.) In this case, we divide the problem into two parts, because for $\tfrac{1}{2} < x < 1$, $y_{\text{upper}} = 1/x$, but for $1 < x < 2$, $y_{\text{upper}} = x^2$.

Thus the required area is given by

$$A = \int_{1/2}^{1} \left(\frac{1}{x} - x^2 \right) dx + \int_{1}^{2} \left(x^2 - \frac{1}{x} \right) dx$$

Answer $\tfrac{64}{3}$.

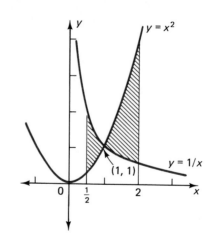

FIGURE 12

$$= \left[\ln x - \frac{x^3}{3} \right]_{1/2}^{1} + \left[\frac{x^3}{3} - \ln x \right]_{1}^{2}$$

$$= \frac{49}{24} \text{ square units.} \quad \text{☞} \ 11$$

☞ 11. Find the area enclosed between $y = x^2 + x + 1$ and $y = x^3 + x^2 + 1$.
[*Hint*: Find the intervals in which $y_1 - y_2 = x - x^3$ is positive and negative.]

We continue by giving an expression for the area bounded by the curve $x = g(y)$, the y-axis, and the horizontal lines $y = c$ and $y = d$. This area (shaded in Figure 13) is given by

$$\int_c^d g(y) \, dy$$

where $d \geq c \geq 0$. We can see this if we redraw the figure with the y-axis horizontal and the x-axis vertical, as shown in Figure 14. The area in question then becomes the area between the curve and the horizontal axis, and is given by the appropriate definite integral. The names of the variables x and y are simply interchanged.

Answer

$$Area = \int_0^1 (x - x^3) \, dx +$$
$$\int_{-1}^0 (x^3 - x) \, dx = \tfrac{1}{2}.$$

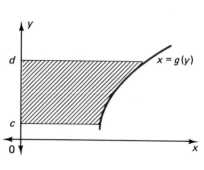

FIGURE 13

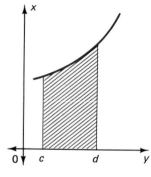

FIGURE 14

12. Sketch the area enclosed between the graphs of $x = y^2$ and $x = y + 2$. Express it as an integral with respect to y, and then evaluate it.

EXAMPLE 6 Find the area bounded by the parabola $y^2 = 4x$, the y-axis, and the horizontal lines $y = 1$ and $y = 3$.

Solution The required area is shown in Figure 15. Here $x = y^2/4$, so that $g(y) = y^2/4$. Thus the required area is

$$\int_1^3 \frac{y^2}{4}\,dy = \left[\frac{1}{4} \cdot \frac{y^3}{3}\right]_{y=1}^3 = \frac{1}{12}(3^3 - 1^3) = \frac{13}{6} \text{ square units.} \quad ☛ \textbf{12}$$

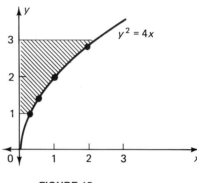

FIGURE 15

Improper Integrals

We sometimes need to evaluate integrals in which the range of integration extends either to $+\infty$ or to $-\infty$ or both. Such integrals are defined by the following:

$$\int_a^\infty f(x)\,dx = \lim_{b \to \infty} \int_a^b f(x)\,dx$$

$$\int_{-\infty}^b f(x)\,dx = \lim_{a \to -\infty} \int_a^b f(x)\,dx$$

$$\int_{-\infty}^\infty f(x)\,dx = \lim_{a \to -\infty} \lim_{b \to \infty} \int_a^b f(x)\,dx$$

provided that the relevant limit exists. Such integrals are called **improper integrals**.* One important application of such integrals occurs in probability theory. (See Section (17-8.)

Answer

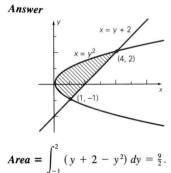

$$\textbf{Area} = \int_{-1}^2 (y + 2 - y^2)\,dy = \tfrac{9}{2}.$$

---

* There is another type of improper integral in which the integrand is unbounded at some point. For example, we define

$$\int_0^1 \frac{1}{\sqrt{x}}\,dx = \lim_{a \to 0+} \int_a^1 \frac{1}{\sqrt{x}}\,dx = \lim_{a \to 0+} [2\sqrt{x}]_a^1 = \lim_{a \to 0+} (2 - 2\sqrt{a}) = 2$$

☛ 13. Evaluate the following improper integrals if they exist:

(a) $\displaystyle\int_1^\infty \frac{1}{\sqrt{x}}\, dx$; (b) $\displaystyle\int_{-\infty}^{-2} \frac{1}{x^3}\, dx$;

(c) $\displaystyle\int_0^\infty e^{-kx}\, dx$ $(k > 0)$.

EXAMPLE 7 Evaluate the following integrals provided they exist.

(a) $\displaystyle\int_1^\infty \frac{1}{x^2}\, dx$ (b) $\displaystyle\int_{-\infty}^0 e^x\, dx$ (c) $\displaystyle\int_{-\infty}^\infty \frac{x}{x^2 + 1}\, dx$

Solutions

(a) $\displaystyle\int_1^\infty x^{-2}\, dx = \lim_{b \to \infty} \int_1^b x^{-2}\, dx = \lim_{b \to \infty} \left[-x^{-1} \right]_1^b$

$\displaystyle\qquad = \lim_{b \to \infty} [1 - b^{-1}] = 1 - 0 = 1.$

(b) $\displaystyle\int_{-\infty}^0 e^x\, dx = \lim_{a \to -\infty} \int_a^0 e^x\, dx = \lim_{a \to -\infty} \left[e^x \right]_a^0$

$\displaystyle\qquad = \lim_{a \to -\infty} [1 - e^a] = 1 - 0 = 1$

(c) $\displaystyle\int_{-\infty}^\infty \frac{x}{x^2 + 1}\, dx = \lim_{a \to -\infty} \lim_{b \to \infty} \int_a^b \frac{x}{x^2 + 1}\, dx$

We make the substitution $x^2 + 1 = u$. Then $2x\, dx = du$, and

$$\int \frac{x}{x^2 + 1}\, dx = \int \frac{1}{2u}\, du = \frac{1}{2} \ln u = \frac{1}{2} \ln (x^2 + 1)$$

(ignoring the constant of integration). Therefore

$$\int_{-\infty}^\infty \frac{x}{x^2 + 1}\, dx = \lim_{a \to -\infty} \lim_{b \to \infty} \left[\frac{1}{2} \ln (b^2 + 1) - \frac{1}{2} \ln(a^2 + 1) \right].$$

Now when $b \to \infty$, the first term on the right becomes infinitely large, so the limit does not exist. Therefore the improper integral in this part does not exist. (Note that we **must not** let $a = -b$ and simply let $b \to \infty$. We must let $a \to -\infty$ as a separate and distinct limit from $b \to \infty$.) ☛ **13**

Answer (a) Does not exist;

(b) $-\dfrac{1}{8}$; (c) $\dfrac{1}{k}$.

EXERCISES 17-2

(1–8) In each of the following exercises, find the area bounded by the curve $y = f(x)$, the x-axis, and the lines $x = a$ and $x = b$.

1. $y = -x^2$; $x = 0, x = 3$

2. $y = 1 - \sqrt{x}$; $x = 1, x = 9$

3. $y = -e^x$; $x = \ln 2, x = \ln 5$

4. $y = x^3$; $x = -1, x = 1$

5. $y = x^2 - 4$; $x = 0, x = 3$

6. $y = x^2 - 3x + 2$; $x = 0, x = 3$

7. $y = 1 - x^2$; $x = 0, x = 2$

8. $y = 2x - 1$; $x = 0, x = 1$

(9–14) Find the area between the following pairs of curves and between the given vertical lines.

9. $y = x^2, y = 3x$; $x = 1, x = 2$

10. $y = x^2, y = 2x - 1$; $x = 0, x = 2$

11. $y = \sqrt{x}, y = x^2$; $x = 0, x = 1$

12. $y = x^2, y = x^3$; $x = 0, x = 2$

700   CH. 17 | THE DEFINITE INTEGRAL

13. $y = e^x$, $y = x^2$; $x = 0$, $x = 1$

14. $y = x^3$, $y = 3x - 2$; $x = 0$, $x = 2$

(15–18) Determine the area of the region enclosed between the following pairs of curves.

15. $y = x^2$, $y = 2 - x^2$

16. $y = x^2$, $y = \sqrt{x}$

17. $y = x^3$, $y = x^2$

18. $y = x^2$, $y = 2x$

(19–20) Find the area bounded by the following curves and lines.

19. $y = x^2$, $y = 0$, $y = 4$, and $x = 0$ (y-axis)

20. $y^2 = x$, $y = 0$, $y = 2$, and $x = 0$

(21–30) Evaluate the following improper integrals if they exist.

21. $\int_{2}^{\infty} \dfrac{1}{x^3}\, dx$

22. $\int_{-\infty}^{0} \dfrac{1}{(x - 2)^2}\, dx$

23. $\int_{-\infty}^{0} (2 - x)^{-3/2}\, dx$

24. $\int_{-1}^{\infty} (2x + 3)^{-4}\, dx$

25. $\int_{0}^{\infty} \dfrac{x}{x^2 + 1}\, dx$

26. $\int_{0}^{\infty} \dfrac{x}{(x^2 + 1)^2}\, dx$

27. $\int_{-\infty}^{2} \dfrac{x - 1}{(x^2 - 2x + 2)^2}\, dx$

28. $\int_{1}^{\infty} \dfrac{2x + 1}{x^2 + x + 1}\, dx$

29. $\int_{-\infty}^{\infty} x\, e^{-x^2}\, dx$

30. $\int_{-\infty}^{\infty} e^{|x|}\, dx$

17-3 APPLICATIONS TO BUSINESS AND ECONOMICS

Coefficient of Inequality for Income Distributions

Let y be the proportion of the total income of a certain population that is received by the proportion x of income recipients whose income is least. For example, suppose that when $x = \frac{1}{2}$ then $y = \frac{1}{4}$. This would mean that the lowest paid 50% of the population receive 25% of the total income. Or if $y = 0.7$ when $x = 0.9$, then the lowest paid 90% of the population receive 70% of the total income. In general, since x and y are fractional parts of a whole, they both lie between 0 and 1 inclusive ($0 \le x \le 1$ and $0 \le y \le 1$) and y is a function of x, that is, $y = f(x)$.

We assume that there is no person who receives zero income, so that $f(0) = 0$. Furthermore, all the income is received by 100% of the recipients, so $f(1) = 1$. The graph of the function $f(x)$ describing the actual income distribution is called a **Lorentz curve.**

Suppose a Lorentz curve is given by the equation $y = \frac{15}{16}x^2 + \frac{1}{16}x$. (See Figure 16.) When $x = 0.2$, we have

$$y = \tfrac{15}{16}(0.2)^2 + \tfrac{1}{16}(0.2) = 0.05.$$

This means that the lowest paid 20% of the people receive only 5% of the total income. Similarly, when $x = 0.5$, we have

$$y = \tfrac{15}{16}(0.5)^2 + \tfrac{1}{16}(0.5) = 0.2656$$

that is, the lowest paid 50% of the people receive only 26.56% of the total income.

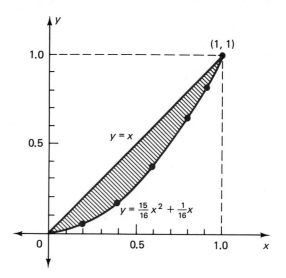

$y = x$

$(1, 1)$

$y = \frac{15}{16}x^2 + \frac{1}{16}x$

FIGURE 16

Perfect equality of income distribution is represented by the line $y = x$. According to this, for example, 10% of the people receive 10% of the total income, 20% of the people receive 20% of the total income, and so on. The deviation of the actual income distribution from perfect equality is measured by the amount by which the actual Lorentz curve departs from the straight line $y = x$. If the Lorentz curve is close to the straight line, income will be distributed approximately uniformly, while a large departure from the line indicates considerable inequality of distribution. We define the **coefficient of inequality** of the Lorentz curve as

$$L = \frac{\text{Area Between the Curve and the Line } y = x}{\text{Area Under the Line } y = x}.$$

Now, the area under the line $y = x$ is a right triangle, so is given by

$$\tfrac{1}{2}(\text{Base}) \times (\text{Height}) = \tfrac{1}{2} \cdot 1 \cdot 1 = \tfrac{1}{2}.$$

Therefore the coefficient of inequality of a Lorentz curve is given by

$$L = 2 \cdot \text{Area Between Lorentz Curve and the Line } y = x$$
$$= 2 \int_0^1 [x - f(x)]\, dx$$

where $y = f(x)$ is the equation of the Lorentz curve.

For example, the coefficient of inequality for the Lorentz curve given by $y = f(x) = \frac{15}{16}x^2 + \frac{1}{16}x$ is

$$L = 2 \int_0^1 \left[x - \left(\frac{15}{16}x^2 + \frac{1}{16}x \right) \right] dx$$

$$= 2 \int_0^1 \left(\frac{15}{16}x - \frac{15}{16}x^2 \right) dx$$

$$= 2 \cdot \frac{15}{16} \int_0^1 (x - x^2)\, dx = \frac{15}{8}\left[\frac{x^2}{2} - \frac{x^3}{3} \right]_0^1$$

$$= \frac{15}{8}\left(\frac{1}{2} - \frac{1}{3} - 0 + 0 \right) = \frac{15}{8} \cdot \frac{1}{6} = \frac{5}{16}.$$

☛ **14.** Calculate the coefficient of inequality for the Lorentz curve given by $y = ax^2 + (1 - a)x$, where a is a constant. Verify the result in the example in the text.

The coefficient of inequality always lies between 0 and 1, as is clear from its geometric definition. When the coefficient is zero, income is distributed perfectly uniformly; the closer the coefficient is to 1, the greater the inequality of distribution of income. ☛ **14**

Learning Curves

In production industry, the management often has to estimate in advance the total number of work-hours that will be required to produce a given number of units of its product. This is required, for example, in order to establish the selling price or the delivery date or to bid for a contract. A tool that is often used for such forecasting is called a *learning curve*.

It is found that a person tends to require less time to perform a particular task if he or she has already done it a number of times before. In other words, the more often a person repeats a task, the more efficient he or she becomes and the less time is required to do it again. Thus as more units are produced on a production run, the time needed to produce each unit goes down.

Let $T = F(x)$ denote the time (for example, in work-hours) necessary for the production of the first x units. An increment Δx in production requires an increment ΔT in time, and the ratio $\Delta T / \Delta x$ is the average time per additional unit produced when the number produced changes from x to $x + \Delta x$. In the limit as $\Delta x \to 0$, this ratio approaches the derivative $dT/dx = F'(x)$, which is the time required per additional unit when there is a small increase in production. As with other marginal rates, this quantity may be identified to a close approximation with the time necessary to produce the next unit, that is, the $(x + 1)$th unit.

If we set $F'(x) = f(x)$, the function commonly used for such a situation is of the form

$$f(x) = ax^b$$

where a and b are constants with $a > 0$ and $-1 \le b < 0$. The choice of ax^b with $-1 \le b < 0$ assures that the time required per unit declines as more and more units are produced. (See Figure 17.) The graph of $f(x)$ is called a **learning curve**. In practice, the constants a and b would be determined from some preliminary production run or from past experience with similar products.

Answer $\frac{1}{3}a$. The example in the text corresponds to $a = \frac{15}{16}$.

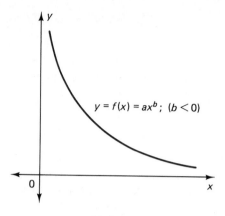

FIGURE 17

● **15.** A learning curve is given by $f(x) = 1 + 3x^{-0.2}$. Calculate the number of work hours needed to produce the first 100 units and the second 100 units.

Provided that the improvement in efficiency or learning is regular enough, the learning curve—once it has been established—can be used to predict the total number of work-hour requirements in advance for future production. *The total number of work-hours ΔT required to produce units numbered $c + 1$ through d is given by*

$$\Delta T = (\text{Work-Hours to Produce } d \text{ Units})$$

$$-(\text{Work-Hours to Produce the First, } c \text{ of Them})$$

$$= F(d) - F(c).$$

That is,

$$\Delta T = \int_c^d f(x)\,dx = \int_c^d ax^b\,dx.$$

EXAMPLE 1 After producing 1000 television sets, a manufacturing firm determines that its assembly plant is following a learning curve of the form

$$f(x) = 20x^{-0.152}$$

where $f(x)$ is the number of work-hours required to assemble the $(x + 1)$th set. Estimate the total number of work-hours required to assemble an additional 4000 television sets.

Solution The total number of work-hours required to assemble an additional 4000 sets after the first 1000 sets is given by

$$\Delta T = \int_{1000}^{5000} f(x)\,dx = \int_{1000}^{5000} 20x^{-0.152}\,dx = \left[20 \cdot \frac{x^{-0.152+1}}{-0.152 + 1} \right]_{1000}^{5000}$$

$$= \frac{20}{0.848}[(5000)^{0.848} - (1000)^{0.848}] = 23.59(1370 - 350)$$

$$= 24{,}060. \quad ● \ 15$$

Answer 249.3 and 210.6.

Profit Maximizing over Time

There exist certain business operations such as mining and oil extraction that turn out to be nonprofitable after a certain period of time. In such operations, the rate of revenue $R'(t)$ (say dollars per month) can be very high at the beginning of the operation but can decrease as time passes because of depletion of the resource. That is, $R'(t)$ eventually becomes a decreasing function of time. On the other hand, the cost rate $C'(t)$ of the operation is small at the beginning but often increases as time passes because of increased maintenance, higher cost of extraction, and many other factors. The cost rate $C'(t)$ is therefore often an increasing function of time. In such operations, there comes a time when the cost rate of running the operation becomes more than the rate of revenue and the operation starts to lose money. The management of such an operation is faced with selecting a time at which to close it down that will result in the maximum profit being obtained.

Let $C(t)$, $R(t)$, and $P(t)$ denote the total cost, total revenue, and the total profit up to the time t (measured from the start of the operation), respectively. Then

$$P(t) = R(t) - C(t) \qquad \text{and} \qquad P'(t) = R'(t) - C'(t).$$

The maximum total profit occurs when

$$P'(t) = 0 \qquad \text{or} \qquad R'(t) = C'(t).$$

In other words, the operation should run until the time t_1, at which $R'(t_1) = C'(t_1)$, that is, until the time at which the rate of revenue and the rate of cost are equal. (See Figure 18.)

The total profit at time t_1 is given by

$$P(t_1) = \int_0^{t_1} P'(t)\, dt = \int_0^{t_1} [R'(t) - C'(t)]\, dt.$$

This is the maximum profit that can be obtained and is clearly the area bounded by the graphs of $R'(t)$ and $C'(t)$ and lying between $t = 0$ and $t = t_1$.

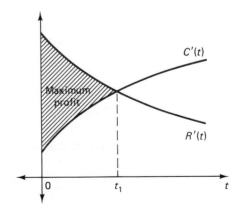

FIGURE 18

Note Since $t = 0$ is the time at which the operation starts production, the total revenue $R(0)$ at that time is zero. In the above analysis, we have also assumed that the total cost $C(0)$ is also zero. This may not be the case in general because of fixed costs (that is, setup costs) which are incurred before production begins. Thus, in practice, we must subtract these fixed costs from the above expression for $P(t_1)$ in order to obtain the actual maximum profit.

EXAMPLE 2 The cost and the revenue rates for a certain mining operation are given by

$$C'(t) = 5 + 2t^{2/3} \qquad \text{and} \qquad R'(t) = 17 - t^{2/3}$$

where C and R are measured in millions of dollars and t is measured in years. Determine how long the operation should continue and find the total profit that can be earned during this period.

Solution The optimal time t_1 that will result in maximum profit is the time at which the two rates (of cost and revenue) are equal. That is,

$$C'(t) = R'(t)$$

$$5 + 2t^{2/3} = 17 - t^{2/3}$$

$$3t^{2/3} = 17 - 5 = 12$$

$$t^{2/3} = 4$$

$$t = 4^{3/2} = 8$$

Thus the operation should continue for $t_1 = 8$ years. The profit that can be earned during this period of 8 years is given by

$$P = \int_0^8 [R'(t) - C'(t)]\, dt$$

$$= \int_0^8 [17 - t^{2/3} - (5 + 2t^{2/3})]\, dt$$

$$= \int_0^8 (12 - 3t^{2/3})\, dt = \left[12t - 3\frac{t^{5/3}}{5/3} \right]_0^8$$

$$= 96 - \tfrac{9}{5}(32) = 38.2 \text{ (millions of dollars)}.$$

Present Value of a Continuous Income

In Section 6-1 we discussed the concept of the present value of a future revenue. In examples like the one just given, where a revenue is spread over a number of years into the future, it is sometimes useful to calculate the present value of this revenue. This can be particularly valuable when a company has to choose between alternative rates of exploiting its resource.

Since the revenue in these cases is obtained continously over a period of time, it is necessary to use continuous discounting to compute the present

16. What is the present value of one cent per minute received continuously over the next 5 years? Assume an annual interest rate of 6%.

value. According to this method, the present value of an income I obtained t years in the future is Ie^{-rt}, where $r = R/100$ and R is the nominal interest rate (see Section 6-1). If $f(t)$ is the rate of profit at time t, then the present value of the total profit obtained between $t = 0$ and $t = T$ is given by

$$\text{Present Value} = \int_0^T f(t)e^{-rt}\, dt \tag{1}$$

Another application of this idea is to the case of an annuity that is paid over a period of time from $t = 0$ to $t = T$. If the annuity is paid frequently, we can regard it at least approximately as being paid continuously. The present value (i.e., the value at $t = 0$) of the annuity is then given by equation (1), where $f(t)$ is the rate of annuity (in dollars per year). ☛ 16

Answer
$$\text{P.V.} = \int_0^5 (60 \cdot 24 \cdot 365)e^{-0.06t}\, dt$$
$$= 2{,}270{,}432 \text{ cents} = \$22{,}704.32.$$

EXAMPLE 3 *(Resource Development Strategy)* A mining company must decide between two strategies for exploiting its resource. By investing $10 million in machinery it will be able to make a net profit of $3 million per year and the resource will last 10 years. Alternatively, the company can invest $15 million in better machinery and then will obtain a net profit of $5 million a year for a period of 7 years. Assuming a nominal discount rate of 10%, which strategy should the company use?

Solution The first strategy has a rate of profit $f(t) = 3$, so its present value is ($r = 0.1$, $T = 10$)

17. A person estimates that his annual income t years from now will be $(60 + 2t)$ thousand dollars. Calculate the present value of this income over the next 20 years, assuming that the income is received continuously and using an annual interest rate of 8%.

$$P_1 = \int_0^{10} 3e^{-0.1t}\, dt - 10$$

$$= \left[-30e^{-0.1t} \right]_0^{10} - 10$$

$$= 30(1 - e^{-1}) - 10 = \$8.964 \text{ million.}$$

(Note that the start-up investment of $10 million can be subtracted from the present value of the profit). Similarly, the present value of the second strategy is

$$P_2 = \int_0^7 5e^{-0.1t}\, dt - 15 = 50(1 - e^{-0.7}) - 15 = \$10.171 \text{ million.}$$

The second strategy is the better one, by about $1.2 million. ☛ 17

Consumers' and Producers' Surplus

Answer
$$\text{P.V.} = \int_0^{20} (60 + 2t)e^{-0.08t}\, dt$$
$$= 747.04 \text{ thousand dollars.}$$

Let the demand curve be $p = f(x)$ for a certain commodity and let the supply curve for the same commodity be given by $p = g(x)$. Here x denotes the amount of the commodity that can be sold or supplied at a price p per unit. In general, the demand function $f(x)$ is a decreasing function indicating that the consumers will buy less if the price increases. On the other hand, the supply function $g(x)$ in general is an increasing function because the producers are

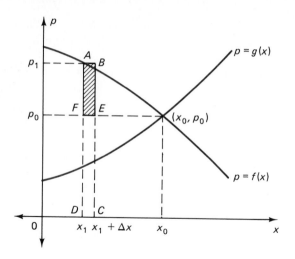

FIGURE 19

willing to supply more if they get higher prices. The market equilibrium (x_0, p_0) is the point of intersection of the demand and supply curves. (See Figure 19.)

From the graph of the demand curve, it is clear that as the price increases, the demand decreases. This implies that there are some consumers who would be willing to buy the commodity at a higher price than the market equilibrium price p_0 that they actually have to pay. These consumers therefore save money as a result of the operation of the free marketplace.

Consider the quantity Δx of units that lie between x_1 and $x_1 + \Delta x$. The area $p_1 \Delta x$ of the rectangle $ABCD$ in the above figure can be interpreted as the total amount of money that the consumers would pay for these Δx units if the price were $p_1 = f(x_1)$ per unit. At the market equilibrium price p_0, the actual amount spent by the consumers on these Δx units is $p_0 \Delta x$. In other words, the consumers save an amount equal to $p_1 \Delta x - p_0 \Delta x = [f(x_1) - p_0]\Delta x$ on these units. This saving is equal to the area of the shaded rectangle $ABEF$ in Figure 19. If we divide the range from $x = 0$ to $x = x_0$ into a large number of intervals of length Δx, we obtain a similar result for each interval: the savings to consumers is equal to the area of a rectangle like $ABEF$ lying between the demand curve and the horizontal line $p = p_0$. Summing up all such savings from $x = 0$ to $x = x_0$, we obtain the total benefit (or savings) to the consumers. This is known as the **consumers' surplus** (C.S.) and is given by the area between the demand curve $p = f(x)$ and the horizontal line $p = p_0$. (See Figure 20.)

The consumers' surplus is given by the definite integral

$$\text{C.S.} = \int_0^{x_0} [f(x) - p_0]\, dx = \int_0^{x_0} f(x)\, dx - p_0 x_0$$

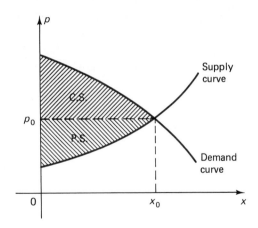

FIGURE 20

Similarly, in a free market there are also producers who would be willing to sell the commodity at a price lower than the market price p_0 that the consumers actually pay. In such a situation, the producers also benefit; this benefit to the producers is called the **producers' surplus** (P.S.).

Using similar reasoning to that above, we can show that the total gain to producers (the P.S.) is given by the area between the supply curve and the horizontal line $p = p_0$. (See Figure 20.) That is,

$$\text{P.S.} = \int_0^{x_0} [p_0 - g(x)] \, dx = p_0 x_0 - \int_0^{x_0} g(x) \, dx$$

where $p = g(x)$ is the supply relation.

EXAMPLE 4 The supply and demand functions for a certain product are given by

$$S: \quad p = g(x) = 52 + 2x \tag{2}$$

$$D: \quad p = f(x) = 100 - x^2. \tag{3}$$

Determine the consumers' and producers' surplus, assuming that market equilibrium has been established.

Solution The equilibrium point (x_0, p_0) is obtained by solving the supply and demand equations simultaneously for x and p. Equating the two values of p in Equations (2) and (3), we have

$$52 + 2x = 100 - x^2$$

$$x^2 + 2x - 48 = 0$$

$$(x - 6)(x + 8) = 0$$

which gives $x = 6$ or $x = -8$. Since the negative value of x is inadmissible, we have $x = 6$. Putting $x = 6$ in Equation (2), we get $p = 52 + 12 = 64$. Thus we have the equilibrium values $x_0 = 6$ and $p_0 = 64$. The consumers' surplus is now given by

$$\text{C.S.} = \int_0^{x_0} [f(x) - p_0]\, dx$$

$$= \int_0^6 [(100 - x^2) - 64]\, dx$$

$$= \left[36x - \frac{x^3}{3} \right]_0^6 = 216 - \frac{216}{3} = 144.$$

The producers' surplus is given by

$$\text{P.S.} = \int_0^{x_0} [p_0 - g(x)]\, dx$$

$$= \int_0^6 [64 - (52 + 2x)\, dx$$

$$= \left[12x - x^2 \right]_0^6 = 72 - 36 = 36. \quad ● \textbf{ 18}$$

EXERCISES 17-3

1. (*Lorentz Curve*) The income distribution of a certain country is given by the Lorentz curve $y = \frac{19}{20}x^2 + \frac{1}{20}x$, where x is the proportion of income recipients and y the proportion of total income received.

  a. What proportion of total income is received by the poorest 20% of the people?

  b. Find the coefficient of inequality for the Lorentz curve.

2. (*Lorentz Curve*) Repeat Exercise 1 for the Lorentz curve $y = 0.94x^2 + 0.06x$.

3. (*Learning Curve*) After painting the first 40 cars, an auto painting shop estimates that the learning curve is of the form $f(x) = 10x^{-0.25}$. Find the total number of work-hours that will be required to paint an additional 60 cars.

4. (*Learning Curve*) X & Y Sound manufactures radios on an assembly line. It is known that the first 100 radios (1 unit) took a total time of 150 work-hours and for each additional unit of 100 radios, less time was required according to the learning curve $f(x) = 150x^{-0.2}$, where $f(x)$

is the number of work-hours required to assemble the $(x + 1)$th unit. How many work-hours will be required to assemble 5 units (that is, 500 radios) after the first 5 units have been assembled?

5. (*Learning Curve*) Murphy Electronics manufactures electronic calculators on its assembly line. The first 50 calculators took 70 hours, and for each additional unit of 50 calculators, less time was required according to the learning curve $f(x) = 70x^{-0.32}$. How much time will be required to assemble 500 calculators after the first 200 calculators have been assembled?

6. (*Learning Curve*) Assuming there is 20% improvement every time the production doubles (for example, the sixth unit requires 80% of the time taken by the third unit, the twentieth unit requires 80% of the time taken by the tenth unit and so on) find the value of the constant b for the learning curve $f(x) = ax^b$.

7. (*Profit Maximization*) The revenue and cost rates for an oil drilling operation are given by

$$R'(t) = 14 - t^{1/2} \quad \text{and} \quad C'(t) = 2 + 3t^{1/2}$$

respectively, where the time t is measured in years and R and C are measured in millions of dollars. How long should the drilling be continued to obtain the maximum profit? What is this maximum profit?

8. *(Profit Maximization)* The revenue and cost rates for a certain mining operation are given by

$$R'(t) = 10 - 2t^{1/3} \quad \text{and} \quad C'(t) = 2 + 2t^{1/3}$$

respectively, where t is measured in years and R and C are measured in millions of dollars. Find how long the operation can be continued for a maximum profit. What is the amount of maximum profit, assuming that the fixed costs for initial operation are $3 million?

(9–14) *(Consumers' and Producers' Surplus)* Find the consumers' and producers' surplus for a product whose demand and supply functions are given below. (Assume that the market equilibrium has been established.)

9. *D:* $p = 15 - 2x$
   S: $p = 3 + x$

10. *D* $p = 17 - 0.5x$
    S: $p = 5 + 0.3x$

11. *D:* $p = 1200 - 1.5x^2$
    S: $p = 200 + x^2$

12. *D:* $p = 120 - x^2$
    S: $p = 32 + 3x$

13. *D:* $p = \dfrac{280}{x + 2}$
    S: $p = 20 + 2.5x$

14. *D:* $p = \dfrac{370}{x + 6}$
    S: $p = 3.8 + 0.2x$

15. *(Investment Decision)* By automating its plant, a company can reduce its labor costs. However, the automated plant requires substantial extra maintenance, which increases with time. The net savings per year after t years are given by $S'(t) = 120 - 4t - (1/2)t^2$ (million dollars per year). Calculate the total savings over the first 8 years. How many years should the automated equipment be kept before the total savings start to decrease? What is the maximum value of the total savings?

16. *(Investment Decision)* A company is considering the purchase of a new machine costing $5000. It is estimated that the machine will save the company money at the rate of $160(5 + t)$ dollars per year at a time t years after its purchase. Will the machine pay for itself during the next 5 years?

17. *(Investment Decision)* To make the correct decision in the preceding exercise, the company ought to compute the present value of its future savings and compare that with the cost of the machine. Calculate the present value of the savings over the first 5 years after the machine's purchase, assuming a nominal interest rate of 8%. Will the machine now pay for itself over this 5-year period?

18. *(Profit Maximization)* The revenue and cost rates for an oil extraction operation are

$$R'(t) = 20 - t, \qquad C'(t) = 4$$

where t is in years and R and C in millions of dollars. Find the number of years the operation should be run to ensure maximum total profit. Calculate the present value of the total profit assuming a nominal discount rate of 10%.

19. *(Profit Maximization)* Repeat the preceding exercise if

$$R'(t) = 50 - 2t, \qquad C'(t) = 20 + t$$

and the nominal discount rate is 12.5%.

20. *(Development Strategy)* A mining company can choose between two strategies for exploiting its resource. The first involves a start-up cost of $25 million and will yield a net profit of $10 million per year for the next 20 years. The second involves an initial cost of $60 million and will yield a net profit of $20 million per year for a 10-year period. Calculate the present value of these two strategies assuming a nominal discount rate of 10%. Which represents the better strategy?

21. *(Development Strategy)* Repeat Exercise 20 in the case when the rate of profit from the first strategy is $P'(t) = (20 - t)$ million dollars and the rate of profit from the second strategy is $P'(t) = (40 - 4t)$ million dollars. Assume the same start-up costs and discount rate.

22. *(Machinery Savings and Costs)* A company has purchased a new machine at a cost of $19,000. They estimate that the machine will save the company money at the rate of $1000(5 + t)$ dollars per year at a time t years after its purchase. However, the cost of operating the machine at this time will be $(1500 + 135t^2)$ dollars per year. Calculate the total net savings to the company during the first t years. Show that after 5 years this net savings has exceeded the purchase price. Determine the number of years the company should keep the machine and the total net savings up to that time.

23. *(Growth of Capital)* If $A(t)$ is the amount of capital stock in a business enterprise at time t and $I(t)$ is the rate of investment, then $dA/dt = I$. Determine the increase in capital stock between $t = 4$ and $t = 9$ if the rate of investment is given by $I(t) = 4 + \sqrt{t}$ (thousands of dollars per year).

(24–26) *(Growth of Capital)* Over the period $0 \le t \le T$, capital is invested continuously in an enterprise at the rate $I(t)$. If the investment grows continuously at the nominal rate of interest R, then capital invested at time t will have increased in value by a factor $e^{r(T-t)}$ by the end of period $(r = R/100)$. The final value of the investment is therefore equal to

$$A(T) = \int_0^T I(t)e^{r(T-t)} \, dt.$$

Calculate the final value if $r = 0.1$ and $T = 10$ in the following cases.

24. $I(t) = I$, constant

25. $I(t) = \begin{cases} 2I & \text{if } 0 \leq t \leq 5 \\ 0 & \text{if } 5 < t \leq 10 \end{cases}$

26. $I(t) = \begin{cases} 0 & \text{if } 0 \leq t \leq 5 \\ 2I & \text{if } 5 < t \leq 10 \end{cases}$

Which of the three strategies in Exercises 24–26 gives the maximum final value? Why?

*27. *(Consumers' and Producers' Surplus)* If the demand curve is $p = f(x)$, show that

$$\left(\frac{d}{dx_0}\right)(\text{C.S.}) = -x_0 f'(x_0).$$

(Hint: If $x_0 \to x_0 + \Delta x_0$, then $\Delta(\text{C.S}) \approx x_0(-\Delta p_0)$.) If the supply curve is $p = g(x)$, show that

$$\left(\frac{d}{dx_0}\right)(\text{P.S.}) = x_0 g'(x_0).$$

Show also that

$$\text{C.S.} = -\int_0^{x_0} xf'(x) \, dx \quad \text{and} \quad \text{P.S.} = \int_0^{x_0} xg'(x) \, dx.$$

Using integration by parts, obtain the expressions for C.S. and P.S. given in the text.

28. Show that

$$\text{P.S.} = \int_0^{p_0} x \, dp \quad \text{and} \quad \text{C.S.} = \int_{p_0}^{p_m} x \, dp$$

where p_m is the price at which the demand falls to zero.

*29. *(Economic Rent)* In a business enterprise in which the capital assets are considered fixed, let $P(x)$ be the dollar value of the output when x work-hours of labor are employed per week. The derivative $P'(x)$ is called the **marginal productivity of labor.** If w is the wage rate (dollars per work-hour), the profit function is given by $P(x) - wx$ (ignoring overhead). Show that it makes sense to hire x_0 work-hours where x_0 is the solution of the equation $P'(x_0) = w$. Show that the profit is then given by

$$\int_0^{x_0} [P'(x) - P'(x_0)] \, dx$$

and interpret this as an appropriate area. This quantity is called the **economic rent** of the given capital assets. Determine the economic rent if the marginal productivity is given by $P'(x) = 120(x + 400)^{-1/2}$, when the wage rate is:

a. \$3 per hour; **b.** \$4 per hour; **c.** \$5 per hour.

◤ 17-4 AVERAGE VALUE OF A FUNCTION

Consider a function $y = f(x)$ defined at the n points $x_1, x_2, x_3, \ldots, x_n$. Then the average value of the n corresponding functional values $f(x_1)$, $f(x_2), \ldots, f(x_n)$ is denoted by $\bar{f}$ or $\bar{y}$ and is given by

$$\bar{y} = \frac{f(x_1) + f(x_2) + \cdots + f(x_n)}{n}$$

This definition can be extended to the case when $f(x)$ is defined and *continuous* at all points of an interval $[a, b]$. Then the average value of $f(x)$ over $[a, b]$ is defined as

$$\bar{f} = \frac{1}{b - a} \int_a^b f(x) \, dx.$$

If $f(x) \geq 0$ on the interval $[a, b]$, then we can interpret $\bar{f}$ geometrically as follows. From the preceding definition of $\bar{f}$, we have

$$\int_a^b f(x)\, dx = \bar{f}(b - a) \tag{1}$$

But $\int_a^b f(x)\, dx$ represents the area between the x-axis, the curve $y = f(x)$, and the vertical lines $x = a$, $x = b$. From Equation (1), this area is equal to $\bar{f}(b - a)$, which is equal to the area of a rectangle of height $\bar{f}$ and width $b - a$, as shown in Figure 21. So $\bar{f}$ is the height of the rectangle that contains the same area as that under the curve.

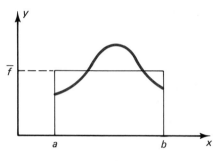

FIGURE 21

EXAMPLE 1 Find the average value of the function $f(x) = x^3$ over the interval $[1, 3]$, and interpret the result geometrically.

Solution We have:

$$\bar{f} = \frac{1}{b - a} \int_a^b f(x)\, dx = \frac{1}{3 - 1} \int_1^3 x^3\, dx = \frac{1}{2}\left[\frac{x^4}{4}\right]_1^3 = 10$$

A rectangle of height 10 and width $b - a = 3 - 1 = 2$ has the same area as that under the curve $y = x^3$ between $x = 1$ and $x = 3$. ☛ **19**

☛ **19.** Calculate **(a)** the average value of e^x over $-1 \leq x \leq 1$; **(b)** the average value of x over $a \leq x \leq b$.

EXAMPLE 2 A dose of 2 milligrams of a certain drug is injected into a person's bloodstream. The amount of the drug left in the blood after t hours is given by $f(t) = 2e^{-0.32t}$. Find the average amount of the drug in the bloodstream during the second hour.

Solution Here we are to find the average value of $f(t)$ in the interval from $t = 1$ to $t = 2$. By definition, we have:

$$\bar{f} = \frac{1}{b - a} \int_a^b f(t)\, dt$$

$$= \frac{1}{2 - 1} \int_1^2 2e^{-0.32t}\, dt = 2\left[\frac{e^{-0.32t}}{-0.32}\right]_1^2$$

$$= \frac{-1}{0.16}(e^{-0.64} - e^{-0.32}) \approx 1.24.$$

Answer
(a) $\frac{1}{2}(e - e^{-1})$;   **(b)** $\frac{1}{2}(a + b)$.

20. In Example 3, calculate the average expected weekly profit during the second year assuming the rate of sales growth remains the same.

EXAMPLE 3 A firm introduces a new product that it prices at \$5. The cost of producing x units per week is $(1000 + 2x)$ dollars. It is projected that during the first year, the weekly sales will rise at a constant rate from 200 to 600 units. Calculate the average expected weekly profit over the first year.

Solution The revenue from x units per week is $5x$ dollars. The weekly profit function is therefore

$$P(x) = \text{Revenue} - \text{Cost} = 5x - (1000 + 2x) = 3x - 1000.$$

The average value of this function over the interval $200 \le x \le 600$ is then

Answer \$1400.

$$\bar{P}(x) = \frac{1}{600 - 200} \int_{200}^{600} (3x - 1000)\, dx = \frac{1}{400} \left[\frac{3}{2}x^2 - 1000x \right]_{200}^{600} = 200.$$

The average profit is therefore \$200 per week during the first year.   ☞ **20**

EXERCISES 17-4

(1–12) Find the average value of the function over the given interval.

1. $f(x) = 3$; $[a, b]$

2. $f(x) = 2x + 5$; $[1, 4]$

3. $f(x) = x^2$; $[0, 2]$

4. $f(x) = 4 - 3x^2$; $[-1, 1]$

5. $f(x) = x^3$; $[0, 2]$

6. $f(x) = 5 - 4x^3$; $[1, 2]$

7. $f(x) = x^2 + 2x$; $[1, 3]$

8. $f(x) = x\sqrt{x^2 + 16}$; $[0, 3]$

9. $f(x) = e^x$; $[0, \ln 2]$

10. $f(x) = (\ln x)/x$; $[1, 5]$

11. $f(x) = 1/x$; $[1, e]$

12. $f(x) = \ln x$; $[1, 3]$

13. *(Average Cost)* The weekly cost C (in dollars) of producing x units of a product is given by

$$C(x) = 5000 + 16x + 0.1x^2$$

The manufacturer estimates that production will be between 200 and 300 units. What is the average cost per week over this interval?

14. *(Average Revenue)* The demand function of a product is $p = 20 - 0.05x$, where x units can be sold at a price of p

each. Find the average revenue over the sales interval $x = 100$ to $x = 200$.

15. *(Average Value of Investment)* If a sum of \$1000 is invested at 6% compounded continuously, then the value V of the investment after t years is $V = 1000e^{0.06t}$. Find the average value of a 5-year investment.

16. *(Average Population Size)* The population of a small town in 1987 was 2000 and was growing according to the formula $p(t) = 2000e^{0.03t}$, where t is measured in years and $t = 0$ corresponds to 1987. Find the average population of the town between the years 1987 and 1997.

17. *(Average Inventory)* A warehouse reorders 100 items from the manufacturer every 4 weeks. During the first 4 weeks the items sell at the rate of 20 items per week, and during the second 4 weeks they sell at 30 items per week. Calculate the average number of items in stock during the 8-week period.

18. *(Average Return)* The income from a mining investment is zero during the first 2 years and then varies according to the formula $R(t) = 5e^{-0.1(t-2)}(t \ge 2)$, where t is time in years. Calculate the average revenue per year over the interval $0 \le t \le 10$.

19. *(Average Population Size)* A population is declining according to the formula $P(t) = 2 \times 10^6/(1 + t)$, where t is time. Find the average population size between $t = 1$ and $t = 3$.

20. *(Average Temperature)* Between 6 A.M. and 6 P.M. on a certain day, the temperature in Vancouver varied according to the formula $T(t) = 13 + 3t - \frac{3}{16}t^2$. (Here t is time in hours with $t = 0$ corresponding to 6 A.M.) Calculate the average temperature:

a. Between 6 A.M. and noon.

b. Between noon and 6 P.M.

21. *(Average Speed)* The speed of an object thrown vertically into the air is given by $V(t) = 64 - 32t$, where t is time in seconds. Calculate the average speed:

a. During the first second.

b. Between $t = 1$ and $t = 3$.

22. *(Average Cost and Learning Curve)* A television manufacturer finds that the learning curve for an assembly line is $f(x) = 20x^{-0.152}$, where $f(x)$ is the number of work-hours required to assemble the $(x + 1)$th set. Calculate the average number of work-hours per set to assemble:

a. The first 1000 sets.

b. Sets 3001 − 4000.

23. *(Average Blood Pressure)* During the course of a tense committee meeting, the chairperson's systolic blood pressure increases according to the formula $P(t) = 140 + 4t + \frac{1}{2}t^2$, where t is time in hours. Calculate the average blood pressure:

a. During the first half-hour.

b. During the third hour.

◢ 17-5 NUMERICAL INTEGRATION (OPTIONAL)

Consider the integral $\int_0^1 \sqrt{1 + x^4}\, dx$. Since $f(x) = \sqrt{1 + x^4}$ is nonnegative and continuous on the interval $0 \le x \le 1$, this integral represents the area under the curve $y = \sqrt{1 + x^4}$ between $x = 0$ and $x = 1$. But we cannot find the antiderivative of $\sqrt{1 + x^4}$ by the methods discussed in this book. In fact this antiderivative cannot be expressed at all in terms of elementary functions. There are, in fact, many such functions whose antiderivatives cannot be found by the known methods of integration. For example, another such function is $f(x) = e^{-x^2/2}$, which is widely used in statistics and whose antiderivative cannot be found in terms of elementary functions. In such cases, we cannot use the fundamental theorem of calculus to evaluate the definite integral. But there do exist methods that allow us to calculate the approximate values of any definite integral and the process is known as **numerical integration.** In this section, we shall describe two such methods of evaluating approximately the definite integral $\int_a^b f(x)\, dx$.

Trapezoidal Rule

Consider the integral $\int_a^b f(x)\, dx$. To derive the trapezoidal rule, we first divide the interval $[a, b]$ into n equal subintervals, each of length h, so that $h = (b - a)/n$. The endpoints of the subintervals are a, $a + h$, $a + 2h$, $a + 3h$, . . . , and we denote the values of $f(x)$ at these endpoints by y_1, y_2, y_3, . . . , y_{n+1}, as shown in Figure 22. In each subinterval, we approximate the area under the curve by the area of the trapezoid which is the four-sided figure with two vertical sides and a top obtained by joining the two points on the graph at the ends of the subinterval. (See Figure 23.) Then the total area under the curve from $x = a$ to $x = b$ is approximately equal to the sum of the area of all the n trapezoids.

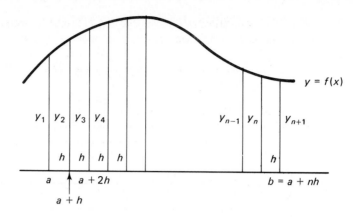

FIGURE 22

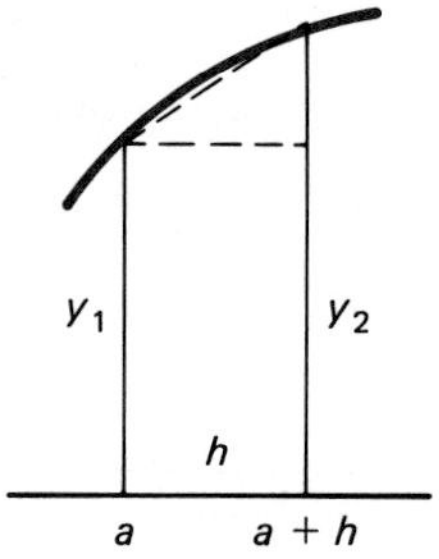

FIGURE 23

Consider the trapezoid in the first subinterval. The area of this trapezoid is equal to the sum of the areas of the rectangle of height y_1 and width h and the triangle of base h and height $(y_2 - y_1)$, as shown in Figure 23. Thus the area of this trapezoid is

$$h \cdot y_1 + \frac{1}{2}h(y_2 - y_1) = \frac{h}{2}(y_1 + y_2).$$

Similarly, the areas of the trapezoids on the other subintervals are

$$\frac{h}{2}(y_2 + y_3), \quad \frac{h}{2}(y_3 + y_4), \quad \ldots, \quad \frac{h}{2}(y_n + y_{n+1}).$$

Therefore,

$$\int_a^b f(x)\, dx \approx \text{Sum of the Areas of the Trapezoids}$$

$$= \frac{h}{2}(y_1 + y_2) + \frac{h}{2}(y_2 + y_3) + \frac{h}{2}(y_3 + y_4)$$

$$+ \cdots + \frac{h}{2}(y_n + y_{n+1})$$

$$= \frac{h}{2}[y_1 + 2(y_2 + y_3 + \cdots + y_n) + y_{n+1}]$$

We can summarize the trapezoidal rule as follows.

Trapezoidal Rule

If $f(x)$ is continuous on $[a, b]$, then

$$\int_a^b f(x)\, dx \approx \frac{h}{2}[y_1 + 2(y_2 + y_3 + \cdots + y_n) + y_{n+1}]$$

where $h = (b - a)/n$ and $y_1, y_2, y_3, \ldots, y_{n+1}$ are the values of $y = f(x)$ at $x = a, a + h, a + 2h, \ldots, a + nh = b$.

It is intuitively obvious that we get a better approximation by increasing the number n of subintervals.

EXAMPLE 1 Find the approximate value of $\int_0^3 e^{-x^2} dx$ by using the trapezoidal rule with $n = 6$.

Solution Here $a = 0$ and $b = 3$, so that

$$h = \frac{b - a}{n} = \frac{3 - 0}{6} = 0.5.$$

The endpoints of the six sub-intervals are then at $x = 0, 0.5, 1, 1.5, 2, 2.5,$ and 3, and the corresponding values of $y = e^{-x^2}$ are given in Table 1. Thus, by the trapezoidal rule, we have

$$\int_0^3 e^{-x^2} dx \approx \frac{h}{2}[(y_1 + y_7) + 2(y_2 + y_3 + y_4 + y_5 + y_6)]$$

$$\approx \frac{0.5}{2}[(1 + 0.0001) + 2(0.7788 + 0.3679$$

$$+ 0.1054 + 0.0183 + 0.0019)]$$

$$\approx 0.8862 \quad \text{☛ 21}$$

☛ **21.** Use the trapezoidal rule to approximate $\int_0^5 x^2 dx$ using

(a) 5 subintervals;
(b) 10 subintervals.
What is the exact value?

TABLE 1

x	0	0.5	1	1.5	2	2.5	3
y	e^0	$e^{-0.25}$	e^{-1}	$e^{-2.25}$	e^{-4}	$e^{-6.25}$	e^{-9}
	1	0.7788	0.3679	0.1054	0.0183	0.0019	0.0001
	y_1	y_2	y_3	y_4	y_5	y_6	y_7

Simpson's Rule

In the approximate evaluation of $\int_a^b f(x) dx$ by the trapezoidal rule, we approximated the curve $y = f(x)$ by a set of straight-line segments. In Simpson's rule, we approximate the curve $y = f(x)$ by a set of parabolic arcs. The resulting formula provides a more accurate approximation to the integral than the trapezoidal rule for the same number n of subintervals.

Simpson's Rule (Statement)

If $y = f(x)$ is continuous on the interval $a \le x \le b$, then

$$\int_a^b f(x) dx \approx \frac{h}{3}[y_1 + y_{n+1} + 2(y_3 + y_5 + \cdots) + 4(y_2 + y_4 + \cdots)]$$

where n is **even**, $h = (b - a)/n$, and $y_1, y_2, y_3, \ldots, y_{n+1}$ are the values of $y = f(x)$ at $x = a, a + h, a + 2h, \ldots, a + nh = b$.

Answer (a) 42.5;   (b) 41.875
(exact value = 41.666 . . .).

The proof of this rule is complicated and is omitted.

> **Practical Form of Simpson's Rule**
>
> $$\int_a^b f(x)\, dx \simeq \frac{h}{3}[X + 2O + 4E]$$
>
> where
>
> $$h = \frac{b-a}{n} = \frac{b-a}{\text{Number of Subintervals (even)}}$$
>
> X = Sum of the **Extreme** Ordinates (i.e., the First and the Last Ordinates)
>
> O = Sum of the **Other Odd** Ordinates (i. e., Omitting the First and the Last Ordinates)
>
> E = Sum of the **Even** Ordinates
>
> **Note** Here the X stands for Ex, the first two letters of the word **extreme**.

EXAMPLE 2 Apply Simpson's rule of approximate integration to approximate $\int_2^{10} \frac{1}{x+1}\, dx$ by taking $n = 8$ equal subintervals. Give the answer correct to three decimal places.

Solution When computing the answer correct to three decimal places, we first compute each term correct to four decimal places (one more) and then round the answer to three places. Here $y = f(x) = 1/(x+1)$, $a = 2$, $b = 10$, and $n = 8$ (even). Therefore $h = (b-a)/n = (10-2)/8 = 1$. Thus the values of x, namely, $a, a+h, a+2h, \ldots, a+8h$ are $2, 3, 4, \ldots, 10$, and the values of $y = f(x)$, namely, $y_1, y_2, y_3, \ldots$, are given by Table 2. Now,

$$X = \text{Sum of the Extreme Ordinates} = y_1 + y_9$$

$$= \frac{1}{3} + \frac{1}{11} = 0.3333 + 0.0909 = 0.4242$$

$$O = \text{Sum of the Other Odd Ordinates (Excluding the First and Last)}$$

$$= y_3 + y_5 + y_7 = \frac{1}{5} + \frac{1}{7} + \frac{1}{9}$$

$$= 0.2000 + 0.1429 + 0.1111 = 0.4540$$

$$E = \text{Sum of the Even Ordinates} = y_2 + y_4 + y_6 + y_8$$

$$= \frac{1}{4} + \frac{1}{6} + \frac{1}{8} + \frac{1}{10}$$

$$= 0.2500 + 0.1667 + 0.1250 + 0.1000 = 0.6417.$$

Therefore

$$\int_a^b f(x)\, dx \simeq \frac{h}{3}[X + 2O + 4E]$$

TABLE 2

x	2	3	4	5	6	7	8	9	10
$y = f(x)$	$\frac{1}{3}$ y_1	$\frac{1}{4}$ y_2	$\frac{1}{5}$ y_3	$\frac{1}{6}$ y_4	$\frac{1}{7}$ y_5	$\frac{1}{8}$ y_6	$\frac{1}{9}$ y_7	$\frac{1}{10}$ y_8	$\frac{1}{11}$ y_9

☛ **22.** Use Simpson's rule to approximate $\int_0^8 x^4 \, dx$ using

(a) 4 subintervals;

(b) 8 subintervals.

What is the exact value?

or

$$\int_2^{10} \frac{1}{x+1} \, dx \approx \frac{1}{3}[0.4242 + 2(0.4540) + 4(0.6417)] \approx 1.300$$

(The actual value is $\ln \frac{11}{3} \approx 1.299$.) ☛ **22**

EXAMPLE 3 Use Simpson's rule to find the approximate area between the x-axis, the lines $x = 2$, $x = 8$, and a continuous curve passing through the points listed in Table 3.

TABLE 3

x	2	3	4	5	6	7	8
y	3.2	3.7	4.1	5	4.3	3.5	3.1

Solution Here $f(x)$ is not given in explicit form. From the given data, the length of each subinterval is $h = 1$ and the values of $y_1, y_2, \ldots$ are given. Note that we have the results shown in Table 4. Thus

$$X = \text{Sum of the Extreme Ordinates} = y_1 + y_7$$
$$= 3.2 + 3.1 = 6.3$$
$$O = \text{Sum of the Other Odd Ordinates} = y_3 + y_5$$
$$= 4.1 + 4.3 = 8.4$$
$$E = \text{Sum of the Even Ordinates} = y_2 + y_4 + y_6$$
$$= 3.7 + 5 + 3.5 = 12.2.$$

TABLE 4

y_1	y_2	y_3	y_4	y_5	y_6	y_7
3.2	3.7	4.1	5	4.3	3.5	3.1

Answer **(a)** 6570.67; **(b)** 6554.67 (exact value = 6553.6).

Therefore, by Simpson's rule, the area is approximately given by

$$\int_2^8 f(x)\, dx \approx \frac{h}{3}[X + 2O + 4E]$$

$$\approx \frac{1}{3}[6.3 + 2(8.4) + 4(12.2)]$$

$$\approx 24.0 \text{ square units} \qquad \text{(rounded to one decimal place).}$$

Formulas for the approximate numerical evaluation of integrals such as those just given are very conveniently evaluated by digital computer. In such cases a very large number n of subintervals can be taken, and extremely accurate values can be obtained for most integrals. If you have taken a programming course you might find it valuable to write programs to compute integrals by either of the two rules given in this section. Test your programs with various values of n for the exercises given in Section 17-1.

EXERCISES 17-5

(1–4) Apply the trapezoidal rule of approximate integration to evaluate the following definite integrals. Round the answer to three decimal places. (In Exercises 1 and 4 check the accuracy of the answer by antidifferentiation.)

1. $\int_1^2 \frac{1}{x}\, dx$ by taking four equal intervals.

2. $\int_0^1 \frac{1}{1 + x^2}\, dx$ by taking four equal intervals.

3. $\int_0^1 \frac{1}{\sqrt{1 + x^2}}\, dx$ by taking five equal intervals.

4. $\int_4^8 \frac{1}{x - 3}\, dx$ by taking eight equal intervals.

(5–8) Apply Simpson's rule to find the approximate values of the following definite integrals (to three decimal places).

5. $\int_4^8 \frac{1}{x}\, dx$ by taking eight equal intervals.

6. $\int_0^1 \sqrt{1 + x^2}\, dx$ by taking four equal intervals.

7. $\int_0^3 \frac{1}{\sqrt{2 + x}}\, dx$ by taking six equal intervals.

8. $\int_0^1 e^{-x^2/2}\, dx$ by taking four equal intervals.

9. Use Simpson's rule to find the approximate value of $\int_{-3}^3 x^4\, dx$ by taking seven equidistant ordinates. Compare it with the exact value.

10. Given that $e^0 = 1$, $e^1 = 2.718$, $e^2 = 7.389$, $e^3 = 20.086$, and $e^4 = 54.598$, use Simpson's rule to find the approximate value of $\int_0^4 e^x\, dx$ and compare it with the exact value found using antidifferentiation.

11. Use both the trapezoidal rule and Simpson's rule to find the approximate area under a continuous curve that passes through the points:

x	1	2	3	4	5	6	7
y	1.82	4.19	6.90	9.21	11.65	14.36	16.72

12. Repeat Exercise 11 for the curve through the points

x	0	0.5	1	1.5	2
y	2	2.03	2.24	2.72	3.46

13. *(Area of Cross Section)* A river is 80 feet wide. The depth d at a distance x feet from one bank is given by the following table:

x	0	10	20	30	40	50	60	70	80
y	0	4	7	9	12	15	14	8	3

Show that the approximate area of cross section is 710 square feet according to Simpson's rule.

14. *(Land Measurements)* The front of a parcel of land is 80 feet long. The widths at intervals of 10 feet are as shown in the figure. Find the approximate area of the land by using:

a. The trapezoidal rule.

b. Simpson's rule.

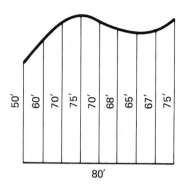

◥ 17-6 DIFFERENTIAL EQUATIONS: AN INTRODUCTION

There are many situations in business and economics where the mathematical formulation of a problem results in an equation that involves the derivative of an unknown function. Consider, for example, the following situation.

An amount of capital A_0 is invested at the nominal rate of interest R percent per annum where the investment is subject to a continuous growth at every instant, that is, the interest on the investment is compounded continuously. (See Section 6-1.) Suppose we wish to determine the total value of the investment $A(t)$ at any time t. We choose $t = 0$ to correspond to the time at which the initial investment is made. In other words. $A(0) = A_0$.

To formulate this problem mathematically, we first calculate the value of the investment $A(t)$ when the interest is compounded n times a year. If Δt denotes the length of each interest period and there are n interest periods in each year, then $n \cdot \Delta t = 1$ or $\Delta t = 1/n$ years. If $A(t)$ and $A(t + \Delta t)$ are the values of the investment at times t and $t + \Delta t$, then the interest earned during the interval of time from t to $t + \Delta t$ is given by the difference

$$A(t + \Delta t) - A(t) = \Delta A.$$

This interest ΔA is earned on the principal sum, which was $A(t)$ at the beginning of the given time interval. But if the nominal annual rate of interest is R percent, with n periods per year, then the percentage interest during one period is R/n. So the interest during the interval Δt is also equal to

$$(\text{Principal}) \times (\text{Percentage Interest})/100 = A(t)(R/100n) = A(t)r\,\Delta t$$

where $r = R/100$ and $\Delta t = 1/n$. Therefore

$$\Delta A = rA\,\Delta t \qquad \text{or} \qquad \frac{\Delta A}{\Delta t} = rA.$$

If the interest is to be compounded continuously, we must increase the number of interest periods in a year indefinitely, that is, we must take the limit as $n \to \infty$. As $n \to \infty$, $\Delta t = 1/n \to 0$ and $\Delta A/\Delta t \to dA/dt$. The above equation then becomes

$$\frac{dA}{dt} = rA. \tag{1}$$

Now dA/dt represents the rate of change in the value of investment at any time t. The above equation therefore states that the *rate of growth of the investment is proportional to the value of the investment at time t when the interest is compounded continuously*.

The value of the investment $A(t)$ at any time t must satisfy Equation (1), which involves the derivative of the unknown function $A(t)$. This equation is an example of what are known as *differential equations*. Let us now give some formal definitions.

DEFINITION Let $y = f(t)$ be a differentiable function of the independent variable t and let y', y'', . . . , $y^{(n)}$ denote the derivatives of y with respect to t of orders up to and including n. Then a **differential equation of order n** for the function y is an equation relating the variables t, y, y', . . . , $y^{(n)}$. The **order n** is the order of the highest derivative that occurs in the differential equation.

EXAMPLE 1

(a) $dy/dt = ry$ is a first-order differential equation. (It is the same as Equation (1).)

(b) $d^2y/dt^2 - e^{ty} = 0$ is a second-order differential equation.

(c) $d^4y/dt^4 - t^2(d^3y/dt^3) = t^2 + 1$ is a fourth-order differential equation.

DEFINITION A differential equation for y, a function of t, is said to be **linear** if the terms in the equation consist of y or one of its derivatives multiplied by a function of t or else of a function of t alone.

EXAMPLE 2

(a) In Example 1, the differential equations in parts (a) and (c) are linear. Part (b) is not linear, however, because y appears in the term e^{ty}, which is not a linear function of y.

(b) $d^2y/dt^2 = 3(dy/dt)^2$ is a nonlinear, second-order differential equation.

(c) $d^2y/dt^2 = 3t^2(dy/dt)$ is a linear, second-order differential equation.

Note that y and its derivatives appear linearly. The fact that the independent variable t occurs as the factor t^2 does not make the equation nonlinear. **23**

DEFINITION A function $y(t)$ is said to be a **solution** of a differential equation if, upon substituting $y(t)$ and its derivatives into the differential equation, this equation is satisfied for all values of t in the domain of $y(t)$.

EXAMPLE 3

(a) The function $y = t^2$ is a solution of the differential equation

$$t \frac{dy}{dt} - 2y = 0.$$

This is so because $dy/dt = 2t$ and so

$$t \frac{dy}{dt} = t \cdot 2t = 2t^2 = 2y.$$

(b) The function $y = e^{kt}$, where k is a constant, is a solution of the differential equation

$$\frac{d^2y}{dt^2} - k^2 y = 0$$

since

$$\frac{dy}{dt} = ke^{kt} \qquad \text{and} \qquad \frac{d^2y}{dt^2} = k^2 e^{kt} = k^2 y.$$

(c) The function $y = 2 \ln t$ is a solution of the differential equation

$$\frac{d^2y}{dt^2} + \frac{1}{2}\left(\frac{dy}{dt}\right)^2 = 0.$$

We have

$$\frac{dy}{dt} = \frac{2}{t} \qquad \text{and} \qquad \frac{d^2y}{dt^2} = -\frac{2}{t^2},$$

> **24.** Show that $y = x^2$ is a solution of the equation $xy\frac{dy}{dx} = y^2 + x^4$.

and so

$$\frac{d^2y}{dt^2} + \frac{1}{2}\left(\frac{dy}{dt}\right)^2 = -\frac{2}{t^2} + \frac{1}{2}\left(\frac{2}{t}\right)^2 = 0. \quad \text{☛ 24}$$

EXAMPLE 4 Solve the differential equation derived earlier for continuous compounding:

$$\frac{dA}{dt} = rA,$$

where r is a constant and $A(0) = A_0$.

Solution The given equation can be written as

$$\frac{dA}{A} = r\, dt$$

where we have multiplied both sides by the differential dt and divided by A. The purpose of doing this is to put all the A's on one side of the equation and all the t's on the other. Integrating both sides, we get

$$\int \frac{1}{A} \, dA = \int r \, dt.$$

Thus

$$\ln A = rt + C_1$$

(because $A > 0$) where C_1 is the constant of integration. Solving for A, we get

$$A = e^{rt+C_1} = e^{C_1} \cdot e^{rt} = Ce^{rt} \tag{2}$$

where $C = e^{C_1}$ is another constant. The value of C can be determined by making use of the additional fact that $A(0) = A_0$. Thus, putting $t = 0$ in Equation (2),

$$A_0 = A(0) = Ce^{r(0)} = C.$$

Thus, substituting $C = A_0$ into Equation (2), we get

$$A(t) = A_0 e^{rt}.$$

In other words, when the interest is being compounded continuously, the investment grows exponentially. This result agrees with that found earlier in Section 6-1.

We can summarize the main result of the last example as follows:

> The differential equation $dy/dt = ky$ where k is a given constant has the solution $y = Ce^{kt}$ where C is an arbitrary constant.

Notice the presence of the arbitrary constant C in the solution. Because of this, $y = Ce^{kt}$ is called the **general solution** of this differential equation. The differential equation does not determine the solution uniquely; the general solution contains an unknown constant.

To determine the value of the constant C we need an additional piece of information beyond the differential equation. For example, in Example 4 we were given the initial value of the investment, $A(0) = A_0$. In general (except for certain irregular cases) the solution of any first-order differential equation contains one arbitrary constant, and additional information is required to determine it. Usually, this information takes the form of the value of the dependent variable being given at one particular value of the independent variable, such as $A = A_0$ at $t = 0$. This type of information is called an **initial condition.**

EXAMPLE 5 (*Population Growth*) Let $P(t)$ denote the size (in millions) of the population of the United States at time t measured in years, with $t = 0$

corresponding to 1900. Assume that this quantity satisfies the differential equation

$$\frac{dP}{dt} = kP$$

where $k = 0.02 \ln 2 \approx 0.01386$. The population was 150 million in 1950. Find an expression for the population at a general time t and use this formula to evaluate the population in 1900 and in 1980.

Solution The differential equation is of the same type as in Example 4. Its general solution is therefore

$$P(t) = Ce^{kt}$$

where C is an arbitrary constant. In order to determine C we use the additional information that $P = 150$ when $t = 50$ (that is, in 1950). Substituting these values into the general solution, we have

$$150 = Ce^{k(50)} = Ce^{(0.02 \ln 2)(50)} = Ce^{\ln 2} = 2C$$

where we have substituted the given value of k and used the fact that $e^{\ln a} = a$ for any positive real number a. Therefore $C = 75$.

Substituting this value of C back into the general solution, we get the following expression for the population at time t,

$$P(t) = 75e^{kt} = 75e^{(0.01386)t}.$$

In 1900 ($t = 0$) the population has the value $P(0) = 75e^{(0.01386)(0)} = 75$ million. In 1980 ($t = 80$) the population is $P(80) = 75e^{(0.01386)(80)} = 75e^{1.109} = 75(3.03) = 227$ million. ☞ **25**

☞ **25.** Find the solution of the differential equation $\frac{dy}{dt} = -2y$ that satisfies the initial condition $y(1) = 3$.

Linear First-Order Equation with Constant Coefficients

We continue by considering the differential equation

$$\frac{dy}{dt} = ky + b \tag{3}$$

where k and b are two given constants. Later in this section, we will show how such a differential equation may be used as a model of population growth when such effects as immigration or harvesting are included. First, however, we shall derive its general solution.

We can write the differential equation in the form

$$\frac{dy}{dt} = k\left(y + \frac{b}{k}\right).$$

Now change the dependent variable to $z = y + b/k$. Then $dz/dt = dy/dt$ and so the differential equation becomes

$$\frac{dz}{dt} = kz.$$

Answer $y = 3e^{-2(t-1)}$.

But we already know from the previous discussion that the general solution of this equation is $z = Ce^{kt}$. Therefore, since $y = z - b/k$ the general solution for y is

$$y = Ce^{kt} - \frac{b}{k}. \tag{4}$$

Again note the presence of the arbitrary constant. We can summarize this result as follows:

> The differential equation $dy/dt = ky + b$, where k and b are given constants, has the general solution $y = Ce^{kt} - b/k$, where C is an arbitrary constant.

EXAMPLE 6 Find the solution of the differential equation

$$\frac{dy}{dt} = 2y + 1$$

that satisfies the initial condition $y(0) = 3$.

Solution We proceed as in the derivation for the general case. First write the differential equation as

$$\frac{dy}{dt} = 2(y + \tfrac{1}{2}).$$

Then transform to the new variable $z = y + \tfrac{1}{2}$. The differential equation becomes $dz/dt = 2z$ and its general solution is $z = Ce^{2t}$. Therefore, since $y = z - \tfrac{1}{2}$, the general solution for y is

$$y = Ce^{2t} - \tfrac{1}{2}.$$

We could of course have obtained this solution by simply substituting $k = 2$ and $b = 1$ in the formula (4) we derived above for the general case.

The constant C is arbitrary and must be determined from the given initial condition that $y(0) = 3$. Setting $t = 0$ and $y = 3$ in the last equation, we get

$$3 = Ce^{2(0)} - \tfrac{1}{2} = C - \tfrac{1}{2} \quad \text{or} \quad C = \tfrac{7}{2}.$$

Thus, substituting for C in the general solution, we get

$$y = \tfrac{7}{2}e^{2t} - \tfrac{1}{2}. \quad \text{☛ 26}$$

☛ **26.** Find the solution of the differential equation $\dfrac{dy}{dt} = 4 - y$ that satisfies the initial condition $y(0) = 3$.

We shall now discuss some applications of the differential equation $dy/dt = ky + b$. Consider first the growth of an investment compounded n times per annum with nominal annual interest rate $R\%$. Let $A(t)$ be the value of the investment at time t and let $\Delta t = 1/n$ denote the time interval between compoundings. Then, as we discussed at the beginning of this section, the increment in A from one compounding to the next is given by

Answer $y = 4 - e^{-t}$.

$$\Delta A = A(t + \Delta t) - A(t) = rA(t) \Delta t$$

where $r = R/100$. Now, suppose that a further total amount I is invested each year in the account in equal amounts just prior to each compounding. Then, each additional investment is $I \div n = I \Delta t$ so that the increment in the value of the account is

$$\Delta A = \text{Interest During } \Delta t + \text{New Investment During } \Delta t$$

$$\Delta A = rA(t) \Delta t + I \Delta t.$$

Thus

$$\frac{\Delta A}{\Delta t} = rA + I.$$

Continuous compounding corresponds to the limit $n \to \infty$, which means $\Delta t \to 0$. In this limit, the above equation becomes the differential equation

$$\frac{dA}{dt} = rA + I$$

which is exactly of the type we have just been discussing.

A much more important application of the differential equation (3) is to population growth. The differential equation $dy/dt = ky$ which corresponds to the special case $b = 0$ can be applied in many cases where a population expands in an environment that places no restriction on its growth. The constant k is called the **specific growth rate** of the population. The differential equation states that the natural growth rate is proportional to the population size. Its solution is an exponentially growing function of t.

The more general equation $dy/dt = ky + b$ can be used for such a population that is expanding not only through its own natural growth but also as a result of a steady immigration of members from outside. The left side of the differential equation gives the total rate of growth of the population size y, the first term on the right is the contribution to that rate of growth from natural expansion, while the second term, b, is the rate of growth due to immigration. If the rate of immigration is constant, we can use the method developed above to find the solution.

The case of a population that is losing members through emigration is similar: the only difference is that the constant b becomes negative, with $-b$ being the rate of emigration. Perhaps the most important case, however, is that of a population that is losing members as a result of hunting or harvesting. Such examples are fundamental for conservation of stocks of certain species that are harvested for human consumption.

EXAMPLE 7 A certain species of fish has an initial population size of 100 units, each unit being 1 million fish, and has a natural specific growth rate of 0.25, with time measured in years. The population is being harvested at the rate of h units per year, so that the size y satisfies the differential equation and initial condition:

$$\frac{dy}{dt} = 0.25y - h \qquad y(0) = 100.$$

Determine y as a function of t in the cases (a) $h = 20$; (b) $h = 25$; (c) $h = 30$. Discuss the significance of the results.

Solution The given differential equation is of the type under discussion with $k = 0.25$ and $b = -h$. The solution can be obtained by following the same procedure as before, or simply by substituting these values of k and b into the general solution (4). We find

$$y = Ce^{0.25t} + 4h.$$

Setting $t = 0$ and $y = 100$, we find the value of C: $C = 100 - 4h$. Thus

$$y = (100 - 4h)e^{0.25t} + 4h.$$

For the three given values of the rate of harvesting, this becomes

$$h = 20: \qquad y = 20e^{0.25t} + 80$$

$$h = 25: \qquad y = 100$$

$$h = 30: \qquad y = 120 - 20e^{0.25t}.$$

The significance of these results is as follows. When the harvesting rate is 25 units per year, the harvesting exactly balances the natural growth of the population and the size remains constant. We have in this case a steady and sustained yield from the harvesting. When h is less than 25, as illustrated by the solution for $h = 20$, the natural growth more than compensates for the harvesting and the population size increases (allowing the prospect of larger harvests in the future). When h is greater than 25, as illustrated by the result for $h = 30$, the population size decreases, since the exponential term has a negative coefficient. Eventually, the population is driven to extinction by this overharvesting. (Check that $y = 0$ when $t = 4 \ln 6 \approx 7.2$) ➡ **27**

EXERCISES 17-6

(1–4) Show that the functions given below satisfy the stated differential equations.

1. $y = t^{-4}$; $t \, dy/dt + 4y = 0$

2. $y = t \ln t$; $t^2 \, d^2y/dt^2 - t \, dy/dt + y = 0$

3. $y = te^{-t}$; $t \, dy/dt + ty = y$

4. $y = t^3 + 2\sqrt{t}$; $2t^2 \, d^2y/dt^2 - 5t \, dy/dt + 3y = 0$

(5–16) Find the general solution of the following differential equations.

5. $dy/dt = t^2 + 1/t$

6. $dy/dx = xe^x$

7. $dy/dt - 4y = 0$

8. $2 \, dy/dt + y = 0$

9. $dy/dt - \sqrt{t} = 0$

10. $2 \, dy/dt + \ln t = 0$

11. $dy/dt = y + 5$

12. $dy/dt = 1 - 3y$

13. $dy/dt - 2y = 1$

14. $3 \, dy/dt + y = 2$

15. $2 \, dy/dt + 2y = 3$

16. $dy/dt - 0.5y + 2 = 0$

(17–22) Find the solutions of the following diferential equations that satisfy the given initial conditions.

17. $dy/dt + 2y = 0$; $y = 1$ when $t = 1$.

18. $2\,dy/dt - y = 0$; $y = 3$ when $t = \frac{1}{4}$.

19. $dy/dt - 2e^t = 0$; $y = 7$ when $t = 0$

20. $dy/dx = xe^{x^2}$; $y = 3$ when $x = 0$

21. $dy/dt = 2y + 3$; $y = 5$ when $t = 0$

22. $dy/dt + 2y = 4$; $y = 3$ when $t = 0$

23. (*Continuous Compounding*) An initial investment of $10,000 is growing continuously at a nominal annual rate of interest of 5%.

   a. Find the value of the investment at any time t.

   b. What is the value of the investment after 8 years?

   c. After how many years will the value of the investment be $20,000?

24. (*Continuous Growth of Stock Value*) A stock of initial value $2000 is growing continuously at a constant rate of 6% growth per annum.

   a. Determine the value of the stock after t years.

   b. After how long will the stock be worth $3000?

25. (*Population Growth*) Assume that the proportional growth rate $y'(t)/y(t)$ of the human population of the earth is constant. The population in 1930 was 2 billion and in 1975 was 4 billion. Taking 1930 to be $t = 0$, determine the population $y(t)$ of the earth at time t. According to this model, what should the population have been in 1960?

26. (*Radioactivity*) Thorium is used to date coral and shells. Its decay satisfies the differential equation $dy/dt = -9.2 \times 10^{-6}y$ when t is measured in years. What is the half-life of radioactive thorium? (See Exercise 38 in Section 6-4.)

27. (*Population Growth with Immigration*) A population has an initial size of 10,000 and a specific growth rate 0.04 (time measured in years). If the population increases due to immigration at the rate of 100 per year, what will be the population size after t years?

28. Repeat Exercise 27 in the case when the population is losing members at a rate of 150 per year due to emigration.

29. (*Spread of Epidemic*) An infectious disease spreads slowly through a large population. Let $p(t)$ be the proportion of the population that has been exposed to the disease within t years of its introduction. If $p'(t) = \frac{1}{3}[1 - p(t)]$

and $p(0) = 0$, find $p(t)$ for $t > 0$. After how many years has the proportion increased to 75%?

30. (*Growth with Immigration*) A population has size of $y(t)$ at time t. The specific growth rate for the population is 0.1 and there is a gain of population at a constant rate r due to immigration.

   a. Write down the differential equation that is satisfied by $y(t)$ and find its general solution.

   b. Find the particular solution in the case when $r = 100$ and the initial size of the population at $t = 0$ is 2000.

31. (*Epidemics*) Consider the spread of a disease that has the property that once an individual becomes infected he always remains infectious. While only a small proportion of the population is infected with the disease, its spread can reasonably be modeled by the differential equation $dy/dt = ky$ (where y is the number of infected individuals at time t). Obtain y as a function of t assuming that at time $t = 0$ there are 587 infected individuals and at time $t = 1$ year there are 831 infected individuals in the population.

*32. (*Pollutant flushing*) A small lake of volume 10^6 cubic meters has accidentally been polluted by 10,000 kilograms of a highly toxic substance. A river flows into and then out of the lake at the rate of 20,000 cubic meters per hour. Assuming that the entering river contains fresh water and that the toxic substance is fully mixed throughout the lake at all times, write a differential equation for the mass of pollutant in the lake. Find the solution and calculate the number of hours for the mass of pollutant to decrease to 100 kilograms.

*33 (*Pollution*) The lake in Exercise 32 eventually recovered from the pollution accident, but then someone built a pulpmill upstream and began dumping mercury into the river at the rate of 0.01 kilogram per hour. Write a differential equation for the mass of mercury in the lake and find its solution. How much mercury will the lake eventually contain?

*34. (*Medicine*) A substance is being infused into a patient's bloodstream at the rate R milligrams per minute and is absorbed from the bloodstream at the rate kM, where k is a constant and M is the number of milligrams in the bloodstream at time t. Write a differential equation for $M(t)$ and find the solution, assuming that the infusion starts at $t = 0$. What is the limiting amount of the substance in the bloodstream?

*35. (*Growth of Capital*) An investment grows according to

the differential equation

$$\frac{dA}{dt} = rA + I(t)$$

where $100r$ is the nominal interest rate and $I(t)$ is the rate of new capital investment. Solve this equation when $I(t)$ is constant and $A(0) = 0$. Compare your answer with Exercise 24 in Section 17-3.

*36. *(Price in Nonequilibrium Market)* For a certain commodity, the supply and demand equations are as follows.

$$D: \quad p + 2x_D = 25$$

$$S: \quad p - 3x_S = 5$$

Assume that if the market is not in equilibrium ($x_D \neq x_S$), then the price changes at a rate proportional to the excess of demand over supply:

$$\frac{dp}{dt} = k(x_D - x_S).$$

Substitute for x_D and x_S and solve the resulting differential equation for $p(t)$. Show that whatever the initial price, the market eventually approaches equilibrium at $p = 17$.

37. *(Newton's Law of Cooling)* The temperature T of a cooling body changes according to the differential equation $dT/dt = k(T_s - T)$, where T_s is the temperature of the surroundings. Find a formula for $T(t)$ in the case when T_s is constant and $T(0) = T_0$.

*38. *(Profit and Advertising)* Assume that a firm's profit P as a function of expenditure on advertising A satisfies the differential equation

$$\frac{dP}{dA} = k(C - A)$$

where k and C are positive constants. By considering the sign of dp/dA for $A < C$ and $A > C$, give the significance of the constant C. Solve the differential equation for $P(A)$ given that $P(0) = P_0$. If $P_0 = 100$, $P(100) = 1100$, $P(200) = 1600$, calculate the optimal expenditure on advertising.

◥ 17-7 SEPARABLE DIFFERENTIAL EQUATIONS

A first-order differential equation is said to be **separable** if it can be expressed in the form

$$\frac{dy}{dt} = f(y)g(t)$$

That is, the right side is the product of a function of y and a function of t. ◄ **28**

A separable equation can be solved by moving all the terms involving y to the left (by dividing by $f(y)$) and by moving all terms involving t to the right (by multiplying by dt):

$$\frac{1}{f(y)} \, dy = g(t) \, dt.$$

The variables are said to have been **separated.** Now both sides can be integrated:

$$\int \frac{1}{f(y)} \, dy = \int g(t) \, dt.$$

In practice, these integrals may be hard, or even impossible, to evaluate, but apart from this difficulty, we can always solve a separable equation this way.

You will recognize that this is precisely the method used in Section 17-6 to obtain the general solution of the differential equation $dy/dt = ky$. We can

◄ 28. Are the following differential equations separable?

(a) $xy\dfrac{dy}{dx} = y + 1$;

(b) $\dfrac{dy}{dx} = x + y$;

(c) $\dfrac{dy}{dx} + 2y = xy.$

Answer **(a)** Yes; **(b)** no; **(c)** yes.

also use this method instead of the method we used earlier to solve the equation $dy/dt = ky + b$. We can separate the variables in this equation by writing it as

$$\frac{1}{ky + b} \, dy = dt.$$

Integrating both sides, we then obtain $\int \frac{1}{ky + b} \, dy = \int dt$, or, assuming that $y + \frac{b}{k} > 0$,

$$\frac{1}{k} \ln\left(y + \frac{b}{k}\right) = t + B$$

where B is an arbitrary constant. Solving this for y, we get

$$y + \frac{b}{k} = e^{kt + kB} = Ce^{kt}$$

where $C = e^{kB}$. This is the same solution as before. We leave it for you to verify that this same form is obtained for the solution if $y + b/k < 0$. ☞ **29**

☞ **29.** Find the general solution of the differential equation
$$xy\frac{dy}{dx} = y + 1 \text{ for the case } y > -1.$$

EXAMPLE 1 Find the solution of the differential equation

$$e^x \frac{dy}{dx} = y^2$$

that satisfies the initial condition that $y = 2$ when $x = 0$.

Solution Note that the independent variable here is x, not t. We can write the given differential equation as

$$\frac{1}{y^2} \, dy = \frac{1}{e^x} \, dx \qquad \text{or} \qquad y^{-2} \, dy = e^{-x} \, dx$$

where we have separated all the terms containing y on the left and all those containing x on the right. Integrating both sides we get

$$\int y^{-2} \, dy = \int e^{-x} \, dx.$$

Therefore

$$\frac{y^{-1}}{-1} = \frac{e^{-x}}{-1} + C \qquad \text{or} \qquad \frac{1}{y} = e^{-x} - C$$

where C is the constant of integration. Solving for y we obtain

$$y = \frac{1}{e^{-x} - C} = \frac{e^x}{1 - Ce^x}.$$

Answer $y - \ln(y + 1) = \ln x + C$, or equivalently, $x(y + 1)e^{-y} = B$.

To determine C we use the initial condition. Setting $y = 2$ and $x = 0$ in the general solution, we have

$$2 = \frac{e^0}{1 - Ce^0} = \frac{1}{1 - C}$$

from which it follows that $C = \frac{1}{2}$. Substituting this into the general solution we get

$$y = \frac{e^x}{1 - \frac{1}{2}e^x}$$

which provides the particular solution for the given initial conditions.

EXAMPLE 2 *(Demand Function)* If the elasticity of demand for a certain commodity is $-\frac{1}{2}$ for all values of its unit price, determine the demand relation.

Solution Let x be the number of units demanded at price p. We know that the elasticity of demand η is given by the formula

$$\eta = \frac{p}{x}\frac{dx}{dp}$$

(See Section 15-3.) Since $\eta = -\frac{1}{2}$ we have the differential equation

$$\frac{p}{x}\frac{dx}{dp} = -\frac{1}{2} \qquad \text{or} \qquad \frac{dx}{dp} = -\frac{x}{2p}.$$

Separating the variables, we get

$$\frac{2}{x}\,dx = -\frac{1}{p}\,dp$$

and integrating both sides, we have

$$\int \frac{2}{x}\,dx = -\int \frac{1}{p}\,dp \qquad \text{or} \qquad 2\ln x = -\ln p + C$$

where C is the constant of integration. Combining the logarithms, we then have $\ln(px^2) = C$. We can write this in exponential form as

$$px^2 = D$$

where $D = e^C$. D is again an arbitrary constant that cannot be determined without additional information. This is the required demand relation.

The Logistic Differential Equation

The differential equation

$$\frac{dy}{dt} = py(m - y) \tag{1}$$

where p and m are constants, is called the **logistic equation.** Its importance arose originally as a model of population growth in a restricted environment, but it has subsequently found a number of other applications. Some of these further applications will be found in the exercises.

The differential equation $dy/dt = ky$ applies to a population when the environment does not restrict its growth. In most cases, however, a stage is reached where further growth of the population is not possible, and the size levels off at some value that is the maximum population that can be supported by the given environment. Let us denote this maximum value by m. Then any differential equation that describes the growth must satisfy the condition that the rate of growth becomes zero as y approaches m; that is,

$$\frac{dy}{dt} \to 0 \quad \text{as} \quad y \to m.$$

Furthermore, if by some chance the size of the population happened to exceed m, then it should decrease; that is,

$$\frac{dy}{dt} < 0 \quad \text{if} \quad y > m.$$

Observe that the differential equation (1) satisfies these requirements.

There is also one further requirement that any reasonable model of population growth must satisfy. If the size of the population is very small, then the restrictions imposed by the environment will have little effect, and the growth will be approximately exponential. In equation (1), if y is much smaller than m, then $m - y \approx m$, and the differential equation becomes approximately

$$\frac{dy}{dt} \approx pmy.$$

This does indeed give approximate exponential growth and the specific growth rate is $k = pm$. The logistic equation (1) is not the only differential equation that satisfies these requirements for restricted growth, but it is the simplest equation that does so.

Now let us turn to the solution of the logistic equation. We shall derive the solution for general constants m and p, but if you have difficulty following this, try going through the argument first with some particular values, such as $m = 2$ and $p = 3$. Separating the variables in (1), we obtain

$$\frac{1}{y(m - y)} \, dy = p \, dt$$

and integrating both sides, we get

$$\int \frac{1}{y(m - y)} \, dy = \int p \, dt.$$

The integral on the left here can be evaluated by using Formula 15 in Appendix II. However, rather than this, we shall show you a useful method for finding such integrals—in fact, this is the way Formula 15 was derived. The trick is to express the integrand in terms of partial fractions. In the case we have, it is easy to see that

$$\frac{1}{y} + \frac{1}{m - y} = \frac{m}{y(m - y)}.$$

(Simply combine the two fractions on the left with their common denominator.) Thus, after multiplying through by m, the integrated equation above becomes

$$\int \left[\frac{1}{y} + \frac{1}{m - y}\right] dy = \int mp \, dt.$$

We can now integrate both sides, and we get

$$\ln y - \ln (m - y) = mpt + B$$

where B is the constant of integration. Here we have assumed that $0 < y < m$ so that the arguments of the logarithms are positive and we do not need to use absolute value signs. Combining the logarithms and setting $k = mp$, we get

$$\ln \left(\frac{y}{m - y}\right) = kt + B.$$

Thus

$$\frac{y}{m - y} = e^{B+kt} = e^B e^{kt} = A^{-1} e^{kt}$$

where we have written $A^{-1} = e^B$. The reason for defining A like this is to make the final answer simpler. Then solving for y, we get

$$Ae^{-kt}y = m - y, \qquad y(1 + Ae^{-kt}) = m, \qquad y = \frac{m}{1 + Ae^{-kt}}. \qquad (2)$$

This is the usual form in which the general solution is given and is often called the logistic function. The constant A is determined as usual from the initial value of y.

We leave it as an exercise for you to verify that the general solution is still given by the formula (2) in the case when $y > m$, the only difference being that the constant A is negative. ☛ **30**

☛ **30.** Find the general solution of the differential equation

$$\frac{dy}{dt} = y(y - 1) \text{ for the case when}$$

$$0 < y < 1.$$

EXAMPLE 3 (*Population Growth*) For a certain rabbit population the growth follows the logistic equation (1) with the constant $k = pm$ having the value 0.25 when time is measured in months. The population is suddenly reduced from its steady value m to a size equal to 1% of m by an epidemic of myxamatosis. How many months does it take the population to recover to 90% of its maximum value? Find an expression for the population size after t months.

Solution The population size $y(t)$ satisfies the differential equation

$$\frac{dy}{dt} = py(m - y) = \frac{0.25}{m}y(m - y)$$

Answer $y = \dfrac{1}{1 + Ae^t}$.

since $p = k/m \doteq 0.25/m$ is given. Separating the variables, we get

$$\frac{m}{y(m-y)} \, dy = 0.25 \, dt$$

and proceeding to integrate both sides using partial fractions on the left as we did above, we reach the result

$$\ln\left(\frac{y}{m-y}\right) = 0.25t + C. \qquad \text{(i)}$$

In the present problem, $y > 0$ and $y < m$, so the argument of the logarithm is positive and we do not require absolute value signs. At $t = 0$, the initial size is $y = 0.01m$ and substituting these values allows C to be determined:

$$\ln\left(\frac{0.01m}{m-0.01m}\right) = 0.25(0) + C$$

or $C = -\ln 99$. Substituting this value of C in (i) and combining the logs, we can write the solution as

$$\ln\left(\frac{99y}{m-y}\right) = 0.25t.$$

The first part of the question can be answered directly from this equation. The population reaches 90% of its ultimate size when $y = 0.9m$, and we get

$$0.25t = \ln\left(\frac{99(0.9m)}{m-0.9m}\right) = \ln 891.$$

Hence, $t = 4 \ln 891 \approx 27.2$. So it takes 27.2 months for the population to recover to 90% of its maximum value.

To complete the solution for y, we write it as

$$\frac{99y}{m-y} = e^{0.25t}$$

and then solve this for y. The result is

$$y = \frac{m}{1 + 99e^{-0.25t}}.$$

EXERCISES 17-7

(1–10) Find the general solution of the following differential equations.

1. $dy/dx = xy$

2. $dy/dx = x + xy$

3. $dy/dt + 2ty^2 = 0$

4. $dy/dt = e^{t+y}$

5. $dy/dt = 3t^2 e^{-y}$

6. $dy/dt + 6t^2 \sqrt{y} = 0$

7. $dy/dt = y(y-1)$

8. $dy/dt + y^2 = 4$

9. $t \, dy/dt + ty = y$

10. $t \, dy/dt - ty = 2y$

(11–18) Find the solutions of the following differential equations that satisfy the given initial conditions.

11. $dy/dx = 2xy;\quad y = 1$ when $x = 0$

12. $dy/dt = y/\sqrt{t}$; $y = e$ when $t = 0$

13. $dy/dt = 3t^2 y$; $y = 2$ when $t = 0$

14. $dy/dx = y(y - 1)$, $y > 1$; $y = 2$ when $x = 0$

15. $dy/dt = 2y(3 - y)$, $0 < y < 3$; $y = 2$ when $t = 0$

16. $2\, dy/dt = y(4 - y)$, $y > 4$; $y = 2$ when $t = 0$

17. $dy/dt = te^{t+y}$; $y = 0$ when $t = 0$

18. $du/dy = e^{u-y}$; $u = 0$ when $y = 0$

19. *(Elasticity)* The elasticity of demand for a certain commodity is $\eta = -\frac{2}{3}$. Determine the demand relation $p = f(x)$ if $p = 2$ when $x = 4$.

20. *(Elasticity)* The elasticity of demand for a certain commodity is given by $\eta = -2$. Determine the demand function $p = f(x)$ if $p = \frac{1}{2}$ when $x = 4$.

21. *(Elasticity)* The elasticity of demand for a certain commodity is given by $\eta = (x - 200)/x$. Determine the demand function $p = f(x)$ if $0 < x < 200$ and $p = 5$ when $x = 190$.

22. *(Elasticity)* The elasticity of demand is $\eta = p/(p - 10)$. Determine the demand function $p = f(x)$ if $0 < p < 10$ and $p = 7$ when $x = 15$.

23. *(Biochemistry)* According to the Michaelis-Menten equation, the rate at which an enzyme reaction takes place is given by

$$\frac{dy}{dt} = -\frac{My}{K + y}$$

where M and K are constant and y is the amount present at time t of the substrate that is being transformed by the enzyme. Find an implicit equation expressing y as a function of t.

24. *(Limited Growth Model)* The limited growth model of

von Bertalanffy can be obtained from the differential equation

$$\frac{dy}{dt} = 3ky^{2/3}(y_m^{1/3} - y^{1/3}).$$

Find an expression for y as a function of t. (*Hint:* Substitute $y^{1/3} = u$ in the integral to be evaluated.)

25. *(Logistic Model)* In a town whose population is 2000, the spread of an epidemic of influenza follows the differential equation

$$\frac{dy}{dt} = py(2000 - y)$$

where y is the number of people infected at time t (t measured in weeks) and $p = 0.002$. If there are initially two people infected, find y as a function of t. How long is it before three-quarters of the population are infected?

26. *(Logistic Model)* We can construct a simple model of the spread of an infection through a population in the following way. Let n be the total number of susceptible (i.e., nonimmune) individuals in the original population. Let $y(t)$ be the number of infected individuals at time t. Then $n - y(t)$ gives the number of remaining uninfected susceptibles. The model consists of setting

$$\frac{dy}{dt} = ky(n - y)$$

where k is a constant. (Note that dy/dt is the rate of spread of the infection.) Find the solution for y as a function of t, and sketch its graph.

27. *(Logistic Model)* A population is growing according to the differential equation $dy/dt = 0.1y(1 - 10^{-6}y)$ when t is measured in years. How many years does it take the population to increase from an initial size of 10^5 to a size of 5×10^5?

◣ 17-8 APPLICATIONS TO PROBABILITY

Probability is concerned with observations or measurements taken in situations in which the outcome is to some degree unpredictable. In such a case we use the term *random variable* to denote a variable whose measured value may vary from one repetition of the observation to another. For example, if a standard die is rolled, the number of spots that show is a random variable; it may turn out to have any one of the values 1, 2, 3, 4, 5, or 6.

In contrast to this, there are situations or observations where the random variable may take any one of a *continuous* set of values that lie in a given interval. For example, if the random variable X denotes the height (in feet) of a randomly selected adult person in New York, then X can take any real number as its value which lies in the interval $3 \le X \le 8$ (assuming that the shortest adult is at least 3 feet tall and the tallest is at most 8 feet in height). In such a case, the random variable is known as a *continuous random variable*.

In the case of a continuous random variable we are usually interested in the probability that the measured value falls within some given interval. For example, we may want to know the probability that an adult in New York is between 6 feet and 6.5 feet in height. (Such questions would be of interest to clothing manufacturers, for example.) In general, if X is a continuous random variable taking values in the interval $a \le X \le b$, we would be interested in the probability that the measured value of X falls between c and d, where $c \le d$ are two numbers between a and b. We write this probability as $P(c \le X \le d)$.

For most continuous random variables there exists a function $f(x)$ called the *probability density function** such that this probability is given by the following definite integral:

$$P(c \le X \le d) = \int_c^d f(x)\, dx. \tag{1}$$

Since the probability on the left must be nonnegative for all values of c and d ($c \le d$), the integrand cannot be negative. That is,

$$f(x) \ge 0 \tag{2}$$

for all values of x where it is defined.

In view of the relationship between definite integrals and areas under curves, we see that $P(c \le X \le d)$ as given by Equation (1) is equal to the area underneath the graph of $y = f(x)$ lying between the vertical lines $x = c$ and $x = d$. (See Figure 24.) It is this association of probabilities with areas under the graph of f that give the density function its usefulness.

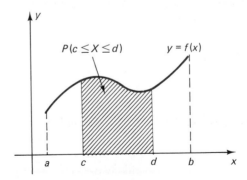

FIGURE 24

---

*Abbreviated p.d.f. The student should beware that many writers use p.d.f. to stand for "probability distribution function," which is different from the probability density function.

The event that X lies in its total interval $[a, b]$ is certain to happen, so has probability 1. That is,

$$P(a \le X \le b) = \int_a^b f(x)\, dx = 1. \tag{3}$$

In other words, the total area under the curve $y = f(x)$ between $x = a$ and $x = b$ must be equal to 1. ☛ **31**

31. For the probability density function $f(x) = 2x$ on the interval $0 \le x \le 1$, calculate the probabilities

(a) $P(0 \le x \le \frac{1}{2})$;

(b) $P(\frac{1}{2} \le x \le 1)$;

(c) $P(\frac{1}{2} \le x \le \frac{3}{4})$.

EXAMPLE 1 Given $f(x) = \frac{1}{4}(2x + 3)$. Determine the constant c so that $f(x)$ represents the p.d.f. of some continuous random variable on the interval $0 \le x \le c$. Find also the probability that this random variable takes a value less than $c/3$.

Solution If $f(x)$ represents a p.d.f. on the interval $0 \le x \le c$, then we must have

$$1 = \int_0^c f(x)\, dx = \int_0^c \tfrac{1}{4}(2x + 3)\, dx = \tfrac{1}{4}\left[x^2 + 3x\right]_0^c = \tfrac{1}{4}(c^2 + 3c).$$

or $c^2 + 3c - 4 = 0$. Therefore,

$$c = \frac{-3 \pm \sqrt{9 + 16}}{2} = \frac{-3 \pm 5}{2} = 1,\ -4.$$

Since c is required to be nonnegative in the problem, the only possible value of c is 1. We still have to verify that $f(x) = \frac{1}{4}(2x + 3)$ is nonnegative on $0 \le x \le 1$. This is true, as can be seen from the graph of $f(x)$ in Figure 25. Hence

$$f(x) = \tfrac{1}{4}(2x + 3) \quad \text{on} \quad 0 \le x \le c$$

represents a p.d.f. provided that $c = 1$.

$$P\left(X < \frac{c}{3}\right) = P(0 \le X < \tfrac{1}{3})$$

$$= \int_0^{1/3} f(x)\, dx = \int_0^{1/3} \tfrac{1}{4}(2x + 3)\, dx$$

$$= \left[\tfrac{1}{4}(x^2 + 3x)\right]_0^{1/3} = \tfrac{1}{4}(\tfrac{1}{9} + 1) = \tfrac{5}{18}. \quad ☛ \ 32$$

Answer **(a)** $\frac{1}{4}$; **(b)** $\frac{3}{4}$; **(c)** $\frac{5}{16}$.

32. Find c such that $f(x) = \frac{5}{6} - \frac{1}{3}x$ is a probability density function of the interval $0 \le x \le c$.

We shall now describe certain widely used probability distributions. The first of these is the so-called *uniform distribution,* which describes a situation or experiment in which the outcomes in the interval $a \le x \le b$ are all equally likely to occur. The p.d.f. in this case is simply the constant function given by

$$f(x) = \begin{cases} \dfrac{1}{b - a} & \text{for } a \le x \le b \\ 0 & \text{otherwise.} \end{cases}$$

Answer $c = 2$
(if $c = 3$, $f(x)$ takes negative values).

The graph of the uniform density function is as shown in Figure 26. $f(x)$ just

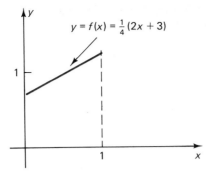

FIGURE 25

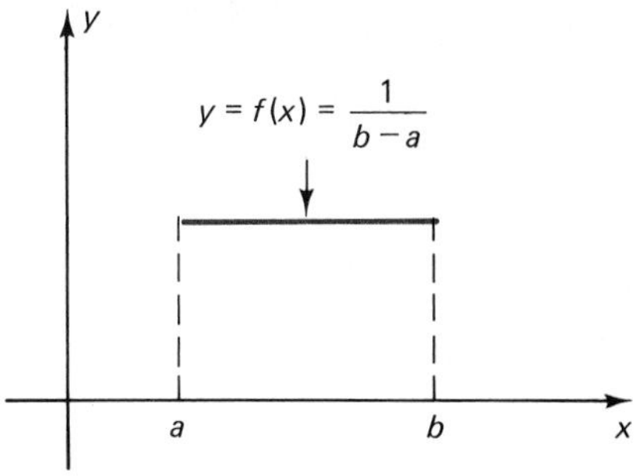

FIGURE 26

defined is clearly a density function, because $f(x) \geq 0$ on $a \leq x \leq b$ (since $b - a > 0$) and

$$\int_a^b f(x)\, dx = \int_a^b \frac{1}{b-a}\, dx = \left[\frac{x}{b-a}\right]_a^b = \frac{b-a}{b-a} = 1.$$

EXAMPLE 2 *(Waiting Time)* The city bus leaves the campus bus stop for downtown every 20 minutes. A student arrives at the bus stop at random to catch the bus. What is the probability that she will have to wait at least 5 minutes before she catches the bus?

Solution The random variable X, which is the time to wait until the next bus arrives, is uniformly distributed in the interval $0 \leq X \leq 20$. Thus the p.d.f. is given by

$$f(x) = \begin{cases} \dfrac{1}{20} & \text{for } 0 \leq x \leq 20 \\ 0 & \text{otherwise.} \end{cases}$$

Therefore

$$\begin{aligned} P(X \geq 5) &= \int_5^{20} f(x)\, dx \\ &= \int_5^{20} \frac{1}{20}\, dx = \left[\frac{x}{20}\right]_5^{20} = \frac{20}{20} - \frac{5}{20} = \frac{3}{4}. \end{aligned}$$ ☛ 33

☛ **33.** Write down the uniform probability density function $f(x)$ on the interval $1 \leq x \leq 9$. Find $P(2 \leq x \leq 5)$.

We often need to consider continuous random variables whose values lie not on a finite interval $a \leq X \leq b$ but on a semi-infinite interval of the type $a \leq X < \infty$ or on the whole infinite interval $-\infty < X < \infty$. In such cases we must allow $b \to \infty$ and (in the second case) $a \to -\infty$; then certain probabilities are given by improper integrals (see Section 17-2). For example, if X takes values in $-\infty < X < \infty$, then the probability that $X \leq d$ is given by

$$P(X \leq d) = \int_{-\infty}^d f(t)\, dt = \lim_{a \to -\infty} \int_a^d f(t)\, dt.$$

Answer $f(x) = \frac{1}{8}; \frac{3}{8}$.

A second probability distribution which has widespread application is the so-called *exponential distribution* and the p.d.f. in this case is given by

$$f(x) = \begin{cases} \dfrac{1}{k} e^{-x/k} & \text{for } 0 \le x < \infty \\ 0 & \text{otherwise.} \end{cases}$$

where k is a certain positive constant. Clearly, $f(x) \ge 0$ and

$$\int_0^\infty f(x)\, dx = \lim_{b \to \infty} \int_0^b \frac{1}{k} e^{-x/k}\, dx = \lim_{b \to \infty} \left[-e^{-x/k} \right]_0^b = \lim_{b \to \infty} (-e^{-b/k} + e^0) = 1$$

because $e^{-b/k} \to 0$ as $b \to \infty$ and $e^0 = 1$. Thus $f(x)$ defined above satisfies both conditions required of a density function. The graph of a typical exponential density function is shown in Figure 27. ☛ 34

☛ **34.** For what value of A is $f(x) = \dfrac{Ax}{(1 + x^2)^2}$ a probability density function on the interval $0 \le x < \infty$? Evaluate $P(0 \le x \le 2)$.

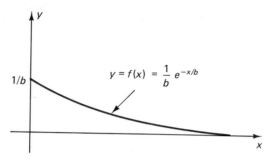

FIGURE 27

EXAMPLE 3 *(Lifetime of Light Bulbs)* The lifetimes of a certain type of light bulbs (in hours) follow an exponential distribution whose density function is given by

$$f(x) = \frac{1}{200} e^{-x/200} \qquad 0 \le x < \infty.$$

Find the probability that a randomly selected light bulb will last: (a) over 100 hours but less than 300 hours; (b) over 200 hours.

Solution If X denotes the lifetime of a randomly selected light bulb, then the probability that the lifetime will lie between two given values c and d is

$$P(c \le X \le d) = \int_c^d f(x)\, dx = \int_c^d \frac{1}{200} e^{-x/200}\, dx$$

$$= \frac{1}{200} \left[-200 e^{-x/200} \right]_c^d = e^{-c/200} - e^{-d/200}.$$

(a) Taking $c = 100$ and $d = 300$, we get

$$P(100 \le X \le 300) = e^{-100/200} - e^{-300/200} = e^{-1/2} - e^{-3/2} \approx 0.38.$$

(b) Taking $c = 200$ and allowing $d \to \infty$, we get

Answer $A = 2$. $P(0 \le x \le 2) = \frac{4}{5}$.

$$P(X \geq 200) = \lim_{d \to \infty} P(200 \leq X < d)$$
$$= \lim_{d \to \infty} (e^{-200/200} - e^{-d/200}) = e^{-1} \approx 0.37.$$

since $e^{-d/200} \to 0$ as $d \to \infty$.

The exponential probability distribution is an important one with many applications. The light bulb example is typical of a range of applications to reliability problems—that is, problems in which we are concerned with the probability of failure of some component or system. Another area of application of this distribution relates to the occurrence of random event in time. For example, we might let the random variable T be the time that will elapse before the next major oil-tanker disaster. Then T will follow an exponential distribution.

In Example 3, we have worked out the probabilities that the lifetime of a randomly selected light bulb will be between 100 and 300 hours or will be over 200 hours. Another question one might ask is: What is the average lifetime of the bulbs? The answer to this question involves the concept of the **expected value** or **mean** of a random variable X which is generally denoted by μ (read as "mu"). This quantity is defined by

$$\mu = \int_a^b x f(x)\, dx$$

where $f(x)$ is the p.d.f. The significance of the mean μ is that it measures the average value of the random variable if a very large number of measurements are made.

EXAMPLE 4 Let X be the lifetime in hours of a randomly selected light bulb of a certain type. The probability density function for X is $f(x) = (1/k)e^{-x/k}$, where k is a certain constant. Find the mean of X, that is, the average lifetime of the light bulbs in question.

Solution

$$\mu = \int_a^b x f(x)\, dx = \int_0^\infty \frac{x}{k} e^{-x/k}\, dx.$$

Observe that the limits on the integration are 0 and ∞ since the lifetime can be any positive real number. From Formula 69 in Appendix II with $a = -1/k$ (or by using integration by parts) we obtain the result that

$$\int_0^\infty xe^{-x/k}\, dx = \lim_{b \to \infty} \left[\int_0^b xe^{-x/k}\, dx \right] = \lim_{b \to \infty} \left[(-kx - k^2)e^{-x/k} \right]_0^b$$
$$= \lim_{b \to \infty} [(-kb - k^2)e^{-b/k} + k^2] = k^2.$$

The value of the integral at the upper limit is zero since $e^{-x/k} \to 0$ and $xe^{-x/k} \to 0$ as $x \to \infty$. (In general, $x^n e^{-cx} \to 0$ as $x \to \infty$ for any positive values of n and c.) Therefore

$$\mu = \frac{1}{k} \int_0^\infty x e^{-x/k}\, dx = k.$$

The constant k occurring in the density function provides the mean lifetime of the bulb. For example, if the density function is $f(x) = (1/200)e^{-x/200}$, the mean lifetime would be 200 hours. ☛ 35

☛ **35.** Find the expected value for
(a) the uniform probability density on the interval $[a, b]$;

(b) the p.d.f. $f(x) = \sqrt{\dfrac{2}{\pi}} e^{-(1/2)x^2}$ on the interval $[0, \infty)$.

Recall that we defined the exponential distribution to correspond to the density function $f(x) = (1/k)e^{-x/k}$. We have now shown that the exponential distribution has the mean $\mu = k$. Because of this the parameter k is often replaced by the symbol μ and the density function is written in the form

$$f(x) = \frac{1}{\mu} e^{-x/\mu}.$$

EXAMPLE 5 Customers arrive at a certain service station with the time between arrivals obeying the exponential distribution with an average of 20 customers per hour. If the attendant leaves his post for a quick smoke and is away for 2 minutes, find the probability that a customer will arrive while he is away.

Solution Since 20 customers arrive on an average each hour, the average time between arrivals is $\frac{1}{20}$ hour or 3 minutes. Thus, taking the random variable to be the time interval until the next customer arrives, X will be exponentially distributed with $\mu = 3$. The p.d.f. is therefore

$$f(x) = \frac{1}{\mu} e^{-x/\mu} = \frac{1}{3} e^{-x/3}.$$

The probability that a customer will arrive in less than 2 minutes is

$$P(X \le 2) = \int_0^2 f(x)\, dx = \int_0^2 \frac{1}{3} e^{-x/3}\, dx = \left[-e^{-x/3} \right]_0^2 = 1 - e^{-2/3} \approx 0.49.$$

So the attendant has a 51% chance of getting away without a customer arriving.

We close this section by describing briefly the most widely used of all distributions, known as the **normal distribution.** The p.d.f. in this case is given by

$$f(x) = \frac{1}{\sqrt{2\pi}\sigma} e^{-(x-\mu)^2/2\sigma^2} \qquad \text{for } -\infty < x < \infty$$

where μ denotes the mean of the normal random variable. The graph of $f(x)$ is the well-known bell-shaped curve which is symmetric about the line $x = \mu$, as shown in Figure 28.

The parameter σ (sigma) which occurs in the density function for the normal random variable is called the *standard deviation*. It provides a measure

Answer **(a)** $\frac{1}{2}(a + b)$; **(b)** $\sqrt{2/\pi}$.

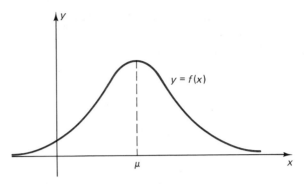

FIGURE 28

of the width of the bell-shaped curve. If σ is very small, the curve is a very tall and narrow bell, which means that the measured values of the random variable will almost always be very close to μ. If σ is large, the curve is flat and wide. In this case the measurements will be widely scattered, often being far away from the mean μ.

The probability that a normal variable X takes any value between c and d is given by the area beneath this curve lying between $x = c$ and $x = d$, that is,

$$P(c \le X \le d) = \int_c^d \frac{1}{\sqrt{2\pi}\sigma} e^{(x-\mu)^2/2\sigma^2} \, dx.$$

The special case $\mu = 0$ and $\sigma = 1$ is called the **standard normal distribution.** Denoting the random variable by Z in this case, we have

$$P(c \le Z \le d) = \frac{1}{\sqrt{2\pi}} \int_c^d e^{-x^2/2} \, dx.$$

This integral cannot be evaluated by elementary methods, but must be evaluated numerically, as in Section 17-5. Its values can be found in any book on introductory statistics.

EXERCISES 17-8

(1–8) In each of the following exercises, determine the constant c so that $f(x)$ is a probability density function on the given interval. Find also the mean μ in each case.

1. $f(x) = cx(3 - x)$ on $0 \le x \le 3$

2. $f(x) = \frac{1}{4}x + c$ on $-1 \le x \le 1$

3. $f(x) = \frac{1}{2}e^{-cx}$ on $0 \le x < \infty$

4. $f(x) = ce^{-3x}$ on $0 \le x < \infty$

5. $f(x) = \frac{2}{3}(x + 1)$ on $0 \le x \le c$

6. $f(x) = \frac{1}{12}(2x + 1)$ on $0 \le x \le c$

7. $f(x) = \dfrac{c}{(1 + x)^4}$ on $0 \le x < \infty$

    (*Hint*: Set $u = 1 + x$ in the integral for μ.)

8. $f(x) = \dfrac{c}{(x - 2)^5}$ on $3 \le x < \infty$

9. Given that $f(x) = 2x - 4$ on $0 \le x \le c$ and that $\int_0^c f(x) \, dx = 1$, determine c. Is $f(x)$ a p.d.f.? If yes, calculate $P(X \le c/3)$.

10. Given that $f(x) = \frac{1}{6}(4x + 1)$ on $0 \le x \le c$ and that

$\int_0^c f(x)\, dx = 1$, determine c. Is $f(x)$ a p.d.f.? If yes, find $P(c/3 \le X \le 2c/3)$.

11. Determine the mean of a uniform distribution defined on the interval $a \le x \le b$.

12. (*Bus Stop Waiting Time*) A person arrives at the nearest bus stop at random to catch a bus for downtown which runs every half an hour.

    a. Find the probability that he will have to wait: (**i**) at most 10 minutes, before he catches the bus; (**ii**) at least 5 minutes before he gets the bus; (**iii**) at least 5 minutes but not over 15 minutes.

    b. What is the average waiting time in this case?

13. (*Airport Waiting Time*) Air service from Montreal to New York is every 1 hour. A person, not knowing the schedule, arrives at the airport at random to catch a flight for New York.

    a. What is the probability that he will have to wait: (**i**) between 10 and 20 minutes; (**ii**) at most 15 minutes; (**iii**) not less than 40 minutes?

    b. What is the expected waiting time in this case?

14. (*Average Travel Time*) Depending on traffic conditions, it takes Susan 25 to 40 minutes to drive from her house to her college. If she leaves home at 9:00 A.M. for her 9:30 A.M. class, what is the probability that she will not be late for her class? What is the average time of travel from her home to college? (Assume a uniform distribution.)

15. (*Uniform Distribution*) A certain machine completes its operation every 20 minutes. What is the probability that a person arriving at random will have to wait for at least 5 minutes for the operation to complete? Evaluate the mean of the waiting time.

16. (*Uniform Distribution*) The weights of medium-grade eggs are uniformly distributed between 38 and 42 grams. What is the probability that a medium-size egg selected at random will weigh over 40 grams? What is the mean weight of these eggs?

17. (*Length of Telephone Calls*) If X denotes the length of telephone calls placed by the employees of a certain firm, then it is known that X follows an exponential distribution with p.d.f.

$$f(x) = 0.4e^{-0.4x}.$$

What is the probability that a random call:

a. Will last for at least 5 minutes?

b. Will not last over 3 minutes?

18. (*Life of Automobiles*) If X is the lifetime (in years) of a certain make of automobile then the density function of X is $\frac{1}{8}e^{-x/8}$. What is the probability that one of these automobiles lasts:

a. less than 5 years?

b. more than 10 years?

19. (*Typist Errors*) The random variable X denotes the number of words a certain typist types before making an error. The density function for X is $c^{-1}e^{-x/c}$, where $c = 1000$. What is the probability that the typist does not make the next error before typing 200 words?

20. (*Insurance Claims*) A large insurance company classifies an accident as "catastrophic" if it results in claims exceeding 10 million dollars. The time interval T (measured in months) between such catastrophes is a continuous random variable whose density function is $\frac{1}{20}e^{-t/20}$. Calculate:

    a. $P(10 \le T \le 20)$

    b. $P(T \ge 15)$.

21. (*Reliability*) The firm Western Electronics Ltd., which manufactures television sets, finds from their previous data that the time when its new television sets need a first major repair can be described by the exponential density function $f(x) = 0.2e^{-0.2x}$ (x is in years).

    a. If the firm guarantees its sets for 2 years, what proportion of sets will come back for a major repair during the guarantee period?

    b. If the firm sold 10,000 sets, how many sets can be expected to come back for a first major repair within 1 year of sale?

22. (*Reliability*) A car manufacturer knows that the time when its new car will need a major repair is described by the exponential density function

$$f(x) = \frac{1}{5}e^{-x/5}.$$

If the manufacturer sold 20,000 cars in a particular year and it gave a 1-year guarantee on any major repair, what number of cars can be expected to need their first major repair during the guarantee period?

23. (*Uniform Distribution*) The weights of medium-grade

eggs are uniformly distributed. If such an egg is selected at random, what is the probability that at least 80% of medium-grade eggs weigh more than the one chosen?

24. *(Income Distribution)* Let X be the income of a randomly chosen family in a certain country (in thousands of dollars). If the density function for X is $\frac{1}{100}xe^{-x/10}$, find the probability that:

a. X lies between 10 and 20.

b. X is greater than 10.

25. *(Sales Volume)* The number of pairs of shoes sold each day by a shoe store is a continuous random variable whose density function is $f(x) = cxe^{-(x/40)^2}$. Find the value of c. Find the probability that more than 50 pairs of shoes are sold on a given day.

26. *(Botany)* The lifespan (measured in days) of a certain species of plant in a given environment is a continuous ran-

dom variable with density function $f(x) = \frac{1}{100}e^{-x/100}$. Determine:

a. The average or expected lifespan of the plants.

b. The probability that a given plant will die within 50 days.

27. *(Digestion Time)* Let T denote the digestion time in hours of a unit of food. Then T is a random variable, and suppose that its probability density function is $f(x) = 9xe^{-3x}$ on the interval $0 \le x < \infty$. Find $P(0 \le T \le x)$ and use this to calculate:

a. The probability that a unit of food is digested within 2 hours.

b. The probability that it is still undigested after 3 hours.

28. The random variable X takes values in the range $0 \le X < \infty$ and $P(0 \le X \le x) = 1 - (1 + x^2)^{-1}$. Find the probability density function. Calculate $P(1 < X < 3)$ and $P(X = 2)$.

CHAPTER REVIEW

Key Terms, Symbols, and Concepts

17.1 Definite integral. Limits of integration, lower limit, upper limit.
Fundamental theorem of calculus.

17.2 Improper integrals,

$$\int_a^\infty f(x)\, dx, \quad \int_{-\infty}^b f(x)\, dx, \quad \int_{-\infty}^\infty f(x)\, dx.$$

17.3 Lorentz curve, coefficient of inequality for income distribution.
Learning curve.
Present value of a continuous income.
Consumers' surplus, producers' surplus.

17.4 Average value of a function.

17.5 Numerical integration. Trapezoidal rule. Simpson's rule.

17.6 Differential equation of order n. Linear differential equation.
Solution of a differential equation. General solution, initial condition.
Specific growth rate.

17.7 Separable differential equation; separation of variables.
Logistic differential equation, logistic function.

17.8 Continuous random variable, probability density function (p.d.f.).
Uniform and exponential probability distributions.
Expected value (mean) of a random variable.

Formulas

$$\int_a^b f(x)\, dx = [F(x)]_a^b = F(b) - F(a), \text{ where}$$
$F'(x) = f(x).$

When $f(x) \ge 0$, the area between $y = f(x)$ and the x-axis from $x = a$ to $x = b$ is equal to $\int_a^b f(x)\, dx$. If $f(x) \le 0$, the area is $\int_a^b -f(x)\, dx$.

Properties of definite integrals:

$$\frac{d}{dx}\int_a^b f(x)\, dx = 0, \quad \frac{d}{dx}\int_a^x f(t)\, dt = f(x)$$

$$\int_a^b \frac{d}{dx} F(x)\, dx = F(b) - F(a)$$

$$\int_a^a f(x)\, dx = 0, \qquad \int_b^a f(x)\, dx = -\int_a^b f(x)\, dx$$

$$\int_a^c f(x)\, dx = \int_a^b f(x)\, dx + \int_b^c f(x)\, dx.$$

Area between two curves from $x = a$ to $x = b$ is
$$\int_a^b (y_{\text{upper}} - y_{\text{lower}})\, dx.$$

Present value $= \displaystyle\int_0^T f(t) e^{-rt}\, dt$, where $r = R/100$.

$$\text{C.S.} = \int_0^{x_0} [f(x) - p_0]\, dx$$

$$\text{P.S.} = \int_0^{x_0} [p_0 - g(x)]\, dx$$

where $\begin{cases} p = f(x) \text{ is the demand relation} \\ p = g(x) \text{ is the supply relation.} \\ (x_0, p_0) \text{ is the market equilibrium point} \end{cases}$

Average value of f: $\quad \bar{f} = \dfrac{1}{b-a} \displaystyle\int_a^b f(x)\, dx.$

Trapezoidal rule:
$$\int_a^b f(x)\, dx \approx \tfrac{1}{2} h\{y_1 + 2(y_2 + y_3 + \cdots + y_n) + y_{n+1}\}$$

Simpson's rule:
$$\int_a^b f(x)\, dx \approx \tfrac{1}{3} h\{y_1 + y_{n+1} + 2(\text{Sum of } y_j \text{ for Odd } j)$$
$$+ 4 (\text{Sum of } y_j \text{ for Even } j)\}$$

The general solution of the differential equation
$$\frac{dy}{dt} = ky \text{ is } y = Ce^{kt}.$$

The general solution of the differential equation
$$\frac{dy}{dt} = ky + b \text{ is } y = Ce^{kt} - \frac{b}{k}.$$

Logistic differential equation: $\dfrac{dy}{dt} = py(m - y).$

Logistic function: $y = \dfrac{m}{1 + Ae^{-kt}} \quad (k = pm).$

$$P(c \le x \le d) = \int_c^d f(x)\, dx, \quad \mu = \int_c^d x f(x)\, dx.$$

REVIEW EXERCISES FOR CHAPTER 17

1. State whether each of the following is true or false. Replace each false statement with a corresponding true statement.

   a. If $f(x)$ is continuous in $a \le x \le b$, then $\int_a^b f(x)\, dx$ represents the area bounded by the curve $y = f(x)$, the x-axis, and the lines $x = a$ and $x = b$.

   b. If $\int_a^b f(x)\, dx = \int_b^a f(x)\, dx$, then $\int_a^b f(x)\, dx = 0$.

   c. $\dfrac{d}{dx} \left[\int_a^x f(t)\, dt \right] = f'(x).$

   d. $\dfrac{d}{dx} \left[\int_a^b f(x)\, dx \right] = \int_a^b \dfrac{d}{dx} [f(x)]\, dx.$

   e. If $F(x)$ is an antiderivative of $f(x)$, then
   $$\frac{d}{dx} \left[\int_a^x f(t)\, dt \right] = F'(x).$$

   f. $\int_a^b f(x)\, dx$ is a function of x.

   g. $\int_a^b f(x)\, dx$ and $\int_a^b f(t)\, dt$ are different from one another.

   h. The function $y = e^{kt}$ is a solution of the differential equation $y(d^2y/dt^2) = (dy/dt)^2$ for all values of k.

   i. The differential equation $dy/dt = y + t$ may be solved by separation of variables.

   j. The differential equation $2(dy/dt)^2 = t^2(dy/dt + 2)$ is of second order.

   k. The solution of the differential equation $dy/dt = yt^2$ may be obtained as follows:
   $$y = \int yt^2\, dt = y \int t^2\, dt = \tfrac{1}{3} yt^3 + C.$$

   Therefore
   $$y(1 - \tfrac{1}{3}t^3) = C \quad \text{or} \quad y = C(1 - t^3/3)^{-1}.$$

l. Any continuous random variable has zero probability of taking any particular value.

m. The total area under the probability density curve is equal to 1.

n. The mean value of a continuous random variable is the value at which the density function is a maximum.

o. A random variable has a probability of 0.5 of being less than its mean.

2. Prove that $\int_a^{ab} \frac{1}{t} \, dt = \int_1^b \frac{1}{t} \, dt$.

(3–5) Find the area bounded by $y = f(x)$, the x-axis, and the lines $x = a$ and $x = b$, where $f(x)$, a, and b have the given values.

3. $f(x) = \ln x$; $a = 1, b = 2$

4. $f(x) = e^x$; $a = -2, b = 2$

5. $f(x) = 1/x^2$; $a = 1, b = 3$

6. Find the area bounded by the two curves $y = x^3$ and $y = \sqrt{x}$.

7. Find the area in the first quadrant bounded by $y = x^3$, $y = 2 - x^2$, and $x = 0$.

8. *(Cost Increment)* The marginal cost of producing the xth unit of a certain commodity is $6 - 0.02x$. Find the change in the total cost of production if the production level changes from 150 to 200.

9. *(Price Increment)* The marginal price of a commodity is given by $p'(x) = 15 - x$. Find the total change in price per unit if the demand increases from $x = 10$ to $x = 15$.

10. *(Lorentz Curve)* Find the coefficient of inequality for the income distribution given by the Lorentz curve $y = \frac{14}{15}x^2 + \frac{1}{15}x$, where x is the cumulative proportion of income recipients and y the cumulative proportion of national income.

11. *(Learning Curve)* After observing the first 400 units of its product, a firm determined that the labor time required to assemble the $(x + 1)$th unit was $f(x) = 500x^{-1/2}$. Find the total number of hours of labor required to produce an additional 500 units.

12. *(Maximum Profit)* The marginal cost and marginal revenue functions of a firm are $C'(x) = 5 + (5 - x)^2$ and $R'(x) = 37 - 4x$, respectively, where x denotes the number of units produced. The fixed costs are 25.

a. Find the level of production which will maximize the profits of the firm.

b. Find the total profit of the firm at this production level.

c. Find the profit if the production level is increased by 2 units beyond the maximum profit level output.

13. *(C.S. and P.S.)* Find the consumers' surplus and producers' surplus if the demand function is $p = 25 - 3x$ and the supply function is $p = 5 + 0.5x^2$.

(14–19) Solve the following differential equations.

14. $dy/dt = 2y(y - 1)$, $y > 1$

15. $dy/dx = xe^{x-y}$

16. $dy/dx = x^2/y$

17. $dy/dx = 4x\sqrt{y} + 2\sqrt{y}e^x$, $x = 0$ when $y = 0$

18. $dy/dt = y/(1 - y)$, $y = \frac{1}{2}$ when $t = 0$

19. $dy/dt + y(2 - y) = 0$, $y > 2$

20. *(Asset Growth)* The proportional growth rate $y'(t)/y(t)$ of a certain firm's asset value is given by $\frac{1}{3}\sqrt[3]{t}$ and its initial net worth at $t = 0$ was \$50,000 (t is measured in years).

a. Determine the net worth $y(t)$ of the firm at any time t.

b. After how many years will the firm be worth \$600,000?

21. *(Continuous Compounding)* An initial investment of P dollars is growing continuously at an annual rate of 6%. If the investment is worth \$26,997 after 5 years, find P.

22. *(Marginal Profit)* The Pacific Dairy Farm finds that its marginal profits are y dollars per liter of milk when the total production is x liters of milk, where

$$y = \frac{5,000,000 \ln (x + 50)}{(x + 50)^3}.$$

The farm breaks even (that is, has zero profit) when $x = 200$. If the farm produces 450 liters of milk each day, calculate the total daily profits of the farm.

23. *(Profit Maximization)* The cost and revenue rates for an oil drilling operation are given by $C'(t) = 9 + 2t^{1/2}$ and $R'(t) = 19 - 3t^{1/2}$, where t is measured in years and R and C are measured in millions of dollars. How long should the drilling be continued? What will be the maximum profit?

24. *(Profit Maximization)* Repeat Exercise 23, when $C'(t) = 4 + 3t^{2/3}$ and $R'(t) = 20 - t^{2/3}$.

25. *(Waiting Time)* A certain machine completes its operation every 15 minutes. What is the probability that a person arriving at random will have to wait at most 5 minutes for the operation to complete? Find the average waiting time.

26. *(Demand)* The demand for a certain commodity follows an exponential distribution whose p.d.f. is given by

$$f(x) = 0.02e^{-0.02x}$$

where x is the number of units of the commodity requested over a 1-week period. What is the probability that during a certain week, the number of units requested will be:

a. Less than 50?

b. Less than 150?

c. More than 200?

27. *(Length of Telephone Calls)* The length in minutes of telephone calls received by the employees of a certain firm follow an exponential distribution with p.d.f.

$$f(x) = 0.3e^{-0.3x}.$$

What is the probability that a random call received by an employee of the firm:

a. Will last for over 5 minutes?

b. Will last for at least 2 minutes?

c. Will not last over 4 minutes?

28. Find the means of the distributions whose p.d.f's are as follow:

a. $f(x) = \frac{1}{10}(3x^2 + 1)$ on $0 \le x \le 2$

b. $f(x) = \frac{6}{125}x(5 - x)$ on $0 \le x \le 5$

29. *(Insecticide Spraying)* Let $y = f(x)$ be the percentage of mosquitoes that survive after aerial spraying with an amount x of insecticide per square mile. Assume that $dy/dx = -ky$, where k is a constant (called the exponential law of survival). If 2000 pounds per square mile of insecticide kill 40% of the mosquitoes, how much insecticide is needed to kill 90% of them?

30. *(Air Travel Time)* Assume that the time of flight between London and Paris is uniformly distributed between 45 and 65 minutes, depending on air traffic density. What is the average flight time? What is the probability that any given flight will take less than an hour?

31. *(Average Demand)* The demand function of a certain product is $p = 30 - 0.1x$, where x units can be sold at a price of p each. Find the average demand over the price interval $p = 1$ to $p = 4$.

32. *(Average Profits)* The demand function of a firm's product is $p = 45 - 0.12x$, where x units can be sold at a price of p each. The cost of producing x units is given by $C(x) = 300 + 5x$. Find the average profit over the sales interval $x = 100$ to $x = 300$.

33. Use both the trapezoidal rule and Simpson's rule to approximate the value of $\displaystyle\int_0^4 \frac{1}{1 + x^2}\, dx$ by taking eight equal intervals. Give the answer to two decimal places.

34. Use both the trapezoidal rule and Simpson's rule to approximate the value of ln 2 by approximating the value of $\displaystyle\int_1^2 \frac{1}{x}\, dx$ by taking eight equal intervals. Give the answer to three decimal places.

CHAPTER 18

Several Variables

Chapter Objectives

18-1 FUNCTIONS AND DOMAINS
(a) Definition of a function of several variables and its domain; computation of function values;
(b) Calculation of the domain of a function of several variables; geometric representation of the domain for two independent variables;
(c) Coordinates in three dimensions;
(d) The graph of a function of two variables;
(e) Contour lines (level curves) for a function of two variables;
(f) Vertical sections for a function of two variables on which one independent variable is constant.

18-2 PARTIAL DERIVATIVES
(a) Definition of partial derivatives; geometrical significance of partial derivatives;
(b) Calculation of partial derivatives; application of the product rule, quotient rule, and chain rule;
(c) Calculation of second- and higher-order partial derivatives; equality of mixed partial derivatives.

18-3 APPLICATIONS TO BUSINESS ANALYSIS
(a) Partial derivative as a marginal rate;
(b) Marginal productivity of labor and capital;
(c) Demand relations; complementary and competing products; elasticity and cross elasticity;
(d) Approximations corresponding to the differentials.

18-4 OPTIMIZATION
(a) Definition of a local maximum and local minimum;
(b) Computation of critical points for a differentiable function of two variables;
(c) The delta test for a local extremum; examples of its application.

18-5 LAGRANGE MULTIPLIERS (OPTIONAL)
(a) Optimization problems involving one or more constraints on the variables, solution by elimination;
(b) Solution of constrained optimization problems using Lagrange multipliers;
(c) Examples and applications of Lagrange multipliers.

18-6 METHOD OF LEAST SQUARES
(a) The mean square error between a set of measurements and the line of best fit;
(b) The principle of least square error; equations for the coefficients;
(c) Example of the calculation of the line of best fit.

CHAPTER REVIEW

So far, we have restricted our attention to cases in which the dependent variable is a function of a single independent variable, $y = f(x)$. However, in many—perhaps most—applications, we come across situations in which one quantity depends not on just one other variable but on several variables.

EXAMPLE 1

(a) Consider a rectangle of length x and width y. Its area A is given by the product of length and width, or

$$A = xy.$$

The variable A depends on the two variables x and y.

(b) The demand, or total sales volume, for a product depends on the price at which it is offered for sale. However, in many cases the sales volume also depends on additional factors such as the amount spent by the manufacturer on advertising the product and the prices of competing products.

(c) The balance of payments of any nation is a function of a large number of variables. Interest rates in the country will affect the amount of foreign investment that flows in. The exchange rate of the national currency will affect the prices of its goods and hence will determine the volume of exports and also of imports. Average wage rates will also affect the prices of exports and therefore the volume. The amount of existing foreign investment in the nation will affect the profits taken out each year. Even the weather can have a powerful effect on the balance of payments if tourism plays a large role in the economy, or if the economy depends substantially on some agricultural crop.

In cases such as these, we need to study functions of several independent variables. For most of this chapter, we shall consider the case of two independent variables, and usually we shall use x and y to stand for them. The generalization to three or more independent variables is in most (but not all) respects quite straightforward. The dependent variable will usually be denoted by z, and we use the notation $z = f(x, y)$ to indicate that z is a function of both x and y.

We first give a formal definition of a function of two variables.

DEFINITION Let D be a set of pairs of real numbers (x, y), and let f be a rule that specifies a unique real number for each pair (x, y) in D. Then we say that f is a **function of the two variables** x and y and the set D is the **domain** of f. The value of f at the pair (x, y) is denoted by $f(x, y)$ and the set of all of these values is called the **range** of f.

EXAMPLE 2 If $f(x, y) = 2x + y$, calculate the value of f at the pair $(1, 2)$. Find the domain of f.

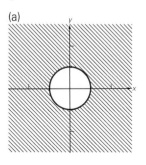
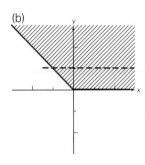
Solution The value is obtained by substituting the pair $x = 1$ and $y = 2$ into the expression for $f(x, y)$:

$$f(1, 2) = 2(1) + 2 = 4.$$

In this case, the value of f is a well-defined real number for all real values of x and y, so that the domain is the set of all pairs (x, y) of real numbers.

The domain D of a function of two variables can be viewed as a subset of points in the xy-plane. In Example 2 all pairs of real numbers (x, y) belong to the domain, so we can say that the domain consists of the whole xy-plane. The range of a function of two variables is a subset of the real numbers, just as it is for a function of one variable.

EXAMPLE 3 Given $f(x, y) = \sqrt{4 - x^2 - y^2}$, calculate $f(0, 2)$, $f(1, -1)$, and $f(1, 2)$. Find the domain of f and represent it graphically.

Solution Substituting the given values of x and y, we obtain the following values.

$$f(0, 2) = \sqrt{4 - 0^2 - 2^2} = \sqrt{0} = 0$$
$$f(1, -1) = \sqrt{4 - 1^2 - (-1)^2} = \sqrt{2}$$
$$f(1, 2) = \sqrt{4 - 1^2 - 2^2} = \sqrt{-1} \quad \text{(not defined)}$$

The pair $(1, 2)$ therefore does not belong to the domain of f. **☛ 1**

In order for $f(x, y)$ to be a well-defined real number, the quantity under the radical sign must be nonnegative. Thus

$$4 - x^2 - y^2 \geq 0$$
$$x^2 + y^2 \leq 4.$$

The domain of f therefore consists of those points (x, y) such that $x^2 + y^2 \leq 4$.

Geometrically, $x^2 + y^2 = 4$ is the equation of a circle centered at the origin with radius 2, and the inequality $x^2 + y^2 \leq 4$ holds at points inside and on this circle. These points form the domain D. (See Figure 1.) The point $(1, 2)$ lies outside the circle, consistent with our earlier finding that $f(1, 2)$ does not exist. **☛ 2**

EXAMPLE 4 (*Cost Function*) A firm produces two products, A and B. The cost of material and labor is \$4 for each unit of product A and \$7 for each unit of B. The fixed costs are \$1500 per week. Express the weekly cost C in terms of the units of A and B produced each week.

Solution If x units of product A and y units of product B are produced each week, then the labor and material costs for the two types of products are $4x$

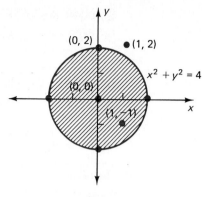

FIGURE 1

and $7y$ dollars, respectively. Thus the cost C (in dollars) is given by

$$C = \text{Labor and Material Costs} + \text{Fixed Costs} = 4x + 7y + 1500.$$

C is a function of x and y.

Now let us consider the generalization of the preceding ideas to functions of more than two variables. When there are three independent variables, it is usual to denote them by x, y, and z, and to denote the dependent variable by w. The function consists of a rule that assigns a real number to each set of three values (x, y, z) of the independent variables. We write the value as $w = f(x, y, z)$.

EXAMPLE 5 If

$$f(x, y, z) = \frac{\sqrt{9 - x^2 - y^2}}{x + z}$$

evaluate $f(1, -1, 4)$ and $f(-1, 2, 1)$. Find the domain of f.

Solution The value of f at any given (x, y, z) is obtained by substituting the given values of x, y, and z into the algebraic expression that defines f.

$$f(1, -1, 4) = \frac{\sqrt{9 - 1^2 - (-1)^2}}{1 + 4} = \frac{\sqrt{7}}{5}$$

$$f(-1, 2, 1) = \frac{\sqrt{9 - (-1)^2 - 2^2}}{-1 + 1} = \frac{\sqrt{4}}{0} \quad \text{(not defined)}$$

We see that the point $(-1, 2, 1)$ does not belong to the domain of f.

For $f(x, y, z)$ to be well defined, it is necessary that the quantity under the radical sign be nonnegative and also that the denominator of f be nonzero. Thus we have the conditions $9 - x^2 - y^2 \geq 0$, or $x^2 + y^2 \leq 9$, and $x + z$

$\neq 0$. Using set notation, we may write the domain as

$$D = \{(x, y, z) \mid x^2 + y^2 \le 9, \quad x + z \neq 0\}.$$

When more than three independent variables occur, it is common to use subscript notation to denote them rather than to introduce new letters. Thus, if there are n independent variables, we would denote them by $x_1, x_2, x_3, \ldots, x_n$. Using z as the dependent variable, we would denote a function of the n variables by $z = f(x_1, x_2, \ldots, x_n)$. Subscript notation is also frequently used for functions of two or three variables; for example, we might write $w = f(x_1, x_2, x_3)$ instead of $w = f(x, y, z)$.

EXAMPLE 6 If
$$z = x_1^2 + e^{x_1 + x_2} + (2x_1 + x_4)^{-1} \sqrt{x_2^2 + x_3^2},$$

evaluate z at the point $(3, -3, 4, -5)$.

Solution By substituting $x_1 = 3$, $x_2 = -3$, $x_3 = 4$, and $x_4 = -5$ into the expression for z, we find

$$z = 3^2 + e^{3+(-3)} + [2(3) + (-5)]^{-1} \sqrt{(-3)^2 + 4^2}$$
$$= 9 + e^0 + (6 - 5)^{-1} \sqrt{9 + 16}$$
$$= 9 + 1 + \tfrac{5}{1} = 15.$$

We have seen how the graph of a function of a single variable helps us to visualize its important features, such as where it is increasing or decreasing, where it is concave up or down, where it is maximum and minimum, and so on. To sketch the graph of $z = f(x, y)$, a function of two variables, we need coordinates in three dimensions, one for each of the variables x, y, and z.

In three dimensions, the x-, y-, and z-axes are constructed at right angles to one another, as shown in Figure 2. It is often convenient to think of the xy-

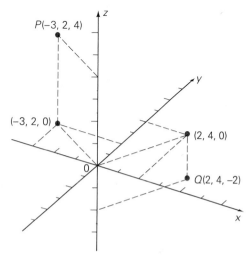

FIGURE 2

plane as being horizontal and the z-axis as pointing vertically upward. The negative z-axis then points downward.

Each pair of axes determines a plane; for example, the x-axis and y-axis determine the xy-plane, the x-axis and z-axis determine the xz-plane, and so on. On the xy-plane, the third coordinate z is equal to zero, and the coordinates x and y are used in the usual way to plot the positions of points in that plane. In Figure 2, the points $(2, 4, 0)$ and $(-3, 2, 0)$ are plotted in order to demonstrate this procedure.

In order to plot the position of a general point (x, y, z) for which $z \neq 0$, we first plot the point $(x, y, 0)$ in the xy-plane and then move from this point parallel to the z-axis according to the given value of the z-coordinate. For example, when plotting $(-3, 2, 4)$, we first plot $(-3, 2, 0)$, as in Figure 2, and then move a distance of 4 units in the direction of the positive z-axis to the point P. In plotting the point $(2, 4, -2)$, we first plot $(2, 4, 0)$ in the xy-plane and then move 2 units parallel to the negative z-axis to the point Q. ☛ 3

All points in the xy-plane satisfy the condition $z = 0$. Correspondingly, all points in the xz-plane satisfy the condition $y = 0$ and all points in the yz-plane satisfy the condition $x = 0$. On the z-axis, both x and y are zero. Correspondingly, on the x-axis, $y = z = 0$, and on the y-axis, $x = z = 0$.

EXAMPLE 7 Plot the points $(0, 2, 4)$, $(3, 0, -2)$, $(0, 0, 5)$, and $(0, -3, 0)$.

Solution The points are plotted in Figure 3. Note that the four points lie, respectively, in the yz-plane, in the xz-plane, on the z-axis, and on the y-axis.

☛ **3.** Plot the points
$(0, 0, -4)$, $(0, -3, 0)$,
$(-3, 4, 3)$, $(2, -3, -2)$ and
$(5, 2, -3)$.

Answer

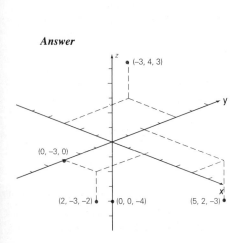

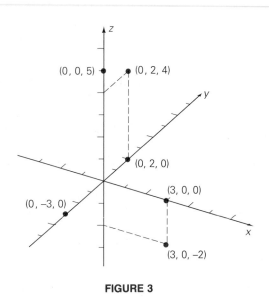

FIGURE 3

4. Draw level curves for the levels $z = 0$, ± 1, and ± 2 for the functions

(a) $z = x + y$; (b) $z = xy$.

Let $z = f(x, y)$ be a function of two variables. Its domain D consists of the set of points in the xy-plane at which the function is defined. For any point (x, y) in D, we can calculate the corresponding value of $z = f(x, y)$ and plot the point (x, y, z) using three-dimensional coordinates. By doing this for every point (x, y) in D, we obtain a set of points (x, y, z) that form a surface in three dimensions. There is one point (x, y, z) on the surface lying above each point of the domain D (or below if $z = f(x, y)$ turns out to be negative). This surface is said to be the **graph** of the function $z = f(x, y)$.

In actual practice, the task of sketching a surface in three dimensions that is the graph of a function $z = f(x, y)$ is by no means as easy as sketching the graph of a function $y = f(x)$ of a single variable. When faced with this task, it is often helpful to examine what are called **sections** of the graph. These are slices made through the graph by specified planes.

Let us consider sections by horizontal planes. A horizontal plane (parallel to the xy-plane) satisfies an equation of the type $z = c$, where c is a constant that gives the height of the plane above the xy-plane (or below, if $c < 0$). So the section of a graph by such a plane consists of points on the graph that lie at a constant height above (or below) the xy-plane. Such a horizontal section can be plotted as a curve in the xy-plane and is called a **contour line,** or **level curve.**

Consider, for example, the function $z = \sqrt{4 - x^2 - y^2}$. The points on the graph of this function that also lie on the horizontal plane $z = c$ satisfy

$$c^2 = 4 - x^2 - y^2.$$

That is,

$$x^2 + y^2 = 4 - c^2.$$

This equation relating x and y is the equation of a circle in the xy-plane centered at $x = y = 0$ and with radius equal to $\sqrt{4 - c^2}$.

For example, let us take $c = 1$, so that we are considering the horizontal slice through the graph by the plane lying 1 unit above the xy-plane. The section is then a circle in the xy-plane with radius $\sqrt{4 - 1^2} = \sqrt{3}$ and center at the point $(0, 0)$. Similarly, if $c = \frac{1}{2}$, the section is a circle of radius $\sqrt{15}/2$, while if $c = \frac{3}{2}$, the section is a circle of radius $\sqrt{7}/2$.

The circles $x^2 + y^2 = 4 - c^2$ corresponding to these three values of c, as well as the outer boundary $x^2 + y^2 = 4$ of the domain, are shown in Figure 4.

The graph of the function $z = \sqrt{4 - x^2 - y^2}$ in three dimensions is a hemisphere centered at the origin, with radius 2 units. The graph is shown in Figure 5, which also shows the contour lines corresponding to $z = 0$, $\frac{1}{2}$, 1, and $\frac{3}{2}$ in their three-dimensional positions. **4**

Another common way of representing a function $z = f(x, y)$ graphically is to hold one of the independent variables, x or y, fixed and to graph z as a function of the remaining variable. By giving the fixed variable several values,

Answer

(a)

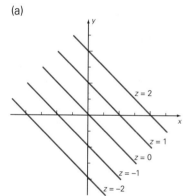

(b)

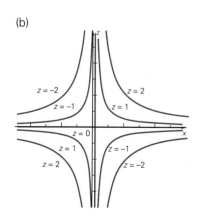

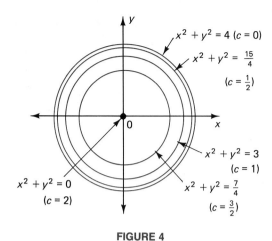

FIGURE 4

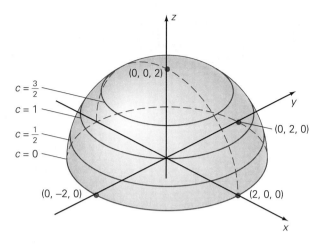

FIGURE 5

a series of such curves are obtained. For example, fixing $y = c$, we obtain $z = f(x, c)$, expressing z as a function of x, and this function can be graphed in the xz-plane. By giving c different values, several such graphs can be drawn.

Now, $y = c$ is the equation of a vertical plane, parallel to the xz-plane, cutting the y-axis at $(0, c, 0)$. Thus the graph of $z = f(x, c)$ is the curve along which this vertical plane intersects the graph of f.

Similarly, setting $x = c$ we get $z = f(c, y)$, whose graph in the yz-plane is the curve of intersection of the graph of f by a vertical plane parallel to the yz-plane.

EXAMPLE 8 Draw vertical sections of the graph of $z = x^2 - y^2$ corresponding to the vertical planes $x = 0$, ± 1, and ± 2 and $y = 0$, ± 1, and 2.

Solution Let us consider first the section on which $x = c$, c a constant. This is the section by the vertical plane that is parallel to the yz-plane and is at a distance c from it. Substituting $x = c$ into the given function $z = x^2 - y^2$, we obtain $z = c^2 - y^2$. This equation in terms of y and z describes a parabola, opening downward, with vertex at the point $y = 0$ and $z = c^2$. For example, if $c = 1$, the vertex of the parabola is at $(y, z) = (0, 1)$. The parabolas corresponding to the values $c = 0$, ± 1, ± 2, drawn in the yz-plane, are shown in Figure 6.

Let us consider the section on which $y = c$, c a constant, which is the section by the vertical plane that is parallel to the xz-plane and is at a distance c from it. Substituting $y = c$ in the given equation, we obtain $z = x^2 - c^2$. This equation in terms of x and z represents a parabola with vertex $x = 0$ and $z = -c^2$; this parabola opens upwards. The parabolas corresponding to $c = 0$, ± 1, ± 2 are shown in Figure 7.

5. For the function $z = x^2 - xy$ draw vertical sections on the xz-plane for $y = 0$ and ± 2 and on the yz-plane for $x = 0$ and ± 1.

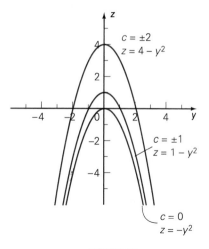

FIGURE 6

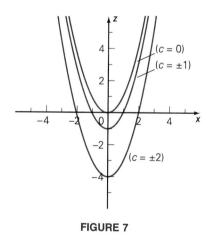

FIGURE 7

Answer

(a)

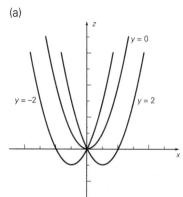

(b)

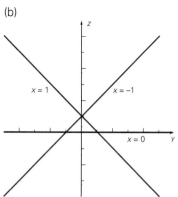

The graph of the function $z = x^2 - y^2$ in three dimensions is given in Figure 8, on which are also shown the vertical sections corresponding to $x = 0$, ± 1, and ± 2 and $y = 0$, -1, and -2. From the figure, we can immediately see the dominant feature of this graph, namely its saddle-like shape near the origin. **☞ 5**

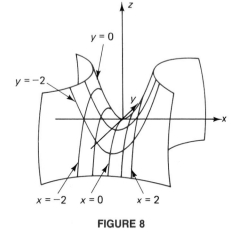

FIGURE 8

EXAMPLE 9 *(Advertising and Sales)* The volume of sales of a particular item depends on its price and also, in many instances, on the amount that the manufacturer spends on promotion and advertising. Let p be the price and A the advertising expenditure per month (both in dollars) and let x be the monthly sales. Then in general, $x = f(p, A)$. Suppose that in a certain case

$$x = 1000(5 - pe^{-kA})$$

where $k = 0.001$. Draw the following vertical sections: graphs of x versus p for $A = 0$, 500, 1000, and 1500 and graphs of x versus A for $p = 1$, 3, 5, and 8.

Solution The required graphs of x versus p are shown in Figure 9. For example, when $A = 0$, $e^{-kA} = e^{-k(0)} = 1$ and so

$$x = 1000(5 - p).$$

The graph of this function is a straight line that intersects the x-axis ($p = 0$) at the value $x = 5000$ and cuts the p-axis ($x = 0$) at $p = 5$. Similarly, when $A = 1000$, $e^{-kA} = e^{-(0.001)(1000)} = e^{-1} = 0.368$ and so $x = 1000(5 - 0.368p)$. This again is a straight line that intersects the x-axis at $x = 5000$ and the p-axis at $p = 5/0.368 \approx 13.6$. This graph represents the demand x as a function of price p when \$1000 per month is spent on advertising.

When p is fixed at the required values, we have the graphs of x versus A shown in Figure 10. For example, when $p = 3$.

$$x = 5000 - 3000e^{-0.001A}.$$

This function provides the volume of sales in terms of the advertising expenditure when the item is priced at \$3.

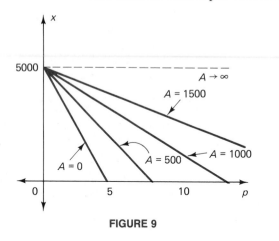

FIGURE 9

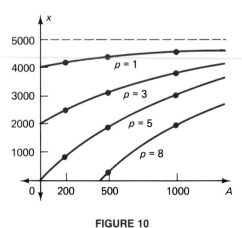

FIGURE 10

EXERCISES 18-1

(1–8) Calculate the values of the given functions at the indicated points.

1. $f(x, y) = x^2 - 2xy + y^2$; $(x, y) = (3, -2)$ and $(-4, -4)$

2. $f(x, y) = \dfrac{(x - 1)(y - 1)}{x + y}$; $(x, y) = (1, -2)$, $(2, -2)$, and $(3, -2)$

3. $f(x, t) = \dfrac{x - t + 1}{x^2 + t^2}$; $(x, t) = (2, 1)$, $(3, \frac{1}{2})$, and $(-\frac{1}{4}, \frac{3}{4})$

4. $f(u, v) = u + \ln |v|$; $(u, v) = (2, 1)$, $(-2, -e)$, and $(0, e^3)$

5. $f(x, y, z) = x^2 + 2y^2 + 3z^2$; $(x, y, z) = (1, 2, 3)$ and $(-2, 1, -4)$

6. $f(x, y, t) = \dfrac{x + y + t}{x + y - t}$; $(x, y, t) = (\frac{1}{2}, -\frac{1}{2}, 1)$ and $(\frac{1}{2}, \frac{1}{2}, -1)$

7. $f(u, v, z) = \dfrac{2u + 3v + 4z}{4u - 3v - z}$; $(u, v, z) = (\frac{1}{2}, 1, 1)$ and $(\frac{1}{4}, -\frac{1}{3}, 2)$

8. $f(a, b, c) = \dfrac{2a^2 + b^2}{\sqrt{c^2 - 4}}$; $(a, b, c) = (1, 2, 3)$ and $(2, 2, -4)$

(9–16) Find the domains of the following functions.

9. $f(x, y) = x^2 + 2xy + y^2$

10. $f(x, y) = \dfrac{x^2}{y^2 - 1}$

11. $f(x, y) = \sqrt{x^2 + y^2 - 9}$

12. $f(x, y) = \sqrt{1 - (x + y)^2}$

13. $f(x, t) = \ln (x - t)$

14. $f(x, y) = e^{y - x}$

15. $f(x, y, z) = x + \sqrt{yz}$

16. $f(u, v, w) = \sqrt{e^{u + v + w} - 1}$

(17–22) Sketch the level curves for each of the following functions corresponding to the given values of z.

17. $z = 2x + 3y$; $z = 0, 1, 2, 3$

18. $z = 3x - y$; $z = 0, 1, -2, 3$

19. $z = \sqrt{16 - x^2 - y^2}$; $z = 0, 1, 2, 3$

20. $z = \sqrt{25 - x^2 - y^2}$; $z = 0, 1, 2, 3$

21. $z = x^2 + y^2$; $z = 1, 2, 3, 4$

22. $z = x^2 - y^2$; $z = 0, \pm 1, \pm 2$

(23–26) Sketch the vertical sections of the graphs of the following functions corresponding to given values of x or y.

23. $z = \sqrt{16 - x^2 - y^2}$; $x = 0, \pm 1, \pm 2$

24. $z = \sqrt{25 - x^2 + y^2}$; $y = 0, \pm 1, \pm 2$

25. $z = x^2 + y^2$; $y = 0, \pm 1, \pm 2, \pm 3$

26. $z = 2x^2 - y^2$; $x = 0, \pm 1, \pm 2, y = 0, \pm 1, \pm 2$

27. (*Cost of a Can*) A cylindrical can has radius r and height h. If the material from which it is made costs \$2 per unit area, express the cost of the can, C, as a function of r and h.

28. (*Cost of a Can*) In Exercise 27, find an expression for C that includes the cost of joining the two ends of the can to the curved side. This cost is \$0.40 per unit length of perimeter of each end.

29. (*Cost of Water Tank*) An open rectangular tank is to be constructed to hold 100 cubic feet of water. The material costs \$5 per square foot for the base and \$3 per square foot for the vertical walls. If C denotes the total cost (in dollars), determine C as a function of the dimensions of the base.

30. (*Cost of Water Tank*) Repeat Exercise 29 if the open tank is of cylindrical shape.

31. (*Cost Function*) A firm produces two products, X and Y. The unit costs of labor and material are \$5 for product X and \$12 for product Y. In addition, the firm also has overhead costs of \$3000 per month. Express the monthly cost C (in dollars) as a function of the units of X and Y produced. What is the total cost of producing 200 units of X and 150 units of Y?

32. (*Cost and Profit Functions*) Western Electronics manufactures two sizes of cassette tapes, 60 minutes and 90 minutes. The unit cost of labor and material for the two sizes are 30¢ and 40¢. In addition, the firm has weekly fixed costs of \$1200.

   a. Give the weekly cost C (in dollars) as a function of the units of the two sizes of tapes produced.

   b. Evaluate the total cost of producing 10,000 tapes of 60-minute size and 8000 tapes of 90-minute size.

   c. If the company sells the two kinds of tapes at 60¢ and 75¢ each, respectively, give the weekly profit as a function of the numbers of units produced and sold each week.

33. (*Arctic Pipeline Cost*) A pipeline is to be constructed from a point A to a point B that lies 500 miles south and 500 miles east of A. From A, the 200 miles to the south consist of tundra, the next 100 miles consist of swamp, and the last 200 miles consist of a dry rocky terrain. The

pipeline costs P dollars per mile over this last terrain, $3P$ dollars per mile over the swamp, and $2P$ dollars per mile over the tundra. Let the pipeline consist of three straight sections, one across each type of terrain, and let it travel x miles to the east across the strip of tundra and a further y miles to the east across the swamp. Express its total cost in terms of x and y.

18-2 PARTIAL DERIVATIVES

We shall now turn to the question of differentiating functions of several variables. In this section, we shall simply concern ourselves with the mechanics of differentiation, but in the following sections, we shall turn to matters of interpretation and use of the resulting derivatives.

Let $z = f(x, y)$ be a function of two independent variables. If the variable y is held fixed at a value $y = y_0$, then the relation $z = f(x, y_0)$ expresses z as a function of the one variable x. This function will have as its graph a curve in the xz-plane, which is in fact the vertical section of the graph of $z = f(x, y)$ by the plane $y = y_0$.

Figure 11 illustrates a typical section $z = f(x, y_0)$. At a general point on this curve, the tangent line can be constructed and its slope can be calculated by differentiating z with respect to x from the relation $z = f(x, y_0)$. This derivative is found in the usual way as a limit according to the following formula:

$$\frac{d}{dx} f(x, y_0) = \lim_{\Delta x \to 0} \frac{f(x + \Delta x, y_0) - f(x, y_0)}{\Delta x}.$$

It is called the *partial derivative of z with respect to x*, and is usually denoted by $\partial z / \partial x$. (Note that we use ∂ not d in this situation. The letter d is reserved for the derivative of a function of a single variable.)

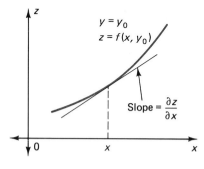

FIGURE 11

DEFINITION Let $z = f(x, y)$ be a function of x and y. Then the **partial derivative of z with respect to x** is defined to be

$$\frac{\partial z}{\partial x} = \lim_{\Delta x \to 0} \frac{f(x + \Delta x, y) - f(x, y)}{\Delta x}.$$

In writing this definition we have dropped the subscript from y_0; we must remember that *when calculating $\partial z/\partial x$, the variable y is held constant*.

Correspondingly, the **partial derivative of z with respect to y** is defined to be

$$\frac{\partial z}{\partial y} = \lim_{\Delta y \to 0} \frac{f(x, y + \Delta y) - f(x, y)}{\Delta y}.$$

In calculating $\partial z/\partial y$, the variable x is held constant and the differentiation is carried out with respect to y only.

At a point $(x_0\ y_0)$ in the domain of the given function $z = f(x, y)$, the partial derivative $\partial z/\partial x$ provides the slope of the vertical section of the graph by the plane $y = y_0$. Correspondingly, the partial derivative $\partial z/\partial y$ provides the slope of the vertical section by the plane $x = x_0$. This latter vertical section has the equation $z = f(x_0, y)$ expressing z as a function of y, and its slope is obtained by differentiating z with respect to y with x set equal to x_0 and held fixed.

These geometric interpretations of partial derivatives are illustrated in Figure 12. Here $A'B'$ represents the vertical section by the plane $y = y_0$ and

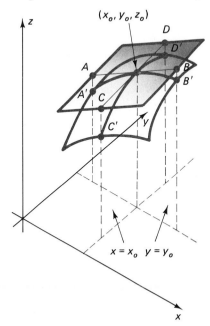

FIGURE 12

the tangent to this curve at $x = x_0$ is the straight line AB. The slope of this line is given by the partial derivative $\partial z/\partial x$ evaluated at (x_0, y_0). Similarly, the straight line CD is the tangent to the vertical section by the plane $x = x_0$. Its slope is $\partial z/\partial y$ evaluated at (x_0, y_0).

Partial derivatives can be evaluated using essentially the same techniques as those used for evaluating ordinary derivatives. *We must simply remember to treat any variable except the one with respect to which we are differentiating as if it were a constant.* Apart from this, the familiar power formula, product and quotient rules, and chain rule can all be used in the usual way.

EXAMPLE 1 Calculate $\partial z/\partial y$ when $z = x^3 + 5xy^2 + 2y^3$.

Solution Treating x as a constant and differentiating with respect to y, we have

$$\frac{\partial z}{\partial y} = 0 + 5x(2y) + 2(3y^2) = 10xy + 6y^2.$$

EXAMPLE 2 Calculate $\partial z/\partial x$ for $z = \sqrt{x^2 + y^2}$.

Solution With y held constant, we use the chain rule.

$$\frac{\partial z}{\partial x} = \frac{\partial}{\partial x}(x^2 + y^2)^{1/2}$$

$$= \frac{1}{2}(x^2 + y^2)^{-1/2}\frac{\partial}{\partial x}(x^2 + y^2)$$

$$= \frac{1}{2}(x^2 + y^2)^{-1/2}(2x + 0)$$

since $\partial(y^2)/\partial x = 0$ as y^2 is held constant. Therefore

$$\frac{\partial z}{\partial x} = x(x^2 + y^2)^{-1/2} = \frac{x}{\sqrt{x^2 + y^2}}.$$

EXAMPLE 3 Calculate $\partial z/\partial x$ and $\partial z/\partial y$ for $z = (x^2 + y^2)/(\ln x)$.

Solution Using the quotient formula, we obtain

$$\frac{\partial z}{\partial x} = \frac{\ln x (\partial/\partial x)(x^2 + y^2) - (x^2 + y^2)(\partial/\partial x)(\ln x)}{(\ln x)^2}$$

$$= \frac{\ln x \cdot (2x) - (x^2 + y^2) \cdot (1/x)}{(\ln x)^2}$$

$$= \frac{2x^2 \ln x - (x^2 + y^2)}{x(\ln x)^2}$$

after multiplying numerator and denominator by x.

● **6.** Calculate $\dfrac{\partial z}{\partial x}$ and $\dfrac{\partial z}{\partial y}$ for
the functions
(a) $z = x^2 y - 3xy^2$;
(b) $z = y \ln (x + y)$.

We do not need to use the quotient formula in order to evaluate $\partial z/\partial y$, since the denominator of the given quotient is a function of x alone, and so is a constant as far as partial differentiation with respect to y is concerned.

$$\frac{\partial z}{\partial y} = \frac{\partial}{\partial y}\left(\frac{x^2 + y^2}{\ln x}\right) = \frac{1}{\ln x}\frac{\partial}{\partial y}(x^2 + y^2)$$

$$= \frac{1}{\ln x}(0 + 2y) = \frac{2y}{\ln x} \qquad ● \mathbf{6}$$

It can be seen from these examples that the calculation of partial derivatives of a function of two variables is essentially no different from differentiating a function of just one variable. We must simply remember that *when finding the partial derivative with respect to one of the variables, we treat the other variable as a constant*, then differentiate in the familiar way.

The partial derivative $\partial z/\partial x$ is itself a function of x and y, and therefore we can construct its partial derivatives with respect to both x and y. These are called **second-order partial derivatives** of z, and the following notation is used:

$$\frac{\partial^2 z}{\partial x^2} = \frac{\partial}{\partial x}\left(\frac{\partial z}{\partial x}\right) \qquad \text{and} \qquad \frac{\partial^2 z}{\partial y\,\partial x} = \frac{\partial}{\partial y}\left(\frac{\partial z}{\partial x}\right).$$

Similarly, $\partial z/\partial y$ may be differentiated with respect to x and y, thus providing two more second-order partial derivatives:

$$\frac{\partial^2 z}{\partial y^2} = \frac{\partial}{\partial y}\left(\frac{\partial z}{\partial y}\right) \qquad \text{and} \qquad \frac{\partial^2 z}{\partial x\,\partial y} = \frac{\partial}{\partial x}\left(\frac{\partial z}{\partial y}\right).$$

The two derivatives $\partial^2 z/\partial x\,\partial y$ and $\partial^2 z/\partial y\,\partial x$ are often called **mixed partial derivatives** of second order. Provided that these mixed partial derivatives are continuous functions of x and y, they are equal to another:

$$\frac{\partial^2 z}{\partial x\,\partial y} = \frac{\partial^2 z}{\partial y\,\partial x}.$$

EXAMPLE 4 Calculate all the second-order derivatives of the function $z = \sqrt{x^2 + y^2}$.

Solution In Example 2, we showed that

$$\frac{\partial z}{\partial x} = \frac{x}{\sqrt{x^2 + y^2}}$$

for this function. It follows in a similar way that

$$\frac{\partial z}{\partial y} = \frac{y}{\sqrt{x^2 + y^2}}.$$

Answer (a) $\dfrac{\partial z}{\partial x} = 2xy - 3y^2$,

$\dfrac{\partial z}{\partial y} = x^2 - 6xy$;

(b) $\dfrac{\partial z}{\partial x} = \dfrac{y}{x + y}$,

$\dfrac{\partial z}{\partial y} = \ln (x + y) + \dfrac{y}{x + y}$.

The two mixed derivatives of second-order are obtained as follows (using the chain rule in the intermediate step).

$$\frac{\partial^2 z}{\partial y \, \partial x} = \frac{\partial}{\partial y}\left(\frac{\partial z}{\partial x}\right) = \frac{\partial}{\partial y}\left(\frac{x}{\sqrt{x^2 + y^2}}\right) = x\frac{\partial}{\partial y}[(x^2 + y^2)^{-1/2}]$$

$$= x(-\tfrac{1}{2})(x^2 + y^2)^{-3/2}(2y)$$

$$= -xy(x^2 + y^2)^{-3/2}$$

$$\frac{\partial^2 z}{\partial x \, \partial y} = \frac{\partial}{\partial x}\left(\frac{\partial z}{\partial y}\right) = \frac{\partial}{\partial x}\left(\frac{y}{\sqrt{x^2 + y^2}}\right) = y\frac{\partial}{\partial x}[(x^2 + y^2)^{-1/2}]$$

$$= -xy(x^2 + y^2)^{-3/2}$$

It can be seen that these two derivatives are equal to one another.
For the remaining two derivatives, the quotient rule must be used.

$$\frac{\partial^2 z}{\partial x^2} = \frac{\partial}{\partial x}\left(\frac{x}{\sqrt{x^2 + y^2}}\right)$$

$$= \frac{\sqrt{x^2 + y^2} \cdot (\partial/\partial x)(x) - x(\partial/\partial x)(\sqrt{x^2 + y^2})}{(\sqrt{x^2 + y^2})^2}$$

$$= \frac{\sqrt{x^2 + y^2} \cdot (1) - x \cdot (x/\sqrt{x^2 + y^2})}{(x^2 + y^2)}$$

$$= \frac{(x^2 + y^2) - x^2}{(x^2 + y^2)\sqrt{x^2 + y^2}} = \frac{y^2}{(x^2 + y^2)^{3/2}}$$

In a similar way, we can show that

$$\frac{\partial^2 z}{\partial y^2} = \frac{x^2}{(x^2 + y^2)^{3/2}}. \qquad \text{☛ 7}$$

☛ **7.** Calculate $\dfrac{\partial^2 z}{\partial x^2}$ and $\dfrac{\partial^2 z}{\partial y \partial x}$ for the functions (a) $z = x^p y^q$; (b) $z = \dfrac{x}{x + y}$.

We may continue this process and calculate partial derivatives of higher orders:

$$\frac{\partial^3 z}{\partial x^3} = \frac{\partial}{\partial x}\left(\frac{\partial^2 z}{\partial x^2}\right), \quad \frac{\partial^3 z}{\partial y \, \partial x^2} = \frac{\partial}{\partial y}\left(\frac{\partial^2 z}{\partial x^2}\right), \quad \frac{\partial^3 z}{\partial x \, \partial y \, \partial x} = \frac{\partial}{\partial x}\left(\frac{\partial^2 z}{\partial y \, \partial z}\right)$$

and so on. Provided that all derivatives of the given order are continuous, the order in which the x and y differentiations are carried out is immaterial. Thus, for example, the following mixed derivatives are all equal.

$$\frac{\partial^3 z}{\partial y \, \partial x^2} = \frac{\partial^3 z}{\partial x \, \partial y \, \partial x} = \frac{\partial^3 z}{\partial x \, \partial x \, \partial y}$$

They are denoted by $\partial^3 z/\partial x^2 \, \partial y$, indicating two differentiations with respect to x and one with respect to y.

Answer

(a) $\dfrac{\partial^2 z}{\partial x^2} = p(p - 1)x^{p-2}y^q$,

$\dfrac{\partial^2 z}{\partial y \partial x} = pqx^{p-1}y^{q-1}$;

(b) $\dfrac{\partial^2 z}{\partial x^2} = \dfrac{-2y}{(x + y)^3}$,

$\dfrac{\partial^2 z}{\partial y \partial x} = \dfrac{x - y}{(x + y)^3}$.

8. Calculate z_x, z_{xx}, and z_{xxy} for the function $z = xe^{2x+y^2}$.

EXAMPLE 5 Calculate $\partial^3 z/\partial x^2 \, \partial y$ and $\partial^4 z/\partial x \, \partial y^3$ for $z = x^3 y^4$.

Solution We have

$$\frac{\partial z}{\partial y} = x^3 \cdot 4y^3 = 4x^3 y^3$$

$$\frac{\partial^2 z}{\partial x \, \partial y} = \frac{\partial}{\partial x}(4x^3 y^3) = 4y^3 \cdot 3x^2 = 12x^2 y^3.$$

Therefore

$$\frac{\partial^3 z}{\partial x^2 \partial y} = \frac{\partial}{\partial x}\left(\frac{\partial^2 z}{\partial x \, \partial y}\right) = \frac{\partial}{\partial x}(12x^2 y^3) = 24xy^3$$

and

$$\frac{\partial^3 z}{\partial x \, \partial y^2} = \frac{\partial}{\partial y}\left(\frac{\partial^2 z}{\partial x \, \partial y}\right) = \frac{\partial}{\partial y}(12x^2 y^3) = 36x^2 y^2.$$

Thus

$$\frac{\partial^4 z}{\partial x \, \partial y^3} = \frac{\partial}{\partial y}\left(\frac{\partial^3 z}{\partial x \, \partial y^2}\right) = \frac{\partial}{\partial y}(36x^2 y^2) = 72x^2 y.$$

As with ordinary derivatives, there are several alternative notations that are used for partial derivatives. The most commonly encountered of these is the use of subscripts to indicate partial derivatives, and we shall use this notation ourselves from time to time. According to this notation, we have the following.

$\dfrac{\partial z}{\partial x}$ is denoted by z_x or $f_x(x, y)$.

$\dfrac{\partial z}{\partial y}$ is denoted by z_y or $f_y(x, y)$.

$\dfrac{\partial^2 z}{\partial x^2}$ is denoted by z_{xx} or $f_{xx}(x, y)$.

$\dfrac{\partial^2 z}{\partial y \, \partial x}$ is denoted by z_{xy} or $f_{xy}(x, y)$.   (Note that $z_{xy} = z_{yx}$ if they are continuous.)

$\dfrac{\partial^4 z}{\partial y^3 \, \partial x}$ is denoted by z_{xyyy} or $f_{xyyy}(x, y)$.

Further partial derivatives are denoted in a similar manner.   **☛ 8**

The notion of partial derivatives extends in a straightforward way to functions $z = f(x_1, x_2, \ldots, x_n)$ of several variables. For example, $\partial z/\partial x_1$ is obtained by differentiating z with respect to x_1 keeping $x_2, \ldots, x_n$ all constant, and so on.

Answer
$z_x = (1 + 2x)e^{2x+y^2}$
$z_{xx} = 4(1 + x)e^{2x+y^2}$
$z_{xxy} = 8y(1 + x)e^{2x+y^2}$

EXAMPLE 6 If $z = f(x_1, x_2, x_3) = x_1^2 + x_1\sqrt{x_2^2 - x_3^2}$, find $\partial z/\partial x_1$, $\partial z/\partial x_2$, and $\partial z/\partial x_3$.

Solution

$$\frac{\partial z}{\partial x_1} = 2x_1 + \sqrt{x_2^2 - x_3^2}$$

$$\frac{\partial z}{\partial x_2} = 0 + x_1(\tfrac{1}{2})(x_2^2 - x_3^2)^{-1/2}(2x_2) = \frac{x_1 x_2}{\sqrt{x_2^2 - x_3^2}}$$

$$\frac{\partial z}{\partial x_3} = 0 + x_1(\tfrac{1}{2})(x_2^2 - x_3^2)^{-1/2}(-2x_3) = -\frac{x_1 x_3}{\sqrt{x_2^2 - x_3^2}} \quad \text{☞ 9}$$

☞ **9.** Calculate w_x, w_y and w_z for the function $w = (x + y + z)$ $\ln (y - z)$.

Answer $w_x = \ln (y - z)$,
$w_y = \ln (y - z) + \dfrac{x + y + z}{y - z}$,
$w_z = \ln (y - z) - \dfrac{x + y + z}{y - z}$.

EXERCISES 18-2

(1–24) Calculate $\partial z/\partial x$ and $\partial z/\partial y$ for the following functions.

1. $z = x^2 + y^2$

2. $z = 3x^3 + 5y^4 + 7$

3. $z = 3e^{2x} - 5 \ln y + 7$

4. $z = xy^2 + x^2 y$ 5. $z = xe^y + ye^{-x}$

6. $z = x \ln y + y^2 \ln x$

7. $z = x^2 + xy + y^2$ 8. $z = xy + \ln (xy)$

9. $z = e^{2x+3y}$ 10. $z = e^{x^2+y^2}$

11. $z = (2x + 3y)^7$ 12. $z = \sqrt{x^2 - y^2}$

13. $z = (x + 2y^3)^{1/3}$ 14. $z = \dfrac{x}{\sqrt{y - x}}$

15. $z = xe^{xy}$

16. $z = (2x + 3y)e^{4x+5y}$

17. $z = \left(\dfrac{x}{y}\right)e^{xy}$ 18. $z = xye^{x/y}$

19. $z = \ln (x^2 + y^2)$

20. $z = (x^2 + y^2) \ln (x + y)$

21. $z = \ln (e^x + xy^3)$ 22. $z = \dfrac{x^2 - y^2}{x^2 + y^2}$

23. $z = \dfrac{y}{y - x}$ 24. $z = xy \sqrt{x^2 + y^2}$

(25–38) Calculate $\partial^2 z/\partial x^2$, $\partial^2 z/\partial y^2$, and $\partial^2 z/\partial x\,\partial y$ for the following functions.

25. $z = x^4 + y^4 + 3x^2 y^3$

26. $z = xe^{-y} + ye^{-x}$

27. $z = xy + \ln (x + y)$

28. $z = x^{3/2} y^{-4}$ 29. $z = x^5 y^{-1/2}$

30. $z = e^{x-2y}$ 31. $z = ye^{xy}$

32. $z = \ln (2x + 3y)$ 33. $z = \ln (x^2 + y^2)$

34. $z = (x^2 + y^2)^5$ 35. $z = \dfrac{x}{x + y}$

36. $z = e^{x^2+y^2}$ 37. $z = \dfrac{xy}{x - y}$

38. $z = (x^2 + y^2)e^{xy}$

39. If $z = e^{y/x}$, show that $xz_x + yz_y = 0$

40. If $z = x^2 e^{-x/y}$, show that $xz_x + yz_y = 2z$.

41. If $z = x^3 + y^3$, show that $xz_x + yz_y = 3z$.

42. If $z = f(ax + by)$, show that $bz_x - az_y = 0$.

43. If $C = ae^{kx+wt}$, show that $\partial C/\partial t = (p/4)(\partial C^2/\partial x^2)$, provided $w = pk^2/4$.

44. If $f(x, y) = xe^{y/x}$, prove that $xf_{xx} + yf_{xy} = 0$.

45. If $C = e^{kx+wt}$, show that $\partial^2 C/\partial t^2 - \partial^2 C/\partial x^2 = 0$, provided $k = \pm w$.

46. *(Microbiology)* In the process of metabolism of a bacterium the rate M at which a chemical substance can be absorbed into the bacterium and distributed throughout its volume is given by $M = aS/V$, where S is the surface area, V is the volume of the bacterium, and a is a constant. For a cylindrical bacterium of radius r and length l, $V = \pi r^2 l$ and $S = 2\pi r l + 2\pi r^2$. Calculate $\partial M/\partial r$ and $\partial M/\partial l$, thus finding how an increase in radius or in length affects the rate of metabolism.

47. *(Zoology)* The rate at which an animal's body loses heat by convection is given by $H = a(T - T_0)v^{1/3}$, where T and T_0 are the temperatures of the animal's body and of the surrounding air, v is the wind velocity, and a is a constant. Calculate $\partial H/\partial T$, $\partial H/\partial T_0$, and $\partial H/\partial v$, and interpret these quantities.

48. *(Diffusion)* If a substance is injected into a vein, the concentration of the substance at any point in the vein will vary with time t and with the distance x from the point of injection. Under certain conditions the concentration can be described by a function of the form

$$C(x, t) = \frac{c}{\sqrt{t}} e^{-x^2/at}$$

where a and c are constants. Show that $C(x, t)$ satisfies the following equation,

$$\frac{\partial C}{\partial t} = \left(\frac{a}{4}\right)\frac{\partial^2 C}{\partial x^2}$$

(This equation is known as the diffusion equation.)

◼ 18-3 APPLICATIONS TO BUSINESS ANALYSIS

The ordinary derivative dy/dx can be regarded as the rate of change of y with respect to x. This interpretation is often useful—for example, the marginal revenue $R'(x)$ gives the rate of change of revenue with respect to the volume of sales, or, approximately, the change in revenue per additional unit sold. Similar interpretations can be made in the case of partial derivatives. For example, if $z = f(x, y)$, then $\partial z/\partial x$ gives the rate of change of z with respect to x when y is constant.

EXAMPLE 1 A new product is launched onto the market. The volume of sales x increases as a function of time t and also depends on the amount A spent on the advertising campaign. If, with t measured in months and A in dollars,

$$x = 200(5 - e^{-0.002A})(1 - e^{-t})$$

calculate $\partial x/\partial t$ and $\partial x/\partial A$. Evaluate these derivatives when $t = 1$ and $A = 400$ and interpret them.

Solution We have

$$\frac{\partial x}{\partial t} = 200(5 - e^{-0.002A})e^{-t}, \qquad \frac{\partial x}{\partial A} = 0.4e^{-0.002A}(1 - e^{-t}).$$

Setting $t = 1$ and $A = 400$, we obtain the values

$$\frac{\partial x}{\partial t} = 200(5 - e^{-0.8})e^{-1} \approx 335, \qquad \frac{\partial x}{\partial A} = 0.4e^{-0.8}(1 - e^{-1}) \approx 0.11.$$

The partial derivative $\partial x/\partial t$ gives the rate of increase in the sales volume with respect to time when the advertising expenditure is maintained fixed. For ex-

▶ **10.** Repeat Example 1 if the sales volume is given by
$$x = 25\frac{1 + (1 + 0.01\sqrt{A})t}{4 + t}.$$

ample, when this expenditure is fixed at \$400, the volume of sales after one month ($t = 1$) is growing at the instantaneous rate of 335 per month.

Similarly, $\partial x/\partial A$ gives the increase in the sales volume at a fixed time that occurs for each additional dollar spent on advertising. At the time $t = 1$, when \$400 is already spent on advertising, an additional dollar so spent will increase the sales volume by 0.11 unit. ▶ **10**

Marginal Productivity

The total output of the product of a business firm depends on a number of factors, which the firm often has some flexibility to change. The two most important such factors are usually the amount of labor employed by the firm and the amount of capital invested in buildings, machinery, and so on. Let L denote the number of units of labor employed by the firm (say in work-hours per year or in dollars per year spent in wages) and let K denote the cost of the capital investment in the firm's productive plant. Then the total output P—for example, the number of units of the firm's product produced per month—is some function of L and K, and we write $P = f(L, K)$. This function is known as the firm's **production function** and the variables L and K are examples of **production input factors**—that is, variables that affect the level of production.

In certain cases, changes in K and L are not independent of one another. For example, if the firm buys an extra machine it must also hire extra labor to operate it. On the other hand, K and L are often to some degree independent variables in the context of the firm's basic production strategy. For example, the firm can choose to invest a large amount of capital in a highly automated plant and so make use of relatively little labor or, on the other hand, it can decide to use less sophisticated machinery and more labor. Thus, in general, K and L can be regarded as independent variables.

The partial derivative $\partial P/\partial L$ is called the **marginal productivity of labor** and $\partial P/\partial K$ is called the **marginal productivity of capital.** $\partial P/\partial L$ measures the increase in production per unit increase in the amount of labor employed when the capital input K is held fixed. Correspondingly, $\partial P/\partial K$ measures the increase in production per unit increase in the capital invested when the labor usage is constant.

EXAMPLE 2 The production function of a certain firm is given by

$$P = 5L + 2L^2 + 3LK + 8K + 3K^2$$

where L is the labor input measured in thousands of work-hours per week, K is the cost of capital investment measured in thousands of dollars per week, and P is the weekly production in hundreds of items. Determine the marginal productivities when $L = 5$ and $K = 12$ and interpret the result.

Solution Given that

$$P = 5L + 2L^2 + 3LK + 8K + 3K^2$$

Answer $x_t = 25\dfrac{3 + 0.04\sqrt{A}}{(4 + t)^2}$,

$x_A = 25\dfrac{0.005t}{\sqrt{A}(4 + t)}.$

When $t = 1$ and $A = 400$, $x_t = 3.8$, $x_A = 0.00125$.

the marginal productivities are

$$\frac{\partial P}{\partial L} = 5 + 4L + 3K \qquad \text{and} \qquad \frac{\partial P}{\partial K} = 3L + 8 + 6K.$$

When $L = 5$ and $K = 12$,

$$\frac{\partial P}{\partial L} = 5 + 4(5) + 3(12) = 61, \qquad \frac{\partial P}{\partial K} = 3(5) + 8 + 6(12) = 95.$$

This means that when $L = 5$ and $K = 12$ (that is, 5000 work-hours per week are used and the cost of capital investment is \$12,000 per week), then P increases by 61 for each unit increase in L and P increases by 95 for each unit increase in K. Thus the production increases by 6100 items per week for each additional 1000 work-hours of employed labor when K is held fixed, and the production increases by 9500 items per week for each additional \$1000 increase in the weekly cost of capital investment when L is held fixed. ☛ 11

☛ **11.** Find the marginal productivities of labor and capital for the production function $P = cK^aL^{1-a}$, where c and a are constants.

The second derivatives of P with respect to K and L also have interpretations as marginal rates of change. The rate at which the marginal productivity $\partial P/\partial K$ increases with respect to changes in the cost of capital is measured by $\partial^2 P/\partial K^2$. Similarly, $\partial^2 P/\partial L^2$ measures the rate at which the marginal productivity $\partial P/\partial L$ increases with respect to changes in the amount of labor employed. Corresponding interpretations can be made for the mixed derivatives $\partial^2 P/\partial K \, \partial L$ and $\partial^2 P/\partial L \, \partial K$.

Demand Relations: Cross Elasticities

Now let us consider a different application—to demand relations. Earlier we supposed that the demand for a commodity depends only on the price per unit of that particular commodity. In practice, this is not always true because the demand for a commodity can be affected by the price of some other related commodity. For example, the demand for beef in the supermarket not only depends on the price per pound of beef but also on the price per pound of pork. Any change in the price of pork will always affect the demand for beef and vice versa, since some customers will be willing to switch from one product to the other.

In general, let A and B be two related commodities such that the change in price of one affects the demand for the other. Let p_A and p_B denote the unit prices for the two commodities. Then, their demands x_A and x_B are assumed to be functions of both the prices p_A and p_B, that is,

$$x_A = f(p_A, p_B) \qquad \text{and} \qquad x_B = g(p_A, p_B).$$

We can calculate four partial first-order derivatives.

$$\frac{\partial x_A}{\partial p_A}, \quad \frac{\partial x_A}{\partial p_B}, \quad \frac{\partial x_B}{\partial p_A}, \quad \frac{\partial x_B}{\partial p_B}$$

Answer
$$\frac{\partial P}{\partial L} = (1 - a)cK^aL^{-a},$$
$$\frac{\partial P}{\partial K} = acK^{a-1}L^{1-a}.$$

The partial derivative $\partial x_A / \partial p_A$ may be interpreted as the **marginal demand for A with respect to p_A.** Similarly, $\partial x_A / \partial p_B$ is the **marginal demand for A with respect to p_B** and measures the amount by which the demand for A increases per unit increase in the price of B. Similar interpretations can be given to the other two partial derivatives.

If the price of commodity B is held fixed, then, in general, an increase in the price of A results in a decrease in the demand x_A for A. In other words, $\partial x_A / \partial p_A < 0$. Similarly, $\partial x_B / \partial p_B < 0$. The partial derivatives $\partial x_A / \partial p_B$ and $\partial x_B / \partial p_A$ can be positive or negative, depending on the particular interaction between the two products. Suppose, for example, the two commodities are beef (A) and pork (B). An increase in the price of A (beef) in general results in an increase in demand for B (pork) when the price of B remains unchanged, since some consumers will switch from A to B. Thus $\partial x_B / \partial p_A > 0$. Similarly, if the price of A (beef) remains unchanged, an increase in the price of B (pork) will result in an increase in demand for A (beef), that is, $\partial x_A / \partial p_B > 0$.

The two commodities A and B are said to be **competitive** if

$$\frac{\partial x_B}{\partial p_A} > 0 \qquad \text{and} \qquad \frac{\partial x_A}{\partial p_B} > 0$$

that is, if an increase in the price of either one of them results in an increase in the demand for the other.

Sometimes an increase in the price of either commodity results in a decrease in the demand for the other (assuming its price remains unaltered). In other words, $\partial x_A / \partial p_B$ and $\partial x_B / \partial p_A$ are both negative. In such a case, the two products A and B are said to be **complementary.** For example, film and cameras are two complementary products. If cameras become costlier, there will be a drop in the demand for film.

EXAMPLE 3 The demands x_A and x_B for the two products A and B are given by the functions

$$x_A = 300 + 5p_B - 7p_A^2 \qquad \text{and} \qquad x_B = 250 - 9p_B + 2p_A$$

where p_A and p_B are the unit prices of A and B, respectively. Determine the four marginal demand functions and find whether the products A and B are competitive or complementary.

Solution The four marginal demand functions are given by the four partial derivatives.

$$\frac{\partial x_A}{\partial p_A} = -14p_A, \qquad \frac{\partial x_A}{\partial p_B} = 5$$

$$\frac{\partial x_B}{\partial p_A} = 2, \qquad \frac{\partial x_B}{\partial p_B} = -9$$

Since $\partial x_A / \partial p_B$ and $\partial x_B / \partial p_A$ are both positive, the two products are competitive.

Consider the demand function for the product A: $x_A = f(p_A, p_B)$ where p_A is the price per unit of A and p_B is the unit price for the related product B. Then the **price elasticity of demand for A** is defined to be

$$\eta_{P_A} = \frac{\partial x_A/\partial p_A}{x_A/p_A} = \frac{p_A}{x_A} \frac{\partial x_A}{\partial p_A}.$$

(See Section 15-3.) The **cross elasticity of demand for A with respect to p_B** is defined to be

$$\eta_{P_B} = \frac{\partial x_A/\partial p_B}{x_A/p_B} = \frac{p_B}{x_A} \frac{\partial x_A}{\partial p_B}.$$

Here, η_{P_A} may be interpreted as the ratio of the percentage change in the demand for A to the percentage change in the price of A when the price of B remains fixed. Similarly, η_{P_B} may be thought of as the ratio of the percentage change in the demand for A to the percentage change in the price of B when the price of A is unchanged.

EXAMPLE 4 The demand function for the product of A is given by

$$x_A = 250 + 0.3p_B - 5p_A^2.$$

Determine η_{P_A} and η_{P_B} when $p_A = 6$ and $p_B = 50$.

Solution In this case, we have

$$\frac{\partial x_A}{\partial p_A} = -10p_A \qquad \text{and} \qquad \frac{\partial x_A}{\partial p_B} = 0.3.$$

When $p_A = 6$ and $p_B = 50$, we have

$$x_A = 250 + 0.3(50) - 5(6^2) = 85$$

$$\frac{\partial x_A}{\partial p_A} = -10(6) = -60 \qquad \text{and} \qquad \frac{\partial x_A}{\partial p_B} = 0.3.$$

Therefore

$$\eta_{P_A} = \frac{\partial x_A/\partial p_A}{x_A/p_A} = \frac{-60}{(85/6)} \approx -4.24 \qquad \text{and}$$

$$\eta_{P_B} = \frac{\partial x_A/\partial p_B}{x_A/p_B} = \frac{0.3}{(85/50)} \approx 0.176.$$

Thus we can say that an increase of 1% in the price of A will result in approximately a 4.24% drop in the demand for this product, while a 1% increase in the price of B will result in a 0.176% increase in the demand for A. **12**

Approximations

In the case of a function $y = f(x)$, we saw in Section 15-1 how the derivative can be used to calculate approximate values of the function at points $x_0 + \Delta x$

● **13.** Given $f(x, y) = x^2 y^3$, approximate $f(-3 + h, 2 + k)$ by a linear expression in h and k.

when $f(x_0)$ is known, provided that Δx is sufficiently small. The approximation is given by

$$f(x_0 + \Delta x) \approx f(x_0) + f'(x_0) \, \Delta x.$$

This approximation formula extends in quite a straightforward way to functions of several variables.

Let $z = f(x, y)$ be a function of two variables that is suitably differentiable. Then, provided that Δx and Δy are small enough,

$$f(x_0 + \Delta x, y_0 + \Delta y) \approx f(x_0, y_0) + f_x(x_0, y_0) \, \Delta x + f_y(x_0, y_0) \, \Delta y.$$

EXAMPLE 5 For $f(x, y) = \sqrt{x + y} + \sqrt{x - y}$, it is readily seen that $f(10, 6) = 6$. Find an approximate expression for $f(10 + h, 6 + k)$ valid for small values of h and k.

Solution Taking $x_0 = 10$ and $y_0 = 6$ in the approximation formula above, we have

$$f(10 + \Delta x, 6 + \Delta y) \approx f(10, 6) + f_x(10, 6) \, \Delta x + f_y(10, 6) \, \Delta y.$$

After partial differentiation, we obtain

$$f_x(x, y) = \tfrac{1}{2}(x + y)^{-1/2} + \tfrac{1}{2}(x - y)^{-1/2}$$
$$f_y(x, y) = \tfrac{1}{2}(x + y)^{-1/2} - \tfrac{1}{2}(x - y)^{-1/2}.$$

At the point $(x_0, y_0) = (10, 6)$, these derivatives have the values

$$f_x(10, 6) = \tfrac{1}{2}(10 + 6)^{-1/2} + \tfrac{1}{2}(10 - 6)^{-1/2} = \tfrac{3}{8}$$
$$f_y(10, 6) = \tfrac{1}{2}(10 + 6)^{-1/2} - \tfrac{1}{2}(10 - 6)^{-1/2} = -\tfrac{1}{8}.$$

Therefore, since $f(10, 6) = 6$,

$$f(10 + \Delta x, 6 + \Delta y) \approx 6 + \tfrac{3}{8} \, \Delta x - \tfrac{1}{8} \, \Delta y.$$

Finally, replacing $\Delta x = h$ and $\Delta y = k$ we obtain the required approximation

$$f(10 + h, 6 + k) \approx 6 + \tfrac{3}{8} h - \tfrac{1}{8} k. \tag{1}$$

For example, take $h = 0.1$ and $k = -0.2$. Then we get

$$f(10.1, 5.8) \approx 6 + \frac{0.3 + 0.2}{8} = 6.0625.$$

For comparison, the exact value of $f(10.1, 5.8)$ is

$$\sqrt{10.1 + 5.8} + \sqrt{10.1 - 5.8} = \sqrt{15.9} + \sqrt{4.3} = 6.0611 \ldots \ldots \quad ● \; \textbf{13}$$

The approximate formula in Equation (1) is a *linear* function of h and k, and consequently it is much easier to work with than the complete expression for $f(10 + h, 6 + k)$. This illustrates the general advantage of this approximation technique in replacing a complicated function by a linear one.

Answer $72 - 48h + 108k$.

14. A firm's revenue R depends on the unit price p it charges for its product and the amount per week it spends on advertising. It is known that when $p = 15$ and $A = 5000$, $R = 25,000$, and the marginal revenue with respect to p is -500 and with respect to A is 4. Calculate the approximate revenue if the price were to be reduced to 12 and the advertising expenditure reduced to 4500.

EXAMPLE 6 By using L units of labor and K units of capital, a firm can produce P units of its product, where $P = f(L, K)$. The firm does not know the precise form of this production function, but it does have the following information.

1. When $L = 64$ and $K = 20$, P is equal to 25,000.

2. When $L = 64$ and $K = 20$, the marginal productivities of labor and capital are $P_L = 270$ and $P_K = 350$.

The firm is contemplating an expansion in its plant that would change L to 69 and K to 24. Find the approximate increase in output which would result.

Solution Taking $L_0 = 64$ and $K_0 = 20$, then for small ΔL and ΔK

$$P = f(L_0 + \Delta L, K_0 + \Delta K) \approx f(L_0, K_0) + f_L(L_0, K_0)\,\Delta L + f_K(L_0, K_0)\,\Delta K$$

$$= 25,000 + 270\,\Delta L + 350\,\Delta K.$$

In the new operation, we would have $\Delta L = 69 - 64 = 5$ and $\Delta K = 24 - 20 = 4$. Therefore

$$P \approx 25,000 + 270(5) + 350(4) = 27,750.$$

The increase in output is therefore $27,750 - 25,000 = 2750.$ ☞ **14**

Answer 24,500.

EXERCISES 18-3

(1–6) *(Marginal Productivities)* For the following production functions $P(L, K)$, find the marginal productivities for the given values of L and K.

1. $P(L, K) = 7L + 5K + 2LK - L^2 - 2K^2$;
   $L = 3, K = 10$

2. $P(L, K) = 18L - 5L^2 + 3LK + 7K - K^2$;
   $L = 4, K = 8$

3. $P(L, K) = 50L + 3L^2 - 4L^3 + 2LK^2 - 3L^2K - 2K^3$;
   $L = 2, K = 5$

4. $P(L, K)$
   $= 25L + 2L^2 - 3L^3 + 5LK^2 - 7L^2K + 2K^2 - K^3$;
   $L = 3, K = 10$

5. $P(L, K) = 100L^{0.3}K^{0.7}$

6. $P(L, K) = 250L^{0.6}K^{0.4}$

7. *(Cobb-Douglass Production Function)* A production function of the form $P(L, K) = cL^aK^b$, where c, a, and b are positive constants and $a + b = 1$, is called a *Cobb-Douglass production function*. Prove that for this produc-

tion function,

$$L\frac{\partial P}{\partial L} + K\frac{\partial P}{\partial K} = P.$$

8. *(Homogeneous Production Function)* A production function $P(L, K)$ is said to be homogeneous of degree n if $L\,\partial P/\partial L + K\,\partial P/\partial K = nP$ for some constant n. Determine whether the production function given by

$$P(L, K) = 5LK + L^2 - 3K^2 + a(L + K)$$

is homogeneous or not. If it is homogeneous, what is the degree of its homogeneity?

(9–12) *(Marginal Demands)* For the following demand functions for the two products A and B, find the four marginal demand functions and determine whether the products A and B are competitive or complementary.

9. $x_A = 20 - 3p_A + p_B$; $x_B = 30 + 2p_A - 5p_B$

10. $x_A = 150 - 0.3p_B^2 - 2p_A^2$; $x_B = 200 - 0.2p_A^2 - 3p_B^2$

11. $x_A = 30\sqrt{p_B}/\sqrt[3]{p_A}$; $x_B = 50p_A/\sqrt[3]{p_B}$

12. $x_A = 200p_B/p_A^2; \quad x_B = 300\sqrt{p_A}/p_B^3$

(13–16) *(Cross Elasticity)* For the following demand functions for the product A, determine η_{p_A} and η_{p_B} at the given price levels for the two related products A and B.

13. $x_A = 250 + 0.3p_B - 2p_A^2; \quad p_A = 5, p_B = 40$

14. $x_A = 127 - 0.2p_B - p_A^2; \quad p_A = 6, p_B = 30$

15. $x_A = 60p_B/\sqrt{p_A}; \quad p_A = 9, p_B = 2$

16. $x_A = 250/(p_A\sqrt{p_B}); \quad p_A = 5, p_B = 4$

17. *(Elasticities of Demand)* The demand function for the product A is given by

$$Q = 327 + 0.2I + 0.5p_B - 2p_A^2$$

where Q is the quantity demanded, I is the consumers' personal disposable income, and p_A and p_B are the unit price of A and the unit price of the related product B, respectively.

a. Compute the price elasticity of demand η_{p_A} when $p_A = 3$, $p_B = 20$, and $I = 200$.

b. Compute the cross elasticity of demand η_{p_B} for A when $p_A = 3$, $p_B = 20$, and $I = 200$.

c. Compute the income elasticity of demand for A,

$$\eta_I = \frac{\partial Q/\partial I}{Q/I} = \frac{I}{Q}\frac{\partial Q}{\partial I}$$

when $p_A = 3$, $p_B = 20$, and $I = 200$.

18. *(Elasticities of Demand)* Repeat Exercise 17 for a product A if the demand is given by the formula

$$Q = 250 + 0.1I + 0.3p_B - 1.5p_A^2.$$

*19. The demand for a certain commodity is given by the function $x = ap^{-b}I^c$, where a, b, and c are constants, p is the price, and I is the consumer's disposable income. Calculate the price elasticity and the income elasticity of demand. (See Exercise 17.) If the supply function for the commodity is $x = rp^s$, where r and s are constants, find the value of p at which market equilibrium is achieved. From this p calculate dp/dI and interpret this derivative.

20. *(Marginal Profits)* The profit per acre from a certain crop of wheat is found to be

$$P = 40L + 5S + 20F - 3L^2 - S^2 - 2F^2 - 4SF$$

where L is the cost of labor, S the cost of seed, and F the cost of fertilizer. Find $\partial P/\partial L$, $\partial P/\partial S$, and $\partial P/\partial F$ and evaluate them when $L = 10$, $S = 3$, and $F = 4$. Interpret these derivatives.

(21–22) If $f(x, y) = \sqrt{x^2 + y^2}$, find the approximate value of each of the following.

21. $f(3.1, 4.1)$ 22. $f(5.1, 11.8)$

(23–24) If $f(x, y) = \sqrt{x^2 - y^2}$, find the approximate value of each of the following.

23. $f(5.2, 2.9)$ 24. $f(25.1, 23.9)$

(25–26) If $f(x, y) = (x - y)/\sqrt{x + y}$, find the approximate value of each of the following.

25. $f(2.1, 1.95)$ 26. $f(4.0, 5.1)$

27. *(Change in Production Level)* A firm can produce P units of its product when it uses L units of labor and K units of capital, where

$$P(L, K) = 100L^{3/4}K^{1/4}.$$

a. Calculate the total output when $L = 81$ and $K = 16$.

b. Approximate the effect of reducing L to 80 and increasing K to 17.

28. *(Change in Production Level)* A firm's production function is given by

$$P(L, K) = 450L^{3/5}K^{2/5}$$

where P represents the output when L units of labor and K units of capital are used.

a. Determine the firm's output when $L = 243$ and $K = 32$.

b. Approximate the effect of increasing the labor to 248 units and decreasing the capital to 31 units.

29. *(Approximate Production)* The production function of a firm is given by

$$P(L, K) = 9L^{2/3}K^{1/3}$$

where P represents the total output when L units of labor and K units of capital are used. Approximate the total output when $L = 1003$ and $K = 28$.

⬙ 18-4 OPTIMIZATION

We saw in Chapter 14 that one of the most important and widely applicable uses of the calculus of functions of a single variable is to calculate maximum and minimum values of functions. The corresponding problem, calculating maxima and minima of functions of several variables, is equally important, and in this section we shall discuss it in the case of functions of two variables.

DEFINITION The function $f(x, y)$ has a **local maximum** at the point (x_0, y_0) if $f(x, y) < f(x_0, y_0)$ for all points (x, y) sufficiently close to (x_0, y_0), except for (x_0, y_0) itself.

The function $f(x, y)$ has a **local minimum** at the point (x_0, y_0) if $f(x, y) > f(x_0, y_0)$ for all points (x, y) sufficiently close to (x_0, y_0) except for (x_0, y_0) itself.

The corresponding value $f(x_0, y_0)$ is called the **local maximum value** (or **local minimum value,** as the case may be) of the function f. The term **extremum** is used to cover both maxima and minima.

In the case of functions of one variable, we discussed two types of extrema, one for which the derivative vanished and the other for which the derivative failed to exist, corresponding to a corner or a spike on the graph of the function. In this section, for the sake of simplicity we shall restrict ourselves to the first type. That is, we shall only consider functions whose graphs are smooth surfaces in three dimensions. This restriction is not too serious since the majority of applications concern functions with smooth graphs.

Let the function $z = f(x, y)$ have a local maximum at (x_0, y_0). Let us construct the vertical section of the graph on which $y = y_0$, that is, the section through the maximum point. This has the equation $z = f(x, y_0)$ and can be represented by a graph in the xz-plane. (See Figure 13.) Since the surface $z = f(x, y)$ has a local maximum when $x = x_0$ and $y = y_0$, this section must

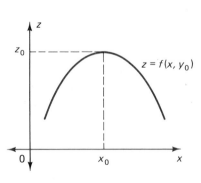

FIGURE 13

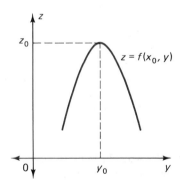

FIGURE 14

15. Find the critical points of the following functions:

(a) $f(x, y) = x^2 + 5xy + y^2$;

(b) $f(x, y) = x^2 + 3xy - 5y^2 - 7x + 4y + 8$.

have a local maximum at $x = x_0$. Therefore the slope of the section, which is given by the derivative $\partial z/\partial x = f_x(x, y_0)$, must be zero when $x = x_0$.

In a similar way, we can consider the section on which $x = x_0$, which consists of a curve in the yz-plane with equation $z = f(x_0, y)$. This curve has a maximum when $y = y_0$, and so the slope $\partial z/\partial y = f_y(x_0, y)$ must be zero when $y = y_0$. (See Figure 14.)

Thus we are led to the following theorem.

THEOREM 1 If $f(x, y)$ has a local maximum or a local minimum at the point (x_0, y_0), then it is necessary that

$$f_x(x_0, y_0) = 0 \quad \text{and} \quad f_y(x_0, y_0) = 0.$$

(The discussion for a local minimum is parallel to that given above for a local maximum.)

DEFINITION A **critical point** of a smooth function $f(x, y)$ is a point (x_0, y_0) at which $f_x(x_0, y_0) = f_y(x_0, y_0) = 0$.

It is clear from the preceding discussion that every local extremum of a smooth function must be a critical point. However, *not every critical point is an extremum*, just as for functions of a single variable. We shall return to this question in a moment.

EXAMPLE 1 Find the critical points of the function

$$f(x, y) = x^3 + x^2y + x - y.$$

Solution We must set the two partial derivatives f_x and f_y equal to zero:

$$f_x(x, y) = 3x^2 + 2xy + 1 = 0$$
$$f_y(x, y) = x^2 - 1 = 0.$$

From the second of these equations, it follows that $x^2 = 1$, or $x = \pm 1$. From the first equation, we then have

$$2xy = -3x^2 - 1 = -3(1) - 1 = -4, \quad \text{that is,} \quad y = \frac{-4}{2x} = \frac{-2}{x}.$$

Thus $y = -2$ when $x = 1$; and when $x = -1$, $y = +2$.

There are therefore two critical points, $(1, -2)$ and $(-1, 2)$. **15**

In the case of a function $f(x)$ of one variable, we saw in Chapter 14 that not every critical point is necessarily a local extremum. A critical point for which $f'(x) = 0$ can be either a local maximum or a local minimum or else a point of inflection, and in Chapter 14 we developed tests to distinguish be-

Answer (a) $(0, 0)$; (b) $(2, 1)$.

tween these possibilities. Similar tests are necessary in the case of a function $f(x, y)$ of two variables, since it is again true that not every critical point need be an extremum. This is illustrated by the function $z = x^2 - y^2$, which was considered in Example 8 of Section 18-1. This function has a critical point at the origin which is neither a local maximum nor a local minimum. The vertical section of its graph by the plane $y = 0$ has a local minimum at the origin, while the vertical section by the plane $x = 0$ has a local maximum at the origin. (See Figures 6 to 8.) A critical point of this type is called a **saddle point.**

If $f(x, y)$ has a local maximum at (x_0, y_0), then it is necessary that the section on which $y = y_0$ should also have a local maximum at $x = x_0$. (This is clear from Figure 13.) This is guaranteed if $f_x(x_0, y_0) = 0$ and $f_{xx}(x_0, y_0) < 0$, by the second derivative test of Section 14-3. Similarly, if $f_y(x_0, y_0) = 0$ and $f_{yy}(x_0, y_0) < 0$, then the section of the graph on which $x = x_0$ is constant must be concave down and so has a local maximum at (x_0, y_0).

The two conditions $f_{xx} < 0$ and $f_{yy} < 0$ at (x_0, y_0) are, however, not sufficient to guarantee that the surface itself has a local maximum at (x_0, y_0). They guarantee only that the vertical sections by the two coordinate planes $x = x_0$ and $y = y_0$ have local maxima at the point (x_0, y_0). It is quite possible for the sections of the graph to have local maxima on these vertical planes, yet to have a local minimum on some other vertical plane through (x_0, y_0). ☛ **16**

It is clear, therefore, that some extra condition is required in order to complete the test for a local maximum or minimum. This is provided by the following theorem, which we shall not prove.

☛ **16.** Let $f(x, y) = (x - y)^2 - 2(x + y)^2$. Show that f has a critical point at $(0, 0)$ at which $f_{xx} = f_{yy} = -2$. However, f does not have a local maximum at the origin: on the vertical plane $y = -x$, $f(x, y) = (x - y)^2 = 4x^2$, which has a local minimum at $x = 0$.

THEOREM 2 Let (x_0, y_0) be a critical point of the function $f(x, y)$ for which $f_x(x_0, y_0) = f_y(x_0, y_0) = 0$. Let

$$\Delta(x, y) = f_{xx}(x, y)f_{yy}(x, y) - [f_{xy}(x, y)]^2.$$

(a) If $f_{xx}(x_0, y_0) < 0$, $f_{yy}(x_0, y_0) < 0$, and $\Delta(x_0, y_0) > 0$, then $f(x, y)$ has a local maximum at (x_0, y_0).

(b) If $f_{xx}(x_0, y_0) > 0$, $f_{yy}(x_0, y_0) > 0$, and $\Delta(x_0, y_0) > 0$, then $f(x, y)$ has a local minimum at (x_0, y_0).

(c) If $\Delta(x_0, y_0) < 0$, then (x_0, y_0) is not a local extremum of $f(x, y)$, but is a saddle point.

Notes

1. If $\Delta(x_0, y_0) = 0$, then this theorem cannot be used to test for a maximum or minimum.

2. If $\Delta(x_0, y_0) > 0$, then f_{xx} and f_{yy} necessarily have the same sign at (x_0, y_0). Therefore in cases (a) and (b) of Theorem 2, the sign of only one of these second derivatives needs to be checked.

EXAMPLE 2 Find the local extrema of the function

$$f(x, y) = x^2 + 2xy + 2y^2 + 2x - 2y.$$

Solution First, let us find the critical points.

$$f_x = 2x + 2y + 2 = 0$$
$$f_y = 2x + 4y - 2 = 0$$

Solving these two simultaneous equations, we obtain $x = -3$ and $y = 2$. Thus $(-3, 2)$ is the only critical point.

Now let us apply Theorem 2 in order to test whether this critical point is a local maximum or a local minimum. We find upon differentiating a second time that

$$f_{xx} = 2, \qquad f_{yy} = 4, \quad \text{and} \quad f_{xy} = 2.$$

☛ **17.** Find the local extrema of the following functions:
(a) $f(x, y) = x^2 + 4xy + 2y^2$;
(b) $f(x, y) = x^2 + 3xy + 4y^2 - x + 2y + 1$.

Therefore, $\Delta = f_{xx}f_{yy} - f_{xy}^2 = (2)(4) - 2^2 = 8 - 4 = 4$. So we see that $f_{xx} > 0$, $f_{yy} > 0$, and $\Delta > 0$, and thus the point $x = -3$, $y = 2$ is a local minimum of f. The local minimum value of f is

$$f(-3, 2) = (-3)^2 + 2(-3)(2) + 2(2)^2 + 2(-3) - 2(2) = -5. \quad \text{☛ 17}$$

EXAMPLE 3 (*Pricing Decision*) The Organic Toothpastes Corporation produces toothpaste in two sizes, 100 milliliter and 150 milliliter. The costs of production for each size tube are 60¢ and 90¢, respectively. The weekly demands x_1 and x_2 (in thousands) for the two sizes are

$$x_1 = 3(p_2 - p_1)$$
$$x_2 = 320 + 3p_1 - 5p_2$$

where p_1 and p_2 are the prices in cents per tube. Determine the prices p_1 and p_2 that will maximize the company's profits.

Solution The profit obtained from each 100-milliliter tube of toothpaste is $(p_1 - 60)$ cents and the profit from each 150-milliliter tube is $(p_2 - 90)$ cents. Therefore the profit P (in thousands of cents, because the demands are in thousands) obtained by selling x_1 tubes of 100-milliliter size and x_2 of 150-milliliter size is given by

$$P = (p_1 - 60)x_1 + (p_2 - 90)x_2$$
$$= 3(p_1 - 60)(p_2 - p_1) + (p_2 - 90)(320 + 3p_1 - 5p_2)$$
$$= -3p_1^2 - 5p_2^2 + 6p_1p_2 - 90p_1 + 590p_2 - 28,800.$$

Therefore,

$$\frac{\partial P}{\partial p_1} = 6p_2 - 6p_1 - 90 \qquad \text{and} \qquad \frac{\partial P}{\partial p_2} = 6p_1 - 10p_2 + 590.$$

Answer (a) Saddle point at $(0, 0)$;
(b) local minimum at $(2, -1)$.

For maximum profit, $\partial P/\partial p_1 = \partial P/\partial p_2 = 0$. That is,

$$6p_2 - 6p_1 - 90 = 0 \qquad \text{and} \qquad 6p_1 - 10p_2 + 590 = 0.$$

Solving these two equations, we get $p_1 = 110$ and $p_2 = 125$. Also $\partial^2 P/\partial p_1^2 = -6$, $\partial^2 P/\partial p_2^2 = -10$, and $\partial^2 P/\partial p_1 \partial p_2 = 6$. Consequently,

$$\Delta = \frac{\partial^2 P}{\partial p_1^2} \cdot \frac{\partial^2 P}{\partial p_2^2} - \left(\frac{\partial^2 P}{\partial p_1 \partial p_2}\right)^2 = (-6)(-10) - 6^2 > 0.$$

Since $\Delta > 0$ and $\partial^2 P/\partial p_1^2$, $\partial^2 P/\partial p_2^2$ are negative, the prices $p_1 = 110\cent$ and $p_2 = 125\cent$ will yield a maximum profit for the company. At these values of p_1 and p_2, the demands are $x_1 = 45$ and $x_2 = 25$ (thousands per week).

Problems also arise in which we are required to find the maximum and minimum values of a function $f(x_1, x_2, \ldots, x_n)$ of several variables. We again solve such problems by setting all of the first partial derivatives equal to zero:

$$\frac{\partial f}{\partial x_1} = \frac{\partial f}{\partial x_2} = \cdots = \frac{\partial f}{\partial x_n} = 0.$$

This provides n equations that must be solved for the variables $x_1, \ldots, x_n$. The resulting point is a critical point of f.

The test that must be applied in order to verify whether the critical point is a local maximum or a local minimum or a saddle point is more complicated than for functions of two variables, and we shall not give it here.*

EXERCISES 18-4

(1–22) Find the critical values of the following functions and test whether each is a local maximum or a local minimum.

1. $f(x, y) = x^2 + y^2 - 2x + 4y + 7$

2. $f(x, y) = 5 + 4x + 6y - x^2 - 3y^2$

3. $f(x, y) = 2x^2 - 3y^2 + 4x + 12y$

4. $f(x, y) = x^2 + 2y^2 - 2x - 2y + 1$

5. $f(x, y) = 2x^2 + xy + 2y^2$

6. $f(x, y) = x^2 + 4xy + y^2$

7. $f(x, y) = 2xy - x^2 - 3y^2 - x - 3y$

8. $f(x, y) = x^2 + 2y^2 - xy - 3x + 5y + 4$

9. $f(x, y) = x^3 + y^2 - 3x - 4y + 7$

10. $f(x, y) = x^3 + y^3 - 12x - 3y$

11. $f(x, y) = x^3 + 3x^2y + y^3 - y$

12. $f(u, v) = u^3 + v^3 - 3uv^2 - 3u + 7$

13. $f(x, y) = 2xy(x + y) + x^2 + 2x$

14. $f(x, y) = xy(x - y) + y^2 - 4y$

15. $f(x, y) = x + y + \dfrac{1}{x} + \dfrac{4}{y}$

16. $f(x, y) = xy - \dfrac{2}{x} - \dfrac{4}{y}$

17. $f(x, y) = (x - 2)(y - 2)(x + y - 3)$

18. $f(x, y) = (x - 1)(y + 2)(x + y - 2)$

19. $f(x, y) = xy + \ln x + y^2$

20. $f(x, y) = 2x^2 + y^2 - \ln (xy^2)$

---

*See, for example, A. E. Taylor and W. R. Mann, *Advanced Calculus*, 2nd ed. (Lexington, Mass.: Xerox College Publishing), p. 230.

21. $f(x, y) = xe^{-x} + ye^{-2y}$

22. $f(p, q) = 25q(1 - e^{-p}) - 50p - q^2$

23. *(Minimum Cost of Production)* A firm produces two types of products, A and B. The total daily cost (in dollars) of producing x units of A and y units of B is given by $C(x, y) = 250 - 4x - 7y + 0.2x^2 + 0.1y^2$. Determine the number of units of A and B that the firm must produce each day to minimize its total cost.

24. *(Maximum Profit)* If the firm in Exercise 23 can sell each unit of A for $20 and each unit of B for $16, find the levels of production of A and B that will maximize the profits for the firm. What is this maximum daily profit?

25. *(Minimum Cost of Production)* Repeat Exercise 23 if

$$C(x, y) = 1500 - 7.5x - 15y - 0.3xy$$
$$+ 0.3x^2 + 0.2y^2.$$

26. *(Optimum Promotion and Production Levels)* If x denotes the firm's output (in hundreds) and y the amount spent (in thousands of dollars) on the promotional efforts to sell the product, then the firm's profit P (in thousands of dollars) is given by $P(x, y) = 16x + 12y + 2xy - x^2 - 2y^2 - 7$. What values of x and y will yield the maximum profit? What is this maximum profit?

27. *(Optimum Labor Usage and Batch-size)* The toal cost C per production run (in thousands of dollars) of a certain industry is given by $C(x, y) = 3x^2 + 4y^2 - 5xy + 3x - 14y + 20$, where x denotes the number of work-hours (in hundreds) and y the number of units (in thousands) of the product produced per run. What values of x and y will result in the minimum total cost per production run?

28. *(Maximum Output)* By using L units of labor and K units of capital, the total weekly output of a firm is given by $P(L, K) = 20K + 32L + 3LK - 2L^2 - 2.5K^2$. Find the number of units of labor and capital that the firm must use to maximize its output.

29. *(Optimal Material Usage)* A manufacturing firm uses two types of raw materials, X and Y, in its product. By using x units of X and y units of Y, the firm can produce P units of the product, where $P = 0.52x + 0.48y + 0.12xy - 0.07x^2 - 0.06y^2$. It costs $5.10 for each unit of X and $1.80 for each unit of Y used, and the firm can sell all the units it produces at $15 each. What amounts of X and Y should be used by the firm to maximize its profits?

30. *(Minimum Cost)* By using L units of labor input and K units of capital input, a firm produces certain output of its product whose total cost T (in millions of dollars) is given by $T = 40 - 5K - 3L - 2KL + 1.5K^2 + L^2$. Find the amount of each input that will make the total cost minimum for the firm. What is this minimum cost?

31. *(Optimal Pricing of Competing Products)* The Western Sweets Candy Company makes candy bars in two sizes at unit costs of 10¢ and 20¢, respectively. The weekly demands (in thousands) x_1 and x_2 for the two sizes are

$$x_1 = p_2 - p_1 \quad \text{and} \quad x_2 = 60 + p_1 - 3p_2$$

where p_1 and p_2 denote the prices in cents of the candy bars of the two size. Determine the prices p_1 and p_2 that will maximize the company's weekly profits.

32. *(Optimal Pricing of Competing Products)* Mack-Oh Toys produces two different types of plastic cars at the cost of 10¢ and 30¢ each. The annual demands x_1 and x_2 (in thousands) are given by

$$x_1 = 30 + 2p_2 - 5p_1 \quad \text{and}$$
$$x_2 = 100 + p_1 - 2p_2$$

where p_1 and p_2 are the unit prices (in cents) of the two types of cars. Determine the prices p_1 and p_2 that the company must charge to maximize its profits.

33. *(Optimal Advertising and Pricing)* It costs a company $2 per unit to manufacture its product. If A dollars per month are spent on advertising, then the number of units per month which will be sold is given by $x = 30(1 - e^{-0.001A})(22 - p)$ where p is the selling price. Find the values of A and p that will maximize the firm's net monthly profit, and calculate the value of this maximum profit.

*34. *(Optimal Advertising and Pricing)* In order to manufacture x items of its product per week, a company's weekly cost function is $C(x) = 50 + \frac{20}{3}x + \frac{1}{60}x^2$. If A dollars per week are spent on advertising, the price p (in dollars) at which the demand will be x items per week is given by

$$p = 20 - \frac{x}{60(1 - e^{-0.001A})}.$$

Find the values of x and A that maximize the weekly profit and calculate this maximum profit.

35. *(Optimum Crop Yield)* The dollar value of a crop of tomatoes produced under artificial heat is given by $V = 25T(1 - e^{-x})$ per unit area of ground. Here T is the

maintained temperature in degrees Celsius above 10°C and x the amount per unit area of fertilizer used. The cost of the fertilizer is $50x$ per unit area and the cost of heating is equal to T^2 per unit area. Find the values of x and T that maximize the profit from the crop. Calculate the maximum profit per unit area.

*36. *(Agriculture)* The average number of apples produced per tree in an orchard in which there are n trees per acre is given by $(A - \alpha n + \beta \sqrt{x})$ where A, α, and β are constants and x is the amount of fertilizer used per acre. The value of each apple is V and the cost per unit of fertilizer is F. Find the values of x and n that make the profit (that is, value of crop of apples less the cost of fertilizer) a maximum.

*37. *(Optimal Fish Stocks)* A lake is to be stocked with two species of fish. When there are x fish of the first species and y fish of the second species in the lake, the average weights of the fish in the two species at the end of the season are $(3 - \alpha x - \beta y)$ pounds and $(4 - \beta x - 2\alpha y)$ pounds, respectively. Find the values of x and y that make the total weight of fish a maximum.

*38. *(Optimal Fish Stocks)* Repeat Exercise 37 for the case when the average weight of the two species of fish are $(5 - 2\alpha x - \beta y)$ pounds and $(3 - 2\beta x - \alpha y)$ pounds, respectively.

39. *(Water Tank Design)* A tank is to be built with width x, length y, and depth z, and it is to be large enough to hold 256 cubic feet of liquid. If the top is open, what dimensions will minimize the total area of the remaining five sides of the tank (and hence minimize the amount of material used in its construction)?

40. *(Medicine)* The reaction to an injection of x units of a certain drug when measured t hours after injection is given by $y = x^2(a - x)te^{-t}$. Calculate $\partial y / \partial x$ and $\partial y / \partial t$ and find the values of x and t that make the reaction maximum.

41. *(Medicine)* If in Exercise 40 the reaction to the drug is given by the formula $y = x(a - x)t^{1/2}e^{-xt}$, calculate:

   a. The value of t which, for fixed x, makes y a maximum.

   b. The values of x and t which together make y maximum.

42. *(Medicine)* Two drugs are used simultenously in the treatment of a certain disease. The reaction R, measured in suitable units to x units of the first drug and y units of the second is

$$R = x^2y^2(a - 2x - y).$$

Find the values of x and y that make R a maximum.

◪ 18-5 LAGRANGE MULTIPLIERS (OPTIONAL)

Sometimes we are faced with minimizing or maximizing a certain function subject to some constraint on the variables involved. Consider the following example.

EXAMPLE 1 A firm wants to construct a rectangular tank to hold 1500 cubic feet of water. The base and the vertical walls have to be made with concrete and the top with steel. If the steel costs twice as much per unit area as the concrete, determine the dimensions of the tank that will minimize the total cost of construction.

Solution Let x, y, and z (in feet) be the length, width, and height of the rectangular tank, respectively. (See Figure 15.) Then,

$$\text{Area of Base} = \text{Area of Top} = xy$$

FIGURE 15

and

$$\text{Area of Four Walls} = 2xz + 2yz.$$

Let the cost of concrete per square foot be p. Then the cost of steel per square foot is $2p$. The cost of constructing the base and four vertical walls with concrete at p per unit area is

$$p(xy + 2xz + 2yz).$$

The cost of constructing the top with steel at $2p$ per unit area is $2pxy$. The total cost C is therefore

$$C = p(xy + 2xz + 2yz) + 2pxy = p(3xy + 2xz + 2yz). \qquad (1)$$

The volume of the box should be 1500 cubic feet. That is,

$$xyz = 1500. \qquad (2)$$

Notice that we have to minimize the function of Equation (1) subject to the condition in Equation (2). We can solve this problem by using the constraint in Equation (2) to remove one of the variables. From Equation (2) $z = 1500/xy$, and substituting this value of z into Equation (1), we get

$$C = p\left(3xy + \frac{3000}{x} + \frac{3000}{y}\right).$$

Now C is a function of two variables that are independent and we can find its minima in the usual way. For a maximum or minimum,

$$C_x = p\left(3y - \frac{3000}{x^2}\right) = 0 \quad \text{or} \quad x^2y = 1000$$

$$C_y = p\left(3x - \frac{3000}{y^2}\right) = 0 \quad \text{or} \quad xy^2 = 1000.$$

It follows, therefore, that $x^2y = xy^2$. Dividing both sides of xy (note that x and y are nonzero), we get $x = y$.

Using $y = x$ in $x^2y = 1000$, we get $x^3 = 1000$ or $x = 10$. Therefore $y = x = 10$.

It is readily verified that when $x = y = 10$, C_{xx}, C_{yy}, and $\Delta = C_{xx}C_{yy} - C_{xy}^2$ are all positive. Therefore the cost C is minimum. When $x = 10$ and $y = 10$, Equation (2) gives $z = 15$. Thus for minimum cost, the dimensions of the tank should be 10 feet by 10 feet by 15 feet.

In Example 1, we eliminated one of the variables (z in this case) from the function C with the help of the constraint equation and then found the critical points of C. But it might happen that we cannot solve the constraint equation for any one of the variables, so that none of them can be eliminated. For example, if the constraint equation is $x^5 - 5x^3y^3 + z^3 + z^5 + 2y^5 + 16 = 0$, we cannot solve for x, y, or z in terms of the other two variables. On the other hand, even if we are able to eliminate one variable by the use of the constraint equation, it may happen that the resulting function to be maximized or minimized is very complicated to handle.

An alternative method—one that avoids the use of elimination—was devised by the French mathematician J. L. Lagrange (1736–1813) and is known as the method of *Lagrange multipliers*. Suppose we are interested in finding the extreme value of the function $f(x, y, z)$ subject to the constraint $g(x, y, z) = 0$. Then we construct an auxiliary function $F(x, y, z, \lambda)$, defined by

$$F(x, y, z, \lambda) = f(x, y, z) - \lambda g(x, y, z).$$

The new variable λ (lambda) is called the **Lagrange multiplier.**

According to the method of Lagrange multipliers if $(x_0, y_0, z_0, \lambda_0)$ is a critical point of $F(x, y, z, \lambda)$ then (x_0, y_0, z_0) is a critical point of $f(x, y, z)$ subject to the constraint $g(x, y, z) = 0$, and conversely. Thus, in order to find the critical points of $f(x, y, z)$ subject to the constraint $g(x, y, z) = 0$, we can instead find the critical points of the auxiliary function $F(x, y, z, \lambda)$. These are given by the conditions

$$F_x = f_x - \lambda g_x = 0$$
$$F_y = f_y - \lambda g_y = 0$$
$$F_z = f_z - \lambda g_z = 0$$
$$F_\lambda = \qquad -g = 0.$$

The last of these equations is nothing but the given constraint equation $g(x, y, z) = 0$. The method of Lagrange multipliers does not indicate directly whether $f(x, y, z)$ will have a maximum, a minimum, or a saddle point at the critical point. In practical problems, we often rely on intuitive reasoning to decide that the critical point gives a maximum or a minimum. There is a test that can be applied, but it is rather a complicated one. ☛ **18**

☛ **18.** Suppose we wish to find the minimum value of $f(x, y) = x^2 + y^2$ subject to the constraint $g(x, y) = 2x + 3y - 12 = 0$. Write down the Lagrange multiplier conditions for the critical point and solve them.

EXAMPLE 2 Let us solve Example 1 again, this time by the method of Lagrange multipliers. We had the function

$$f(x, y, z) = C = p(3xy + 2yz + 2zx)$$

and the constraint $xyz = 1500$. This constraint can be written in the form

$$g(x, y, z) = xyz - 1500 = 0.$$

The auxiliary function in this case is

$$F(x, y, z, \lambda) = f(x, y, z) - \lambda g(x, y, z)$$
$$= p(3xy + 2yz + 2zx) - \lambda(xyz - 1500).$$

The critical points of F are given by the following conditions

$$F_x = p(3y + 2z) - \lambda yz = 0$$
$$F_y = p(3x + 2z) - \lambda xz = 0$$
$$F_z = p(2x + 2y) - \lambda xy = 0$$

Answer $2x - \lambda \cdot 2 = 0$,
$2y - \lambda \cdot 3 = 0$,
$2x + 3y - 12 = 0$.
Solution is $x = \frac{24}{13}$, $y = \frac{36}{13}$.

19. Use the Lagrange multiplier method to find the minimum value of $f(x, y) = xy + 2yz + 3zx$ subject to the constraint $= xyz - 6000 = 0$.

and

$$F_\lambda = -xyz + 1500 = 0.$$

From the first three equations, we have

$$\frac{\lambda}{p} = \frac{3y + 2z}{yz} = \frac{3}{z} + \frac{2}{y}$$

$$\frac{\lambda}{p} = \frac{3x + 2z}{xz} = \frac{3}{z} + \frac{2}{x}$$

$$\frac{\lambda}{p} = \frac{2x + 2y}{xy} = \frac{2}{x} + \frac{2}{y}.$$

From the first and second values of λ/p, we have

$$\frac{3}{z} + \frac{2}{y} = \frac{3}{z} + \frac{2}{x} \qquad \text{or} \qquad \frac{2}{y} = \frac{2}{x}$$

from which it follows that $x = y$. From the second and third values of λ/p,

$$\frac{3}{z} + \frac{2}{x} = \frac{2}{x} + \frac{2}{y} \qquad \text{or} \qquad \frac{3}{z} = \frac{2}{y}.$$

Therefore $z = 3y/2$. Substituting $x = y$ and $z = 3y/2$ into the expression for F_λ, we have

$$-y \cdot y \cdot \tfrac{3}{2}y + 1500 = 0 \qquad \text{or} \qquad y^3 = 1000.$$

Hence $y = 10$. Therefore $x = y = 10$ and $z = \tfrac{3}{2}y = 15$.

The critical point of $C(x, y, z)$ subject to constraint $xyz = 1500$ is thus given by $x = 10$, $y = 10$, and $x = 15$, as before. **☛ 19**

EXAMPLE 3 *(Labor and Capital Investment Decision)* By using L units of labor and K units of capital, a firm can produce P units of its product, where

$$P(L, K) = 50L^{2/3}K^{1/3}.$$

It costs the firm $100 for each unit of labor and $300 for each unit of capital used. The firm has a sum of $45,000 available for production purposes.

(a) Use the method of Lagrange multipliers to determine the units of labor and capital that the firm should use to maximize its production.

(b) Show that at this maximum level of output, the ratio of the marginal productivities of labor and capital is equal to the ratio of their unit costs.

(c) Show that if $1 additional is available for production at this maximum level of production, the firm can produce approximately λ extra units of its product, where λ is the Lagrange multiplier. In other words, λ can be interpreted as the *marginal productivity of money*.

Answer
$x = 20$, $y = 30$, $z = 10$,
$f_{min} = 1800$.

Solution

(a) Here the function to be maximized is

$$P(L, K) = 50L^{2/3}K^{1/3}.$$

The cost of using L units of labor at \$100 each and K units of capital at \$300 each is $(100L + 300K)$ dollars. Since we want to make use of the whole available sum of \$45,000, we must have

$$100L + 300K = 45,000.$$

We wish to maximize $P(L, K)$ subject to this constraint.

The auxiliary function is

$$F(L, K, \lambda) = 50L^{2/3}K^{1/3} - \lambda(100L + 300K - 45,000).$$

To obtain a maximum of $P(L, K)$, we must have

$$F_L = \tfrac{100}{3}L^{-1/3}K^{1/3} - 100\lambda = 0 \qquad (3)$$

$$F_K = \tfrac{50}{3}L^{2/3}K^{-2/3} - 300\lambda = 0 \qquad (4)$$

$$F_\lambda = -(100L + 300K - 45,000) = 0.$$

Solving the first two equations for λ, we get

$$\lambda = \tfrac{1}{3}L^{-1/3}K^{1/3} \qquad \text{and} \qquad \lambda = \tfrac{1}{18}L^{2/3}K^{-2/3}. \qquad (5)$$

We then equate the two values of λ.

$$\tfrac{1}{3}L^{-1/3}K^{1/3} = \tfrac{1}{18}L^{2/3}K^{-2/3}$$

Multiplying both sides by $L^{1/3}K^{2/3}$, we get

$$\tfrac{1}{3}K = \tfrac{1}{18}L \qquad \text{or} \qquad L = 6K.$$

Using this in the value of F_λ we get

$$600K + 300K - 45,000 = 0 \qquad \text{or} \qquad K = 50.$$

Therefore $L = 6K = 300$ and the firm can maximize its output if it uses 300 units of labor and 50 units of capital.

(b) The marginal productivities of labor and capital are given by

$$P_L = \tfrac{100}{3}L^{-1/3}K^{1/3}, \qquad P_K = \tfrac{50}{3}L^{2/3}K^{-2/3}.$$

At the maximum level of output, from Equations (3) and (4) we have

$$P_L = 100\lambda \qquad \text{and} \qquad P_K = 300\lambda. \qquad (6)$$

Therefore

$$\frac{\text{Marginal Productivity of Labor}}{\text{Marginal Productivity of Capital}} = \frac{P_L}{P_K} = \frac{100\lambda}{300\lambda} = \frac{1}{3}.$$

But,

$$\frac{\text{Unit Cost of Labor}}{\text{Unit Cost of Capital}} = \frac{100}{300} = \frac{1}{3}.$$

Thus, at the maximum level of output, the ratio of the marginal productivities of labor and capital is equal to the ratio of the unit costs of labor and capital.

(c) At the maximum level of production, when $L = 300$ and $K = 50$, we have two ways of calculating λ (from Equations (5)):

$$\lambda = \tfrac{1}{3}(300)^{-1/3}(50)^{1/3} = 0.1835$$

$$\lambda = \tfrac{1}{18}(300)^{2/3}(50)^{-2/3} = 0.1835.$$

Suppose we can buy ΔL units of labor and ΔK units of capital with the extra $1 available. Then

$$100\,\Delta L + 300\,\Delta K = 1. \tag{7}$$

The increase in production when labor is increased from 300 to $300 + \Delta L$ and capital is increased from 50 to $50 + \Delta K$ is given by

$$\Delta P = P(300 + \Delta L, 50 + \Delta K) - P(300, 50)$$

$$\approx P_L(300, 50) \cdot \Delta l + P_K(300, 50) \cdot \Delta K.$$

Now from Equation (6) it follows that at the maximum, $P_L(300, 50) = 100\lambda$ and $P_K(300, 50) = 300\lambda$. Therefore the increase in production is approximately given by

$$\Delta P \approx 100\lambda\,\Delta L + 300\lambda\,\Delta K = \lambda(100\,\Delta L + 300\,\Delta K) = \lambda$$

where we used Equation (7). Thus an extra dollar available for production will increase the production by an approximate amount $\lambda = 0.1835$ unit. In other words, λ represents the marginal productivity of money.

EXAMPLE 4 *(Production Decision)* A company can use its plant to produce two types of products, A and B. It makes a profit of $4 per unit of A and $6 per unit of B. The numbers of units of the two types that can be produced by the plant are restricted by the product transformation equation, which is

$$x^2 + y^2 + 2x + 4y - 4 = 0$$

where x and y are the numbers of units (in thousands) of A and B, respectively, produced per week. Find the amounts of each type that should be produced in order to maximize the profit.

Solution We wish to maximize the profit P, which is given by

$$P(x, y) = 4x + 6y$$

(in thousands of dollars per week). Here x and y are subject to the constraint

$$g(x, y) = x^2 + y^2 + 2x + 4y - 4 = 0. \tag{8}$$

☛ **20.** Re-solve Example 4 if the product transformation equation is $x^2 + 2y^2 + x + y = \frac{7}{4}$.

Using the method of Lagrange multipliers, we construct the function

$$F(x, y, \lambda) = P(x, y) - \lambda g(x, y).$$

Then the critical points are given by

$$F_x = P_x - \lambda g_x = 4 - \lambda(2x + 2) = 0$$

$$F_y = P_y - \lambda g_y = 6 - \lambda(2y + 4) = 0$$

$$F_\lambda = -g = 0.$$

This value of F_λ duplicates the given constraint equation. From the values of F_x and F_y,

$$\lambda = \frac{2}{x + 1} = \frac{3}{y + 2}.$$

Therefore $2(y + 2) = 3(x + 1)$ or $y = (3x - 1)/2$. Substituting this into Equation (8), we obtain an equation for x alone.

$$x^2 + \left(\frac{3x - 1}{2}\right)^2 + 2x + 4\left(\frac{3x - 1}{2}\right) - 4 = 0$$

After simplification, this reduces to $13x^2 + 26x - 23 = 0$. From the quadratic formula, we find the roots

$$x = -1 \pm \frac{6\sqrt{13}}{13} \approx 0.664 \qquad \text{or} \qquad -2.664.$$

Of course, only the positive root $x = 0.664$ is meaningful. With this value of x, we have

$$y = \frac{3x - 1}{2} = \frac{3(0.664) - 1}{2} = 0.496.$$

Thus the optimum production levels are 664 units of A and 496 units of B per week. The maximum profit is

$$P = 4(0.664) + 6(0.496) = 5.63$$

that is, $5630 per week. ☛ **20**

_____

The method of Lagrange multipliers can also be used when there are more than one constraint. If $f(x, y, z)$ is to be maximized or minimized subject to two constraints $g_1(x, y, z) = 0$ and $g_2(x, y, z) = 0$, then we construct the auxiliary function F as follows.

$$F(x, y, z, \lambda_1, \lambda_2) = f(x, y, z) - \lambda_1 g_1(x, y, z) - \lambda_2 g_2(x, y, z)$$

The critical points are then obtained by solving the equations

$$F_x = F_y = F_z = F_{\lambda_1} = F_{\lambda_2} = 0.$$

Answer $x = y = 0.5$.

(1–10) Use the method of Lagrange multipliers to determine the critical points of f subject to the given constraints.

1. $f(x, y) = x^2 + y^2$; $2x + 3y = 7$

2. $f(x, y) = x^2 + y^2 - 3xy$; $2x + 3y = 31$

3. $f(x, y) = 3x + 2y$; $x^2 + y^2 = 13$

4. $f(x, y) = 2x^2 + 3y^2$; $xy = \sqrt{6}$

5. $f(x, y, z) = x^2 + y^2 + z^2$; $2x + 3y + 4z = 29$

6. $f(x, y, z) = xyz$; $xy + yz + 2zx = 24$ $(xyz \neq 0)$

7. $f(x, y, z, u) = x^2 + 2y^2 - 3z^2 + 4u^2$;
   $2x - 3y + 4z + 6u = 73$

8. $f(u, v, w, x) = 3u^2 - v^2 + 2w^2 + x^2$;
   $3u + v - 2w + 4x = 20$

9. $f(x, y, z) = x^2 + 2y^2 - 3z^2$; $x + 2y - 3z = 5$,
   $2x - 3y + 6z = -1$

10. $f(u, v, w) = uv + vw + wu$;
    $3u - v + 2w + 13 = 0$, $2u + 3v - w = 0$

11. *(Minimum Production Costs)* The cost of producing x regular models and y deluxe models of a firm's product is given by the joint cost function $C(x, y) = x^2 + 1.5y^2 + 300$. How many units of each type should be produced to minimize total costs if the firm decides to produce a total of 200 units?

12. *(Minimum Production Costs)* A firm can manufacture its product at two of its plants. The cost of manufacturing x units at the first plant and y units at the second plant is given by the joint cost function $C(x, y) = x^2 + 2y^2 + 5xy + 700$. If the firm has a supply order of 500 units, how many should be manufactured at each plant to minimize the total cost?

13. *(Optimal Use of Capital and Labor)* The production function for a firm is $P(L, K) = 80L^{3/4}K^{1/4}$, where L and K represent the numbers of units of labor and capital used and P the number of units of the product produced. Each unit of labor costs $60 and each unit of capital costs $200, and the firm has $40,000 available for production purposes.

a. Use the method of Lagrange multipliers to determine the number of labor and capital units that the firm must use to obtain a maximum output.

b. Show that when the labor and capital are at their maximum levels, the ratio of their marginal productivities is equal to the ratio of their unit costs.

c. At this maximum level of output, determine the increase in production if $1 extra is available for production. Show that it is approximately equal to the Lagrange multiplier.

14. *(Optimal Use of Capital and Labor)* Repeat Exercise 13 for

$$P(L, K) = 800\sqrt{3L^2 + 1.5K^2}.$$

The unit costs of labor and capital are $250 and $50, and the firm has $6750 to spend for production.

15. *(Optimal Use of Capital and Labor)* Repeat Exercise 13 if

$$P(L, K) = 113L + 15K + 3LK - L^2 - 2K^2$$

and the unit costs for labor and capital are $60 and $100, respectively. The firm has a budget constraint of $7200 for production.

16. *(Optimal Use of Capital and Labor)* Repeat Exercise 13 for

$$P(L, K) = 72L + 30K + 5LK - 2L^2 - 3K^2.$$

The unit costs for labor and capital are $80 and $150, respectively. The budget constraint is $5640.

17. *(Optimal Use of Capital and Labor)* By using L units of labor and K units of capital, a firm can produce P units of its product, where $P(L, K) = 60L^{2/3}K^{1/3}$. The costs for labor and capital are $64 and $108 per unit. Suppose the firm decides to produce 2160 units of its product.

a. Use the method of Lagrange multipliers to find the number of labor and capital inputs that must be used to minimize the total costs.

b. Show that at this level of production, the ratio of marginal costs of labor and capital is equal to the ratio of their unit costs.

18. *(Optimal Use of Capital and Labor)* Repeat Exercise 17 if

$$P(L, K) = 60\sqrt{5(L^2 + K^2)}$$

and the unit costs for labor and capital are $200 and $100, respectively. The firm decides to produce 4500 units.

19. *(Investments)* An investment of p dollars in the four investments A, B, C, and D results in an annual return of $\sqrt{p}$, $\sqrt{1.2p}$, $\sqrt{1.3p}$, and $\sqrt{1.5p}$ dollars, respectively. A person has \$12,000 to invest in these four enterprises. How much should be invested in each to maximize the annual return?

20. *(Optimal Advertising)* If a firm spends x thousand dollars on advertising in city A, its potential sales (in thousands of dollars) in that city are given by $300x/(x + 10)$. If it spends x thousand dollars advertising in city B, its poten-

tial sales (in thousands of dollars) in that city are given by $500x/(x + 13.5)$. If the profit is 25% of its sales and the firm has \$16,500 for advertising budget in the two cities, how much should be spent in each city to maximize the firm's net profit?

21. *(Physics)* Three spheres are to be made of materials of densities 1, 2, and 3 grams per cubic centimeter such that their total weight is 10 grams. Find the radii of the spheres for which the sum of their three surface areas is a minimum.

▧ 18-6 METHOD OF LEAST SQUARES

On many occasions in the course of this book, we have quoted formulas for such things as the demand relation for a particular product, the cost of manufacturing x items of a product, the sales volume as a function of advertising expenditure, production functions, and so on. In writing a textbook, we are in the fortunate position of being able to invent our own examples of these functions. However, in real situations, a firm is not able to invent its own cost function, for example, but instead must determine this function from observation of its operations.

In these real cases, we usually do not have a mathematical formula expressing the relationship in question; what we do have are certain data collected from measurements made in the past. Sometimes these data arise in the course of the normal operations of the firm and sometimes they arise as a result of deliberate experimentation. For example, in order to test the effectiveness of its advertising, a company might conduct comparative tests in different towns, varying the advertising expenditure from one town to another.

The measured data can be plotted as a series of points on a graph. In order to obtain an approximation to the complete graph of the relationship, a smooth curve is drawn passing as close as possible to these data points. Usually, the curve which is drawn will not pass through every one of the data points, because to make it do so would reduce its smoothness. In fact, we often approximate the relationship by drawing the graph as a straight line passing as closely as possible to the plotted points. Consider the following example.

Suppose the management of the Pacific Rubber Company is faced with the problem of forecasting its tire sales in the coming years. From past experience, it is known that tire sales increase with the number of cars on the road. The firm has available the data in Table 1, collected in the past. If we plot the number of cars along the x-axis and the tire sales along the y-axis, we obtain the set of points shown in Figure 16.

Looking at the points, we can reasonably conclude that the relationship between x and y is approximately linear, and on this basis we can draw the straight line that lies closest to the set of points. (See Figure 17.) Even though

TABLE 1

Number of Cars (in millions)	18	18.3	18.9	19.4	19.8	20.3
Tire Sales (in thousands)	40	44	52	59	67	77

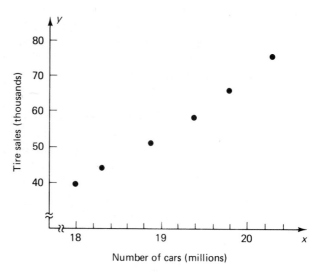

FIGURE 16

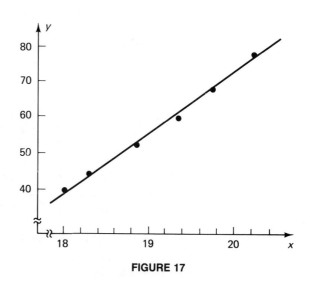

FIGURE 17

all the points do not lie precisely on the straight line, the line does approximate the observed data quite well. This line can be used to predict future tire sales if the management of the firm has some estimate of the number of cars that will be on the road in future years.

Drawing such a line by eye is not objective in the sense that it may be possible to draw another line that appears to fit the set of data points just as well or even better than the one drawn. What is needed is some objective criterion for deciding on the particular straight line that provides the best fit to the observed points. Such a criterion is provided by the **method of least squares.**

Let us suppose that there are n observed data that are plotted as a sequence of points $(x_1 \; y_1)$, $(x_2, \; y_2)$, . . . , $(x_n, \; y_n)$ on the xy-plane. We shall seek the straight line that in a certain sense lies closest to these points.

Let the equation of the straight line of best fit through the given n points be

$$y = ax + b \tag{1}$$

where a and b are constants. Our purpose is to determine a and b, which will then fix the straight line. When $x = x_i$, the observed value of y is y_i; however, the "correct" value is $ax_i + b$, obtained by replacing $x = x_i$ in Equation (1).

The **error** in the value y_i is equal to the difference $y_i - (ax_i + b)$ between the observed value and the theoretical value of y. (See Figure 18.) The **square error** is defined to be simply $(y_i - ax_i - b)^2$. Then the **mean square error,** E, is defined as the average of all the square errors. That is,

$$E = \frac{1}{n}[(y_1 - ax_1 - b)^2 + (y_2 - ax_2 - b)^2 + \cdots + (y_n - ax_n - b)^2].$$

Thus, in calculating E, we determine the square error from each individual data point, add these together for all n data points, and then divide by n.

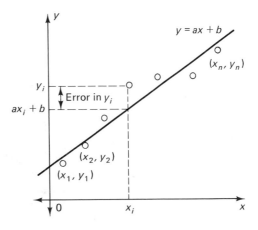

FIGURE 18

Of course, we cannot calculate E yet, since we do not know the values of the constants a and b. According to the method of least squares, what we must do is to choose a and b in such a way that E is minimized. Necessary conditions for this are that the two partial derivatives $\partial E/\partial a$ and $\partial E/\partial b$ be zero.

$$\frac{\partial E}{\partial a} = \frac{\partial}{\partial a} \frac{1}{n}[(y_1 - ax_1 - b)^2 + \cdots + (y_n - ax_n - b)^2]$$

$$= \frac{1}{n}[2(y_1 - ax_1 - b)(-x_1) + \cdots + 2(y_n - ax_n - b)(-x_n)]$$

$$= \frac{2}{n}[(-x_1 y_1 + ax_1^2 + bx_1) + \cdots + (-x_n y_n + ax_n^2 + bx_n)]$$

$$= \frac{2}{n}[a(x_1^2 + x_2^2 + \cdots + x_n^2) + b(x_1 + x_2 + \cdots + x_n)$$

$$-(x_1 y_1 + x_2 y_2 + \cdots + x_n y_n)]$$

and

$$\frac{\partial E}{\partial b} = \frac{\partial}{\partial b} \frac{1}{n}[(y_1 - ax_1 - b)^2 + \cdots + (y_n - ax_n - b)^2]$$

$$= \frac{1}{n}[2(y_1 - ax_1 - b)(-1) + \cdots + 2(y_n - ax_n - b)(-1)]$$

$$= \frac{2}{n}[a(x_1 + x_2 + \cdots + x_n) + nb - (y_1 + y_2 + \cdots + y_n)]$$

Therefore, setting these two derivatives equal to zero, we obtain the following two equations.

$$a(x_1^2 + x_2^2 + \cdots + x_n^2) + b(x_1 + x_2 + \cdots + x_n)$$
$$= (x_1 y_1 + x_2 y_2 + \cdots + x_n y_n)$$
$$a(x_1 + x_2 + \cdots + x_n) + nb = (y_1 + y_2 + \cdots + y_n)$$

These equations form a pair of simultaneous linear equations for the unknown constants a and b, and they may be solved in the usual way. Having calculated a and b, the best straight line through the given data points has equation $y = ax + b$.

Let us find the equation of the best straight line through the data points given for the Pacific Rubber Company. If x denotes the number of cars (in millions) on the road and y the number of tire sales (in thousands) of the company, then we have the data given in Table 2.

When using the method of least squares, it is convenient to make a table as illustrated in Table 3. In the four columns of the table, the values of x_i, y_i, x_i^2, and $x_i y_i$ are listed for each data point. Then the columns are summed. In

☞ 21. Find the equations for a and b for the following three data points, and hence find the straight line that best fits this data.

x	1	2	3
y	5	4	2

TABLE 2

x	18.0	18.3	18.9	19.4	19.8	20.3
y	40	44	52	59	67	77

TABLE 3

x_i	y_i	x_i^2	$x_i y_i$
18.0	40	324.00	720.0
18.3	44	334.89	805.2
18.9	52	357.21	982.8
19.4	59	376.36	1144.6
19.8	67	392.04	1326.6
20.3	77	412.09	1563.1
114.7	339	2196.59	6542.3

this case, we have

$$x_1 + x_2 + \cdots = 114.7$$

$$y_1 + y_2 + \cdots = 339$$

$$x_1^2 + x_2^2 + \cdots = 2196.59$$

$$x_1 y_1 + x_2 y_2 + \cdots = 6542.3.$$

In addition, we have six data points and so $n = 6$. The constants a and b are now given by

$$a(x_1^2 + x_2^2 + \cdots) + b(x_1 + x_2 + \cdots) = x_1 y_1 + x_2 y_2 + \cdots$$

$$a(x_1 + x_2 + \cdots) + nb = y_1 + y_2 + \cdots.$$

That is,

$$2196.59a + 114.7b = 6542.3$$

$$114.7a + 6b = 339.$$

Solving these two equation, we obtain

$$a = 15.8 \quad \text{and} \quad b = -246.$$

Thus the best straight line through the given data points has the equation

$$y = 15.8x - 246. \quad ☞ \ 21$$

Answer $14a + 6b = 19$, $6a + 3b = 11$. $y = -\frac{3}{2}x + \frac{20}{3}$.

As an example of the way in which this result could be used, let us suppose that the government forecast that the number of cars on the road next year would be 22.3 million. Then the Pacific Rubber Company can estimate

22. If you studied Section 7.5, rewrite the equations for a and b using sigma notation.

Answer

$$a \sum_{i=1}^{n} x_i^2 + b \sum_{i=1}^{n} x_i = \sum_{i=1}^{n} x_i y_i,$$

$$a \sum_{i=1}^{n} x_i + nb = \sum_{i=1}^{n} y_i.$$

that its tire sales will be (in thousands)

$$(15.8)(22.3) - 246 = 106.$$

It can be shown without too much difficulty that $\partial^2 E / \partial a^2 > 0$, $\partial^2 E / \partial b^2 > 0$, and (with a little more difficulty) that

$$\Delta = \frac{\partial^2 E}{\partial a^2} \cdot \frac{\partial^2 E}{\partial b^2} - \left(\frac{\partial^2 E}{\partial a \, \partial b} \right)^2 > 0$$

so that the values of a and b found by setting $\partial E / \partial a = \partial E / \partial b = 0$ do correspond to a minimum of E.

It is worth mentioning that the method of least squares is not limited to fitting best straight lines but can be extended to many types of curves. For example, it is often used to fit the best polynomial function to a set of data points. (See Exercise 33 in the Review Exercises for an illustration.) ☛ **22**

EXERCISES 18-6

(1–4) Using the method of least squares find the best straight line through the following sets of data points.

1.

x	2	3	5	6	9	12
y	3	4	6	5	7	8

2.

x	3	4	5	6	7	8
y	0.7	1.9	2.1	2.5	3.4	4.5

3.

x	0	1	2	3
y	1	1.5	2.5	3

4.

x	2	3	4	5	6
y	2	4	3.5	5	6.5

5. *(Growth of Sales)* A department store finds that the trend of sales of a new electric razor is as given in Table 4. Find the equation of the straight line that best fits the data.

TABLE 4

Week Number (x)	1	2	3	4	5	6
Units Sold (y)	20	24	28	33	35	39

6. *(Profits and Advertising)* A firm finds that its net profit increases with increase in the amount spent on advertising the product. The firm has available from its past records the data given in Table 5.

TABLE 5

Advertising Expenditure (x) (thousands of dollars)	10	11	12.3	13.5	15
Net Profits (y) (thousands of dollars)	50	63	68	73	75

a. Find the equation of the straight line that best fits the data.

b. Estimate the money that should be spent on advertising to obtain a net profit of $80,000.

7. *(Demand Curve)* A firm is trying to determine the demand curve for its product. It sells the product in different cities at different prices and determines the volume of sales. After one month, the data given in Table 6 are obtained.

a. Determine the straight line that best fits the data.

TABLE 6

Price (p) (dollars)	2	2.25	2.50	2.75	2.90
Volume Sales (x)	300	290	270	230	200

b. Use the demand curve of part (a) to determine the volume of sales if the price is $3.

c. Use the result of part (a) to determine the price that maximizes the monthly revenue.

8. *(Profit and Production Level)* The production level and the profits of a certain firm for the past few years are shown in Table 7.

TABLE 7

Production (x) (thousands of units)	40	47	55	70	90	100
Profits (y) (thousands of dollars)	32	34	43	54	72	85

a. Determine the equation of the straight line that best fits the data.

b. Estimate the profits when the production level is increased to 120 thousand units.

9. *(Sales and TV Commercials)* A marketing firm wishes to determine the effect of television commercials on the sales of a certain product. The firm has feedback from six large cities as given in Table 8. Determine the equation of the straight line that best fits the data. Estimate the volume of sales that would result from 24 commercials.

10. *(Sales Volume and Commissions)* The sales commissions paid and the sales volume at seven branches of a large chain store for the last year were as given in Table 9. Determine the equation of the straight line that best fits the data.

11. *(Growth of GNP)* Four-year averages for the gross national product (GNP) of a certain country are given in billions of dollars by Table 10. Determine the equation of the straight line that best fits the data. Estimate the GNP for 1980.

TABLE 10

Year (x)	1956	1960	1964	1968	1972	1976
GNP (y)	453	562	691	862	1054	1310

12. *(Agriculture)* The average yield y in bushels of corn per acre in the United States varies from one year to another. The value for the period 1960–1971 are given in Table 11, in which t denotes the date starting with $t = 0$ in

TABLE 8

City	A	B	C	D	E	F
No. of commercials (x)	10	12	15	20	18	21
Sales (y) (hundreds)	40	45	56	68	67	70

TABLE 9

Store	1	2	3	4	5	6	7
Commissions (x) (thousands of dollars)	37.2	45.3	80.5	56.4	67.2	74.6	62.7
Sales (y) (hundreds of thousands of dollars)	4.3	5.1	7.9	5.4	7.1	7.2	6.5

TABLE 11

t	0	1	2	3	4	5	6	7	8	9	10	11
y	54	63	65	67	70	73	72	80	79	87	83	88

1960 and increasing to $t = 11$ in 1971. Show that during this period, a linear equation of the form $y = at + b$ fits this data quite well, and determine the values of a and b.

13. *(Epidemic)* During the growth period of a certain cholera outbreak, the numbers of new cases (y) on successive days were as given in the following table (x denotes the day in question). Find the best straight line through these data points and use it to predict how many new cases will arise on days 6 and 7. (This method of prediction is called *linear extrapolation*.)

TABLE 12

x	1	2	3	4	5
y	6	7	10	12	15

14. *(Entomology)* A certain species of insect is gradually extending its habitat in a northerly direction. The following table gives the most northerly latitude y at which the in-

TABLE 13

x	1955	1960	1965	1970	1975
y	30°N	35°N	38°N	42°N	45°N

sect was normally found during the year x. Find the best straight line through these data points and use it to predict when the insect will reach latitude 49°N.

15. *(Physical Chemistry)* The maximum amount y of a certain substance that will dissolve in 1 liter of water depends on the temperature T. The following experimental results are obtained. Find the best straight-line fit to these data.

TABLE 14

T (°C):	10	20	30	40	50	60
y (grams)	120	132	142	155	169	179

CHAPTER REVIEW

Key Terms, Symbols, and Concepts

18.1 Function of two (or more) variables, domain, range.
Coordinates in three dimensions; x-, y-, and z-axes.
Graph of a function $z = f(x, y)$.
Contour line (or level curve). Vertical section.

18.2 Partial derivatives; $\dfrac{\partial z}{\partial x}$, $\dfrac{\partial z}{\partial y}$.

Second order partial derivatives;

$$\frac{\partial^2 z}{\partial x^2}, \quad \frac{\partial^2 z}{\partial y^2}, \quad \frac{\partial^2 z}{\partial x\, \partial y}, \quad \frac{\partial^2 z}{\partial y\, \partial x}.$$

Notation: $z_{xx}, \quad z_{yy}, \quad z_{xy}$ or

$$f_{xx}(x, y), \quad f_{yy}(x, y), \quad f_{xy}(x, y).$$

18.3 Production function, production input factors.
Marginal productivity of capital and labor.
Competitive and complementary products.
Cross elasticity of demand.

18.4 Local maximum and local minimum for a function of two variables.
Saddle point. Δ-test.

18.5 Constraint. Lagrange multipliers.
Marginal productivity of money.

18.6 Method of least squares. Mean square error.

Formulas

$$\frac{\partial z}{\partial x} = \lim_{\Delta x \to 0} \frac{f(x + \Delta x, y) - f(x, y)}{\Delta x} \quad [\,= f_x(x, y)].$$

$$\frac{\partial z}{\partial y} = \lim_{\Delta y \to 0} \frac{f(x, y + \Delta y) - f(x, y)}{\Delta y} \quad [= f_y(x, y)].$$

$$\frac{\partial^2 z}{\partial x\, \partial y} = \frac{\partial^2 z}{\partial y\, \partial x}.$$

$$f(x_0 + \Delta x,\ y_0 + \Delta y) \approx$$
$$f(x_0, y_0) + f_x(x_0, y_0)\, \Delta x + f_y(x_0, y_0)\, \Delta y.$$

Critical point when f differentiable:

$$f_x(x_0, y_0) = 0, \quad f_y(x_0, y_0) = 0.$$

Δ-test: $\Delta(x, y) = f_{xx}(x, y) f_{yy}(x, y) - [f_{xy}(x, y)]^2.$

If $f_{xx} < 0$, $f_{yy} < 0$, and $\Delta > 0$, critical point is local maximum.

If $f_{xx} > 0$, $f_{yy} > 0$, and $\Delta > 0$, critical point is local minimum.

If $\Delta < 0$, critical point is saddle point.

Method of least squares: $y = ax + b$

$$a(x_1^2 + x_2^2 + \cdots + x_n^2) + b(x_1 + x_2 + \cdots + x_n)$$
$$= (x_1 y_1 + x_2 y_2 + \cdots + x_n y_n)$$

$$a(x_1 + x_2 + \cdots + x_n) + nb = (y_1 + y_2 + \cdots + y_n)$$

REVIEW EXERCISES FOR CHAPTER 18

1. State whether each of the following is true or false. Replace each false statement with a corresponding true statement.

   a. The range of a function $z = f(x, y)$ is the region in the xy-plane where the function takes its values.

   b. The domain of $f(x, y) = (x + y)/(x^2 + y^2 + 1)$ is the set of all real numbers.

   c. In three dimensions, the x-coordinate is zero on the x-axis.

   d. In three dimensions, $y = 0$ and $z = 0$ on the yz-plane.

   e. If $z = f(x + y)$ is any function in which x and y occur only in the combination $(x + y)$, such as $z = (x + y)^2 \ln (x + y)$, then $\partial z/\partial x = \partial z/\partial y$.

   f. If $z = f(x, y)$ and if $\partial z/\partial x = 0$, then z is independent of x, and we can write $z = g(y)$, a function of y only.

   g. If the second-order partial derivatives of $f(x, y)$ are all continuous, then

   $$\frac{\partial^3 f}{\partial x^2\, \partial y} = \frac{\partial^3 f}{\partial y\, \partial x^2}.$$

   h. $\dfrac{\partial}{\partial x}(x^3 y^2) = \dfrac{x^4}{4} y^2$

   i. $\dfrac{\partial}{\partial y}\left(\dfrac{x^2}{y}\right) = x^2 \ln y$

   j. If $f(x, y)$ has a local maximum at (a, b), then $f_x(a, b) = f_y(a, b) = 0$.

   k. If (a, b) is a critical point of $f(x, y)$ and f_{xx}, f_{yy}, are negative at (a, b), then (a, b) is a local maximum point for $f(x, y)$.

   l. If f_{xx} and f_{yy} are of opposite sign at a critical point (a, b) of $f(x, y)$, then f has a saddlepoint at (a, b).

   m. If $f_x(a, b) = 0$ and $f_y(a, b) = 0$, then $f(x, y)$ has a local extremum at (a, b).

   n. If f_{xx} and f_{yy} are positive at the point (a, b), then (a, b) is a local minimum point for $f(x, y)$.

   o. The straight line that best fits a given set of data points must pass through at least two of the data points.

   p. Given only two data points, the straight line obtained by the method of least squares passes through both points.

   q. The critical points of $x^2 + y^2$ subject to the condition $x - y = \ln (x + y)$ can be obtained by finding the critical points of the function $F(x, y, \lambda) = x^2 + y^2 - \lambda(y + \ln (x + y) - x)$.

(2–5) Give the domains of the following functions.

2. $z = \sqrt{x(y + 1)}$

3. $z = y^{-1} \ln (x + y)$

4. $z = \sqrt{1 - x_1^2 - x_2^2 - x_3^2}$

5. $z = \dfrac{\ln (x_1 + x_2 + x_3)}{\sqrt{x_1 - x_2}}$

(6–9) Evaluate $\partial z/\partial x$, $\partial z/\partial y$, $\partial^2 z/\partial x\, \partial y$, and $\partial^2 z/\partial y^2$ for the following functions.

6. $z = \sqrt{x}/y$

7. $z = xy(x + y)$

8. $z = xe^{x+y}$

9. $z = \ln(x^2 + y^2) + e^{y\sqrt{x}}$

10. If $z = f(ax + by) + g(ax - by)$, show that $a^2 z_{yy} - b^2 z_{xx} = 0$.

11. *(Marginal Profits)* Burnaby Mountain Service Station has total annual profits P given by

$$P(x, y, z) = 3xy + 5yz + zx - x^2 - 2y^2 - 1.5z^2$$

where x is the number of employees, y is the number of gas pumps, and z is the value of other supplies (such as oil or tires) carried in inventory. Calculate P_x, P_y, and P_z and interpret P_y.

12. *(Marginal Productivities)* A company's production function is

$$P(L, K, T) = 30L + 25K + 4T - \tfrac{1}{2}L^2 - \tfrac{1}{3}K^2 - \tfrac{1}{20}T^2$$

where P represents the total output and L, K, and T are production input factors. Find the marginal productivity function for each of the factors of production.

13. *(Elasticity)* The demand x_A for a product A is given by

$$x_A = 100\sqrt{p_A p_B}$$

where p_A is the price per unit of A and p_B is the price per unit of a related product B.

a. Show that $p_A \dfrac{\partial x_A}{\partial p_A} + p_B \dfrac{\partial x_A}{\partial p_B} = x_A$.

b. Calculate the elasticities of demand η_{p_A} and η_{p_B} and evaluate their sum.

14. *(Marginal Costs)* If L units of labor and K units of capital are used to produce P items of a certain product, the firm's total cost of producing P units is given by

$$C = 2L^2 + 3K^2 - 5K - 4L \quad \text{(millions of dollars)}.$$

Find $\partial C/\partial L$ and $\partial C/\partial K$. Interpret $\partial C/\partial L$.

15. *(Marginal Costs)* A monopolist faces the demand functions for its two products A and B given by

$$x_A = 3 - p_A + 0.2p_B, \quad x_B = 5 + 0.3p_A - 2p_B.$$

The joint cost function is given by

$$C = x_A^2 + x_B^2 - x_A x_B.$$

Find $\partial C/\partial p_A$ and $\partial C/\partial p_B$. What is the meaning of these derivatives?

(16–19) Determine the local maxima and minima of the following functions.

16. $f(x, y) = 7 + 2xy + 5x - x^2 - 2y^2$

17. $g(u, v) = uv + \dfrac{1}{u} - \dfrac{8}{v} + 7$

18. $F(x, y) = 2xy - \sqrt{x^2 + 4y^2} - 31$

19. $G(x, y) = \dfrac{x}{y} - 2y^2 + \dfrac{8}{3}x^3$

20. *(Optimal Pricing of Competing Products)* The Fresh Meats Company can sell x pounds of pork at p cents per pound and y pounds of beef at q cents per pound, where

$$x = 600 + 4q - 10p, \quad y = 120 + 5p - 3q.$$

What prices per pound for pork and beef should the company charge to maximize its revenue?

21. *(Optimal Pricing of Competing Products)* A meat company can sell x pounds of beef at a price of p cents per pound and y pounds of pork at a price of q cents per pound, where

$$p = 40.5 - 0.03x - 0.004y$$

$$q = 25.4 - 0.005x - 0.02y.$$

How much of each kind of meat should the company sell to maximize its revenue?

22. *(Optimal Use of Capital and Labor)* By using L units of labor and M units of material, a firm can produce T units of its product, where

$$T = 180L + 150M + 3LM - L^2 - 4M^2.$$

a. Determine the number of units of labor and material that must be used to maximize the firm's output.

b. Suppose the firm can sell all it produces at \$5 per unit. It costs the firm \$40 per unit of labor and \$10 for each unit of material. How many units of each should be used to maximize the profits? What is the maximum profit?

c. The unit costs of labor and material are as in part (b). The firm has \$12,660 available for both labor and material. Determine the optimal allocation of these funds.

23. *(Optimal Use of Labor and Advertising)* The monthly profits (in dollars) of a service organization are given by

$$P(x, y) = 448x + 233y + xy - 2x^2 - 3y^2 - 15,000$$

where x denotes the number of workers and y the number of times the firm advertises.

a. Determine x and y that will maximize the organization's monthly profits. What are these maximum profits?

b. The workers get a wage of $500 per month and the advertising costs $100 each time. The working capital is such that an expenditure of $53,100 is to be made on labor and advertisement. Determine the values of x and y that make the profit a maximum.

24. *(Optimal Use of Raw Materials)* A firm uses two types of raw materials, A and B, to manufacture its product. By using x units of A and y units of B, the firm can produce T units of its product, where

$$T = 70x + 240y + 3xy - 4x^2 - 5y^2.$$

a. How many units of each raw material should the firm use in order to maximize its output? What is this maximum output?

b. If it costs the firm $5 for each unit of A and $7 for each unit of B and the firm can sell all it produces at $10 per unit, what amounts of A and B will maximize the firm's profits?

c. The costs for raw materials are as in part (b). If the firm has $250 available for raw materials, what amounts of each will maximize the firm's production?

d. In part (c), when the amounts of the two raw materials used are at the optimum levels, show that the ratio of their marginal productivities is equal to the ratio of their unit costs.

e. Approximate the increase in production when the amount available for raw materials in part (c) is increased by $1.

*25. *(Optimal Use of Labor and Capital)* By using L units of labor and K units of capital, a firm can produce $P(L, K)$ units of its product. The unit costs of labor and capital are p and q dollars, respectively. The firm has a budget constraint of C dollars.

a. Use the method of Lagrange multipliers to show that at the maximum level of production, the ratio of the marginal productivities of labor and capital is equal to the ratio of their unit costs.

b. Show that the firm can produce approximately λ extra units of its product if $1 extra is available at this maxi-

mum level of output, where λ is the Lagrange multiplier.

*26. *(Optimal Use of Labor and Capital)* When L units of labor and K units of capital are used, a firm's output is given by $P(L, K)$. The unit costs of labor and capital are a and b dollars, respectively. Suppose the firm decides to produce P_0 units of its product and the combination of labor and capital is used that produces these units at minimum cost. Show that the ratio of the marginal productivities of labor and capital is equal to the ratio of their unit costs.

27. *(Physiology)* The amount of heat generated by a human infant is sometimes represented by the formula $H = 0.0128LW^{2/3}$, where L is the length of the infant in centimeters and W the weight in kilograms. For an infant of length 60 centimeters and weight 3.8 kilograms, who continues to grow in such a way that $W = kL^3$, where k is a constant, calculate dH/dt if W increases at the rate of 0.2 kilograms per week.

28. *(Physiology)* The surface area of the average human body of weight W kilograms and height H meters is given in square meters by

$$S = 0.2W^{0.4}H^{0.7}.$$

Calculate $\partial S/\partial W$ and $\partial S/\partial H$ when $W = 90$ and $H = 1.75$. If W is changed to 85 and H to 1.8, calculate the approximate increment in S.

*29. *(Diffusion)* The model equation in Exercise 48 of Section 18-2 for the diffusion of a substance through the bloodstream does not take account of the drift due to the motion of the blood. A better equation is

$$C(x, t) = \frac{c}{\sqrt{t}} e^{-[(x-vt)^2/at]}$$

where v is the velocity of the blood. Show that for this equation

$$\frac{\partial C}{\partial t} = \frac{a}{4} \frac{\partial^2 C}{\partial x^2} - v \frac{\partial C}{\partial x}.$$

30. *(Zoology)* In an experiment the weight W of the antlers of deer was measured for a number of deer of different ages. The results are given in Table 15, W being in kilograms and the age A in months. Show that the data fit closely with the linear relation $W = mA + b$, and determine the constants m and b.

TABLE 15

A	20	22	30	34	42	43	46	54	56	68	70
W	0.08	0.10	0.15	0.20	0.27	0.26	0.31	0.36	0.40	0.49	0.49

31. *(Sales and Advertising)* The annual advertising expenditures and sales of six brand-name soaps for a particular year are given by the data in Table 16. Determine the equation of the straight line that best fits the data.

TABLE 16

Advertising Expenditure (x) (hundreds of thousands of dollars)	9	7	11	8	4	6
Sales (y) (millions of dollars)	13	10	15	11	5	8

32. *(Income and Car Ownership)* A study made on the median income in thousands of dollars of families in seven large cities and the percentage of families who own at least one car gave the results (obtained from census data) shown in Table 17. Determine the equation of the straight line that best fits this data.

TABLE 17

Median Income (x)	7.1	4.3	6.5	3.7	5.2	7.8	5.9
Percentage of Car Owners (y)	67	48	62	49	54	72	60

*33. *(Least Squares for Quadratic Function)* It is desired to fit a set of experimental data $\{(x_i, y_i)\}$ with a quadratic function $y = ax^2 + bx + c$. We define the mean square error to be

$$E = \frac{1}{n}[(y_i - ax_1^2 - bx_1 - c)^2$$
$$+ (y_2 - ax_2^2 - bx_2 - c)^2 + \cdots$$
$$+ (y_n - ax_n^2 - bx_n - c)^2].$$

By setting $\partial E/\partial a$, $\partial E/\partial b$, and $\partial E/\partial c$ equal to zero, find three equations for a, b, and c. Evaluate a, b, and c from these equations for the data given in Table 18.

TABLE 18

x	0	1	2	3	4
y	9.9	8.2	6.1	3.7	2.4

APPENDIX ONE

Table of Standard Derivatives

I. BASIC FORMULAS

1. The derivative of a constant is zero.

2. For any constant c, $\dfrac{d}{dx}[cf(x)] = cf'(x)$.

3. $\dfrac{d}{dx}[f(x) \cdot g(x)] = f(x)g'(x) + g(x)f'(x)$.   —Product Formula

4. $\dfrac{d}{dx}\left(\dfrac{f(x)}{g(x)}\right) = \dfrac{g(x)f'(x) - f(x)g'(x)}{[g(x)]^2}$.   —Quotient Formula

5. If $y = f(u)$ and $u = g(x)$ then

$$\frac{dy}{dx} = \frac{dy}{du} \cdot \frac{du}{dx},$$ —Chain Rule

    or

$$\frac{d}{dx}(f[g(x)]) = f'[g(x)] \cdot g'(x)$$ —Chain Rule

6. $\dfrac{d}{dx}(x^n) = nx^{n-1}$.   —Power Formula

II. EXPONENTIAL AND LOGARITHMIC FUNCTIONS

$\dfrac{d}{dx}(e^x) = e^x$.

$\dfrac{d}{dx}(a^x) = a^x \ln a$.

$\dfrac{d}{dx}(\ln x) = \dfrac{1}{x}$.

$\dfrac{d}{dx}(\log_a x) = \dfrac{1}{x} \log_a e = \dfrac{1}{x \ln a}$.

APPENDIX TWO

Table of Integrals

Note In every integral the constant of integration is omitted and should be supplied by the reader.

SOME FUNDAMENTAL FORMULAS

1. $\int [f(x) \pm g(x)] \, dx = \int f(x) \, dx \pm \int g(x) \, dx.$

2. $\int cf(x) \, dx = c \int f(x) \, dx.$

3. $\int f[g(x)]g'(x) \, dx = \int f(u) \, du$ where $u = g(x).$

4. $\int f(x)g(x) \, dx = f(x) \int g(x) \, dx - \int f'(x) \left[\int g(x) \, dx \right] dx.$

RATIONAL INTEGRANDS INVOLVING $(ax + b)$

5. $\int (ax + b)^n \, dx = \dfrac{(ax + b)^{n+1}}{a(n + 1)} \quad (n \neq -1).$

6. $\int (ax + b)^{-1} \, dx = \dfrac{1}{a} \ln |ax + b|.$

7. $\int x(ax + b)^n \, dx = \dfrac{1}{a^2}(ax + b)^{n+1} \left[\dfrac{ax + b}{n + 2} - \dfrac{b}{n + 1} \right] \quad (n \neq -1, -2).$

8. $\int x(ax + b)^{-1} \, dx = \dfrac{x}{a} - \dfrac{b}{a^2} \ln |ax + b|.$

9. $\int x(ax + b)^{-2} \, dx = \dfrac{1}{a^2} \left[\ln |ax + b| + \dfrac{b}{ax + b} \right].$

10. $\int \dfrac{x^2}{ax + b} \, dx = \dfrac{1}{a^3} \left[\dfrac{1}{2}(ax + b)^2 - 2b(ax + b) + b^2 \ln |ax + b| \right].$

11. $\int \dfrac{x^2}{(ax + b)^2} \, dx = \dfrac{1}{a^3} \left[ax + b - \dfrac{b^2}{ax + b} - 2b \ln |ax + b| \right].$

12. $\int \dfrac{1}{x(ax + b)} \, dx = \dfrac{1}{b} \ln \left| \dfrac{x}{ax + b} \right| \quad (b \neq 0).$

13. $\int \dfrac{1}{x^2(ax + b)} \, dx = -\dfrac{1}{bx} + \dfrac{a}{b^2} \ln \left| \dfrac{ax + b}{x} \right| \quad (b \neq 0).$

14. $\int \dfrac{1}{x(ax + b)^2} \, dx = \dfrac{1}{b(ax + b)} - \dfrac{1}{b^2} \ln \left| \dfrac{ax + b}{x} \right| \quad (b \neq 0).$

15. $\displaystyle\int \frac{1}{(ax + b)(cx + d)}\, dx = \frac{1}{bc - ad} \ln \left| \frac{cx + d}{ax + b} \right| \quad (bc - ad \neq 0).$

16. $\displaystyle\int \frac{x}{(ax + b)(cx + d)}\, dx = \frac{1}{bc - ad} \left[\frac{b}{a} \ln |ax + b| - \frac{d}{c} \ln |cx + d| \right]$
$$(bc - ad \neq 0).$$

17. $\displaystyle\int \frac{1}{(ax + b)^2(cx + d)}\, dx = \frac{1}{bc - ad} \left[\frac{1}{ax + b} + \frac{c}{bc - ad} \ln \left| \frac{cx + d}{ax + b} \right| \right]$
$$(bc - ad \neq 0).$$

18. $\displaystyle\int \frac{x}{(ax + b)^2(cx + d)}\, dx = -\frac{1}{bc - ad} \left[\frac{b}{a(ax + b)} + \frac{d}{bc - ad} \ln \left| \frac{cx + d}{ax + b} \right| \right]$
$$(bc - ad \neq 0).$$

INTEGRALS CONTAINING $\sqrt{ax + b}$

19. $\displaystyle\int x\sqrt{ax + b}\, dx = \frac{2}{a^2} \left[\frac{(ax + b)^{5/2}}{5} - \frac{b(ax + b)^{3/2}}{3} \right].$

20. $\displaystyle\int x^2\sqrt{ax + b}\, dx = \frac{2}{a^3} \left[\frac{(ax + b)^{7/2}}{7} - \frac{2b(ax + b)^{5/2}}{5} + \frac{b^2(ax + b)^{3/2}}{3} \right].$

21. $\displaystyle\int \frac{x}{\sqrt{ax + b}}\, dx = \frac{2ax - 4b}{3a^2} \sqrt{ax + b}.$

22. $\displaystyle\int \frac{1}{x\sqrt{ax + b}}\, dx = \frac{1}{\sqrt{b}} \ln \left| \frac{\sqrt{ax + b} - \sqrt{b}}{\sqrt{ax + b} + \sqrt{b}} \right| \quad (b > 0).$

23. $\displaystyle\int \frac{1}{x^n\sqrt{ax + b}}\, dx = -\frac{1}{b(n - 1)} \frac{\sqrt{ax + b}}{x^{n-1}} - \frac{(2n - 3)a}{(2n - 2)b} \int \frac{1}{x^{n-1}\sqrt{ax + b}}\, dx$
$$(n \neq 1).$$

24. $\displaystyle\int \frac{\sqrt{ax + b}}{x}\, dx = 2\sqrt{ax + b} + b \int \frac{1}{x\sqrt{ax + b}}\, dx \quad \text{(see 22)}.$

25. $\displaystyle\int \frac{\sqrt{ax + b}}{x^2}\, dx = -\frac{\sqrt{ax + b}}{x} + \frac{a}{2} \int \frac{1}{x\sqrt{ax + b}}\, dx \quad \text{(see 22)}.$

INTEGRALS CONTAINING $a^2 \pm x^2$

26. $\displaystyle\int \frac{1}{a^2 - x^2}\, dx = \frac{1}{2a} \ln \left| \frac{x + a}{x - a} \right|.$

27. $\displaystyle\int \frac{1}{(a^2 - x^2)^2}\, dx = \frac{x}{2a^2(a^2 - x^2)} + \frac{1}{4a^3} \ln \left| \frac{x + a}{x - a} \right|.$

28. $\displaystyle\int \frac{x}{a^2 \pm x^2}\, dx = \pm\frac{1}{2} \ln |a^2 \pm x^2|.$

29. $\int \dfrac{1}{x(a^2 \pm x^2)}\, dx = \dfrac{1}{2a^2} \ln \left| \dfrac{x^2}{a^2 \pm x^2} \right|.$

INTEGRALS CONTAINING $\sqrt{a^2 - x^2}$

30. $\int \dfrac{x}{\sqrt{a^2 - x^2}}\, dx = -\sqrt{a^2 - x^2}.$

31. $\int \dfrac{1}{x\sqrt{a^2 - x^2}}\, dx = -\dfrac{1}{a} \ln \left| \dfrac{a + \sqrt{a^2 - x^2}}{x} \right|.$

32. $\int \dfrac{1}{x^2\sqrt{a^2 - x^2}}\, dx = -\dfrac{\sqrt{a^2 - x^2}}{a^2 x}.$

33. $\int \dfrac{1}{(a^2 - x^2)^{3/2}}\, dx = \dfrac{1}{a^2} \dfrac{x}{\sqrt{a^2 - x^2}}.$

34. $\int \dfrac{x}{(a^2 - x^2)^{3/2}}\, dx = \dfrac{1}{\sqrt{a^2 - x^2}}.$

35. $\int \dfrac{1}{x(a^2 - x^2)^{3/2}}\, dx = \dfrac{1}{a^2\sqrt{a^2 - x^2}} - \dfrac{1}{a^3} \ln \left| \dfrac{a + \sqrt{a^2 - x^2}}{x} \right|.$

36. $\int \dfrac{1}{x^2(a^2 - x^2)^{3/2}}\, dx = \dfrac{1}{a^4} \left[-\dfrac{\sqrt{a^2 - x^2}}{x} + \dfrac{x}{\sqrt{a^2 - x^2}} \right].$

37. $\int x\sqrt{a^2 - x^2}\, dx = -\tfrac{1}{3}(a^2 - x^2)^{3/2}.$

38. $\int \dfrac{\sqrt{a^2 - x^2}}{x}\, dx = \sqrt{a^2 - x^2} - a \ln \left| \dfrac{a + \sqrt{a^2 - x^2}}{x} \right|.$

39. $\int x(a^2 - x^2)^{3/2}\, dx = -\tfrac{1}{5}(a^2 - x^2)^{5/2}.$

40. $\int \dfrac{(a^2 - x^2)^{3/2}}{x}\, dx = \dfrac{(a^2 - x^2)^{3/2}}{3} + a^2\sqrt{a^2 - x^2} - a^3 \ln \left| \dfrac{a + \sqrt{a^2 - x^2}}{x} \right|.$

41. $\int x^n(a^2 - x^2)^{3/2}\, dx = \dfrac{1}{n + 1} x^{n+1}(a^2 - x^2)^{3/2} + \dfrac{3}{n + 1} \int x^{n+2}\sqrt{a^2 - x^2}\, dx$

$(n \neq -1).$

42. $\int x^n\sqrt{a^2 - x^2}\, dx = -\dfrac{1}{n + 2} x^{n-1}(a^2 - x^2)^{3/2} + \dfrac{a^2(n - 1)}{n + 2} \int x^{n-2}\sqrt{a^2 - x^2}\, dx$

$(n \neq -2).$

INTEGRALS CONTAINING $\sqrt{x^2 \pm a^2}$

43. $\int \dfrac{1}{\sqrt{x^2 \pm a^2}}\, dx = \ln |x + \sqrt{x^2 \pm a^2}|.$

44. $\int \dfrac{x}{\sqrt{x^2 \pm a^2}}\, dx = \sqrt{x^2 \pm a^2}.$

45. $\int \dfrac{x^2}{\sqrt{x^2 \pm a^2}}\, dx = \dfrac{1}{2}x\sqrt{x^2 \pm a^2} \mp \dfrac{1}{2}a^2 \ln|x + \sqrt{x^2 \pm a^2}|.$

46. $\int \dfrac{1}{x\sqrt{x^2 + a^2}}\, dx = -\dfrac{1}{a} \ln\left|\dfrac{a + \sqrt{x^2 + a^2}}{x}\right|.$

47. $\int \dfrac{1}{x^2\sqrt{x^2 \pm a^2}}\, dx = \mp \dfrac{\sqrt{x^2 \pm a^2}}{a^2 x}.$

48. $\int \dfrac{1}{(x^2 \pm a^2)^{3/2}}\, dx = \pm \dfrac{1}{a^2} \dfrac{x}{\sqrt{x^2 \pm a^2}}.$

49. $\int \dfrac{x}{(x^2 \pm a^2)^{3/2}}\, dx = -\dfrac{1}{\sqrt{x^2 \pm a^2}}.$

50. $\int \dfrac{x^2}{(x^2 \pm a^2)^{3/2}}\, dx = -\dfrac{x}{\sqrt{x^2 \pm a^2}} + \ln|x + \sqrt{x^2 \pm a^2}|.$

51. $\int \dfrac{1}{x(x^2 + a^2)^{3/2}}\, dx = \pm\dfrac{1}{a^2}\left[\dfrac{1}{\sqrt{x^2 + a^2}} + \int \dfrac{1}{x\sqrt{x^2 + a^2}}\, dx\right]$ (see 46).

52. $\int \dfrac{1}{x^2(x^2 \pm a^2)^{3/2}}\, dx = -\dfrac{1}{a^4}\left[\dfrac{\sqrt{x^2 \pm a^2}}{x} + \dfrac{x}{\sqrt{x^2 \pm a^2}}\right].$

53. $\int \dfrac{1}{(x^2 \pm a^2)^{5/2}}\, dx = \dfrac{1}{a^4}\left[\dfrac{x}{\sqrt{x^2 \pm a^2}} - \dfrac{1}{3}\dfrac{x^3}{(x^2 \pm a^2)^{3/2}}\right].$

54. $\int \dfrac{x}{(x^2 \pm a^2)^{5/2}}\, dx = -\dfrac{1}{3}\dfrac{1}{(x^2 \pm a^2)^{3/2}}.$

55. $\int \dfrac{x^2}{(x^2 \pm a^2)^{5/2}}\, dx = \pm\dfrac{1}{3a^2}\dfrac{x^3}{(x^2 \pm a^2)^{3/2}}.$

56. $\int \sqrt{x^2 \pm a^2}\, dx = \tfrac{1}{2}x\sqrt{x^2 \pm a^2} \pm \tfrac{1}{2}a^2 \ln|x + \sqrt{x^2 \pm a^2}|.$

57. $\int x\sqrt{x^2 \pm a^2}\, dx = \tfrac{1}{3}(x^2 \pm a^2)^{3/2}.$

58. $\int x^2\sqrt{x^2 \pm a^2}\, dx = \dfrac{x}{4}(x^2 \pm a^2)^{3/2} \pm \dfrac{1}{8}a^2 x\sqrt{x^2 \pm a^2} - \dfrac{1}{8}a^4 \ln|x + \sqrt{x^2 \pm a^2}|.$

59. $\int \dfrac{\sqrt{x^2 + a^2}}{x}\, dx = \sqrt{x^2 + a^2} - a \ln\left|\dfrac{a + \sqrt{x^2 + a^2}}{x}\right|.$

60. $\int \dfrac{\sqrt{x^2 \pm a^2}}{x^2}\, dx = -\dfrac{\sqrt{x^2 \pm a^2}}{x} + \ln|x + \sqrt{x^2 \pm a^2}|.$

61. $\int (x^2 \pm a^2)^{3/2}\, dx = \dfrac{x}{4}(x^2 \pm a^2)^{3/2} \pm \dfrac{3}{8}a^2 x\sqrt{x^2 \pm a^2} + \dfrac{3}{8}a^4 \ln|x + \sqrt{x^2 \pm a^2}|.$

62. $\int x(x^2 \pm a^2)^{3/2}\, dx = \tfrac{1}{5}(x^2 \pm a^2)^{5/2}.$

63. $\int \dfrac{(x^2 \pm a^2)^{3/2}}{x}\, dx = \dfrac{1}{3}(x^2 \pm a^2)^{3/2} \pm a^2 \int \dfrac{\sqrt{x^2 \pm a^2}}{x}\, dx.$

64. $\displaystyle\int x^n(x^2 \pm a^2)^{3/2}\,dx = \frac{1}{n+1}x^{n+1}(x^2 \pm a^2)^{3/2} - \frac{3}{n+1}\int x^{n+2}\sqrt{x^2 \pm a^2}\,dx$

$$(n \neq -1).$$

65. $\displaystyle\int x^n\sqrt{x^2 \pm a^2}\,dx = \frac{1}{n+2}x^{n-1}(x^2 \pm a^2)^{3/2} \pm \frac{a^2(n-1)}{n+2}\int x^{n-2}\sqrt{x^2 \pm a^2}\,dx$

$$(n \neq -2).$$

INTEGRALS CONTAINING $ax^2 + bx + c$

66. $\displaystyle\int \frac{1}{ax^2 + bx + c}\,dx = \frac{1}{\sqrt{b^2 - 4ac}}\ln\left|\frac{2ax + b - \sqrt{b^2 - 4ac}}{2ax + b + \sqrt{b^2 - 4ac}}\right| \quad (b^2 - 4ac > 0)$

INTEGRALS CONTAINING EXPONENTIALS AND LOGARITHMS

67. $\displaystyle\int e^{ax}\,dx = \frac{1}{a}e^{ax}.$

68. $\displaystyle\int a^x\,dx = \frac{1}{\ln a}a^x.$

69. $\displaystyle\int xe^{ax}\,dx = \frac{1}{a^2}(ax - 1)e^{ax}.$

70. $\displaystyle\int x^n e^{ax}\,dx = \frac{1}{a}x^n e^{ax} - \frac{n}{a}\int x^{n-1}e^{ax}\,dx.$

71. $\displaystyle\int \frac{1}{b + ce^{ax}}\,dx = \frac{1}{ab}[ax - \ln(b + ce^{ax})] \quad (ab \neq 0).$

72. $\displaystyle\int \ln|x|\,dx = x\ln|x| - x.$

73. $\displaystyle\int x\ln|x|\,dx = \tfrac{1}{2}x^2\ln|x| - \tfrac{1}{4}x^2.$

74. $\displaystyle\int x^n \ln|x|\,dx = \frac{1}{n+1}x^{n+1}\left[\ln|x| - \frac{1}{n+1}\right] \quad (n \neq -1).$

75. $\displaystyle\int \frac{1}{x\ln|x|}\,dx = \ln|\ln|x||.$

76. $\displaystyle\int \ln^n|x|\,dx = x\ln^n|x| - n\int \ln^{n-1}|x|\,dx.$

77. $\displaystyle\int x^m \ln^n|x|\,dx = \frac{1}{m+1}\left\{x^{m+1}\ln^n|x| - n\int x^m \ln^{n-1}|x|\,dx\right\} \quad (m \neq -1).$

78. $\displaystyle\int \frac{\ln^n|x|}{x}\,dx = \frac{1}{n+1}\ln^{n+1}|x|.$

MISCELLANEOUS INTEGRALS

79. $\displaystyle\int \frac{1}{x(ax^n + b)}\, dx = \frac{1}{nb} \ln \left| \frac{x^n}{ax^n + b} \right| \quad (n \neq 0,\ b \neq 0).$

80. $\displaystyle\int \frac{1}{x\sqrt{ax^n + b}}\, dx = \frac{1}{n\sqrt{b}} \ln \left| \frac{\sqrt{ax^n + b} - \sqrt{b}}{\sqrt{ax^n + b} + \sqrt{b}} \right| \quad (b > 0).$

81. $\displaystyle\int \sqrt{\frac{x + a}{x + b}}\, dx = \sqrt{x + b}\,\sqrt{x + a} + (a - b)\ln \left| \sqrt{x + b} + \sqrt{x + a} \right|.$

APPENDIX THREE

NUMERICAL TABLES
TABLE A.3.1   Four Place Common Logarithms.

N	0	1	2	3	4	5	6	7	8	9
1.0	.0000	.0043	.0086	.0128	.0170	.0212	.0253	.0294	.0334	.0374
1.1	.0414	.0453	.0492	.0531	.0569	.0607	.0645	.0682	.0719	.0755
1.2	.0792	.0828	.0864	.0899	.0934	.0969	.1004	.1038	.1072	.1106
1.3	.1139	.1173	.1206	.1239	.1271	.1303	.1335	.1367	.1399	.1430
1.4	.1461	.1492	.1523	.1553	.1584	.1614	.1644	.1673	.1703	.1732
1.5	.1761	.1790	.1818	.1847	.1875	.1903	.1931	.1959	.1987	.2014
1.6	.2041	.2068	.2095	.2122	.2148	.2175	.2201	.2227	.2253	.2279
1.7	.2304	.2330	.2355	.2380	.2405	.2430	.2455	.2480	.2504	.2529
1.8	.2553	.2577	.2601	.2625	.2648	.2672	.2695	.2718	.2742	.2765
1.9	.2788	.2810	.2833	.2856	.2878	.2900	.2923	.2945	.2967	.2989
2.0	.3010	.3032	.3054	.3075	.3096	.3118	.3139	.3160	.3181	.3201
2.1	.3222	.3243	.3263	.3284	.3304	.3324	.3345	.3365	.3385	.3404
2.2	.3424	.3444	.3464	.3483	.3502	.3522	.3541	.3560	.3579	.3598
2.3	.3617	.3636	.3655	.3674	.3692	.3711	.3729	.3747	.3766	.3784
2.4	.3802	.3820	.3838	.3856	.3874	.3892	.3909	.3927	.3945	.3962
2.5	.3979	.3997	.4014	.4031	.4048	.4065	.4082	.4099	.4116	.4133
2.6	.4150	.4166	.4183	.4200	.4216	.4232	.4249	.4265	.4281	.4298
2.7	.4314	.4330	.4346	.4362	.4378	.4393	.4409	.4425	.4440	.4456
2.8	.4472	.4487	.4502	.4518	.4533	.4548	.4564	.4579	.4594	.4609
2.9	.4624	.4639	.4654	.4669	.4683	.4698	.4713	.4728	.4742	.4757
3.0	.4771	.4786	.4800	.4814	.4829	.4843	.4857	.4871	.4886	.4900
3.1	.4914	.4928	.4942	.4955	.4969	.4983	.4997	.5011	.5024	.5038
3.2	.5051	.5065	.5079	.5092	.5105	.5119	.5132	.5145	.5159	.5172
3.3	.5185	.5198	.5211	.5224	.5237	.5250	.5263	.5276	.5289	.5302
3.4	.5315	.5328	.5340	.5353	.5366	.5378	.5391	.5403	.5416	.5428
3.5	.5441	.5453	.5465	.5478	.5490	.5502	.5514	.5527	.5539	.5551
3.6	.5563	.5575	.5587	.5599	.5611	.5623	.5635	.5647	.5658	.5670
3.7	.5682	.5694	.5705	.5717	.5729	.5740	.5752	.5763	.5775	.5786
3.8	.5798	.5809	.5821	.5832	.5843	.5855	.5866	.5877	.5888	.5899
3.9	.5911	.5922	.5933	.5944	.5955	.5966	.5977	.5988	.5999	.6010
4.0	.6021	.6031	.6042	.6053	.6064	.6075	.6085	.6096	.6107	.6117
4.1	.6128	.6138	.6149	.6160	.6170	.6180	.6191	.6201	.6212	.6222
4.2	.6232	.6243	.6253	.6263	.6274	.6284	.6294	.6304	.6314	.6325
4.3	.6335	.6345	.6355	.6365	.6375	.6385	.6395	.6405	.6415	.6425
4.4	.6435	.6444	.6454	.6464	.6474	.6484	.6493	.6503	.6513	.6522
4.5	.6532	.6542	.6551	.6561	.6571	.6580	.6590	.6599	.6609	.6618
4.6	.6628	.6637	.6646	.6656	.6665	.6675	.6684	.6693	.6702	.6712
4.7	.6721	.6730	.6739	.6749	.6758	.6767	.6776	.6785	.6794	.6803
4.8	.6812	.6821	.6830	.6839	.6848	.6857	.6866	.6875	.6884	.6893
4.9	.6902	.6911	.6920	.6928	.6937	.6946	.6955	.6964	.6972	.6981
5.0	.6990	.6998	.7007	.7016	.7024	.7033	.7042	.7050	.7059	.7067
5.1	.7076	.7084	.7093	.7101	.7110	.7118	.7126	.7135	.7143	.7152
5.2	.7160	.7168	.7177	.7185	.7193	.7202	.7210	.7218	.7226	.7235
5.3	.7243	.7251	.7259	.7267	.7275	.7284	.7292	.7300	.7308	.7316
5.4	.7324	.7332	.7340	.7348	.7356	.7364	.7372	.7380	.7388	.7396
N	0	1	2	3	4	5	6	7	8	9

TABLE A.3.1   Four Place Logarithms (Continued).

N	0	1	2	3	4	5	6	7	8	9
5.5	.7404	.7412	.7419	.7427	.7435	.7443	.7451	.7459	.7466	.7474
5.6	.7482	.7490	.7497	.7505	.7513	.7520	.7528	.7536	.7543	.7551
5.7	.7559	.7566	.7574	.7582	.7589	.7597	.7604	.7612	.7619	.7627
5.8	.7634	.7642	.7649	.7657	.7664	.7672	.7679	.7686	.7694	.7701
5.9	.7709	.7716	.7723	.7731	.7738	.7745	.7752	.7760	.7767	.7774
6.0	.7782	.7789	.7796	.7803	.7810	.7818	.7825	.7832	.7839	.7846
6.1	.7853	.7860	.7868	.7875	.7882	.7889	.7896	.7903	.7910	.7917
6.2	.7924	.7931	.7938	.7945	.7952	.7959	.7966	.7973	.7980	.7987
6.3	.7993	.8000	.8007	.8014	.8021	.8028	.8035	.8041	.8048	.8055
6.4	.8062	.8069	.8075	.8082	.8089	.8096	.8102	.8109	.8116	.8122
6.5	.8129	.8136	.8142	.8149	.8156	.8162	.8169	.8176	.8182	.8189
6.6	.8195	.8202	.8209	.8215	.8222	.8228	.8235	.8241	.8248	.8254
6.7	.8261	.8267	.8274	.8280	.8287	.8293	.8299	.8306	.8312	.8319
6.8	.8325	.8331	.8338	.8344	.8351	.8357	.8363	.8370	.8376	.8382
6.9	.8388	.8395	.8401	.8407	.8414	.8420	.8426	.8432	.8439	.8445
7.0	.8451	.8457	.8463	.8470	.8476	.8482	.8488	.8494	.8500	.8506
7.1	.8513	.8519	.8525	.8531	.8537	.8543	.8549	.8555	.8561	.8567
7.2	.8573	.8579	.8585	.8591	.8597	.8603	.8609	.8615	.8621	.8627
7.3	.8633	.8639	.8645	.8651	.8657	.8663	.8669	.8675	.8681	.8686
7.4	.8692	.8698	.8704	.8710	.8716	.8722	.8727	.8733	.8739	.8745
7.5	.8751	.8756	.8762	.8768	.8774	.8779	.8785	.8791	.8797	.8802
7.6	.8808	.8814	.8820	.8825	.8831	.8837	.8842	.8848	.8854	.8859
7.7	.8865	.8871	.8876	.8882	.8887	.8893	.8899	.8904	.8910	.8915
7.8	.8921	.8927	.8932	.8938	.8943	.8949	.8954	.8960	.8965	.8971
7.9	.8976	.8982	.8987	.8993	.8998	.9004	.9009	.9015	.9020	.9025
8.0	.9031	.9036	.9042	.9047	.9053	.9058	.9063	.9069	.9074	.9079
8.1	.9085	.9090	.9096	.9101	.9106	.9112	.9117	.9122	.9128	.9133
8.2	.9138	.9143	.9149	.9154	.9159	.9165	.9170	.9175	.9180	.9186
8.3	.9191	.9196	.9201	.9206	.9212	.9217	.9222	.9227	.9232	.9238
8.4	.9243	.9248	.9253	.9258	.9263	.9269	.9274	.9279	.9284	.9289
8.5	.9294	.9299	.9304	.9309	.9315	.9320	.9325	.9330	.9335	.9340
8.6	.9345	.9350	.9355	.9360	.9365	.9370	.9375	.9380	.9385	.9390
8.7	.9395	.9400	.9405	.9410	.9415	.9420	.9425	.9430	.9435	.9440
8.8	.9445	.9450	.9455	.9460	.9465	.9469	.9474	.9479	.9484	.9489
8.9	.9494	.9499	.9504	.9509	.9513	.9518	.9523	.9528	.9533	.9538
9.0	.9542	.9547	.9552	.9557	.9562	.9566	.9571	.9576	.9581	.9586
9.1	.9590	.9595	.9600	.9605	.9609	.9614	.9619	.9624	.9628	.9633
9.2	.9638	.9643	.9647	.9652	.9657	.9661	.9666	.9671	.9675	.9680
9.3	.9685	.9689	.9694	.9699	.9703	.9708	.9713	.9717	.9722	.9727
9.4	.9731	.9736	.9741	.9745	.9750	.9754	.9759	.9763	.9768	.9773
9.5	.9777	.9782	.9786	.9791	.9795	.9800	.9805	.9809	.9814	.9818
9.6	.9823	.9827	.9832	.9836	.9841	.9845	.9850	.9854	.9859	.9863
9.7	.9868	.9872	.9877	.9881	.9886	.9890	.9894	.9899	.9903	.9908
9.8	.9912	.9917	.9921	.9926	.9930	.9934	.9939	.9943	.9948	.9952
9.9	.9956	.9961	.9965	.9969	.9974	.9978	.9983	.9987	.9991	.9996
N	0	1	2	3	4	5	6	7	8	9

TABLE A.3.2   Natural Logarithms

	0.00	0.01	0.02	0.03	0.04	0.05	0.06	0.07	0.08	0.09
1.0	0.0000	0.0100	0.0198	0.0296	0.0392	0.0488	0.0583	0.0677	0.0770	0.0862
1.1	0.0953	0.1044	0.1133	0.1222	0.1310	0.1398	0.1484	0.1570	0.1655	0.1740
1.2	0.1823	0.1906	0.1989	0.2070	0.2151	0.2231	0.2311	0.2390	0.2469	0.2546
1.3	0.2624	0.2700	0.2776	0.2852	0.2927	0.3001	0.3075	0.3148	0.3221	0.3293
1.4	0.3365	0.3436	0.3507	0.3577	0.3646	0.3716	0.3784	0.3853	0.3920	0.3988
1.5	0.4055	0.4121	0.4187	0.4253	0.4318	0.4383	0.4447	0.4511	0.4574	0.4637
1.6	0.4700	0.4762	0.4824	0.4886	0.4947	0.5008	0.5068	0.5128	0.5188	0.5247
1.7	0.5306	0.5365	0.5423	0.5481	0.5539	0.5596	0.5653	0.5710	0.5766	0.5822
1.8	0.5878	0.5933	0.5988	0.6043	0.6098	0.6152	0.6206	0.6259	0.6313	0.6366
1.9	0.6419	0.6471	0.6523	0.6575	0.6627	0.6678	0.6729	0.6780	0.6831	0.6881
2.0	0.6931	0.6981	0.7031	0.7080	0.7130	0.7178	0.7227	0.7275	0.7324	0.7372
2.1	0.7419	0.7467	0.7514	0.7561	0.7608	0.7655	0.7701	0.7747	0.7793	0.7839
2.2	0.7885	0.7930	0.7975	0.8020	0.8065	0.8109	0.8154	0.8198	0.8242	0.8286
2.3	0.8329	0.8372	0.8416	0.8459	0.8502	0.8544	0.8587	0.8629	0.8671	0.8713
2.4	0.8755	0.8796	0.8838	0.8879	0.8920	0.8961	0.9002	0.9042	0.9083	0.9123
2.5	0.9163	0.9203	0.9243	0.9282	0.9322	0.9361	0.9400	0.9439	0.9478	0.9517
2.6	0.9555	0.9594	0.9632	0.9670	0.9708	0.9746	0.9783	0.9821	0.9858	0.9895
2.7	0.9933	0.9969	1.0006	1.0043	1.0080	1.0116	1.0152	1.0188	1.0225	1.0260
2.8	1.0296	1.0332	1.0367	1.0403	1.0438	1.0473	1.0508	1.0543	1.0578	1.0613
2.9	1.0647	1.0682	1.0716	1.0750	1.0784	1.0818	1.0852	1.0886	1.0919	1.0953
3.0	1.0986	1.1019	1.1053	1.1086	1.1119	1.1151	1.1184	1.1217	1.1249	1.1282
3.1	1.1314	1.1346	1.1378	1.1410	1.1442	1.1474	1.1506	1.1537	1.1569	1.1600
3.2	1.1632	1.1663	1.1694	1.1725	1.1756	1.1787	1.1817	1.1848	1.1878	1.1909
3.3	1.1939	1.1970	1.2000	1.2030	1.2060	1.2090	1.2119	1.2149	1.2179	1.2208
3.4	1.2238	1.2267	1.2296	1.2326	1.2355	1.2384	1.2413	1.2442	1.2470	1.2499
3.5	1.2528	1.2556	1.2585	1.2613	1.2641	1.2669	1.2698	1.2726	1.2754	1.2782
3.6	1.2809	1.2837	1.2865	1.2892	1.2920	1.2947	1.2975	1.3002	1.3029	1.3056
3.7	1.3083	1.3110	1.3137	1.3164	1.3191	1.3218	1.3244	1.3271	1.3297	1.3324
3.8	1.3350	1.3376	1.3403	1.3429	1.3455	1.3481	1.3507	1.3533	1.3558	1.3584
3.9	1.3610	1.3635	1.3661	1.3686	1.3712	1.3737	1.3762	1.3788	1.3813	1.3838
4.0	1.3863	1.3888	1.3913	1.3938	1.3962	1.3987	1.4012	1.4036	1.4061	1.4085
4.1	1.4110	1.4134	1.4159	1.4183	1.4207	1.4231	1.4255	1.4279	1.4303	1.4327
4.2	1.4351	1.4375	1.4398	1.4422	1.4446	1.4469	1.4493	1.4516	1.4540	1.4563
4.3	1.4586	1.4609	1.4633	1.4656	1.4679	1.4702	1.4725	1.4748	1.4770	1.4793
4.4	1.4816	1.4839	1.4861	1.4884	1.4907	1.4929	1.4952	1.4974	1.4996	1.5019
4.5	1.5041	1.5063	1.5085	1.5107	1.5129	1.5151	1.5173	1.5195	1.5217	1.5239
4.6	1.5261	1.5282	1.5304	1.5326	1.5347	1.5369	1.5390	1.5412	1.5433	1.5454
4.7	1.5476	1.5497	1.5518	1.5539	1.5560	1.5581	1.5602	1.5623	1.5644	1.5665
4.8	1.5686	1.5707	1.5728	1.5748	1.5769	1.5790	1.5810	1.5831	1.5851	1.5872
4.9	1.5892	1.5913	1.5933	1.5953	1.5974	1.5994	1.6014	1.6034	1.6054	1.6074
5.0	1.6094	1.6114	1.6134	1.6154	1.6174	1.6194	1.6214	1.6233	1.6253	1.6273
5.1	1.6292	1.6312	1.6332	1.6351	1.6371	1.6390	1.6409	1.6429	1.6448	1.6467
5.2	1.6487	1.6506	1.6525	1.6544	1.6563	1.6582	1.6601	1.6620	1.6639	1.6658
5.3	1.6677	1.6696	1.6715	1.6734	1.6752	1.6771	1.6790	1.6808	1.6827	1.6845
5.4	1.6864	1.6882	1.6901	1.6919	1.6938	1.6956	1.6974	1.6993	1.7011	1.7029

$$\ln (N \cdot 10^m) = \ln N + m \ln 10, \qquad \ln 10 = 2.3026$$

TABLE A.3.2   Natural Logarithms (*Continued*).

	0.00	0.01	0.02	0.03	0.04	0.05	0.06	0.07	0.08	0.09
5.5	1.7047	1.7066	1.7084	1.7102	1.7120	1.7138	1.7156	1.7174	1.7192	1.7210
5.6	1.7228	1.7246	1.7263	1.7281	1.7299	1.7317	1.7334	1.7352	1.7370	1.7387
5.7	1.7405	1.7422	1.7440	1.7457	1.7475	1.7492	1.7509	1.7527	1.7544	1.7561
5.8	1.7579	1.7596	1.7613	1.7630	1.7647	1.7664	1.7682	1.7699	1.7716	1.7733
5.9	1.7750	1.7766	1.7783	1.7800	1.7817	1.7834	1.7851	1.7867	1.7884	1.7901
6.0	1.7918	1.7934	1.7951	1.7967	1.7984	1.8001	1.8017	1.8034	1.8050	1.8066
6.1	1.8083	1.8099	1.8116	1.8132	1.8148	1.8165	1.8181	1.8197	1.8213	1.8229
6.2	1.8245	1.8262	1.8278	1.8294	1.8310	1.8326	1.8342	1.8358	1.8374	1.8390
6.3	1.8406	1.8421	1.8437	1.8453	1.8469	1.8485	1.8500	1.8516	1.8532	1.8547
6.4	1.8563	1.8579	1.8594	1.8610	1.8625	1.8641	1.8656	1.8672	1.8687	1.8703
6.5	1.8718	1.8733	1.8749	1.8764	1.8779	1.8795	1.8810	1.8825	1.8840	1.8856
6.6	1.8871	1.8886	1.8901	1.8916	1.8931	1.8946	1.8961	1.8976	1.8991	1.9006
6.7	1.9021	1.9036	1.9051	1.9066	1.9081	1.9095	1.9110	1.9125	1.9140	1.9155
6.8	1.9169	1.9184	1.9199	1.9213	1.9228	1.9242	1.9257	1.9272	1.9286	1.9301
6.9	1.9315	1.9330	1.9344	1.9359	1.9373	1.9387	1.9402	1.9416	1.9430	1.9445
7.0	1.9459	1.9473	1.9488	1.9502	1.9516	1.9530	1.9544	1.9559	1.9573	1.9587
7.1	1.9601	1.9615	1.9629	1.9643	1.9657	1.9671	1.9685	1.9699	1.9713	1.9727
7.2	1.9741	1.9755	1.9769	1.9782	1.9796	1.9810	1.9824	1.9838	1.9851	1.9865
7.3	1.9879	1.9892	1.9906	1.9920	1.9933	1.9947	1.9961	1.9974	1.9988	2.0001
7.4	2.0015	2.0028	2.0042	2.0055	2.0069	2.0082	2.0096	2.0109	2.0122	2.0136
7.5	2.0149	2.0162	2.0176	2.0189	2.0202	2.0215	2.0229	2.0242	2.0255	2.0268
7.6	2.0282	2.0295	2.0308	2.0321	2.0334	2.0347	2.0360	2.0373	2.0386	2.0399
7.7	2.0412	2.0425	2.0438	2.0451	2.0464	2.0477	2.0490	2.0503	2.0516	2.0528
7.8	2.0541	2.0554	2.0567	2.0580	2.0592	2.0605	2.0618	2.0631	2.0643	2.0656
7.9	2.0669	2.0681	2.0694	2.0707	2.0719	2.0732	2.0744	2.0757	2.0769	2.0782
8.0	2.0794	2.0807	2.0819	2.0832	2.0844	2.0857	2.0869	2.0882	2.0894	2.0906
8.1	2.0919	2.0931	2.0943	2.0956	2.0968	2.0980	2.0992	2.1005	2.1017	2.1029
8.2	2.1041	2.1054	2.1066	2.1078	2.1090	2.1102	2.1114	2.1126	2.1138	2.1150
8.3	2.1163	2.1175	2.1187	2.1190	2.1211	2.1223	2.1235	2.1247	2.1258	2.1270
8.4	2.1282	2.1294	2.1306	2.1318	2.1330	2.1342	2.1353	2.1365	2.1377	2.1389
8.5	2.1401	2.1412	2.1424	2.1436	2.1448	2.1459	2.1471	2.1483	2.1494	2.1506
8.6	2.1518	2.1529	2.1541	2.1552	2.1564	2.1576	2.1587	2.1599	2.1610	2.1622
8.7	2.1633	2.1645	2.1656	2.1668	2.1679	2.1691	2.1702	2.1713	2.1725	2.1736
8.8	2.1748	2.1759	2.1770	2.1782	2.1793	2.1804	2.1815	2.1827	2.1838	2.1849
8.9	2.1861	2.1872	2.1883	2.1894	2.1905	2.1917	2.1928	2.1939	2.1950	2.1961
9.0	2.1972	2.1983	2.1994	2.2006	2.2017	2.2028	2.2039	2.2050	2.2061	2.2072
9.1	2.2083	2.2094	2.2105	2.2116	2.2127	2.2138	2.2148	2.2159	2.2170	2.2181
9.2	2.2192	2.2203	2.2214	2.2225	2.2235	2.2246	2.2257	2.2268	2.2279	2.2289
9.3	2.2300	2.2311	2.2322	2.2332	2.2343	2.2354	2.2364	2.2375	2.2386	2.2396
9.4	2.2407	2.2418	2.2428	2.2439	2.2450	2.2460	2.2471	2.2481	2.2492	2.2502
9.5	2.2513	2.2523	2.2534	2.2544	2.2555	2.2565	2.2576	2.2586	2.2597	2.2607
9.6	2.2618	2.2628	2.2638	2.2649	2.2659	2.2670	2.2680	2.2690	2.2701	2.2711
9.7	2.2721	2.2732	2.2742	2.2752	2.2762	2.2773	2.2783	2.2793	2.2803	2.2814
9.8	2.2824	2.2834	2.2844	2.2854	2.2865	2.2875	2.2885	2.2895	2.2905	2.2915
9.9	2.2925	2.2935	2.2946	2.2956	2.2966	2.2976	2.2986	2.2996	2.3006	2.3016

TABLE A.3.3 Exponential Functions.

x	e^x	e^{-x}		x	e^x	e^{-x}
0.00	1.0000	1.0000		0.45	1.5683	0.6376
0.01	1.0101	0.9900		0.46	1.5841	0.6313
0.02	1.0202	0.9802		0.47	1.6000	0.6250
0.03	1.0305	0.9704		0.48	1.6161	0.6188
0.04	1.0408	0.9608		0.49	1.6323	0.6126
0.05	1.0513	0.9512		0.50	1.6487	0.6065
0.06	1.0618	0.9418		0.51	1.6653	0.6005
0.07	1.0725	0.9324		0.52	1.6820	0.5945
0.08	1.0833	0.9231		0.53	1.6989	0.5886
0.09	1.0942	0.9139		0.54	1.7160	0.5827
0.10	1.1052	0.9048		0.55	1.7333	0.5769
0.11	1.1163	0.8958		0.56	1.7507	0.5712
0.12	1.1275	0.8869		0.57	1.7683	0.5655
0.13	1.1388	0.8781		0.58	1.7860	0.5599
0.14	1.1503	0.9694		0.59	1.8040	0.5543
0.15	1.1618	0.8607		0.60	1.8221	0.5488
0.16	1.1735	0.8521		0.61	1.8044	0.5434
0.17	1.1853	0.8437		0.62	1.8589	0.5379
0.18	1.1972	0.8353		0.63	1.8776	0.5326
0.19	1.2092	0.8270		0.64	1.8965	0.5273
0.2(	1.2214	0.8187		0.65	1.9155	0.5220
0.21	1.2337	0.8106		0.66	1.9348	0.5169
0.22	1.2461	0.8025		0.67	1.9542	0.5117
0.23	1.2586	0.7945		0.68	1.9739	0.5066
0.24	1.2712	0.7866		0.69	1.9937	0.5016
0.25	1.2840	0.7788		0.70	2.0138	0.4966
0.26	1.2969	0.7711		0.71	2.0340	0.4916
0.27	1.3100	0.7634		0.72	2.0544	0.4868
0.28	1.3231	0.7558		0.73	2.0751	0.4819
0.29	1.3364	0.7483		0.74	2.0959	0.4771
0.30	1.3499	0.7408		0.75	2.1170	0.4724
0.31	1.3634	0.7334		0.76	2.1383	0.4677
0.32	1.3771	0.7261		0.77	2.1598	0.4630
0.33	1.3910	0.7189		0.78	2.1815	0.4584
0.34	1.4049	0.7118		0.79	2.2034	0.4538
0.35	1.4191	0.7047		0.80	2.2255	0.4493
0.36	1.4333	0.6977		0.81	2.2479	0.4449
0.37	1.4477	0.6907		0.82	2.2705	0.4404
0.38	1.4623	0.6839		0.83	2.2933	0.4360
0.39	1.4770	0.6771		0.84	2.3164	0.4317
0.40	1.4918	0.6703		0.85	2.3396	0.4274
0.41	1.5068	0.6637		0.86	2.3632	0.4232
0.42	1.5220	0.6570		0.87	2.3869	0.4190
0.43	1.5373	0.6505		0.88	2.4109	0.4148
0.44	1.5527	0.6440		0.89	2.4351	0.4107

TABLE A.3.3 Exponential Functions (*Continued*).

x	e^x	e^{-x}	x	e^x	e^{-x}
0.90	2.4596	0.4066	2.75	15.643	0.0639
0.91	2.4843	0.4025	2.80	16.445	0.0608
0.92	2.5093	0.3985	2.85	17.288	0.0578
0.93	2.5345	0.3946	2.90	18.174	0.0550
0.94	2.5600	0.3906	2.95	19.106	0.0523
0.95	2.5857	0.3867	3.00	20.086	0.0498
0.96	2.6117	0.3829	3.05	21.115	0.0474
0.97	2.6379	0.3791	3.10	22.198	0.0450
0.98	2.6645	0.3753	3.15	23.336	0.0429
0.99	2.6912	0.3716	3.20	24.533	0.0408
1.00	2.7183	0.3679	3.25	25.790	0.0388
1.05	2.8577	0.3499	3.30	27.113	0.0369
1.10	3.0042	0.3329	3.35	28.503	0.0351
1.15	3.1582	0.3166	3.40	29.964	0.0334
1.20	3.3201	0.3012	3.45	31.500	0.0317
1.25	3.4903	0.2865	3.50	33.115	0.0302
1.30	3.6693	0.2725	3.55	34.813	0.0287
1.35	3.8574	0.2592	3.60	36.598	0.0273
1.40	4.0552	0.2466	3.65	38.475	0.0260
1.45	4.2631	0.2346	3.70	40.447	0.0247
1.50	4.4817	0.2231	3.75	42.521	0.0235
1.55	4.7115	0.2122	3.80	44.701	0.0224
1.60	4.9530	0.2019	3.85	46.993	0.0213
1.65	5.2070	0.1920	3.90	49.402	0.0202
1.70	5.4739	0.1827	3.95	51.935	0.0193
1.75	5.7546	0.1738	4.00	54.598	0.0183
1.80	6.0496	0.1653	4.10	60.340	0.0166
1.85	6.3598	0.1572	4.20	66.686	0.0150
1.90	6.6859	0.1496	4.30	73.700	0.0136
1.95	7.0287	0.1423	4.40	81.451	0.0123
2.00	7.3891	0.1353	4.50	90.017	0.0111
2.05	7.7679	0.1287	4.60	99.484	0.0101
2.10	8.1662	0.1225	4.70	109.95	0.0091
2.15	8.5849	0.1165	4.80	121.51	0.0082
2.20	9.0250	0.1108	4.90	134.29	0.0074
2.25	9.4877	0.1054	5.00	148.41	0.0067
2.30	9.9742	0.1003	5.20	181.27	0.0055
2.35	10.486	0.0954	5.40	221.41	0.0045
2.40	11.023	0.0907	5.60	270.43	0.0037
2.45	11.588	0.0863	5.80	330.30	0.0030
2.50	12.182	0.0821	6.00	403.43	0.0025
2.55	12.807	0.0781	7.00	1096.6	0.0009
2.60	13.464	0.0743	8.00	2981.0	0.0003
2.65	14.154	0.0707	9.00	8103.1	0.0001
2.70	14.880	0.0672	10.00	22026.	0.00005

TABLE A.3.4 Compound Interest Tables.

	$i = \frac{1}{2}\%$ ($i = 0.005$)				$i = \frac{3}{4}\%$ ($i = 0.0075$)		
n	$(1+i)^n$	$a_{\overline{n}|i}$	$s_{\overline{n}|i}$	n	$(1+i)^n$	$a_{\overline{n}|i}$	$s_{\overline{n}|i}$
1	1.005000	0.995025	1.000000	1	1.007500	0.992556	1.000000
2	1.010025	1.985099	2.005000	2	1.015056	1.977723	2.007500
3	1.015075	2.970248	3.015025	3	1.022669	2.955556	3.022556
4	1.020151	3.950496	4.030100	4	1.030339	3.926110	4.045225
5	1.025251	4.925866	5.050251	5	1.038067	4.889440	5.075565
6	1.030378	5.896384	6.075502	6	1.045852	5.845598	6.113631
7	1.035529	6.862074	7.105879	7	1.053696	6.794638	7.159484
8	1.040707	7.822959	8.141409	8	1.061599	7.736613	8.213180
9	1.045911	8.779064	9.182116	9	1.069561	8.671576	9.274779
10	1.051140	9.730412	10.228026	10	1.077583	9.599580	10.344339
11	1.056396	10.677027	11.279167	11	1.085664	10.520675	11.421922
12	1.061678	11.618932	12.335562	12	1.093807	11.434913	12.507586
13	1.066986	12.556151	13.397240	13	1.102010	12.342345	13.601393
14	1.072321	13.488708	14.464226	14	1.110276	13.243022	14.703404
15	1.077683	14.416625	15.536548	15	1.118603	14.136995	15.813679
16	1.083071	15.339925	16.614230	16	1.126992	15.024313	16.932282
17	1.088487	16.258632	17.697301	17	1.135445	15.905025	18.059274
18	1.093929	17.172768	18.785788	18	1.143960	16.779181	19.194718
19	1.099399	18.082356	19.879717	19	1.152540	17.646830	20.338679
20	1.104896	18.987419	20.979115	20	1.161184	18.508020	21.491219
21	1.110420	19.887979	22.084011	21	1.169893	19.362799	22.652403
22	1.115972	20.784059	23.194431	22	1.178667	20.211215	23.822296
23	1.121552	21.675681	24.310403	23	1.187507	21.053315	25.000963
24	1.127160	22.562866	25.431955	24	1.196414	21.889146	26.188471
25	1.132796	23.445638	26.559115	25	1.205387	22.718755	27.384884
26	1.138460	24.324018	27.691911	26	1.214427	23.542189	28.590271
27	1.144152	25.198028	28.830370	27	1.223535	24.359493	29.804698
28	1.149873	26.067689	29.974522	28	1.232712	25.170713	31.028233
29	1.155622	26.933024	31.124395	29	1.241957	25.975893	32.260945
30	1.161400	27.794054	32.280017	30	1.251272	26.775080	33.502902
31	1.167207	28.650800	33.441417	31	1.260656	27.568318	34.754174
32	1.173043	29.503284	34.608624	32	1.270111	28.355650	36.014830
33	1.178908	30.351526	35.781667	33	1.279637	29.137122	37.284941
34	1.184803	31.195548	36.960575	34	1.289234	29.912776	38.564578
35	1.190727	32.035371	38.145378	35	1.298904	30.682656	39.853813
36	1.196681	32.871016	39.336105	36	1.308645	31.446805	41.152716
37	1.202664	33.702504	40.532785	37	1.318460	32.205266	42.461361
38	1.208677	34.529854	41.735449	38	1.328349	32.958080	43.779822
39	1.214721	35.353089	42.944127	39	1.338311	33.705290	45.108170
40	1.220794	36.172228	44.158847	40	1.348349	34.446938	46.446482
41	1.226898	36.987291	45.379642	41	1.358461	35.183065	47.794830
42	1.233033	37.798300	46.606540	42	1.368650	35.913713	49.153291
43	1.239198	38.605274	47.839572	43	1.378915	36.638921	50.521941
44	1.245394	39.408232	49.078770	44	1.389256	37.358730	51.900856
45	1.251621	40.207196	50.324164	45	1.399676	38.073181	53.290112
46	1.257879	41.002185	51.575785	46	1.410173	38.782314	54.689788
47	1.264168	41.793219	52.833664	47	1.420750	39.486168	56.099961
48	1.270489	42.580318	54.097832	48	1.431405	40.184782	57.520711
49	1.276842	43.363500	55.368321	49	1.442141	40.878195	58.952116
50	1.283226	44.142786	56.645163	50	1.452957	41.566447	60.394257

TABLE A.3.4 Compound Interest Tables (*Continued*).

	$i = 1\%$ ($i = 0.01$)				$i = 2\%$ ($i = 0.02$)		
n	$(1 + i)^n$	$a_{\overline{n}\,\rvert i}$	$s_{\overline{n}\,\rvert i}$	n	$(1 + i)^n$	$a_{\overline{n}\,\rvert i}$	$s_{\overline{n}\,\rvert i}$
1	1.010000	0.990099	1.000000	1	1.020000	0.980392	1.000000
2	1.020100	1.970395	2.010000	2	1.040400	1.941561	2.020000
3	1.030301	2.940985	3.030100	3	1.061208	2.883883	3.060400
4	1.040604	3.901966	4.060401	4	1.082432	3.807729	4.121608
5	1.051010	4.853431	5.101005	5	1.104081	4.713460	5.204040
6	1.061520	5.795476	6.152015	6	1.126162	5.601431	6.308121
7	1.072135	6.728195	7.213535	7	1.148686	6.471991	7.434283
8	1.082857	7.651678	8.285671	8	1.171659	7.325481	8.582969
9	1.093685	8.566018	9.368527	9	1.195093	8.162237	9.754628
10	1.104622	9.471305	10.462213	10	1.218994	8.982585	10.949721
11	1.115668	10.367628	11.566835	11	1.243374	9.786848	12.168715
12	1.126825	11.255077	12.682503	12	1.268242	10.575341	13.412090
13	1.138093	12.133740	13.809328	13	1.293607	11.348374	14.680332
14	1.149474	13.003703	14.947421	14	1.319479	12.106249	15.973938
15	1.160969	13.865053	16.096896	15	1.345868	12.849264	17.293417
16	1.172579	14.717874	17.257864	16	1.372786	13.577709	18.639285
17	1.184304	15.562251	18.430443	17	1.400241	14.291872	20.012071
18	1.196147	16.398269	19.614748	18	1.428246	14.992031	21.412312
19	1.208109	17.226008	20.810895	19	1.456811	15.678462	22.840559
20	1.220190	18.045553	22.019004	20	1.485947	16.351433	24.297370
21	1.232392	18.856983	23.239194	21	1.515666	17.011209	25.783317
22	1.244716	19.660379	24.471586	22	1.545980	17.658048	27.298984
23	1.257163	20.455821	25.716302	23	1.576899	18.292204	28.844963
24	1.269735	21.243387	26.973465	24	1.608437	18.913926	30.421862
25	1.282432	22.023156	28.243200	25	1.640606	19.523456	32.030300
26	1.295256	22.795204	29.525631	26	1.673418	20.121036	33.670906
27	1.308209	23.559608	30.820888	27	1.706886	20.706898	35.344324
28	1.321291	24.316443	32.129097	28	1.741024	21.281272	37.051210
29	1.334504	25.065785	33.450388	29	1.775845	21.844385	38.792235
30	1.347849	25.807708	34.784892	30	1.811362	22.396456	40.568079
31	1.361327	26.542285	36.132740	31	1.847589	22.937702	42.379441
32	1.374941	27.269589	37.494068	32	1.884541	23.468335	44.227030
33	1.388690	27.989693	38.869009	33	1.922231	23.988564	46.111570
34	1.402577	28.702666	40.257699	34	1.960676	24.498592	48.033802
35	1.416603	29.408580	41.660276	35	1.999890	24.998619	49.994478
36	1.430769	30.107505	43.076878	36	2.039887	25.488842	51.994367
37	1.445076	30.799510	44.507647	37	2.080685	25.969453	54.034255
38	1.459527	31.484663	45.952724	38	2.122299	26.440641	56.114940
39	1.474123	32.163033	47.412251	39	2.164745	26.902589	58.237238
40	1.488864	32.834686	48.886373	40	2.208040	27.355479	60.401983
41	1.503752	33.499689	50.375237	41	2.252200	27.799489	62.610023
42	1.518790	34.158108	51.878989	42	2.297244	28.234794	64.862223
43	1.533978	34.810008	53.397779	43	2.343189	28.661562	67.159468
44	1.549318	35.455454	54.931757	44	2.390053	29.079963	69.502657
45	1.564811	36.094508	56.481075	45	2.437854	29.490160	71.892710
46	1.580459	36.727236	58.045885	46	2.486611	29.892314	74.330564
47	1.596263	37.353699	59.626344	47	2.536344	30.286582	76.817176
48	1.612226	37.973959	61.222608	48	2.587070	30.673120	79.353519
49	1.628348	38.588079	62.834834	49	2.638812	31.052078	81.940590
50	1.644632	39.196118	64.463182	50	2.691588	31.423606	84.579401

TABLE A.3.4   Compound Interest Tables (*Continued*).

$i = 3\%$ ($i = 0.03$) $i = 4\%$ ($i = 0.04$)

n	$(1 + i)^n$	$a_{\overline{n}\|i}$	$s_{\overline{n}\|i}$	n	$(1 + i)^n$	$a_{\overline{n}\|i}$	$s_{\overline{n}\|i}$
1	1.030000	0.970874	1.000000	1	1.040000	0.961538	1.000000
2	1.060900	1.913470	2.030000	2	1.081600	1.886095	2.040000
3	1.092727	2.828611	3.090900	3	1.124864	2.775091	3.121600
4	1.125509	3.717098	4.183627	4	1.169859	3.629895	4.246464
5	1.159274	4.579707	5.309136	5	4.216653	4.451822	5.416323
6	1.194052	5.417191	6.468410	6	1.265319	5.242137	6.632975
7	1.229874	6.230283	7.662462	7	1.315932	6.002055	7.898294
8	1.266770	7.019692	8.892336	8	1.368569	6.732745	9.214226
9	1.304773	7.786109	10.159106	9	1.423312	7.435332	10.582795
10	1.343916	8.530203	11.463879	10	1.480244	8.110896	12.006107
11	1.384234	9.252624	12.807796	11	1.539454	8.760477	13.486351
12	1.425761	9.954004	14.192030	12	1.601032	9.385074	15.025805
13	1.468534	10.634955	15.617790	13	1.665074	9.985648	16.626838
14	1.512590	11.296073	17.086324	14	1.731676	10.563123	18.291911
15	1.557967	11.937935	18.598914	15	1.800944	11.118387	20.023588
16	1.604706	12.561102	20.156881	16	1.872981	11.652296	21.824531
17	1.652848	13.166118	21.761588	17	1.947900	12.165669	23.697512
18	1.702433	13.753513	23.414435	18	2.025817	12.659297	25.645413
19	1.753506	14.323799	25.116868	19	2.106849	13.133939	27.671229
20	1.806111	14.877475	26.870374	20	2.191123	13.590326	29.778079
21	1.860295	15.415024	28.676486	21	2.278768	14.029160	31.969202
22	1.916103	15.936917	30.536780	22	2.369919	14.451115	34.247970
23	1.973587	16.443608	32.452884	23	2.464716	14.856842	36.617889
24	2.032794	16.935542	34.426470	24	2.563304	15.246963	39.082604
25	2.093778	17.413148	36.459264	25	2.665836	15.622080	41.645908
26	2.156591	17.876842	38.553042	26	2.772470	15.982769	44.311745
27	2.221289	18.327031	40.709634	27	2.883369	16.329586	47.084214
28	2.287928	18.764108	42.930923	28	2.998703	16.663063	49.967583
29	2.356566	19.188455	45.218850	29	3.118651	16.983715	52.966286
30	2.427262	19.600441	47.575416	30	3.243398	17.292033	56.084938
31	2.500080	20.000428	50.002678	31	3.373133	17.588494	59.328335
32	2.575083	20.388766	52.502759	32	3.508059	17.873551	62.701469
33	2.652335	20.765792	55.077841	33	3.648381	18.147646	66.209527
34	2.731905	21.131837	57.730177	34	3.794316	18.411198	69.857909
35	2.813862	21.487220	60.462082	35	3.946089	18.664613	73.652225
36	2.898278	21.832252	63.275944	36	4.103933	18.908282	77.598314
37	2.985227	22.167235	66.174223	37	4.268090	19.142579	81.702246
38	3.074783	22.492462	69.159449	38	4.438813	19.367864	85.970336
39	3.167027	22.808215	72.234233	39	4.616366	19.584485	90.409150
40	3.262038	23.114772	75.401260	40	4.801021	19.792774	95.025516
41	3.359899	23.412400	78.663298	41	4.993061	19.993052	99.826536
42	3.460696	23.701359	82.023196	42	5.192784	20.185627	104.819598
43	3.564517	23.981902	85.483892	43	5.400495	20.370795	110.012382
44	3.671452	24.254274	89.048409	44	5.616515	20.548841	115.412877
45	3.781596	24.518713	92.719861	45	5.841176	20.720040	121.029392
46	3.895044	24.775449	96.501457	46	6.074823	20.884654	126.870568
47	4.011895	25.024708	100.396501	47	6.317816	21.042936	132.945390
48	4.132252	25.266707	104.408396	48	6.570528	21.195131	139.263206
49	4.256219	25.501657	108.540648	49	6.833349	21.341472	145.833734
50	4.383906	25.729764	112.796867	50	7.106683	21.482185	152.667084

TABLE A.3.4 Compound Interest Tables (*Continued*).

	$i = 5\%\ (i = 0.05)$				$i = 6\%\ (i = 0.06)$		
n	$(1+i)^n$	$a_{\overline{n}|i}$	$s_{\overline{n}|i}$	n	$(1+i)^n$	$a_{\overline{n}|i}$	$s_{\overline{n}|i}$
1	1.050000	0.952381	1.000000	1	1.060000	0.943396	1.000000
2	1.102500	1.859410	2.050000	2	1.123600	1.833393	2.060000
3	1.157625	2.723248	3.152500	3	1.191016	2.673012	3.183600
4	1.215506	3.545951	4.310125	4	1.262477	3.465106	4.374616
5	1.276282	4.329477	5.525631	5	1.338226	4.212364	5.637093
6	1.340096	5.075692	6.801913	6	1.418519	4.917324	6.975319
7	1.407100	5.786373	8.142008	7	1.503630	5.582381	8.393838
8	1.477455	6.463213	3.549109	8	1.593848	6.209794	9.897468
9	1.551328	7.107822	11.026564	9	1.689479	6.801692	11.491316
10	1.628895	7.721735	12.577893	10	1.790848	7.360087	13.180795
11	1.710339	8.306414	14.206787	11	1.898299	7.886875	14.971643
12	1.795856	8.863252	15.917127	12	2.012196	8.383844	16.869941
13	1.885649	9.393573	17.712983	13	2.132928	8.852683	18.882138
14	1.979932	9.898641	19.598632	14	2.260904	9.294984	21.015066
15	2.078928	10.379658	21.578564	15	2.396558	9.712249	23.275970
16	2.182875	10.837770	23.657492	16	2.540352	10.105895	25.672528
17	2.292018	11.274066	25.840366	17	2.692773	10.477260	28.212880
18	2.406619	11.689587	28.132385	18	2.854339	10.827603	30.905653
19	2.526950	12.085321	30.539004	19	3.025600	11.158116	33.759992
20	2.653298	12.462210	33.065954	20	3.207135	11.469921	36.785591
21	2.785963	12.821153	35.719252	21	3.399564	11.764077	39.992727
22	2.925261	13.163003	38.505214	22	3.603537	12.041582	43.392290
23	3.071524	13.488574	41.430475	23	3.819750	12.303379	46.995828
24	3.225100	13.798642	44.501999	24	4.048935	12.550358	50.815577
25	3.386355	14.093945	47.727099	25	4.291871	12.783356	54.864512
26	3.555673	14.375185	51.113454	26	4.549383	13.003166	59.156383
27	3.733456	14.643034	54.669126	27	4.822346	13.210534	63.705766
28	3.920129	14.898127	58.402583	28	5.111687	13.406164	68.528112
29	4.116136	15.141074	62.322712	29	5.418388	13.590721	73.639798
30	4.321942	15.372451	66.438848	30	5.743491	13.764831	79.058186
31	4.538039	15.592811	70.760790	31	6.088101	13.929086	84.801677
32	4.764941	15.802677	75.298829	32	6.453387	14.084043	90.889778
33	5.003189	16.002549	80.063771	33	6.840590	14.230230	97.343165
34	5.253348	16.192904	85.066959	34	7.251025	14.368141	104.183755
35	5.516015	16.374194	90.320307	35	7.686087	14.498246	111.434780
36	5.791816	16.546852	95.836323	36	8.147252	14.620987	119.120867
37	6.081407	16.711287	101.628139	37	8.636087	14.736780	127.268119
38	6.385477	16.867893	107.709546	38	9.154252	14.846019	135.904206
39	6.704751	17.017041	114.095023	39	9.703507	14.949075	145.058458
40	7.039989	17.159086	120.799774	40	10.285718	15.046297	154.761966
41	7.391988	17.294368	127.839763	41	10.902861	15.138016	165.047684
42	7.761588	17.423208	135.231751	42	11.557033	15.224543	175.950545
43	8.149667	17.545912	142.993339	43	12.250455	15.306173	187.507577
44	8.557150	17.662773	151.143006	44	12.985482	15.383182	199.758032
45	8.985008	17.774070	159.700156	45	13.764611	15.455832	212.743514
46	9.434258	17.880066	168.685164	46	14.590487	15.524370	226.508125
47	9.905971	17.981016	178.119422	47	15.465917	15.589028	241.098612
48	10.401270	18.077158	188.025393	48	16.393872	15.650027	256.564529
49	10.921333	18.168722	198.426663	49	17.377504	15.707572	272.958401
50	11.467400	18.255925	209.347996	50	18.420154	15.761861	290.335905

TABLE A.3.4   Compound Interest Tables (*Continued*).

| | | $i = 7\%$ ($i = 0.07$) | | | | $i = 8\%$ ($i = 0.08$) | |
n	$(1 + i)^n$	$a_{\overline{n}|i}$	$s_{\overline{n}|i}$	n	$(1 + i)^n$	$a_{\overline{n}|i}$	$s_{\overline{n}|i}$
1	1.070000	0.934579	1.000000	1	1.080000	0.925926	1.000000
2	1.144900	1.808018	2.070000	2	1.166400	1.783265	2.080000
3	1.225043	2.624316	3.214900	3	1.259712	2.577097	3.246400
4	1.310796	3.387211	4.439943	4	1.360489	3.312127	4.506112
5	1.402552	4.100197	5.750739	5	1.469328	3.992710	5.866601
6	1.500730	4.766540	7.153291	6	1.586874	4.622880	7.335929
7	1.605781	5.389289	8.654021	7	1.713824	5.206370	8.922803
8	1.718186	5.971299	10.259803	8	1.850930	5.746639	10.636628
9	1.838459	6.515232	11.977989	9	1.999005	6.246888	12.487558
10	1.967151	7.023582	13.816448	10	2.158925	6.710081	14.486562
11	2.104852	7.498674	15.783599	11	2.331639	7.138964	16.645487
12	2.252192	7.942686	17.888451	12	2.518170	7.536078	18.977126
13	2.409845	8.357651	20.140643	13	2.719624	7.903776	21.495297
14	2.578534	8.745468	22.550488	14	2.937194	8.244237	24.214920
15	2.759032	9.107914	25.129022	15	3.172169	8.559479	27.152114
16	2.952164	9.446649	27.888054	16	3.425943	8.851369	30.324283
17	3.158815	9.763223	30.840217	17	3.700018	9.121638	33.750226
18	3.379932	10.059087	33.999033	18	3.996019	9.371887	37.450244
19	3.616528	10.335595	37.378965	19	4.315701	9.603599	41.446263
20	3.869684	10.594014	40.995492	20	4.660957	9.818147	45.761964
21	4.140562	10.835527	44.865177	21	5.033834	10.016803	50.422921
22	4.430402	11.061240	49.005739	22	5.436540	10.200744	55.456755
23	4.740530	11.272187	53.436141	23	5.871464	10.371059	60.893296
24	5.072367	11.469334	58.176671	24	6.341181	10.528758	66.764759
25	5.427433	11.653583	63.249038	25	6.848475	10.674776	73.105940
26	5.807353	11.825779	68.676470	26	7.396353	10.809978	79.954415
27	6.213868	11.986709	74.483823	27	7.988061	10.935165	87.350768
28	6.648838	12.137111	80.697691	28	8.627106	11.051078	95.338830
29	7.114257	12.277674	87.346529	29	9.317275	11.158406	103.965936
30	7.612255	12.409041	94.460786	30	10.062657	11.257783	113.283211
31	8.145113	12.531814	102.073041	31	10.867669	11.349799	123.345868
32	8.715271	12.646555	110.218154	32	11.737083	11.434999	134.213537
33	9.325340	12.753790	118.933425	33	12.676050	11.513888	145.950620
34	9.978114	12.854009	128.258765	34	13.690134	11.586934	158.626670
35	10.676581	12.947672	138.236878	35	14.785344	11.654568	172.316804
36	11.423942	13.035208	148.913460	36	15.968172	11.717193	187.102148
37	12.223618	13.117017	160.337402	37	17.245626	11.775179	203.070320
38	13.079271	13.193473	172.561020	38	18.625276	11.828869	220.315945
39	13.994820	13.264928	185.640292	39	20.115298	11.878582	238.941221
40	14.974458	13.331709	199.635112	40	21.724521	11.924613	259.056519
41	16.022670	13.394120	214.609570	41	23.462483	11.967235	280.781040
42	17.144257	13.452449	230.632240	42	25.339482	12.006699	304.243523
43	18.344355	13.506962	247.776496	43	27.366640	12.043240	329.583005
44	19.628460	13.557908	266.120851	44	29.555972	12.077074	356.949646
45	21.002452	13.605522	285.749311	45	31.920449	12.108402	386.505617
46	22.472623	13.650020	306.751763	46	34.474085	12.137409	418.426067
47	24.045707	13.691608	329.224386	47	37.232012	12.164267	452.900152
48	25.728907	13.730474	353.270093	48	40.210573	12.189136	490.132164
49	27.529930	13.766799	378.999000	49	43.427419	12.212163	530.342737
50	29.457025	13.800746	406.528929	50	46.901613	12.233485	573.770156

Answers to Odd-Numbered Exercises

CHAPTER 1

Exercises 1-1

1. a. True   **b.** False; $(3x)(4y) = 12x^2$
   c. False; $2(5 - 4y) = 10 - 8y$
   d. False; $-(x + y) = -x - y$
   e. False; $5x - (2 - 3x) = 5x - 2 + 3x = 8x - 2$
   f. False; $5 - 2x \neq 3x$, $5 - 2x$ cannot be simplified.
   g. False; $-3(x - 2y) = -3x + 6y$
   h. False; $(-a)(-b)(-c) \div (-d) = abc \div d$
   i. True   **j.** True
   k. False; $(-x)(-y) = xy$
   l. True
   m. False; true only if $x \neq 0$.
3. -4 **5.** 21 **7.** 3 **9.** 7 **11.** -30
13. -10 **15.** 12 **17.** $3x + 6y$ **19.** $4x - 2y$
21. $-x + 6$ **23.** $3x - 12$ **25.** $2x + 4$ **27.** $-xy + 6x$
29. $6x - 2y$ **31.** $-3x + 4z$ **33.** $5x - 3y$
35. $43x - 22y$ **37.** xyz **39.** $2x^2 + 6x$
41. $-6a + 2a^2$ **43.** $2x^2 + 8x$ **45.** $-x^2 + 4x - 2$
47. 1 **49.** $x - 2y$ **51.** $8x - 4$ **53.** 0 **55.** $1 + 2/x$
57. $-3/2 + 1/2x$ **59.** $1/y + 1/x$

Exercises 1-2

1. a. True   **b.** False; $x/3 + x/4 = 7x/12$
   c. False; $a/b + c/d = (ad + bc)/bd$
   d. True   **e.** True
   f. False; $(a/b) \div [(c/d) \div (e/f)] = ade/bcf$
   g. False; $1/a + 1/b = (b + a)/ab$
   h. False; $x/(x + y) = 1/(1 + yx^{-1})$
   i. False; $(\frac{6}{7})(\frac{8}{9}) = (6 \cdot 8)/(7 \cdot 9) = \frac{48}{63}$
   j. True
3. 10 **5.** $\frac{2}{7}$ **7.** $35x/36$ **9.** $10x^2/3$ **11.** $\frac{35}{3}$

13. $\frac{9}{25}$ **15.** $10/3y$ **17.** $45/32x^2$ **19.** $45/2x$ **21.** x/y
23. $y/6x$ **25.** $4t$ **27.** $16t^3/9$ **29.** $1/6$ **31.** $3/2x$
33. $(3y + 2)/6x$ **35.** $7a/18b$ **37.** $(9y - 5x)/30x^2$
39. $(x^2z + xy^2 + yz^2)/xyz$ **41.** $(x^2 + 12y^2)/3y$
43. $(3x^2 + x - 12)/6x$ **45.** $(x^2 + 4)/12$ **47.** $-1/2$
49. $10/27$ **51.** $5/2$ **53.** $19x/44y$ **55.** $23a/31b$
57. $40a/87b$

Exercises 1-3

1. $2^{10} = 1024$ **3.** a^{21} **5.** $-x^{10}$ **7.** y^7
9. $1/a^2$ **11.** $9/x^5$ **13.** $32/x$ **15.** $x^{10}y^7z^3$
17. x^4/y^2 **19.** x^2y **21.** 16 **23.** $3^6 = 729$
25. x^7 **27.** x^2 **29.** 1 **31.** $-x^9$ **33.** $1/x^8y^5$
35. $-8y^2$ **37.** -3 **39.** $4b/a^{11}$ **41.** $x^6 - 2x^3$
43. $2x^6 + 6$ **45.** $2x^6 - x^5 - 3x^2$ **47.** $2x/(x + 2)$
49. $1/(x + y)$ **51.** $15/4x^2$ **53.** $(9y^2 + 4x^2)/30x^3y$
55. $5x^2/6$ **57.** $3x^4/8y^2$ **59.** $6/y^2$ **61.** $(x^2 + 1)/x^2$

Exercises 1-4

1. $10/3$ **3.** $-2/3$ **5.** $1/8$ **7.** 9 **9.** $5/4$ **11.** -2
13. 3 **15.** $1/27$ **17.** 2.5 **19.** 4 **21.** $2/3$ **23.** 4
25. $1/18$ **27.** $8x^3$ **29.** $2x/y^2$ **31.** $2\sqrt{x}$ **33.** $\sqrt[3]{x}$
35. $x^{4/7}y^{1/5}$ **37.** p^4q^8 **39.** $6x^{11/6}/y^{7/20}$ **41.** $11\sqrt{5}$
43. $2\sqrt{2}$ **45.** $14\sqrt{7}$ **47.** $2\sqrt{5}$ **49.** $a^{-5/6}$ **51.** 1
53. 1 **55.** 3^{3n}
57. a. False   **b.** True   **c.** True   **d.** True   **e.** False
   f. False   **g.** False   **h.** False   **i.** False   **j.** False
   k. True

Exercises 1-5

1. $3a + 10b + 6$ **3.** $5\sqrt{a} + 3\sqrt{b}$
5. $t^3 + 12t^2 + 3t - 5$ **7.** $\sqrt{x} + 3\sqrt{2y}$
9. $14x + 22y$ **11.** $9x + 3y$

13. $2x^3 - 2x^2y + 3xy^2 - y^3$ **15.** $xy + 2x - 3y - 6$

17. $6xy - 8x + 3y - 4$ **19.** $3a^2 + 2a - 8$

21. $2x^3 + x^2 - 8x + 21$ **23.** $x^2 - 16$ **25.** $4t^2 - 25x^2$

27. $x - 9y$ **29.** $x^2 + y^2 + 2xy - z^2$

31. $x^5 - x^3 + 2x^2 - 2$ **33.** $x^5 + 2x^3 - x^2 - 2$

35. $y^2 + 12y + 36$ **37.** $4x^2 + 12xy + 9y^2$

39. $2x^2 - 2x\sqrt{6y} + 3y$ **41.** $8x^2 + 18y^2$

43. $2x^3y + 2xy^3$ **45.** $3x^2 + 135x - 90$

47. $2a^3 + 8a$ **49.** $2x^2 - 3x/2$

51. $x + 7 - \dfrac{5}{x} + \dfrac{4}{x^2}$ **53.** $t^{3/2} - 2\sqrt{t} + 7/\sqrt{t}$

55. $4x - 2y$ **57.** $x - 3$ **59.** $t + 1 + \dfrac{2}{t - 1}$

61. $x^2 + 1 + \dfrac{3}{x + 2}$ **63.** $x^2 - 2x + 3 + \dfrac{3}{2x + 1}$

Exercises 1-6

1. $3(a + 2b)$ **3.** $2y(2x - 3z)$ **5.** $(2 - a)(u - v)$

7. $(x - 2)(y + 4)$ **9.** $(3 + p)(x - y)$

11. $2(3x + 2y)(z - 4)$

13. $(x - 4)(x + 4)$ **15.** $3(t - 6a)(t + 6a)$

17. $xy(x - 5y)(x + 5y)$ **19.** $(x + 1)(x + 2)$

21. $(x + 2)(x - 1)$ **23.** $(x - 2)(x + 1)$

25. $(x - 6)(x - 9)$ **27.** $(x - 11)(x - 1)$

29. $2(x - 2)(x + 3)$ **31.** $5y^2(y + 7)(y - 2)$

33. $(2x + 3)(x + 1)$ **35.** $(2x + 3)^2$

37. $(x - 3)(5x - 2)$ **39.** $(5x + 2)(2x - 3)$

41. $(q + 4)(3q + 8)$ **43.** $2xy(x - 1)(3x + 5)$

45. $(x + y)(x + 5y)$ **47.** $(p - 5q)(p + 4q)$

49. $(2t - 3u)(t + 2u)$ **51.** $(2a - 3b)(3a + 5b)$

53. $(x - 3)(x^2 + 3x + 9)$ **55.** $(3u + 2v)(9u^2 - 6uv + 4v^2)$

57. $xy^2(4x - 3y)(16x^2 + 12xy + 9y^2)$

59. $(x - 3)(x + 3)(y - 2)(y + 2)$

61. $(x - 2)(x + 2)(x^2 + z^2)$

63. $(x + y)(x^2 + y^2)$ **65.** $5x(x + y)^3(3x - 2y)^3$

67. $(x + y + 1)(x + y + 2)$ (*Hint:* Put $x + y = u$.)

69. $(3a - 3b + 2)(a - b + 1)$

71. $(x^n + 2)(3x^n + 1)$ (*Hint:* Put $x^n = u$.)

73. $(x - \sqrt{2}y)(x + \sqrt{2}y)(x^2 + 2y^2 - \sqrt{2}xy)(x^2 + 2y^2 + \sqrt{2}xy)$

75. $(\sqrt{3}x - \sqrt{2})(\sqrt{3}x + \sqrt{2})$

77. $(x^2 + 2y^2 - 2xy)(x^2 + 2y^2 + 2xy)$

79. $(x + y)(x^4 - x^3y + x^2y^2 - xy^3 + y^4)$

Exercises 1-7

1. 2   **3.** $x - 2$ **5.** $\dfrac{5x + 7}{x + 2}$ **7.** $\dfrac{2(x^2 + x + 3)}{(x + 2)(2x - 1)}$

9. $\dfrac{3 + 4x - 3x^2}{x^2 - 1}$ **11.** $\dfrac{2 - x}{(2x - 1)(x + 1)}$

13. $\dfrac{2}{(x - 1)(x - 2)(x - 3)}$ **15.** $\dfrac{x^2 + 3}{(x - 1)^2(x + 3)}$

17. $\dfrac{10 + 4x - 2x^2}{(x + 3)(x + 1)(x - 1)}$ **19.** $(x - 1)(x + 2)$

21. $\dfrac{-2(x + 1)}{3}$ **23.** $\dfrac{(x + 2)(2x - 1)}{(x - 2)(2x + 1)}$ **25.** 3   **27.** $\dfrac{2}{x - 1}$

29. $\dfrac{(x - 1)(2x - 1)}{(x + 1)^2}$ **31.** $\dfrac{x^2 - 1}{x - 2}$ **33.** $\dfrac{(x - 1)(x + 2)}{(x - 2)(x - 5)}$

35. $\dfrac{(x + y)^2}{xy}$ **37.** $\dfrac{-(x^2 + y^2)}{(x + y)^2}$ **39.** $\dfrac{-1}{x(x + h)}$

41. $\dfrac{3 - \sqrt{7}}{2}$ **43.** $\left(\dfrac{1}{2}\right)(\sqrt{5} + \sqrt{10} - \sqrt{3} - \sqrt{6})$

45. $\dfrac{3 - \sqrt{3}}{2}$ **47.** $\dfrac{\sqrt{x} + \sqrt{y}}{x - y}$ **49.** $\sqrt{x + 2} + \sqrt{2}$

51. $(-2/3)(\sqrt{x + 3} + 2\sqrt{x})$ **53.** $11/(5 + \sqrt{3})$

55. $1/(\sqrt{x + h} + \sqrt{x})$

Review Exercises for Chapter 1

1. a. False; $a^m b^n$ cannot be simplified by the laws of exponents.

b. False; $a^m + b^m \neq (a + b)^m$. For example, $(a + b)^2 = a^2 + 2ab + b^2$ and not $a^2 + b^2$.

c. True   **d.** False; $(a - b)^2 = a^2 - 2ab + b^2$

e. False; $-2(a + b) = -2a - 2b$ **f.** False;

g. False; $\sqrt{a - b} \neq \sqrt{a} - \sqrt{b}$. For example, if $a = 25$, $b = 9$, then $\sqrt{a - b} = \sqrt{25 - 9} = 4$ and $\sqrt{a} - \sqrt{b} = 5 - 3 = 2$; clearly $4 \neq 2$.

h. False; $(a + 2b)/a = 1 + 2b/a$ **i.** True

j. False; $(1/a) - (1/b) = (b - a)/ab$

k. False; $\dfrac{a/b}{c} = \dfrac{a}{b} \cdot \dfrac{1}{c}$ **l.** False; $(2a)^5 = 2^5 a^5 = 32a^5$

m. True   **n.** True   **o.** True   **p.** True   **q.** True

r. False; a rational number can be expressed either as a terminating decimal or a repeating decimal.

3. $3/2^{10}$ **5.** $\dfrac{3a^{11}}{8b^7}$ **7.** $(-27/2)x^{13}y^6$ **9.** $1/x$

11. $6x^{11/6}/y^{7/20}$ **13.** $x^{2ab - 2bc}$

15. $\dfrac{3 + 2x^2}{x^2 + 1}$ **17.** $\dfrac{4x^2 + 15x - 1}{(x + 1)^2(x + 3)(x - 2)}$

19. $\dfrac{(x + 2)(y + 2)}{(x + 3)(y + 3)}$ **21.** $\dfrac{(a + b)(x - 3)}{(a - b)(x + 3)}$

23. $\dfrac{(a + 2)(a + 1)(a - 3)}{a}$ **25.** $3(x - 5y)(x + 5y)$

27. $(3x - 5)(2x + 3)$ **29.** $(x - 3)(x + 4)$

31. $(k + 5)(k - 4)$ **33.** $(4x - 3)(2x - 3)$

35. $(y - 5)(y + 2)$ **37.** $3(a + 3)(a - 1)$ **39.** $-3(4x + 7)$

41. $\left(x + \dfrac{2}{x}\right)\left(x^2 - 2 + \dfrac{4}{x^2}\right)$

CHAPTER 2

Exercises 2-1

1. Yes   **3.** No   **5.** 2 is a solution and 5 is not.

7. No   **9.** No   **11.** $10x^2 - x - 7 = 0$; degree 2

13. $y - 6 = 0$; degree 1   **15.** 1   **17.** -2 **19.** 4

21. $\frac{17}{5}$ **23.** -1 **25.** -10 **27.** $-\frac{19}{7}$ **29.** $-\frac{13}{50}$

31. -2 **33.** 3   **35.** $\frac{4}{3}$ **37.** 2   **39.** 1

41. a. $x = (cz - by)/a$ **b.** $b = (cz - ax)/y$

43. a. $x = ty/(y - t)$ **b.** $t = xy/(x + y)$

Exercises 2-2

1. $x + 4$ **3.** $2 + x/2$ **5.** $x - 1$

7. 18   **9.** 15   **11.** 25 years

13. 5 quarters and 10 dimes

15. $52,000 at 8% and $8000 at 10.5%

17. $3000 at 10% and $5000 at 8%

19. $2.50   **21.** $55.00   **23.** 15 ounces   **25.** 30 ounces

27. 2:1   **29.** 10,000   **31.** $2200 and $700

Exercises 2-3

1. $-2, -3$ **3.** $-2, -7$ **5.** -2 **7.** $3, 4$

9. $1, -1$ **11.** $0, 8$ **13.** $-\frac{1}{6}, -\frac{1}{4}$ **15.** $-\frac{3}{2}, -1$

17. $\frac{1}{2}, -\frac{2}{3}$ **19.** $0, 1$ **21.** $1, -1, 2, -2$

23. $\dfrac{-3 \pm \sqrt{5}}{2}$ **25.** $\dfrac{-3 \pm \sqrt{41}}{4}$ **27.** $\dfrac{-1 \pm \sqrt{13}}{2}$

29. $-\frac{5}{2}$ **31.** $\dfrac{-5 \pm \sqrt{10}}{5}$ **33.** $3 \pm 2\sqrt{2}$

35. $-3 \pm \sqrt{10}$ **37.** $\dfrac{3 \pm \sqrt{13}}{2}$ **39.** $1 \pm \dfrac{\sqrt{7}}{2}$

41. $-1 \pm \sqrt{5}$ **43.** $\dfrac{-9 \pm \sqrt{17}}{4}$ **45.** $\pm\sqrt{\dfrac{11}{6}}$

47. $0, \frac{11}{6}$ **49.** $4, -\frac{4}{3}$ **51.** $\frac{1}{3}, \frac{3}{2}$

53. $-1 \pm \sqrt{5}$ **55.** $\dfrac{4 \pm \sqrt{10}}{2}$ **57.** $6, -\frac{1}{2}$

59. $\dfrac{3 \pm \sqrt{17}}{4}$ **61.** $\dfrac{5 \pm \sqrt{-11}}{6}$ (no real solution)

63. $-3, -1$ **65.** $2, -2$ **67.** $-1, 1/8$ (Put $x^{1/3} = u$.)

69. $t = \dfrac{-u \pm \sqrt{u^2 + 2gs}}{g}$

71. $R = \dfrac{-\pi H \pm \sqrt{\pi^2 H^2 + 2\pi A}}{2\pi}$ **73.** 1

75. a. $x = y \pm \sqrt{4y^2 - 1}$ **b.** $y = \dfrac{-x \pm \sqrt{4x^2 + 3}}{3}$

Exercises 2-4

1. 4, 11   **3.** -12 and -11 or 11 and 12

5. 5 cm and 12 cm   **7.** 4 in. and 6 in.

9. 3 in.

11. a. 4, 1 second   **b.** 5 seconds   **c.** 100 feet

13. $50   **15.** 5%   **17.** $195

19. $(38 \pm 2\sqrt{21})$ dollars   **21.** 4% and 8%

23. a. 50 or 70 units

b. $300   **c.** 45 or 60 units   **d.** $325

25. $5 or $7; $8 or $6

Review Exercises for Chapter 2

1. a. True; provided that the constant is nonzero.

b. True; provided that the expression is well-defined for all values of x.

c. False; multiplying both sides of an equation by an expression containing the variable may result in new roots that are not the roots of the original equation.

d. False; for example, if we square $x = 2$, we get $x^2 = 4$ whose roots are 2 and -2 which are not the same as the roots of $x = 2$.

e. False; if $px = q$, then $x = q/p$.

f. False; $ax^2 + bx + c = 0$ is a quadratic equation, provided that $a \neq 0$.

g. False; the solution of $x^2 = 4$ is given by $x = 2$ or -2.

h. False; the roots of the equation $ax^2 + bx + c = 0 (a \neq 0)$ are given by $x = (-b \pm \sqrt{b^2 - 4ac})/2a$.

i. True

j. False; a quadratic equation may have two identical roots or no real roots at all.

k. False; a linear equation will always have exactly one root.

l. True

3. 1/3   **5.** No solution   **7.** No solution   **9.** 2, 5/3

11. abc, provided that $a + b + c \neq 0$ **13.** -1

15. 5   **17.** 1, -4 **19.** 7

21. pqr, provided that $pq + qr + pr \neq 0$ **23.** 3, 3/2

25. 4   **27.** 2   **29.** 9/5

31. **a.** $r = (a - S)/(l - S)$ **b.** $l = (a + rS - S)/r$

33. \$75,000 at 8% and \$25,000 at 10%   **35.** 1600   **37.** \$4 or \$5

39. **a.** $P = \$(2400 - 600C)$ **b.** \$600

CHAPTER 3

Exercises 3-1

1. $\{-1, 0, 1, 2, 3, 4\}$ **3.** $\{4, 3, 2, 1, 0, -1, -2, \ldots\}$

5. $\{2, 3, 5, 7, 11, 13, 17, 19\}$ **7.** $\{2, 3\}$

9. $\{x \mid x$ is an even number, $0 < x < 100\}$ or $\{x \mid x = 2n$, n is a natural number; $1 < n < 49\}$

11. $\{x \mid x$ is an odd number, $0 < x < 20\}$ or $\{x \mid x = 2n + 1$, n is an integer and $0 < n \leq 9\}$

13. $\{x \mid x$ is a natural number divisible by 3$\}$ or $\{x \mid x = 3n$, n is a natural number$\}$

15. $\{x \mid x$ is a real number: $-1 \leq x \leq 1\}$

17. $[3, 8]$ **19.** $(-7, -3)$ **21.** $2 \leq x < 5$ **23.** $x < 3$

25. **a.** True   **b.** False; $3 \in \{1, 2, 3, 4\}$

c. False; $4 \notin \{1, 2, 5, 7\}$ **d.** True

e. False; $\varnothing$ is a set, whereas 0 is a number. A set cannot be equal to a number.

f. False; $\varnothing$ is an empty set with no element, whereas the set $\{0\}$ contains the element 0.

g. False; $\varnothing$ does not contain any element, so 0 cannot be in $\varnothing$.

h. False; the empty set $\varnothing$ is not a member of $\{0\}$.

i. True   **j.** True

k. False; $2 \notin \{x \mid (x - 2)^2/(x - 2) = 0\}$, whereas $2 \in \{x \mid x - 2 = 0\}$.

l. True   **m.** True

n. False; The set of all squares in a plane is a subset of the set of all rectangles in a plane.

o. True   **p.** True

q. False; $\{x \mid 2 \leq x \leq 3\} \subset \{y \mid 1 \leq y \leq 5\}$.

r. True

29. No, it contains -2.

Exercises 3-2

1. $x < 2$ **3.** $u \geq -\frac{17}{3}$ **5.** $x > 2$ **7.** $x > 1/8$

9. $y < -\frac{7}{5}$ **11.** $t \leq 13$ **13.** $-1 < x < 3$

15. $x > -\frac{1}{2}$ **17.** No solution   **19.** $-\frac{2}{3} < x < \frac{2}{3}$

21. $x \geq 2$ **23.** No solution   **25.** \$5000

27. 1501 or more   **29.** More than 1875   **31.** At least 1600

Exercises 3-3

1. $2 < x < 5$ **3.** $x \leq -3$ or $x \geq \frac{5}{2}$ **5.** $3 \leq x \leq 4$

7. $-2 < x < 1$ **9.** $y < -2$ or $y > \frac{3}{2}$

11. $x < -2\sqrt{2}$ or $x > 2\sqrt{2}$ **13.** $x \leq -2$ or $x \geq 2$

15. All x **17.** 3   **19.** All $x \neq -1$ **21.** No solution

23. All x **25.** No solution   **27.** 60 units

29. $45 \leq x \leq 60$ **31.** $x > 150$

33. If x yards is the length of one side, then $30 \leq x \leq 70$.

35. At most 3 feet   **37.** $80 < n < 120$.

39. $20¢ \leq p \leq 30¢$

Exercises 3-4

1. $7\sqrt{2}$ **3.** $3 - \pi$ **5.** 1, $-\frac{1}{7}$

7. 1/2   **9.** $-1, \frac{3}{2}$ **11.** No solution

13. No solution   **15.** $\frac{27}{17}$ or $\frac{33}{19}$ **17.** $-1, \frac{1}{7}$

19. $-\frac{11}{3} < x < -1$ or $(-\frac{11}{3}, -1)$

21. $x \leq -\frac{1}{5}$ or $x \geq 1$; $(-\infty, -\frac{1}{5}]$ or $[1, \infty)$

23. $1 < x < 2$; $(1, 2)$ **25.** No solution

27. All real numbers or $(-\infty, \infty)$ **29.** No solution

31. All real numbers   **33.** No solution

35. $x > \frac{5}{2}$, i.e., $(\frac{5}{2}, \infty)$

37. **a.** $|x - 3| < 5$; $x \in (-2, 8)$

b. $|y - 7| \leq 4$; $y \in [3, 11]$

c. $|t - 5| = 3$; $t \in \{2, 8\}$

d. $|z - \mu| < \sigma$; $z \in (\mu - \sigma, \mu + \sigma)$

e. $|x - 4| > 3$; $x \in (-\infty, 1)$ or $x \in (7, \infty)$

f. $|\bar{x} - \mu| < 5$; $x \in (\mu - 5, \mu + 5)$

39. $|p - 22| \leq 5$.

Review Exercises for Chapter 3

1. **a.** True

b. False; when two sides of an inequality are multiplied by a *positive* constant, the direction of inequality is preserved.

c. False; a quadratic inequality has either no solution or one solution or an *infinite* number of solutions.

d. False; if a negative number is subtracted from both sides of an inequality, the direction of inequality must be preserved.

e. False; the statement is true only if $a \geq 0$.

f. True   **g.** True   **h.** True   **i.** True

j. False; for example, if $x = 2$ and $y = -7$, then $x > y$, whereas $|x| < |y|$ because $2 < 7$.

k. True　　l. True

3. $x \geq \frac{2}{5}$　　5. $x > -6$　　7. $x < \frac{2}{9}$

9. $x > \frac{1}{6}$　　11. No solution　　13. $-1 < x \leq \frac{5}{2}$

15. $x < \frac{2}{3}$ or $x > 1$　　17. $x \leq -\frac{1}{2}$ or $x \geq 5$

19. $-3 < x < \frac{5}{2}$　　21. All x　　23. No solution

25. $x \leq -3$ or $0 \leq x \leq 5$　　27. $2 < x < 6$

29. $x \leq -\frac{1}{2}$ or $x \geq 2$　　31. $x \leq -2$ or $x \geq 5$

33. $x \leq 1$ or $x \geq \frac{5}{2}$　　35. $x < -4$ or $x > 10$

37. No solution　　39. All x　　41. $\frac{7}{2}, \frac{3}{4}$

43. $0, -\frac{8}{5}$　　45. $1, 2$

47. a. At least 120,000　　b. At least 220,000

49. $\$6 \leq p \leq \$8$　　51. \$35,000　　53. $20 \leq x \leq 40$

55. $20 \leq x \leq 28$　　57. $\$30 \leq p \leq \$38$

CHAPTER 4

Exercises 4-1

1.

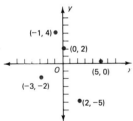

3. $\sqrt{5}$　　5. $\frac{1}{2}\sqrt{29}$　　7. 0 or -6　　9. 19 or -5

11. $y = -3$ or 1　　13. $x^2 + y^2 + 2x - 6y + 1 = 0$

15. $2(x^2 + y^2 + x - y) = 1$

17.

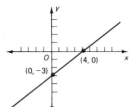

19.

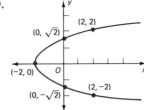

21.

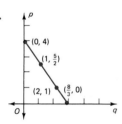

23.

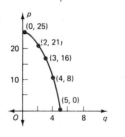

Exercises 4-2

1. 2　　3. 0　　5. No slope　　7. $y = 5x - 9$

9. $y = 4$　　11. $y = 6x - 19$　　13. $x = 3$

15. $y = \frac{1}{3}x - 4$　　17. $y = -3x + 5$　　19. $y = x - 1$

21. $y = 4$　　23. $y = -x - 1$　　25. $\frac{3}{2}; -3$

27. $2; -3$　　29. $0; \frac{3}{2}$　　31. Perpendicular　　33. Neither

35. Parallel　　37. Perpendicular　　39. $(3, 0), (0, 3)$.

Exercises 4-3

1. $y_c = 7x + 150$; \$850

3. a. $y_c = 3x + 45$　　b. \$105　　c. \$3 and \$45

5. $y_c = 5.5x + 300$; \$465　　7. $y_c = 10x + 150$

9. $p = 1.7 - 0.00005x$　　11. $p = 0.00025x + 0.5$

13. $p = (1/600)x + 8$

15. $V = 10,000 - 1200t$; \$4000　　17. $V = 800 - 120t$; \$80

19. a. $2x + 5y = 280$

b. $m = -\frac{2}{5}$; it indicates that every increase of 5 units of the first type will be at the cost of 2 units of the second type.

c. 40

21. a. $y = 650 - 25t$　　b. 26 days

c. At the end of 21st day

23. $m = 7/800, c = -9/80$; $A = 70$ months

25. $5x + 8y = 100$

Exercises 4-4

1. $x = -1, y = -2$　　3. $x = 1, y = 6$

5. $x = -1, t = 3$　　7. $x = 4, y = 3$

9. $x = 40, y = 60$　　11. No solution

13. Coordinates of any point on the line $x + 2y = 4$.

15. $x = 1, y = 2, z = 3$.　　17. $x = 1, y = 2, z = 3$.

19. $x_1 = 1, x_2 = 2, x_3 = -1$.　　21. $x = 3, y = -2, z = 1$

23. $x = 5, y = -2$ or $x = -2, y = 5$

25. 1600 lb of I and 600 lb of II.

27. 100 units of X and 40 units of Y.

29. 60 gal of 25% acid solution and 140 gal of 15% acid solution.

31. 18 and 35

33. 45/14 tons of type A, 65/14 tons of type B, and 23/14 tons of type C.

35. 2:2:1

Exercises 4-5

1. 800　　3. a. 500　　b. \$4.14

5. Yes, because he will have to sell fewer items to break even.

7. $x = 40$ or 20 **9.** $x = 30$, $p = 5$ **11.** $x = 10$,
$p = 10$ **13.** $x = 3$, $p = 4$

15. a. $p = 42 - 0.06x$ **b.** $x = 150$, $p = 33$

   c. $p_1 = \$33.90$, $x_1 = 135$; price increases by $\$0.90$ and the demand decreases by 15 units.

   d. $\$5.44$ **e.** $\$4.08$

17. $x = -\frac{3}{7}$, $p = \frac{32}{7}$; since x cannot be negative, market equilibrium occurs at $x = 0$ (that is, no goods are being produced
and sold).

19. a. $p = 14 - x/1000$;

   b. $x = 7500$, $p = \$6.50$; $\$48,750$

21. $x = 9.28$, $p = 0.539$; $x = 0.718$, $p = 6.96$

23. In Exercises 20 and 21, stable at the higher equilibrium price
and unstable at the lower equilibrium price. In Exercise 22, stable at $p = \frac{7}{8}$, $p = 2$ and unstable at $p = \frac{3}{2}$.

Review Exercises for Chapter 4

1. a. False; if a point lies on the x-axis, its *ordinate* is zero.

   b. False; each point on the y-axis has its x-coordinate zero.

   c. False; if a point is in the first quadrant, then $x > 0$, $y > 0$.

   d. False; the origin $(0, 0)$ does not lie in any quadrant.

   e. False; a horizontal line has *zero* slope

   f. True   **g.** False; a vertical line has *no* slope.

   h. False; the distance of the point (a, b) from the origin is
$\sqrt{a^2 + b^2}$.

   i. False; slope $= (y_2 - y_1)/(x_2 - x_1)$, provided $x_2 \neq x_1$.

   j. False; the equation $Ax + By + C = 0$ represents a straight
line, provided the constants A and B are not both zero.

   k. False; the slope of the line given by $x = my + b$ is $1/m$,
provided $m \neq 0$.

   l. True   **m.** True   **n.** True   **o.** True   **p.** True

3. $y = 2$ **5.** $x = 2$ **7.** $2x - 3y = 11$

9. $4x - 6y = 3$; $(3,0)$, $(0,-2)$ **11.** $x = 1$, $y = 1$

13. $x = y = \frac{1}{6}$ **15.** $u = 1$, $v = -1$ **17.** $u = \frac{1}{6}$, $v = \frac{1}{9}$

19. $x = 3$, $y = -1$ **21.** $x = 3$, $y = -2$, $z = 1$

23. $x = 6$, $y = 3$ or $x = 3$, $y = 6$

25. If x units of A and y units of B are produced, then
$5x + 8y = 640$.

27. The cost y_c of producing x units is given by $y_c = 25x + 3000$;
$\$5500$.

29. $\$2.50$ and $\$3.00$ **31.** $\$15,000$; 6%

33. a. S: $90p = x + 1200$; D: $10p + x = 5000$

   b. $x = 4380$, $p = \$62$ **c.** $\$1.80$; 18 tons;   **d.** $\$6.11$

35. a. $y_c = 4x + 2500$ **b.** 2000 units   **c.** 2160 units

37. Over 5000 belts

39. a. $p = 20$, $x = 30$;   **b.** $\$600,000$ **41.** 11,500

43. D: $p = 1700 - x/2$; S: $p = 500 + x/4$; $p = \$900$, $x = 1600$

45. a. Second   **b.** First   **c.** Second

47. a. $p = -0.625x + 4.125$ **b.** 3.4 million rolls

49. a. $p = 1300n - 260,000$ **b.** $n = 200$ **c.** $n = 400$

CHAPTER 5

Exercises 5-1

1. 5; -4; $3x^2 + 2$; $3(x + h) + 2$

3. 12; -8; $5c + 7$; $12 + 5c$; $19 + 5c$

5. 9; 4; a^2, x, $(x + h)^2$ **7.** 3; 3; 3; **3**

9. 2; $|x|$; $\sqrt{a^2 + h^2}$

11. 7; $(3 - 5t + 7t^2)/t^2$; $3(c^2 + h^2) - 5(c + h) + 14$;
$3(c + h)^2 - 5(c + h) + 7$

13. a. 6   **b.** 11   **c.** 12

   d. $f(5 + h) = 7 + 2h$; $f(5 - h) = 3h - 9$

17. 10   **19.** 2   **21.** 0   **23.** $2x - 3 + h$

25. All real numbers   **27.** All $x \neq 2$ **29.** All $x \neq 2, 1$

31. All u **33.** All $y \geq \frac{2}{3}$ **35.** All $u < \frac{3}{2}$

37. All $x \neq 5$ **39.** All $x \neq 2, 3$ **41.** All $x \neq 4, \frac{1}{2}$

43. 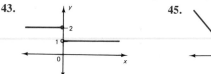   **45.**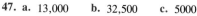

47. a. 13,000   **b.** 32,500   **c.** 5000

49. a. 3.9   **b.** 4.0

   c. $3.775 \approx 3.8$ rounded to one decimal place.

51. $C(x) = 15x + 3000$; $R(x) = 25x$; $P(x) = 10x - 3000$

53. If x denotes one of the sides, then $A = x(100 - x)$.

55. $x =$ side of square base; $C(x) = 5.5x^2 + 1800/x$.

57. $C(x) = \begin{cases} 25x & \text{if } x \leq 50 \\ 20x & \text{if } x > 50 \end{cases}$

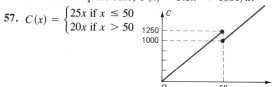

59. a. $R = (200 + 5x)(70 - x)$

   b. $R = p\left[70 - \dfrac{p - 200}{5}\right] = \left(\dfrac{p}{5}\right)(550 - p)$

61. $C(x) = \begin{cases} 500 & \text{if } 6 \leq x \leq 12 \\ 450 + \dfrac{600}{x} & \text{if } x > 12 \end{cases}$

63. $A = f(x) = x(10 - x)$; $D_f = \{x \mid 0 < x < 10\}$

65. $V = f(x) = x(16 - 2x)(20 - 2x); D_f = \{x \mid 0 < x < 8\}$

67. No   **69.** No   **71.** Yes

Exercises 5-2

1. $(0, -3)$ **3.** $(-1, 1)$ **5.** $(-\frac{1}{4}, \frac{17}{8})$

7.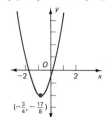

9. $(-\frac{1}{6}, \frac{37}{12})$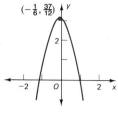

11. $f_{min} = -\frac{9}{4}$ **13.** $f_{max} = \frac{5}{4}$ **15.** $600; R_{max} = \$3600$

17. a. $C(x) = 25x + 2000$ **b.** $3000; R_{max} = \$90,000$
   c. $1750; P_{max} = \$28,625$

19. 15,625 sq. yd   **21.** $x = 10$

23. $\$22.50; R_{max} = \$202,500$

25. $\$175; R_{max} = \$6125$

Exercises 5-3

1. $D_f = \{x \mid -2 \leq x \leq 2\}$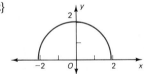

3. $D_f = \{x \mid x \leq 3\}$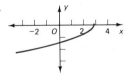

5. $D_f = \{x \mid x \neq 0\}$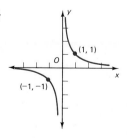

7. $D_f = \{$all real numbers$\} = \{x \mid x$ is real$\}$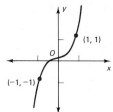

9. $D_f = \{x \mid x$ is real$\}$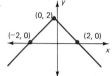

11. $D_f = \{x \mid x$ is real$\}$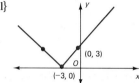

13. $D_f = \{x \mid x \neq 3\}$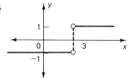

15. a. Yes; $y = \sqrt{9 - x^2}$ **b.** No   **c.** No   **d.** Yes;
   $y = -\sqrt{4 - x^2}$.

17. $x^2 + y^2 - 4x - 10y + 20 = 0$

19. $x^2 + y^2 = 49$

21. $x^2 + y^2 + 4x - 2y - 8 = 0$

23. $x^2 + y^2 + 6x - 6y + 9 = 0$

25. Yes; center: $(2, 4)$, radius $= 4$

27. Yes; center: $(\frac{5}{4}, -1)$, radius $= \frac{7}{4}$ **29.** No

31. 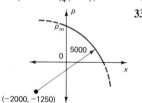   **33.**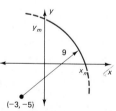

$\rho_m = \$3332.57$

Max. Coronado $(x_m) = 4.48$
Max. Eastern Star $(y_m) = 3.49$

35. $g(x) = \begin{cases} 200 - x & \text{if } 0 \leq x < 200 \\ 700 - x & \text{if } 200 \leq x < 700 \\ 1000 - x & \text{if } 700 \leq x \leq 1000 \end{cases}$

37. $f_{min} = 2$ occurs at $x = -\frac{1}{2}$; no maximum

39. $f_{max} = 1; f_{min} = -1$

41. $f_{min.} = f(\pm\frac{1}{3}) = -2; f_{max} = f(0) = -1$

43. $f_{max} = f(0) = 1; f_{min} = f(\pm 4) = -3$

45. $f_{min} = f(1) = 1$; no maximum

Exercises 5-4

1. $(f \pm g)(x) = x^2 \pm 1/(x - 1)$; $(fg)(x) = x^2/(x - 1)$;
$(f/g)(x) = x^2(x - 1)$; $(g/f)(x) = 1/x^2(x - 1)$;
$D_{f+g} = D_{f-g} = D_{fg} = D_{f/g} = \{x \mid x \neq 1\}$;
$D_{g/f} = \{x \mid x \neq 0, 1\}$

3. $(f \pm g)(x) = \sqrt{x - 1} \pm 1/(x + 2)$;
$(fg)(x) = \sqrt{x - 1}/(x + 2)$; $(f/g)(x) = (x + 2)\sqrt{x - 1}$;
$(g/f)(x) = 1/(x + 2)\sqrt{x - 1}$;
$D_{f+g} = D_{f-g} = D_{fg} = D_{f/g} = \{x \mid x \geq 1\}$; $D_{g/f} = \{x \mid x > 1\}$

5. $(f \pm g)(x) = (x + 1)^2 \pm 1/(x^2 - 1)$;
$(fg)(x) = (x + 1)/(x - 1)$;
$(f/g)(x) = (x + 1)^3(x - 1)$; $(g/f)(x) = 1/(x + 1)^3(x - 1)$;
$D_{f+g} = D_{f-g} = D_{fg} = D_{f/g} = D_{g/f} = \{x \mid x \neq \pm 1\}$

7. $\sqrt{8}$ 9. $\sqrt{3}$ 11. Not defined   13. 0

15. Not defined   17. $-\frac{1}{3}$ 19. $-\frac{1}{2}$ 21. Not defined

23. $f \circ g(x) = |x| + 1$; $g \circ f(x) = (\sqrt{x} + 1)^2$

25. $f \circ g(x) = 2 + |x - 2|$; $g \circ f(x) = x$

27. $f \circ g(x) = \sqrt{x^2} = |x|$ (all x); $g \circ f(x) = x$ (all $x \geq 0$)

29. $f \circ g(x) = (1 - x)/x$; $g \circ f(x) = (x - 1)^{-1}$

31. $f \circ g(x) = 3$; $g \circ f(x) = 7$

33. $f(x) = x^3$, $g(x) = x^2 + 1$ is the simplest answer.

35. $f(x) = 1/x$, $g(x) = x^2 + 7$ is the simplest answer.

37. $R = x(2000 - x)/15$

39. $R = f(t) = 243(t + 1)^5 + 3^{3/2}(t + 1)^{1/2}$; $9 + 3^{10}$

41. $3a + 4b = 7$

43. $R = \dfrac{25(t + 24)^2}{50(t + 24)^2 + 8(t + 72)^2}$; 240,200 units

Exercises 5-5

1. $y = 3 - 3x/4$ 3. $y = x/(1 - x)$

5. $y = -x$ or $y = x + 1$

7. $y = -x + 2$ or $y = -x - 2$

9. $y = \frac{2}{3}\sqrt{9 - x^2}$, $y = -\frac{2}{3}\sqrt{9 - x^2}$

11. $y = (1 - \sqrt{x})^2$ 13. $y = -x$, $y = 1/x$

15. $x = f^{-1}(y) = -(y + 4)/3$

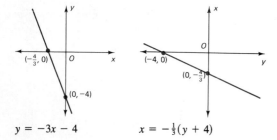

$y = -3x - 4$ $x = -\frac{1}{3}(y + 4)$

17. $x = f^{-1}(p) = 10 - 5p/2$

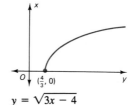   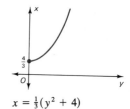

$p = 4 - \frac{2}{5}x$ $x = 10 - \frac{5}{2}p$

19. $x = f^{-1}(y) = (y^2 + 4)/3$ ($y \geq 0$)

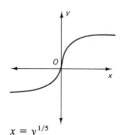

$y = \sqrt{3x - 4}$ $x = \frac{1}{3}(y^2 + 4)$

21. $x = f^{-1}(y) = y^{1/5}$

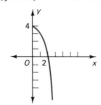

$y = x^5$ $x = y^{1/5}$

23. $x = f^{-1}(y) = 4 - y^2$ ($y \geq 0$)

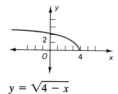

$y = \sqrt{4 - x}$

$x = 4 - y^2$ ($y \geq 0$)

25. $x = f^{-1}(y) = -1 + \sqrt{y}$, $x \geq -1$;
$x = f^{-1}(y) = -1 - \sqrt{y}$, $x \leq -1$

27. $x = f^{-1}(y) = y^{3/2}$, if $x \geq 0$; $x = f^{-1}(y) = -y^{3/2}$, if $x \leq 0$

29. $x = f^{-1}(y) = 1 + y$ if $x \geq 1$; $x = f^{-1}(y) = 1 - y$ if $x \leq 1$

Review Exercises for Chapter 5

1. a. False; the domain is the set of all real numbers.
   b. True   c. True
   d. False; $|x^2 - 9|/(x - 3) = x + 3$ if $x^2 > 9$, that is, if
   $x < -3$ or $x > 3$; $|x^2 - 9|/(x - 3) = -(x + 3)$ if
   $x^2 < 9$, that is, if $-3 < x < 3$

e. False; a curve is the graph of a function if any vertical line meets the graph in *at most* one point.

f. False; a function is a rule which assigns to each value in the domain *only* one value in the range.

g. True   h. False; true only if $a \neq 0$.

i. False; the domain of a polynomial function is the set of all *real numbers*.

j. False; The domain of f/g may differ from that of $f + g$, $f - g$, and fg.

k. False; In general, $f \circ g \neq g \circ f$.

l. False; The vertex need not be at the origin.

m. False; In $F(x, y) = 0$, x and y are not both independent variables; either one can be the independent variable, but not both.   n. True

o. False; The graph of f^{-1} is the reflection of the graph of f in the line $y = x$.

3. $F(x)$ and $G(x)$ are equal to $f(x)$; $g(x)$ and $h(x)$ are not.

5. $p = -3$ or 5 7. $x^2 + y^2 - 6x - 6y + 9 = 0$

9. Yes, center $(-2, -2)$, radius $= 3$; $(-2 \pm \sqrt{5}, 0)$ and $(0, -2 \pm \sqrt{5})$ 11. No

13. Yes, center $(-5, 2)$, radius $= 5$; $(-5 \pm \sqrt{21}, 0)$ and $(0, 2)$

15. D_f = set of all real numbers

17. $f \circ g(x) = |1 - x^2|$; $g \circ f(x) = 1 - x^2$

19. $g(x) = \sqrt{x}$, $f(x) = x/(x + 1)$ is the simplest answer.

21. $E = \begin{cases} 1000 & \text{if } 0 \leq x \leq \$6000 \\ 520 + 0.08x & \text{if } x \geq \$6000; \end{cases}$
Domain: all $x \geq 0$ a. $1000   b. $1160

23. $R = 160p - 40p^2$

25. a. $R(p) = 108p - 0.4p^2$, p = rent in dollars per suite
    b. $R(x) = 270x - 2.5x^2$, x = number of suites occupied. A rent of $135 per suite will yield the max. revenue.

27. $250; 18,820 units   29. a. 250 units   b. 180 units

31.

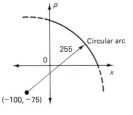

highest price $= \$159.57$

33. $y = \pm\sqrt{x - 1}$ 35. $x = f^{-1}(y) = 4 - y^2$

39. $p(0) = 1000$; $p(1) = 2000$; $p(2) = 2600$
    $t = f^{-1}(p) = \sqrt{(p - 1000)/(3000 - p)}$

CHAPTER 6

Exercises 6-1

1. $2524.95   3. $146.93   5. $2000(1.03)^4 = \$2251.02$
7. $2000(1.06)^8 = \$3187.70$ 9. 6.09%   11. 12.68%
13. 8%   15. 9.57%   17. Semiannual compounding
19. Quarterly compounding
21. 7.18%   23. 14.61%   25. $R = 6$, $P = \$50,000$
27. 1.5068   29. 2981.0   31. 0.5066   33. $5986.09
35. $1822.12   37. $207.51
39. a. 8.66 years   b. 13.73 years
41. 10.14%   43. 13.73%   45. 4.6%
47. Annual compounding at 5.2%
49. Quarterly compounding at 8.2%
51. 3.92%   53. $2577.10.   55. $1000 now is better.
57. $2000 now is better.

Exercises 6-2

5. D_f = all x, R_f = all $y > 0$
7. D_f = all t, R_f = all $y < 0$ 9. D_g = all x, R_g = all $y > 5$
11. D_f = all x, R_f = all $y < \frac{1}{3}$

13.

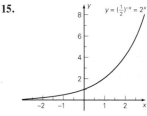

15.

17.

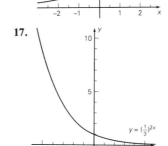

19.

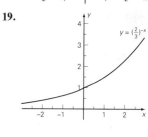

21.

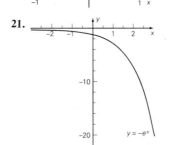

23.

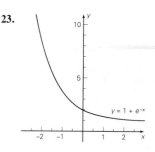

25.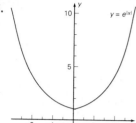

27.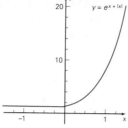

29.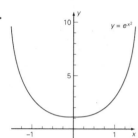

31.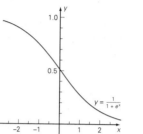

33. $y = 200(2^{3t})$

35. a. 7812

    b. $y = 1{,}000{,}000(2^{-t/3})$, where t is measured in days

37. a. 409,400   **b.** 274,400

39. 1.98%. It is constant and does not depend on time.

41. a. \$2019   **b.** 18.1%

Exercises 6-3

1. $\log_{27}\left(\frac{1}{81}\right) = -\frac{4}{3}$ **3.** $\log_{125} 25 = \frac{2}{3}$

5. $\log_{8/27}\left(\frac{3}{2}\right) = -\frac{1}{3}$ **7.** $3^3 = 27$

9. $4^{-1/2} = \frac{1}{2}$ **11.** 9   **13.** 8   **15.** -3

17. 100   **19.** $p/2$ **21.** $\frac{1}{3}$

23. 0.3010   **25.** 1.0791   **27.** 1.4771

29. $\log\left(\dfrac{x + 1}{x}\right)$ **31.** $\ln\left(\dfrac{x^2 t^4}{y^3}\right)$ **33.** $\ln\left(\dfrac{x^2 \cdot 3^x}{\sqrt{x + 1}}\right)$

35. $\log\left(\dfrac{xy^2}{1000}\right)$ **37.** $-\frac{5}{2}$ **39.** 2, 3

41. No solution, because the base of log cannot be 1 or a negative number.

43. $\frac{1}{4}$ **45.** $\frac{3}{2}$ **47.** 2   **49.** 6

51. $x = -y^2/(1 - y)$ **55.** 1.2267

57. 4.4332   **59.** -1.0759

63. All $x > 2$ **65.** $-2 < x < 2$ **67.** All $x > 0$

69. All $x > 0$, $\neq 1$ **71.** All $x \neq 0$ **73.** All $x > 0$, $\neq 1/e$

75.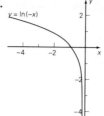

77.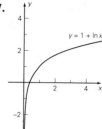

79.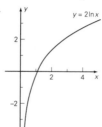

81.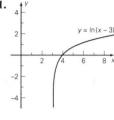

83. a. 15.41   **b.** 2.81   **c.** 1.69

85. 22.58; second design is cheaper for larger x.

Exercises 6-4

1. 1.3979   **3.** 0.4362   **5.** 0.7737   **7.** 2.155

9. $(\log c)(\log a - \log b)^{-1}$ **11.** 46.3 years after 1976

13. 38.4 years after 1976   **15.** After 23.2 months

17. 14.27 years

23. $y = e^{(0.6931)t}$ **25.** $y = 5e^{(0.0392)t}$

27. $y = 4e^{(0.0198)t}$ billion

29. a. 10.13%   **b.** $I = 121e^{(0.0965)t}$

    c. $t = 7.52$ years after January 1975

31. $100(e^k - 1)$ percent; $(1/k) \ln 2$; $(1/k) \ln 3$

33. 1.55 months   **35.** 11.33%   **37.** 9.95; $t = 8.8$

39. $k = 4.36 \times 10^{-4}$; 6.47 grams

41. Approximately 251 times

43. 1580; 3.16×10^6; 10^{14} **45.** 126.3 minutes

47. 51.7 minutes

49. $A = 20$; $k = 0.691$; 18.75 nuts per 5-minute period

51. 0.5; 0.8; 1.143; 1.684

Review Exercises for Chapter 6

1. a. False; True only if $a > 0$, $\neq 1$.

    b. False; If $a^x = y$, then $\log_a y = x$ holds only if $a > 0$ and $\neq 1$.

    c. True   **d.** False; $\log (78) = \log (7.8) + 1$

    e. True   **f.** True   **g.** False; $\log_a (xy) = \log_a x + \log_a y$

h. False; If $\ln x > 1$, then $x > e$.　　**i.** True　　**j.** True

k. True　　**l.** False; $\log (x^n) = n \log x$

m. False; $(\ln x^3)/(\ln x^2) = (3 \ln x)/(2 \ln x) = \frac{3}{2}$

3. $-1/4$　　**5.** $(3 - 4x)/2$　　**7.** $(1 - x)/(1 - 2x)$

9. 0.5204　　**11.** 1　　**13.** $11/17$

15. $\frac{1}{2}, \frac{3}{2}$　　**19.** $y = x/(x - 1)$

21. All x　　**23.** All x　　**25.** All x in $-3 < x < 3$

27.

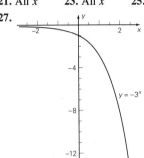

29. $R = px = \$200xe^{-x/50}$; $\$3032.65$

31. a. 2980 units　　**b.** $\$58.70$

33. $\$802.89$　　**35.** $\$576.19$　　**37.** 7.18%　　**39.** Bank B

41. 1.86%　　**43.** 10.99%

47. a. 6.09%　　**b.** 6.14%　　**c.** 6.17%　　**d.** 6.18%

49. 1.961; 12.88; 52.20

51. 15; 27.97; 30. The rate of production reaches 30 units in the long run.

53. a. 8.204%　　**b.** $0.5e^{0.07885t}$　　**c.** 13.93 yrs.

CHAPTER 7

Exercises 7-1

1. 39; 59　　**3.** $74 - 2r$　　**5.** 54　　**7.** 27th

9. 1335　　**11.** $n(5n - 1)/2$　　**13.** 414　　**15.** 12

19. $\$120$　　**21.** $\$201$　　**23. a.** 30　　**b.** $\$241$

25. $\$1275$　　**27.** $\$1122$　　**29.** 9

31. a. $\$(200 + 20n)$　　**b.** $\$300$

35. $\$150,000$; $\$100,000$; $\$50,000$

Exercises 7-2

1. 768　　**3.** $(\frac{2}{9})(-\frac{3}{2})^{n-1}$　　**5.** 10th

7. 16, 24, 36, 54, 81, . . . ; $19,683/32$　　**9.** 0.16

11. $(3^{12} - 1) = 531,440$　　**13.** $(2^n - 1)$　　**15.** 2

17. $2\sqrt{2}/3$　　**21.** $\sqrt{2}$　　**23.** 6 yr

25. $\$4997.91$　　**27.** $\$4502.02$

29. a. $\$4246.71$　　**b.** $\$5081.96$

31. $\$49,045.42$　　**33.** $\$126,217.96$

35. $\$7,144.57$　　**37.** $\$11,392.19$; $\$18,458.31$; $\$22,841.15$

Exercises 7-3

1. 7.335929　　**3.** 41.152716　　**5.** 9.471305

7. 18.987419　　**9.** $\$14,486.56$　　**11.** $\$4607.11$

13. $\$8615.38$　　**15.** $\$2044.73$　　**17.** $\$1073.00$

19. 25th　　**21. a.** $\$11,733.20$　　**b.** $\$25,000[(1.08)^n - 1]$

23. $\$162.83$　　**25.** $\$8957.21$　　**27.** $\$3511.79$

29. $\$7931.51$　　**31.** $\$4831.08$　　**33.** $\$77,698,00$

35. a. $\$79,426.86$　　**b.** $\$88,632.52$　　**37.** $\$1358.68$

39. $\$50,053.40$　　**41.** $\$304.91$　　**43.** $\$335.68$

45. a. $\$37,308.98$　　**b.** $\$26,499.03$

47. $\$71,295.10$　　**49.** $\$63.73$

Exercises 7-4

1. 1　　**3.** 4　　**5.** 5

15.

n	0	1	2	3	4	5	6	7	8	9	10
y_n	10	15	22	30	39	47	53	57	59	59	60

17. $y_n = 1.1y_{n-1}$

n	0	1	2	3	4	5
y_n	10,000	11,000	12,100	13,310	14,641	16,105
n	6	7	8	9		10
y_n	17,716	19,487	21,436	23,579	25,937	

19. $y_n = ca^n = c2^n$　　**21.** $y_n = 5 \cdot 3^n$　　**23.** $y_n = c \cdot 2^n - 1$

25. $y_n = c(-1)^n + 1$　　**27.** $y_n = 8$　　**29.** $y_n = 6 + 4n$

31. $y_n = 20(0.2)^n + 5$　　**33.** $y_n = 4(-1)^n - 3$

35. 6 years

37. $y_n = 1.06y_{n-1}$, $y_0 = 5000$;
$y_n = 5000(1.06)^n$;
$y_8 = 5000(1.06)^8 = \$7969.24$

39. $y_n = y_{n-1} + 160$, $y_n = 2000 + 160n$ and $y_{10} = \$3600$

41. a. $y_n = 1.15y_{n-1} - 900$, $y_0 = 5000$;
$y_n = 6000 - 1000(1.15)^n$
b. $y_{10} = \$1954.44$

43. a. $y_n = 1.01y_{n-1} + 200$, $y_1 = 200$;
$y_n = 20{,}000(1.01)^n - 20{,}000$
b. $y_{30} = \$6956.98$

45. a. $y_n = 1.0125y_{n-1} - 200$, $y_{36} = 0$
$y_n = 16{,}000[1 - (1.0125)^{n-36}]$
b. $y_0 = \$5769.45$ **c.** $y_{20} = \$2884.06$

47. a. $y_n = 1.01y_{n-1} - P$, $y_0 = 10{,}000$;
$y_n = [10{,}000 - 100P](1.01)^n + 100P$
b. $P = \$263.34$

49. a. $y_n = 1.01y_{n-1} - 500$, $y_{120} = 0$;
$y_n = 50{,}000[1 - (1.01)^{n-120}]$
b. $y_0 = \$34{,}850.26$

Exercises 7-5

1. 8.   **3.** 64.   **5.** $\frac{29}{6}$.   **7.** $\frac{3}{4}$.

9. n^2.   **11.** $\frac{1}{3}n(n^2 + 3n + 5)$
13. $\frac{1}{6}n(4n^2 + 9n - 1)$
15. $\frac{1}{4}n(n^3 + 2n^2 + 15n + 10)$
17. $\frac{1}{3}(20)(20^2 + 12 \cdot 20 - 7) = 4220$
19. $\frac{1}{6}(25)(2 \cdot 25^2 + 15 \cdot 25 + 31) = 6900$

21. 42540   **25. a.** 11   **b.** 151
27. a. 9   **b.** 496.

Review Exercises for Chapter 7

1. a. False; $T_n = a + (n - 1)d$ is the nth term only if the sequence is in A.P.
b. False; the given formula is the sum to n terms of a G.P.
c. False; the formula only applies when $-1 < r < 1$.
d. False; the pth term of a G.P. is ar^{p-1}.
e. True   **f.** True   **g.** True   **h.** True; unless $r = 1$.
i. True   **j.** True
k. False; the sequence $1, x, x^2, x^3, \ldots,$ is in G.P.
3. $\frac{2}{3}, \frac{6}{3}, \frac{10}{3}, \frac{14}{3}, \frac{18}{3}, \frac{22}{3}$ **5.** 3   **7.** 8, 4, 2, 1, 0.5, 0.25
9. 820   **11.** $3(2^n - 1)$ **13.** 10.8
15. Number of terms = 8; last term = 4   **17.** 15 months
19. $\$155; \$5$ **21.** nth payment = $(242 - 2n); \$420$
23. a. $\$137{,}639.05$ **b.** $\$117{,}817.77$
25. $\$131.67$ **27.** $\$318.00$
31. $y_n = c(-1/4)^n$ **33.** $y_n = 3$
35. $y_n = 1 - 3n$
37. $A_n = 1.12A_{n-1}$, $A_0 = 10{,}000$; 4 years

CHAPTER 8

Exercises 8-1

1. $\{A, B, C, D, E\}$ where $A, \ldots, E$ are the names of the five candidates
3. $\{H1, H2, H3, H4, H5, H6, T1, T2, T3, T4, T5, T6\}$
5. $\{ABC, ACB, BCA, BAC, CAB, CBA\}$ where A, B, C are the names of the salespeople.
7. $\{BBB, BBW, BWW\}$ where B = brown, W = white, and the order of B's and W's is of no significance.
9. The sample space consists of 13 points, one for each spade card in the deck.
11. $\{(6, 4), (5, 5), (4, 6)\}$
13. $\{(6, 1) (5, 2), (4, 3), (3, 4), (2, 5), (1, 6)\}$
15. $E_1 \cup E_2 = \{x \mid x$ is a black card or a heart$\}$ is the event that the card drawn is either a black card or a heart.
17. $E_1 \cup E_3 = \{x \mid x$ is a heart or is of denomination less than 7, or both$\}$ is the event that the drawn card is either a heart or is of denomination less than 7.
19. $E_3 \cap E_4 = \{x \mid x$ is an ace$\} = E_4$ is the event that the drawn card is an ace and is of denomination less than 7.
21. $E_2' = \{x \mid x$ is a red card$\}$ is the event that the drawn card is not black.
23. The following pairs are mutually exclusive: E_1 and E_4; E_3 and E_4; E_3 and E_5; E_6 and E_k where $k = 1, 2, 3, 4, 5$.
25. The given events are *not* mutually exclusive.
27. The given two events *are* mutually exclusive.
29. a. $F' \cap U$ **b.** $U' \cap E'$
c. $(U \cap E') \cap F$ **d.** $(U' \cap E) \cap F'$
31. a. The applicant has experience and no degree and is over 40.
b. The applicant has experience, has a degree, and is over 40.
c. The applicant has no degree and is under 40.
d. The applicant has a degree but no experience.
e. The applicant has no experience and in addition either has a degree and is over 40 or else has no degree and is under 40.

Exercises 8-2

1. $\frac{1}{4}$ **3.** $\frac{3}{4}$ **5.** $\frac{19}{52}$ **7.** $\frac{3}{8}$ **9.** $\frac{7}{8}$ **11.** 1   **13.** $\frac{5}{18}$
15. $\frac{11}{36}$
17. a. 0.55   **b.** 0.75   **c.** 0
d. 0.3   **e.** 0.25   **f.** 1
19. a. E_2' = person earns less than \$20,000; $P(E_2') = 0.65$.
b. $E_2' \cup E_1$ = either person earns less than \$20,000 or has been employee for 10 yr or more, or both; $P(E_2' \cup E_1) = 0.85$.

c. $E_1' \cap E_2$ = person has been employed for less than 10 yr and earns \$20,000 or more; $P(E_1' \cap E_2) = 0.15$.

d. $(E_1' \cap E_2')' = E_1 \cup E_2$ = the person is not in the category of being both an employee for less than 10 yr and earning less than \$20,000; $P[(E_1' \cap E_2')'] = 0.4$.

21. 0.85 **23. a.** $\frac{4}{25}$ **b.** $\frac{16}{25}$ **c.** $\frac{61}{125}$

25. a. 0.6 **b.** 0.3 **c.** 0.1

d. Insufficient information to calculate $P(7)$. **27.** $\frac{14}{60}$

29. a.

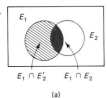

(a)

b.

E_1 $E_1 \cap E_2'$ $E_1' \cap E_2'$ E_2

(b)

31. 0.55

33.

	C	C'	
T	0.68	0.22	0.90
T'	0.07	0.03	0.10
	0.75	0.25	1

a. $P(T' \cap C') = 0.03$
b. $P(C \cap T) = 0.68$
c. $P(T \cap C') = 0.22$
d. $P(T \cup C') = P(T) + P(C') - P(T \cap C')$
$\qquad = 0.90 + 0.25 - 0.22 = 0.93$
e. $P(T' \cup C) = 0.78$
f. $P(C' \cup T') = 0.32$

Exercises 8-3

1. $\frac{1}{2}$ **3.** No, $P(A) = \frac{3}{4}$, $P(B) = \frac{1}{2}$,
$P(A \cap B) = \frac{1}{4} \neq P(A)P(B)$.

5. $\frac{1}{2}$ **7.** $\frac{2}{6}$ **9.** $\frac{20}{26}$

11. $\frac{3}{6}$ **13.** $\frac{3}{4}$ **15.** 0.1 **17. a.** $\left(\frac{150}{500}\right)\left(\frac{149}{499}\right)$ **b.** $\left(\frac{350}{500}\right)\left(\frac{349}{499}\right)$

19. 3% **a.** 0.065 **b.** $\frac{7}{13}$

21. $\frac{3}{7}$ **23.** $\frac{6}{13}$ **25.** No

27. $\frac{5}{18} + \left(\frac{13}{18}\right)\left(\frac{5}{18}\right) = \frac{155}{324}$

Exercises 8-4

1. $\frac{7}{15}$ **3.** 0.9755 **5.** $\frac{5}{8}$ **7.** $\frac{20}{47}$
9. $\frac{3}{12}, \frac{4}{12}, \frac{5}{12}$

11. Yes, provided there is a zero probability of a two-tailed penny.

13. $\frac{50}{89}$ **15. a.** $\frac{6}{43}$ **b.** $\frac{15}{43}$ **c.** $\frac{18}{43}$

Exercises 8-5

1. 90 **3.** 120 **5.** $n(n-1)$
7. 45 **9.** 1 **11.** 1225 **13.** 380 **15.** 7
17. 19 **19.** 8 **21.** $1/13^3 = 1/2197$
23. a. $\frac{2}{21}$ **b.** $\frac{10}{21}$ **25.** $\frac{969}{2639} = 0.37$
27. 20 outcomes
29. 40 (assuming no distinction between which driver drives which cab).
31. 72 **33.** $\frac{28}{33}$ **35.** $2^{10} = 1024$
37. 26^{-11} **39.** $\frac{1}{2}$
41. a. $\frac{5}{42}$ **b.** $\frac{10}{21}$ **c.** $\frac{1}{21}$
43. a. $\frac{4}{91}$ **b.** $\frac{24}{91}$ **c.** $\frac{34}{455}$

Exercises 8-6

1. $\frac{1}{4}$ **3.** $10(5^3/6^5) = 625/3888$
5. $\frac{128}{625}$ **7.** $1 - \left(\frac{4}{5}\right)^6 = 11,529/15,625$
9. $1 - 61(3^6/4^8) = 0.32$ **11. a.** $\frac{135}{4096}$ **b.** $\frac{1}{4096}$
13. $\frac{63}{256} = 0.25$ **15.** $252(4^5/5^{10}) = 0.026$
17. a. $\left(\frac{4}{5}\right)^8$ **b.** $1 - 3\left(\frac{4}{5}\right)^8$
19. $1 - 90(4^6/5^8) = 0.056$ **21.** $\frac{25}{36}; \frac{171}{1296}$
23. a. 0.251 **b.** 0.215 **c.** 0.201
25. $45(0.2)^2(0.98)^8 = 0.0153$
27. a. 0.251 **b.** 0.215 **c.** 0.201 **29.** 0.679

Exercises 8-7

1. $a^7 + 7a^6b + 21a^5b^2 + 35a^4b^3 + 35a^3b^4 + 21a^2b^5$
$\qquad\qquad\qquad\qquad\qquad + 7ab^6 + b^7$
3. $32x^5 + 40x^4y + 20x^3y^2 + 5x^2y^3 + (5/8)xy^4 + y^5/32$
5. $32x^5/243 + 40x^3/27 + 20x/3 + 15x^{-1} + 135x^{-3}/8$
$\qquad\qquad\qquad\qquad\qquad + 243x^{-5}/32$
7. $a^6 - 12a^5b + 60a^4b^2 - 160a^3b^3 + 240a^2b^4 - 192ab^5$
$\qquad\qquad\qquad\qquad\qquad + 64b^6$
9. $32p^5 - 240p^4q + 720p^3q^2 - 1080p^2q^3 + 810pq^4 - 243q^5$
11. $p^{10} - 5p^7q + 10p^4q^2 - 10pq^3 + 5q^4/p^2 - q^5/p^5$
13. $567x^3y^6/16$ **15.** $(-1)^{n-1}\binom{3n}{n-1}x^3$
17. $252x^2$ **19.** -252 **21.** $189x^{17}/8, -21x^{19}/16$
25. $n = 8, a = 1/2$ **27.** $140\sqrt{2}$

Review Exercises for Chapter 8

1. a. False; If A and B are mutually exclusive, then
$P(A \mid B) = P(B \mid A) = 0$.

b. True

c. False; If E_1 and E_2 are independent, then
$P(E_1 \cap E_2) = P(E_1)P(E_2)$.

d. True

e. False; In general, the outcomes have unequal probabilities,
though in many cases they are equally likely.

f. False; The sum of the probabilities of complementary events
is 1.

g. False; If E_1 and E_2 are independent,
$P(E_1 \cap E_2) = P(E_1)P(E_2)$. If they are mutually exclusive,
$P(E_1 \cap E_2) = 0$.

h. True   **i.** True

j. False; $P(A \mid B) = P(B \mid A)[P(A)/P(B)]$

k. True   **l.** True

m. False; $P(A \mid A') = 0$.

n. False; If $E_2 \subseteq E_1$, then $P(E_1 \mid E_2) = 1$.

o. False; $P(E_1 \mid E_2) = P(E_2 \mid E_1)[P(E_1)/P(E_2)]$.

p. False; If E_1 and E_2 are independent, then
$P(E_1 \mid E_2) = P(E_1)$.

q. True   **r.** True

s. False; For example, take $B = C$. Then B and C are certainly
not independent. Alternatively, let A be the event that a coin
lands heads, B the event that a card drawn from a deck is a
spade, and C the event that this same card is a red card.

3. 72   **5.** 28   **7.** $\frac{1}{2}$

9. $\frac{15}{4}$ **11.** 8   **13.** 281

17. a. $\frac{2}{15}$ **b.** $\frac{1}{3}$ **c.** $\frac{2}{3}$

19. Not independent   **21.** 12%   **25.** $\frac{93}{256}$

27. $\frac{5}{12}$; 10.5%   **29.** $\frac{16}{31}$ **31.** $\frac{3}{5}$; $\frac{9}{20}$ **33.** $_{36}P_{30}$; $_{30}P_{26}$

35. $\frac{2}{5}, \frac{8}{15}$ **37.** $\frac{8}{81}$ **39.** $\frac{1}{17}$ **41.** $\frac{5}{13}$

CHAPTER 9

Exercises 9-1

1. A: 2×2; **B:** 2×3; **C:** 3×1; **D:** 3×3; **E:** 2×3; **F:** 2×2;
G: 1×3; **H:** 1×1

3. $A = \begin{bmatrix} 0 & 1 \\ 1 & 2 \end{bmatrix}$ **5.** Any matrix of the form $\begin{bmatrix} 0 & x & y \\ -x & 0 & z \\ -y & -z & 0 \end{bmatrix}$

7. $\begin{bmatrix} 6 & 12 \\ 3 & 9 \end{bmatrix}$

9. $\begin{bmatrix} 2 & 0 & 5 \\ 0 & 6 & -1 \end{bmatrix}$ **11.** $\begin{bmatrix} -4 & 13 \\ 1 & 6 \end{bmatrix}$

13. $\begin{bmatrix} 2 & 1 & 12 \\ 13 & 4 & -12 \\ 5 & 10 & 21 \end{bmatrix}$

15. $x = 1, y = 4$ **17.** $x = 1, y = 5, z = -2$

19. $x = 6, y = 5, z = 2, t = -2, u = 2, v = 3$

21. $x = 1, y = 2, z = 3, t = 4, u = -1, v = 2, w = 5$

23. $x = 0, y = 1, z = 2, t = -1, u = -2, v = 3, w = 0$

25. a. $\begin{bmatrix} 11 & 13 & 16 \\ 14 & 11 & 13 \\ 9 & 16 & 7 \\ 17 & 10 & 11 \end{bmatrix}$ **b.** $\begin{bmatrix} 12 & 14.4 & 18 \\ 15.6 & 12 & 14.4 \\ 9.6 & 18 & 7.2 \\ 19.2 & 10.8 & 12 \end{bmatrix}$

27. a. $\begin{bmatrix} 0 & 33 & 39 \\ 35 & 0 & 38 \\ 45 & 30 & 0 \end{bmatrix}$ **b.** $\begin{bmatrix} 0 & 165 & 195 \\ 175 & 0 & 190 \\ 225 & 150 & 0 \end{bmatrix}$

29. a. $\begin{bmatrix} 65 & 64 & 46 \\ 97 & 45 & 34 \\ 37 & 50 & 57 \end{bmatrix}$ **b.** $\begin{bmatrix} 88.75 & 88.5 & 62.5 \\ 132.5 & 61.25 & 46.5 \\ 49.75 & 69 & 77.5 \end{bmatrix}$

Exercises 9-2

1. 3×3 **3.** 2×4 **5.** 2×5 **7.** [23]

9. $\begin{bmatrix} 18 \\ 28 \end{bmatrix}$ **11.** $\begin{bmatrix} 1 & 4 \\ 5 & -1 \\ -4 & 5 \end{bmatrix}$ **13.** $\begin{bmatrix} 11 \\ -6 \\ 32 \end{bmatrix}$

15. $\begin{bmatrix} -25 & 14 \\ -58 & 32 \end{bmatrix}$ **17.** $\begin{bmatrix} 8 & 33 \\ 5 & -22 \end{bmatrix}$ **19.** $\begin{bmatrix} 4 & 12 \\ 12 & 16 \end{bmatrix}$

21. a. $\begin{bmatrix} 9 & 5 \\ 0 & 4 \end{bmatrix}$ **b.** $\begin{bmatrix} 6 & 0 \\ 3 & 7 \end{bmatrix}$

c. $(A + B)^2 \neq A^2 + 2AB + B^2$ **23.** $p = 1, q = 4$

25. $A = [3, -1]$ **27.** $A = \begin{bmatrix} 3 & 0 \\ 0 & 1 \end{bmatrix}$

29. $\begin{bmatrix} 2 & 3 \\ 1 & 4 \end{bmatrix}\begin{bmatrix} x \\ y \end{bmatrix} = \begin{bmatrix} 7 \\ 5 \end{bmatrix}$

31. $\begin{bmatrix} 1 & 2 & 3 \\ 2 & -1 & 4 \\ 0 & 3 & -2 \end{bmatrix}\begin{bmatrix} x \\ y \\ z \end{bmatrix} = \begin{bmatrix} 8 \\ 13 \\ 5 \end{bmatrix}$

33. $\begin{bmatrix} 2 & 1 & 0 & -1 \\ 0 & 3 & 2 & 4 \\ 1 & -2 & 4 & 1 \end{bmatrix}\begin{bmatrix} x \\ y \\ z \\ u \end{bmatrix} = \begin{bmatrix} 0 \\ 5 \\ 12 \end{bmatrix}$

35. $\begin{bmatrix} -2x & -2y \\ x & y \end{bmatrix}$, $(x, y$ arbitrary)   **37.** $A^n = \begin{bmatrix} 1 & 0 \\ 0 & 1 \end{bmatrix}$ for all n

39. $A^n = \begin{bmatrix} 1 & 0 \\ 1 - 2^{-n} & 2^{-n} \end{bmatrix}$

$$41. \quad \begin{bmatrix} 5 & 8 & 4 & 10 \end{bmatrix} \begin{bmatrix} 650 \\ 550 \\ 500 \\ 300 \end{bmatrix} = 12{,}650$$

43. Let $\mathbf{A} = \begin{bmatrix} 3 & 2 & 4 \\ 4 & 1 & 3 \end{bmatrix}$, $\mathbf{B} = \begin{bmatrix} 20 & 30 \end{bmatrix}$, $\mathbf{C} = \begin{bmatrix} 6 \\ 10 \\ 12 \end{bmatrix}$

    a. $\mathbf{BA} = \begin{bmatrix} 180 & 70 & 170 \end{bmatrix}$ b. $\mathbf{AC} = \begin{bmatrix} 86 \\ 70 \end{bmatrix}$

    c. $\mathbf{BAC} = \begin{bmatrix} 3820 \end{bmatrix}$

45. a. $\mathbf{A} = \begin{bmatrix} 0 & 1 & 0 & 0 \\ 1 & 0 & 1 & 1 \\ 0 & 1 & 0 & 1 \\ 0 & 1 & 1 & 0 \end{bmatrix}$, $\mathbf{A}^2 = \begin{bmatrix} 1 & 0 & 1 & 1 \\ 0 & 3 & 1 & 1 \\ 1 & 1 & 2 & 1 \\ 1 & 1 & 1 & 2 \end{bmatrix}$

    b.

$$\mathbf{A} = \begin{bmatrix} 0 & 1 & 0 & 0 & 1 \\ 1 & 0 & 1 & 1 & 0 \\ 0 & 1 & 0 & 1 & 0 \\ 0 & 1 & 1 & 0 & 1 \\ 1 & 0 & 0 & 1 & 0 \end{bmatrix}, \mathbf{A}^2 = \begin{bmatrix} 2 & 0 & 1 & 2 & 0 \\ 0 & 3 & 1 & 1 & 2 \\ 1 & 1 & 2 & 1 & 1 \\ 2 & 1 & 1 & 3 & 0 \\ 0 & 2 & 1 & 0 & 2 \end{bmatrix}$$

$\mathbf{A}^3$ gives the numbers of routes passing through two vertices.

Exercises 9-3

1. $x = 2, y = 1$ 3. $u = 4, v = -1$
5. $x = 1, y = 2, z = 3$ 7. $x_1 = 1, x_2 = 2, x_3 = -1$
9. $p = -1, q = 2, r = 2$
11. $x = -1, y = 3, z = -3, t = 2$
13. $x = y = 1, z = -1, w = 2$ 15. $x = 2, y = 1, z = 3$
17. $x = 4, p = 17$ 19. $x = \frac{91}{4}, p_1 = \frac{115}{4}$
21. 100, 150, and 200 units of A, B, and C
23. \$6000 at 6%, \$7200 at 10%, \$6800 at 8%
25. $p = 3, x = -1$. In practice, there will be no market (i.e., $x = 0$).

Exercises 9-4

1. $x = 2 + z, y = 3 - 2z$ 3. No solution
5. No solution   7. $x = -4 + 6z, y = 10 - 11z$
9. No solution   11. $u = 1, v = 2, w = -1$
13. No solution   15. No solution
17. $x = (3 - z)/5, y = (7 + 6z)/5$
19. If x, y, and z denote, respectively, the number of houses of first, second, and third type which can be made, then $(x, y, z) = (50, 0, 0), (42, 2, 5), (34, 4, 10), (26, 6, 15), (18, 8, 20), (10, 10, 25),$ and $(2, 12, 30)$.
21. No solution

Review Exercises for Chapter 9

1. a. False; The array is not rectangular.
   b. False; $\mathbf{A}$ and $\mathbf{B}$ have different sizes, so $\mathbf{A} + \mathbf{B}$ cannot be formed.
   c. True
   d. True
   e. False; The number of *columns* in $\mathbf{A}$ must equal the number of *rows* in $\mathbf{B}$.
   f. False; Unless $\mathbf{A}$ and $\mathbf{B}$ are both *square* matrices of the same size.
   g. False; Usually $\mathbf{AB} \neq \mathbf{BA}$ but not always. For example, take $\mathbf{A}$ to be square and $\mathbf{B} = \mathbf{I}$, then $\mathbf{BA} = \mathbf{AB}$.
   h. False; $\mathbf{AB}$ and $\mathbf{BA}$ are both defined if $\mathbf{A}$ is $m \times n$ and $\mathbf{B}$ is $n \times m$. Then $\mathbf{AB}$ is of size $m \times m$ and $\mathbf{BA}$ is $n \times n$.
   i. True. (Note that if $\mathbf{A}$ is $m \times n$, then $\mathbf{I}$ is of size $n \times n$ in the product $\mathbf{AI}$ and is of size $m \times m$ in the product $\mathbf{IA}$.)
   j. True
   k. True
   l. False; For example, take $\mathbf{A} = \begin{bmatrix} 1 & 0 \end{bmatrix}$ and $\mathbf{B} = \begin{bmatrix} 0 \\ 1 \end{bmatrix}$.
   m. False; The system may have no solutions, one unique solution, or an infinite number of solutions.
   n. False; If a system with more variables than equations is *consistent* then it must have infinitely many solutions.
   o. False; A system is consistent if it has one or more solutions.

3. $\begin{bmatrix} 1 & -11 \\ 20 & 10 \end{bmatrix}$ 5. $\begin{bmatrix} 7 & 1 \\ 11 & 9 \end{bmatrix}$ 7. $\begin{bmatrix} 9 & 1 \\ -5 & -9 \end{bmatrix}$

9. $x = 1, y = 2$

11. $x = 1, y = -1$ 13. $x = -1, y = 2$ 15. No solution

17. $\mathbf{X} = \begin{bmatrix} 1 & -1 \\ 0 & 1 \end{bmatrix}$

19. $\mathbf{X} = \begin{bmatrix} 2 & -1 \\ 1 & 2 \end{bmatrix}$

21. $\mathbf{A} = \begin{bmatrix} 2 & 1 \\ -3 & -1 \end{bmatrix}$, $\mathbf{B} = \begin{bmatrix} 0 & -1 \\ 3 & 3 \end{bmatrix}$, for example

23. $\begin{bmatrix} 517 & 345 & 189 \\ 257 & 284 & 408 \end{bmatrix}$ or $\begin{bmatrix} 517 & 257 \\ 345 & 284 \\ 189 & 408 \end{bmatrix}$

25. a. $\begin{bmatrix} 33 & 59 & 33 \\ 47 & 38 & 54 \end{bmatrix}$ b. $\begin{bmatrix} 35.6 & 61.4 & 36 \\ 49.4 & 40.8 & 58.8 \end{bmatrix}$

27. $\begin{bmatrix} 3 & 5 & 2 & 3 \end{bmatrix} \begin{bmatrix} 12 \\ 5 \\ 3 \\ 20 \end{bmatrix} = 127$

29. $\mathbf{PA} = \begin{bmatrix} 114 & 84.5 & 92 & 84 \end{bmatrix}$; Elements are the earnings in weeks I–IV.

CHAPTER 10

Exercises 10-1

1. $\begin{bmatrix} -\frac{4}{7} & \frac{5}{7} \\ \frac{3}{7} & -\frac{2}{7} \end{bmatrix}$ **3.** $\begin{bmatrix} -2 & -1 \\ -\frac{3}{2} & -\frac{1}{2} \end{bmatrix}$

5. No inverse

7. $\frac{1}{11}\begin{bmatrix} -1 & 2 & 6 \\ -2 & 4 & 1 \\ 6 & -1 & -3 \end{bmatrix}$

9. $\begin{bmatrix} -2 & -\frac{1}{3} & \frac{4}{3} \\ 1 & \frac{2}{3} & -\frac{2}{3} \\ \frac{1}{2} & -\frac{1}{3} & -\frac{1}{6} \end{bmatrix}$ **11.** No inverse

13. $\begin{bmatrix} \frac{3}{14} & \frac{1}{2} & \frac{1}{14} \\ \frac{1}{14} & -\frac{1}{2} & \frac{5}{14} \\ -\frac{5}{14} & -\frac{1}{2} & \frac{3}{14} \end{bmatrix}$

15. $\begin{bmatrix} -6 & \frac{7}{2} & \frac{9}{2} & -\frac{3}{2} \\ 3 & -2 & -2 & 1 \\ -4 & \frac{5}{2} & \frac{7}{2} & -\frac{3}{2} \\ 7 & -4 & -5 & 2 \end{bmatrix}$

17. $x = 2, y = 1$

19. $u = 1, v = 2$ **21.** $x = 1, y = 2, z = -1$

23. $u = 1, v = 0, w = 3$ **25.** 1600 lb of P and 600 lb of Q

Exercises 10-2

1. a. $\begin{bmatrix} 0.2 & 0.7 \\ 0.5 & 0.1 \end{bmatrix}$

b. 250 and 180 units for I and II, respectively.

c. 75 and 36 units for I and II, respectively.

3. a. $\begin{bmatrix} 0.3 & 0.5 \\ 0.4 & 0.2 \end{bmatrix}$ **b.** $\begin{bmatrix} 470 \\ 450 \end{bmatrix}$

c. 141 and 135 units for P and Q, respectively.

5. Final demands; 80 units for P, 70 for Q. (Total outputs: 275 and 225 units.)

7. a. $A = \begin{bmatrix} 0.1 & 0.4 & 0.4 \\ 0.4 & 0.2 & 0.2 \\ 0.3 & 0.3 & 0.3 \end{bmatrix}$ **b.** $\begin{bmatrix} 360 \\ 323 \\ 317 \end{bmatrix}$

c. 72, 32.3 and 31.7 units

Exercises 10-3

1. Yes   **3.** No (first row does not sum to 1).

5. No (not a square matrix).

7. Yes; $P^2 = \begin{bmatrix} \frac{1}{3} & \frac{2}{3} \\ \frac{2}{9} & \frac{7}{9} \end{bmatrix}$ has nonzero elements.

9. No; $P^n = \begin{bmatrix} 1 & 0 \\ 0 & 1 \end{bmatrix}$ for all n, so always has zero elements.

11. Yes; P^2 has all nonzero elements.

13. Yes   **15.** No (elements do not sum to 1).

17. a. Probability of a transition from state 2 to state 1 $= \frac{1}{4}$.

b, c. State matrix is $[\frac{19}{36} \quad \frac{17}{36}]$ **d.** $[\frac{3}{7} \quad \frac{4}{7}]$

19. a. $[0.22 \quad 0.47 \quad 0.31]$ **b.** $[0.21 \quad 0.44 \quad 0.35]$

c. $[\frac{17}{67} \quad \frac{27}{67} \quad \frac{23}{67}]$

21. a. $\frac{3}{20}$ **b.** $\frac{5}{16}$ **c.** $\frac{29}{150}$ **d.** $[\frac{18}{55} \quad \frac{32}{55} \quad \frac{1}{11}]$
Party X will on average be in power $\frac{18}{55}$ of the time, party Y $\frac{32}{55}$ of the time, and party Z $\frac{1}{11}$ of the time.

23. a. $\frac{27}{50}$ **b.** $\frac{27}{80}$ **c.** $[\frac{8}{13} \quad \frac{5}{13}]$ (state 1 is "tall," 2 is "short")
In the long run $\frac{5}{13}$ of persons in the population will be tall and $\frac{8}{13}$ will be short.

25. a. 35%, 38%, and 27%   **b.** 38.3%, 36.6%, and 25.1%

c. 44.4%, 33.3%, and 22.2%

27. $P = \begin{bmatrix} 0.85 & 0.15 \\ 0.35 & 0.65 \end{bmatrix}$; $[0.7 \quad 0.3]$
In the long run, 70% of the energy is produced from oil, gas, or coal and only 30% from atomic power.

Exercises 10-4

1. $(an - lc)$ **3.** $-(am - lb)$ **5.** 25   **7.** -38
9. $3a + 2b$ **11.** 32   **13.** -192 **15.** $(ac - b^2)$
17. -21 **19.** 35   **21.** 0   **23.** 6   **25.** adf
27. 10   **29.** $x = 3$ **31.** $x = 6; -1$
33. $x = 1, y = -1$ **35.** $x = 1, y = 2$ **37.** $x = 6, y = 10$
39. No solution   **41.** $x = 1, y = -1, z = -1$
43. $x = -1, y = -1, z = 3$
45. $x = 3, y = 1, z = 2$ **47.** No solution
49. $x = -1, y = 2, z = 1$

Exercises 10-5

1. $\begin{bmatrix} 2 & 3 \\ 5 & -7 \end{bmatrix}$ **3.** $\begin{bmatrix} a_1 & b_1 \\ a_2 & b_2 \\ a_3 & b_3 \end{bmatrix}$ **5.** $\begin{bmatrix} 1 & 0 \\ 0 & 1 \end{bmatrix}$

7. $\begin{bmatrix} \frac{1}{5} & -\frac{2}{5} \\ \frac{1}{5} & \frac{3}{5} \end{bmatrix}$ **9.** $\begin{bmatrix} 2 & 3 \\ 4 & 5 \end{bmatrix}$

11. $\begin{bmatrix} -\frac{1}{6} & \frac{1}{2} & -\frac{5}{6} \\ \frac{5}{6} & -\frac{1}{2} & \frac{7}{6} \\ -\frac{1}{2} & \frac{1}{2} & -\frac{1}{2} \end{bmatrix}$

13. $\frac{1}{9}\begin{bmatrix} 1 & -2 & 4 \\ -2 & 4 & 1 \\ 4 & 1 & -2 \end{bmatrix}$ 15. No inverse exists

17. a. $\begin{bmatrix} 0.2 & 0.2 & 0.1 \\ 0.3 & 0.1 & 0.3 \\ 0.4 & 0.5 & 0.2 \end{bmatrix}$

b. 623, 1007, and 1466 units for I, II, and III, respectively.

c. 62.3, 201.4, and 586.4 units, respectively.

19. a. $\begin{bmatrix} 0.3 & 0.1 & 0.4 \\ 0.3 & 0.3 & 0.1 \\ 0.2 & 0.2 & 0.3 \end{bmatrix}$

b. 246, 227, and 250 units for A, B, and C, respectively.

33. $\begin{bmatrix} 1 & 0 & 0 & 0 \\ 0 & 1 & 0 & 0 \\ 0 & 0 & 1 & 0 \\ 0 & 0 & 0 & 1 \end{bmatrix}$ 35. $\begin{bmatrix} 0 & \frac{1}{3} & 0 & \frac{2}{3} \\ \frac{2}{3} & 0 & \frac{1}{3} & 0 \\ 0 & \frac{2}{3} & 0 & \frac{1}{3} \\ \frac{1}{3} & 0 & \frac{2}{3} & 0 \end{bmatrix}$

37. $\begin{bmatrix} 0.8 & 0.2 \\ 0.4 & 0.6 \end{bmatrix}$ (Tom = 1; Dick = 2).

39. a. $\begin{bmatrix} 0.8 & 0.2 \\ 0.1 & 0.9 \end{bmatrix}$ b. 48.7%   c. 33.3%

Review Exercises for Chapter 10

1. a. True   b. True   c. False; The zero matrix is *not* invertible.

d. False; The statement is true if **A** is invertible but not for a general **A**.

e. False; For example, the identity matrix is invertible.

f. False; For example, $\begin{bmatrix} 1 & 1 \\ 1 & 1 \end{bmatrix}$ is not invertible.

g. True   h. True

i. False; The determinant of $k\mathbf{A}$ is equal to $k^n|\mathbf{A}|$.   j. True

k. False; The quantity defined is the minor, not the cofactor.

l. False; The cofactor and minor are equal in absolute value. They sometimes have the same sign, sometimes opposite signs.

m. False; $(\mathbf{AB})^{-1} = \mathbf{B}^{-1}\mathbf{A}^{-1}$

n. True   o. True (but the proof requires care)

p. True   q. True

3. $\frac{1}{11}\begin{bmatrix} 5 & 3 \\ -2 & 1 \end{bmatrix}$ 5. $[1/(a^2 + b^2)]\begin{bmatrix} a & -b \\ b & a \end{bmatrix}$

7. $\begin{bmatrix} 1 & 0 & 0 \\ 0 & \frac{1}{2} & 0 \\ 0 & 0 & \frac{1}{3} \end{bmatrix}$ 9. $\frac{1}{3}\begin{bmatrix} -2 & 1 & 1 \\ -8 & 7 & -2 \\ 7 & -5 & 1 \end{bmatrix}$

11. Not invertible   13. Not invertible (not a square matrix).

15. $x = 1$, $y = 1$ 17. $x = 1$, $y = 1$, $z = -1$

19. $x = -\frac{1}{2}$, $y = \frac{1}{2}$ 21. $x = 1$, $y = 2$, $z = 3$

23. No solution   25. $(a + 4)(a - 4)$

27. $(x + 8)(x - 5)$ 29. $x(1 - x)$

31. a. $\begin{bmatrix} 0.1 & 0.4 \\ 0.7 & 0.2 \end{bmatrix}$

b. 177.27 and 253.86 units for I and II, respectively.

c. 35.45 and 101.54 units, respectively.

CHAPTER 11

Exercises 11-1

1.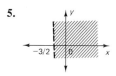

3.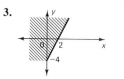

5.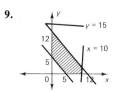

7.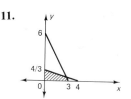

9.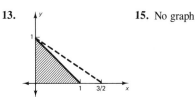

11.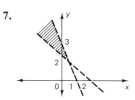

13.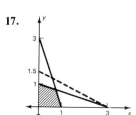

15. No graph

17.

19. $0 \le x \le 70$, $0 \le y \le 90$;   **21.** $x - 2y \le 40$
$40 \le x + y \le 100$

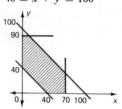

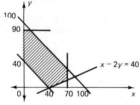

23.

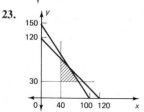

25. $2x + 3y \ge 110$

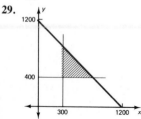

27. $15x + 8y \le 100$; $110x + 83y < 900$; $7x + 7y \ge 60$; $x, y \ge 0$

29.

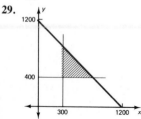

31. $7x + 3y \ge 50$ (x ounces meat, y ounces soybean)

Exercises 11-2

1. $Z = 15$ at $(5, 0)$ **3.** $Z = 7$ at $(1, 2)$ **5.** $Z = \frac{35}{3}$ at $(\frac{7}{3}, 0)$
7. No solution (feasible region is empty)   **9.** $Z = 4$ at $(3, 1)$
11. No solution (feasible region is unbounded)
13. $Z = 8$ at $(4, 1)$ **15.** $Z = 5$ at $(3, 1)$ **17.** $x = 2000$,
$y = 3000$, $Z = \$28{,}000$
19. $x = 10$, $y = 20$, $Z = \$1700$
21. 50 of A, 20 of B; $Z = \$18{,}500$
23. $(500/3)$ dollars per unit of B
25. 30 and 60 acres in crops I & II; $Z = 21{,}000$ dollars
27. 2.7 oz of A and 2.2 oz of B
29. $x = 3000$, $y = 0$; Min. cost is $\$225{,}000$.

Exercises 11-3

1.
$$\begin{array}{c} \\ t \end{array} \begin{array}{cccc} x & y & t \\ [1 & 1 & 1 \mid 5] \end{array}$$

3.
$$\begin{array}{c} \\ t \\ u \end{array} \begin{array}{cccc} x & y & t & u \\ \begin{bmatrix} 2 & 1 & 1 & 0 \\ 1 & 2 & 0 & 1 \end{bmatrix} & \begin{matrix} 4 \\ 5 \end{matrix} \end{array}$$

5.
$$\begin{array}{c} \\ t \\ u \\ v \end{array} \begin{array}{ccccc} x & y & t & u & v \\ \begin{bmatrix} 3 & 1 & 1 & 0 & 0 \\ 1 & 1 & 0 & 1 & 0 \\ 1 & 2 & 0 & 0 & 1 \end{bmatrix} & \begin{matrix} 7 \\ 3 \\ 5 \end{matrix} \end{array}$$

7.
$$\begin{array}{c} \\ t \\ u \end{array} \begin{array}{cccc} x & y & t & u \\ \begin{bmatrix} 3 & 2 & 1 & 0 \\ 3 & 4 & 0 & 1 \end{bmatrix} & \begin{matrix} 12{,}000 \\ 18{,}000 \end{matrix} \end{array}$$

9.
$$\begin{array}{c} \\ t \\ u \end{array} \begin{array}{cccc} x & y & t & u \\ \begin{bmatrix} 2 & 4 & 1 & 0 \\ 5 & 3 & 0 & 1 \end{bmatrix} & \begin{matrix} 100 \\ 110 \end{matrix} \end{array}$$

11.
$$\begin{array}{c} \\ t \\ u \\ v \end{array} \begin{array}{ccccc} x & y & t & u & v \\ \begin{bmatrix} 2 & 5 & 1 & 0 & 0 \\ 4 & 1 & 0 & 1 & 0 \\ 3 & 2 & 0 & 0 & 1 \end{bmatrix} & \begin{matrix} 200 \\ 240 \\ 190 \end{matrix} \end{array}$$

13.
$$\begin{array}{c} \\ t \\ u \end{array} \begin{array}{ccccc} x & y & z & t & u \\ \begin{bmatrix} 2 & 1 & 1 & 1 & 0 \\ 1 & 2 & 1 & 0 & 1 \end{bmatrix} & \begin{matrix} 5 \\ 4 \end{matrix} \end{array}$$

15.
$$\begin{array}{c} \\ y \\ u \end{array} \begin{array}{cccc} x & y & t & u \\ \begin{bmatrix} \frac{2}{3} & 1 & \frac{1}{3} & 0 \\ 3 & 0 & -2 & 1 \end{bmatrix} & \begin{matrix} \frac{8}{3} \\ 3 \end{matrix} \end{array} \rightarrow \begin{array}{c} \\ y \\ x \end{array} \begin{array}{cccc} x & y & t & u \\ \begin{bmatrix} 0 & 1 & \frac{7}{9} & -\frac{2}{9} \\ 1 & 0 & -\frac{2}{3} & \frac{1}{3} \end{bmatrix} & \begin{matrix} 2 \\ 1 \end{matrix} \end{array}$$

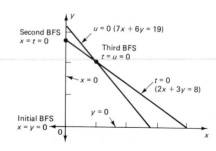

17.
$$\begin{array}{c} \\ y \\ u \end{array} \begin{array}{cccccc} x & y & z & t & u \\ \begin{bmatrix} \frac{1}{2} & 1 & \frac{1}{2} & \frac{1}{2} & 0 \\ 2 & 0 & 3 & -1 & 1 \end{bmatrix} & \begin{matrix} \frac{5}{2} \\ 11 \end{matrix} \end{array} \rightarrow$$

$$\begin{array}{c} \\ y \\ x \end{array} \begin{array}{ccccc} x & y & z & t & u \\ \begin{bmatrix} 0 & 1 & -\frac{1}{4} & \frac{3}{4} & -\frac{1}{4} \\ 1 & 0 & \frac{3}{2} & -\frac{1}{2} & \frac{1}{2} \end{bmatrix} & \begin{matrix} -\frac{1}{4} \\ \frac{11}{2} \end{matrix} \end{array}$$

The second solution here is infeasible.

19.
$$\begin{array}{c} \\ s \\ x \\ u \end{array} \begin{array}{cccccc} x & y & s & t & u \\ \begin{bmatrix} 0 & -\frac{17}{2} & 1 & -\frac{3}{2} & 0 \\ 1 & \frac{9}{4} & 0 & \frac{1}{4} & 0 \\ 0 & -\frac{3}{2} & 0 & -\frac{1}{2} & 1 \end{bmatrix} & \begin{matrix} -\frac{17}{2} \\ \frac{17}{4} \\ -\frac{5}{2} \end{matrix} \end{array}$$

$$\begin{array}{c} \\ s \\ x \\ y \end{array} \begin{array}{ccccc} x & y & s & t & u \\ \begin{bmatrix} 0 & 0 & 1 & \frac{4}{3} & -\frac{17}{3} \\ 1 & 0 & 0 & -\frac{1}{2} & \frac{3}{2} \\ 0 & 1 & 0 & \frac{1}{3} & -\frac{2}{3} \end{bmatrix} & \begin{matrix} \frac{17}{3} \\ \frac{1}{2} \\ \frac{5}{3} \end{matrix} \end{array}$$

$$\begin{array}{c} u \\ x \\ y \end{array} \begin{bmatrix} 0 & 0 & -\frac{3}{17} & -\frac{4}{17} & 1 & -1 \\ 1 & 0 & \frac{9}{34} & -\frac{5}{34} & 0 & 2 \\ 0 & 1 & -\frac{2}{17} & \frac{3}{17} & 0 & 1 \end{bmatrix}$$

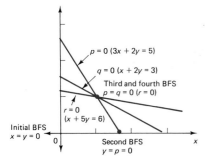

21.

$$\begin{array}{c} \\ x \\ q \\ r \end{array} \begin{array}{cccccc} x & y & p & q & r & \\ \begin{bmatrix} 1 & \frac{2}{3} & \frac{1}{3} & 0 & 0 & \frac{5}{3} \\ 0 & \frac{4}{3} & -\frac{1}{3} & 1 & 0 & \frac{4}{3} \\ 0 & \frac{13}{3} & -\frac{1}{3} & 0 & 1 & \frac{13}{3} \end{bmatrix} \end{array}$$

$$\begin{array}{c} x \\ y \\ r \end{array} \begin{bmatrix} 1 & 0 & \frac{1}{2} & -\frac{1}{2} & 0 & 1 \\ 0 & 1 & -\frac{1}{4} & \frac{3}{4} & 0 & 1 \\ 0 & 0 & \frac{3}{4} & -\frac{13}{4} & 1 & 0 \end{bmatrix}$$

$$\begin{array}{c} x \\ y \\ q \end{array} \begin{bmatrix} 1 & 0 & \frac{7}{8} & 0 & -\frac{13}{8} & 1 \\ 0 & 1 & -\frac{13}{16} & 0 & \frac{39}{16} & 1 \\ 0 & 0 & \frac{3}{4} & 1 & -\frac{13}{4} & 0 \end{bmatrix}$$

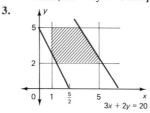

Exercises 11-4

Answers to Exercises 1–13 are given in the answers to Exercises 11-2 above.

15. $Z_{max} = 8$ at $(0, 0, 4)$

17. 2000, 3000, and 4000 lb of regular, super, and deluxe per day.

19. $x = \frac{11}{7}, y = \frac{6}{7}, z = 0; Z = \frac{23}{7}$

21. Maximum $Z = 4$ is attained at the two vertices $x = 1, y = 1, z = 2$ and $x = 3, y = 1, z = 0$ and therefore at all points on the edge joining them.

23. $x = 4, y = 3; Z = 31$

25. $x = y = 0, z = 1; Z = 2$

27. $Z_{min} = 4$ when $x = 3, y = 1$,**29.** $Z_{min} = \frac{49}{100}$ when $x = \frac{27}{10}, y = \frac{11}{5}$

Review Exercises for Chapter 11

1. a. False; The graph of a linear inequality in two variables is a *region* in the *xy*-plane that is bounded by a dashed line if the inequality is *strict* and by a solid line if it is weak.

b. False; If $y - 2x \geq 1$, then $2x - y \leq -1$.

c. False; If $y - 3x \leq 2$, then $3x - y \geq -2$.

d. False; If $y > a$ and $x > b$, then $y + x > a + b$. **e.** True

f. False; We cannot conclude separate inequalities for x and y from the single joint inequality $y - x > a - b$.

g. True

h. False; $4x - 2y > 6$ is equivalent to $-2x + y < -3$.

3.

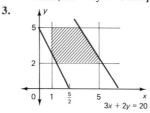

5. $Z_{max} = 19$ when $x = 1, y = 2$.

7. $Z_{max} = \frac{7}{3}$ when $x = \frac{7}{3}, y = 0$.
$Z_{min} = -\frac{12}{5}$ when $x = 0, y = \frac{12}{5}$.

9. $Z_{max} = 15$ when $x = 5, y = 0$.

11. $Z_{min} = 0$ when $x = y = 0$.

13. $Z_{max} = 14$ when $x = 0, y = 2, z = 2$.

15. $Z_{max} = 13$ when $x = 0, y = 7, z = 1$.
$Z_{min} = -3$ when $x = 0, y = 0, z = 3$.

17. $x = 0, y = 90$; min. cost is $1600.

19. Number of S is $\frac{300}{7}$, number of $T = \frac{1200}{7}$. max. weight is $\frac{3300}{7}$ lb.

CHAPTER 12

Exercises 12-1

1. 0.40

3. $g(x)$ is not defined over the whole interval from x to $x + \Delta x$.

5. -50 **7.** $\Delta x - 2\Delta x / x(x + \Delta x)$

9. -7 **11.** 1   **13.** 0.1613

15. $3a^2 + 3ah + h^2 + 1$

17. a. 40   **b.** 160   **c.** 220   **d.** 250
e. $1000 - 240t - 120\Delta t$

19. 10.1   **21. a.** $452.38   **b.** $201.06

23. a. $\Delta p = e^{-0.3} - e^{-0.6} = 0.1920$ **b.** $\Delta p / \Delta t = 0.064$

25. a. 20   **b.** -28 (descending)   **c.** $100 - 32t - 16\Delta t$

27. 7

Exercises 12-2

1. 25   **3.** 4   **5.** 0   **7.** 4   **9.** 1

11. $-\frac{1}{2}$ **13.** 0   **15.** -3 **17.** -6

19. $\frac{3}{2}$ **21.** $\frac{1}{192}$ **23.** No limit   **25.** $\frac{1}{4}$ **27.** $\frac{1}{8}$

29. 2   **31.** 2   **33.** 4   **35.** 2

37. 7   **39.** 0   **41.** $4x + 5$

43. a. 8 feet per second   **b.** -24 feet per second

47. Limit is 1

Exercises 12-3

1. 2   **3.** 0   **5.** $2x$ **7.** $2u + 1$

9. $-6x$ **11.** $-1/(x + 1)^2$ **13.** $-2/(2t + 3)^2$

15. a. $-4x$ **b.** 3   **17. a.** $1/2\sqrt{y}$ **b.** $-(y + 2)/y^3$

19. -2 **21.** 3   **23.** 7

25. Slope $= 12$; $y = 12x - 16$ **27.** Slope $= -\frac{1}{9}$; $y = \frac{2}{3} - \frac{x}{9}$

29. Slope $= -1$; $y = 3 - x$ **31.** Slope $= -2$; $y = 7 - 2x$

33. a. 2000   **b.** 400   **c.** -1200.

35. $m'(t) = -3 + t/2$; $m(0) = 9$, $m'(0) = -3$; $m(6) = 0$, $m'(6) = 0$; $-m'(t)$ is the rate at which compound A is being broken down.

Exercises 12-4

1. $5x^4$ **3.** $-3/t^4$ **5.** $-1/u^6$ **7.** $(-2/3)x^{-5/3}$

9. $12x^2 - 6x$ **11.** $12x^3 - 21x^2 + 10x$ **13.** $6u - 6/u^3$

15. $1.2x^{0.2} - 0.6x^{-1.6}$ **17.** $x^{-1/2} - x^{-3/2}$

19. $3\sqrt{x} - 3x^{-5/2}$ **21.** $3x^{1/2} + 5x^{1/4}$

23. $12x^3 + 8x - 4$ **25.** $4x - 23$ **27.** $4u + 3$

29. $27t^2 + 6t - 5$ **31.** $3x^2 + 12x + 12$

33. $-3(x^2 + 2x + 1)/x^4$ **35.** $-48/y^3$

37. $1 - 1/x^2$ **39.** $1/2\sqrt{t} - (9/2)t^{-5/2}$ **41.** 0

43. $1/\sqrt{2y} - 1/3y^2$ **45.** $6t^{-1/4} + (\frac{3}{32})t^{-7/4}$

47. $3x^2 - 3/x^4$

49. $3u^2 - 10u - 14/(3u^3)$ **53.** $y = 3 - x$

55. $y = -2 - x/2$ **57.** (2, 5)

59. a. $6t^2 - 1/2\sqrt{t}$ **b.** 95.75

61. a. 0.1   **b.** 0.3   **c.** 0.5

63. $p'(r) = -3\sqrt{r_0}/(8r^{3/2}) - r_0/4r^2$; $p'(2r_0) = -(3 + \sqrt{2})/(16\sqrt{2}r_0)$ **65.** $(k/2)\sqrt{ca/t}$

67. $m'(t) = 6 + 6t$; $m(2) = 26$; $m'(2) = 18$; $m(2)$ represents the mass at time 2, and $m'(2)$ represents the rate at which the mass is increasing at time $t = 2$.

69. $A(15) = 916.42\pi$; $A'(15) = 32.05\pi$; $A(30) = 1322.38\pi$; $A'(30) = 23.54\pi$

Exercises 12-5

1. 2   **3.** $0.0003x^2 - 0.18x + 20$ **5.** $1 - 0.02x$

7. $0.1 - (2 \times 10^{-3})x - (2.5 \times 10^{-5})x^{3/2}$

9. $R'(x) = 25 - 0.5x$ **11.** 10

13. $20 - 0.5x$ **15. a.** 1.23   **b.** 6.25

17. $x = 40$; $P(40) = 300$, $p = 15$ **19.** $R'(x) = 9 - 0.1x$; $p = \$4.50$

23. 0.54; 0.46

Exercises 12-6

1. a. 1   **b.** 2   **c.** Does not exist

3. a. 0   **b.** 0   **c.** 0   **5.** 0

7. Limit does not exist.

9. 1   **11.** $\frac{1}{6}$ **13.** $+\infty$ **15.** $+\infty$

17.

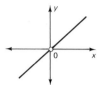

Discontinuous at $x = 0$

19.

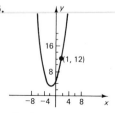

Discontinuous at $x = 0$

21.

Discontinuous at $x = 0$

23.

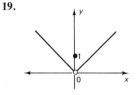

Continuous at $x = 1$

25.

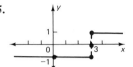

Discontinuous at $x = 3$

27.

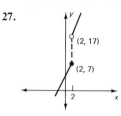

Discontinuous at $x = 2$

29. 2   **31.** 3, -2 **33.** No x **35.** 2, 3

37. $h = 0$ **39.** $x = 0$ **41.** $x = 1$

43. $f(x) = \begin{cases} 6x + 20 & \text{if } 0 \le x \le 35 \\ 9x - 85 & \text{if } 35 < x \le 45 \\ 12x - 220 & \text{if } x > 45 \end{cases}$

$f(x)$ is continuous for all $x \geq 0$ but is not differentiable at $x = 35, 45$.

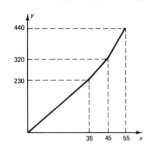

$$45.\ T_2 - T_1 = \begin{cases} -0.1x & \text{if } 0 < x \leq 2000 \\ 0.05x - 300 & \text{if } 2000 < x \leq 6000 \\ 0.1x - 600 & \text{if } x > 6000 \end{cases}$$

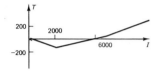

47.
$$f(x) = \begin{cases} 12 & \text{if } 0 < x \leq 1 \\ 24 & \text{if } 1 < x \leq 2 \\ 36 & \text{if } 2 < x \leq 3 \\ \vdots & \vdots \\ 96 & \text{if } 7 < x \leq 8 \end{cases}$$

$f(x)$ is discontinuous and nondifferentiable at $x = 0$, 1, 2, 3, 4, 5, 6, 7, 8.

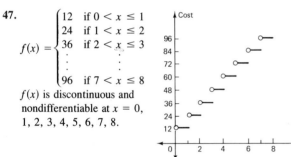

Review Exercises for Chapter 12

1. a. False; Increment in the independent variable can be positive or negative.
  b. True
  c. False; For example, $f(x) = (x^2 - 4)/(x - 2)$ is *not* defined at $x = 2$ but $\lim_{x \to 2} f(x)$ exists and is equal to 4.
  d. False; $x/x = 1$ only if $x \neq 0$.
  e. True **f.** True
  g. False; The derivative of y with respect to x represents the *instantaneous* rate of change of y with respect to x.
  h. False; $f(x)$ is continuous at $x = c$ if $\lim_{x \to c} f(x) = f(c)$.
  i. False; For example, $f(x) = |x|$ is continuous at $x = 0$ but is *not* differentiable at $x = 0$.
  j. True **k.** False; If $f(x) = |x|$, then $f'(0)$ does not exist.

3. 640; 32 **5.** 0 **7.** No limit **9.** 0 **11.** 0
13. 0 **15.** $1/2\sqrt{x}$ **17.** $-2(x + 1)^{-3}$
19. $\frac{3}{2}x^{1/2} + 3x^{-5/2}$
21. $(\frac{7}{2})x^{5/2}$ **23.** $(\frac{1}{6})x^{-5/6}$ **25.** $(\frac{5}{6})x^{-1/6}$
27. $9t^2 - 4t + 3$ **29.** $1 - 3/(2y^2) + 3/y^4$
31. $20x$ **33.** $R'(x) = 50 - 0.1x$ **35.** $40 - 0.1x$
37. -62
39. -4; as the price rises from $p = 2$ to $p = 3$, the demand falls by approximately 4 units
41. No **43.** $h = 2$
$$45.\ \bar{C}(x) = \begin{cases} 2 & \text{if } 0 < x \leq 20 \\ 1.75 + 5/x & \text{if } 20 < x \leq 40 \\ 1.5 + 15/x & \text{if } x > 40 \end{cases}$$
continuous for all $x > 0$ but not differentiable at $x = 20, 40$.

CHAPTER 13

Exercises 13-1

1. $4x^3 + 3x^2 + 3$
3. $11 - 42x$ **5.** $6x^2 - 14x - 13$ **7.** $9x^2 + 2x - 11$
9. $3y^2 - 2 + 15/y^2$ **11.** $24x^3 - 33x^2 + 18x - 11$
13. $R'(x) = 500 - x$ **15.** $R'(x) = 160,000 - 160x + 0.03x^2$
17. $6200 + 520t + 21t^2 + 0.4t^3$ **19.** $-6/(2x + 7)^2$
21. $1/(u + 1)^2$ **23.** $-3(x - 1)^{-2}$
25. $(t^2 - 10t + 35)(t - 5)^{-2}$
27. $-u^{-1/2}(\sqrt{u} - 1)^{-2}$ **29.** $-2x(x^2 + 1)^{-2}$
31. $(12x^3 + 3x^2 - 6x - 11)/(3x - 1)^2$
33. $(18u^6 + 20u^4 - 30u^2 + 112u)/(u^2 + 1)^2$
35. $y = 4x + 14$ **37.** $y = -x + 6$ **39.** $(0, 1)$
41. $(3, 0), (-9, 2)$ **43.** $\bar{C}'(x) = -a/x^2$
45. $(0.6 - 0.12t - 0.003t^2)(4 + 0.1t + 0.01t^2)^{-2}$
49. $8t^3 - 6t^2 + 14t - 3$ **51.** $1/[2\sqrt{t}(\sqrt{t} + 1)^2]$

Exercises 13-2

1. $21(3x + 5)^6$ **3.** $6x(2x^2 + 1)^{1/2}$
5. $-8x(x^2 + 1)^{-5}$ **7.** $t(t^2 + a^2)^{-1/2}$
9. $-t^2(t^3 + 1)^{-4/3}$ **11.** $10(t^2 + 1/t^2)^4(t - 1/t^3)$
13. $1.2x(x^2 + 1)^{-0.4}$ **15.** $(t^2 + 1/t^4)(t^3 - 1/t^3)^{-2/3}$
17. $(\frac{5}{3})(x^2 + 1)^{-5/6}$ **19.** $2(u^2 + 1)^2(7u^2 + 3u + 1)$
21. $(x + 1)^2(2x + 1)^3(14x + 11)$ **23.** $x^2(x^2 + 1)^6(17x^2 + 3)$
25. $4[(x + 1)(x + 2) + 3]^3(2x + 3)$
27. $-35(3x + 2)^6/(x - 1)^8$
29. $3(u^2 + 1)^2(u^2 + 2u - 1)/(u + 1)^4$
31. $(x^2 + 1)(3x^2 + 4x - 1)(x + 1)^{-2}$

33. $-(x^2 - 1)^{-3/2}$ **35.** $(t^3 + 8t)(t^2 + 4)^{-3/2}$ **37.** 28

39. $y = (4/5)x + (9/5)$ **41.** $y = 6x - 11$

43. $x/\sqrt{100 + x^2}$ **45.** $-100x^{-2}(x^2 + 100)^{-1/2}$

47. $(100 - 0.15x - (2 \times 10^{-4})x^2)(100 - 0.1x - 10^{-4}x^2)^{-1/2}$

49. $dC/dt = 100$ per month

51. \$220   **53.** 0.955 kilometer per hour

55. \$22   **57.** $5/[2(t + 2)^{3/2}\sqrt{3t + 1}]$

59. a. $dN/dt = -1/160$ **b.** 0.0051

Exercises 13-3

1. $7e^x$ **3.** $3e^{3x}$ **5.** $2xe^{x^2}$ **7.** $e^{\sqrt{x}}/2\sqrt{x}$

9. $(x + 1)e^x$ **11.** $(2x - x^2)e^{-x}$ **13.** $-xe^{-x}$

15. $(2x - 1)e^{x^2 - x}$

17. $(x + 1)e^x/(x + 2)^2$ **19.** $e^x/(e^x + 1)^2$ **21.** $\frac{1}{7}x$

23. 0   **25.** $1/(x \ln 7)$ **27.** $2x/(x^2 + 5)$ **29.** $5x^{-1}(\ln x)^4$

31. $-1/[x(\ln x)^2]$ **33.** $-1/[2x(\ln x)^{3/2}]$

35. $2(1 + \ln x)$ **37.** $2x^3(x^2 + 1)^{-1} + 2x \ln (x^2 + 1)$

39. $e^x(x^{-1} + \ln x)$ **41.** $(1 - \ln x)/x^2$

43. $[1 - \ln (x + 1)]/(x + 1)^2$ **45.** $2x \ln 3$

47. $3x^2 \log e$ **49.** $1/\ln 3$ **51.** $2x(x^2 + 1)^{-1} - (x + 1)^{-1}$

53. $[2(x + 1)]^{-1} - 2x(x^2 + 4)^{-1}$ **55.** $a^x \ln a$

57. $1/x \ln a$ **59.** 0   **61.** 0

63. $[(x + 1)^{-1} \ln x - x^{-1} \ln (x + 1)]/(\ln x)^2$

65. $x/(\ln 10) + 2x \log x$ **67.** e

69. $y = (-\frac{1}{2})x$ **71.** $y = 0$ **73.** $5 - (1 + 0.1x)e^{0.1x}$

75. $\ln (2 - 0.001x) - 0.001x/(2 - 0.001x)$

77. $C'(x) = 1 - 0.5e^{-0.5x}$;
$\overline{C}'(x) = -100x^{-2} - (1 + 0.5x)x^{-2}e^{-0.5x}$

79. $10e^{-0.005A} - 1$ **81.** $C = c(0) - c_s$

83. -4.343×10^3; -4.343×10^6; -4.343×10^9

85. $dy/dt = 3Cky_m e^{-kt}(1 - Ce^{-kt})^2$

Exercises 13-4

1. $dy/dx = 15x^4 + 21x^2 - 8x$; $d^2y/dx^2 = 60x^3 + 42x - 8$;
$d^3y/dx^3 = 180x^2 + 42$; $d^4y/dx^4 = 360x$; $d^5y/dx^5 = 360$;
$d^n y/dx^n = 0$ for $n \geq 6$.

3. $f'(x) = 3x^2 - 12x + 9$; $f''(x) = 6x - 12$; $f'''(x) = 6$;
$f^{(n)}(x) = 0$ for $n \geq 4$.

5. $y'' = 2(1 - 3x^2)/(1 + x^2)^3$ **7.** $g^{(4)}(u) = 1944(3u + 1)^{-5}$

9. $2(3x^2 - 1)(x^2 + 1)^{-3}$ **11.** $2x^{-3}$ **13.** $(x + 4)e^x$

15. $-(x + 1)^{-2} - (x + 2)^{-2}$ **17.** $(x - 1)e^{-x}$

21. a. vel. $= 9 + 32t$; acc. $= 32$
b. vel. $= 9t^2 + 14t - 5$; acc. $= 18t + 14$

23. $C'(x) = 30 - 0.2x + 0.006x^2$; $C''(x) = -0.2 + 0.012x$

27. $-Ck^2 y_m e^{-kt}(1 - Ce^{-kt})(1 + Ce^{-kt})^{-3}$

Review Exercises for Chapter 13

1. a. False; $(d/dx)(uv) = uv' + vu'$.
b. True; $(u/v)' = (u \cdot 1/v)' = u'(1/v) + u(1/v)'$ (product rule).
c. False; $(d/dx)[u(x)]^n = n[u(x)]^{n-1}u'(x)$.
d. True   **e.** True
f. False; The second derivative of any quadratic function is a *constant*.
g. False; If accleration is zero, then velocity is constant, not necessarily zero.
h. False; $(d^2/dx^2)[u(x)]^n = nu^{n-2}[(n - 1)(u')^2 + uu'']$.
i. False; $(d/dx)(e^{x^2}) = 2xe^{x^2}$.
j. False; $(d/dx) \ln (x^2 + 1) = 2x/(x^2 + 1)$.
k. False; $(d/dx)(\ln 2) = 0$, because $\ln 2$ is a constant.
l. False; $(d/dx)(e^x) = e^x$.
m. False; $(d/dx)(1/x^3) = (d/dx)(x^{-3}) = -3x^{-4}$.

3. $3x(1 + 2 \ln x)$ **5.** $x^2(3 - x)e^{-x}$

7. $e^x/(x + 3) + e^x \ln (x + 3)$

9. $6(2x + 1)^2(3x - 1)^3(7x + 1)$

11. $(x^2 + 1)^2(2x^2 - 6x - 4)(x - 1)^{-5}$

13. $(2x^2 + 4)(x^2 + 4)^{-1/2}$ **15.** $(\frac{1}{6})(x + 1)^{-5/6}$

17. $x^{\sqrt{2} - 1}(1 + \sqrt{2} \ln x)$

19. $\ln 2 - (3/2)x^2 (x^3 + 1)^{-1}$ **21.** $1/x + 2x - 3/\{2(3x + 5)\}$

23. $x^x(1 + \ln x)$ **25.** $y = x$ **27.** $y = (3/8)x + (1/4)$

29. $30(3x - 7)^4(x + 1)^2(27x^2 - 18x - 5)$

31. $-(1 + \ln x)(x \ln x)^{-2}$ **33.** $R'(x) = (1 - x/b)e^{(a-x)/b}$

35. $C'(x) = 0.5 + 0.01(x + 1)e^x$; $\overline{C}'(x) = -100/x^2 + 0.01e^x$

37. $-e^{50-x}$

39. $-4/3$; if the price rises from \$40 to \$41, the demand falls approximately by $\frac{4}{3}$ units.

41. $R'(x) = 15(20 - x)e^{-x/20}$; $P'(x) = 15(20 - x)e^{-x/20} - 20$

43. $(9t + 7)/2\sqrt{t + 1}$ **45.** $t = p$

CHAPTER 14

Exercises 14-1

1. a. $x > 3$ **b.** $x < 3$

3. a. $x > 1$ or $x < -1$ **b.** $-1 < x < 1$

5. **a.** $x > 1$ or $x < -1$ **b.** $-1 < x < 0, 0 < x < 1$

7. **a.** All $x \neq -1$ **b.** No x

9. **a.** $x > 0$ **b.** No x

11. **a.** $x > 1/e$ **b.** $0 < x < 1/e$

13. **a.** $x < 0$ or $x > 4$ **b.** $0 < x < 4$

15. **a.** $x > 2$ **b.** $x < 2$

17. **a.** $x > 1$ **b.** $x < 1$

19. **a.** $x > 0$ **b.** $x < 0$

21. **a.** $x > 0$ **b.** No x

23. **a.** No x **b.** All $x \neq 0$

25. **a.** Increasing for all $x > 0$.

   b. Increasing for $0 < x < 100$; decreasing for $x > 100$.

   c. Increasing for $0 < x < 90$; decreasing for $x > 90$.

27. **a.** Always increasing.

   b. Increasing for $0 < x < a/2b$; decreasing for $x > a/2b$.

   c. Increasing for $0 < x < (a - k)/2b$; decreasing for $x > (a - k)/2b$.

29. **a.** $x > \frac{1}{2}$ **b.** $x < \frac{1}{2}$

31. **a.** No x **b.** All $x > 0$

Exercises 14-2

1. $\frac{3}{2}$ 3. $1, -1$ 5. $0, 1, -1$ 7. $0, 1, \frac{2}{5}$ 9. $\frac{1}{3}, -\frac{1}{3}$

11. $0, 2$ 13. $0, \pm 1$ 15. $1, \frac{3}{2}$ 17. $e^{-1/3}$ 19. 3

21. Minimum at $x = 6$.

23. Minimum at $x = 4$; maximum at $x = 0$.

25. Maximum at $x = 1$; minimum at $x = 2$.

27. Maximum at $x = 4$; minimum at $x = 8$.

29. Maximum at $x = 1$; minimum at $x = 3$;
no extrema at $x = 0$.

31. Maximum at $x = \frac{3}{5}$; minimum at $x = 1$;
no extrema at $x = 0$.

33. Minimum at $x = 0$.

35. Minimum at $x = 1/e$.

37. Maximum value 5 when $x = -2$;
minimum value -22 when $x = 1$.

39. Minimum value $-1/e$ when $x = -1$.

41. Minimum value 0 when $x = 1$;
maximum value $(\frac{9}{11})^3(\frac{2}{11})^{2/3}$ when $x = \frac{9}{11}$;
no extrema at $x = 0$.

43. Maximum value $1/2e$ when $x = \sqrt{e}$.

45. Minimum value 0 when $x = 1$.

47. Maximum value $\frac{1}{4}$ when $x = \frac{3}{2}$;
minimum value 0 when $x = 1, 2$.

49. Minimum value 1 when $x = 0$; no local maximum

51. Minimum value 0 when $x = 2$

Exercises 14-3

1. **a.** All x **b.** No x; no point of inflection

3. **a.** $x > 0$ **b.** $x < 0$; $x = 0$ is the point of inflection.

5. **a.** $x < -\sqrt{3}$ or $x > \sqrt{3}$

   b. $-\sqrt{3} < x < \sqrt{3}$; $x = \sqrt{3}$ and $x = -\sqrt{3}$ are the points of inflection.

7. **a.** $x > 0$

   b. $x < 0$; no point of inflection

9. **a.** No x

   b. $x > 5$; no point of inflection.

11. **a.** $x > 4$

   b. $x < 4$; $x = 4$ is the point of inflection.

13. **a.** $x > \frac{5}{2}$ **b.** $x < \frac{5}{2}$ **c.** All x

   d. No x; no point of inflection

15. **a.** $x > 3$ **b.** $x < 0$ and $0 < x < 3$

   c. $x < 0$ and $x > 2$

   d. $0 < x < 2$; $x = 0$ and $x = 2$ are the points of inflection.

17. **a.** $x > 1$ **b.** $0 < x < 1$ **c.** $x > 0$

   d. No x (Note that y is *not* defined for $x < 0$); no point of inflection

19. **a.** $x > 3$ **b.** $0 < x < 3$ **c.** $x > 0$

   d. No x; no point of inflection

21. Not concave up or down

23. Concave up for $x > \frac{25}{3}$ and concave down for $0 < x < \frac{25}{3}$

25. Minimum value $= -22$ at $x = 5$;
no maximum

27. Maximum value $= 51$ at $x = -2$;
minimum value $= -74$ at $x = 3$

29. Maximum value $= 3$ at $x = 0$;
minimum value $= 0$ at $x = 1$ and -125 at $x = -4$

31. Minimum value $= -1/e$ at $x = -1$;
no maximum

33. Maximum value $= 1$ at $x = 0$;
no minimum

35. Minimum value $= (1 + \ln 2)/2$ at $x = 1/\sqrt{2}$;
no maximum

37. Minimum value $= -1/e$ at $x = 1/e$;
no maximum

39. Maximum value $= 3^3 \cdot 4^4/7^7$ at $x = 10/7$;
minimum value $= 0$ at $x = 2$ (no extrema at $x = 1$)

41. Minimum value = 0 at $x = 1$; no maximum

Exercises 14-4

1.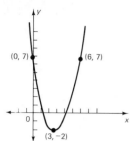

3.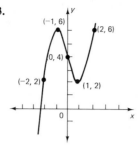

5.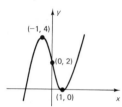

7.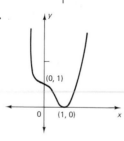

9.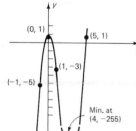

11.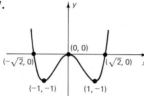

13.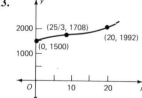

Exercises 14-5

1. 5; 5   **3.** 50; 25

7. Maximum area a^2 occurs when the length of rectangle is twice the width $(a/\sqrt{2})$.

9. $30\sqrt{2}$ yd $\times$ $15\sqrt{2}$ yd   **11.** 6 ft $\times$ 6 ft $\times$ 9 ft

13. Minimum value of $\overline{C}$ is 41 when $x = 2$.   **15.** $x = 2000$

17. a. 10   **b.** 15

19. $x = 3; p = 15/e$

21. a. $P(x) = 2.7x - 0.001x^2 - 50$ **b.** 1350
   c. \$1772.50

23. $x = 2500$; \$2250   **25.** $x = 2500$; $x = 2000$; maximum profit is \$1200.

27. $x = 50$

29. b. 2000   **31.** 1000   **33.** 10,000 m²

35. $n = 6.9$ (Since n must be an integer, $n = 7$ gives smaller loss than $n = 6$.)

37. $x = 2$ cm; $y = 12$ cm   **39.** $x = 50$

41. $x = 200(4 - t)/3$; $t = 2$ **43.** $n = \sqrt{a/c}$

45. $(r^{-1} - 5)$ years   **47.** Radius = 9.27 ft; height = 37.07 ft

49. R_{max} when $x = 350$; P_{max} when $x = 225$; $x = 200$

51. $x = 50 - 5\sqrt{2} \approx 42.9$ km

53. First drug

Exercises 14-6

1. Absolute maximum 7 at $x = 6$; absolute minimum -2 at $x = 3$.

3. Absolute maximum 75 at $x = -1$; absolute minimum -249 at $x = 5$.

5. Absolute maximum 56 at $x = 2$; absolute minimum -79 at $x = -1$.

7. Absolute maximum -3 at $x = 2$; absolute minimum -33 at $x = \frac{1}{2}$.

9. Absolute maximum e^{-1} at $x = 1$; absolute minimum $2e^{-2}$ at $x = 2$.

11. Absolute maximum $(e - 1)$ at $x = e$; absolute minimum 1 at $x = 1$.

13. Absolute maximum 0 at $x = 0$; absolute minimum $-1/e$ at $x = 1/e$.

15. $y_{max} = 6000$ at $t = 0$; $y_{min} = 2000$ at $t = 20$.

17. 3000; 2000   **19. a.** 45   **b.** 40

21. Base = 4.64 ft; height = 2.32 ft   **23.** 6 times

25. Minimum rate $-0.25a$ at $t = \ln 2$; maximum rate $-0.168a$ at $t = 2$

Exercises 14-7

1. 1   **3.** $\frac{1}{2}$ **5.** $-\frac{2}{3}$ **7.** 1

9. $-\frac{2}{3}$ **11.** 0   **13.** Does not exist

15. Does not exist   **17.** 0   **19.** No limit

21. 1   **23.** 1   **25.** 1

27. $\frac{2}{3}$ **29.** $\frac{3}{2}$ **31.** 5   **33.** $-\frac{2}{3}$ **35.** 0

37. a. $+\infty$ **b.** $-\infty$

39. a. $-\infty$ **b.** $+\infty$ **41. a.** $+\infty$ **b.** $+\infty$

43. a. $-\infty$ **b.** $+\infty$

45. a. $+\infty$ **b.** Does not exist

47.

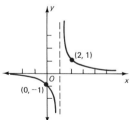

Horizontal asymptote: $y = 0$;
Vertical asymptote: $x = 1$

49.

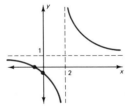

Horizontal asymptote: $y = 1$;
Vertical asymptote: $x = 2$

51.

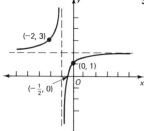

Horizontal asymptote: $y = 2$;
Vertical asymptote: $x = -1$

53.

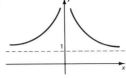

Horizontal asymptote: $y = 1$;
Vertical asymptote: $x = 0$

55.

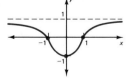

Horizontal Asymptote: $y = 1$;
No vertical asymptote

57.

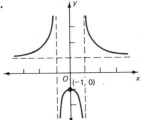

Horizontal asymptote: $y = 1$;
Vertical asymptote: $x = \pm 1$

59.

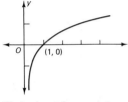

No horizontal asymptote;
Vertical asymptote: $x = 0$

61.

No horizontal asymptote;
Vertical asymptote: $x = 0$

63.

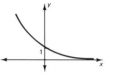

Horizontal asymptote: $y = 0$;
No vertical asymptote

65.

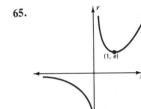

Horizontal asymptote: $y = 0$;
Vertical asymptote: $x = 0$

67.

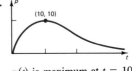

$p(t)$ is maximum at $t = 10$

69.

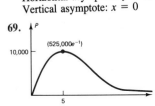

71.

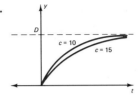

Horizontal asymptote: $y = D$;
No vertical asymptote

73.

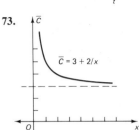

75.

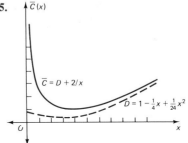

77.

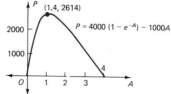

Review Exercises for Chapter 14

1. a. False; $f(x)$ is increasing for values of x at which $f'(x) > 0$ and is decreasing for values of x at which $f'(x) < 0$.

b. False; For example $f(x) = x^3$ is increasing for all x, but $f'(0) = 0$.

c. True

d. True

e. False; At a point of inflection, $f''(x)$ may be undefined.

f. True

g. False; if $f(x)$ has an extremum at $x = c$, then either $f'(c) = 0$ or $f'(c)$ fails to exist.

h. False; if $f'(c) = 0$, then $f(x)$ could have a point of inflection at $x = c$.

i. False; for example $f(x) = |x|$ has a local minimum at $x = 0$, but the tangent cannot be drawn at this point on the graph.

j. False; the tangent at the point of inflection need not be horizontal.

k. False; a local maximum value of a function can be *less* than a local minimum value. This is the case for $f(x) = x + 1/x$ for example.

l. True

m. False; a cubic function may have two or no local extrema.

n. False; maximum revenue need not lead to a maximum profit.   **o.** True

p. False; above a certain level, the cost of further advertising outweighs the extra revenue it generates.

q. True

r. False; for example, $\displaystyle\lim_{x \to +\infty} \frac{x + 2}{\sqrt{x^2 + 4}} = 1$, whereas $\displaystyle\lim_{x \to -\infty} \frac{x + 2}{\sqrt{x^2 + 4}} = -1$.

s. False; $\displaystyle\lim_{x \to -\infty} \frac{1}{\sqrt{x}}$ does not exist, because $\sqrt{x}$ is not defined for *negative* values of x.   **t.** True; A vertical line cannot cross the graph of a function more than once.

3. a. $x < 2$ or $x > 4$

b. $2 < x < 4$

c. $x > 3$

d. $x < 3$

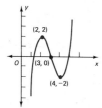

5. a. $-2 < x < 0$ or $x > 2$

b. $x < -2$ or $0 < x < 2$

c. $x < -\sqrt{\frac{12}{5}}$ or $x > \sqrt{\frac{12}{5}}$

d. $-\sqrt{\frac{12}{5}} < x < \sqrt{\frac{12}{5}}$

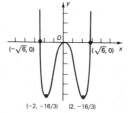

7. a. $x > -1$

b. $x < -1$

c. $x > -2$

d. $x < -2$

9. $x > 60$

11. a. increasing

b. $t = (100/7) \ln [215Y/(375 - Y)]$

13. Local max. at $t = 0$, local min. at $t = 1$.

15. Local min. at $x = \frac{1}{2}$.

17. Local max. at $x = \frac{1}{4}$; local min. at $x = 0$, $x = 1$.

19. a. $k = \frac{3}{2}$ **b.** $k = -48$ **c.** $k = -32$

21. $A > 0$

23. Absolute min. value 0 at $x = 0$, 2 and absolute max. $8\sqrt[3]{2}$ at $x = -2$.

25. $x = 100$; $P_{\max} = 50$; $P(120) = 44$ **27.** $x = 6$

31. b. $Q = 2000$; $T_{\min} = \$10,250$ **c.** $\$10,256.25$

33. 2341 pairs of men's shoes; 3746 pairs of women's shoes.

35. a. $x = (4 - t)/10$; $p = (104 + 4t)/10$

b. $(4 - t)^2/20$ **c.** $t = 2$

37. a. $F = 10$ **b.** 49   **39.** 9 months

41.

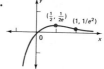

$p(0) = 100$; $p(2) = 63.03$; $\displaystyle\lim_{t \to \infty} p(t) = 20$

43. $A =$ probability of success in the long run; $p'(0) = A/B$

45. $n = 70$

47. a. dy/dt represents the rate at which the proportion of people infected is increasing.

b. $t = 2$

c. Increasing for $0 < t < 2$ and decreasing for $t > 2$

49. 9 ft $\times$ 9 ft $\times$ 6 ft

51.

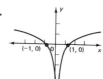

53.

55.

No horizontal asymptote;
Vertical asymptote: $x = 0$

39. a. $-\infty$ **b.** $+\infty$ **41. a.** $+\infty$ **b.** $+\infty$

43. a. $-\infty$ **b.** $+\infty$

45. a. $+\infty$ **b.** Does not exist

47.

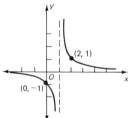

Horizontal asymptote: $y = 0$;
Vertical asymptote: $x = 1$

49.

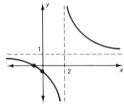

Horizontal asymptote: $y = 1$;
Vertical asymptote: $x = 2$

51.

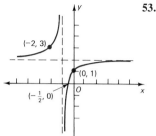

Horizontal asymptote: $y = 2$;
Vertical asymptote: $x = -1$

53.

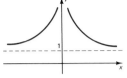

Horizontal asymptote: $y = 1$;
Vertical asymptote: $x = 0$

55.

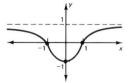

Horizontal Asymptote: $y = 1$;
No vertical asymptote

57.

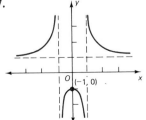

Horizontal asymptote: $y = 1$;
Vertical asymptote: $x = \pm 1$

59.

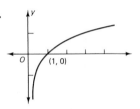

No horizontal asymptote;
Vertical asymptote: $x = 0$

61.

No horizontal asymptote;
Vertical asymptote: $x = 0$

63.

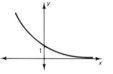

Horizontal asymptote: $y = 0$;
No vertical asymptote

65.

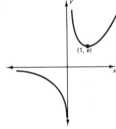

Horizontal asymptote: $y = 0$;
Vertical asymptote: $x = 0$

67.

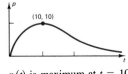

$p(t)$ is maximum at $t = 10$

69.

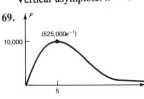

71.

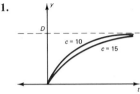

Horizontal asymptote: $y = D$;
No vertical asymptote

73.

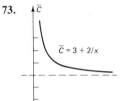

75.

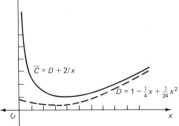

77.

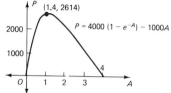

Review Exercises for Chapter 14

1. a. False; $f(x)$ is increasing for values of x at which $f'(x) > 0$ and is decreasing for values of x at which $f'(x) < 0$.

b. False; For example $f(x) = x^3$ is increasing for all x, but $f'(0) = 0$.

c. True

d. True

e. False; At a point of inflection, $f''(x)$ may be undefined.

f. True

g. False; if $f(x)$ has an extremum at $x = c$, then either $f'(c) = 0$ or $f'(c)$ fails to exist.

h. False; if $f'(c) = 0$, then $f(x)$ could have a point of inflection at $x = c$.

i. False; for example $f(x) = |x|$ has a local minimum at $x = 0$, but the tangent cannot be drawn at this point on the graph.

j. False; the tangent at the point of inflection need not be horizontal.

k. False; a local maximum value of a function can be *less* than a local minimum value. This is the case for $f(x) = x + 1/x$ for example.

l. True

m. False; a cubic function may have two or no local extrema.

n. False; maximum revenue need not lead to a maximum profit.   **o.** True

p. False; above a certain level, the cost of further advertising outweighs the extra revenue it generates.

q. True

r. False; for example, $\lim\limits_{x \to +\infty} \dfrac{x + 2}{\sqrt{x^2 + 4}} = 1$, whereas $\lim\limits_{x \to -\infty} \dfrac{x + 2}{\sqrt{x^2 + 4}} = -1$.

s. False; $\lim\limits_{x \to -\infty} \dfrac{1}{\sqrt{x}}$ does not exist, because $\sqrt{x}$ is not defined for *negative* values of x.   **t.** True; A vertical line cannot cross the graph of a function more than once.

3. a. $x < 2$ or $x > 4$
b. $2 < x < 4$
c. $x > 3$
d. $x < 3$

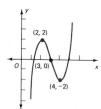

5. a. $-2 < x < 0$ or $x > 2$
b. $x < -2$ or $0 < x < 2$
c. $x < -\sqrt{\frac{12}{5}}$ or $x > \sqrt{\frac{12}{5}}$
d. $-\sqrt{\frac{12}{5}} < x < \sqrt{\frac{12}{5}}$

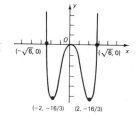

7. a. $x > -1$
b. $x < -1$
c. $x > -2$
d. $x < -2$

9. $x > 60$

11. a. increasing
b. $t = (100/7) \ln [215Y/(375 - Y)]$

13. Local max. at $t = 0$, local min. at $t = 1$.

15. Local min. at $x = \frac{1}{2}$.

17. Local max. at $x = \frac{1}{4}$; local min. at $x = 0$, $x = 1$.

19. a. $k = \frac{3}{2}$ **b.** $k = -48$ **c.** $k = -32$

21. $A > 0$

23. Absolute min. value 0 at $x = 0$, 2 and absolute max. $8\sqrt[3]{2}$ at $x = -2$.

25. $x = 100$; $P_{\max} = 50$; $P(120) = 44$ **27.** $x = 6$

31. b. $Q = 2000$; $T_{\min} = \$10,250$ **c.** $\$10,256.25$

33. 2341 pairs of men's shoes; 3746 pairs of women's shoes.

35. a. $x = (4 - t)/10$; $p = (104 + 4t)/10$
b. $(4 - t)^2/20$ **c.** $t = 2$

37. a. $F = 10$ **b.** 49   **39.** 9 months

41.

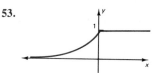

$p(0) = 100$; $p(2) = 63.03$; $\lim\limits_{t \to \infty} p(t) = 20$

43. $A = $ probability of success in the long run; $p'(0) = A/B$

45. $n = 70$

47. a. dy/dt represents the rate at which the proportion of people infected is increasing.
b. $t = 2$
c. Increasing for $0 < t < 2$ and decreasing for $t > 2$

49. 9 ft $\times$ 9 ft $\times$ 6 ft

51.

53.

55.

No horizontal asymptote;
Vertical asymptote: $x = 0$

CHAPTER 15

Exercises 15-1

1. $(2x + 7)\,dx$ **3.** $(1 + \ln t)\,dt$

5. $2z(z^2 + 1)^{-1}\,dz$ **7.** $ue^u(u + 1)^{-2}\,du$

9. $[(2x - 3)/2\sqrt{x^2 - 3x}]\,dx$ **11.** $2\,dx$

13. 0   **15.** 0.12   **17.** 0.003

19. $dy = 0.12,\ \Delta y = 0.1203$

21. $dy = 0.02,\ \Delta y = \ln(1.02) = 0.0198$

23. 2.0833   **25.** 1.9875

27. $\pm 0.512\pi$ cm^3 **29.** $\frac{5}{3}\%$

31. $p = 1.96$

33. $x = 1000;\ C = 1200$; error in $x = -50$; error in $C = 10$

35. $C(x) = 5000 + 20(x - 200) = 20x + 1000$

37. $P(x) = 15x + 1250$

39. $R(x) = -30x + 8000$

Exercises 15-2

1. $-x/(y + 1)$ **3.** $-x^2/y^2$

5. $(4x - y)/(x + 2y)$ **7.** $-y(2x + y)/x(2y + x)$

9. $(x^4 - y)/(x - y^4)$ **11.** $-y/(e^y + x)$

13. $-y/x$ **15.** $-5t/3x$

17. $(x - 3y^2)/(3x^2 - y)$ **19.** $3y = x + 5$

21. $y = -x/2$

23. a. $(1, 5);\ (1, -1)$

 b. $(4, 2);\ (-2, 2)$

25. a. $(2, 4);\ (-2, -4)$

 b. $(4, 2)$ and $(-4, -2)$

27. 0   **29.** $-\frac{16}{27}$

31. $x[(y + 1)^2 + (x + 1)^2]/y^2(x + 1)^3$ **33.** $[-y/(x + 2y)]\,dx$

35. $[y(z - 1)/z(1 - y)]\,dz$ **37.** $-\dfrac{\sqrt{100 - 9x^2}}{9x}$

39. $-2(2e^x + 7)/(4pe^x - 3e^{x/2})$

41. $p = 3 + \sqrt{9 - P}\ (p \ge 3);\ p = 3 - \sqrt{9 - P}\ (0 \le p \le 3)$

43. $-[x(t^2 + 1) - h + tk]/[y(t^2 + 1) - (k + th)]$

Exercises 15-3

1. $(x^2 + 1)(x - 1)^{1/2}[2x/(x^2 + 1) + 1/2(x - 1)]$

3. $(x^2 - 2)(2x^2 + 1)(x - 3)^2[2x/(x^2 - 2) + 4x/(2x^2 + 1)$
$+ 2/(x - 3)]$

5. $(x^2 + 1)^{1/3}(x^2 + 2)^{-1}[2x/3(x^2 + 1) - 2x/(x^2 + 2)]$

7. $[(2x^2 + 5)/(2x + 5)]^{1/3}[4x/3(2x^2 + 5) - 2/3(2x + 5)]$

9. $x^x[x + 2x \ln x]$ **11.** $e^{(x + e^x)}$

13. $x^{\ln x}(2 \ln x)/x$ **15.** $x^x(1 + \ln x) + x^{-2 + 1/x}(1 - \ln x)$

17. $-n$ **19.** $\sqrt{p}/(2\sqrt{p} - 8)$

21. a. $-\frac{1}{3}$ **b.** -1 **c.** -3

23. a. $3 < p < 6$ **b.** $0 < p < 3$

25. a. $\frac{16}{9} < p < 4$ **b.** $0 < p < \frac{16}{9}$

27. $\eta = -\frac{36}{17};\ -18\%$

33. a. Revenue decreases

 b. Revenue increases

Review Exercises for Chapter 15

1. a. False; The differential of x^2 is $2x\,dx$.

 b. True   **c.** True (provided $dx \ne 0$).

 d. False; In $f(x, y) = 0$, one of the variables is independent and the other is a dependent variable.

 e. False; $(dy/dx)(dx/dy) = 1$ provided both derivatives are defined.

 f. False; If y is an increasing function of x, then x is also an increasing function of y.   **g.** True

 h. False; The logarithmic derivative of x with respect to y is $(1/x)(dx/dy)$ and the logarithmic derivative of y with respect to x is $(1/y)(dy/dx)$. If the statement is to be true, then we must have $(1/x)\,dx/dy = [(1/y)(dy/dx)]^{-1}$, that is, $(dx/dy)(dy/dx) = yx$, or $xy = 1$, which is not, in general, true.

 i. False; The logarithmic derivative of x^n is n/x.

 j. True   **k.** True   **l.** True   **m.** True

3. $-\frac{1}{4}$ **5.** 34

7. $-[1 + x(2t + x)(x + t)][1 + t(2x + t)(x + t)]^{-1}$

9. dy is undefined, because at $x = -1,\ y = 0,\ dy/dx$ does not exist.

11. $-2/e$

13. $[(x^4 - 4)/x(x + 1)^2]^{1/3}[4x^3/3(x^4 - 4) - 1/3x - 2/3(x + 1)]$

15. 0   **17. a.** $p = 6$ **b.** $p = 8$ **c.** $p = 3$

19. a. $-9/16$ **b.** $-16/9$

21. \$50   **25.** $(\ln x - 1)\,dx/(\ln x)^2$ or $(y/x - y^2/x^2)\,dx$

CHAPTER 16

Exercises 16-1

1. $x^8/8 + C$ **3.** $-1/(2x^2) + C$

5. $7x^2/2 + C$ **7.** $e^3 \ln|x| + C$

9. $\ln|x|/\ln 2 + C$ **11.** $e \ln|x| + x^2/2e + C$

13. $(e^2 - 2^e)e^x + C$ **15.** $-(\ln 2)/x + C$

17. $x^8/8 + 7x^2/2 + 7 \ln|x| + 7x + C$

19. $7x^3/3 - 3x^2/2 + 8x + \ln|x| - 2/x + C$

21. $x^3/3 + 5x^2/2 + 6x + C$　　**23.** $x^3 + x^2/2 - 2x + C$

25. $x^3/3 + 2x^2 + 4x + C$　　**27.** $x^3/3 + 2x - 1/x + C$

29. $4x^3/3 - 12x - 9/x + C$　　**31.** $x^4/4 + (4x^{5/2})/5 + C$

33. $x^6/6 + 3x^5/5 + x^4/2 + C$

35. $x^3 + 3x^2 - x - 2\ln|x| + C$

37. $3x/2 + C$　　**39.** $2x + C$

41. $e^x/\ln 2 + C$　　**43.** $x^3/3 + x + C$

45. $(1/2)x^2 + 4x^{3/2} + 9x + C$　　**47.** $x^3 - 6x + C$

49. $x^3/3 + x^2/2 + C$　　**51.** $(4/3)x^{3/2} - 6x + C$

53. $x^4 + x^3 + x^2 + x + \ln|x| - x^{-2}/2 + C$

55. $\frac{2}{7}u^{7/2} + \frac{6}{5}u^{5/2} + \frac{14}{3}u^{3/2} + C$

57. $\frac{4}{7}x^{7/2} + \frac{2}{5}x^{5/2} - \frac{2}{3}x^{3/2} + C$

59. $-1/x + 3\ln|x| + 7x - x^2 + C$

61. $\theta^3 - 3\theta^2 + 9\ln|\theta| + 4e^\theta + C$

63. $f(x) = \dfrac{2x^3}{3} + \dfrac{x^2}{2} - 6x + 7$

65. $s = t^3/3 + (4/5)t^{5/2} + t^2/2$

67. a. $C(x) = 2000 + 30x + 0.025x^2$　　**b.** $7062.50

　　c. 500 units

69. $1320; $500

71. a. $R(x) = 4x - 0.005x^2$　　**b.** $p = 4 - 0.005x$

73. $P(x) = 5x - 0.001x^2 - 180$

75. 1680; 750

Exercises 16-2

1. $\frac{1}{16}(2x + 1)^8 + C$　　**3.** $\frac{1}{5}(2 - 5t)^{-1} + C$

5. $\frac{1}{2}\ln|2y - 1| + C$　　**7.** $\frac{1}{2}\ln|2u + 1| + C$

9. $\frac{1}{3}e^{3x+2} + C$　　**11.** $-e^{5-x} + C$

13. $\frac{1}{3}e^{3x+2} + C$　　**15.** $\frac{1}{5}(x^2 + 7x + 3)^5 + C$

17. $-\frac{1}{2}(x^2 + 3x + 1)^{-2} + C$　　**19.** $\frac{1}{3}(x^2 + 1)^{3/2} + C$

21. $\frac{1}{2}\ln(x^2 + 1) + C$　　**23.** $\frac{1}{2}(t^3 + 8)^{2/3} + C$

25. $\frac{1}{3}(\sqrt{x} + 7)^6 + C$　　**27.** $\frac{1}{9}(2 + x\sqrt{x})^6 + C$

29. $\frac{1}{2}e^{t^2} + C$

31. $(1/n)e^{x^n} + C$　　**33.** $-\frac{1}{3}e^{3/x} + C$

35. $-\frac{2}{3}e^{-x\sqrt{x}} + C$　　**37.** $\frac{1}{3}e^{x^3} + C$

39. $e^{x^2-x} + C$　　**41.** $-1/(e^x + 1) + C$

43. $-\frac{1}{3}\ln|3 - e^{3x}| + C$

45. $\ln|e^x - e^{-x}| + C$　　**47.** $\frac{1}{2}(\ln x)^2 + C$

49. $\frac{2}{3}(\ln x)^{3/2} + C$　　**51.** $\frac{1}{4}(\ln x)^4 + C$

53. $\ln|1 + \ln x| + C$　　**55.** $\ln|t^3 + t| + C$

57. $\frac{1}{3}(x^2 + 4x + 1)^{3/2} + C$　　**59.** $(\ln 2)\ln|x| + (1/2)(\ln x)^2 + C$

61. $t^2/2 + t + \ln|t - 1| + C$

63. $(\frac{2}{5})(x + 1)^{5/2} - (2/3)(x + 1)^{3/2} + C$

65. $g(x) = 1 + \sqrt{x^2 + 1}$　　**67.** $(\frac{1}{3})f(3x) + C$

69. $2f(\sqrt{x}) + C$　　**71.** $f(\ln x) + C$

73. $C(x) = \dfrac{1}{3000}(x^2 + 2500)^{3/2} + 175/3$

75. a. 4.92　　**b.** 5.52

77. $P(t) = 60,000 - 2,400,000/\sqrt{t + 1600}$; 60,000 barrels

79. $t = 3; 5\ln 2$

Exercises 16-3

1. $(1/\sqrt{5})\ln|(2x - 3 - \sqrt{5})/(2x - 3 + \sqrt{5})| + C$

　　(Formula 66)

3. $\frac{1}{4}[\ln|2x - 3| - 3/(2x - 3)] + C$　(Formula 9)

5. $2\sqrt{3x + 1} + \ln|(\sqrt{3x + 1} - 1)/(\sqrt{3x + 1} + 1)| + C$

　　(Formula 22, 24)

7. $-\frac{1}{4}\ln|(4 + \sqrt{t^2 + 16})/t| + C$　(Formula 46)

9. $\frac{1}{2}y\sqrt{y^2 - 9} + \frac{9}{2}\ln|y + \sqrt{y^2 - 9}| + C$　(Formula 45)

11. $\frac{1}{2}\ln|(\sqrt{3x + 4} - 2)/(\sqrt{3x + 4} + 2)| + C$　(Formula 22)

13. $1/3(2x + 3) - \frac{1}{9}\ln|(2x + 3)/x| + C$　(Formula 14)

15. $(x^3/6 - x/8)(x^2 - 1)^{3/2} + \frac{1}{16}x(x^2 - 1)^{1/2}$

　　$+ \frac{1}{16}\ln|x + (x^2 - 1)^{1/2}| + C$

　　(Formulas 58, 64, 65)

17. $\frac{1}{8}(4x^3 - 6x^2 + 6x - 3)e^{2x} + C$　(Formulas 70, 69)

19. $\frac{1}{4}\sqrt{2x + 3}\sqrt{4x - 1} +$

　　$(7/4\sqrt{2})\ln|\sqrt{4x - 1} + \sqrt{4x + 6}| + C$　(Formula 81)

21. $\ln|(1 - e^x)/(2 - 3e^x)| + C$　(Formula 15)

23. $-\frac{1}{2}\ln|(2x^2 + 3)/(x^2 + 1)| + C$　(Formula 15)

25. $\frac{1}{2}\ln x - \frac{3}{4}\ln|3 + 2\ln x| + C$　(Formula 8)

Exercises 16-4

1. $(x^2/2)\ln x - x^2/4 + C$

3. $[x^{n+1}/(n + 1)]\ln x - x^{n+1}/(n + 1)^2 + C$

5. $x\ln x - x + C$　　**7.** $\frac{2}{9}x^{3/2}(3\ln x - 2) + C$

9. $\sqrt{x}(\ln x - 2) + C$

11. $\frac{1}{9}(x + 1)^3[3\ln(x + 1) - 1] + C$

13. $x\ln x + C$　　**15.** $(\frac{1}{9})x^3(2 + 3\ln x) + C$

17. $(\ln 10)^{-1}(x\ln x - x) + C$

19. $(\ln 10)^{-1}(x^2/4)(2\ln x - 1) + C$　　**21.** $xe^x - e^x + C$

23. $(x/m)e^{mx} - (1/m^2)e^{mx} + C$　　**25.** $\frac{1}{3}(2x + 1)e^{3x} - \frac{2}{9}e^{3x} + C$

27. $(x^2/2)\ln x - x^2/4 + C$　　**29.** $(x^2 - 2x + 2)e^x + C$

31. $\frac{1}{2}(x^2 - 1)e^{x^2} + C$　　**33.** $(x^3/3)(\ln x - \frac{1}{3}) + C$

37. $C(x) = 2250 + 250\ln 20 - 5000(x + 20)^{-1}[1 + \ln(x + 20)]$

Review Exercises for Chapter 16

1. a. False; the antiderivative contains an arbitrary constant.

  b. True

  c. False; The integral of the product of two functions can often be obtained by integration by parts.

  d. False; $\int (d/dx)[f(x)]\, dx = f(x) + C.$

  e. False; $(d/dt)[\int f(t)\, dt] = f(t)$

  f. False; If $f'(x) = g'(x)$, then $f(x) - g(x)$ is constant, not necessarily zero.

  g. False; $\int (1/x)\, dx = \ln |x| + C.$

  h. False; $\int e^x\, dx = e^x + C.$

  i. False; $\int(1/e^t)\, dt = \int e^{-t}\, dt = -e^{-t} + C.$

  j. False; $\int [f(x)]^n f'(x)\, dx = [f(x)]^{n+1}/(n + 1) + C$, $n \neq -1$.

  k. False; $\int x^n\, dx = x^{n+1}/(n + 1) + C$ only if $n \neq -1$.

  l. False; Integration by parts gives:
    $\int xf(x)\, dx = x \int f(x)\, dx - \int(\int f(x)\, dx)\, dx.$

  m. False; $\int(1/x^2)\, dx = (-1/x) + C.$

  n. False; $\int e^{x^2}\, dx$ cannot be expressed in terms of elementary functions.

  o. False; $\int e^t\, dt = e^t + C.$

3. $t^3/3 + 4 \ln |t| + (8/3)t^{3/2} + C$ **5.** $e^2x + C$

7. $(2/3)x^{3/2} + C$ **9.** $x/\ln 10 + C$

11. $x(\log e/\ln 3) + C$, because $\log e/\ln 3$ is simply a constant.

13. $\ln |1 + \ln x| + C$

15. $2e^{\sqrt{x+1}} + C$ **17.** $\frac{1}{2}(1 + x^3)^{2/3} + C$

19. $2\sqrt{2 + \ln x} + C$

21. $\frac{2}{3} \ln |x - 1| + \frac{7}{3} \ln |x + 2| + C$ (Formulas 15, 16)

23. $\frac{1}{81}[-\sqrt{9 - x^2}/x + x/\sqrt{9 - x^2}] + C$ (Formula 36)

25. $\frac{1}{2}t\sqrt{25t^2 + 9} + \frac{9}{10} \ln |5t + \sqrt{25t^2 + 9}| + C$ (Formula 56)

27. $\frac{1}{3}[3x - \ln (1 + 2e^{3x})] + C$ (Formula 71)

29. $\frac{1}{2}x^6 + C$ (*Note:* $\log_x x^3 + 3 \log_x x = 3 \cdot 1 = 3.$)

31. $\frac{1}{27}x^3[9 (\ln x)^2 - 6 \ln x + 2] + C$ (Formulas 77, 74)

33. $\frac{1}{3} \ln |(x^3 - 1)/x^3| + C$ (Multiply and divide the integrand by x^2 and then use the substitution $x^3 = y$.) (Formula 12)

35. $\frac{1}{25} \ln |(e^x + 1)/(e^x - 4)| - \frac{1}{5}(e^x - 4)^{-1} + C$ (Formula 17; set $e^x = t$).

37. $-\frac{1}{2}e^{(1/x^2)} + C$ **39.** $7(x \ln x - x - x \ln 2) + C$

41. $f(e) = 2 - 2/e$

43. a. $R(x) = 12x - 0.1x^2 - 0.01x^3$

  b. 120 **c.** $p = 12 - 0.1x - 0.01x^2$

  d. $5(\sqrt{37} - 1) = 25.4$

45. 184; 384 **47.** $P(x) = \sqrt{x^2 + 900} - 2x - 100$

49. $R = 200xe^{-x/20}$; $p = 200e^{-x/20}$

51. $C(I) = 0.25I + 0.45I^{2/3} + 4.2$ **53.** $\rho = 21,100$

55. $R(t) = (10t^2 - t^3/3)$ million dollars; $t = 20$; $R_{max} = \$\frac{4000}{3}$ million

57. a. 2250 million barrels **b.** 1666 million barrels

  c. 7191 million barrels

59. $f(x) = x^4 - 2x^{3/2} + 1$

CHAPTER 17

Exercises 17-1

1. $\frac{1}{3}$ **3.** 0 **5.** 12 **7.** $\frac{13}{2}$ **9.** $-2 \ln 2$

11. $\frac{1}{6}$ **13.** $(2\sqrt{2} - 1)/3$ **15.** $e^{-1} - e$ **17.** $\frac{3}{2}$

19. $\frac{1}{2} \ln (\frac{5}{2})$ **21.** 0 **23.** $(e + 1)/(3 + \ln 2)$

25. 0 **27.** 16 **29.** $\frac{16}{3}$ **31.** $\frac{81}{4}$ **33.** 4.5

35. $\ln 2 \approx 0.693$ **37.** 1 **39.** 1 **41.** $(e^x \ln x)/(1 + x^2)$

43. 0 **45.** $4e^{\sqrt{2}} \ln 2$ **47.** 950

49. 500 **51.** \$156; \$524

Exercises 17-2

1. 9 **3.** 3 **5.** $\frac{23}{3}$ **7.** 2 **9.** $\frac{13}{6}$ **11.** $\frac{1}{3}$

13. $e - \frac{4}{3}$ **15.** $\frac{8}{3}$ **17.** $\frac{1}{12}$ **19.** $\frac{16}{3}$ **21.** $\frac{1}{8}$

23. $\sqrt{2}$ **25.** Does not exist **27.** $-\frac{1}{4}$ **29.** 0

Exercises 17-3

1. a. 4.8% **b.** $\frac{19}{60}$ **3.** 210 **5.** 356

7. 9 years; \$36 million **9.** C.S. = 16; P.S. = 8

11. C.S. = 8000; P.S. = 16,000/3

13. C.S. = 178.16; P.S. = 45

15. \$746.67 million; 12 years; \$864 million **17.** \$4835.60; No

19. 10 years; \$103.01 million

21. \$88.53 million; \$87.15 million; First strategy is better.

23. $\$\frac{98}{3}$ thousand **25.** 21.39I

29. a. \$1200 **b.** \$400 **c.** \$80

Exercises 17-4

1. 3 **3.** $\frac{4}{3}$ **5.** 2 **7.** $\frac{25}{3}$ **9.** $1/\ln 2$ **11.** $1/(e - 1)$

13. \$15,333.33 **15.** \$1166.20 **17.** 60

19. (ln 2) million **21. a.** 48 **b.** 0

23. a. 141.0 units **b.** 153.2 units

Exercises 17-5

1. 0.697 (Correct result is 0.693.)   **3.** 0.880

5. 0.693   **7.** 1.644   **9.** 98; Exact value is 486/5.

11. 55.58 square units; 55.56 square units

Exercises 17-6

5. $y = \frac{1}{3}t^3 + \ln|t| + C$ **7.** $y = ce^{4t}$ **9.** $y = \frac{2}{3}t^{3/2} + C$

11. $y = ce^t - 5$ **13.** $y = ce^{2t} - 0.5$ **15.** $y = ce^{-t} + 3/2$

17. $y = e^{2-2t}$ **19.** $y = 5 + 2e^t$ **21.** $y = 6.5e^{2t} - 1.5$

23. a. $10,000e^{0.05t}$ **b.** \$14,918.25   **c.** 13.86 years

25. $y = 2e^{kt}$ billion, where $k = (\ln 2)/45 = 0.01540$; in 1960,
$y = 2e^{30k} = 3.17$ billion

27. $y = 2500(5e^{0.04t} - 1)$

29. $p(t) = 1 - e^{-t/5}$; $t = 5 \ln 4 \approx 6.9$ years

31. $y = 587e^{0.3476t}$

33. $dm/dt = 0.01 - 0.02m$; $m = 0.5(1 - e^{-0.02t})$; 0.5 kg

35. $A = (I/r)(e^{rt} - 1)$ **37.** $T = T_s - (T_s - T_0)e^{-kt}$

Exercises 17-7

1. $y = ce^{(1/2)x^2}$ **3.** $y = 1/(t^2 + C)$ **5.** $y = \ln(t^3 + C)$

7. $y = (1 - ce^t)^{-1}$

9. $y = cte^{-t}$ **11.** $y = e^{x^2}$

13. $y = 2e^{t^3}$ **15.** $y = 3/(1 + 0.5e^{-6t})$

17. $y = -t - \ln(1 - t)$ **19.** $p = 16x^{-3/2}$

21. $p = 100 - 0.5x$ **23.** $K \ln y + y = -Mt + K \ln y_0 + y_0$

25. $y = 4000(2 + 1998e^{-4t})^{-1}$; 2.0w   **27.** 22.0

Exercises 17-8

1. $c = \frac{2}{9}$; $\mu = \frac{27}{4}$ **3.** $c = \frac{1}{2}$; $\mu = 2$

5. $c = 1$; $\mu = \frac{5}{9}$ **7.** $c = 3$; $\mu = \frac{1}{2}$

9. $c = 2 + \sqrt{5}$; $f(x)$ is *not* a p.d.f., because $f(x)$ is not ≥ 0 on
the interval $0 < x < c$. Note that $f(1) = 2 - 4 = -2 < 0$.

11. $\mu = (b + a)/2$

13. a. (i) $\frac{1}{6}$ **(ii)** $\frac{1}{4}$ **(iii)** $\frac{1}{3}$ **b.** 30 min

15. $\frac{3}{4}$; 10 min

17. a. $e^{-2} \approx 0.1353$

   b. $1 - e^{-1.2} \approx 0.6988$

19. $e^{-0.2} \approx 0.8187$ **21. a.** 33%   **b.** 1813 sets

23. 0.2   **25.** $c = \frac{1}{800}$; 0.2096

27. $P(0 \leq T \leq x) = 1 - (1 + 3x)e^{-3x}$;

   a. 0.983;   **b.** 0.001

Review Exercises for Chapter 17

1. a. False; The statement is true only if $f(x) \geq 0$ in $a \leq x \leq b$.

   b. True

   c. False; $\left(\dfrac{d}{dx}\right)\left[\displaystyle\int_a^x f(t)\, dt\right] = f(x)$.

   d. False; $\left(\dfrac{d}{dx}\right)\left[\displaystyle\int_a^b f(x)\, dx\right] = 0$ and

   $\displaystyle\int_a^b \left(\dfrac{d}{dx}\right)[f(x)]\, dx = f(b) - f(a)$ **e.** True

   f. False; $\displaystyle\int_a^b f(x)\, dx$ is always some real number.

   g. False; $\displaystyle\int_a^b f(x)\, dx = \int_a^b f(t)\, dt$.   **h.** True

   i. False; The given equation cannot be expressed in the form $f(y)dy = g(t)\, dt$.

   j. False; The given equation is of *first* order as it involves derivatives only of first order.

   k. False; It is incorrect to write $\int yt^2\, dt = y \int t^2\, dt$ since y is a function of t and may not be taken outside the integral.

   l. True, provided that the random variable has a probability density function.

   m. True

   n. False; The density function may not have a maximum value at the mean value of the continuous random variable.

   o. False; The statement is true if the graph is symmetrical about the mean, but not in general.

3. $2 \ln 2 - 1$ **5.** $\frac{2}{3}$ **7.** $\frac{17}{12}$

9. \$12.50   **11.** 10,000   **13.** C.S. $= 24$, P.S. $= \frac{64}{3}$

15. $y = \ln[c + (x - 1)e^x]$ **17.** $y = (x^2 + e^x - 1)^2$

19. $y = 2(1 - ce^{2t})^{-1}$ **21.** \$20,000

23. $t = 4$ yr; \$$\frac{40}{3}$ million   **25.** $\frac{1}{3}$; $\mu = 7.5$ min

27. a. $e^{-1.5} \approx 0.223$ **b.** $e^{-0.6} \approx 0.5488$ **c.** 0.698

29. 9015.15 lb/mi²   **31.** 275   **33.** 1.33; 1.32

CHAPTER 18

Exercises 18-1

1. $f(3, -2) = 25$; $f(-4, -4) = 0$

3. $f(2, 1) = \frac{2}{3}$; $f(3, \frac{1}{2}) = \frac{14}{37}$; $f(-\frac{1}{4}, \frac{3}{4}) = 0$

5. $f(1, 2, 3) = 36$; $f(-2, 1, -4) = 54$

7. $f(\frac{1}{2}, 1, 1) = -4$; $f(\frac{1}{4}, -\frac{1}{3}, 2)$ is undefined.

9. $D = $ whole xy-plane   **11.** $D = \{(x, y)\,|\,x^2 + y^2 \geq 9\}$

13. $D = \{(x, t)\,|\,x - t > 0\}$ **15.** $D = \{(x, y, z)\,|\,yz \geq 0\}$

17.

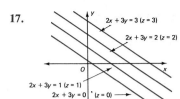

2x + 3y = 3 (z = 3)
2x + 3y = 2 (z = 2)
2x + 3y = 1 (z = 1)
2x + 3y = 0 (z = 0)

19.

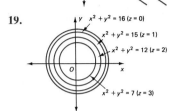

$x^2 + y^2 = 16$ (z = 0)
$x^2 + y^2 = 15$ (z = 1)
$x^2 + y^2 = 12$ (z = 2)
$x^2 + y^2 = 7$ (z = 3)

21.

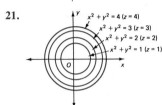

$x^2 + y^2 = 4$ (z = 4)
$x^2 + y^2 = 3$ (z = 3)
$x^2 + y^2 = 2$ (z = 2)
$x^2 + y^2 = 1$ (z = 1)

23.

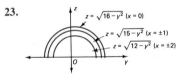

$z = \sqrt{16 - y^2}$ (x = 0)
$z = \sqrt{15 - y^2}$ (x = ±1)
$z = \sqrt{12 - y^2}$ (x = ±2)

25.

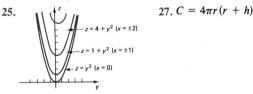

$z = 4 + y^2$ (x = ±2)
$z = 1 + y^2$ (x = ±1)
$z = y^2$ (x = 0)

27. $C = 4\pi r(r + h)$

29. Let x and y denote the dimensions of the base (in feet). Then
$C = 5xy + 600(1/x + 1/y)$.

31. If x units of X and y units of Y are produced, then
$C = 3000 + 5x + 12y$; $C(200, 150) = \$5800$.

33. $C(x, y) = P[2\sqrt{200^2 + x^2} + 3\sqrt{100^2 + y^2}$
$+ \sqrt{200^2 + (500 - x - y)^2}]$

Exercises 18-2

1. $2x$; $2y$ **3.** $6e^{2x}$; $-5/y$

5. $e^y - ye^{-x}$; $xe^y + e^{-x}$ **7.** $2x + y$; $x + 2y$

9. $2e^{2x+3y}$; $3e^{2x+3y}$ **11.** $14(2x + 3y)^6$; $21(2x + 3y)^6$

13. $\frac{1}{3}(x + 2y^3)^{-2/3}$; $2y^2(x + 2y^3)^{-2/3}$

15. $(xy + 1)e^{xy}$; x^2e^{xy}

17. $(x + 1/y)e^{xy}$; $(x^2/y - x/y^2)e^{xy}$

19. $2x/(x^2 + y^2)$; $2y/(x^2 + y^2)$

21. $(e^x + y^3)/(e^x + xy^3)$; $3xy^2/(e^x + xy^3)$

23. $y/(y - x)^2$; $-x/(y - x)^2$

25. $12x^2 + 6y^3$; $12y^2 + 18x^2y$; $18xy^2$

27. $-1/(x + y)^2$; $-1/(x + y)^2$; $1 - (x + y)^{-2}$

29. $20x^3y^{-1/2}$; $(3/4)x^5y^{-5/2}$; $(-5/2)x^4y^{-3/2}$

31. y^3e^{xy}; $x(xy + 2)e^{xy}$; $(xy^2 + 2y)e^{xy}$

33. $2(y^2 - x^2)/(x^2 + y^2)^2$; $2(x^2 - y^2)/(x^2 + y^2)^2$;
$-4xy/(x^2 + y^2)^2$

35. $-2y/(x + y)^3$; $2x/(x + y)^3$; $(x - y)/(x + y)^3$

37. $2y^2/(x - y)^3$; $2x^2/(x - y)^3$; $-2xy/(x - y)^3$

47. $\dfrac{\partial H}{\partial T} = av^{1/3} \approx$ increase in heat loss per degree increase in body temperature.

$\dfrac{\partial H}{\partial T_0} = -av^{1/3} \approx$ increase in heat loss per degree increase in temperature of surroundings.

$\dfrac{\partial H}{\partial v} = (a/3)(T - T_0)v^{-2/3} \approx$ increase in heat loss per unit increase in wind velocity.

Exercises 18-3

1. $P_L(3, 10) = 21$; $P_K(3, 10) = -29$

3. $P_L(2, 5) = 4$; $P_K(2, 5) = -122$

5. $P_L = 30(K/L)^{0.7}$; $P_K = 70(L/K)^{0.3}$

9. $\partial x_A/\partial p_A = -3$, $\partial x_A/\partial p_B = 1$; $\partial x_B/\partial p_A = 2$; $\partial x_B/\partial p_B = -5$; Competitive.

11. $\partial x_A/\partial p_A = -20p_B^{1/2}p_A^{-5/3}$; $\partial x_A/\partial p_B = 15p_B^{-1/2}p_A^{-2/3}$; $\partial x_B/\partial p_A = 50p_B^{-1/3}$; $\partial x_B/\partial p_B = (-50/3)p_A p_B^{-4/3}$; competitive.

13. $-\frac{25}{53}$; $\frac{3}{53}$ **15.** $-\frac{1}{2}$; 1

17. a. $-\frac{36}{359}$ **b.** $\frac{10}{359}$ **c.** $\frac{40}{359}$

19. $\eta_p = -b$, $\eta_I = c$; $p = (aI^c/r)^{1/(b+s)}$; $dp/dI = [c/(s + b)I](aI^c/r)^{1/(b+s)}$; This provides the rate at which the equilibrium price increases with respect to increase in consumers' income.

21. 5.14   **23.** 4.325   **25.** 0.075

27. a. 5400   **b.** 5434.375   **29.** 2738.73

Exercises 18-4

1. Local minimum at $(1, -2)$

3. Saddle point at $(-1, 2)$; no extrema

5. Local minimum at $(0, 0)$ 7. Local maximum at $(-\frac{3}{2}, -1)$

9. Local minimum at $(1, 2)$; saddle point at $(-1, 2)$

11. Local minimum at $(0, 1/\sqrt{3})$; local maximum at $(0, -1/\sqrt{3})$; saddle points at $(\pm 2/\sqrt{15}, \mp 1/\sqrt{15})$

13. Saddle points at $(2, -1)$ and $(-\frac{2}{3}, \frac{1}{3})$; no extrema

15. Local minimum at $(1, 2)$; local maximum at $(-1, -2)$; saddle points at $(\pm 1, \mp 2)$

17. Local maximum at $(\frac{5}{3}, \frac{5}{3})$; saddle points at $(2, 2)$, $(1, 2)$ and $(2, 1)$

19. Saddle points at $(\sqrt{2}, -1/\sqrt{2})$; no extrema (Note that f is defined only for $x > 0$.)

21. Local maximum at $(1, 1/2)$

23. $x = 10$, $y = 35$ 25. $x = 50$, $y = 75$

27. $x = 2$, $y = 3$

29. 27 units of X and 30 units of Y.   31. $p_1 = 20¢$, $p_2 = 25¢$

33. $p = 12$, $A = 1000 \ln 3$, $P_{max} = 1000(2 - \ln 3)$.

35. $x = \ln 5$, $T = 10$; $P_{max} = \$50\,(2 - \ln 5) = \$19.53$

37. Total weight is maximum when $x = (3\alpha - 2\beta)/(2\alpha^2 - \beta^2)$, $y = (4\alpha - 3\beta)/(4\alpha^2 - 2\beta^2)$, provided that $4\alpha \geq 3\beta$. (We need $\Delta = 2\alpha^2 - \beta^2 > 0$ and $x, y \geq 0$.)

39. $x = y = 8$ ft, $z = 4$ ft

41. **a.** $t = 1/2x$ **b.** $x = a/3$, $t = 3/2a$

Exercises 18-5

1. $(\frac{14}{13}, \frac{21}{13})$ 3. $(3, 2)$, $(-3, -2)$

5. $(2, 3, 4)$ 7. $(12, -9, -8, 9)$

9. $(\frac{5}{2}, -1, -\frac{3}{2})$ 11. $x = 120$, $y = 80$

13. **a.** $L = 500$, $K = 50$ **c.** $\lambda = 0.5623$

15. **a.** $L = 70$, $K = 30$ **c.** $\lambda = 1.05$ 17. $L = 54$, $K = 16$

19. \$2400, \$2880, \$3120, and \$3600

21. $3a$, $1.5a$, a, where $a = (10/49\pi)^{1/3}$

Exercises 18-6

1. $y = 0.47x + 2.58$ 3. $y = 0.7x + 0.95$

5. $y = 3.8x + 16.53$

7. **a.** $x = 533.45 - 111.07p$ **b.** 200   **c.** \$2.40

9. **a.** $y = 2.86x + 11.95$ **b.** 80.6

11. $y = 42.37x + 398.3$; 1415 ($x = 0$ corresponds to year 1956).

13. $y = 2.3x + 3.1$; Predictions are therefore $y = 16.9$ (or 17) when $x = 6$ and $y = 19.2$ (or 19) when $x = 7$.

15. $y = 1.197T + 107.6$

Review Exercises for Chapter 18

1. **a.** False; the range of a function $f(x, y)$ is the set of values that the function takes. It is a subset of the real numbers, not of the xy-plane.

   b. False; the domain is the whole xy-plane.

   c. False; on x-axis, $y = z = 0$.

   d. False; on yz-plane, $x = 0$.   **e.** True   **f.** True

   g. False; if the *third* order partial derivatives of f are continuous, then it follows that $\partial^3 f/\partial x^2 \partial y = \partial^3 f/\partial y \partial x^2$.

   h. False; $(\partial/\partial x)(x^3 y^2) = 3x^2 y^2$.

   i. False; $(\partial/\partial y)(x^2/y) = -x^2/y^2$.

   j. True; provided f is differentiable at (a, b).

   k. False; the conditions stated are necessary in order that $f(x, y)$ have a local maximum, but are not sufficient. Sufficient conditions are obtained by adding the condition $f_{xx}f_{yy} - f_{xy}^2 > 0$.

   l. True

   m. False; if $f_x(a, b) = f_y(a, b) = 0$, then (a, b) could be a saddle point for $f(x, y)$.

   n. False; for (a, b) to be a local minimum point for $f(x, y)$, we should have at (a, b), $f_x = f_y = 0$, $f_{xx} > 0$, $f_{yy} > 0$, and $f_{xx}f_{yy} - f_{xy}^2 > 0$.

   o. False; the line of best fit need not pass through any data points.

   p. True   **q.** True

3. $D = \{(x, y) \mid x + y > 0 \text{ and } y \neq 0\}$

5. $D = \{(x_1, x_2, x_3) \mid x_1 + x_2 + x_3 > 0 \text{ and } x_1 > x_2\}$

7. $2xy + y^2$; $x^2 + 2xy$; $2x + 2y$; $2x$

9. $2x/(x^2 + y^2) + (y/2\sqrt{x})e^{y\sqrt{x}}$; $2y/(x^2 + y^2) + \sqrt{x}e^{y\sqrt{x}}$; $-4xy/(x^2 + y^2)^2 + [(1 + y\sqrt{x})/2\sqrt{x}]e^{y\sqrt{x}}$; $2(x^2 - y^2)/(x^2 + y^2)^2 + xe^{y\sqrt{x}}$

11. $P_x = 3y + z - 2x$; $P_y = 3x + 5z - 4y$; $P_z = 5y + x - 3z$; P_y represents the increase in profits when an additional gas pump is installed without changing the number of employees or inventory.

13. **b.** $\eta_{P_A} = \eta_{P_B} = \frac{1}{2}$; sum $= 1$

15. $\partial C/\partial p_A = 1.1 + 2.78p_A - 3.66p_B$; $\partial C/\partial p_B = 13.8 - 3.66p_A + 8.88p_B$; these derivatives represent the rate of change of manufacturing costs with respect to increases in the prices of the two products.

17. Local maximum at $(-\frac{1}{2}, 4)$.   19. Saddle point at $(\frac{1}{2}, -\frac{1}{2})$.

21. 600 lb of beef and 500 lb of pork.

23. **a.** $x = 127$, $y = 60$; maximum profits $= \$20,438$

   b. $x = 96$, $y = 51$

27. 0.0984   31. $y = 1.46x - 0.60$

33. $a(x_1^4 + x_2^4 + \cdots + x_n^4) + b(x_1^3 + x_2^3 + \cdots + x_n^3)$
$\qquad + c(x_1^2 + \cdots + x_n^2) = (x_1^2 y_1 + x_2^2 y_2 + \cdots + x_n^2 y_n)$
$a(x_1^3 + x_2^3 + \cdots + x_n^3) + b(x_1^2 + x_2^2 + \cdots + x_n^2)$
$\qquad + c(x_1 + x_2 + \cdots + x_n) = (x_1 y_1 + x_2 y_2 + \cdots + x_n y_n)$
$a(x_1^2 + x_2^2 + \cdots + x_n^2) + b(x_1 + x_2 + \cdots + x_n) + nc$
$\qquad\qquad = (y_1 + y_2 + \cdots + y_n)$

For the given data, these equations reduce to
$354a + 100b + 30c = 104.3$
$100a + 30b + 10c = 41.1$
$30a + 10b + 5c = 30.3$
Then $a = 0.0357$, $b = -2.09$, $c = 10.03$.

Index

A

Abscissa, 119
Absolute maximum, 608
Absolute minimum, 608
Absolute value, 108
Acceleration, 559
Addition
  fractions, 12, 49
  matrix, 367
Additive inverse, 7
Adjoint, 433
Algebraic expression, 29
Algebraic operations, 29–37
Amortization, 280, 294
Annuity, 277, 296
  future value of, 277
  present value of, 280
Antiderivative, 655
Approximations, 632, 771
Area between curves, 695
Area under curves, 685, 692
Argument, 170
Arithmetic progression (A.P.), 260
Artificial variable, 478
Associative properties, 4, 375
Asymptotes, 613, 621
  horizontal, 191, 616
  vertical, 190, 619
Augmented matrix, 382
Average cost, 517
Average rate of change, 486
Axes, coordinate, 118

B

Base
  change formula, 246, 247
  of an exponent, 18, 228
  of a logarithm, 232

Basic feasible solution, 464
Basis, 465
Bayes' theorem, 337
Bernoulli trials, 350
B.F.S. (*see* Basic feasible solution)
Binomial, 29
  coefficients, 357
  expansion, 357
  probabilities, 351
  theorem, 356
  Break-even analysis, 154
  Break-even point, 155

C

Calculus
  differential, 655
  integral, 655
Cancellation, 11
Cartesian
  coordinates, 118
  plain, 118
Certain event, 312
Chain rule, 543
Circles, 193
Closed intervals, 93
Coefficient, 29
Coefficient of inequality for income distributions, 702
Coefficient matrix, 377
Cofactor, 426
Column matrix, 365, 371
Column vector, 365
Combinations, 344
Common
  difference, 260
  factor, 11
  logarithms, 240
  ratio, 268

Commutative properties, 4
Competitive products, 770
Complementary products, 770
Complement of an event, 315
Composite function, 202
Compound interest, 219–222
  continuous compounding, 219–222
Concave down, 582
Concave up, 581
Conditional probability, 329
Consistent (system of equations), 392
  test for, 393
Constant function, 178
Constant of integration, 655
Constant profit line, 453
Constraints, 456
Consumer's surplus, 708
Continuous function, 495, 526
Contour line, 755
Coordinate axes, 118
Costs, 81, 136
  average, 517
  fixed, 136
  marginal, 515
  total, 136
  variable, 136
Cramer's rules, 430
Critical points, 574, 711
Cross elasticity, 771
Cube root, 24
Curves
  area between, 638
  concavity, 581–82
  demand, 139
  learning, 703
  level, 755
  Lorentz, 701
  product transformation, 195
  sketching, 589, 591, 621
  supply, 140

D

Data points, 791
Decaying exponential, 227
Decreasing function, 567
Definite integral, 683
Degree of a polynomial function, 179
Demand curve, 139
Demand elasticity, 646
Demand equation, 139

Departing variable, 465, 470
Derivatives, 502
  of composite function, 543
  of exponential functions, 552
  higher, 560
  of inverse functions, 641
  logarithmic, 645
  mixed partial, 763
  partial, 761
  of power functions, 508
  of products and quotients, 536, 538
  second order partial, 763
Determinants, 525, 526, 529
  expansion, of 526, 527
  inverse by, 434
Diagonal elements, 375
Difference equations, 285
  applications in finance, 291
  linear first order, 289
  order of, 285
  solution of, 285, 287
Differential, 630
Differential coefficient, 502
Differential equation, 722
  general solution of, 724, 726, 734
  linear, 722
  logistic, 732
  order of, 722
  separable variable type, 730
  solution of, 623
Differentiation
  implicit, 636
  logarithmic, 643
Discontinuity, jump, 526
Discontinuous function, 495, 526
Discount rate, 223
Discriminant, 77
Distance formula, 120
Distribution (probability), 737
  expected value of, 741
  exponential, 740
  mean of, 741
  uniform, 738
Distributive properties, 4
Dividend, 35
Division
  of fractions, 10, 52
  long, 35
  by zero, 9
Divisor, 35
Domain of a function, 169

E

e, 221, 228
Economic lot size, 604, 626
Economic order quantity, 604
Economic rent, 712
Effective rate, 217, 222, 246
Elasticity
   cross, 771
   of demand, 646
   income, 774
   unit, 648
Element of a matrix, 365
Element of a set, 89
Empty set, 91
End points of an interval, 93
Entering variable, 465, 470
Entry of a matrix, 365
Equally likely, 319
Equations
   applications of, 66–70. 79–84
   difference, 285
   differential, 722
   general linear, 131
   graph of, 122, 133
   input–output, 409
   linear, 62, 131, 133
   matrix, 377
   in one variable, 58
   polynomial, 61
   quadratic, 62, 71
   roots of, 58
   solution of, 58, 145
   systems of linear, 145, 376, 382
Equilibrium price, 158
Equilibrium quantity, 158
Equivalent system, 382
Error, 634
   mean-square, 791
   percentage, 634
   relative, 634
   square, 791
Events, 311
   certain, 312
   complements of, 315
   impossible, 312
   independent, 332
   intersection of two, 314
   mutually exclusive, 315
   union of two, 313
Exponential form, 234
Exponential function, 227

Exponents, 18
   base of, 18
   fractional, 23
   laws of, 19–21, 26
Expressions
   addition and subtraction of, 30
   division of, 34
   multiplication of, 31
   Extrema, 573, 608, 775
   test for local, 576, 586, 777

F

Factorial, *n*, 343
Factoring, 38–45
Factors, 38
Feasible solution, 451
   basic, 464
Final demand, 408
Finite sequence, 260
First-degree equation (*see* Linear equations)
First derivative, 560
First derivative test, 576
Fixed costs, 136
Fractions, 10
Functions, 169
   absolute value, 197
   algebraic, 180
   argument of, 170
   average rate of change of, 486
   average value of, 712
   Cobb–Douglas production, 773
   combination of, 201
   composite, 202
   constant, 178
   continuous, 495, 526
   cost, 182
   cubic, 179
   decaying exponential, 227
   decreasing, 567
   differentiable, 502, 527
   domain of, 169, 750
   explicit, 205
   exponential, 227
   graph of, 171, 567, 755
   growing exponential, 227
   implicit, 206
   increasing, 567
   inverse, 207, 210
   joint-cost, 798
   linear, 179

Functions (*cont.*)
  logarithmic, 233
  natural exponential, 229
  objective, 456
  polynomial, 179
  power, 189
  probability density, 737
  production, 768
  profit, 181
  quadratic, 179, 183
  range of, 169, 750
  rational, 179
  revenue, 181
  transcendental, 180
  of two variable, 750
  value of, 169
Fundamental theorem, 685, 689

G

General linear equation, 131
General term, 260
Geometric progressions (G.P.), 268
Graph of an equation, 122
Graph of a function, 171, 755
Graph of inequality, 443
Graph theory, 380
Graphing linear equations, 133
Grouping method, 40

H

Half-plane, 444
Higher derivatives, 560
Horizontal asymptote, 191, 616
Horizontal line
  equations of, 130
  slope of, 128
Horizontal plane, 754

I

Identity elements, 6
Identity matrix, 375
Implicit differentiation, 636
Inplicit relation, 205
Impossible event, 312
Increasing function, 567
Increment, 483
Indefinite integral, 655
Independent variable. 170
Indicator, 470
Indifference line, 453

Inequalities, 89
  linear, 95, 443–44
  quadratic, 102
  strict, 89
Inequality symbol, 89
Infinite G.P., sum of, 272
Inflexion point, 584
Initial condition, 724
Input–output analysis, 407–10
Input–output matrix, 409
Instantaneous speed, 491–91
Integers, 2
Integrals, 655, 683
  definite, 683
  improper, 699
  indefinite, 655
  table of, 667, 670
Integral sign, 656
Integrand, 656
Integration, 655
  constant of, 655
  limits of, 683
  numerical, 715
  by parts, 674–77
  power formula for, 656
  by substitution, 663–68
  by using table, 670–73
  variable of, 656
Intercepts, 130
Intersection of events, 314
Interest
  compound, 215–17, 270
  compounded continuously, 219–21
  effective rate of, 217, 222, 246
  nominal rate of, 216
  simple, 262
Interval
  closed, 93
  end-points of, 93
  open, 93
Inventory cost model, 602
Inverse of a matrix, 400
Invertible matrix, 401
Irrational numbers, 2

J

Jump discontinuity, 526

L

Langrange multipliers, 783
Learning curve, 703

Least common denominator (L.C.D.),
  13
Least squares method, 789–94
Leontief, 407
Level curve, 755
Like terms, 30
Limiting value, 493, 614
Limits, 493
  at infinity, 614
  of integration, 683
  lower, 683
  one-sided, 523
  theorems, 495–97
  upper, 683
Line
  general equation of, 131
  indifference, 453
  parallel and perpendicular, 133
  point-slope formula of, 129
  slope of, 127
  slope-intercept formula of, 130
Linear cost model, 136
Linear equations, 62, 131
Linear function, 179
Linear inequalities, 95, 443–44
Linear models, 633
Linear optimization
  by geometric approach, 450
  by simplex method, 470
Linear programming problem, 450
Literal part, 29
Local maximum or minimum, 572, 775
Logarithmic differentiation, 643
Logarithmic form, 234
Logarithmic function, 232–33
Logarithms, 232
  applications of, 243–45
  common, 240
  natural, 237
  properties of, 235–36
Logistic model, 249, 732
Long division, 35
Lorentz curve, 701
Lower limit, 683

M

Marginal
  analysis, 514
  average cost, 539
  cost, 515
  cost rate, 563
  demand, 770

Marginal (*cont*.)
  physical productivity, 565
  price, 537
  productivity, 520
  productivity of capital, 768
  productivity of labor, 520, 768
  profit, 519
  propensity to save and consume, 521
  revenue, 518
  tax rate, 521, 530
  yield, 521
Market equilibrium, 158
Markov chains, 414–25
Matrix, 364
  addition and subtraction of, 367
  adjoint of, 433
  augmented, 382
  coefficient, 377
  cofactor, 433
  column, 365, 371
  demand, 409
  diagonal element of, 375
  elements of, 365
  identity, 375
  input–output, 409
  inverse of, 400
  invertible, 401
  multiplication of, 371, 373
  non-singular, 401
  output, 409
  production, 370
  reduced, 384, 393
  regular, 422
  row, 365, 371
  scalar multiplication of, 366
  singular, 401
  size of, 365
  square, 366
  state, 418
  steady state, 422
  transition, 415
  transpose, 433
  zero, 365
Maxima and minima
  absolute, 608
  applications, 594–607
  local, 572–73, 775
  tests for, 576, 586, 777
Minor, 426
Model
  input–output, 407, 412, 435
  inventory cost, 602

Model (*cont.*)
  linear cost, 136
  logistic, 249, 732
Monomial, 29
Multinomial, 29
Multiplicative inverse, 7
Mutually exclusive, 315

N

$n!$ defined, 343
Natural logarithms, 237
Natural number, 2
Nominal rate of interest, 216
Non-singular matrix, 401
Null set, 91
Number line, 3
Numbers
  complex, 78
  imaginary, 78
  irrational, 2
  natural, 2
  properties of real, 4, 6
  rational, 2
  real, 3
Numerical coefficient, 29
Numerical iteration, 287

O

Objective function, 456
One-sided limits, 523
Open interval, 93
Optimization, 594–607, 775–79
Ordered pair, 119
Ordinate, 119
Origin, 118
Outcome, 310
Output matrix, 409

P

Parabolas, 183
Parallel and perpendicular lines, 133
Partial derivatives, 761
Partial listing, 90
Pascal triangle, 355
Penalty method, 478
Permutations, 342
Pivot element, 471
Pivoting, 464
Point-slope formula of a line, 129
Polynomials, 35, 179
Power formula, 508, 656

Present value, 222, 280, 706
Primary inputs, 408
Principal nth root, 24
Probability, 317–18
  binomial, 351
  conditional, 329
  formulas, 321–24, 330
  multiplication rule of, 332
  posterior, 338
  prior, 338
  transition, 415
Producer's surplus, 709
Production function, 768
Production input factors, 768
Product rule, 536
Product transformation curve, 195
Profit, 81
Progressions
  arithmetic, 260
  geometric, 268

Q

Quadrants, 119
Quadratic
  equations, 71
  formula, 73, 77
  function, 179, 183
Quotient, 35
Quotient rule, 538

R

Radical, 24
Radical sign, 24
Random process, 414
Random variable, 736–37
Range of a function, 169
Rational function, 179
Rationalizing the denominator, 47
Rational number, 2
Real number, 3
Reciprocal, 6
Reduced matrix, 384, 393
Related rates, 547
Remainder, 35
Revenue, 81
Richter scale, 254
Root
  cube, 24
  of an equation, 58
  principal nth, 24
  square, 24

Row matrix, 365, 371
Row operations, 383, 384
Row reduction method, 384, 393

S

Saddle point, 777
Sample point, 310
Sample space, 310, 311
Savings plans, 275, 292
Scrap value, 261
Second derivative, 560, 763
Second derivative test, 586
Semiopen interval, 93
Separation of variables, 730
Sets
    elements of, 89
    empty, 91
    finite, 91
    infinite, 91
    member of, 89
    null, 91
    subset of, 92
    void, 91
Simplex method, 470
Simplex tableau, 465, 470
Simpson's rule, 717–18
Singular matrix, 401
Sinking fund, 274
Slack variable, 461, 464
Slope-intercept formula, 130
Slope of a line, 127
Solution of an equation, 58, 145
Specific decay rate, 246
Specific growth rate, 246
Square matrix, 366
Standard form, 462
Stellar magnitudes, 254
Structural variable, 461
Stochastic process, 414
Summation notation, 299
Supply curve, 140
Systems
    of linear equations, 145, 376, 382
    consistent, 392
    inconsistent, 392
    singular, 390–94

T

Tangent line, 505
Third derivative, 560

Transpose of a matrix, 433
Trapezoidal rule, 716
Trinomial, 29

U

Uniqueness, test for, 393
Unit elasticity, 648
Union of events, 313
Upper limit, 683

V

Variable cost, 136
Variables
    artificial, 478
    departing, 465, 470
    dependent, 170
    entering, 465, 470
    independent, 170
    of integration, 656
    random, 736–37
    slack, 461, 464
Vector
    column, 365
    row, 365
    state, 418
    value, 377
    variable, 377
Venn diagram, 312
Vertex of a parabola, 183
Vertical asymptote, 190, 619
Vertical line
    equation of, 131
    slope of, 128
Vertical line test, 173

X

x-axis, 118, 753
x-coordinate, 119
xy-plane, 118, 753
xz-plane, 754

Y

y-axis, 118, 753
y-coordinate, 119
y-intercept, 130
yz-plane, 754

Z

z-axis, 753
Zero matrix, 365